International Conference on Strongly Correlated Electronic Systems 2010

(SCES 2010)

AA008903

Journal of Physics: Conference Series Volume 273

Santa Fe, New Mexico, USA
27 June-2 July 2010

ISBN: 978-1-61782-658-0
ISSN: 1742-6588

Printed from e-media with permission by:

Curran Associates, Inc.
57 Morehouse Lane
Red Hook, NY 12571

Some format issues inherent in the e-media version may also appear in this print version.

Copyright© (2010) by the Institute of Physics
All rights reserved.

Printed by Curran Associates, Inc. (2011)

For permission requests, please contact the Institute of Physics
at the address below.

Institute of Physics
Dirac House, Temple Back
Bristol BS1 6BE UK

Phone: 44 1 17 929 7481
Fax: 44 1 17 920 0979

techtracking@iop.org

Additional copies of this publication are available from:

Curran Associates, Inc.
57 Morehouse Lane
Red Hook, NY 12571 USA
Phone: 845-758-0400
Fax: 845-758-2634
Email: curran@proceedings.com
Web: www.proceedings.com

TABLE OF CONTENTS

PLENARY

QUANTUM PHASE TRANSITIONS INCLUDING NON-FERMI LIQUIDS AND EXOTIC QUANTUM PHASES

PHYSICS OF TRANSURANICS

HEAVY FERMIONS INCLUDING KONDO IMPURITY AND KONDO LATTICE SYSTEMS

UNCONVENTIONAL SUPERCONDUCTORS

HIGH-TC SUPERCONDUCTIVITY INCLUDING CUPRATES AND PNICTIDES

NOVEL MAGNETS AND NEW MATERIALS

QUANTUM MAGNETISM AND FRUSTRATED MAGNETS

COUPLED SPIN, CHARGE AND ORBITAL DEGREES OF FREEDOM

MOTT–HUBBARD SYSTEMS

CORRELATED ELECTRONS IN REDUCED DIMENSIONS AND CONFINED GEOMETRIES

CONCLUDING REMARKS

PREFACE

Preface

The 2010 International Conference on Strongly Correlated Electron Systems was held in Santa Fe, New Mexico, a reunion of sorts of the 1989 International Conference on the Physics of Highly Correlated Electron Systems that also convened in Santa Fe. SCES 2010 – co-chaired by John Sarrao and Joe Thompson – followed the tradition of earlier conferences in this century, hosted by Buzios ('08), Houston ('07), Vienna ('05), Karlsruhe ('04), Krakow ('02) and Ann Arbor ('01). Every three years since 1997, SCES has joined the International Conference on Magnetism (ICM), held in Recife ('00), Rome ('03), Kyoto ('06) and Karlsruhe ('09). Like its predecessors, SCES 2010 topics included strongly correlated f- and d-electron systems, heavy-fermion behaviors, quantum-phase transitions, non-Fermi liquid phenomena, unconventional superconductivity, and emergent states that arise from electronic correlations. Recent developments from studies of quantum magnetism and cold atoms complemented the traditional subjects and were included in SCES 2010.

2010 celebrated the 400th anniversary of Santa Fe as well as the birth of astronomy. So what is the connection to SCES? The Dutch invention of the first practical telescope and its use by Galileo in 1610 and subsequent years overturned the dogma that the sun revolved about the earth. This revolutionary, and at the time heretical, conclusion required innovative combinations of new instrumentation, observation and mathematics. These same combinations are just as important 400 years later and are the foundation of scientific discoveries that were discussed during SCES 2010. As we learned, past dogmas about strongly correlated materials and phenomena must be re-examined with an open and inquisitive mind. A glimpse at the excitement this conference had to offer can be seen in the SCES 2010 proceedings, which are published simultaneously by the Institute of Physics Publishing in issues of the *Journal of Physics: Conference Series* (contributed papers) and *Journal of Physics: Condensed Matter* (invited papers).

This conference would not have been possible without the hard work of the SCES 2010 Program Committee, International and National Advisory Committees, Local Committee and conference organizers, the New Mexico Consortium. We thank them as well as those organizations that generously provided financial support: ICAM-I2CAM, Quantum Design, Lakeshore, the National High Magnetic Field Laboratory and the Department of Energy National Laboratories at Argonne, Berkeley, Brookhaven, Los Alamos and Oak Ridge. Of course, we especially thank the participants for bringing new ideas and new results, without which SCES 2010 would not have been possible.

All papers published in this volume of *Journal of Physics: Conference Series* have been peer reviewed through processes administered by the proceedings Editors. Reviews were conducted by expert referees to the professional and scientific standards expected of a proceedings journal published by IOP Publishing.

This is a co-publication with *Journal of Physics: Condensed Matter.* The bulk of the papers, after peer review, are published in *Journal of Physics: Conference Series*. However a selection of papers, are published separately in a special issue of *Journal of Physics: Condensed Matter*.

Trends in Heavy Fermion Matter

J. Flouquet[1], D. Aoki[1], F. Bourdarot[1], F. Hardy[2], E. Hassinger[1], G. Knebel[1], T. D. Matsuda[1,3], C. Meingast[2], C. Paulsen[4], V. Taufour[1]

[1] CEA-Grenoble, INAC/SPSMS, 17 rue des Martyrs, 38054 Grenoble cedex 9, France.
[2] KIT, Institut für Festkörperphysik, 76021 Karlsruhe, Germany.
[3] Advanced Science Research Center, JAEA Tokai 319-1195, Japan.
[4] Institut Néel, CNRS/UJF Grenoble, 38042 Grenoble cedex 9, France.

E-mail: `jacques.flouquet@cea.fr`

Abstract.
A brief review on major advances in heavy fermion physics is presented including the Ce metal phase diagram, the huge effective mass detected in $CeAl_3$, and the successive discoveries of unconventional superconductivity in $CeCu_2Si_2$ and three U based compounds, UBe_{13}, UPt_3 and URu_2Si_2. In order to track the origin of the huge effective mass, the case of intermediate valence compounds is discussed with emphasis on the differences between Yb and Ce materials. The formation of the effective mass is analyzed by two regular- and singular-part contributions. Examples are given for both, antiferromagnetic ($CeRu_2Si_2$ series) and ferromagnetic tricriticalities (UGe_2). Pressure and magnetic-field studies on the ferromagnetic superconductor URhGe illustrate the role of the singular effective mass enhancement on the superconducting pairing. The discovery of the Ce-115 material gives the opportunity to study deeply the interplay of antiferromagnetism and superconductivity. This is clearly demonstrated by field re-entrant AF inside the SC phase just below the superconducting upper critical field (H_{c2}) for $CeCoIn_5$ or on both sides of H_{c2} within a restricted pressure window for $CeRhIn_5$. The present status of the search for the hidden-order parameter of URu_2Si_2 is given and we emphasize that it may correspond to a lattice unit-cell doubling which leads to a drastic change in the band structure and spin dynamics, with the possibility of competition between multipolar ordering and antiferromagnetism.

Published under licence by IOP Publishing Ltd

1. Introduction

Heavy fermion compounds (HFC) are of special interest as a large variety of ground states can be achieved and a strong interplay between them occurs. Due to the weakness of their corresponding characteristic temperature a change from one state to another can be easily realized by moderated temperature (T), pressure (P), or magnetic field (H) tunings. Heavy-fermion quasiparticles are the result of strong magnetic and valence fluctuations.

Heavy fermion physics started with the discovery of the Ce metal phase diagram in which the occupation number (n_f) of the 4f trivalent Ce^{3+} configuration is a key parameter [1]. A boost in the field was the discovery in $CeAl_3$ [2] that even with n_f close to unity the establishment of long range magnetic ordering can be avoided with the benefit of the formation of huge heavy quasiparticles. The Sommerfeld coefficient $\gamma \sim 1\,\mathrm{J\,mole^{-1}K^{-2}}$ surpasses that of a free electron value by three order of magnitude. The description from single Kondo impurity to a regular array of Kondo centers (the Kondo lattice) is still the subject of debate [3]. At least, now it is well established that heavy quasiparticles move along the trajectories of the Fermi Surface [4, 5].

A major breakthrough was the discovery of superconductivity (SC) in $CeCu_2Si_2$ [6] which opened the era of unconventional SC. This new research field was reinforced by the concomitant discoveries of SC in three uranium HFC namely UBe_{13} [7], UPt_3 [8], and URu_2Si_2 [9, 10]. Unconventional SC with a sign-reversing order parameter (OP) was directly proven by the observation of multiple SC phases in UPt_3 [11]. New features of unconventional SC, e.g. the unitary treatment of impurities [12, 13], were rapidly derived from HFC studies [14, 15], whereas it took, for instance, a decade before serious considerations of unconventional SC were put forward in organic conductors.

The achievement of a huge effective mass was rapidly related to the proximity to a magnetic instability, but it took over a decade before the advent of quantitative studies focusing on quantum criticality such as the switch from paramagnetic (PM) to antiferromagnetic (AF) ground states. Since this point is highly discussed by other authors [16, 17], we will focus here on different facets of HFC going from the main discoveries to the present status.

In this article special attention is given to: (i) the (H, P, T) phase diagram ranging from intermediate valence compounds to examples of tricriticality for both antiferromagnetic (AF) and ferromagnetic (FM) materials, (ii) recent investigations on the interplay between effective mass enhancement and SC, and (iii) the resolution of the puzzle of the hidden order (HO) phase of URu_2Si_2.

2. Intermediate valence compounds

De facto, HFC belong to the general class of the anomalous 4f or 5f compounds. These were recognized six decades ago via the discovery of the high pressure phase diagram of Ce metal (Fig. 1). The first order transition at $T_{\gamma\alpha}$ from a PM γ phase to a PM

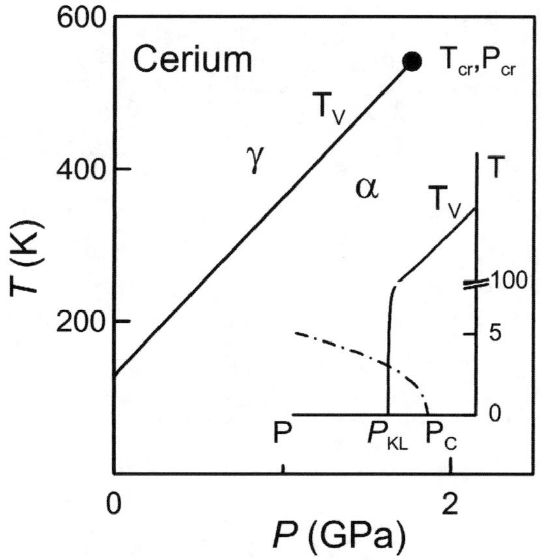

Figure 1. Pressure-temperature phase diagram of Ce metal (T, P) with the first order transition $T_V(P)$ between the γ and the α phase and its critical end point. If the volume can be expanded: (i) $T_V(P)$ will collapse abruptly at $T = 0\,\mathrm{K}$ for $P = P_{KL}$ and (ii) magnetic ordering (dashed dotted line) will collapse at the critical pressure P_c[18]

.

α phase is isostructural. It ends up at a critical end point (CEP) at $T_{cr} \sim 600\,\mathrm{K}$ and $P_{cr} \sim 2\,\mathrm{GPa}$. From high energy spectroscopy, n_f in the γ phase is near unity while even in the α phase n_f remains high ($n_f \sim 0.9$). Thus the transition at $T_{\gamma\alpha}$ is connected to an abrupt change in the valence and $T_{\gamma\alpha} = T_V$ the temperature of the valence transition. At ambient pressure, $T_V(0) \sim 100\,\mathrm{K}$, such a high value prevents the establishment of long range magnetic ordering. However, if $T_{\gamma\alpha}$ collapses we must consider on equal footing the interplay of valence and magnetic instabilities [18]. If a negative pressure could be realized, AF for example would appear below a Néel temperature (T_N) up to a pressure P_c. An interesting problem is the consequence of an intercept between the valence transition line ($T_V(P)$) which must collapse at some pressure P_{KL} at $T \to 0\,\mathrm{K}$ and the AF–PM boundary. Thus in addition to the magnetic instability at P_c a possible Fermi surface instability may occur. Furthermore, if T_V becomes comparable to any characteristic temperature of a specific bare ground state like T_N for AF, T_{Curie} for FM, or T_{SC} for SC, the valence transition or its fluctuations near its CEP must be considered in addition to the spin fluctuations. The interplay between the different phase diagrams reported Fig. 2 may lead to new coexisting matter such as the AF + SC phase of $CeRhIn_5$ discussed later. A discussion on valence fluctuations is given in the contribution of K. Miyake [19]. Strikingly, a first order valence transition similar to that in Ce metal has almost never been observed in any Ce HFC. The main reason is that the repulsion term U_{fc} between the 4f electron and the light electron plays a critical

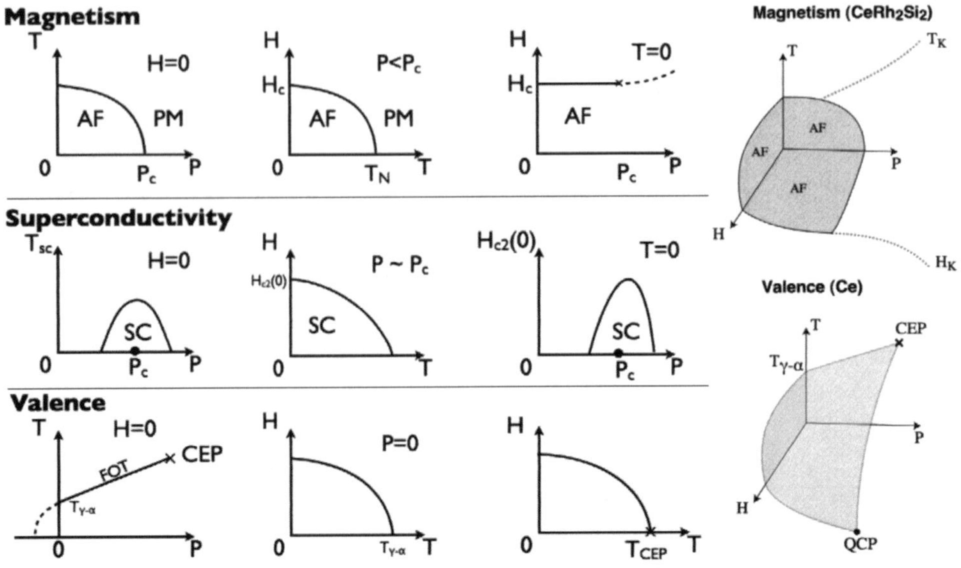

Figure 2. Interplay between (H, T, P) phase diagram for the magnetic, superconducting and valence transition. H re-entrance of AF results from the interplay between magnetic and superconducting phase diagram; in Yb HFC, our guess is that there is a strong interplay between magnetic and valence phase diagram[21].

role in controlling the position of the zero temperature CEP as a function of the relative ε_f of the 4f level with respect to the Fermi level [20]. Thus, very often the valence fluctuations can only be felt and the precise location of the valence quantum critical point is a "ghost" centered around $P = P_V$ but will react to magnetic field sweeps. For Ce HFC, there are clear examples, e.g. $CeCu_2Si_2$ or $CeRu_2Si_2$, where P_c and P_V are well separated ($P_V - P_c \sim 4\,GPa$) but in most cases $P_c \sim P_V$. Furthermore P_V coincides with the crossover pressure P^*_{CF} where the Kondo energy $k_B T_K$ surpasses the crystal field splitting C_{CF} and the full degeneracy $2J + 1$ of the 4f level is recovered [21].

Often the trivalent Yb ions with 13 4f electrons are considered as the 4f hole analog of trivalent Ce. However, we want to point out two major differences [21, 22]: (i) the deeper localization of the 4f electron of Yb with respect to Ce implies that the width Δ of the 4f virtual band states is one order of magnitude smaller in Yb HFC than in Ce HFC, (ii) the larger strength of the spin-orbit coupling λ between the $j = l - 1/2$ and $j = l + 1/2$ individual configuration of the angular momentum $l = 3$ for Yb than Ce. This leads to the hierarchy $\Delta > \lambda > C_{CF}$ for Ce and $\lambda > \Delta \sim C_{CF}$ for Yb. Consequently, in Ce HFC, the variation of n_f is restricted between 1 to 0.84. Contrary, for Yb centers, n_f can vary from 0 to 1 i.e. with a valence changing from 2 to 3. These hand waiving arguments are clearly illustrated in band structure calculations (for example see in ref.[22] the contrast between $CeRh_2Si_2$ and $YbRh_2Si_2$).

Another difference is that the pressure P^*_{CF} can be located deep inside the intermediate valence phase (Fig. 3) as pointed out two decades ago for the analysis of

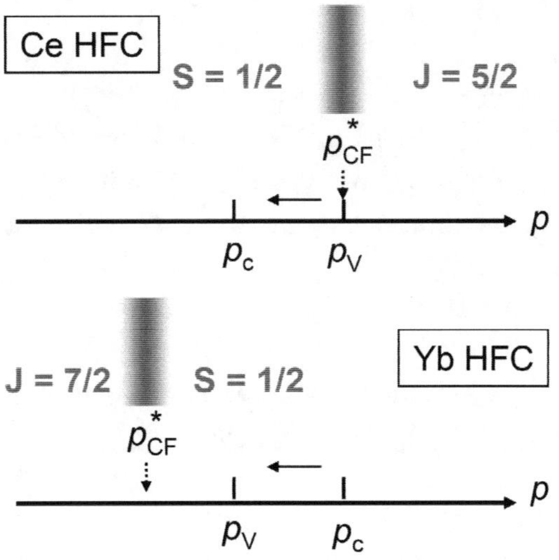

Figure 3. Schematic difference between Ce and Yb HFC on the hierarchy between their characteristic pressure for the crystal field wipe out (P^*_{CF}); for the valence transition (P_{V}) and for the magnetic instability (P_{c})[21].

Mössbauer experiments on $YbCu_2Si_2$ [23] and experimentally observed for $YbInCu_4$ [24]. The latter compound represents a beautiful example of a first order valence transition with the strong interplay between valence and magnetic fluctuations (Fig. 4). In agreement with recent theoretical developments [19] we want to stress that at least for Yb HFC the interplay between valence and magnetic fluctuations is strong. A nice

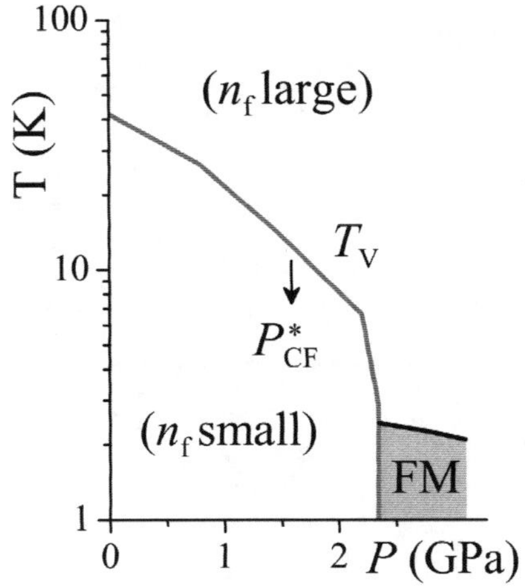

Figure 4. Schematic (T, P) phase diagram of $YbInCu_4$. T_{V} has been determined by resistivity and NQR, T_{Curie} by susceptibility and NQR. P^*_{CF} indicates the crossover pressure for the mergence of crystal field effects[21, 24]

illustration for this is the work on β-$YbAlB_4$ [25] presented by S. Nakatsuji at this conference.

Before discussing the formation of the huge effective mass, let us stress that the metal-insulator transition, which can occur in anomalous rare earth compounds, remains a quite puzzling problem. Four decades ago, systems like SmS [26], SmB_6 [27], YbB_{12} [28], where the valence mixing is directly linked to the formation of an extra carrier via a relation like $Sm^{2+} \leftrightarrow Sm^{3+} + 5d$, surprisingly end up in an insulating ground state, despite the fact that the occupation number of the divalent configuration $1 - n_f$ is only about 0.3. Paradoxically, the low temperature properties seem to be renormalized to the divalent non-magnetic state even if the occupation number of the trivalent configuration $n_f = 0.7$ is higher than the divalent one. New sets of experiments for SmS [29, 30] and SmB_6 [31] have clearly shown (Fig. 5) that this unconventional insulating phase is rather robust, but the ground states changes through a first order transition to a metallic long range ordered magnetic phase under pressure when n_f approaches closely to unity. We believe that these new data require new theoretical insights as well as a better understanding on their differences with the examples of so-called Kondo insulators such as CeNiSn [32] and $CeRu_4Sn_6$ [33].

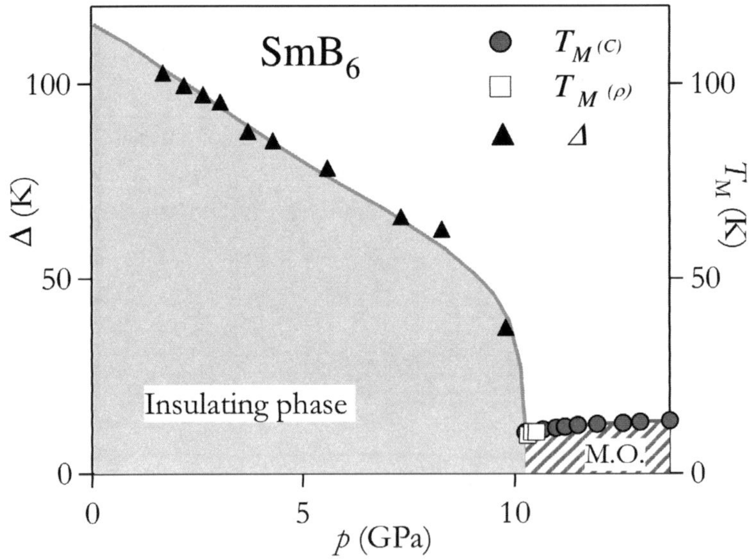

Figure 5. (T, P) phase diagram of SmB_6. At low pressure, the ground state is PM and insulating. Above $P \sim 10$ GPa, it switches to metallic and long range magnetic ordered through a first order transition [31].

3. Origin of heavy fermion quasiparticle

The huge effective mass $m^* \sim 100\, m_0$ is the result of two additional effects: (i) the renormalization of the bare band mass m_B and (ii) an additional contribution m^{**} due to the development of magnetic or valence correlation. Thus $m^* = m_B + m^{**}$. The last term m^{**} is directly linked to the possible occurrence of unconventional SC. In the strong coupling theory of SC the coupling constant λ is equal to m^{**}/m_B. However, fits to the temperature dependence of the superconducting upper critical field $H_{c2}(T)$ indicate that λ usually never reaches a very high value and even near a magnetic quantum critical point ($1 < \lambda < 2$) [34]. Thus, a large part of the effective mass enhancement comes from the renormalization of the band mass m_B.

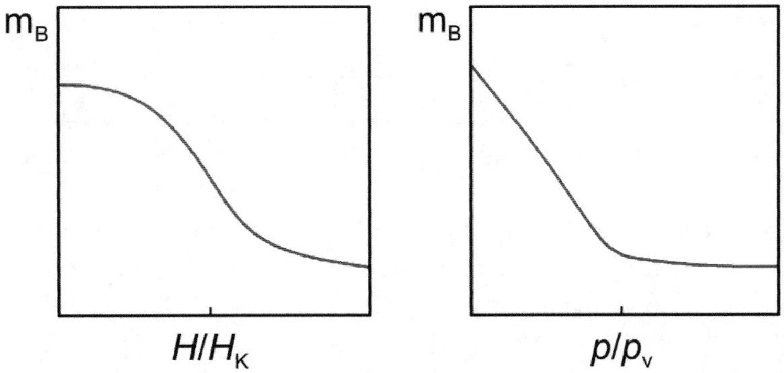

Figure 6. Schematic representations of the field and pressure variation of the regular renormalized band effective mass.

Figure 6 mimics the H and P variation of the regular part (m_B) in a scheme where the renormalized band mass m_B will be mainly given by local Kondo fluctuations. Thus m_B will depend on the strength of the Zeeman energy in comparison to the Kondo energy leading to a crossover field $H_K \sim k_B T_K / g\mu_B$ and on the proximity to the valence transition (P_V). Magnetic fields can induce rapidly a strong magnetic polarization (often 10 % in a field of 10 T) that can decouple spin up and spin down sub-bands with different Fermi wave vector ($k_{F\uparrow}$, $k_{F\downarrow}$) and effective masses. In the frame of global criticality, it is assumed that the regular part m_B has a weaker H and P dependence than the singular part m^{**} close to the quantum singularity.

Figure 7 represents the qualitative P and H dependence of the dynamical susceptibility of an AF system at its hot spot at Q and the corresponding variation of m^{**}_Q of an antiferromagnet which goes from an AF to PM phase through a metamagnetic transition. As shown below for $CeRu_2Si_2$, for $P > P_c$ the metamagnetic field H_c is replaced by pseudo-metamagnetic phenomena at H_M which can be considered as a continuation of the metamagnetic critical end point H^*_c. Furthermore, the induced magnetization leads to the development of FM fluctuations that originate from the AF correlations with enhanced fluctuations at H_c or H_M. Hence an extra contribution of m^* (m^{**}_Q) will appear in magnetic field. It has been recently demonstrated that for AF systems both singular contributions m^{**}_Q and m^{**}_0 are comparable close to P_c [35].

Before discussing in more detail tricriticality phenomena for two selected examples, the $CeRu_2Si_2$ series (AF) and of UGe_2 (FM), we point out that there is no consensus about the possibility of a divergence of m^* in HFC. For the global scenario often referred to as the conventional spin-fluctuation approach [36], there is no divergence of m_Q^{**} at the AF singularity, even for a second order quantum critical point. For FM materials a possible divergence of m_0^{**} at the FM quantum critical point is totally wiped out by the first-order nature of the singularity [37]. In the local quantum criticality scenario, a divergence of effective mass is often invoked at its magnetic quantum critical field H_C notably for $YbRh_2Si_2$ [38, 39]. We consider there is no evidence of such a divergence [22]. Effectively a maximum of m^* occurs at H_c with a very sharp decrease of m^* for $H > H_c$. This situation is quite comparable to that of the polarized phase of $CeRu_2Si_2$ which is discussed hereafter and in contrast with a smooth H increase of m^* going from $H = 0$ to $H = H_c$. It is obvious that a new set of experiments is required to settle this point. Possible candidates to study are cage compounds like the recently investigated YbT_2Zn_{20} systems [40, 41]. In these system a γ term as large as $10\,\mathrm{J\,mole^{-1}K^{-2}}$ has been reported with strong magnetic field effects and they open new possibilities to study the formation of the effective mass. Due to the insertion of Yb ions in a cage, one may

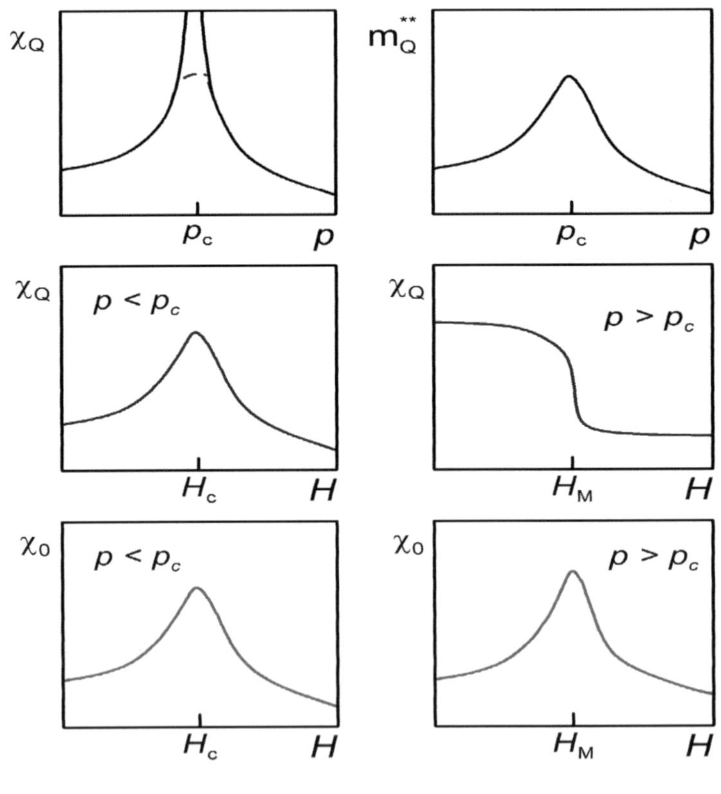

Figure 7. (a) Pressure variation of the AF dynamical susceptibiliy χ_Q and (b) corresponding contribution to the effective mass enhancement m_Q^{**}. (c,d) Magnetic field dependence of the AF dynamical susceptibility χ_Q and (e,f) of the FM dynamical response χ_0 arising from the AF pseudogap on both side of the ctitical pressure p_c.

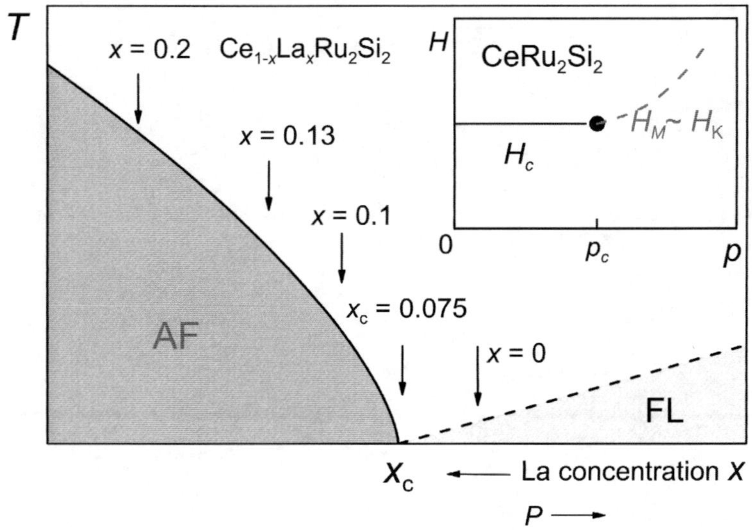

Figure 8. Magnetic phase diagram of $Ce_{1-x}La_xRu_2Si_2$ as function of the La concentration x. The inset shows for $CeRu_2Si_2$ the evolution of the metamagnetic field H_c as function of pressure. H_c ends at the metamagnetic critical field end points H_c^*. Just above the quantum critical point, the metamagnetic phenomena is replaced by a sharp pseudo-metamagnetic crossover at H_M[18].

suspect a very high value of m_B which will furthermore highly vary with the magnetic field.

4. Tricriticality: the AF $CeRu_2Si_2$ series — the FM UGe_2 case

$CeRu_2Si_2$ has been highly studied as it is located close to the AF instability on the PM side ($P_c \sim -0.3$ GPa). It is also one of the few HFC where the Fermi surface is fully determined [42] and thus allows a real proof on the itinerant 4f electron description. Furthermore, extensive inelastic neutron experiments show that the AF correlations develop up to $T_{corr} = 60$ K i. e. at a temperature higher than the single Kondo temperature $T_K \sim 25$ K [43]. Clearly, there is no evidence of an intermediate temperature regime where the effective mass enhancement will be dominated by m_B. This is well demonstrated by the continuous increase of the electronic Grüneisen parameter ($\Omega(T)$) on cooling [44, 45] since if, in a finite T range, the free energy will be controlled by a main parameter T^*, a constant Grüneisen regime must emerge. The large value of the ratio T_{corr}/T_K agrees with the conclusion that in HFC the 4f centers behave as single site Kondo centers only at high temperatures [46].

Expanding the lattice volume of the $CeRu_2Si_2$ by La doping, induces AF order above a critical doping level (x_c) near 7% La (Fig. 8). The Ising character of the localized spin leads to clear first order metamagnetic phenomena at a field $H_c \sim 4$ T. The value of H_c is weakly pressure dependent leading to a critical end point at ~ 4 T for $P = P_c$

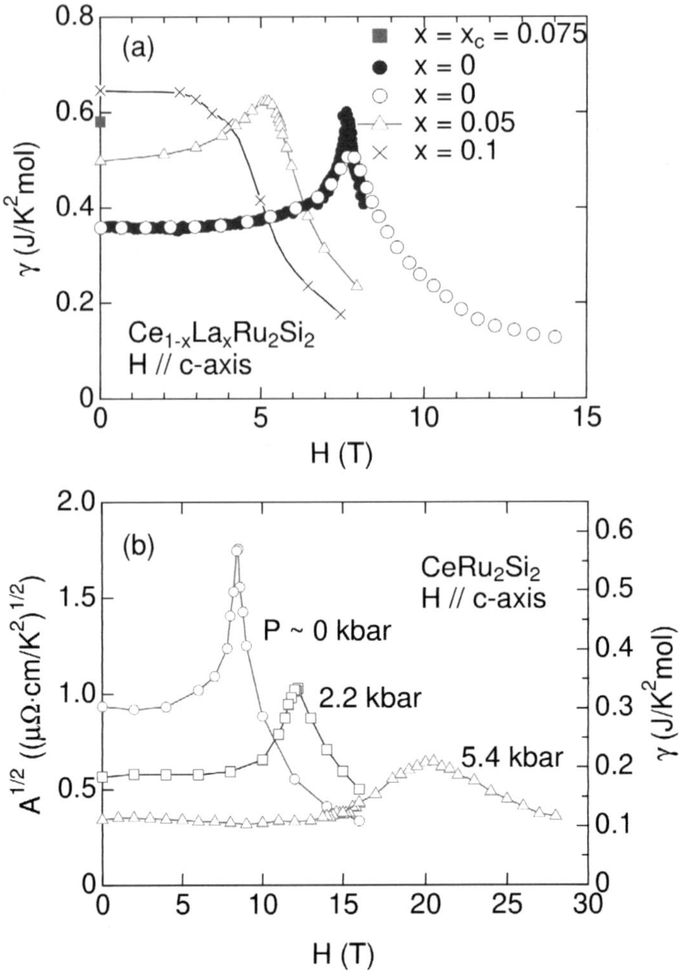

Figure 9. (a) Field variation of the Sommerfeld coefficient of $Ce_{1-x}La_xRu_2Si_2$ for $x = 0$, $x = 0.05$ which is on the AF side ($x < x_c \sim 0.075$), and $x = 0.1$ which are located on the paramagnetic side. At H_c or H_M close to P_c, $\gamma(P_c)$ i. e. the value at x_c is almost the values at H_c or H_M ($\gamma(H_c \,\text{or}\, H_M)$) [51]. Lower panel (b) shows the variation of $\gamma(H)$ at ambient pressure compared to the field dependence of $\sqrt{A}$ under pressure recently deduced from resistivity experiments [50].

or $x = x_c$. Entering in the PM regime above P_c, the metamagnetic transition ($P < P_c$, $H = H_c$) changes to a sharp pseudo-metamagnetic crossover at a field H_M. Furthermore the characteristic field H_c, or H_M, corresponds to a critical value $M_c \sim 0.4\,\mu_B$ of the magnetization. This points out that the increase of the volume of the majority spin band is the source of the metamagnetic and pseudo-metamagnetic phenomena. Microscopic inelastic neutron scattering experiments show [47, 48] that, for $CeRu_2Si_2$, a drastic softening of the FM fluctuation occurs at H_M while the intensity of the AF correlation collapses at H_M with almost a persistence of the same energy for the AF fluctuation

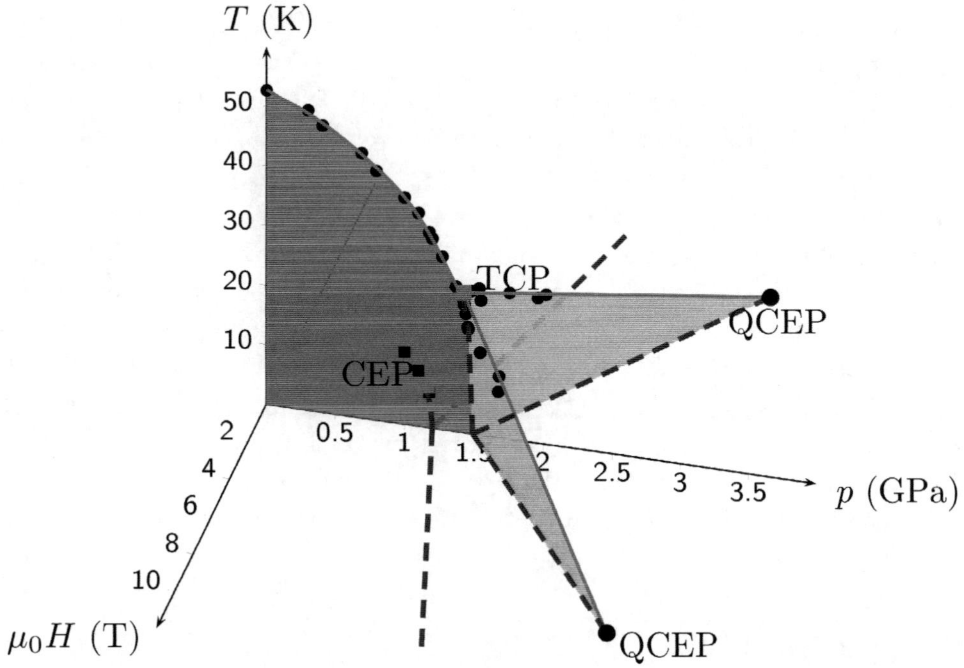

Figure 10. UGe_2 wings with the tricritical pressure P_c at $T = 0\,K$ and the collapse of the first order metamagnetic transition at $T = 0$ for $P_{QCP} > 3\,GPa$ and $H_{QCP} > 16\,T$[54]

itself [49]. It was recently stressed [50, 51] that close to P_c a quasi-convergence between the effective mass m^* (H_c or H_M) at H_c or H_M and the effective mass $m^*(P_c)$ at P_c in zero magnetic field exists: $m^*(P_c) \sim m^*(H_c)$ as illustrated in Fig. 9. For $x = 0.1$ with a Néel temperature $T_N \approx 4\,K$ the Sommerfeld coefficient at $H = 0$ is about $\gamma = 0.65\,J\,mole^{-1}K^{-2}$ and decreases monotonously with increasing field. At the critical point $x_c \sim 0.075$ we find $\gamma \approx 0.6\,J\,mole^{-1}K^{-2}$, quite similar to the value for $CeRu_2Si_2$ at H_M. Escaping further from P_c by applying a pressure on $CeRu_2Si_2$, the γ term at $H = 0$ decreases strongly in good agreement with the large magnitude of the Grüneisen parameter $\Omega(0) \sim +200$. The relative enhancement of $\gamma(H_M)/\gamma(H = 0)$ is at least at low pressure weakly pressure dependent [50, 51]. Under pressure, correlations are still efficient but form a quite smoother pseudogap than the one built at $P = 0$; the FM interaction is created as the spin up and spin down sub-bands decouple when a critical value of the magnetization is reached [18]. Approaching the valence fluctuation regime, near $P = 4\,GPa$, such a picture will certainly fail. It is worthwhile to point out that in the research on quantum criticality in HFC, most of the results are concentrated on the PM side ($P > P_c$ for Ce HFC); in Grenoble attempts are made now to look more carefully to the AF side ($P < P_c$) and thus to make quantitative comparisons with the recent developments on tricriticality.

It is well known that a FM state can be induced by magnetic field at H_c through a first order transition from a PM ground state at a pressure P higher than the critical pressure P_c of the FM–PM instability at $H = 0$ [52]. A nice example is UGe_2 [53] as, at the first order FM critical pressure $P_c \sim 1.49\,\mathrm{GPa}$, the sublattice magnetization jumps from the FM phase to the PM one from $M_0 = 0.9\,\mu_B$ to zero. In zero field, the transition line $T_{\mathrm{Curie}}(P)$ between FM and PM changes from second order to first order at a tricriticality point $T_{\mathrm{TCP}} \sim 24\,\mathrm{K}$, $P_{\mathrm{TCP}} \sim 1.42\,\mathrm{GPa}$ [54, 55]. The domain of the first order transition at $H = 0$ is quite narrow $(P_c - P_{\mathrm{TCP}})/P_{\mathrm{TCP}} \sim 0.05$, but recent resistivity measurements [54] as well as thermal expansions [55] data show that FM wings [56] appear under magnetic field, and a first order transition will occur. They will only collapse at $T = 0\,\mathrm{K}$ for $P > P_{\mathrm{QCEP}} \sim 3\,\mathrm{GPa}$ i. e. $2P_c$ around $H_{\mathrm{QCEP}} \sim 10\,\mathrm{T}$. For UGe_2, a large pressure is needed $(2P_c)$ to reach a PM ground state at any magnetic field (Fig. 10). When the size of the FM sublattice magnetization close to P_c decreases, as for UCoGe ($M_0 \sim 0.07\,\mu_B$) [57], no detection of FM wings has been reported. Thus the large separation between P_{QCEP} and P_c in UGe_2 is clearly linked to the large jump of M_0 at P_c.

5. Evidence of m_B and m^{**} in FM superconductors

Three U based FM-SC have been discovered, namely UGe_2 [58], URhGe [59], and UCoGe [57]. The respective values of M_0 at $P = 0$ are $1.4\,\mu_B$, $0.4\,\mu_B$ and $0.07\,\mu_B$ and their Curie temperatures $T_{\mathrm{Curie}} = 50\,\mathrm{K}$, $9.5\,\mathrm{K}$, and $2.8\,\mathrm{K}$. In UGe_2, SC appears only under pressure with a maximum of the critical SC temperature $T_{\mathrm{SC,max}} \sim 0.7\,\mathrm{K}$ reached close to the pressure P_x where the system switches from the ferromagnetic FM2 phase to another ferromagnetic FM1 phase [53]. The discoveries of SC in URhGe ($T_{\mathrm{SC}} \sim 0.25\,\mathrm{K}$) and in UCoGe ($T_{\mathrm{SC}} \sim 0.7\,\mathrm{K}$) already at $P = 0$ open the possibility of careful measurements even at ambient pressure. UGe_2 and UCoGe have a PM ground state above respectively $P_c \sim 1.49\,\mathrm{GPa}$ [53] and $P_c \sim 1.2\,\mathrm{GPa}$ [60, 61, 62], while in URhGe an initial P increase of T_{Curie} is observed, at least up to $P = 13\,\mathrm{GPa}$ [63]. In contrast to UGe_2, where SC exists only on the FM side, in UCoGe, SC persists also in the PM regime above P_c. SC in URhGe is suppressed above $p \approx 4\,\mathrm{GPa}$ [64]. For the low T_{Curie} materials (URhGe [65] and UCoGe [66]), the measurements of the specific heat (see Fig. 11 for URhGe) show clearly the occurrence of a term linear in temperature above T_{Curie} which can be associated with the formation of the renormalized band mass m_B already at T_{Curie}. Furthermore applying a magnetic field along the easy c axis of magnetization leads to quench the FM fluctuation (thus to the collapse of m_0^{**}) and to restore the sole m_B contribution. By contrast, in the high T_{Curie} ($\sim 50\,\mathrm{K}$) material like UGe_2 [67] there is no mark of renormalized band mass contribution to the specific heat above T_{Curie} at $P = 0$ as its characteristic temperature T_B may be quite similar to T_{Curie}. The "hidden" character of m_B is quite similar to the previous discussion made for $CeRu_2Si_2$.

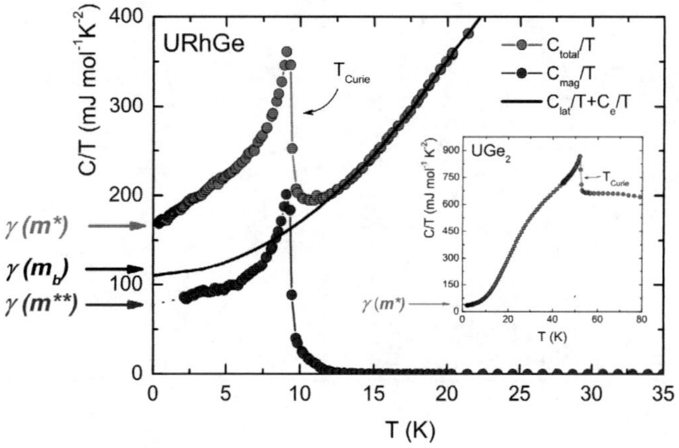

Figure 11. Evidence of the m_B strength at ambient pressure for URhGe from the temperature dependence of C/T. The inset shows $C/T(T)$ for UGe_2.

6. Link between m^{**} and T_{SC} in the ferromagnetic superconductor URhGe

URhGe is a very nice example to illustrate the link between the strength of m^{**} and its feedback on SC. A remarkable discovery was that the application of a magnetic field along the b hard-axis induces the re-orientation of the magnetization from the easy c axis to the b axis at a field $H_R \sim 12\,$T and this re-orientation is directly associated with a H re-entrance of SC (Fig. 12) [68]. To verify that the driving phenomena is an enhancement of m^{**}, careful resistivity experiments as a function of pressure and magnetic field have been realized [64, 69]. Furthermore, this coupling was verified by detailed magnetization (M) studies. Figure 13 compares the field enhancement of $\Delta m^*/m^*$ detected by resistivity (ρ) and magnetization [70]. Taking into account that magnetization measurements are limited down to $T = 1.5$ K while resistivity has been realized down to $70\,mK$, the agreement can be considered excellent. Assuming that the H enhancement of m^* is that of m_0^{**}, via the McMillan-type formula $T_{SC}(m^*) = T_0 \exp(-m^*/m^{**})$, the field re-entrance (RSC) of SC as well as its collapse at a pressure $P_{RSC} \sim 2\,$GPa, which is two times smaller than the pressure $P_{LFSC} \sim 4\,$GPa where SC would collapse in a low field, were quantitatively explained. To our knowledge, it is the first case where a link between the mass enhancement and the appearance of SC has been shown successfully on a given material between the (P, H) variation of m^{**} and of $T_{SC}(m^{**})$.

An anomalous temperature dependence of the upper critical field $H_{c2}(T)$ has also been reported in UCoGe when the field is applied perpendicular to the easy c axis [71]. The common point with URhGe is that in this Ising type ferromagnet when H is applied along the hard magnetization axis (see [70]) the Curie temperature decreases with H, as proven theoretically [72], which will obviously lead to field mass enhancement. In UCoGe, there is no evidence that the anomalies of H_{c2} for field perpendicular to the c

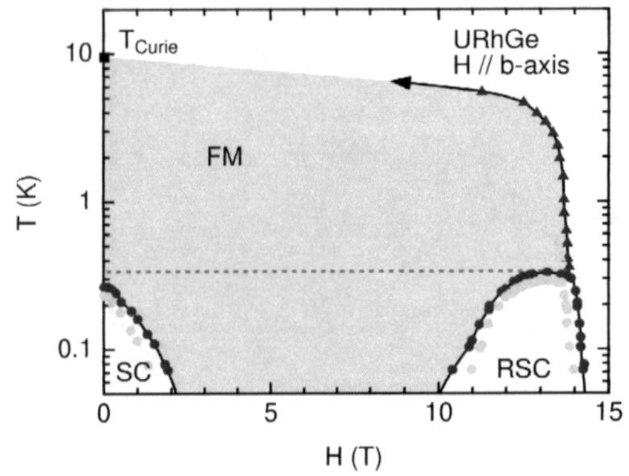

Figure 12. H, T phase diagram of URhGe at ambient pressure with the occurrence of low field SC and H reentrant superconductivity at H_R which is directly associated with the field mass enhancement (see [68, 69]).

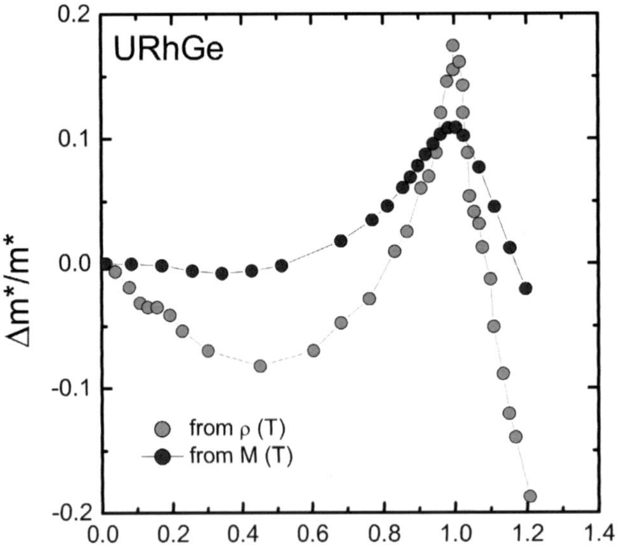

Figure 13. H enhancement of the effective mass as deduced from the A coefficient of the resistivity measurements down to 80 mK and from the temperature dependence of magnetization measurements down 1.5 K using the Maxwell relation [65].

axis is correlated with a spin re-orientation. The key parameter may be the size of the field induced moment for example $\chi_b H$ (χ_b is the susceptibility along the hard b axis) by comparison to M_0. It will drive the spin re-orientation if χ_b is greater than χ_c as it is the case for URhGe. Any induced supplementary component along the c axis via an angular misalignment of the b axis versus the magnetic field direction will wipe out rapidly the phenomena as an extra magnetic component along c leads to the opposite to a decrease of the strength of the spin fluctuations. Recently a theory on field induced Ising spin fluctuations has been developed [73].

7. Interplay of antiferromagnetism and superconductivity: before and after the discovery of 115 compounds

Careful studies on the interplay between AF and SC were quite limited before the appearance of the Ce-115 HFC as the SC dome near P_c occurs at low temperature ($T = 0.6\,\mathrm{K}$ or lower) far below the maximum of the Néel temperature ($T_{\mathrm{N,max}}$) of the AF transition. The discovery of SC in Ce-115 compounds (discussed by T. Park at this conference) opens new opportunities to study in detail the interplay of AF and SC, notably in $CeRhIn_5$, via resistivity, microcalorimetry, and neutron scattering as both transition temperatures are comparable. Figure 14 represents the (T, P, H) phase diagram of $CeRhIn_5$ determined in Grenoble [22, 74].

The main point is that at $H = 0$, above $P = P_c^* \approx 2\,\mathrm{GPa}$ where $T_{\mathrm{SC}} = T_{\mathrm{N}}(P = P_c^*)$, the opening of a large SC gap leads to the rapid suppression of antiferromagnetism at $P = P_c^* + \varepsilon$. However, in absence of SC, AF will survive up to $P_c \sim 2.5\,\mathrm{GPa}$. From macroscopic measurements [22, 74, 75], the coexisting domain of AF + SC is small with the onset of SC at $P_{\mathrm{S}}^+ \sim 1.5\,\mathrm{GPa}$ and a disappearance of AF close to P_c^*. Looking only to the specific heat anomaly at the phase transition from the AF phase to the AF + SC state indicates that the transition is not at all BCS like. NMR experiments give two pieces of evidence [76] that (i) the ground state appear homogeneous i. e. with no phase separation between AF and SC components, (ii) the magnetic structure switches from incommensurate to commensurate.

The striking new phenomena in magnetic field is the lowering of the strength of SC and the creation of vortices. Both phenomena lead to the H re-entrance of AF to occur up to the pressure P_c [22, 74, 75]. Up to now, no microscopic studies have been realized in the high magnetic field AF + SC phase. As indicated on Fig. 14, two clear limits exist: (i) the low pressure $P < P_s^+ = 1.5\,\mathrm{GPa}$ regime, where only AF occurs, and the Fermi surface corresponds to a local behavior of the 4f electron, (ii) the high pressure limit with $P > P_c$, where a pure SC phase exists up to H_{c2} and the Fermi surface implies an itinerant character of the 4f electrons [77]. Between P_s^+ and P_c for $H < H_{c2}$, the FS must evolve in pressure and magnetic field between the AF + SC phase and the pure SC phase.

The $(H, P, T = 0\,\mathrm{K})$ phase diagram of $CeRhIn_5$ is quite analogous to that proposed for the high T_c superconductors (e.g. $YBa_2Cu_3O_{7-x}$) with the change from spin density wave insulator to SC + spin density wave with small Fermi surface pockets and then ending to a pure SC phase with a "normal" large Fermi surface [78] as a function of the hole concentration. Thanks to the recent discoveries of new SC materials such as the pnictides, e.g.(see [79]) there are now many examples of an interplay between AF and SC with the occurrence of either a homogeneous phase or phase separation. Precise experiments can be realized in the future to reach a clearer view.

$CeCoIn_5$ corresponds to the case of a pure SC phase at ambient pressure. A consequence of the previous remarks that high magnetic polarization can be reached plus the weakness of the Pauli limit of $H_{c2}(0)$ with respect to the orbital one pushes the interpretation of the appearance of a new high magnetic field low temperature phase (HFLT) to be a FFLO phase [80, 81] as predicted four decades ago by Fulde-Ferrel [82] and Larkin-Ovchinnikov [83]. However recent NMR [84, 85] and neutron scattering experiments [86, 87] for $H \parallel a$ prove clearly that the HFLT phase has a long range magnetic component and the wave vector does not change with the direction of the magnetic field. Here, we will not discuss in detail the physics of the HFLT, but recently we have realized a new set of magnetization experiments with $H \parallel c$ in order to clarify whether exactly at $H_{c2}(0)$ a quantum singularity exists since the $H_{c2}(0)$ point is not associated with a magnetic quantum criticality [88]. Our magnetization ($M(H)$) measurements performed down to 80 mK are mainly an improvement by a fine zoom in the $(H - H_{c2})/H_{c2}$ window of previous published data [89]. Figure 15 shows at constant field value, the temperature variation of $M(H,T)$. The spectacular result is that, below $H_{c2}(0)$, $M(T)$ approaches a T^2 dependence with a T decrease of $M(H)$ on cooling while above $H_{c2}(0)$, $M(H,T)$ does not obey a T^2 Fermi liquid law but a quasi linear T dependence with an increase of $M(H)$ on cooling. Using Maxwell relations it can be clearly shown that coming from the SC phase a sharp maxima of the γ term of the specific heat will occur at $H_{c2}(0)$. The extrapolation of the suspected divergence

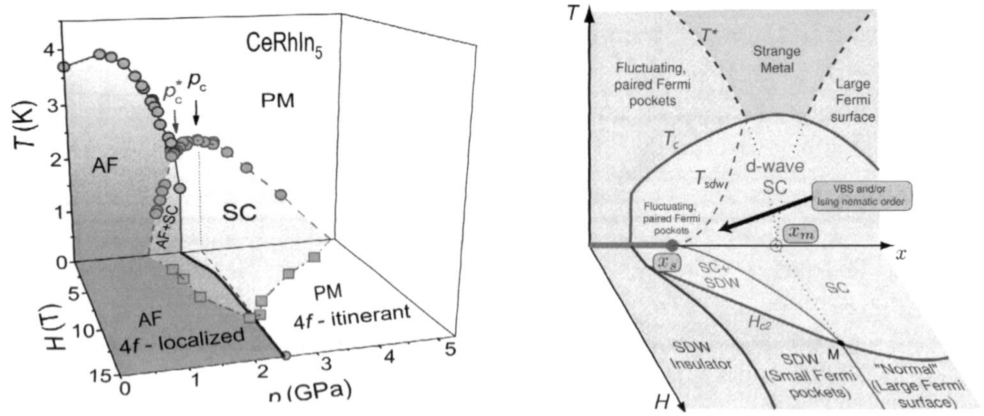

Figure 14. (H,T,P) phase diagram of $CeRhIn_5$ compared to the one proposed for the high T_c superconductor [77, 78].

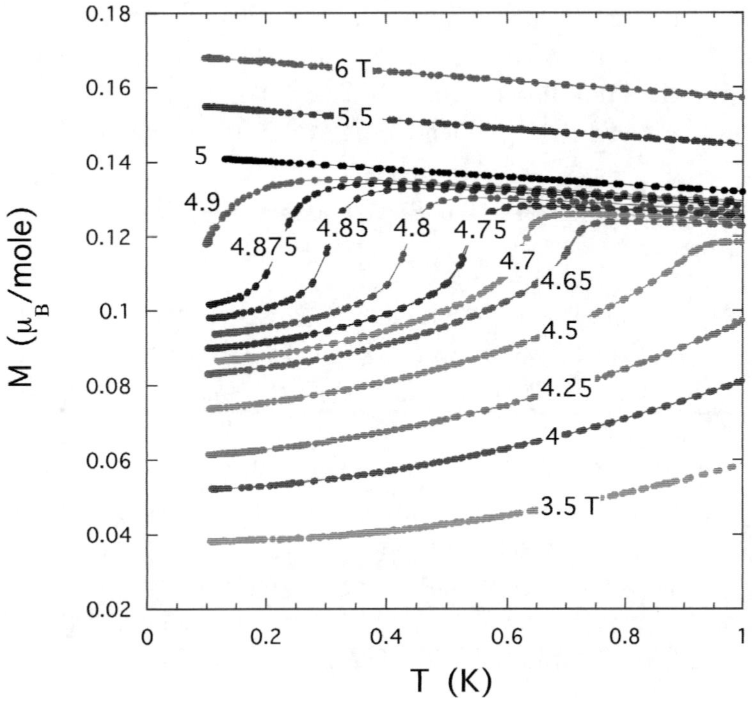

Figure 15. Magnetization as function of temperature at constant field $H \parallel c$ of $CeCoIn_5$ showing clearly the singular magnetic behaviour right at H_{c2}.

of the effective mass from the field dependence of the A coefficient of the resistivity measured at fields above $H_{c2}(0)$ at $H_{QCP} \sim 4.3$ T [90] is not marked by any change of the field dependence of the $\gamma(H)$ at H_{QCP}. Experimentally it should be possible to push down the magnetization measurements below 20 mK where a Fermi liquid temperature dependence of M may be obeyed on both sides of $H_{c2}(0)$ and to establish if at $H_{c2}(0)$ the field dependence of γ has no discontinuity linked to the first order nature of the transition. The unusual behavior, notably the quasi linear T variation of M above $H_{c2}(0)$ at least down to 80 mK is a consequence of the interference between magnetic fluctuations and superconducting correlations as also discussed recently in organic and pnictide superconductors [91]. Thus, there is clear evidence that $H_{c2}(0)$ is a quantum singularity caused by the interplay of normal and superconducting properties. It was proven that AF criticality can be induced by the onset of superconductivity around $H_{c2}(0)$ [92]. The exciting phenomena in $CeCoIn_5$ is that magnetism is glued to SC as if the H weakening of the SC gaps leads to a restoration of a pseudogap structure favorable to AF. Even more the first order magnetization jump is quite similar macroscopically to the metamagnetic phenomena described previously for the $CeRu_2Si_2$ series. It has been proposed that the tendency of field induced AF ordering is linked to strong Pauli depairing (as can be viewed for pseudometamagnetism) and strongly favored by unconventional SC gap node along the AF modulation [93]. The location of a magnetic quantum critical point in the (H, T, P) phase diagram of $CeCoIn_5$ in the absence of SC is still an open question. At least, the absence of H re-entrant AF above $H_{c2}(0)$ for any field direction implies that the magnetic critical field at $P = 0$ will be lower than $H_{c2}(0)$ and already at ambient pressure $CeCoIn_5$ is above $P_c^\star$ [34].

8. The hidden order phase of URu_2Si_2: feedback between FS instability and spin dynamics

Attempts to resolve the nature of the hidden order phase of URu_2Si_2 boosts an intense experimental and theoretical activity over the last decade (see [94]). As several invited contributions (J. C. Davis, P. M. Oppeneer, and H. Harima (see [95, 96, 97]) have treated this problem, we will stress only a few points. One of the paradox of URu_2Si_2 represented in Fig. 16 is that above its ordering temperature $T_0 \sim 17.5\,K$, it looks as a classical intermediate valence compound with e.g. a broad maximum of C/T as observed for $CeSn_3$ [98]. Knowing the entropy involved at low temperature, without any phase transition at T_0, its residual γ term would be around $100\,mJ\,mole^{-1}K^{-2}$ and a smooth continuous decrease of C/T will occur on cooling. However the phase transition at T_0 is associated with a drastic change leading to a marked increase of C/T on cooling as observed for usual HFC [99]. Low energy excitations must be involved as the temperature variation of the Grüneisen coefficient is large on cooling [100].

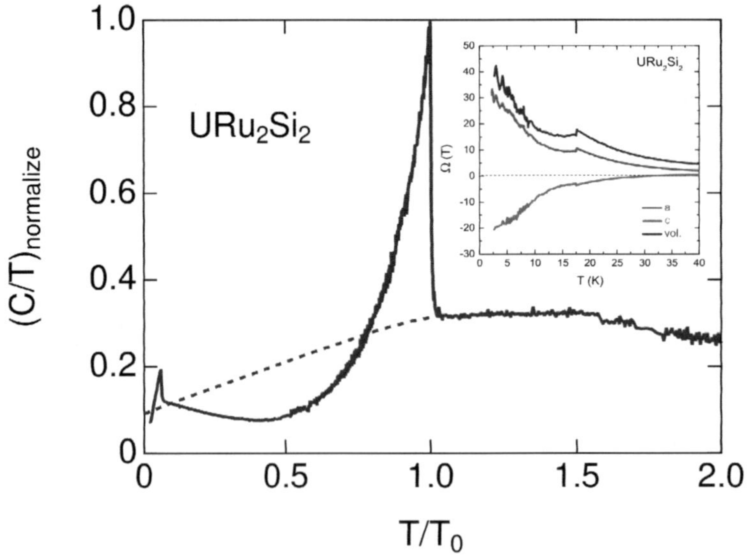

Figure 16. Temperature dependence of normalized specific heat for URu_2Si_2 (full line). The dashed line indicates an extrapolation below T_0 of C/T assuming the survival of PM state. The insert shows the T dependence of the Grüneisen coefficient [98].

It is now well established by transport measurements (Hall effect, Nernst effect, thermoelectric power) (see [94, 98]), by NMR [101], high energy spectroscopy [102]

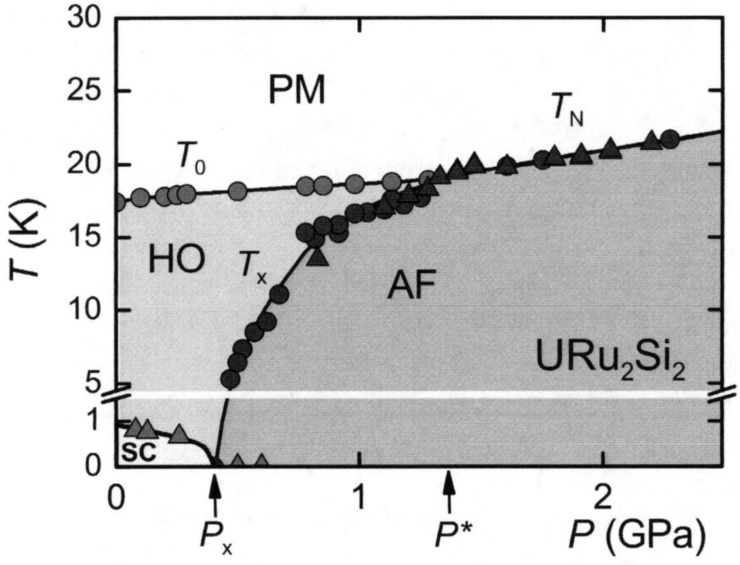

Figure 17. (T, P) phase diagram of URu_2Si_2 with the location of the PM, HO and HF boundaries. SC exists only in the HO phase i. e. below P_x. AF, HO and PM meets at the initial pressure P^* [104].

and recently by two scanning tunneling microscope observations [95, 103] that a Fermi surface reconstruction must occur below T_0 going from PM to HO states. The (P, H, T) phase diagram of URu_2Si_2 (Fig. 17) [104, 105] is now well established despite qualitative differences coming from the high sensitivity of the system to pressure inhomogeneity. The observation of an inelastic excitation at wave vector $Q_0 = (1, 0, 0)$ by inelastic neutron scattering experiments in the HO phase at the same wave vector [106] where AF is detected at high pressure via strong elastic neutron reflection and the strong similarity [107, 108] in the dHvA frequencies detected for both HO and AF ground states are in strong support of the idea that both phases must have the same wave vector $(0, 0, 1)$ implying a doubling of the lattice along c. This conclusion is reinforced by recent low energy ARPES measurements reported in this conference [109]. The idea is that at T_0, a Fermi Surface reconstruction occurs going from body center tetragonal space group #139 above T_0 to a simple tetragonal either #136 or #123 (see reference [97, 96]). The persistence inside the tetragonal symmetry is in full agreement with STM data. The change of tetragonal space group is associated with a gap opening of characteristic energy Δ_G; the drop of carrier density has a direct feedback on the spin dynamics (development of sharp excitation at $Q_0 = (1, 0, 0)$ and $Q_1 = (0.6, 0, 0)$) which may open the possibility that the U centers even in an intermediate valence phase may have their multipolar properties renormalized to the U^{4+} configuration see [96]). Thus quadrupole [97] or hexadecapole [110] are invoked for the OP of URu_2Si_2. It is even proposed that the HO phase may be still a PM state with a lattice switch from space group #139 to #123 [96] with a gap opening in the HO phase generated by the strong dynamical longitudinal Ising type magnetic fluctuation induced along the c axis. An incommensurate spin density wave was proposed in association with the existence of sharp excitations at $Q_1 = (0.6, 0, 0)$ [111], however there is no evidence of any associated detectable OP in neutron scatterings X-ray or STM. Despite many attempts the puzzle of the HO is not resolved. For example, up to now all aims to detect quadrupole or hexadecapole ordering by X-ray methods at synchrotron beam lines have failed. Recently, careful designed uniaxial strain neutron scattering experiments on URu_2Si_2 [112] indicate that HO and AF cannot be mixed when the strain is applied in the basal plane. This result supports strongly the absence of time reversal breaking in HO opposite to the case of AF. Definitively, strong progress has been made but the final solution on the hidden order is not given. Thus still different theoretical proposals are made and experimentalists dream to find the solution.

9. The Future: Material, Instrumentation, Experiments

9.1. Material

The main new trends are often associated with the discovery of unexpected effects on new material. In the last decade, we can quote for HFC:

- the ferromagnetic superconductor,
- the appearance of Ce115 series,
- the SC in compounds with non-centrosymmetric crystal structure (see E. Bauer this conference),
- the criticality of Yb compounds ($YbRh_2Si_2$, $\beta YbAlB_4$) and recently YbT_2Zn_{20} series.

Even on well defined physical problems such as SC of FM HFC it is crucial to find a case at ambient pressure where clean crystals can be grown enabling deep studies on up-up and down-down spin component. For Yb HFC, the priority is to find an example with high critical field singularities (H_c) allowing unambiguously the determination of the Fermi Surface instability.

9.2. Instrumentation

Major progresses have been in experiments in extreme conditions (very low temperature, high pressure, high magnetic fields) which allow large scans for the (H, P, T) determination of phase diagrams. Important breakthroughs have been made in the determination of basic quantities such as specific heat, thermal expansion (via strain gauge, Larmor precession), and NMR. High energy spectroscopy can now be realized at low temperature ($T \sim 4\,\mathrm{K}$) with excellent resolution. STM will certainly be able to give new insights on electronic structures related either with intrinsic properties or extrinsic properties correlated with specific defects. One hope is that fancy superstructures or intrinsic defects maybe discovered associated with the strong pressure and uniaxial strain dependence of materials near quantum singularities. At least, it is clear that all this improvements already play an important role in determining the HO state of URu_2Si_2.

9.3. Physical problems

Precise determination of a (H, P, T) phase diagram is a first key goal as studies on Ce-115 HFC or URu_2Si_2 have shown. For example, in ferromagnetic superconductors such as UGe_2, it is not clear what is the thermodynamic boundary of SC inside the FM phases and notably, if SC can occur homogeneously inside the low pressure large sublattice magnetization phase FM2. To give a definite answer the combination of experiments with different probes are absolutely necessary.

An interesting field is the polarized paramagnetic phase which can be obtained through a metamagnetic or pseudo-metamagnetic transition. However the full experimental determination of the Fermi surface has never been achieved. For example, for $CeRu_2Si_2$ above H_M, a large part of the Fermi surface is still missing. To resolve this enigma is a challenging goal. A new emerging phenomena is the occurrence of 2.5 Lifshiftz instability directly associated with the large magnetic polarization induced by the magnetic field (two cases have been pointed out $CeIn_3$ [113] and URu_2Si_2 [114].

The detection of other exotic phases such as multipolar ordering [115], Pomeranchuk instability, nematic phases, Kondo topological insulator [116] requires the exploration of a large domain. Systematic studies on skutterudite materials have pointed out quite new cases such as $PrFe_4P_{12}$ [117]. In unconventional SC, we have already stressed the interest to precisely determine the properties of the AF + SC phases. For FM-SC, it is quite obvious that spontaneous vortices must occur as the internal field produced by the magnetization surpasses often the first superconducting critical field $H_{c1}(0)$. However, up to now a clear observation of spontaneous vortices have not been achieved [118]. Another interesting question concerns (e.g. UCoGe) the possible occurrence of SC on both side of P_c may lead to a change in the phases near P_c [119]. In most HFC-SC, the origin of SC is considered to be mediated through magnetic, valence fluctuations or mixed effects of both phenomena. In exotic cases like URu_2Si_2 or $PrOs_4Sb_{12}$ [120], SC is associated to the HO phase and to the proximity to a multipolar instability, respectively. To clarify their SC singularities with respect to the previous examples is also a stimulating goal.

To stress that many problems remain to be revisited (see [18]), let us indicate that for example, the classification of $CeAl_3$ as a canonical example of PM Kondo lattice is quite controversial as (i) in single crystals a clear AF anomaly has been observed [121, 122] and (ii) microscopic measurements (as muon spectroscopy presented here [123]) confirm that a long range magnetic component exists at $P = 0$. These points agree with a negative sign of the thermal expansion indicating that quantum singularity must occur at a slightly higher pressure ($P_c \sim 0.2\,\mathrm{GPa}$, $H_c \sim 2\,\mathrm{T}$) than at ambient pressure. For the archetype strong coupling HFC UBe_{13}, there is strong evidence that P and H are carrier "pumps" [18] and thus lead to a continuous change of the zero field SC reference. One can speculate that a magnetic field breaks the cubic symmetry leading to a drastic change of the FS with H. This hypothesis has to be tested by modeling band structure calculations with magnetostriction data. Even for the double transition of UPt_3, the origin of the postulated symmetry breaking field by AF remains unclear as the origin of the tiny ordered moment observed only in neutron scattering experiments is far from being fully understood.

Thus, depending on the level of understanding, the HFC are challenging materials with properties which have a strong impact on other strongly correlated electronic materials going from organic conductors, high T_c SC, recent pnictides, and 3d transition metals.

References

[1] Jayaraman A 1965 *Phys. Rev.* **137** A179–A182
[2] Andres K, Graebner J E and Ott H R 1975 *Phys. Rev. Lett.* **43** 1779
[3] Varma C 2006 *Physica B* **378-380** 17
[4] Taillefer L and Lonzarich G G 1988 *Phys. Rev. Lett.* **60** 1570
[5] Settai R, Takeuchi T and Ōnuki Y 2007 *J. Phys. Soc. Jpn.* **76** 051003
[6] Steglich F, Aarts J, Bredl C D, Lieke W, Meschede D, Franz W and Schäfer H 1979 *Phys. Rev. Lett.* **43** 1892
[7] Ott H R, Rudigier H, Fisk Z and Smith J L 1983 *Phys. Rev. Lett.* **50** 1595
[8] Stewart G R, Fisk Z, Willis J O and Smith J L 1984 *Phys. Rev. Lett.* **52** 679–682
[9] Schlabitz W *et al.* 1986 *Z. Phys. B* **62** 171
[10] Palstra T T M, Menovsky A A, Berg J v d, Dirkmaat A J, Kes P H, Nieuwenhuys G J and Mydosh J A 1985 *Phys. Rev. Lett.* **55** 2727.
[11] Fisher R A *et al.* 1989 *Phys. Rev. Lett.* **62** 1411
[12] Coffey L, Rice T M and Ueda K 1985 *J. Phys. C* **18** L813
[13] Pethick C J and Pines D 1986 *Phys. Rev. Lett.* **57** 118
[14] Ravex A, Flouquet J, Tholence J L, Jaccard D and Meyer A 1987 *J. Magn. Magn. Mater.* **63-64** 400
[15] Jaccard D *et al.* 1987 *J. Appl. Phys.* **26** 517.
[16] Coleman P 2006 *Physica B* **378-380** 1160
[17] Löhneysen v H, Rosch A, Vojta M and Wölfle P 2007 *Rev. Mod. Phys.* **79** 1015
[18] Flouquet J 2005 *Progress in Low Temperature Physics* **15** 139
[19] Miyake K 2007 *J. Phys.: Condens. Matter* **19** 12501
[20] Watanabe S, Tsuruta A, Miyake K and Flouquet J 2009 *J. Phys. Soc. Jpn.* **78** 104706
[21] Flouquet J and Harima H *arXiv:0910.3110*
[22] Knebel G, Aoki D, Braithwaite D, Salce B and Flouquet J 2006 *Phys. Rev. B* **74** 020501
[23] Zevin V, Zwicknagl G and Fulde P 1988 *Phys. Rev. Lett.* **60** 2331
[24] Sarrao J *et al* 1996 *Phys. Rev. B* **54** 12207
[25] Okawa M, Matsunami M, Ishizaka K, Eguchi R, Taguchi M, Chainani A, Takata Y, Yabashi M, Tamasaku K, Nishino Y, Ishikawa T, Kuga K, Horie N, Nakatsuji S and Shin S 2010 *Phys. Rev. Lett.* **104** 247201
[26] Lapierre F *et al.* 1981 *Solid State Commun.* **40** 347
[27] Allen J W, Batlogg B and Wachter P 1979 *Phys. Rev. B* **20** 4807
[28] Okamura H, Kimura S, Shinozaki H, Nanba T, Iga F, Shimizu N and Takabatake T 1998 *Phys. Rev. B* **58** R7496
[29] Barla A *et al.* 2004 *Phys. Rev. Lett.* **92** 066401

[30] Imura K, Matsubayashi K, Suzuki H S, Kabeya N, Deguchi K and Sato N K 2009 *J. Phys. Soc. Jpn.* **78** 104602
[31] Derr J *et al* 2008 *Phys. Rev. B* **77** 193197
[32] Takabatake T, Teshima F, Fujii H, Nishigori S, Suzuki T, Fujita T, Yamaguchi Y, Sakurai J and Jaccard D 1990 *Phys. Rev. B* **41** 9607
[33] Paschen S, Winkler H, Nezu T, Kriegisch M, Hilscher G, Custers J and Prokofiev A 2010 *J. Phys.: Conf. Ser.* **200** 012156
[34] Knebel G, Aoki D, Brison J P and Flouquet J 2008 *J. Phys. Soc. Jpn.* **77** 114704
[35] Misawa T, Yamaji Y and Imada M 2009 *J. Phys. Soc. Jpn.* **78** 084707
[36] Moriya T and Takimoto T 1995 *J. Phys. Soc. Jpn.* **64** 960-
[37] Belitz D, Kirkpatrick T R and Vojta T 2005 *Rev. Mod. Phys.* **77** 579
[38] Gegenwart P, Si Q and Steglich F 2008 *Nature Physics* **4** 186
[39] Friedemann S, Westerkamp T, Brando M, Oeschler N, Wirth S, Gegenwart P, Krellner C, Geibel C and Steglich F 2009 *Nature Physics* **5** 465
[40] Jia S, Bud'ko S L, Samolyuk G D and Canfield P C 2007 *Nature Physics* **3** 334
[41] Honda F, Yasui S, Yoshiuchi S, Takeuchi T, Settai R and Ōnuki Y 2010 *J. Phys. Soc. Jpn.* **79** 083709
[42] Suzuki M T and Harima H 2010 *J. Phys. Soc. Jpn.* **79** 024705
[43] Rossat-Mignot, Regnault L P, Jacoud J L, Vettier C, Lejay P, Flouquet J, Walker E, Jaccard D and Amato A 1988 *J. Magn. Magn. Mater.* **76-77** 376
[44] Lacerda A, de Visser A, Puech L, Lejay P, Haen P, Flouquet J, Voiron J and Okhawa F J 1989 *Phys. Rev. B* **40** 11429
[45] Weickert F, Brando M, Steglich F, Gegenwart P and Garst M 2010 *Phys. Rev. B* **81** 134438
[46] feng Yang Y, Curro N J, Fisk Z, Pines D and Thompson J D *arXiv:1005.5184*
[47] Flouquet J *et al* 2004 *J. Magn. Magn. Mater.* **272-276** 27
[48] Sato M, Koike Y, Katano S, Metoki N, Kadowaki H and Kawarazaki S 2001 *J .Phys. Soc. Jpn.* **Suppl. A 70** 118
[49] Raymond S *et al.* 1998 *J. Phys.: Condens. Matter* **10** 2363
[50] Matsuda T D *et al.* to be published
[51] Flouquet J to be published in J. Low Temp. Phys.
[52] Shimizu M 1964 *Proc. Phys. Soc.* **84** 397
[53] Huxley A, Sheikin I, Ressouche E, Kernavanois N, Braithwaite D, Calemczuk R and Flouquet J 2001 *Phys. Rev. B* **63** 144519
[54] Taufour V *et al.* to be published
[55] Kabeya N, Iijima R, Osaki E, Ban S, Imura K, Deguchi K, Aso N, Homma Y, Shiokawa Y and Sato N K 2010 *J. Phys.: Conf. Ser.* **200** 032028
[56] Belitz D, Kirkpatrick T R and Vojta T 1999 *Phys. Rev. Lett.* **82** 4707–4710
[57] Huy N T, Gasparini A, de Nijs D E, Huang Y, Klaasse J C P, Gortenmulder T, de Visser A, Hamann A, Görlach T and v Löhneysen H 2007 *Phys. Rev. Lett.* **99** 067006
[58] Saxena S S, Agarwal P, Ahilan K, Grosche F M, Haselwimmer R K W, Steiner M J, Pugh E, Walker I R, Julian S R, Monthoux P, Lonzarich G G, Huxley A, Sheikin I, Braithwaite D and Flouquet J 2000 *Nature* **406** 587
[59] Aoki D, Huxley A, Ressouche E, Braithwaite D, Flouquet J, Brison J P, Lhotel E and Paulsen C 2001 *Nature* **413** 613
[60] Hassinger E, Aoki D, Knebel G and Flouquet J 2008 *J. Phys. Soc. Jpn.* **77** 073703
[61] Hassinger E, Aoki D, Knebel G and Flouquet J 2010 *J. Phys.: Conf. Ser.* **200** 012055
[62] Slooten E, Naka T, Gasparini A, Huang Y K and de Visser A 2009 *Phys. Rev. Lett.* **103** 097003
[63] Hardy F, Huxley A, Flouquet J, Salce B, Knebel G, Braithwaite D, Aoki D, Uhlarz M and Pfleiderer C 2005 *Physica B* **359** 1111
[64] Miyake A, Aoki D and Flouquet J 2009 *J. Phys. Soc. Jpn.* **78** 063703
[65] Hardy F *et al.* to be published

[66] Aoki D *et al.* to be published
[67] Hardy F, Meingast C, Taufour V, Flouquet J, v Löhneysen H, Fisher R A, Phillips N E, Huxley A and Lashley J C 2009 *Phys. Rev. B* **80** 174521
[68] Lévy F, Sheikin I, Grenier B and Huxley A D 2005 *Science* **309** 1343
[69] Miyake A, Aoki D and Flouquet J 2008 *J. Phys. Soc. Jpn.* **77** 094709
[70] Hardy F *et al.* to be published
[71] Aoki D, Matsuda T D, Taufour V, Hassinger E, Knebel G and Flouquet J 2009 *J. Phys. Soc. Jpn.* **78** 113709
[72] Mineev V P 2010 *Phys. Rev. B* **81** 180504
[73] Tada Y and Fujimoto S to be published
[74] Knebel G *et al.* 2004 *J. Phys.: Condens. Matter* **16** 8905
[75] Park T, Ronning F, Yuan H Q, Salamon M B, Movshovich R, Sarrao J L and Thompson J D 2006 *Nature* **440** 65
[76] Yashima M, Kawasaki S, Mukuda H, Kitaoka Y, Shishido H, Settai R and Ōnuki Y 2007 *Phys. Rev. B* **76** 020509
[77] Knebel G, Aoki D and Flouquet J *arXiv:0911.5223*
[78] Sachdev S *arXiv:0910.0846*
[79] Ishida K, Nakai Y and Hosono H 2009 *J. Phys. Soc. Jpn.* **78** 062001
[80] Bianchi A, Movshovich R, Oeschler N, Gegenwart P, Steglich F, Thompson J D, Pagliuso P G and Sarrao J L 2002 *Phys. Rev. Lett.* **89** 137002
[81] Matsuda Y and Shimahara H 2007 *J. Phys. Soc. Jpn.* **76** 051005
[82] Fulde P and Ferrell R A 1964 *Phys. Rev.* **135** A550
[83] Larkin A I and Ovchinnikov Y N 1965 *Sov. Phys. JETP* **20** 762
[84] Young B L, Urbano R R, Curro N J, Thompson J D, Sarrao J L, Vorontsov A B and Graf M J 2007 *Phys. Rev. Lett.* **98** 036402
[85] Koutroulakis G, Stewart M D, Mitrović V F, Horvatić M, Berthier C, Lapertot G and Flouquet J 2010 *Phys. Rev. Lett.* **104** 087001
[86] Kenzelmann M, Strässle T, Niedermayer C, Sigrist M, Padmanabhan B, Zolliker M, Bianchi A D, Movshovich R, Bauer E D, Sarrao J L and Thompson J D 2008 *Science* **321** 1652
[87] Kenzelmann M, Gerber S, Egetenmeyer N, Gavilano J L, Strässle Th, Bianchi A D, Ressouche E, Movshovich R, Bauer E D, Sarrao J L and Thompson J D 2010 *Phys. Rev. Lett.* **104** 127001
[88] Ronning F, Capan C, Bauer E D, Thompson J D, Sarrao J L and Movshovich R 2006 *Phys. Rev. B* **73** 064519
[89] Tayama T, Harita A, Sakakibara T, Haga Y, Shishido H, Settai R and Ōnuki Y 2002 *Phys. Rev. B* **65** 180504
[90] Howald L, Seyfarth G, Knebel G, Lapertot G, Aoki D and Brison J-P 2010 arxiv 1009.4894
[91] Doiron-Leyraud N, Auban-Senzier P, de Cotret S R, Bourbonnais C, Jérome D, Bechgaard K and Taillefer L 2010 *Phys. Rev. B* **80** 214531
[92] Fujimoto Y, Tsuruta A and Miyake K 2008 *J. Phys. Cond. Matter* **20** 325226
[93] Ikeda R, Hatayema Y and Aoyama K 2010 *Phys. Rev. B* **82** 060510
[94] Amitsuka H, Matsuda K, Kawasaki I, Tenya K, Yokoyama M, Sekine C, Tateiwa N, Kobayashi T C, Kawarazaki S and Yoshizawa H 2007 *J. Magn. Magn. Mater.* **310** 214
[95] Schmidt A R, Hamidian M H, Wahl P, Meier F, Balatsky A V, Garrett J D, Williams T J, Luke G M and Davis J C 2010 *Nature* **465** 570
[96] Elgazzar S, Rusz J, Amft M, Oppeneer P M and Mydosh J A 2009 *Nature Materials* **8** 337
[97] Harima H, Miyake K and Flouquet J 2010 *J. Phys. Soc. Jpn.* **79** 033705
[98] Hassinger E, Derr J, Levallois J, Aoki D, Behnia K, Bourdarot F, Knebel G, Proust C and Flouquet J 2008 *J. Phys. Soc. Jpn. Suppl. A* **77** 172
[99] Fisher R A *et al.* 1990 *Physica B* **163** 419
[100] Hardy F *et al.* to be published
[101] Matsuda K, Kohori Y, Kohara T, Kuwahara K and Amitsuka H 2001 *Phys. Rev. Lett.* **87** 087203

[102] Santander-Syro A F, Klein M, Boariu F L, Nuber A, Lejay P and Reinert F 2009 *Nature Physics* **5** 637
[103] Aynajian P, da Silva Neto E H, Parker C V, Huang Y, Pasupathy A, Mydosh J and Yazdani A 2010 *PNAS* **107** 10383
[104] Hassinger E, Knebel G, Izawa K, Lejay P, Salce B and Flouquet J 2008 *Phys. Rev. B* **77** 115117
[105] Niklowitz P G, Pfleiderer C, Keller T, Vojta M, Huang Y K and Mydosh J A 2010 *Phys. Rev. Lett.* **104** 106406
[106] Villaume A, Bourdarot F, Hassinger E, Raymond S, Taufour V, Aoki D and Flouquet J 2008 *Phys. Rev. B* **78** 012504
[107] Nakashima M, Ohkuni H, Inada Y, Settai R, Haga Y, Yamamoto E and Ōnuki Y 2003 *J. Phys.: Condens. Matter* **15** S2011
[108] Hassinger E *et al.* to be published
[109] Yoshida R *et al.* to be published and this conference
[110] Haule K and Kotliar G 2009 *Nature Physics* **5** 796
[111] Wiebe C R, Janik J A, Macdougall G J, Luke G M, Garrett J D, Zhou H D, Jo Y J, Balicas L, Qiu Y, Copley J R D, Yamani Z and Buyers W J L 2007 *Nature Physics* **3** 96
[112] Bourdarot F *et al.* to be published
[113] Gor'kov L P and Grigoriev P D 2006 *Phys. Rev. B* **73** 060401
[114] Shishido H, Hashimoto K, Shibauchi T, Sasaki T, Oizumi H, Kobayashi N, Takamasu T, Takehana K, Imanaka Y, Matsuda T D, Haga Y, Onuki Y and Matsuda Y 2009 *Phys. Rev. Lett.* **102** 156403
[115] Kuramoto Y, Kusunose H and Kiss A 2009 *J. Phys. Soc. Jpn.* **78** 072001
[116] Dzero M, Sun K, Galitski V and Coleman P 2010 *Phys. Rev. Lett.* **104** 106408
[117] **76** 043705
[118] Ohta T, Hattori T, Ishida K, Nakai Y, Osaki E, Deguchi K, Sato N K and Satoh I 2010 *J. Phys. Soc. Jpn.* **79** 023707
[119] Mineev V P *arXiv:0812.2171*
[120] Bauer E D, Frederick N A, Ho P C, Zapf V S and Maple M B 2002 *Phys. Rev. B* **65** 100506
[121] Jaccard D, Cibin R, Bezinge A, Sierro J, Matho K, Flouquet J 1988 *J. Magn. Magn. Mater.* **76** 255
[122] Lapertot G, Calemczuk R, Marcenat C, Henry J Y, Boucherle J X, Flouquet J, Hammann J, Cibin R, Cors J, Jaccard D, Sierro J 1993 *Physica B* **188** 454
[123] Gavilano G *et al.* to be published and this conference

Journal of Physics: Conference Series **273** (2011) 012002 doi:10.1088/1742-6596/273/1/012002

Quadrupole Ordering and Rattling in Clathrate Compound $Pr_3Pd_{20}Ge_6$

G Ano[1], K Matsuo[1], K Araki[1], Y Tachikawa[1], K Mitsumoto[1], M Akatsu[1], Y Nemoto[1], T Goto[1], N Takeda[2], A Kikkawa[3], A Dönni[3] and H Kitazawa[3]

[1]Graduate School of Science and Technology, Niigata University, Niigata 950-2181, Japan
[2]Faculty of Engineering, Niigata University, Niigata 950-2181, Japan
[3]National Institute for Materials Science, Tsukuba 305-0044, Japan

E-mail: ano@phys.sc.niigata-u.ac.jp

Abstract. We have studied quadrupole ordering and rattling in clathrate compound $Pr_3Pd_{20}Ge_6$ using ultrasonic measurements. The softening were observed in temperature dependence of elastic constants C_{11}, $(C_{11} - C_{12})/2$ and C_{44}. The C_{11} and $(C_{11} - C_{12})/2$ show minimums at 250 mK indicating a transition point to an antiferroquadrupole ordered phase due to a Γ_3 doublet ground state at 8c site. Furthermore, the C_{11}, $(C_{11} - C_{12})/2$ and C_{44} exhibit discontinuitive anomalies at 60 mK revealing a ferroquadrupole ordering due to a Γ_5 triplet ground state at 4a site. Two distinctive ultrasonic dispersions caused by thermally activated Γ_5-type rattling were observed in C_{44} around 15 K and 25 K.

1. Introduction

The clathrate compounds $R_3Pd_{20}X_6$ (R = rare-earth, X = Si or Ge) have been intensively investigated because of various interesting properties consisting of quadrupole ordering caused by orbital degrees of freedom of 4f electrons and rattling motions of a guest rare-earth atom in an oversized cage. $R_3Pd_{20}X_6$ has a cubic $Cr_{23}C_6$-type structure with a space group Fm$\bar{3}$m. There are 116 atoms consisting of 12 R atoms, 80 Pd atoms and 24 X atoms in a conventional unit cell. The R atoms occupy two different crystallographic sites of 4a site with O_h symmetry and 8c site with T_d symmetry [1]. The Ce-based compound $Ce_3Pd_{20}Ge_6$ with Γ_8 ground states at 4a and 8c sites shows a O_u-type ferroquadrupole (FQ) ordering accompanied by a spontaneous tetragonal distortion at $T_Q = 1.25$ K and an antiferromagnetic ordering at $T_N = 0.75$ K [2, 3]. The elastic constant C_{44} of $Ce_3Pd_{20}Ge_6$ shows a frequency dependent shoulder-like anomaly around 10 K accompanying considerable ultrasonic attenuation, which is caused by the thermally activated rattling motion of the guest atoms of the 4a site [3, 4]. Similarly, we found ultrasonic dispersion in the series of $R_3Pd_{20}Ge_6$ (R = La, Pr, Nd) [4, 5]. In the isomorphic compounds of $Ce_3Pd_{20}Si_6$ and $Pr_3Pd_{20}Si_6$, however, no elastic anomaly of C_{44} associated with the rattling was observed. These compounds show antiferroquadrupole (AFQ) ordering with appreciable magnetic anisotropy at very low temperatures [6, 7].

Inelastic neutron scatterings and magnetization experiments on $Pr_3Pd_{20}Ge_6$ revealed that the ground state of the Pr^{3+} ion at the 8c site is a Γ_3 doublet and the one at 4a site is Γ_5 triplet in the crystalline electric fields (CEF) [8]. Magnetic susceptibility of $Pr_3Pd_{20}Ge_6$ has no anomaly

Published under licence by IOP Publishing Ltd

down to 80 mK. Therefore, $Pr_3Pd_{20}Ge_6$ is expected to undergo quadrupole ordering free from magnetic moments at low temperatures. In this paper, we present the study of quadrupole ordering and rattling in $Pr_3Pd_{20}Ge_6$ by using ultrasonic method down to 20 mK.

2. Experiment

We prepared single crystal of $Pr_3Pd_{20}Ge_6$ grown by a floating zone method. The specimen for the ultrasonic measurements was cut to $3.64\times3.05\times2.80$ mm^3 in dimension. Ultrasonic waves were generated and detected by piezoelectric $LiNbO_3$ transducers bonded on parallel surfaces of the sample. The change of ultrasound velocity v was measured by using a phase difference method and the attenuation of the ultrasonic echo signals were measured by a digital storage oscilloscope. The elastic constant $C = \rho v^2$ was determined by the ultrasound velocity v and the mass density $\rho = 10.29$ g/cm^3 for $Pr_3Pd_{20}Ge_6$ with a lattice parameter $a = 1.2445$ nm [9]. The low-temperature measurements were carried out by a homemade ^{3}He refrigerator down to 0.45 K and a top-loading type dilution refrigerator equipped with a superconducting magnet down to 20 mK in magnetic fields up to 14 T.

3. Results and Discussion

Figure 1 shows the temperature dependence of the elastic constants C_{11}, $(C_{11} - C_{12})/2$ and C_{44} of $Pr_3Pd_{20}Ge_6$ at zero magnetic field. All of the elastic constants exhibit monotonous increase with decreasing temperature from 170 K down to 50 K. Sizable elastic softening were observed below about 10 K. The softening is described in term of quadrupole susceptibility dominated by the Curie term proportional to reciprocal temperature.

The C_{11} and $(C_{11} - C_{12})/2$ show minimums at 290 mK and discontinuitive increases at $T_{Q1} = 250$ mK with decreasing temperature and further anomalies at $T_{Q2} = 60$ mK. The C_{44} shows the upturn at $T_{Q2} = 60$ mK, while it shows little anomaly at T_{Q1}. Another notable feature is that the ultrasonic echo signals of all elastic modes are considerably attenuated below 0.1 K, not shown here. It is important to insist that the softening of C_{44} below T_{Q1} indicates quadrupole degeneracy remains for the Γ_5 triplet ground state at the 4a site down to T_{Q2}. Consequently, we suggest that the doublet Γ_3 at 8c site orders first at T_{Q1}, and then the triplet Γ_5 orders at the 4a site at T_{Q2}.

The elastic softening of $(C_{11} - C_{12})/2$ and C_{44} in $Pr_3Pd_{20}Ge_6$ are described in terms of quadrupole susceptibility of 4f electrons. The solid lines for $(C_{11} - C_{12})/2$ and C_{44} in Fig. 1 are calculations of quadrupole susceptibility based on the CEF level schemes at 4a and 8c sites, which were obtained from inelastic neutron scatterings and magnetization experiments [8]. The fits well reproduce the experimental results of the softening in $(C_{11} - C_{12})/2$ and C_{44} below 50 K down to 0.8 K. From this analysis, we determined quadrupole-strain coupling constant g_Γ and the intersite quadrupole coupling constant g'_Γ to be $|g_{\Gamma_3[4a]}| = 30$ K, $|g_{\Gamma_3[8c]}| = 90$ K, $g'_{\Gamma_3[4a]} = 0$ K, $g'_{\Gamma_3[8c]} = -36$ mK for the $(C_{11} - C_{12})/2$ and $|g_{\Gamma_5[4a]}| = 31$ K, $|g_{\Gamma_5[8c]}| = 13$ K, $g'_{\Gamma_5[4a]} = 13$ mK, $g'_{\Gamma_5[8c]} = -30$ mK for the C_{44}.

In order to elucidate the low-temperature magnetic phase diagram in $Pr_3Pd_{20}Ge_6$, we have carried out ultrasonic measurements of $(C_{11} - C_{12})/2$ down to 20 mK under magnetic fields along the [110] direction. As shown in Fig. 2, we determined the $H - T$ phase diagram of $Pr_3Pd_{20}Ge_6$. We specify the high temperature phase I to be a paramagnetic phase, the ordered phase II appearing below T_{Q1} and the ordered phase III below T_{Q2}. The I-II phase transition point T_{Q1} increases with increasing magnetic fields up to 0.5 T. Above 0.5 T, T_{Q1} decreases as the field increases and crosses about 1.5 T at absolute zero. The phase II accompanied by the lifting of Γ_3 doublet ground state at the 8c site is expected to be quadrupole ordering because magnetic susceptibility has a little peak above 80 mK [8]. The magnetic fields stabilized phase II in $Pr_3Pd_{20}Ge_6$, which resembles the AFQ phase of CeB_6. Furthermore, the coupling constant

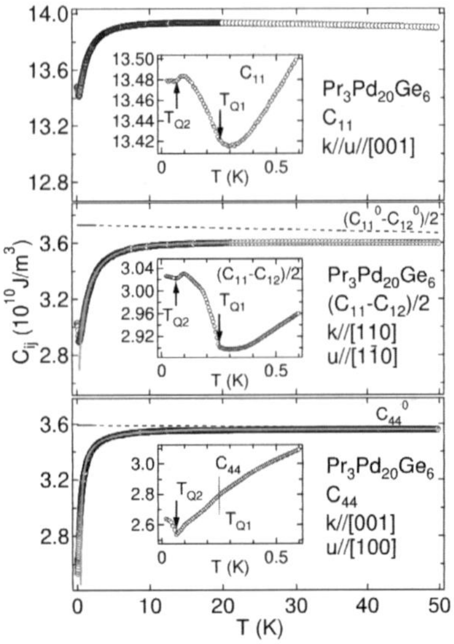

Figure 1. Temperature dependence of the elastic constants C_{11}, $(C_{11} - C_{12})/2$ and C_{44} of $Pr_3Pd_{20}Ge_6$ at zero magnetic field. The insets show the expanded view of C_{11}, $(C_{11} - C_{12})/2$ and C_{44} below 0.6 K.

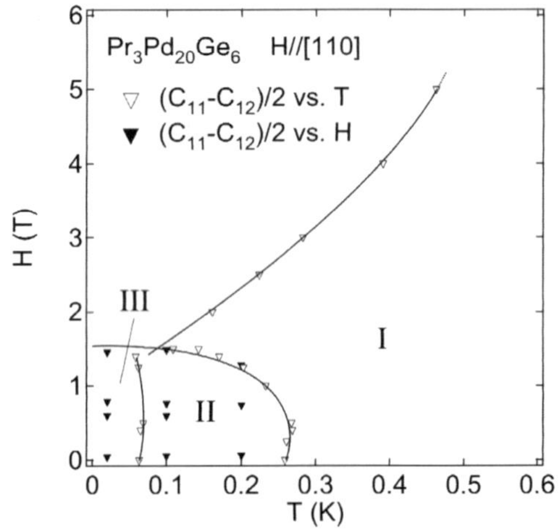

Figure 2. $H - T$ phase diagram of $Pr_3Pd_{20}Ge_6$ determined by low-temperature ultrasonic measurements of $(C_{11} - C_{12})/2$ under magnetic fields along [110] direction.

of $g'_{\Gamma_3[8c]}$ in $Pr_3Pd_{20}Ge_6$ possesses negative sign, which indicates the antiferro-type inter-site quadrupole interaction among 8c sites. These results suggest that the phase II is AFQ ordering at the 8c site. The II-III phase transition point T_{Q2} shows little shift with increasing fields up to 1.4 T. Moreover, T_{Q2} shifts to higher temperatures as the field increases from 1.5 T to 5 T. The phase III is probably FQ ordering state at the 4a site because the coupling constant of $g'_{\Gamma_5[4a]}$ possesses positive sign, which indicates the ferro-type intersite quadrupole interaction among 4a sites.

Figures 3(a) and (b) show frequency dependence of the elastic constant C_{44} and the attenuation coefficient α_{44} in $Pr_3Pd_{20}Ge_6$. We performed ultrasonic experiments with frequencies from 15 to 151 MHz. It should be noted that we newly discover two distinctive shoulder-like anomalies in C_{44} accompanying considerable ultrasonic attenuation around 15 K and 25 K. These elastic anomalies and the attenuation peaks shift to higher temperatures with increasing frequencies, which obeys thermally activated rattling processes. The C_{11} and $(C_{11} - C_{12})/2$ mode in relation to Γ_3-type quadrupole fluctuation, however, exhibits no sign of ultrasonic attenuation. The results of these ultrasonic dispersions support the thermal rattling of Pr^{3+} ion in a cage having Γ_5-type quadrupole fluctuation as similar as the ultrasonic dispersions in C_{44} for $Ce_3Pd_{20}Ge_6$ [3]. These ultrasonic dispersions are well fitted by the Debye-type formula of elastic constant $C(\omega) = C_\infty - \frac{C_\infty - C_0}{1+\omega^2\tau^2}$ and attenuation coefficient $\alpha(\omega) = \frac{C_\infty - C_0}{2\rho v^3}\frac{\omega^2\tau}{1+\omega^2\tau^2}$. From the fitting with Arrenius-type relaxation time $\tau = \tau_0 \exp(E/k_B T)$, we determined the activation energies of $E_{(L)} = 102$ K, $E_{(H)} = 198$ K and the attempt time $\tau_{0(L)} = 4.2 \times 10^{-11}$ s, $\tau_{0(H)} = 1.1 \times 10^{-11}$ s of the rattling in $Pr_3Pd_{20}Ge_6$. Here, $E_{(L)}$ and $\tau_{0(L)}$ were estimated from the dispersion around 15 K, and $E_{(H)}$ and $\tau_{0(H)}$ were from that around 25 K.

We carried out the low-temperature ultrasonic measurements in the clathrate compound

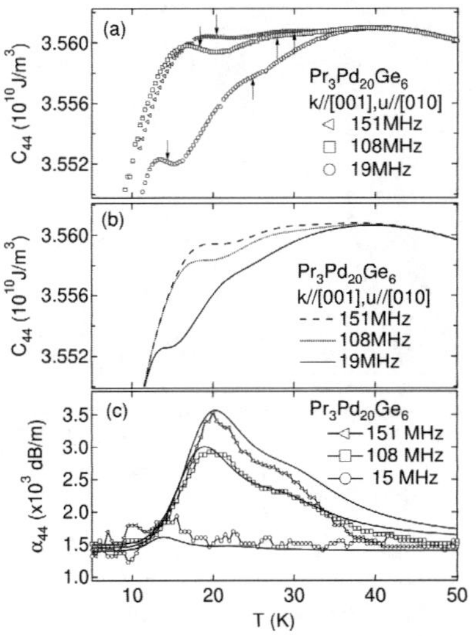

Figure 3. (a) The elastic constant C_{44} as a function of temperature at frequencies of 19 MHz, 108 MHz and 151 MHz. (b) The lines show calculation of C_{44} using the Debye type formula. (c) Temperature dependence of ultrasonic attenuation coefficient α_{44} at frequencies of 15 MHz, 108 MHz and 151 MHz. Solid lines are calculation of α_{44} using the Debye fits.

$Pr_3Pd_{20}Ge_6$ to investigate the quadrupole ordered phases and rattling. The elastic constants C_{11}, $(C_{11} - C_{12})/2$ and C_{44} shows the elastic softening associated with the Γ_3 doublet ground state at 8c site and the Γ_5 triplet ground state at 4a site. The analysis of the elastic softenings $(C_{11} - C_{12})/2$ and C_{44} in terms of quadrupole susceptibility suggest that the intersite interaction among 8c site is antiferro-type and that among 4a site is ferro-type. Furthermore, the anomalies originated with phase transition at T_{Q1} =250 mK and T_{Q2} =60 mK have been found. Therefore, we conclude that the phase II is AFQ ordering state and the phase III is FQ ordering state. The $H - T$ phase diagram under magnetic fields along the [110] direction was determined. We found two different ultrasonic dispersions around 15 K and 25 K in C_{44} due to the thermally activated Γ_5-type rattling motion in $Pr_3Pd_{20}Ge_6$.

Acknowledgments

This work was supported by a Grant-in-Aid for Specially Promoted Research (No.18002008) "Strongly correlated quantum phases associated with charge fluctuations" from the Ministry of Education, Culture, Sports and Technology of Japan.

References

[1] Gribanov A. V, Seropegin Yu. D and Bodak O. I 1994 *J.Alloys.Compd.* 204 L9-L11
[2] Kitagawa J, Takeda N and Ishikawa M 1996 *Phys.Rev.* **B53** 5101-5103
[3] Nemoto Y, Yamaguchi T, Horino T, Akatsu M, Yanagisawa T, Goto T, Suzuki O, Dönni A and Komatsubara T 2003 *Phys.Rev.* **B68**, 184109
[4] Goto T, Nemoto Y, Yamaguchi T, Akatsu M, Yanagisawa T, Suzuki O and Kitazawa H 2004 *Phys.Rev.* **B70** 184126
[5] Nemoto Y, Yanagisawa T, Yasumoto Y, Kobayashi H, Yamaguchi A, Tsuduku S, Goto T, Takeda N, Ochiai A, Sugawara H, Sato H and Kitazawa H 2008 *J.Phys.Soc.Jpn.* **77** 153-158
[6] Goto T, Watanabe T, Tsuduku S, Kobayashi H, Nemoto Y, Yanagisawa T, Akatsu M, Ano G, Suzuki O, Takeda N, Dönni A and Kitazawa H 2009 *J.Phys.Soc.Jpn.* **78** 024716
[7] Kobayashi H, Tsuduku S, Suzuki O, Yanagisawa T, Nemoto Y, Goto T, Dönni A, Kitazawa H and Takeda N 2008 *J.Phys.Soc.Jpn.* **77** Suppl. A, 263-265
[8] Keller L, Dönni A, Zolliker M and Komatsubara T 1999 *PhysicaB* **259 – 261** 336-337, Suzuki O, Kitazawa H, Kido G, Yamaguchi T, Nemoto Y and Goto T 2006 *J.Alloys.Compd.* **408 – 412** 107-109
[9] Kitagawa J 1999 *DoctorThesisTokyoUniversity*

Journal of Physics: Conference Series **273** (2011) 012003 doi:10.1088/1742-6596/273/1/012003

Metamagnetic Transition in Heavy Fermion Compounds YbT_2Zn_{20} (T : Co, Rh, Ir)

Yusuke Hirose[1], Masatoshi Toda[1], Shingo Yoshiuchi[1], Shinichi Yasui[1], Kiyohiro Sugiyama[1], Fuminori Honda[1], Masayuki Hagiwara[2], Koichi Kindo[3], Rikio Settai[1] and Yoshichika Ōnuki[1]

[1]Graduate School of Science, Osaka University, Toyonaka Osaka 560-0043, Japan
[2]KYOKUGEN, Osaka University, Toyonaka, Osaka 560-8531, Japan
[3]ISSP, University of Tokyo, Kashiwanoha, 5-1-5, Kashiwa, Chiba 277-8581, Japan

E-mail: yhirose@crystal.phys.sci.osaka-u.ac.jp

Abstract. We measured the magnetization in high magnetic fields up to 500kOe, together with the magnetic susceptibility, ac-susceptibility and magnetoresistance for heavy fermion compounds YbT_2Zn_{20} (T : Co, Rh, Ir). The metamagnetic behaviour or an abrupt nonlinear increase of magnetization was observed at the magnetic field H_m at temperatures lower than a characteristic temperature $T_{\chi_{max}}$ below which the magnetic susceptibility becomes almost constant : $T_{\chi_{max}} = 7.4$K and $H_m = 97$kOe in $YbIr_2Zn_{20}$, $T_{\chi_{max}} = 5.3$K and $H_m = 64$kOe in $YbRh_2Zn_{20}$, and $T_{\chi_{max}} = 0.32$K and $H_m = 6$kOe in $YbCo_2Zn_{20}$. From the present data and the data in several Ce and U heavy fermion compounds, a simple relation between $T_{\chi_{max}}$ and H_m was obtained : $H_m(\text{kOe}) = 15T_{\chi_{max}}(\text{K})$.

1. Introduction

Most of Ce and U compounds order antiferromagnetically. Some compounds such as $CeCu_6$, $CeRu_2Si_2$ and UPt_3 exhibit no long-range magnetic ordering[1]. The magnetic susceptibility of these compounds shows a maximum at a characteristic temperature $T_{\chi_{max}}$. Below $T_{\chi_{max}}$, the susceptibility becomes almost temperature-independent, and an f-electron nature is changed into a new electronic state, called the heavy fermion state.Here, $T_{\chi_{max}}$ approximately corresponds to the Kondo temperature T_K. One of the characteristic properties in the heavy fermion compounds is the metamagnetic behaviour or an abrupt nonlinear increase of magnetization at the magnetic field H_m at temperatures lower than $T_{\chi_{max}}$. The metamagnetic behaviour appears at $H_m = 77$kOe in $CeRu_2Si_2$ and $H_m = 200$kOe in UPt_3, for example[2],[3].

Very recently, we found the metamagnetic behaviour at $H_m = 97$kOe for H // $\langle 100 \rangle$ in a new heavy fermion compound $YbIr_2Zn_{20}$ with the cubic cage-structure[4]. Here we report the metamagnetic behaviour in $YbIr_2Zn_{20}$ and the similar compounds $YbRh_2Zn_{20}$ and $YbCo_2Zn_{20}$ by measuring the magnetization, magnetic susceptibility, ac-susceptibility and magnetoresistance, and propose a relation between $T_{\chi_{max}}$ and H_m in the heavy fermion compounds.

Published under licence by IOP Publishing Ltd

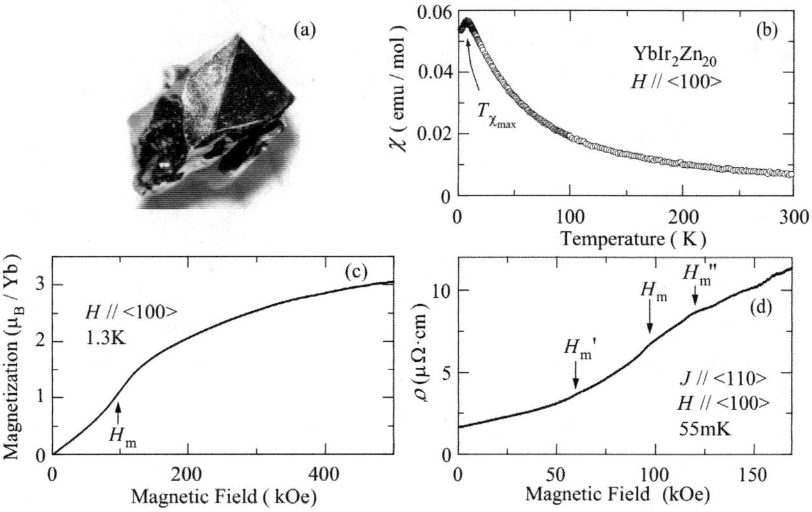

Figure 1. (a) Single crystal ingot, (b) temperature dependence of the magnetic susceptibility, (c) magnetization and (d) transverse magnetoresistance in $YbIr_2Zn_{20}$.

2. Experimental results

Single crystal ingots, grown by the the Zn-flux method, are pyramidal in shape, as shown in Figures 1(a) and 3(a) for $YbIr_2Zn_{20}$ and $YbCo_2Zn_{20}$, respectively, where flat planes correspond to the {111} planes, reflecting the diamond structure of Yb atoms. Yb and T atoms are surrounded by Zn atoms, forming a cubic cage-structure.

Figure 1(b) shows the temperature dependence of the magnetic susceptibility χ for $H//\langle 100\rangle$. The susceptibility follows the Curie-Weiss law of $\chi = C/(T + \theta_P) + \chi_0$: the Curie term with an effective magnetic moment $\mu_{eff} = 4.54\mu_B/\text{Yb}$, close to $\mu_{eff} = 4.57\mu_B/\text{Yb}$ of Yb^{+3}, a paramagnetic Curie temperature θ_P = -27K and χ_0 = -1.05×10^{-3}emu/mol. The susceptibility possesses a broad peak at $T_{\chi_{max}}$ = 7.4K, which is characteristic in the heavy fermion compounds. The present result is the same as the previous one[5].

The high-field magnetization for $H//\langle 100\rangle$ is shown in Figure 1(c), revealing a metamagnetic behaviour at H_m = 97kOe. The corresponding magnetoresistance is also shown in Figure 1(d). A change of the magnetoresistance was observed at H_m = 97kOe, as shown by an arrow, together with small change at H'_m = 60kOe and H''_m = 120kOe. The metamgnetic transition field is slightly anisotropic : H_m = 120kOe for $H//\langle 110\rangle$.

The similar measurements were carried out for $YbRh_2Zn_{20}$, as shown in Figure 2. The magnetic susceptibility has a maximum at $T_{\chi_{max}}$ = 5.3K, as shown in Figure 2. The metamagnetic behviour was observed at H_m = 64kOe in the magnetization, as shown in Figure 2(b) and another anomaly was furthermore observed at H'_m = 84kOe in the magnetoresistance, as shown in Figure 2(c)

In $YbCo_2Zn_{20}$, the metamagnetic behaviour was observed at a very small magnetic field H_m = 6kOe at temperatures lower than 0.3K. The susceptibility increases with decreasing temperature down to 1.8K, following the Curie-Weiss law, as shown in Figure 3(b). We therefore measured the ac-susceptibility below 1K, and obtained a peak at $T_{\chi_{max}}$ = 0.32K, as shown in inset of Figure 3(b). The magnetization at 1.3K didn't reveal the metamagnetic behaviour, as shown in Figure 3(c). The metamagnetic behaviour at H_m = 6kOe is reflected in the ac-susceptibility at 60mK, as shown in inset of Figure 3(c), and in the magnetoresistance at 100mK, as shown

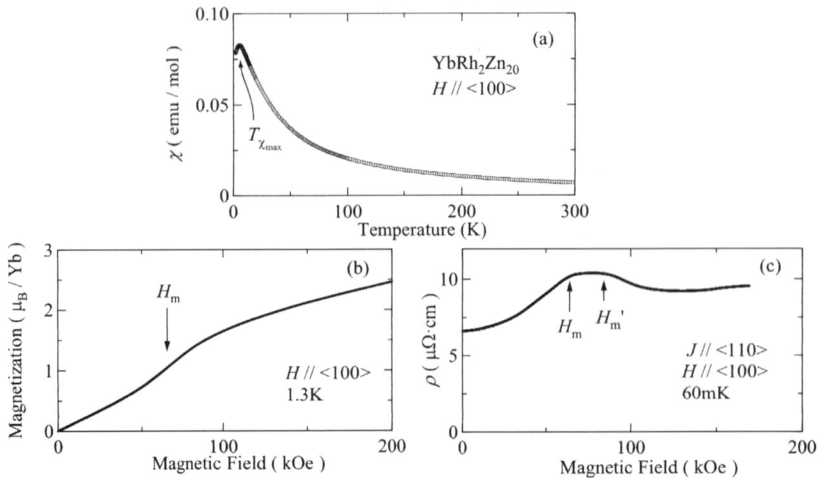

Figure 2. (a) Temperature dependence of the magnetic susceptibility, (b) magnetization and (c) transverse magnetoresistance in $YbRh_2Zn_{20}$.

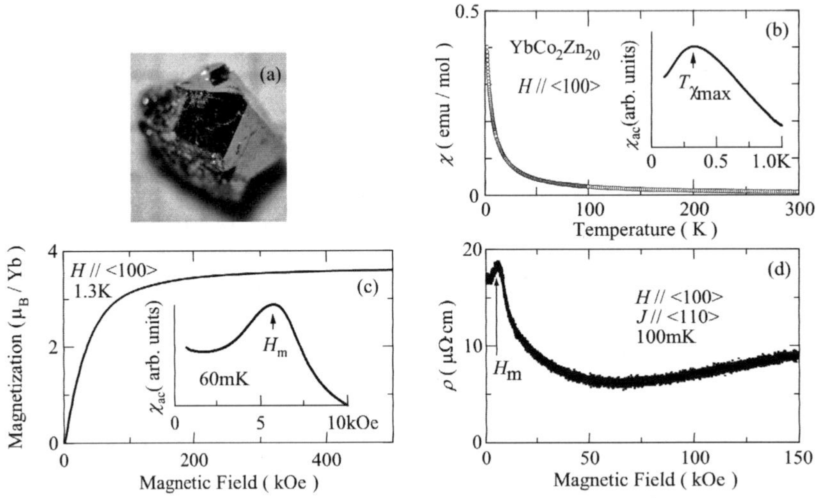

Figure 3. (a) Single crystal ingot, (b) temperature dependence of the magnetic susceptibility, (c) magnetization and (d) transverse magnetoresistance in $YbCo_2Zn_{20}$.

in Figure 3(d). $YbCo_2Zn_{20}$ is therefore located in the vicinity of the quantum critical point. In fact, the electronic specific heat coefficient γ is very large, $8000 mJ/K^2{\cdot}mol$[5].

From the present results of the metamagnetic behaviour, together with those of Ce and U compounds, we constructed the relation between $T_{\chi_{max}}$ and H_m, as shown in Figure 4. A solid line in Figure 4 indicates a simple relation of $H_m(kOe) = 15T_{\chi_{max}}(K)$.

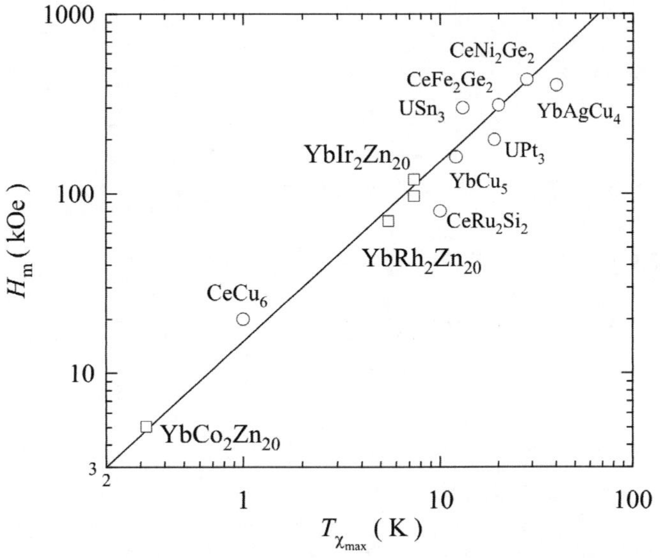

Figure 4. Relation between H_m and $T_{\chi_{max}}$ in heavy fermion compounds.

3. Conclusion

A simple relation between the characteristic temperature $T_{\chi_{max}}$ and the metamagnetic field H_m was obtained experimentally for heavy fermion compounds YbT_2Zn_{20} (T : Ir, Rh, and Co) : $H_m = 15\ T_{\chi_{max}}$(K). In $YbCo_2Zn_{20}$, an electronic state with a very small Kondo temperature $T_{\chi_{max}} = 0.32$K is realized, which corresponds to a very large γ value of 8000mJ/K$^2\cdot$ mol. This might be mainly due to a long distance between Yb atoms, 6Å, together with the cage structure.

Acknowledgments

We are very grateful to Profs. K. Miyake and S. Watanabe for fruitful discussions. This work was supported by Grant-in-Aid for Scientific Research on Specially Promoted Research (No. 20001004), Osaka University: Global COE Program "Core Research and Engineering of Advanced Materials-Interdisciplinary Education Center for Materials Science" (No. G10), Grant-in-Aid for Scientific Research on Innovative Areas "Heavy Electrons" (20102002, 21102511), Priority Areas of New Materials Science Using Regulated Nano Spaces (20045008, 22013009), Scientific Research (B)(21340102, 21740256) from the Ministry of Education, Culture, Sports, Science and Technology, Japan.

References

[1] Y. Ōnuki, R. Settai, K. Sugiyama, T. Takeuchi, T.C. Kobayashi, Y. Haga and E. Yamamoto : J. Phys. Soc. Jpn. **73** (2004) 769.

[2] P. Haen, J. Flouquet, F. Lapierre, and P. Lejay : J. Low Temp. Phys. **67** (1987) 391.

[3] K. Sugiyama, M. Nakashima, D. Aoki, K. Kindo, N. Kimura, H. Aoki, T. Komatsubara, S. Uji, Y. Haga, E. Yamamoto, H. Harima and Y. Ōnuki Phys. Rev. B **60** (1999) 9248.

[4] T. Takeuchi, S. Yasui, M. Toda, M. Matsushita, S. Yoshiuchi, M. Ohya, K. Katayama, Y. Hirose, N. Yoshitani, F. Honda, K. Sugiyama, M. Hagiwara, K. Kindo, E. Yamamoto, Y. Haga, T. Tanaka, Y. Kubo, R. Settai, and Y. Ōnuki J. Phys. Soc. Jpn. **79** (2010) 064609.

[5] M. S. Torikachvili, S. Jia, E. D. Mun, S. T. Hannahs, R. C. Black, W. K. Neils, D. Martien, S. L. Bud'ko, and P. C. Canfield: Proc. Natl. Acad. Sci. U.S.A. **104** (2007) 9960.

Evidence for power-law Griffiths singularities in a layered Heisenberg magnet

Fawaz Hrahsheh[1], Hatem Barghathi[1], Priyanka Mohan[2], Rajesh Narayanan[2] and Thomas Vojta[1]

[1] Department of Physics, Missouri University of Science and Technology, Rolla MO 65409, USA

[2] Department of Physics, Indian Institute of Technology Madras, Chennai 600036, India

E-mail: `vojtat@mst.edu`

Abstract. We study the ferromagnetic phase transition in a randomly layered Heisenberg model. A recent strong-disorder renormalization group approach [Phys. Rev. B **81**, 144407 (2010)] predicted that the critical point in this system is of exotic infinite-randomness type and is accompanied by strong power-law Griffiths singularities. Here, we report results of Monte-Carlo simulations that provide numerical evidence in support of these predictions. Specifically, we investigate the finite-size scaling behavior of the magnetic susceptibility which is characterized by a non-universal power-law divergence in the Griffiths phase. In addition, we calculate the time autocorrelation function of the spins. It features a very slow decay in the Griffiths phase, following a non-universal power law in time.

1. Introduction

Impurities, defects, and other types of quenched disorder influence zero-temperature quantum phase transitions much more strongly than generic classical phase transitions that occur at nonzero temperatures. The interplay of quantum and disorder fluctuations produces unconventional phenomena such as power-law quantum Griffiths singularities [1, 2, 3], infinite-randomness critical points featuring exponential instead of power-law scaling [4, 5], and smeared phase transitions [6, 7]. A recent review of these phenomena can be found in Ref. [8], while Ref. [9] focuses on metallic systems and also discusses experiments.

Disorder effects are stronger at quantum phase transitions than at classical transitions because the quenched disorder is perfectly correlated in the *imaginary time* direction. Imaginary time acts as an extra dimension at a quantum phase transition and becomes infinitely extended at zero temperature. Therefore, the impurities and defects are effectively "infinitely large" in this extra dimension, which makes them much harder to average out than finite-size defects.

This argument suggests that strong disorder effects should also occur at a classical thermal phase transition provided that the disorder is perfectly correlated in one or more *space* dimensions. An example occurs in the McCoy-Wu model, a disordered classical two-dimensional Ising model having perfect disorder correlations in one of the two dimensions. McCoy and Wu [10, 11, 12, 13] showed that this model exhibits an unusual phase transition featuring a smooth specific heat while the susceptibility is infinite over an entire temperature range. Fisher [4, 5] achieved an essentially complete understanding of this phase transition with the help of a strong-disorder renormalization group approach (using the equivalence between the McCoy-Wu model

Published under licence by IOP Publishing Ltd

and the random transverse-field Ising chain). He determined that the critical point is of exotic infinite-randomness type and is accompanied by power-law Griffiths singularities.

Recently, some of us investigated another classical system with perfectly correlated disorder, the randomly layered Heisenberg magnet, by means of a strong-disorder renormalization group [14]. This theory predicts that the (thermal) critical point in the randomly layered Heisenberg magnet is of infinite-randomness type as well. Moreover, it is in the same universality class as the quantum critical point of the random transverse-field Ising chain. In this paper, we present the results of Monte-Carlo simulations of the randomly layered Heisenberg model. They provide numerical evidence in support of the above renormalization group predictions.

2. Model and renormalization group predictions

We consider a ferromagnet consisting of a random sequence of layers made up of two different ferromagnetic materials. Its Hamiltonian, a classical Heisenberg model on a three-dimensional lattice of perpendicular size $L_\perp$ (in z direction) and in-plane size $L_\parallel$ (in the x and y directions) is given by

$$H = -\sum_{\mathbf{r}} J_z^\parallel \left(\mathbf{S_r}\cdot\mathbf{S}_{\mathbf{r}+\hat{\mathbf{x}}} + \mathbf{S_r}\cdot\mathbf{S}_{\mathbf{r}+\hat{\mathbf{y}}}\right) - \sum_{\mathbf{r}} J_z^\perp\, \mathbf{S_r}\cdot\mathbf{S}_{\mathbf{r}+\hat{\mathbf{z}}}. \tag{1}$$

Here, $\mathbf{S_r}$ is a three-component unit vector on lattice site $\mathbf{r}$, and $\hat{\mathbf{x}}$, $\hat{\mathbf{y}}$, and $\hat{\mathbf{z}}$ are the unit vectors in the coordinate directions. The interactions within the layers, $J_z^\parallel$, and between the layers, $J_z^\perp$, are both positive and independent random functions of the perpendicular coordinate z. In the following, we take all $J_z^\perp$ to be identical, $J_z^\perp \equiv J^\perp$, while the $J_z^\parallel$ are drawn from a binary probability distribution $P(J^\parallel) = (1-p)\,\delta(J^\parallel - J_u) + p\,\delta(J^\parallel - J_l)$ with $J_u > J_l$. Here, p is the concentration of the "weak" layers.

The qualitative behavior of (1) is easily explained. At sufficiently high temperatures, the model is in a conventional paramagnetic phase. Below a temperature T_u (the transition temperature of a hypothetical system having $J_z^\parallel \equiv J_u$ for all z) but above the actual critical temperature T_c, rare thick slabs of strong layers develop local order while the bulk system is still nonmagnetic. This is the *paramagnetic* Griffiths phase (or Griffiths region). In the *ferromagnetic* Griffiths phase, located between T_c and a temperature T_l (the transition temperature of a hypothetical system having $J_z^\parallel \equiv J_l$ for all z), bulk magnetism coexists with rare nonmagnetic slabs. Finally, below T_l, the system is in a conventional ferromagnetic phase.

In Ref. [14], the behavior in both Griffiths phases and at criticality has been derived within a strong-disorder renormalization group calculation. Here, we simply motivate and summarize the results. The probability of finding a slab of L_{RR} consecutive strong layers is given by simple combinatorics; it reads $w(L_{RR}) \sim (1-p)^{L_{RR}} = e^{-\tilde{p}L_{RR}}$ with $\tilde{p} = -\ln(1-p)$. Each such slab is equivalent to a two-dimensional Heisenberg model with an effective interaction $L_{RR}J_u$. Because the two-dimensional Heisenberg model is exactly at its lower critical dimension, the renormalized distance from criticality, ϵ, of such a slab decreases exponentially with its thickness, $\epsilon(L_{RR}) \sim e^{-bL_{RR}}$ [8, 15]. Combining the two exponentials gives a power law spectrum of locally ordered slabs,

$$\rho(\epsilon) \sim \epsilon^{\tilde{p}/b-1} = \epsilon^{1/z-1} \tag{2}$$

where the second equality defines the conventionally used dynamical exponent, z. It increases with decreasing temperature throughout the Griffiths phase and diverges as $z \sim 1/|T - T_c|$ at the actual critical point.

Many important observables follow from appropriate integrals of (2). The susceptibility can be estimated by $\chi \sim \int d\epsilon\, \rho(\epsilon)/\epsilon$. In an infinite system, the lower bound of the integral is 0; therefore, the susceptibility diverges in the entire temperature region where $z > 1$. A finite system size $L_\parallel$ in the in-plane directions introduces a nonzero lower bound $\epsilon_{\min} \sim L_\parallel^{-2}$. Thus,

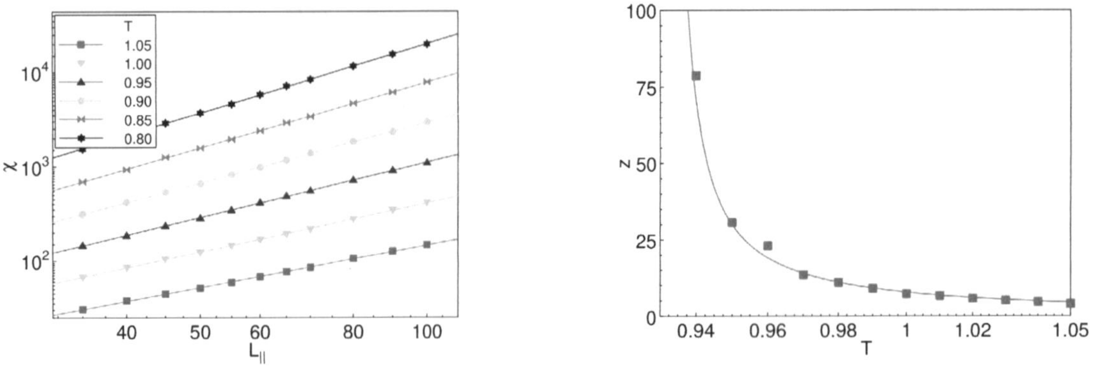

Figure 1. Left: Susceptibility χ as a function of in-plane system size $L_\parallel$ for several temperatures in the Griffiths region. The perpendicular size is $L_\perp = 800$; the data are averages over 160 disorder configurations. The solid lines are fits to the power law (3). Right: Resulting Griffiths dynamical exponent z as a function of temperature T in the weakly disordered Griffiths region.

the susceptibility in the Griffiths region diverges as

$$\chi(L_\parallel) \sim L_\parallel^{2-2/z} \tag{3}$$

for $z > 1$. The behavior of the time autocorrelation function $C(t) = (1/N)\sum_{\mathbf{r}}\langle \mathbf{S}(\mathbf{r},t)\cdot\mathbf{S}(\mathbf{r},0)\rangle$ within model A dynamics [16] can be determined analogously (N is the total number of spins). As the correlation time of a single, locally ordered slab is inversely proportional to its renormalized distance from criticality, we obtain, in the Griffiths phase,

$$C(t) \sim \int d\epsilon\, \rho(\epsilon) e^{-\epsilon t} \sim t^{-1/z} \ . \tag{4}$$

3. Monte-Carlo simulations

We have carried out Monte-Carlo simulations of the Hamiltonian (1) with system sizes up to $L_\perp = 800$ and $L_\parallel = 150$. While studying the thermodynamics, we have used the efficient Wolff cluster algorithm [17] to eliminate critical slowing down. To measure the time autocorrelation function, we have equilibrated the system using the Wolff algorithm but then propagated the dynamics by means of the Metropolis algorithm [18] which implements model A dynamics. All results reported below are for $J_z^\perp \equiv 1.00$ and a concentration $p = 0.8$ of weak layers with $J_u = 1.00$ and $J_l = 0.25$. The data are averages over a large number (20 to 160) of disorder realizations.

To test the finite-size behavior (3) of the susceptibility, one needs to consider samples having sizes $L_\perp \gg L_\parallel$ such that $L_\perp$ is effectively infinite. We have used system sizes $L_\perp = 800$ and $L_\parallel = 35$ to 100. Figure 1 shows the susceptibility χ as a function of $L_\parallel$ for several temperatures in the Griffiths region below $T_u \approx 1.443$. χ follows a nonuniversal power law in $L_\parallel$ with a temperature-dependent exponent. Simulations for many more temperature values yield analogous results. The value of the exponent z extracted from fits to (3) is shown in the right panel of figure 1 for the paramagnetic side of the Griffiths region. z can be fitted to the predicted power law $z \sim 1/|T - T_c|$, as discussed after (2), giving the estimate $T_c \approx 0.933$

Figure 2 shows the autocorrelation function $C(t)$ of the spins as a function of time t for several temperatures in the Griffiths phase. In agreement with the prediction (4), the long-time behavior of $C(t)$ follows a nonuniversal power law. We have also studied (not shown) the behavior at criticality. It displays the logarithmic time-dependence predicted in Ref. [14].

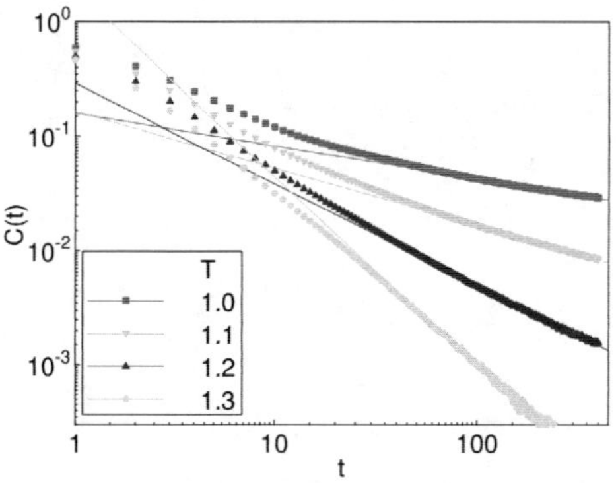

Figure 2. Time autocorrelation function $C(t)$ for several temperatures in the Griffiths phase. The system sizes are $L_\perp = 400$ and $L_\parallel = 100$. The data are averages over 80 disorder configurations. The solid lines are fits to (4).

4. Conclusions

In summary, we have performed Monte-Carlo simulations of a randomly layered three-dimensional Heisenberg model. Our numerical results for the magnetic susceptibility and the time autocorrelation function in the paramagnetic Griffiths phase support the infinite-randomness critical point scenario that arises from the strong-disorder renormalization group approach [14]. A full confirmation of the theoretical predictions requires performing a scaling analysis of the infinite-randomness critical point itself, and measuring all three independent critical exponents. This work is in progress.

Experimental verifications of infinite-randomness critical behavior and the accompanying power-law Griffiths singularities have been hard to come by, in particular in higher-dimensional systems. Only very recently, promising measurements have been reported [19, 20] of the quantum phase transitions in $CePd_{1-x}Rh_x$ and $Ni_{1-x}V_x$. The randomly layered Heisenberg magnet considered here provides an alternative realization of an infinite-randomness critical point. It may be more easily realizable in experiment because the critical point is classical, and samples can be produced by depositing random layers of two different ferromagnetic materials.

This work has been supported in part by the NSF under grant nos. DMR-0339147 and DMR-0906566 and by Research Corporation.

References

[1] Thill M and Huse D A 1995 *Physica A* **214** 321
[2] Guo M, Bhatt R N and Huse D A 1996 *Phys. Rev. B* **54** 3336
[3] Rieger H and Young A P 1996 *Phys. Rev. B* **54** 3328
[4] Fisher D S 1992 *Phys. Rev. Lett.* **69** 534
[5] Fisher D S 1995 *Phys. Rev. B* **51** 6411
[6] Vojta T 2003 *Phys. Rev. Lett.* **90** 107202
[7] Hoyos J A and Vojta T 2008 *Phys. Rev. Lett.* **100** 240601
[8] Vojta T 2006 *J. Phys. A* **39** R143
[9] Vojta T 2010 (*Preprint* `arxiv:1005.2707`)
[10] McCoy B M and Wu T T 1968 *Phys. Rev. Lett.* **21** 549
[11] McCoy B M and Wu T T 1968 *Phys. Rev.* **176** 631
[12] McCoy B M and Wu T T 1969 *Phys. Rev.* **188** 982
[13] McCoy B M 1969 *Phys. Rev. Lett.* **23** 383
[14] Mohan P, Narayanan R and Vojta T 2010 *Phys. Rev. B* **81** 144407
[15] Vojta T and Schmalian J 2005 *Phys. Rev. B* **72** 045438
[16] Hohenberg P C and Halperin B I 1977 *Rev. Mod. Phys.* **49** 435
[17] Wolff U 1989 *Phys. Rev. Lett.* **62** 361
[18] Metropolis N, Rosenbluth A, Rosenbluth M and Teller A 1953 *J. Chem. Phys.* **21** 1087
[19] Westerkamp T, Deppe M, Küchler R, Brando M, Geibel C, Gegenwart P, Pikul A P and Steglich F 2009 *Phys. Rev. Lett.* **102** 206404
[20] Ubaid-Kassis S, Vojta T and Schroeder A 2010 *Phys. Rev. Lett.* **104** 066402

Low-temperature thermal transport coefficients of heavy fermion β-$YbAlB_4$

Y. Machida[1], C. Ogura[1], K. Izawa[1], K. Kuga[2] and S. Nakatsuji[2]

[1]Department of Physics, Tokyo Institute of Technology, 2-12-1 O-Okayama, Meguro, Tokyo 152-8551, Japan

[2]Institute for Solid State Physics, University of Tokyo, 5-1-5 Kashiwanoha, Kashiwa, Chiba 277-8581, Japan

E-mail: machida@ap.titech.ac.jp

Abstract. The first Yb-based heavy fermion superconductor β-$YbAlB_4$ with T_c = 80 mK is stoichiometrically quantum critical under ambient pressure and under zero magnetic field. We report the low-temperature thermoelectric power $S(T)$ on a high-quality single crystal of β-$YbAlB_4$ down to 40 mK. In zero field, $S(T)/T$ exhibits a dramatic enhancement and takes large negative values at low temperatures, followed by a fast drop below 80 mK as an indication of the superconducting transition. Under the field of B = 25 mT ($\geq B_{c2}$), $-S(T)/T$ is found to continuously increase down to the lowest temperature, which is highly in contrast to what is expected for the Fermi liquids ($S/T \sim$ const.).

1. Introduction

Heavy fermion compounds have been intensively studied for their ability to use the pressures or the magnetic fields to tune continuously from a magnetic phase into a Fermi liquid through a quantum critical point (QCP). One consequence of the QCP is the vanishing characteristic energy scale of the Fermi liquid. So far, this behavior has been indicated mainly by the divergence of the electronic specific heat [1]. In some cases, however, the dominant nuclear contributions at low temperatures prevent the detailed studies on the QCP by means of the specific heat. Recently, new insight on the QCP via the thermoelectric coefficients, e.g., the thermoelectric power and the Nernst coefficient, has been examined as a promising probe to unravel the nature of the QCP because of their sensitivity to the low-energy excitations [2,3].

Here, we present the results of the low-temperature thermoelectric power of β-$YbAlB_4$ using a high-quality single crystal. β-$YbAlB_4$ is the first Yb-based heavy fermion superconductor with T_c = 80 mK [4]. Moreover, the system is stoichiometrically quantum critical under ambient pressure and under zero magnetic field. This enable us to make detailed investigations to elucidate the nature of quantum criticality.

2. Experiments

Single crystals were grown by using the Al-flux method as described in the literature [4]. Thermal transport coefficient was measured by employing a steady-state method in a dilution refrigerator. The heat current q was injected parallel to the ab plane, and the magnetic field was applied parallel to the c axis on the sample with a rectangular shape ($500 \times 100 \times 7\mu$m). The thermal

Published under licence by IOP Publishing Ltd

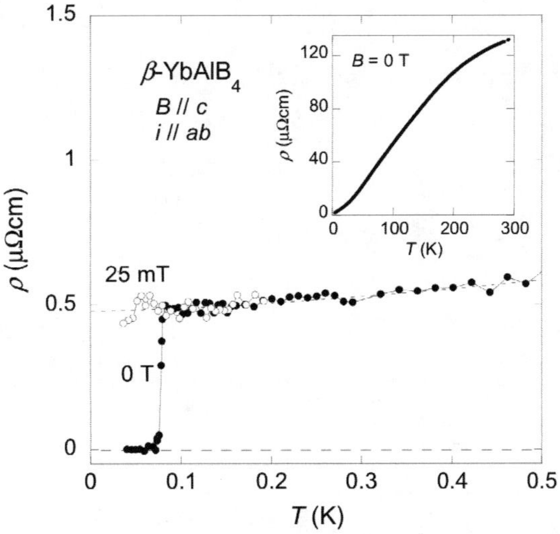

Figure 1. Temperature dependence of the electric resistivity $\rho(T)$ below 0.5 K under zero field and 25 mT. Inset: zero-field ρ vs. T plot in the temperature range of 4.2-300 K. A dotted line represents a fit to the data at $B = 25$ mT by the power-law form $\rho(T) = \rho_0 + AT^{1.5}$.

contacts with resistance of ~10 mΩ at room temperature were made to the sample by using a spot welding technique. The same contacts were used to measure the resistivity by a standard four-contact method.

3. Results and Discussions

First, we demonstrate the high-quality of our sample from the resistivity measurements. Figure 1 displays the temperature dependence of the electrical resistivity $\rho(T)$ under zero field and 25 mT. The inset of Fig. 1 shows the high-temperature part of zero-field resistivity. Under zero field, $\rho(T)$ steeply decreases on cooling and shows a sharp drop at 80 mK associated with the superconducting transition. As denoted by a dotted line in Fig. 1, the normal state $\rho(T)$ is well described by the power-law form $\rho(T) = \rho_0 + AT^{1.5}$, consistent with the previous report [4]. The low residual resistivity $\rho_0 = 0.48$ $\mu\Omega$cm with a high residual resistivity ratio $RRR \sim 270$ indicates high-quality of the sample. By applying the field of 25 mT, which is just above the upper critical field B_{c2}^c along the c axis [5], the superconducting transition is entirely suppressed.

Next, we concentrate on the thermal transport properties of β-$YbAlB_4$. Figure 2 shows the temperature dependence of the thermoelectric power divided by temperature $-S(T)/T$ under zero field and 25 mT. Above 0.2 K, $-S(T)/T$ exhibits a dramatic enhancement with a nearly logarithmic form and reaches a large negative value $-S/T \sim 6$ μV/K^2, which is two order of magnitude larger than the values for simple metals such as Cu ($S/T \sim -10$ nV/K^2 at 1 K) [6]. The large S/T is also found in other heavy-fermion compounds [2,3,7] and are considered to be related to the strongly enhanced specific heat. Indeed, the dimensionless ratio ($q = \frac{SN_Ae}{C}$, N_A and e represent the Avogadro number and the charge of electron, respectively) linking the thermoelectric power and the specific heat C is found to be of the order of unity in a wide range of systems [8]. On further decreasing the temperature, $-S(T)/T$ under zero field shows steep increase even below 0.2 K but with a smaller slope followed by a sudden drop at 80 mK due to the superconducting transition. By contrast, under 25 mT ($> B_{c2}$), the drop disappears and $-S(T)/T$ is found to continuously increase down to the lowest temperature, incompatible

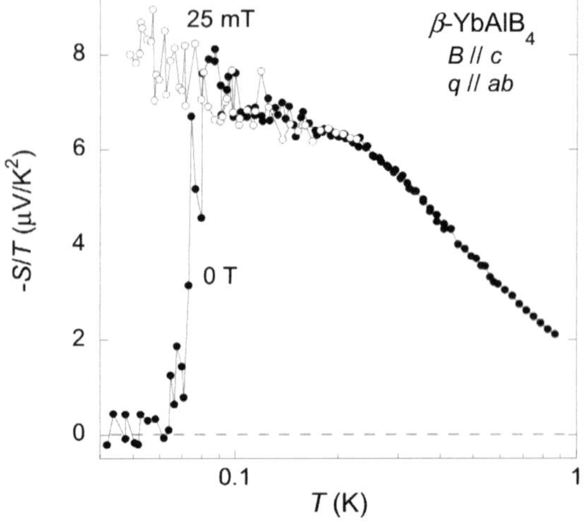

Figure 2. Temperature dependence of the thermoelectric power divided by temperature $-S(T)/T$ under zero field and 25 mT.

with the Fermi liquid behavior in which $S(T)/T$ is expected to be constant [8]. Moreover, the pronounce increase near the QCP is in contrast to the other quantum critical heavy-fermion compounds, e.g., $CeCoIn_5$ [2] and $YbRh_2Si_2$ [3,9], in which the downturn of $|S/T|$ was observed around the field-induced QCP.

Finally, let us examine the dimensionless ratio q of the thermoelectric power S and the specific heat C for β-$YbAlB_4$. Figure 3 displays the temperature dependence of q for β-$YbAlB_4$ together with the results in the vicinity of the QCP for $CeCoIn_5$ [2] and $YbRh_2Si_2$ [9]. For β-$YbAlB_4$, we obtained q by combining the specific heat data reported in Ref. [4] with our thermopower data. Here, the specific heat data below 0.4 K is estimated by assuming that it increases logarithmically down to the lowest temperature. It can be clearly seen that q for all the three systems takes the values of the order of unity at high temperatures. Nevertheless, q for β-$YbAlB_4$ is almost temperature-independent below 0.2 K, while the one for $CeCoIn_5$ and $YbRh_2Si_2$ decreases upon cooling. The apparently different behavior of $q(T)$ among the quantum critical materials would be useful to distinguish the nature of the quantum criticality. Theoretically, it is argued that q decreases considerably toward antiferromagnetic QCP while it remains essentially unchanged for the ferromagnetic QCP or QCP due to a local criticality [10]. To further elucidate the quantum criticality of β-$YbAlB_4$, detailed investigations are highly required.

4. Summary

Our results of the low-temperature thermoelectric power S measurements using the high-quality single crystal of β-$YbAlB_4$ reveal the dramatic enhancement of S/T toward the QCP, giving rise to the apparently different behavior of the dimensionless ratio q compared with $CeCoIn_5$ and $YbRh_2Si_2$.

Acknowledgments

We thank K. Miyake and S. Watanabe for stimulating discussions. This work is supported by Grant-in-Aids (Nos. 20684016, 21740250) for Scientific Research from the Japanese Society for the Promotion of Science (JSPS), a Grant-in-Aid for Scientific Research on Innovative Areas "Heavy Electrons" (No. 20102006A) from the Ministry of Education, Culture, Sports, Science,

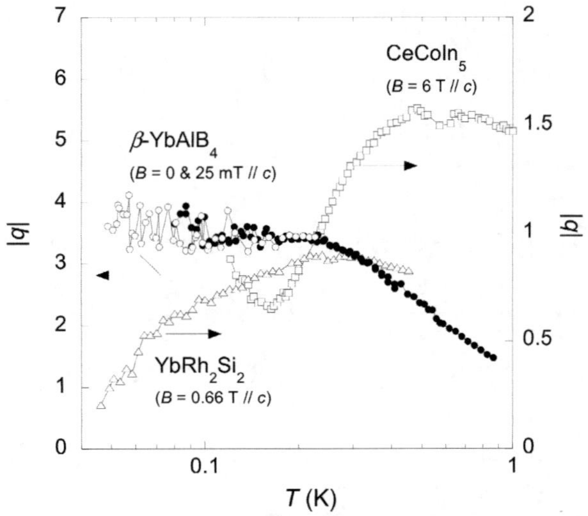

Figure 3. Temperature dependence of the dimensionless ratio $|q|$ of the thermoelectric power and the specific heat for β-$YbAlB_4$ (0 T, closed circles and 25 mT, open circles), $CeCoIn_5$ (squares) [2] and $YbRh_2Si_2$ (triangles) [9] in the vicinity of the QCP.

and Technology of Japan (MEXT), and a Grant-in-Aid for the Global COE Program from MEXT through the Nanoscience and Quantum Physics Project of the Tokyo Institute of Technology.

References

[1] Custers J, Gegenwart P, Wilhelm H, Neumaier K, Tokiwa Y, Trovarelli O, Geibel C, Steglich F, Pépin C and Coleman P 2003 *Nature* **424** 524
[2] Izawa K, Behnia K, Matsuda Y, Shishido H, Settai R, Onuki Y and Flouquet J 2007 *Phys. Rev. Lett.* **99** 147005
[3] Hartmann S, Oeschler N, Krellner C, Geibel C, Paschen S and Steglich F 2010 *Phys. Rev. Lett.* **104** 096401
[4] Nakatsuji S, Kuga K, Machida Y, Tayama T, Sakakibara T, Karaki Y, Ishimoto H, Yonezawa S, Maeno Y, Pearson E, Lonzarich G G, Balicas L, Lee H and Fisk Z 2008 *Nat. Phys.* **4** 603
[5] Kuga K, Karaki Y, Matsumoto Y, Machida Y and Nakatsuji S 2008 *Phys. Rev. Lett.* **101** 137004
[6] Rumbo E R 1976 *J. Phys. F: Met. Phys.* **6** 85
[7] Benz J, Pfleiderer C, Stockert O and Löhneysen H v 1999 *Physica B* **259-261** 380
[8] Behnia K, Jaccard D and Flouquet J 2004 *J. Phys.: Condens. Matter* **16** 5187
[9] Machida Y, Tomokuni K, Izawa K, Lapertot G, Knebel G, Brison J-P and Flouquet J unpublished
[10] Miyake K and Kohno H 2005 *J. Phys. Soc. Jpn.* **74** 254

International Conference on Strongly Correlated Electron Systems (SCES 2010) IOP Publishing
Journal of Physics: Conference Series **273** (2011) 012006 doi:10.1088/1742-6596/273/1/012006

Local moment behaviors of the valence fluctuating systems β-YbAlB$_4$ and α-YbAlB$_4$

Y. Matsumoto, K. Kuga, N. Horie and S. Nakatsuji

Institute for Solid State Physics, University of Tokyo, Kashiwa 277-8581, Japan

E-mail: `matsumoto@issp.u-tokyo.ac.jp, satoru@issp.u-tokyo.ac.jp`

Abstract. -YbAlB$_4$ is the first example of an Yb-based heavy Fermion superconductor with $T_c = 80$ mK and exhibits pronounced non-Fermi-liquid behavior above T_c. On the other hand, recent hard x-ray photoemission spectroscopy measurements have revealed strong intermediate valence of both -YbAlB$_4$ and its locally isostructural polymorph α-YbAlB$_4$. Here we present the results of the specific heat and magnetization measurements of -YbAlB$_4$ and α-YbAlB$_4$. The results indicate a Fermi liquid ground state for α-YbAlB$_4$ in contrast to the quantum criticality found in -YbAlB$_4$. Interestingly, both systems exhibit Kondo lattice behavior with large Wilson ratio below T^* 8 K, through the renormalization of a high valence fluctuation scale 200 K.

1. Introduction

$4f$ based heavy Fermion (HF) systems have provided prototypical systems to study interesting phenomena in the vicinity of quantum critical points, such as unconventional superconductivity and non-Fermi-liquid (NFL) phenomena. While a number of unconventional superconductors including CeCu$_2$Si$_2$, CeCoIn$_5$, and CeIrIn$_5$, have been found in Ce ($4f^1$) based intermetallic HF systems near a quantum critical point, much attention has also been paid to the unconventional quantum criticality, which does not follow the standard SDW description, observed in systems suchs as CeCu$_{6-x}$Au$_x$ and YbRh$_2$Si$_2$ [1, 2].

Our recent studies have found the first Yb ($4f^{13}$) based HF superconductivity with the transition temperature $T_c = 80$ mK in the new compound -YbAlB$_4$ [3, 4]. Pronounced non-Fermi-liquid behavior above T_c and its magnetic field dependence indicate that the system is a rare example of a pure metal that displays quantum criticality at ambient pressure and close to zero magnetic field [3]. Furthermore, the T/B scaling found in our recent high-precision magnetization measurements clarifies its unconventional zero-field quantum criticality without tuning[5]. Interestingly, hard x-ray photoemission spectroscopy (HXPES) measurements have revealed intermediate valence of Yb$^{+2.75}$[6], providing the first unique example of quantum criticality in a mixed valent system. Whether the valence fluctuation is relevant for the mechanism of the quantum criticality and superconductivity is an interesting open question.

In this paper, we present the results of the specific heat and magnetization measurements of -YbAlB$_4$ and α-YbAlB$_4$. Both phases are locally isostructural with different arrangement of distorted hexagons made of Yb atoms (space group: *Cmmm*(-YbAlB$_4$), *Pbam*(α-YbAlB$_4$)) [7]. α-YbAlB$_4$ also has an intermediate valence of Yb$^{+2.73}$[6]. The results indicate a Fermi liquid ground state for α-YbAlB$_4$ in contrast to the unconventional quantum criticality in

Published under licence by IOP Publishing Ltd

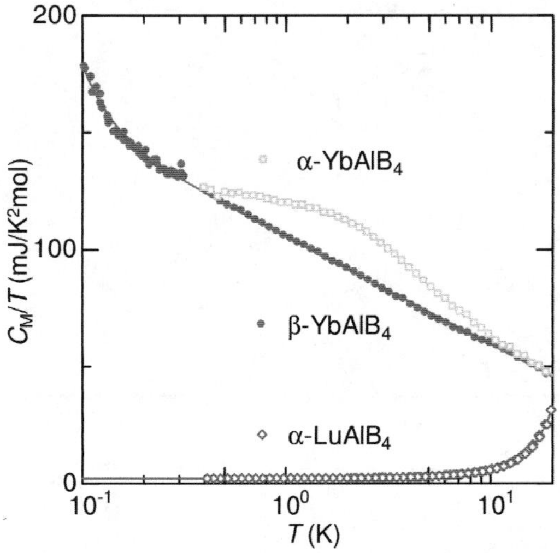

Figure 1. Magnetic part (f-electron contribution) of the specific heat C_M plotted as C_M/T versus T for both - (●) and α-$YbAlB_4$ (□) under zero field. C_M/T for the phase shows a $\ln T$ dependence for 0.2 K $< T <$ 20 K. $T_0 \sim 180 \pm 10$ K and $S_0 = 3.7 \pm 0.1$ J/molK are determined from the fit of the results to $C_M/T = S_0/T_0 \ln(T_0/T)$. The upturn in the lowest T may contain a nuclear contribution. C/T of α-$LuAlB_4$ is also shown(◇).

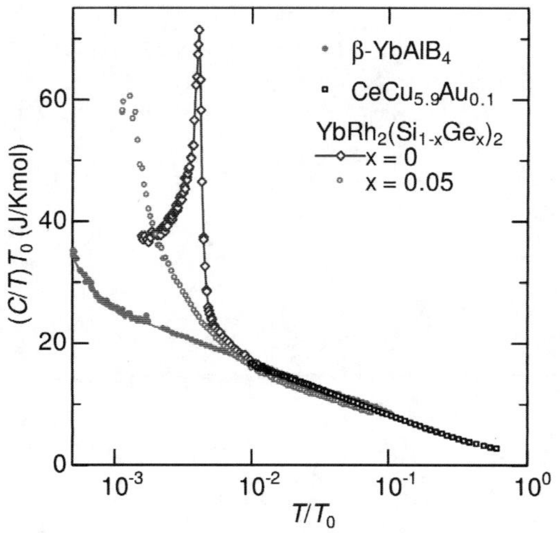

Figure 2. C_M/T of -$YbAlB_4$ scaled by T_0 compared with other quantum critical systems $CeCu_{5.9}Au_{0.1}$ (T_0 = 6.2 K)[8] and $YbRh_2(Si_{1-x}Ge_x)_2$ (T_0 = 24 K) [2, 9]. The $\ln T$ dependence of the three quantum critical materials collapses on top of each other using nearly the same coefficient $S_0 \sim 4$ J/mole K.

-$YbAlB_4$. The temperature dependence of the specific heat and susceptibility indicates that both systems have large valence fluctuation scale of $\sim$ 200 K and a small renormalized Kondo lattice scale of $\sim$ 8 K regardless of its ground state. Large Wison ratios found in the both systems indicate the importance of ferromagnetic intersite coupling for the emergence of the small Kondo lattice scale in the mixed valent compounds.

2. Experimental

High purity single crystals of -$YbAlB_4$ were grown by a flux method [7]. Energy dispersive x-ray analysis found no impurity phases, no inhomogeneities and a ratio Yb:Al of 1:1. Surface impurities were carefully removed with dilute nitric acid before measurements.

The magnetization M was measured by a commercial SQUID magnetometer using pure single crystals of 2.0 mg. The typical residual resistivity ratio (RRR) of these samples is 140.

The specific heat C was measured by a relaxation method. For the measurements above (below) 0.4 K, high-purity single crystals of 0.8 mg (2.2 mg) with RRR > 200 (RRR > 140) were measured. The magnetic part of the specific heat C_M was obtained by subtracting the specific heat of α-$LuAlB_4$, which is the non-magnetic isostructural counterpart of α-$YbAlB_4$.

3. Results and discussion

First, we present the magnetic part of the specific heat C_M devided by temperature in Fig. 1. In both - and α-$YbAlB_4$, C_M/T values are strongly enhanced to be larger than 130 mJ/molK^2 in

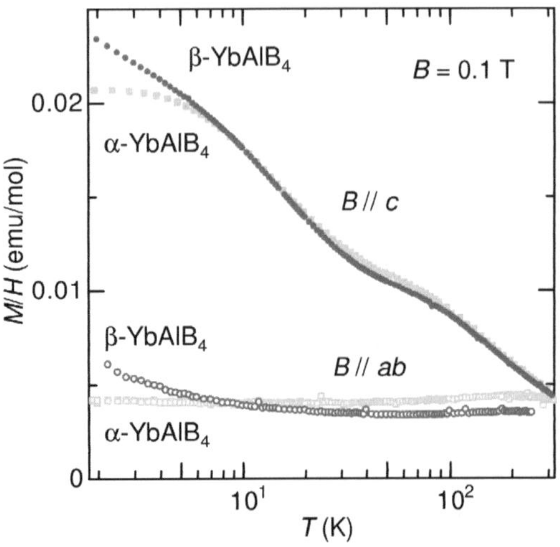

Figure 3. Temperature dependence of the d.c. susceptibility M/H measured in a field of 0.1 T along the ab plane and c axis for both - (red) and α-$YbAlB_4$ (green symbols).

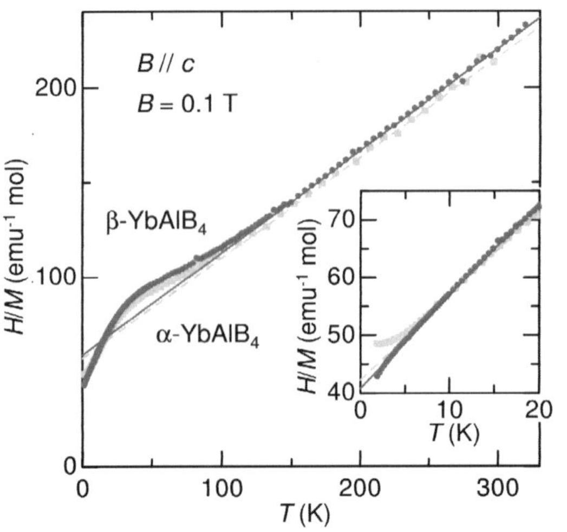

Figure 4. Temperature dependence of inverse susceptibility H/M in $B = 0.1$ T along the c axis. Solid and broken lines are Curie-Weiss fits above 150 K for α-(green) and -$YbAlB_4$ (red symbols), respectively. Inset shows low temperature part of H/M. Lines are Curie-Weiss fits below 20 K (see text).

the low T limit, which is two orders magnitude larger than the band calculation estimates ($\sim$ 6 mJ/molK^2) [10, 11]. While clear ln$T$ divergent behavior is observed in -$YbAlB_4$ in the temperature range of 0.2 K $< T <$ 20 K, C_M/T of α-$YbAlB_4$ nearly saturates in the low temperatures below 1 K, indicating a Fermi liquid ground state. On the other hand, at higher temperatures above 10 K, C_M/T of α-$YbAlB_4$ merges to the one of -$YbAlB_4$. Fitting the ln$T$ behavior of -$YbAlB_4$ to $C_M/T = S_0/T_0 \ln(T_0/T)$ yields $T_0 = 180 \pm 10$ K and $S_0 = 3.7 \pm 0.1$ J/molK. Here, T_0 sets a characteristic hybridization scale for the system and is close to the coherence temperature 250 K at the peak of the resistivity[3]. Note that T_0 in -$YbAlB_4$ is much higher than other quantum critical Kondo lattice systems such as $CeCu_{5.9}Au_{0.1}$ (T_0 = 6.2 K)[8] and $YbRh_2(Si_{1-x}Ge_x)_2$ (T_0 = 24 K) [2, 9]. This is consistent with the intermediate valence in -$YbAlB_4$ because mixed-valent compounds are typically characterized by a much higher value of T_0 than Kondo lattice systems. Interestingly however, if we plot C_M/T scaled by T_0 vs. T/T_0, the $-\ln T$ dependence in these three quantum critical materials collapses on top of each other, indicating common meaning of T_0 as the T scale below which $\sim$ 70 % of the ground doublet entropy, $R\ln 2$, is released.

The temperature dependences of the d.c. magnetic susceptibility $\chi = M/H$ are shown in Fig. 3. Both systems exhibit strong Ising anisotropy with moments aligned along the c axis, and almost T independent χ along the ab plane. Broad peaks found around 200 K in χ_{ab} for both systems are close to the T_0 scale obtained from C_M and the coherence temperature of the resistivity. χ along the c axis for these two systems shows almost the same temperature dependence for both α and phases down to $T \sim 8$ K, where it exhibits a shoulder. Below $T \lesssim 8$ K, on the other hand, these two systems show contrasting behavior *i.e.* while α-$YbAlB_4$ shows saturating behavior, consistent with the Fermi liquid ground state, -$YbAlB_4$ continues to diverge due to the quantum criticality [5]. The Curie-Weiss behavior, $\chi_c = C/(T + \Theta_W)$, is observed in the T range larger than 150 K with $\Theta_W = 110 \pm 2$, 108 ± 5 K for α and phase, respectively (Fig. 4). Ising moments $I_z = 2.22,\ 2.24\ \ _\mathrm{B}$ for α and phases are deduced from the Curie constant $C = N_\mathrm{A} I_z^2/k_\mathrm{B}$ where N_A and k_B are Avogadro and Boltzmann constant, respectively. Furthermore, at the lower temperatures below 20 K, another Curie-Weiss behavior is observed (Fig. 4 inset). If we fit the data in T range of $6 \lesssim T \lesssim 15$ K, $\Theta_\mathrm{W} = 29,\ 25$ K and $I_z = 1.4,\ 1.3\ \ _\mathrm{B}$ are obtained for the α and phases, respectively.

These observations indicate an existence of local moments far below $T_0 \sim 200$ K at least down to ~ 8 K. This Kondo lattice behavior with a low temperature scale $T^* \sim 8$ K is striking compared with ordinary valence fluctuating materials, such as $\mathrm{YbAl_3}$ [12], where Pauli paramagnetism is normally expected. A possible origin of this behavior may lie in the presence of ferromagnetic (FM) interactions between Yb $4f$-electron spins. Indeed, the Wilson ratio $R_\mathrm{W} = (\pi^2 k_B^2/\ \ _0 I_z^2)(\chi/\gamma) \sim 7$ is obtained for both α and phases by using χ_c at $B = 0.1$ T, $T = 0.4$ K, $\gamma = C_\mathrm{M}/T$ at $B = 0$, $T = 0.4$ K and I_z obtained from the high temperature Curie-Weiss fit. The R_W values are considerablly large compared with the normal value 2 expected for the Kondo lattice. If we use I_z obtained from the low temperature Curie-Weiss fit, then the Wilson ratio becomes $R_\mathrm{W} \sim 25$ for the both systems. This significantly high value strongly suggests the existence of FM correlations. The observations of the ESR signal [13] is also consistent with the presence of FM correlations [14]. Such FM interactions are known to give rise to *Kondo resonance narrowing* [15] in d-electron systems where Hund's coupling causes a large downward renormalization of the Kondo temperature [15]. In our systems, however, the FM interactions driving the narrowing presumably derive from intersite RKKY coupling. Neutron scattering measurements and studies of Lu dilution effect in $\mathrm{Yb_{1-x}Lu_xAlB_4}$ systems are important future works to clarify the origin of the ferromagnetic intersite coupling between Yb $4f$-electrons and the low temperature scale of $T^* \sim 8$ K in the mixed valence compounds.

Acknowledgments

We thank P. Coleman, A. H. Nevidomskyy, D.E. MacLaughlin, Q. Si, T. Tomita, K. Ueda, and T. Sakakibara for useful discussions. This work is partially supported by Grants-in-Aid (No.21684019) from JSPS, by Grants-in-Aids for Scientific Research on Innovative Areas "Heavy Electrons" and Global COE Program "the Physical Sciences Frontier" of MEXT, Japan.

References

[1] v Löhneysen H, Rosch A, Vojta M and Wölfle P 2007 *Rev. Mod. Phys.* **79** 1015–1075
[2] Gegenwart P, Si Q and Steglich F 2008 *Nature Phys.* **4** 186–197
[3] Nakatsuji S, Kuga K, Machida Y, Tayama T, Sakakibara T, Karaki Y, Ishimoto H, Yonezawa S, Maeno Y, Pearson E, Lonzarich G G, Balicas L, Lee H and Fisk Z 2008 *Nature Phys.* **4** 603–607
[4] Kuga K, Karaki Y, Matsumoto Y, Machida Y and Nakatsuji S 2008 *Phys. Rev. Lett.* **101** 137004
[5] Matsumoto Y, Nakatsuji1 S, Kuga K, Karaki Y, Horie N, Shimura Y, Sakakibara T, Nevidomskyy A H and Coleman P 2010 preprint
[6] Okawa M, Matsunami M, Ishizaka K, Eguchi R, Taguchi M, Chainani A, Takata Y, Yabashi M, Tamasaku K, Nishino Y, Ishikawa T, Kuga K, Horie N, Nakatsuji S and Shin S 2010 *Phys. Rev. Lett.* **104** 247201
[7] Macaluso R T, Nakatsuji S, Kuga K, Thomas E L, Machida Y, Maeno Y, Fisk Z and Chan J Y 2007 *Chem. Mater.* **19** 1918–1922
[8] v Löhneysen H 1996 *J. Phys.: Condens. Matter* **8** 9689–9706
[9] Custers J *et al.* 2003 *Nature* **424** 524–527
[10] Nevidomskyy A H and Coleman P 2009 *Phys. Rev. Lett.* **102** 077202
[11] O'Farrell E C T, Tompsett D A, Sebastian S E, Harrison N, Capan C, Balicas L, Kuga K, Matsuo A, Kindo K, Tokunaga M, Nakatsuji S, Csányi G, Fisk Z and Sutherland M L 2009 *Phys. Rev. Lett.* **102** 216402
[12] Cornelius A L, Lawrence J M, Ebihara T, Riseborough P S, Booth C H, Hundley M F, Pagliuso P G, Thompson J L S J D, Jung M H, Lacerda A H and Kwei G H 2002 *Phys. Rev. Lett.* **88** 117201
[13] Holanda L M, Vargas J M, Rettori C, Nakatsuji S, Kuga K, Fisk Z, Oseroff S and Pagliuso P G 2009 preprint available at http://arxiv.org/abs/0908.0044v1
[14] Krellner C, Förster T, Jeevan H, Geibel C and Sichelschmidt J 2008 *Phys. Rev. Lett.* **100** 066401
[15] Nevidomskyy A H and Coleman P 2009 *Phys. Rev. Lett.* **103** 147205

Quantum dynamical phase transition in the soliton nucleation model of density wave transport

J H Miller, Jr.
Department of Physics and Texas Center for Superconductivity,
University of Houston, 4800 Calhoun, Houston, Texas 77204-5005, USA

E-mail: jhmiller@uh.edu

Abstract. In the sine-Gordon model of a pinned density wave (DW), the electrostatic energy generated by charged solitons (S) and antisolitons ($\hat{S}$) leads to a Coulomb blockade threshold field E_T for quantum S-$\hat{S}$ pair creation. This field can be far smaller than the classical depinning field, since the quantum instability occurs as soon as the formerly lowest energy potential well rises to become a metastable well, or "false vacuum." The analogy to time-correlated single electron tunnelling, as well as broader implications of the proposed tunnelling process, are briefly discussed.

1. Introduction

The charge density wave (CDW) is a highly anisotropic correlated electron-phonon system in which quantum interference between right- and left-moving electron states leads to a divergent $2k_F$ Lindhard response and softening of the $2k_F$ phonon mode [1]. Below the Peierls transition, this results in a lattice distortion and charge modulation $\rho(x,t) = \rho_0(x,t) + \rho_1 \cos[2k_F x - \phi(x,t)]$ along the main axis of the needle-like crystal. Here $\rho_0(x,t)$ contains the background charge of the condensed electrons and any excess or deficiency of charge $\propto \partial\phi/\partial x$. A Peierls gap $2\Delta(T)$ opens up at the Fermi surface, and its temperature dependence resembles that of the BCS gap in a superconductor. The electron-hole condensate is coupled to a macroscopic occupation of $2k_F$ phonons, which increases the electron effective mass by the Fröhlich mass ratio, $\mu = M/m$. A spin density wave (SDW) is equivalent to two out-of-phase CDWs for the spin-up and spin-down subbands, and is thought to be mediated by electron-electron interactions. In either case, the DW wavelength is given by $\lambda_{DW} = \pi\hbar/p_F = \pi/k_F$.

The Peierls gaps at $\pm k_F$ can be displaced in momentum space in an unpinned DW, resulting in a DW current $\propto \partial\phi/\partial t$. Its fermionic representation resembles a Luttinger liquid [2] of dressed electrons (quantum solitons), except the Fermi velocity gets replaced by the phason velocity $c_0 = \mu^{-1/2} v_F$ [3] and a real DW comprises many parallel chains. When pinned by impurities, a DW can still transport a current provided the applied electric field exceeds a threshold E_T. Displacing the DW by one wavelength (advancing ϕ by 2π) returns the system to its original state, except for charge displaced between contacts, so the pinning energy is a periodic function of phase. Observations of narrow-band noise and coherent oscillations [1] suggest that a sine-Gordon model can provide a suitable idealized framework, provided electrostatic interactions between charged kinks are taken into account.

Published under licence by IOP Publishing Ltd

2. Coulomb blockade threshold field and time-correlated soliton tunnelling

The pivotal experiments on $NbSe_3$ by Monceau et al [4] revealed Zener-like behaviour of the form: $J_{cdw} \sim E\exp[-E_0/E]$. The characteristic field E_0 proved far too small to represent Zener tunnelling of normal electrons across the Peierls gap. This observation motivated Bardeen [5] to propose coherent Zener tunnelling of condensed electrons through a tiny "pinning gap" created by impurity scattering which, unlike the Peierls gaps, would be fixed in momentum space at $\pm k_F$. An instanton picture, proposed by Maki [6], interprets the Zener-like behavior in terms of quantum nucleation of soliton-antisoliton pairs. Later experiments [7] revealed a sharp threshold field E_T for CDW transport, which Bardeen interpreted [8] in terms of a maximum tunnelling distance. Impedance and mixing experiments [9-11] showing agreement with photon-assisted tunnelling theory, and $h/2e$ quantum interference in the CDW magnetoconductance of $NbSe_3$ crystals [12] with columnar defects and TaS_3 rings [13], provide compelling evidence of their quantum character.

Another interpretation of the threshold field [14], as an S-$\hat{S}$ pair-creation threshold due to Coulomb blockade [15], is motivated by Coleman's paper on the massive Schwinger model [16]. A pair of soliton (S) and antisoliton ($\hat{S}$) domain walls [17] carry charges $\pm Q_0 = \pm 2Ne$ (in a fully condensed system counting both spins) where N is the number of parallel chains. Like a parallel plate capacitor, these produce an internal electric field of magnitude, $E^* = Q_0/\epsilon A$, where ϵ is the dielectric constant and A is the cross-sectional area, as shown in figure 1. When an external field E is applied, the difference in electrostatic energies of a state with a pair of separation l and that of the 'vacuum' with no pair is: $\Delta U = \frac{1}{2}\epsilon Al[(E \pm E^*)^2 - E^2] = Q_0 l[\frac{1}{2}E^* \pm E]$, which is positive when $|E| < \frac{1}{2}E^*$. Conservation of energy thus forbids pair production for fields less than the Coulomb blockade threshold, $E_T \equiv \frac{1}{2}E^* = Q_0/2\epsilon A \sim Ne/\epsilon A$, which can be substantially smaller than the classical depinning field. The experimentally observed scaling law [1], $\epsilon E_T \sim e n_{ch}$ (for a fully condensed system), where $n_{ch} = N/A$, arises naturally in this model.

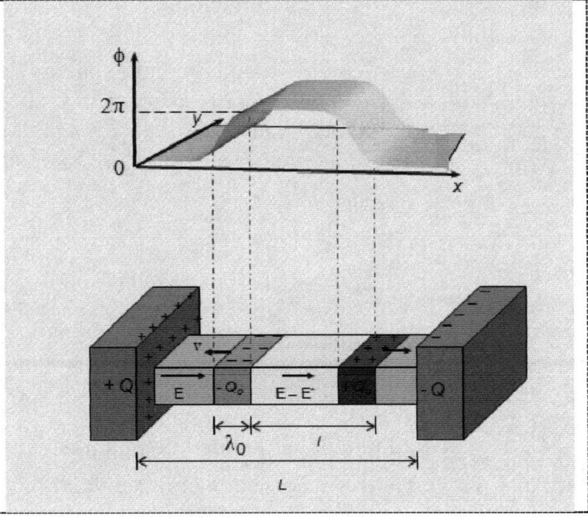

Figure 1. Top. Density wave phase *vs.* position, showing the production of a soliton-antisoliton domain wall pair.

Bottom. Model of density wave capacitance, in which the nucleated domain walls are moving toward the contacts. The applied field E partially or completely cancels the internal field E^*. The distance l, soliton width $\lambda_0 = c_0/\omega_0$, and crystal thickness are greatly exaggerated for clarity.

If the DW phase is fixed at *zero* at the contacts or at $\pm\infty$, advancing the phase by ϕ in the middle creates charged kinks that produce an internal electric field: $E_\phi = (E^*/2\pi)\phi$. E^* can also be written as $E^* = 2e/\epsilon A_{ch}$ (counting both spins), where A_{ch} is the cross-sectional area of a single chain, since $A = NA_{ch}$. If an applied field, represented by a dimensionless parameter $\theta = 2\pi(E/E^*) = \pi(E/E_T)$, partially cancels the internal field E_ϕ, this yields an electrostatic energy proportional to $(\theta - \phi)^2$. The Hamiltonian for a single chain can then be written as:

Journal of Physics: Conference Series **273** (2011) 012007 doi:10.1088/1742-6596/273/1/012007

$$H = \int_{-L/2}^{L/2} dx \left\{ \sum_{\sigma=\uparrow,\downarrow} \left[\frac{1}{2D}\Pi_\sigma^2(x) + \frac{1}{2}Dc_0^2\left(\frac{\partial\phi_\sigma(x)}{\partial x}\right)^2 + D\omega_0^2[1-\cos\phi_\sigma(x)] \right] \mp U\cos[\phi_\uparrow(x) - \phi_\downarrow(x)] + u_E\left(\theta - \phi(x)\right)^2 \right\} \quad (1)$$

where ϕ_σ represent phases corresponding to the two spin sub-bands, $\phi \equiv (\phi_\uparrow + \phi_\downarrow)/2$, $D = \frac{\mu\hbar}{4\pi v_F}$, ω_0 is the pinning frequency, $\Pi_\sigma(x) = D\frac{\partial\phi_\sigma(x)}{\partial t}$ is the canonical momentum density, and $u_E = \frac{e^2}{2\pi^2\epsilon A}$. The minus sign is appropriate for a CDW in the interaction term, $\mp U\cos[\phi_\uparrow - \phi_\downarrow]$, since this tends to keep the two spin sub-bands in-phase, whereas the + sign applies to an SDW. Defining $u_p \equiv D\omega_0^2$ and assuming $U \gg u_p$, we see that the interaction term tends to constrain the phases along a saddle point in a plot of potential energy *vs.* $\phi_\uparrow$ & $\phi_\downarrow$ (focusing for the moment on the CDW) such that $\phi_\uparrow \approx \phi_\downarrow \approx \phi$, which simplifies the remaining potential energy contributions to:

$$U[\phi] = \int_{-L/2}^{L/2} dx\left\{2u_p[1-\cos\phi(x)] + u_E\left(\theta - \phi(x)\right)^2\right\}. \quad (2)$$

The usual linear coupling $\sim\theta\phi$ is implicitly contained in the quadratic energy term, which also includes electrostatic contributions $\sim\phi^2$ and θ^2. The inclusion of both pinning and electrostatic contributions clearly delineates a quantum threshold, which is much smaller than the classical depinning threshold when $u_E \ll u_p$. Setting $F \propto -\partial u/\partial\phi = 0$, in the limit $\alpha \equiv u_E/u_p \ll 1$ we find (using $\sin\phi \approx \phi$) that $\phi \cong [u_E/(u_E + u_p)]\theta \approx \alpha\theta \sim 0$ when $\theta < \pi$. When $\theta > \pi$, the state with $\phi \sim 2\pi$ becomes the lowest energy state, as illustrated in figure 2 (left). Thus, $\theta = \pi$ demarcates the boundary above which a dynamical phase transition via quantum decay of the "false vacuum" becomes possible.

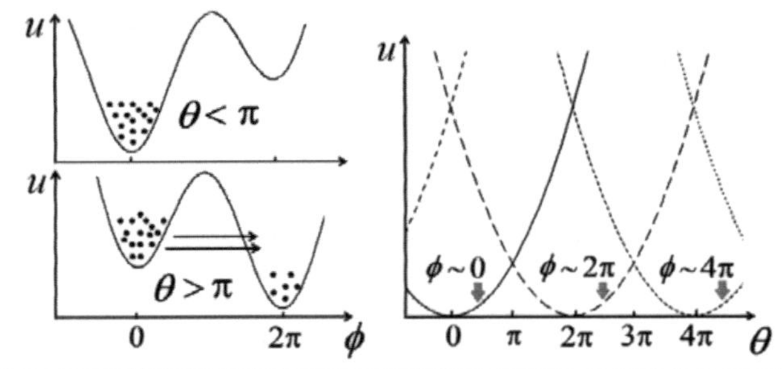

Figure 2. Left. Potential energy *vs.* ϕ for two values of θ, in which the many spatial degrees of freedom are illustrated schematically. Tunnelling can only occur if $\theta > \pi$, when parts of the system tunnel into the adjacent well. **Right.** Potential energy vs. θ, in which the phase is sitting in various potential minima, $\phi \sim 2\pi n$.

The right side of figure 2 shows plots of u *vs.* θ, obtained by minimizing the energy for $\phi \sim 2\pi n$. An applied field induces a displacement charge Q across the contacts, which is related to θ by: $\theta = 2\pi(Q/Q_0)$. Suppose the system is initially sitting in the $\phi \sim 0$ potential minimum. When θ exceeds π, what was once the ground state (true vacuum) now becomes a metastable state (false vacuum) as the solid line in the right side of figure 2 crosses above the dashed line. One or more bubbles of true vacuum then nucleate, in which ϕ advances by 2π with soliton-antisoliton domain walls at the boundaries, and each bubble rapidly expands as the electric field drives the oppositely charged soliton domain walls toward the contacts (figure 1). The total charging energy, $\propto (\theta - 2\pi n)^2$ in figure 2, can be re-expressed (for N parallel chains) in terms of total displacement charge as $U_C = (Q - nQ_0)^2/2C$, showing a clear analogy to time-correlated single-electron tunnelling in Coulomb blockade tunnel junctions. The origin of the narrow-band noise, coherent voltage or current oscillations, and mode locking with an ac source thus becomes clear in this context.

3. Discussion

Observations of flat, essentially bias voltage-independent dielectric and other ac response data below threshold [11] suggest that many DW samples show little phase displacement below threshold and are thus in the region where $\alpha = u_E/u_p \ll 1$ (recalling that $\phi \approx \alpha\theta$). This is consistent with the interpretation that, for these samples, the Coulomb blockade threshold for soliton pair production is well below the classical depinning field. Dias and Lemos [18] propose that the probability of nucleating a soliton domain wall pair by tunnelling is determined by the mass-energy (or action) density of a domain wall rather than the total mass-energy, and is therefore scale invariant – i.e. independent of the transverse dimensions. An alternative argument [3,17] is that each domain wall can be viewed as a condensate of microscopic quantum solitons that tunnel coherently within a system with only one thermal degree of freedom, as in Josephson tunnelling. This approach suggests that, rather than computing the total Euclidean action or "bounce," the tunnelling Hamiltonian matrix element concept should be extended to other systems, such as quantum solitons. A growing body of literature proposes that the entire universe came into existence through a quantum nucleation process, and subsequently tunnelled hundreds or thousands of times into successively lower potential wells, in order to explain why the cosmological constant is small and positive (e.g. see [19]). This being the case, it would make sense to determine whether large scale quantum tunnelling can occur in the laboratory. Moreover, any possibility of observing quantum interference in rings of NbS_3, which undergoes a Peierls transition above room temperature, could be of practical importance for biomedical and other applications.

4. References

[1] Grüner G 1994 *Density Waves in Solids* (Reading, MA: Addison-Wesley)
[2] Artemenko S N and Remisov S V 2005 *Phys. Rev. B* **72** 125118
[3] Maiti A and Miller J H Jr 1991 *Phys. Rev. B* **43** 12205
[4] Monceau P, Ong N P, Portis A M, Meershaut A and Rouxel J 1976 *Phys. Rev. Lett.* **37** 602
[5] Bardeen J 1979 *Phys. Rev. Lett.* **42** 1498
[6] Maki K 1977 *Phys. Rev. Lett.* **39** 46
[7] Fleming R M and Grimes C C 1979 *Phys. Rev. Lett.* **42** 1423
[8] Bardeen J 1980 *Phys. Rev. Lett.* **45** 1978
[9] Miller J H Jr, Richard J, Tucker J R, and Bardeen J 1983 *Phys. Rev. Lett.* **51** 1592
[10] Miller J H Jr, Thorne R E, Lyons W G, Lyding J W, Tucker J R and Bardeen J 1985 *Phys. Rev. B* **31** 5229
[11] Miller J H Jr 1985 *Quantum Tunneling of Charge Density Waves in Quasi-One-Dimensional Conductors* (PhD Dissertation, University of Illinois at Urbana-Champaign)
[12] Latyshev Yu I, Laborde O, Monceau P and Klaumünzer S 1997 *Phys. Rev. Lett.* **78** 919
[13] Tsubota M, Inagaki K and Tanda S 2009 *Physica B* **404** 416
[14] Krive I V and Rozhavsky A S 1985 *Solid State Commun.* **55** 691
[15] Miller J H Jr, Ordóñez C and Prodan E 2000 *Phys. Rev. Lett.* **84** 1555
[16] Coleman S 1976 *Ann. Phys., NY* **101** 239
[17] Miller J H Jr, Cárdenas G, García-Perez A, More W and Beckwith A W 2003 *J. Phys. A: Math. Gen.* **36** 9209
[18] Dias O J C and Lemos J P S 2001 *J. Math. Phys.* **42** 3292
[19] Steinhardt P J and Turok N 2006 *Science* **312** 1180

Acknowledgments

The author gratefully acknowledges the assistance of Divya Pamaraj in preparing figure 2 and financial support from NIH (NCI & NHLBI) R21CA133153, NIH (NCI) ARRA supplement 3R21CA133153-03S, and the State of Texas through the Texas Center for Superconductivity at the University of Houston and the Norman Hackerman Advanced Research Program.

Journal of Physics: Conference Series **273** (2011) 012008 doi:10.1088/1742-6596/273/1/012008

Color superfluidity and trionic state of three-component lattice fermionic atoms

Shin-ya Miyatake[1], **Kensuke Inaba**[2,3], **and Sei-ichiro Suga**[4]

[1] Department of Applied Physics, Osaka University, Suita, Osaka 565-0871, Japan
[2] NTT Basic Research Laboratories, NTT Corporation, Atsugi 243-0198, Japan
[3] JST, CREST, Sanbancho, Chiyoda-ku, Tokyo 102-0075, Japan
[4] Department of Materials Science and Chemistry, University of Hyogo, Himeji 671-2280, Japan

E-mail: suga@eng.u-hyogo.ac.jp

Abstract. We investigate three-component (colors) fermionic atoms with anisotropic attractive interactions in optical lattices at half filling using the dynamical mean field theory. In the weakly interacting region the color superfluid (CSF) state emerges, where atoms with two of three colors form the Cooper pairs. As the interaction is increased, a first-order quantum phase transition to the trionic state occurs, where singlet bound states of three different color atoms are formed. We show that the trionic state survives up to a fairly large anisotropic region. The phase diagram for the CSF and trionic states is obtained.

1. Introduction

Studies of ultracold fermionic atoms have attracted considerable attention of both theorists and experimentalists. Feshbach resonances provide the means for controlling both the strength of the interaction between fermionic atoms and its sign [1]. By loading fermionic atoms into optical lattices, diverse interaction configurations can be introduced. Using these techniques, fascinating phenomena caused by many-body effects have been observed. Recently, a balanced mixture of three different hyperfine states of ^{6}Li fermionic atoms was succeeded in creating [2, 3]. The enhanced three-body loss coefficient was observed at a certain magnetic field. Its origin was discussed in connection with the formation of the trionic state which is singlet bound states of three fermionic atoms with different internal degrees of freedom (colors). For the three-component fermionic atoms in optical lattices, it was shown theoretically that for the small attractive interaction a color superfluid (CSF) emerges, where atoms with two of three colors form the Cooper pairs and the third ones remain a Fermi liquid [4, 5, 6, 7]. As the attractive interaction becomes strong, a quantum phase transition from the CSF to the trionic state occurs [6, 7]. In experiments, it is difficult to realize the isotropic interactions between atoms with different colors. In fact, at the magnetic field where the enhanced three-body loss coefficient was observed, two scattering lengths are nearly the same, while the remaining one takes a different enhanced value [2]. These results mean the anisotropy of the attractive interactions between atoms with three colors. These findings motivated us to investigate the stability of the CSF and trionic state in the three-component fermionic atoms in optical lattices with anisotropic interactions.

Published under licence by IOP Publishing Ltd

2. Model and method

In this paper, we investigate the effects of anisotropic attractive interactions in the three-component fermionic atoms in optical lattices. The system is assumed to keep a balanced population of fermionic atoms with three colors. The fermionic optical lattice systems can be well described by the following Hubbard-type Hamiltonian [8]:

$$\mathcal{H} = -t \sum_{\langle i,j\rangle,\alpha} c_{i\alpha}^\dagger c_{j\alpha} + \sum_{i,\alpha\neq\beta} \frac{U_{\alpha\beta}}{2} n_{i\alpha} n_{i\beta} - \sum_{i,\alpha} \mu_\alpha n_{i\alpha}, \tag{1}$$

where $c_{i\alpha}$ annihilates a fermionic atom with color $\alpha(=1,2,3)$ at the ith lattice site, $n_{i\alpha} = c_{i\alpha}^\dagger c_{i\alpha}$, the subscript $\langle i,j\rangle$ indicates the sum over the nearest-neighbor sites, and $U_{\alpha\beta}(<0)$ is the attractive interaction between two atoms with different colors α and β. We set $U_{12} \equiv U$ and $U_{23} = U_{31} \equiv U'$ with $U'/U < 1$.

We make use of a dynamical mean field theory (DMFT). In the DMFT, the lattice model is mapped onto a single impurity model connected dynamically to a bath. The Green's function is obtained via the self-consistent solution of this impurity problem. This retains nontrivial local quantum fluctuations missing in conventional mean-field theories. We apply here the two-site DMFT method [9], which allows us to study quantum phase transitions of lattice fermions qualitatively. To study the color superfluid of lattice fermions, we extend this method to the case when the superfluid order exists [10]. It is expected that the fermionic atoms for the strongest attractive interaction are paired to form the s-wave Cooper pair with an order parameter $\Delta_{\alpha\beta} \equiv \langle c_{i\alpha} c_{i\beta}\rangle$. We thus choose the gauge $\Delta_{12} = \Delta$ and $\Delta_{23} = \Delta_{31} = 0$. The order parameter of the color superfluid is thus Δ. We calculate the quasiparticle weights Z_α, which are defined by $Z_\alpha = [1 - d\mathrm{Re}[\Sigma_\alpha(\omega)]/d\omega|_{\omega=0}]^{-1}$ in terms of the self-energy $\Sigma_\alpha(\omega)$ of the respective color. In our case, $Z_1 = Z_2 \equiv Z$ and $Z_3 \equiv Z'$. The quasiparticle weight Z represents the coherent spectral weight of the Bogoliubov quasiparticle for $\alpha = 1,2$. We also calculate the single-particle excitation spectra (SPES) $\rho_\alpha(\omega)$. The density of states for an infinite-dimensional Bethe lattice is adopted, which is independent of color. The chemical potentials are set to be $\mu_1 = \mu_2 = (U+U')/2$ and $\mu_3 = U'$ so that particle-hole symmetry can be satisfied. In this case, the system is half filling. We take no account of the possibility of the density wave state. The hopping integral t is used in units of energy.

3. Numerical results

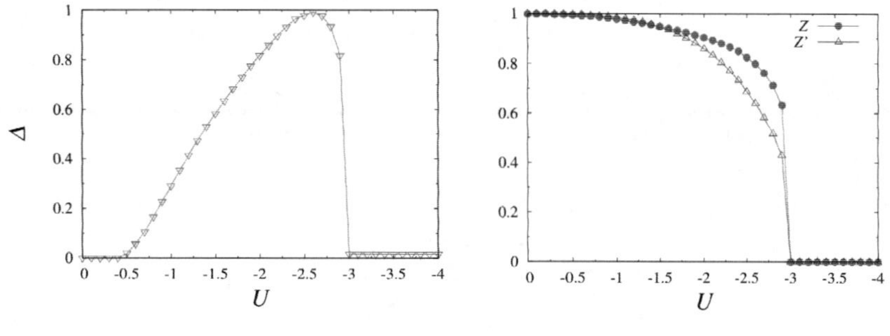

Figure 1. (Color Online) (a) The CSF order parameter Δ and (b) the quasiparticle weights Z and Z' as a function of U for $U'/U = 0.6$.

In Fig. 1, we show the order parameter Δ, and the quasiparticle weights Z and Z' for the fixed ratio $U'/U = 0.6$ as a function of U. For the weak interaction, Δ takes a very small value and increases with increasing $|U|$. The quasiparticle weights Z and Z' significantly decrease from unity in different ways, reflecting the difference in the renormalization of the attractive interaction. As U becomes strong, the quasiparticle weights jump to 0 with vanishing Δ simultaneously. These results indicate that a discontinuous quantum phase transition occurs at $U = -3.0$. In correlated lattice fermion systems, the zero quasiparticle weight means a transition to the Mott insulating state. For the strong attractive interaction, it is considered that atoms with different three colors form the trion. In the DMFT, the trion is caused by local correlation effects and thus localized on a site, yielding a Mott insulator.

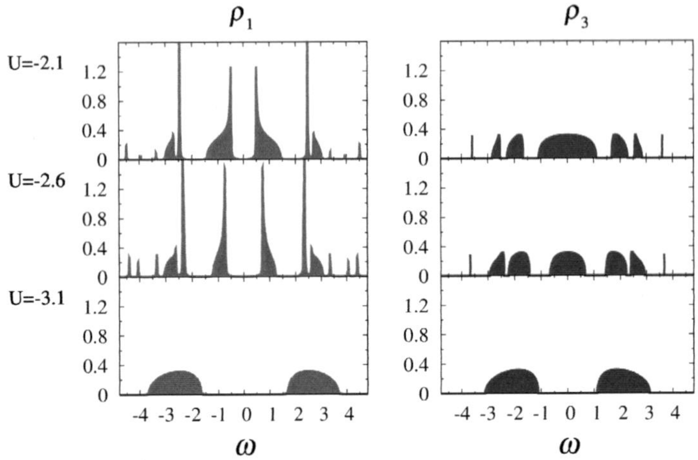

Figure 2. (Color Online) The single-particle excitation spectra for $U = -2.1, -2.6$ and -3.1 with $U'/U = 0.6$. (a) $\rho_1(\omega)(=\rho_2(\omega))$ and (b) $\rho_3(\omega)$.

We investigate the quantum phase transition in a spectral point of view. In Fig. 2, the SPES are shown for several values of U with $U'/U = 0.6$. For the CSF phase ($U = -2.1, -2.6$), we observe the increase of the superfluid gap around $\omega = 0$ in $\rho_1(\omega)(=\rho_2(\omega))$ and the narrowing of the coherent peaks, as the interaction becomes strong. These results mark the significant renormalization of the Bogoliubov quasiparticle in agreement with the decrease in Z. In $\rho_3(\omega)$, the quasiparticle peak at $\omega = 0$ is seen, indicating the Fermi liquid. The results demonstrate that atoms with colors 1 and 2 form the Cooper pair, while atoms with color 3 remain the Fermi liquid. For $U = -3.1$, the SPES show qualitatively the same profile which is typical of the Mott insulator. Actually, the spectral gap of the trionic state measured between the two peaks is ~ 4.96 (~ 3.72) for $\rho_1(\omega)$ ($\rho_3(\omega)$). This value is the same as $|U + U'| = 4.96$ ($2|U'| = 3.72$) which is the energy needed to take one atom of color 1 (3) away from the trionic bound state. The results for the quasipartricle weights, the CSF order parameter, and the SPES are consistent with each other. In this way, we have confirmed that the CSF-trion quantum phase transition occurs even in the anisotropic interaction system.

Performing the same two-site DMFT calculations for various values of U'/U and U, we obtain the phase diagram at half filling as shown in Fig. 3. Note that the CSF-trion transition is of the first order irrespective of U'/U. We find that the trionic state is robust against the anisotropy of the attractive interaction and survives in a fairly large anisotropic region.

4. Discussions and Summary

In this study, we have shown that the CSF-trion transition is of the first order. This result is in contrast with that in Refs. [6, 7], where the CSF-trion transition below half filling is of

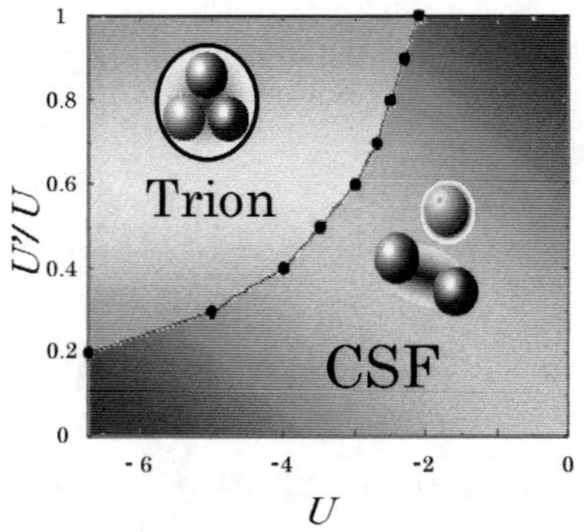

Figure 3. Phase diagram of the three-component attractive fermionic atoms in optical lattices at half lling.

the second order. Recently, two of the present authors have investigated the zero- and nite-temperatures properties of the present system using the self-energy functional approach [11]. In Ref. [11], it was shown that for half lling the CSF-trion transition is of the rst order. The discrepancy may originate in the di erence of the accuracy for the calculation of correlation e ects. In Refs. [6, 7], the trionic ground-state energy was estimated as $3Un$, where n is the lling. In the present calculation and also in Ref [11], we obtain the lower trionic ground-state energy than $3Un$. When the ground state energy in our calculations is expanded in terms of $1/U$, the zeroth order corresponds to $3Un$. The higher order corrections lower the ground state energy, which may yield a discontinuous change of the ground state energy between the CSF and the trionic state, resulting in the rst-order quantum phase transition.

In summary, we have investigated the CSF-trion quantum phase transition of the three-component fermionic atoms in optical lattices at half lling using the DMFT. We have shown that the trionic state is robust against the anisotropy of the attractive interactions.

Acknowledgments

We thank T. Higashiyama and M. Yamashita for useful comments and valuable discussions. Numerical computations were carried out at the Supercomputer Center, the Institute for Solid State Physics, University of Tokyo. This work was supported by Grants-in-Aid for Scienti c Research (C) (No. 20540390) from the Japan Society for the Promotion of Science and on Innovative Areas (No. 21104514) from the Ministry of Education, Culture, Sports, Science and Technology.

References

[1] For a review, see Chen Q, Stajic J, Tan S and Levin K 2005 *Phys. Rep.* **412** 1
[2] Ottenstein T B, Lompe T, Kohnen M, Wenz A N and Jochim S 2008 *Phys. Rev. Lett.* **101** 203202
[3] Huckans J H, Williams J R, Hazlett E L, Stites R W and O'Hara K M 2009, *Phys. Rev. Lett.* **102** 165302
[4] Honerkamp C and Hoftetter W 2004 *Phys. Rev. Lett.* **92** 170403
[5] Honerkamp C and Hoftetter W 2004 *Phys. Rev.* B **70** 094521
[6] Rapp A, Zaránd G, Honerkamp C and Hoftetter W 2007 *Phys. Rev. Lett.* **98** 160405
[7] Rapp A, Hoftetter W and Zaránd G 2008 *Phys. Rev.* B. **77** 144520
[8] Jaksch D, Bruder C, Cirac J I, Gardiner C W and Zoller P 1998 *Phys. Rev. Lett.* **81** 3108
[9] Pottho M 2001 *Phys. Rev.* B **64** 165114
[10] Higashiyama T, Inaba K and Suga S 2008, *Phys. Rev.* A **77** 043624
[11] Inaba K and Suga S 2009 *Phys. Rev.* A **80** 041602(R)

Journal of Physics: Conference Series **273** (2011) 012009 doi:10.1088/1742-6596/273/1/012009

Low pressure phase diagram of $CeCoGe_3$

Masakazu Mizoo, Takashi Nishioka, Harukazu Kato, Masahiro Matsumura

Graduate School of Integrated Arts and Sciences, Kochi University, Kochi 780-8520, Japan

E-mail: nisioka@kochi-u.ac.jp

Abstract. We have performed measurements of the Hall magnetization and electrical resistivity on a pressure-induced antiferromagnetic heavy fermion superconductor $CeCoGe_3$ at pressures up to ~1.7 GPa, which crystallizes in the tetragonal $BaNiSn_3$-type structure without inversion symmetry. We have clarified that five successive phase transitions appear at ambient pressure. As pressure is applied, these transitions approach one another until ~1.2 GPa, and then separate again. We have also found another phase transition emerges above ~0.5 GPa, which is consistent with a recent Co-NQR study. From these results, we propose a low pressure *P-T* phase diagram, which contains at least eight ordered phases below ~2 GPa. This phase diagram is consistent with a recent reported high pressure phase diagram.

$CeCoGe_3$ crystallizes in the $BaNiSn_3$-type structure without inversion symmetry, and becomes a superconductor below 0.7 K and at ~5.5 GPa [1]. In addition, $CeCoGe_3$ undergoes successive phase transitions despite only one Ce crystallographic site. This was first reported by Pecharsky *et al.* from the macroscopic measurements of a polycrystalline sample [2]. They claimed that $CeCoGe_3$ shows a ferrimagnetic transition at 21 K and an antiferromagnetic transition at 18 K. Contrary to this, Eom *et al.* insisted that the transition at 21 K is an antiferromagnetic one and that at 18 K is a ferrimagnetic one from the macroscopic measurements of a polycrystalline sample [3]. Das and Nigam reported two phase transitions at 21 K and 16 K in $CeCoGe_3$ from their precise electrical resistivity measurements on a polycrystalline sample [3]. Single crystal study of $CeCoGe_3$ was first performed by Thamizhavel *et al.*, and they proposed three antiferromagnetic transitions at T_{N1}=21 K, T_{N2}=12 K and T_{N3}=8 K [5]. From the microscopic point of view, Matsumura *et al.* clarified that at least two phase transitions exist from Co-NQR study [6]. The neutron diffraction study revealed the complex magnetic structure [7, 8]. Thus, how many phase transitions occur in $CeCoGe_3$ is still controversial. Recently, Knebel *el al.* performed high pressure specific measurements on a single crystalline $CeCoGe_3$ and constructed a *P-T* phase diagram [9]. They clarified six ordered phases. However, the relationship between the transitions at ambient pressure and those at high pressures is not clear due to the lack of low pressure measurement especially below 1 GPa. In addition, their phase diagram is based on only specific heat measurements and does not involve magnetic information. They also found the anomalous superconductivity with extremely high upper critical field.

In order to understand the anomalous superconductivity, the physical properties of the normal state should be clarified. We examine the following two queries in this paper. The first one is how many true phase transitions exist at ambient pressure. The second one is how relate with the transitions at ambient pressure and those at high pressures. In order to resolve these, we

Published under licence by IOP Publishing Ltd

performed detailed measurements of the Hall magnetization [10] up to ~1.0 GPa and electrical resistivity up to ~1.7 GPa, and propose a low pressure *P-T* phase diagram of $CeCoGe_3$, which is consistent with a high pressure *P-T* phase diagram [9] and fills the insufficiency of the high pressure phase diagram.

The polycrystalline sample was prepared by melting the constituents in an arc furnace under an argon atmosphere, after which it was annealed at 1000 °C for one week. The X-ray analysis confirmed almost single phase formation of sample. The magnetization was measured using Hall sensor at a field of 15 Oe (Hall magnetometer [10]). The value V_H was evaluated by averaging the absolute values of positive and negative fields to subtract the unbalanced voltage of the Hall sensor, so that V_H gives similar information to the ac susceptibility χ_{ac}. The electrical resistivity was measured using the conventional ac four terminal method using the commercial ac resistance bridge (LakeShore 370). Pressure was generated by a hybrid piston-cylinder cell and the pressure transmitting medium was Daphne 7373 oil. The low temperature pressure was determined by the superconducting transition temperature of indium measured by χ_{ac}. Low temperature was generated by Kochi university method using GM refrigerator [11].

Figure 1 shows the temperature dependence of V_H at measured pressures. At ambient pressure, $V_H(T)$ is similar to $\chi_{ac}(T)$ (— · —) as mentioned above. Some discrepancy probably comes from the difference of their driving field. The value of V_H suddenly increases below T_{M1}=19.2 K and shows a broad peak at T_{M2}=16.2 K. Several small jagged anomalies observed below T_{M2} are due to experimental problems. Zero (geographic) field magnetization (······) measured by a SQUID magnetometer suddenly decreases below T_{M3}=12.5 K and shows a small hump at T_{M4}. At 0.13 GPa, zero-field-cooled V_H shows distinct anomalies at T_{M3} and T_{M4}, and V_H shows a clear anomaly at T_{M2}. These anomalies become clearer at 0.37 GPa. As pressure is increased, these anomalies approach gradually one another and turns into one sharp peak at ~1 GPa. Further increasing pressure, the peak magnitude is strongly suppressed and we can see only a tiny hump at 1.05 GPa in $V_H(T)$.

Figure 2 shows the temperature dependence of the electrical resitivity ρ at several pressures. As can be seen from the figure, the shape of $\rho(T)$ is very complicated. We define the characterestic temperatures as $T_{R1} \sim T_{R7}$, where T_{R1} is the temperature at which $d\rho/dT$ indicates the first increase when temperature decreases, and $T_{R2} \sim T_{R7}$ are the temperature of the maximums of $d\rho/dT$. This definition is the identical with that by Das and Nigam [4]. These anomalies also approach one another up to ~1 GPa like Fig. 2 and separate again at higher fields.

Figure 3 shows a phase diagram obtained from this study. Previous reported transition temperatures estimated from the magnetization *M* (□) [5] and the specific heat *C* (○) [9] are also plotted. The difference at ambient pressure between T_{R1} and T_{R2} is small, while it increases under pressure. Since high temperature anomaly in *C* occurs at T_{R1}, the anomaly in *C* and a sudden increase in the magnetization occur at different temperatures. Although Thamizhavel *et al.* regarded these two anomalies occurred at the same temperature T_{N1}, there is slight difference between the transition temperature from *M* and that from *C* (see the temperatures indicated by arrows in Fig. 5 for *M* and Fig. 11 for *C* in Ref. [5]). From these facts, $CeCoGe_3$ shows antiferromagnetic transition at T_{R1} and ferrimagnetic-like one at T_{M1}, which has been pointed out by Eom *et al.* [3]. We consider another phase transition exists at ~16 K, because not only V_H and $d\rho/dT$ in this study but also *C* in literatures [2, 5] show anomalies at around this temperature. In fact, Pecharsky *et al.* [2] and Das and Nigam *et al.* [4] adopted these anomalies as transition temperatures. From the above consideration, we suggest that five successive phase transitions appear in $CeCoGe_3$ at ambient pressure. Four transitions except for the highest temperature one approach gradually one another and seems to merge into one anomaly at ~1 GPa. We can see another new phase transition appears above ~0.5 GPa. Recently, Matsumura *et al.* reported the magnetic structures at 0 GPa and 0.77 GPa are completely different from

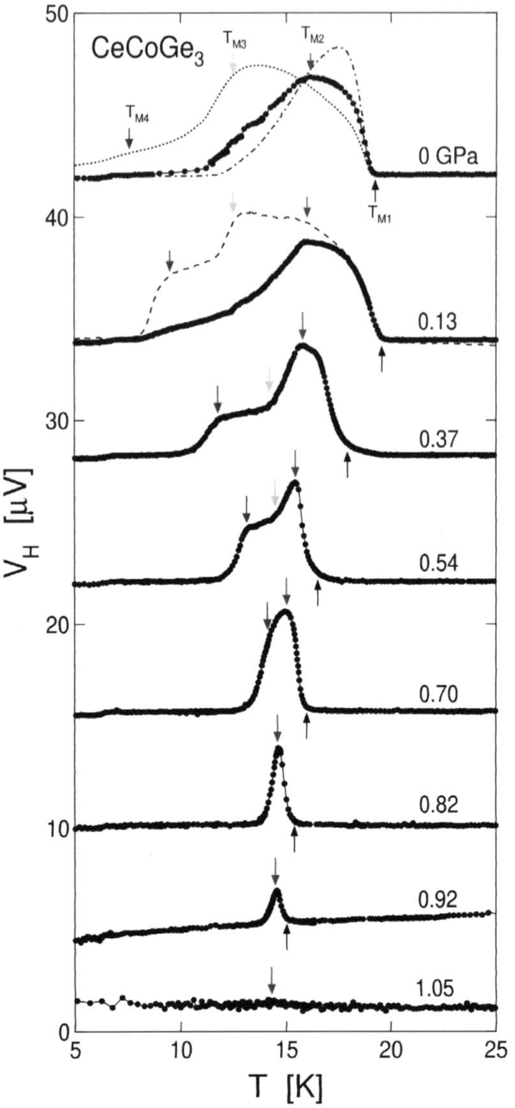

Figure 1. Temperature dependence of the Hall magnetization of $CeCoGe_3$ at a field of 15 Oe under pressure. The longitudinal zero point is shifted for clarity. The arrows indicate the temperatures where some anomaly appears in V_H. The dotted line and the dashed-dotted line at 0 GPa are the results of magnetization at zero field (earth field) and ac susceptibility at a driving field of 1.0 Oe, respectively. The dashed line are the results of zero-field-cooled (ZFC) Hall magnetization (- - - -) at 15 Oe.

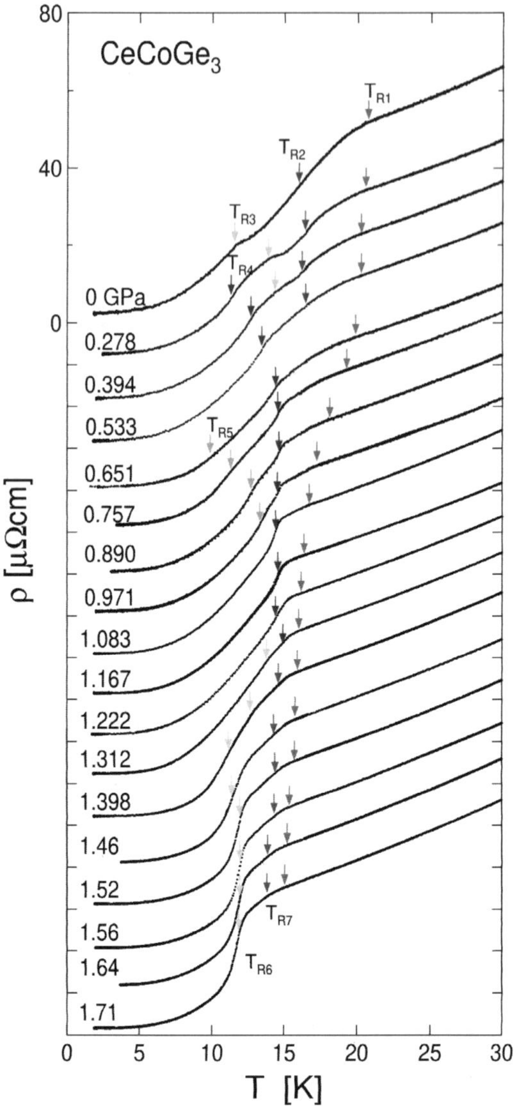

Figure 2. Temperature dependence of the electrical resistivity ρ under pressure. The arrows indicate the temperatures where some anomaly appears in ρ. T_{R1} is defined as a temperature where $d\rho/dT$ increases rapidly and $T_{R2} \sim T_{R7}$ are defined as peak temperatures of $d\rho/dT$.

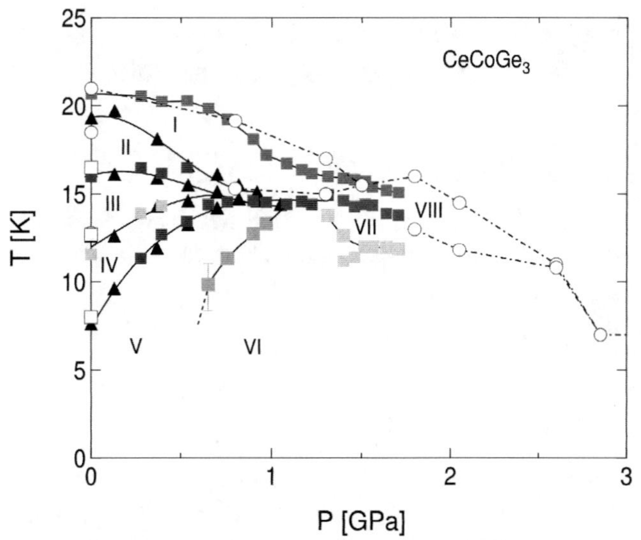

Figure 3. Proposed *P*-*T* phase diagram at low pressures for $CeCoGe_3$. The closed symbols are obtained from this study (Hall magnetization (●), electrical resistivity (■). The open circles are from literatures (specific heat (○) [2, 9], electrical resistivity (□) [5].

Co-NQR study [6], therefore, this phase transition should exist. The phase transitions gradually approach one another at ~1 GPa, and separate again above ~1.2 GPa. We can see another phase transition seems to appear about 1.4 GPa, which requires further research. It should be noted that this phase diagram is consistent with a high pressure phase diagram [9], and complements it. From Fig. 3, $CeCoGe_3$ contains eight ordered phases below ~2 GPa, which may correlate with the lack of inversion symmetry in the crystal structure of $CeCoGe_3$.

This work was partially supported by a Grant-in-Aid for Scientific Research (C) of the Ministry of Education, Culture, Sports, Science and Technology, Japan and the Comprehensive Support Programs for Creation of Regional Innovation of Japan Science and Technology Agency (JST).

References

[1] Settai R, Sugitani I, Okuda I, Thamizhavel A, Nakashima M, Ōnuki Y and Harima H 2007 *J. Magn. Magn. Mat.* **310** 844.
[2] Pecharsky V N, Hyun O -B and Gschneidner Jr K A 1993 *Phys. Rev. B* **47** 11839.
[3] Eom D, Ishikawa M, Kitagawa G and Takeda N 1998 *J. Phys. Soc. Jpn.* **67** 2495.
[4] Das A and Nigam A K 2000 *J. Phys.: Cond. Mat.* **12** 1315.
[5] Thamizhavel A, Takeuchi T, Matsuda T D, Haga Y, Sugiyama K, Settai R and Ōnuki Y 2005 *J. Phys. Soc. Jpn.* **74** 1858.
[6] Matsumura M, Sato Y, Mizoo M, Kawamura Y, Kato Hand Nishioka T (SCES 2010).
[7] Das A, Kremer R K, R. Pöttgen and Ouladdiaf B 2006 *Physica B* **378 - 380** 837.
[8] Kaneko K, Metoki N, Takeuchi T, Matsuda T D, Haga Y, Thamizhavel A, Settai R and Ōnuki Y 2009 *J. Phys.: Conf. Ser.* **150** 042082.
[9] Knebel G, Aoki D, Lapertot G, Salce B, Flouquet J, Kawai T, Muranaka H, Settai R and Ōnuki Y 2009 *J. Phys. Soc. Jpn.* **78** 074714.
[10] Mizoo M, Kawamura Y, Nishioka T, kato H and Matsumura M 2010 *J. Phys.: Conf. Ser.* **200** 112007.
[11] Nishioka T, Sumida T, Takesaka T, Kawamura Y, Kato H and Matsumura M 2009 *J. Phys.: Conf. Ser.* **150** 012030.

Journal of Physics: Conference Series **273** (2011) 012010 doi:10.1088/1742-6596/273/1/012010

Transport and magnetic behavior under pressure and high-resolution photoemission studies of $Ce_2Rh_{0.7}Co_{0.3}Si_3$, an alloy on the verge of quantum critical point

K. Mukherjee, Kartik K Iyer, Swapnil Patil, K. Maiti, and E.V. Sampathkumaran

Department of Condensed Matter Physics and Materials Science, Tata Institute of Fundamental Research, Homi Bhabha Road, Mumbai-400005, India.
E-mail: kmukherjee@tifr.res.in
E-mail: sampath@tifr.res.in

Abstract. We report the influence of external pressure on the temperature dependence of magnetization and electrical resistivity as well as high-resolution photoemission studies for an alloy, $Ce_2Rh_{0.7}Co_{0.3}Si_3$, ordering magnetically below 3 K. It is found that external pressure has the same effect as that induced by (further) Co substitution for Rh in the series, $Ce_2Rh_{1-x}Co_xSi_3$, resulting in qualitative changes in the features in the magnetic and transport data, with a suppression of magnetic ordering followed by quantum critical point effect. The high-resolution photoemission spectra reveal signature of Kondo feature even at ambient pressure. These findings support the validity of spin-density-wave picture in this series.

1. Introduction

There is a considerable debate in the current literature [1-3] on how Fermi surface evolves as one gradually transforms a magnetically ordered Ce/U intermetallic to non-magnetic phase. That is, there are some materials where magnetically ordered state before 'quantum critical point' (QCP) can be viewed as a spin-density-wave (SDW) state due to itinerancy of *f*-electrons (associated with the Kondo effect), while some other experiments support 'localized picture'. Recently, we addressed [4] this question in the solid solution, $Ce_2Rh_{1-x}Co_xSi_3$, forming in an AlB_2-derived hexagonal structure. Ce_2RhSi_3 [5] and Ce_2CoSi_3 [6] have been known to be antiferromagnetic (near T_N = 7 K) and non-magnetic respectively, QCP near x= 0.6. The electrical resistivity (ρ) data showed distinct qualitative changes below T_N with increasing Co content. In particular, apart from other features, there is a distinct upturn appearing in $\rho(T)$ below T_N for x= 0.3 which disappeared with increasing x near QCP (see figure 1). This was taken as an evidence of gradual transformation of Fermi surface as QCP is approached that supports 'spin-density-wave picture' rather than 'local moment picture'. In order to support this line of argument, we considered it worthwhile to probe how external pressure influences the transport behavior for such a composition. Therefore, we carried out high pressure (< 15 kbar) *ac* resistivity (ρ) and *dc* magnetization *(M)* studies on the alloy, $Ce_2Rh_{0.7}Co_{0.3}Si_3$. In addition, we have performed high-resolution (< 5 meV) photoemission studies as a function of temperature *(T)* to see the evolution of Kondo features. Our results support the views based on spin density wave scenario as discussed earlier for this family [4].

2. Experimental details

The polycrystalline specimen studied here is the same as reported earlier [4]. $M(T)/H$ studies under pressure (≤10 kbar) were carried out (1.8-300 K) in a magnetic field of 5 kOe with a SQUID magnetometer (Quantum Design) using a pressure cell procured from

Published under licence by IOP Publishing Ltd

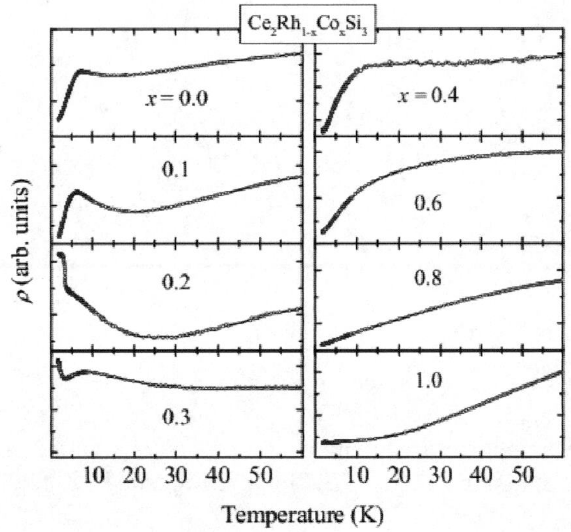

Figure 1. Electrical resistivity (linear scale) at low temperatures for the alloys, $Ce_2Rh_{1-x}Co_xSi_3$. The data were taken from Ref. 4.

EasyLab Technologies Ltd (UK) in a hydrostatic pressure medium of daphne oil. The $\rho(T)$ data in zero field and in field (<140 kOe) were carried out (1.8-300 K) with a physical property measurements system (Quantum Design) in a hydrostatic medium of a mixture of pentane and iso-pentane up to $\leq$15 kbar. The electrical contacts of the leads to the specimen were made by a conducting silver paste. Photoemission measurements were carried out on scraped sample surface using an electron analyzer SES2002 and monochromatic photon sources from Gammadata Scienta at a base pressure of 3×10^{-11} torr and an energy resolution set to 5 meV.

3. Results and discussion

The results of magnetization measurements measured in a field of 5 kOe at low temperatures (20 K) are shown in figure 2(*a*) in the form of a plot of *M*/*H versus T* to enable us to draw qualitative conclusions on the magnetic ordering behavior. The feature due to magnetic ordering appearing as a change in slope at 3 K for ambient pressure conditions persists at 5 kbar as well, though it is prominently seen in the form of a peak for this pressure. For a slight increase of pressure to 6 kbar, interestingly, this maximum disappears. For an application of 10 kbar, it appears that the Curie Weiss variation persists down to 1.8 K, as though the magnetic ordering is suppressed. In figure 2(*b*) and 2(*c*), we show the $\rho(T)$ behavior. It is clear that the upturn below 3 K (due to possible SDW gap) seen at ambient pressure vanishes at 5 kbar resulting in a positive $d\rho/dT$ at very low temperatures, as though there is a change in the Fermi level position with respect to the gap. However, the feature (a drop) near 8 K attributable to the Kondo coherence effect is enhanced marginally by a few degrees, which appears to smear out the double-peaked structure (characterizing Kondo lattices) in $\rho(T)$. This smearing is very significant for higher pressures (10 and 15 kbar) resulting in a single broad peak

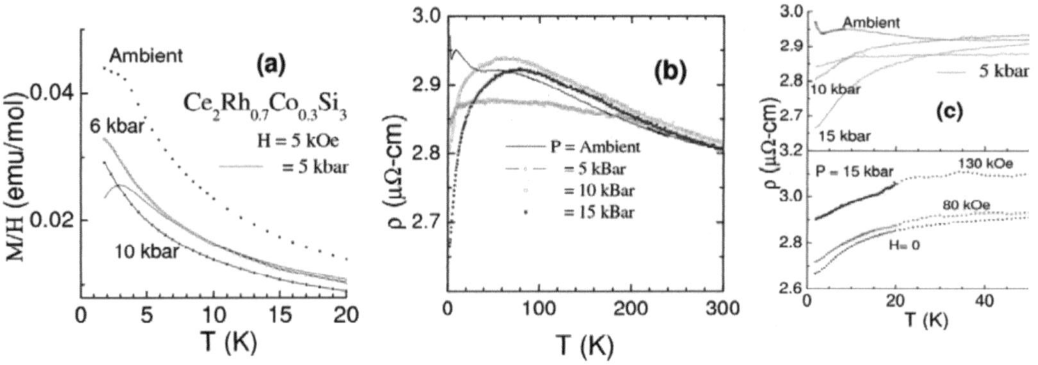

Figure 2. (*a*) Temperature dependence of magnetization divided by magnetic field, under external pressure of the alloy, $Ce_2Rh_{0.7}Co_{0.3}Si_3$. (*b*) Electrical resistivity as a function of temperature for the same specimen under pressures is plotted. (*c*) The ρ data under pressure as well as in magnetic field at low temperatures are plotted.

(around 60 and 70 K respectively). It is well-known that an increase of this (Kondo coherence) peak with temperature is the characteristic influence of pressure in the Ce-based Kondo lattices. A noteworthy feature is that ρ appears to vary linearly below about 8 K for P = 10 kbar (the pressure at which magnetic ordering appears to get suppressed), as though at this pressure-induced QCP, non-Fermi-liquid behavior develops. For the higher pressure (15 kbar), quasi-linearity with a higher exponent ($\sim T^{1.2}$) is distinctly observed. All these features are qualitatively the same as that seen for gradual replacement of Rh by Co, thereby endorsing the arguments in Ref. 4. An additional feature we stress is that the quasi-linear behavior seen in $\rho(T)$ at 15 kbar interestingly becomes linear if ρ is measured in the presence of a large field (say, e.g., 130 kOe) as shown in figure 2(*c*), as though there is a tendency to move away from Fermi-liquid behavior at very high fields for the pressurized specimen.

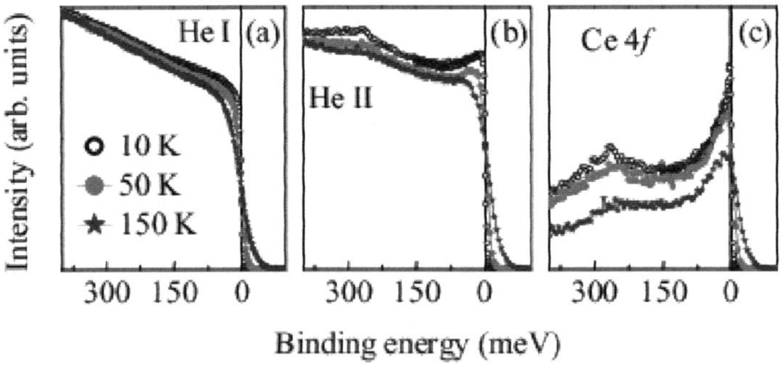

Figure 3. (*a*) He I and (b) He II spectra of $Ce_2Rh_{0.7}Co_{0.3}Si_3$ at different temperatures. (c) (He II - He I) spectra presenting the Ce 4f signal at different temperatures.

In Fig. 3, we show the high resolution He I and He II spectra collected at different temperatures. He I spectra shown in Fig. 3(a) exhibit typical temperature induced change at the Fermi level following closely the Fermi-Dirac distribution function; correlation induced features in such strongly correlated systems containing late transition metals appear at higher energies [7,8]. Interestingly, the He II spectra in Fig. 3(b) reveal two distinct features: one at the Fermi level and the other around 280 meV binding energies. Relative photoemission cross section of Ce 4*f* states is significantly larger in He II spectra compared to that in He I spectra. Thus, the features in the He II spectra can be attributed to the signature of Ce 4*f* states (the feature at 280 meV is spin-orbit split Ce 4*f* signal [8]) that can be delineated by subtracting He I spectra from the He II spectra. The subtracted spectra shown in figure 3(***c***) exhibit distinct signatures of the growth of Ce 4*f* features characteristic of Kondo resonance. The observation of this feature at ambient pressure for an antiferromagnetic alloy is interesting and demonstrates the persistence of Kondo compensation effect within the magnetically ordered region in Doniach's magnetic phase diagram, thereby supporting SDW scenario.

4. Conclusion

We have reported how external pressure influences magnetic and transport behavior of a magnetically ordered alloy, $x = 0.3$. In addition, high-resolution photoemission spectra down to low temperatures were also measured which show the signature of Kondo resonance feature at ambient pressure. These findings clearly support the validity of SDW picture in this family of materials.

References

[1] Paschen S, Lűhmann T, Wirth S, Gegenwart P, Trovarelli O, Geibel C, Steglich F, Colemann P, and Si Q, 2004 *Nature* (London) **432** 881
[2] Schrőder A, Aeppli G, Coldea R, Adams M, Stockert O, Lőhneyson H V, Bucher E, Ramazashvilli R, and Coleman P, 2000 *Nature* (London) **407** 351
[3] Kadowaki H, Tabata Y, Sato M, Aso N, Raymond S and Kawarazaki S, 2006 *Phys. Rev. Lett.* **96** 016401
[4] Patil S, Iyer K K, Maiti K and Sampathkumaran E V, 2008 *Phys. Rev.* B **77** 094443
[5] Das I and Sampathkumaran E V, *J. Magn. Magn. Mater* 1994 **137,** L239; Nakano T, Sengupta K, Rayaprol S, Hedo M, Uwatoko Y and Sampathkumaran E V, *J. Phys.: Condens. Matter* 2007 **19** 326205
[6] Majumdar S and Sampathkumaran E V, 2000 *Phys. Rev.* B **62**, 8959
[7] Maiti K, Mahadevan P and Sarma D D, 1999 *Phys. Rev. B* **59**, 12457 (1999); Maiti K and Sarma D D, 1998 *Phys. Rev. B* **58**, 9746.
[8] Patil S, Pandey S K, Medicherla V R R, Singh R S, Bindu R, Sampathkumaran E V and Maiti K, 2010 *J. Phys.: Condens. Matter* **22** 255602

Ultrasonic investigation of a heavy fermion compound YbAgGe

Y Nakanishi[1], R. Kashiwazaki[1], K. Deto[1], F. Shichinomiya[1], M. Nakamura[1], H. Kubo[2] K. Umeo[2], T. Onimaru[2], T. Takabatake[2] and M. Yoshizawa[1]

[1] Graduate School of Engineering, Iwate University, Morioka 020-8551, Japan
[1] Department of Quantum Matter, ADSM, Hiroshima University, Higashihiroshima 739-8530, Japan

E-mail: yoshiki@iwate-u.ac.jp

Abstract. We have performed ultrasonic measurements on a heavy fermion compound with geometrical frustration YbAgGe which undergoes a complex series of phase transitions of $T_{\rm M1}$ = 0.8 K, and $T_{\rm M2}$ = 0.65 K in zero eld. A pronounced elastic softening toward the transition temperature $T_{\rm M2}$ was observed in the temperature dependence of longitudinal elastic constants C_{11} and C_{33}, and transverse one C_{44}, while no softening was observed in transverse one C_{66}. Furthermore, a sharp drop was observed at $T_{\rm M2}$ in all the elastic constants. The data is difficult to explain only by crystalline electric eld effect proposed previously. Alternately, the deformation potential approximation can reproduce the experimental data reasonably. It is found that the bandwidth W of quasi-particles formed by hybridization between conduction electrons and $4f$ localized ones derived from Yb ion increases gradually with increasing the applied magnetic eld, indicating an increase of Kondo temperature $T_{\rm K}$. A sudden change of the W appears in a eld of 5 T, suggesting that the application of eld suppresses in part the magnetic frustration, and enhances the Kondo temperature $T_{\rm K}$. We discuss our results with a model where the main contribution to the elastic softening arises from the deformation coupling between quasi-particles and elastic waves, and with an effect attributed to order-parameter uctuations near $T_{\rm M2}$.

1. Introduction

Over the past few years, a considerable number of studies have been conducted on the Ytterbium compounds since they display a rich variety of intriguing physical properties such as quantum critical behavior, non-Fermi liquid behavior, multipolar ordering and so on.[1] The unusual behavior is attributed to two energetically similar electronic configurations, trivalent $4f^{13}$ and divalent $4f^{14}$ of Yb ions, following the Hund's rule tendency to fill the $4f$ shell. As a result, a quantum mechanical admixture of these states leads to intermediate valence character. Furthermore, additionally geometrical frustration can cause sufficiently large quantum spin fluctuations leading to suppression of a long-ranged magnetic order as to drive a system into interesting types of intrinsically quantum-mechanical ground states. YbAgGe was recognized to be one of the few stoichiometric Yb-based heavy fermion compounds which ground state would be determined by the competition between the Kondo effect and the Ruderman-Kittel-Kasuya-Yoshida (RKKY) interaction derived from the exchange interaction between $4f$ electrons and

Published under licence by IOP Publishing Ltd

conduction ones. In addition, a high degree of geometrical frustration plays also an important role in this compound.

YbAgGe, crystallizing in the hexagonal ZrNiAl-type structure, undergoes successive two magnetic phase transitions at $T_{M1} = 0.8$ K (second order) and $T_{M2} = 0.65$ K (first order), which disappear with the application of a magnetic field.[2-3] At the same time, non-Fermi liquid behavior is observed in specific-heat and resistivity measurements.[4] Furthermore, YbAgGe is recognized to be the moderate heavy-fermion (HF) compound with a Kondo temperature $T_K \sim 20$ K and high-temperature linear coefficient of the specific heat $C_p/T = 150$ mJ/mol·K^2. It is worthwhile also to mention that the magnetic ordering temperatures are well separated from the Kondo temperature $T_K \sim 20$ K that is itself well below the estimated first excited crystal electric field (CEF) level $T_{CEF} = 60 \sim 100$ K.[5] As a result, most likely, the low temperature property would be governed by Kramers doublet including the competition between the Kondo effect and the RKKY interaction, and the geometrical frustration effect.

In this way YbAgGe really gives unique possibility to explore the possible interplay and the role of the Kondo effect, quantum criticality, intermediate valence, RKKY interaction and CEF effect using ultrasonic measurements. Ultrasound measurement has been used for the last three decades as a powerful method to investigate the elastic properties of Heavy Fermion HF compound. In particular, one can obtain detailed information concerning a ground state multiplet of rare-earth ions split mainly by the CEF effect and the character of a quadrupolar moment active in low-lying energy levels. We present in this communication the first ultrasonic investigation on the Yb-based moderate HF compound YbAgGe.

2. Experiment

Single crystal of YbAgGe was grown using a Ag-Ge rich self-flux method. The three Yb^{3+} ions per unit cell lie on a quasi-Kagome triangular lattice in the basal plane of the hexagonal non-centrosymmetric structure (space group $P\bar{6}2m$, $a = 7.05$, $c = 4.14$ $\AA$) Specimen used for the present ultrasonic measurement was cut into a rectangular shape with two axes along the a-and c-axis.

The sound velocity, as the elastic constant was measured by an ultrasonic apparatus based on a phase comparison method at temperatures down to 0.5 K in magnetic field up to 10 T. Plates of $LiNbO_3$ was used for the piezoelectric transducer. The fundamental resonance frequency of $LiNbO_3$ transducer is 10 - 30 MHz. The transducer was glued on the parallel planes of the sample by an elastic polymer Thiokol. The absolute value of the sound velocity was obtained by measuring the delay time between the ultrasonic echo signals with an accuracy of a few percent. The elastic constant was calculated as $C = \ v^2$ by using the sound velocity v and the density of the crystal. The lattice constant of YbAgGe at room temperature was used for the estimation of the density $= 9.801$ g/cm^3.

3. Deformation potential approximation

There are various couplings of lattice distortions and electronic states. In the case of intermetallic rare earth, especially for HF compounds a deformation potential coupling to conduction electrons plays an important role in their elastic property in addition to the quadrupolar response mentioned above. A large density of states of quasi-particles in the vicinity of Fermi level formed by a strong hybridization, a coupling constant d between the quasi-particles and a relevant elastic strain associated with sound waves becomes significantly strong. [6-7] This situation often brings about an elastic anomaly. For the simplicity, we employed here a rectangular density of states with a total band width W. According to the deformation potential approximation with the rectangular density of states, the elastic constant C_Γ can be written by

$$C_\Gamma(T) = C_\Gamma^{(0)}(T) - N\frac{4d^2}{W}\tanh(\frac{W}{4}) \quad (1)$$

where, $C_\Gamma{}^{(0)}$, N, d and W denote a background elastic constant, number of ions in a unit volume, coupling constant and total band width, respectively. denotes $1/T$.

Table 1. Elastic constants at 4.2 K and fitting parameters in zero field in formula (1) of YbAgGe.

C_{11}	C_{33}	C_{44}	C_{66}	d	W
61.8 GPa	39.7 GPa	34.4 GPa	25 GPa	2.8 K	6.0 K

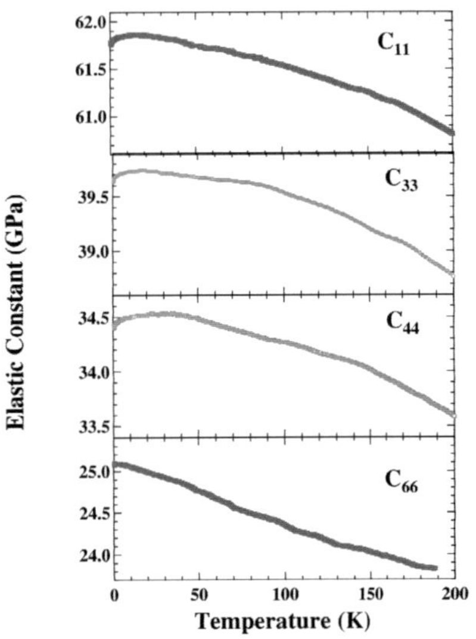

Figure 1. Temperature dependence of longitudinal elastic constants C_{11}, C_{33} and transverse ones C_{44}, C_{66} of YbAgGe.

4. Experimental Results and Discussions

Figure 1 depicts an overview of the longitudinal (C_{11}, C_{33}) and transverse (C_{44}, C_{66}) elastic constants as a function of temperature (T) for YbAgGe. The absolute values of each elastic constants are listed Table I. These data show normal behavior at higher temperature; a stiffening with decreasing T. However, a softening was found below ~20 K in the T dependence of C_{11}, C_{33}. Figure 2(a) depicts the T dependence of C_{11} at low temperatures under the selected magnetic fields (H) along a-axis, offset for clarity in ascending order bottom to top. The softening below around 10 K becomes suppressed gradually with increasing the H, and finally disappear above 4 T. The solid blue lines in Fig. 2(a) show the calculated results based on the formula (1). These fits yield the important values: background $C_\Gamma{}^{(0)}=a+bT$, deformation coupling constant d, and band width W as summarized in Table I. Figure 2(b) depicts the H dependence of W and d under the selected H along a-axis. It is found that the both values increase dramatically around 4 T with increasing the H.

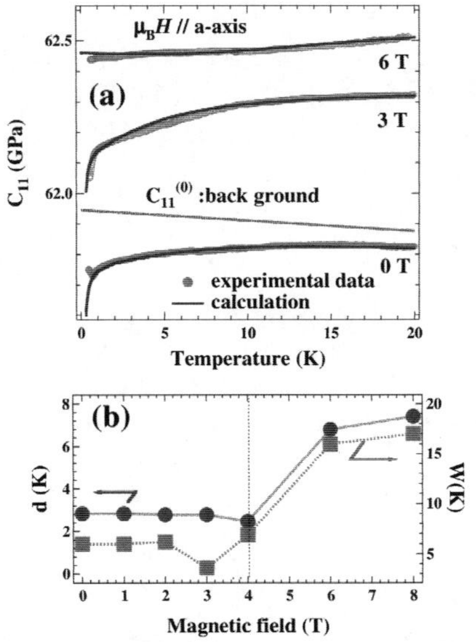

Figure 2. (a) Temperature dependence of elastic constant C_{11} under the selected H. A red circle denotes the experimental result. A blue line denotes the caluculation result based on the formula (1). A dotted line is a background $C_{\Gamma}{}^{(0)}$. (b)Blue circles and red squares denote the magnetic field dependence of the coupling constant d and band width W, respectively.

Here, we discuss the low temperature property of YbAgGe. The softening observed is difficult to reproduce only by the CEF effect if the Kramers doublet is the ground state in YbAgGe as proposed by the experimental results of neutron inelastic scattering measurement.[5] Clearly, the quadrupolar degrees of freedom are not active on the same energy scale as the correlation or any exchange energy. Another factor need to be considered when understanding the strong softening observed. Now we examine the origin from a different angle. One can obtain better agreement if one applies the deformation potential approximation as stated above. This indicates that the coupling between quasiparticles and the relevant elastic strain associated with sound waves is of great importance and plays a crucial role at low temperatures. An increase in the band width W with increasing H suggests a tendency toward the broadening of $4f$ level to gain the Zeeman energy, leading to the suppressed HF behavior or/and suppression of a magnetic fluctuation due to the geometrical frustration. This point remains as a matter to be discussed further. We will discuss them more in detail in another article as space is limited.[8] Nevertheless, it might be reconfirmed, from this work, that the possible interplay and the role of the Kondo effect, quantum criticality, RKKY interaction and geometrical frustration effect would give rise to the novel ground state that the magnetic ordered state to the NFL one, being tuned by a H.

5. Summary

We have performed ultrasonic measurements on YbAgGe single-crystalline sample. Experimental results and the calculated fitting demonstrate that the coupling between quasiparticles and the relevant elastic strain associated with sound waves is of great importance, and also that the band width W becomes broader with the application of a magnetic field when crossing the critical field $H_c{\sim}4$ T. It seems that this notable feature would be associated with the field-induced non-Fermi-liquid behavior. The deeper investigation is necessary to understand its origin and to interpret properly the corresponding experimental data in the future. For a further discussion, the same measurements for the Au-substitution system $YbAg_{1-x}Au_xGe$ are currently in progress.

6. Acknowledgement

The measurements have been performed in the Cryogenic Division of the Center for Instrumental Analysis, Iwate University. This work was supported by a Grant-in-Aid for Scientific Research on Innovative Areas "Heavy Electrons"(Grant No. 20102007) of The Ministry of Education, Culture, Sports, Science, and Technology, Japan. One of the authors (Y.N) has been supported financially in partial from *Takahashi Industrial and Economic Research Foundation* and *The Murata Science Foundation*.

7. References

[1] For example, Torikachvili M S *et al.* 2007 *Proc. Natl. Acad. Sci. U. S. A.* **104** 9960
[2] Gibson B *et al.* 1996 *J. Alloys Compd.* **239** 34
[3] Umeo K *et al.* 2004 *J. Phys. Soc. Jpn.* **73** 537
[4] Yokiwa Y *et al.* 2006 *Phys. Rev* B **73** 094435
[5] Matsumura T *et al.* 2004 *J. Phys. Soc. Jpn.* **73** 2967
[6] Luthi B 1985 *J. Phys. Chem. Solids.* **52** 70
[7] Thalmeier P 1988 *J. Magn. Magn. Mater.* **76 & 77** 299
[8] Nakanishi Y *et al.* 2010 (in preparation)

Pair density wave instability and Cooper pair insulators in gapped fermion systems

P Nikolić[1,2], A A Burkov[3] and A Paramekanti[4]

[1] Department of Physics and Astronomy, George Mason University, Fairfax, VA 22030, USA
[2] Department of Physics and Astronomy, Johns Hopkins University, Baltimore, MD 21218, USA
[3] Department of Physics and Astronomy, University of Waterloo, Waterloo, Ontario N2L 3G1, Canada
[4] Department of Physics, University of Toronto, Toronto, Ontario M5S 1A7, Canada

E-mail: `pnikolic@gmu.edu`

Abstract. By analyzing simple models of fermions in lattice potentials we argue that the zero-temperature pairing instability of any ideal band-insulator occurs at a finite momentum. The resulting supersolid state is known as "pair density wave". The pairing momentum at the onset of instability is generally incommensurate as a result of phase-space restrictions and relative strengths of interband and intraband pairing. However, commensurate pairing occurs in the strong-coupling limit and becomes a Cooper-channel analogue of the Halperin-Rice exciton condensation instability in indirect bandgap semiconductors. The exceptional sensitivity of incommensurate pairing to quantum fluctuations can lead to a strongly-correlated insulating regime and a non-BCS transition, even in the case of weak coupling as shown by an exact renormalization group analysis.

1. Introduction

One of the central challenges in the theory of high temperature superconductors is how to reconcile the mean-field character of Cooper pairing which is pivotal in any fermionic superfluid, and the seemingly fluctuation-driven superconducting transition in underdoped cuprates [1]. Motivated by this problem, we consider s-wave pairing instability in generic models of band-insulators and argue that certain phenomena familiar from cuprates, such as density-wave patterns [2] and fluctuation-driven transitions, are possible in these simple systems. The most direct physical realization of our models can be found in ultra-cold gases of alkali atoms tuned near a broad Feshbach resonance [3].

The simplest generic band-insulator is obtained at zero temperature by placing an even number of fermionic particles in each unit-cell of a periodic potential $V_{\boldsymbol{r}}$:

$$H = \int \mathrm{d}^3 r \left[c_\sigma^\dagger \left(-\frac{\boldsymbol{\nabla}^2}{2m} - \mu + V_{\boldsymbol{r}} \right) c_\sigma - U c_\uparrow^\dagger c_\downarrow^\dagger c_\downarrow c_\uparrow \right] . \quad (1)$$

We require that the chemical potential μ sit in a band-gap, but allow changes of μ, lattice amplitude V and attractive interaction strength U to drive phase transitions (we use $\hbar = 1$). In order to stay focused on universal aspects of unconventional pairing, we choose to work in

Published under licence by IOP Publishing Ltd

the unitarity limit which defines a strongly-interacting quantum critical point at which the only relevant interaction, included in (1), is related to the two-body vacuum scattering length a through a cut-off scale implicit in the following integral:

$$\frac{1}{U} = -\frac{m}{4\pi a} + \sum_{\boldsymbol{n}} \int \frac{\mathrm{d}^3 k}{(2\pi)^3} \frac{1}{2\epsilon_{\boldsymbol{nk}}}\Big|_{V=0} . \tag{2}$$

Here $\boldsymbol{n}$ and $\boldsymbol{k}$ are band-index and crystal momentum quantum numbers of quasiparticles, and $\epsilon_{\boldsymbol{nk}}$ are their bare energies.

2. Pair density wave instability

In our recent work [4] we argued that the pairing instability in the continuum model (1) generally occurs at a *finite incommensurate* wavevector $\boldsymbol{Q}$. Before subjecting this conclusion to more scrutiny, we outline the robust reasons for finite-momentum pairing. The mean-field inverse pairing susceptibility matrix Π at frequency $\Omega = 0$ in the insulating state is:

$$\Pi_{\boldsymbol{Gq};\boldsymbol{G'q'}} = \sum_{\boldsymbol{n}_1 \boldsymbol{n}_2} \int \frac{\mathrm{d}^3 k_1}{(2\pi)^3} \frac{\mathrm{d}^3 k_2}{(2\pi)^3} \frac{f(\xi_{\boldsymbol{n}_1 \boldsymbol{k}_1}) - f(-\xi_{\boldsymbol{n}_2 \boldsymbol{k}_2})}{\xi_{\boldsymbol{n}_1 \boldsymbol{k}_1} + \xi_{\boldsymbol{n}_2 \boldsymbol{k}_2}} \Gamma^{\boldsymbol{Gq}*}_{\boldsymbol{n}_1 \boldsymbol{k}_1; \boldsymbol{n}_2 \boldsymbol{k}_2} \Gamma^{\boldsymbol{G'q'}}_{\boldsymbol{n}_1 \boldsymbol{k}_1; \boldsymbol{n}_2 \boldsymbol{k}_2} + \frac{(2\pi)^3}{U} \delta(\boldsymbol{q}-\boldsymbol{q}') \delta_{\boldsymbol{GG'}} \tag{3}$$

where $\boldsymbol{G}$ are reciprocal lattice vectors, $\xi_{\boldsymbol{nk}} = \epsilon_{\boldsymbol{nk}} - \mu$, and $f(\xi)$ is Fermi-Dirac distribution function. The conservation of crystal momentum implies $\Pi_{\boldsymbol{Gq};\boldsymbol{G'q'}} = \Pi_{\boldsymbol{GG'}}(\boldsymbol{q}) \times (2\pi)^3 \delta(\boldsymbol{q} - \boldsymbol{q}')$. This expression is obtained by integrating out the internal Matsubara frequency in the bubble Feynman diagram, where the fermion propagators are written in the basis of Bloch states [4]. Such a basis choice introduces non-trivial momentum and band dependent vertex functions:

$$\Gamma^{\boldsymbol{Gq}}_{\boldsymbol{n}_1 \boldsymbol{k}_1; \boldsymbol{n}_2 \boldsymbol{k}_2} = \int \mathrm{d}^3 r \, e^{-i(\boldsymbol{q}+\boldsymbol{G})\boldsymbol{r}} \psi_{\boldsymbol{n}_1 \boldsymbol{k}_1}(\boldsymbol{r}) \psi_{\boldsymbol{n}_2 \boldsymbol{k}_2}(\boldsymbol{r}) \propto \delta(\boldsymbol{k}_1 + \boldsymbol{k}_2 - \boldsymbol{q}) . \tag{4}$$

Typical plots of the lowest eigenvalue $\Pi(\boldsymbol{q})$ of (3) are shown in Fig.1(a). These numerical calculations differ by cutoffs N_B (the number of bands) and N_O (the number of order parameter plane-wave harmonics at different $\boldsymbol{G}$), and also illustrate the dependence on the orientation of $\boldsymbol{q}$. By keeping the density at two fermions per site we invariably find that $\Pi(q)$ is linear for small q and reaches its minimum at a finite value of $q = |\boldsymbol{Q}|$ in the direction $\hat{\boldsymbol{q}} = (1,1,1)/\sqrt{3}$ for the simple cubic cosine potential. It can be seen from (3) that by varying the interaction strength U all eigenvalues of the matrix Π are shifted by the same amount. Pairing occurs when the smallest eigenvalue $\Pi(\boldsymbol{q})$ becomes negative, but this happens at a finite momentum $\boldsymbol{Q}$.

The momentum dependence of $\Pi(\boldsymbol{q})$ enters through both the denominator in (3) and the vertex functions. The former can lead to *commensurate* pairing instability if the interband pairing channels are strong enough, a particle-particle analogue of Halperin-Rice exciton condensation in indirect gap semiconductors [5]. The vertex functions, however, contain linear terms in $\boldsymbol{q}$ which respect lattice symmetries (such as $\boldsymbol{q}\boldsymbol{n}_1$, etc.), arising from phase-space restrictions for intraband and interband pairing. Their contributions are not canceled out in (3) thanks to the non-linear factor involving $\xi_{\boldsymbol{nk}}$, and make $\Pi(\boldsymbol{q}) \sim A_{\hat{\boldsymbol{q}}}|\boldsymbol{q}|$ for small q leading to an *incommensurate* pairing instability. A negative value $A_{\hat{\boldsymbol{q}}} < 0$ is required in order to secure a local minimum at finite $\boldsymbol{Q}$. We never found a positive $A_{\hat{\boldsymbol{q}}}$ in our calculations, which would imply a uniform pairing instability with an unphysical dispersion for Goldstone modes.

Fluctuation corrections can be systematically included by promoting the theory to a simplectic $Sp(2N)$ symmetry group and performing a perturbative $1/N$ expansion [6]. The corrections still involve the same vertex functions, so the linear-q features are unlikely to be accidentally canceled. Therefore, we can expect incommensurate finite-momentum pairing instability to all orders of $1/N$.

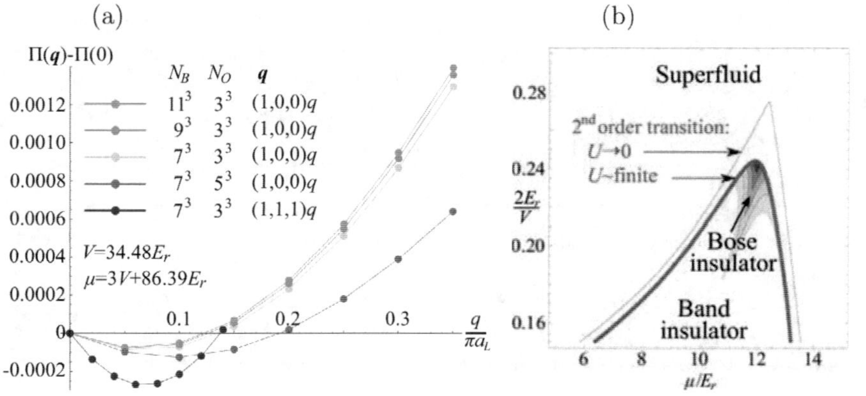

Figure 1. (a) $\Pi(\boldsymbol{q})$ in $d = 3$ for the potential $V_r = V\sum_{i=1}^{d}\cos(2\pi x_i/a_L)$; $E_r = \pi^2\hbar^2/2ma_L^2$, N_B is the number of bands, N_O is the number of order parameters. (b) A $d = 2$ phase diagram at $T = 0$.

However, an equilibrium superfluid phase with incommensurate pairing is not possible. The equilibrium order parameter must be a standing wave involving all symmetry-related wavevectors $\boldsymbol{Q}$ at which pairing occurs most readily. The obtained state is called a pair-density-wave (PDW) and breaks translational symmetry both with its global U(1) phase and density. An incommensurate density modulation is very frustrated in a lattice potential and thus not pinned to the lattice. Quantum fluctuations must remove this frustration by stabilizing some commensurate order, at least at $\boldsymbol{Q} = 0$. The arguments above suggest that the linear-q dependence of $\Pi(q)$ can hardly be completely removed, but fluctuations can modify $\Pi(q)$ to push its quadratically-dispersing minimum to a finite commensurate wavevector.

3. Fluctuation effects

The emergence of a fluctuation-dominated regime can be clearly seen in renormalization group (RG) analysis. The theory (1) contains a number of interacting fixed points associated with a diverging "vacuum" scattering length a at zero temperature. Any band-insulator can be viewed as a "vacuum" of particles or holes and taken to the unitarity limit by adjusting the chemical potential to its nearest band-edge and tuning U to a critical value U^* appropriate for the given lattice potential. The universal theory near unitarity generally contains a single fermion species, either particles or holes, whose low-energy dynamics is entirely captured by an effective mass; fermionic excitations in other bands are gapped and thus irrelevant near the fixed point:

$$S' = \int \mathcal{D}k\, f_{k,\alpha}^{\dagger}\left(-i\omega + E(\boldsymbol{k})\right) f_{k,\alpha} + U \int \mathcal{D}k_1 \mathcal{D}k_2 \mathcal{D}q\, f_{k_1,\alpha}^{\dagger} f_{k_1+q,\alpha} f_{k_2,\beta}^{\dagger} f_{k_2-q,\beta} ,$$

where $k = (\omega, \boldsymbol{k})$, $\mathcal{D}k = \mathrm{d}\omega \mathrm{d}^d\boldsymbol{k}/(2\pi)^{d+1}$, and $E(\boldsymbol{k}) = E_0 + \frac{k^2}{2m^*}$. Since the only fermion field f in this theory lives in a vacuum state, the fermion propagator acquires no renormalization at all orders of perturbation theory (all Feynman loop diagrams are zero). Only the interaction vertex is renormalized by high energy fluctuations, from an exactly summable geometric series of ladder diagrams. This allows deriving the exact RG equations to all orders of perturbation theory [7]:

$$\frac{\mathrm{d}E_g}{\mathrm{d}l} = 2E_g \quad , \quad \frac{\mathrm{d}U}{\mathrm{d}l} = (2-d)U - \Pi U^2 , \tag{5}$$

where $E_g = E_0 + U$ is the effective band-gap and Π is a positive cutoff-dependent constant. The unitarity fixed point is $E_g^* = 0$, $U^* = -\epsilon\Pi^{-1}$ in $d = 2 + \epsilon \geq 2$ dimensions. Having the exact RG equations, we find the flow of couplings with the scale parameter:

$$U(l) = \left\{ \begin{array}{ll} \frac{U(0)}{1+\Pi U(0) l} & , \quad d=2 \\ \frac{U(0)}{[1+\Pi U(0)]e^l - \Pi U(0)} & , \quad d=3 \end{array} \right\} \tag{6}$$

In two dimensions any initially negative U flows toward $-\infty$, while in three dimensions only sufficiently large attractive U undergoes a run-away flow. In both cases, however, the run-away flow escalates at a finite l where $U(l) \to -\infty$ has a vertical asymptote. This indicates a breakdown of the RG. The working assumption behind the RG was that high-energy fermions are not paired. When $U \to -\infty$, even the high-energy fermions must form Cooper pairs, and this happens at a finite scale l. At very large attractive U one must abandon the theory (5) and continue the RG using a theory which contains bosonic fields at all energy scales. This clearly affects the universality class of the superconducting transition, as order parameter fluctuations become qualitatively important.

The bosonic effective theory in question can be derived by rewriting the model (1) in the basis of Wannier states. Assuming that fermions live only at "high energies", either due to their band-gap or pairing, we can integrate them out and arrive at a lattice boson model:

$$S_{\text{eff}}^{(2)} = \int \mathrm{d}\tau \left[\sum_{ij} b_i^\dagger K_{ij} \left(\frac{\partial}{\partial \tau} \right) b_j + \sum_{ijkl} \mathcal{U}_{ij}^{kl} b_i^\dagger b_j^\dagger b_k b_l + \cdots \right] . \tag{7}$$

Here $K_{ij}(x)$ is a function analytic at $x = 0$, and the bosonic fields b_i live on lattice sites and have multiple orbital states, both labeled by indices i, j. On the other side of the superfluid transition is a bosonic Mott insulator with gapped but long-lived bosonic excitations.

This theory is valid up to the energy scales at which the boson excitations cannot decay into free fermion pairs. A bosonic insulator can be accommodated as long as the gap for collective excitations Δ_b lies below the threshold for two fermion creation $2\Delta_f$. This threshold can be estimated in the mean-field theory from the frequency dependence of the bosonic Green's function $G(\boldsymbol{q}, \Omega) = -\Pi^{-1}(\boldsymbol{q}, \Omega)$. The Fig.1(b) illustrates a Bose-insulator regime in two dimensions at $T = 0$ in which the lowest excitations are infinite-lifetime gapped bosonic modes (darker shades correspond to larger positive $2\Delta_f - \Delta_b$). The superconducting transition occurring from this bosonic insulator is fluctuation-driven. The transition is effectively in the BEC regime, even though the required attractive interaction U can be very small.

The most familiar Mott insulators have an integer number of bosons localized at each site due to strong repulsive forces among them [8]. However, Mott insulators can be stabilized at other commensurate boson densities leading to translational symmetry breaking [9]. In the present model the density of bosons is an integer per site since they originate from pairing of fermions in a fully populated band. However, the model is inescapably a multi-orbital one, and there is no constraint on the density of bosons in any particular orbital. The multi-orbital character of the effective theory (7) is tied to the same properties of the vertex functions (4) which led to the PDW pairing instability. Therefore, we conjecture that the microscopic models such as (1) and the corresponding effective theories (7) can contain stable bosonic Mott insulating phases, adjacent to the PDW superfluid, which break translational symmetry.

4. Acknowledgments

We acknowledge support from NIST (PN, grant 70NANB7H6138, Am 001), NSERC of Canada (AAB and AP) and the Sloan Foundation (AP).

[1] Emery V J and Kivelson S A 1995 *Nature* **374** 434–437
[2] Hanaguri T, Lupien C, Kohsaka Y, Lee D H, Azuma M, Takano M, Takagi H and Davis J C 2004 *Nature* **430** 1001
[3] Chin J K, Miller D E, Liu Y, Stan C, Setiawan W, Sanner C, Xu K and Ketterle W 2006 *Nature* **443** 961–964
[4] Nikolic P, Burkov A A and Paramekanti A 2010 *Physical Review B* **81** 012504
[5] Halperin B I and Rice T M 1968 *Reviews of Modern Physics* **40** 755
[6] Nikolic P and Sachdev S 2007 *Physical Review A* **75** 033608 (pages 14)
[7] Sachdev S 1999 *Quantum Phase Transitions* (Cambridge, UK: Cambridge University Press)
[8] Fisher M P A, Weichman P B, Grinstein G and Fisher D S 1989 *Physical Review B* **40** 546
[9] Balents L, Bartosch L, Burkov A, Sachdev S and Sengupta K 2005 *Physical Review B* **71** 144508

International Conference on Strongly Correlated Electron Systems (SCES 2010) IOP Publishing
Journal of Physics: Conference Series **273** (2011) 012013 doi:10.1088/1742-6596/273/1/012013

Relation between Metamagnetic Transition and Quantum Critical Point in Heavy Fermion Compound $YbIr_2Zn_{20}$

Y. Ōnuki[1,2], S. Yasui[1], S. Yoshiuchi[1], M. Ohya[1], M. Matsushita[1], Y. Hirose[1], T. Takeuchi[3], F. Honda[1], R. Settai[1], K. Sugiyama[1], E. Yamamoto[2], and Y. Haga[2]

[1]Graduate School of Science, Osaka University, Toyonaka, Osaka 560-0043, Japan
[2]ASRC, Japan Atomic Energy Agency, Tokai, Osaka 560-0043, Japan
[3]Low Temperature Center, Osaka University, Toyonaka, Osaka 560-0043, Japan

E-mail: onuki@phys.sci.osaka-u.ac.jp

Abstract. We measured the electrical resistivity in magnetic eld H under pressure P for a heavy fermion compound $YbIr_2Zn_{20}$. The quantum critical point is found to be de ned as an electronic state where the metamagnetic transition eld H_m becomes zero. This is realized at $P_c \simeq 5.2$ GPa. The most characteristic feature in the quantum critical pressure region is an anomalously large A value of the electrical resistivity $\rho = \rho_0 + AT^2$, reaching $A = 380$ $\mu\Omega{\cdot}cm/K^2$ at 5.0 GPa, which is, however, strongly reduced with increasing the magnetic eld: $A = 1.45$ $\mu\Omega{\cdot}cm/K^2$ at 80 kOe. From the result of the de Haas-van Alphen experiment at ambient pressure, the corresponding cyclotron effective mass of conduction electrons at 5.0 GPa is 400 - 500 m_0 (m_0: rest mass of an electron).

1. Introduction

The electronic states in Ce and U compounds can be tuned by applying pressure P and/or magnetic field H. For example, the Néel temperature $T_N = 5$ K in $CeIrSi_3$ becomes a guiding parameter to tune the electronic state under pressure. The quantum critical point is therefore defined as the electronic state at which the Néel temperature becomes zero at the critical pressure P_c: $T_N \to 0$ at $P \to P_c$ (= 2.25 GPa).[1] In a wide pressure region around P_c, unconventional superconductivity is found in $CeIrSi_3$.

On the other hand, a non-magnetic Yb compound of $YbCu_2Si_2$ is known to become a magnetically ordered compound above 8 GPa, for example.[2] What is a guiding parameter in the non-magnetic Yb compound? In this paper, we introduce a new guiding parameter of the metamagnetic transition field H_m in a heavy fermion compound $YbIr_2Zn_{20}$ with the cubic cage structure. The quantum critical point is realized at $P_c \simeq 5.2$ GPa, where H_m =97 kOe at ambient pressure for $H \parallel$ <100> becomes zero.

2. Experimental Results

One of the characteristic properties in the heavy fermion compound is the metamagnetic behavior or an abrupt nonlinear increase of magnetization at the magnetic field H_m. $YbIr_2Zn_{20}$ belongs to

Published under licence by IOP Publishing Ltd

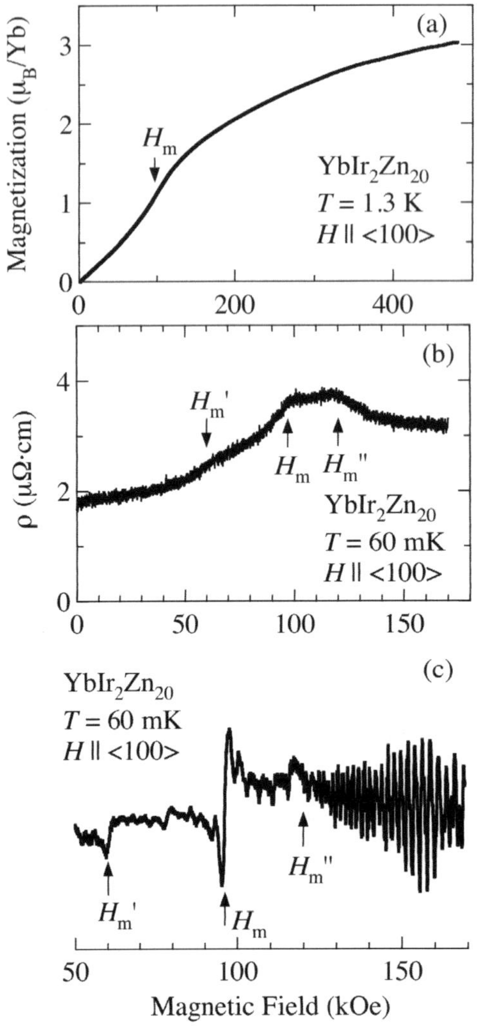

Figure 1. (a)Magnetization curve at 1.3 K and (b) magnetoresistance at 60 mK and (c) de Haas-van Alphen oscillation for H || <100> in $YbIr_2Zn_{20}$.

the heavy fermion compound with the electronic specific heat coefficient $\gamma = 540$ mJ/(K^2·mol).[3] In fact, we detected the large cyclotron effective mass of 10 - 30 m_0 (m_0: rest mass of an electron) from the de Haas-van Alphen (dHvA) effect.[4, 5]

Figure 1(a) shows the magnetization curve at 1.3 K for H || <100>. The metamagnetic transition is observed at $H_m = 97$ kOe. The present metamagnetic transition is also found in the longitudinal magnetoresistance, as shown in Fig. 1(b). A small change of the magnetoresistance is also observed at $H_m{}' = 60$ kOe and $H_m{}'' = 120$ kOe. Almost the same results are observed in the de Haas-van Alphen (dHvA) oscillations, as shown in Fig. 1(c). Later we neglect the change at $H_m{}'$ and $H_m{}''$ because it is not observed under pressure.

We measured the transverse magnetoresistance at 0.1 K under several pressures, as shown in Fig. 2(a). With increasing pressure, the transition field H_m, shown by arrows, shifts to lower

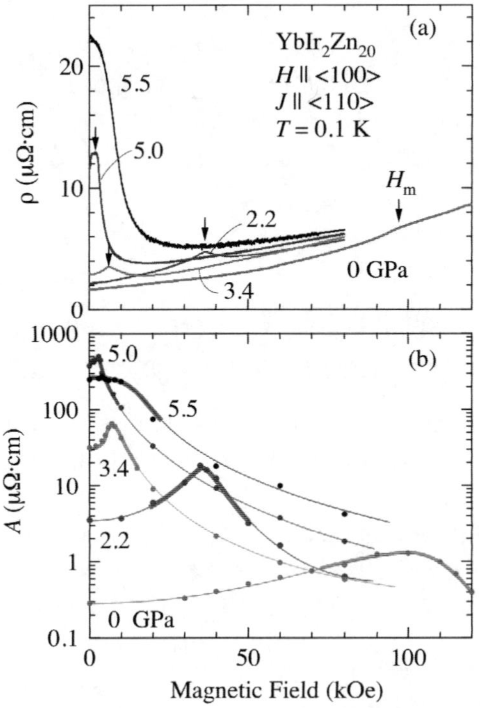

Figure 2. Magnetic field dependence of (a) electrical resistivity at 0.1 K for $H \parallel$ <100> and (b) A-value in the Fermi liquid relation under several pressures in $YbIr_2Zn_{20}$.

magnetic fields, and the resistivity anomaly at H_m becomes sharp and distinct, revealing a peak structure at 2.2 and 3.4 GPa. Moreover, the resistivity $_0$ is steeply enhanced at 5.0 and 5.5 GPa, but is reduced strongly in magnetic fields. It is noticed that the metamagnetic transition is observed at $H_m = 1.4$ kOe at 5.0 GPa, but is not observed at 5.5 GPa. It is thus concluded that the critical pressure is $P_c \simeq 5.2$ GPa. The metamagnetic transition field is thus a good tuning parameter to reach the quantum critical point.

We measured the temperature dependence of the electrical resistivity below 0.8 K under magnetic field and pressure. The Fermi liquid relation of $\ = \ _0 + AT^2$ is satisfied and the obtained A value is shown in Fig. 2(b) as a function of magnetic field. A broad peak at H_m = 97 kOe at ambient pressure in the A value is changed into a distinct peak at 2.2 and 3.4 GPa, together with an anomalous enhancement of A value at high pressures, $A = 380\ \ \cdot \mathrm{cm/K}^2$ at 0 kOe under 5.0 GPa, for example.

We studied the relation between $\sqrt{A}$ and the cyclotron effective mass m_c^* obtained from the dHvA experiment. Figure 3 shows the field dependence of the cyclotron mass for the dHvA frequency $F = 1.1\ \ 10^7$ Oe in the magnetic field $H \parallel$ <100> at ambient pressure, shown by open squares. [4, 5] The cyclotron mass increases slightly with increasing magnetic field up to $H_m = 97$ kOe, but decreases gradually in much higher fields. The cyclotron mass well correlates with the $\sqrt{A}$ value at 0 GPa. From the experimental results of $\sqrt{A}$ at 0 and 5.0 GPa in Fig. 3, we can estimate the cyclotron mass at 0 kOe to be about 450 m_0 at 5.0 GPa, revealing an extremely large cyclotron mass. The detected cyclotron mass at ambient pressure is in the range from 10 to 30 m_0, depending on the field direction and the corresponding dHvA branch. Much larger cyclotron masses are expected to exist at 5.0 GPa. These heavy cyclotron masses of 400 - 500

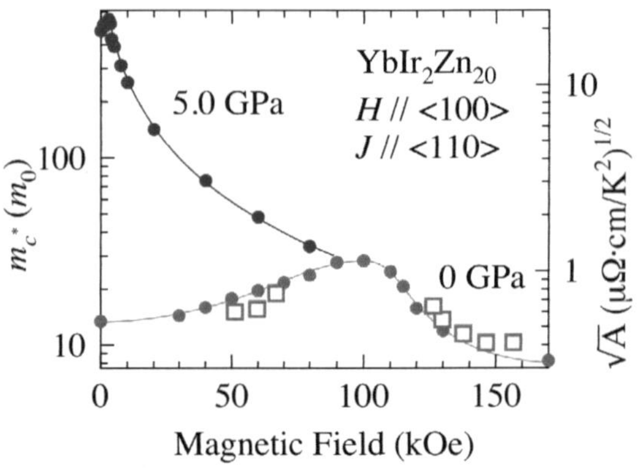

Figure 3. Field dependence of the cyclotron effective mass m_c^* at 0 GPa, shown by open squares and the $\sqrt{A}$ value under 0 and 5.0 GPa, shown by circles in $\mathrm{YbIr_2Zn_{20}}$.

m_0 are strongly reduced in magnetic fields.

3. Conclusion

We found the metamagnetic transition at $H_{\mathrm{m}} = 97$ kOe for $H \parallel$ <100> in a heavy fermion compound $\mathrm{YbIr_2Zn_{20}}$. The transition field shifts to lower magnetic fields with increasing pressure and becomes zero at $P_{\mathrm{c}} \simeq 5.2$ GPa, reaching the quantum critical point. The metamagnetic transition field is found to be a new guiding parameter. Correspondingly, the A value of the electrical resistivity $= {}_0 + AT^2$ increases extremely in magnitude from $A = 0.29$ $\cdot$ cm/K^2 at ambient pressure to 380 $\cdot$ cm/K^2 at 5.0 GPa under 0 kOe, which is strongly reduced to 1.45 $\cdot$ cm/K^2 in magnetic field of 80 kOe.

Acknowledgment

We are very grateful to Profs. K. Miyake and S. Watanabe for fruitful discussions. This work was supported by Grant-in-Aid for Scientific Research on Specially Promoted Research (No. 20001004), Osaka University: Global COE Program "Core Research and Engineering of Advanced Materials-Interdisciplinary Education Center for Materials Science" (No. G10), Grant-in-Aid for Scientific Research on Innovative Areas "Heavy Electrons" (20102002, 21102511), Priority Areas of New Materials Science Using Regulated Nano Spaces (20045008, 22013009), Scientific Research (B)(21340102, 21740256) from the Ministry of Education, Culture, Sports, Science and Technology, Japan.

References

[1] R. Settai, Y. Miyauchi, T. Takeuchi, F. Levy, I. Sheikin, and Y. Ōnuki: J. Phys. Soc. Jpn. **77** (2008) 073705.
[2] H. Winkelmann, M. M. Abd-Elmeguid, H. Micklitz, J. P. Sanchez, P. Vulliet, K. Alami-Yadri, and D. Jaccard: Phys. Rev. B **60** (1999) 3324.
[3] M. S. Torikachvili, S. Jia, E. D. Mun, S. T. Hannahs, R. C. Black, W. K. Neils, D. Martien, S. L. Bud'ko, and P. C. Can eld: Proc. Natl. Acad. Sci. U.S.A. **104** (2007) 9960.
[4] S. Yoshiuchi, M. Toda, M. Matsushita, S. Yasui, Y. Hirose, M. Ohya, K. Katayama, F. Honda, K. Sugiyama, M. Hagiwara, K. Kindo, T. Takeuchi, E. Yamamoto, Y. Haga, R. Settai, T. Tanaka, Y. Kubo, Y. Ōnuki: J. Phys. Soc. Jpn. **78** (2009) 123711.
[5] T. Takeuchi, S. Yasui, M. Toda, M. Matsushita, S. Yoshiuchi, M. Ohya, K. Katayama, Y. Hirose, N. Yoshitani, F. Honda, K. Sugiyama, M. Hagiwara, K. Kindo, E. Yamamoto, Y. Haga, T. Tanaka, Y. Kubo, R. Settai, and Y. Ōnuki: J. Phys. Soc. Jpn. (2010) in press.

Journal of Physics: Conference Series **273** (2011) 012014 doi:10.1088/1742-6596/273/1/012014

Quenching of ferromagnetism in β–UB_2C and $UNiSi_2$ at high pressure.

V A Sidorov[1, 2], P H Tobash[1], C Wang[1], B L Scott[1], Tuson Park[1, 3], E D Bauer[1], F Ronning[1], J D Thompson[1] and Z Fisk[4]

[1] Los Alamos National Laboratory, Los Alamos, New Mexico, 87545, USA
[2] Institute for High Pressure Physics, Russian Academy of Sciences, 142190 Troitsk, Russia
[3] Department of Physics, Sungkyunkwan University, Suwon 440-746, Korea
[4] Department of Physics and Astronomy, University of California, Irvine, California, 92697, USA

E-mail: vs_hppi@mail.ru

Abstract. β-UB_2C exhibits itinerant ferromagnetism below T_c = 75 K; whereas, $UNiSi_2$ exhibits ferromagnetism of localized uranium moments below T_c = 95 K and Kondo lattice behaviour at higher temperatures. We have found that ferromagnetism in both compounds is quenched at high pressure. In β-UB_2C the Curie temperature continuously approaches zero at 2.95 GPa, where resistivity and specific heat reveal the behaviour similar to that observed in Ce-based antiferromagnets near quantum critical points. In $UNiSi_2$ the Curie temperature decreases gradually to 25 K at 5.34 GPa. Above this pressure hysteretic phenomena appear in the temperature dependences of resistivity between 5 and 17 K, signalling a change in the magnetic order. No signature of magnetism was found above 5.5 GPa down to 1.15 K. Kondo lattice behaviour of resistivity with the enhanced residual resistivity is observed in $UNiSi_2$ above 5.5 GPa.

1. Introduction

Strongly correlated f-electron systems exhibit a variety of interesting properties at low temperatures. Magnetic order, which is the ground state of many lanthanides and actinides, can be destroyed simply by raising temperature, but quenching of magnetism also may be accomplished at zero temperature by adjusting a non-thermal control parameter. For example, pressure can suppress a magnetic ordering temperature to zero and produce a change of the ground state at a quantum-critical point (QCP), where magnetic order is destroyed by quantum and not thermal fluctuations. Ferromagnetic QCP's are rare because ferromagnetic order is relatively uncommon for cerium and uranium compounds. Nevertheless, superconductivity has been found near zero-temperature ferromagnetic instabilities in UGe_2 [1] and in UIr [2] under high pressure. Recently, it was suggested [3] that the itinerant ferromagnet β-UB_2C, by analogy with UGe_2 and UIr, might exhibit a pressure-induced QCP and superconductivity. $UNiSi_2$ on the other hand is a nearly localized ferromagnet [4] and the trends in $UFeSi_2$-$UCoSi_2$-$UNiSi_2$ series allow one to expect the suppression of magnetism in this compound at high pressure. The aim of our experiments has been to establish the evolution of transport and

Published under licence by IOP Publishing Ltd

thermodynamic properties of these two uranium compounds in a wide range of pressures and temperatures and to clarify the effect of pressure on magnetism.

2. Experimental details

We produced β-UB_2C by argon arc melting a stoichiometric mixture of constituents on a water-cooled copper hearth in a Zr-gettered high-purity argon atmosphere, a method similar to that used by Tran et al. [3]. Powder x-ray diffraction of the product revealed the main phase to be rhombohedral β-UB_2C (a = 0.6537(2) nm, c = 1.0784(2) nm) which is formed during rapid cooling of the arc-melted boule. Orthorhombic UBC also was found as an ~10% impurity phase in the powder pattern. Single crystals of $UNiSi_2$ with orthorhombic $CeNiSi_2$-type structure were grown in gallium flux. The lattice parameters (a = 0.4016(3) nm, b = 1.6014(1) nm, c = 0.4016(3) nm) are close to those reported for polycrystalline samples [4]. Magnetic properties and the specific heat of both materials were characterized with Quantum Design MPMS and PPMS instruments at atmospheric pressure. Electrical resistivity, ac-susceptibility and ac-specific heat measurements at high hydrostatic pressure to 6 GPa were performed with a clamped toroidal anvil device. Samples and Pb pressure sensor were placed in a Teflon ampoule 2 mm in diameter and height filled with a glycerol-water (3:2) liquid.

3. Results and discussion

3.1. β-UB_2C

As shown in the left panel of Figure 1, a kink in ρ(T) at ambient pressure signals the establishment of magnetic order at the Curie temperature T_c ~75 K. For comparison, the temperature-dependent resistivity of single-crystalline UBC at ambient pressure also is plotted. Pressures to 6 GPa do not produce substantial changes in the overall shape or magnitude of ρ(T) of β-UB_2C above 75 K; however, the resistive anomaly at T_c(P) shifts to lower temperatures and broadens with increasing pressure. Differentiation of ρ(T) reveals a hump at a characteristic temperature T* in addition to a sharp increase and a peak around T_c. Representative curves of dρ(T)/dT are depicted in the right panel of Figure 1. At ambient pressure, the hump, located at T* ≈ 40 K, reflects excitations of the ordered state [3]. Application of pressure also shifts T* to lower temperatures. It is still visible at 3.6 K and 2.78 GPa, but both T*(P) and T_c(P) tend to zero at a critical pressure P_c just below 3 GPa. At higher pressures no signature of magnetic ordering is found in ρ(T), and there is only a broad maximum in the dρ(T)/dT at T_{max}, which increases gradually between 3 and 6 GPa and is distinct from T*.

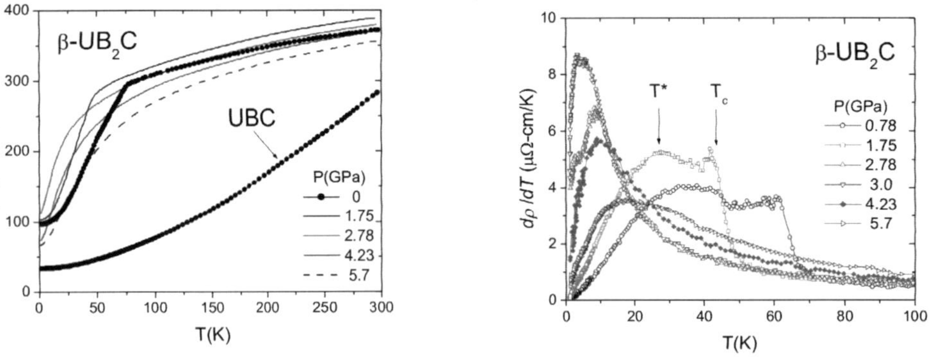

Figure 1. Temperature dependences of resistivity ρ(T) of β-UB_2C and dρ(T)/dT at some representative pressures.

From fits of the low-temperature resistivity to $\rho(T) = \rho_0 + AT^n$, we extract the pressure dependence of ρ_0, A, and n, which is displayed in the left panel of Figure 2. The coefficient A exhibits a pronounced peak at P_c. The exponent n is ~2.5 in the magnetically ordered state at low and modest pressures, above which it decreases sharply to $n \sim 1.2$ near P_c. For $P > P_c$, n gradually approaches the Fermi-liquid value $n = 2$ at the highest pressures. The residual resistivity ρ_0, which already is large at ambient pressure, increases by ~ 20 μΩ-cm as P approaches P_c from below and then decreases by over 40 μΩ-cm for pressures well above P_c.

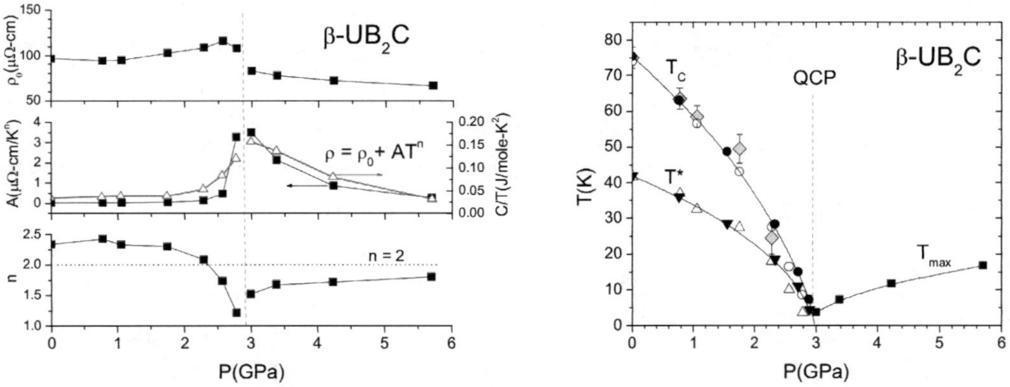

Figure 2. Pressure dependences of ρ_0, A, and n extracted from fits of low temperature resistivity and pressure dependence of C/T at 1.12 K (left panel). P-T diagram of β-UB_2C (right panel) .

In ac-calorimetry measurements, a peak in specific heat due to magnetic order is well visible below 2 GPa (diamonds in the P-T diagram,). A broad anomaly is still visible at 2.29 GPa, but close to P_c a calorimetric signature of T_c vanishes. The value of C(P)/T at 1.12 K exhibits a peak at P_c (Figure 2). The absolute value of C/T was normalized to the ambient pressure value 0.035 J/mole-K^2 determined in a PPMS. All these measurements suggest a ferromagnetic QCP at P_c ~3 GPa.

3.2. $UNiSi_2$

On cooling the electrical resistivity ρ(T) of $UNiSi_2$ increases in a Kondo-like (–lnT) manner, as was found earlier for polycrystalline samples [4], passes through a maximum and drops at the ferromagnetic ordering temperature T_c. At high pressure T_c decreases and a qualitative change in ρ(T) takes place between 5 and 5.6 GPa (Figure3, left panel).

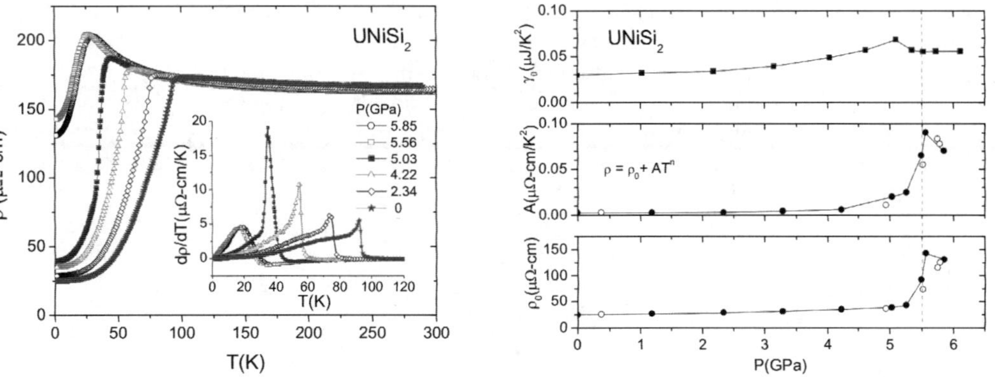

Figure 3. Temperature dependences of ρ(T) at different pressures (left panel) and pressure dependences of ρ_0 and A extracted from fits of low temperature resistivity in the form $\rho(T) = \rho_0 + AT^n$ (right panel). n was found to be 2±0.05 at all pressures. C/T from ac-calorimetry (T=0 extrapolation) also is shown in the upper right panel. Open circles – measurements made on downloading.

In this pressure range the peak of magnetic ac-susceptibility accompanying the transition to the ferromagnetic state decreases dramatically and vanishes. The specific heat C(T) anomaly observed at T_c decreases and disappears above 5.1 GPa (Figure 4, left panel). All these experiments show that ferromagnetism of nearly localized U f-electrons is quenched above 5.5 GPa. In the pressure region around this critical pressure the appearance of another – weak ferromagnetism – was found. It is visible in an upturn in ac-susceptibility below ~15 K where a very weak peak of susceptibility due to the main transition is still present at 26 K. Signatures of this weak ferromagnetism also are visible in ρ(T) and C(T) behavior in the region centered near 5.2 GPa at the P-T diagram (Figure 4, right panel).

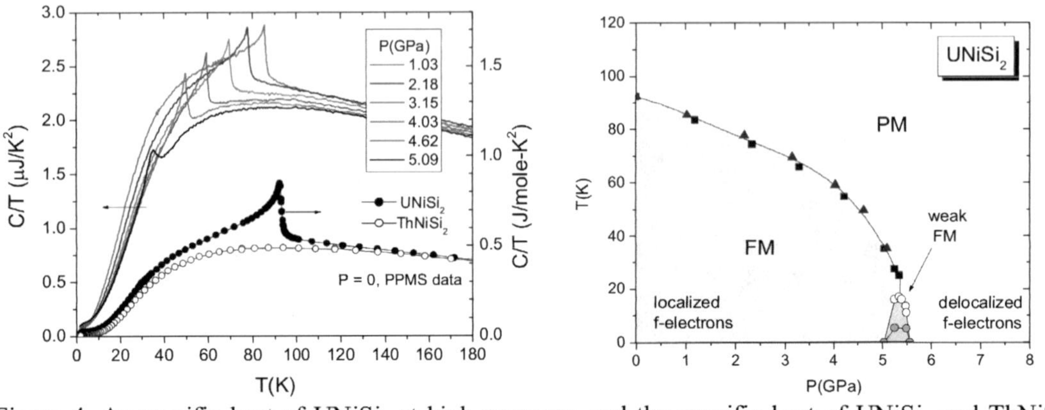

Figure 4. Ac-specific heat of $UNiSi_2$ at high pressures and the specific heat of $UNiSi_2$ and $ThNiSi_2$ measured at P = 0 in PPMS (left panel). Right panel shows the P-T diagram of $UNiSi_2$.

Uranium compounds are of special interest due to a dual nature of their f-electrons: localized and itinerant. For ferromagnetism of localized electrons the magnetic entropy is equal to Rln(2J+1), where J is the total moment. In the case of fully localized U^{3+} moments one should expect magnetic entropy of the transition ~Rln10 = 19.1 J/mole-K. For $UNiSi_2$ the magnetic entropy is 11.3 J/mole-K as estimated from Figure 4. It means that f-electrons in this material are mostly localized as concluded earlier [4]. On applying pressure the magnetic entropy is strongly reduced and vanishes above 5.5 GPa. At this pressure the ground state of $UNiSi_2$ changes to nonmagnetic with enhanced values of γ_0 and ρ_0 (Figure 3). We suggest the delocalization of f-electrons of $UNiSi_2$ at a critical pressure 5.5 GPa with the formation of a hybridized f-d band of heavy electrons. The magnetic entropy of localized U moments is now transferred to these heavy charge carriers. The enhanced value of ρ_0 may be explained by a combined effect of lower mobility of carriers in this band and strong inter-band scattering. We stress that the high value of ρ_0 in the high-pressure state of $UNiSi_2$ is not due to formation of defects in the lattice. Upon releasing pressure, the sample recovers its ambient pressure RRR within a few percent of the initial value. In contrast to β-UB_2C no non-Fermi-Liquid behaviour of resistivity was found around the critical pressure in $UNiSi_2$ and we can not regard this localized-to-itinerant transition as a QCP.

Work at Los Alamos was performed under the auspices of the U.S. Department of Energy, Office of Basic Energy Sciences. VAS acknowledges a support by grant RFBR-09-02-00336. ZF acknowledges a support by grant NSF-DMR-0801253.

References

[1] Saxena S S *et al.* 2004 *Nature* **406** 587
[2] Akazawa T *et al.* 2004 *J.Phys. Soc. Japan* **73** 3129
[3] Tran V H, Rogl P, André G, Bourée F 2006 *J. Phys.: Condens. Matter* **18** 703
[4] Kaczorowski D 1996 *Solid State Commun.* **99** 949

International Conference on Strongly Correlated Electron Systems (SCES 2010) IOP Publishing
Journal of Physics: Conference Series **273** (2011) 012015 doi:10.1088/1742-6596/273/1/012015

The Hall effect in quantum critical $CeAuSb_2$

A. M. Strydom‡

Physics Department, University of Johannesburg, PO Box 524, Auckland Park 2006, South Africa

E-mail: amstrydom@uj.ac.za

A. Thamizhavel

Department of Condensed Matter Physics and Materials Science, Tata Institute of Fundamental Research, Mumbai 400005, India.

Abstract. $CeAuSb_2$ forms in a two-dimensional structure type with plates of $CeSb_2$ separated by layers of Au and Sb atoms respectively. Kondo effect and moderate heavy fermion characteristics are evident. Antiferromagnetic order occurs at low temperature with magnetic moments along the interplanar [001] direction. We have established $T_N = 5.6$ K from $B = 0.01$ T, $\mathbf{B} \parallel \mathbf{c}$ Squid magnetometric measurements. The same field orientation presents furthermore metamagnetic behaviour at $B_{meta1} = 2.6$ T and at $B_{meta2} \simeq B = 5.4$ T, as was also previously reported in the literature. It is known that continuous suppression of the magnetic order at T_N by applied field leads to a quantum critical point at a critical field $B_c = 5.4$ T. Here we present and discuss results of Hall effect measurements up to applied fields of 14 T on single-crystal $CeAuSb_2$, a property that offers the possibility to distinguish between types of quantum criticality. Analysis of the field-dependent Hall effect show a jump in the Hall resistivity close to B_c, the temperature dependence of which suggests that for $CeAuSb_2$ the jump evolves into a discontinuity at $T = 0$ K, $B = B_c$. Our results are interpreted in terms of local-moment quantum criticality in $CeAuSb_2$.

PACS numbers: 71.27.+a, 75.20.Mb, 73.50.Jt, 64.70.Tg

‡ A major part of the experimental work was done at the Max Planck Institute for Chemical Physics of Solids, Dresden

Published under licence by IOP Publishing Ltd

Journal of Physics: Conference Series **273** (2011) 012015 doi:10.1088/1742-6596/273/1/012015

1. Introduction

A large volume of experimental and theoretical work has been devoted to gaining a deeper understanding of the physics at work in the quantum critical (QC) state [1]. The magnetic order parameter appears to have been especially amenable as a scale with which to measure the approach to a quantum phase transition which is, by definition, located at $T = 0$ K. The QC state is manifested in physical properties close to the phase transition at finite temperatures however, and since the work of Coleman *et al* [2] in which the modality of change in the Fermi surface at a QC point was connected with the quasiparticle behaviour across this point, the Hall effect has been of special interest. In contrast to for instance the low-temperature logarithmic or power-law temperature dependencies observed in electrical resistivity, specific heat or magnetic susceptibility and which are taken as fingerprints of quantum criticality, the enduring feature of Hall effect studies of heavy-fermion systems [3] is the possibility it offers to distinguish between the local-moment ordering picture involving the breakdown of heavy quasiparticles at a QC point, and the spin-density wave scenario in which heavy quasiparticles rather exist on *both* sides of the QC point. The scope of details found in the non-Fermi-liquid behaviour of systems close to a QC point has led to the notion of non-universality of the QC state, but the case of universality has been made for the quasiparticle linewidth close to the QC state [4]. It is therefore of interest to broaden the scope of experimental parameters with which to study the QC state.
In this work we focus on Hall effect measurements on $CeAuSb_2$. It adopts the tetragonal crystal structure of space group $P4/nmm$ [5]. The strong two-dimenionality of the structure is conducive to highly anisotropic electronic transport and magnetic properties [6]. A Kondo interaction on the Ce $4f-$electron moment is evident in the $\mathbf{j} \parallel \mathbf{c}$ electrical resistivity in the higher-temperature paramagnetic region. The ground state is magnetic: antiferromagnetic (AF) order sets in at $T_N = 5.6$ K and the $c-$axis, which is also the crystallographic stacking direction, is the easy axis of magnetization [6]. The AF ordering can be suppressed in a continuous manner to zero by applied magnetic fields [7] and electronic and magnetic properties convert into non-Fermi-liquid temperature dependencies in the process. It is of special interest that a metamagnetic transion is seen in the easy-axis magnetization of $CeAuSb_2$ well within the ordered region at 2 K, in a field of about 2.5 T [6, 7]. The AF transition at T_N, as well as this metamagnetic feature are both inconspicuous in the basal plane magnetization [6].

2. Experimental and Results

Procedures for growing single-crystal specimens of $CeAuSb_2$ have been described elsewhere [6]. A platelet was selected for Hall effect measurements in the ac-transport option of a PPMS station (Quantum Design, San Diego). Hall voltage measurements were collected as function of field at a number of temperatures, with field reversal in

order to eliminate non-orthogonal voltages.
Figure 1 illustrates field scans of the Hall resistivity $\rho_H(B)$ of $CeAuSb_2$ at a number of temperatures close to T_N. The magnetic field was applied along the easy axis of magnetization, $\mathbf{B} \parallel \mathbf{c}$, and excitation current was applied within the basal plane. The sample was demagnetized thermally in zero field between successive scans. Starting from a zero Hall signal at zero field, small ρ_H values are evident in small fields up to about 2.6 T. Our magnetization measurements (not shown) confirmed this as the field value of the lower one of two metamagnetic transitions known [6, 7] to occur in $CeAuSb_2$. The high-field regions for all isotherms are not only linear-in-B, but they also exhibit the same slope which means that here the Hall coefficient $R_H = \Delta\rho_H/\Delta B$ is field– and temperature *in*dependent. Taking profit from this and to isolate the anomaly in the transition region near B_c, we subtracted the high-field background line from each of the isotherms to yield a bare Hall resistivity, plotted for selected temperatures in Fig. 2 (left).

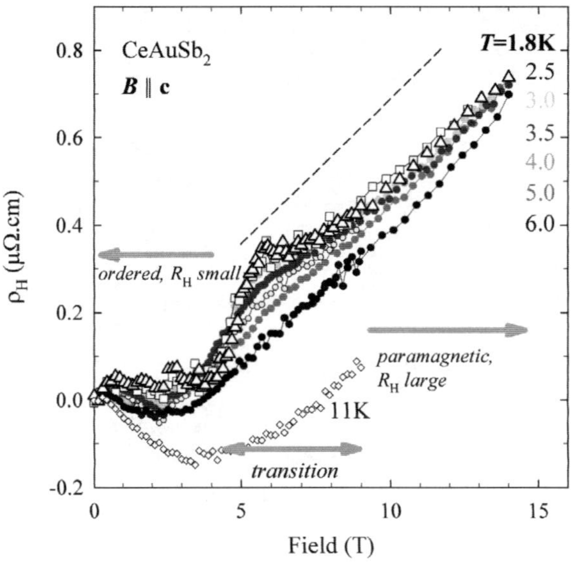

Figure 1. Field dependent Hall resistivity $\rho_H(B)$ of $CeAuSb_2$ at a number of temperatures spanning the $T_N = 5.6$ K AF order. The dashed line emphasizes the high-field $R_H = \Delta\rho_H/\Delta B =$ *constant* behaviour. This is used as a background and subtracted from the $\rho_H(B)$ curves - resulting difference curves are plotted in Fig. 2. A transition region separates small R_H values at the low-field magnetically ordered region from the high-field, large Hall coefficient, paramagnetic region. Note that equally large R_H values are found (although of negative sign) in the low-field 11 K paramagnetic isotherm.

In the conventional view of heavy fermions, a heavy Fermi liquid is formed by hybridization between the conduction- and f-electrons which is characteristic of the Kondo effect. The heavy Fermi liquid exists on both sides of a QC point, and the Fermi surface volume which accommodates both types of electrons, needs to be suitably large. The more unconventional local QC point [8] occurs when a metal undergoes a phase transition from the local f-electron magnetically ordered state into a paramagnetic phase with itinerant f-electrons. In this case the localization of f-electrons in the ordered phase removes them from the quasiparticle fluid. A reduced and smaller Fermi surface now accommodates just the conduction electrons. Our results can be reconciled with the latter view, i.e. of a *local quantum critical point* in $CeAuSb_2$. Comparing the Hall coefficient values $R_H = \Delta\rho_H/\Delta B$ obtainable from Fig. 2, the reduced Hall coefficient

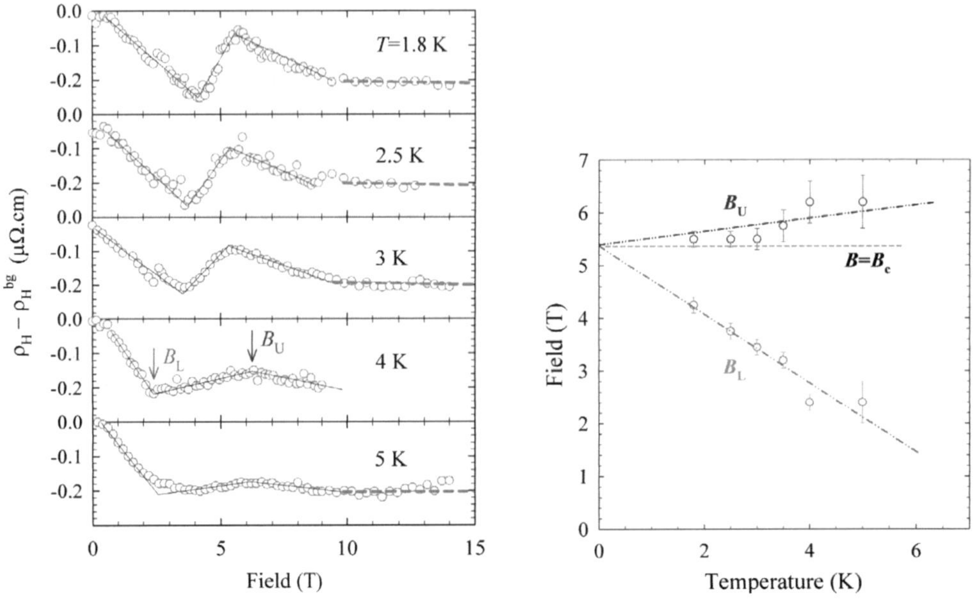

Figure 2. (left) Bare Hall resistivity $\rho_H - \rho_H^{bg}(B)$ of $CeAuSb_2$ obtained by subtracting a linear-in-B background (see Fig. 1). The high-field regions (upward of $\sim$ 9 T) are flat, which is a consequence of a field *in*dependent Hall coefficient at high B. At lower B, two successive distinct turning points are observed; a minimum at B_L and a maximum at B_U. The turning points sharpen up considerably and close in on each other with lowering temperatures. (right) Temperature dependence of the minima B_L and maxima B_U found in the bare Hall resistivity curves. Lines are guides to the eye and interesect in the point ($T = 0, B = B_c = 5.4$ T), suggesting a discontinuity in the Hall coefficient at this point.

values from the low-field, $T < T_N$ isotherms are associated with the magnetically ordered state in which ordered local moments are not accommodated in the Fermi surface volume. The high-field region on the other hand, as well as the low-field, $T > T_N$ ranges exhibit large Hall coefficients which portray a large Fermi surface volume in which the composite $f - c$ quasiparticles need to be counted. AMS acknowledges the SA-NRF (2072956) and UJ-URC for financial assistance.

References

[1] see e.g. *Proc. Int. Conf. Quantum Criticality and Novel Phases 2009* ed Frank Steglich, Oliver Stockert, and Steffen Wirth *Phys. Stat. Sol. B* **247**.
[2] Coleman P, Pépin C, Si Q, and Ramazashvili R 2001 *J. Phys.: Condens. Matter* **13** R723
[3] Paschen S 2006 *Physica B* **378-380** 28
[4] Schlottmann P 2006 *Phys. Rev. B* **73** 085110
[5] Sologub O, Hiebl K, Rogl P, Noël H, and Bodak O 1994 *J. Alloys Comp* **210** 153
[6] Thamizhavel A *et al* 2003 *Phys. Rev. B* **68** 054427
[7] Balicas L *et al* 2005 *Phys. Rev. B* **72** 064422
[8] Si Q 2006 *Physica B* **378-380** 23

International Conference on Strongly Correlated Electron Systems (SCES 2010) IOP Publishing
Journal of Physics: Conference Series **273** (2011) 012016 doi:10.1088/1742-6596/273/1/012016

Three-component Repulsive Fermionic Atoms in Optical Lattices at finite temperatures

Sei-ichiro Suga[1] **and Kensuke Inaba**[2,3]

[1] Department of Materials Science and Chemistry, University of Hyogo, Himeji 671-2280, Japan
[2] NTT Basic Research Laboratories, NTT Corporation, Atsugi 243-0198, Japan
[3] JST, CREST, Sanbancho, Chiyoda-ku, Tokyo 102-0075, Japan

E-mail: suga@eng.u-hyogo.ac.jp

Abstract. We investigate finite-temperature properties of three-component (color) repulsive fermionic atoms in optical lattices at half filling using a self-energy functional approach. For a certain anisotropy of the interactions, we find a color density-wave (DW) state at low temperatures. In the weakly interacting region, the second-order phase transition from the color DW state to the Fermi-liquid state occurs with increasing temperature. In the strongly interacting region, by contrast, the first-order transition from the color DW state to a color selective Mott transition (CSMT) state or a Mott insulating state occurs. As the interaction increases above the critical temperatures, we observe a crossover from the CSMT state to the Mott insulating state.

1. Introduction

Recent signi cant progress in cold fermionic atoms trapped in optical lattices provides us with a way to investigate the fascinating aspects of quantum many-body e ects. For example, the Mott transition was successfully demonstrated in ^{40}K fermionic atoms [1, 2]. Quite recently, the Mott insulating states of six-component ^{173}Yb formionic atoms in an optical lattice were observed [3]. The ^{173}Yb system can be a candidate for realizing three-component repulsive lattice fermionic atoms. The three-component system attracts much interest of both theorists and experimentalists. It was shown theoretically that its ground state for the isotropic interactions is a color density-wave (DW) state [4], and that the Mott transition for the isotropic interactions occurs at commensurate 1/3 and 2/3 llings but never occur at incommensurate half lling [5].

In a recent study, we investigated the ground state of three-component repulsive fermionic atoms in optical lattices at zero temperature [6]. We showed that, depending on the anisotropy of the interactions, two kinds of ordered states appear at half lling and that both states obey a rst-order quantum phase transition at the isotropic interaction point. When uctuation e ects are taken into account, a color-selective Mott transition (CSMT) state appears, where two di erent-color atoms are localized at the di erent sites and the third color atoms are itinerant throughout the system. In the strongly interacting region, a Mott insulating state appears. Both states have slightly higher energies than those of the ordered ground states. By analogy with strongly correlated electron systems [7, 8], the CSMT state and the Mott insulating state are considered to extend at nite temperatures, and thus they are expected to be observed in experiments. These motivate us to investigate the ordered states and the Mott transition states

Published under licence by IOP Publishing Ltd

at nite temperatures. In this paper, we investigate three-component repulsive fermionic atoms in optical lattices at zero and nite temperatures using a self-energy functional approach (SFA) [9, 10, 11].

2. Model and method

In accordance with the conventional model for cold atoms in optical lattices [12], we set the nearest-neighbor hopping and the on-site interaction between the atoms with di erent colors. Low-energy properties can be described by the repulsive three-component Hubbard model,

$$\mathcal{H} = \sum_{<i,j>} \sum_{=1}^{3} (t + \mu \ \delta_{i,j}) c_i^\dagger \ c_j \ + \frac{1}{2} \sum_{i} \sum_{\neq \beta} U \ _\beta n_i \ n_{i\beta}, \tag{1}$$

where $c_i^\dagger$ (c_i) is the fermionic creation (annihilation) operator for the state with color ($=$ 1, 2, 3) in the ith site and $n_i \ = c_i^\dagger \ c_i$. Here, t denotes the nearest-neighbor hopping integral, μ is the chemical potential for the atom with color , and $U \ _\beta (> 0)$ is the on-site repulsive interaction between two atoms with colors and β. The condition $\mu \ = (1/2)(U \ _\beta + U \ )$ is imposed to maintain a half- lled system with a balanced population of three-component atoms. We assume a homogeneous optical lattice and neglect the con nement potential for a rst approximation.

The SFA allows us to deal e ciently with zero- and nite-temperature properties concerning the phase transition driven by correlation e ects [9, 10]. This method is based on the Luttinger-Ward variational method. The framework of this variational approach enables us to introduce a proper reference system which has to include the same interaction term as that of the original Hamiltonian (1). Other parameters, $\mathbf{t}$, of the reference system correspond to variational parameters, where $\mathbf{t}$ is the parameter matrix of the noninteracting part of the reference Hamiltonian. In the SFA [9, 10], we calculate the grand potential Ω as a function of the self-energy of the reference system $\Sigma_\mathbf{t}$. Then, from the condition $\partial\Omega/\partial\mathbf{t} = (\partial\Sigma_\mathbf{t}/\partial\mathbf{t})(\partial\Omega/\partial\Sigma_\mathbf{t}) =$ 0, we obtain the reference self-energy $\Sigma_\mathbf{t}$ which properly describes the original correlated system. Here, we apply the two-site Anderson impurity model as the reference system. This approximation is in some sense regarded as a nite-temperature extension of the two-site DMFT [13, 6]. To examine the staggered ordered state, we divide the bipartite lattice into two sublattices [7, 14]. We then calculate the order parameter of the staggered state M , which is de ned by the di erence in the atom numbers ($n_{A(B),}$), $M \ = n_{A,} \quad n_{B,}$, between the sublattices A and B with color per site. We use a semicircular density of states for an in nite dimensional Bethe lattice, which is independent of .

We investigate the anisotropically interacting system. For this purpose, we set $U_{12} = U$, $U_{13} = U'$, and $U_{23} = U''$, and then set $U'/U = 0.6$ and $U''/U = 2$. In the following, the hopping integral t is used in units of energy.

3. Numerical results

In Fig. 1(a), we show M for $T = 0$. As U/t increases, M_1 and M_3 increase towards unity, and M_2 decrease towards minus unity. The results indicate that sites doubly occupied by color-1 and 3 atoms and sites solely occupied by color-2 atoms appear alternately. This state is a color DW state. In this case, color-2 and 3 atoms are localized at the di erent sites rigidly, while color-1 atoms occupy the site more exibly. In fact, M_1 approaches unity slowly compared with M_3. From this result, it is expected that color-1 atoms may become itinerant at the lower critical temperature than the critical temperature above which color-2 and 3 atoms become itinerant. In this case, a CSAF state emerges between these two critical temperatures. In Fig. 1(b), we show M for the weak interaction $U/t = 0.8$ as a function of T/t. We nd that M approach

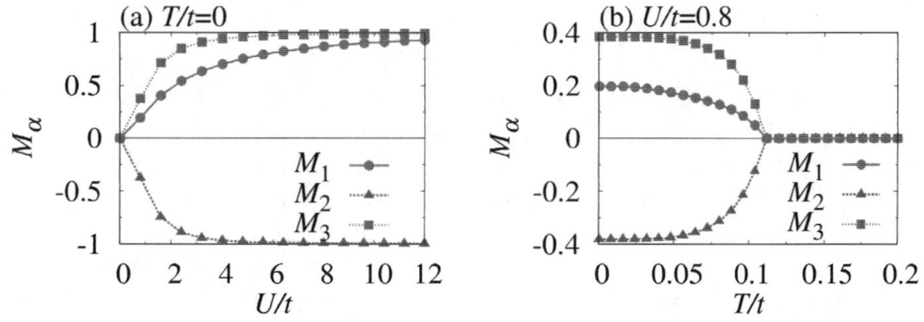

Figure 1. (Color Online) (a) The order parameter of the staggered state as a function of U/t for $T = 0$. (b) The order parameter of the staggered state as a function of T/t for $U/t = 0.8$.

zero continuously at the same critical temperature $T/t = 0.11$. The results indicate that in this weakly interacting region the CSAF state does not appear and the direct second-order phase transition from the color DW state to the Fermi liquid state occurs.

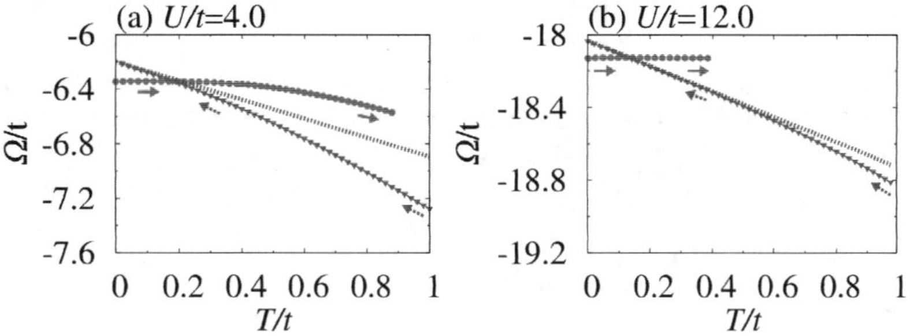

Figure 2. (Color Online) The grand potentials with the right-arrows denote the staggered ordered states which appear by increasing temperature from $T = 0$, while the grand potentials with the left-arrows denote the unordered states which appear by decreasing temperature from the high temperature region. The broken grey lines express $\Omega/t = \quad (\ln 2)T/t + \text{const.}$

We investigate the phase transition from the color DW state for the stronger U/t. It is considered that, as temperature is increased, the ordered state disappears because of thermal uctuations and Mott transition states are expected to appear. In Fig. 2, we show the grand potentials. From the analysis by the SFA, we nd that the grand potential with the right-arrows denotes the color DW state which appears by increasing temperature from $T = 0$. In fact, the grand potentials with the right-arrows are almost independent of T/t in $T/t < 0.4$, which is a manifestation of the ordered state because of $S \quad 0$. On the other hand, the grand potential with the left-arrows denotes the unordered state which appears by decreasing temperature from the high temperature region. As a reference, we depict $\Omega/t = \quad (\ln 2)T/t + \text{const.}$, which shows the temperature dependence of the Mott transition state.

For $U/t = 4.0$, both grand potentials cross at $T/t \quad 0.18$, which indicates that the rst-order phase transition from the color DW state occurs there. In $T/t > 0.18$, the lower grand potential obeys the temperature dependence $\Omega/t \quad a(T/t)^2 \quad (\ln 2)T/t$ with a positive constant a. The entropy is thus $S \quad 2aT/t + \ln 2$. The rst term stems from the contribution of the itinerant atoms and the second term stem from the localized two-color atoms which are distributed randomly. Therefore, in $T/t > 0.18$, color-2 and 3 atoms with the strongest U'' are localized at the di erent sites randomly as a result of the Mott transition, while color-1 atoms are itinerant throughout the system. This state is a CSMT state.

For $U/t = 12.0$, both grand potentials cross at $T/t \quad 0.13$. In $0.13 < T/t < 0.4$, the lower grand potential obeys the temperature dependence $\Omega/t \quad (\ln 2)T/t$, leading to $S \quad \ln 2$. This state is the Mott insulating state, where color-1 and 3 atoms are localized at the same sites and the color-1 atoms are localized solely at the di erent sites. The stronger repulsions U and U'' push the color -1 and 3 atoms with the weakest U' localized at the same sites. The paired sites and solely localized sites appear randomly. In $T/t > 0.4$, the lower grand potential obeys $\Omega/t \quad b(T/t)^2 \quad (\ln 2)T/t$ with a positive constant b and thus $S \quad 2bT/t + \ln 2$. This state is the CSMT state as mentioned above. From the numerical results, we nd that at $T/t \quad 0.13$ the rst-order phase transition occurs from the color DW state to the Mott insulating state with increasing temperature. When temperature is further increased, a crossover occurs from the Mott insulating state to the CSMT state around $T/t \quad 0.4$.

4. Summary

We have investigated nite-temperature properties of three-component repulsive fermionic atoms in optical lattices at half lling using the SFA. In the present anisotropy of the interactions, we have shown that the color density-wave state survives in the low temperature region. In the high temperature region, the Fermi liquid state, the color-selective Mott transition state, and the Mott insulating state appear. These high temperature states are considered to be observed in experiments. Finding a way to detect these three states is our important future issue.

Acknowledgments

We thank Y. Takahashi and M. Yamashita for useful comments and valuable discussions. Some of the numerical computations were undertaken at the Supercomputer Center at ISSP, University of Tokyo. This work was supported by Grants-in-Aid for Scienti c Research (C) (No. 20540390) from the Japan Society for the Promotion of Science and on Innovative Areas (No. 21104514) from the Ministry of Education, Culture, Sports, Science and Technology.

References

[1] Jördens R, Strohmaier N, Günter K, Moritz H and Esslinger T 2008 *Nature* **455** 204
[2] Schneider U, Hackermüller L, Will S, Best Th, Bloch I, Costi T A, Helmes R W, Rasch D and Rosch A 2008 *Science* **322** 1520
[3] Taie S, Sugawa S, Yamazaki R and Takahashi Y march 2010 *in Talk of International Symposium on Physics of New Quantum Phases in Superclean Materials, Yokohama*
[4] Honerkamp C and Hofstetter W 2004 *Phys. Rev. Lett.* **92** 170403
[5] Gorelik E V and Blümer N 2009 *Phys. Rev.* A **80** 051602(R)
[6] Miyatake S, Inaba K and Suga S 2010 *Phys. Rev.* A **81** 021603(R)
[7] Georges A, Kotliar G, Krauth W and Rozenberg M J 1996 *Rev. Mod. Phys.* **68** 13
[8] Kotliar G and Vollhardt D 2004 *Physics Today* **March 2004** 53
[9] Pottho M 2003 *Eur. Phys. J. B.* **32** 429
[10] Pottho M 2003 *Eur. Phys. J. B.* **36** 335
[11] Koga A, Inaba K and Kawakami N 2005 *Prog. Theor. Phys.* **160** 253
[12] Jaksch D, Bruder C, Cirac J I, Gardiner C W and Zoller P 1998 *Phys. Rev. Lett.* **81** 3108
[13] Pottho M 2001 *Phys. Rev.* B **64** 165114
[14] Chitra R and Kotliar G 1999 *Phys. Rev. Lett.* **83** 2386 (1999)

International Conference on Strongly Correlated Electron Systems (SCES 2010) IOP Publishing
Journal of Physics: Conference Series **273** (2011) 012017 doi:10.1088/1742-6596/273/1/012017

Magnetic field evolution of critical end point in UGe_2

Valentin Taufour, Alain Villaume, Dai Aoki, Georg Knebel, Jacques Flouquet

INAC/SPSMS, CEA-Grenoble, 17 rue des Martyrs, 38054 Grenoble, France

E-mail: `valentin.taufour@cea.fr`

Abstract. New measurements on the ferromagnetic heavy-fermion superconductor UGe_2 are reported. Thermal expansion measurements via strain gage allow us to locate precisely the critical end point (CEP) of the first order boundary between the large moment ferromagnetic phase FM2 ($M_0 = 1.5\ \mu_B$) and the small moment phase FM1 ($M_0 = 0.9\ \mu_B$). The field dependence of the CEP is compared with that of the paramagnetic-FM1 transition above the tricritical point.

1. Introduction

UGe_2 is a ferromagnet below 53 K at ambient pressure. The Curie temperature can be driven to 0 K with applying pressure : $p_c \approx 1.49$ GPa. Pressure studies have revealed two ferromagnetic ground states : FM1 and FM2. Both phases have the magnetization along the a-axis of the orthorhombic structure (space group $Cmmm$) but different ordered moments : $M_0 = 1.5\ \mu_B$/U in FM2 and 0.9 μ_B/U in FM1. Figure 1 shows the boundary between these phases in the p-T phase diagram. FM2 exists only below p_x and the transition line, which is of first order [1, 2], terminates at a critical end point (CEP). This means that below $p_{\rm CEP}$, a crossover regime exists between FM1 and FM2. With magnetic field applied along the a-axis, the critical pressures p_x and p_c are shifted to higher pressure [2]. In this paper, we report on the field evolution of $p_{\rm CEP}$ and $T_{\rm CEP}$. Interest of the studies of the FM1-FM2 transition has been enhanced by the discovery of superconductivity in the ferromagnetic state (below p_c) with $T_{\rm sc_{max}}$ around p_x [3].

This study is quite complementary with our recent results on the FM1-PM boundary which revealed the tricritical point (TCP) of UGe_2 [4] by resistivity measurements. The FM1-PM transition changes from second order to first order at a TCP (see fig. 1 : $T_{\rm TCP} \approx 24$ K and $p_{\rm TCP} \approx 1.42$ GPa). Under magnetic field, the first order transition line occurs at higher pressure, which leads to a plane of first order transition in the p-T-H space. At a given field, or pressure, the first order transition ends up at a critical point which can be driven with pressure, or field, to 0 K at a quantum critical end point (QCEP) [5]. Between 1.42 GPa and 1.85 GPa, we found that the critical point decreases from 24 K at 0 T to 16.5 K at 2.5 T, so that we roughly estimate $p_{\rm QCEP_{PM\text{-}FM1}} \approx 3$ GPa and $H_{\rm QCEP_{PM\text{-}FM1}} \approx 9$ T for the PM-FM1 transition [4]. By symmetry, there are two first order planes for positive and negative magnetic field, so that they are called wings [6]. The same wing structure is expected for the FM1-FM2 transition studied here : we will show that the CEP is found at higher magnetic field with increasing pressure, but that $T_{\rm CEP}$ does not decrease significantly up to 8 T.

Published under licence by IOP Publishing Ltd

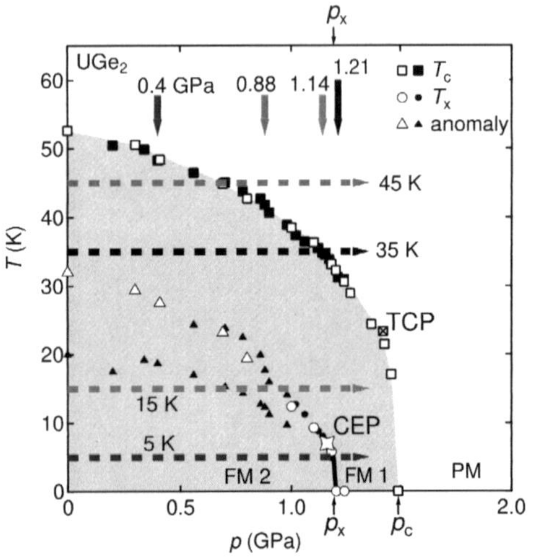

Figure 1. (Color online) Phase diagram of UGe_2 determined by our resistivity (open symbol) and thermal expansion measurements (full symbol). Vertical arrows show the temperature scans presented in fig. 2(a) and horizontal dashed lines show the pressure scans of fig. 2(b). The superconducting phase is not shown here.

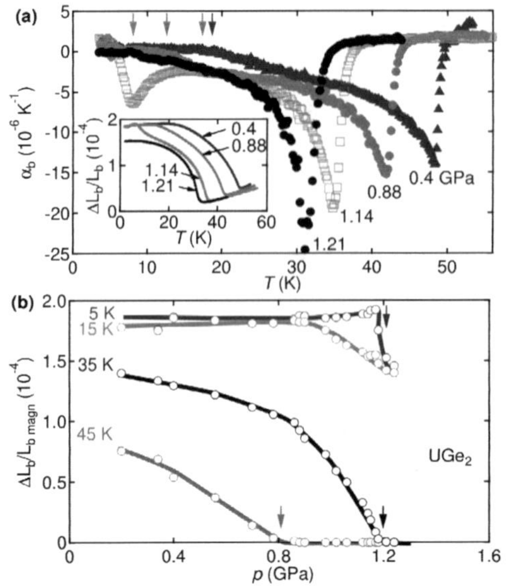

Figure 2. (Color online) (a) Temperature dependence of the thermal expansion coefficient along the b-axis at different pressures. The inset show the temperature dependence of the corresponding relative elongation. (b) Pressure dependence of the isothermal spontaneous magnetostriction.

2. Experimental technique

For the determination of the CEP of the first order line (T_x, p_x) of the FM1-FM2 transition, thermal expansion experiments via strain gage have been used. And for the location of the FM1-FM2 wings $T_{\text{CEP}}(H)$ under magnetic field, resistivity measurements have been performed.

The single crystal of UGe_2 was grown by the Czochralski pulling method in a tetra-arc furnace. The residual resistivity ratio (RRR) is higher than 300. For the thermal expansion measurements, the pressure was applied via a CuBe piston cylinder cell with Daphne 7373 as a pressure-transmitting medium. Thermal expansion was measured along the b-axis via strain gage. Silicon was used as a reference material. Another sample was used for resistivity measurements. Pressure was applied via a NiCrAl-CuBe hybrid piston-cylinder cell with Daphne 7373. Electrical resistivity was measured down to 2 K and at high fields up to 9 T, by the 4-probe AC-method with current parallel to the a-axis. Magnetic field was applied along the easy magnetization a-axis.

3. At $H = 0$ T : first order nature of the FM1-FM2 transition, CEP and crossover

The inset of figure 2(a) shows the temperature dependence of the relative elongation $\Delta L_b/L_b$ at different pressures. The expansion coefficient α_b corresponding to the temperature derivative is shown in figure 2(a). A sharp anomaly corresponding to the Curie temperature T_{C} is observed. Such feature in α_b is observed up to 1.24 GPa, and thus confirms that the PM-FM1 transition is second order at least up to 1.24 GPa [7]. Further experiments located the tricritical point at $p_{\text{TCP}} \approx 1.42$ GPa [4].

In addition, below T_{C}, we observe a clear anomaly at 1.14 GPa at T_x which disappears completely at 1.21 GPa, above $p_x \approx 1.19$ GPa. At low pressure, only a large hump can be recorded confirming that the switch from FM1 to FM2 corresponds to a broad crossover regime at ambient pressure [8] and thus the existence of the CEP which terminates the first order line between FM1 and FM2.

The non magnetic contribution to the relative elongation can be determined by fitting $\Delta L_b/L_b$ (inset of fig. 2(a)) in the paramagnetic regime by T^2 [9]. Then, we obtain the spontaneous magnetostriction $\Delta L_b/L_{b\mathrm{magn}}$ by substracting the non magnetic part to $\Delta L_b/L_b$. Fine tuning of the pressure allows us to draw $\Delta L_b/L_{b\mathrm{magn}}$ as a function of pressure in fig. 2(b). The anomaly at 45 K and 35 K is associated to the PM-FM1 transition. By comparison to the continuity of the spontaneous magnetostriction at that second order transition, we observe a discontinuity at low temperature at the FM2-FM1 transition. This discontinuity changes to a continuity above 7 K (curve at 15 K for fig. 2(b)). Thus, the FM1-FM2 transition is first order below 7 K, and the transition line ends up at a CEP.

For pressure lower than p_{CEP}, anomalies can be detected as one inflection point or two (0.4 and 0.88 GPa in fig. 2(a)). Closer to the CEP, a minimum is also visible in α_b similar to the minimum observed above the CEP at the first order transition. The CEP coordinates have been determined when the minimum in the thermal expansion coefficient is the deepest : $p_{\mathrm{CEP}} \approx 1.16$ GPa, $T_{\mathrm{CEP}} \approx 7$ K. As we will show, our resistivity measurements confirm these values. In fig. 1, the anomalies corresponding to the crossover between FM1 and FM2 are reported. Our resistivity data are in good agreement with previously reported resistivity measurements [10, 11] but does not perfectly agree with the thermal expansion data. Above p_{CEP}, at the genuine FM1-FM2 transition, the agreement is very good, as well as for the PM-FM1 transition.

4. Evolution of T_{CEP} in magnetic field : the FM1-FM2 wings

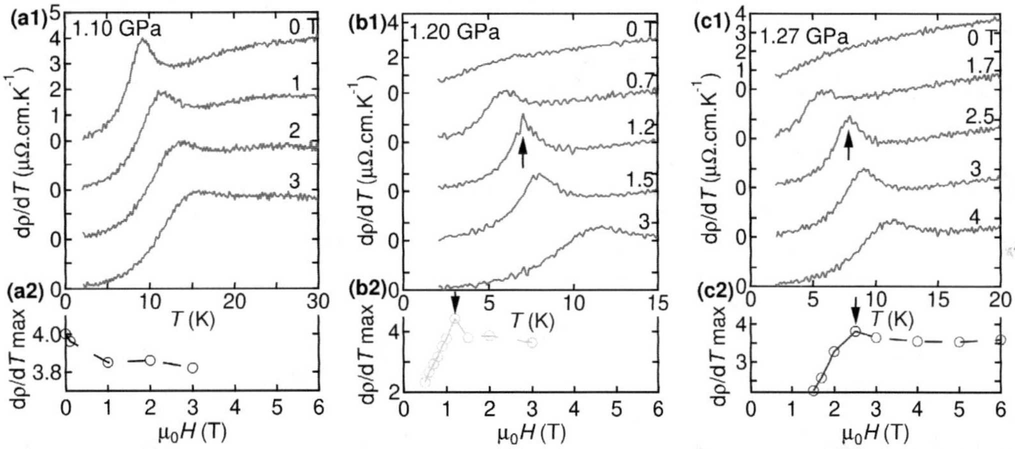

Figure 3. (Color online) Temperature derivative of resistivity $d\rho/dT$ for three typical pressures : (a) below p_{CEP} ; (b) near p_x ; (c) above p_x. (a1) The anomaly smears out with magnetic field. (a2) The value of $d\rho/dT$ at the maximum (i.e. at T_x) decreases. (b1) Up to 1.2 Tesla, the FM1-FM2 transition is sharper. At higher field, the anomaly is broadened and disappears. (b2) Below 1.2 T, $d\rho/dT_{\mathrm{max}}$ increases with field (full line) which is interpreted here as the first order FM1-FM2 transition and then decreases (dashed line) as in a crossover regime. (c1) Below 1.5 Tesla, there is no anomaly (FM1 state). From 1.5 to 2.5 Tesla, anomalies are detected. Above 2.5 Tesla, the anomaly is broadened. (c2) The feature is the same as for 1.2 GPa but shifted to higher fields.

To determine the field evolution of the FM1-FM2 boundary and CEP, accurate resistivity measurements were realized. Figure 3 shows the temperature variation of $d\rho/dT$ (temperature derivative of the resistivity ρ) at different magnetic fields for three pressures : at 1.1 GPa just below $p_{\mathrm{CEP}} \approx 1.14$ GPa, at 1.2 GPa just above $p_x = 1.19$ GPa and at 1.27 GPa between p_x and $p_c = 1.49$ GPa. Two different behavior are observed : the peak of $d\rho/dT$, characteristic of the FM1-FM2 transition, is either sharpened with magnetic field or broadened. The first case is interpreted as the first order FM1-FM2 transition and the second as the crossover regime. The magnetic field value between these two behavior is thus defined as H_{CEP}. The corresponding $T_{\mathrm{CEP}}(H)$ is obtained as explained in fig. 4. Clearly, the field dependence of T_{CEP} must be very small up to 8 T.

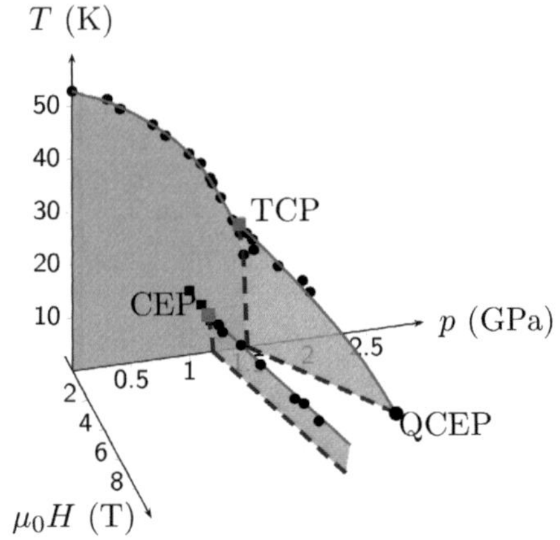

Figure 4. (Color online) Temperature-pressure-magnetic field phase diagram of UGe_2 drawn from resistivity measurements. Gray planes are planes of first order transition. Full red lines are second order lines. For the magnetic field dependence of T_{CEP}, each point is determined as follow (see arrows in fig.3(b) and (c)) : first H_{CEP} is determined by taking the maximum position of $d\rho/dT_{\mathrm{max}}$ (for example 1.2 T from the data in fig. 3(b2) at 1.2 GPa), and then T_{CEP} is obtained as the position of the peak on the $d\rho/dT$ curve at H_{CEP} (7 K in the data of fig. 3(b1) at 1.2 GPa and 1.2 T). Details of the field dependence of the TCP will be presented in ref. [4]

The fact that T_{CEP} of the FM1-FM2 transition does not decrease significantly up to 8 T is in contrast with the reduction from 24 K to 16.5 K at 2.5 T of the critical point of the PM-FM1 transition [4]. These results agree with the theoretical prediction that H_{QCEP} varies as m_0^3 where m_0 is a microscopic magnetization parameter [5].

In summary, we have performed thermal expansion and resistivity measurements under pressure. The magnetic field evolution of the CEP has been determined up to 8 T and 1.6 GPa. Within the resolution of the field steps of the experiment, the CEP seems to remain at roughly the same temperature.

Acknowledgments

This work was financially supported by ANR-CORMAT and SINUS.

References

[1] Pfleiderer C and Huxley A 2002 *Phys. Rev. Lett.* **89**
[2] Huxley A, Sheikin I and Braithwaite D 2000 *Physica B* **284** 1277–1278
[3] Saxena S S *et al.* 2000 *Nature* **406** 587–592
[4] Taufour V *et al.* to be published
[5] Belitz D, Kirkpatrick T and Rollbuhler J 2005 *Phys. Rev. Lett.* **94**
[6] Griffiths R B 1973 *Phys. Rev. B* **7** 545–551
[7] Nishimura K, Oomi G, Yun S and Onuki Y 1994 *J. Alloys Comp.* **213-214** 383–386
[8] Hardy F, Meingast C, Taufour V, Flouquet J, Loehneysen H v, Fisher R A, Phillips N E, Huxley A and Lashley J C 2009 *Phys. Rev. B* **80**
[9] Pfleiderer C, Boni P, Keller T, Rossler U K and Rosch A 2007 *Science* **316** 1871–1874
[10] Bauer E, Dickey R, Zapf V and Maple M 2001 *J. Phys.-Condes. Matter* **13** L759–L770
[11] Tateiwa N, Kobayashi T, Hanazono K, Amaya K, Haga Y, Settai R and Onuki Y 2001 *J. Phys.-Condes. Matter* **13** L17–L23

Quantum critical behavior in the heavy Fermion single crystal $Ce(Ni_{0.935}Pd_{0.065})_2Ge_2$

C. H. Wang[1,2,3], J. M. Lawrence[1], A. D. Christianson[3], S. Chang[4], K. Gofryk[2], E. D. Bauer[2], F. Ronning[2], J. D. Thompson[2], K. J. McClellan[2], J. A. Rodriguez-Rivera[4,5], J. W. Lynn[4]

[1]University of California, Irvine, California 92697
[2]Los Alamos National Laboratory, Los Alamos, NM 87545
[3]Neutron Scattering Science Division, Oak Ridge National Laboratory, Oak Ridge, TN, 37831
[4]NCNR, National Institute of Standards and Technology, Gaithersburg, MD 20899-6102
[5]Department of Materials Science and Engineering, University of Maryland, College Park, MD 20742

E-mail: `wangc1@ornl.gov`

Abstract.
We have performed magnetic susceptibility, specific heat, resistivity, and inelastic neutron scattering measurements on a single crystal of the heavy Fermion compound $Ce(Ni_{0.935}Pd_{0.065})_2Ge_2$, which is believed to be close to a quantum critical point (QCP) at T = 0. At lowest temperature(1.8-3.5 K), the magnetic susceptibility behaves as $\chi(T) - \chi(0) \propto T^{-1/6}$ with $\chi(0) = 0.032 \times 10^{-6}$ m^3/mole (0.0025 emu/mole). For $T < 1$ K, the specific heat can be fit to the formula $\Delta C/T = \gamma_0 - T^{1/2}$ with γ_0 of order 700 mJ/mole-K^2. The resistivity behaves as $\rho = \rho_0 + AT^{3/2}$ for temperatures below 2 K. This low temperature behavior for $\gamma(T)$ and $\rho(T)$ is in accord with the SCR theory of Moriya and Takimoto[1]. The inelastic neutron scattering spectra show a broad peak near 1.5 meV that appears to be independent of Q; we interpret this as Kondo scattering with $T_K = 17$ K. In addition, the scattering is enhanced near Q=(1/2, 1/2, 0) with maximum scattering at $\Delta E = 0.45$ meV; we interpret this as scattering from antiferromagnetic fluctuations near the antiferromagnetic QCP.

In strongly correlated electron systems, a quantum critical point (QCP) separates an antiferromagnetic (AFM) or ferromagnetic(FM) state from a nonmagnetic Fermi liquid state at $T = 0$ K. This QCP can be tuned by adjusting a control parameter such as doping parameter x, external pressure P, or applied magnetic field H. In the vicinity of the QCP, the critical fluctuations are quantum in nature and induce unique behavior. The nature of this quantum ground state phase transition poses one of the most significant challenges in condensed matter physics. Heavy Fermion (HF) compounds are very good candidates for studying the QCP. These compounds behave as Fermi Liquids (FL), with large values for the specific heat coefficient $\gamma = C/T$ and susceptibility $\chi(0)$, and with the resistivity varying as $\Delta\rho \propto T^2$ at low temperature. When such systems are tuned close to a QCP, fluctuations of the nearby magnetically ordered state affect the thermodynamic behavior and lead to Non-Fermi Liquid (NFL) behavior such as $\Delta\rho \propto T^\alpha$ with $\alpha < 2$ and $C/T \propto ln(T_0/T)$ at low temperature[2].

The approach to a QCP has been attained in HF systems by doping in the tetragonal compounds $Ce(Ru_{1-x}Rh_x)_2Si_2$ (x=0.03)[3, 4] and $Ce_{1-x}La_xRu_2Si_2$ (x=0.075)[5, 6] as well as

Published under licence by IOP Publishing Ltd

in the orthorhombic compound $CeCu_{6-x}Au_x$ (x=0.1)[7, 8, 9], by application of a magnetic field in $YbRh_2Si_2$[10], and with no additional control parameter in the compound β-$YbAlB_4$[11].

A conventional spin fluctuation theory has been established to explain the non-Fermi liquid behavior. This theory, implemented through a renormalization-group approach[12, 13] or through the self-consistent renormalization(SCR) method[1], has successfully explained various aspects of the non-Fermi liquid behavior. However, the experimental results for some systems do not follow this spin fluctuation theory. A locally critical phase transition has been invoked to explain the behavior of a few systems[14]. Hence, measurements in other systems are needed to understand the behavior near the QCP in HF materials.

Below 2 K, the tetragonal compound $CeNi_2Ge_2$ exhibits behavior for the resistivity and specific heat that obeys the predictions of the SCR theory for a 3D AFM QCP[1]: $\Delta\rho \propto T^{3/2}$ and $C/T \propto \gamma(0) - A\sqrt{T}$ [15, 16]. Inelastic neutron scattering shows two low energy features[17, 18, 19]. A broad peak at 4 meV that is only weakly **Q**-dependent corresponds to Kondo scattering with $T_K = 46$ K[20]. A peak at 0.7 meV that is highly **Q**-dependent and shows a maximum intensity at $\mathbf{Q}_N$= (1/2 1/2 0) corresponds to scattering from antiferromagnetic fluctuations. Although this compound is clearly close to a QCP, when T decreases below 0.4 K or B increases above 2 T, the system enters the FL state and C/T shows saturation[15, 16, 21]. The compound can be brought closer to the QCP by alloying with Pd. According to the phase diagram proposed by Knebel et al[22], the QCP occurs in $Ce(Ni_{2-x}Pd_x)_2Ge_2$ when x =0.065. Fukuhara et al found[23] that when x = 0.10, the ordering wavevector remains $\mathbf{Q}_N$= (1/2 1/2 0); for larger x it changes to $\mathbf{Q}_N$= (1/2 1/2 1/6).

Here we report measurements of the magnetic susceptibility, the specific heat for a series of magnetic fields, the resistivity, and the inelastic neutron scattering of a single crystal of $Ce(Ni_{0.935}Pd_{0.065})_2Ge_2$ grown with 58-Ni to reduce the incoherent scattering. For this alloy, we report the inelastic neutron spectra for the first time.

Single crystals were grown using the Czochralski method. The magnetization was measured in a commercial superconducting quantum interference device (SQUID) magnetometer. The specific heat measurements were performed in a commercial physical properties measurement system (PPMS). The electrical resistivity was also measured in the PPMS using the four wire method. The inelastic neutron experiment were performed on the MACS spectrometer[24] at the NIST center for neutron research (NCNR).

The magnetic susceptibility $\chi(T)$ is shown in figure 1(a). At high temperatures the data follow Curie-Weiss (CW) behavior $\chi(T) = C_{eff}/(T-\theta)$; in the range 150-300 K, the CW fit yields a moment 2.84 μ_B and θ = -118 K. Below 100 K, $\chi(T)$ is enhanced over the CW value but no long range magnetic order is observed down to 1.8 K. In the traditional spin fluctuation theory[1, 2, 12, 13] for a three dimensional system near a QCP, $\chi(T) \propto T^{-3/2}$ is expected for an AFM while $\chi(T) \propto T^{-4/3}$ is expected for a FM. In figure 1 (b), our data for $Ce(Ni_{0.935}Pd_{0.065})_2Ge_2$ follow the behavior of $(\chi(T) - \chi(0)) \propto T^{-1/6}$. Exponents for the susceptibility that are smaller than 1 are also observed for β-$YbAlB_4$[11] where $\chi(T) \propto T^{-1/3}$ when B=0.05 T and for $YbRh_2(Si_{0.95}Ge_{0.05})_2$ where $(\chi(T) - \chi(0)) \propto T^{-0.6}$[25].

In figure 2(a) we plot the specific heat C/T. The zero field data are very similar to that reported by Kuwai et al for $Ce(Ni_{0.936}Pd_{0.064})_2Ge_2$[26]. The data are logarithmic with temperature in the range 1 K to 5 K. When an external field is applied, the data deviate from the ln$T$ behavior and show saturation. As for $CeNi_2Ge_2$, this is because the magnetic field forces the system to enter the FL state. However, in figure 2(b), the low temperature (T <1 K) data can be seen to deviate slightly from the ln$T$ behavior; in this temperature range the data can be fit(figure 2(c)) to the form $\gamma_0 - aT^{1/2}$ (with γ_0 of order 695 mJ/mole-K^2), which as mentioned is the expected behavior in the SCR theory.

In figure 3(a) the resistivity $\rho(T)$ data are seen to be roughly linear in temperature over a wide range 0.4 K to 12 K, indicating NFL behavior. On the expanded scale of figure 3(b), $\rho(T)$

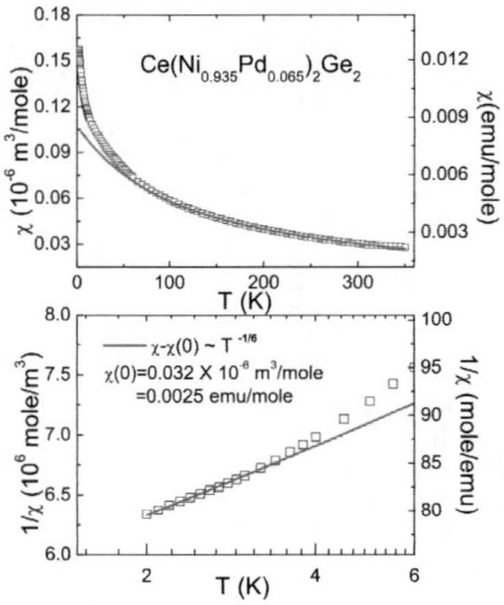

Figure 1. (a): Magnetic susceptibility of $Ce(Ni_{0.935}Pd_{0.065})_2Ge_2$. The solid line is the high temperature Curie-Weiss fit. (b): Low temperature inverse susceptibility. The solid line is $\chi(T) - \chi(0) \propto T^{-1/6}$.

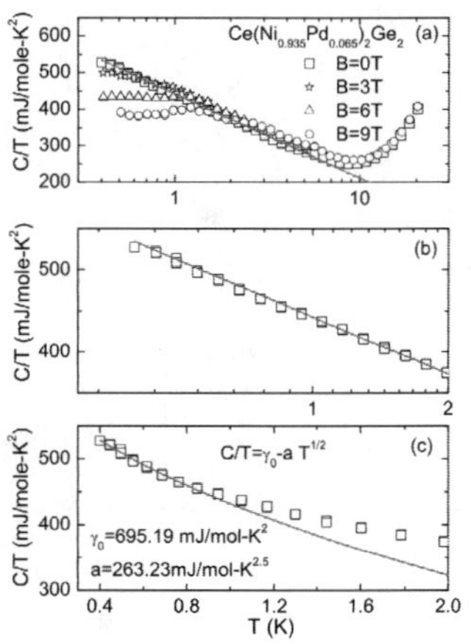

Figure 2. (a): C/T in applied fields of B=0, 3 T, 6 T and 9 T. (b): Low temperature zero field C/T curve in a logarithmic temperature scale. (c) Low temperature zero field C/T data. The solid line is $\gamma_0 - aT^{1/2}$.

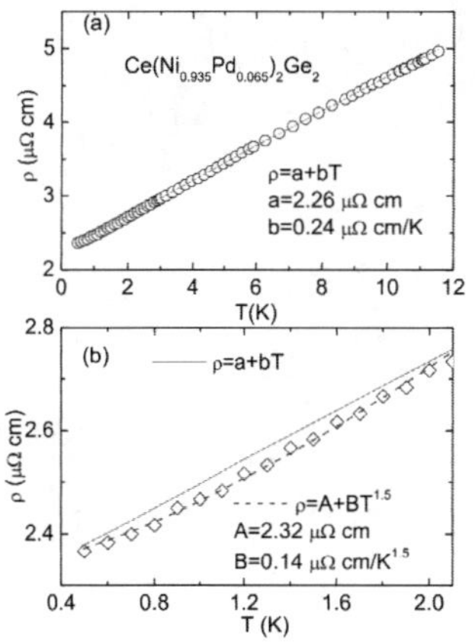

Figure 3. Resistivity for $Ce(Ni_{0.935}Pd_{0.065})_2Ge_2$. The solid line is the linear fit for the temperature range 0.4 K to 12 K while the dashed line represents $T^{3/2}$ dependence in the temperature range 0.4 K to 2 K.

deviates from T-linear, following a $T^{3/2}$ dependence. Again, this is the expected behavior in the SCR theory.

Hence, at temperatures higher than 2 K, this alloy exhibits typical NFL behavior, with the resistivity varying as $\rho(T) \propto T$, the specific heat varying as $C/T \propto \ln\frac{T_0}{T}$, and the susceptibility varying as $(\chi(T) - \chi(0)) \propto T^{-1/6}$. At temperatures below 1 K, however, $\rho(T) \propto T^{3/2}$ and $C/T \propto \gamma_0 - a\sqrt{T}$, thereby following the expected behavior of the SCR theory for a three dimensional AFM QCP spin fluctuation system.

In figure 4, we plot the INS spectra for $Ce(Ni_{0.936}Pd_{0.065})_2Ge_2$. A broad peak near 1.5 meV appears to be only weakly Q-dependent. As for the 4 meV peak in $CeNi_2Ge_2$, we identify this as representing Kondo scattering. Extra intensity is observed when $\Delta E < 1$ meV in the vicinity of $\mathbf{Q} = (1/2\ 1/2\ 0)$. This is the wavevector of the antiferromagnetic fluctuations seen for $CeNi_2Ge_2$ near 0.7 meV, hence we identify it as the critical wavevector $\mathbf{Q}_N$ for the QCP.

We first fit these spectra to the sum of a quasi-elastic Lorentzian (peak width Γ_0) and an inelastic Lorentzian (peak position E_1 and width Γ_1) (Fig. 4(a)). For $\mathbf{Q} = (1/2\ 1/2\ 0)$ we find E_1=1.51 meV, Γ_1=0.90 meV and Γ_0=0.35 meV; for $\mathbf{Q}$=(3/4 3/4 0), we find E_1=1.35 meV and Γ_1=0.88 meV. In figure 4(b), we use a second approach to obtain the AFM fluctuation spectra: we subtract the spectra for $\mathbf{Q} = (3/4\ 3/4\ 0)$ from that measured at (1/2 1/2 0). A fit of the difference to a quasielastic Lorentzian peak gives the AFM fluctuation energy scale as $\Gamma_0 = 0.44$ meV.

The value of E_1 is around 1.35-1.51 meV represents a Kondo temperature in the range 15.7-17.5 K. We note that this is smaller than the value 46 K seen in $CeNi_2Ge_2$, so that the approach to the QCP involves a reduction of T_K. We note in addition that T_K does not vanish at the QCP. A finite Kondo temperature is expected for a spin density wave system at a QCP.

The value of the AFM fluctuation energy Γ_0 is around 0.35-0.44 meV appears to be smaller

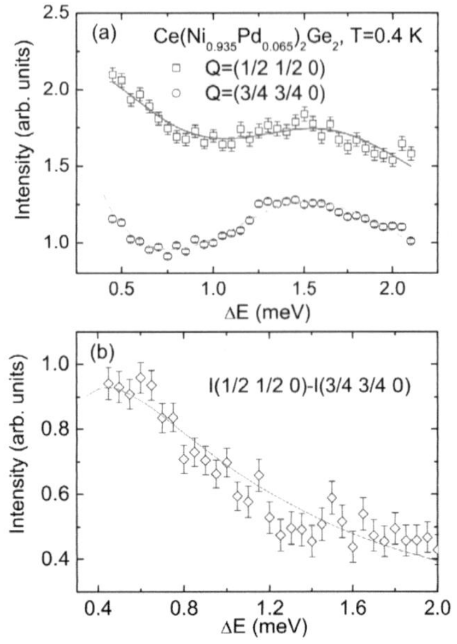

Figure 4. Inelastic neutron scattering spectra for $Ce(Ni_{0.935}Pd_{0.065})_2Ge_2$. The data were collected on MACS. In (a), the scattering at the critical wavevector $\mathbf{Q}_N = (1/2\ 1/2\ 0)$ is compared to the scattering at (3/4 3/4 0). The lines represent fits to the sum of a quasi-elastic and an inelastic Lorentzian peak, as described in the text. In (b) the difference between the data for **Q**= (1/2 1/2 0) and (3/4 3/4 0) is compared to a quasielastic Lorentzian (solid line).

than the fluctuation energy scale of pure $CeNi_2Ge_2$ where Γ=0.7 meV [17]. This represents "critical slowing down" of the AFM spin fluctutaions. However, despite the fact that this system is believed to sit at the QCP, the lifetime of the AFM spin fluctuations is finite, i.e. it does not diverge as expected at the critical point. Such a "saturation" of the spin fluctuation lifetime has been observed for other HF systems where the QCP is attained by alloying [4, 6] and may arise from the fact that the QCP occurs in a disordered environment.

Research at UC Irvine was supported by the U.S. Department of Energy, Office of Basic Energy Sciences, Division of Materials Sciences and Engineering under Award DE-FG02-03ER46036. Work at Los Alamos National Laboratory was performed under the auspices of the U.S. DOE/Office of Science. Work at ORNL was sponsored by the Laboratory Directed Research and Development Program of ORNL, managed by UT-Battelle, LLC, for the U. S. DOE, and was supported by the Scientific User Facilities Division, Office of Basic Energy Sciences, DOE. We acknowledge the support of the National Institute of Standards and Technology, U. S. Department of Commerce, in providing the neutron research facilities used in this work. The identification of commercial products does not imply endorsement or recommendation by the National Institute of Standards and Technology.

References

[1] Moriya T and Takimoto T 1995 *J. Phys. Soc. Jpn.* **64** 960
[2] Stewart G R 2001 *Rev. Mod. Phys.* **73** 797
[3] Sekine C, Sakakibara T, Amitsuka H, Miyako Y and Goto T, 1992 *J. Phys. Soc. Jpn* **61** 4536
[4] Kadowaki H, Tabata Y, Sato M, Aso N, Raymond S and Kawarazaki S 2006 *Phys. Rev.* Lett. **96** 016401
[5] Kambe S, Raymond S, Regnault L.-P, Flouquet J, Lejay P and Haen P 1996 *J. Phys. Soc. Jpn* **65** 3294
[6] Knafo W, Raymond S, Flouquet J, Fåk B, Adams M A, Haen P, Lapierre F, Yates S, and Lejay P 2004 *Phys. Rev.* B **70** 174401
[7] Stockert O, Löneysen H v, Rosch A, Pyka N, and Loewenhaupt M 1998 *Phys. Rev.* Lett. **80** 5627
[8] Schröder A, Aeppli G, Bucher E, Ramazashvili R, and Coleman P 1998 *Phys. Rev.* Lett. **80** 5623
[9] Schröder A, Aeppli G, Coldea R, Adams M, Stockert O, Löhneysen H v, Bucher E, Ramazashvilil R and Coleman P 2000 *Nature* **407** 351

[10] Custers J, Gegenwart P, Wilhelm H, Neumaier K, Tokiwa Y, Trovarelli O, Geibell C, Steglich F, Péin C and Coleman P, 2003 *Nature* **423** 524
[11] Nakatsuji S, Kuga K, Machida Y, Tayama T, Sakakibara T, Karaki Y, Ishimoto H, Yonezawa S, Maeno Y, Pearson E, Lonzarich G G, Balicas L, Lee H and Fisk Z 2008 *Nature* Phys. **4** 603
[12] Hertz J A, 1976 *Phys. Rev.* B **14** 1165
[13] Millis A J, 1993 *Phys. Rev.* B **48** 7183
[14] Si Q, Rabello S, Ingersent K, and Smith J L 2001 *Nature* **413** 804.
[15] Steglich F, Buschinger B, Gegenwart P, Lohmann M, Helfrich R, Langhammer C, Hellmann P, Donnevert L, Thomas S, Link A, Geibel C, Lang M, Sparn G and Assmus W 1996 *J. Phys: Condens. Matter* **8** 9909
[16] Aoki Y, Urakawa J, Sugawara H, Hideyuki S, Fukuhara T and Maezawa K 1997 *J. Phys. Soc. Jpn* **66** 2993
[17] Kadowaki H, Fåk B, Fukuhara T, Maezawa K, Nakajima K, Adams M A, Raymond S, and Flouquet J 2003 *Phys. Rev.* B **68** 140402
[18] Knopp G , Loidl A, Caspary R, Gottwick U, Bredl C D, Spille H, Steglich F and Murani A P 1998 *J. Magn. Magn. Mater.* **74**, 341
[19] Fåk B, Flouquet J, Lapertot G, Fukuhara T and Kadowaki H 2000 *J. Phys: Condens. Matter* **12** 5423
[20] A fit to the susceptibility of $CeNi_2Ge_2$ suggested a crystal field excitation at 4 meV; however, the same model predicted the excitation would not be observable in neutron scattering, due to a vanishing matrix element. See Frost C D, Rainford B D, Carter F V and Saxena S S 2000 *Physica* B **276-278** 290
[21] Koerner S, Scheidt E -W, Schreiner T, Heuser K, and Stewart G R 2000 *J. Low Temp. Phys.* **119** 147
[22] Knebel G, Brando M, Hemberger J, Nicklas M, Trinkl W, and Loidl A 1999 *Phys. Rev.* B **59** 12390
[23] Fukuhara T, Hiroaki K, Naofumi A, Takeuchi T, Kuwai T, Sakurai J and Maezawa K 2002 *J. Phys. Soc. Jpn.* **71** 1360
[24] MACS: a new high intensity cold neutron spectrometer at NIST, Rodriguez J A, Adler D M, Brand P C, Broholm C, Cook J C, Brocker C, Hammond R, Huang Z, Hundertmark P, Lynn J W, Maliszewskyj N C, Moyer J, Orndorff J, Pierce D, Pike T D, Scharfstein G, Smee S A and Vilaseca R, 2008 *Meas. Sci. Technol.* **19** 034023
[25] Gegenwart P, Custers J, Tokiwa Y, Geibel C, and Steglich F, 2005 *Phys. Rev.* Lett. **94** 076402
[26] Kuwai T, Takeuchia M, Tsuchiya T, Isikawa Y, Fukuhara T, 2004 *J. Mag. Mag. Mater.* **272** 68

Entropy accumulation near quantum critical points: effects beyond hyperscaling

Jianda Wu[1], Lijun Zhu[2] and Qimiao Si[1]

[1]Department of Physics & Astronomy, Rice University, Houston, Texas 77005, USA
[2]Theoretical Division and Center for Nonlinear Studies, Los Alamos National Laboratory, Los Alamos, New Mexico 87545, USA

E-mail: jw5@rice.edu

Abstract. Entropy accumulation near a quantum critical point was expected based on general scaling arguments, and has recently been explicitly observed. We explore this issue further in two canonical models for quantum criticality, with particular attention paid to the potential effects beyond hyperscaling. In the case of a one-dimensional transverse field Ising model, we derive the specific scaling form of the free energy. It follows from this scaling form that the singular temperature dependence at the critical field has a vanishing prefactor but the singular field dependence at zero temperature is realized. For the spin-density-wave model above its upper critical dimension, we show that the dangerously irrelevant quartic coupling comes into the free energy in a delicate way but in the end yields only subleading contributions beyond hyperscaling. We conclude that entropy accumulation near quantum critical point is a robust property of both models.

1. Introduction

Quantum critical points (QCPs) have been extensively studied in heavy fermion metals and related systems [1]. They occur as a non-thermal control parameter is tuned to a second-order phase transition at zero temperature. For thermodynamics, it was shown [2] based on scaling considerations that the thermal expansion [$\alpha = (1/V)(\partial V/\partial T)_{p,N} \propto \partial S/\partial p$, the variation of entropy S with pressure p] is more singular than the specific heat [$c_p = (T/N)(\partial S/\partial T)_p$], in the general case when the tuning parameter is linearly coupled to pressure. Correspondingly, the Grüneisen ratio, $\Gamma = \alpha/c_p$, diverges at the QCP, and the entropy is maximized. The same conclusions apply to a field-tuned QCP, where the magnetic Grüneisen ratio is the magnetocaloric effect. The predicted divergence of the Grüneisen ratio is by now widely observed in quantum critical heavy-fermion metals [3, 4], and the entropy enhancement has recently been explicitly observed in a quantum critical ruthenate [5].

The scaling arguments proceed as follows. Near a QCP, the critical part of the free energy takes the hyperscaling form $F = F_0 T^{d/z+1} f(r/T^{1/(\nu z)})$, where $r = p - p_c$, d is the spatial dimension and ν, z are respectively the correlation-length and dynamic exponents. The universal function $f(x)$ has different asymptotic behaviors in the $x \to 0$ and $x \to \pm\infty$ limits, respectively corresponding to the quantum critical and quantum disordered/renormalized classical regimes. The divergence of Γ can be readily derived, in universal forms as $1/r$ for $|r| \gg T$ and as $1/T^{1/\nu z}$ for $|r| \ll T$ [2]. The $1/r$ divergence in the low-temperature limit also amounts to a sign change of Γ across the QCP. As seen in Fig. 1, this corresponds to an entropy maximization at the QCP

Published under licence by IOP Publishing Ltd

in the low-temperature limit. The extension of this sign change to the finite temperature phase transitions has also been discussed [6].

The hyperscaling form for the free energy is based on the existence of a single critical energy scale, *i.e.*, the gap for the quantum critical excitations $\Delta \sim r^{\nu z}$. In this paper we explore how the scaling form of the free energy arises in some specific models for QCPs, paying particular attention to the possible effects that go beyond hyperscaling.

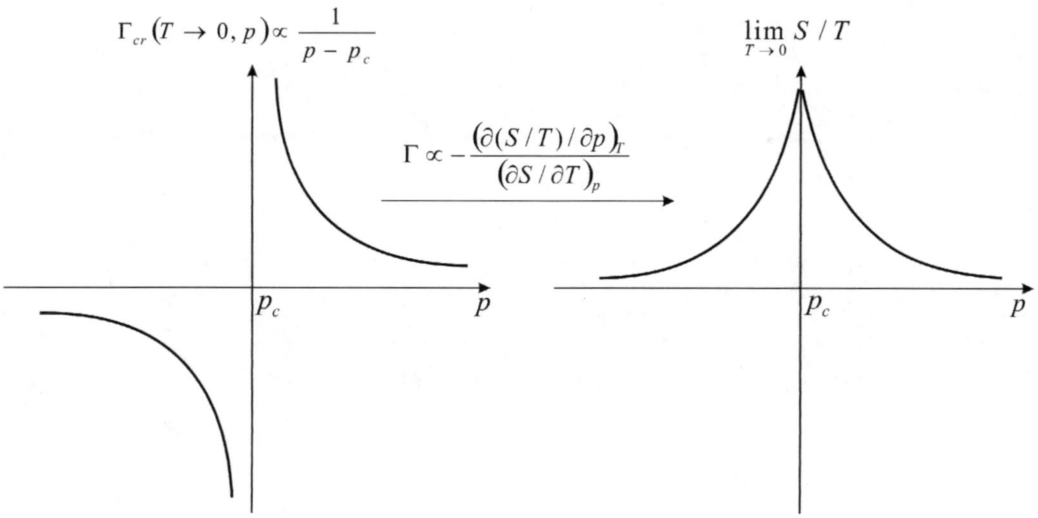

Figure 1. Divergence of the Grüneisen ratio and the accumulation of entropy near QCPs.

2. One-dimensional transverse field Ising model

Consider first the one-dimensional transverse field Ising model (1D TFIM), defined by the Hamiltonian [7]:

$$H_I = -J \sum_i \left(g\hat{\sigma}_i^x + \hat{\sigma}_i^z \hat{\sigma}_{i+1}^z \right), \tag{1}$$

where $J > 0$ is the ferromagnetic Ising coupling between the neighboring spins and gJ is the transverse magnetic field. With the tuning of g, there exists a quantum phase transition from a ferromagnet to quantum paramagnet at $g_c = 1$ [7, 8]. Using a Jordan-Wigner transformation, we can represent the Hamiltonian in the thermodynamic limit in terms of free fermions and the free energy can be written as $F = -Nk_BT \left[\ln 2 + \frac{1}{\pi} \int_0^\pi d\, k \ln \cosh(\varepsilon_k / 2k_BT) \right]$, with the dispersion $\varepsilon_k = 2J \left(1 + g^2 - 2g \cos k \right)^{1/2}$. The normalized free energy density is

$$f \equiv \frac{F}{NJk_B} = f_0(g) - \frac{t}{\pi} \int_0^\pi dk \ln \left(1 + e^{-2A}\right); A = \sqrt{(1-g)^2/t^2 + 4(g/t^2)\sin^2(k/2)} \tag{2}$$

where $t = k_BT/J$, and $f_0(g)$ is the ground state energy. Our focus will be on the T-dependent part, $f_1(g, T) = f(g, T) - f_0(g)$.

Consider first the quantum critical regime, where $|g - 1|/t \ll 1$. At temperatures small compared to J, we can introduce a momentum cutoff $k_c \approx t$, below which $|\sin(k/2)|/t \ll 1$ and therefore $A \ll 1$. The free energy can be approximated by the contributions from $k < k_c$, which

Journal of Physics: Conference Series **273** (2011) 012019 doi:10.1088/1742-6596/273/1/012019

is

$$f_1 \approx -\frac{t}{\pi}\left[\frac{1}{2}\left(\frac{1-g}{t}\right)^2 k_c + \frac{g}{t^2}\frac{k_c^3}{6} + k_c \ln 2\right] \overset{k_c \approx t}{\approx} -\frac{t^2}{\pi}\left(\left(\frac{g}{6} + \ln 2\right) + \frac{1}{2}\left(\frac{1-g}{t}\right)^2\right) . \quad (3)$$

While Eq. (3) contains approximations for the regular pieces, it reveals an important point. Compared with the general hyperscaling form, this expression is particular in that the linear term $|1-g|/t$ in the expansion vanishes. As a result, the magnetic Grüneisen ratio as a function of temperature at the QCP stays finite.

Consider next the low-temperature regions where $|1-g|/t \gg 1$. Here $A \gg 1$ and we can expand $\ln(1+e^{-2A})$ as e^{-2A} and obtain

$$f_1 \approx -\frac{t^2}{\sqrt{2\pi g}}\left(\frac{|1-g|}{t}\right)^{1/2} e^{-\frac{2}{t}|1-g|} . \quad (4)$$

We can further replace $1/\sqrt{g}$ by $1/\sqrt{g_c}$ without missing any singularity. The free energy has the exact hyperscaling form with $z = 1$. We obtain the magnetic Grüneisen ratio to be

$$\Gamma = \frac{1}{g-1} , \quad (5)$$

which is indeed divergent at the QCP, $g_c = 1$. The accumulation of entropy immediately follows (Fig. 1).

3. Spin-density-wave quantum critical point

Quantum phase transitions in itinerant magnets are traditionally described in terms of a $T = 0$ spin-density-wave (SDW) transition and formulated as a quantum Ginzburg-Landau theory with dynamic exponent $z > 1$ [9]. For concreteness, we take the 3D antiferromagnetic case as an example, *i.e.*, with $d = 3$ and $z = 2$. The effective dimension of the Ginzburg-Landau theory is $d+z$, which is above the upper critical dimension 4. Correspondingly, the quartic coupling u, appearing in the action as $u\int \phi^4$, is irrelevant in the renormalization-group (RG) sense. Under the RG transformation, u renormalizes to zero as the fixed point is reached, leaving only the Gaussian (quadratic) part. However, u is dangerously irrelevant [10, 11]. For instance, in the quantum critical regime, the correlation length is determine as $\xi^{-2} \sim r(T) = r + cuT^{3/2}$ (where c is a constant, $4(n+2)(1/3 - \ln 2\sqrt{2}/(3\pi^3))$, with n being the number of components of the order parameter). One can consider u as introducing a new energy scale, and the scaling functions for generic physical quantities are expressed in terms of two variables $f(r/T, uT^{3/2}/r)$.

Within the RG approach [10], it is natural to focus on the free energy associated with the Gaussian term, F_G. The expression for F_G in the RG approach can be shown to be equal to a Gaussian free energy with r replaced by renormalized $r(T)$. We will focus on the quantum critical regime, and write $F_G = F_G^{(1)} + F_G^{(2)}$, where

$$F_G^{(1)} = -nV \int_0^{\Lambda} \frac{d^3q}{(2\pi)^3} \int_0^{\Gamma_q} \frac{d\varepsilon}{2\pi} \left(\coth\frac{\varepsilon}{2T} - 1\right) \tan^{-1}\frac{\varepsilon/\Gamma_q}{r(T) + (q/\Lambda)^2} , \quad (6)$$

and

$$F_G^{(2)} = -nV \int_0^{\Lambda} \frac{d^3q}{(2\pi)^3} \int_0^{\Gamma_q} \frac{d\varepsilon}{2\pi} \tan^{-1}\frac{\varepsilon/\Gamma_q}{r(T) + (q/\Lambda)^2} . \quad (7)$$

Here, $\Gamma_q = \Gamma_0 q^{z-2}$, while Λ and Γ_0 are respectively the ultraviolet momentum and energy cutoffs. The leading temperature-dependent contribution from $F_G^{(2)}$ turns out to be $\frac{nV\Lambda^3\Gamma_0}{4\pi^3}\left(\frac{\sqrt{2\pi}}{4} + \ln 2 - \frac{\sqrt{2}}{2}\ln(\sqrt{2}+1)\right) c(u/\Gamma_0^{3/2})T^{3/2}$, which is more singular than that from $F_G^{(1)}$.

This Gaussian form, however, is incomplete. The quartic coupling also introduces an explicitly linear in u term ($t \equiv T/\Gamma_0$), which takes the form

$$F_u = -2n(n+2)N\Gamma_0 u \int_0^\infty e^{-6x} \left(\int_0^1 \frac{d^3q}{(2\pi)^3} \int_0^1 \frac{d\varepsilon}{\pi} \coth \frac{\varepsilon}{2te^{2x}} \frac{\varepsilon}{\varepsilon^2 + (re^{2x}+q^2)^2} \right) f_u \, dx \,,$$
$$f_u \equiv \frac{1}{2\pi^2} \int_0^1 \frac{d\varepsilon}{\pi} \coth \frac{\varepsilon}{2te^{2x}} \frac{\varepsilon}{\varepsilon^2 + (re^{2x}+1)^2} + \frac{2}{\pi} \int \frac{d^3q}{(2\pi)^3} \coth \frac{1}{2te^{2x}} \frac{1}{1 + (re^{2x}+q^2)^2} \,. \quad (8)$$

We find that, to the linear order in u, the leading temperature-dependent contribution from F_u exactly cancels that of $F_G^{(2)}$, leaving the singular contribution to the total free energy to be

$$F_{tot} \approx F_G^{(1)} \,, \quad (9)$$

which is given by

$$F_{tot} \sim -2\frac{T^2}{\Gamma_0^2} + \frac{2\pi T^{5/2}}{\Gamma_0^{5/2}} \left(1 + \frac{9}{2} \frac{r + cuT^{3/2}/\Gamma_0^{3/2}}{T/\Gamma_0} \right) . \quad (10)$$

We can trace the singular terms included in Eq. (6) to the contributions of the low-energy degrees of freedom, and those included in Eq. (7) to a unphysical origin from the degress of freedom at high energies and short distances near the cutoff scales. Eqs. (6,9) are therefore expected to be the regularization prescription that is valid to all orders in u. The explicit result given in Eq. (10) is consistent with those often quoted in the literature. The T^2 term represents the background Fermi liquid contribution. The critical part of the free energy is consistent with the scaling form $T^{5/2}(1+r/T)$. The contribution of the renormalized quartic coupling $uT^{3/2}$ to the free energy is subleading; the dangerously irrelevant parameter does not modify the leading singular part of the free energy from its hyperscaling form, thereby preserving the divergence of the Grüneisen ratio and the accompanied entropy accumulation.

4. Summary

We have considered the thermodynamic properties of two models of quantum criticality. We have analyzed the potential effects beyond the most general hyperscaling forms in both models, and established that entropy accumulation is a robust property of both models.

Acknowledgments

We acknowledge the support of NSF Grant No. DMR-1006985, the Robert A. Welch Foundation Grant No. C-1411 (JW and QS), and the U.S. DOE through the LDRD program at LANL (LZ).

References

[1] Si Q, Steglich F 2010 *Science* **329** 1161
[2] Zhu L et al. 2003 *Phys. Rev. Lett* **91** 066404
[3] Küchler R et al. 2003 *Phys. Rev. Lett.* **91** 066405
[4] Gegenwart P. 2010 *J. Low Temp. Phys.* **161** 117
[5] Rost A W et al. 2009 *Science* **325** 1360
[6] Garst M et al. 2005 *Phys. Rev.* B**72** 205129
[7] Pfeuty P 1970 *Ann. Phys. NY* **57** 79
[8] Chakrabarti B K, Duta A and Sen P 1996 *Quantum Ising Phase and Transitions in Transverse Ising Models* (Springer).
[9] Hertz J A 1976 *Phys. Rev.* B**14** 1165
[10] Millis A J 1993*Phys. Rev.* B**48** 7183; Zülicke U and Millis A J 1995 *Phys. Rev.* B**51** 8996
[11] Moriya T 1985 *Spin Fluctuations in Itinerant Electron Magnets* (Berlin: Springer-Verlag)

Stability of superflow in supersolid phases of lattice bosons with dipole-dipole interaction

Daisuke Yamamoto[1] **and Ippei Danshita**[2]

[1] Department of Physics, Waseda University, Okubo, Shinjuku-ku, Tokyo 169-8555, Japan
[2] Computational Condensed Matter Physics Laboratory, RIKEN, Wako, Saitama 351-0198, Japan

E-mail: yamamoto@kh.phys.waseda.ac.jp

Abstract. We investigate the stability of superflow of bosons with isotropic dipole-dipole interactions in a two-dimensional optical lattice. We perform linear stability analyses for the dipolar Bose-Hubbard model in the hardcore boson limit, and show that the superflow can exist in a supersolid phase unless the velocity exceeds a certain critical value and that the critical value is remarkably smaller than that in the standard superfluid phase. Additionally, it is found that there exists a parameter range in which the SS phases are stabilized by a finite superflow. We also discuss the influence of quantum fluctuations on these results within the cluster mean-field approximation.

Recently, the physics of systems with strong dipole-dipole interactions has received considerable interest because of the successful experimental preparation of ultracold dipolar atoms [1] and polar molecules [2]. The long-range nature and anisotropy of the dipole-dipole interaction greatly enrich the variety of phenomena which can be observed in experiments with cold atoms and molecules. Recent theoretical studies predicted that various quantum phases, such as fermionic p-wave superfluids (SF) [3] and Haldane-Bose insulators [4], can be realized by controlling appropriately the strength of the dipole-dipole interaction and the orientation of the dipole moments. Furthermore, the systems of ultracold dipolar gases have the potential to answer the long-standing question of whether a solid can exhibit superfluidity [5–8]. The coexisting phase of solid order and superfluidity is usually called the "supersolid (SS)" phase. Recent quantum Monte Carlo (QMC) simulations for a two-dimensional Bose-Hubbard system with the isotropic long-range interactions have shown that an off-diagonal long-range order with a finite SF stiffness coexists with a crystalline long-range order in a wide range of parameters [7]. Hence, it is highly likely that the SS phases will be experimentally observed in the context of dipolar Bose gases loaded into optical lattices in the near future.

In order to prove the existence of SS phases in experiments, one has to verify that both the crystalline order and superfluidity are simultaneously present. While the existence of the crystalline order can be identified by using the Bragg scattering techniques in experiments of ultracold gases, it is diffecult to measure directly the SF fraction. Instead, the superfluidity of the gases can be revealed by measuring the critical velocity above which superflow breaks down. In the experimental demonstrations of superfluidity of weakly- and strongly-interacting Bose gases [9, 10], and fermionic SF across the BEC-BCS crossover [11], the critical velocity of the superflow has been measured by using a moving optical lattice. In this work, we investigate the stability of superflow of dipolar Bose gases loaded into a two-dimensional moving optical

Published under licence by IOP Publishing Ltd

lattice, and propose that the superfluidity of the SS phases can be also identified by measuring the critical velocity of superflow.

Assuming that the dipole moments are polarized to the direction perpendicular to the lattice plane, we consider here the Bose Hubbard model with isotropic dipole-dipole interactions [5],

$$\hat{H} = -J\sum_{\langle j,l\rangle}(\hat{a}_j^\dagger \hat{a}_l + \text{h.c.}) + \frac{U}{2}\sum_j \hat{n}_j(\hat{n}_j - 1) + \sum_{j<l} V_{jl}\hat{n}_j\hat{n}_l - \mu\sum_j \hat{n}_j, \tag{1}$$

where $\hat{a}_j^\dagger$ is the boson creation operator at site j of a square lattice, $\hat{n}_j = \hat{a}_j^\dagger \hat{a}_j$, J is the hopping between nearest-neighbor sites, and U is the onsite interaction. The long-range part of the dipole-dipole interaction can be expressed by $V_{jl} = Vd^3/|\mathbf{r}_j - \mathbf{r}_l|^3$. Here, d is the lattice spacing and $\mathbf{r}_j = (j_x d, j_y d)$ is a lattice vector, where j_x and j_y are integers. The average density of bosons, $n \equiv \sum_j \langle \hat{n}_j \rangle / M$, where M is the total number of lattice sites, is controlled by the chemical potential μ. In the hard-core boson limit, $U \to \infty$, only the occupation numbers $\hat{n}_j = 0$ or 1 are allowed, and thereby Eq. (1) can be mapped onto the following quantum spin-1/2 Hamiltonian:

$$\hat{H}_{\text{s}} = -J\sum_{\langle j,l\rangle}\left(\hat{S}_j^+ \hat{S}_l^- + \text{h.c.}\right) + \sum_{j<l} V_{jl}\hat{S}_j^z \hat{S}_l^z - h\sum_j \hat{S}_j^z \tag{2}$$

with $\hat{S}_j^- = \hat{a}_j$ and $\hat{S}_j^z = \hat{n}_j - 1/2$. The shifted chemical potential $h = \mu - 1/2\sum_{l(\neq j)} V_{jl}$ plays the role of the *magnetic field* acting on the spins. In the spin language, the density ordering and Bose-Einstein condensation are expressed by the spatial modulation of the expectation values of the longitudinal spin components and existence of non-zero transverse spin components, respectively [12].

The Weiss molecular-field approximation, or simply the mean-field (MF) approximation, is widely used to understand qualitatively the properties and behaviors of the spin systems. At zero temperature, the MF energy is obtained by replacing the spin operators at each lattice site by the classical vectors of length $S = 1/2$; $\hat{\mathbf{S}}_j \to \mathbf{S}_j^{\text{cl}} = S(\cos\varphi_j \sin\theta_j, \sin\varphi_j \sin\theta_j, \cos\theta_j)$. For the case that the system does not have superflow, we can take $\varphi_j = 0$ without loss of generality. Minimizing the MF energy with respect to $\{\theta_j\}$ within an ansatz for some possible sublattice structures, we obtain the ground-state phase diagram in the $(V/J, h/J)$-plane as shown in Fig. 1. One can see that in addition to the usual SF and Mott insulator (MI) phases, there exist the checkerboard solid (CS) and checkerboard supersolid (CSS) phases for $V/J > 3.024$. These features of the phase diagram, including the presence of stable CSS phases, are in qualitative agreement with the QMC results [7], which is surprising because the mean-field prediction of the existence of SS phases is not correct in the hard-core bosonic Hubbard model with nearest- and next-nearest-neighbor interactions [13]. This is attributed to the fact that the long-range nature of the dipole-dipole interactions sufficiently suppresses the quantum fluctuations. Other solid and SS phases with a more complicated structure appear in the region of large V/J. Since our main purpose here is to investigate superfluidity of the SF and CSS phases, in the figure we only show the region $V/J < 5.2$, where an additional SS phase, named SS2, is found.

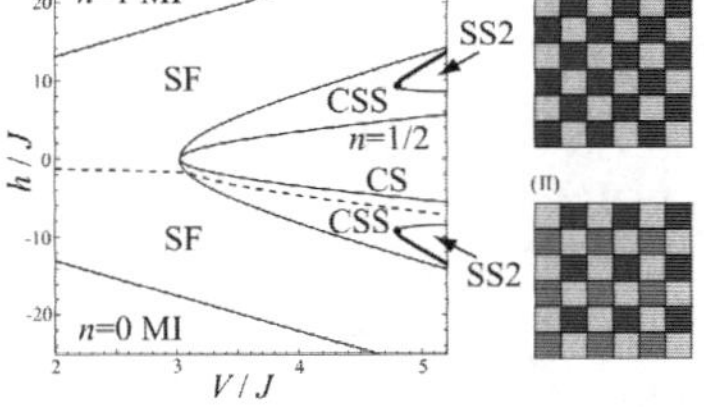

Figure 1. The ground-state MF phase diagram of the dipolar Bose-Hubbard model in the hard-core boson limit. Second and first-order phase transitions are indicated by thin and thick solid lines, respectively. In the CS and CSS phases, the system has a checkerboard-type two-sublattice structure sketched in (I). The symmetry of SS2 state is sketched in (II). The dashed line represents the contour of $n = 0.45$.

Now let us investigate the stability of superflow in the SF and CSS phases (see also Ref. 8). We consider here that the optical lattice confining hardcore dipolar bosons is moving at a constant velocity $\mathbf{v}$, and thereby in the frame where the lattice is at rest, the SF component of the hardcore bosons is flowing with quasi-momentum $\mathbf{K} = -m\mathbf{v}$ [9, 10], where m is the particle mass. Our stability analysis of superflow is based on the linear spin-wave (LSW) theory. First, minimizing the MF energy with $\varphi_j = -\mathbf{K} \cdot \mathbf{r}_j$, we derive the current-carrying MF solutions of $\{\theta_j\}$ for the SF and CSS phases. Next, we perform local rotations of the spin reference frame, so that at each site the new spin quantization axis is oriented along the direction of the classical spin vector. Introducing bosons via the standard Holstein-Primakoff transformation, we diagonalize the quadratic part of the resulting boson Hamiltonian by a Bogoliubov transformation. From the behavior of the excitation spectra obtained in this way, we examine the stability of superflow.

Hereafter we assume $\mathbf{K} = (K, 0)$, namely that the supercurrent flows along the x direction in the two-dimentional lattice. In Fig. 2, we show the excitation spectra $\omega(\mathbf{q})$, where $\mathbf{q}$ is the quasi-momentum of the excitation, for $V = 3.5J$, $n = 0.45$, and the different values of K. When $K = 0$, the system is in the stable CSS phase, and its excitation has two branches, reflecting the two-sublattice structure. The dispersion curves gradually deform (or *tilt*) with increasing K. When K reaches a certain critical value, $K^{\mathrm{L}}_{\mathrm{CSS}}$, the lower branch touches zero as shown in Figs. 2(b) and (d). When $K > K^{\mathrm{L}}_{\mathrm{CSS}}$, a part of the excitation spectra has negative excitation energy, which signals the Landau instability (LI). Yet, since the LI is effective only in the presence of the dissipation, it cannot destabilize the system at (or near) zero temperature [14]. The existence of excitation modes with a complex frequency causes another type of instability, the dynamical instability (DI), which means exponential growth of the fluctuations in time and is relevant even at zero temperature. When K increases further and exceeds a critical value $K^{\mathrm{D}}_{\mathrm{CSS}}$, long-wavelength phonons cause the DI, as shown in Figs. 2(c), (e), and (f), which reflects the fact that the effective mass in the x-direction is negative, resulting in the imaginary sound speed in the x direction.

The stability phase diagrams in the (n, Kd)- and $(V/J, Kd)$-planes are shown in Figs. 3(a) and (b). One can see that in Fig. 3(a), the critical quasi-momentum for DI in the CSS phase is remarkably smaller than that in the SF phase and independent on the density n within the MF and LSW calculations. The V/J-dependence of the critical quasi-momentum for DI exhibits a characteristic minimum structure near the phase boundary between SF and CSS as shown in Fig. 3(b). Additionally, it is found that there exists a parameter range in which the CSS phase is stabilized by a finite superflow.

Finally, let us discuss the influence of quantum fluctuations. To this end, we extend the cluster mean-field (CMF) theory [15] to the case where the system has finite superflow. Our calculations are based on a cluster of 2×2 sites. The interactions between spins (hardcore bosons) in a cluster are treated exactly, while the influence of outside spins is replaced by effective fields whose values are determined self-consistently from the expectation values of inside spins in the

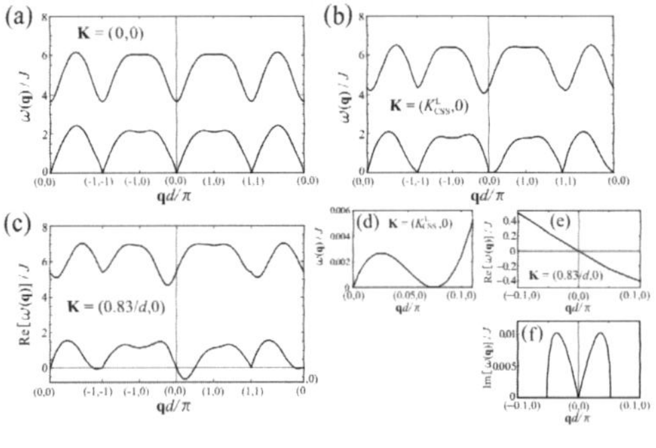

Figure 2. Excitation spectra $\omega(\mathbf{q})$ in the CSS phase for $V = 3.5J$, $n = 0.45$, and different values of K, where $K = 0$ (a), $K = K^{\mathrm{L}}_{\mathrm{CSS}}$ (b), and $K = 0.83/d > K^{\mathrm{D}}_{\mathrm{CSS}}$ (c). (d) and (e) are magnification of (b) and (c) focused on the region of $|q_x| \ll 1/d$ and $q_y = 0$. In (f), the imaginary part of $\omega(\mathbf{q})$ corresponding to (e) is shown.

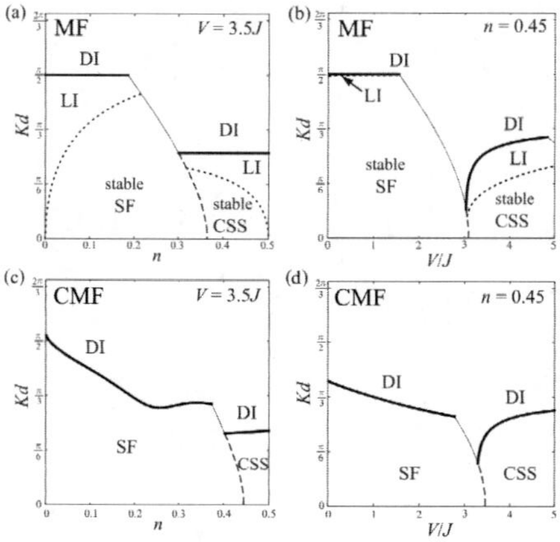

Figure 3. Stability phase diagrams of hardcore dipolar bosons flowing in a two-dimentinal lattice with quasi-momentum $\mathbf{K} = (K, 0)$ for $V = 3.5J$ in the (n, Kd)-plane [(a) and (c)] and for $n = 0.4$ in the $(V/J, Kd)$-plane [(b) and (d)]. The upper (lower) two figures show the results of the MF and LSW (CMF) calculations. The regions of SF, CSS, Landau instability (LI), and dynamical instability (DI) are located. The dotted, thick solid, and thin solid lines represent the critical quasi-momenta for LI, DI caused by phonons, and DI caused by roton-like excitations, respectively. The dashed line indicates the location of the flow-induced phase transition from SF to CSS state.

usual manner. We determine the critical momenta for DI by the condition that the inverse effective mass becomes zero. As shown in Figs. 3(c) and (d), the overall values of the critical momenta for DI are smaller than those of the MF calculations, as a result of including the effects of the quantum fluctuations. However, the main qualitative features, e.g., the minimum structure in the $(V/J, Kd)$-plane and the existence of the flow-induced CSS phase, are not changed.

In conclusion, we have investigated the stability of superflow of dipolar Bose gases in a two-dimensional moving optical lattice. We have derived the mean-field phase diagram of the dipolar Bose-Hubbard model in the hardcore boson limit and have calculated the critical quasi-momenta for Landau and dynamical instabilities in the superfluid and checkerboard supersolid phases. It is expected that the superfluidity of the supersolid phase can be identified by measuring the critical quasi-momenta above which superflow breaks down.

One of the authors (D.Y.) is supported by a Grant-in-Aid from JSPS.

References

[1] Griesmaier A, Werner J, Hensler S, Stuhler J and Pfau T 2005 *Phys. Rev. Lett.* **94** 160401; Lahaye T, Koch T, Froehlich B, Fattori M, Metz J, Griesmaier A, Giovanazzi S and Pfau T 2007 *Nature (London)* **448** 672.
[2] Ni K-K, Ospelkaus S, de Miranda M H G, Pe'er A, Neyenhuis B, Zirbel J J, Kotochigova S, Julienne P S, Jin D S and Ye J 2008 *Science* **322** 231.
[3] Baranov M A 2008 *Phys. Rep.* **464** 71; Bruun G M and Taylor E 2008 *Phys. Rev. Lett.* **101** 245301.
[4] Dalla Torre E G, Berg E and Altman E 2006 *Phys. Rev. Lett.* **97** 260401.
[5] Goral K, Santos L and Lewenstein M 2002 *Phys. Rev. Lett.* **88** 170406.
[6] Yi S, Li T and Sun C P 2007 *Phys. Rev. Lett.* **98** 260405; Danshita I and Sá de Melo C A R 2009 *Phys. Rev. Lett.* **103** 225301.
[7] Capogrosso-Sansone B, Trefzger C, Lewenstein M, Zoller P and Pupillo G 2010 *Phys. Rev. Lett.* **104** 125301.
[8] Danshita I and Yamamoto D 2010 *Phys. Rev.* A **82** 013645 .
[9] De Sarlo L, Fallani L, Lye J E, Modugno M, Saers R, Fort C and Inguscio M 2005 *Phys. Rev.* A **72** 013603.
[10] Mun J, Medley P, Campbell G K, Marcassa L G, Pritchard D E and Ketterle W 2007 *Phys. Rev. Lett.* **99** 150604.
[11] Miller D E, Chin J K, Stan C A, Liu Y, Setiawan W, Sanner C and Ketterle W 2007 *Phys. Rev. Lett.* **99** 070402.
[12] Matsuda H and Tsuneto T 1970 *Suppl. Prog. Theor. Phys.* **46** 411.
[13] Batrouni G G and Scalettar R T 2000 *Phys. Rev. Lett.* *84* 1599.
[14] Konabe S and Nikuni T 2006 *J. Phys.* B **39** S101; Iigaya K, Konabe S, Danshita I and Nikuni T 2006 *Phys. Rev.* A **74** 053611.
[15] Oguchi T 1955 *Prog. Theor. Phys.* **13** 148; Hassan S R, de Medici L and Tremblay A-M S 2007 *Phys. Rev.* B **76** 144420; Yamamoto D 2009 *Phys. Rev.* B **79** 144427.

Journal of Physics: Conference Series **273** (2011) 012021 doi:10.1088/1742-6596/273/1/012021

Electronic structure of $U(Ru_{1-x}Rh_x)_2Si_2$ studied by laser angle-resolved photoemission spectroscopy

R Yoshida[1], Y Nakamura[1], M Fukui[1], Y Haga[2], E Yamamoto[2], Y Ōnuki[2,3], M Okawa[4], S Shin[4], M Hirai[1], Y Muraoka[1] and T Yokoya[1]

[1] The Graduate School of Natural Science and Technology, Okayama Univeresity, Okayama, 700-8530, Japan

[2] Advanced Science Research Center, Japan Atomic Energy Agency, Tokai, Ibaraki 319-1195 , Japan

[3] Graduate School of Science, Osaka University, Toyonaka, Osaka 560-0043, Japan

[4] Institute for Solid State Physics, The University of Tokyo, Kashiwa, Chiba 277-8581, Japan

E-mail: dns421103@s.okayama-u.ac.jp

Abstract. We have studied electronic structures of $U(Ru_{1-x}Rh_x)_2Si_2$ employing ultrahigh-resolution laser angle-resolved photoemission spectroscopy to understand the effect of Rh substitution. A hole-like dispersive feature, which presumably has Rh *d*-band character, was observed in $U(Ru_{1-x}Rh_x)_2Si_2$ for both x = 0 and x = 0.03. However, although a heavy quasiparticle band appears in the hidden-order state of URu_2Si_2 (x = 0), it was not observed for x = 0.03 over the temperature range studied. In addition, it was found that energy-distribution curves at the Fermi vector of the hole-like band for x = 0.03 behave similarly to the Fermi-Dirac function. We also present the mapping of Fermi surfaces formed by the hole-like band, which stay nearly unaffected by Rh substitution.

1. Introduction

A heavy fermion superconductor URu_2Si_2 shows an anomalous second-order phase transition at 17.5 K, whose order parameter is still unknown [1, 2]. Although its nature is a big mystery, the 'hidden-order' state below 17.5 K seems to be closely related to the antiferromagnetic phase, which appears as a bulk property under pressure [3]. Thus, the comparison between these two phases looks beneficial to understand the underlying nature of the hidden order. However, it is not always possible to carry out such comparative studies because some experimental techniques are inherently impossible to be performed under pressure.

The above problem may be compensated by the technique of chemical substitution, which are intensively carried out in the recent studies [4, 5]. In URu_2Si_2 system, the effect similar to the pressure effect can be obtained by substituting Rh for Ru, which induces the antiferromagnetic phase coexisting with the hidden order. In fact, this behavior helps various experimental techniques, such as angle-resolved photoemission spectroscopy (ARPES), to carry out comparative studies in URu_2Si_2 as a function of Rh concentration.

Motivated by recent advancements in both photoemission spectroscopy [6] and the substitution studies on URu_2Si_2 [4, 5], we have carried out ARPES experiments on $U(Ru_{1-x}Rh_x)_2Si_2$ (x = 0 and 0.03) employing a state-of-art laser photoemission spectrometer. We observed a hole-like dispersive feature, which presumably holds Rh *d*-band character, in both

Published under licence by IOP Publishing Ltd

Rh concentrations. However, we observed a heavy quasiparticle band only for x = 0 over the temperature range studied. In addition, the temperature dependence of the hole-like band for x = 0.03 follows the typical metallic behavior. We also present the mapping of Fermi surfaces formed by the hole-like band, which seems nearly unaffected by Rh substitution. These findings suggest that the emergence of the narrow band is related to the hidden-order state while some electronic states stay nearly unmodified.

2. Experimental

Single crystals of $U(Ru_{1-x}Rh_x)_2Si_2$ (x = 0 and 0.03) were prepared by the Czochralski method employing a tetra-arc furnace where the detail of synthesis has been described elsewhere [7, 8]. ARPES spectra were obtained employing laser ARPES system at the Institute for Solid State Physics, The University of Tokyo, where the system consists of a vacuum ultraviolet laser (photon energy of 6.994 eV) and a Gammadata-Scienta R4000WAL electron analyzer. We used circularly-polarized light unless otherwise noted. The (001) surfaces were obtained by *in situ* cleaving in ultrahigh vacuum where the base pressure of the main chamber was kept better than 4×10^{-11} Torr throughout the experiment. Temperature dependence of ARPES spectra were studied in the temperature range of 7 - 26 K. Binding energies of spectra and sample temperature were calibrated in reference to the Fermi edge of a gold film evaporated near the sample, and the total energy resolution was set at 2 meV.

3. Result and Discussion

3.1. ARPES intensity plots

Figure 1 and Figure 2 illustrate ARPES data obtained for $U(Ru_{1-x}Rh_x)_2Si_2$ along (100) direction. Note that *s*-polarization was employed for these data, except for the one in Figure 1(c). In ARPES intensity plots at 7 K (Figure 1(a)) and corresponding energy-distribution curves (EDCs) (Figure 1(b)) for x = 0, we can observe a heavy band appearing near the Fermi level (E_F). We can also see a hole-like dispersive feature, which is also confirmed to intersect E_F at 25 K with the Fermi vector (k_F) $\sim$ 0.12 Å (Figure 1(c)). In the data for x = 0.03 taken at 7 K and 25 K along (100) direction, however, we can only observe a hole-like band whose k_F $\sim$ 0.14 Å (Figure 2). We note that these k_F values are close to the ones recently reported [9]. These observations show that although the existence of hole-like band is almost unvaried, the narrow band is extremely sensitive to the substitution of Rh.

3.2. Temperature dependence of energy-distribution curves

Figure 3(a) and 3(b) show the temperature dependence of EDCs for the same Rh concentrations where the cuts have been taken at the k_F of the hole-like band in each dataset. In Figure 3(a) for x = 0, one can see the development of the sharp peak with decreasing temperature. Moreover, the peak shifts toward higher binding-energy side as temperature is lowered. In Figure 3(b) for x = 0.03, however, spectra near E_F behave similarly to Fermi-Dirac function, and we are unable to detect the opening of gap. Being consistent with the observations seen in the ARPES intensity plots, remarkable difference, coming from the emergence of the narrow band, appears in energy-distribution curves.

3.3. Fermi Surface Topology

In Figure 4(a), we show the result of Fermi surface mapping performed for x = 0 at 25 K (paramagnetic phase). We observed a small hole-like Fermi surface where the positions of k_F, obtained from fitting MDCs, are superimposed in the same figure. The Fermi surface may have small anisotropy in shape, but significant deviation from spherical shape is absent. An estimated cross-sectional area is $\sim$ 0.71 $Å^{-2}$, which is close to the hole-like Fermi surface observed in the

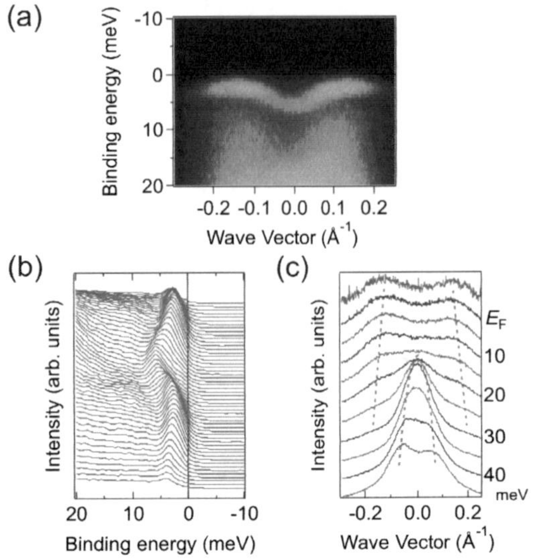

Figure 1. ARPES intensity plot along (100) direction measured for x = 0 sample: (a) Intensity map at 7 K, (b) EDCs at 7 K, and (c) MDCs at 25 K (where curves have been normalized to the area). Note that a concave-up dispersive feature, which does not cross the Fermi level, appears in the MDCs, and this feature is known as a surface state [9].

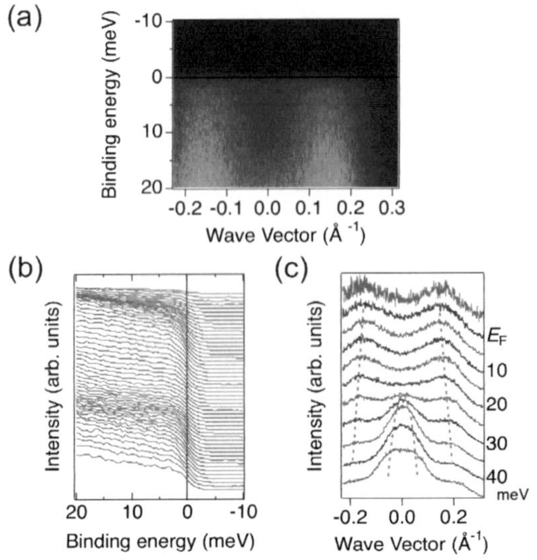

Figure 2. ARPES intensity plot along (100) direction measured for x = 0.03 sample: (a) Intensity map at 7 K, (b) EDCs at 7 K, and (c) MDCs at 25 K (where curves have been normalized to the area). Note that a concave-up dispersive feature, which is similar to the surface state shown in Figure 1(c), also appears in the MDCs for x = 0.03.

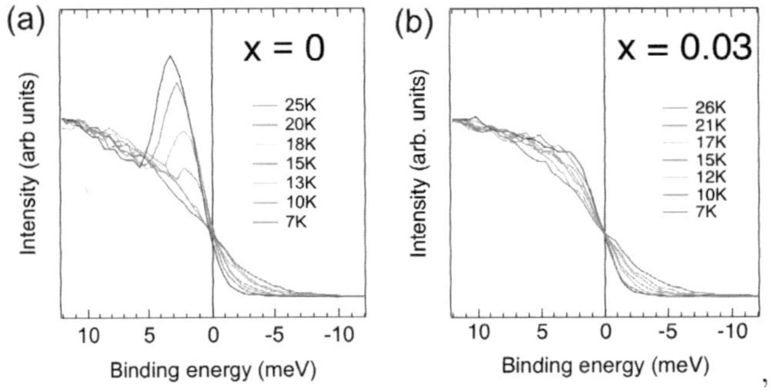

Figure 3. Temperature dependence of energy-distribution curves at the k_F of hole-like band; (a) for x = 0, and (b) for x = 0.03. Note that spectra have been normalized to the intensity at E_B = 12 meV.

dHvA study [7]. Figure 4(b) illustrates the Fermi surface mapped for x = 0.03 at 26 K where *s*-polarization was employed. In the same way, we observed a hole-like Fermi surface, which size and shape anisotropy are comparable to the one for x =0 within the experimental errors. This is consistent with the view that Rh substitution has small e ect on the hole-like band.

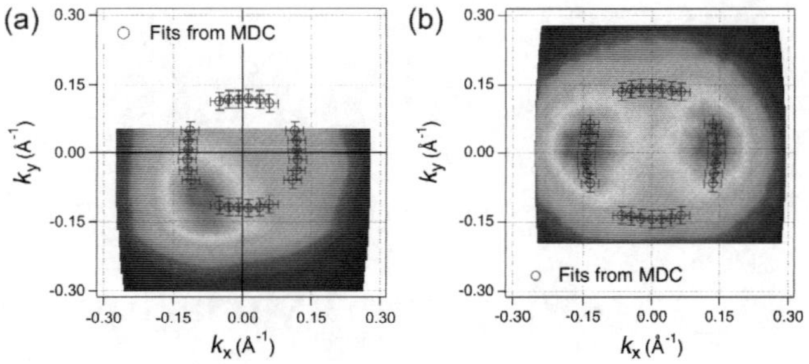

Figure 4. Fermi surface mapping performed (a) for x = 0 at 25 K, and (b) for x = 0.03 at 26 K. Note that *s*-polarization was employed for (b), and the positions of k_F obtained from tting MDCs have been symmetrized and superimposed on the gures.

4. Conclusion

In this paper, we present the ARPES data on the system of $U(Ru_{1\,x}Rh_x)_2Si_2$ and discuss the e ect of Rh substitution. We show that substitution of Rh signi cantly a ects the emergence of narrow band observed in the hidden-order state of URu_2Si_2. However, the e ect of Rh substitution on hole-like bands appears rather small. These experimental data can be a good starting point to di erentiate hidden-order state from others and would promote further systematic studies of the system of $U(Ru_{1\,x}Rh_x)_2Si_2$.

Acknowledgments

We are grateful to I. Kawasaki, H. Yamagami, H. Harima, K. Miyake, T. Shimojima, K. Okazaki, Y. Ishida, W. Malaeb, and K. Machida for valuable comments. This work is supported by a Grant-in-Aid for Scienti c Research on Innovative Areas "Heavy Electrons" (No. 20102002 and No. 20102003) and Grant-in-Aid for Scienti c Research (S)(No. 20224015) of MEXT. Photoemission experiments were carried out under the Visiting Researcher's Program of the Institute for Solid State Physics, the University of Tokyo. R. Y. acknowledges the nancial support from JSPS.

References

[1] Palstra T T M, Menovsky A A, Berg J V D, Dirkmaat A J, Kes P H, Nieuwenhuys G J and Mydosh J A 1985 *Phys. Rev. Lett.* **55** 2727
[2] Chandra P, Coleman P, Mydosh J A and Tripathi V 2002 *Nature* **417** 831
[3] Amitsuka H, Sato M, Metoki N, Yokoyama M, Kuwahara K, Sakakibara T, Morimoto H, Karawazaki S, Miyako Y and Mydosh J A 1999 *Phys. Rev. Lett.* **83** 5114
[4] Yokoyama M, Amitsuka H, Itoh S, Kawasaki I, Tenya K and Yoshizawa H 2004 *J. Phys. Soc. Jpn.* **73** 545
[5] Baek S-H, Graf M J, Balatsky A V, Bauer E D, Cooley J C, Smith J L and Curro N J 2010 *Phys. Rev. B* **81** 132404
[6] Hufner S 2007 *Very High Resolution Photoelectron Spectroscopy* (Berlin: Springer)
[7] Ohkuni H *et al.* 1999 *Phil. Mag. B* **79** 1045
[8] Haga Y, Sakai H and Kambe S 2007 *J. Phys. Soc. Jpn.* **76** 051012
[9] Santander-Syro A F, Klein M, Boariu F L, Nuber A, Lejay P and Reinert F 2009 *Nature Phys.* **5** 637

Journal of Physics: Conference Series **273** (2011) 012022 doi:10.1088/1742-6596/273/1/012022

Unusual electric field gradient variation in the $NpPd_2Ge_2$ compound

P. Gaczyński, E. Colineau, F. Wastin, P. Boulet, J. Rebizant, R. Caciuffo

European Commission, Joint Research Centre, Institute for Transuranium Elements; Postfach 2340, 76125 Karlsruhe, Germany

E-mail: `Piotr.Gaczynski@ec.europa.eu`

Abstract. $NpPd_2Ge_2$ has been studied by ^{237}Np Mössbauer spectroscopy between 4.2 K and 80 K. The value of the isomer shift suggests a Np^{3+} electronic configuration. The low temperature spectra provide a value of 2.10 μ_B for the magnetic ordered moment on the Np site. These values suggest a rather localized character of the $5f$ electrons in this system. The magnetic transition around 60 K is accompanied by a large increase of the quadrupolar interaction that is evidence of a redistribution of the electronic charges around the neptunium nucleus.

1. Introduction

The family of ternary intermetallics systems AT_2X_2 (where A = rare earth or actinide atoms, T = transition metals and X= Si or Ge) is well known to exhibit a wide variety of electronic and magnetic properties [1, 2] ranging from simple Pauli paramagnetism to Kondo lattice/heavy fermion behaviour and complex magnetic ordering. Among analogue compounds with stochiometry 1:2:2, it is worth mentioning $CeCu_2Si_2$ [3, 4] which presents a superconducting phase embedded in a magnetic phase, the antiferromagnetic $CePd_2Si_2$ [5] that becomes superconducting under high pressures, $CePd_2Ge_2$ [6] showing coexistence of magnetic order and superconducting state at ambient pressure and URu_2Si_2 [7] where a still mysterious "hidden order" occurs below 17.5 K.

All Np-based NpT_2X_2 intermetallics order magnetically [8] and the transition atoms do not carry any localized magnetic moment except Mn in $NpMn_2Si_2$ and $NpMn_2Ge_2$ compounds [8, 9] which have ordering temperature around room temperature. Neutron diffraction experiments on the $NpCo_2Si_2$ compound [10] indicate an antiferromagnetic ordering of the Np magnetic moments (μ_{Np}) canted by 52° away from the c - axis. In $NpRu_2Si_2$, incommensurate and modulated antiferromagnetic ordering of μ_{Np} was estabilished [11].

The AT_2X_2 compounds crystallize either in the tetragonal $ThCr_2Si_2$ type structure (space group I4/mmm) or, more seldom, in the $CaBe_2Ge_2$ type structure (space group P4/nmm) [12]. $NpPd_2Ge_2$ belongs to the first group.

Magnetisation measurements on the $NpPd_2Ge_2$ compound shows the antiferromagnetic nature of the magnetic ordering below 60 K [13]. Here we present results of a ^{237}Np Mössbauer effect study on $NpPd_2Ge_2$.

Published under licence by IOP Publishing Ltd

2. Experimental details

The $NpPd_2Ge_2$ samples were prepared by arc melting the pure constituents in stoichiometric amounts under a purified argon atmosphere. X-ray powder diffraction confirms the tetragonal $ThCr_2Si_2$-type structure with lattice parameters $a = 4.252$ Å and $c = 10.110$ Å and fractional atomic coordinate $z = 0.3757$. The ^{237}Np Mössbauer investigations were performed on a powder absorber with a thickness of 100 mg of Np/cm^2 using a ^{241}Am source of 100 mCi kept around 4.2 K and a transmission spectrometer calibrated with the standard absorber $NpAl_2$. The spectra were analyzed using Lorentzian lineshapes. The position and relative intensities of the absorption lines of Np nuclei were calculated by solving the complete Hamiltonian for the hyperfine interactions in both the excited and ground nuclear state of ^{237}Np.

3. Results and Discussion

Typical ^{237}Np Mössbauer spectra taken at different temperatures between 4.2 K and 80 K are shown in fig. 1.

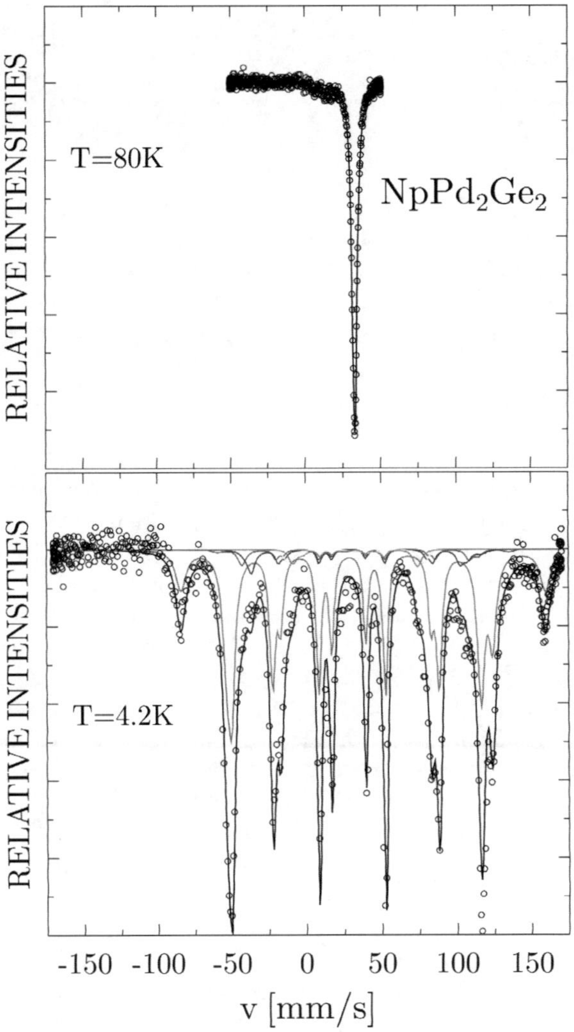

Figure 1. ^{237}Np Mössbauer spectra of $NpPd_2Ge_2$ taken at 4.2 and 80 K. The solid black curve represents the best fit to the data. The colored curves represent the subspectra (see text).

At 80 K, in the paramagnetic region, a single neptunium site is observed, in agreement with the crystallographic structure. The spectrum consists of a single line, barely broadened by a vanishing quadrupolar splitting ($|\frac{e^2qQ}{2}| = 1.76$ mm/s). Because of the axial symmetry at the Np site, the asymmetry parameter η is equal to zero. The isomer shift value of 19.48 mm/s (versus $NpAl_2$) falls into the range expected for Np^{3+}, corresponding to an electronic configuration $5f^4$ ($^5I_4$) [14]. It is interesting to note that this value is notably higher than in most other NpT_2X_2 analogues, which can be explained by the more localized character of the $5f$ electrons in $NpPd_2Ge_2$.

The spectrum recorded at 4.2 K reveals a large magnetic splitting due to the ordering of the neptunium magnetic moments. It shows a well resolved pattern, but a careful analysis shows that it cannot be reproduced by a single static site. The spectrum shown on figure 1 was fitted assuming dynamic effects (relaxation). This model used for fitting spectra gives good $\chi^2 = 4.29*10^{-6}$ parameter and contains relatively low number of free parameters. However, it is possible to use an alternative model which takes into account a distribution of hyperfine field that could arise from a complex magnetic structure with a modulation of the magnetic moment μ_{Np} [15] ($\chi^2 = 1.47*10^{-5}$). Both models gives close value of magnetic hyperfine filed B_{hf}, QS and *IS* parametrs. The Dunlap-Lander relation [16] gives the estimated value of 2.10 μ_B at Np site. Although this value is lower than that expected for free Np^{3+} ion [14] (2.47 μ_B), it is significantly higher than in all other NpM_2T_2 compounds, at the exception of $NpRh_2Ge_2$ (2.3 μ_B) [15], confirming the $5f$ electrons are more localized in $NpPd_2Ge_2$ than in the other NpT_2X_2 compounds.

The quadrupole interaction measured in the paramagnetic state ($|\frac{e^2qQ}{2}| = 1.76$ mm/s at 80 K) depends mostly on the aspherical distribution of the valence electron density close to the nucleus [17]. Below the magnetic ordering temperature, the quadrupole interaction reaches a value of -7.85 mm/s indicating a redistribution of the electron density around Np atoms. Because the magnitude and sign of ΔQS the simple explanation namely the variation of μ_{Np} vs. electric field gradient (EFG) cannot explain this phenomenon. The increasing magnitude of QS could be associated with the lowering of local symmetry around Np. The sudden jump in the quadrupole parameter at a given temperature is usually connected with the structural transitions. The displacement of the atoms influences the electronic properties of the material. Changes in the electronic structure of $NpPd_2Ge_2$ as seen by the QS variation should also affect the isomer shift parameter. As the magnitude of the *IS* does not change between paramagnetic and magnetically ordered states as well as η, we suggest that the electron density redistribution around the Np atoms is not due to a structural distortion. In NpO_2, crystallizing in the CaF_2 - type of structure above 25 K, a symmetry lowering from $Fm\bar{3}m$ to $Pn\bar{3}m$ as a consequence of reduced point symmetry at Np site without atoms displacement, leading to the multipolar order below 25 K, was reported and discused in many papers [18–21]. The Mössbauer parameters *IS* and QS behave in similar way in both NpO_2 [22] and $NpPd_2Ge_2$ compounds. Long-range order of electric quadrupoles has also been observed in UO_2 below 30.8 K. In that case, however, quadrupole order is accompanied by an ordering of magnetic dipoles, and associated with an internal distortion of the oxygen sub-lattice [23]. A similar scenario is not excluded in $NpPd_2Ge_2$. We cannot, however, directly prove the multipolar model from ^{237}Np Mössbauer data alone. Further experimental and theoretical investigations are necessary.

4. Conclusions

We have investigated $NpPd_2Ge_2$ by ^{237}Np Mössbauer spectroscopy and showed unusual change in QS parameter between paramagnetic and magnetically ordered state. ΔQS can not be explained like a function of μ_B direction vs. principal axis of EFG. So, to explain this phenomenon multipolar ordering accompanied by magnetic ordering below 60 K was proposed. The magnetic moment carried by neptunium amounts to 2.1 μ_B, is significantly larger than the values observed in most other NpM_2T_2 compounds. This, together with the isomer shift value, suggests a rather localized character of the $5f$ electrons in this system. Further studies are under way to better elucidate the magnetic and electronic properties of $NpPd_2Ge_2$.

4.1. Acknowledgments

PG acknowledges the European Commission for support in the frame of the "Training and Mobility of Researchers" programme. The high purity Np metals required for the fabrication of the compound were made available through a loan agreement between Lawrence Livermore National Laboratory and ITU, in the frame of a collaboration involving LLNL, Los Alamos National Laboratory and the US Department of Energy.

References

[1] Żolnierek Z, Mulak J 1995 *J. Magn. Magn. Mater.* **140-144** 1393
[2] Leciejewicz J, Szytuła A 1987 *Magnetic ordering in Ternary Intermetallics MT_2X_2 Systems* (Państwowe Wydawnictwo Naukowe, Warsaw - Cracow)
[3] Steglich F, Aarts J, Bredl C D, Lieke W, Meschede D, Franz W, Schäfer H 1979 *Phys. Rev. Lett.* **43** 1892
[4] Sheikin I, Braithwaite D, Brison J P, Buzdin A I, Assmus W 1998 *J. Phys.:Condens. Matter* **10** L749-L755
[5] Grosche F M, Julian S R, Mathur N D, Lonzarich G G 1996 *Physica B* **223&224** 50-52
[6] Mathur N D, Grosche F M, Julian S R, Walker I R, Freye D M, Haselwimmer R K W, Lonzarich G G 1998 *Nature* **394** 39
[7] Schmidt A R, Hamidian M H, Wahl P, Meier F, Balatsky A V, Garrett J D, Williams T J, Luke G M, Davis J C 2010 *Nature* **465** 570 and refs. therein
[8] Gal J, Kroupp M, Hadari Z 1977 *Phys. Rev. B* **16** 3913
[9] Schäfer W, Will G, Litterst F J, Potzel W, Kalvius G M, Fredo S, Gal J 1987 *Inorg. Chim. Acta* **140** 83
[10] de Novion C H, Gal J, Buevoz J L 1980 *J. Mag. Mag. Mat.* **21** 85
[11] Bogé M, Bonnisseau D, Burlet P, Fournier J M, Pleska E, Quezel S, Rebizant J, Rossat-Mignod J, Spirlet J C, Wulff M 1989 *J. Nuc . Mat.* **166** 77
[12] Wastin F, Rebizant J, Spirlet J C, Sari C, Walker C T, Fuger J 1993 *J. Alloys Compd.* **196** 87
[13] Colineau E, Wastin F, Boulet P, Rebizant J 2001 31th *Journèes des Actinides* Saint-Malo, France
[14] Kalvius G M, Gal J, Asch L, Potzel W 1992 *Hyp. Int.* **72** 77
[15] Wastin F 1991 PhD Thesis, Université de Liège
[16] Dunlap B D, Lander G H 1974 *Phys. Rev. Lett.* **33** 1046
[17] Blaha P, Schwarz K, Dederichs P H 1988 *Phys. Rev. B* **37** 2792
[18] Faber J, Lander G H 1976 *Phys. Rev. B* **14** 1151
[19] Mannix D, Lander G H, Rebizant J, Caciuffo R, Bernhoeft N, Lidström E, Vettier C 1999 *Phys. Rev. B* **60** 15187
[20] Paixão J A, Detlefs C, Longfield M J, Caciuffo R, Santini P, Bernhoeft N, Rebizant J, Lander G H 2002 *Phys. Rev. Lett.* **89** 187202
[21] Santini P, Carretta S, Amoretti G, Caciuffo R, Magnani N, Lander G H 2009 *Rev. Mod. Phys.* **81**, 807
[22] Friedt J M, Litterst F J, Rebizant J 1985 *Phys. Rev. B* **32** 257
[23] Wilkins S B, Caciuffo R, Detlefs C, Rebizant J, Colineau E, Wastin F, Lander G H 2006 *Phys. Rev. B* **73** 060406

International Conference on Strongly Correlated Electron Systems (SCES 2010) IOP Publishing
Journal of Physics: Conference Series **273** (2011) 012023 doi:10.1088/1742-6596/273/1/012023

Pu Electronic Structure and Photoelectron Spectroscopy

J.J. Joyce[1], T. Durakiewicz[1], K.S. Graham[1], E.D. Bauer[1], D.P. Moore[1], J.N. Mitchell[1], J.A. Kennison[1], R.L. Martin[1], L.E. Roy[1], and G.E. Scuseria[2]

[1] Los Alamos National Laboratory, P.O. Box 1663, Los Alamos, NM 87544, U.S.A.
[2] Department of Chemistry, Rice University, Houston, TX 77005, U.S.A.

E-mail: jjoyce@lanl.gov

Abstract. The electronic structure of $PuCoGa_5$, Pu metal, and PuO_2 is explored using photoelectron spectroscopy. Ground state electronic properties are inferred from temperature dependent photoemission near the Fermi energy for Pu metal. Angle-resolved photoemission details the energy vs. crystal momentum landscape near the Fermi energy for $PuCoGa_5$which shows significant dispersion in the quasiparticle peak near the Fermi energy. For the Mott insulators AnO_2(An = U, Pu) the photoemission results are compared against hybrid functional calculations and the model prediction of a crossover from ionic to covalent bonding is found to be reasonable.

1. Introduction

The properties of Pu materials are defined in large part by the 5f electrons which are at the boundary between localized and itinerant character. In the actinide elements, Pu is the transition point for the 5f electrons moving from bonding character in the lighter actinides (Np and lower) to localized character in the heavier actinides (Am and higher). Moreover, this crossover point in 5f character at Pu in the actinide elements is also found in many Pu compounds as well. The three materials δ-Pu metal, $PuCoGa_5$, and PuO_2 serve as examples. In Pu metal there are six solid state allotropes with a 21% volume change between the alpha and delta phases. The 5f character shifts from more itinerant in the high density alpha phase toward a more localized character in the lower density delta phase. [1, 2] The 18.5 K superconductor $PuCoGa_5$ [3] has a T_C an order of magnitude higher than its Ce counterparts while $UCoGa_5$ shows no evidence of strongly correlated electron characteristics, properties again pointing toward the degree of interaction for the f-electrons in bonding. Within the Mott insulator family of actinide dioxides, [4] there is a crossover between ionic and covalent bonding at PuO_2. Central to the understanding of these electronic structure properties is an accurate picture for the characteristics of Pu 5f electrons. Photoelectron spectroscopy (PES) in various forms including temperature dependent PES and angle-resolved photoemission (ARPES) is used along with advanced computational methods to determine the electronic structure of these complex Pu materials.

Due to the aforementioned crossover point in 5f character, a full range of Pu materials and associated 5f characteristics may be observed covering localized 5fs in PuTe [5] and PuSb [6] to rather itinerant 5fs in α-Pu [2]. The Pu materials presented below δ-Pu, $PuCoGa_5$, and PuO_2 show a somewhat intermediate 5f character and experimental signatures associated with hybridized electronic structure. For single crystal $PuCoGa_5$, the appearance of periodic and dispersive 5f states provides compelling evidence that there is a strong component of itinerant character to the electronic structure

Published under licence by IOP Publishing Ltd

near the Fermi energy. In δ-Pu, the binding energy, lineshape, and temperature dependence of the spectral feature at the Fermi energy help determine the ground state electronic properties.

2. Results

Photoemission data were collected using the Laser Plasma Light Source taking advantage of the high energy resolution available from the He lamp combined with a low temperature capability. The Pu metal samples were polycrystalline, Ga stabilized, δ- (fcc) phase. In Figure 1a, the energy resolution is 35 meV, at a photon energy of 40.8 eV and a sample temperature of 15 K. The energy resolution and position were determined by a Cu Fermi level reference spectrum. In Figure 1b, the resolution is relaxed to 60 meV in order to collect data at two temperatures before any significant sample deterioration occurs.

Photoemission data for δ-Pu is show in Figure 1a with two important details; 1) the lineshape of the spectral feature may be described as a symmetric oscillator function cut by the low temperature Fermi function, 2) the position of the Fermi energy high on the spectral feature indicates the energy position of the peak is at or very near the Fermi energy and that this is a ground state energy position. In the b) and c) frames of Figure 1, the temperature dependence of the PES data and a simple model for a narrow peak are presented. In the b) frame, we show the PES data taken at 77 and 300K. While there is a significant change with temperature in the PES data, the c) frame shows a simple model for temperature dependence that is consistent with the PES data based on a Lorentzian lineshape 200 meV in width at the Fermi energy. The model temperature dependence illustrates the role of the Fermi function acting on a ground state feature. Combining this Pu high resolution and temperature dependent data with previously published photon energy dependence [7] makes a strong case that the peak at the Fermi energy is a narrow band, of mixed 5f-conduction band character, centered at or very near the Fermi energy in the ground state of δ-Pu.

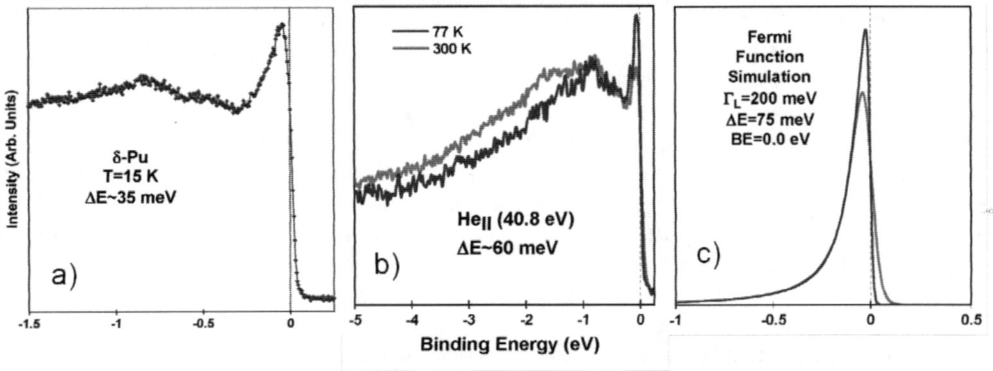

Figure 1. a) High resolution PES for δ-Pu with 35 meV energy resolution at a temperature of 15 K, **b)** temperature dependence of Pu PES between 77 and 300 K, **c)** simulation of temperature dependence for a narrow peak at the Fermi energy as a function of temperature (77 and 300 K after b)).

ARPES data for $PuCoGa_5$ is presented in Figure 2. The original photoemission research on single crystal, flux grown samples [8] cleaned by laser ablation showed a narrow peak at the Fermi energy consistent with both the modest enhanced mass value for $PuCoGa_5$ as well as a range of other Pu materials showing both a localized feature well below the Fermi energy, and a hybridized 5f feature at the Fermi energy. In Figure 2, we present ARPES data for single crystal $PuCoGa_5$, cleaved at 10K to expose a clean, stoichiometric surface. The results from the new, cleaved single crystal data are consistent with the the original single crystal results. The new data presented in Figure 2, show clear evidence of a dispersive peak, with some Pu 5f character, moving through the Fermi energy.

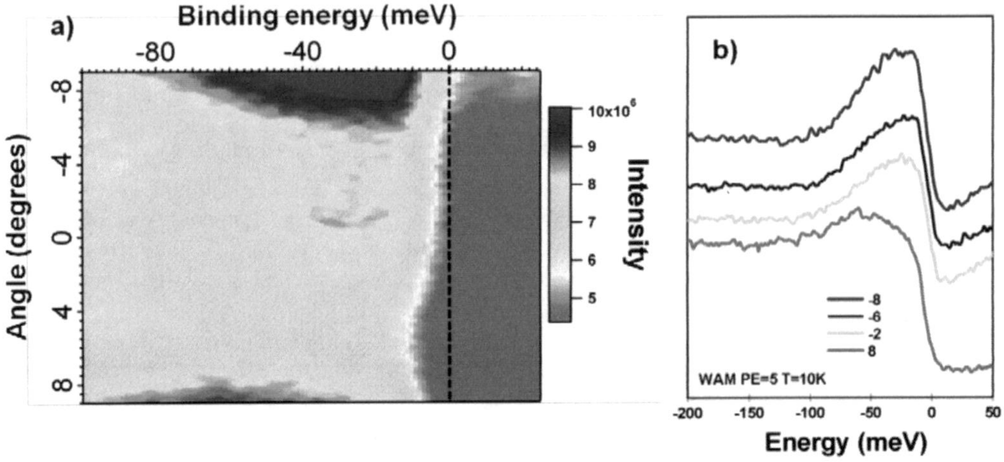

Figure 2. a) Angle-resolved PES for $PuCoGa_5$ at hν=21.2 eV showing 120 meV near the Fermi energy (dotted black line) in the horizontal plane and 18 degrees of angle in the vertical plane. The 18 degrees of angle corresponds to 90% of the crystal momentum space along Γ-X using 10 eV for the inner potential. Color represents relative intensity. b) Energy distribution curves at four different angles showing the dispersion of the peak through the Fermi level.

The position in reciprocal space for the ARPES data was assigned by comparison to both $UCoGa_5$ and $CeCoIn_5$ data and their associated inner potentials, then projecting an averaged inner potential of 10 eV for $PuCoGa_5$. With a 10 eV inner potential and a photon energy of 21.2 eV, one may calculate the reciprocal lattice location to be just below a Γ-point in the Γ-Z direction. Transforming angles into crystal momentum at this energy shows the in-plane Γ-X direction requires about 20 degrees of angular acceptance to cover the zone center to edge. We use angles in degrees rather than reciprocal lattice dimensions in figure 2 since the inner potential is estimated by comparison to known, closely related materials rather than experimentally determined. Comparing to the original photoemission on $PuCoGa_5$ [8], the linewidth of this peak near the Fermi energy is less than 100 meV and the dispersion is at least 50 meV within the portion of reciprocal space defined by Figure 2. This is consistent with the reported linewidth and energy position, as the original data was angle-integrated PES, which would be a superposition of the narrower but dispersive peaks in the ARPES data set.

For the AnO_2 family of materials, the competition between kinetic energy and onsite Coulomb repulsion leads to Mott insulating character. To accommodate the competition between kinetic energy and Coulomb repulsion the hybrid functional model is applied to the AnO_2 family. [4,9] The hybrid functional predicted UO_2 as an ionic solid with small orbital character overlap between the O 2p and the U 5f levels. The energy separation of the 5f and 2p bands was observed experimentally by ARPES and a small amount of 5f-2p intermixing was observed in the U 5f levels as 200 meV of dispersion when the U-U interatomic separation would preclude any wavefunction overlap.[10]

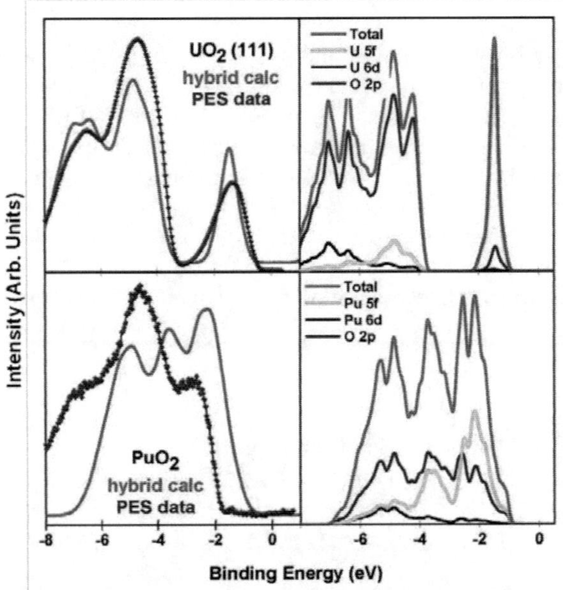

Figure 3. PES and hybrid functional calculations for UO_2 and PuO_2. The left frames show the PES (blue) plotted against the calculations (red - adjusted for lifetime and instrument resolution). The right frames show the orbital (p, d, f) character for U, Pu and O within the model calculations. The agreement between PES and the calculation is good for both UO_2 and PuO_2, showing the separation between U 5f and O 2p states in the top frames and the hybridization between Pu 5f and O 2p states in the bottom frames. The energetic overlap in states for PuO_2 leads to the covalent bonding whereas the separation into two manifolds of states in UO_2 results in primarily ionic bonding.

In Figure 3 we show the hybrid functional calculations for UO_2 and PuO_2 compared with the PES data. The hybrid functional calculations predicted that for PuO_2, the separation between the O 2p levels and the Pu 5f levels would be eliminated leading to strong hybridization and thus a significant amount of covalency in PuO_2 whereas UO_2 had clear ionic character. In order to quantify the covalent character of PuO_2, recent ARPES data has demonstrated the strong intermixing between Pu 5f and O 2p levels. [11] The combination of the ARPES in references 10 and 11 with the angle-integrated photoemission in Figure 3 provides strong evidence for hybridization in PuO_2. Similarly, the UO_2 data is consistent with limited hybridization and energetically separated regions of 5f and 2p character. The photoemission results thus support the predictions of the hybrid functional model for the AnO_2 family of materials, which show strong mixing of the Pu 5f and the O 2p states, a few eV below the Fermi energy. The systematic trend in the AnO_2 series showing convergence of the O 2p levels and the actinide 5f levels as the series moves from Th, U, Np to Pu is completely consistent with the PES results shown here for UO_2 and PuO_2.

For the metals δ-Pu and $PuCoGa_5$, the first peak in the photoemission data is associated with a 5f-conduction state hybridized bulk feature. The identification of this peak as a bulk electronic structure feature rather than surface state is based on; 1) comparison to several different Pu materials all showing similar spectral features (δ-Pu, $PuCoGa_5$, Pu_2RhGa_8, $PuIn_3$, $PuSb_2$), 2) the photon energy dependence of the peak [1], and 3) the observation of sample degradation with vacuum contaminants. These three points argue for a bulk electronic structure feature, the possibility of a minor component in this spectral feature which could be associated with a surface feature is open to future research. Additional experiments using normal emission ARPES and controlled dosing of adsorbates would enhance the points of differentiation between bulk vs. surface.

3. Summary

Both of the strongly correlated metals, δ-Pu and $PuCoGa_5$ show two regions of 5f character, one near the Fermi energy and one removed by 1 eV or more from the Fermi energy. These two regions have been associated with hybridized Pu 5f character and localized Pu 5f character respectively. Temperature dependent and high resolution PES data for δ-Pu metal give strong evidence that there is a narrow spectral feature centered very near the Fermi energy in the ground state. The position of the Fermi level high on the peak further supports the premise of a narrow feature at the Fermi energy. In the case of $PuCoGa_5$, the ARPES data provides direct experimental evidence that this first peak near the Fermi energy is a dispersive quasiparticle state which moves through the Fermi energy as a function of crystal momentum. The $PuCoGa_5$ data show a peak less than 100 meV in width and dispersing at least 50 meV. For $PuCoGa_5$, direct observation of a dispersive peak in the ARPES data demonstrates the significant role of the periodic potential in understanding the electronic structure. For the AnO_2 Mott insulators, PES results show limited but non-zero wavefunction intermixing for the U 5f and O 2p states while there was strong hybridization between the Pu 5f and the O 2p levels consistent with the prediction of ionic bonding in UO_2 with covalent bonding for PuO_2. Overall, a combination of ARPES results and temperature dependent PES in addition to conventional PES provides insight on the electronic structure and ground state properties for a range of Pu materials from strongly correlated metals to Mott insulators.

Acknowledgments
Research supported by the U.S. Department of Energy, Office of Basic Energy Sciences, Division of Materials Sciences and Engineering, the LANL LDRD program, and Campaign II.

References
[1] J.M. Wills, *et al.*, J. Elect. Spectr. and Related Phenom. **135**, 163 (2004).
[2] A.J. Arko, *et al.*, Phys. Rev. B **62**, 1773 (2000).
[3] J.L. Sarrao, *et al.*, Nature **420**, 297 (2002).
[4] I.D. Prodan, *et al.*, Phys. Rev. B **76**, 33101 (2007).
[5] T. Durakiewicz, *et al.*, Phys. Rev. B **70**, 205103 (2004).
[6] T. Gouder *et al.*, Phys. Rev. Lett. **84**, 3378 (2000).
[7] J.J. Joyce, *et al.*, Mat. Res. Soc. Symp. Proc. **986**, 35 (2007).
[8] Joyce, *et al.*, Phys. Rev. Lett. **91,** 176401 (2003).
[9] I.D. Prodan, *et al.*, Phys. Rev. B **73**, 045104 (2006).
[10] L.E. Roy, *et al.*, J. of Computational Chemistry, **29** (13), 2288 (2008).
[11] J.J. Joyce, *et al.*, Mat. Res. Soc. Symp. Proc. accepted (2010).

Crystal structure and physical properties of $NpRh_2Sn$, a new Np-based ternary compound

T Klimczuk[1,2], J-C Griveau[1], P Gaczy ski[1], R Eloirdi[1], E Colineau[1], and R Caciuffo[1]

[1]European Commission, Joint Research Center, Institute for Transuranium Elements, Postfach 2340, Karlsruhe, D-76125 Germany
[2]Faculty of Applied Physics and Mathematics, Gdansk University of Technology, Narutowicza 11/12, 80-233 Gdansk, Poland

E-mail: Tomasz.Klimczuk@ec.europa.eu

Abstract. We report on the synthesis and physical characterization of a new compound, $NpRh_2Sn$, prepared by arc-melting under argon atmosphere. Rietveld analysis of the powder x-ray diffraction pattern reveals an orthorhombic, Fe_3C-type structure, with lattice parameters that are the shortest among the AnT$_2$Sn series (An = U, Np, Pu, and T = Pd, Rh). A fit of the high-temperature magnetic susceptibility curve, $\chi(T)$, gives a Curie-Weiss temperature Θ_{CW} = -29K, and an effective magnetic moment $\mu_{eff} = 2.42\mu_B$. The stabilization of the antiferromagnetic order below T_N = 34 K is revealed by a cusp in $\chi(T)$ and by a small anomaly in the specific heat curve, $C_p(T)$. An enhanced value of the Sommerfeld coefficient, $\gamma \approx 107$ mJ mol^{-1} K^{-2}, indicates a moderate heavy-fermion state. $NpRh_2Sn$ is the first member of the AnRh$_2$Sn family, and a rare representative of heavy-fermion systems amidst Np- intermetallics.

1. Introduction

The large family of the ternary actinide compounds AnT_2M, where An is an actinide element, T is a transition element, and M is a metalloid, forms mainly in the orthorhombic Fe_3C - type crystal structure (Pnma, s.g. 62). The uranium based alloys (UT_2M) have attracted considerable attention due to a wide range of physical properties, which originates from the sensitive nature of the uranium $5f$-electrons. Very few compounds with An other than uranium have been reported, and most of them contain palladium as the transition element.

Here, we present the crystal structure and physical properties of a new intermetallic $NpRh_2Sn$ compound, which is the first member of the $AnRh_2M$ family.

2. Experimental

Polycrystalline $NpRh_2Sn$ sample was synthesized by arc-melting stoichiometric amounts of elements under a zirconium gettered ultra pure argon atmosphere. The purity of the product was verified by powder X-ray diffraction (Bruker D8 Focus, CuK_α radiation, graphite monochromator). Data were collected in the 2θ range of 10° – 100° with a step size of 0.01°. The FullProf Suite package was used for Rietveld structure refinement [1]. The magnetic susceptibility was measured using a Magnetic Property Measurement System (MPMS, Quantum Design), whereas the specific heat and electrical

Published under licence by IOP Publishing Ltd

resistivity were determined using a Physical Properties Measurement System (PPMS, Quantum Design).

3. Results and discussion

The as-cast sample extracted from the batch was studied by x-ray powder diffraction and the result is presented in Figure 1. The known crystal structure of URh_2Sn was employed as a starting structural model [2]. As shown in Fig. 1, there is good agreement between the model and the data, suggesting that $NpRh_2Sn$ is isostructural to the U analogue. The lattice parameters for $NpRh_2Sn$ were calculated to be a=9.7208(6) Å, b=4.4221(3) Å and c=6.9043(4) Å and are smaller than the reported ones for $NpPd_2Sn$ [3]. It was found that 8% of Np substitutes the Rh(1) site and 8% of Rh are located on the Np site.

Selected physical parameters for $NpRh_2Sn$ and other members of the AnT_2M series are given in Table 1.

Table 1. Lattice parameters, An-An distance (d_{An-An}) , Néel temperature (T_N) and Curie-Weiss temperature (Θ_{CW}) for 5 members of AnT_2Sn family

	a (Å)	b (Å)	c (Å)	d_{An-An} (Å)	T_N (K)	Θ_{CW} (K)
$NpRh_2Sn$	9.7208(6)	4.4221(3)	6.9043(4)	3.9702	34	-29
$NpPd_2Sn$ [3]	10.004(3)	4.535(2)	6.961(1)	4.0606	15	-80
$PuPd_2Sn$ [4]	10.053(9)	4.502(4)	7.065(6)	---	11	-30
URh_2Sn [2,5]	9.7923(2)	4.37474(7)	6.9639(1)	3.935	---	-200
UPd_2Sn [6]	9.9787(1)	4.58843(5)	6.89166(8)	4.1817	---	-77

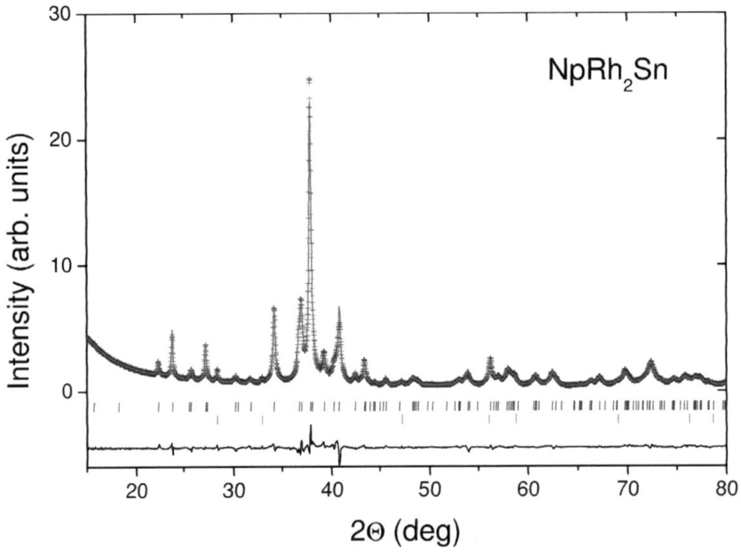

Figure 1. Rietveld refinement of room temperature the x-ray diffraction pattern for $NpRh_2Sn$. Crosses represent observed data, red solid line is the calculated intensity. The black tick marks correspond to $NpRh_2Sn$ and green set refers to the NpO_2 impurity (~2% wt.). Figure of merits: R_p = 4.04, R_{wp} = 5.97, R_{exp} = 2.60 and χ^2 = 5.28.

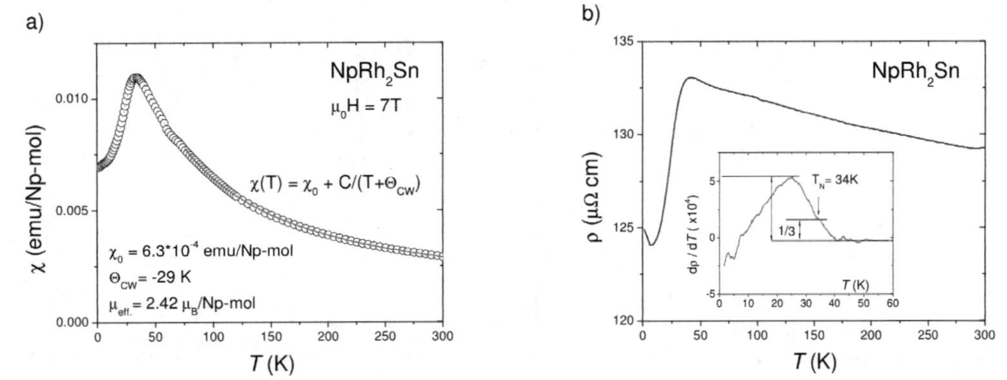

Figure 2. The temperature dependence of the a) magnetic susceptibility ($\chi(T)$) and b) electrical resistivity ($\rho(T)$) of $NpRh_2Sn$. The solid red line in panel a) is a fit by the modified Curie-Weiss law. The inset in panel b) shows the derivative of the electrical resistivity ($d\rho/dT$) vs. temperature. The Néel temperature, defined as shown in the inset, is 34K.

In Figure 2 we show the temperature dependence of magnetic susceptibility (a) and electrical resistivity (b) of $NpRh_2Sn$. The Curie-Weiss fit of the $\chi(T)$ data above 75 K (red solid line) gives a Curie-Weiss temperature Θ_{CW} = -29 K, and an effective magnetic moment μ_{eff} = 2.42 μ_B/Np-mol. This is close to the value expected for Np^{+3} (2.68 μ_B). The measurement indicates an antiferromagnetic anomaly at T_N around 35 K.

Electrical resistivity as a function of temperature (main panel) and the temperature derivative of the resistivity (inset) for $NpRh_2Sn$ are plotted in Figure 2b. At room temperature the resistivity is 129 μΩ cm which is very close to what is observed for $PuPd_2Sn$ [4] and $NpPd_2Sn$ [3]. With decreasing temperature the electrical resistivity slightly increases and at 50 K reaches a value of 133 μΩ cm. Interestingly the non-metallic behavior in the high temperature region was also reported for $NpPd_2Sn$ whereas $PuPd_2Sn$ exhibit metallicity in the whole temperature range. Below 40 K a rapid drop of resistivity, and a pronounced positive peak of $d\rho/dT$, are visible. This behavior is caused by the antiferromagnetic transition. Using the derivative of the electrical resistivity we can determine the ordering temperature (1/3 of the $d\rho/dT$ peak). We found T_N = 34 K, which is very close to the magnetic ordering temperature.

The temperature dependence of the specific heat (C_p) and specific heat over temperature (C_p/T) of $NpRh_2Sn$ are shown in Figure 3. At room temperature C_p reaches a value that corresponds to the Dulong-Petit law: C_p=3nR=99.77 J mol^{-1} K^{-1}, where n=4 and R=8.314 J mol^{-1} K^{-1}. The magnetic phase transition discussed above is not well pronounced in the specific heat measurement. Close to T=30K only a slight change of slope, as shown in the inset of Fig. 3a, is observed on the specific heat curve $C_p(T)$. Similar result, notably a marginal evidence of a specific heat anomaly was reported also for $UPd_{2-x}Sn$ (x=0.05 and 0.15), as discussed in ref. [7].

Figure 3b shows C_p/T versus T^2 (only low temperature region) where the Debye temperature (Θ_D) and Sommerfeld parameter (γ) can be estimated from the linear fit: $C_p/T = C_{el}/T + C_{ph}/T = \gamma + \beta T^2$, where $\beta = \frac{12\pi^4}{5}\frac{nR}{\Theta_D^3}$. Although the experimental data were fitted only up to 10K, which is three times lower than T_N, we are aware of the magnetic contribution (C_{mag}) to the specific heat. Subtracting C_{mag} should not change the value of γ, but might slightly change β, and consequently the Debye temperature. The calculated Sommerfeld parameter for $NpRh_2Sn$, γ = 107 mJ mol^{-1} K^{-2} indicates

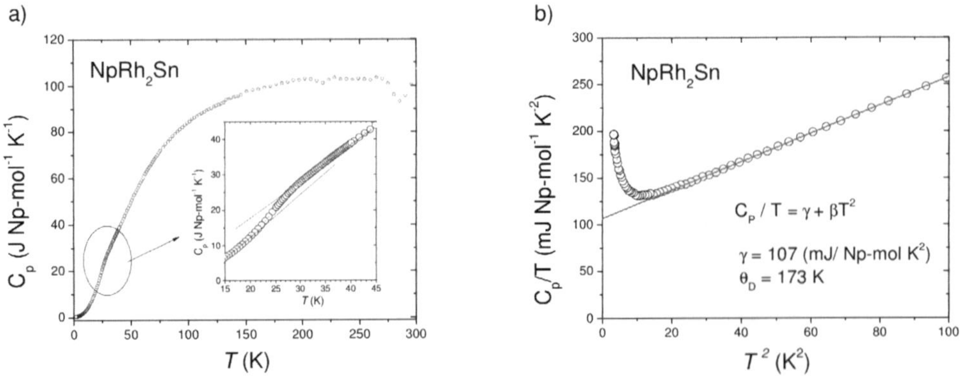

Figure 3. Temperature dependent specific heat (a) and specific heat over temperature (b) of $NpRh_2Sn$. The inset of the figure a) emphasises the magnetic transition region.

moderately heavy fermion state, and is of the same order of magnitude as reported for $PuPd_2Sn$ and $NpPd_2Sn$, which are 180 mJ mol^{-1} K^{-2}, and 400 mJ mol^{-1} K^{-2} respectively.

To summarize, we have synthesized and studied a new Np-based ternary $NpRh_2Sn$ compound. Its properties are similar to those of $NpPd_2Sn$, although a shorter Np-Np distance causes an increase of the Néel temperature. $NpRh_2Sn$ is a rare representative of heavy-fermion systems amidst Np-intermetallics.

Acknowledgements
Np metal required for the fabrication of the compound was made available through a loan agreement between Lawrence Livermore National Laboratory and ITU, in the frame of a collaboration involving Lawrence Livermore National Laboratory, Los Alamos National Laboratory, and the US Department of Energy. T.K. acknowledges the European Commission for financial support.

References

[1] J. Rodriguez-Carvajal 1993 *Physica* B **192** 55
[2] P. Gravereau, H. Guengard, F. Mirambet, L. Trut, J. Grannec, B. Chevalier, A. Tressaud, J. Etourneau 1994 *Mat. Sci. Forum* **166-169** 671
[3] D. Kaczorowski, K. Gofryk, P. Boulet, J. Rebizant, P. Javorský, E. Colineau, F. Wastin, G.H. Lander 2005 *Physica* B **359-361** 1102
[4] K. Gofryk, D. Kaczorowski, J-C. Griveau, N. Magnani, R. Jardin, E. Colineau, J. Rebizant, , F. Wastin and R. Caciuffo 2008 *Phys. Rev.* B **77** 014431
[5] F. Mirambet, B. Chevalier, P. Gravereau, L. Fournès, J. Etourneau 1995 *J. Magn. Magn. Mater.* **140-144** 1361
[6] M. Marezio, D.E. Cox, C. Rossel, M.B. Maple 1988 *Solid State Commun.* **67** 831
[7] I. Maksimov, F. J. Litterst, H. Rechenberg, M. A. C. de Melo, R. Feyerherm, R. W. A. Hendrikx, T. J. Gortenmulder, J. A. Mydosh, and S. Süllow 2003 *Phys. Rev.* B **67** 104405.

Magnetic phase transitions in CePtSn under high pressure

A M Alsmadi[1], M S Torikachvili[2], M K Kothapalli[3], F Nasreen[3], and H. Nakotte[3]

[1]Physics Department, The Hashemite University, 13115 Zarqa, Jordan
[2]Physics Department, San Diego State University, San Diego, CA 92182, USA
[3]Physics Department, New Mexico State University, Las Cruces, NM 88003, USA

E-mail: alsmadi_abd@yahoo.com

Abstract. CePtSn crystallizes in the orthorhombic TiNiSi-type structure. The compound undergoes two antiferromagnetic transitions at T_N = 7.5 K and T_M = 5 K, and it exhibits large magnetic and transport anisotropy. We report on measurements of the electrical resistivity of single-crystalline CePtSn as a function of temperature and magnetic field (*i* || *c* axis, *H* || *a* axis) performed under hydrostatic pressure up to 20 kbar. At ambient pressure, the overall temperature dependence indicates Kondo-type behavior in the presence of the crystal field. The application of pressure causes a reduction of the magnetoresistance effect. We find that the maximum in $\Delta\rho / \rho_{B=0}$ is only 0.05 at 20 kbar, while it is about 0.15 at ambient pressure. On the other hand, application of hydrostatic pressure in CePtSn has little effect on both T_N and T_M. It initially increases both temperatures, but then reduces them again above 10 kbar. The results are discussed in terms of moment fluctuations and the variation due to increased 4*f*-ligand hybridization.

1. Introduction

CePtSn crystallizes in the orthorhombic TiNiSi-type structure, and it is known as an anisotropic dense Kondo compound. CePtSn orders antiferromagnetically (AF) below T_N = 7.5 K, and it undergoes an additional magnetic phase transition at T_M = 5K [1]. Previously, several attempts to determine the magnetic structure of CePtSn were performed. Neutron-diffraction studies [2] were originally interpreted in terms of two sinusoidally modulated incommensurate antiferromagnetic structures characterized by the propagation vector **q** = (0, δ, 0) with δ = 0.466 at temperatures $T < T_M$ and δ = 0.418 for $T_M < T < T_N$. However, this scenario was found to be inconsistent with the results of muon spin relaxiation (μSR) experiments [3] that indicate locally commensurate ordering. A single-frequency μSR spectrum was seen for $T_M < T < T_N$, while three discrete frequencies were found below T_M. To reconcile the results of the two experiments, a spin-slip model had been proposed based on irreducible representation analysis [4]. In this picture the occurrence of periodic spin-slips of an essentially commensurate, **q** = (0, 1/2, 0) phase was proposed. Since μSR is sensitive to the local moment configuration and neutron diffraction is sensitive to the average long-range correlations, such arrangement of magnetic moments can satisfactorily explain both types of data.

Recent magnetoresistance measurements performed on CePtSn at 1.8 K in fields applied along the *a* or *b* axis revealed sharp magnetoresistive (MR) anomalies at 12.5 T for B || *a*, and 11 T for B || *b* respectively. Furthermore, the *b* axis magnetoresistance at 1.8 K shows a sharp drop at 3.5 T, which

[1] To whom any correspondence should be addressed.

Published under licence by IOP Publishing Ltd

appears to be irreversible upon field removal [5]. With the anomalies from magnetization, magnetoresistance and specific heat, H-T magnetic phase diagrams of CePtSn for fields along the *a* and *b* axis had been proposed [5,6]. For B || *a*, the two-phase boundaries associated with T_N and T_M appear to merge around 8 T. In zero field, one can distinguish two antiferromagnetic phases AF_1 and AF_2, in addition to the paramagnetic phase above T_N. For fields applied along the *a* axis, a third field-induced antiferromagnetic phase appears. Also for B || *b*, two zero field antiferromagnetic phases can be distinguished, but now the AF_1 phase persists into high fields and two field-induced phases can be distinguished at low temperatures.

A spin-slip structure is consistent with the magnetoresistance data, for which irreversibility was observed upon application of a magnetic field (H). The intermediate field induced phase AF_3 is associated with the irreversible magnetoresistance transition. This phase is induced upon application of a magnetic field from a zero-field cooled state, but persists down to zero field upon field removal. Within the spin-slip model, this may be interpreted as a change in the periodicity of the spin-slip discommensurations by the applied field, or perhaps its suppression altogether. It is reasonable to expect that such a change would have a much large effect on the electrical resistivity (ρ) than on the magnetic susceptibility. In addition, irreversibility may be expected since the original spin slip configuration is unlikely to be recovered after the removal of the magnetic field.

Previous magnetic susceptibility measurements on CePtSn under hydrostatic pressures up to 8.2 kbar show very weak effect [7] leaving T_N and T_M nearly unaffected. This is rather unexpected because the volume contraction is possibly the main reason for the change from the antiferromagnetic Kondo metal behavior with $T_N = 7.5$ K in CePtSn to a non-magnetic Kondo semi-metal CeNiSn [8].

Here, we present the effect of hydrostatic pressure (P) on the electrical resistivity and magnetoresistance for fields applied along the *a* axis of single-crystalline CePtSn. The data supplement previously published data for fields applied along the other two directions [9,10].

2. Experimental

The CePtSn sample was cut from the same Czochralski grown crystal of Ref. 6. The ρ(T) measurements in pressures up to 20 kbar were performed in a 9 T superconducting magnet system at San Diego State University, using a Cu-Be pressure cell with a core of hardened NiCrAl alloy. The pressure-transmitting medium was a 50:50 mixture of Fluorinert FC72:FC84, and the pressure was determined from the superconducting transition of a lead manometer.

3. Results and discussion

Electrical resistivity and magnetoresistance studies were performed in order to determine the pressure dependence of T_N and T_M. The behavior of ρ(T) under different pressures is shown in Fig. 1. The behavior of ρ(T) at P = 0 is reminiscent of other Kondo-type materials with crystal field splitting [11,12], with ρ(T) being characterized by a broad peak around 67 K followed by a minimum near 22 K and a sharp drop below 7 K. A CEF level scheme based on inelastic neutron scattering measurements was proposed for CePtSn (Ce^{3+} is a $^2F_{5/2}$ ion), yielding a ground-state doublet and two excited doublets as $\Delta_1 = 273$ K and $\Delta_2 = 273$ K, respectively [12], all with Γ_5 symmetry.

The minimum in ρ(T) shifts to higher values with pressure while the Kondo temperature remains almost constant up to 20 kbar. Application of hydrostatic pressure on CePtSn increases the absolute value of the resistivity. The value of two antiferromagnetic transitions at $T_N = 7.5$ K and $T_M = 5$ K can be extracted from the derivative of the resisitivity (the inset of Fig. 1 shows the derivative of the resisitivity at ambient pressure). The effect of pressure on both T_N and T_M is minimal as shown in the inset of Fig. 2. Pressure increases both T_N and T_M initially but then reduces them again above 10 kbar. Even in an applied magnetic field, we observed similar pressure effect on the transition temperatures. A comparison of dT_N/dT and dT_M/dT with values for other Ce compounds [13], suggests that the 4*f* electrons the CePtSn seem to be fairly localized. For example, application of pressure causes a reduction of the ordering temperature in CePdAl at an approximate rate of –0.32 K/kbar and a rough estimate of the critical pressure to suppress the ordering in CePdAl is found to be around 10 kbar. In addition, the magnetic susceptibility (χ) in the CePtSn follows Curie–Weiss law very closely with an effective magnetic moment of 2.33 μ_B that is only slightly reduced from that of the Ce^{3+} ion (2.45 μ_B), indicating a well-localized character of the Ce 4*f* electrons [14]. A deviation from the Curie– Weiss behavior was observed on going from CePtSn to CeRhAs, were the unit-cell volume decreases, suggesting an increase of the hybridization of 4*f* electron states with the conduction band [14].

For Ce compounds, the key ingredient for the formation of 4*f* magnetic moments and the long-range magnetic ordering is the degree of the 4*f*-ligand hybridization and the competition between the on-site Kondo interaction (with the quenching the 4*f* localized magnetic moments) and the intersite RKKY interaction between these moments. Hydrostatic pressure usually shortens the interatomic distances and promotes hybridization. It seems that the 4*f*-ligand hybridization is not enhanced enough in 20 kbar to allow a change in balance between the Kondo and RKKY interactions.

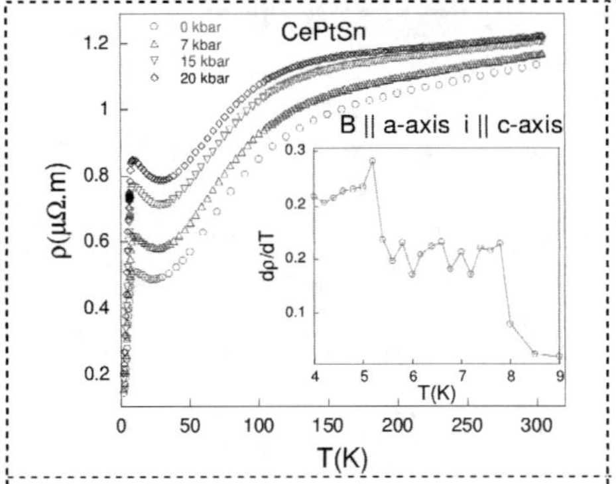

Figure 1. Temperature dependence of the normalized electrical resistivity of CePtSn at various pressures with *i* || *c* axis. The inset shows the temperature dependence of the derivative electrical resistivity at ambient pressure.

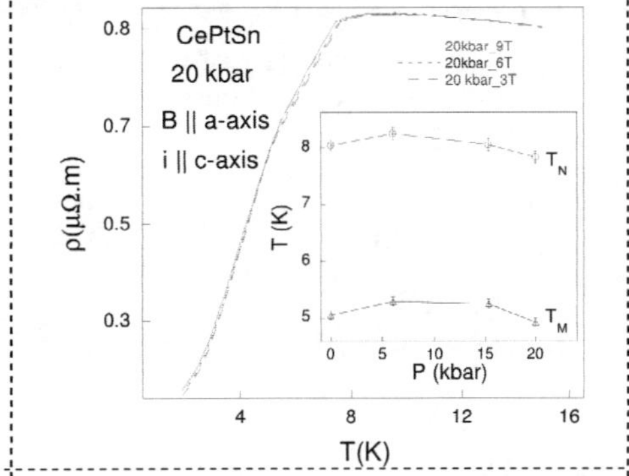

Figure 2 Temperature dependence of the magnetoresistance of CePtSn under 20.3 kbar for various fields (*i* || *c* axis and *H* || *a* axis). The inset shows the pressure dependence of the critical temperatures in zero field.

Fig. 2 shows the ρ(T) data for CePtSn for P = 20 kbar, for various fields parallel to the *a* axis. Similar to the ambient pressure data, both magnetic phase transitions are accompanied by anomalies in ρ(T). The transition associated with T_N slowly moves to lower temperature with increasing fields, while the contrary is true for the transition associated with T_M, in good agreement with the previously published data [5,6].

In Fig. 3, the normalized magnetoresistance is plotted as a function of the magnetic field at T = 2.1 K, for P = 0 and 20 kbar. The application of pressure causes a reduction in magnetoresistance. We find that the maximum in $\Delta\rho/\rho_{B=0}$ is only 0.05 at 20 kbar, while it amounts to 0.15 at ambient pressure. Unfortunately, the superconducting magnet of this experiment was limited to 9 T, which is insufficient to probe the effect of pressure on the metamagnetic transition field (~12 T at 4.2 K for *H* || *a* axis). The anomalies in the ρ(T) data permit delineating the magnetic phase diagram for field applied along the *a*-axis at ambient and under hydrostatic pressure of 20 kbar. The results are displayed in Fig. 4, which also includes the ambient pressure data taken from Ref. 5.

4. Conclusion

In conclusion, we studied the effect of hydrostatic pressure up to 20 kbar on the magnetic ordering temperatures of CePtSn by means of measurements of ρ(T) in H • 9 T and P • 20 kbar, with up to 9 T. The field applied along the *a* axis The application of hydrostatic pressure on CePtSn has little effect on both T_N and T_M. Initially, pressure leads to small increase of both temperatures, T_N and T_M, followed by a reduction after 10 kbar. The application of pressure causes a reduction in magnetoresistance. Measurements of ρ(T, P, H) in higher fields to probe the stability of the spin-slip phase are in order.

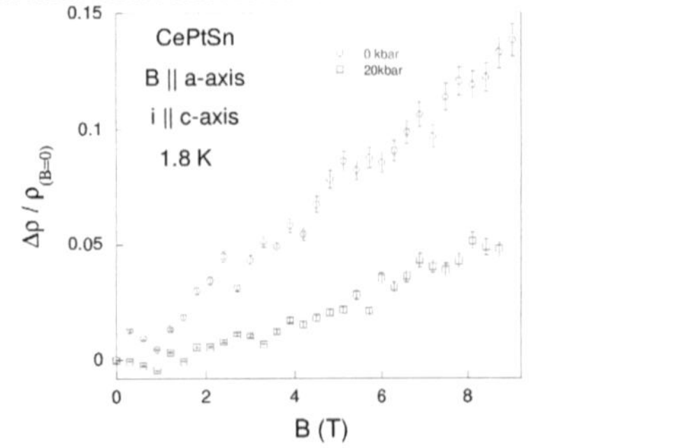

Figure 3. Field dependence of the normalized electrical resistivity of CePtSn at 1.8 K at ambient pressure and under 20 kbar ($i \parallel c$ axis and $H \parallel a$ axis).

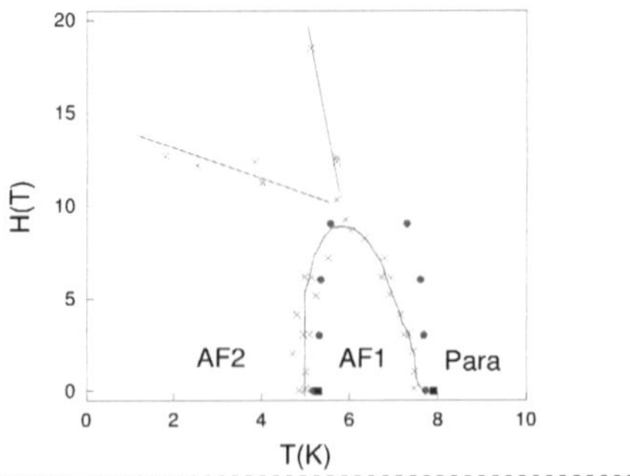

Figure 4. Magnetic H-T phase diagram of CePtSn for $H \parallel a$ axis. Solid squares are taken from the anomalies in ρ(T) at ambient pressure while crosses are taken from Ref. 5 The solid circles are taken from the anomalies in ρ(T) at 20 kbar.

References

[1] Takabatake T, Iwasaki H, Nakamoto G, Fujii H, Nakotte H, de Boer F R, and Sechovsky V 1993 Physica B **183** 108

[2] Kadowaki H, Ekino T, Iwasaki H, Takabatake T, Fujii H, and Sakurai J1993 J. Phys. Soc. Jpn. **62** 4426

[3] Kalvius G M, Kratzer A, Munch K H, Wagner F E, Zwirner S, Kobayashi H, Takabatake T, Nakamoto G, Fujii H, Kreitzman S R, and Kiefl R 1993 Physica B **186-188** 412

[4] Kadowaki H 1998 J. Phys. Soc. Jpn **67** 3261

[5] Nakotte H, Chang S, Torikachvili M S , Bordallo H N , Lacerda A H, and Takabatake T 2000 Physica B **281-282** 103

[6] Khmelevska T, Svoboda P, Janousova B, Sechovsky V, Chang S, Nakotte H, and Torikachivili M S 2001 J. Appl. Phys. **89** 7189

[7] Janousova B, Sechovsky V, Sushko Yu V, and Komatsubara T 2003 Acta Physica Polonica **34** 1039

[8] Kalvius G M, Kratzer A, Grosse G, Noakes D R, Wäppling R, Löhneysen H v, Takabatake T, and Echizen Y 2000 Physica B **289-290** 256

[9] Prokleška J, Mišek M, Sechovsky V, and Javorsky P 2009 J. Appl. Phys. **105** 07E106

[10] Mišek M, Prokleška J, Javorsky P, and Sechovsky V 2009 J. Alloys Compd. **480**,147

[11] Geibel C, Kammerer C, Seidel B, Bredl C D, Grauel A, and Steglich F 1992 J. Magn. Magn. Mater. **108** 207

[12] Diviš M, Nakotte H, de Boer F R, de Chatel P F and Sechovesky V 1994 J. Phys.: condens matter **6** 6895

[13] Alsmadi A M, Nakotte H, Sarrao J L, Jung M H, Lacerda A H 2003 J. Appl. Phys. **93** 8343

[14] Takabatake T, Sasakawa T, Kitagawa J, Suemitsy T, Echizen Y, Umeo K, and Sera M 2003 Physica B **328** 53

Acknowledgments

The support from NSF grants DMR-0805335 for the work at SDSU, and DMR-0804032 for the work at NMSU is gratefully acknowledge.

Flux distribution in vortex lattice states in d-wave and paramagnetic superconductors

Kazushi Aoyama[1,2] **and Ryusuke Ikeda**[1]

[1] Department of Physics, Kyoto University, Kyoto 606-8502, Japan
[2] Young Researcher Development Center, Kyoto University, Kyoto 606-8302, Japan

E-mail: bonn@scphys.kyoto-u.ac.jp

Abstract. Roles of field-induced antiferromagnetic (AFM) fluctuation in the vortex lattice in d-wave superconductors with strong Pauli paramagnetic depairing (PD) are studied with a focus on the heavy-fermion superconductor $CeCoIn_5$. Based on the Ginzburg-Landau approach, we calculate the field dependence of the vortex lattice form factor (VLFF) which reflects the magnetic flux distribution in the vortex lattice. It is found that, in a magnetic field perpendicular to the two-dimensional plane ($\mathbf{H} \parallel c$), the strong AFM fluctuation near $H_{c2}(0)$ enhances the VLFF in high fields and that this enhancement is remarkable when the AFM moment is perpendicular to $\mathbf{H}$ ($\mathbf{m} \perp \mathbf{H}$). The obtained result in $\mathbf{m} \perp \mathbf{H}$ well explains an anomalous field dependence of the VLFF obtained in neutron scattering experiments on $CeCoIn_5$, which suggests the necessity of taking account of the AFM fluctuation and the possibility that the configuration $\mathbf{m} \perp \mathbf{H}$ is realized in $\mathbf{H} \parallel c$.

1. Introduction

Interplay of magnetism and superconductivity has been one of the most interesting issues in condensed matter physics and heavy-fermion materials have attracted interest as a good field to investigate this problem. In recently discovered heavy-fermion superconductors, such as $CeCoIn_5$, $CeRhIn_5$, and $NpPd_5Al_2$, Pauli paramagnetic effect is known to be strong. In these materials, the antiferromagnetic (AFM) quantum critical behavior is observed in transport phenomena around $H_{c2}(T{=}0)$ [1, 2, 3]. Considering this intriguing property peculiar to paramagnetic superconductors, we study roles of AFM fluctuation in superconductors with strong Pauli paramagnetic depairing (PD).

$CeCoIn_5$ is a typical example of paramagnetic superconductors with a two-dimensional electronic structure and its pairing symmetry is identified with *d*-wave. In a magnetic field perpendicular to the two-dimensional plane ($\mathbf{H} \parallel c$), the superconducting state in this material has eminent properties discussed as a consequence of strong PD, such as the high field and low temperature superconducting phase where the Fulde-Ferrell-Larkin-Ovchinnikov vortex lattice state can be realized [4], the reentrant structural transition of the vortex lattice [5, 6]. Another fascinating property, the field-induced AFM quantum critical phenomenon just near the $H_{c2}(0)$, is also explained as a consequence of strong PD [7]. In ref.[7], it is mentioned that an AFM fluctuation or ordering is enhanced by PD in the high field *d*-wave superconducting phase. This theoretical result suggests that, in the high field region in d-wave and paramagnetic superconductors, the AFM fluctuation should be commonly taken into account. As a phenomenon which seems to be strongly relevant to the field-induced AFM

Published under licence by IOP Publishing Ltd

fluctuation, we focus on the magnetic flux distribution in the vortex lattice which is reflected by the vortex lattice form factor (VLFF) obtained in neutron scattering experiments. Experimental data on $CeCoIn_5$ in $\mathbf{H} \parallel c$ case, show that the VLFF increases with increasing field and enhances abruptly near the $H_{c2}(T)$ in low temperatures, which is in contrast to the typical field dependence of the VLFF in type II superconductors [5]. Although such an anomalous behavior is qualitatively explained as a result of strong PD [8], the origin of the abrupt enhancement in the high field region is not sufficiently explained. In this report, we discuss the possibility that this remarkable increase of the VLFF is attributed to the strong AFM fluctuation near $H_{c2}(0)$.

2. Theoretical Framework

We consider the case with a magnetic field $\mathbf{H} = H\hat{z} \parallel c$ which corresponds to the situation in the neutron scattering experiment. In the vortex lattice state, the gap function is expressed as $|\Delta(\hat{\mathbf{p}}, \mathbf{r})|^2 = |w_{\mathbf{p}}|^2 \sum_{\mathbf{K}} T_{\mathbf{K}} \exp(i\,\mathbf{K}\cdot\mathbf{r})$, where $\mathbf{K}$ is a reciprocal lattice vector and $w_{\mathbf{p}}$ denotes a pairing symmetry. Then, the longitudinal flux distribution $B_z(\mathbf{r})$ is also expressed with $\mathbf{K}$,

$$B_z(\mathbf{r}) = \sum_{\mathbf{K}} F_{\mathbf{K}} \exp(i\,\mathbf{K}\cdot\mathbf{r}). \tag{1}$$

The Fourier component $|F_{\mathbf{K}}|$ with the shortest $|\mathbf{K}| \neq 0$ corresponds to the VLFF $|F|$ which is, in the neutron scattering experiment, obtained from the integrated intensity of a diffraction peak [5]. The purpose of this paper is to obtain the field dependence of $|F_{\mathbf{K}}|$.

We derive $F_{\mathbf{K}}$ from a BCS-like electronic Hamiltonian which is written in zero field case as

$$H_\Delta - \mu N = \sum_{\mathbf{k},\alpha,\beta} \hat{c}^\dagger_{\mathbf{k},\alpha}\,\xi(\mathbf{k})\,\delta_{\alpha,\beta}\,\hat{c}_{\mathbf{k},\beta} - \sum_{\mathbf{q},\alpha,\beta}\Big(\Delta(\mathbf{q})\hat{\Psi}^\dagger(\mathbf{q}) + m_\nu(\mathbf{q})\,\hat{S}^\dagger_\nu(\mathbf{q}) + \text{h.c.}\Big),$$

where $\hat{\Psi}(\mathbf{q}) = -i\,(\sigma_y)_{\alpha,\beta}\sum_{\mathbf{k}} w_{\mathbf{k}}\,\hat{c}_{-\mathbf{k}+\mathbf{q}/2,\alpha}\hat{c}_{\mathbf{k}+\mathbf{q}/2,\beta}/2$, $\hat{S}_\nu(\mathbf{q}) = (\sigma_\nu)_{\alpha,\beta}\sum_{\mathbf{k}}\hat{c}^\dagger_{\mathbf{k}-\mathbf{q},\alpha}\,\hat{c}_{\mathbf{k}+\mathbf{Q},\beta}/2$, $\hat{c}_{\mathbf{k},\alpha}$ creates a quasiparticle with spin index α and momentum $\mathbf{k}$, σ_ν are the Pauli matrices, μ is the chemical potential, and $\mathbf{m}$ represents the AFM moment. The dispersion $\xi(\mathbf{k})$ satisfies $\xi(\mathbf{k}) = -\xi(\mathbf{k}+\mathbf{Q})$ for the commensurate AFM modulation and $w_{\mathbf{p}}$ satisfies $w_{\mathbf{p}} = -w_{\mathbf{p}+\mathbf{Q}}$ for the $d_{x^2-y^2}$ pairing, where $\mathbf{Q}$ is the commensurate AFM wave vector and is (π,π) for the $d_{x^2-y^2}$ pairing of our interest. In the case with a nonzero magnetic field, the PD effect is introduced by adding the Zeeman term $\gamma_B H(\sigma_z)_{\alpha,\beta}$ to $\xi(\mathbf{k})\delta_{\alpha,\beta}$ and the orbital depairing effect is taken into consideration by introducing the vector potential $\mathbf{A}$ which satisfies $\nabla\times\mathbf{A} = \mathbf{B}$. From the saddle point equation of the total free energy with respect to $\mathbf{A}$, the Maxwell equation is derived,

$$-\frac{1}{4\pi}\nabla\times\mathbf{B} = \frac{|e|}{2m_e\,i}\lim_{\mathbf{r}'\to\mathbf{r}} T\sum_{\varepsilon,\sigma}[\nabla-\nabla']\mathcal{G}_{\varepsilon,\sigma}(\mathbf{r},\mathbf{r}') - \nabla\times\Big(T\sum_{\epsilon,\sigma}\gamma_B\sigma\,\mathcal{G}_{\varepsilon,\sigma}(\mathbf{r},\mathbf{r})\Big)\hat{z}, \tag{2}$$

where $\mathcal{G}_{\varepsilon,\sigma}(\mathbf{r},\mathbf{r}')$ is a Green's function of a quasiparticle with Matsubara frequency ε and spin σ. Below, we consider two cases, $\mathbf{m} \parallel H\hat{z}$ ($\nu = z$) and $\mathbf{m} \perp H\hat{z}$ ($\nu = x$). The difference between two cases appears in the spin $\overline{\sigma}$ of a quasiparticle with the momentum $\mathbf{k}+\mathbf{Q}$ which is coupled with one with the spin σ and the momentum $\mathbf{k}$ through the $(\sigma_\nu)_{\sigma,\overline{\sigma}}$ term in $\hat{S}_\nu$, so that $\overline{\sigma} = \sigma$ in $\mathbf{m} \parallel \mathbf{H}$ and $\overline{\sigma} = -\sigma$ in $\mathbf{m} \perp \mathbf{H}$. Based on the Ginzburg-Landau (GL) approach, we expand $\mathcal{G}_{\varepsilon,\sigma}(\mathbf{r},\mathbf{r}')$ with respect to $\Delta(\hat{\mathbf{p}},\mathbf{r})$ and calculate Feynman diagrams shown in Fig. 1 (a)-(c). Since the calculation method based on the GL approach taking account of both the orbital depairing and the PD is given in ref. [4, 9], we only note the relation for the normal Green's function described by a solid line in Fig. 1, $\mathcal{G}^N_{\varepsilon,\overline{\sigma}}(\mathbf{p}+\mathbf{Q}) = -\mathcal{G}^N_{-\varepsilon,-\overline{\sigma}}(\mathbf{p})$, which is used in calculating Fig. 1 (b) and (c). The AFM fluctuation $|m_\nu(\mathbf{q})|^2$ which is expressed by a wavy line in Fig. 1, can be taken as $T\sum_\Omega X(\mathbf{q},\Omega)$ by using the familiar phenomenological form of $X(\mathbf{q},\Omega)$,

$$X(\mathbf{q},\Omega) = \frac{1}{N(0)}\frac{\xi^2(h,t)[\xi_0^{(\mathrm{N})}]^{-2}}{1+\xi^2(h,t)|\mathbf{q}\times\hat{z}|^2 + |\Omega|/(\Gamma_0\xi^{-2}(h,t))}, \tag{3}$$

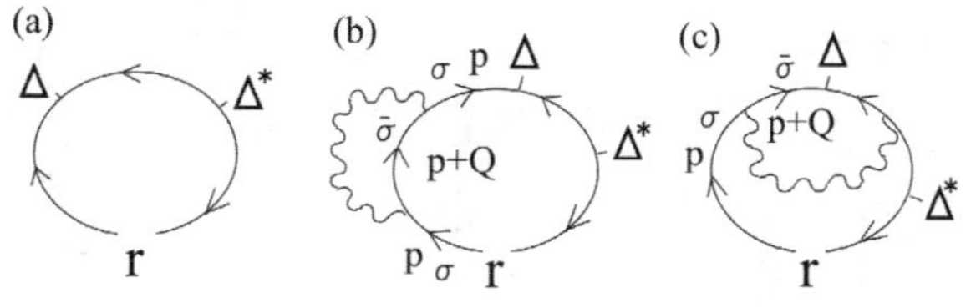

Figure 1. Feynman diagrams which give the local flux distribution. (a) is the second order term in Δ with no AFM fluctuation. (b) and (c) describe contirbutions from the quasiparticle self energy and the vertex correction, respectively. A wavy line denotes the AFM fluctuation.

where $\xi(h,t) = \xi_0^{(N)}(t+1-h/h_{QCP})^{-1/2}$. $\xi_0^{(N)}$ is a characteristic length scale of the AFM fluctuation and h_{QCP} represents an AFM quantum critical point close to $H_{c2}(0)$. In this report, we set $\xi_0^{(N)} = 0.6\,\xi_0$ where ξ_0 is the superconducting coherence length at $T=0$. Fig. 1 (b) properly gives the real part of the quasiparticle self energy Σ brought by the AFM fluctuation. The imaginary part of the self energy, namely, the quasiparticle damping which suppresses PD [9], can be calculated straightforwardly and is expressed as $\mathrm{Im}\Sigma = \pi T\xi(h,t)/\left(4T_c p_F[\xi_0^{(N)}]^2\right)$. Fig. 1 (c) describing the vertex correction changes its sign depending on the pairing symmetry since this term is proportional to $w_{\mathbf{p}}w_{\mathbf{p+Q}}$. Below, we assume the triangular lattice and calculate the lowest order Fourier component $F_{\mathbf{K}}$ which is, hereafter, denoted by F.

3. Result

The field dependence of the VLFF at $t \equiv T/T_c = 0.06$ is shown in Fig. 2 (a) where a dotted curve corresponds to the case with no AF contribution except the quasiparticle damping, and a dashed (solid) curve corresponds to the case with the AF fluctuation in $\mathbf{m} \parallel \mathbf{H}$ ($\mathbf{m} \perp \mathbf{H}$). Green and orange curves denote the results with and without the vertex correction contribution, respectively. The parameters used in Fig. 2 are α_M^c (Maki parameter in $\mathbf{H} \parallel c$) $\equiv \sqrt{2}H_{\mathrm{orb}}^c(0)/(1.25T_c/\gamma_B) = 6.5$, $p_F\xi_0 = 7.0$, $\Gamma_0 = 2\pi T_c$, and $h_{QCP} = 0.942$, where $H_{\mathrm{orb}}^c(0)$ is the orbital limit and h represents a magnetic field normalized by $H_{c2}(0)$ in this system. In the case with no AFM contribution (dotted curve) where the fourth order term in Δ is incorporated, the VLFF increases with increasing field, suggesting the concentration of the magnetic flux in the vortex core, in other words, the enhancement of the magnetic screening far away from the vortex core [8]. When only the self energy contribution is taken into account (orange curves in Fig. 2 (a)), the AFM fluctuation enhances the VLFF in $\mathbf{m} \parallel \mathbf{H}$, while it suppresses the VLFF in $\mathbf{m} \perp \mathbf{H}$. However, when all the AFM contributions are considered (green curves in Fig. 2 (a)), the VLFF is enhanced near the H_{c2} by the AFM fluctuation in both two cases, $\mathbf{m} \parallel \mathbf{H}$ and $\mathbf{m} \perp \mathbf{H}$. Comparing the green curves with the orange ones, we find that what plays an important role in enhancing the VLFF is the self energy contribution in $\mathbf{m} \parallel \mathbf{H}$ and the vertex correction contribution in $\mathbf{m} \perp \mathbf{H}$. We note that the enhancement of the VLFF in $\mathbf{m} \perp \mathbf{H}$ is due to the d-wave pairing symmetry since the sign of the vertex correction term is determined by the d-wave character, $w_{\mathbf{p}}w_{\mathbf{p+Q}} < 0$.

Fig. 2 (b) shows the field dependence of the VLFF at $t = 0.02$ (red), $t = 0.04$ (blue), and $t = 0.06$ (green). Dotted, dashed, and solid curves in Fig. 2 (b) correspond to the same cases as those in Fig. 2 (a). In $\mathbf{m} \perp \mathbf{H}$ case, the enhancement of the VLFF due to the AFM fluctuation becomes more remarkable on cooling and exceeds the enhancement in $\mathbf{m} \parallel \mathbf{H}$ at $t = 0.02$. The obtained field dependence of the VLFF in lowering temperature in $\mathbf{m} \perp \mathbf{H}$ is consistent with the experimental data [10]. This fact suggests that the abrupt increase of the VLFF in the high field region is explained as a result of the strong AFM fluctuation induced by PD in d-

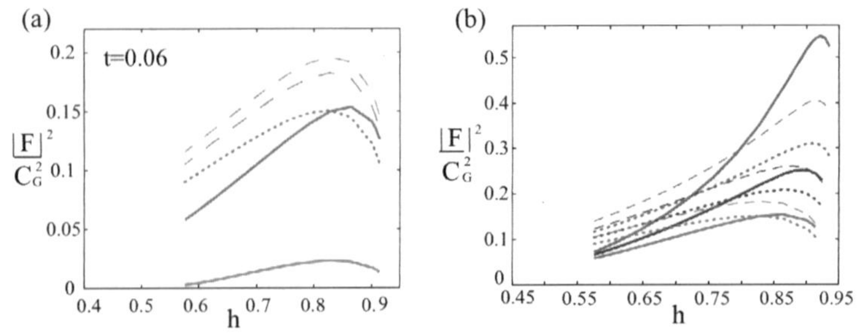

Figure 2. (Color online) Field dependence of the VLFF normalized by the value near T_c, C_G, in the case with no AF contribution (dotted curves), in the case with the AF fluctuation in $\mathbf{m} \parallel \mathbf{H}$ (dashed ones), and the corresponding $\mathbf{m} \perp \mathbf{H}$ case (solid ones) (a) at $t = 0.06$ and (b) at $t = 0.02$ (highest (red)), $t = 0.04$ (middle (blue)), and $t = 0.06$ (lowest (green)). In (a), in order to see each contribution of the Feynman diagrams shown in Fig. 1, results without the vertex correction contribution (Fig. 1 (c)) are shown in two cases, $\mathbf{m} \parallel \mathbf{H}$ (higher (orange) dashed curve) and $\mathbf{m} \perp \mathbf{H}$ (lower (orange) solid one).

wave superconductors and that the AFM moment may be perpendicular to the magnetic field ($\mathbf{m} \perp \mathbf{H}$), as is satisfied in $\mathbf{H} \perp c$ [11].

4. Conclusion

The contribution of the field-induced AFM fluctuation to the vortex lattice form factor (VLFF) in d-wave and paramagnetic superconductors in $\mathbf{H} \parallel c$ is calculated based on the Ginzburg-Landau approach. In addition to the PD-induced increase of the VLFF with increasing field, the strong AFM fluctuation near $H_{c2}(0)$ enhances the VLFF in high fields. The temperature dependence of this enhancement is remarkable in $\mathbf{m} \perp \mathbf{H}$ where the vertex correction plays an important role. The obtained field dependence of the VLFF in $\mathbf{m} \perp \mathbf{H}$ is consistent with the experimental result in $CeCoIn_5$, which suggests that the anomalous field dependence of the VLFF in $CeCoIn_5$ is explained as a result of the strong AFM fluctuation induced by PD and that the AFM moment may be perpendicular to the magnetic field ($\mathbf{m} \perp \mathbf{H}$).

Acknowledgments

This work was financially supported by Grant-in-Aid for Scientific Research [No. 20102008 and 21540360] from MEXT, Japan.

References

[1] Singh S et al., 2007 *Phys. Rev. Lett.* **98**, 057001; F. Ronning et al., 2005 *Phys. Rev.* B **71**, 104528.
[2] Park T et al., 2006 *Nature* **440**, 65 and arXiv: 0910.2287.
[3] Honda F et al., 2008 *J. Phys. Soc. Jpn. Suppl.* A **77**, 339.
[4] Ikeda R, 2007 *Phys. Rev.* B **76**, 134504 .
[5] Bianchi A et al., 2008 *Science* **319**, 177; Eskildsen R M, private communication.
[6] N. Hiasa and R. Ikeda, 2008 *Phys. Rev. Lett.* **101**, 027001.
[7] Ikeda R, Hatakeyama Y, and Aoyama K, 2010 *Phys. Rev.* B **82** 060510(R).
[8] Ichioka M and Machida K, 2007 *Phys. Rev.* B **76** 064502.
[9] Adachi H and Ikeda R, 2003 *Phys. Rev.* B **68**, 184510.
[10] White S J et al., 2010 *New J. Phys.* **12**, 023026.
[11] Kenzelmann M et al., 2008 *Science* **321**, 1652.

Journal of Physics: Conference Series **273** (2011) 012027 doi:10.1088/1742-6596/273/1/012027

Superconducting condensation energy in a diluted Kondo lattice

C Capan[1], G Seyfarth[1,2,3], D Hurt[1], A D Bianchi[1,2], Z Fisk[1]

[1]Department of Physics and Astronomy, University of California Irvine, Irvine, CA 92697-4575
[2]Department de Physique, Universite de Montreal, Montreal H3C 3J7 Canada
[3]Department of Condensed Matter Physics, University of Geneva, Switzerland

E-mail: ccapan@uci.edu

Abstract. Heavy fermion compounds are host to many interesting but poorly understood strong correlation effects, leading in exceptional cases to unconventional superconductivity. $CeCoIn_5$ is such an example of an ambient pressure heavy fermion superconductor with a T_c of 2.3K. Such compounds are best described as a Kondo lattice, where Ce, Yb or U ions form a periodic lattice of magnetic moments coupled to the conduction electrons via antiferromagnetic exchange interaction. Dilution, or substitution, of the Kondo ion with a lanthanide, actinide or alkali-earth element that does not participate in the Kondo lattice, results in the so-called "Kondo hole". The electron scattering off the Kondo hole leads to significant changes in both the normal and superconducting properties. We have investigated the suppression of superconductivity in $CeCoIn_5$ diluted with La, Yb, Y, Ca, Th, Gd, Er, Eu and Lu via specific heat measurements on well characterized single crystals. The suppression of the superconducting condensation energy exhibits a remarkable departure from the Abrikosov-Gorkov impurity pair breaking picture, emphasizing that the Kondo holes are no ordinary pair-breakers.

$CeCoIn_5$ is an ambient pressure[1], d-wave[2], superconductor with the highest T_c (2.3 K) among the Ce-based heavy fermion superconductors. Together with the parent $PuCoGa_5$[3] ($T_c = 18$ K), superconductivity in this system forms out of a lattice of incompletely quenched f-electron local moments. This leads to the suggestion that the pairing mechanism directly involves the Kondo interaction[4] in addition to, or rather than antiferromagnetic spin fluctuations. Superconductivity in $CeCoIn_5$ also displays an unusual interplay with antiferromagnetism. Although no long range magnetic order is reported at H=0 for the pure $CeCoIn_5$, a resonance peak is observed below T_c in inelastic neutron scattering[5] along the commensurate wavevector $Q = (1/2, 1/2, 1/2)$, corresponding to undamped magnon-like excitations within the superconducting state[6]. A long range antiferromagnetic state is stabilized, with the same commensurate structure[7], when In is substituted by very small amounts of Cd[8] or Hg[9]. When superconductivity is suppressed with magnetic fields, strong deviations from Fermi liquid behavior are observed in the specific heat[10] and resistivity measurements[11], signaling the proximity of a zero temperature magnetic instability, also known as a quantum critical point(QCP).

La substitution for Ce in this compound has been investigated by various groups and the emerging picture is rather counter-intuitive. First, the Kondo coherence temperature is weakly doping-dependent up to about 40%La beyond which it vanishes abruptly[12]. This is hard to reconcile with the idea that the coherence in the Kondo lattice results from intersite coupling

Published under licence by IOP Publishing Ltd

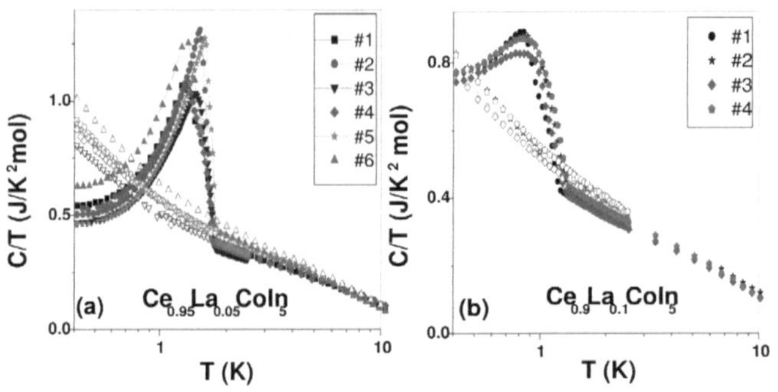

Figure 1. Electronic specific heat vs Temperature at $H = 0$(closed symbols) and $H = 5T$(open symbols) in single crystals of $Ce_{1-x}La_xCoIn_5$ for (a) $x = 0.05$ and (b) $x = 0.1$.

among neighboring Ce's. Moreover, the residual thermal conductivity in the superconducting state at $T = 0$ does not have a universal, doping independent, value[13], as expected in the presence of nodal quasiparticles in a d-wave superconductor. This has been interpreted[13] in an extreme multiband scenario in which there remains unpaired electrons on the 3D pockets of the Fermi surface[13], an interpretation that has been contested both on theoretical[14] and experimental[15] grounds. Motivated by these results, we have investigated the effect of Ce-site substitution on superconducting properties of $CeCoIn_5$ for an extensive range of dopants and concentrations.

Single crystals of $Ce_{1-x}R_xCoIn_5$ were grown from excess In flux with $R = La^{3+}$, Yb^{2+}, Y^{3+}, Th^{4+}, Lu^{3+}, Ca^{2+}, Eu^{2+}, Er^{3+}, Gd^{3+} as described elsewhere[1]. Samples were etched in dilute HCl to remove excess In and carefully screened for free-In inclusions. Energy dispersive X-ray spectroscopy (EDS) shows that Ce has been effectively substituted by R, although an accurate determination of the actual concentration was not possible due to the non-uniform distribution of dopants. Figure 1 shows the electronic specific heat, measured in In-free single crystals of $Ce_{1-x}La_xCoIn_5$ with nominal La concentrations of $x = 0.05$ (1a) and $x = 0.1$ (1b). The electronic specific heat is obtained by subtracting the lattice contribution, determined from $LaCoIn_5$. At zero field, the specific heat transition has the same onset temperature of $T_c = 1.8$K ($x = 0.05$) and 1.3K ($x = 0.1$) for all samples from a given batch, whereas the size of the specific heat anomaly, as well as the $T = 0$ intercept of $\frac{C}{T}$, γ_0, is sample-to-sample dependent. As also shown in Fig. 1, a field of $H = 5T$ applied along [001] suppresses the SC transition and reveals a divergent $\frac{C}{T}$ in all samples, similar to the non-Fermi Liquid behavior reported in pure $CeCoIn_5$[10]. Similar observations hold when Ce is substituted by $R = Yb^{2+}$, Th^{4+}, Y^{3+}, Lu^{3+}, Eu^{2+}, Er^{3+}, Gd^{3+} and Ca^{2+}, as seen in Figure 2: the specific heat jump broadens and is suppressed as the T_c decreases, while γ_0 systematically increases.

Figure 3 summarizes our main results. The specific heat jump ΔC is shown in Fig. 3a as a function of the critical temperature T_c, both quantities normalized with the values of 3.544 J/K mol and 2.25 K found in the pure compound. Regardless of the nature of R (magnetic or not, isovalent or not), the data for all $Ce_{1-x}R_xCoIn_5$ compounds fall on a single line which is clearly distinct from the prediction of Abrikosov-Gorkov (AG) theory[17], also shown in Fig. 3a as a solid line. In contrast to Ce-site substitution, In-site substitution by Sn[16] do follow the AG theory fairly well. The difference between Ce vs In site substitution is a compelling evidence that the deviation we observe is due to the Kondo hole scattering. Fig. 3b shows the doping dependence of T_c. The doping x is obtained via EDS analysis and the non-uniform spatial distribution of doping within samples prevented an accurate determination of x. We attribute

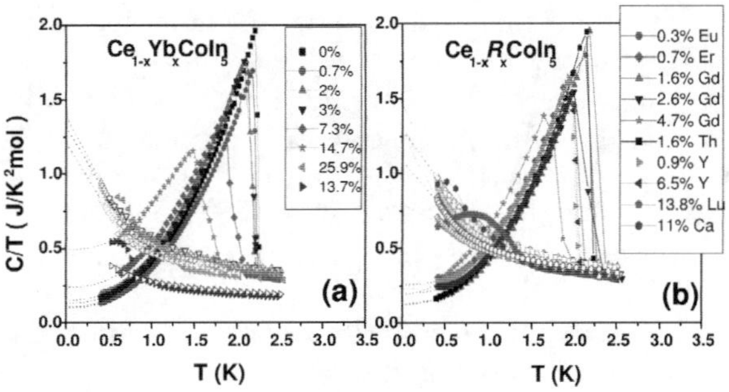

Figure 2. Electronic specific heat vs Temperature at $H = 0$(closed symbols) and $H = 5T$(open symbols) in single crystals of (a) $Ce_{1-x}Yb_xCoIn_5$ and (b) $Ce_{1-x}R_xCoIn_5$ (R =Th,Lu,Eu,Er,Gd,Ca,Y) for the indicated concentrations. Dashed lines correspond to power law (linear) fits to the $H = 0 (H = 5\text{T})$ data.

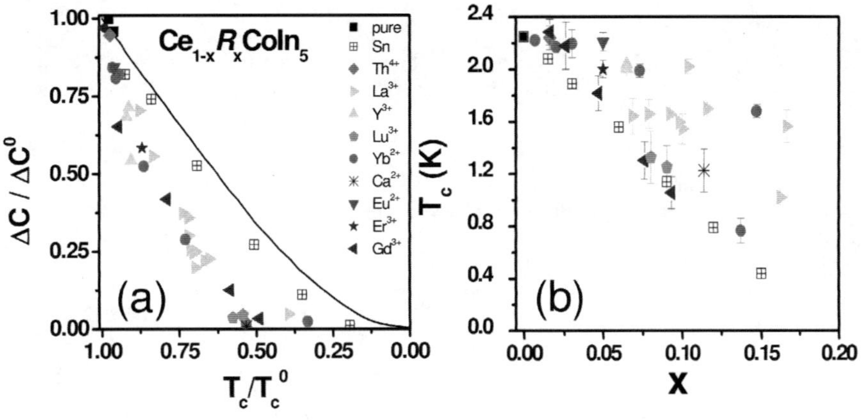

Figure 3. Normalized specific heat jump vs normalized T_c at $H = 0$ in single crystals of $Ce_{1-x}R_xCoIn_5$ with $R = La^{3+}$, Yb^{2+}, Y^{3+}, Th^{4+}, Lu^{3+}, Ca^{2+}, Eu^{2+}, Er^{3+}, Gd^{3+}, as indicated. The values of $\Delta C = 3.544$J/K mol and $T_c = 2.25$K for pure $CeCoIn_5$ were used for normalization. For comparison, results of Sn substitution for In are also shown[16]. Solid line corresponds to the prediction based on Abrikosov-Gorkov theory[17].(b) Doping dependence of T_c in the same crystals.

the spread in the data to the inaccurate determination of x via EDS. A previous investigation of rare-earth substitution, via resistivity, claimed a universal suppression of T_c with increasing residual resistivity, irrespective of the magnetic nature of the substituent[18]. Nevertheless, there is some spread in T_c vs ρ_0 as well. Taken together, these results suggest that the pair-breaking effect due to the Ce-site substituent is not sensitive to the nature of this substituent.

The term Kondo Hole has been coined[19] to describe the creation of a vacancy in the Kondo lattice by chemical substitution of Ce with an element that does not have a local moment which could couple to the conduction band via Kondo effect. While a marked upturn in resistivity observed in $CePd_3$ doped with as little as 3% La has been attributed to Kondo Hole scattering[20], the transport and thermodynamic signatures of Kondo Holes have remained elusive and hard to disentangle from the characteristic properties of the heavy fermion metal in the normal state. Heavy fermion superconductivity, on the other hand, offers an unequivocal playground to investigate the physics of the Kondo Hole, as the non s-wave pairing is known to be extremely sensitive to any source of scattering. Previous thermodynamic investigations of the suppression of superconductivity upon Ce- or U-site dilution[19, 21, 22, 23, 24] have revealed deviations from the Abrikosov-Gorkov theory[17] in the doping dependence of both T_c and the specific heat jump. However, the universality we observe in $Ce_{1-x}R_xCoIn_5$ with respect to the magnetic or electronic nature of R has been missed in earlier dilution studies in other heavy fermion superconductors.

In conclusion, we have investigated the suppression of the superconductivity via specific heat measurements in well-characterized single crystals of $Ce_{1-x}R_xCoIn_5$ with $R = La^{3+}$, Yb^{2+}, Y^{3+}, Th^{4+}, Lu^{3+}, Ca^{2+}, Eu^{2+}, Er^{3+}, Gd^{3+}. The suppression of the specific heat anomaly at T_c is universal, independent of the valence or the magnetic nature of the substituent. Moreover, we found a striking deviation from the Abrikosov-Gorkov theory for the Ce-site substitution, in contrast to In-substitution, giving compelling evidence for the Kondo hole scattering. More theoretical work is needed in order to account for the suppression of superconductivity in presence of Kondo holes in heavy fermion superconductors.

Acknowledgments
Work at UC Irvine was supported through NSF Grant No. NSF-DMR-0801253.

References

[1] Petrovic C, Pagliuso P G, Hundley M F, Movshovich R, Sarrao J L, Thompson J D, Fisk Z and Monthoux P 2001 *J. Phys. : Condens. Matter* **13** L337
[2] Izawa K, Yamaguchi H, Matsuda Y, Shishido H, Settai R and Onuki Y 2001 *Phys. Rev. Lett.* **87** 057002
[3] Sarrao J L, Morales L A, Thompson J D, Scott B L, Stewart G R, Wastin F, Rebizant J, Boulet P, Colineau E and Lander G H 2003 *Nature* **420** 297
[4] Flint R, Dzero M and Coleman P 2008 *Nat. Phys.* **4**
[5] Stock C, Broholm C, Hudis J, Kang H J and Petrovic C 2008 *Phys. Rev. Lett.* **100** 087001
[6] Chubukov A V and Gor'kov L P 2008 *Phys. Rev. Lett.* **101** 147004
[7] Nicklas M, Stockert O, Park T, Habicht K, Kiefer K, Pham L D, Thompson J D, Fisk Z and Steglich F 2007 *Phys. Rev. B* **76** 052401
[8] Pham L D, Park T, Maquilon S, Thompson J D and Fisk Z 2006 *Phys. Rev. Lett.* **97** 056404
[9] Bauer E, Ronning F, Maquilon S, Pham L, Thompson J and Fisk Z 2008 *Physica B* **403** 1135 – 1137
[10] Bianchi A, Movshovich R, Vekhter I, Pagliuso P G and Sarrao J L 2003 *Phys. Rev. Lett.* **91** 257001
[11] Paglione J, Tanatar M A, Hawthorn D G, Boaknin E, Hill R W, Ronning F, Sutherland M, Taillefer L, Petrovic C and Canfield P C 2003 *Phys. Rev. Lett.* **91**
[12] Nakatsuji S, Yeo S, Balicas L, Fisk Z, Schlottmann P, Pagliuso P G, Moreno N O, Sarrao J L and Thompson J D 2002 *Phys. Rev. Lett.* **89** 106402
[13] Tanatar M A, Paglione J, Nakatsuji S, Hawthorn D G, Boaknin E, Hill R W, Ronning F, Sutherland M, Taillefer L, Petrovic C, Canfield P C and Fisk Z 2005 *Phys. Rev. Lett.* **95** 067002
[14] Barzykin V and Gor'kov L P 2007 *Phys. Rev. B* **76** 014509
[15] Seyfarth G, Brison J P, Knebel G, Aoki D, Lapertot G and Flouquet J 2008 *Phys. Rev. Lett.* **101** 046401
[16] Bauer E D, *et al Phys. Rev. B* **73** 245109
[17] Skalski S, Betbeder-Matibet O and Weiss P R 1964 *Phys. Rev.* **136**
[18] Paglione J, Sayles T A, Ho P C, Jeffries J R and Maple M B 2007 *Nat. Phys.* **3** 706
[19] Steglich F, Ahlheim U, Rauchschwalbe U and Spille H 1987 *Physica B* **148** 6
[20] Lawrence J M, Thompson J D and Chen Y Y 1985 *Phys. Rev. Lett.* **54** 2537
[21] Smith J L, Fisk Z, Willis J O, Ott H R, Lambert S E, Dalichaouch Y and Maple M B 1987 *J. Magn. and Magn. Materials* **63-64** 464 – 466
[22] Dalichaouch Y, de Andrade M C, Gajewski D A, Chau R, Visani P and Maple M B 1995 *Phys. Rev. Lett.* **75** 3938–3941
[23] de la Torre A L, Visani P, Dalichaouch Y, Lee B W and Maple M B 1992 *Physica B* **179** 208 – 214
[24] Aronson M C, Vorenkamp T, Koziol Z, de Visser A and Bakker K 1991 *J. Appl. Phys.* **69** 5487–5489

The S=1 Underscreened Anderson Lattice model for Uranium compounds

C. Thomas[1,2], A. S. R. Simões[1], J. R. Iglesias[1], C. Lacroix[2], N. B. Perkins[3] and B. Coqblin[4]

[1] Inst. de Física, Universidade Federal do Rio Grande do Sul, 91501-970 Porto Alegre, Brazil
[2] Institut Néel, CNRS-UJF, BP 166, 38042 Grenoble Cedex 9, France
[3] Department of Physics, University of Wisconsin-Madison, Madison, WI 53706, USA
[4] L. P. S., Université Paris-Sud, UMR-8502 CNRS, 91405 Orsay, France

Abstract. Magnetic properties of uranium and neptunium compounds showing coexistence of the Kondo effect and ferromagnetic order are investigated within the degenerate Anderson Lattice Hamiltonian, describing a $5f^2$ electronic configuration with $S = 1$ spins. Through the Schrieffer-Wolff transformation, both an exchange Kondo interaction for the $S = 1$ f-spins and an effective f-band term are obtained, allowing to describe the coexistence of Kondo effect and ferromagnetic ordering and a weak delocalization of the $5f$-electrons. We calculate the Kondo and Curie temperatures and we can account for the pressure dependence of the Curie temperature of UTe.

1. INTRODUCTION

The interplay between Kondo effect and magnetism is very important in cerium, ytterbium, uranium or other anomalous rare-earth and actinide systems [1, 2]. In contrast to $4f$-electrons, the $5f$-electrons in actinide compounds can be either localized or itinerant or in-between, depending on the studied system [3, 4, 5, 6, 7].

The difference between the $4f$- and $5f$-electrons leads to different magnetic properties. In the case of cerium Kondo compounds, the competition between the Kondo effect on each Ce atom and the magnetic ordering is described by the "Doniach diagram" [8], which gives rather low ordering temperatures, typically of order 5 to 10 K. On the other hand, some uranium compounds, like UTe [9], $UCu_{0.9}Sb_2$ [10] or $UCo_{0.5}Sb_2$ [11] present a ferromagnetic order with large Curie temperatures (equal respectively to 102 K, 113 K and 64.5 K) and also exhibit a logarithmic Kondo-type decrease of the resistivity above T_C. Recently, a similar behavior has been observed in two neptunium compounds $NpNiSi_2$ [12] and Np_2PdGa_3 [13], with T_C equal to respectively 51.5 K and 62.5 K.

Thus, it is clear that Kondo effect and ferromagnetic order coexist in some uranium and neptunium compounds. However, the localization of the $5f$-electrons is a difficult question and in fact the values of the magnetic moments observed for example in UTe are substantially smaller than the free-ion values for either the $5f^2$ or the $5f^3$ configurations [9]. On the same side, in the series of uranium monochalcogenides, US lies closest to the itinerant side for the $5f$-electrons, USe is in the middle and the $5f$-electrons are more localized in UTe, as evidenced by magnetization measurements [9]. Moreover, the Curie temperature of UTe is passing through a maximum and is then decreasing with applied pressure, which is interpreted

Published under licence by IOP Publishing Ltd

as a weak delocalization of the $5f$-electrons under pressure [4, 9]. The dual nature of the $5f$-electrons, assuming two localized $5f$-electrons and one delocalized one, has been also introduced to account for the properties of some uranium compounds [5] .

The coexistence between the Kondo effect and the ferromagnetism in some uranium compounds was described by the Underscreened Kondo Lattice (UKL) model which considers localized f-spins $S = 1$ without any f-band width [14]. Here we will present a recent work in which a finite $5f$-bandwidth is introduced within the Underscreened Anderson Lattice (UAL) model, in order to describe a weak delocalization of the $5f$-electrons and to account for the pressure dependence of the Curie temperature of UTe [15, 16]. In contrast with $S = 1/2$ case, Kondo screening in Kondo $S = 1$ is not complete [17] and this allows coexistence of Kondo effect and magnetism.

2. THE EFFECTIVE $S = 1$ HAMILTONIAN.

We start here from the UAL Hamiltonian with two f-electrons per site (in different orbitals α=1,2) forming a $S = 1$ spin. The UAL model, which is explicitely given in Ref. [15], is composed of a conduction band of energy $\epsilon_{\mathbf{k}}$, a two-fold degenerate f-level at the energy E^f, a $c - f$ hybridization term, and Coulomb interactions between f-electrons: U and U' among electrons in respectively the same and different orbitals and Hund's coupling J. Then, we use the Schrieffer-Wolff (SW) transformation [18] and we obtain an effective Hamiltonian: $H = H_c + H_{cf} + H_W$. The first term describes the conduction electrons, the second one is the usual Kondo exchange Hamiltonian, but here with f-spins $S = 1$, and it is given by:

$$H_{cf} = \frac{1}{2}\sum_{i\mathbf{k}\mathbf{k}'} J_K \left[c^\dagger_{\mathbf{k}'\uparrow} c_{\mathbf{k}\downarrow} S^-_i + c^\dagger_{\mathbf{k}'\downarrow} c_{\mathbf{k}\uparrow} S^+_i + (c^\dagger_{\mathbf{k}'\uparrow} c_{\mathbf{k}\uparrow} - c^\dagger_{\mathbf{k}'\downarrow} c_{\mathbf{k}\downarrow}) S^z_i \right] \tag{1}$$

with the different components of the spin $S = 1$ given by: $S^z_i = n^f_{i1\uparrow} n^f_{i2\uparrow} - n^f_{i1\downarrow} n^f_{i2\downarrow}$, $S^+_i = n^f_{i1} f^\dagger_{i2\uparrow} f_{i2\downarrow} + f^\dagger_{i1\uparrow} f_{i1\downarrow} n^f_{i2}$ and a similar expression for S^-_i. J_K is the usual Kondo interaction integral. The third term H_W of the new derived Hamiltonian yields an effective f-band term, with a bandwidth proportional to J_K. Detailed calculations can be found in Refs [15] and [16].

Besides these two terms H_{cf} and H_W, we add an exchange interaction J_H between the localized spins to account for the effective intersite ferromagnetic exchange. In fact, the exchange interaction J_K can in principle yield both the Kondo effect and magnetic order, but the addition of the exchange interaction J_H is necessary to really describe the magnetic order, as already well established theoretically [2, 14, 19]. Here J_H is considered as a ferromagnetic interaction between nearest neighbor f-spins.

Thus, the total mean field Hamiltonian is :

$$H_{tot} = \sum_{\mathbf{k}\sigma} \epsilon^c_{\mathbf{k}\sigma} n^c_{\mathbf{k}\sigma} + \sum_{i\sigma\alpha} E^f_{0\sigma} n^f_{i\alpha\sigma} + \sum_{\mathbf{k}\sigma\alpha} \Lambda_\sigma (c^\dagger_{\mathbf{k}\sigma} f_{\mathbf{k}\alpha\sigma} + h.c.) + \sum_{\mathbf{k}\sigma\alpha} A_{\mathbf{k}\sigma} f^\dagger_{\mathbf{k}\alpha\sigma} f_{\mathbf{k}\alpha\sigma} + C \tag{2}$$

The effective f-hopping term $A_{\mathbf{k}\sigma}$ is given by :

$$A_{\mathbf{k}\sigma} = -\epsilon_{\mathbf{k}} \frac{J_K}{2} \left[(n^f_\sigma)^2 + \frac{1}{2} n^f_\sigma n^f_{\bar\sigma} + \frac{1}{4} (n^f_{\bar\sigma})^2 \right] , \tag{3}$$

while the effective $c - f$ hybridization is equal to $\Lambda_\sigma = -\frac{J_K}{4}(\lambda_\sigma + \lambda_{\bar\sigma})$, with $\epsilon^c_{\mathbf{k}\sigma} = \epsilon_{\mathbf{k}} + \Delta_\sigma$ and $\Delta_\sigma = J_K \sigma M^f$.

$E^f_{0\sigma}$ is the effective position of the f-level given by :

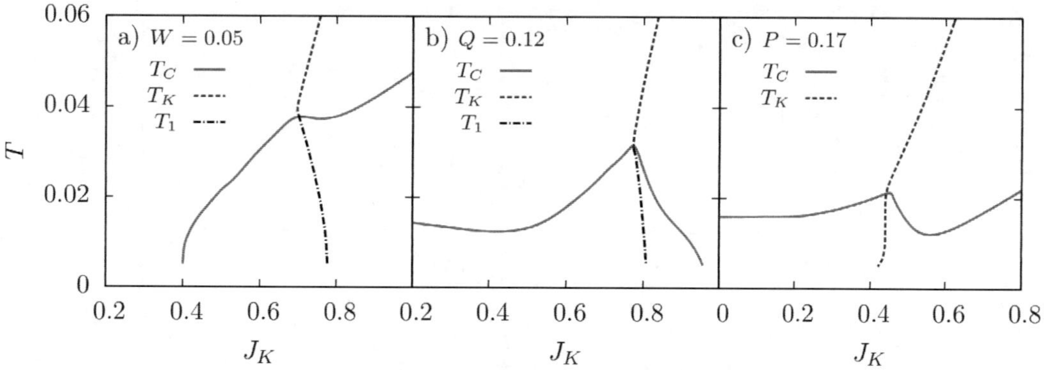

Figure 1. Plots of T_C, T_K and T_1 versus J_K for the three cases (a), (b) and (c), with $J_H = -0.01$, $n^c = 0.8$ and $n^f_{\mathrm{tot}} = 2$ and the values of the different parameters given in the figures (all the temperatures and the energies are expressed in units of D).

$$E^f_{0\sigma} = E^f + U'n^f_{\bar{\sigma}} + (U' - J)n^f_{\sigma} + J_K\sigma m^c - \tfrac{J_K}{8}(\lambda_\uparrow + \lambda_\downarrow)^2 + J_H z\sigma M^f \ , \tag{4}$$

and C is equal to :

$$C = -2U'Nn^f_\uparrow n^f_\downarrow - (U' - J)N[(n^f_\uparrow)^2 + (n^f_\downarrow)^2] + \tfrac{J_K}{2}N(\lambda_\uparrow + \lambda_\downarrow)^2 - \tfrac{J_H}{2}zN(M^f)^2 - J_K N m^c M^f \ , \tag{5}$$

with $\sigma = \pm\frac{1}{2}$, $M^f = n^f_\uparrow - n^f_\downarrow$ and $m^c = \frac{1}{2}(n^c_\uparrow - n^c_\downarrow)$.

$n^f_{i\alpha\sigma}$ is the number of f-electrons per site, spin and orbital α, $n^f_\sigma =< n^f_{i\alpha\sigma} >$ and n^c the total number of conduction electrons. The f and c magnetizations $M^f = n^f_\uparrow - n^f_\downarrow$, $m^c = \frac{1}{2}(n^c_\uparrow - n^c_\downarrow)$ and Kondo parameter $\lambda_\sigma = \langle c^\dagger_{\mathbf{k}\sigma} f_{i\alpha\sigma}\rangle$ should be calculated selfconsistently [14, 16].

The diagonalization of the Hamiltonian gives one pure f-band with dispersion given by: $E^{f-band}_{\mathbf{k}\sigma} = E^f_{0\sigma} + A_{\mathbf{k}\sigma}$

and two hybridized $f - c$ bands $E^\pm_{\mathbf{k}\sigma}$ given by :
$E^\pm_{\mathbf{k}\sigma} = \frac{1}{2}[\epsilon_{\mathbf{k}} + A_{\mathbf{k}\sigma} + E^f_{0\sigma} + \Delta_\sigma \pm S_{\mathbf{k}\sigma}]$,

with:
$S_{\mathbf{k}\sigma} = \sqrt{[\epsilon_{\mathbf{k}} - A_{\mathbf{k}\sigma} - E^f_{0\sigma} + \Delta_\sigma]^2 + 8(\Lambda_\sigma)^2}$.

3. RESULTS and CONCLUSIONS.

We present here the main results obtained by the present calculations. The Kondo (T_K) or Curie (T_C) temperatures are defined in the mean field approach as the temperatures at which respectively the λ_σ parameters or the magnetizations vanish. T_K depends essentially on J_K,

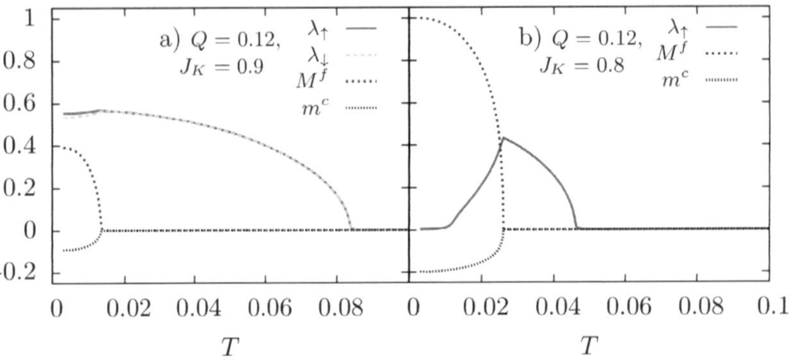

Figure 2. M^f, m^c and λ_σ versus temperature (expressed in units of D) for the case (b), with $Q = 0.12$, $J_H = -0.01$, $n^c = 0.8$, $n^f_{\text{tot}} = 2$, and respectively two values of $J_K = 0.9$ (Figure 2.a) and $J_K = 0.8$ (Figure 2.b).

while T_C depends obviously on both J_K and J_H. We consider a constant density of states for the conduction electron band with a band width $2D$, and in the following all energies and temperatures are expressed in units of D.

Our basic interest is here to study the effect of the finite f-band width W on the magnetic properties. Thus, we have considered the three following cases:
- case (a) : W is taken constant,
- case (b) : W is taken proportional to the exchange Kondo integral, i.e. $W = QJ_K$,
- case (c) : W is given by our present derivation of the SW transformation, i.e. $W = 2PA_{\mathbf{k}\sigma}/\epsilon_{\mathbf{k}}$.

Figure 1 gives both T_K and T_C, as a function of the Kondo exchange integral J_K, for the three cases (a), (b) and (c) defined above, with respectively the following parameters: $W = 0.05$, $Q = 0.12$ and $P = 0.17$. For small J_K values, there is no Kondo effect and T_C increases continuously with J_K, while for large J_K values, T_K increases rapidly with J_K. Two effects have to be pointed out: (1) in some cases T_C decreases for large J_K; (2) there is a "peculiar behavior" of T_K for the two cases (a) and (b): in a certain J_K range, λ is non-zero only between two temperatures T_K and T_1 and vanishes below T_1.

Then, we present on Figure 2 two plots of M^f, m^c and λ_σ versus temperature for $n^c = 0.8$, $J_H = -0.01$ and for the case (b) with $Q = 0.12$, but for two different values of J_K equal to 0.9 (Figure 2a) and 0.8 (Figure 2b). Figure 2a shows clearly that both the Kondo effect and ferromagnetism exist together for $J_K = 0.9$ down to the lowest temperatures, with T_K larger than T_C. Figure 2b corresponding to the "peculiar behavior" shows that for $J_K = 0.8$, with decreasing temperature, Kondo effect occurs at T_K, then there is a coexistence of Kondo and ferromagnetism, and finally the Kondo effect disappears at T_1 to yield only a strong ferromagnetism at very low temperatures. This behavior can be interpreted in the following way: Kondo effect for a spin $S = 1$ cannot be complete when there is only one screening channel [17]. Thus if exchange is large enough, the ordering of the remaining f-moments occurs in the Kondo phase. However, at lower temperature, when these ordered magnetic moments are large, they act as an internal magnetic field and Kondo effect is destroyed by this effective magnetic field. One should notice that there is at present no experimental evidence in favor or in contrast of such an effect in actinide compounds at very low temperatures.

Finally, the decrease of the Curie temperature observed in the cases (b) and (c) of Figure 1 for large J_K values above the intersection point with T_K can probably be considered as resulting from a possible effect of "delocalization" of the $5f$-electrons. Let us also remark that J_K increases with increasing pressure and that the two figures corresponding to the cases (b) and

Journal of Physics: Conference Series **273** (2011) 012028 doi:10.1088/1742-6596/273/1/012028

(c) can give a good description of the experimentally observed variation of T_C with pressure in UTe compound, which is passing through a maximum and then decreasing with applied pressure [4, 9].

We can conclude that the present calculation improves the previous $S = 1$ UKL model of Ref. [14]. The model described in Ref. [14] started from a localized description of the $5f$-electrons in the $5f^2$ configuration, while here we started from an Anderson Hamiltonian and derived by the Schrieffer-Wolff transformation a new Hamiltonian with a non zero f-band width, in addition to the usual Kondo exchange term. This new Hamiltonian can describe the onset of the "delocalization" of the $5f$-electrons and the new curves giving a maximum of T_C as a function of J_K can account for the pressure dependence of T_C in UTe. Thus, the present work yields an improvement with respect to the UKL model of Ref. [14] for the description of the $5f$-electrons in some uranium compounds.

References

[1] Hewson A C, *The Kondo problem to Heavy Fermions*, Cambridge University Press (1993).
[2] Coqblin B, AIP Conference Procedings, volume **846**, pp. 3-93 (2006).
[3] Coqblin B, Iglesias-Sicardi J R and Jullien R 1978 *Contemporary Physics* **19** 327
[4] Sheng Q G and Cooper B R 1996 *J. Magn. Magn. Mater.* **164** 335
[5] Zwicknagl G, Yaresko A N and Fulde P 2003 *Phys. Rev.* B **68** 052508
[6] Moore K T and van der Laan G 2009 *Rev. Mod. Phys.* **81** 235
[7] Santini P, Carretta S, Amoretti G, Caciuffo R, Magnani N and Lander G H 2009 *Rev. Mod. Phys.* **81** 807
[8] Doniach S 1976 *Proc. of the "Int. Conf. on Valence Instabilities and Related Narrow-band Phenomena"* ed R D Parks (Plenum Press) p 168
[9] Schoenes J, Vogt O, Lohle J, Hulliger F and Mattenberger K 1996 *Phys. Rev.* B **53** 14987
[10] Bukowski Z, Troc R, Stepien-Damm J, Sulkowski C and Tran V H 2005 *J. Alloys and Compounds* **403** 65
[11] Tran V H, Troc R, Bukowski Z, Badurski D and Sulkowski C 2005 *Phys. Rev.* B **71** 094428
[12] Colineau E, Wastin F, Sanchez J P and Rebizant J 2008 *J. Phys.: Cond. Matter* **20** 075207
[13] Tran V H, Griveau J C, Eloirdi R, Miiller W and Colineau E March 2010, *presented at the 40 emes Journees des Actinides*, Geneva, Switzerland
[14] Perkins N B, Núñez-Regueiro M D, Coqblin B and Iglesias J R 2007 *Phys. Rev.* B **76** 125101.
[15] Thomas C, Simões A S R, Lacroix C, Iglesias J R and Coqblin B 2009 *Physica* B **404** 3008.
[16] Thomas C, Simões A S R, Iglesias J R, Lacroix C, Perkins N B and Coqblin B, *to be published*
[17] Nozieres P and Blandin A 1980 *J. de Physique* **41** 193
[18] Schrieffer J R and Wolff P A 1966 *Phys. Rev.* **149** 491
[19] Iglesias J R, Lacroix C and Coqblin B 1997 *Phys. Rev.* B **56** 11820

Journal of Physics: Conference Series **273** (2011) 012029 doi:10.1088/1742-6596/273/1/012029

Band renormalization effects in correlated f-electron systems

T. Durakiewicz[1], J.J. Joyce[1], Y. Li[1], P. S. Riseborough[2], P.M. Oppeneer[3], E.D. Bauer[1], K.S. Graham[1]

[1]Los Alamos National Laboratory, PO Box 1663, Los Alamos, USA; [2]Physics Department, Temple University, Philadelphia; [3]Department of Physics and Astronomy, Uppsala University, SE-75120 Uppsala, Sweden.

E-mail: tomasz@lanl.gov

Abstract. Two band renormalization effects, novel in the context of f-electron systems, are presented here. First, the interband scattering mechanism leads to the point-like Fermi surface renormalization along k_x/k_y directions in the Brillouin zone. Second, in the normal direction the renormalization leads to reduction in the number of Fermi sheets and in creation of additional periodicity in band dispersion. It is shown that such renormalization effects can lead to significant differences between calculated and measured electronic structures, and need to be considered in f-electron systems.

1. Introduction

Band renormalization effects in strongly correlated systems lead to significant differences between measured and calculated electronic structures. Numerous examples from d-electron arena, including high temperature superconductors, show differences in dispersion, effective mass, and energy between calculated and measured band structures. This discrepancy is brought about by many-body interactions present in nature, but not reproduced by theory. The motivation for this work was to investigate the possibility of existence and the nature of the f-band renormalization effects similar to those found in d-electron systems. When such effects are found, one has to consider their influence on binding energy and dispersion of bands, and even on the very number of Fermi sheets; all facts of significance in both the experimental practice and theoretical approaches to f-electron systems.

2. Results

Our system of choice is USb_2, investigated by Angle Resolved Photoemission (ARPES) method in the Synchrotron Radiation Center. The Brillouin zone of USb_2 is shown in Fig. 1, and the Fermi surface comprises several cylindrical surfaces along z direction. The two band renormalization effects discussed here are seen along k_x/k_y and k_z direction in the Brillouin zone, respectively.

First, the existence of a kink structure was found in the in-plane ARPES data [1-4] collected at single excitation energy and variable emission angle. The kink structure in the dispersion of a hole-like band was observed at around 21meV below the Fermi level, and is a consequence of band renormalization due to interband scattering. A single hole-like parabolic band is predicted by LSDA calculation to touch the Fermi level at Gamma point of high symmetry. This band, due to interband scattering process involving a boson, is renormalized below the Fermi surface in a way that induces flattening of the band and shift in the Fermi energy. Two combined effects result in a formation of a

Published under licence by IOP Publishing Ltd

kink structure at energy scale of 21meV and a gap of 17meV, while the Luttinger's condition of S1 = S2 is maintained, where S1 and S2 represent the Fermi volume before and after renormalization.

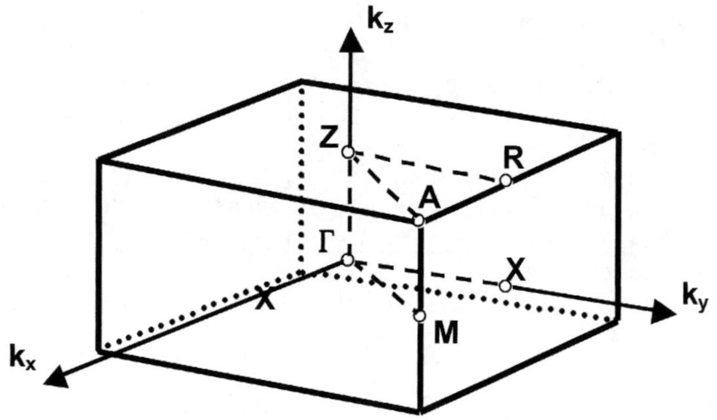

Fig. 1. Brillouin zone of USb_2. The two band renormalization effects discussed here are measured in plane (Gamma – X) and in normal direction (Gamma-Z), respectively.

This observation provided the first evidence for a kink structure in band dispersion in the in f-electron system and allowed us to transplant the self-energy approach to band structure developed for d-electron systems into the world of f-electrons. Details of the model are given elsewhere [1-4]. Since the point-like surface was now renormalized below the Fermi level, it was interesting to see if this effect coincides with a reduction in the very number of Fermi sheets.

Evidence in support of such a possibility came from normal emission experiments, where the excitation energy is adjusted, and emission angle is always normal [5]. Dispersion in the perpendicular momentum component, or in the Gamma-Z direction, can be measured in such an experiment.

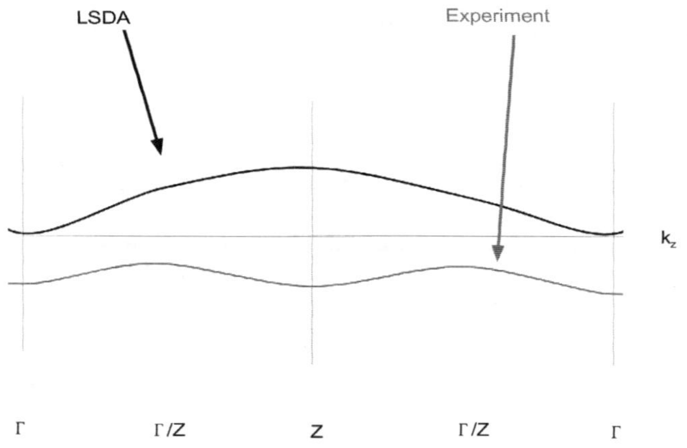

Fig. 2. Renormalization of the near-Fermi-level band in USb_2, as seen in normal emission ARPES experiment. Calculated LSDA band is shown in black, while measured band dispersion is shown in red.

The Fermi surface of uranium diantimonide consists of several uniaxial cylindrical sheets. We have found that the bare-band LSDA calculation over-counts the number of sheets by one, while one of the calculated surfaces shrinks below the Fermi level, as shown in Fig. 2. The bare LSDA bands can be renormalized numerically using a low-order self-energy expansion in three-band inter-band scattering model, and very good fit between the renormalized calculation and measured dispersion is obtained. We conclude that inter-band scattering in USb_2 influences the fermiology of this system in terms of changing the shape and number of Fermi sheets.

All experiments are performed in AF state, at temperatures around 10K. In addition to the above discussed renormalization, the effect of doubling the periodicity of the already AF-folded band is seen in AF state. This doubling on top of the AF folding comes into play solely due to interband scattering, without any additional magnetic ordering.

We propose that similar band renormalization effects may significantly influence the band structure of other f-electron systems, and need to be considered.

Acknowledgements
Work supported by the U.S. Department of Energy, Office of Basic Energy Sciences, Division of Materials Sciences and Engineering, and the LANL LDRD program. SRC supported by NSF.

References
[1] T. DURAKIEWICZ et al., Europhysics Letters 84, 37003 (2008).
[2] T. DURAKIEWICZ et al., Acta Physica Polonica A, 117 264 (2010).
[3] X. YANG et al., J. Phys:Conf. Ser. 200 012164, 1 (2010).
[4] X. YANG et al., Phil. Mag. 89 1893 (2009).
[5] work prepared for publication

Journal of Physics: Conference Series **273** (2011) 012030 doi:10.1088/1742-6596/273/1/012030

Low temperature phase of YbPd investigated by Raman scattering

T. Hasegawa[1], **N. Ogita**[1], **M. Sugishima**[2], **A. Mitsuda**[2], **H. Wada**[2] **and M. Udagawa**[1,3]

[1] Graduate School of Integrated Arts & Sciences, Hiroshima University, Higashi Hiroshima 739-8521, Japan.

[2] Department of Physics, Kyushu University, Fukuoka 812-8581, Japan.

[3] Institute for Advanced Material Research, Hiroshima University, Higashi-Hiroshima 739-8530, Japan.

E-mail: hase@hiroshima-u.ac.jp

Abstract. The possibility of the structural change in YbPd has been investigated by Raman scattering in the temperature range from 4 K to 300 K. One peak was clearly observed around 91 cm^{-1} below 132 K. This is a clear evidence of a structural change at the transition temperature 132 K from the CsCl structure. The polarization dependence of the 91 cm^{-1} peak shows that the cubic and tetragonal symmetries are ruled out for the crystal structure at 4 K.

1. Introduction

YbPd is an intermediate valence compound and four transitions at 0.5, 1.9, 105, and 125 K have been reported from the measurements of specific heat, thermal expansion, resistivity, and magnetic susceptibility [1]. The transition at 1.9 K is a magnetic order [2]. The magnetic ordered state has attracted attention, since it is realized accompanying valence fluctuation of magnetic ions. The valence of Yb is reported as 2.8+ by X-ray absorption spectrum of Yb L_{III} edge [1] and as 2.75+ by photoemission spectrum, eliminating surface contribution [3]. At low temperatures, neutron scattering [4] and Mössbauer [2] measurements have shown that a half of Yb are Yb^{3+} and have magnetic moment.

Since the Yb state in the intermediate valence compound is affected by a change of its surroundings, the transitions at 105 K and 125 K should be elucidated. The thermal expansion coefficient shows volume expansion around 105 K and 125 K with decreasing temperature. Reflecting the expansion, the transition temperature decreases under pressures, and the transition at 105 K vanishes at 1.5 GPa [5]. Although the expansion suggests decrease of Yb valence, the valence did not show any change at these transitions [1]. Therefore, it is reasonable to consider that the anomalous thermal expansion is originated by successive structural transitions from the room temperature CsCl structure. However, any structural distortion has not been detected by neutron scattering [4] and Mössbauer measurement of ^{170}Yb [2] which has a quadrupole moment in its excited nuclear state. These results suggest that structural distortion is, if any, quite small despite relatively high transition temperature. Such a small distortion would indicate that the transitions are not caused by structural instability.

Published under licence by IOP Publishing Ltd

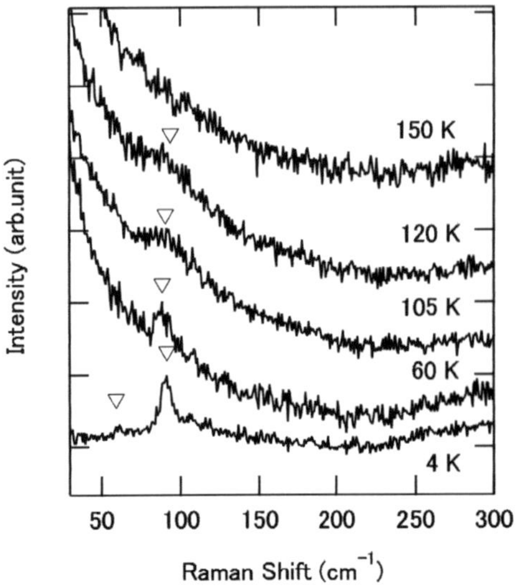

Figure 1. Temperature dependence of (x, x) Raman spectra.

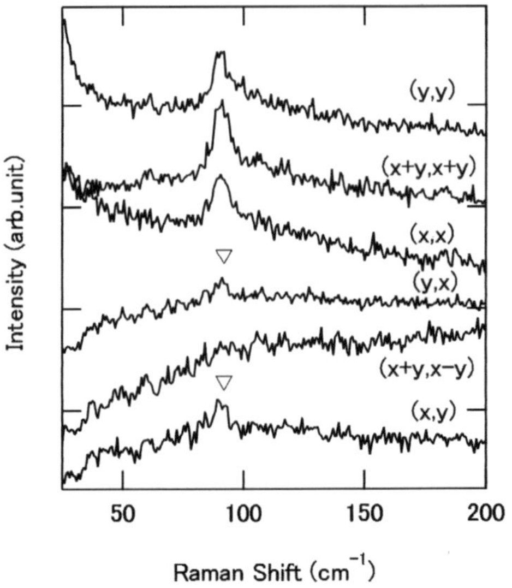

Figure 2. Polarization dependence of Raman spectra on (100) surface at 4 K.

The transitions at 105 K and 125 K are still unclear. It is important whether or not the structure changes at the transitions, even for the magnetic ordered state below 1.9 K. Raman scattering measurement is one of the most sensitive tools to investigate changes of structure and symmetry. This paper reports the Raman scattering results of YbPd at low temperatures.

2. Experimental

The single crystal of YbPd was grown by a self-flux method under Ar atmosphere. Detailed method has been reported elsewhere. The crystal has a cubic shape with (100) surfaces. The transition temperatures at 105 K and 125 K became 110 K and 132 K, respectively, in this crystal. The reason for the increase of the transition temperatures is unknown. But the single crystal exhibits sharper transitions than the previous polycrystal sample does. The energy dispersive X-ray spectroscopy confirms that composition of the present single crystal is more stoichiometric. Thus, the quality of the single crystal is good.

Raman scattering spectra were measured using a triple-monochromator (JASCO NR-1800) with a liquid-N_2 cooled CCD detector (Princeton Instruments Inc. LN/CCD-1100PB). Excitation light was 514.5nm line of an Ar^+ laser and its power was 10 mW before a sample surface. For the measurements at low temperatures, a cryostat cooled by a GM cryocooler (SHI SRDK-205) was used.

We use the conventional description of Raman scattering geometry, such as (x, y). (x, y) means that polarization of the excitation light along x is irradiated on the sample, and that polarization of scattered light along y is measured. The propagation directions of light, z or $\bar{z}$, are omitted, because the back-scattering geometry was used for all measurements.

3. Result and discussion

Evaluation of structural distortion using Raman scattering method is very simple in this case. YbPd has the CsCl structure at room temperature. Since the CsCl structure has no Raman active mode, if we observe any peak of phonon at low temperatures, we can conclude the

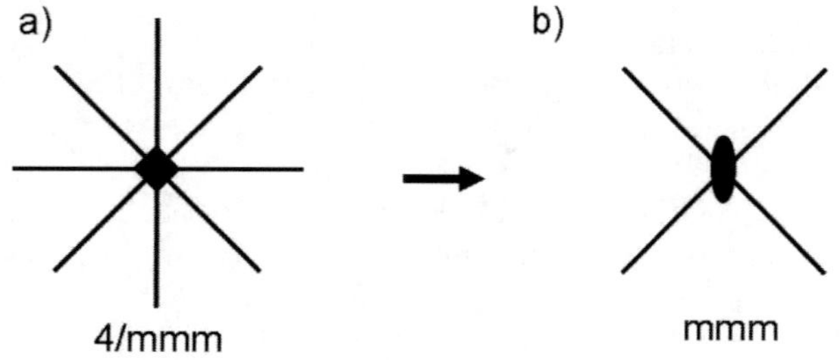

Figure 3. Symmetries on the (100) surface: a) at room temperature, b) at 4 K.

structural distortion. Figure 1 shows temperature dependence of Raman spectra for (x, x) geometry. As expected, there is no peak above 132 K, while at 4 K, two peaks are clearly observed at 60 cm^{-1} and 91 cm^{-1}. In this energy region, possible elementary excitations are phonon or crystal field excitations of Yb^{3+}. The crystal field excitations have been reported at 38 cm^{-1} and 99 cm^{-1} by neutron inelastic scattering measurement[4]. These energies are different from ones observed by present study. Moreover, the polarization dependence of the 91 cm^{-1} peak does not consist with polarization dependence expected for the crystal field excitations. Therefore, the observed peaks are phonon excitations. This result is a clear evidence of structural distortion from the CsCl structure. Since the peak at 90 cm^{-1} was observed at least up to 120 K as shown in Fig. 1, the structural change occurs at the transition of 132 K. The peak at 60 cm^{-1} appears at the lowest temperature. This appearance suggests the additional distortion occurs at a lower temperature.

Figure 2 shows polarization dependence of Raman spectra at 4 K. We focus on the 90 cm^{-1} peak. This mode shows strong intensity whenever the polarization directions of the incident and scattered lights are parallel. This is the characteristic of the fully-symmetric representation Γ_1 among the irreducible representations. On the other hand, this mode was observed weakly in (x, y) spectra, while it was not observed in $(x + y, x - y)$ spectra. From these results, we can write the Raman tensor R in the two dimensional space on the (100) surface as,

$$R = \begin{pmatrix} a & c \\ c & b \end{pmatrix}. \tag{1}$$

The intensity of the peak is described by a^2 for (x, x), b^2 for (y, y), c^2 for (x, y), and $(a - b)^2/4$ for $(x + y, x - y)$. From the experimental results, we obtain $a = b$ and $c \neq 0$. From the form of the tensor R, the cubic symmetry is ruled out, since the tensor of Γ_1 is proportional to the unit matrix in all cubic symmetries, including $m\bar{3}$. In addition, the four-fold axis normal to the (100) surface is lost, since if the four-fold axis exists, c is exactly zero for Γ_1. The fact that the peak was not observed in the $(x + y, x - y)$ spectra means that the tensor R becomes diagonal when principal axes are taken along the [110] directions. Therefore, principal axes on the surface become the [110] directions from the [100] ones. Since the four-fold axes do not lie along the [110] directions, all four-fold axes have vanished at 4 K. Consequently, the crystal symmetry is neither cubic nor tetragonal at 4 K. The possible symmetries are orthorhombic, rhombohedral, or their subgroups. Figure 3 summarizes the possible symmetries on the (100) surface at 4 K. Since the three-fold axes along [111] don't lie on the (100) surface, we can not discuss their existence from the present study.

4. Conclusion

We have measured Raman spectra of YbPd single crystal in the temperature range from 4 K to 300 K. At 4 K, two peaks were clearly observed at 60 cm^{-1} and 91 cm^{-1}, while no peak was found at 300 K. These peaks are clear evidence of structural distortion from the CsCl structure in YbPd. Since the peak at 90 cm^{-1} has been observed up to 120 K, the structural distortion occurs at the 132 K transition. From the detailed measurements of polarization dependence on the (100) surface at 4 K, it is found that all four-fold axes vanish at low temperatures. Therefore, the crystal symmetry at 4 K is neither cubic nor tetragonal.

Acknowledgments

This work is supported by Grant-in-Aids for Scientific Research on Innovative Areas "Heavy Electrons" (No. 20102005) of the Ministry of Education, Culture, Sports, Science, and Technology, Japan, and for Young Scientists (B) (No. 21740527) of Japan Society for the Promotion of Science, and also supported by N-BARD and IAMR of Hiroshima University.

References

[1] Pott R, Boksch W, Leson G, Politt B, Schmidt H, Freimuth A, Keulerz K, Langen J, Neumann G, Oster F, Röhler J, Walter U, Weidner P and Wohlleben D 1985 *Phys. Rev. Lett.* **54** 481

[2] Bonville P, Hammann J, Hodges J A, Imbert P and Jéhanno G J 1986 *Phys. Rev. Lett.* **57** 2733

[3] Domke M, Laubschat C, Sampathkumaran E V, Prietsch M, Mandel T, Kaindl G and Middelmann H U 1985 *Phys. Rev.* B **32** 8002

[4] Walter U and Wohlleben D 1987 *Phys. Rev.* B **35** 3576

[5] Mitsuda A, Yamada K, Sugishima M and Wada H 2009 *Physica* B **404** 3002

Inelastic contribution of the resistivity in the hidden order in URu_2Si_2

E Hassinger, T D Matsuda[1], G Knebel, V Taufour, D Aoki and J Flouquet

INAC-SPSMS, CEA Grenoble, 17 rue des Martyrs, 38054 Grenoble, France

E-mail: georg.knebel@cea.fr

Abstract. In the hidden order of URu_2Si_2 the resistivity at very low temperature shows no T^2 behavior above the transition to superconductivity. However, when entering the antiferromagnetic phase, the Fermi liquid behavior is recovered. We discuss the change of the inelastic term when entering the AF phase with pressure considering the temperature dependence of the Grüneisen parameter at ambient pressure and the influence of superconductivity by an extrapolation of high field data.

1. Introduction

URu_2Si_2 has attracted the interest of researchers for a long time, because it has an ordered phase below $T_0 = 17.8\,\mathrm{K}$ with huge anomalies in macroscopic quantities at the transition but unknown order parameter, which coexists with unconventional superconductivity below $T_{SC} = 1.5\,\mathrm{K}$ [1, 2]. Its resistivity is anisotropic and by a factor of about two larger in the basal plane of the tetragonal crystal structure than along the crystallographic c axis. At the transition to the so-called hidden order (HO) at T_0, the resistivity shows a nesting like behavior with a jump due to partial gapping of the Fermi surface and a loss of carriers by a factor of 3-10 [2, 3] which has also been determined by band structure calculations [4]. By angular resolved photoemission spectroscopy it was directly shown that when entering the HO, some bands drop below the Fermi energy [5]. There are signs of different kinds of gaps (spin gap and charge gap) in many properties observable by an exponential behavior of macroscopic quantities [1, 2], gap in optical conductivity [6] and sharply gapped Ising-like longitudinal magnetic excitations in neutron scattering [7, 8]. The small ordered magnetic moment below T_0 was proven to be extrinsic [9]. Recent theories tend to describe the hidden order by a local multipole order parameter [10, 11], but have to take into account the change of hybridization of the conduction electrons [12]. Certainly, both effects are mixed and play a role in the formation of the HO phase. Under pressure, the ground state changes from HO to an antiferromagnetic phase (AF) at $P_x = 0.5\,\mathrm{GPa}$ and superconductivity is suppressed at the same time [13, 14, 15]. However, the critical pressure P_x seems to depend strongly on the sample and pressure homogeneity conditions [16, 17]. The resistivity anomaly does not change at the transition to either phase below T_0 or T_N, indicating that the rearrangement of the Fermi surface is similar in both phases. Below T_0

[1] Present address: Advanced Science Research Center, Japan Atomic Energy Agency, Ibaraki 319-1195, Japan

Published under licence by IOP Publishing Ltd

or T_N, the resistivity can be described quite well by the following formular [18]

$$\rho(T) = \rho_0 + AT^2 + B\frac{T}{\Delta}(1 + 2\frac{2T}{\Delta})\exp(-\frac{\Delta}{T})$$

which describes the resistivity of a Fermi liquid interacting with gapped spin waves in an antiferromagnet [19]. Obviously, this is not the right model in the non-magnetic HO state as described in Ref. [15]. Nevertheless, it allows to determine the value of the anisotropic gap Δ, which increases by 20 % when entering the AF phase [20]. Already McElfresh *et al.* [21] have remarked, that at low temperature one never finds a T^2 behavior, even for a very small temperature range just above the superconducting transition. Recently, an anisotropic inelastic term of the resistivity in the HO state was found [22] which could indicate an anisotropic scattering mechanism due to the order parameter in this phase. In Ref.[23] it has been shown that a quadrupolar charge order and its excitation can cause this anisotropy. However, a detailed study on crystals with different quality should clarify the influence of sample quality [24]. If this anisotropy is a signature of the special order parameter of the HO state, the temperature behavior of the resistivity should change under pressure when entering the antiferromagnetic state. Here we present a high pressure resistivity experiments to address this question.

2. Experimental Details

The single crystal was grown by Czochralsky pulling method in a tetra arc furnace. We have carried out resistivity measurements with current along the a-axis and magnetic field along $c$axis under pressure. Due to the non-quadratic temperature dependence of the resistivity at low temperature just above the superconducting transition [21, 15], the extrapolation to $T = 0$ is ambiguous and can lead to a nearly zero or even negative residual resistivity in very good crystals. Therefore, we defined the residual resistivity ratio RRR as $\rho(300K)/\rho(2K)$ above the onset of the superconducting transition. In this study, the crystal had RRR = 160. We have carried out standard four point resistivity measurements with a current of $I \approx 10\,\mu$A through the sample. The signal was amplified by a low temperature transformer by factor of 1000 keeping the noise level very low. The measurement frequency of 71 Hz was chosen because of its bandpass. The signal was further amplified by a room temperature pre-amplifier and detected with a lock-in amplifier. Low temperatures were created with a dilution refrigerator with base temperature of about 20 mK. Pressure was applied with a copper beryllium piston cylinder pressure cell. For the determination of the upper critical field H_{c2}, field sweeps at constant temperature were carried out at low temperatures. H_{c2} is defined where the resistivity starts to rise from zero.

3. Results and Discussion

The temperature dependence of the resistivity for the measured pressures is shown in Fig. 1 as a function of T (top) and T^2 (bottom). One clearly sees, that the resistivity does not show a T^2 dependence over an extended temperature range for the low pressures whereas for $P = 1.55$ GPa a nice T^2 up to 3.5 K is observed. Only for the three lowest pressures a complete superconducting transition with zero resistivity appears whereas only an onset of the superconducting transition if observed for the highest pressures.

In Fig. 2 we plotted the pressure dependence of some parameters one can extract from these data. One striking feature in almost all of them is the discontinuous pressure dependence between 0.85 GPa and 1.32 GPa. It evidently shows, that the transition between the HO and the pressure induced AF state appears between these two pressures, i.e. 0.85 GPa$< P_x <$ 1.32 GPa. Magnetoresistance measurements on the same sample also show a clear change between the lower three and the higher two pressures [25]. In Fig. 2a, we show the A coefficient of a forced $\rho = \rho_0 + AT^2$ fit at $H = 0$ between 1.8 K and 2.5 K. It decreases strongly with pressure. However, in the present measurements, the pressure steps were too large to detect an discontinuity of A

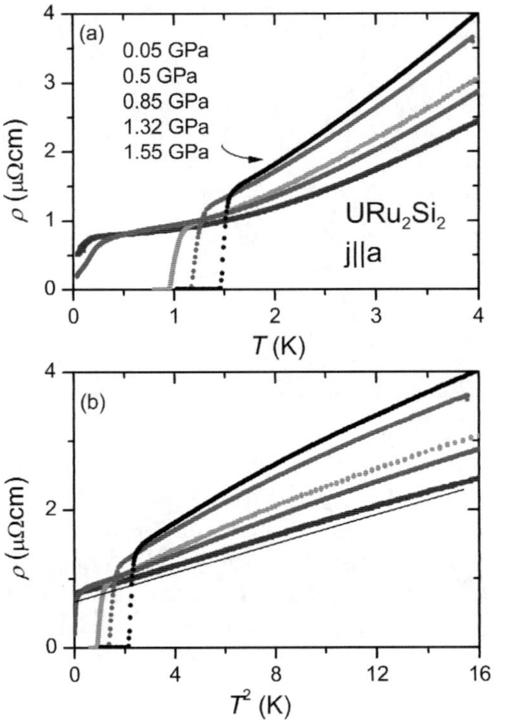

Figure 1. (Color online) (a) Resistivity as a function of temperature in zero magnetic field for several pressures. (b) The same data as a function of T^2. The straight line is a guide to the eye.

at the critical pressure P_x [15]. As $A \propto (m^\star)^2$ we have plotted the normalized square of the effective band mass of the β branch determined in recent Shubnikov-de Haas measurements [25] (blue stars) as well as the normalized square of the initial slope of the upper critical field H_{c2} near T_{SC} (small triangles), which are both proportional to the effective mass as a function of pressure. The pressure dependence of all these parameters is consistent.

As expected from a bare look at the data, this T^2 fit does not reproduce the data well at low pressures. Therefore, we fitted the data also with a $\rho = \rho_0 + AT^x$ fit for temperatures between 1.8 K and 3.5 K (Fig. 2b). The exponent is around 1.5 in the HO state and a clear T^2 behavior is found only deep inside the AF phase. The superconducting parameters T_{SC} and H_{c2} (Fig. 2c and d) decrease linearly with pressure and then disappear abruptly above P_x. The critical pressure where superconductivity would disappear extrapolated from the low pressure data is at around 2 GPa. In earlier pressure studies of the resistivity on samples with lower quality, the decrease $T_c(P)$ was linear [15, 26, 21] up to the pressure where the superconducting phase disappeared. The superconducting transition was found up to quite high pressures (far above 1 GPa for the above mentioned references). In contrast, in magnetic susceptibility and especially specific heat measurements, which are sensitive to bulk transitions, the superconducting state disappeared at the critical pressure [14, 15]. The explanation for this difference is that in the same way as droplets of the AF phase exist within the HO state ($P < P_x$) near lattice imperfections (and are responsible for the small moment observed in neutron scattering) droplets of the superconducting HO state exist under pressure ($P > P_x$) in the non-superconducting AF phase. In resistivity, a shortcut through the sample by inhomogeneous superconducting droplets can give the zero resistivity signal.

The non-Fermi liquid behavior in resistivity is observed precisely in the HO phase and disappears in the AF phase. The same happens with superconductivity. It is tempting to claim, that the component of linear temperature behavior is a signature of the HO state or even stronger, that superconductivity is linked to a linear resistivity term [27] as was claimed in high

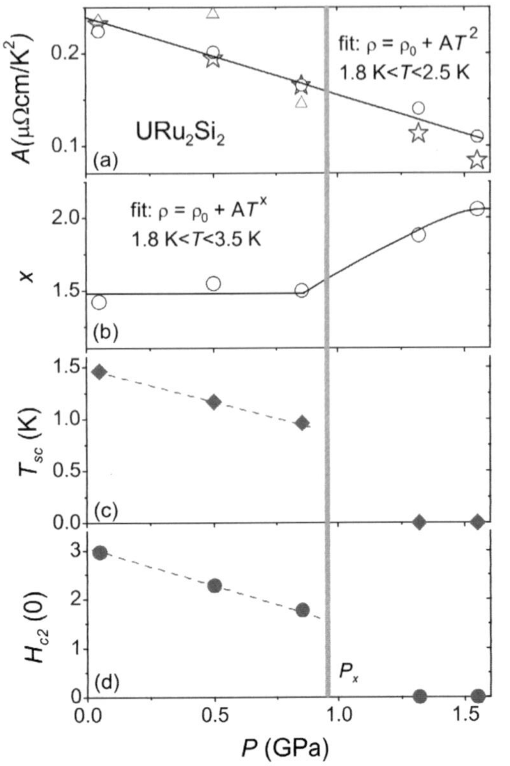

Figure 2. Pressure dependence of several parameters determined from resistivity measurements at zero field (a-c) and with $H \parallel c$. In (a) we further show the pressure dependence of the normalized square of the effective mass of the heavy β branch (blue stars) and the normalized square of the initial slpope of the upper critical field (triangles). The critical pressure P_x is indicated by the gray line. Lines are to guide the eye.

temperature and organic superconductors. However, in URu_2Si_2 one should be careful with claims like this. A recent study of the Grüneisen parameter gives a saturation only at very low temperature [28]. Accordingly, the Fermi liquid behavior in resistivity is maybe hidden by the onset of superconductivity. Zhu *et al.* have studied the temperature behavior of the resistivity in field with a longitudinal configuration ($H \parallel I$) to avoid transverse magnetoresistance. If the resistivity is expressed as $\rho_0 + AT^x$, they report $x < 2$ for the zero field measurements in both crystal directions. With field, a Fermi liquid behavior is recovered above $H_{c2} = 3\,\mathrm{T}$ for a current and field along the c axis whereas a linear temperature dependence for current and field along a remains up to the maximum field of $H = 12\,\mathrm{T}$. However this field is not high enough to suppress superconductivity along the a axis completely. To overcome this problem we have performed magnetoresistance measurements as function of field for different constant temperatures [24]. Extrapolating the magnetoresistance from the normal state at high fields above H_{c2} to zero field at constant temperatures $T < T_{SC}$, the resistivity in the normal state in absence of superconductivity has been extrapolated. From this extrapolations a T^2 behavior [24] with a coefficient A in good agreement with our result at zero pressure has been observed.

Knowing the Grüneisen parameter $\Gamma = 40$ [28] and the compressibility $\kappa \approx 0.5 \cdot 10^{-6}\,\mathrm{bar}^{-1}$ [29] allows a prediction for the relative change of A with pressure: As $A \propto (m^\star)^2$ with the characteristic temperature $T^\star \propto 1/m^\star$, then $dT^\star/T^\star = dm^\star/m^\star = 1/2dA/A$. With the definition of the Grüneisen parameter

$$\Gamma = -d\ln T^\star/d\ln V = \kappa^{-1} d\ln T^\star/dp$$

we find for a pressure difference of $\Delta p = 0.5\,\mathrm{GPa}$ a relative change of $\Delta A/A \approx 0.2$ in agreement with our data.

4. Conclusion

To summarize, we show the change of the temperature dependence of the electrical resistivity in URu_2Si_2 with pressure. As expected from the Grüneisen parameter, Fermi liquid behavior appears at very low temperatures $T < T_{SC}$ only. Superconductivity has to be suppressed to be able to observe a T^2 temperature dependence. However, we cannot exclude peculiar scattering in the HO state, but it is difficult to find a decisive answer.

Acknowledgments

We acknowledge funding from the French ANR within the projects CORMAT, DELICE, and SINUS.

References

[1] Palstra T T M, Menovsky A A, Vandenberg J, Dirkmaat A J, Kes P H, Nieuwenhuys G J and Mydosh J A 1985 *Phys. Rev. Lett.* **55** 2727–2730
[2] Maple M B, Chen J W, Dalichaouch Y, Kohara T, Rossel C, Torikachvili M S, McElfresh M W and Thompson J D 1986 *Phys. Rev. Lett.* **56** 185–188
[3] Dawson A L, Datars W R, Garrett J D and Razavi F S 1989 *J. Phys.: Condens. Matter* **1** 6817–6828
[4] Harima H *private communication*
[5] Santander-Syro A F, Klein M, Boariu F L, Nuber A, Lejay P and Reinert F 2009 *Nature Physics* **5** 637–641
[6] Bonn D A, Garrett J D and Timusk T 1988 *Phys. Rev. Lett.* **61** 1305–1308
[7] Broholm C, Kjems J K, Buyers W J L, Matthews P, Palstra T T M, Menovsky A A and Mydosh J A 1987 *Phys. Rev. Lett.* **58** 1467–1470
[8] Broholm C, Lin H, Matthews P T, Mason T E, Buyers W J L, Collins M F, Menovsky A A, Mydosh J A and Kjems J K 1991 *Phys. Rev. B* **43** 12809–12822
[9] Matsuda K, Kohori Y, Kohara T, Amitsuka H, Kuwahara K and Matsumoto T 2003 *J. Phys.: Condens. Matter* **15** 2363–2373
[10] Haule K and Kotliar G 2009 *Nature Physics* **5** 796–799
[11] Harima H, Miyake K and Flouquet J 2010 *J. Phys. Soc. Jpn.* **79** 033705
[12] Schmidt A R, Hamidian M H, Wahl P, Meier F, Balatsky A V, Garrett J D, Williams T J, Luke G M and Davis J C 2010 *Nature* **465** 570–576
[13] Amitsuka H, Sato M, Metoki N, Yokoyama M, Kuwahara K, Sakakibara T, Morimoto H, Kawarazaki S, Miyako Y and Mydosh J A 1999 *Phys. Rev. Lett.* **83** 5114–5117
[14] Amitsuka H, Matsuda K, Kawasaki I, Tenya K and Yokoyama M 2007 *J. Magn. Magn. Mater.* **310** 214–220
[15] Hassinger E, Knebel G, Izawa K, Lejay P, Salce B and Flouquet J 2008 *Phys. Rev. B* **77** 115117
[16] Motoyama G, Nishioka T and Sato N K 2003 *Phys. Rev. Lett.* **90** 166402
[17] Amitsuka H, Matsuda K, Yokoyama M, Kawasaki I, Takayama S, Ishihara Y, Tenya K, Tatelwa N, Kobayashi T C and Yoshizawa H 2008 *Physica B: Condensed Matter* **403** 925–927
[18] Palstra T T M, Menovski A A and Mydosh J A 1986 *Phys. Rev. B* **33** 6527–6530
[19] Hessel Andersen N 1980 *Crystalline Field and Structural Effects in f-Electron Systems* (Plenum, New York) p 373
[20] Aoki D, Bourdarot F, Hassinger E, Knebel G, Miyake A, Raymond S, Taufour V and Flouquet J 2009 *J. Phys. Soc. Jpn.* **78** 053701
[21] McElfresh M W, Thompson J D, Willis J O, Maple M B, Kohara T and Torikachvili M S 1987 *Physical Review B* **35** 43–47
[22] Zhu Z W, Hassinger E, Xu Z A, Aoki D, Flouquet J and Behnia K 2009 *Physical Review B* **80** 172501
[23] Miyake K and Flouquet J 2010 *J. Phys. Soc. Jpn.* **79** 035001
[24] Matsuda T D 2010 *et al. to be published*
[25] Hassinger E 2010 *to be published*
[26] Jeffries J R, Butch N P, Yukich B T and Maple M B 2008 *J. Phys.: Condens. Matter* **20** 095225
[27] Doiron-Leyraud N, Auban-Senzier P, Ren de Cotret S, Bourbonnais C, Jrome D, Bechgaard K and Taillefer L 2009 *Phys. Rev. B* **80** 214531–
[28] Hardy F 2010 *to be published*
[29] Jeffries J R 2010 *arXiv:1002.2245v1*

Journal of Physics: Conference Series **273** (2011) 012032 doi:10.1088/1742-6596/273/1/012032

Pressure-induced bicritical point in UPd_2Si_2

H. Hidaka[1], A. Tanaka[1], S. Takahashi[1], T. Yanagisawa[2], and H. Amitsuka[1]

[1]Department of Physics, Hokkaido University, Sapporo 060-0810, Japan,
[2]Creative Research Institution, Hokkaido University, Sapporo 001-0021, Japan

E-mail: hidaka@phys.sci.hokudai.ac.jp

Abstract. We have studied the effects of pressure on the incommensurate (IC)-commensurate (C) antiferromagnetic (AFM) phase transition of UPd_2Si_2 by measuring the electrical resistivity up to 4.21 GPa. We observed that the low-temperature C-AFM (type-I) phase expands and prevails over the IC-AFM phase as pressure is increased. The obtained pressure-temperature phase diagram strongly suggests the existence of a novel bicritical point, more specifically, the so-called Lifshitz point at $P_c \sim 3.4$ GPa, where magnetic modulation continuously changes from IC to C.

1. Introduction

UPd_2Si_2 has the $ThCr_2Si_2$-type body-centered tetragonal structure [1]. This compound shows an incommensurate-antiferromagnetic (IC-AFM) ordering with a propagation vector $\boldsymbol{Q} = 0.73\boldsymbol{c}^*$ at 135 K ($= T_{\rm Nh}$), and successively undergoes a simple commensurate (C) type-I AFM ordering at 108 K ($= T_{\rm NI}$) [1, 2]. It is known from the specific heat that the phase transitions are of second order at $T_{\rm Nh}$ and of first order at $T_{\rm NI}$ [2]. In addition, an anti-phase ordering state ($\boldsymbol{Q} = 2/3\boldsymbol{c}^*$) becomes stable in high magnetic field applied along the c-axis [2, 3]. The magnetism of UPd_2Si_2 is phenomenologically well explained in terms of the axial-next-nearest-neighbor-Ising (ANNNI) model assuming localized $5f$ spins, whereas Fermi liquid behavior with a moderately large electric-specific-heat coefficient $\gamma \sim 30$ mJ/molK^2 is also seen at low temperature [2]. Such a duality between the localized and the itinerant features of the $5f$-electronic states in U compounds is one of the interesting issues in the physics of strongly correlated electron systems. Hydrostatic pressure can be a useful tool both for investigating the role of electronic correlations in f-electron compounds and for probing an exotic electronic behavior emerging due to the competition of various interactions. High-pressure studies on UPd_2Si_2 have been carried out so far up to 1.5 GPa [4, 5]. They have revealed that $T_{\rm Nh}$ is insensitive to pressure and $T_{\rm NI}$ increases linearly at a rate of 5.8 K/GPa by applying pressure [4, 5]. These results predict a presence of a pressure-induced multicritical point near 3 GPa. In the present study, we have measured the electrical resistivity of UPd_2Si_2 along the c-axis by extending the pressure range up to 4.21 GPa in order to clarify the pressure variation of the ordered phases and understand the $5f$-electronic states of this compound.

2. Experimental Details

A single crystal of UPd_2Si_2 was grown by the Czochralsky method as described in Ref. 2. Electrical resistivity for current $J \parallel c$-axis was measured by means of a conventional four

Published under licence by IOP Publishing Ltd

probe method in the temperature range 1.8 K – 300 K. Hydrostatic pressure was applied by an indenter cell with Daphne oil 7373 as a pressure-transmitting medium [6]. The pressure value was determined by the superconducting transition temperature of lead. The resistivity at room temperature (ρ_{RT}) at ambient pressure was scaled to the value of the previous work [2] because of the uncertainty in the value of the resistivity estimated for a tiny sample in the pressure cell.

3. Results and Discussion

The electrical resistivity $\rho(T)$ measured for high pressures up to 4.21 GPa is shown in Fig. 1. The temperature dependence in a paramagnetic (PM) state and below 100 K is almost unchanged with pressure, except for a continuous increase of ρ_{RT} and an enhancement of the residual resistivity above 3.6 GPa.

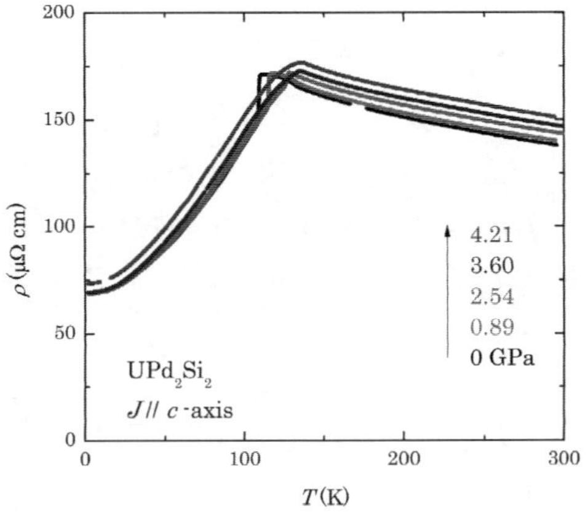

Figure 1. Temperature dependence of the electrical resistivity of UPd_2Si_2 with $J \parallel c$-axis for high pressures up to 4.21 GPa.

Figures 2 (a) and (b) show $\rho(T)$ near the transition temperatures below and above 3.3 GPa, respectively. At ambient pressure, the $\rho(T)$ shows an upturn at T_{Nh} and an abrupt drop with a hysteresis at T_{Nl} indicated lower arrows, which are consistent with the data reported previously [2]. T_{Nh} is almost constant for pressures below $\sim$ 3 GPa. Above this pressure, however, the anomaly becomes indistinct and we cannot determine T_{Nh} from the resistivity. On the other hand, T_{Nl} increases linearly at a rate of +7.0 K/GPa with increasing pressure up to 3.28 GPa, where both the abrupt drop and the hysteresis gradually become indistinct.

Above 3.60 GPa, only one peak can be observed as shown in Fig. 2 (b). Here, we newly define the temperature of the resistivity maximum above 3.60 GPa as T_{N}. With increasing pressure, T_{N} increases monotonously. The T_{N} should be different from both T_{Nh} and T_{Nl}, because $\rho(T)$ just below T_{N} shows a convex-upward shape. No hysteresis is observed near T_{N}, and the $\rho(T)$ curve below T_{N} is quite similar to that below T_{Nl} for $P \leq 3.28$ GPa as shown in Fig. 1. It is thus considered that T_{N} corresponds to the second-order transition temperature from the PM phase to the type-I C-AFM phase.

Figure 3 shows an obtained pressure-temperature (P–T) phase diagram of UPd_2Si_2. The behavior below 1.5 GPa is almost consistent with the previous data [4, 5]. Our experimental results suggest that the type-I AFM phase is stabilized by applying pressure, and predominates over the IC-AFM phase above $\sim$ 3.4 GPa ($= P_{\mathrm{c}}$). Consequently, it is considered that two second-order phase-boundary lines and a first-order one merge at P_{c}, forming a pressure-induced

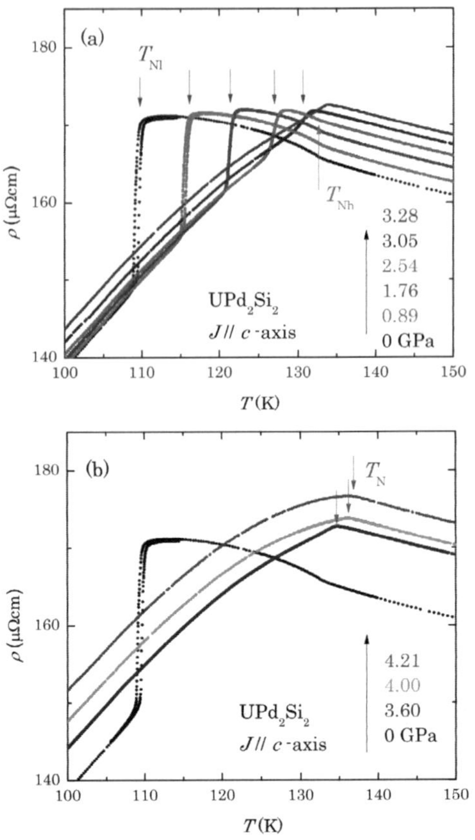

Figure 2. Temperature dependence of the resistivity between 100 K and 150 K of UPd_2Si_2 (a) below 3.3 GPa and (b) above 3.3 GPa. Red, blue and green arrows represent T_{Nh}, T_{Nl} and T_N, respectively.

bicritical point, although it is difficult to assure whether the transitions near the critical point are of first or second order from the present data.

This pressure-induced bicritical point on UPd_2Si_2 might be characterized as a novel Lifshitz point. The Lifshitz point is defined as a bicritical point which separates a second-order phase-boundary line into two parts such that on one part the order parameter is characterized by a fixed wave vector $\boldsymbol{Q} = \boldsymbol{Q_0}$ $(= 1\boldsymbol{c}^*$ in the present case), while on the second part $\boldsymbol{Q}$ varies continuously from $\boldsymbol{Q_0}$ by changing pressure, magnetic field and so on. The physical properties around the Lifshitz point have been attracting much attention due to its novel criticality [7, 8]. A well-known example for a magnetic substance will be a magnetic-field-induced Lifshitz point in MnP [9]; however, there are few other examples to our knowledge. UPd_2Si_2 might be a rare magnetic substance that has a Lifshitz point in zero magnetic field. We may thus expect to be able to study the critical phenomena associated with the Lifshitz point in detail using this material. To prove the presence of the Lifshitz point in UPd_2Si_2, however, a continuous shift of the $\boldsymbol{Q}$ vector from IC to C at T_{Nh} must be confirmed. Further experiments, such as neutron scattering and magnetization under high pressure, will be required. It is also interesting and important to investigate the consistency between the characteristics of the obtained P-T phase diagram and the localized-spin model applied for the ambient-pressure data [2]. To understand

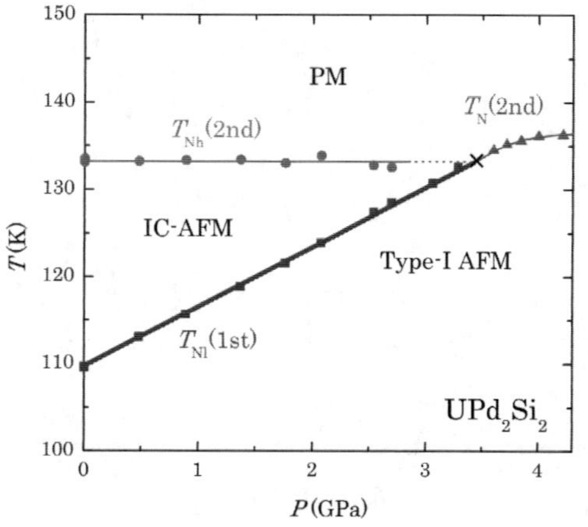

Figure 3. Pressure-temperature phase diagram of UPd_2Si_2 up to 4.2 GPa. The lines are guides to eyes: the bold and fine lines represent the first- and second-order phase transitions, respectively.

the $5f$-electronic states of this compound, the numerical calculations using the ANNNI model and the detailed analyses on the low-temperature Fermi-liquid behavior are now in progress.

4. Summary

We have measured the electrical resistivity of UPd_2Si_2 under high pressure using an indenter cell, and constructed a P–T phase diagram in a pressure range extended up to 4.21 GPa. T_{Nl} increases linearly with increasing pressure, and seems to meet T_{Nh} at $P_c \sim 3.4$ GPa. Above P_c, the system is considered to behave as a simple type-I antiferromagnet. UPd_2Si_2 might be a novel magnetic substance which has a pressure-induced Lifshitz point at zero field. In addition to the interest in the critical phenomena, future studies on the pressure change in the magnetic structure and the characteristic energy scales including that for the low-temperature Fermi-liquid behavior would provide useful information for the basic feature of the $5f$-electron magnetism.

5. Acknowledgments

This work was supported by Grant-in-Aid for Scientific Research B (19340086) and S (20224015) from the Ministry of Education, Culture, Sports, Science and Technology (MEXT), Japan.

References

[1] Shemirani B, Lin H, Collins M F, Stager C V and Garrett J D 1993 *Phys. Rev. B* **47** 8672
[2] Honma T, Amitsuka H, Yasunami S, Tenya K, Sakakibara T, Mitamura H, Goto T, Kido G, Kawarazaki S, Miyako Y, Sugiyama K and Date M 1998 *J. Phys. Soc. Jpn.* **67** 1017
[3] Collins M F, Shemirani B, Stager C V, Garrett J D, Lin H, Buyers W J L and Tun Z 1993 *Phys. Rev. B* **48** 16500
[4] Quirion G, Razavi F S, Plumer M L and Garrett J D 1998 *Phys. Rev. B* **57** 5220
[5] Sekine C, Kihou K, Inaba T, Togashi T, Shirotani I and Honma T 1999 *Muroran−IT. Academic Resources Archive* **49** 105
[6] Kobayashi T C, Hidaka H, Kotegawa H, Fujiwara K and Eremets M I 2007 *Rev. Sci. Instrum.* **78** 023909
[7] Hornreich R M, Luban M and Shtrikman S 1975 *Phys. Rev. Lett.* **35** 1678
[8] Michelson A 1977 *Phys. Rev. B* **16** 577
[9] Becerra C C, Bindilatti V and Oliveira Jr. N F 2000 *Phys. Rev. B* **62** 8965

Magnetic and transport properties of YbT_2Al$_{20}$ (T = Ti, V and Cr)

Ryuji Higashinaka, Akihiro Nakama, Makoto Ando, Makoto Watanabe, Yuji Aoki and Hideyuki Sato

Graduate School of Science, Tokyo Metropolitan University, Hachioji, Tokyo 192-0397, Japan

E-mail: `higashinaka@phys.metro-u.ac.jp`

Abstract. The magnetic and transport properties of single crystalline YbT_2Al$_{20}$ (T = Ti, V and Cr) grown by Al self- ux method were investigated. All compounds show weakly-temperature dependent Pauli paramagnetic behavior in the DC susceptibility and show normal metallic behavior without any noticeable shoulders and peaks in the temperature dependence of resistivity. From these measurements, the valence state of Yb ions is expected to be close to divalent.

The physics of cage-like compounds such as filled skutterudites and clathrate compounds has attracted much attention in the past 10 years because of their wide variety of strongly correlated electronic properties attributed to their crystal structures[1, 2, 3]. In these systems, because the rare earth ion with $4f$-electrons is surrounded by the large number of cage ions, a strong hybridization between the conduction electrons and 4f electrons of the rare earth ion is expected. Also, because this site has a high cubic point symmetry, small crystal electric field (CEF) effects are expected. These characteristics induce interesting properties of strongly correlated electron system such as novel superconductivity, heavy fermion behavior, multipolar ordering and so on.

Among the cage-like systems, RT_2X_{20} (R = rare earth, T = transition metal, X = Zn, Al) is a new cadidate. In these compounds, similarly to other cage-like compounds, the R ions occupy their own single unique crystallographic site with the polyhedron of 16 X atoms surrounding it (Fig.1(a))[4, 5, 6, 7, 8]. In addition to such cage-like structure, T site atoms constitute a network of corner-shared tetrahedra (the pyrochlore lattice) with strong geometrical frustration (Fig.1(a)). In this sense, RT_2X_{20} is a rare system in which two characteristic crystal structures coexist. Although the systems with cage-like or geometrically frustrated structure show interesting physical properties, there has been reported little studies for the system containing both structures.

Since the discovery of RT_2X_{20} compounds, X = Zn compounds have been intensively studied[9, 10, 11]. One of the most important results of the recent studies on these compounds is the especially enormous Sommerfeld coefficient of 7.9 J/mol K^2 of YbCo$_2$Zn$_{20}$[9]. In order to investigate the origin of the enormous γ, systematic and comparative studies of magnetic, transport and structural properties of the Zn- and Al-based compounds are essential. In contrast to the Zn-based compound, although RT_2Al$_{20}$ compounds had been discovered earlier by Niemann and Jeitschko[5], only limited works have been reported on them mainly about the crystal growth technique and structural determination. On the Yb-compounds, only the lattice

Published under licence by IOP Publishing Ltd

constant of $YbTi_2Al_{20}$ was reported in ref. [5] and there are no other reports on the physical properties of the Yb-compound. In order to investigate the nature of YbT_2Al_{20} compounds in more detail, further investigation specially into the magnetic and transport properties in the transition metal series of the Yb-compound are important. In this work, we prepared single crystals of YbT_2Al_{20} (T = Ti, V, Cr) and investigated magnetic and transport properties of these materials.

Single crystals of $Yb(Ti, V, Cr)_2Al_{20}$ have been prepared by the Al self-flux method from the

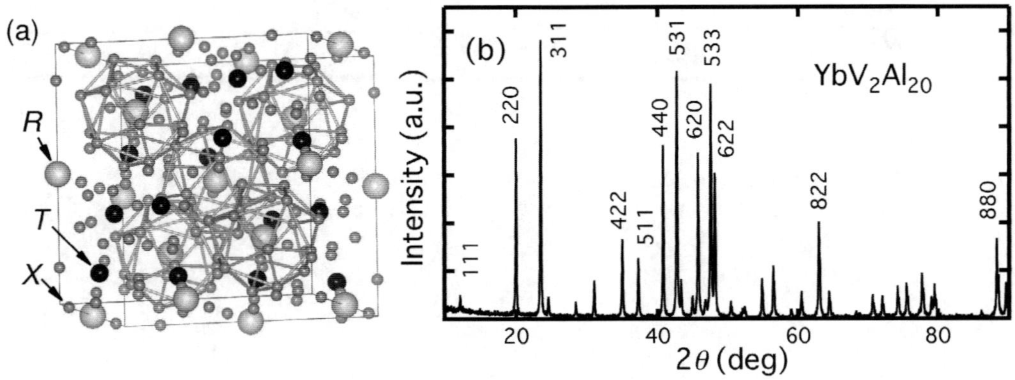

Figure 1. (a) Crystal structure of RT_2X_{20} compound. R ion is surrounding by 16 X ions. T ions form the corner shared tetrahedral network named pyrochlore latteice. (b) X-ray powder diffraction pattern of YbV_2Al_{20} using Co-Kα radiation. For main peaks, the Miller's indices of cubic $Fd\bar{3}m$ symmetry for main peaks are shown.

starting elements, chips of Yb(3N), powders of Ti(5N), V(4N) or Cr (4N) and grains of Al(4N) at the atomic ratio of 1:2:40. These mixtures were placed in an alumina crucible and they were sealed in an evacuated quartz tube. These sealed tubes were heated up to 1050°C, kept for 2 hours, then cooled down to 750°C for 45 hours. The samples were removed from the furnace and the excess Al was spun off in a centrifuge. Small amount of residual Al-flux on the crystals were removed by etching with 3 vol% HCl or NaOH in H_2O for 2 hours in a ultrasonic bath. The quality of the samples was checked using a powder x-ray diffractometer with Co-Kα radiation (Fig. 1(b)). All of the peak are indexed by the Miller's indices of the $CeCr_2Al_{20}$ type structure with cubic $Fd\bar{3}m$ symmetry and absence of any impurity phases was confirmed. From these results, the lattice parameters a for YbT_2Al_{20} (T = Ti, V, Cr) are evaluated to be 14.6841(2)Å, 14.5334(8)Å and 14.4466(3)Å, respectively. The parameter a of $YbTi_2Al_{20}$ is consistent to the previous report of polycrystals [5]. The crystallographic orientations of the single crystals were determined by the back Laue x-ray scattering technique. Typical size and mass of measured single crystals were about 0.5 mm and about 3 mg, respectively. The DC magnetization was measured using a Quantum Design MPMS down to 2 K and up to 7 T along the [110] direction. The resistivity was measured using a homemade cryostat down to 4 K.

In Fig. 2(a), we show the temperature dependence of the DC susceptibility, $\chi(= M/H)$, for YbT_2Al_{20} (T = Ti, V and Cr) at 1 T. All the compounds exhibit Pauli paramagnetic susceptibility. This behavior indicates that for all compounds the valence of Yb ion is close to divalent. However, because there are weak temperature dependence in $\chi(T)$, it is possible that a small part of the 4f levels are unoccupied and Yb^{3+} contributions remain to some extent. This

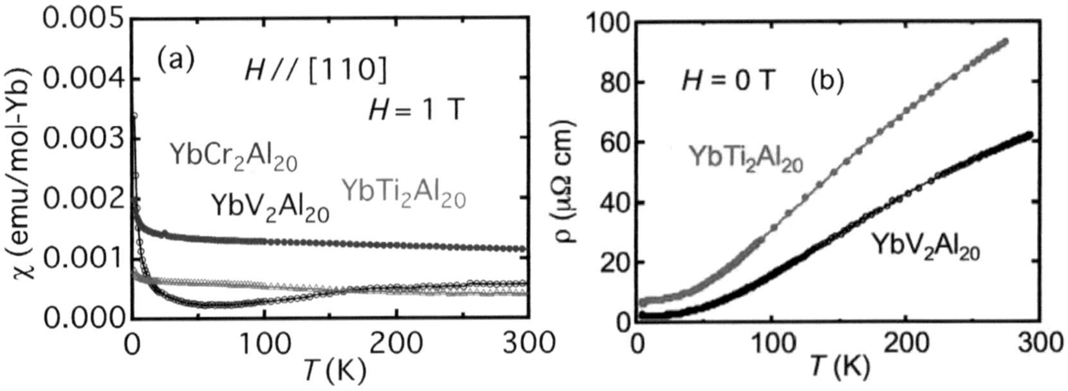

Figure 2. (a) Temperature dependence of DC susceptibility χ of YbT_2Al$_{20}$ (T = Ti, V and Cr) for the fields of 1 T applied parallel to the [110] direction. (b) Temperature dependence of resistivity ρ of YbT_2Al$_{20}$ (T = Ti and V) at 0 T.

may indicate the possibility of valence fluctuation of Yb ions. However, it is difficult to estimate the degree of c-f hybridization only by our present experimental results. The upturn at lower temperatures in all susceptibility data is expected to be the contributions from paramagnetic impurity phases because its value is very small. The values of Pauli paramagnetic susceptibility extrapolated to 0 K, $\chi(0)$ ($0.5 \sim 1.0 \times 10^{-3}$ emu/mol) are larger than those of normal metals ($\sim 10^{-5}$ emu/mol) and almost comparable to those of La compounds without $4f$ electrons ($\sim 0.5 \times 10^{-3}$ emu/mol: data not shown). These magnetic properties are very different from those of the Zn-based compounds that show strongly correlated heavy fermion behavior.

In Fig. 2(b), we show the temperature dependence of the resistivity for YbT_2Al$_{20}$ (T = Ti and V) at 0 T. Their resistivity at room temperature ($60 \sim 100$ $\mu\Omega$cm) and the residual resistivity ratio $\rho(300\mathrm{K})/\rho(4.2\mathrm{K})$ (~ 20) are similar to those of the Zn-based compounds[9]. However, the temperature dependence is qualitatively different. For $YbTi_2Al_{20}$ and YbV_2Al_{20}, ρ monotonously decreases with decreasing temperature down to the lowest investigated temperature without any shoulder or peak structures at low temperatures which are observed in the Zn-based compound. These normal metallic behaviors are consistent with the Pauli paramagnetic susceptibility, and are also an indication of almost Yb^{2+} state in the Al-based compounds. Usually, the cell volume with Yb^{2+} ions is larger than that with Yb^{3+} ions. Therefore, the positive deviation of the cell volume for $YbTi_2Al_{20}$ in the RTi$_2$Al$_{20}$ series reported in Ref.[5] is consistently understood based on the lanthanoid contraction. Similarly, the lattice parameters of the other two compounds evaluated from our powder X-ray diffraction patterns also show the positive deviation from the lanthanoid contraction line of the RT_2Al$_{20}$ series.

In summary, we has studied the magnetic and transport properties of single crystalline YbT_2Al$_{20}$ (T = Ti, V, and Cr). From their nonmagnetic metallic behavior, we concluded that the valence of Yb ions is close to 2+ in these compounds. The weakly-temperature dependent susceptibility indicates the possibility of valence fluctuation of Yb ions. Considering the cage-like structure of the present system, this feature may originate from hybridization between 4f electron and conduction electrons, leading to a Kondo temperature of the order of a few hundred Kelvin corresponding to the weak T-dep. structure in $\chi(T)$. In order to examine this scenario, electronic structure studies (by PES and others) are needed.

This work was supported by a Grant-in-Aid for Young Scientist (B) (No.21740261) and

Scientific Research (B) (No.20340094) from the Japan Society for the Promotion of Science, and a Grant-in-Aid for Scientific Research on Priority Area "Ubiquitous" (No.20045015), and Innovative Areas "Heavy Electrons" (No.20102007) from the Ministry of Education, Culture, Sports, Science and Technology (MEXT) of Japan.

References

[1] M. B. Maple, Z. Henkie, R. E. Baumbach, T. A. Sayles, N. P. Butch, P.-C. Ho, T. Yanagisawa, W. M. Yuhasz, R. Wawryk, T. Cichorek, and A. Pietraszko: J. Phys. Soc. Jpn. **77** (2008) Supl.A. 7.

[2] H. Sato, H. Sugawara, Y. Aoki and H. Harima: in *Handbook of Magnetic Materials*, ed. K.H.J. Buschow. (Elsevier, Amsterdam, 2009) Vol.18, pp. 1-110.

[3] B. C. Sales: in *Handbook on the Physics and Chemistry of Rare Earth*, ed. K.A. Gschneidner, Jr., J.-C. G. Bunzli and V. K. Pecharsky (Elsevier Science, Amsterdam, 2003) Vol. 33, p.1-34.

[4] P.I. Krypyakevych and O.S. Zarechnyuk: Dopov. Akad. Nauk Ukr. RSR Ser. A: Fiz. Tekh **30** (1968) 364.

[5] S. Niemann and W. Jeitschko: J. Solid State Chem. **114** (1995) 337.

[6] T. Nasch, W. Jeitschko and U. C. Rodewald: Z. Natuurforsch., B: Chem. Sci. **52** (1997) 1023.

[7] O. Moze, L. D. Tung, J. J. M. Franse and K. H. J. Buschow: J. Alloys Compd **268** (1998) 39.

[8] V. M. T. Thiede, W. Jeitschko, S Niemann and T. Ebel: J. Alloys Compd. **267** (1998) 23.

[9] M.S. Torikachvili, S. Jia, E.D. Mun, S.T. Hannahs, R.C. Black, W.K. Neils, D. Martien, S.L. Bud'ko and P.C. Can eld: Proc. Natl. Acad. Sci. U.S.A. **104** (2007) 9960.

[10] Y. Saiga, K. Matsubayashi, T. Fujiwara, M. Kosaka, S. Katano, M. Hedo, T. Matsumoto and Y. Uwatoko: J. Phys. Soc. Jpn. **77** (2008) 053710.

[11] T. Onimaru, K. T. Matsumoto, Y. F. Inoue, K. Umeo, Y. Saiga, Y. Matsushita, R. Tamura, K. Nishimoto, I. Ishii, T. Suzuki and T. Takabatake: J. Phys. Soc. Jpn. **79** (2010) 033704.

Staggered Order with Itinerant and Localized Singlets in Generalized Kondo Lattice

S Hoshino[†], **J Otsuki**, **Y Kuramoto**
Department of Physics, Tohoku University, Sendai 980-8578, Japan
E-mail: [†]hoshino@cmpt.phys.tohoku.ac.jp

Abstract. The generalized Kondo lattice where even number of localized electrons are coupled with two conduction bands is investigated using the dynamical mean-field theory combined with continuous-time quantum Monte Carlo method. The crystalline electric field (CEF) singlet-triplet states are taken for localized states. With one conduction electron per site, a staggered order with Kondo and CEF singlets arises. Nature of the electronic order is studied in terms of the local susceptibility. The e ective CEF splitting in one sublattice remains almost the same as the original CEF splitting, while the other sublattice looks like Kondo singlets. Hence, this electronic order is interpreted as alternating sites of itinerant and localized singlet states.

1. Introduction

Distinction between itinerant and localized characters of electrons is one of the most fundamental problems in strongly correlated electron systems. The typical systems are Lanthanide and Actinide compounds which have nearly localized f electrons. In the system with one f electrons per site as in Ce compounds, the competition between the Kondo e ect and the Ruderman-Kittel-Kasuya-Yosida (RKKY) interaction determines the behavior of the system. As a result of the competition, the quantum phase transition between magnetic and nonmagnetic ground states arises. The overall behavior of this system is now well understood. In the system with two f electrons per site as in some Pr and U compounds, on the other hand, the problem is qualitatively di erent from f^1 systems because of the absence of the Kramers degeneracy.

Some f^2 systems can form the crystalline electric eld (CEF) singlet ground state in the localized limit of f electrons. If f electrons strongly interact with conduction electrons, the system forms a collective Kondo singlet. In this case, f electrons acquire the itinerancy because of the Kondo e ect. In the competing region between Kondo and CEF e ects, or itinerant and localized states of f electrons, an exotic ordering is expected. This picture may be relevant to real systems such as $PrFe_4P_{12}$ and URu_2Si_2 [1, 2]. In our previous work, we have demonstrated the existence of the staggered order with the Kondo and CEF singlets (staggered Kondo-CEF singlet order) using the model where the CEF singlet-triplet states interact with the two-band conduction electrons [3].

In the staggered Kondo-CEF singlet ordered phase, the conduction electrons accumulate more on the Kondo-singlet site, and form the charge density wave. The driving force for this order is the e ective inter-site repulsion between Kondo singlets, which is naturally understood from the strong coupling limit with respect to the Kondo exchange interaction [4, 5]. In this regard, the staggered Kondo-CEF singlet order is in contrast with the ordinary multipole order

Published under licence by IOP Publishing Ltd

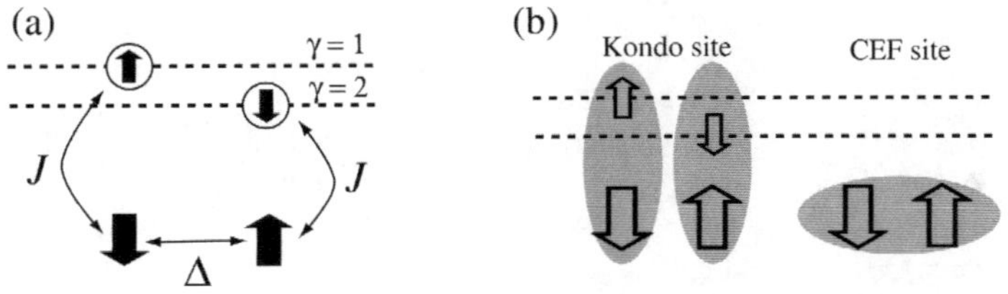

Figure 1. (a) Schematic on-site picture of the 2BSTKLM, and (b) illustration of the staggered Kondo-CEF singlet order. The upper part on the lines shows the two conduction bands, and the lower part shows two localized spins.

by the RKKY interaction which is understood from the weak coupling limit. In this paper, we investigate the properties of this electronic order in terms of itinerant and localized characters of f electrons using local susceptibilities.

2. Model and Methods

We take the generalized Kondo lattice model as follows [6]:

$$\mathcal{H} = \sum_{\boldsymbol{k}\gamma\sigma}(\varepsilon_{\boldsymbol{k}\gamma} - \mu)c^{\dagger}_{\boldsymbol{k}\gamma\sigma}c_{\boldsymbol{k}\gamma\sigma} + J\sum_{i\gamma}\boldsymbol{S}_{\gamma i}\cdot\boldsymbol{s}_{\mathrm{c}\gamma i} + \Delta\sum_{i}\boldsymbol{S}_{1i}\cdot\boldsymbol{S}_{2i}, \tag{1}$$

where σ (γ) denotes the spin (orbital) index. $\boldsymbol{s}_{\mathrm{c}\gamma i}$ denotes the conduction spin at site i with orbital γ. $\boldsymbol{S}_{1i}$ and $\boldsymbol{S}_{2i}$ are the localized spins which represent f-electron states. As shown in the second term, the conduction electrons with the orbital index γ interact with localized spins with the same index by the Kondo exchange interaction J. The third term is the interaction Δ between localized spins at the same site, which tends to form the local pair singlet. This model is called the "two-band singlet-triplet Kondo lattice model" (2BSTKLM) in the following. Figure 1(a) schematically shows this model. Note that the 2BSTKLM with $\Delta = 0$ corresponds to the two-independent Kondo lattice models. Since the localized states can be regarded as the CEF singlet-triplet states with splitting Δ, the word 'CEF singlet' is used to represent the local pair singlet in the rest of this paper.

We employ the dynamical mean-field theory (DMFT) for our analysis [8]. The staggered ordered phase can be investigated by extending the DMFT to the two-sublattice system. As the impurity solver, we use the continuous-time quantum Monte Carlo method (CT-QMC).[9, 10, 11] . Since the present CT-QMC is free from numerical difficulties such as the negative sign problem, our calculations are highly accurate in the framework of the DMFT. In this paper, we fix the number n_{c} of conduction electrons per site as $n_{\mathrm{c}} = 1$, which corresponds to the quarter filling with two bands. The conduction bands are taken as $\varepsilon_{\boldsymbol{k}1} = \varepsilon_{\boldsymbol{k}2} = \varepsilon_{\boldsymbol{k}}$ for simplicity, although the realistic bands are not symmetric. Each conduction electron forms the hypercubic lattice whose density of states is given by

$$\rho(\varepsilon) = \frac{1}{D}\sqrt{\frac{2}{\pi}}\exp\left(-\frac{2\varepsilon^2}{D^2}\right), \tag{2}$$

where we take $D = 1$ as a unit of energy. We choose $J = 0.8$ and $T = 0.01$ in this paper.

In the 2BSTKLM at quarter filling, the staggered Kondo-CEF singlet order arises at low temperatures as illustrated in Fig. 1 (b) [3]. The system becomes insulating under the present condition. Since the $\Delta = 0$ case coincides with the two-independent Kondo lattice models, this electronic order connects to the charge density wave found in the Kondo lattice model at

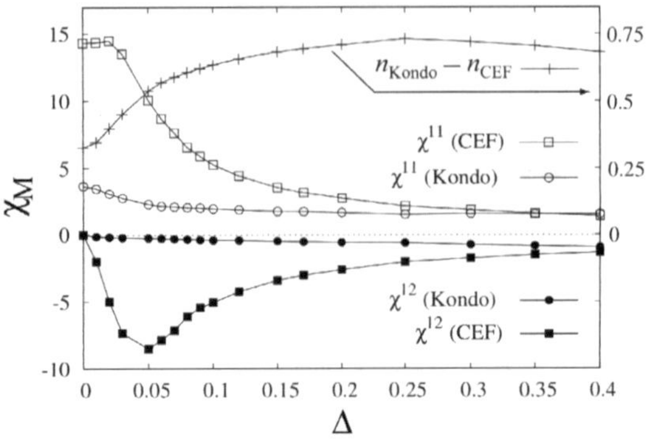

Figure 2. Local magnetic susceptibilities on Kondo- and CEF-singlet sites (left scale), and order parameter (right scale) as a function of Δ at $T = 0.01$.

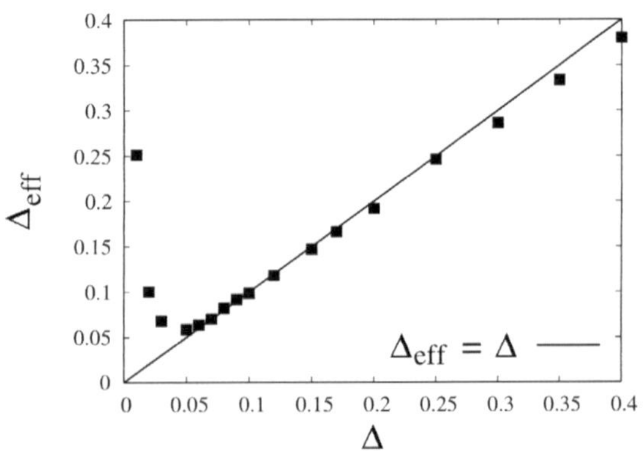

Figure 3. Effective CEF splitting as a function of Δ. The straight line shows the relation $\Delta_{\text{eff}} = \Delta$.

quarter filling [7]. As Δ increases, the ordered phase breaks down by a first-order transition to the homogeneous phase. At $T = 0.01$, the transition occurs at $\Delta \sim 0.42$. We note that the present order survives as long as the number of conduction electrons is close to the quarter filling. For $J = 0.8$ and $\Delta = 0.2$, this range is $0.8 \lesssim n_c \lesssim 1.3$.

3. Magnetic Susceptibilities and Effective CEF Splitting

Let us investigate basic characteristics of the staggered Kondo-CEF singlet ordered phase using local magnetic susceptibilities defined by

$$\chi_{\text{M}}^{\gamma\gamma'} = \int_0^\beta \left[\langle T_\tau S_\gamma^z(\tau) S_{\gamma'}^z \rangle - \langle S_\gamma^z \rangle \langle S_{\gamma'}^z \rangle \right] \mathrm{d}\tau. \tag{3}$$

Under the present condition, the susceptibility has two independent components $\chi_{\text{M}}^{11} = \chi_{\text{M}}^{22}$ and $\chi_{\text{M}}^{12} = \chi_{\text{M}}^{21}$. In the limit of large Δ, we obtain $\chi_{\text{M}}^{11} = -\chi_{\text{M}}^{12} = 1/(2\Delta)$. In the opposite limit of $\Delta = 0$ with Kondo temperature T_{K}, we have $\chi_{\text{M}}^{11} = 1/(2T_{\text{K}})$, while $\chi_{\text{M}}^{12} = 0$ because each spin independently makes the singlet with conduction electrons.

Figure 2 shows the local magnetic susceptibilities computed at $T = 0.01$ as a function of Δ. The order parameter, which corresponds to the difference of the number of conduction electrons between Kondo- and CEF-singlet sites, is also shown as $n_{\text{Kondo}} - n_{\text{CEF}}$. In one sublattice, identified as the Kondo-singlet site, $|\chi_{\text{M}}^{12}|$ is always smaller than χ_{M}^{11}. Furthermore, the value of χ_{M}^{11} is almost independent of Δ. In the other sublattice, identified as the CEF-singlet site, the computed results for χ_{M}^{11} and χ_{M}^{12} are well understood in terms of the effective Hamiltonian $\mathcal{H}_{\text{eff}} = \Delta_{\text{eff}} \boldsymbol{S}_1 \cdot \boldsymbol{S}_2$ on the CEF site. Here Δ_{eff} is the effective CEF splitting. With use of the expression

$$\chi_{\text{M}}^{12}(T = 0) = -\frac{1}{2\Delta_{\text{eff}}}, \tag{4}$$

the effective CEF splitting can be estimated from the local magnetic susceptibility at low temperature. Figure 3 shows Δ_{eff} estimated from χ_{M}^{12} at $T = 0.01$ using (4). Here the straight line represents $\Delta_{\text{eff}} = \Delta$. For $\Delta > 0.05$, we observe good correspondence between Δ_{eff} and the original CEF splitting Δ. Hence on the CEF-singlet site, the interaction should be weak between the localized states and conduction electrons.

On the other hand, the f electrons at the Kondo-singlet site acquire the itinerancy because of the Kondo effect. Hence, the staggered Kondo-CEF singlet order can be regarded as the staggered order with itinerant and localized singlet states. As is clear from the formation of the CEF-singlet site, this order is a characteristic for the f^2 systems. In the strong coupling limit, the conduction electrons accumulate solely on the Kondo sublattice. With finite coupling, the Kondo cloud extends to the CEF sublattice, but charge density wave (CDW) of conduction electrons still remains. Namely, the conduction electrons at the CEF-singlet site is a part of spatially extended Kondo cloud at the Kondo-singlet site, and do not significantly affect the bare CEF character.

This staggered order is responsible for the insulating behavior of the quarter-filled bands. This staggered order is responsible for the insulating behavior of the quarter-filled bands. Figure 1(b) illustrates the electron configurations for the Kondo site and the CEF site in the staggered order. In the doubled unit cell, the total number of itinerant electrons is four, and the two renormalized bands are completely filled. Note that the CEF site does not participate in the conduction. In the absence of the staggered order, the number of electrons is odd per unit cell, and the system becomes metallic.

Finally we consider the deviation from the bare CEF splitting in the region with $\Delta < 0.05$ shown in Fig. 3. Since the case with $\Delta = 0$ corresponds to the two-independent Kondo lattices, $\chi_{\rm M}^{12}$ should vanish in this limit. Thus $\Delta_{\rm eff}$ should diverge as $\Delta \to 0$. On the other hand, the difference of $\chi_{\rm M}^{11}$ between Kondo- and CEF-singlet sites at $\Delta = 0$ comes from the difference of charge density of conduction electrons. The Kondo-singlet site has more conduction electrons, and has thus smaller $\chi_{\rm M}^{11}$.

4. Summary

The basic characteristic of the staggered Kondo-CEF singlet order at low temperature has been investigated in terms of the local magnetic susceptibilities. Our calculation using the DMFT+CT-QMC successfully characterizes this novel electronic order. The calculated effective CEF splitting at the CEF-singlet site shows good agreement with the original CEF splitting. Hence, the CEF-singlet site is almost decoupled from conduction electrons. This fact allows us to interpret the staggered Kondo-CEF singlet order as the alternating sites of itinerant and localized singlet states.

One of the authors (S. H.) is supported by the global COE program of The Ministry of Education, Culture, Sports, Science, and Technology (MEXT), Japan. This work is partly supported by a Grant-in-Aid for Scientific Research on Innovative Areas "Heavy Electrons" (No 20102008) of MEXT, Japan.

References

[1] For a review, see Y. Kuramoto, H. Kusunose, and A. Kiss: J. Phys. Soc. Jpn. **78** (2009) 072001.
[2] E. Hassinger *et al*: J. Phys. Soc. Jpn. 77 (2008) Suppl. A, p. 172.
[3] S. Hoshino, J. Otsuki, and Y. Kuramoto: J. Phys. Soc. Jpn. **79** (2010) 074720.
[4] J. E. Hirsch: Phys. Rev. B **30** (1984) 5383.
[5] M. Sigrist *et al*: Phys. Rev. B **46** (1992) 13838.
[6] S. Watanabe *et al*: J. Phys. Soc. Jpn. **68** (1999) 159.
[7] J. Otsuki, H. Kusunose, and Y. Kuramoto: J. Phys. Soc. Jpn. **78** (2009) 034719.
[8] For a review, see A. Georges *et al*: Rev. Mod. Phys. **68** (1996) 13.
[9] A. N. Rubtsov, V. V. Savkin, and A. I. Lichtenstein: Phys. Rev. B **72** (2005) 035122.
[10] P. Werner and A. J. Millis: Phys. Rev. B **74** (2006) 155107.
[11] J. Otsuki *et al*: J. Phys. Soc. Jpn. **76** (2007) 114707.

Journal of Physics: Conference Series **273** (2011) 012035 doi:10.1088/1742-6596/273/1/012035

Electronic hybridization effects in dense intermetallics measured by electron spin resonance

V. A. Ivanshin[1], **T. O. Litvinova,**
Kazan State University, Kremlevskaya str. 18, 420008 Kazan, Russia
A.A. Sukhanov
Zavoisky Physical-Technical Institute, Sibirsky trakt 10/7, 420029 Kazan, Russia

E-mail: Vladimir.Ivanshin@ksu.ru

Abstract. Recent theoretical studies associate the unexpected well resolved electron spin resonance (ESR) signals below Kondo temperature in several undoped heavy-fermion (HF) compounds with hybridization effects between *4f* and conduction electrons (CE) in the presence of ferromagnetic (FM) fluctuations. We analyze ESR experiments in different concentrated Yb-, Ce-, and Eu-based intermetallic systems. We believe that the exotic ESR absorption here is caused by a novel type of ESR excitations – *hybridized electronic states,* which can be created in some strongly correlated electronic systems due to hybridization between the *4f*-orbitals and the wave functions of the CE of the outer electronic shells in conjunction with FM RKKY interaction.

1. Introduction

Recently, much attention has been focused on observation of well-defined ESR signals in several concentrated intermetallic systems [1-6] where the HF behaviour, evolution of the Kondo state, and hybridization mechanisms can be studied directly from the ESR measurements in contrast to earlier investigations in some Kondo lattices doped with different rare-earth ions as a spin probes [7]. According to several theories developed during two last years [8-12], an effective hybridization between localized *f* electrons and conduction bands and strongly anisotropic Kondo-ion interactions in conjunction with FM correlations which reduce usually a very effective spin-lattice relaxation are discussed as a possible origins of the narrowed ESR lines below the Kondo temperature in these experiments. In the present contribution, we report the orientation-dependent ESR spectra in $YbCo_2Zn_{20}$ at room temperature and compare the results with those of all known ESR studies in dense intermetallic compounds.

2. Experimental results

Details of synthesis and characterization of the $YbCo_2Zn_{20}$ single crystals have been described previously [13]. The single crystal specimen used in our study has a size 5×4×4 mm^3. ESR spectra of $YbCo_2Zn_{20}$ were recorded on an EMX/plus Bruker spectrometer operating at X-band. The strongly anisotropic ESR spectra with the resonance fields of the components up to 10 kOe have been detected at room temperature. Angular variations of the ESR lines are shown for two different orientations with the *dc* external magnetic field ***H*** rotated either in the $(10\bar{1})$ plane (Fig.1, top) or in the (011) plane (Fig.1, bottom). For the latter orientation two resonances with different resonance field positions (at g

[1] To whom any correspondence should be addressed.

Published under licence by IOP Publishing Ltd

~ 2.2 and g ~ 4.3) and line widths have been clearly resolved. Figure 2 illustrates the very different angular dependences of the effective ESR g-factor of both signals at this orientation.

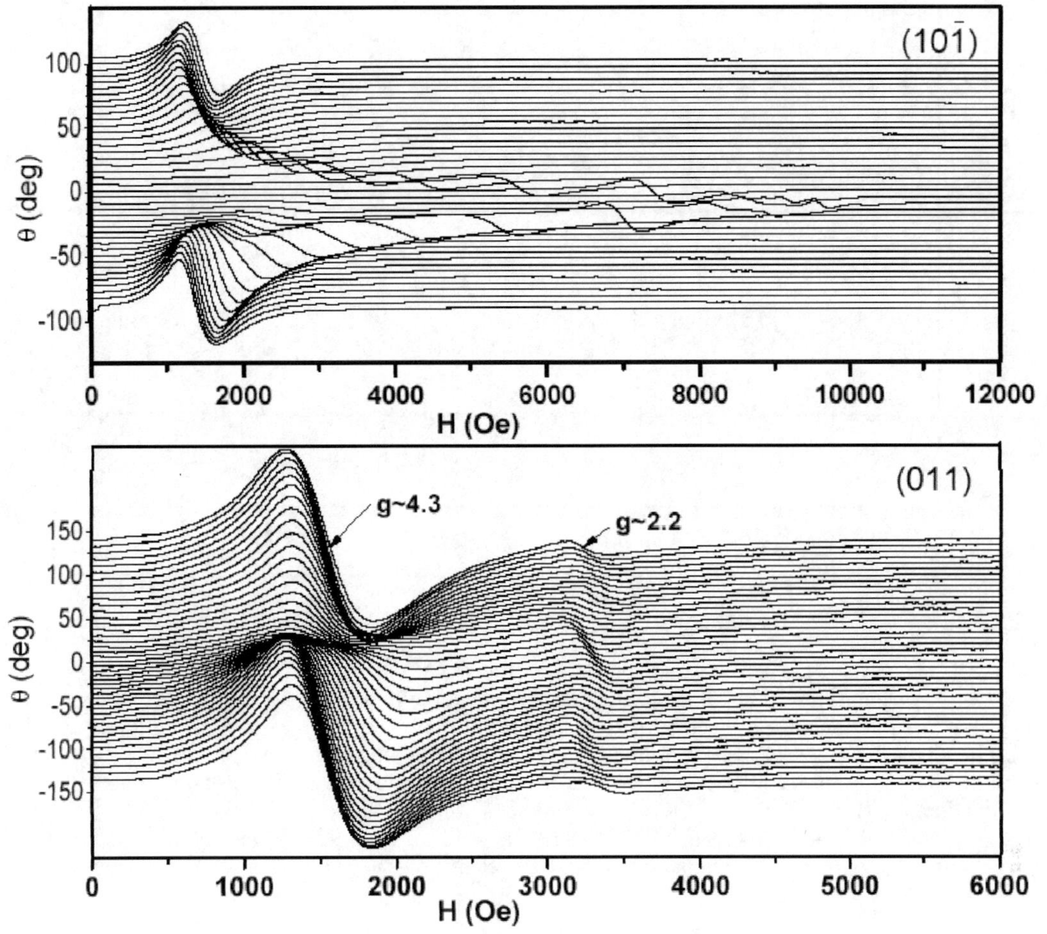

Figure 1. Angular dependences of the ESR spectra at room temperature in the $YbCo_2Zn_{20}$ single crystal at 9.4 GHz for the constant magnetic field rotated in the $(10\bar{1})$ plane (top) and in the (011) plane (bottom).

3. Discussion

A coexistence of two ESR lines in $YbCo_2Zn_{20}$ reminds the results of recent ESR experiments in the first Yb-based HF superconductor β-$YbAlB_4$ where the single ESR response at g ~ 2.3 also at room temperature has been related to the CE spin resonance (CESR). At T = 4.2 K it reaches g ~ 3.0 and was attributed with the g-value of the Yb^{3+} Kramers doublet [14]. Although the signal at g ~2.2 in our studies can be ascribed to CESR too, the values of g ~ 4.3 for the low-field signal in $YbCo_2Zn_{20}$ and g ~ 7.33 in the iso-structural $YbFe_2Zn_{20}$ alloy [5] significantly differ from the corresponding quantities for the Yb^{3+} ion. The average ESR g-factors and a possible hybridization types of all concentrated intermetallic systems in which the ESR signals were observed are listed in the Table 1. For most compounds, these g-values are very close to those found for the corresponding *4f*-ions Yb, Ce or Eu in different crystal symmetries.

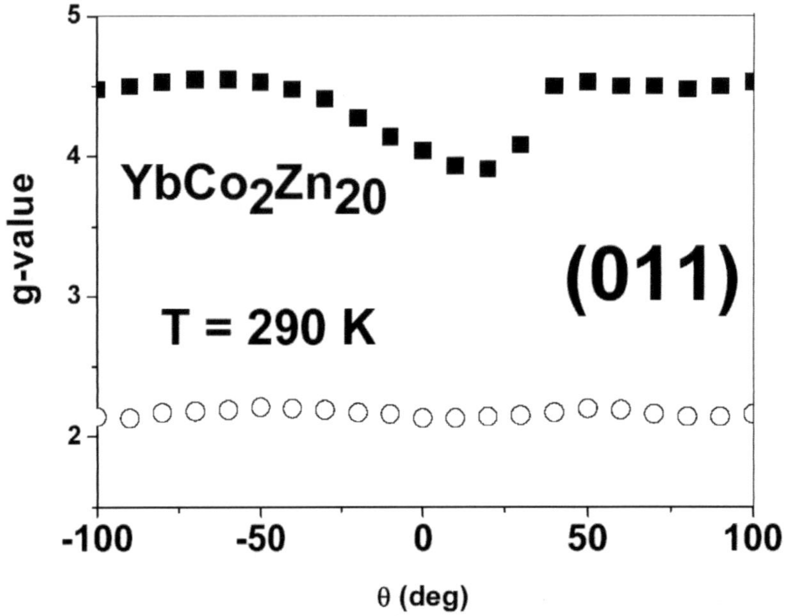

Figure 2. The angular dependences for lines with g ~ 2.2 (full squares) and g ~ 4.3 (empty circles) of the ESR spectrum in the $YbCo_2Zn_{20}$ single crystal at room temperature. The external magnetic field has been rotated in the (011) plane.

In contrast, in a case of the *4f-3d* hybridization, i.e., for YbCuAl [15] and YbT_2Zn_{20} (T = Co, Fe) the average ESR g-values can be attributed with those of the corresponding *3d*-ions Cu, Co, and Fe. This means that the fraction of the *4f*-electrons that hybridize with the CE and become itinerant can be much higher for the *4f-3d* hybridization than for other types. At the same time, a part of *3d*-electrons can be localized and the ESR *g*-tensor is given by the crystalline field scheme of the *d* states. Very recently, such coexistence of the *3d* local moments and itinerant electrons in iron pnictides [17] and interplay of *3d* and *4f* magnetism in CeCoPO [18] have been discussed. In the latter compound the main contribution to magnetism results from the *3d*-electrons of Co because the Ce *4f*-electrons are

Table 1 Average ESR g-factors and possible hybridization types in some undoped intermetallic compounds investigated by ESR

Compound	Average ESR g-factor	Hybridization type
YbCuAl	1.94[15]	4f-3d3s3p
$YbCo_2Zn_{20}$	2.2; 4.33[5]	4f-3d4s
$YbFe_2Zn_{20}$	7.33[5]	4f-3d4s
$YbRh_2Si_2$	3.55[1,2]	4f-4d3p
YbRh	2.55[4]	4f-4d5s
$YbRh_2Pb$	3.75[7]	4f-4d6s6p
CeRuPO	2.6[4]	4f-4d2s2p
$YbIr_2Si_2$	3.357[3]	4f-5d3p
YbBiPt	3.75[5]	4f-5d6s6p
β-$YbAlB_4$	2.3 (295 K); 3.0 (4.2 K)[14]	4f-2s2p
EuB_6	2.0[16]	4f-2s2p

screened due to pronounced Kondo-interactions. The exotic ESR response in dense intermetallics has been called as "HF ESR" [8, 19] or "Kondo coupled resonant modes" [9]. However, these ESR signals are not special to the HF systems or to the Kondo effect because they were also detected in the common alloys such as $YbRh_2Pb$ [6] and EuB_6 [16] or in the non-Kondo lattice YbRh [4]. Up to now is not clear that either heavy mass only or/and the anisotropic part of the strong electron-electron interactions cause an observable ESR [11, 19]. Therefore, this novel type of ESR excitations can be treated as *hybridized electronic states* which arise in some strongly correlated electron systems in conjunction with FM correlations and lead to a narrow collective spin mode of coupled localized and itinerant electrons. Usually it is assumed that based solely on the ESR g values and their anisotropy, we cannot distinguish between localized and conducting states resonating [9, 12]. Our orientation-dependent ESR measurements in $YbCo_2Zn_{20}$ provide probably the first experimental evidence of such separation.

In summary, we have discussed recent investigations on the ESR in dense intermetallic materials. A coexistence of CESR and local moment ESR signals has been observed in the strongly anisotropic ESR spectra of the $YbCo_2Zn_{20}$ crystals. The observed ESR behavior originates from the Co *3d-* electrons rather than the Yb *4f*-electrons due to very effective *4f-3d* hybridization.

This work was supported in part by RNP 2.1.1/6183. We wish to thank S.L. Bud'ko for furnishing the $YbCo_2Zn_{20}$ single crystals and G.R. Bulka for fruitful discussions.

4. References

[1] Ivanshin VA, Aminov LK, Kurkin IN, Sichelschmidt J, Stockert O, Ferstl J and Geibel C 2003 *JETP Lett.* **77** 526
[2] Sichelschmidt J, Ivanshin VA, Ferstl J, Geibel C, Steglich F 2003 *Phys. Rev. Lett.* **91** 156401
[3] Sichelschmidt J, Wykhoff J, Krug von Nidda H-A, Fazlishanov II, Hossain Z, Krellner C, Geibel C and Steglich F 2007 *J. Phys.: Condens. Matter* **19** 016211
[4] Krellner C, Förster T, Jeevan H, Geibel C, Sichelschmidt J 2008 *Phys. Rev. Lett.* **100** 066401.
[5] Ivanshin VA, Sokolov DA, Aronson MC, Jia S, Bud'ko SL and Canfield PC 2009 *J. Alloys and. Compds.* **480** 126
[6] Ivanshin VA, Litvinova TO, Sukhanov AA, Sokolov DA, Aronson MC 2009 *JETP Lett.* 2009 **90** 116
[7] Krug von Nidda H-A, Heinrich M, Loidl A 2003 *Relaxation phenomena: liquid crystals, magnetic systems, polymers, high-T_c superconductors, metallic glasses*, ed W Haase and S Wrobel (Berlin: Springer) p 112
[8] Abrahams E, Wölfle P 2008 *Phys. Rev. B* **78** 104423
[9] Schlottmann P 2009 *Phys. Rev. B* **79** 045104
[10] Zvyagin AA, Kataev V, Büchner B **80** 2009 024412
[11] Huber DL 2009 *J. Phys.: Condens. Matter* **21** 322203
[12] Wölfle P, Abrahams E 2009 *Phys. Rev. B* **80** 235112
[13] Torikachvili MC, Jia S, Mun ED, Hannahs ST, Black RC, Neils WK, Martien D, Bud'ko SL and Canfield PC 2007 *PNAS* **104** 9960.
[14] Holanda LM, Vargas JM, Rettori C, Nakatsuji S, Kuga K, Fisk Z, Oseroff SB, Pagliuso PG 2009 *cond-mat/arXiv: 0908.0044v1*
[15] Tien C, Yu J-T, Duh H-M 1993 *Jpn. J. Appl. Phys.* **32** 2658
[16] Glushkov VV, Semeno AV, Sluchanko NE, Dukhnenko AV, Fillippov VB, Demishev SV 2008 *Physica B* **403** 932
[17] Kou, S-P, Weng, Z-Y 2010 *cond-mat/arXiv: 0811.4111v4*
[18] Krellner C, Burkhardt U, Geibel C, *Physica* B 2009 **404** 3206
[19] Schaufuβ U, Kataev V, Zvyagin AA, Büchner B, Sichelschmidt J, Wykhoff J, Krellner C, Geibel C, Steglich F 2009 *Phys. Rev. Lett.* **102** 076405

International Conference on Strongly Correlated Electron Systems (SCES 2010) IOP Publishing
Journal of Physics: Conference Series **273** (2011) 012036 doi:10.1088/1742-6596/273/1/012036

Magnetic frustration effects in uranium intermetallics

Yu Jiang[1], C H Booth[1], P H Tobash[2], K Gofryk[2], M A Torrez[2], F Ronning[2], E D Bauer[2] and J D Thompson[2]

[1] Chemical Sciences Division, Lawrence Berkeley National Laboratory, Berkeley, California 94720, USA

[2] Materials Physics and Applications Division, Los Alamos National Laboratory, Los Alamos, New Mexico 87545, USA

E-mail: chbooth@lbl.gov

Abstract. The effect of geometrical frustration on the development of the heavy-fermion state and quantum criticality is studied in $UAuCu_4$, $UAuPt_4$, UAu_3Ni_2 samples through measurements of their magnetic susceptibility, heat capacity, and electrical resistivity. In addition, since lattice disorder can play a large role in defining magnetic properties in frustrated systems, extended X-ray absorption fine structure (EXAFS) data have also been obtained. The local structure results show a strong correlation with the magnetic properties in these samples.

Geometrical frustration often leads to a variety of interesting states of matter, such as spin-ice, spin-liquid, and spin-glass states [1, 2], which have been widely studied in pyrochlores [3] and Mott insulators [4]. In contrast, the effect of frustration on the development of the heavy-fermion (HF) state [5] or quantum criticality in intermetallic compounds [6] has received much less attention. As the class of strongly correlated intermetallics demonstrating strikingly different behavior from those of normal Fermi Liquids grows, it becomes increasingly important to account for the role of frustration in generating such non-Fermi-liquid (NFL) behavior, including the role of spin fluctuations and disorder in the quantum critical region of HF systems [7, 8, 9].

To study the effects of magnetic-order suppression and quantum criticality from frustration in HF compounds, we have studied a $UM_{5-x}X_x$ series of intermetallic compounds with the $AuBe_5$ structure: $UAuCu_4$, $UAuPt_4$ and UAu_3Ni_2. Since these *fcc* lattices have antiferromagnetic (AF) interactions, their spins are geometrically frustrated, although more weakly than that of, say, the triangular Kagomé lattice [2]. Previous measurements of the parent compounds UPt_5 and UNi_5 show standard HF behavior at low T: the linear specific heat coefficients, γ, are of order 100 mJ/mol·K^2, both the specific heats over temperature (C/T) and the magnetic susceptibilities, $\chi(T)$, go to a constant, and the dc resistivities, $\rho(T)$, follow a T^2 law [10, 11]. UCu_5 displays similar behavior, except for a magnetic transition that occurs near 1 K [12, 13]. Varying M and X atomic species should alter the electronic density and conduction-band hybridization of the f electronic states. In addition, X/M site interchange can occur, where some X and M atoms then occupy the same symmetry site, as in $UPdCu_4$ [14]. Such chemical disorder can generate random bond length disorder, leading to magnetic disorder, the effects of which would be reflected in transport and magnetic properties. In this paper, we report magnetic and transport properties for this family of intermetallics, correlated with local structural measurements using the extended X-ray absorption fine-structure (EXAFS) technique.

Samples of $UAuCu_4$, $UAuPt_4$, and UAu_3Ni_2 were arc-melted on a water-cooled copper hearth

Published under licence by IOP Publishing Ltd

Table 1. (Top) Summary of magnetic, specific heat data, and EXAFS fit results for three compounds. R_{nom} is the shortest U-M pair distance from the nominal crystal structure at room temperature. Note: in the fit for γ of $UAuCu_4$, we only used the data below T = 30 K.

	γ (J/mol·K^2)	μ_{eff} (μ_B)	T_N or T_f (K)	θ_{CW} (K)	f	f_{4c}^M
$UAuCu_4$	0.1	3.19	$\sim$ 30	-161	5	0
$UAuPt_4$	0.26	3.20	$\leq$0.15	-154	>1000	5±3%
UAu_3Ni_2	0.27	3.50	$\sim$ 3.6	-150	42	20±10%

U-M(16e)	R / R_{nom} (Å)	σ^2_{stat}($\times 10^{-4}$ Å^2)	θ_{cD}(K)
U-Cu	2.94 / 2.93	-1(1)	284(4)
U-Pt	3.09 / 3.09	6(2)	186(6)
U-Ni	2.95 / 3.02	139(2)	242(4)

with a Zr-getterered UHP Ar atmosphere, and χ, C/T and ρ were measured (figure 1). $\chi(T)$ data were fit with a Curie-Weiss law, $\chi = \frac{N_A \mu_{eff}^2}{3k_B(T-\theta_{CW})}$, above ~150 K yielding the values of μ_{eff} and θ_{CW} listed in table 1. The C/T data were fit with $C/T = \gamma + \beta T^2$ to obtain the specific heat Sommerfeld coefficient γ, with fit ranges between 2 - 30 K ($\leq T_N$) for $UAuCu_4$ and between 6 - 13 K (10 - 20 K) for $UAuPt_4$ (UAu_3Ni_2). $\rho(T)$ data at low-T were fit with a power law, $\rho = \rho_0 + AT^\alpha$, to obtain the temperature exponent, α. The frustration parameter ($f = \theta_{CW}/T_N$ or T_f, defined below) then quantifies the degree of frustration in these compounds. These properties are summarized in table 1.

X-ray absorption data were collected at U L_{III}, Au L_{III}, Pt L_{III}, Cu K and Ni K edges between 30 and 300 K on beamline 4-1 of the Stanford Synchrotron Radiation Lightsource (SSRL), using a half-tuned double crystal Si(220) monochromator with a slit height of 0.7 mm. Data were reduced using standard procedures [15, 16]. A similar set of constraints were employed to fit the data as used previously [14], which include the site-interchange model. From the fit, we get the pair distance, R, the percentage of 4c sites occupied by M, f_{4c}^M, and the Debye-Waller factor, $\sigma^2(T)$, which is then fit with the correlated-Debye model [17] using a static bond distribution width offset, σ^2_{stat}, and correlated-Debye temperature, θ_{cD} (table 1). Examples of the EXAFS data and fits are shown in figure 1d. $UAuCu_4$ data are similar to $UAuPt_4$.

$UAuCu_4$ is AF below $T_N \sim$ 30 K and has a much larger $\theta_{CW} \sim$ -161 K [18]. The $\chi(T)$ and C/T data are similar to those of both its parent compound, UCu_5 ($T_N \sim$ 15 K, $\theta_{CW} \sim$ -180 K) and the Ag-substituted compound, $UAgCu_4$ ($T_N \sim$ 18 K, $\theta_{CW} \sim$ 160 K) [18]. However, their θ_{CW} to T_N ratio are larger than for typical HF metals, such as $YbAgCu_4$ ($\theta_{CW} \sim$ -16 K and $T_K \sim$ 150 K [19]), suggesting that the large θ_{CW} in the present compounds may be influenced by a large interatomic AF coupling. As $T \to 0$ K, χ and $C/T \to$ constant, consistent with Fermi-liquid behavior. The EXAFS results indicate similar local and average (nominal) structures, and the local bond length disorder, σ^2_{stat}, is consistent with zero. These data also indicate no Au/Cu site-interchange, and hence, no chemical disorder occurs, in contrast to $UPdCu_4$ [14]. All of these data are consistent with an ordered lattice and moderate AF frustration.

$UAuPt_4$, however, has quite different magnetic and transport properties. For instance, γ is more than twice that of $UAuCu_4$, $\chi(T)$ is logarithmic with T below $T \sim$ 20 K, and $\rho(T) = \rho_0 + AT^{1.5}$ between T = 0.4 and 4 K (figure 1c inset). Even more surprisingly, C/T at $T \leq 6$ K increases linearly up to 0.65 J/mol·K^2 at 0.4 K. This behavior, given the small field dependence of C/T (not shown), cannot be described with a nuclear Schottky term or by the Hertz-Millis [20] or any other known NFL theory [21]. It is unlikely that this unusual behavior arises from a putative af transition as C/T does not show any magnetic order down to 0.15 K

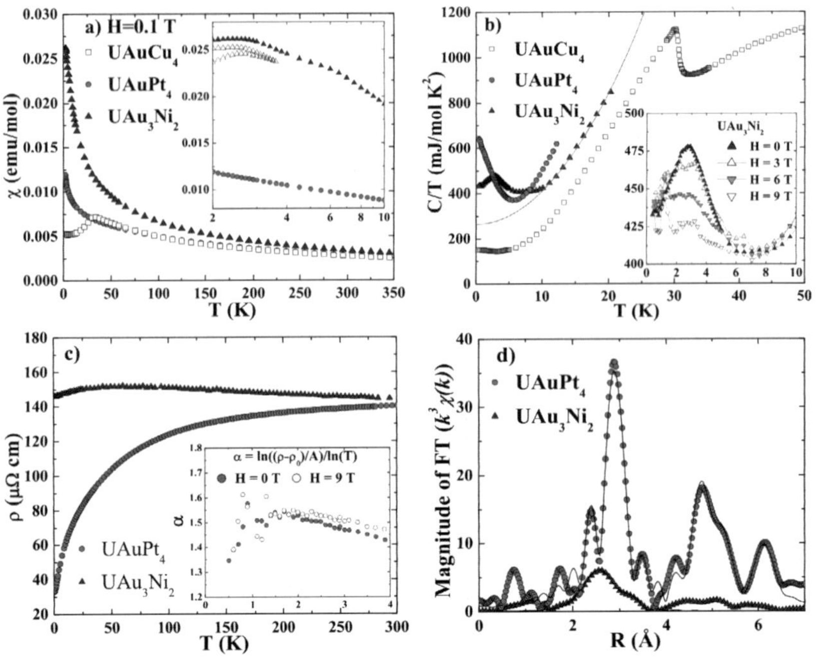

Figure 1. a) $\chi(T)$ at $H = 0.1$ T for $UAuCu_4$, $UAuPt_4$, and UAu_3Ni_2; inset shows the data below $T = 20$ K on a logarithmic temperature scale, in which the field-cooled (FC, open triangle, upper) and zero-field cooled (ZFC, open triangle, lower) data at $H = 0.01$ T for UAu_3Ni_2 are also displayed. b) C/T for three samples; inset shows the H-dependence of UAu_3Ni_2. c) ρ for $UAuPt_4$ and UAu_3Ni_2; inset shows resistivity exponent α vs. T at $H = 0$ T and $H = 9$ T for $UAuPt_4$ (open circle). d) Magnitude of Fourier transform (FT) of $k^3\chi(k)$ of U L_{III}-edge for $UAuPt_4$ ($T = 50$ K, transformed from 3.0-15.0 Å^{-1}, and Gaussian broadened by 0.3 Å^{-1}), and UAu_3Ni_2 ($T = 30$ K, tranformed from 3.0-11.5 Å^{-1}, and Gaussian broadened by 0.3 Å^{-1}); the r-space fit (solid lines) to $UAuPt_4$ data is between 2.2-6.3 Å, and for UAu_3Ni_2 data is between 1.9-3.6 Å. Note: $\chi(k)$ represents the EXAFS oscillations, not the magnetic susceptibility χ.

[10]. All of these properties represent deviations from normal Fermi-Liquid behavior, suggesting a NFL ground state of some kind. Although $UAuPt_4$ has a large frustration parameter, the EXAFS data are consistent with a well-ordered lattice structure (table 1). Hence, $UAuPt_4$ appears to be a good candidate for a frustrated NFL metal.

UAu_3Ni_2, on the other hand, shows a logarithmic T-dependence of $\chi(T)$ below 20 K. At about $T_f = 3.6$ K, the zero-field-cooled (ZFC) and field-cooled (FC) $\chi(T)$ data start to diverge, indicative of a spin glass (SG). C/T of UAu_3Ni_2 has a broad peak at $T \sim T_f$ which decreases with external magnetic field, and is almost suppressed at $T \sim 9$ T. The $5f$ entropy at 6 K, $S_{5f} \sim$ 3.1 J/mol·K, is smaller than $R \ln 2$. This behavior is consistent with the generic behavior of a classic SG, such as CuMn 0.3 at.% and $Eu_{0.4}Sr_{0.6}S$ [1]. $\rho(T) = \rho_0 + AT^{1.5}$ at low T; however, ρ_0 is much larger than that of $UAuPt_4$, possibly reflecting both the spin disorder of the SG state and a significant amount of lattice disorder/distortion due to random site occupancies. U/Au L_{III}- and Ni K-edge EXAFS data indicate a surprisingly large fraction of the $4c$ site occupied by Ni, $f_{4c}^{Ni} \sim 20\% \pm 10\%$, suggesting a nearly random distribution of the Au and Ni on both $4c$ and $16e$ sites (in the nominal structure, $f_{4c}^{Ni} \sim 0$, while in a random distribution model, $f_{4c}^{Ni} \sim 40\%$).

In addition, we find a large bond static bond length disorder for the nearest-neighbor atom pair, $\sigma^2_{stat} \sim 0.0139$ Å^2. The average bond lengths also deviate from the nominal crystal structure, e.g., the distortion for the shortest U-16e pairs is R_{U-Ni} - $R_{U-Au'} \geq 0.1$ Å. The significant measured disorder is consistent with the large ρ_0 and SG-like behavior. Together with $\chi(T)$ and C/T data, we consider UAu_3Ni_2 to be a frustrated SG.

It is possible to qualitatively understand these $UM_{5-x}X_x$ intermetallics within a magnetic frustration (Q) vs. Kondo coupling (K) phase diagram, as in reference [22]. $UAuCu_4$ is well within the AF metal region with moderate frustration and weak Kondo coupling strength. Isostructural $UAuPt_4$ has very different properties without any apparent magnetic order. Since χ, C/T and ρ data all show NFL behavior above $T = 0.15$ K [10], it appears to be in the spin liquid (SL) phase. In addition, α from $\rho \propto T^\alpha$ changes very little in applied fields up to 9 T, indicative of a much larger separation from the heavy Fermi liquid (HFL) phase than $YbRh_2Si_2$ [6, 23]. However, the low-T slope in C/T decreases slightly at $H = 9$ T (not shown). The H-dependence of C/T and ρ therefore suggests that $UAuPt_4$ is near the AF and SL boundary, far away from the HFL phase, and hence close to the quantum critical point between AF and SL. UAu_3Ni_2, however, is not easily included in the two-dimensional (2D) Q vs. K phase diagram since it is also strongly disordered.

Acknowledgments

We thank John Mydosh and Cristian Batista for useful comments and discussions. Work at Lawrence Berkeley National Laboratory was supported by the U.S. Department of Energy (DOE), Office of Basic Energy Sciences (OBES) under Contract No. DE-AC02-05CH11231. Work at Los Alamos National Laboratory (LANL) was performed under the auspices of the U.S. DOE OBES and was funded in part by the LANL Directed Research and Development program. X-ray absorption data were collected at SSRL, a national user facility operated by Stanford University on behalf of the DOE/OBES.

References

[1] Mydosh J A 1993 *Spin Glasses: An Experimental Introduction* 1st ed (London: CRC Press)
[2] Ramirez A P 2001 *Handbook on Magnetic Materials* vol 13 (Amsterdam: Elsevier) chap 4, p 423
[3] Gardner J S, Gingras M J and Greedan J E 2010 *Rev. Mod. Phys.* **82** 53
[4] Balents L 2010 *Nature* **464** 199
[5] Shiga M, Fujisawa K and Wada H 1993 *J. Phys. Soc. Japn.* **62** 1329
[6] Custers J *et al.* 2003 *Nature* **424** 524
[7] Burdin S, Grempel D R and Georges A 2002 *Phys. Rev. B* **66** 045111
[8] Senthil T, Vojta M and Sachdev S 2004 *Phys. Rev. Lett.* **69** 035111
[9] Coleman P 1999 *Physica B* **259-261** 353
[10] Ott H R, Rudigier H, Felder E, Fisk Z and Thompson J D 1987 *Phys. Rev. B* **35** R1452
[11] H J van Daal, K H J Buschow, P B van Aken and M H van Maaren 1975 *Phys. Rev. Lett.* **23** 1457
[12] Ott H R, Rudigier H, Felder E, Fisk Z and Batlogg B 1985 *Phys. Rev. Lett.* **55** 1595
[13] Nakamura H, Kitaoka Y, Asayama K, Onuki Y and Shiga M 1994 *J. Phys.: Condens. Matter* **6** 10567
[14] Bauer E D, Booth C H, Kwei G H, Chau R and Maple M B 2001 *Phys. Rev.* B **65** 245114
[15] Hayes T M and Boyce J B 1982 *Solid State Physics* vol 37 ed Ehrenreich H, Seitz F and Turnbull D (New York: Academic) p 173
[16] Li G G, Bridges F and Booth C H 1995 *Phys. Rev.* B **52** 6332
[17] Crozier E D, Rehr J J and Ingalls R 1988 *X-Ray Absorption: Principles, Applications, Techniques of EXAFS, SEXAFS, XANES* ed Konigsberger D and Prins R (New York: Wiley) p 373
[18] Umarji A M *et al.* 1987 *J. Phys. F: Met. Phys.* **17** L25
[19] Sarrao J L *et al.* 1999 *Phys. Rev. B* **59** 6855
[20] Millis A J 1993 *Phys. Rev. B* **48** 7183
[21] Stewart G R 2001 *Rev. Mod. Phys.* **73** 797
[22] Custers J, Gegenwart P, Geibel C, Steglich F, Coleman P and Paschen S 2010 *Phys. Rev. Lett* **104** 186402
[23] Friedemann S *et al.* 2009 *Nature Phys.* **5** 465

A NQR study of $CeOs_2Al_{10}$

Harukazu KATO[1], Tomoaki TAKESAKA[1], Riki Kobayashi[1], Takashi NISHIOKA[1], Masahiro MATSUMURA[1], Yo TOKUNAGA[2], and Shinsaku KAMBE[2]

[1] Faculty of Science, Kochi University, Kochi 780-8520, Japan
[2] Advanced Science Research Center, Japan Atomic Energy Agency, Tokai, Ibaraki 319-1195, Japan

E-mail: katoharu@kochi-u.ac.jp

Abstract. NQR measurements have been carried out for the Al nuclei in $CeOs_2Al_{10}$, which shows a mysterious transition at T_0 = 29 K. Some of the NQR lines split into two lines below T_0, reflecting by symmetry lowering which occurs in the ordered state. The spin-lattice relaxation rate $1/T_1$ does not show a linear temperature dependence in the ordered state, but reaches a constant value with the temperature down to zero kelvin. This implies that the present compound has no density of states at the Fermi energy. These results are compared with those reported on $CeRu_2Al_{10}$.

1. Introduction

CeT_2Al_{10} (T = Ru, Os) crystallizes in the $YbFe_2Al_{10}$-type structure with the orthorhombic *Cmcm* space group[1, 2], where the Ce-Ce distance is rather large (~ 5.2 Å). Because of the long Ce distance, usual RKKY-type magnetic interaction between the f moments would be expected as weak. In fact, other Ce compounds with similar Ce-Ce distances show no transition or, if any, magnetic ordering only at a temperature less than 2 K[3]. On the other hand, CeT_2Al_{10} shows a phase transition at a remarkable high temperature of T_0 = 27 K (Ru), 29 K (Os)[2, 4, 5, 6, 7], which might be hardly ascribed as typical RKKY-interaction-induced magnetic ordering of the f-moment. In addition, the transition is characterized by the decrease of the magnetic susceptibility along *all* of the $a-$, $b-$, and $c-$ axes, as well as by the *increase* of the electric resistivity just below T_0[4, 5, 6], which are not seen in an ordinary magnetic transition.

Nuclear quadrupole resonance (NQR) is a powerful tool to investigate local electric and magnetic environments at the nuclear site. CeT_2Al_{10} has five inequivalent Al sites, e.g., Al(1) and Al(2) with the different $8g$ sites in the *Cmcm* space group, Al(3) and Al(4) with the different $8f$, and Al(5) with the $8e$. NQR for the Al nucleus would be reflected by symmetry changes through the transition, if any, in points of both the crystallographic and magnetic views. Our Al-NQR measurements carried out on $CeRu_2Al_{10}$ indicate that the NQR lines for Al(1)-Al(4) are significantly affected in the ordered state, and that the Al(5) remains unchanged[8]. By fine analyses, those changes have been identified as owing to a crystallographic distortion which breaks the two-fold axis along the a-axis, not owing to appearance of internal field at the ligand site.

Following the NQR measurements, many experiments have been carried out on $CeRu_2Al_{10}$. Based on the NQR results as well as the transport and magnetic properties in the ordered state, a possible scheme to deal with the transition has been suggested, which is a spin-singlet state formation of the f-electron[9, 10]. It would be accompanied with appearance of a gap, which has been in fact evidenced by an inelastic neutron experiment[11]. It is known that spin-singlet formation often couples with the lattice deformation, as known in a spin-Peierls system, which might arise in the present system[12]. However,

Published under licence by IOP Publishing Ltd

very recently, μSR measurements [13, 14] as well as neutron diffraction studies[14] have proposed that long-ranged ordered moment appears below T_0. In this context, the "simple" spin-singlet model is clearly inappropriate. It is to be still noted that, although the transition releases a large entropy of ~ 0.7 Rln2[4], magnitude of the ordered moment estimated from the neutron experiments is as, at most, $\sim 0.3\ \mu_B$ [14], which might *not* be enough to explain quenched degree of freedom of the f electron. The nature of the ordered state might be controversial.

In the context, the ordered state of $CeOs_2Al_{10}$ also come into question. Here we comment on the difference in physical behaviors between $CeRu_2Al_{10}$ and $CeOs_2Al_{10}$: although they commonly show an increase in electric resistivity just at T_0, the former (Ru) shows a sudden drop in resistivity with decreasing temperature far below T_0, and the latter (Os) is semiconductive down to zero temperature[2, 4, 5]. In this paper, we report NQR measurements on $CeOs_2Al_{10}$. The spectrum change through the transition in the present compound resembles that in the Ru compound, which suggests that $CeOs_2Al_{10}$ shows a similar symmetry breaking to $CeRu_2Al_{10}$ below T_0. The ordered state of $CeOs_2Al_{10}$ and $CeRu_2Al_{10}$ might have the same origin. The spin-lattice relaxation rate $1/T_1$ is also presented. $1/T_1$ in the ordered state goes to a constant value with decreasing temperature, implying no density of states remains at the Fermi energy. We compare the results with those of $CeRu_2Al_{10}$[8].

2. Experiments

A powder sample of $CeOs_2Al_{10}$, which was prepared by crushing single crystals, was used for our experiments. The single crystals were obtained by an Al self-flux method[4]. NQR measurements presented were carried out in zero field. Since an Al nucleus has a nuclear spin I of 5/2, each Al site with a non-cubic crystallographic environment gives two NQR lines, corresponding to the $\pm 5/2 \rightleftarrows \pm 3/2$ and $\pm 3/2 \rightleftarrows \pm 1/2$ transitions. Hereafter we call the former the higher frequency (HF) line, and the latter the lower frequency (LF). We also define $\nu_Q \equiv 3e^2qQ/[2hI(2I-1)]$ and $\eta \equiv (V_{XX} - V_{YY})/V_{ZZ}$, where V_{XX}, V_{YY} and V_{ZZ} are the principal values of the electric field gradient tensor, $eq = V_{ZZ}$, and eQ the quadrupole moment of the nucleus. All signals were detected by a spin-echo method with a homemade phase coherent spectrometer. Spectra in the paramagnetic and ordered states were taken by collecting the integrated intensities of the echo signal as the frequency is swept. Corrections on the signal intensities in the spectra were not taken. Relaxation time measurements were done with an inversion recovery method. In this experiments, we treat $1/T_1$ for the HF lines from the Al(5) and Al(4) sites, both of which has $\eta \sim 0$ as discussed below. In such cases, The nuclear magnetization $M(t)$ after the inversion pulse obeys a function of $1 - M(t)/M(\infty) \propto 0.427\exp(-3t/T_1) + 0.573\exp(-10t/T_1)$, with which we estimated $1/T_1$ as a fitting parameter.

3. Results and discussion

Figure 1(a) shows the NQR spectrum in the paramagnetic state. Combined with the NMR results (not shown here) in an external field, these lines are successfully assigned as shown in the figure. The HF lines for Al(5)-Al(3) and Al(1) are clearly found, although the corresponding LF lines are considerably weak in intensity, whose reason has been unclear. (The lines for Al(2) were missing in our frequency range measured, probably because the site has a smaller ν_Q value) From the observed frequencies of the HF lines, we then estimate $\nu_Q = 2.61$ MHz (Al(5)), 1.82 MHz (Al(4)), 1.44 MHz (Al(3)), and 1.78 MHz (Al(1)), where η is assumed as ~ 0 for Al(5), Al(4), and Al(3), and as 0.77 for Al(1) based on the NMR spectrum. These ν_Q values estimated are similar to those in $CeRu_2Al_{10}$[8]. In the ordered state, the NQR spectrum is changed as shown in Fig. 1(b). Here we could ascribe the observed lines as composed of the unaffected Al(5)-HF line and a set of the two lines split from each of the Al(4)-, Al(3)-, and Al(1)-HF lines. In fact, with increasing temperature, the frequency positions of the split lines move toward that of the corresponding line in the paramagnetic state (not shown here). The qualitatively trend of the splitting in the ordered state coincides with that in $CeRu_2Al_{10}$. This strongly suggested that symmetry breaking arising in the ordered state is the same between $CeOs_2Al_{10}$ and $CeRu_2Al_{10}$. Possibly, the transition seen in $CeOs_2Al_{10}$ and $CeRu_2Al_{10}$ has the same origin.

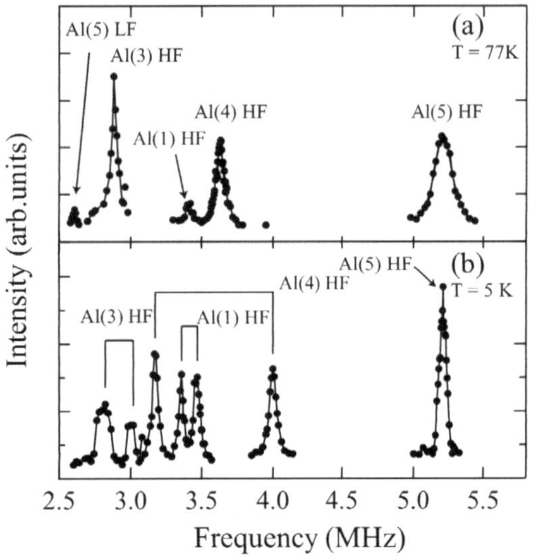

Figure 1. Zero-external-field-applied NQR spectra in the paramagnetic state (a) and in the ordered state (b). An assignment of the lines is also presented. Corrections for the signal intensities have not been done.

We have previously proposed on $CeRu_2Al_{10}$ that the lattice deformation which breaks the two-fold axis symmetry along the a axis should occur in the ordered state[8]. At present, we infer the statement might be also valid on $CeOs_2Al_{10}$. Here, it should be noted that the transition for these compounds accompanies a small but finite ordered magnetic moment of $\sim 0.3\,\mu_B$, as evidenced by the recent neutron diffraction experiments on $CeRu_2Al_{10}$[14] and on $CeOs_2Al_{10}$ where the report on the latter compound will be published elsewhere: the long-ranged ordered moments at Ce would give rise to, more or less, internal field at the ligand sites, which might be an alternative origin of splitting of the NQR lines. However, we confirm that, as far as assuming the proposed magnetic structure[14] and dipole field from the ordered moments, it seems still difficult to make a fully explanation for the observed alternation of the NQR lines only in terms of magnetic origin. In order to interpret the NQR and the neutron results consistently, more careful analyses would be desired.

Figure 2 shows the temperature dependence of $1/T_1$ measured on the HF lines of the Al(5) and Al(4) sites. In measuring $1/T_1$ of Al(4) in the ordered state, we use the line observed at higher frequency between the split two. $1/T_1$ on both the site shows a kink at T_0 = 29 K. It is noted that any divergent increase of $1/T_1$ is not seen for Al(5) and Al(4), even though the latter site causes the splitting of the corresponding NQR lines. In an ordinary magnetic transition, critical slow down of fluctuating magnetic moments often causes a divergent increase of $1/T_1$ at the transition temperature. The absence found in the present compound might be reflected by smallness of the ordered magnetic moments. With decreasing temperature in the ordered state, $1/T_1$ in the present compound monotonically decreases and goes toward a constant value at zero temperature. This behavior is clearly different from that seen in $CeRu_2Al_{10}$[8] (plotted as open circles in the figure); $1/T_1$ in $CeRu_2Al_{10}$ shows a behavior of $1/T_1 \propto T$ at low temperatures. In general, linear temperature dependence of $1/T_1$, which is known as the Korringa-law, originates from fluctuations of electrons on the Fermi surface, and the slope (the $1/T_1T$ values) is proportional to the density of states (DOS) at the Fermi energy. The discrepancy in the low-temperature behavior of $1/T_1$ between $CeOs_2Al_{10}$ and $CeRu_2Al_{10}$ could be explained in terms of different properties of the Fermi surface, i.e., the present compounds has no DOS at the Fermi energy. In fact, the electric resistivity of $CeOs_2Al_{10}$ shows a semiconductive behavior, although that of $CeRu_2Al_{10}$ far below T_0 shows a positive slope against temperature[2, 4, 5, 6, 7].

A possibility that the ordered state accompanies formation of a pseudo gap around the Fermi energy has been proposed on both $CeOs_2Al_{10}$ and $CeRu_2Al_{10}$[4, 5]. In this framework, the difference between the two compounds is whether or not the residual DOS remains at the Fermi energy in opening a pseudo

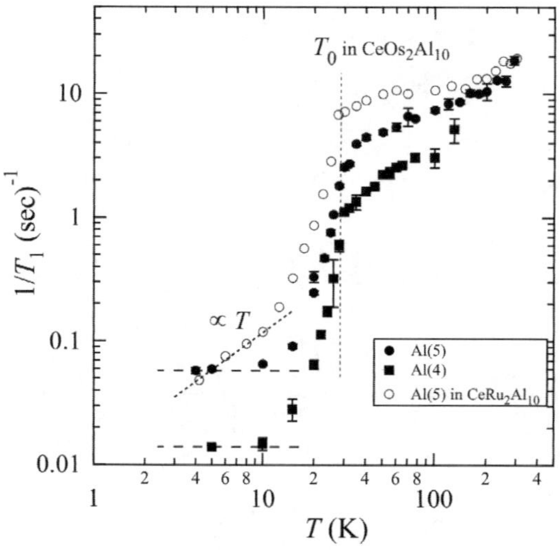

Figure 2. Nuclear-lattice relaxation rate, $1/T_1$, plotted against temperature in a log-log scale. The solid circle and solid square indicate $1/T_1$ measured for the Al(5)- and Al(4)-HF lines, respectively. For comparison, $1/T_1$ measured on the Al(5)-HF lines in $CeRu_2Al_{10}$ is also shown, which is denoted as the open circle.

gap. At present, we have no idea of the origin of the discrepancy. However, we still point out that the transition temperature is controlled by external pressure and that the physical properties of $CeRu_2Al_{10}$ with a pressure of 2 GPa resembles that of the Os compound[4]. This might imply that the residual DOS seen in $CeRu_2Al_{10}$ is well controlled by some external parameters, that is, it is not only owing to extrinsic effect such as impurities. An unified picture to involve the DOS alternation would be desired.

acknowledgements
We would like to thank Professor Yasuoka for valuable discussion. This work is partially supported by KAKENHI for Young-Scientists (B), 20740200. Some of the authors (YT and SK) are supported by a Grant-in-Aid for Scientific Research on Innovative Areas "Heavy Electrons" of the Ministry of Education, Culture, Sports, Science and Technology of Japan.

References

[1] Thiede V M T, Ebel T, and Jeitschko W, 1998 *J. Mat. Chem.* **8** 125-130
[2] Strydom A M, 2009 *Phys. B* **404** 2981-2984
[3] see for example, Sereni J G, 1991 *Handbook on the Physics and Chemistry of Rare Earths*, vol. 23 ed Gschneider K A Jr and Eyring L (Amsterdam:Elsevier) p 1
[4] Nishioka T, Kawamura Y, Takesaka T, Kobayashi R, Kato H, Matsumura M, Kodama K, Matsubayashi K, and Uwatoko Y, 2009 *J. Phys. Soc. Jpn.* **78** 123705
[5] Muro Y, Kajino J, Umeo K, Nishimoto K, Tamura R, and Takabatake T, 2010 *Phys. Rev.B* **81** 214401
[6] Taksesaka T, Oe K, Kobayashi R, Kawamura Y, Nishioka T, Kato H, Matsumura M, and Kodama K, 2010 *J. Phys. Conf. Series* **200** 012201
[7] Muro Y, Motoya K, Saiga Y, and Takabatake T, 2010 *J. Phys. Conf. Series* **200** 012136
[8] Matsumura M, Kawamura Y, Edamoto S, Takesaka T, Kato H, Nishioka T, Tokunaga Y, Kambe S, and Yasuoka H, 2009 *J. Phys. Soc. Jpn.* **78** 123713
[9] Tanida H, Tanaka D, Sera M, Moriyoshi C, Kuroiwa Y, Takesaka T, Nishioka T, Kato H, and Matsumura M, 2010 *J. Phys. Soc. Jpn.* **79** 043708
[10] K. Hanzawa, 2010 *J. Phys. Soc. Jpn.* **79** 047310
[11] Robert J, Mignot J M, Andre G, Nishioka T, Kobayashi R, Matsumura M, Tanida H, Tanaka D, and Sera M, *Preprint* cond-mat/1003.4933
[12] Hanzawa K, *Preprint* cond-mat/1005.5017
[13] Kambe S, Chudo H, Tokunaga Y, Koyama T, Sakai H, Ito T U, Ninomiya K, Higemoto W, Takesaka T, Nishioka T, and Miyake Y, 2010 *J. Phys. Soc. Jpn.* **79** 053708
[14] Khalyavin D D et al., *Preprint* cond-mat/1006.1618

International Conference on Strongly Correlated Electron Systems (SCES 2010) IOP Publishing
Journal of Physics: Conference Series **273** (2011) 012038 doi:10.1088/1742-6596/273/1/012038

Magnetic and transport properties of CeT_2Al$_{10}$ (T = Fe, Ru, Os) under pressure

Yukihiro Kawamura[1], Yuta Ogane[1], Takashi Nishioka[1], Harukazu Kato[1], Masahiro Matsumura[1], Kazuyuki Matsubayashi[2] and Yoshiya Uwatoko[2]

[1]Graduate School of Integrated Arts and Sciences, Kochi University, Kochi 780-8520, Japan
[2]Institute for Solid State Physics, University of Tokyo, Kashiwa 305-0047, Japan

E-mail: b08d6a02@s.kochi-u.ac.jp

Abstract. The results of electrical resistivity ρ measurements for CeT_2Al$_{10}$ (T = Fe, Ru, Os) under pressure are reported. The abrupt increase was observed at $T_0 \sim 27$ K for CeRu$_2$Al$_{10}$ and 29 K for CeOs$_2$Al$_{10}$ at ambient pressure. By applying pressure, T_0 for CeRu$_2$Al$_{10}$ and CeOs$_2$Al$_{10}$ suddenly disappeared at 4 GPa and 2 GPa, respectively. Although the ground state of CeRu$_2$Al$_{10}$ is metal at ambient pressure, ρ at low temperatures increased and the ground state became semiconductor at 2 GPa. The increase in ρ suppressed above 3 GPa and the ground state became metal above 5 GPa. Magnetic contribution to CeRu$_2$Al$_{10}$ above 4 GPa shows maximum, which seems to be attributed to Kondo coherence. These systematical change in $\rho(T)$ was also observed in CeFe$_2$Al$_{10}$ and CeOs$_2$Al$_{10}$. From the pressure dependence of the temperature of the maximum, we suggests that CeFe$_2$Al$_{10}$ and CeOs$_2$Al$_{10}$ is corresponding to 3.2 GPa and 1.3 GPa of CeRu$_2$Al$_{10}$.

1. Introduction

Heavy fermion compounds CeT_2Al$_{10}$ (T = Fe, Ru, Os) crystallizes in the YbFe$_2$Al$_{10}$-type crystal structure(orthorhombic, space group $Cmcm$)[1, 2]. These compounds is characteristic of their new phase transition, which occurred at $T_0 \sim 27$ K on CeRu$_2$Al$_{10}$[3] and 29 K on CeOs$_2$Al$_{10}$[4], and for Kondo semiconducting behavior on CeFe$_2$Al$_{10}$[5]. At first the new phase transition is reported to be attributed to antiferromagnetic order[3]. However, ^{27}Al-NQR study for CeRu$_2$Al$_{10}$ denies it because the splitting of line spectrum below T_0 is different from antiferromagnetic order[6]. Besides, 27 K is too high to order antiferromagnet in this small concentration of magnetic system, where the nearest-neighbor of Ce exceeds 5Å. This is why this new phase transition is attracted much attention. Recently, μ^+SR and neutron diffraction experiments revealed that this transition is due to long-range magnetic ordering of the Ce sub-lattice with a reduced moment of 0.34 μ_B[7]. However, the origin of very high T_0 and low value of the ordered moments has not clarified yet. In addition, the relation between Kondo semiconducting behavior and this phase transition is not clarified. The scaling behavior against pressure among isostructural CeT_2Al$_{10}$ (T=Ru, Os and Fe) may be one of the key aspects to reveal these issues.

The purpose of this study is to clarify the relationship between each CeT_2Al$_{10}$. In our previous study, we reported overall behavior of $\rho(T)$ for CeT_2Al$_{10}$ under pressure. In this paper, we report

Published under licence by IOP Publishing Ltd

$\rho(T)$ for CeT_2Al$_{10}$ in detail. In addition, we discuss pressure dependence of each characteristic temperature.

2. Experimental

Single crystals of CeT_2Al$_{10}$ (T = Fe, Ru, Os) were grown by using Al self-flux method. The starting composition of CeT_2Al$_{70}$ was placed in an Al_2O_3 crucible, sealed in a quartz ampule in vacuum and heated to 1000 °C and then cooled to 720 °C at a rate of 5 °C/h and the Al flux was removed by centrifugation. The remaining flux was removed by sinking 5 mol/L NaOH solution for 24 hours. The resulting single crystals were confirmed as single phase by powder X-ray diffraction. The electrical resistivity under 2 GPa was measured by the ac 4 terminal method with the frequency of 40 Hz and effective current is 3 mA. The pressure was applied by using piston cylinder type pressure cell with Daphne 7373 oil as a pressure transmitting oil. The pressure was calibrated by the temperature of the superconductivity on In. The electrical resistivity above 2 GPa was measured by the conventional dc 4 terminal method with effective current of 3 mA. The pressure was applied by cubic anvil cell with the mixture of Fluorinert FC70:FC77=1:1 as a pressure transmitting oil.

3. Results and Discussion

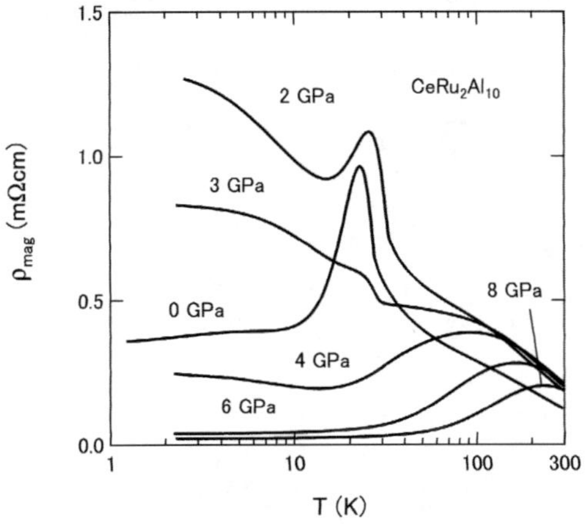

Figure 1. Temperature dependence of the magnetic contribution to the electrical resistivity ρ_{mag} (see text in detail) for $CeRu_2Al_{10}$ under pressure.

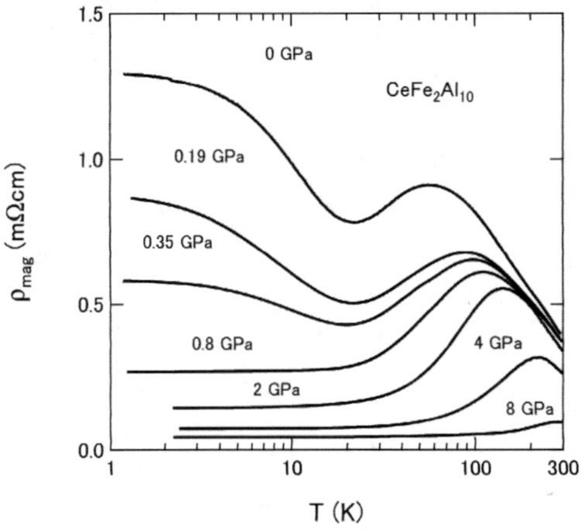

Figure 2. ρ_{mag} vs T for $CeFe_2Al_{10}$ under pressure.

Fig. 1 shows temperature dependence of the magnetic contribution to the electrical resistivity ρ_{mag} for $CeRu_2Al_{10}$ under pressure. We have evaluated the ρ_{mag} from the experiment by subtracting the phonon contribution ρ_{ph} of the appropriate La compound. At ambient pressure, $\rho(T)$ increases with decreasing temperature from room temperature to 27 K. At 27 K, $\rho(T)$ increases abruptly reflecting the appearance of the new phase transition. It exhibits a maximum at 23 K and then shows decreasing gap-like below 23 K. At 2 GPa, T_0 increases to 33 K. The value of ρ at low temperatures increases. This increase in ρ at low temperatures indicates ground state of this compound becomes semiconductor from metal. At 3 GPa, T_0 decreases to 29 K. With decreasing T_0, $\rho(T)$ at low temperatures suppresses. The broad peak, which seems to be attributed to high density Kondo effect, is observed over 4 GPa, where the novel phase transition disappears. The temperature of this maximum denotes T_{max}. The T_{max} increases monotonically with pressure.

Fig. 2 shows the temperature dependence of ρ_{mag} for $CeFe_2Al_{10}$. At ambient pressure, ρ_{mag} exhibits a maximum at $T_{max} \sim 70$ K and shows increasing at low temperatures. The former seems to be attributed to dense Kondo effect and the latter seems to be attributed to Kondo semiconducting gap. This behavior is similar to ρ_{mag} for $CeRu_2Al_{10}$ at 4 GPa. However, the rate of the increase in ρ_{mag} for $CeFe_2Al_{10}$ is higher than that for $CeRu_2Al_{10}$. On the other hand, ρ_{mag} for $CeFe_2Al_{10}$ does not show phase transition, which one can see that for $CeRu_2Al_{10}$ at 3 GPa. Thus, ρ_{mag} for $CeFe_2Al_{10}$ is considered to be attributed to that for $CeRu_2Al_{10}$ at the pressure between 3 GPa and 4 GPa. Increasing at low temperatures suppresses and T_{max} increases with pressure. Although semiconducting behavior at 0 GPa suddenly suppresses at 2 GPa in our previous study, we confirmed that semiconducting behavior gradually suppresses with pressure in this study.

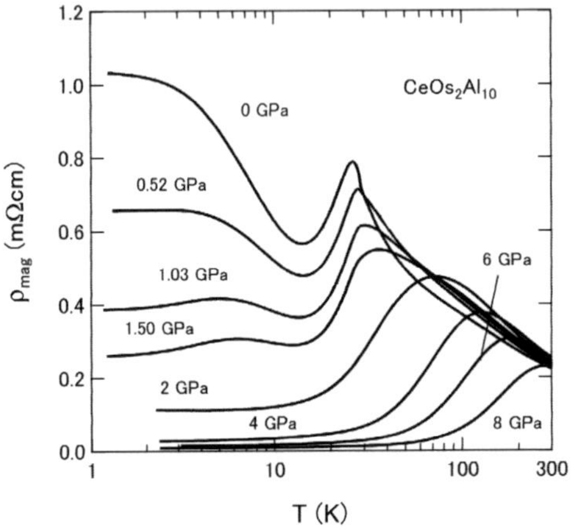

Figure 3. ρ_{mag} vs T for $CeOs_2Al_{10}$ under pressure.

Fig. 3 shows the temperature dependence of ρ_{mag} for $CeOs_2Al_{10}$. At ambient pressure, $\rho(T)$ in $CeOs_2Al_{10}$ increases abruptly at $T_0 \sim 29$ K and exhibit a peak at 26 K followed by a minimum and increasing with decreasing temperature. The behavior of ρ_{mag} for $CeOs_2Al_{10}$ is similar to that for $CeRu_2Al_{10}$ at 2 GPa, although the value of a peak after T_0 is smaller than that for $CeRu_2Al_{10}$. At 0.52 GPa, the abrupt increase in ρ_{mag} below T_0 is suppressed. At 1.03 GPa, the kink appears instead of the abrupt increase. This kink decreases with increasing pressure. On the other hand, the increase at low temperature is gradually suppressed as the pressure increases. The ground state become metal above 2 GPa. Above 3 GPa, T_{max} gradually increases with pressure.

Fig. 4(a) shows the pressure dependence for T_0. T_0 for $CeRu_2Al_{10}$ increases slightly with pressure at first and exhibit a maximum at around 2 GPa. At 4 GPa, T_0 disappears suddenly like first order transition. On the other hand, T_0 for $CeOs_2Al_{10}$ decreases slightly with pressure and suddenly disappears at 2 GPa. Fig. 4(b) shows the pressure dependence for T_{max}. T_{max} increases monotonically with pressure in each sample. The dotted line indicates least square fitting to the each data. T_{max} obtained from the data of pressure measurements using piston cylinder cell is considered to be higher than actual value, because the pressure is calibrated at low temperatures and expected to be higher at high temperatures. Thus, we used the data of ambient pressure measurements and of pressure measurements using cubic anvil cell as fitting data. The dT_{max}/dP is similar to each other. The pressure of the extrapolation to the x axis for $CeFe_2Al_{10}$, $CeRu_2Al_{10}$ and $CeOs_2Al_{10}$ is -1.9 GPa, 1.3 GPa and 0 GPa, respectively. This result suggests that $CeFe_2Al_{10}$ and $CeOs_2Al_{10}$ is corresponding to 3.2 GPa and 1.3 GPa of $CeRu_2Al_{10}$. Since the lattice constant of $CeFe_2Al_{10}$ is smaller than that of $CeRu_2Al_{10}$, this result is consistent with chemical pressure. However the lattice constant of $CeRu_2Al_{10}$ is almost same as but is smaller than that of $CeOs_2Al_{10}$. Although it contradicts to present results, the reason is not clarified.

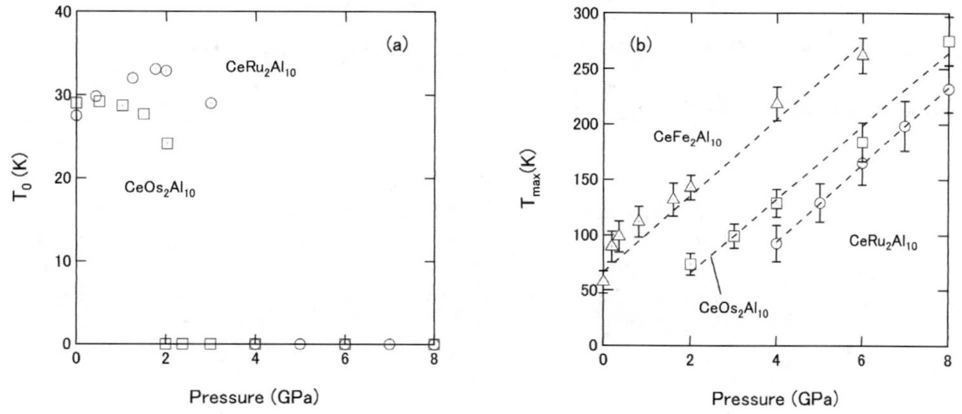

Figure 4. (a)T_0 vs P for $CeRu_2Al_{10}$(open circle) and for $CeOs_2Al_{10}$ (open square). The data pointed at 0 K indicates that there is not any phase transition above 2 K. (b) T_{max} vs P for $CeFe_2Al_{10}$ (open triangle), $CeRu_2Al_{10}$ (open circle) and $CeOs_2Al_{10}$(open square). The dotted line indicates least square fitting to the each data(see text in detail).

4. Summary

$\rho(T)$ for $CeRu_2Al_{10}$ increases abruptly at T_0 and exhibit a maximum at 23 K and then decreases with temperature. By applying pressure, T_0 slightly increases and $\rho(T)$ at low temperatures enhances. This behavior is similar to $\rho(T)$ for $CeOs_2Al_{10}$ at ambient pressure. At 4 GPa, Kondo semiconducting like behavior is observed, which is similar to $\rho(T)$ for $CeFe_2Al_{10}$ at ambient pressure. The broad peak, which is seems to be attributed to high density Kondo effect, is observed over 4 GPa. The temperature of this broad peak increases monotonically with pressure, which is also observed in $\rho(T)$ for $CeOs_2Al_{10}$ and $CeFe_2Al_{10}$. From these behaviors, we suggests that $CeFe_2Al_{10}$ and $CeOs_2Al_{10}$ is corresponding to 3.2 GPa and 1.3 GPa of $CeRu_2Al_{10}$.

References

[1] A. I. Tursina, S. N. Nesterenko, E. V. Murashova, I. V. Chernyshev, H. Noel and Y. D. Seropegin: Acta Cryst. E **61** (2005) i12.
[2] V. M. T. Thiede, T. Ebel, W. Jeitschoko: J. Mater. Chem. **8** (1998) 125.
[3] A. M. Strydom: Physica B **404** (2009) 2981.
[4] T. Nishioka, Y. Kawamura, T. Takesaka, R. Kobayashi, H. Kato, M. Matsumura, K. Kodama, K. Matsubayashi and Y. Uwatoko: J. Phys. Soc. Jpn. **78** (2009) 123705.
[5] Y. Muro, K. Motoya, Y. Saiga, and T. Takabatake: J. Phys. Soc. Jpn. **78** (2009) 083707.
[6] M. Matsumura, Y. Kawamura, S. Edamoto, T. Takesaka, H. Kato, T. Nishioka, Y. Tokunaga, S. Kambe and H. Yasuoka: J. Phys. Soc. Jpn. **78** (2009) 123713.
[7] D.D. Khalyavin et al., : cond-mat. /1006.1618v1.

Electronic structure of URu_2Si_2 in paramagnetic phase studied by soft x-ray photoemission spectroscopy

I. Kawasaki[1], S.-i. Fujimori[1], Y. Takeda[1], T. Okane[1], A. Yasui[1], Y. Saitoh[1], H. Yamagami[1,2], Y. Haga[3], E. Yamamoto[3] and Y. Ōnuki[3,4]

[1] Condensed Matter Science Division, Japan Atomic Energy Agency, Sayo, Hyogo 679-5148, Japan
[2] Department of Physics, Faculty of Science, Kyoto Sangyo University, Kyoto 603-8555, Japan
[3] Advanced Science Research Center, Japan Atomic Energy Agency, Tokai, Ibaraki 319-1195, Japan
[4] Graduate School of Science, Osaka University, Toyonaka, Osaka 560-0043, Japan

E-mail: kawasaki.ikuto@jaea.go.jp

Abstract. We have performed soft x-ray photoemission spectroscopy study on the heavy-fermion compound URu_2Si_2. We have found that the U 5f states are distributed in the binding energy range between 0.6 eV and the Fermi level. In angle-resolved photoemission spectra, we have observed the U 5f-Ru 4d hybridized band. Our results strongly suggest that the U 5f states form itinerant energy bands in this system.

1. Introduction

The heavy-fermion compound URu_2Si_2 has attracted much attention over the past two decades. It undergoes a puzzling phase transition into the "hidden order" (HO) state at T_0 = 17.5 K. Although intensive experimental and theoretical studies have been performed so far [1], the true order parameter of the HO phase has not been directly detected experimentally, and none of the proposed theories can fully account for all the experimental observations. According to various experiments such as specific heat [2], thermal conductivity [3, 4] and Hall effect [5], an energy gap opens in a large portion of Fermi surface below T_0, leading to a large reduction of carrier density. This implies a drastic reconstruction of the electronic state at T_0. Therefore it is essential to elucidate the electronic structure in order to clarify the nature of the HO transition. Photoemission spectroscopy (PES) provides a direct probe of the band structure of solids. To investigate the electronic structure of URu_2Si_2, angle-resolved photoemission spectroscopy (ARPES) studies have been performed in photon energy region of $h\nu$ = 21-156 eV [6, 7, 8]. According to these ARPES studies, the U 5f states are distributed in a narrow energy range within 0.2 eV below the Fermi level (E_F) and form less dispersive bands. The bandwidth of the observed U 5f bands is much narrower than that of the band structure calculation based on the local density approximation, suggesting an existence of a strong renormalization effect. However, these studies were performed in the lower photon energy range, and it was difficult to relate the observed band structure to the bulk properties because of the high surface sensitivity of those experiments.

Published under licence by IOP Publishing Ltd

Recently, PES experiments in soft x-ray region (SX-PES) have been recognized as a useful experimental technique in probing bulk electronic structures of strongly correlated materials [9, 10]. In the present study, we have performed the soft x-ray angle-integrated and angle-resolved PES experiments to elucidate the electronic structure of URu_2Si_2.

2. Experimental methods

Single crystal samples of URu_2Si_2 were prepared by the Czochralski pulling method in a tetra-arc furnace. The samples were cleaved *in situ* by the top post method with the surface parallel to (001) plane just before the measurements. SX-PES measurements were performed at BL23SU in SPring-8 using a Scienta SES 2002 electron energy analyzer. The base pressure was typically 4×10^{-8} Pa. The energy resolutions were about 80 and 180 meV for $h\nu = 403$ and 760 eV, respectively. The sample temperature was kept at 20 K above T_0 in order to avoid a reconstruction of Brillouin zone.

3. Results and discussion

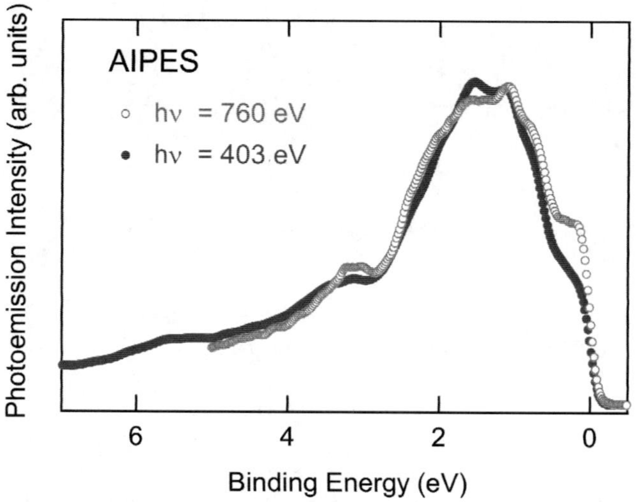

Figure 1. AIPES spectra of URu_2Si_2 measured at $h\nu = 403$ and 760 eV.

We first show the angle-integrated photoemission (AIPES) spectra. Figure 1 shows the AIPES spectra of URu_2Si_2 measured at $h\nu = 403$ and 760 eV. These photon energies were chosen to trace the cuts including the Γ point in momentum space. The solid angle determined by the angular acceptance of the electron analyzer corresponds to the two dimensional plane which covers the first and part of the neighboring Brillouin zones (BZs). Both spectra were normalized to match the tail of valence band states below 2 eV. The photoionization cross-section calculations for atomic orbitals [11] suggest that the contributions from the U 5f and the Ru 4d states are about one order of magnitude larger than those of other states such as Si 3p and U 6d in this photon energy range. The relative photoionization cross section of the U 5f states with respect to the Ru 4d states increases by a factor of about 2 by increasing photon energy $h\nu$ from 403 to 760 eV. As shown in Fig. 1, the intensity in the energy region from 0.6 eV to E_F enhances by increasing photon energy $h\nu$, suggesting the existence of the U 5f states in this energy region.

To investigate the U 5f states in more detail, we have measured ARPES spectra at $h\nu = 760$ eV. This photon energy corresponds to the scan along the Γ-X-Z plane in the body-centered tetragonal BZ (Fig. 2(a)) by assuming the inner potential of photoelectron final states to be 12 eV [6]. Figure 2(b) shows the ARPES spectra of URu_2Si_2 along the Z-Y-X line. As seen

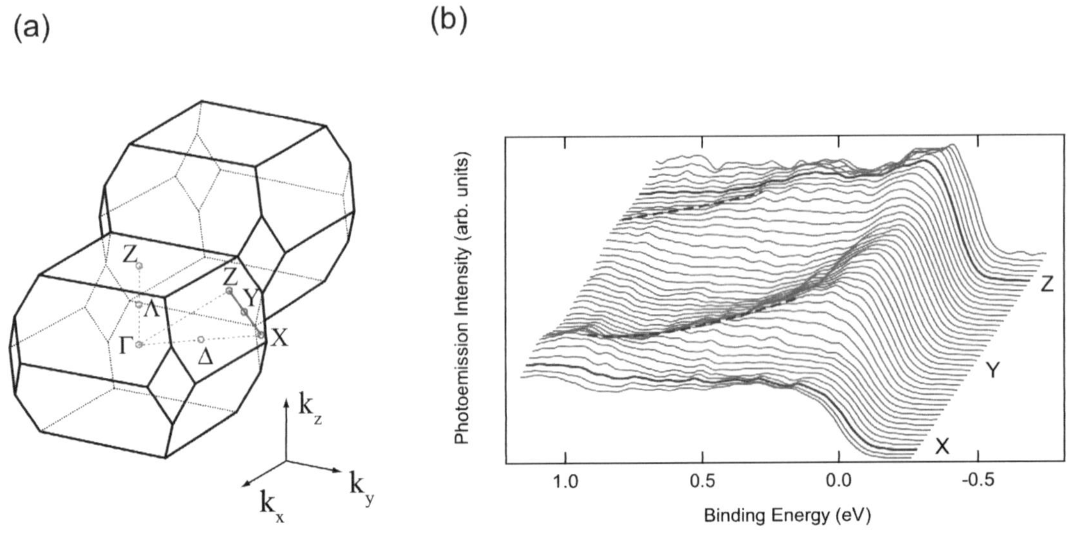

Figure 2. (a) Brillouin zone of URu_2Si_2. (b) ARPES spectra of URu_2Si_2 measured along the Z-Y-X line. The green broken lines are guides to the eyes.

in Fig. 2(b), a strongly dispersive hole-like band with a band maximum at around E_F was observed around the Z point. In addition, we have observed another dispersive band along the Y-X. These bands are considered to be contributions mainly from the Ru 4d states since these bands disperse to higher binding energies. In contrast, in the vicinity of E_F, we have observed a less dispersive band between the Z point and the Y point, suggesting a strong U 5f character. This band strongly hybridized with the rapidly dispersing band in the Y-X, suggesting that the U 5f states are strongly hybridized with the Ru 4d states. The existence of the f-d hybridized band is in contrast to the case of U 5f localized compound UPd_3 [12] where the hybridization between the U 5f states and the valence bands is weak and the U 5f states form non-dispersive bands in higher binding energy. Thus this result suggests that the U 5f states contribute to the energy band formation and have an itinerant character in this compound.

We compare the present results with previous ARPES studies. Denlinger *et al.* [6] and Ito *et al.* [8] have also revealed that the U 5f states are distributed in the vicinity of E_F. However, the energy distributions of the U 5f states in these studies are within 0.2 eV from E_F, which is more than three times narrower. In addition, Ito *et al.* [8] have observed a U 5f derived band in the vicinity of E_F along Z-Y-X line. The energy dispersion of the reported U 5f band is quite different from that of present study and is much narrower. Since the escape depths of photoelectrons in the present experiment are much larger than those of previous studies, those differences are considered to be originated from U 5f electronic states between the surface and the bulk electronic structures.

4. Conclusion

We have studied the electronic structure of URu_2Si_2 by means of SX-PES. In the AIPES spectra, we have revealed that the U 5f states are distributed over the binding energy region between 0.6 eV and E_F. In the ARPES spectra, we have found that the U 5f states are strongly hybridized with the rapidly dispersing Ru 4d band, suggesting that the U 5f states contribute to the energy-band formation in this compound and have an itinerant character.

Acknowledgments

This work was performed under the Proposal No. 2009A3821 at SPring-8 BL23SU and was financially supported by a Grant-in-Aid for Scientific Research on Innovative Areas "Heavy Electrons" (No. 20102003) from the Ministry of Education, Culture, Sports, Science and Technology, Japan.

References

[1] Shah N, Chandra P, Coleman P and Mydosh J A 2000 *Phys. Rev.* B **61** 564

[2] Maple M B, Chen J W, Dalichaouch Y, Kohara T, Rossel C, Torikachvili M S, McElfresh M W and Thompson J D 1986 *Phys. Rev. Lett.* **56** 185

[3] Behnia K, Bel R, Kasahara Y, Nakajima Y, Jin H, Aubin H, Izawa K, Matsuda Y, Flouquet J, Haga Y, Ōnuki Y and Lejay P 2005 *Phys. Rev. Lett.* **94** 156405

[4] Sharma P A, Harrison N, Jaime M, Oh Y S, Kim K H, Batista C D, Amitsuka H and Mydosh J A 2006 *Phys. Rev. Lett.* **97** 156401

[5] Schoenes J, Schonenberger C, Franse J J M and Menovsky A A 1987 *Phys. Rev.* B **35** 5375

[6] Denlinger J D, Gweon G H, Allen J W, Olson C G, Maple M B, Sarrao J L, Armstrong P E, Fisk Z and Yamagami H 2001 *J. Electron Spectrosc. Relat. Phenom.* **117-118** 347 (2001).

[7] Denlinger J D, Gweon G H, Allen J W, Olson C G, Dalichaouch Y, Lee B W, Maple M B, Fisk Z, Can eld P C and Armstrong P E 2000 *Physica* B **281&282** 716

[8] Ito T, Kumigashira H, Takahashi T, Haga Y, Yamamoto E, Honma T, Ohkuni H and Ōnuki Y 1999 *Phys. Rev.* B **60** 13390

[9] Sekiyama A. Iwasaki T, Matsuda K, Saitoh Y, Ōnuki Y and Suga S 2000 *Nature* **403** 396

[10] Ohkochi T, Fujimori S i, Yamagami H, Okane T, Saitoh Y, Fujimori A, Haga Y, Yamamoto E and Ōnuki Y 2008 *Phys. Rev.* B **78** 165110

[11] Yeh J J and Lindau I 1985 *At. Data Nucl. Data Tables* **32** 1

[12] Ito T, Kumigashira H, Souma S, Takahashi T, Haga Y, Tokiwa Y and Ōnuki Y 2002 *Phys. Rev.* B **66** 245110

NMR studies in the half-Heusler type compound YbPtSb

T Koyama[1], M Abe[1], T Mito[1], K Ueda[1], T Kohara[1], H S Suzuki[2]

[1] Graduate School of Material Science, University of Hyogo, Kamigori, Hyogo 678-1297, Japan
[2] National Institute for Materials Science, Tsukuba 305-0047, Japan

E-mail: t-koyama@sci.u-hyogo.ac.jp

Abstract. ^{121}Sb and ^{195}Pt nuclear magnetic resonance (NMR) has been studied in the half-Heusler type compound YbPtSb to obtain information on local magnetic behavior. The characteristics of the localized $4f$ spins are observed in the Cuire-Weiss type behavior of the Knight shifts K for both ^{121}Sb and ^{195}Pt. From the slope of K-χ plots we estimated hyperfine coupling constants of -3.8 and -4.6 kOe/μ_B at Sb and Pt sites, respectively. It was found that the spin-echo decay rate $1/T_2$ of ^{121}Sb shows a clear peaks at 10 K. Similar tendency was also observed in case of ^{195}Pt. However, static properties do not show any anomalies near 10 K.

1. Introduction

The study of Yb-based compounds is an active area of research in strongly correlated electron systems. The $4f$ shell of trivalent Yb ion is one electron short of full occupation and it is often considered as the hole analog of trivalent Ce ion in which a lone electron resides in the $4f$ shell.

The huge liner-temperature specific heat coefficient, γ, in Yb-based compounds is one of the important issues in strongly correlated f electron systems. The extremely large γ was first discovered in YbPtBi with γ=8 J /mol K^2 [1]. This value is an order magnitude more than in typical heavy-fermion materials. $YbCo_2Zn_{20}$ is also known to have a huge γ=8 J /mol K^2 [2] and the pressure-induced magnetic transition was suggested [3]. The important question is whether the massive electronic state observed in these Yb-base compounds can be associated with the existence of very heavy electrons or not.

YbPtSb forms the half-Heusler MgAgAs structure (space group $F\bar{4}3m$), which is isostructual to YbPtBi. The magnetic susceptibility of YbPtSb obeys the Cuire-Weiss law with the effective magnetic moment, $\mu_{\rm eff}$=4.4$\mu_{\rm B}$, which is close to the free Yb^+ value (Yb^{3+} : $\mu_{\rm eff}$=4.54$\mu_{\rm B}$). The Weiss temperature, Θ, derived from between 5 and 300 K is estimated to be Θ=-5.3 K [4]. The electrical resistivity shows the metallic behavior. The value of $C_{\rm el}/T$ exceeds 6 J /mol K^2 at low temperatures [4, 5]. Here $C_{\rm el}$ is the electronic part of the specific heat. Low temperature specific heat and resistivity measurements suggested the magnetic ordering at 0.4 K [6]. It was also reported that an elastic softening of the longitudinal elastic constant C_{11} below around 20 K and its strong field dependence [7].

2. Experimental

Single crystals of YbPtSb were grown from Sb-flux method. For performing the nuclear magnetic resonance (NMR) experiment, several single crystals were crushed into powder. NMR

Published under licence by IOP Publishing Ltd

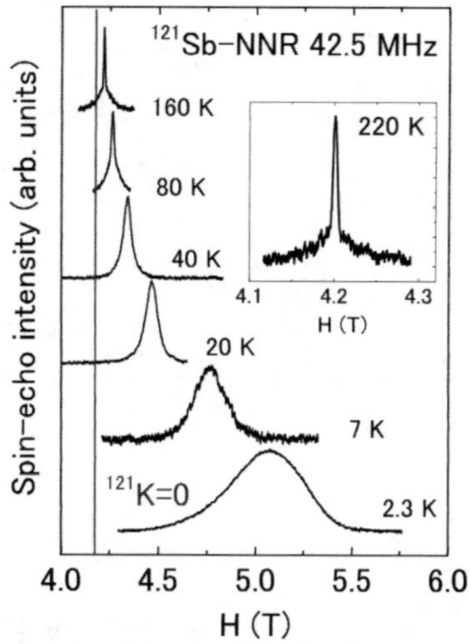

Figure 1. Temperature dependence of ^{121}Sb NMR spectra at a constant frequency of 42.5 MHz. $^{121}K=0$ indicates the zero shift. Inset shows the ^{121}Sb spectrum at 220 K.

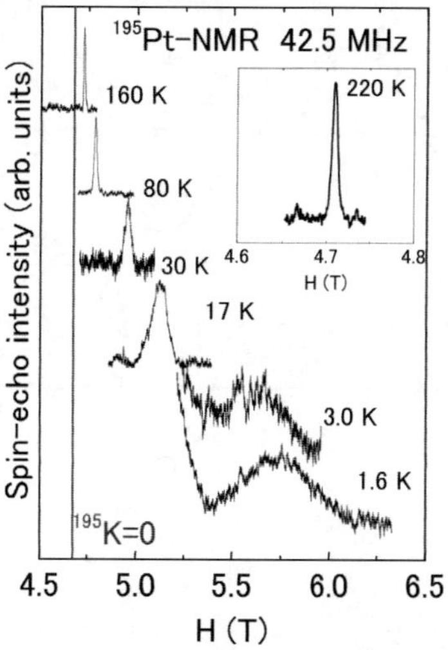

Figure 2. Temperature dependence of ^{195}Pt NMR spectra. $^{195}K=0$ indicates the zero shift. The large signals around 5 T at 3.0 K and 1.6 K originate from ^{121}Sb. Inset shows the ^{195}Pt spectrum at 220 K.

measurements were carried out using a conventional phase-coherent pulse spectrometer in the temperature range between 1.5 and 250 K. The magnetic susceptibility was measured using a superconducting quantum interference device magnetometer (Quantum Design, MPMS-5) at applied magnetic field of 4.5 T.

3. Results and discussion

Figure 1 and 2 show the temperature dependence of the ^{121}Sb (nuclear spin $I=5/2$) and ^{195}Pt ($I=1/2$) NMR spectra obtained by sweeping external field at a fixed frequency of 42.5 MHz. At high temperatures ^{195}Pt NMR spectrum is a single Gaussian-type. On the other hand we observed two spectra overlap with a narrow peak and a broad one for ^{121}Sb. These results are clearly seen in the inset of Fig. 1 and 2. We measured the magnetic field, H, dependence of ^{121}Sb NMR spectra (not shown). The line width of the narrow peak is proportional to H, whereas that of the broad one is independent of H. These results suggest that former is due to the inhomogeneous distribution of the Knight shift and latter is due to the inhomogeneous distribution of the electric field gradient (EFG). We suppose that two spectra overlap for ^{121}Sb originates from the local off-stoichiometry of Pt, as reported in other half-Heusler compounds [8].

The half-Heusler compound YbPtSb crystallizes in the face-centered cubic structure. Yb and Sb atoms are located at $4a(0, 0, 0)$ and $4b(1/2, 1/2, 1/2)$ positions forming rock salt structure arrangement and Pt atoms are located at one of the cube center positions $4c(1/4, 1/4, 1/4)$ leaving the other $4d(3/4, 3/4, 3/4)$ empty. However, it is likely that there are some voids at $4c$ positions or interstitial occupancies at $4d$ position in the present sample. The narrow NMR

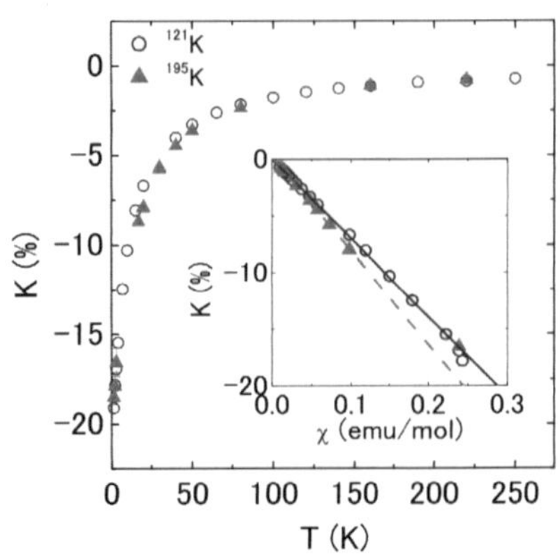

Figure 3. Temperature dependence of Knight shift K of ^{121}Sb (open circle) and ^{195}Pt (closed triangle). Inset shows K vs χ plots with temperature as an implicit parameter. Solid and dashed lines indicate the result of liner fitting to the data. χ is defined as M/H where H=4.5 T.

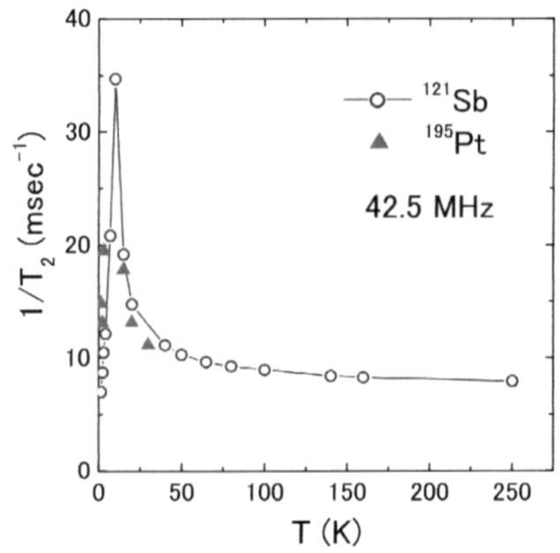

Figure 4. Temperature dependence of $1/T_2$ for ^{121}Sb (open circle) and ^{195}Pt (closed triangle). T_2 were measured at the peak position of the resonance lines (see Fig. 1 and Fig. 2). $1/T_2$ for ^{195}Pt around 10 K was not obtained due to the low NMR intensity.

spectrum of ^{121}Sb originates from 4*b* sites where the neighboring 4*c* position is occupied by Pt atom. Consequently the line width of ^{121}Sb should be narrow reflecting the cubic local symmetry of Sb site. On the other hand, there is local EFG experienced by those nuclei in the vicinity of crystalline defects (such as vacancies or interstitial occupancies), which break the local cubic symmetry of the Sb atoms. As a result, Sb NMR spectra become broad. ^{195}Pt spectra are not affected by local EFG owing to the lack of the nuclear quadrupole moment.

The NMR resonance lines of ^{121}Sb and ^{195}Pt shift to higher field side and broaden on cooling. We determined the ^{121}Sb and ^{195}Pt Knight shift (^{121}K and ^{195}K, respectively) from the peak position of the spectrum. Figure 3 shows the temperature dependence of ^{121}K and ^{195}K. Both ^{121}K and ^{195}K are large, negative and follow the Curie-Weiss behavior of the bulk susceptibility, χ. The inset of Fig. 3 displays the Knight shift vs χ plot with the temperature as an implicit parameter. ^{121}K and ^{195}K are proportional to χ, indicating that the Knight shift is mainly due to a transferred hyperfine field from localized Yb 4f-moments. Note that we used M/H as χ measured at 4.5 T, which is close value of magnetic field H to the present NMR measurement. In fact H dependence of the magnetization, M, does not show the linear relation at low temperatures. The slopes of the Knight shift vs χ yield hyperfine coupling constants of ^{121}A=-3.8 kOe/μ_B and ^{195}A=-4.6 kOe/μ_B for ^{121}Sb and ^{195}Pt, respectively.

As temperature approaching $T^* \equiv 10$ K the ^{121}Sb NMR lines lose intensity (Fig. 1) due to highly enhanced spin-echo decay rate $1/T_2$. Figure 4 shows the temperature dependence of $1/T_2$. T_2 was determined by fitting the single exponential function to the spin-echo decay curve. Although $1/T_2$ of ^{121}Sb is almost independent of temperature above 50 K, it exhibits a strong peak at 10 K. Unfortunately reliable $1/T_2$ values of ^{195}Pt could not be obtained near 10 K due to the poor signal-to-noise ratio. Each value of $1/T_2$ and T^* is unaffected by H. In order to obtain more information about this anomaly at 10 K, we have checked the static physical properties in YbPtSb. In spite of careful measurement of M, any anomalies were not detected around T^* in our experimental accuracy. Moreover, specific heat and resistivity measurements under H do not show anomalous behavior [5]. The charge glass-like state is suggested by an elastic softening of the longitudinal elastic constant and its logarithmic temperature dependence below around 20 K [7]. However, the NMR Knight shift, following the Curie-Weiss law, indicates the trivalent Yb ions. The relevance of the elastic softening to the anomalous $1/T_2$ behavior is not clear now.

^{121}Sb has both a quadrupole and magnetic moments while ^{195}Pt has only magnetic moment. Thus anomalous $1/T_2$ behavior observed in both nuclei can be understood to be the magnetic origin. In this case, $1/T_2$ is affected by the magnetic field fluctuation both parallel and perpendicular to the static magnetic field direction around observed nuclei. On the other hand, the spin-lattice relaxation rate $1/T_1$ is affected by the fluctuation only perpendicular to the static magnetic field. $1/T_1$ measurement is in progress to clarify an anisotropy of the fluctuations.

The isostructural compound YbPtBi exhibit the drastic enhancement of $1/T_2$ at low temperatures, accompanying the similar behavior of the spin-lattice relaxation rate $1/T_1$, but the peak of $1/T_2$ was not reported down to 35 K [9]. It is a open question whether the rapid increase of $1/T_2$ at low temperatures is inherent in the formation of the massive electron state in the Yb-based half-Heusler type compound or not.

4. Conclusion

We have presented a NMR investigation of YbPtSb which exhibits a sharp upturn of $C_{\rm el}/T$ at low temperatures. ^{121}K and ^{195}K decrease with temperature. The Knight shifts show the linear relation with χ, indicating that the K is mainly due to a transferred hyperfine field from localized Yb $4f$-moments. It appeares that, in contrast to the temperature independent behavior above 50 K, $1/T_2$ of ^{121}Sb shows the divergent increase around 10 K and decreases at low temperatures. However, no anomalies were detected in the temperature dependence of the NMR line width, χ and $C_{\rm el}$ at present. To clarify the origin of the $1/T_2$, further experiment will be needed.

Acknowlegements

This work was supported by a Grant-in-Aid for Scientific Research of The Ministry of Education, Culture, Sports, Science, and Technology, Japan (No. 21540378 and 21102522).

References

[1] Fisk Z, Canfield P C, Beyermann W P, Thompson J D, Hundley M F, Ott H R, Felder E, Maple M B, Lopez de la Torre M A, Visani P and Seaman C L, 1991 *Phys. Rev. Lett* **67** 3310.
[2] Torikachvili M S, Mun S J E D, Hannahs S T, Black R C, Neils W K, Martien D, Bud'ko S L and Canfield P C, 2007 *PNAS* **104** 9960.
[3] Saiga Y, Matsubayashi K, Fujiwara T, Kosaka M, Katano S, Hedo M, Matsumoto T and Uwatoko Y, 2008 *J. Phys. Soc. Jpn.* **77** 053710.
[4] Kasaya M, Suzuki H, Yamaguvhi T and Katoh K, 1992 *J. Phys. Soc Jpn* **61** 4187.
[5] K. Matsubayashi, private communication.
[6] Munakata K, Matsubayashi K, Suzuki H S, Yamada A, Matsumoto T and Uwatoko Y, unpublished.
[7] Nakanishi Y, Ito K, Kamiyama T, Kashiwazaki R, Deto K, Shichinomiya F, Nakamura M, Yhoshizawa M and Suzuki H S, 2010 *Phys. Status Solidi B* **247** 754.
[8] Harmening T, Eckert H and Pöttgen R, 2009 *Solid State Sciences* **11** 900.
[9] Reyes A P, Le A P, Heffner R H, Ahrens E T, Fisk Z and Canfield P C, 1995 *Physica B* **206 & 207** 332.

Size effects on mixed valence $CePd_3$

Y. H. Lin[1,*], C. R. Wang[2], C. L. Dong[3], M. N. Ou[1], Y. Y. Chen[1,]**
[1] Institute of Physics, Academia Sinica, Taipei, Taiwan
[2] Department of Physics, Tunghai University, Taichung, Taiwan
[3] National Synchrotron Radiation Research Center, Hsinchu, Taiwan

*E-mail: linyh@phys.sinica.edu.tw; **cheny2@phys.sinica.edu.tw

Abstract. To study the size effects on mixed-valence state of $CePd_3$, nanoparticles of $CePd_3$, sizes ranging from 5.2 to 9.5 nm, were prepared. The mixed valence increased from 3.3 to 3.5 as particle size reduced from bulk to 5.2 nm. This consequence was illustrated by the enhancements of valence fluctuations and 4f electron hybridization with conduction band through size reduction. Another interesting finding in the nanoparticles is that a certain fraction ~ 25% of the sample becomes trivalent which is evident from the increase of magnetic susceptibility at low temperatures.

1. Introduction

$CePd_3$ is a mixed valence compound due to the mixing interaction between trivalent Ce^{3+}: $|4f^1 5d^1 6s^2>$ and tetravalent Ce^{4+}: $|4f^0 5d^2 6s^2>$. The valence of $CePd_3$ is intermediate with valence ~ 3.5 [1]. Its resistivity exhibits a maximum at a temperature T_{max} = 124 K. This value corresponds to the inelastic magnetic neutron linewidth Γ; thus, one can use $2T_{max}$ to estimate its spin-fluctuation temperature T_{sf} ~ 248 K [2]. The result is fairly consistent with that estimated from the temperature of χ_{max} in magnetic susceptibility, based on a single ion model [2, 3]. Previous studies on the Ce L_{III} x-ray absorption edges (XANES) in $CeAl_2$ films suggested a trivalent electronic configuration in cerium, with a mixture of trivalent and tetravalent cerium ions on the film surfaces [4, 5]. In this study, we investigate size effects on the mixed valence state of $CePd_3$. A series of $CePd_3$ nanoparticles was prepared with various sizes. The valence variation was then examined through the measurements of susceptibility and Ce L_3-edge XANES.

2. Experimental

The sample of $CePd_3$ was synthesised through mixing element Ce (99.99% purity) and Pd (99.99% purity) in 1:3 atomic ratio thoroughly, followed by arc melting under a flowing atmosphere of high purity argon. To exclude the possible impurity phase and mechanical stress, the obtained $CePd_3$ was subsequently annealed at 570 ℃ for 7 days. Using this as target, nanoparticles with sizes d = 5.2, 6.2, 9.5 nm were fabricated by ArF excimer laser ablation. The nanoparticles were collected on a liquid nitrogen-cooled trap. X-ray diffraction (XRD) patterns reveal the cubic structure of $CePd_3$ bulk and nanoparticles (Fig. 1.). All reflection peaks of bulk and nanoparticles were correctly indexed to that of $CePd_3$; the lattice parameter a = 4.126 Å of bulk is in good agreement with those reported in literature.

Published under licence by IOP Publishing Ltd

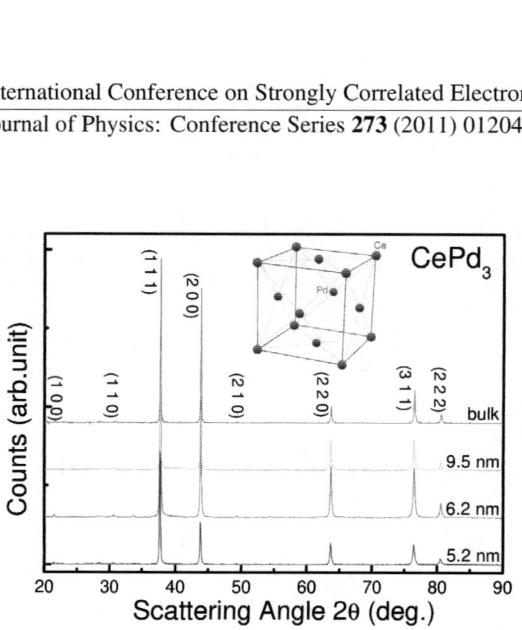

Figure. 1. The x-ray diffraction patterns of $CePd_3$ bulk, 9-, 6-, and 5-nm nanoparticles at room temperature.

The morphology of nanoparticles was directly observed by high resolution transmission electron microscopy (HRTEM) (Insets in Fig. 2). The average sizes of nanoparticles were estimated by TEM images and fits the full width at half maximum (FWHM) of the (200) peak to theoretical simulation. The lattice parameters of nanoparticles are about the same as that of the bulk based on GSAS analysis.

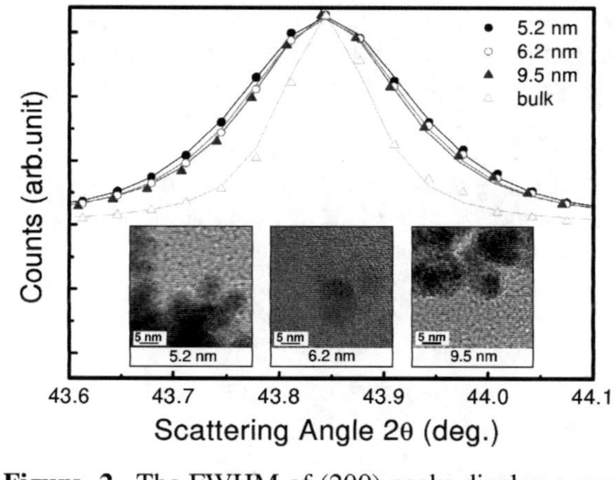

Figure. 2. The FWHM of (200) peaks display a measure of the nanoparticle sizes for $CePd_3$ bulk, 9-, 6-, and 5-nm nanoparticles with theoretical simulation (lines). Insets: TEM images of nanoparticles for various sizes.

2. Results and Discussion

Ce L_{III}-edge XANES spectroscopies were performed at the National Synchrotron Radiation Research Center (NSRRC), Taiwan. The experiments of Ce L_{III}-edge XANES spectra were carried out at Wiggler 17C beam-line using fluorescent mode at room temperature. The Ce L_{III}-edge XANES spectra exhibit the mixed valence nature for both bulk and nanoparticles. Detailed valence information was evaluated by the spectroscopic analysis. The normalization procedure was conducted in the energy range between 5705 and 5755 eV. After evaluating the area contributed from trivalent and tetravalent Ce ions, peaked at 5726 and 5735 eV respectively, the results revealed the valence z of Ce ion increases with size reduction. The valence z is estimated to be 3.28, 3.31, 3.34 and 3.37 for bulk, 9.5, 6.2 and 5.2 nm nanoparticles respectively (Fig. 3.). Although the absolute values of valence calculated from Ce L_{III}-edge XANES spectra significantly depends on the fitting and background subtraction, nevertheless, the trend in the increase of mix valence with size reduction is clearly observed in nanoparticles.

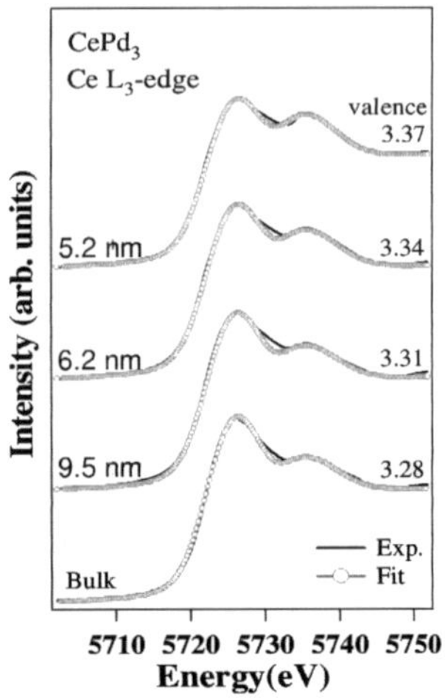

Figure. 3. Ce L_{III}-edge XANES spectra of $CePd_3$ bulk and nanoparticles specimens.

The temperature dependent susceptibility $\chi(T)$ of $CePd_3$ exhibits a paramagnetic behaviour at temperatures above 300 K, followed by a round shaped transition at 100-300 K (inset (a) in Fig. 4). This is consistent with single-ion scaling at spin-fluctuation temperature T_{sf} ~ 250 K [2]. The rise at T< 50 K may be correlated to the coherence observed in low temperature resistivity data. For nanoparticles, an enhancement of magnetic susceptibility at low temperatures appeared and smeared out the round shaped transition near 125 K in the bulk specimen (Fig. 4 and inset (b)). Magnetic susceptibility increased at low temperatures with size reduction, this indicates a certain fraction x of Ce ions in the sample becomes trivalent. In order to estimate the value of fraction x, Curie constant was estimated from the magnetic susceptibility data.

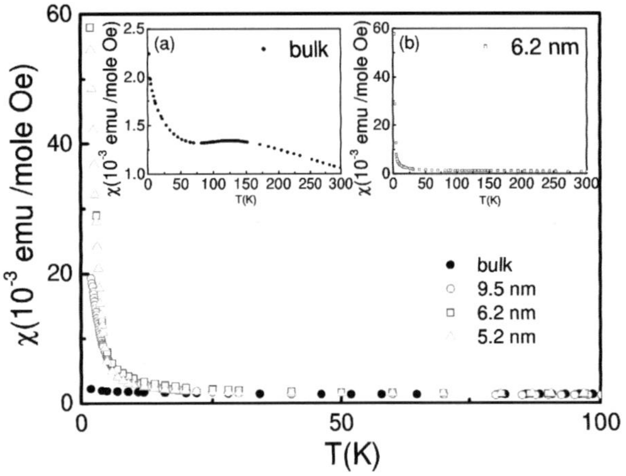

Figure. 4. The temperature dependence of magnetic susceptibility χ for $CePd_3$ bulk and nanoparticles. Insets: susceptibility χ for bulk and 6.2 nm $CePd_3$.

The temperature dependence of inversed magnetic susceptibility $1/\chi$ for $CePd_3$ bulk and nanoparticles is shown in Fig. 5. Since the spin-fluctuation temperature T_{sf} is ~ 250 K, it is difficult to extract the values of Curie constant from the data at high temperatures T = 200-300 K. For low temperature region, the susceptibility was fitted to the Curie-Weiss law $\chi(T) = \chi_0 + C/(T-\Theta)$ for T = 2-10 K, where χ_0 is a constant, and Θ is Curie-Weiss temperature. From Curie constant, the molar percentage of trivalent cerium ions is obtained. For $CePd_3$ bulk, the Curie constant C at low temperature is 0.007 emu/K f.u. Based on the theoretical values 0.204 emu/K f.u. for J = 1/2 ground state of magnetic Ce ions, the percentages of magnetic trivalent cerium in $CePd_3$ bulk is estimated to be near 3% [6].

Assuming the rest of the 97% Ce ions are mixed valence with valence *V*, the actual *V* can then be estimated from the equation, $z = 3x + (1-x)V$: *z* being the average valence obtained from Ce L_{III}-edge XANES, and *x* the fraction of trivalent cerium estimated from Curie constant. For the bulk, the mixed valence *V* is estimated to about 3.3. For nanoparticles, the value of Curie constant *C* increases as particle size decreases. The values of *x* calculated from *C* are 0.262, 0.265, and 0.255 for 9.5, 6.2 and 5.2 nm specimens respectively. The new fraction of trivalent cerium was conjectured from the sample surfaces of nanoparticles. Following similar data analysis as those of the bulk, the valence of mixed valence states of Ce in nanoparticles are obtained as 3.42, 3.46 and 3.50 for 9.5, 6.2 and 5.2 nm specimens respectively.

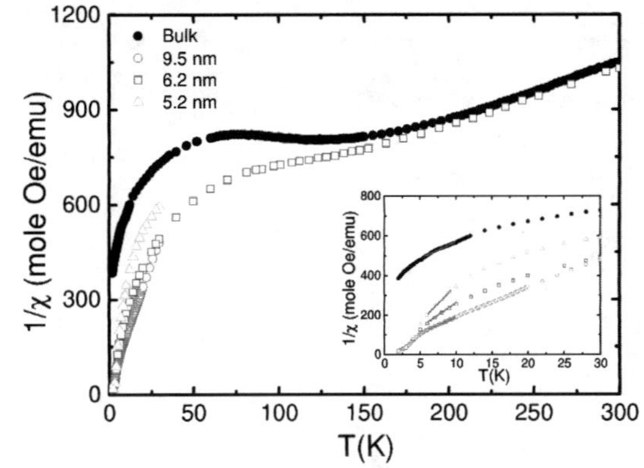

Figure. 5. The temperature dependence of the inverse susceptibility $1/\chi$ for $CePd_3$ bulk and nanoparticles. Inset: $1/\chi$ versus T at low temperatures.

In conclusion, the size effect on valence of Ce was presented in $CePd_3$ nanoparticles by means of magnetic susceptibility and Ce L_{III}-edge XANES spectra. As the particle size decreased to the nanoscale, about 25 % of cerium became trivalent from magnetic susceptibility data. Meanwhile, the valence of mixed valence cerium increases from 3.3 to 3.5. The trend in the increase of mixed valence with size reduction indicates that 4f electron hybridization with conduction band in $CePd_3$ nanoparticles was enhanced as particle size decreases. The measurement of magnetic susceptibility and Ce L_{III}-edge XANES spectra provides only an average valence of the specimens. Therefore, whether the valence change occurred in the whole particle or just on the surface of nanoparticles is unclear. To answer this question, studies with bulk-surface resolved measurements must be performed.

References

[1] S. K. Malik et al., Solid State Commun. 43, 243-245 (1982)
[2] J. M. Lawrence et al., Phys. Rev. Lett. 54, 23 (1985)
[3] J. M. Lawrence, P. S. Riseborough, and R. D. Parks, Rep. Prog. Phys. 44, 1 (1981)
[4] C.L. Dong et al., J. Electron Spectrosc. Relat. Phenom. 152, 1-5 (2006)
[5] C. R. Wang et al., J. Magn. Magn. Mater. 239, 524-526 (2002)
[6] Y.Y. Chen et al., Phys. Rev. Lett. 98, 157206 (2007)

Journal of Physics: Conference Series **273** (2011) 012042 doi:10.1088/1742-6596/273/1/012042

Observation of Kondo resonance in rare-earth hexaborides using high resolution photoemission spectroscopy

Kalobaran Maiti, Swapnil Patil and Ganesh Adhikary

Department of Condensed Matter Physics and Materials Science, Tata Institute of Fundamental Research, Homi Bhabha Road, Colaba, Mumbai 400005, India

Geetha Balakrishnan

Department of Physics, University of Warwick, Coventry, CV4 7AL, UK.

E-mail: `kbmaiti@tifr.res.in`

Abstract.

We studied the electronic structure of rare earth hexaborides, CeB_6, PrB_6 and NdB_6 using *state-of-the-art* high resolution photoemission spectroscopy. CeB_6 is a dense Kondo system. PrB_6 and NdB_6 are antiferromagnetic (Neel temperature $\sim$7 K), known to be stable moment systems and do not exhibit Kondo effect. Photoemission spectra exhibit distinct signature of surface and bulk electronic structures of these compounds. The energy position of the surface feature is not influenced by the $4f$ density of states. High resolution spectra of CeB_6 reveal multiple Kondo resonance features in the bulk spectra due to various photoemission final states. Interestingly, high resolution photoemission spectra of antiferromagnetic PrB_6 also exhibit a sharp feature at the Fermi level that shows temperature dependence similar to the Kondo resonance features.

PACS numbers: 75.20.Hr, 71.27.+a, 71.28.+d, 71.15.Mb

Published under licence by IOP Publishing Ltd

Magnetic impurity in a metal often leads to a logarithmic enhancement in resistivity at lower temperatures, which is known as Kondo effect. At very low temperatures, the conduction electrons couples to the impurity moment antiferromagnetically forming a singlet state (non-magnetic state), termed as Kondo singlet. Such electronic states manifest as a sharp feature in the spectral density of states at the Fermi level, ϵ_F called Kondo resonance feature (KRF), which depends on the proximity of the impurity band to ϵ_F and the hybridization of the impurity states with the conduction electronic states. Numerous efforts are put forward during past several decades in identifying KRF in the electronic spectra using various forms of experimental techniques [1, 2]. However, experimental identification of KRF delineated from the uncompensated moment contributions is still a puzzle. Hexaborides are an important class of materials used extensively in technology and possess plethora of interesting properties making them well suited to address the issue. For example, LaB_6 is a good metal [3], CeB_6 is a dense Kondo system [4], PrB_6 & NdB_6 are local moment antiferromagnets [5], SmB_6 is a Kondo insulator [6], EuB_6 is a ferromagnetic semiconductor [7].

Such wide varieties of properties appear due to the interaction of $4f$ electrons with the conduction electrons of B $2s2p$ character [8, 9]. It is predicted that the bulk non-$4f$ electronic density of states are very similar across the series[10]. Thus, by changing rare-earth element, one can tune the $4f$ binding energy, ϵ_{4f} and hence the coupling of $4f$ moment with the conduction electrons. Here, we studied the electronic structure of RB_6 (R = Ce, Pr and Nd) employing high resolution photoemission spectroscopy that revealed distinct features corresponding to surface and bulk electronic structure. CeB_6 spectra exhibit multiple Kondo resonance features. Interestingly, the signature of Kondo resonance feature could also be observed in the photoemission spectra of PrB_6.

Large single crystals of all the compounds were produced at Warwick by the floating zone technique using a four mirror Xe arc lamp image furnace. The growths were carried out in a flow of argon gas and growth speeds of 10-18 mm/hr. Cylindrical boules of ~6 mm diameter and ~50 - 60 mm length were obtained by this technique [11]. Photoemission measurements were performed using monochromatic He I ($h\nu$ = 21.2 eV) and He II ($h\nu$ = 40.8 eV) photons, and Gammadata Scienta SES2002 analyzer operating at an energy resolution of 2 meV. The sample surfaces were prepared by cleaving the single crystals *in situ.* It was noticed that the angle integrated spectra from cleaved surface and scraped surface using a diamond file are identical.

In Fig. 1(a), we show the He II spectra of NdB_6 and CeB_6 collected at 10 K from freshly cleaved samples. The NdB_6 spectrum exhibit a sharp feature at about 1.7 eV binding energy as observed earlier [12]. This feature gradually weakens in intensity with time delay and vanished completely in about 3 hours after cleaving. Since the surface electronic structure is associated to only top few layers [13], the vanishing of the peak with time characterizes it to be a surface feature and can be attributed to dangling B $2s2p$ states. Earlier studies of LaB_6 also showed such a surface feature [3, 14]. In both, NdB_6 and LaB_6, the binding energy of $4f$ band is different from 2 eV and hence may have small influence on surface states.

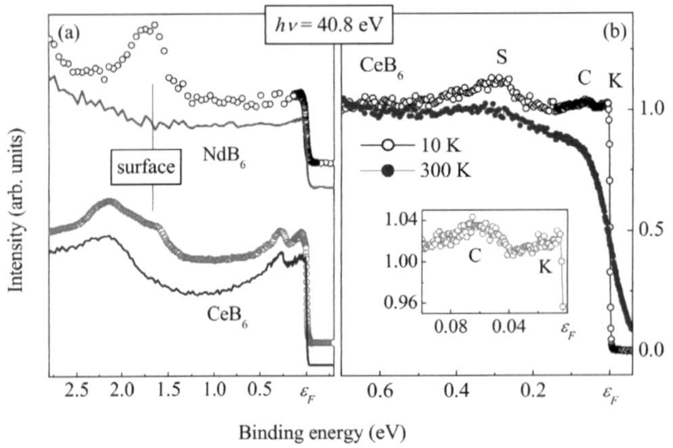

Figure 1. (a) He II spectra of NdB_6 and CeB_6 at 10 K from freshly cleaved surface (symbols) and after a delay (lines). (b) He II spectra of CeB_6 at 10 K and 300 K. The inset shows the near ϵ_F region of 10 K spectrum in an expanded scale.

In order to probe the influence of $4f$ states on the surface feature, we study the He II spectrum of CeB_6 in the same figure, where $4f$ signal appear close to ϵ_F. Ultra high resolution employed in the He II spectroscopy helped to reveal four distinct features at 2.1 eV, 1.7 eV, 0.28 eV and 0.05 eV binding energies. The feature around 1.7 eV disappears gradually with the increase in delay time and completely vanishes in about 2.5 hours after cleaving. This again indicates the surface character of this feature. The identical energy position in both the cases establish that the surface feature in hexaborides are not influenced by the $4f$ electronic states. Evidently, the other features can be attributed to the bulk electronic structure.

The feature around 2.1 eV corresponds to the unscreened final state having f^0 electronic configuration [15]. The well screened features having f^1 electronic configuration appear near ϵ_F and represent the signature of Kondo resonance feature [2]. In order to reveal the details of these features, we show the He II spectrum at 10 K and 300 K in an expanded scale in Fig. 1(b). There are three distinct features marked S, C and K could be viewed in the spectrum at 10 K and can be attributed to the photoemission final states, $|f^1_{7/2}\underline{C}>$, $|f_{5/2}(\Gamma_7)^1\underline{C}>$ and $|f_{5/2}(\Gamma_8)^1\underline{C}>$, respectively ($\underline{C}$ represents a hole in the conduction band). These features almost vanished in the 300 K spectrum characterizing them as the signature of Kondo resonance features.

In Fig. 2(a), we compare the He I and He II spectra of PrB_6 at 10 K. While the line shape of the features look very similar in the whole energy range shown, a small mismatch is observed near ϵ_F. In order to bring this out clearly, we show the subtracted (He II - He I) spectrum in Fig. 2(c). Such subtraction provides the Pr $4f$ contributions as the relative photoemission cross section of $4f$ states increases significantly in He II energy compared to that in He I energy. Thus, the sharp feature visible at ϵ_F in the figure possess $4f$ character. Interestingly, the intensity of this feature vanishes at 104

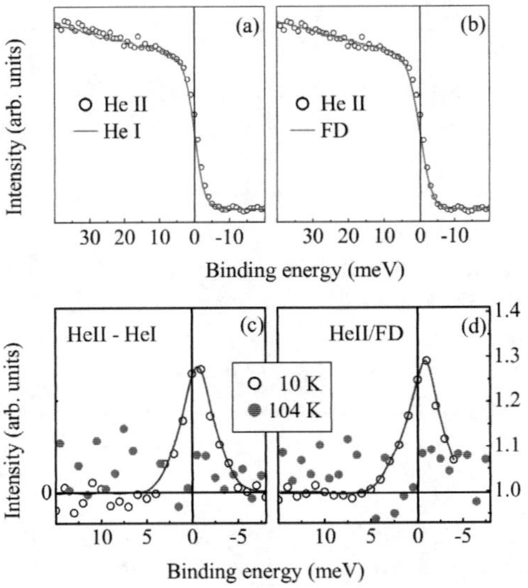

Figure 2. (a) He I (line) and He II (open circles) spectra of PrB_6 at 10 K. (b) He II (open circles) spectrum is compared with the corresponding Fermi-Dirac (FD) distribution function (line) at 10 K. (c) (He II - He I) and (d) He II/FD spectra at 10 K and 104 K.

K.

Spectral density of states (SDOS) can directly be obtained by dividing the resolution broadened Fermi-Dirac (FD) distribution function. Thus obtained SDOS exhibit a feature similar to that in Fig. 2(c) that vanishes at 104 K - the comparison of the raw data and the FD function is shown in Fig. 2(b). Sharp peaks near ϵ_F can appear from bare band structure and photoemission cross section effects and/or photoemission final state effects. Therefore, the identification of Kondo resonance feature has been a long standing issue since the birth of Kondo physics. The key difference appears in the temperature evolution - in all the above cases, the feature remains unchanged with temperature while the Kondo peak vanishes at higher temperature. The observation of vanishing of the feature at 104 K in Figs. 2(c) and 2(d), thus, attribute it to be a Kondo resonance feature. This is not unusual as the Pr $4f$ levels are significantly hybridized with the conduction electrons. The absence of such effect in the bulk measurements may be related to the dominance of other bulk effects.

In summary, we investigated the evolution of the electronic structure in RB_6 (R = Ce, Pr and Nd) using high resolution photoemission spectroscopy and good quality single crystals. We observe that the surface states in these compounds are generic and are not influenced by the $4f$ electronic states. High resolution spectra exhibit Kondo resonance feature in CeB_6 - a dense Kondo system. Interestingly, high resolution spectra of PrB_6 at 10 K also show a sharp feature at the Fermi level that vanishes at higher temperature indicating it to be signature of Kondo resonance in this compound.

S.P. thanks the Council of Scientific and Industrial Research, Government of India for financial support. G.B. wishes to thank EPSRC, UK, for financial support for the crystal growth activities.

References

[1] Patthey F *et al.* 1987 *Phys. Rev. Lett.* **58**, 2810; Garnier M *et al.* 1997 *Phys. Rev. Lett.* **78**, 4127.
[2] Ehm D *et al.* 2007 *Phys. Rev. B* **76**, 045117; Patil S *et al.* 2010 *J. Phys. Condens. Matter* **22**, 255602; Patil S *et al.* 2010 *Phys. Rev. B appearing.*
[3] R. Nishitani *et al.*, 1980 *Surface Science* **93**, 535.
[4] K. Samwer and K. Winzer 1976 *Z. Phys. B* **25**, 269.
[5] Geballe T H *et al.* 1968 *Science* **160**, 1443
[6] Nyhus P *et al.* 1997 *Phys. Rev. B* **55**, 12488
[7] Guy C N *et al.* 1980 *Solid State Commun.* **33**, 1055
[8] Medicherla V R R *et al.* 2007 *Appl. Phys. Lett.* **90**, 062507.
[9] Maiti K *et al.* 2007 *Phys. Rev. Lett.* **99**, 266401; Maiti K 2008 *Europhys. Lett.* **82**, 67006.
[10] Kimura S *et al.* 1992 *Phys. Rev. B* **46**, 12196.
[11] Balakrishnan G, Lees M R and Paul D 2003 *J. Crystal Growth* **256**, 206; *ibid* 2004 *J. Magn. Magn. Mater.* **272-276**, 601
[12] Tanaka A *et al.* 1997 *Phys. Rev. B* **56**, 7660; Patil S *et al.* 2010 *Appl. Phys. Lett.* **96**, 092106.
[13] Maiti K *et al.*, 2001 *Europhys. Lett.* **55**, 246; *ibid.* 2006 *Phys. Rev. B* **73**, 052508; *ibid.* 2004 *Phys. Rev. B* **70**, 195112; Maiti K and Sarma D D, 2000 *Phys. Rev. B* **61**, 2525.
[14] Aono M *et al.* 1977 *Phys. Rev. B* **16**, 3489; Morimoto O, Kunii S and Kakizaki A 2006 *Phys. Rev. B* **73**, 245423.
[15] Chiaia G *et al.* 1997 *Phys. Rev. B* **55**, 9207.

International Conference on Strongly Correlated Electron Systems (SCES 2010) IOP Publishing
Journal of Physics: Conference Series **273** (2011) 012043 doi:10.1088/1742-6596/273/1/012043

Structural instability of un lled skutterudite compounds TSb$_3$ (T=Co, Rh and Ir) under high pressure

K. Matsui*, **J. Hayashi, K. Akahira, K. Ito, Y. Fukushi, K. Takeda and C. Sekine**†

Graduate School of Engineering, Muroran Institute of Technology, 27-1 Mizumoto-cho, Muroran, Hokkaido 050-8585, Japan

E-mail: *s1724073@mmm.muroran-it.ac.jp, †sekine@mmm.muroran-it.ac.jp

Abstract. Filled skutterudite compounds attracted many researchers' interest as strongly correlated electron systems. The skutterudite family includes binary, un lled skutterudite compounds. We have studied the powder x-ray diffraction of un lled skutterudite compounds TSb$_3$ (T=Rh and Ir) under high pressure up to around 40 GPa, using synchrotron radiation. Pressure-induced irreversible structural changes of TSb$_3$ (T=Rh and Ir) have been observed. Structural instability of the un lled skutterudite compounds TSb$_3$ (T=Co, Rh and Ir) under high pressure is discussed.

1. Introduction

Filled skutterudite compounds have attracted much attention of their wide variety of strongly correlated electron behaviors, such as unconventional superconductivity [1], multipole ordering [2], and so on. Furthermore, the compounds have also received considerable attention owing to their potential as improved, next generation thermoelectric materials [3, 4, 5]. The skutterudite family includes binary, unfilled skutterudite compounds. The compounds crystallize in a body centered cubic structure of space group Im$\bar{3}$ (T$^5{}_h$, No. 204) and have a general formula TX_3 (T=Co, Rh and Ir, occupying site 8c, X=P, As and Sb, site 24g) or $\square T_4X_{12}$ where the symbol $\square$ represents a vacancy (site 2a) [6]. These materials show excellent thermoelectric properties (large Seebeck coefficients, S and high hole mobilities) [7]. The figure of merit, Z, of thermoelectric materials is given by a combination of electrical and thermal transport coefficients, Z=S^2 /κ, where S is the Seebeck coefficient, is the electrical conductivity, and κ is the thermal conductivity, which has both electronic and lattice components, κ_{el} and κ_{lat}, respectively. Although the power factor, S^2 , for TSb$_3$ (T=Co, Rh and Ir) is quite large, the lattice thermal conductivity, κ_{lat} is too high. However, the binary skutterudite compounds have large voids which can be filled by rare-earth ions etc. The filling ions inside the oversized cages are believed to show random motion ("rattling") and produce a large phonon scattering. Actually, such the filling can reduce κ_{lat} greatly [5]. In spite of the presence of large voids in the structure, earlier studies under high pressure indicate that the binary skutterudites are quite stable under high pressure. It has been reported that the binary skutterudite compound IrSb$_3$ is stable in the structure up to 42 GPa [8]. However, recently the structural change of CoSb$_3$

Published under licence by IOP Publishing Ltd

in high pressure has been reported by Kraemer *et al.* [9]. They reported that above 20 GPa $CoSb_3$ undergoes an irreversible isosymmetric transition to another phase that, upon pressure release, exhibits a volume greater than that of pristine $CoSb_3$. This anomalous behavior could be interpreted as a pressure-induced self-insertion reaction of $CoSb_3$, in which antimony atoms from the compound framework partially fill the $2a$ site. This phenomenon of $CoSb_3$ has been verified by synchrotron powder x-ray diffraction [10] and Raman scattering study [11]. In this paper, we have studied the synchrotron powder x-ray diffraction of isostructural families $RhSb_3$ and $IrSb_3$ under high pressure to investigate the structural instability of these compounds. Our preliminary results of $RhSb_3$ and $IrSb_3$ have been reported [10].

2. Experimental

High quality polycrystal samples of the skutterudite compounds $T\mathrm{Sb}_3$ (T=Rh and Ir) were prepared under high pressure [10]. The x-ray diffraction experiments under high pressure, using synchrotron radiation, were performed at the beam line BL-18C in KEK-Photon Factory (Tsukuba, Japan). Powder x-ray diffraction patterns of $T\mathrm{Sb}_3$ (T=Rh and Ir) were measured with a diamond-anvil cell (DAC) and an imaging plate up to around 40 GPa at room temperature. Pressure in the DAC was determined from a pressure shift in the sharp R-line fluorescence spectrum of ruby. A 4:1 methanol-ethanol solution was used as pressure medium.

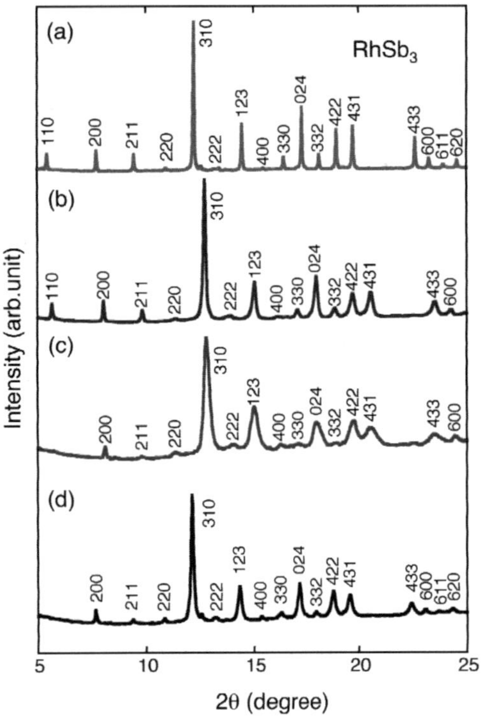

Figure 1. X-ray diffraction patterns of $RhSb_3$ at (a) 0.5 GPa, (b) 19.4 GPa, (c) 27.9 GPa and (d) 0.5 GPa (at decreasing pressure).

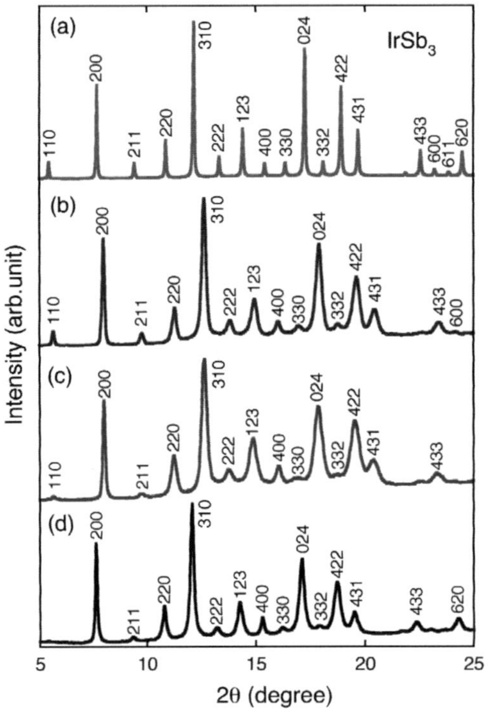

Figure 2. X-ray diffraction patterns of $IrSb_3$ at (a) 0.5 GPa, (b) 26.5 GPa, (c) 29.4 GPa and (d) 0.2 GPa (at decreasing pressure).

3. Results and discussion

Figures 1 and 2 show x-ray diffraction patterns of $RhSb_3$ and $IrSb_3$ at several pressures (λ = 0.619 Å), respectively. The appearance of any new peak was not observed in the profiles for both compounds under pressures. This result indicates that there is no change in crystal structure type within this pressure range for both compounds. However, the intensity of 110, 200 and 211 Bragg peaks was drastically suppressed at 27.9 GPa (Fig. 1(c)) for $RhSb_3$. Furthermore, the suppressed intensity of these peaks was maintained at the diffraction pattern after decreasing pressures (Fig. 1(d)). A similar behavior has been observed for $CoSb_3$[10]. Figure 4 shows a simulation of x-ray diffraction patterns of $Sb_xRh_4Sb_{12-x}$ based on the pressure-induced self-insertion reaction model. The significant reduction of the intensity of 110, 200 and 211 peaks agrees with the observed diffraction patterns. For $IrSb_3$, similar reduction of the intensity of the peaks at low-angle side was observed. Figure 3 shows the relative unit cell volume (V/V_0) of $RhSb_3$ (a) and $IrSb_3$ (b) versus pressure. V_0 is the volume at ambient pressure. Solid and open circles represent increasing and decreasing pressure runs, respectively. The dotted lines show the compression curves of the bulk modulus $B_0 = 95$ ($RhSb_3$) and 105 GPa ($IrSb_3$), which were obtained by fitting of a Birch equation of state to data below 10 GPa. The first derivative of B_0 with respect to pressure, B_0' was fixed at a value of 6, which is the value for the curve fitting of $CoSb_3$ in ref [12]. Both compounds reveal a saturation behavior of the volume reduction with pressure. We define the critical pressure P_c as the pressure that the volume reduction with pressure becomes to be saturated. Further, upon pressure release, a hysteresis was observed in which the cell volume becomes greater than that observed at increasing pressure for both compounds. The samples recovered at ambient pressure exhibit volumes about 2.2% and 1.8-2.2% greater than those of pristine $RhSb_3$ and $IrSb_3$, respectively. The bends around 10 GPa are due to solidification of the 4:1 methanol-ethanol solution used as pressure medium.

Each parameters of TSb_3 (T=Co, Rh and Ir) are summarized in Table 1, which includes the reported values of ref. 10. The nonlinear relation between P_c and B_0 of TSb_3 (T=Co, Rh and Ir) was found. This could be due to the size of guest free space at site 2a and the rigidity. We define the guest free distance r_{GFD} as $r_{GFD} = r_{\Box-Sb}$ - r_{Sb3+} - r_{Sb} where $r_{\Box-Sb}$ is the distance between Sb at site 2a and Sb at site 24g, r_{Sb3+} is the effective ionic radius of Sb^{3+} (0.76Å), and r_{Sb} is the covalent radius of Sb (1.38Å) [13]. While $RhSb_3$ has smaller B_0 than $IrSb_3$, r_{GFD} of $RhSb_3$ and $IrSb_3$ are almost the same. Therefore, $RhSb_3$ seems to be easier to give rise to the pressure-induced self-insertion reaction than $IrSb_3$.

4. Summary

We have observed that $RhSb_3$ and $IrSb_3$ reveal the same pressure-induced irreversible structural change as $CoSb_3$. The self-insertion reaction seems to be valid, however, other possibilities should be also examine such as structural phase transition with a collapse of the cage structure under high pressure. If the self-insertion reaction actually occurs, a small guest such as Sb^{3+} vibrates quite largely in the cage. Therefore, a markable rattling effect is expected for $Sb_xT_4Sb_{12-x}$ (T = Co, Rh and Ir).

Table 1. Cell volume V, bulk modulus B_0, critical pressure P_c, increased volume rate y and guest free distance r_{GFD} for TSb_3 (T=Co, Rh, and Ir).

Compound	V (Å^3)	B_0 (GPa)	P_c (GPa)	y(%)	r_{GFD}(Å)
$CoSb_3$	737.47	84	28	2.3	1.21
$RhSb_3$	785.18	95	20	2.2	1.31
$IrSb_3$	793.10	105	30	1.8-2.2	1.32

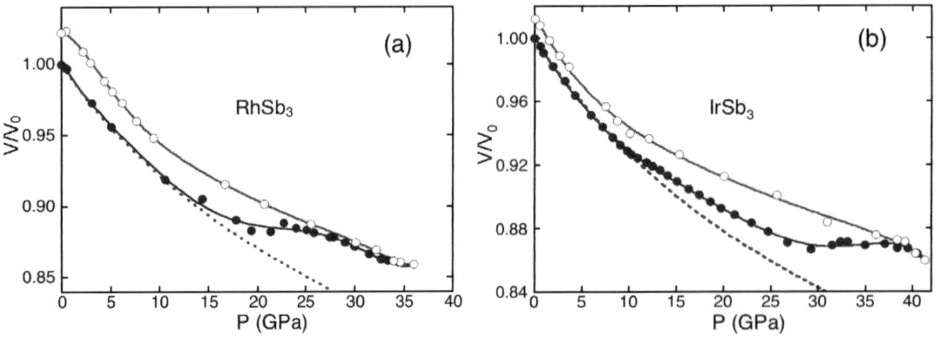

Figure 3. Pressure dependence of the relative unit cell volume for $RhSb_3$ (a) and $IrSb_3$ (b). Solid and open circles represent increasing and decreasing pressure runs, respectively. The dotted lines show the compression curves of the bulk modulus B_0 = 95 ($RhSb_3$) and 105 GPa ($IrSb_3$), which were obtained by fitting of a Birch equation of state to data below 10 GPa.

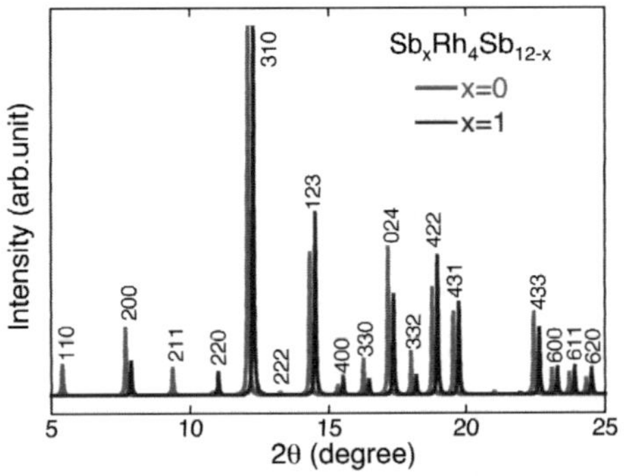

Figure 4. A simulation of x-ray diffraction patterns of $Sb_xRh_4Sb_{12-x}$. The intensity is normalized by that of 310 peak.

Acknowledgments
This work was supported by a Grant-in-Aid for Scientific Research on Innovative Areas "Heavy Electrons" (No. 20102004) of The Ministry of Education, Culture, Sports, Science, and Technology, Japan and a Grant-in Aid for Scientific Research (C) (No. 20540339).

References
[1] Aoki Y, Tayama T, Sakakibara T, Kuwahara K, Iwasa K, Kohgi M, Higemoto W, MacLaughlin D E, Sugawara H and Sato H 2007 *J. Phys. Soc. Jpn.* **76** 051006
[2] Yoshizawa M, Nakanishi Y, Oikawa M, Sekine C, Shirotani I, Saha S R, Sugawara H and Sato H 2005 *J. Phys. Soc. Jpn.* **74** 2141
[3] Morelli D T and Meisner G P 1995 *J. Appl. Phys.* **77** 3777
[4] Sales B C, Mandrus D, Chakoumakos B C, Keppens V and Thompson J R 1997 *Phys. Rev.* B **56** 15081
[5] Nolas G S, Cohn J L and Slack G A 1998 *Phys. Rev.* B **58** 164
[6] Zhuravlev N N and Zhdanov G S 1956 *Sov. Phys. Crystallogr.* **1** 404
[7] Slack G A and Tsoukala V G 1994 *J. Appl. Phys.* **76** 1662
[8] Snider T S, Badding J V, Schujman S B and Slack G A 2000 *Chemistry of Materials* **12** 697
[9] Kraemer A C, Gallas M R, da Jornada J A H and Perottoni C A 2007 *Phys. Rev.* B **75** 024105
[10] Matsui K, Hayashi J, Akahira K, Ito K, Takeda K and Sekine C 2010 *J. Phys.: Conf. Ser.* **215** 012005
[11] Komura M and Kume T, private communication
[12] Shirotani I, Noro T, Hayashi J, Sekine C, Giri R and Kikegawa T 2004 *J. Phys. : Condens. Mtter* **16** 7853
[13] Shannon R 1976 *Acta Crystallogr., Sect. A* **32** 751

International Conference on Strongly Correlated Electron Systems (SCES 2010) IOP Publishing
Journal of Physics: Conference Series **273** (2011) 012044 doi:10.1088/1742-6596/273/1/012044

Successive magnetic transition in non-centrosymmetric $CeCoGe_3$ probed by Co-NQR

M. Matsumura, Y. Sato, M. Mizoo, Y. Kawamura, H. Kato and T. Nishioka

Graduate School of Integrated Arts and Sciences, Kochi University, Kochi 780-8520, Japan

E-mail: matumura@cc.kochi-u.ac.jp

Abstract. Co-NQR measurements for non-centrosymmetric superconductor $CeCoGe_3$ have been performed in ambient and under pressure up to 1.5 GPa to investigate the magnetic structures for the successive magnetic transition at T_{N1}=21 K, T_{N2}=12 K and T_{N3}=8 K. In ambient pressure, the crystallographically equivalent one Co site becomes three sites at $T_{N1} > T > T_{N2}$, and further becomes four sites at $T_{N2} > T$. Applying pressure at 4.2 K, the four Co sites becomes two sites above 0.3 Gpa. These Co sites acquire slightly different internal field parallel to the c-axis, consistent with the anisotropic magnetization with easy c-axis. The T-dependences of the internal fields and the spectral weights for respective sites are estimated. The magnetic structure in each phase is discussed referring to the recent results of neutron scattering.

Recently a great deal of interests in superconductivity in non-centrosymmetric heavy Fermion compounds has been attracted because of their specific Cooper pairing symmetry.[1] The present compound $CeCoGe_3$ with tetragonal $BaNiSn_3$ type (space group: I4mm) is found to be one of the candidates since the discovery of the superconductivity below 0.6 K under pressure ~5.5 GPa near neighbor to the quantum critical point (QCP).[2]

In addition to the interest of the unconventional superconductivity, the magnetic ordered phases in approaching to QCP in $CeCoGe_3$ are also intriguing because of an exhibition of the complex magnetic order in spite of one crystallographically equivalent Ce site. The investigation of the magnetic phase from ambient pressure to near neighbor to QCP under pressure is required to clarify what magnetic background the superconductivity emerges. The magnetic order in ambient pressure occurs successively at T_{N1}=21 K, T_{N2}=12 K and T_{N3}=8 K[3, 4] accompanying by the step-like decrease of the spontaneous magnetization along the c-axis.[4] The complex magnetic phase diagram against pressure is revealed by the magnetization and the specific heat measurements.[5] In this paper, we report the Co-NQR results in ambient pressure and also report the NQR spectral change in the ground state ($T < T_{N3}$) under pressure up to 1.5 GPa.

The polycrystalline sample used in the present experiment was prepared by arc-melting the constituent metals with stoichiometric quantity. The ingot was annealed at 800°C for one week in vacuum and subsequently crushed into powder below 53 μm for NQR measurements. No impurity phases were detected on the powder X-ray diffraction. The NQR measurements were performed using phase coherent spin echo method. The corrections of so called ν^2 and spin echo decay were done for the NQR spectrum. Taking account for the possible 1st order transition at T_{N2} and T_{N3}, all the measurements were performed with rising temperature after cooling to the lowest temperature, 4.2 K.[3] The piston-cylinder cell was used in the NQR measurement

Published under licence by IOP Publishing Ltd

Journal of Physics: Conference Series **273** (2011) 012044 doi:10.1088/1742-6596/273/1/012044

under pressure. The pressure media is the mixture (1:1) of fluorinert FC-77 and FC-70 below 1 GPa and Daphne 7373 oil above 1 GPa. The pressure was determined by the superconducting transition temperature of In metal put in the sample.

Figure 1 exhibits the ^{59}Co(I=7/2)-NQR spectra at representative temperatures. No signals were detected above 9 MHz. The appearance of the NQR signal and almost the same nuclear spin-lattice relaxation rate $1/T_1$ as that in nonmagnetic $LaCoGe_3$ at 300 K suggest that Co atoms retain no magnetic moment in $CeCoGe_3$. The details of $1/T_1$ associated with the dynamics of Ce moments will be described elsewhere. Above 21 K, the three NQR lines corresponding to $\pm1/2 \leftrightarrow \pm3/2$, $\pm3/2 \leftrightarrow \pm5/2$ and $\pm5/2 \leftrightarrow \pm7/2$ with equal frequency interval indicates the asymmetry parameter of the electric field gradient(EFG) at Co site η=0, being consistent with one equivalent Co site with the axial symmetry along the c-axis. The quadrupole frequency ν_Q=1.791±0.005 MHz and the linewidth of about 20 KHz are evaluated at 25 K. The monotonous decrease of ν_Q (not shown) and the no deviation from η=0 with rising temperature imply that no structural transitions occur except for volume expansion.

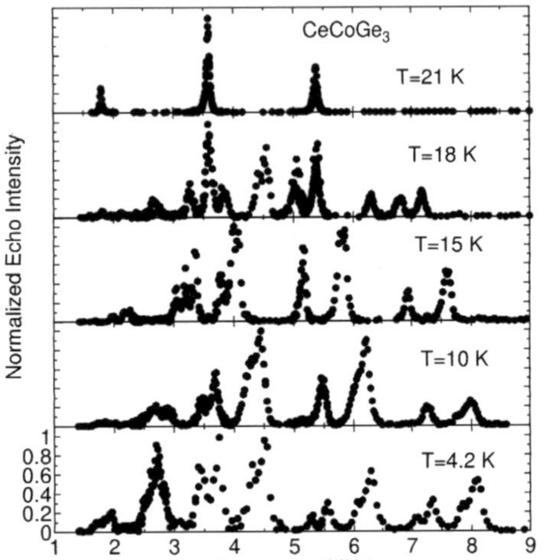

Figure 1. T-dependence of the ^{59}Co NQR spectrum, acrossing the successive magnetic transitions.

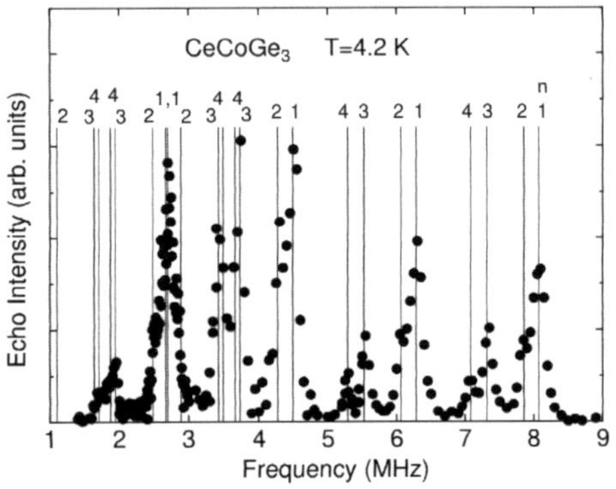

Figure 2. Comparison of the NQR spectrum at 4.2 K with the calculated resonance positions denoted by the solid lines for four kinds (n=1 ~ 4) of Co sites (see text).

Below 21 K, the NQR spectrum spreads over 1 to 9 MHz with the fine structures as shown in Fig. 1, indicating the internal field acted on Co site due to the commensurate magnetic order. One Co site at $T > T_{N1}$ becomes four or three ones with different internal fields below T_{N1}. It is found that all the NQR lines are reproduced for the case that the Co NQR is perturbed by the four or three kinds of internal fields parallel to the Z axis of EFG (coincident to the crystal c-axis). In this case, the nuclear spin Hamiltonian is written for the n-th site (n=1, 2, 3 or 4) in the frame of principal coordinate of EFG,

$$H = H_Q + H_Z = \frac{1}{6}h\nu_Q(3I_z^2 - I(I+1)) + h\nu_L^{(n)}I_Z,$$

where $\nu_L^{(n)}$ is the nuclear Zeeman frequency for the n-th site, relating to the internal field $H_i^{(n)}$ as $\nu_L^{(n)}=(\gamma_{Co}/2\pi)H_i^{(n)}$ using the nuclear gyromagnetic ratio γ_{Co} ($\gamma_{Co}/2\pi$=10.054 MHz/10kOe). The resonances peaks associated with ν_Q, $2\nu_Q$ and $3\nu_Q$ NQR lines rigorously split into ($\nu_L^{(n)}$,

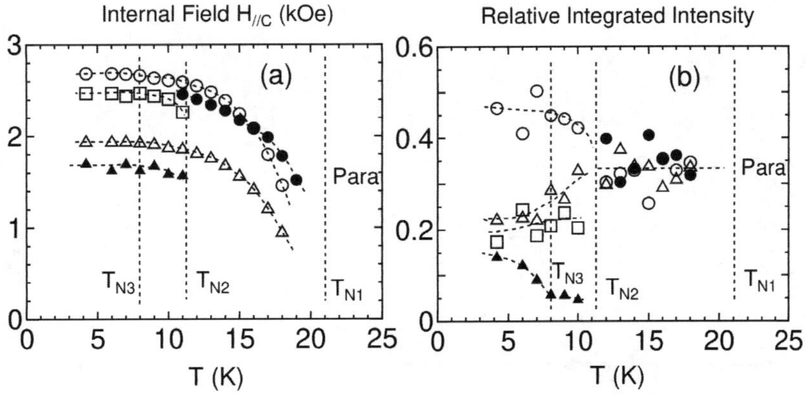

Figure 3. (a) T-dependence of $H_i^{(n)}$ for respective Co sites. (b) T-dependence of the relative I·I for respective Co sites. The dotted lines are guide for eyes.

$|\nu_Q \pm \nu_L^{(n)}|$), $2\nu_Q \pm \nu_L^{(n)}$ and $3\nu_Q \pm \nu_L^{(n)}$, respectively. The consistency between the NQR peaks and the calculation with n=1 ~ 4 is demonstrated in Fig. 2 for the spectrum at 4.2 K. The magnitude of the internal fields are $H_i^{(1)}$ =2.7, $H_i^{(2)}$ =2.4, $H_i^{(3)}$ =1.95 and $H_i^{(4)}$ =1.8 kOe. Here, the values ν_Q and η in the magnetic ordered state are the same as those at 21 K, assuming their T-independences. The reasonable agreement implies that the successive magnetic transition is not related to lattice distortion and/or charge redistribution. The spectra at the other temperatures are also well explained along the same scheme except that n is reduced to 1 ~ 3 above 12 K.

Since the internal field $H_i^{(n)}$ and the spectral weight (relative integrated intensity ($I \cdot I$)) at the respective Co sites can be determined definitely from the part of the spectrum from 6.5 to 9 MHz including $3\nu_Q + \nu_L^{(n)}$ lines, we focus on this part of the spectrum. The T-dependence of $H_i^{(n)}$ and the relative $I \cdot I$ for respective sites are presented in Fig. 3. Above 12 K, the three kinds of Co sites (n=1 ~ 3) with nearly equal quantity are noticed. While, below 11 K, the number of Co sites changes into four (n=1 ~ 4) with unequal spectral weight amounted to be 45, 20, 20, and 15 % at 4.2 K. This spectral change indicates that the magnetic ordered structure is rearranged at the boundary temperature around 11 K, which is consistent with T_{N2}=12 K in the macroscopic measurements.[4] The transition at T_{N3} =8 K found in the macroscopic measurement is not definitely evidenced in the T-dependence of $H_i^{(n)}$ except that the change of the relative $I \cdot I$ becomes prominent above about 8 K toward T_{N2}.

Figure 4 shows the pressure dependence of the NQR spectrum up to 1.52 GPa for $3\nu_Q + \nu_L^{(n)}$ lines at 4.2 K. The drastic changes of NQR spectrum below 0.3 GPa indicate that the magnetic ordered structure is rearranged under low pressure even in $T < T_{N3}$, probably relating to the complex magnetic phase toward to QCP.[5] The four site splitting becomes two site one above about 0.3 GPa with $I \cdot I$ of about 2 : 1 for low : high frequency spectra above 0.7 GPa.

We discuss the magnetic structures for respective phases referring to the neutron scattering results. According to the neutron results, the Ce moments of about 0.5 μ_B are parallel to the c-axis and the basic ordering wave vector $\mathbf{q_1}$=(0,0,1/2) is accompanied by the parasitic component $\mathbf{q_2}$=(0,0,3/4) in the ground AF state at 2.9 K.[6] The magnetic structure for the $\mathbf{q_1}$ mode is proposed as *uudd*, where *u* and *d* denote the moment direction of *up* and *down* within the subsequent c-planes. On the other hand, since the NQR is a local probe, the internal field is determined by the Ce moments around Co site, namely, the four nearest neighbor (nn) Ce at the corner and the one next nearest neighbor (nnn) Ce at the body center in the tetragonal unit

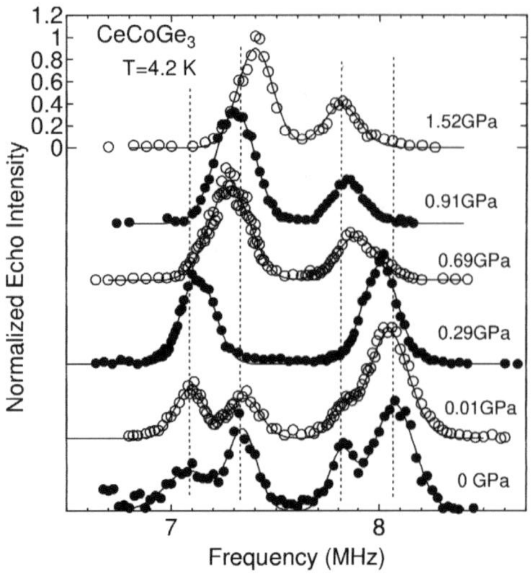

Figure 4. Pressure dependence of the NQR spectrum for $3\nu_Q + \nu_L^{(n)}$ lines at 4.2 K.

cell. In *uudd* structure, only the two Co sites coming from (up 4nn Ce and up 1nnn Ce) and (up 4nn Ce and down 1nnn Ce) with the relative $I \cdot I$ of 1 : 1 are expected as long as all the Ce moment size is equal and the moment direction is confined to be along the c-axis.

At $T < T_{N2}$, the four Co sites and their $I \cdot I$ in NQR are inconsistent with the expectation from the *uudd* structure. At least, two Ce sites with different moment are required. Further study is needed if the combined mode $\mathbf{q_1} + \mathbf{q_2}$, which predicts more than two Ce moment sizes, might explain the NQR results.[6] Note that relative $I \cdot I$=2 : 1 under $p > 0.7$ GPa is also inconsistent with the expectation from the *uudd* structure despite of the consistency with two Co sites. A long period modulation over two times unit cell might be considered.

In the phase $T_{N2} < T < T_{N1}$, where the spontaneous magnetization of 0.13 μ_B/Ce is observed,[4] the internal field for one Co site exhibit different T-dependence from the other two as shown in Fig. 3(a). The one Co site may be ascribed to the ferromagnetic region and the other two Co sites may assigned to the long period antiferromagnetic region possessing no net magnetization. If this is the case and the Ce moment size is assumed to be 0.5 μ_B as in $T < T_{N2}$, the ferromagnetic magnetization is 0.5/3 μ_B/Ce being roughly agreement with the spontaneous magnetization of 0.13 μ_B/Ce. In the range $T_{N3} < T < T_{N2}$, some rearrangement between modes may occur through a flip of moments along the c-axis. For further discussions, more detailed neutron scattering in these temperature ranges is highly desired at present.

This work was partially supported by a Grant-in-Aid for Scientific Research (C) of the Ministry of Education, Culture, Sports, Science and Technology, Japan.

References

[1] Gorkov L P and Rashba E I 2001 *Phys. Rev. Lett.* **87** 037004
[2] Settai R, Sugitani I, Okuda Y, Thamizhavel A, Nakashima M, Onuki Y and Harima H 2007 *J. Magn. Magn. Mater.* **310** 844.
[3] Pecharsky V K, Hyun O -B and Gshneidner Jr K A 1993 *Phys. Rev.* B **47** 11839.
[4] Thamizhavel A, Takeuchi T, Matsuda T D, Haga Y, Sugiyama K, Settai R and Onuki Y 2005 *J. Phys. Soc. Jpn.* **74** 1858.
[5] Knebel G, Aoki D, Lapertot G, Salce B, Flouquet J, Kawai T, Muranaka H, Settai R and Onuki Y 2009 *J. Phys. Soc. Jpn.* **78** 074714.
[6] Kaneko K, Metoki N, Takeuchi T, Matsuda T D, Haga Y, Thamizhavel A, Settai R and Onuki Y 2009 *J. Physcs: Conf. Series* **150** 042082.

Thermal and magnetic properties of a low-temperature antiferromagnet $Ce_4Pt_{12}Sn_{25}$.

R. Movshovich[1], N. Kurita[1,2], H.-O. Lee[1], Pei-Chun Ho[3], M. Brian Maple[4], Yoshifumi Tokiwa[1],5 Corneliu F. Miclea[1], Eric D. Bauer[1], Filip Ronning[1], Pinaki Sengupta[1], Ilya Vekhter[6], Zachary Fisk[7], J. D. Thompson[1],

[1] Los Alamos National Laboratory, Los Alamos, New Mexico 87545, USA
[2] National Institute for Materials Science, Tsukuba, Ibaraki 305-0003, Japan
[3] Department of Physics, California State University, Fresno, California 93740, USA
[4] Department of Physics, University of California, San Diego, California 92093, USA
[5] Department of Physics, University of Goettinghen, Goettinghen 37077, Germany
[6] Louisiana State University, Baton Rouge, Louisiana 70803, USA
[7] Department of Physics and Astronomy, University of California, Irvine, California 92697, USA

E-mail: roman@lanl.gov

Abstract. We report specific heat (C) and magnetization (M) of single crystalline $Ce_4Pt_{12}Sn_{25}$ at temperature down to $\approx$ 50 mK and in fields up to 3 T. C/T exhibits a sharp anomaly at 180 mK, with a large jump in a Sommerfeld coefficient $\gamma = C/T$ of $\Delta\gamma$ = 30 J/molK^2-Ce, which, together with corresponding cusp-like magnetization anomaly, indicate antiferromagnetic (AFM) ground state with Nel temperature $T_N = 0.18$ K. Numerical calculations based on Heisenberg model reproduce well zero field specific heat data, and point to a very small Kondo scale T_K, clearly placing $Ce_4Pt_{12}Sn_{25}$ in the weak exchange coupling $J < J_c$ limit of the Doniac diagram. Magnetic field suppresses AFM state at $H^* \approx 0.7$ T, much more rapidly than indicated by theoretical calculations.

1. Introduction

Competition between magnetically ordered and heavy fermion states is ubiquitous in compounds with f-electrons. Formation of the heavy electron states is widely believed to be due to Kondo screening of the local f-moments [1], while the magnetism is driven by the Ruderman-Kittel-Kasuya-Yoshida (RKKY) [2, 3, 4] interaction between the f-moments, mediated by the conduction (c) electrons. This competition was originally addressed by Doniach in the 1D-chain model [5, 6, 7]. At low f-c exchange coupling J the long-range AFM ground state of f-ions is stabilized, while beyond some critical coupling J_c Kondo screened state is stabilized.

$Ce_4Pt_{12}Sn_{25}$ presents an example of a metallic $4f$-dilute Ce-based system [8], with Ce - Ce interatomic distance d_{Ce-Ce}=6.14Å. It therefore appears to be a good system to study the Kondo-limit of the Doniach model. $Ce_4Pt_{12}Sn_{25}$ is a cubic compound with a rather complicated unit cell with a lack of 4 - fold point symmetry, three inequivalent Sn sites, and a lattice constant a=12.38Å. Previous specific heat, resistivity, and ac-susceptibility measurements uncovered a phase transition at 0.18 K. [9] In this paper, we report the low-temperature specific heat

Published under licence by IOP Publishing Ltd

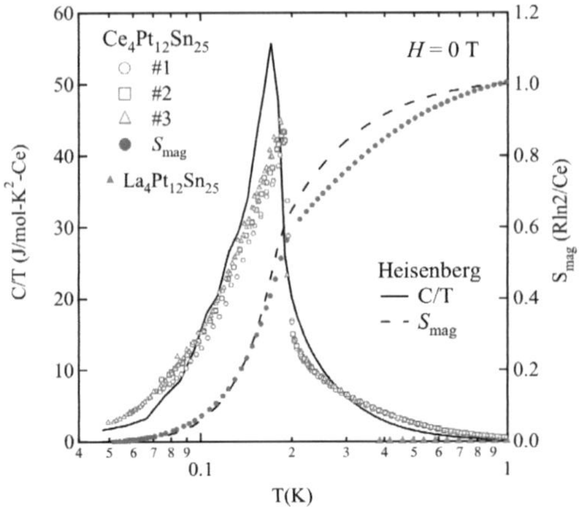

Figure 1. (Color online) (a) Temperature dependence of C/T and entropy S of $Ce_4Pt_{12}Sn_{25}$ for samples #1, #2 and #3, and of the La-analog $La_4Pt_{12}Sn_{25}$ in zero field. Solid dots represent the temperature dependence of magnetic entropy S_{mag} in units of $R\ln 2$. Solid and dashed lines represent specific heat and entropy, respectively, from the Heisenberg model calculations described in the text.

measurements in magnetic field, and compare them to the results of numerical calculation based on Heisenberg model. We find that the zero-field signatures of the magnetic transition are reasonably close to those described by a spin-1/2 Heisenberg model. Notable differences, however, arise with increasing magnetic field.

2. Experimental details

Single crystals of $Ce_4Pt_{12}Sn_{25}$ were grown by Sn self-flux method. The details of the sample growth and the physical properties are described in Ref. [9]. Specific heat was measured in a SHE dilution refrigerator with 9 T superconducting magnet by means of a quasi-adiabatic heat pulse method with a RuO_2 thermometer. We used three samples from different batches labeled in the figures as samples #1 (7.18 mg), #2 (2.58 mg) and #3 (3.70 mg).

3. Results and Discussion

Figure 1 shows the temperature dependence of C/T of $Ce_4Pt_{12}Sn_{25}$ (50 mK - 3 K) for samples #1, #2 and #3, and that of a non-magnetic analogue $La_4Pt_{12}Sn_{25}$ (0.4 K - 3 K) in zero field. C/T of $La_4Pt_{12}Sn_{25}$ is negligible compared to that of $Ce_4Pt_{12}Sn_{25}$, and the specific heat of $Ce_4Pt_{12}Sn_{25}$ is of purely magnetic origin C_{mag}. With temperature decreasing from 3 K, C/T monotonically increases down to 0.18 K, at which point it exhibits a sharp anomaly with the magnitude of the jump $\Delta C/T \sim 30$ J/mol K^2-Ce. We identify the anomaly at 0.18 K as due to antiferromagnetic (AFM) ordering by means of magnetization measurements, not shown. Solid circles represents the temperature dependence of the magnetic entropy $S_{mag}(T)$ of $Ce_4Pt_{12}Sn_{25}$. Entropy gain is $\approx 0.5\,R\ln 2$ at T_N and reaches the value of $1.0\,R\ln 2$ at 3 K. This indicates that the crystal electric field (CEF) ground state of Ce^{3+} in $Ce_4Pt_{12}Sn_{25}$ is a Γ_7 doublet. One remarkable feature of the specific heat of $Ce_4Pt_{12}Sn_{25}$ is the long tail above T_N. There is a number of possible origins of such tail feature, including the onset of the Kondo screening, the development of short-range magnetic correlations in the lattice of Ce moments above the bulk transition temperature T_N, or frustration of spin-spin interactions. Similar C/T behavior was reported in the structurally frustrated system Yb_2Pt_2Pb [10], for example.

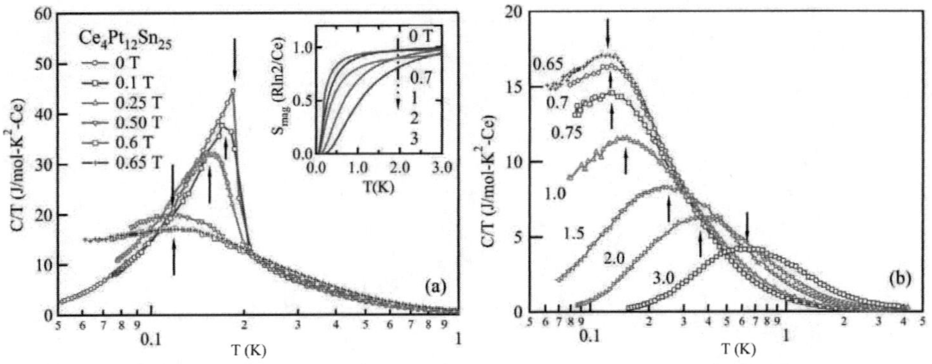

Figure 2. (Color online) Temperature dependence of C/T of $Ce_4Pt_{12}Sn_{25}$ in the magnetic fields, (a) 0-0.65 T, (b) 0.65-3 T. The arrows indicate the temperature where C/T has maximum. The inset (a): $S_{\text{mag}}(T)$ in several fields up to 3 T.

To elucidate the origin of this behavior we performed Quantum Monte Carlo simulations of the spin-1/2 three-dimensional Heisenberg model in zero and applied magnetic field. The exchange interaction was fixed to have the AFM ordering temperature of $T_N = 0.18$ K for $H = 0$, while the g-factor of the spins was determined from the saturation magnetization at high fields. The results for zero field specific heat and entropy are displayed in Fig. 1 as solid and dashed curves, respectively. The agreement between the model calculations and experimental data is rather good. In particular, the behavior of the entropy below T_N is very similar, and the entropy at T_N for model calculations is about $0.55\,R\ln 2$, about 10% larger than the experimental value. The most notable difference is that experimental entropy is lower than the calculated values above the transition, indicating that some degrees of freedom remain "locked". This discrepancy is also seen from the somewhat longer tail of the measured specific heat in the lower panel of Fig. 1 relative to the Heisenberg model.

Such contribution may still be due to remnant Kondo physics, although it would require a rather low temperature scale $T_K < T_N$. The primary factor responsible for the low ordering temperature is the large distance between Ce ions of $Ce_4Pt_{12}Sn_{25}$. In spite of that, RKKY interaction still dominates the Kondo interactions, and the system orders magnetically below $T_N = 0.18$ K. $Ce_4Pt_{12}Sn_{25}$ is therefore in the low $J < J_c$ limit of the Doniach phase diagram. Note that Ce-based skutterudite compounds and YbT_2Zn_{20}, whose Ce-Ce and Yb-Yb distances are similar to that in $Ce_4Pt_{12}Sn_{25}$, do not show magnetic ordering, and are in the $J > J_c$ limit. It should also be noted that there is no significant sample dependence for different samples #1, #2, and #3, with respect to the ordering temperature, the magnitude of the anomaly in C/T, and its width.

Figure 2 shows C/T for magnetic field (a) $0 \leq B \leq 0.65$ T and (b) $0.65 \leq B \leq 3.0$ T. As shown in Fig. 2(a), the steep jump in C/T associated with AFM ordering is quickly suppressed, and T_{max}, where C/T has its maximum value, gradually shifts to lower temperature, as indicated by arrows, reflecting suppression of the AFM ordering with field. The feature associated with the AFM order is suppressed entirely around 0.7 T. With further increase of magnetic fields, T_{max} shifts to higher temperature, while the peak hight is continuously suppressed, as shown in Fig. 2(b). This is due the to increase of magnetic field Zeeman splitting of the crystal electric field (CEF) doublet ground state. As seen in the inset of Fig. 2(a), S_{mag} reaches $1.0\,R\ln 2$ independently of applied fields up to 3 T, although the ground state and the shape of the specific heat anomaly both vary strongly with applied field, reflecting a two-fold degeneracy of the CEF

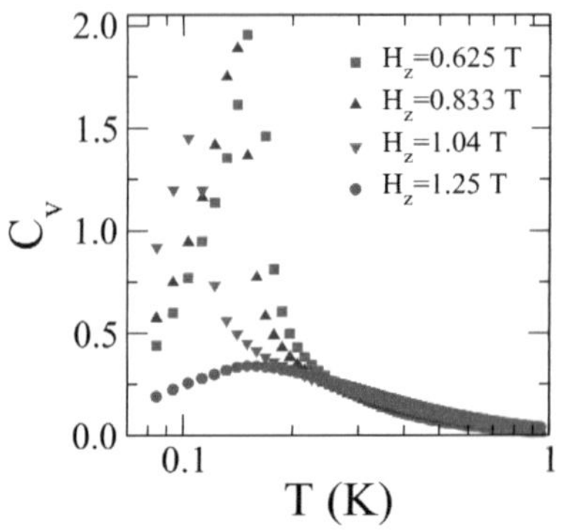

Figure 3. (Color online) Numerical results for C/T (in arbitrary units) for the spin-1/2 Heisenberg model as described in the text in the high field region.

doublet.

4. Theoretical modeling in field

Some of the results of our numerical calculations with magnetic field included are shown in Fig 3. At low fields (not shown) the anomaly is very robust, and does not change up to $H = 0.4$ T. At higher field the discrepancy between the experimental data and numerical modeling is even greater, since experimentally the AFM anomaly is suppressed completely for $H \approx 0.5T$, whereas theoretical calculations reveal a strong anomaly even at $H = 1.04$ T. This is an important difference, and can be due to a variety of factors absent in the model, including next nearest neighbors interactions, frustration, remaining Kondo physics, or RKKY coupling strength changing in magnetic field, and is an important issue to address in the future investigations of RKKY-Kondo competition.

5. Conclusion

In conclusion, we have performed low-temperature field-dependent specific heat to elucidate the ground state properties of $Ce_4Pt_{12}Sn_{25}$. T_N is suppressed with an initial increase of field up to 0.6 T, whereas T_{max} of the maximum in $C(T)$ begins to move to higher temperature above 0.7 T. This latter evolution is ascribed to an electronic Schottky contribution from the Zeeman-split ground state Γ_7 doublet. Therefore, the ground state of $Ce_4Pt_{12}Sn_{25}$ changes above 0.6 T from AFM to paramagnetic.

The model calculations, based on a CEF scheme with a Γ_7 ground state, reproduce rather well experimental high field specific heat and magnetization data. Lower field data is described very well by the numerical calculation based on the Heisenberg model. Small deviation between experimental data and numerical results can be ascribed to Kondo screening with a low characteristic temperature $T_K \leq T_N = 0.18$ K. This indicates that electronic spins on Ce^{3+} are not screened substantially by the conduction electrons, and places $Ce_4Pt_{12}Sn_{25}$ in the $J \ll J_c$ of the Doniach phase diagram. The Kondo temperature T_K is exponentially small in this regime, and a J^2 dependence of the RKKY interaction stabilizes the AFM ground state *in spite* of a large distance between the Ce ions. $Ce_4Pt_{12}Sn_{25}$ therefore presents us with a *counter example* to an expectation that dilute f-electron compounds will likely fall into the Kondo screened regime, and therefore should be good candidates to study Kondo physics. Weak $c - f$ hybridization in $Ce_4Pt_{12}Sn_{25}$ may be due to large distances between Ce and surrounding Sn and Pt atoms [9].

We would like to thank Hironori Sakai for useful discussions. Work at Los Alamos National Laboratory was performed under the auspices of the US Department of Energy. I. V. was supported in part by the US DOE Grant No. DE-FG02-08ER46492.

References

[1] J. Kondo, Prog. Theor. Phys. **32**, 37 (1964).
[2] M. A. Ruderman and C. Kittel, Phys. Rev. **96**, 99 (1954).
[3] T. Kasuya, Prog. Theor. Phys. (Kyoto) **16**, 45 (1956).
[4] K. Yosida, Phys. Rev. **106**, 893 (1957).
[5] S. Doniach, Physica B, **91**, 231 (1977).
[6] S. Doniach, in *Valence Instabitilies and Related Narrow Band Phenomena, edited by R. D. Parks (Plenum, New York, 1977)*, p. 169.
[7] Y.-F. Yang, Z. Fisk, H.-O. Lee, J. D. Thompson, and D. Pines, Nature **454**, 611 (2008).
[8] B. Chafik El Idrissi, G. Venturini and B. Malaman, Mat. Res. Bull., **25**, 807 (1990).
[9] H.-O. Lee, N. Kurita, P.-C. Ho, C. L. Condron, P. Klavins, S. M. Kauzlarich, M. B. Maple, R. Movshovich, J. D. Thompson, Z. Fisk, J. Phys.: Condens. Matter **22**, 065601 (2010).
[10] M. S. Kim, M. C. Bennett, and M. C. Aronson, Phys. Rev. B **77**, 144425 (2008)

Journal of Physics: Conference Series **273** (2011) 012046 doi:10.1088/1742-6596/273/1/012046

Magnetic properties of $Ce(Ru_{1-x}Fe_x)_2Al_{10}$

Takashi Nishioka[1], Daishi Hirai[1], Yukihiro Kawamura[1], Harukazu Kato[1], Masahiro Matsumurta[1], H. Tanida[2], M. Sera[2], Kazuyuki Matsubayashi[3], Yoshiya Uwatoko[3]

[1]Graduate School of Integrated Arts and Sciences, Kochi University, Kochi 780-8520, Japan
[2]Institute for Solid State Physics, University of Tokyo, Kashiwa 277-8581, Japan
[3]Department of Quantum Matter, ADSM, Hiroshima University, Higashi-Hiroshima, Hiroshima 739-8530, Japan

E-mail: nisioka@kochi-u.ac.jp

Abstract. $YbFe_2Al_{10}$-type $CeRu_2Al_{10}$ exhibits a mysterious phase transition at 27 K. The isostructural $CeFe_2Al_{10}$ is a Kondo semiconductor, whose macroscopic properties are very similar to those of $CeRu_2Al_{10}$ at a pressure of ~4 GPa. In order to clarify the relationship between the physical and chemical pressure effect on $CeRu_2Al_{10}$, we have performed measurements of the electrical resistivity, magnetic susceptibility and specific heat on pseudo-ternary $Ce(Ru_{1-x}Fe_x)_2Al_{10}$ single crystals. These results suggest that $Ce(Ru_{1-x}Fe_x)_2Al_{10}$ gives almost the same information as $CeRu_2Al_{10}$ under pressure, and furthermore the primary origin of the phase transition is not simple magnetic one.

1. Introduction

Very recently, there has been a growing interest in orthorhombic $YbFe_2Al_{10}$-type CeT_2Al_{10} (T=eight group elements: Fe, Ru and Os) [1], because they exhibit quite peculiar properties compared with conventional Ce compounds. Among them, the key material is $CeRu_2Al_{10}$ that shows a phase transition at T_0~27 K. This was first reported by Strydom on a polycrystalline sample, and he proposed this transition is due to magnetic origin [2]. The most important feature of $CeRu_2Al_{10}$ is that the content of Ce ions is very low, so that the RKKY interaction, which leads to magnetic orderings, should be strongly suppressed, and thus high temperature magnetic ordering is expected to be hardly realized. A typical magnetic ordering temperature for such "dilute" Ce compound is ~1 K [3]. Moreover, $CeRu_2Al_{10}$ lies in the valence fluctuation regime, because the cell volume deviates from that expected from the Lanthanoid contraction [1]. From such a point of view, we synthesized the single crystal for the first time and performed the precise macroscopic measurements, suggesting partial gap opens over the Fermi surface below T_0 [4, 5]. Subsequently, Matsumura *et al.* asserted that the phase transition is not due to magnetic origin but structural one from ^{27}Al NQR/NMR measurements [6]. On the other hand, a quite recent neutron diffraction measurement has revealed the antiferromagnetic ordered moment of 0.34 μ_B/Ce [7]. However, the proposed magnetic structure is completely inconsistent with the NQR spectra by Matsumura *et al.* [6]. Thus, the true nature of this phase transition is still controversial.

The pressure study of $CeRu_2Al_{10}$ gives important information on the origin of the mysterious phase transition. As pressure is applied to $CeRu_2Al_{10}$, T_0 disappears suddenly at around 4 GPa,

Published under licence by IOP Publishing Ltd

and it shows Kondo semiconducting behavior. Further increasing pressure, $CeRu_2Al_{10}$ shows heavy fermion behavior. The temperature dependences of electrical resistivity of $CeOs_2Al_{10}$ and $CeFe_2Al_{10}$ are very similar to those of $CeRu_2Al_{10}$ at ~2 GPa and ~4 GPa, respectively [5]. $CeFe_2Al_{10}$ is a Kondo semiconductor, which was discovered by Muro *et al.* [8]. Their pressure dependences are also expected from those of $CeRu_2Al_{10}$. Precise measurements are so difficult under pressure that research at ambient pressure is highly desirable. One effective method is the chemical pressure effect.

The purpose of this study is to examine the relationship between the physical and chemical pressure effect on $CeRu_2Al_{10}$. We synthesized the single crystals of $Ce(Ru_{1-x}Fe_x)_2Al_{10}$, in which Ru is replaced by Fe with smaller ionic radius, and have measured the electrical resistivity, magnetic susceptibility and specific heat. These results resemble those of $CeRu_2Al_{10}$ under pressure, and therefore $Ce(Ru_{1-x}Fe_x)_2Al_{10}$ realizes $CeRu_2Al_{10}$ under pressure.

2. Experimental

The single crystalline samples of $Ce(Ru_{1-x}Fe_x)_2Al_{10}$ (x=0.125, 0.250, 0.375, 0.500, 0.625, 0.750 and 0.875) were grown by the Al self-flux method. The X-ray analysis confirmed single phase formation of the sample. The lattice constants do not obey the Vegard's law, indicating the valence of Ce varies with x. The electrical resistivity was measured using the conventional ac four terminal method using the commercial ac resistance bridge (LakeShore 370). The magnetic susceptibility was measured by a SQUID magnetometer (Quantum Design: MPMS XL5). The specific heat was measured by the relaxation method (Quantum Design: PPMS).

3. Results and Discussion

Figure 1 shows the temperature dependence of the normalized electrical resistivity ρ/ρ_{300K}. The behavior of $\rho(T)$ at x=0 is characterized by a broad peak just below T_0, a shallow minimum at around 10 K and a small hump below 5 K [5]. As x increases up to 0.5, the broad peak gradually shrinks and the hump largely develops, thereby the shallow minimum becomes clear. The shape

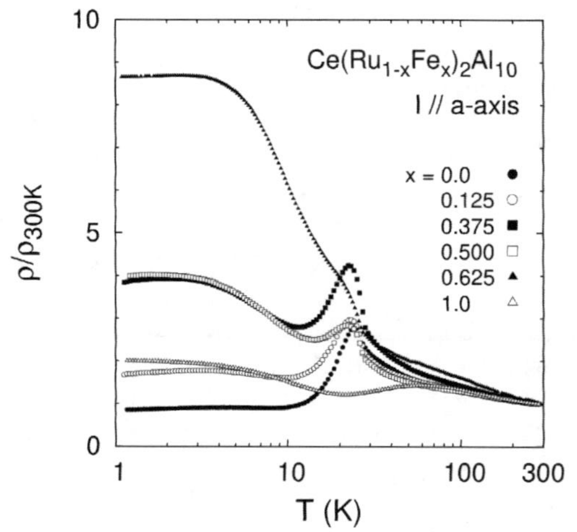

Figure 1. Temperature dependence of the normalized electrical resistivity ρ/ρ_{300K} of $Ce(Ru_{1-x}Fe_x)_2Al_{10}$ along the a-axis.

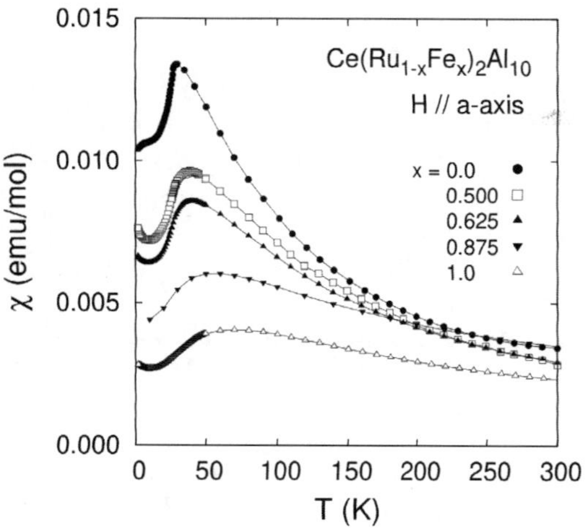

Figure 2. Temperature dependence of the magnetic susceptibility χ of $Ce(Ru_{1-x}Fe_x)_2Al_{10}$ along the a-axis. The applied fields are 1 T for x=0.375, 0.625, 0.875 and 5 T for x=0.0, 1.0.

Journal of Physics: Conference Series **273** (2011) 012046 doi:10.1088/1742-6596/273/1/012046

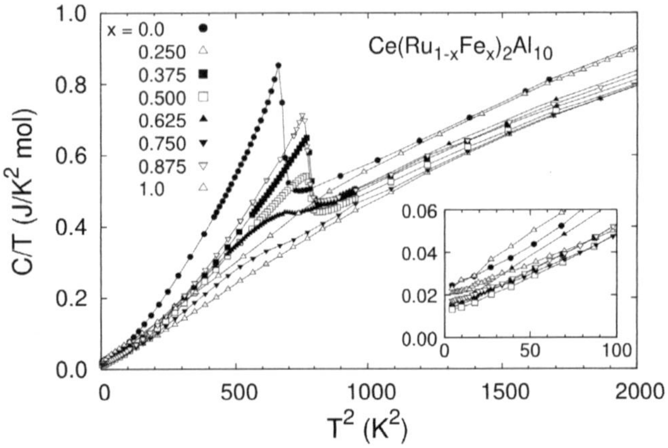

Figure 3. C/T vs T^2 plot for $Ce(Ru_{1-x}Fe_x)_2Al_{10}$. The inset shows an expanded view of the low-temperature region.

of $\rho(T)$ at $x = 0.5$ is almost the same as that of $CeOs_2Al_{10}$. At x=0.625, the hump becomes so large that both the local minimum and maximum anomalies turn into merely inflections, which is similar to $\rho(T)$ of $CeRu_2Al_{10}$ at 3 GPa. The x dependence of $\rho(T)$ for $Ce(Ru_{1-x}Fe_x)_2Al_{10}$ is quite similar to the pressure dependence of $\rho(T)$ in $CeRu_2Al_{10}$.

Figure 2 shows the temperature dependence of χ along the easy a-axis. At x=0, $\chi(T)$ follows the Curie-Weiss like law above ~50 K, and shows a broad peak at $T_{\rm max}$=30 K, followed by a sharp drop at T_0=27 K. $T_{\rm max}$ would correspond to the Kondo temperature. Although the difference between the temperature of $T_{\rm max}$ and $T_{\rm min}$ is very small at $x = 0$, it increases with x. The magnitude of χ becomes small with increasing x. The $\chi(T)$ at $x = 0.5$ is almost the same as that of $CeOs_2Al_{10}$. The overall behavior is qualitatively the same as our preliminary high pressure susceptibility measurements of $CeRu_2Al_{10}$.

Figure 3 shows the specific heat C of $Ce(Ru_{1-x}Fe_x)_2Al_{10}$ in a plot of C/T vs T^2. At x =0, C shows a sharp λ-type anomaly, and the temperature dependence follows the activated type form with a gap magnitude of ~100 K. As shown in the inset, the electronic specific coefficient γ_0=23 mJ/K^{2}mol, which is considerably smaller than the extrapolated value of γ_1=230 mJ/K^{2}mol to T=0 K above T_0. This reduction should be attributed to the reduction of the Fermi surface below T_0. The λ-type anomaly becomes smaller with increasing x, and the shape at x=0.5 is very similar to that of $CeOs_2Al_{10}$. Further increasing x, the λ-type anomaly becomes broad and only a small shoulder appears at 0.75. There seems to disappear the phase transition at x=0.875. The γ_0 value shows a minimum at x=0.5, whose value is nearly the same as that of $CeOs_2Al_{10}$.

Figure 4 summarizes the x dependence of T_0 from this study. As is easily seen, the values of T_0 obtained from three different measurements almost coincide. The x dependence is very similar to that of $CeRu_2Al_{10}$ under pressure. As pointed out above, these three measurements suggest that the behaviors of $Ce(Ru_{1-x}Fe_x)_2Al_{10}$ at x=0.5 and $CeOs_2Al_{10}$ are nearly identical. Furthermore, x dependences of these measurements are almost the same as the pressure dependence of $CeRu_2Al_{10}$. Accordingly, the chemical pressure effect of $CeRu_2Al_{10}$ almost gives the similar information of the physical pressure effect. From Fig. 3, T_0 suddenly disappears between x=0.75 and 0.875, where observed T_0 is always above 25 K. On the other hand, the entropy associated with the phase transition is strongly reduced with x, which means the ordered moment is strongly reduced with increasing x. If the transition is due to a normal RKKY-type magnetic transition, the ordered temperature should be proportional to $(g_J - 1)^2 J(J + 1)$, and then T_0 gradually decreases with increasing x. Nevertheless, the observed T_0 is always above 25 K, and disappears

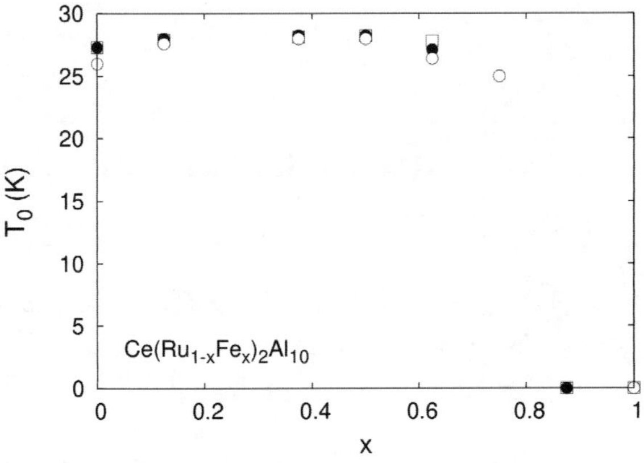

Figure 4. x dependence of T_0 for $Ce(Ru_{1-x}Fe_x)_2Al_{10}$. The symbols indicate as follows: C (○) , ρ (●), χ (□).

suddenly. This suggests that the primary origin of the phase transition is due to something unknown one, and the observed magnetic moment by the neutron diffraction might be induced by this unknown one. The inconsistency between NQR and neutron diffraction measurements may be related with the unknown origin. The detailed analysis will be published elsewhere.

4. Summary

Orthorhombic $YbFe_2Al_{10}$-type $CeRu_2Al_{10}$ shows a mysterious phase transition at $T_0 \sim 27$ K. An isostructural compound $CeFe_2Al_{10}$ is a Kondo semiconductor. The $\rho(T)$ of $CeRu_2Al_{10}$ at ~4 GPa is very similar to that of $CeFe_2Al_{10}$. We have grown single crystals of $Ce(Ru_{1-x}Fe_x)_2Al_{10}$ by the Al-self flux method and measured ρ, χ and C in order to investigate the relation between physical and chemical pressure. These macroscopic properties are very similar to those of $CeRu_2Al_{10}$ under pressure, and therefore we can conclude that $Ce(Ru_{1-x}Fe_x)_2Al_{10}$ gives almost the same information as $CeRu_2Al_{10}$ under pressure. T_0 suddenly disappears between x=0.75 and 0.875. We also found that the magnitude of χ and the anomaly in C at T_0 reduces with x. These results suggest that the primary origin of the mysterious phase transition is not simple magnetic one.

Acknowledgments

TN expresses his sincere thanks to Prof. Takabatake for continuous interest. This work was partially supported by a Grant-in-Aid for Scientific Research (C) of the Ministry of Education, Culture, Sports, Science and Technology.

References

[1] Thieda V M T, Ebel T and Jeitschko W 1998 *J. Mater. Chem.* **8** 2981.
[2] Strydom A M 2009 *Physica B* **404** 2981.
[3] See for example, Sereni J G 1991 *Handbook on the Physics and Chemistry of Rare Earths* vol 23, ed K A Gschneidner Jr and L Eyring (Amsterdam: Elsevier) p.1
[4] Takesaka T, Oe K, Kobayashi R, Kawamura Y, Nishioka T, Kato H, Matsumura M and Kodama K 2010 *J. Phys.: Conf. Ser.* **200** 012201.
[5] Nishioka T, Kawamura Y, Oe K, Kobayashi R, Kato H, Matsumura M, Kodama K, Matsubayashi K and Uwatoko Y 2009 *J. Phys. Soc. Jpn.* **78** 123705.
[6] Matsumura M, Kawamura Y, Edamoto S, Takesaka T, Kato H, Tokunaga Y, Kambe S and Yasuoka H 2009 *J. Phys. Soc. Jpn.* **78** 123713.
[7] Khalyavin D D, Hillier A D, Adroja D T, Strydom A M, Manuel P, Chapon L C, Peratheepan P, Knight K, Deen P, Ritter C, Muro Y and Takabatake T 2010 asXiV:1006.1618v1.
[8] Muro Y, Motoya K, Saiga Y and Takabatake T 2009 *J. Phys. Soc. Jpn.* **78** 083707.

Magnetically robust non-fermi liquid behavior due to the competition between crystalline-electric field singlet and Kondo-Yosida singlet in f^2-based heavy fermion systems

Shinya Nishiyama, Hiroyasu Matsuura and Kazumasa Miyake

Division of Materials Physics, Department of Materials Engineering Science, Graduate School of Engineering Science, Osaka University, Toyonaka, Osaka 560-8531, Japan

E-mail: `nishiyama@blade.mp.es.osaka-u.ac.jp`

Abstract. In f^2-based heavy fermion systems with a tetragonal symmetry, we investigate the magnetic field dependence of a non-fermi liquid (NFL) which arises related to the quantum critical point (QCP) due to the competition between the crystalline-electric field (CEF) singlet and the Kondo-Yosida singlet states. On the basis of the Wilson numerical renormalization group method, we find that the magnetic field less than a characteristic magnetic field H_z^* does not affect the characteristic temperature T_F^* at which the specific heat takes a maximum value. Since such H_z^* increases as the deviation from the QCP increases, slightly off the QCP, there are parameter regions where NFL behaviors are robust at an observable temperature range $T > T_F^*$ against a magnetic field of up to H_z^* which is far larger than T_F^*. Our result suggests that such robust NFL behaviors can arise also in systems with other CEF symmetries; e.g., magnetically robust NFL behaviors observed in UBe_{13} may be understood on this basis.

1. Introduction

The competition between the crystalline-electric field (CEF) singlet and the Kondo-Yosida (K-Y) singlet states in f^2-based heavy fermion systems has attracted attention. The former one is caused by a localized character of the f-electron, while the latter one is caused by the itinerant character of the f-electron due to the hybridization between the f-electron and conduction electrons. Because of these conflicting characters of the f-electron, the competition of these singlet states causes the quantum critical point (QCP) where the system exhibits non-Fermi liquid (NFL) behaviors similar to the two-channel Kondo model (TCKM) [1]. It is known that properties of $Th_{1-x}U_xRu_2Si_2$ [2], and some Pr-based filled skutterudite compounds can be explained on the basis of this NFL [3, 4]. However, the magnetic field effect on this NFL has not been fully studied so far.

In this paper, we investigate the magnetic field dependence of this NFL in a tetragonal symmetry on the basis of the Wilson numerical renormalization group method (NRG) [5, 6, 7]. As in [2], its competition suppresses the characteristic temperature T_F^* where the specific heat takes a maximum value compared to the energy scale of each singlet state, T_{K2} (the lower Kondo temperature) and Δ (the CEF splitting for first excited state). In this case, NFL behaviors appear at $T_F^* < T < \min(T_{K2}, \Delta)$ as in the case of the TCKM. We demonstrate that T_F^* is not affected by the magnetic field less than the characteristic magnetic field H_z^* which depends on the distance from the QCP. In such a case, there is little change in NFL behaviors under

Published under licence by IOP Publishing Ltd

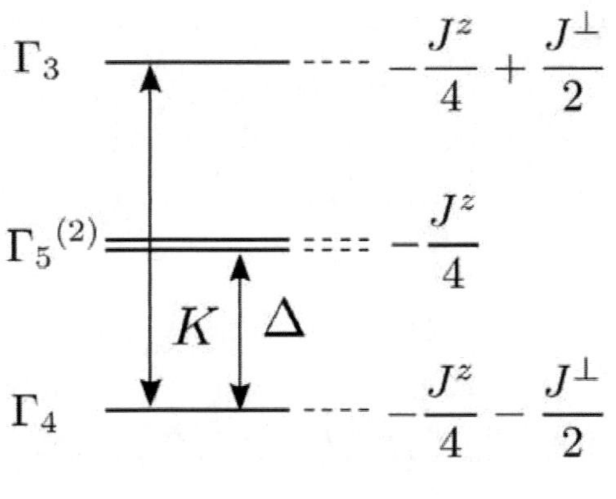

Figure 1. f^2-CEF level scheme. K and Δ is related to coupling constants in the Hamiltonian (1) as $J_\perp = K$ and $J_z = 2\Delta - K$, respectively.

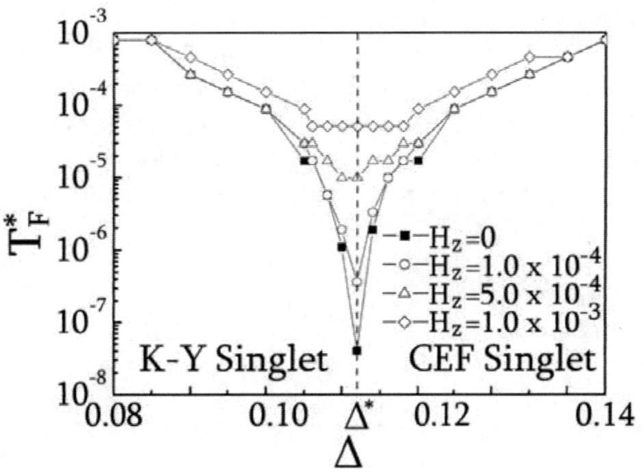

Figure 2. Δ dependence of T_F^* for various magnetic fields, H_z. The dashed line represents the critical point where two ground states exchange.

a magnetic field of up to H_z^*, i.e., NFL behaviors are robust against the magnetic field. Our result can be applicable to systems with other CEF symmetries: e.g., magnetically robust NFL behaviors observed in UBe_{13} [8] may be understood on this basis.

2. Numerical Results and Discussion

We use the two-orbital impurity Anderson model as an appropriate one for treating the present problem. It has already been known that the f^2-CEF level scheme in the tetragonal symmetry can be reproduced by introducing an anisotropic "antiferromagnetic Hund's-rule coupling" [2]. Then, the Hamiltonian is given as follows:

$$\mathcal{H} = \sum_{m\vec{k}\sigma} \varepsilon_{\vec{k}m} c^\dagger_{\vec{k}m\sigma} c_{\vec{k}m\sigma} + \sum_{m\sigma} E_{fm} f^\dagger_{m\sigma} f_{m\sigma} + \sum_{m} U_m f^\dagger_{m\uparrow} f^\dagger_{m\downarrow} f_{m\downarrow} f_{m\uparrow} + \sum_{m\vec{k}\sigma} \left[V_{m\vec{k}} f^\dagger_{m\sigma} c_{m\vec{k}\sigma} + \text{h.c.} \right] + \frac{J_\perp}{2} \left[S_1^+ S_2^- + S_1^- S_2^+ \right] + J_z S_1^z S_2^z, \quad (1)$$

where $f^\dagger_{m\sigma}$ is a creation operator of the f-electron on the orbital $m(=1,2)$ with the energy E_{fm}, and its pseudospin operator is $\vec{S}_m$. $c^\dagger_{m\vec{k}\sigma}$ is a creation operator of the conduction electron with the wave vector $\vec{k}$ hybridizing with the f-electron of the orbital m with the strength $V_{m\vec{k}}$. We explicitly consider the on-site intra-orbital Coulomb repulsion U_m, while other Coulomb repulsion terms are implicitly included in the "antiferromagnetic Hund's-rule coupling" of which coupling constants are related to f^2-CEF level splittings as shown in Fig.1. The hybridizations are set to be isotropic in the momentum space (i.e., $V_{m\vec{k}} \equiv V_m$), and the conduction bands to be symmetric in the energy space (with an extent from $-D$ to D) about a Fermi level. We perform NRG calculations by taking the discretization parameter as $\Lambda = 3.0$, and keeping states up to 1500 states in each iteration step.

The system described by the Hamiltonian (1) shows the competition between the K-Y singlet and the f^2-CEF singlet states. There exists two stable fixed points corresponding to these singlet states, and an unstable fixed point between those around where NFL behaviors appear. In the present paper, we take the parameter set in the Hamiltonian (1) as $E_{f1} = E_{f2} = -0.4$, $U_1 = U_2 = 1.0$, $V_1 = 0.45$, $V_2 = 0.3$, $K = 0.16$ measured in unit of D, and change the CEF splitting Δ. The energy scale of the K-Y singlet state is given by the lower Kondo temperature $T_{K2} = 6.01 \times 10^{-3}$ which is obtained by a Wilson's definition, $4T_K\chi_{imp}(T = 0) = 0.413$, in the case of $K = \Delta = 0$, although, precisely speaking, the actual lower Kondo temperature is different from T_{K2} due to the CEF effect. On the other hand, the energy scale of the f^2-CEF singlet state is given by $\min(K, \Delta)$. The competition between the K-Y singlet and the CEF singlet states arises around the point where these two characteristic energies are about the same.

First, we focus on the magnetic field dependence of the characteristic temperature T_F^* at which the specific heat of f-electrons, $C_{imp} \equiv \partial S_{imp}(T)/\partial(\log T)$, takes a maximum value just before the entropy, $S_{imp}(T)$, approaching 0 as $T \to 0$. Here, the effect of the magnetic field is taken into account through the Zeeman term, $\mathcal{H}_Z = -g\mu_B H_z j_z$ for f^1 states with $j = 5/2$ and $g = 6/7$. That on f^2 states also arises by Zeeman term for each f^1 orbitals composing f^2-states, and the off-diagonal matrix element between Γ_3 and Γ_4 singlet states. Fig.2 shows the Δ dependence of T_F^* under various magnetic fields, H_z. For $H_z = 0$, at a critical point $\Delta \equiv \Delta^* \simeq 0.112$, one can see that T_F^* is sharply suppressed. This is because T_F^* is given by the energy splitting between two singlet states which are nearly degenerated, not by the characteristic energies of each singlet state, T_{K2} and Δ, near $\Delta \sim \Delta^*$. As the distance from Δ^* increases, T_F^* asymptotically approaches those of each singlet state: i.e., $T_F^* \simeq T_{K2}$ for $\Delta \ll \Delta^*$, and $T_F^* \simeq K$ for $\Delta \gg \Delta^*$. On the other hand, the magnetic field dependence of $T_F^*(H_z)$ is hard to see from Fig.2.

Second, in order to see this point, we show the magnetic field dependence of T_F^* near the critical point for $\Delta < \Delta^*$ (the K-Y singlet ground state region) in Fig.3. It is noted that we only show results for $\Delta < \Delta^*$ in this paper, although similar behaviors are observed for $\Delta > \Delta^*$ i.e., in the CEF singlet ground state region. As one can see, $T_F^*(H_z)$ remains constant for H_z less than the characteristic magnetic field H_z^* (as shown by arrows in Fig.3) which is defined approximately as that where T_F^* starts to increase as H_z increases. As the distance from the critical point increases, H_z^* increases. Therefore, H_z^* is considered to be determined by two effects which break the criticality of the TCKM-type NFL fixed point. One is the distance of Δ from Δ^*, and the other is the magnetic field which breaks the degeneracy of the two singlet

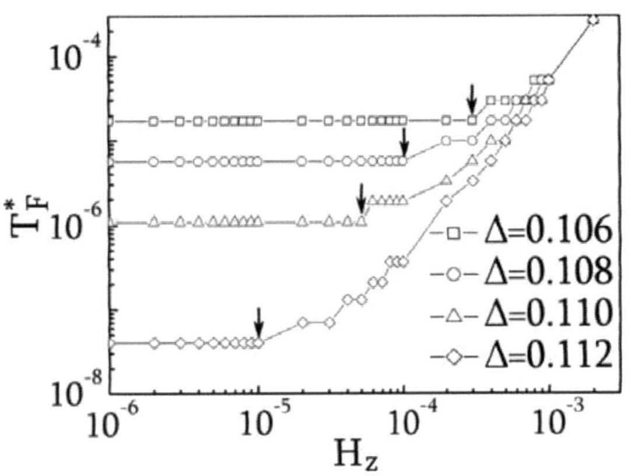

Figure 3. Magnetic field dependence of T_F^* around the critical point in the case of the K-Y singlet ground state. Arrows indicate the characteristic magnetic field H_z^*.

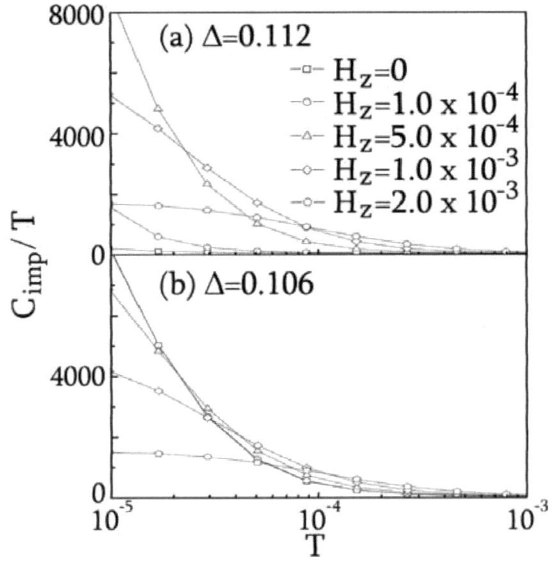

Figure 4. Sommerfeld coefficient C_{imp}/T vs T under various magnetic fields, for (a)$\Delta = 0.112$ and (b)0.106.

states which gives rise to the residual entropy $S_{\rm imp}(T=0)=0.5\log 2$ as in the case of the NFL state in TCKM. In other words, H_z^* corresponds to the energy scale characterizing a crossover from the TCKM-type NFL fixed point to the polarized local Fermi liquid fixed point.

It is remarked that the magnetic field affects two singlet states. However, since $H_z(<H_z^*)$ is much smaller than $T_{\rm K2}$ and Δ, the magnetic field is essentially considered not to break each singlet state. This means that H_z^* is not determined by the magnetic field dependence of two singlet states in this case.

Finally, in Fig.4, we show the temperature dependence of the Sommerfeld coefficient $\gamma_{\rm imp}(T)\equiv C_{\rm imp}/T$ for (a)$\Delta=0.112\simeq\Delta^*$ and (b)$\Delta=0.106$ under various magnetic fields. In the case of (a), extremely close to the critical point, $\gamma_{\rm imp}(T)$ increases as H_z increases. This is attributed to the drastic increase of $T_{\rm F}^*(H_z)$, especially in the 10^{-4} to 10^{-3} range of the magnetic field. As a result, $\gamma_{\rm imp}(T)$ logarithmically increases as T decreases down to 10^{-5}. On the other hand, in the case of (b), slightly off the critical point, the logarithmic divergence of $\gamma_{\rm imp}(T)$ is robust against a magnetic field of up to $H_z\sim 1.0\times10^{-3}$ for $T>3.0\times10^{-5}$. However, that is suppressed for $H_z=2.0\times10^{-3}$. As shown in Fig.3, $T_{\rm F}^*$ remains constant of up to $H_z=H_z^*\sim 2\times10^{-4}$, and increases rather moderately as H_z increases below $H_z\sim10^{-3}$. This is why the temperature dependence of $\gamma_{\rm imp}(T)$ remains the same as that at $H_z=0$ for $T>3\times10^{-5}\sim T_{\rm F}^*$ up to $H_z\sim10^{-3}$.

3. Conclusion and Discussion

In conclusion, we have investigated the effect of the magnetic field effect on NFL behaviors due to the competition between the K-Y singlet and the CEF singlet states in f^2-based heavy fermion systems with the tetragonal symmetry. Near the critical point, $T_{\rm F}^*$ has been shown not to be affected by the magnetic field less than the characteristic magnetic field H_z^*. We demonstrated that, in the vicinity of the QCP, there are parameter regions where the NFL behavior of the Sommerfeld coefficient is robust against the magnetic field of up to $H_z\sim H_z^*$. This is because such H_z^* increases as the system deviating from the QCP.

We expect that such an effect of the competition also causes the NFL behavior similar to the present one in systems with other CEF symmetries. UBe_{13}, a cubic heavy fermion system, is considered to be one of its candidates because of the experimental result indicating the competition between the K-Y singlet and the CEF singlets [9]. A detailed report will be published elsewhere. Although all the results in this paper are for the f^2-impurity problem, the effect of the competition can play an important role also in lattice systems to exhibit such a NFL robust against the magnetic field because the contribution to the Sommerfeld coefficient from the competition is known to be large [2, 4].

Acknowledgments

We would like to thank K. Hattori and S. Yotsuhashi for valuable conversation and discussion. This work is supported by a Grant-in-Aid for Scientific Research on Innovative Areas "Heavy Electron" (No.20102008). S.N. and H.M. are supported by the Global COE program (G10) from the Japan Society for the Promotion of Science.

References

[1] Cox D L and Zawadowski A 1998 *Adv. Phys.* **47** 599
[2] Yotsuhashi S, Miyake K and Kusunose H 2002 *J. Phys. Soc. Jpn* **71** 389
[3] Otsuki J, Kusunose H and Kuramoto Y 2005 *J. Phys. Soc. Jpn* **74** 2082
[4] Hattori K and Miyake K 2005 *J. Phys. Soc. Jpn* **74** 2193
[5] Wilson K G 1975 *Rev. Mod. Phys* **47** 773
[6] Krishna-murthy H R, Wilkins J W and Wilson K G 1980 *Phys. Rev.* B **21** 1003
[7] Krishna-murthy H R, Wilkins J W and Wilson K G 1980 *Phys. Rev.* B **21** 1044
[8] Gegenwart P, Langhammer C, Helfrich R, Oeschler N, Land M, Kim J S, Stewart G R and Steglich F 2004 *Physica* C **408** 157
[9] Kim J S, Andraka B, Jee C S, Roy S B and Stewart G R 1990 *Phys. Rev.* B **41** 11073

Journal of Physics: Conference Series **273** (2011) 012048 doi:10.1088/1742-6596/273/1/012048

Transport and magnetic properties of new heavy-fermion antiferromagnet $YbNi_3Al_9$

S Ohara, T Yamashita, Y Mori and I Sakamoto

Department of Engineering Physics, Electronics and Mechanics, Graduate School of Engineering, Nagoya Institute of Technology, Nagoya 466-8555, Japan

E-mail: ohara.shigeo@nitech.ac.jp

Abstract. We have synthesized a new Yb-based Kondo-lattice compound $YbNi_3Al_9$. This compound crystallizes in a trigonal $ErNi_3Al_9$-type structure (space group R32), in which the Yb-ion is arranged in a two-dimensional honey-comb lattice perpendicular to the c-axis. We report the first measurements of electrical resistivity and magnetization for single-crystalline samples of $YbNi_3Al_9$. The electrical resistivity of $YbNi_3Al_9$ is characteristic of the typical properties of heavy-fermion antiferromagnets with a Neel temperature of T_N = 3.4K. The transport and magnetic properties exhibit large anisotropy in the low-temperature region owing to an interplay among the crystalline-electric-field effect, the Ruderman-Kittel-Kasuya-Yoshida interaction, and the Kondo effect. Below T_N, the metamagnetic transition is observed at a very low magnetic field of around 1 kOe with the field applied along the a-axis. The magnetic structure of $YbNi_3Al_9$ is highly sensitive to the applied magnetic field.

1. Introduction

Many Ce- and Yb-based intermetallic compounds have attracted considerable interest owing to their various physical properties such as unusual superconductivity and heavy-fermion behavior and the phenomena they exhibit, such as valence fluctuations and anomalous magnetic transitions[1]. These properties and phenomena arise from the interplay among the crystalline-electric-field (CEF) effect, the Ruderman-Kittel-Kasuya-Yoshida (RKKY) interaction, and the Kondo effect. Numerous studies have been conducted on heavy-fermion Ce-based compounds. However, relatively few studies have been conducted on heavy-fermion Yb-based compounds. This is mainly due to difficulties in the sample preparation of Yb-based compounds, which is due to the high vapor pressure of Yb[2] and also because most of the known Yb-based intermetallic compounds are found to be usual mixed-valent or non-magnetic metals.

Recently, we have succeeded in synthesizing single crystals of $YbNi_3Al_9$ through an investigation of the Al-rich parts of the phase diagrams of Yb-T-Al systems, where T is a transition metal. The crystal structures of ternary aluminides RNi_3Al_9 (R = Y, Gd, Dy, Er) were reported by Gladyshevskii[3]. However, to the best of our knowledge, the physical properties of $YbNi_3Al_9$ have not yet been investigated in detail.

We have found that $YbNi_3Al_9$ crystallizes in the trigonal $ErNi_3Al_9$-type structure (space group R32)[3]. In this crystal structure, three types of layers are observed to pile up along the c-axis: Al_3-layers with a triangular mesh, Ni_3-layers with a triangular mesh, and Yb_2Al_3-layers. In the Yb_2Al_3-layer, the Yb ion is arranged in a two-dimensional honey-comb lattice site and at the center of the Al_3 triangle. The Yb ion occupies a site equivalent to 6c. The stacking sequence

Published under licence by IOP Publishing Ltd

of these layers is repeated three times along the c-axis: $Al_3(B)$-$Ni_3(C)$-$Al_3(A)$-$Al_3(C)$-$Al_3(A)$-$Ni_3(C)$-$Al_3(B)$-$Yb_2Al_3(A)$[3]. Here, we use the notation of the site A, B and C that is introduced for the stacking of the closed-packed layers. The unit cell contains 78 atoms. The shortest Yb-Yb distance of $a/\sqrt{3} = 0.42$ nm in the Yb_2Al_3 layer is less than half of the Yb-Yb distance of $c/3 = 0.91$ nm along the c-axis. We expect that the unique two-dimensional magnetic honey-comb lattice structure will give rise to unusual magnetic properties. Here, we report the first results of electrical resistivity and magnetization measurements on single crystalline $YbNi_3Al_9$.

2. Experimental

Single crystals of $YbNi_3Al_9$ are synthesized in Al flux[4]. The excess Al surrounding the grown crystals is removed using centrifugal force and then corroded away using HCl solution. Crystallographic characterization is performed by means of X-ray powder diffraction. Since the X-ray powder diffraction pattern of $YbNi_3Al_9$ is in good agreement with that of $ErNi_3Al_9$, we conclude that the crystal structure of $YbNi_3Al_9$ is trigonal $ErNi_3Al_9$-type[3]. Additional reflections of impurities are not observed. The obtained lattice parameters are $a = 0.7270(5)$ nm and $c = 2.736(4)$ nm. The single crystalline nature is confirmed using the Laue method. The magnetization M and electrical resistivity ρ are measured in the temperature range of 1.8-300 K. Magnetization measurements are carried out using a SQUID magnetometer (Quantum Design) in magnetic fields H of up to 50 kOe. The electrical resistivity measurements are performed using the usual dc four-probe method.

3. Results and Discussion

Temperature dependence of electrical resistivity for $YbNi_3Al_9$ and $LuNi_3Al_9$ with the current $J \| a$-axis (ρ_a) and $J \| c$-axis (ρ_c) is shown in Fig. 1. The resistivity of $LuNi_3Al_9$ shows typical metallic behavior. Slight anisotropy is observed between ρ_a and ρ_c. In both cases, the residual resistivity is less than $1\mu\Omega$cm and the residual resistivity ratio is approximately 30. These values indicate that the crystals of $LuNi_3Al_9$ are of good quality. The electrical resistivity of $YbNi_3Al_9$ corresponds to the typical properties of heavy-fermion antiferromagnets with a magnetic ordering temperature of 3.4 K. The low value of resistivity at 1.5 K (approximately 5 $\mu\Omega$cm) indicates minimal impurity scattering.

The magnetic contribution of $YbNi_3Al_9$ to resistivity, ρ_m, is obtained by subtracting the resistivity of $LuNi_3Al_9$ from that of $YbNi_3Al_9$; this magnetic contribution is plotted against $\log T$ in Fig. 2. ρ_m along the a-axis is proportional to $-\log T$ between 100 and 30 K. The $-\log T$

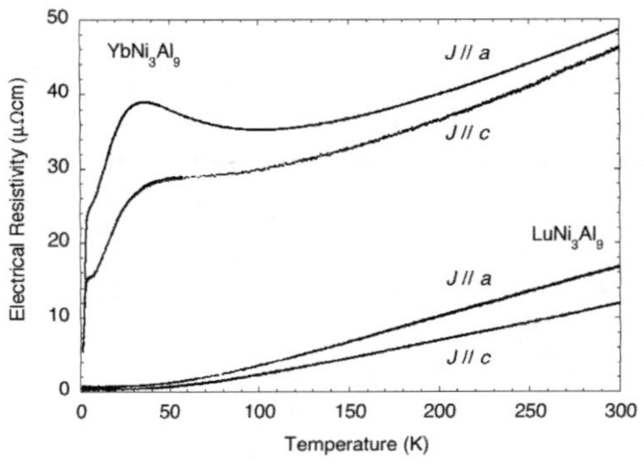

Figure 1. Temperature dependence of electrical resistivity for $YbNi_3Al_9$ and $LuNi_3Al_9$.

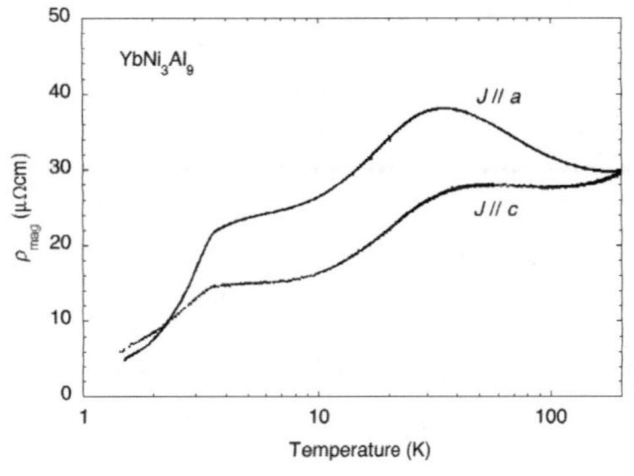

Figure 2. Magnetic contribution to the electrical resistivity ρ_m for $YbNi_3Al_9$ as a function of log T.

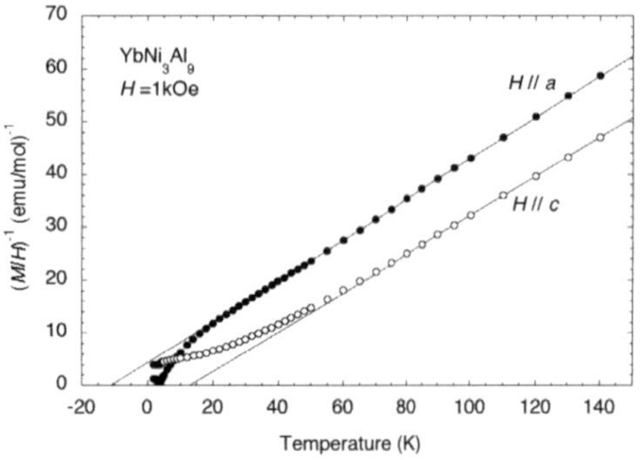

Figure 3. Temperature dependence of reciprocal magnetic susceptibilities χ^{-1} = $(M/H)^{-1}$ of $YbNi_3Al_9$ measured in a magnetic field of H= 1kOe

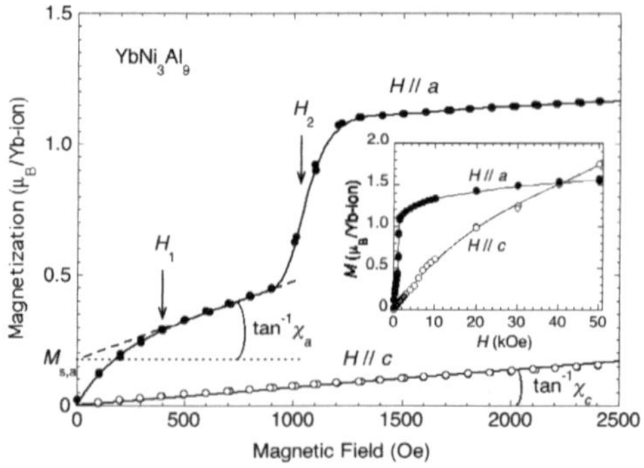

Figure 4. Magnetization curves of $YbNi_3Al_9$ for both $H\|a$- and $H\|c$-axis at 2K.

dependence is characteristic of the Kondo scattering. Along the c-axis, the –log T dependence is relatively weak. This may be due to a characteristic two-dimensional array of magnetic Yb ions perpendicular to the c-axis in $YbNi_3Al_9$. The broad hump around 30 K is attributed to the competition between the Kondo effect and the CEF effect which is expected to divide the J = 7/2 multiplet in four Kramer's doublets. The strong reduction of the magnetic scattering at T = 3.4 K is attributed to the magnetic ordering.

Figure 3 shows the temperature dependence of reciprocal magnetic susceptibilities χ^{-1} = $(M/H)^{-1}$ of $YbNi_3Al_9$ measured in a magnetic field of H= 1kOe for both $H\|a$- and $H\|c$-axis. In both magnetic field directions, the susceptibilities of $YbNi_3Al_9$ show Curie-Weiss behavior above 80 K. Below 30 K, a large deviation from the Curie-Weiss law are observed. This behavior may be ascribed to the CEF effect. The effective magnetic moment μ_{eff} and the paramagnetic Curie temperatures Θ_p are estimated to be $4.55\mu_B$/Yb-ion and –12 K for the a-axis and $4.65\mu_B$/Yb-ion and +12K for the c-axis, respectively. These values of μ_{eff} agree with those expected in case of free Yb^{3+} ions (i.e., $\mu_{eff} = 4.54\mu_B$), accounting for the experimental accuracy. Since $LuNi_3Al_9$, in which the rare-earth element Lu is not expected to carry magnetic moment, shows diamagnetism, the Ni ion in the compounds has no magnetic moment. The room-temperature susceptibility value of $LuNi_3Al_9$ is -3×10^{-4}emu/mol.

The magnetization curves for both $H\|a$- and $H\|c$-axis in a low magnetic field region at 2K are shown in Fig. 4. With increase in the magnetic field, the magnetization along the a-axis shows a rapid increase up to H_1, followed by a linear increase, and then a metamagnetic transition at around H_2 = 1kOe. It should be noted that these field-induced anomalies occur at very low fields. Since these anomalies disappear at the magnetic transition temperature (Fig. 6), they must be an intrinsic property of $YbNi_3Al_9$. Above H_2, the magnetization for a-axis is gradually increased and reaches $1.60\mu_B$ at 50 kOe, as shown in the inset of the figure. In contrast, the magnetization along the c-axis increases smoothly and reaches $1.75\mu_B$ at 50 kOe. These values are consistent with the magnitude of the magnetic moment of the ground state Kramer's doublet induced by CEF. Residual magnetization and hysteresis are not observed along both directions, within experimental accuracy.

For obtaining the temperature dependence of magnetic susceptibility in case of $YbNi_3Al_9$ at around the magnetic ordering temperature of 3.4 K, we measured the magnetic field dependence of magnetization along the a-axis in the temperature range from 2 to 4 K in steps of 0.2 K. As shown in Fig. 4, we determine χ_a and χ_c as the slopes of M-H curves in the low field region.

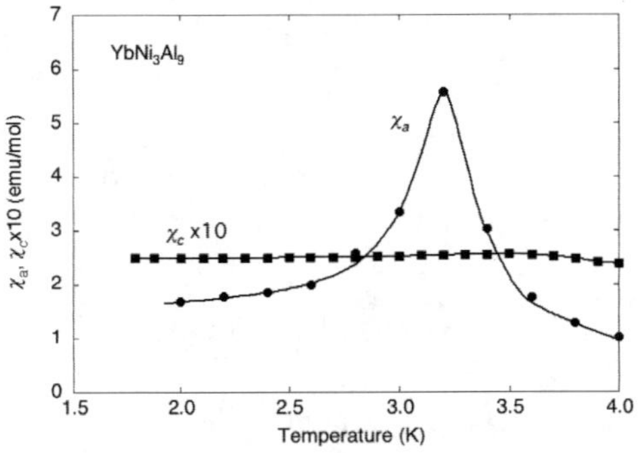

Figure 5. Temperature dependence of magnetic susceptibility of $YbNi_3Al_9$ at a low temperature region, obtained from the slope of *M*-*H* curves.

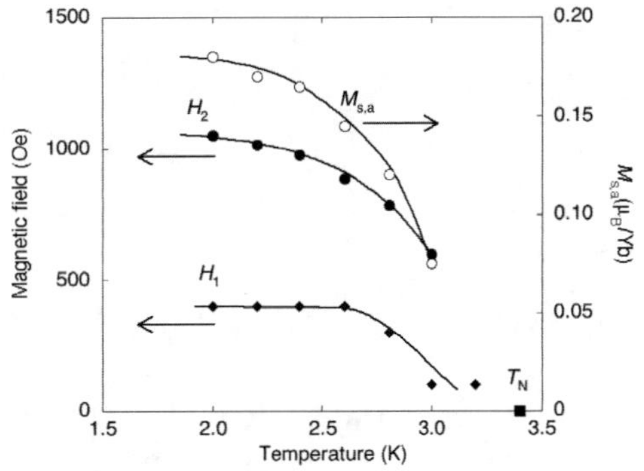

Figure 6. Magnetic field-temperature phase diagram of $YbNi_3Al_9$ for the magnetic field along the *a*-axis.

The temperature dependence of χ_a and χ_c is shown in Fig. 5. The different susceptibility scales between χ_a and χ_c are noteworthy. There is a large anisotropy in the magnitude of susceptibility along two crystallographic directions. χ_a shows a peak at 3.2 K, while χ_c is almost constant below 3.4 K. These characteristic features of susceptibility can be ascribed to the antiferromagnetic ordering with the magnetic easy plane perpendicular to the *c*-axis. This result is consistent with the observation of metamagnetic transition along the $H \| a$-axis.

We have estimated the field-induced spontaneous magnetization $M_{s,a}$ using the intercept of linear extrapolation of the *M*-*H* curve, as shown in Fig. 4. The value of $M_{s,a}$=0.18μ_B at 2K amounts to approximately 10% of that expected for the Kramer's doublet ground state of the Yb^{3+} ion. The field-induced spontaneous magnetization may be considered to arise from a canted antiferromagnetic structure.

Finally, we show the temporary magnetic field-temperature phase diagram of $YbNi_3Al_9$ for the magnetic field along the *a*-axis in Fig. 6. The magnetic structure of $YbNi_3Al_9$ is highly sensitive to the magnetic field. A more detailed study is required to understand the magnetic structure of $YbNi_3Al_9$.

4. Summary

We have synthesized single crystals of $YbNi_3Al_9$. $YbNi_3Al_9$ is a heavy-fermion antiferromagnet with $T_N = 3.4$K. The magnetic structure of $YbNi_3Al_9$ is highly sensitive to the magnetic field. To obtain more detailed information about the magnetic phase diagram and the magnetic structure of $YbNi_3Al_9$, specific heat and neutron diffraction experiments under magnetic fields are being conducted. A more detailed analysis of the magnetic properties of $YbNi_3Al_9$ will be reported elsewhere.

Acknowledgments

The authors acknowledge Dr. Y. Uwatoko, Dr. Y. Aoki, and Dr. K. Mibu for their helpful discussions.

References

[1] See e.g., Doniach 1977 *Valence Instabilities and Relation Narrow Band Phenomena* (New York: Plenum)
[2] Fisk Z., Maple M. B. 1992 *J. Alloys Compd.* **183** 303
[3] Gladyshevskii R. E., Cenzual K., Flack H. D., and Parthe E. 1993 *Acta Cryst.* **49** 468
[4] Ohara S., Chen G. F., Sakamoto I. 2001 *J. Alloys Compd.* **323-324** 632

Journal of Physics: Conference Series **273** (2011) 012049 doi:10.1088/1742-6596/273/1/012049

Competing magnetic interactions in $CeNi_{9-x}Co_xGe_4$

L Peyker[1], C Gold[1], W Scherer[1], H Michor[2] and E-W Scheidt[1]

[1] CPM, Institut für Physik, Universität Augsburg, 86159 Augsburg, Germany

[2] Institut für Festkörperphysik, Technische Universität Wien, 1040 Wien, Austria

E-mail: ludwig.peyker@physik.uni-augsburg.de

Abstract. $CeNi_9Ge_4$ exhibits outstanding heavy fermion features with remarkable non-Fermi-liquid behavior which is mainly driven by single-ion effects. The substitution of Ni by Cu causes a reduction of both, the RKKY coupling and Kondo interaction, coming along with a dramatic change of the crystal field (CF) splitting. Thereby a quasi-quartet ground state observed in $CeNi_9Ge_4$ reduces to a two-fold degenerate one in $CeNi_8CuGe_4$. This leads to a modification of the effective spin degeneracy of the Kondo lattice ground state and to the appearance of antiferromagnetic (AFM) order. To obtain a better understanding of consequences resulting from a reduction of the effective spin degeneracy, we stepwise replaced Ni by Co. Thereby an increase of the Kondo and RKKY interactions through the reduction of the effective d-electron count is expected. Accordingly, a paramagnetic Fermi liquid ground state should arise. Our experimental studies, however, reveal AFM order already for small Co concentrations, which becomes even more pronounced with increasing Co content x. Thereby the modification of the effective spin degeneracy seems to play a crucial role in this system.

1. Introduction

Due to the wide variety of their ground states, Ce based ternary intermetallic compounds have attracted large attention during the last years. Inherent to these materials is the existence of localized $4f$ electrons (Ce^{3+}) at high temperatures. In the low temperature limit, the competition between Kondo and RKKY interactions results in different ground states depending on their relative magnitudes [1]: The formation of long-range magnetic order depends quadratically on the dimensionless coupling parameter $N(E_F)J_0$, while the development of a local Kondo ground state depends exponentially on $N(E_F)J_0$ [2]. The dimensionless effective exchange coupling parameter $N(E_F)J_0$ correlates the exchange interaction, J_0, between $4f$ localized magnetic moments and conduction electrons with the electronic density of states at the Fermi level, $N(E_F)$. Accordingly, small values of $N(E_F)J_0$ favor long range magnetic order, while for large values of $N(E_F)J_0$ a paramagnetic Kondo-screened ground state is anticipated. At the borderline between these two regimes non Fermi-liquid (NFL) behavior has been observed.

Apart from this classical scenario other mechanisms, e.g., via a change of the effective spin degeneracy N may occur. The latter was analyzed by Coleman [3]. This additional mechanism seems to be relevant for the ground state of the heavy fermion system $CeNi_9Ge_4$ [4]. In the substitution series $CeNi_{9-x}Cu_xGe_4$ a reduction from a quasi-quartet ground state in $CeNi_9Ge_4$ to a two-fold degenerated one in $CeNi_8CuGe_4$ leads to antiferromagnetic (AFM) order[4]. This is also benefitted by a reduction of the effective exchange coupling $N(E_F)J_0$ due to an enhanced effective d-electron count.

Published under licence by IOP Publishing Ltd

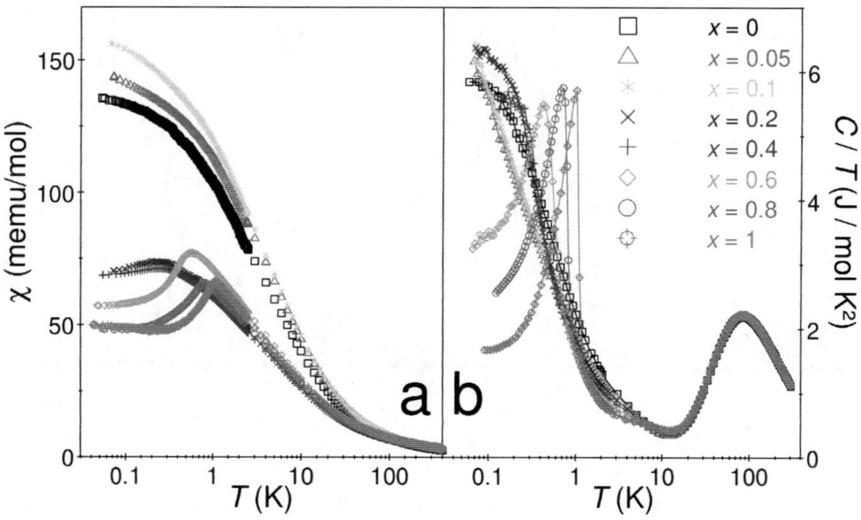

Figure 1. (a) The magnetic susceptibility χ and (b) the specific heat divided by temperature C/T of $CeNi_{9-x}Co_xGe_4$ in semi-logarithmic plots. AFM transitions are evident for $x \geq 0.2$.

In the present work, the hybridization strength in $CeNi_9Ge_4$ is tuned contrary through non-isoelectronic Ni/Co substitution to study the influences of the effective spin degeneracy in more detail.

2. Sample preparation and characterization

For the synthesis of the samples of $CeNi_{9-x}Co_xGe_4$ materials of high purity were used: Ce: 4N; La: 3N8 (Ames MPC); Ni: 4N5; Co: 4N8; Ge: 5N. They were prepared by arc melting under argon atmosphere followed by an annealing process at 950° C for two weeks inside an evacuated quartz tube. Less than 0.5 % weight loss occurred during the melting process. Optical emission spectroscopy in an inductively coupled plasma (ICP-OES) and energy dispersive X-ray spectroscopy (EDX) analysis indicated that the samples used in this work are essentially single phase. X-ray powder diffraction experiments revealed that all samples crystallize in the $LaFe_9Si_4$-type structure (tetragonal spacegroup $I4/mcm$). For the whole concentration range, the replacement of Ni by Co does not lead to any significant changes in the lattice parameters and therefore also not in the unit cell volume. Thus, volume effects in $CeNi_{9-x}Co_xGe_4$ due to the substitution, i.e. chemical pressure, are hardly relevant.

3. Experimental results

Figure 1 a presents the temperature dependent magnetic susceptibility χ of $CeNi_{9-x}Co_xGe_4$. The magnetic behavior of the parent compound $CeNi_9Ge_4$ is discussed in more detail in [5]. The Co substituted samples follow a simple modified Curie-Weiss-type law, $\chi(T) = C/(T - \Theta) + \chi_0$, above 100 K . From a least-square fit the following parameters were obtained: i) the paramagnetic Curie-Weiss temperature Θ around -40 K, ii) the temperature independent susceptibility contribution χ_0 between 0.9 memu mol^{-1} for $x = 0.05$ and 1.6 memu mol^{-1} for $x = 1$, iii) the Curie constant C corresponding to an effective paramagnetic moment between 2.5 μ_B and 2.6 μ_B, which is in line with the theoretical value of 2.54 μ_B for a Ce^{3+} ion. An itinerant paramagnetic

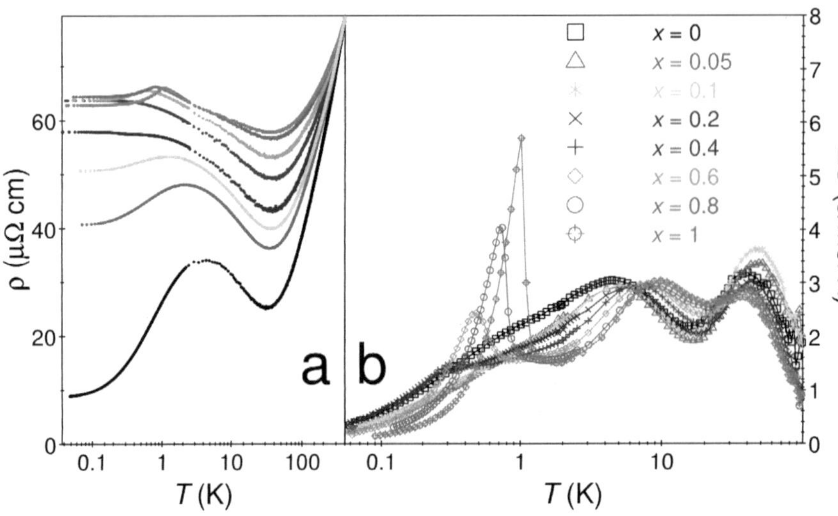

Figure 2. (a) The electrical resistivity $\rho(T)$ of $CeNi_{9-x}Co_xGe_4$ normalized at 300 K to that of $LaNi_9Ge_4$ in a semi-logarithmic plot. (b) Temperature dependence of the magnetic specific heat ΔC of $CeNi_{9-x}Co_xGe_4$ in semi-logarithmic representation.

Co sublattice as observed in La- and $CeCo_9Ge_4$ [6] is not observed. For smallest Co concentrations $x \leq 0.1$ the substitution of Ni by Co causes a moderate increase of the low temperature susceptibility as compared to the parent compound $CeNi_9Ge_4$. This may be caused by the influence of substitutional disorder leading to partial reduction of the AFM intersite coupling or to some ferromagnetic correlations. Between $x = 0.1$ and $x = 0.2$ the low temperature susceptibility decreases dramatically. This huge change in the magnetic properties can not only be caused by a marked increase of the Kondo screening but also by a reduction of the effective spin degeneracy. On further increasing x the development of sharp cusps in $\chi(T)$ finally indicates phase transitions towards AFM order. $CeNi_8CoGe_4$ exhibits magnetic ordering below $T_N \approx 1$ K. These observations are corroborated by specific heat results. Figure 1 b shows the specific heat divided by temperature C/T of $CeNi_{9-x}Co_xGe_4$ between 60 mK and 300 K on a semi-log scale. Already a small amount of Co ($x \leq 0.1$) causes a $C/T \propto -\ln(T)$ divergence of the Sommerfeld coefficient which holds over more than one decade in temperature down to $T = 70$ mK. This reveals a temperature dependent Sommerfeld-Wilson ratio, $R \propto \chi_0/\gamma$, which has been discussed in terms of crystal field effects and a fourfold degenerated ground state in the case of $CeNi_9Ge_4$ [7; 8]. In addition, the divergence of the Sommerfeld coefficient indicates the vicinity to a QCP transition which may occur around $x = 0.15$. The existence of a QCP transition is supported by a linear dependence of the Néel-temperature T_N for $x \geq 0.6$. At higher Co concentrations susceptibility and specific heat reveal, initially a flattening ($x = 0.2; 0.4$) and then a peak $x \geq 0.6$ indicating an AFM transition. Additionally, the low temperature limit of C/T decreases with higher x values, indicating an enhancement of the Kondo coupling.
In Figure 2 a the electrical resistivity of $CeNi_{9-x}Co_xGe_4$ is displayed. The classical Kondo lattice behavior of $CeNi_9Ge_4$ which is observed, is explicit illustrated in [4]. The two samples with a small Co concentration ($x = 0.05$ and $x = 0.1$) exhibit a similar behavior but with enhanced residual resistances at low temperature. Additionally, with increasing Co content, substitutional disorder shifts the Kondo coherence maximum to lower temperatures. On further increase of x the Kondo lattice behavior disappears and the residual resistance increases further. The cusps

in the electrical resistivity for $x \geq 0.6$ indicate the AFM transition, already observed in magnetic susceptibility and specific heat data.
In order to study the crossover from single-ion Kondo behavior towards long range AFM order, the magnetic contribution to the specific heat ΔC (Fig 2 b) was extracted by subtracting the total specific heat of the corresponding La compounds with unoccupied $4f$ states. For $CeNi_9Ge_4$ two pronounced maxima around 5 und 35 K are obtained. The first Schottky-like anomaly is associated with a quasi-quartet ground state characterized by a subtle splitting into two doublets $\Gamma_7^{(1)}$ and $\Gamma_7^{(2)}$ which interacts with the Kondo screening on the same energy scale [5]. The second maximum is associated with the Γ_6 CF doublet. With increasing Co concentration the upper maximum remains almost constant, while the lower maximum shifts to higher temperatures (10 K for $x = 1$). This refers to a reduction of the effective spin degeneracy N like it was found in $CeNi_{9-x}Cu_xGe_4$ [4] and appears to be the driving force for the AFM ordering. For $x \geq 0.2$ the AFM phase transition at low temperatures is observed.

4. Conclusions

Studying the ground state evolution in $CeNi_{9-x}Co_xGe_4$, we found already for small Co concentrations $x \leq 0.1$ a logarithmic divergence of the Sommerfeld coefficient $\gamma \simeq C/T$, as it is typical for a QCP transition. In contrast, the magnetic susceptibility of these compounds shows a flattening instead of the logarithmic divergence which would be expected in case of a temperature independent Sommerfeld-Wilson ratio χ_0/γ. A further increase of the Co content $x \geq 0.2$ leads to a significant pronounced increase of the Kondo temperature but nonetheless AFM order appears in this system. The latter is counter-intuitive to the usual Doniach picture and seems to result from a reduction of the effective spin degeneracy N. The fact that a similar scenario is observed also in $CeNi_{9-x}Cu_xGe_4$ [4], where the $3d$ electron count changes in the opposite way, clearly suggests that the reduction of the effective spin degeneracy away from the paramagnetic quasi-quartet ground state in $CeNi_9Ge_4$ plays a crucial role in the formation of long range AFM order in the system $CeNi_{9-x}TM_xGe_4$ (TM = Cu, Co).

5. Acknowledgments

This work was supported by the Deutsche Forschungsgemeinschaft (DFG) under Contract No. SCHE487/7-1/2.

References

[1] Stewart G R 2001 *Rev. Mod. Phys.* **73** 797 and
Stewart G R 2006 *Rev. Mod. Phys.* **78** 743

[2] Doniach S 1977 *Physica B & C* **91** 231

[3] Coleman P 1983 *Phys. Rev. B* **28** 5255

[4] Peyker L, Gold C, Scheidt E-W, Scherer W, Donath J G, Gegenwart P, Mayr F, Unruh T, Eyert V, Bauer E and Michor H 2009 *J. Phys.: Condens. Matter* **21** 235604

[5] Michor H, Adroja D T, Bauer E, Bewley R, Dobozanov D, Hillier A D, Hilscher G, Killer U, Koza M, Manalo S *et al.* 2006 *Physica B* **378** 640

[6] Gold C, Peyker L, Scheidt E-W, Michor H and Scherer W 2010 *J. Phys.: Conf. Ser.* **200** 012049

[7] Scheidt E-W, Mayr F, Killer U, Scherer W, Michor H, Bauer E, Kehrein S, Pruschke T and Anders F 2006 *Physica B* **378** 154

[8] Anders F B and Pruschke T 2006 *Phys. Rev. Lett.* **96** 86404

Journal of Physics: Conference Series **273** (2011) 012050 doi:10.1088/1742-6596/273/1/012050

Continuous-Time Monte Carlo study of the pseudogap Bose-Fermi Kondo model

J. H. Pixley[1], Stefan Kirchner[2,3], M. T. Glossop[1] and Qimiao Si[1]

[1] Department of Physics & Astronomy, Rice University, Houston, Texas, 77005, USA

[2] Max Planck Institute for the Physics of Complex Systems, 01187 Dresden, Germany

[3] Max Planck Institute for Chemical Physics of Solids, 01187 Dresden, Germany

E-mail: jp11@rice.edu

Abstract. We study the pseudogap Bose-Fermi Anderson model with a continuous-time quantum Monte Carlo (CT-QMC) method. We discuss some delicate aspects of the transformation from this model to the Bose-Fermi Kondo model. We show that the CT-QMC method can be used at sufficiently low temperatures to access the quantum critical properties of these models.

Over the past decade intermetallic compounds have served as model systems to study instabilities of metallic magnets near zero temperature [1–3]. In particular the quantum critical properties found in several heavy fermion compounds seem to be beyond the Ginzburg-Landau paradigm of criticality [4–7]. A natural explanation for the new type of criticality invokes inherently-quantum critical modes in addition to the gapless order parameter fluctuations. In the case of heavy fermion compounds, the additional critical mode has been identified with the destruction of the Kondo effect [8–10]. Since the destruction of the Kondo effect is local in space it can be systematically studied in simplified, local models. In these quantum impurity systems the critical state only appears at the (spatial) boundary of suitable hosts.

In the present article, we study the Kondo-destroying quantum critical point in the pseudogap Bose-Fermi Kondo model (PBFKM) with Ising anisotropy, defined as

$$H_{\text{PBFKM}} = \sum_{k,\sigma} \epsilon_k c^\dagger_{k,\sigma} c_{k,\sigma} + \sum_q \omega_q \phi^\dagger_q \phi_q + J\mathbf{S} \cdot \sum_{k,k',\sigma,\sigma'} c^\dagger_{k,\sigma} \frac{\vec{\sigma}}{2} c_{k',\sigma'} + g S_z \sum_q (\phi^\dagger_q + \phi_{-q}). \quad (1)$$

Where ϵ_k, ω_q are the fermonic and bosonic bath dispersions, J is the Kondo coupling between the spin of the conduction electrons and the spin of the impurity, $\vec{\sigma}$ is a vector of Pauli spin matrices and g is the coupling between the z-component of the impurity spin and the bosonic bath. We take a pseudogap density of states (DOS) for the conduction electrons, $\rho_c(\epsilon) \propto |\epsilon|^r$ for $|\epsilon| < D$ and $0 < r < 1/2$ where we have taken the Fermi energy to be zero, and a sub-ohmic density of states for the bosons, $\rho_B(\omega) = \sum_q [\delta(\omega - \omega_q) - \delta(\omega + \omega_q)] \propto \text{sgn}(\omega)|\omega|^\alpha$ up to a cutoff Λ. A perturbative renormalization group study has been used carried out for the Bose-Fermi Kondo model with non-zero conduction-electron DOS ($r = 0$) in both the Ising and continuous spin symmetry cases [11]. For the pseudogapped case ($r \neq 0$), such a study has been possible only for the continuous spin symmetry case [12]. The numerical renormalization group method, on the other hand, has been used to study the model in the Ising limit [13]. Eq. (1) has as particular

Published under licence by IOP Publishing Ltd

limits the pseudogap Kondo and the Bose-Fermi Kondo models, which have recently been studied using related methods [14, 15]. Despite the seeming simplicity of Eq. (1), it is hard to study the quantum critical properties of the PBFKM directly. In particular, the quantum-relaxational regime ($\hbar\omega < k_B T$) is difficult to address theoretically. The quantum-relaxational regime of the pseudogap Kondo model has been studied previously [15], by applying a CT-QMC algorithm [16–18] to the pseudogap Anderson model whose low-energy sector in the local moment regime can be mapped onto the pseudogap Kondo model via a Schrieffer-Wolff (SW) transformation. Since the Anderson model only involves fermions, standard perturbative expansion methods can be applied to it. In the CT-QMC approach, the full perturbative expansion in terms of the hybridization between conduction and localized fermions is then sampled stochastically using a Monte Carlo algorithm [16–18]. The corresponding Bose-Fermi Anderson model is

$$\begin{aligned} H &= \sum_{k,\sigma} \epsilon_k c^\dagger_{k,\sigma} c_{k,\sigma} + \epsilon_d (n_\uparrow + n_\downarrow) + U n_\uparrow n_\downarrow + \sum_{k,\sigma} (V_k d^\dagger_\sigma c_{k,\sigma} + V^*_k c^\dagger_{k,\sigma} d_\sigma) \\ &+ \sum_q \omega_q \phi^\dagger_q \phi_q + g \frac{(n_\uparrow - n_\downarrow)}{2} \sum_q (\phi^\dagger_q + \phi_{-q}), \end{aligned} \tag{2}$$

where ϵ_d is the energy level of the impurity, U is the on-site interaction , $n_\sigma = d^\dagger_\sigma d_\sigma$, V_k is the hybridization of the impurity with the conduction electrons, and $S_z = \frac{1}{2}(n_\uparrow - n_\downarrow)$. The CT-QMC method has been extended to treat ohmic bosonic baths coupled to the charge of the impurity by invoking a Firsov-Lang (FL) transformation [19, 20]. Here we extend this approach to treat a sub-ohmic bosonic bath that couples to the spin of the impurity and explore the possibility of using it to access the quantum critical properties. As it turns out, the generators of the FL and SW transformations do not commute, raising the important question as to which is the proper order of applying the two transformations.

We focus on the particle-hole symmetric case, $U = -2\epsilon_d$. First, we perform the FL transformation to eliminate the term linear in ϕ exactly. We choose a generator $S_{FL} = g S_z \sum_q \frac{1}{\omega_q}(\phi^\dagger_q - \phi_{-q})$ and the transformed Hamiltonian $\tilde{H} = e^{S_{FL}} H e^{-S_{FL}}$ is

$$\tilde{H} = \sum_{k,\sigma} \epsilon_k c^\dagger_{k,\sigma} c_{k,\sigma} + \tilde{\epsilon_d}(n_\uparrow + n_\downarrow) + \tilde{U} n_\uparrow n_\downarrow + \sum_{k,\sigma} (V_k \tilde{d}^\dagger_\sigma c_{k,\sigma} + V^*_k c^\dagger_{k,\sigma} \tilde{d}_\sigma) + \sum_q \omega_q \phi^\dagger_q \phi_q, \tag{3}$$

where $\tilde{d}^\dagger_\sigma = d^\dagger_\sigma \exp(\frac{\sigma g}{2} \sum_q \frac{1}{\omega_q}(\Phi^\dagger_q - \Phi_{-q}))$, $\sigma = \pm 1$ for $\uparrow/\downarrow$, $\tilde{U} = U + \frac{1}{2} g^2 \sum_q \frac{1}{\omega_q}$, and $\tilde{\epsilon}_d = -\frac{\tilde{U}}{2}$. Note that $\tilde{n}_\sigma = n_\sigma$ and the transformation does not destroy particle-hole symmetry.

A modified SW transformation [21] is used to eliminate the hybridization term in Eq. (3). We only consider the Kondo limit, and for simplicity neglect the k dependence of $V_k = V$. The generator S_{SW} is standard but with $U, \epsilon_d, d, d^\dagger$ replaced by $\tilde{U}, \tilde{\epsilon_d}, \tilde{d}, \tilde{d}^\dagger$, namely $S_{SW} = \sum_{k,\sigma} V\Big(\frac{1-n_{-\sigma}}{\tilde{\epsilon_d}-\epsilon_k} + \frac{n_{-\sigma}}{\tilde{\epsilon_d}+\tilde{U}-\epsilon_k}\Big)(\tilde{d}^\dagger_\sigma c_{k,\sigma} - c^\dagger_{k,\sigma}\tilde{d}_\sigma)$. Writing the Hamiltonian in Eq. (3) as $\tilde{H} = H_0 + H_b + \tilde{H}_h$ where $H_0 = \tilde{\epsilon_d}(n_\uparrow + n_\downarrow) + \tilde{U} n_\uparrow n_\downarrow + \sum_{k,\sigma} \epsilon_k c^\dagger_{k,\sigma} c_{k,\sigma}$, $H_b = \sum_q \omega_q \phi^\dagger_q \phi_q$ and $\tilde{H}_h = \sum_{k,\sigma} V(\tilde{d}^\dagger_\sigma c_{k,\sigma} + c^\dagger_{k,\sigma}\tilde{d}_\sigma)$, we have $H' = e^{S_{SW}} \tilde{H} e^{-S_{SW}} \approx H_0 + H_b + [S_{SW}, H_b] + \frac{1}{2}[S_{SW}, \tilde{H}_h]$. Projecting out unoccupied and doubly occupied states we arrive at

$$\begin{aligned} H' &= \sum_{k,\sigma} \epsilon_k c^\dagger_{k,\sigma} c_{k,\sigma} + \sum_q \omega_q \phi^\dagger_q \phi_q + \sum_{k,k',\sigma} (\frac{1}{2}\tilde{W}_{k,k'} + \frac{1}{4}\tilde{J}_{k,k'}) c^\dagger_{k,\sigma} c_{k',\sigma} \\ &- \sum_{k,k'} \tilde{J}_{k,k'} (\frac{1}{2}(s^+_{k,k'} \tilde{S}^- + s^-_{k,k'} \tilde{S}^+) + s^z_{k,k'} S^z), \end{aligned} \tag{4}$$

where $\tilde{W}_{k,k'} = V^2(\frac{1}{\epsilon_k - \tilde{\epsilon}_d} + \frac{1}{\epsilon_{k'} - \tilde{\epsilon}_d})$, $\tilde{J}_{k,k'} = V^2(\frac{1}{\epsilon_k - \tilde{\epsilon}_d - \tilde{U}} + \frac{1}{\epsilon_{k'} - \tilde{\epsilon}_d - \tilde{U}} - \frac{1}{\epsilon_k - \tilde{\epsilon}_d} - \frac{1}{\epsilon_{k'} - \tilde{\epsilon}_d})$ is the Kondo coupling, $\tilde{S}^+ = S^+ \exp(g \sum_q \frac{1}{\omega_q}(\Phi^\dagger_q - \Phi_{-q}))$, $\tilde{S}^- = S^- \exp(-g \sum_q \frac{1}{\omega_q}(\Phi^\dagger_q - \Phi_{-q}))$,

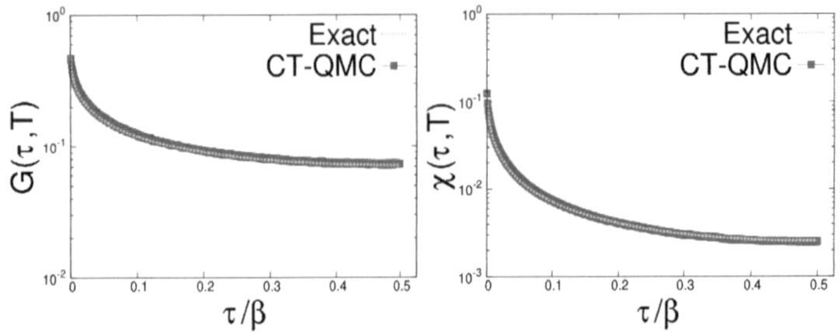

Figure 1. The single particle Green function $G(\tau, T)$ and the spin susceptibility $\chi(\tau, T)$ for $U = g = 0$, $\Gamma = 0.1D$ and $r = 0.4$ at $T = 6.6 \times 10^{-4}D$. Comparing the exact calculation and the CT-QMC result we see agreement for both G and χ.

$\vec{S} = \frac{1}{2}\sum_{\alpha,\beta} d^\dagger_\alpha \vec{\sigma}_{\alpha,\beta} d_\beta$ and $\vec{s}_{k,k'} = \frac{1}{2}\sum_{\alpha,\beta} c^\dagger_{k,\alpha}\vec{\sigma}_{\alpha,\beta}c_{k'\beta}$. The third term in equation (4) represents a potential scattering of the conduction electrons, and the last is the Kondo term, but with renormalized impurity spin flip operators due to the presence of the bosonic bath. $\tilde{J}_{k,k'}$ differs from the standard expression in that U and ϵ_d is replaced by $\tilde{U}$ and $\tilde{\epsilon}_d$.

We will now discuss the opposite order of transformations, namely, $H'' = e^{S_{FL}}e^{S_{SW}}He^{-S_{SW}}e^{-S_{FL}}$. Applying the SW transformation, projecting out charge fluctuations and then applying the FL transformation arrives at equation (4), however with $\tilde{W}_{k,k'}$, $\tilde{J}_{k,k'}$ replaced by $W_{k,k'}$, $J_{k,k'}$. We see that applying first the SW transformation, which is not exact, completely ignores the bosonic baths' influence on the charge degrees of freedom of the impurity. Whereas applying the FL transformation first, which is exact, correctly captures the bosonic baths' influence on the Anderson model which lowers the Kondo coupling. The non-commutativity of the two transformations quantitatively affects the effective Kondo scale at the quantum critical point, but it does not change the universal scaling behavior of the quantum critical properties because the critical value of the Kondo coupling is not universal.

As a check on the CT-QMC approach we first compare the single particle Green function and the spin susceptibility for the numerical result with $U = g = 0$ to the analytic result. For $g = 0$ the bosonic bath decouples from the problem and can be ignored, and taking $U = 0$ reduces the Hamiltonian in Eq. (2) to the resonant level model with a pseudogap. The impurity single particle Green function, $\langle \hat{T}_\tau d(\tau) d^\dagger(0)\rangle$ is then $G(\omega) = (\omega - \epsilon_d - \sum_k \frac{|V_k|^2}{\omega - \epsilon_k})^{-1}$. Using $V_k = V$ and taking the infinite bandwidth limit we can perform the sum over k [22]. We obtain $G(\omega) = (\omega - \epsilon_d - \Sigma(\omega))^{-1}$, where $\mathrm{Re}\Sigma(\omega) = -\Gamma(\omega)\tan(\frac{\pi r}{2})\mathrm{sgn}(\omega)$, $\mathrm{Im}\Sigma(\omega) = -\Gamma(\omega)$ and we have defined the dynamic hybridization function to be $\Gamma(\omega) = \pi|V|^2\rho_0|\omega|^r$. The imaginary time Green function can then be obtain by Fourier transform, $G(\tau, \beta) = \int_{-\infty}^{\infty} \frac{d\omega}{\pi} \frac{e^{-\tau\omega}}{e^{-\beta\omega}+1} \mathrm{Im}(G(\omega + i0^+))$ and the local spin susceptibility can be constructed $\chi(\tau, \beta) = -\frac{1}{2}G(\tau, \beta)G(-\tau, \beta)$. As seen in figure 1 we obtain quantitative agreement, within numerical accuracy, in the long time behavior for both $G(\tau, \beta)$ and $\chi(\tau, \beta)$.

We now turn to the quantum-critical properties of the PBFKM defined in Eq. (2) by measuring the static spin susceptibility, described below. In what follows we fix $r = 0.4$ and $\alpha = 0.6$. After the (FL) transformation, the Hamiltonian in Eq. (3) can be expanded in the hybridization term. We use the CT-QMC algorithm to calculate the partition function, the single particle Green function and the local spin susceptibility. Fixing $U = 0.025D$ and varying g we can tune the model to a quantum critical point (QCP). Within the CT-QMC approach we measure the local spin susceptibility $\chi(\tau, \beta) = \langle T_\tau S_z(\tau) S_z(0)\rangle$ and then calculate the static susceptibility $\chi_{\mathrm{stat}}(\beta) = \int_0^\beta d\tau\, \chi(\tau, \beta)$, where we have set the Lande g-factor and Bohr magneton to unity. For small g, the finite U gives rise to a Kondo screened local moment; the static susceptibility approaches a constant for temperatures well below the Kondo temperature, T_K. For large g, the impurity spin decouples from the conduction band and follows the fluctuations of the bosonic bath; the static susceptibility takes the Curie-Weiss form, $\chi_{\mathrm{stat}}(T) \sim T^{-1}$. At the QCP, the bosonic bath acts to decohere and destroy the Kondo effect [14]. Consequently, at the

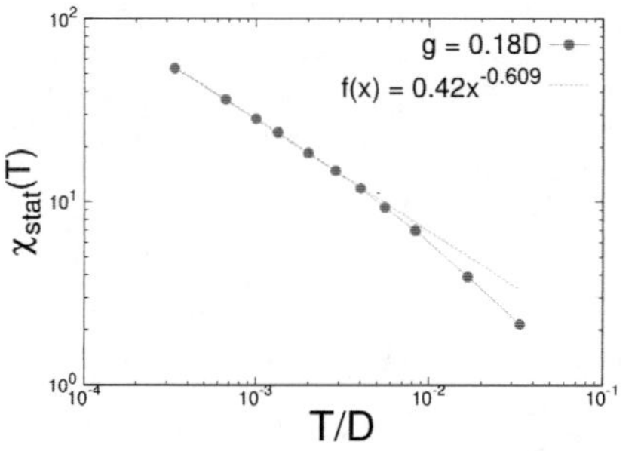

Figure 2. Static susceptibility in the vicinity of the quantum critical point, $g_c \approx 0.18D$. For $r = 0$ the Kondo temperature is, $T_K \approx 0.029D$. Well below T_K we see divergence of the static susceptibility as $\chi_{\mathrm{stat}} \sim T^{-0.609}$.

QCP the scaling of $\chi_{\mathrm{stat}}(T)$ acquires an anomalous exponent $\chi_{\mathrm{stat}}(T) \sim T^{-\alpha}$ for temperatures well below T_K. As seen in figure 2, using the CT-QMC approach we obtain $g_c \approx 0.18D$ and in the vicinity of the quantum critical point, $\chi_{\mathrm{stat}}(T) \sim T^{-x}$ with $x = 0.609$. Our calculated exponent agrees with the numerical renormalization group result within numerical accuracy [13]; the same exponent is also expected in related pseudogap Bose-Fermi Kondo model with continuous spin symmetry [12] or the Bose-Fermi Kondo model with Ising symmetry but with $r = 0$ [11].

In conclusion, we have shown that the low energy properties of the pseudogap Bose-Fermi Kondo model can be addressed within a continuous-time quantum Monte Carlo approach. We have demonstrated that this approach correctly reproduces the exactly solvable limit of the pseudogap resonant level model, and been able to determine the critical behavior of the static local spin susceptibility in an interacting case.

This work has been supported by NSF (Grant No. DMR-1006985), the Robert A. Welch Foundation (Grant No. C-1411), and the W. M. Keck Foundation. The calculations were performed on the Rice Computational Research Cluster funded by the NSF and a partnership between Rice University, AMD and Cray.

References

[1] Löhneysen H v, Rosch A, Vojta M and Wölfe P 2007 *Rev. Mod. Phys.* **79**, 1015
[2] Si Q and Steglich F 2010 *Science* **329**, 1161
[3] Sachdev S 1999 *Quantum Phase Transitions* (Cambridge: Cambridge University Press)
[4] Aronson M C, Osborn R, Robinson R A, Lynn J W, Chau R, Seaman C L and Maple M B 1995 *Phys. Rev. Lett.* **75**, 725
[5] Schröder A, Aeppli G, Coldea R, Adams M, Stockert O, Löhneysen H v, Bucher E Ramazashvili R and Coleman P 2000 *Nature* **407**, 351
[6] Paschen S, Lühmann T, Wirth S, Gegenwart P, Trovarelli O, Geibel C, Steglich F, Coleman P and Si Q 2004 *Nature* **432**, 881
[7] Friedemann S, Oeschler N, Wirth S, Krellner C, Geibel C, Steglich F, Paschen S, Kirchner S and Si Q 2010 *Proc. Natl. Acad. Sci. USA* **107**, 14547-14551
[8] Coleman P, Pépin C, Si Q and Ramazashvili R 2001 *J. Phys. Cond. Matt.* **13**, R723
[9] Si Q, Rabello S, Ingersent K and Smith J L 2001 *Nature* **413**, 804
[10] Si Q, Rabello S, Ingersent K and Smith J L 2003 *Phys. Rev. B* **68**, 115103
[11] Zhu L and Si Q 2002 *Phys. Rev. B* **66**, 024426
[12] Kirćan M and Vojta M 2004 *Phys. Rev. B* **69**, 174421
[13] Glossop M T, Khoshkhou N and Ingersent K 2008 *Physica B* **403**, 1303
[14] Kirchner S and Si Q 2008 *Phys. Rev. Lett.* **100**, 026403
[15] Glossop M T, Kirchner S, Pixley J H and Si Q 2009 *Preprint* arXiv:0912.4521v1
[16] Werner P, Comanac A, Medici L d, Troyer M and Millis A J 2006 *Phys. Rev. Lett.* **97**, 076405
[17] Prokof'ev N V, Svistunov B and Tupitsyn I 1998 *Sov. Phys. JETP* **87**, 310
[18] Rubtsov A N, Savkin V V and Lichtenstein A I 2005 *Phys. Rev. B* **72**, 035122
[19] Werner P and Millis A J 2007 *Phys. Rev. Lett.* **99**, 146404
[20] Werner P and Millis A J 2010 *Phys. Rev. Lett.* **104**, 146401
[21] Schrieffer J R and Wolff P A 1966 *Phys. Rev.* **149**, 491
[22] Gonzalez-Buxton C and Ingersent K 1998 *Phys. Rev. B* **57**, 14254

Heat capacity and magnetic phase diagram of the hydride CeRuSiH

J. Rodríguez Fernández[1], D. P. Rojas[1], E. Gaudin[2], B. Chevalier[2]
[1]CITIMAC, Facultad de Ciencias, Universidad de Cantabria, 39005 Santander, Spain
[2]CNRS, Universit de Bordeaux, ICMCB, 87 Avenue Dr. A. Schweitzer, 33608 Pessac, France
E-mail: rodrigufj@unican.es

Abstract. The hydride CeRuSiH exhibits antiferromagnetic order with two transitions at T_{N1} = 7.5 and T_{N2} = 3.1 K. Furthermore, magnetization measurements M (H) up to H = 45 kOe shows a metamagnetic double transition at low temperatures, suggesting a complex magnetic phase diagram. Here, we present magnetization measurements and heat capacity data up to 90 kOe, which allows as to complete the magnetic phase diagram. In addition, from the analysis of the heat capacity, we propose a model for the crystal field splitting.

1. Introduction

Among the rare earth intermetallic, the Ce based one are characterized for having a strong dependence of their magnetic properties on the interatomic distance. In this sense, many studies have been performed using chemical substitution and/or hydrostatic pressure. Another important way to modify the atomic volume and then the magnetic interactions is hydrogen absorption [1, 2]. Recent hydrogenation studies on the non-magnetic heavy fermion ternary silicide CeRuSi, have discovered that CeRuSiH exhibits antiferromagnetic order with two transitions at T_{N1} = 7.5 and T_{N2} = 3.1 K [3]. In addition, magnetization measurements M (H) up to H = 45 kOe revealed a metamagnetic double transition at low temperatures, suggesting a complex magnetic phase diagram [3]. Neutron powder diffraction experiments evidenced an incommensurate magnetic structure below T_{N1} that transforms into a commensurate one at T_{N2} [4]. Also, the influence of pressure on the magnetic properties of CeRuSiH has also been analyzed up to 10 kbar [5], confirming that the increase of the unit cell volume in the sequence CeRuSi $\longrightarrow$ CeRuSiH is the main responsible of the induced antiferromagnetic order in the hydride. Now, we present a detailed investigation of the M(T) and M(H) measurements up to H = 90 kOe. In addition, we report on the magnetic field dependence of the heat capacity of CeRuSiH. These new results confirm the previous picture obtained up to H = 45 kOe, allowing us to extend the (H, T) magnetic phase diagram up to 90 kOe.

2. Experimental details

A polycrystalline CeRuSi sample was heated at 523 K under vacuum for 12 hours and then exposed for 2 days under 4 MPa of hydrogen gas at the same temperature. More details of sample preparation can be found in reference [3]. Magnetization measurements have been performed in a PPMS system in magnetic fields up to 90 kOe. Heat capacity measurements were carried

Published under licence by IOP Publishing Ltd

out by a relaxation method, also using the PPMS system. The sample was a plate of 0.4 mm thickness and 7 mg weight, obtained by compressing the original powder.

3. Results and discussion

Susceptibility curves, measured cooling the sample in the presence of applied magnetic fields, are depicted in Figure 1. Below 40 kOe, all the curves shows a maximum around $T_{N1} = 7.5$ K. When the temperature decreases, the behavior changes with applied field, defined a second transition at lower temperatures in excellent agreement with reference [3], in which a detailed analysis of this low field range can be found. At 50 kOe, the susceptibility shows a broad maximum centred at 5 K. At higher fields, the maximum disappears, and the susceptibility tends to saturate when the temperature decreases, approaching to the behavior expected in ferromagnetic material. In this field range (H $\geq$ 50 kOe), the transition temperature has been defined at the inflexion point of χ(T) (temperature at which the dχ/dT curve exhibits a minimum).

The magnetic field dependence of the isothermal magnetization, displayed in Figure 2, has been measured first increasing the field up to 90 kOe and then, decreasing the field down to 0 kOe. At T = 2 K, the curve shows two metamagnetic transitions at H_{C1} = 6.7 kOe and H_{C2} = 48 kOe. When the temperature increases, H_{C1} shifts to higher fields, whereas H_{C2} moves in opposite direction. This tendency is followed up to T = 7 K, where the two transitions join at $H_C \approx 30$ kOe. At the critical field H_{C1} the magnetization exhibit a clear irreversibility, which is better observed in the inset of Figure 2. At T = 2 K, the irreversibility has a field width of $\Delta H_{C1} \approx 0.6$ kOe; this value remains nearly constant below 7 K.

The heat capacity (Cp) measured at zero magnetic field between 2 and 300 K is shown in Figure 3. The increase of the Cp above the magnetic transition is due to the electronic and mainly phonon contributions. In addition, we should have a crystal field contribution. In the tetragonal symmetry of CeRuSiH, it is expected that the sixfold degenerated 4f state of Ce^{3+} ions split into three doublets, with the excited stated separated from the ground state by energy

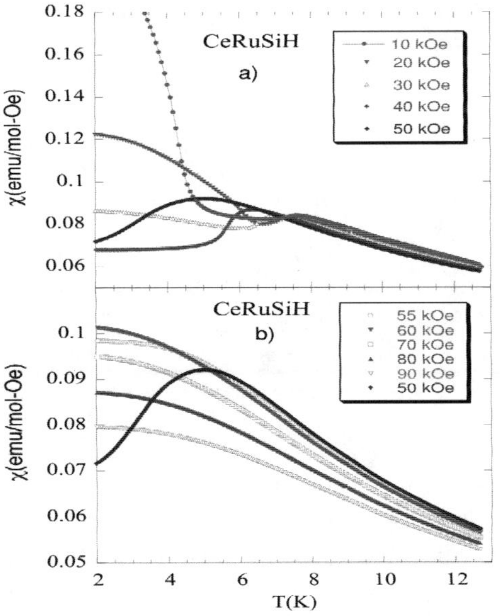

Figure 1. Temperature dependence of the susceptibility of CeRuSiH measured under different applied fields.

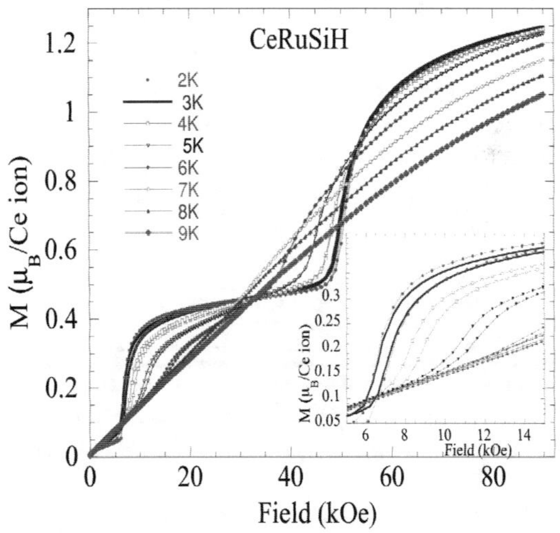

Figure 2. Field dependence of the magnetization of CeRuSiH measured at different temperatures. The inset shows an enlargement around the first metamagnetic transition.

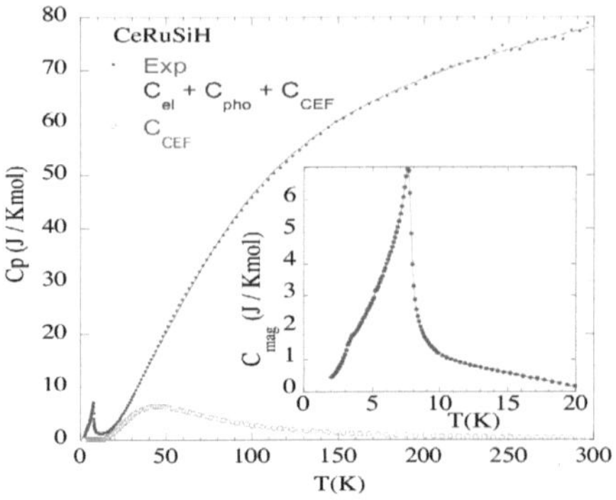

Figure 3. Temperature dependence of the specific heat of CeRuSiH in a zero applied magnetic field. Experimental data (•), fitting curve (solid line), and contribution from the CEF excitations (∘). The inset shows the magnetic contribution.

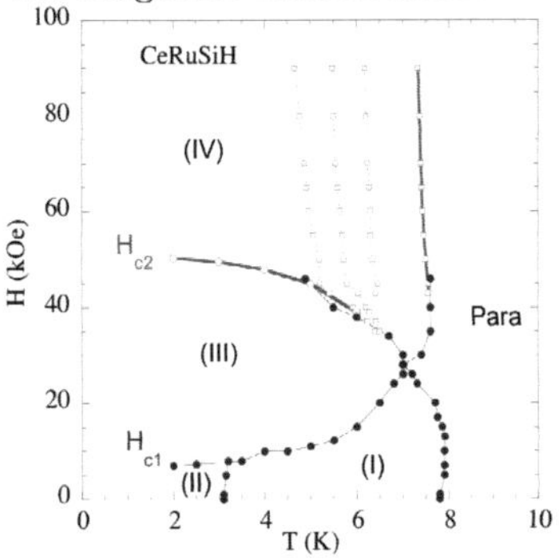

Figure 5. Field-temperature phase diagram of the CeRuSiH deduced from the magnetization and Cp measurements.

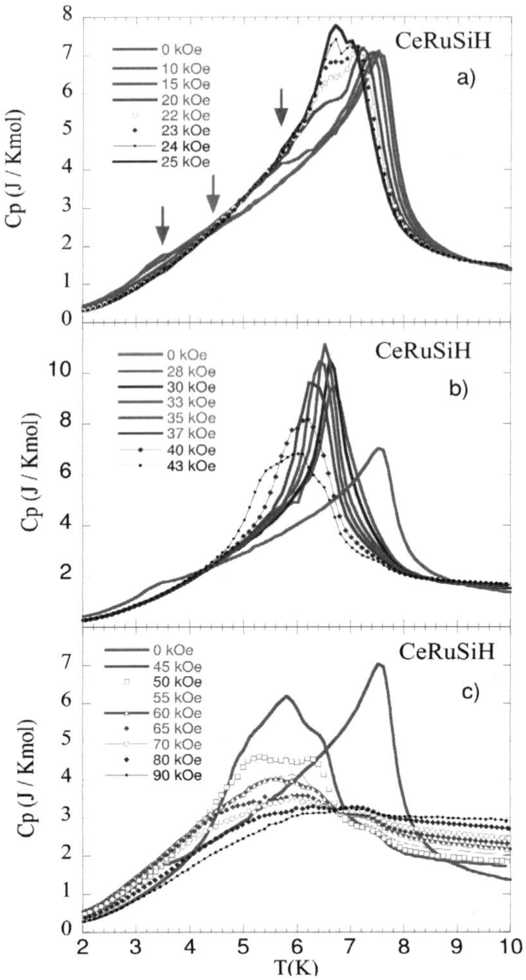

Figure 4. Temperature dependence of the heat capacity of CeRuSiH measured under different applied fields.

gaps Δ_1 and Δ_2. In order to determine the ground multiplet, the experimental data was fitted to the addition of these three contributions. In particular, for the phonon contribution, due to the absence of an isomorphous non magnetic compound, and the important different weight of the constituents, we use a Debye model assuming the existence of two Debye temperatures: Cph $= n_1$Cph $(\Theta_{D1}) + n_2$Cph (Θ_{D2}) where $n_1+n_2 = 4$ was fixed. Furthermore, we use an electronic coefficient $\gamma = 26$ mJ/molK^2 which was obtained from the low temperature analysis [3]. More details about this procedure can be found in reference [6], where a similar method was successfully used in $CeNiGaH_{1.1}$. The best fitting corresponds to $n_1 = 2.9$, $\Theta_{D1} = 358$ K, $\Theta_{D2} = 2300$ K, $\Delta_1 = 127.2$ K and $\Delta_2 = 127.8$ K. The energy of the crystal field splitting is in good agreement with the values suggested from of resistivity measurements [3]. The high value of Θ_{D2} could be attributed to the high energy phonon spectra associated to the light H atoms, which are the main responsibles of the feature that the experimental data still have an important slope at 300 K. In the inset of Figure 3 it is depicted the magnetic contribution obtained as $C_{mag} =$

Cp - C_{pho} - C_{el} - C_{CEF}; the peak and its shoulder at lower temperatures define clearly the two magnetic transitions at T_{N1} = 7.5 and T_{N2} = 3.5 K respectively, in agreement with the previous results.

The effect of the magnetic field on the heat capacity, depicted in Figure 4, provide us a valuable information about the magnetic phase diagram, and in particular on its extension up to 90 kOe. In Figure 4a it is observed that when the field starts to increase, the peak shifts to lower temperatures, as it is expected in antiferromagnetic compounds. Moreover, the low temperature shoulder shifts to higher temperatures, as is indicated by arrows for some selected fields. Above 28 kOe, Figure 4b, the most significant features are i) the two transitions collapse and ii) the peak becomes narrow and its height increases. The last one is quite important, because usually the effects of magnetic field on Cp is the opposite: the peak broadens and decreases in height. The unusual field dependence found in CeRuSiH is related with the incommensurate magnetic structure, which evolves towards commensurate one with field. Therefore, for the magnetic field corresponding to the maximum height (H = 33 kOe), it is expected that the system is directly ordered from the paramagnetic state into a commensurate magnetic structure. At H = 45 kOe, Figure 4c, a small bump appears around 7.4 K and then, by decreasing the temperature, emerges a peak surrounded by two shoulders. By increasing the magnetic field, this multi-structured peak becomes less intense whereas the bump at 7.4 K transforms into a maximum at 90 kOe.

Using all the previous data, it is possible to put forward the magnetic phase diagram shown in Figure 5. Below 45 kOe, the new data coincide with the previous one (blue lines) [3], in which phases (I) and (II) correspond to antiferromagnetic arrangement, whereas phase (III) shows a ferrimagnetic character. The new information obtained in this paper is represented in red. First, it is observed that the increase of the critical filed H_{C2} by decreasing the temperature, tends to saturate at about 50 kOe. Second, between the paramagnetic and the ferromagnetic phase (IV) there are three intermediate phases, probably of ferrimagnetic nature. These phases, although give rise to a broad increase of the magnetization in the M(T) curves, are clearly detected in the heat capacity. This phase diagram resembles those found in other Ce intermetallics as $CeRu_2(Si_{1-x}Ge_x)_2$ [7] or $Ce_{0.8}La_{0.2}Ru_2Si_2$ [8]. In these alloys, the antiferromagnetic order is induced by the expansion of the lattice due to the substitution of a small amount of Ge for Si or La for Ce respectively, and in both cases, the magnetic structure is also incommensurate.

Summarizing, χ (T), M(H) and Cp data up to 90 kOe allowed us to complete the complex magnetic phase diagram of CeRuSiH, in which the incommensurate magnetic structure evolves towards a commensurate order by increasing the magnetic field. Furthermore, from the analysis of the Cp between 2 and 300 K, a crystal field splitting is proposed.

This work was financially supported by the European Science Foundation (ECOMCOST action P16) and by the Spanish Project MAT2008-06542-C04

References

[1] B. Chevalier, M. Pasturel, J-L Bobet, R . Decourt, J. Etourneau, O. Isnard, J. Sanchez Marcos, J. Rodríguez Fernández 2004 *J. Alloys Compounds* **383** 4

[2] B. Chevalier, S. F. Matar, M. Ménetrier, J. Sanchez Marcos and J. Rodríguez Fernández 2006 *J. Phys.: Condens. Matter* **18** 6045

[3] B. Chevalier, E. Gaudin, S. Tencé, B. Malaman, J. Rodríguez Fernández, G. André and B. Coqblin 2008 *Phys. Rev. B* **77** 014414

[4] S. Tencé, G. André, E. Gaudin and B. Chevalier 2008 *J. Phys.: Condens. Matter* **20** 255239

[5] J. Rodríguez Fernández, D.P. Rojas, J.C. Gómez Sal, S. Tencé and B. Chevalier 2010 *J. Phys. Conf. Series* **200** 012167

[6] B. Chevalier, J. Sanchez Marcos, J. Rodríguez Fernández, M. Pasturel, and F. Weill 2005 *Phys. Rev. B* **71** 214437

[7] H. P. Haen, H. Bioud and T. Fukuhara 1999 *Physica B* **259-261** 179

[8] J. M. Mignot, L. P Regnault, J. L Jacoud, J Rossat-Mignod, P. Haen and P. Lejay 1991 *Physica B* **171** 357

Nernst effect in $CeCu_2Si_2$

A-S Rüetschi, K Sengupta, G Seyfarth and D Jaccard
DPMC, University of Geneva, 24 Quai E.-Ansermet, 1211 Genève 4, Switzerland
E-mail: Didier.Jaccard@unige.ch

Abstract.
We report on an ongoing multiprobe experiment on a very high puritiy $CeCu_2Si_2$ single crystal at ambient and high pressure. The Nernst signal, which has not been measured in this compound so far, exhibits quite uncommon features and seems to be coupled to the antiferromagnetic phase, rather than to the vortex motion within the superconducting mixed state.

Although the superconducting state of $CeCu_2Si_2$ has been discovered thirty years ago [1], its nature is not yet fully understood and a lot of activity is still going on in order to understand the complex low temperature $H - T - p$ phase diagram of this compound [2–4]. At $p = 0$ the ground state of $CeCu_2Si_2$ depends strongly on the exact stoichiometry of each sample [3; 5; 6]. Silicon rich samples show an antiferromagnetic ground state (A-type samples), copper rich samples have a superconducting ground state (S-type samples) and in high quality 1:2:2 samples antiferromagnetism and superconductivity compete (A/S-type samples). Under a pressure of about 0.2-0.3 GPa the antiferromagnetic phase disappears [7; 8] and an increase of the superconducting transition temperature T_c from about 0.7 K up to over 2 K is observed at $p > 2$ GPa [9]. Antiferromagnetic fluctuations at ambient pressure [10] and valence fluctuations around the maximum T_c (4.5 GPa) [11] are commonly thought to provide the superconducting pairing glue.

The main motivation of our work is to better understand the nature of the superconducting state of $CeCu_2Si_2$ by performing different magneto-transport measurements on the same sample at ambient and high pressure. Particular attention is put on the Nernst effect, a probe already proven to be useful for studying exotic states in correlated electron systems [12; 13]. Here we present brand-new results of the Nernst effect at ambient pressure.

A multiprobe set-up has been designed, which allows us to measure the electric resistivity ρ, the thermopower S, the transverse Hall resistivity ρ_{xy}, the Nernst coefficient ν and the ac specific heat C_p on a single sample. However, its geometry has been optimized for measuring ν and ρ_{xy}. The ambient pressure set-up can be transferred without modification into a Bridgman-type pressure cell. High pressure measurements are currently under way. The current is injected along the a-b plane of the sample and the magnetic field is applied perpendicular to that plane. Measurements are performed in a He^3/He^4 dilution fridge between about 100 mK and 4.5 K. A high purity single crystal of A/S-type $CeCu_2Si_2$ and with a size of $600(l) \times 576(b) \times 25(w)\ \mu m^3$ is used for the present study. The sample comes from a batch already used earlier by our group [14]. At ambient pressure, it has a T_c of 780 mK and a large residual resistivity ratio of 18. Throughout this communication T_c is determined by the vanishing of ρ. The Néel temperature T_N, as determined by the ac specific heat measurements, roughly coincides with T_c. Under field,

Published under licence by IOP Publishing Ltd

superconductivity is completely suppressed at around 2.5 T and T_N decreases to about 0.6 K at 5 T, in accordance with previous reports [7; 15].

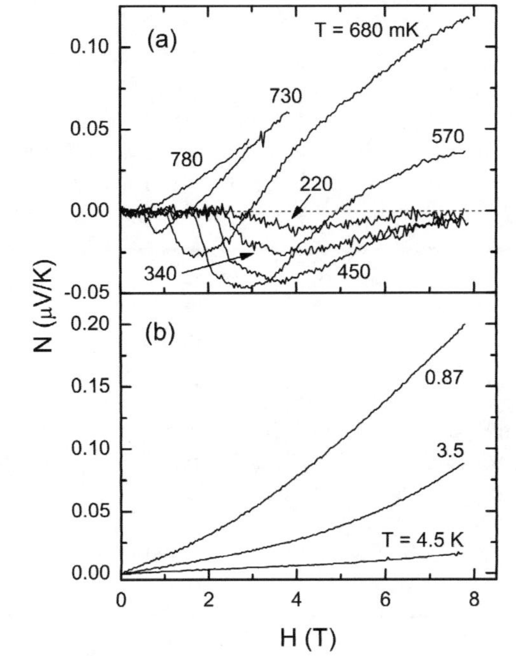

Figure 1. *The Nernst signal N as a function of the applied magnetic field.* ***a*** *Measurements below the zero field T_c,* ***b*** *measurements above the zero field T_c.*

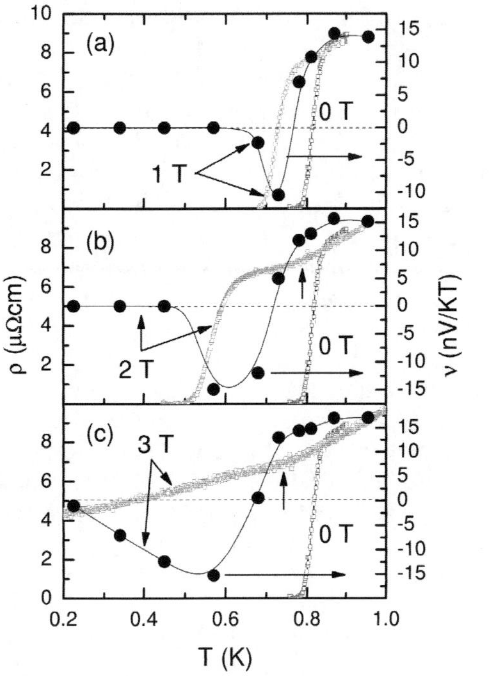

Figure 2. *The Nernst coefficient $\nu = N/B$ and resistivity ρ as a function of temperature at different magnetic fields. In all plots, the zero field resisitivity curve is shown as a reference. The thin black line is a guide to the eye and the horizontal dashed line indicates the zero of ν. The vertical arrows in* ***b*** *and* ***c*** *show T_N.*

Fig. 1 shows the Nernst signal N as a function of an external magnetic field H for various temperatures. N is the transverse electric field generated by a longitudinal temperature gradient $\partial_x T$ ($N = E_y/|\partial_x T|$). The Nernst signal is antisymmetrised with respect to H and its sign is defined according to the so-called vortex convention, which is based on the Josephson equation $E \parallel (H \times (-\nabla T))$ for moving vortices [16]. In this convention the Nernst signal generated by moving vortices is positive. The Nernst signal of $CeCu_2Si_2$ exhibits rather uncommon features. A first look at Fig. 1 shows that the overall magnitude of N is quite small ($N \leq 200\,nV/K$), at least compared with another heavy fermion compound, $CeCoIn_5$ [13; 17]. As expected, N is zero below the corresponding superconducting critical fields (Fig. 1 a). Above the critical fields, N is negative, and with further increasing field, N goes through a minimum before changing sign. At the lowest temperatures, N stays negative up to the highest fields. Above T_c (Fig. 1 b) N is positive throughout the whole explored field range and exhibits a positive curvature. It is worth noting that the curvature of N changes to negative for lower temperatures and fields higher than the minimum of N (Fig. 1 a).

Fig. 2 shows the temperature dependence of the Nernst coefficient $\nu = N/H$ and of the resistivity ρ for 1, 2 and 3 T below 1 K. The Nernst coefficient has a prominent maximum around

900 mK before dropping at higher temperatures (not shown, as can be inferred from Fig. 1 b). On decreasing the temperature, it changes sign, goes through a field dependent minimum and becomes zero in the superconducting state. As seen in Fig. 2 c, above the resistive upper critical field $H_{c2}(0)$ of about 2.5 T, ν diminishes to small but non zero values at low temperatures. One also observes a large broadening of the negative peak of ν with increasing field.

In order to shed light on this complex Nernst signal, a closer look at the H-T phase diagram of $CeCu_2Si_2$ is necessary [6; 15]. Below T_c the superconducting phase exists up to a field of roughly 2.5 T. On further increasing field one enters the antiferromagnetic A-phase, which extends up to about 7 T. Now, a close look at Fig. 2 suggests that the Nernst signal has two main contributions: a positive one, which peaks around 900 mK, and a negative one which becomes dominant below that temperature. Interestingly, the signal below T_c may be best described by being related to the A-phase rather than to vortices or superconducting fluctuations [18]. It is well known that in the superconducting mixed phase moving vortices can generate a large Nernst signal [19]. The sign of the Nernst signal generated by such moving vortices is well defined and would be positive in the sign convention used in this experiment. However in $CeCu_2Si_2$ the sign of the Nernst signal below T_c is negative, a clear indication that the vortex contribution to N is either absent or small compared to the contribution arising from the A-phase. Fig. 2 c corroborates this conclusion. At 3 T, e.g. above the upper critical field and well inside the A-phase, the negative contribution to N is still present, of the same order of magnitude as at lower fields, and appears already at the lowest temperatures since it is not shunted by the superconducting condensate. Furthermore, the temperature dependence of ν at 2 T (Fig 2 b) clearly demonstrates that the sign change in $\nu(T)$ is much better linked to T_N (indicated by the small vertical arrow) than to the superconducting transition. This correspondence holds for all temperatures, where $N(H)$ has been measured. Finally, the A-phase is readily destroyed by pressure [7; 8] and therefore, if the assumption of a Nernst signal arising from the A-phase is correct, one would expect a drastic change in the behavior of $N(T, H)$ with pressure. Preliminary results at 2 GPa do indeed show a more "conventional" Nernst signal, which is positive throughout the whole investigated temperature and field range.

In conclusion, the first low temperature Nernst data at ambient pressure in $CeCu_2Si_2$ reveal unexpectedly a predominant negative contribution from the A-phase, rather than a signature of the vortex state. Not yet understood is the behavior of the positive Nernst signal in the paramagnetic phase, in particular its maximum and subsequent drop above 1 K leading to a rather small value at 4 K. Measurements at higher temperatures and in other Ce compounds are necessary in order to elucidate the uncommon features of the Nernst signal further, especially in the light of a generic Nernst behavior in heavy fermion systems.

References

[1] Steglich F, Aarts J, Bredl C D, Lieke W, Meschede D and Franz W 1979 *Phys. Rev. Lett.* **43** 1892

[2] Faulhaber E, Stockert O, Jeevan H S, Prokes K, Deppe M, Geibel C, Steglich F and Loewenhaupt M 2005 *Physica B* **359** 357

[3] Stockert O, Arndt J, Schneidewind A, Schneider H, Jeevan H S, Geibel C, Steglich F and Loewenhaupt M 2008 *Physica B* **403** 973

[4] Fujiwara K, Hata Y, Kobayashi K, Miyoshi K, Takeuchi J, Shimaoka Y, Kotegawa H, Kobayashi T C, Geibel C and Steglich F 2008 *J. Phys. Soc. Jpn.* **77** 123711

[5] Steglich F, Gegenwart P, Geibel C, Helfrich R, Hellmann P, Lang M, Link A, Modler R, Sparn G, Büttgen N and Loidl A 1996 *Physica B* **223-224** 1–8

[6] Steglich F, Geibel C, Modler R, Lang M, Hellmann P and Gegenwart P 1995 *J. Low Temp. Phys.* **99** 267

[7] Sheikin I, Braithwaite D, Brison J P, Buzdin A and Assmus W 1999 *Physica B* **259** 683

[8] Lengyel E, Nicklas M, Jeevan H S, Sparn G, Geibel C, Steglich F, Yoshioka Y and Miyake K 2009 *Phys. Rev. B* **80** 140513

[9] Bellarbi B, Benoit A, Jaccard D, Mignot J M and Braun H 1984 *Phys. Rev. B* **30** 1182

[10] Miyake K, Schmitt-Rink S and Varma C M 1986 *Phy. Rev. B* **34** 6554

[11] Holmes A T, Jaccard D and Miyake K 2004 *Phys. Rev. B* **69** 024508

[12] Xu Z A, Ong N P, Wang Y, Kakeshita T and Uchida S 2000 *Nature* **406** 486

[13] Bel R, Behnia K, Nakajima Y, K I, Matsuda Y, Shishido H, Settai R and Onuki Y 2004 *Phys. Rev. Lett.* **92** 217002

[14] Vargoz E, Jaccard D, Genoud J Y, Brison J P and Flouquet J 1998 *Solid State Commun.* **106** 631

[15] Bruls G, Wolf D, Finsterbusch D, Thalmeier P, Kouroudis I, Sun W, Assmus W, Lüthi B, Lang M, Gloos K, Steglich F and Modler R 1994 *Phys. Rev. Lett.* **72** 1754

[16] Wang Y, Li L and Ong N P 2006 *Phys. Rev. B* **73** 024510

[17] Onose Y, Li L, Petrovic C and Ong N P 2007 *Eur. Phys. Lett.* **79** 17006

[18] Pourret A, Spathis P, Aubin H and Behnia K 2009 *New Journal of Physics* **11** 055071

[19] Hagen S J, Lobb C J, Greene R L, Forrester M G and Talvacchio J 1990 *Phys. Rev. B* **42** 6777

Journal of Physics: Conference Series **273** (2011) 012053 doi:10.1088/1742-6596/273/1/012053

NMR spectral study under zero external field in pure and diluted $CeRhIn_5$ by La substitutions

Hironori Sakai[1,2], Seung-Ho Baek, Filip Ronning, Eric D. Bauer and Joe D. Thompson

[1] Los Alamos National Laboratory, Los Alamos, NM 87545, USA.

[2] Advanced Science Research Center, Japan Atomic Energy Agency, Tokai, Ibaraki, 319-1195, Japan.

E-mail: sakai.hironori@jaea.go.jp

Abstract. $CeRhIn_5$ is an incommensurate antiferromagnet with T_N=3.8 K. The NMR spectra for tetragonal In(1) sites in $CeRhIn_5$ have been investigated under zero applied field in order to distinguish between spin density wave or spiral order. The ^{115}In(1)-NMR spectra around $2\nu_Q$ under zero field below T_N provide a good indication of spiral order. The La substitutions of 1 % and 2 % indicate a decrease of the internal field on In(1) sites due to the reduction of hyperfine couplings by La substitutions.

1. Introduction

$CeRhIn_5$ belongs to the so-called 115 family which has a tetragonal $HoCoGa_5$ type structure. At ambient pressure, $CeRhIn_5$ is an antiferromagnet below T_N=3.8 K [1]. Unconventional superconductivity emerges under applied pressure, reaching a maximum transition temperature T_c=2.1 K near the critical pressure P_c =2.5 GPa where the antiferromagnetic transition extrapolates to $T = 0$. Recent de Haas-van Alphen (dHvA) measurements reveal a divergence of the e ective mass close to P_c, consistent with proximity to an antiferromagnetic (AFM) quantum critical point (QCP) [2]. The dHvA frequencies of principal orbits increase sharply, as if signaling a transition from small-to-large Fermi volume, which is interpreted as a localized/itinerant crossover of the $4f$ electrons at the QCP. At ambient pressure, the incommensurate AFM order of Ce moments with propagation of $\boldsymbol{Q} = (1/2, 1/2, 0.297)$ has been reported by neutron di raction experiments [3]. Zero- eld ^{115}In NMR measurements on a powder sample of ground single crystals [4] have proposed that such an incommensurate AFM order originates from a spiral arrangement of Ce moments, not from spin-density wave (SDW) modulation of Ce moments. There are two crystallographically inequivalent In sites denoted In(1) ($1c$ site) and In(2) ($4i$ site). The In(1) sites have axial symmetry in the Ce-In planes, but the In(2) sites have two Ce and two Rh neighbors in the a planes and, thus, lower symmetry. As discussed below, key data necessary to conclude spiral order is ^{115}In(1)-NMR spectra under zero eld; although, In(2)-NMR has deduced clear evidence about the incommensurability. As for the previous In(1) spectra in the AFM state [4, 5], data points were scarce and insu cient to distinguish unambiguously between SDW or spiral cases. In this proceedings paper, we look more carefully into the zero- eld ^{115}In(1)-NMR spectra and report the change of internal elds on In(1) sites by dilute La substitutions of 1 % and 2 %.

Published under licence by IOP Publishing Ltd

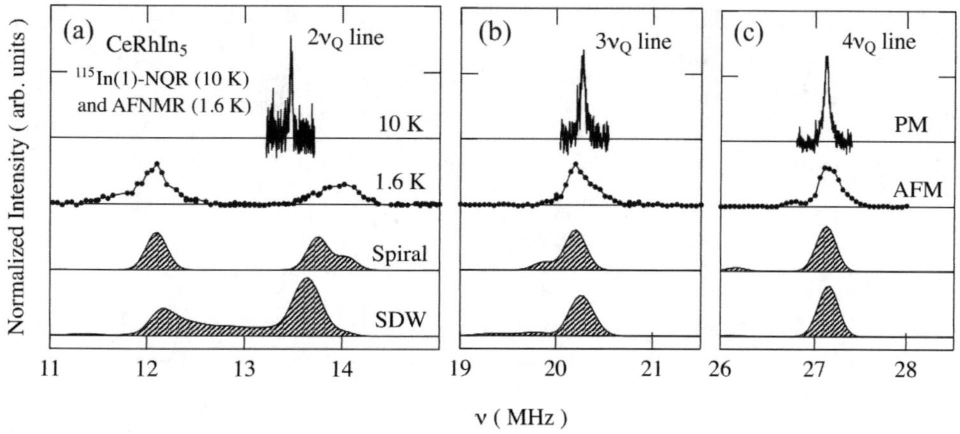

Figure 1. ^{115}In(1)-NQR spectra at 10 K and NMR spectra at 1.6 K for pure $CeRhIn_5$ under zero applied eld. Panels (a), (b) and (c) show spectra taken around 2 $_{Q}$, 3 $_{Q}$ and 4 $_{Q}$ lines of ^{115}In(1) sites, respectively. The lower shaded spectra are simulated in the cases of spiral and spin-density-wave (SDW) ordering of Ce moments.

2. Experimental

Single crystals of $Ce_{1\ x}La_xRhIn_5$ ($x = 0.0, 0.01, 0.02$) were grown by a self- ux method. NQR and NMR measurements in the temperature range 1.5-100 K were performed using a conventional phase-coherent pulsed NMR spectrometer. Frequency-swept spectra under zero eld were measured in the frequency range of 10-30 MHz. One piece of single crystal was mounted into an rf-coil in order to avoid decompositions and/or surface strains by crushing the crystals. This is important because powdering or crushing $CeRhIn_5$ in uences physical properties, especially, the volume fraction of incommensurate/commensurate antiferromagnetism and parasitic superconductivity [6, 7]. Using conventional notation, the quadrupole frequency parameter is de ned as $_{Q} \equiv \frac{3e^2qQ}{2I(2I\ \ 1)h}$, where eQ is the nuclear quadrupolar moment, I is the nuclear spin quantum number, and $eq \equiv V_{ZZ}$ is the principal component of the electric eld gradient (EFG) tensor. Here, V_{ii} denotes EFG tensor components in the principal coordinate system, such that $|V_{XX}| \leq |V_{YY}| \leq |V_{ZZ}|$ for each ionic site.

3. Results and discussions

Figure 1 shows the In(1) NQR spectra at 10 K (above T_N) and NMR spectra at 1.6 K (below T_N) for $CeRhIn_5$ under zero eld. In the paramagnetic state, the NQR spectra show narrow line widths of 50 kHz at each of the transition lines: 2 $_{Q}$, 3 $_{Q}$, and 4 $_{Q}$. These sharp lines indicate negligible lattice disorder in this high quality crystal. The $_{Q}$ is determined to be 6.778 MHz at 10 K for ^{115}In(1) sites, which is consistent with previously reported values [4, 5]. Generally, in an AFM ordered state nuclei experience site-speci c internal elds and NMR spectra with zero applied eld can be observed that exhibit characteristic features of the ordered state. As already reported for $CeRhIn_5$ in its orderd state [4], zero- eld NMR spectra for In(2) show a double-horn type pattern, like a powder-pattern, which is known to be characteristic of incommensurate order. The internal eld on the In(2) sites can be expressed as $H_{\rm int} = A|\boldsymbol{M}(n)|$, where A is the transferred hyper ne coupling constant and $M(n)$ is the Ce staggered moment in the nc plane. Neutron di raction has revealed that Ce moments lie in the c-plane and that the propagation vector is $\boldsymbol{Q} = (1/2, 1/2, \delta \equiv 0.297)$. In the case of SDW order, $|M(n)| \propto \sqrt{1 + \cos(4\ \ n\delta)}$, that is, the magnitude of Ce moments modulates along the c axis.

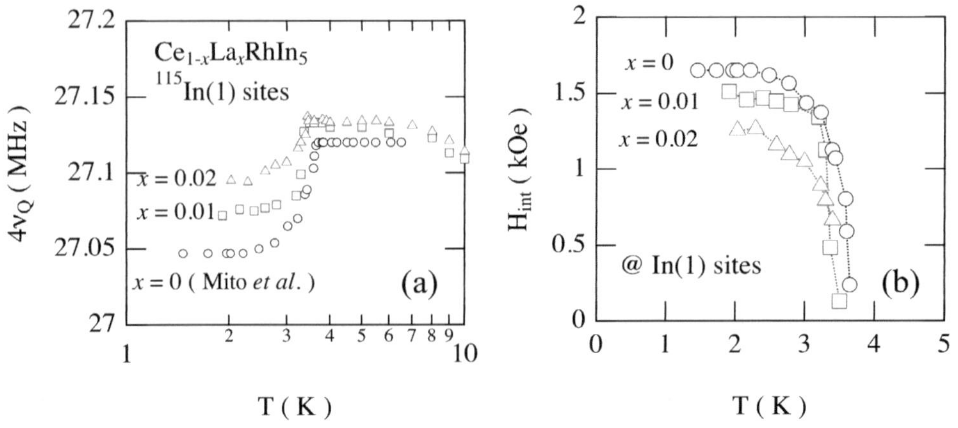

Figure 2. (a) Temperature dependence of $4\nu_Q$ for ^{115}In(1) sites in $Ce_{1-x}La_xRhIn_5$ (x=0, 0.01, 0.02). The data for $CeRhIn_5$ are taken from Ref. [5]. (b) Temperature dependence of H_{int} on the In(1) sites estimated from the $4\nu_Q$ position.

On the other hand, in the case of spiral order, $|\boldsymbol{M}(n)| = M$ is constant, while $\boldsymbol{M}(n)$ points spirally as $\boldsymbol{M}(n) = M(\cos(2\pi n\delta), \sin(2\pi n\delta), 0)$. In the both cases of SDW and helical order, the internal field H_{int} on In(2) sites is parallel to the c axis and oscillates as $\propto \cos(2\pi n\delta)$, that is, we can not distinguish whether it is SDW or spiral from In(2) spectra.

As for In(1) sites, H_{int} on In(1) sites is independent of n in the case of spiral order, but H_{int} oscillates as $\propto \cos(2\pi n\delta)$ in the case of SDW order. The direction of H_{int} is perpendicular to the c axis in the both cases. The simulated spectra for In(1) sites are shown in Fig. 1. These spectra were calculated by numerical diagonalization of the Hamiltonian matrix, including Zeeman ($\gamma_N h\boldsymbol{I} \cdot \boldsymbol{H}_{int}$) and quadrupolar terms. These simulated spectra also are convoluted by a Gaussian function with a natural width of 50 kHz. Values of $\nu_Q = 6.778$ MHz and H_{int}=1.59 kOe are used for the simulations. As shown in Fig. 1(b) and (c), the simulated spectra are similar for both cases around $3\nu_Q$ and $4\nu_Q$ lines; whereas, the spectra around $2\nu_Q$ become distinct in appearance for SDW and spiral ordering. It is noteworhty that finite spectral weight should be observed around 13 MHz in the SDW case and that the spectra should have finite weight even below 12 MHz. As shown in Fig. 1(a),however, spectral weight in the measured spectrum around 13 MHz is zero within the experimental accuracy, and there is no observed intensity below 11.5 MHz (down to 10 MHz in this experiment). Therefore, we can conclude that spiral order in $CeRhIn_5$ is highly plausible as proposed in Ref. [4]. A definite amplitude of Ce moments in such a spiral order also indicates the localized nature of $4f$ electrons in $CeRhIn_5$.

Next, we have investigated the change of H_{int} by La substitutions of 1 % and 2 %. Temperature variations of the $4\nu_Q$ line for $CeRhIn_5$ and lightly La-substituted $CeRhIn_5$ are shown in Fig. 2(a). Using similar calculations as discussed above, $4\nu_Q$ NMR in the ordered state also can be interpreted as H_{int} on In(1) sites. Estimates of H_{int} on In(1) sites are shown in Fig. 2(b). H_{int} decreases from 1.59 kOe in pure $CeRhIn_5$, to 1.48 kOe in 1 %-La, and 1.26 kOe in 2 %-La substituted $CeRhIn_5$, which corresponds to a reduction of H_{int} by 20 % with only 2 % La substitution. Although H_{int} is reduced, neutron diffraction measurements have revealed that La substitution of 1 % does not affect the size of the ordered Ce moment nor the incommensurate propagation δ [8]. So, the change of H_{int} probably originates mainly from a change of transferred hyperfine coupling constant A, not from the change of Ce moment M. Indeed, we also have detected a change in the diagonal term of the hyperfine coupling tensor in the paramagnetic state. The a component A_a for $CeRhIn_5$, $Ce_{0.99}La_{0.01}RhIn_5$, and $Ce_{0.98}La_{0.02}RhIn_5$ is estimated to be 18.3, 15.8, and 10.4 kOe/μ_B, respectively. Here, A_a is de ned from the slope in a plot of Knight shift versus static susceptibility with temperature as an implicit parameter. Because the hyper ne coupling is often changed between ordered and paramagnetic states, it is di cult to estimate quantitatively the corresponding moment size on Ce ions in the ordered state. Moreover, the existence of o -diagonal hyper ne coupling in the ordered state of $CeRhIn_5$, discussed in Ref. [9], would impose di culty in estimating M from NMR experiments. We can say, however, that the transferred coupling on In(1) sites is sensitive to dilute La substitutions. The reduction of T_N by La substitutions [10] may come from the reduction of exchange couplings, which has been suggested microscopically from nuclear spin-lattice relaxation rates $1/T_1$ (not shown). Detailed results of $1/T_1$ will be presented elsewhere. In this regard, we also note that the reported pressure-induced change of H_{int} [5] seems to be due to the change of A, because neutron di raction could not observe a corresponding reduction of M by pressure [11].

4. Conclusion

We have investigated the antiferromagnetically ordered state of $CeRhIn_5$ by NMR/NQR techniques. Speci cally, we have focused on In(1) NMR spectra under zero applied eld to conclude spiral order. As a result, AFM spiral order is highly plausible as suggested previously. A uniform moment on Ce ions is consistent with a localized nature of $4f$ electrons in $CeRhIn_5$. Lanthanum substitutions of 1 % and 2 % reduce the internal eld on In(1) sites, coming from the change in transferred hyper ne coupling via p-f mixing.

5. Acknowledgments

We thank T. Park, R. Movshovich, Y. Tokunaga and S. Kambe for useful discussions. H. S. wishes to acknowledge the hospitality of Los Alamos National Laboratory. Work at Los Alamos National Laboratory was performed under the auspices of U. S. Department of Energy, O ce of Basic Energy Sciences, Division of Materials Sciences and Engineering.

References

[1] Hegger H, Petrovic C, Moshopoulou E G, Hundley M F, Sarrao J L, Fisk Z and Thompson J D 2000 *Phys. Rev. Lett.* **84** 4986
[2] Shishido H, Settai R, Harima H and Onuki Y 2005 *J. Phys. Soc. Jpn.* **74** 1103
[3] Bao W, Pagliuso P G, Sarrao J L, Thompson J D, Fisk Z, Lynn J W and Erwin R W 2000 *Phys. Rev. B* **62** R14621
[4] Curro N J, Hammel P C, Pagliuso P G, Sarrao J L, Thompson J D and Fisk Z 2000 *Phys. Rev. B* **62** R6100
[5] Mito T, Kawasaki S, q Zheng G, Kawasaki Y, Ishida K, Kitaoka Y, Aoki D, Haga Y and Onuki Y 2001 *Phys. Rev. B* **63** 220507
[6] Chen G F, Matsubayashi K, Ban S, Deguchi K and Sato N K 2006 *Phys. Rev. Lett.* **97** 017005
[7] Yashima M, Mukuda H, Kitaoka Y, Shishido H, Settai R and Onuki Y 2009 *Phys. Rev. B* **79** 214528
[8] Bao W, Christianson A D, Pagliuso P G, Sarrao J L, Thompson J D, Lacerda A H and Lynn J W 2002 *Physica B* **312-313** 120
[9] Curro N J 2006 *New Journal of Physics* **8** 173
[10] Pagliuso P G, Moreno N O, Curro N J, Thompson J D, Hundley M F, Sarrao J L, Fisk Z, Christianson A D, Lacerda A H, Light B E and Cornelius A L 2002 *Phys. Rev. B* **66** 054433
[11] Llobet A, Gardner J S, Moshopoulou E G, Mignot J M, Nicklas M, Bao W, Moreno N O, Pagliuso P G, Goncharenko I N, Sarrao J L and Thompson J D 2004 *Phys. Rev. B* **69** 024403

Low-temperature magnetic properties of $Pr(Cu,Ga)_{12.85}$

Y. Shimura[1], T. Sakakibara[1], K. Kuga[1], J.Y. Cho[2] and J.Y. Chan[2]

[1] Institute for Solid State Physics, University of Tokyo, Kashiwa 277-8581, Japan
[2] Department of Chemistry, Louisiana State University, Baton Rouge, Louisiana 70803

E-mail: sakaki@issp.u-tokyo.ac.jp

Abstract. Magnetization of a cubic compound $Pr(Cu,Ga)_{12.85}$ was examined at low temperatures down to 0.07 K in fields up to 14.5 T. The magnetic susceptibility $\chi(T)$ exhibits Curie-Weiss behavior down to 1 K. Field variation of the magnetization $M(H)$ at 0.12 K shows a steep initial rise to 1.3 μ_B/Pr at 2 T, followed by a gradual increase in higher fields reaching 1.9 μ_B/Pr at 14.5 T. The results indicate that the crystalline-electric-field ground state of Pr^{3+} is a Γ_5 magnetic triplet. On cooling below 1 K, $\chi(T)$ tends to saturate and provides no evidence of a phase transition down to ~0.3 K, suggesting onsite as well as intersite fluctuations exist. Below 0.25 K, a weak history dependence is observed in $\chi(T)$, which can be attributed to a spin-glass-like freezing of the Pr moment. A weak magnetic field of ~0.2 T suppresses this spin-glass-like state, and non-Fermi liquid like behavior remains in $M(T)$ down to 0.08 K.

1. Introduction

There is a growing interest in a heavy fermion state in Pr-based compounds. In ordinary Ce-based heavy fermion (HF) compounds, the local magnetic moment of a $4f^1$ configuration couples with conduction electrons to form a Kondo singlet state, leading to a large electronic specific heat coefficient γ at low temperatures. Unlike the case of Ce compounds, however, the ground state degeneracy of the crystalline electric field (CEF) levels in Pr-based compounds ($4f^2$) is not protected by time reversal symmetry. High crystal symmetry is therefore required so that the CEF ground state has degeneracy that plays a key role in the HF formation. Up to present, Pr-based skutterudites (cubic), $PrFe_4P_{12}$ [1] and $PrOs_4Sb_{12}$ [2] are confirmed to be the HFs, the latter being a superconductor below $T_c = 1.7$ K. The cubic compounds $PrAg_2In$ [3] and $PrMg_3$ [4, 5] have also been reported to show the HF behavior in the specific heat $C(T)$ and the magnetic susceptibility $\chi(T)$. Pr based HFs known to date are, however, still not large in number, and the mechanism of the HF formation is not known well. Clearly, further effort is needed to search for the new Pr-based HF systems.

A new member of the cubic compounds, $Pr(Cu,Ga)_{12.85}$, has recently been synthesized and the magnetic properties have been studied [6]. This compound crystallizes in a $NaZn_{13}$-type structure, isostructural to the heavy fermion superconductor UBe_{13}. In this structure, each Pr atom is surrounded by 24 neighboring atoms, thereby a large hybridization effect can be expected. $\chi(T)$ of $Pr(Cu,Ga)_{12.85}$ shows Curie-Weiss behavior down to 2 K with an effective moment very close to the value of the free Pr^{3+} ion. No evidence of a phase transition is obtained in $\chi(T)$, $C(T)$ and the electrical resistivity $\rho(T)$ measured down to the lowest temperature of

Published under licence by IOP Publishing Ltd

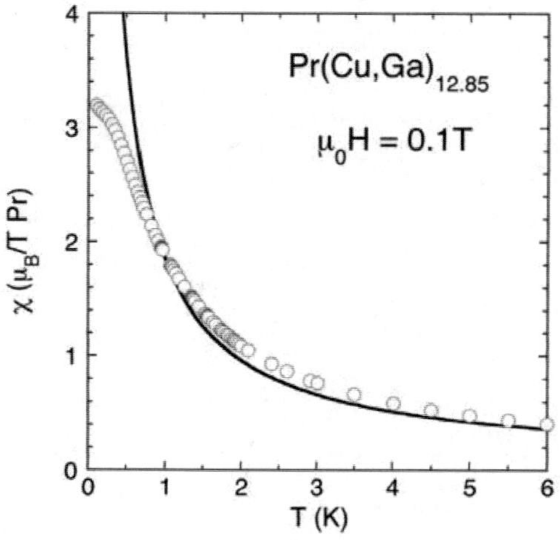

Figure 1. Magnetic susceptibility of $Pr(Cu,Ga)_{12.85}$ below 6 K. A field of 0.1 T was applied parallel to the [100] direction. The data points below 1 K were obtained in a field-cooled condition. Solid line is the calculated result based on the CEF scheme Γ_5(0 K)-Γ_1(20 K)-Γ_4(30 K)-Γ_3(37 K).

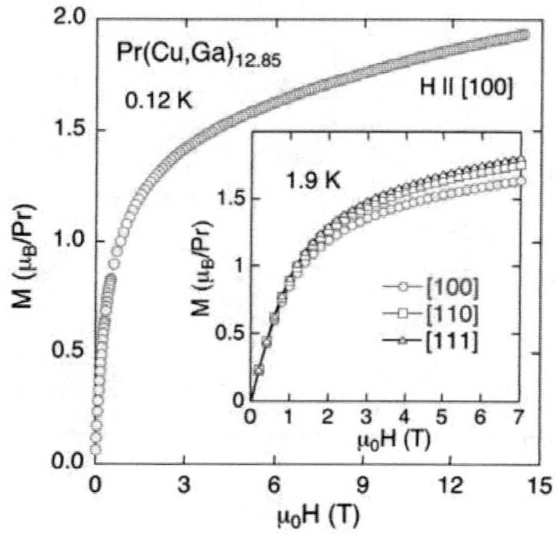

Figure 2. Field dependence of the magnetization of $Pr(Cu,Ga)_{12.85}$ for $H \parallel$ [100] obtained at 0.12 K. The inset shows the results at 1.9 K for H applied to the three principal directions [100] (circles), [110] (squares) and [111] (triangles).

2 K. Very interestingly, $\rho(T)$ shows the AT^2 dependence at low T with a large A coefficient of 0.0727 $\mu\Omega\,\mathrm{cm/K^2}$, and a relatively large C/T value of 97 $\mathrm{mJ/mol{\cdot}K^2}$ is obtained around 2 K. These results suggest that $Pr(Cu,Ga)_{12.85}$ is a HF system. In fact, a Kadowaki-Woods ratio, A/γ^2, of $0.727\times10^{-5}\mu\Omega\,\mathrm{cm{\cdot}mol^2K^2mJ^{-2}}$ is in the order of the expected range for many HFs. In order to examine the ground state magnetic properties of $Pr(Cu,Ga)_{12.85}$, we measured the magnetization down to below 0.1 K.

2. Experimental

Single crystalline samples of $Pr(Cu,Ga)_{12.85}$ were grown by the Ga-flux method. Ga atoms mix into the crystal since the $NaZn_{13}$ structure is only stabilized in the presence of Ga [6]. The atomic ratio of Ga and Cu, as determined from the electron probe microanalysis, is 7.0:6.3 [6]. DC magnetization measurements below 1.9 K in magnetic fields up to 14.5 T were done using a capacitive Faraday magnetometer with a dilution refrigerator [7]. Magnetization above 1.9 K was measured by a SQUID magnetometer (MPMS, Quantum Design).

3. Results and Discussion

Figure 1 shows the magnetic susceptibility $\chi(T)$ of $Pr(Cu,Ga)_{12.85}$ below 6 K down to the base temperature of 0.08 K. It has been reported in the previous paper that $\chi(T)$ above 2 K well follows a Curie-Weiss law, with an effective moment of 3.32 μ_B/Pr and a small negative Weiss temperature of -1.27 K [6]. Our data shows that $\chi(T)$ continues to increase down to below 1 K. This fact strongly indicates that a magnetic triplet is the ground state or in the very low-lying states of the CEF scheme. In this figure, the data points below 1 K were obtained in a field-cooled (FC) condition. Well below 1 K, χ(FC) tends to saturate to a finite value. No evidence of a phase transition is obtained in χ(FC) down to 0.08 K.

Figure 2 shows the field dependence of the magnetization $M(H)$ of $Pr(Cu,Ga)_{12.85}$ for $H \parallel$ [100] measured at 0.12 K. $M(H)$ in low fields shows a steep rise to the value 1.3 μ_B/Pr at

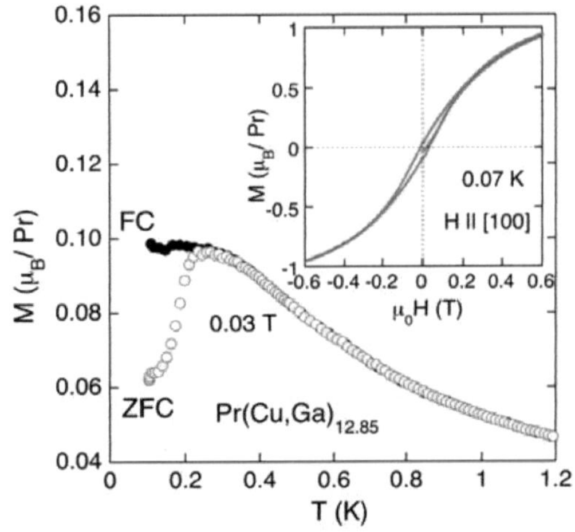

Figure 3. Thermal variation of the magnetization of Pr(Cu,Ga)$_{12.85}$ in a field of 0.03 T (‖[100]) obtained in field-cooled (FC) and zero-field-cooled (ZFC) conditions. Inset shows the full-loop magnetization to ±0.6 T measured at 0.07 K.

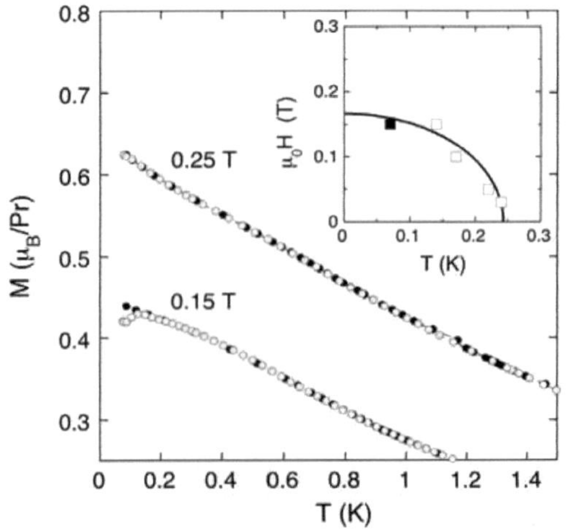

Figure 4. Thermal variation of the magnetization of Pr(Cu,Ga)$_{12.85}$ in fields of 0.15 and 0.25 T (‖[100]). Open (closed) circles are the ZFC (FC) data. Inset shows the $H-T$ diagram for the irreversibility in $M(T,H)$.

2 T, followed by a gradual increase at higher fields. The moment value reaches 1.9 μ_B/Pr at 14.5 T. In the $M(H)$ data, no anomaly that could be related to a phase transition can be seen up to 14.5 T, suggesting that the system is in the paramagnetic state in this field range. The inset shows the $M(H)$ data measured at 1.9 K for the three principal directions $H \parallel [100]$, [110] and [111]. $M(H)$ is isotropic in low fields below 1 T as expected for cubic symmetry. At higher fields, a small but distinct anisotropy develops in the magnetization: $M_{[100]} < M_{[110]} < M_{[111]}$.

Now we discuss the CEF scheme of Pr(Cu,Ga)$_{12.85}$. In a cubic environment, the $J = 4$ multiplet of Pr^{3+} splits into four states: Γ_1 singlet, Γ_3 non-magnetic doublet, and two magnetic triplets Γ_4 and Γ_5. The results of the magnetic measurements in Figs. 1 and 2 indicate that a magnetic triplet is involved in the CEF ground state. The Γ_4 triplet becomes the low-lying state only in a very narrow CEF parameter region where it degenerates with a Γ_1 singlet and a Γ_3 doublet [8]. In this situation, it can be shown that the magnetization becomes strongly anisotropic with a [100] easy axis, $M_{[100]}/M_{[111]} \sim 1.6$ for $H > 3$ T, contrary to the experimental results in Fig. 2. By contrast, the Γ_5 triplet makes an isolated CEF ground state in a much wider parameter range [8], and gives an isotropic moment of 2 μ_B/Pr close to the value observed in Fig. 2. The weak anisotropy observed in high fields can be well explained by admixing with a Γ_3 doublet in CEF excited states [9]. From these, we consider that CEF ground state of this system is a Γ_5 triplet. While it is not easy to fully determine the CEF scheme from the magnetization data alone, we tentatively adopt Γ_5(0 K)-Γ_1(20 K)-Γ_4(30 K)-Γ_3(37 K) to reproduce $\chi(T)$ as well as the anisotropy. The solid line in Fig. 1 shows the calculated susceptibility $\chi_{\mathrm{CEF}}(T)$ assuming the level scheme above, which reproduces the experimental result above 1 K reasonably well. More details of the CEF analysis of this compound will be published elsewhere. Below 1 K, however, χ(FC) significantly deviates from $\chi_{\mathrm{CEF}}(T)$ and levels off. Moreover, $M(H)$ at low T reaches the Γ_5 moment value only in very high fields. These observations imply that the Γ_5 moment at low T is significantly suppressed below the full value, possibly due to onsite as well as intersite fluctuations of the Γ_5 moment which could be related with the heavy fermion behavior observed in this system [6].

While no clear evidence of a phase transition is obtained in χ(FC), we found that a weak history dependence develops in the magnetization in a very low $T - H$ region below 0.25 K and 0.2 T as shown in Fig. 3. Here the field-cooled magnetization M(FC) taken in a field of 0.03 T is compared with the magnetization M(ZFC) obtained in a zero-field-cooled condition. A clear difference appears between M(ZFC) and M(FC) below 0.25 K; while M(FC) is nearly T-independent below 0.25 K, M(ZFC) shows a slight decrease. The inset shows the full loop of $M(H)$ to ± 0.6 T measured at 0.07 K after zero-field cooling the sample to the temperature. A small hysteresis can be seen in $M(H)$ at low fields $|H| < 0.2$ T. These results give strong evidence that the CEF ground state of this compound indeed has a magnetic moment which undergoes a freezing in a spin-glass-like state at low T. This state is easily destroyed by applying a small field $H_c \sim 0.18$ T. At this field, the induced moment value is only $\sim 25\%$ of the saturation value. The inset of Fig. 4 shows the phase diagram for the spin-glass-like state, where open (closed) squares denote the position where the irreversibility disappears in the $M(T)$ ($M(H)$) measurements. Outside the border (solid line), the system is in the paramagnetic state. In Fig. 4, we show $M(T)$ in the paramagnetic region near H_c. Interestingly, we found that the magnetization does not stay constant but shows nearly T-linear increase on cooling down to below 0.1 K (Fig. 4), which could be related to quantum critical behavior.

To summarize, low temperature magnetization measurements on Pr(Cu,Ga)$_{12.85}$ indicate that the CEF ground state is a Γ_5 triplet carrying a magnetic moment, which undergoes a spin-glass-like freezing below 0.25 K. Under a small field of 0.2 T, the system remains in a paramagnetic state down to the lowest temperature of 0.07 K with unusual temperature variation of the magnetization. How the entropy $R\ln 3$ of the Γ_5 state is released in this region is highly of interest, and we are carrying out the specific heat measurements whose results will be published elsewhere.

Acknowledgments

This work has been supported by a Grant-in-Aid for Scientific Research on Innovative Areas "Heavy Electrons" (No. 20102007) of the Ministry of Education, Culture, Sports and Technology, Japan and a Grant-in-Aid for Scientific Research (A) (No. 20244053) from the Japan Society for the Promotion of Science.

References

[1] Sugawara H, Matsuda T D, Abe K, Aoki Y, Sato H, Nojiri S, Inada Y, Settai R and Onuki Y: Phys. Rev. B **66** (2002) 134411.
[2] Bauer E D, Frederick N A, Ho P -C, Zapf V S and Maple M B: Phys. Rev. B **65** (2002) 100506(R).
[3] Yatskar A, Beyermann W P, Movshovic R and Canfield P C 1996 Phys. Rev. Lett. **77** 3637.
[4] Tanida H, Suzuki H.S, Takagi S, Onodera H and Tanigaki K. 2006 J. Phys. Soc. Jpn. **75** 073705.
[5] Morie T, Sakakibara T, Suzuki H S, Tanida H and Takagi S 2009 J. Phys. Soc. Jpn. **78** 033705.
[6] Cho J Y, Thomas E L , Nambu Y, Capan C, Karki A B, Young D P, Kuga K, Nakatsuji S and Chan J Y: Chem. Mater. **21** (2009) 3072.
[7] Sakakibara T, Mitamura H, Tayama T and Amitsuka H: Jpn. J. Appl. Phys. **33** (1994) 5067.
[8] Lea K, Leask M and Wolf W: J. Phys. Chem. Solids. **23** (1962) 1381.
[9] Aoki Y, Namiki T, Matsuda T D, Abe K, Sugawara H and Sato H: Phys. Rev. B **65** (2002) 064446.

What makes a Kondo insulator/semiconductor?

Jozef Spałek[1] **and Andrzej Slebarski**[2]
[1]Marian Smoluchwski Institute of Physics, Jagiellonian University, Reymonta 4, 30-059 Kraków, Poland;
[2]Institute of Physics, University of Silesia, Uniwersytecka 4, 40-007 Katowice, Poland

E-mail: ufspalek@if.uj.edu.pl or andrzej.slebarski@us.edu.pl

Abstract. We provide a brief critical assessment of the concept of Kondo insulator (semiconductor) in view of the experimental results and a simple mean-field-like picture of correlated states within the Anderson-lattice model. A brief distinction between the *Kondo semiconductors* and either ordinary- or magnetic- (including Mott-Hubbard) -semiconductors is discussed.

1. Introduction

The Kondo insulators (KI) belong to the class of strongly correlated materials forming a group of either nonmagnetic semiconductors [1] with the narrowest gap known (i.e. in the conductivity gap Δ in the Kelvin range) or to semimetals [2], both with a heavy-fermion metallic state setting in at elevated temperature $T \geq \Delta$. Their nature is known to a lesser extent than that of the stable heavy-Fermi-liquid metals [3]. CeRhSb is an example of the full-gap semiconductor with $\Delta \approx 7.6$ K [1], whereas CeNiSn is an anisotropic semimetallic system [2]. The understanding of these materials is partly obscured by the circumstance that the relatively small values of magnetic susceptibility and the gap when $T \to 0$ lead to their properties alternation by a minute content of magnetic or other impurities [4]. In effect, it is often difficult to differentiate between intrinsic and extrinsic features of these systems. The situation is, in some sense, analogous to that in the early-stage research of semiconductors until the role of nonintentional impurities in, e.g., Ge or Si has been established. Also, a clear distinction from either ordinary, magnetic, and Mott-Hubbard semiconductors should be singled out. The purpose of this critical brief paper is to propose criteria defining the *Kondo* insulator/semiconductor and then address in an elementary manner the question about the nature of FeSi [5], probably the first system of this type containing 3d electrons.

2. A canonical theoretical interpretation: Kondo semiconductors

The present critical assessment of these Kondo-lattice materials arises from our earlier research on $CeRhSb_{1-x}Sn_x$ [6] and $CeNi_{1-\delta}Sn_{1+\delta-x}Sb_x$ [7] systems, as well as that on $Ce_3Bi_4Pt_3$ [8] and FeSi [5]. The interpretation is based on the earlier saddle-point solution of the Anderson-lattice model [9]. From this accumulated material it follows that the Kondo insulating state can be characterized by the following features:

---

[1] Jozef Spałek.

[2] Andrzej Slebarski

Published under licence by IOP Publishing Ltd

(*i*) *The number of valence electrons is an even number*. For example, in case of CeRhSb it is 18 electrons (including one 4f electron due to Ce). So has CeNiSn, whereas $Ce_3Bi_4Pt_3$ has 54 such electrons and FeSi has 12. From that one can see that a single magnetic moment of either Ce^{3+} or Fe^{4+} can be screened to a substantial extent within a single chemical formula if the coupling of the surrounding screening carrier cloud is of antiferromagnetic character. The remaining washing out of the magnetic moment can be achieved by autoscreening, i.e., the intersite antiferromagnetic correlations between the relevant 4f or 3d electrons; as is explicitly seen for the Anderson-lattice model [9].

(*ii*) *The resistivity* $\rho(T)$ *is often thermally activated in the limited temperature range* with a small (or even extremely small [6]) activation gap. A small value of the gap is sometimes not easy to detect as the "very shallow" impurity in-gap states can form impurity bands overlapping with the upper and/or the lower hybridization bands, thus leading to a semimetallic behaviour. The semimetallic state can also be caused by the absence of an intraatomic part of the hybridization and vanishing intersite hybridization in some directions in **k** space. For example, the intraatomic part of hybridization between 4f and either 5d or 6s states of Ce vanishes as they have different parity. So does that between 3d states Fe and 3p states of Si, but in this case it is possible between 3d and 4s states, so FeSi from this perspective should be a full-gap system. Also, a residual intraatomic part of hybridization between 4f states due to Ce and 5p states due to Sb can produce a full-gap state. However, this simple rationalization does not work for CeNiSn, which in this view should also be regarded as being in the same category, as CeRhSb.

(*iii*) *The magnetic susceptibility* $\chi(T)$ *is strongly temperature dependent and diminishes remarkably as T is reduced below a characteristic temperature* and the system *does not* order magnetically (provided the impurity contribution is identified and properly subtracted).

(*iv*) The resistivity and magnetic susceptibility *obey* [6] *a simple scaling law* $\chi(T)\rho(T)$=const, particularly in the regime $k_BT < \Delta$, i.e., when the number of excited carriers is small.

(*v*) The semiconducting (or semimetallic) state transforms gradually into a moderately heavy-electron metallic state in the region $k_BT > \Delta$ and/or into a system of localized f or d electrons if either $k_BT >> \Delta$ or Δ is large (10^2 K or larger).

One should note that the features (*i*), (*ii*) and (*iv*) may also be obeyed by intrinsic semiconductors if the susceptibility of excited carriers (after substituting the diamagnetism of filled bands) is detectable. Obviously, neither Mott-Hubbard insulators nor magnetic semiconductors do not fall in this category, as they order magnetically (no screening effect of the carriers; they are also closer to ionic systems). Thus the specific features for the Kondo insulators/semiconductors are those listed under the headings (*iii*)-(*v*). Additionally, those systems exhibit a quantum critical behaviour upon electron doping [6], but these topics will not be discussed here.

3. Universal scaling law: $\chi(T)\rho(T)$ = const

We summarize first evidence for the Kondo insulating/semiconducting character of the CeRhSb system and then discuss a situation in $Ce_3Bi_4Pt_3$ and FeSi. If the scaling $\chi(T)\rho(T)$=const is characteristic and universal for the Kondo insulators, it should be obeyed for the all known Kondo insulators. In figure 1a-c we have collected the main properties of the full-gap semiconductor $CeRhSb_{1-x}Sn_x$, with the conductivity gap Δ = 7.6 K (a). The scaling law $\chi(T)\rho(T)$=const is clearly obeyed for $x \leq 0.08$ (b) [6], and the magnetic susceptibility (after subtracting by the impurity effect) $\chi \to 0$ when $T \to 0$. In figure 2 the scaling law $\chi \sim \rho^{-1}$ is illustrated for $Ce_3Bi_4Pt_3$. The susceptibility $\chi(T)$ and resistivity $\rho(T)$ data obtained for the single crystalline sample are taken from Ref. [10]. In the original $\chi(T)$ plot, $Ce_3Bi_4Pt_3$ exhibits a sharp upturn in the susceptibility at the low temperatures T<13 K.This impurity upturn is described by the presence of Ce interstitial impurities with the concentration of about 1%. In effect, with correction for impurities, the scaling χ versus ρ^{-1} is clearly observed in the temperature region 4.5 K < T < 100 K for $Ce_3Bi_4Pt_3$ (for $Ce_3Bi_4Pt_3$ the gap $\Delta \approx 60$ K, therefore the

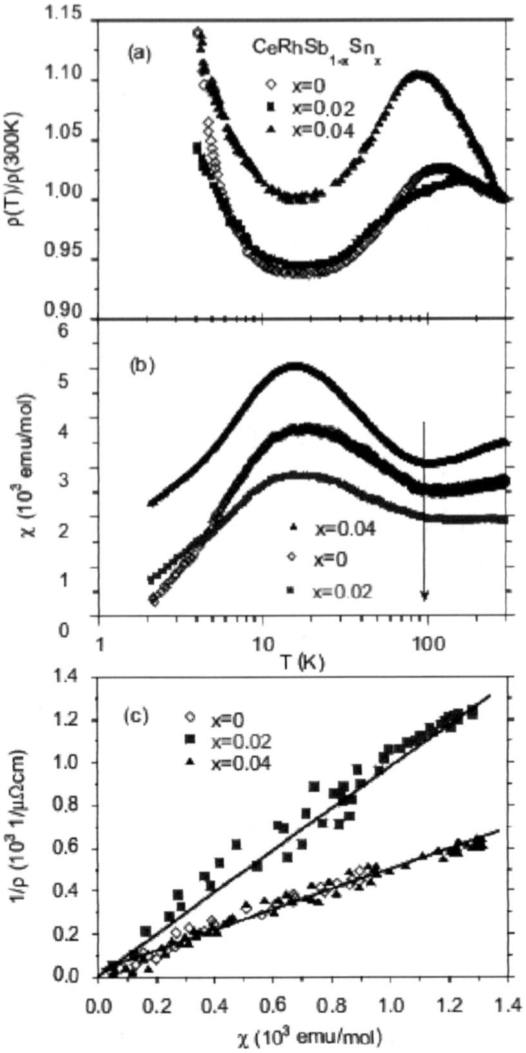

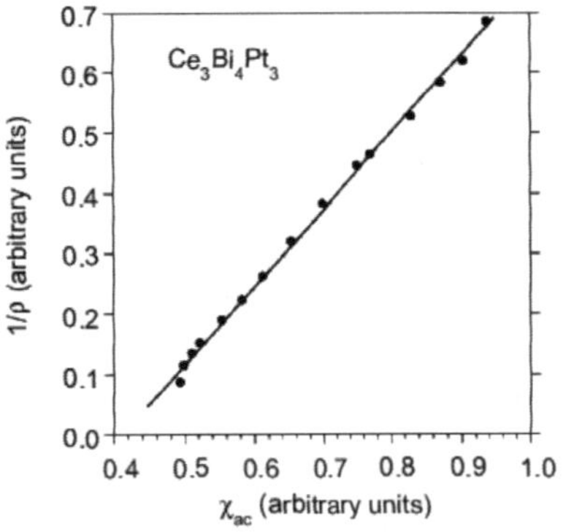

Figure 2. Linear scaling law between the inverse resistivity ρ and the susceptibility χ for $Ce_3Bi_4Pt_3$ for 4.5 K<T<100K. The data are taken from [10].

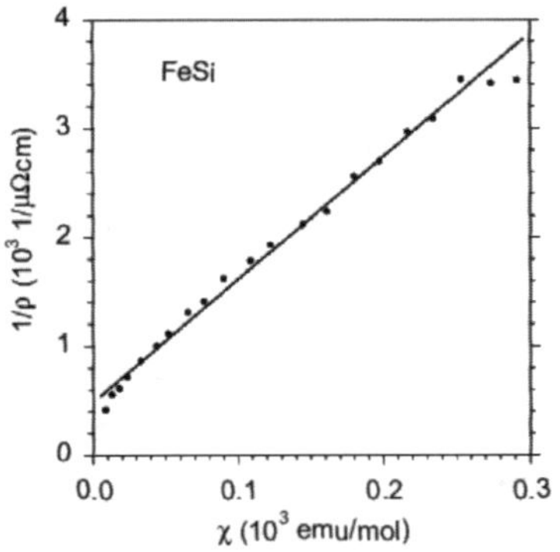

Figure 2. FeSi; ρ^{-1} vs. χ for 130 K < T <300 K. The χ and ρ data are from [11]. The impurity effect is small for T > 130 K.

impurity effect does not dominate in the temperature region T > 10K).

We discuss next evidence for the Kondo insulating character of FeSi which is a nonmagnetic narrow gap semiconductor [5] with unusual features it shares with a class of Ce-Kondo insulators. Very interesting feature, which one encounters in this system, is the fact that the gap ($\Delta \approx$ 300 K) disappears gradually at T_0 with increasing temperature T [12,13] which has been ascribed theoretically as a typical behaviour for a Kondo semiconductor. The temperature T_0 [13] is not identified with the Kondo temperature T_K , but represents an instability point of the paramagnetic Kondo semiconductor. Particularly important is the linear [$\Delta(T) \sim (T_0 - T)$] dependence of the gap when approaching T_0 (for FeSi, $T_0 \approx 200$ K). A very similar $\Delta(T)$ behaviour was also experimentally observed for the Kondo insulator CeRhSb and CeNiSn semimetal [14].
In contradiction to these data, it has been also reported [15] that FeSi is an itinerant semiconductor whose properties can be explained without a local Kondo-like interaction. We therefore elaborate now more detail about our criteria for FeSi. We should note that the features (*i*), (*ii*), and (*iii*) are obeyed [11]. Figure 3 also illustrates the scaling behaviour $\chi(T)\rho(T)$ = const in the temperature region 130 K < T < 300 K (i.e., in the temperature range where the impurity contribution to ρ and χ is negligible). Within our criteria a Kondo insulator description seems to be appropriate also for FeSi.

4. Acknowledgments

The work was supported by the Ministry of Science and Higher Education, Grants Nos. N N202 128736 and N N202 032137.

References:

[1] Malik S K and Adroja D T 1991 *Phys. Rev.* B **43** 6277
[2] Takabatake T, Teshima F, Fujii H, Nishigori S, Suzuki T, Fujita T, Yamaguchi Y, Sakurai J and Jaccard D 1990 *Phys. Rev.* B **41** 9607
[3] Stewart G R 1984 Rev. Mod. Phys. **56** 755
[4] Takabatake T, Nakazawa Y, Ishikawa M, Sakakibara T, Koga K, and Oguro I 1988 J. Magn. Magn. Mater. **77-76** 87
[5] Jaccarino V, Wertheim G K, Wernick J H, Walker L R, and Sigurds A 1967 Phys. Rev. **160** 476; Aeppli G and Fisk Z 1992 Comments Condens. Matter Phys. **16** 155
[6] Slebarski A and Spałek J 2005 Phys. Rev. Lett. **95** 046402; Spałek J, Slebarski A, Goraus J, Spałek L, Tomala K, Zarzycki A, and Hackemer A 2005 Phys. Rev. B **72** 155112
[7] Slebarski A and Spałek J 2009 Phil. Mag. **89** 1845
[8] Hundley M F, Canfield P C, Thompson J D, and Fisk Z 1990 Phys. Rev. B **42** 6842
[9] Doradzi ski R and Spałek J 1997 Phys. Rev. B **56** R14239; Doradzi ski R and Spałek J 1998 Phys. Rev. B **58** 3293
[10] Pietrus T, vLöhneysen H, and Schlottmann P 2008 Phys. Rev. B **77** 115134
[11] Schlesinger Z, Fisk Z, Zhang H T, Maple M B, DiTusa J F, and Aeppli G 1993 Phys. Rev. Lett. **71** 1748
[12] Fäth M, Aarts J, Menovsky A A, Nieuwenhuys G J, and Mydosh J A 1998 Phys. Rev. B **58** 15483; Menzel D, Popovich P, Kovaleva N N, Schoenes J, Doll K, and Boris A V 2009 Phys. Rev. B **79** 165111
[13] Spałek J 2000 Acta Phys. Polon. A **97** 71
[14] Ekino T, Takabatake T, Tanaka H, and Fujii H 1995 Phys. Rev. Lett. **75** 4262
[15] Breuer K, Messerli S, Purdie D, Garnier M, Hengsberger M, and Baer Y 1997 Phys. Rev. B **56** R7061; Klein M, Zur D, Menzel D, Schoenes J, Doll K, Röder J, and Reinert F 2008 Phys. Rev. Lett. **101** 046406

International Conference on Strongly Correlated Electron Systems (SCES 2010) IOP Publishing
Journal of Physics: Conference Series **273** (2011) 012056 doi:10.1088/1742-6596/273/1/012056

Elastic properties of MnSi, FeSi and CoSi

Alla E. Petrova[1], Vladimir N. Krasnorussky[1], William M. Yuhasz[2], Thomas A. Lograsso[2], and Sergei M.Stishov[1]

[1]Institute for High Pressure Physics of RAS, Troitsk 142190, Moscow Region, Russia
[2]Ames Laboratory, Iowa State University, Ames, IA 50011, USA

E-mail: `sergei@hppi.troitsk.ru`

Abstract. Measurements of sound velocities in single crystals of MnSi, FeSi and CoSi were performed in the temperature range 2.5-300 K and elastic constants were calculated. The temperature dependence of the elastic constants reveal nontrivial features, reflecting specifics of the magnetic and electronic subsystems in these materials.

The intermetallic compounds MnSi, FeSi and CoSi with a cubic B20 crystal structure have attracted much attention over decades due to their remarkable magnetic and electrical properties. MnSi is an itinerant helical magnet with a Curie point of 29 K. FeSi, a strongly correlated semiconductor, having a small energy gap of about 0.05 eV at low temperature, reveals metallic properties above 100 K [1, 2, 3, 4, 5]. CoSi is a semimetal with very high residual resistivity, implying existence of a fairly large number of defects [6]. Although these compounds display a range of properties their lattice parameters change very little consistent with the trend in metallic radii for Mn, Fe and Co. Together these facts tell us that the total energies of these systems are not sensitive to the details of their electronic and magnetic structures. However, as is shown below the specifics of the macro properties of these silicides can be observed through the second derivatives of the total energy with respect to deformation, i.e. their elastic properties.

We report here results of ultrasonic studies of single crystals of MnSi, FeSi and CoSi. The measurements were performed using a digital pulse-echo technique (see details in Ref. [7]) in the temperature range 4-300 K (4-150 K for MnSi). The single crystals were grown by the Bridgman (MnSi, CoSi) and the Czokhralski (FeSi) methods. The lattice parameters of the crystals, determined by X-ray diffraction, correspond well to literature values (see Table. 1). For the ultrasonic studies, samples of MnSi, FeSi and CoSi of about 2 - 4 mm thicknesses and with orientations along [100], [110] and [111] were cut from big single crystals. The corresponding surfaces of the samples were made optically flat and parallel. The 36° Y (P-wave) and 41° X (S-wave) cut $LiNbO_3$ transducers were bonded to the samples with various adhesives, including silicon greases and super glue. Temperature was measured by a calibrated Cernox sensor with an accuracy of 0.02 K. The speed of sound and elastic constants are calculated using the known thickness and density of the samples and the relationship $c_{ij} = \rho V^2$. The precision of the sound velocity measurements is about 0.1%, though the absolute accuracy may be of order 1% mainly due to uncertainty connected with a phase shift at the transducers sample bond interface. Results of the measurements and calculations are shown in Figures 1 to 6 and Table 1. Some our data for MnSi have been already published in Ref. [7]. They are partly reproduced here for the comparison purpose. Low resolution ultrasonic data for a number of transition metal silicides were published in [8]. The temperature dependence of the elastic moduli of FeSi was measured

Published under licence by IOP Publishing Ltd

previously using the resonant ultrasound spectroscopy technique [9]. The data from Ref. [9] generally agrees with our results, where they overlap at temperatures ($T > 77K$), although slight differences in temperature calibration are present.

Now we turn to the analysis of the current experimental data. As is seen from Figures 1- 3, the numerical values of the elastic constants in the row MnSi-FeSi-CoSi change according to positions of the metallic atoms in the Mendeleev periodic table or, in the other words, according to the number of 3d valence electrons in these compounds. A slight difference in this characteristic of the compounds evidently results in subtle variation of the cohesion energies and lattice parameters, that caused quite an observable difference in the elastic constants (see Figures 1- 4). It needs to be emphasized that regularity in the variation of the elastic constants variation exist despite the specifics electronic and magnetic structures of the compounds. Indeed it is instructive to look at the situation at low temperatures where CoSi is a semimetal with a good metallic value of the Sommerfeld constant $\gamma = 1.1mJ/moleK^2$ [10], FeSi is a narrow gap semiconductor, and MnSi is a metal with helical magnetic structure and a very large $\gamma = 36mJ/moleK^2$ [11]. The values of c_{11} and c_{12} elastic constants are highest for CoSi, then follow FeSi and MnSi (Fig. 1, 2). The situation is different in case of the shear elastic constant c_{44} (Fig. 3) where FeSi has now the highest value, and CoSi the smallest.

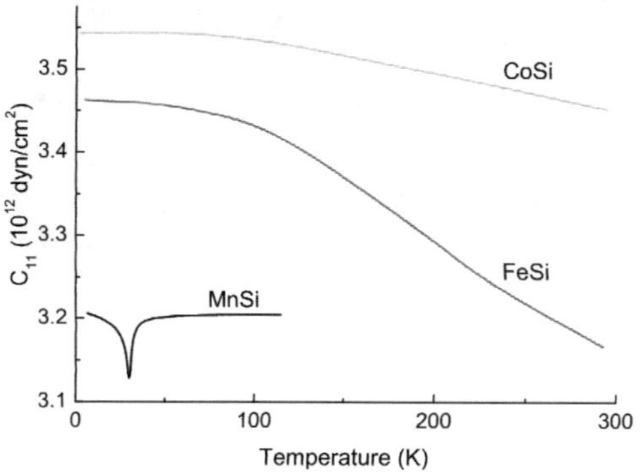

Figure 1. (Color online) Elastic constants c_{11} of MnSi, FeSi and CoSi

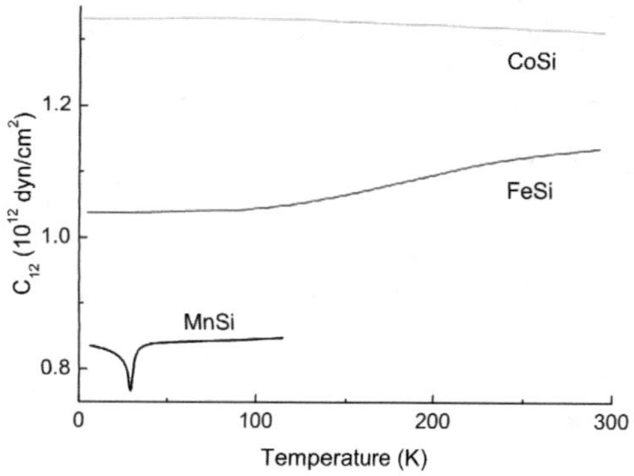

Figure 2. (Color online) Elastic constants c_{12} of MnSi, FeSi and CoSi

It seems quite natural that CoSi with the highest electron concentration, has the highest bulk modulus $K = (c_{11} + 2c_{12})/3$. At the same time, it is not quite clear why CoSi should have the smallest shear moduli c_{44} and $(c_{11} - c_{12})/2$.

The temperature variations of the elastic properties of the three compounds are far from triviality. Though in the case of CoSi there are no significant features to be seen on the scale of Figs. 1- 3. However, if we look closely at the temperature dependence of CoSi below 60 K Fig. 5, we can clearly see an unusual softening of the elastic constants. The elastic constants c_{11} and c_{44} of FeSi strongly decrease at high temperature as compared with CoSi; whereas, c_{12} grows anomalously over the entire range of temperature. As a consequence the shear module $(c_{11}-c_{12})/2$ of FeSi drops off precipitously above 100 K (Fig. 4). This behavior is consistent with the gap closing as indicated in [12, 13]. The MnSi elastic constants that control propagation of longitudinal waves reveal significant softening at $\sim$ 30 K and small discontinuities at 28.8 K, which corresponds to the magnetic phase transition in MnSi. In contrast, the shear elastic moduli, do not show any softening at all and only respond to small volume deformations caused by the magneto-volume effect [11].

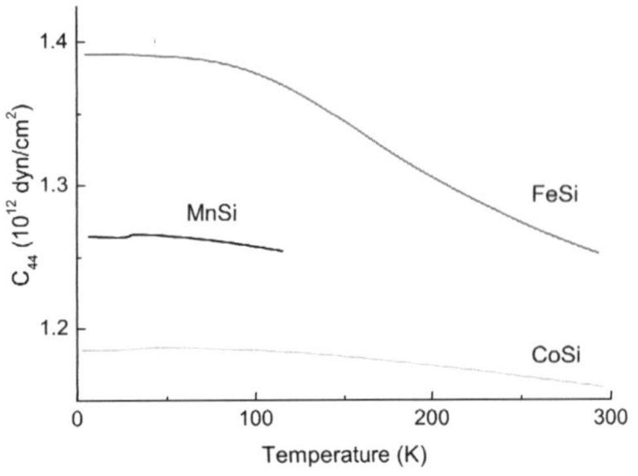

Figure 3. (Color online) Elastic constants c_{44} of MnSi, FeSi and CoSi

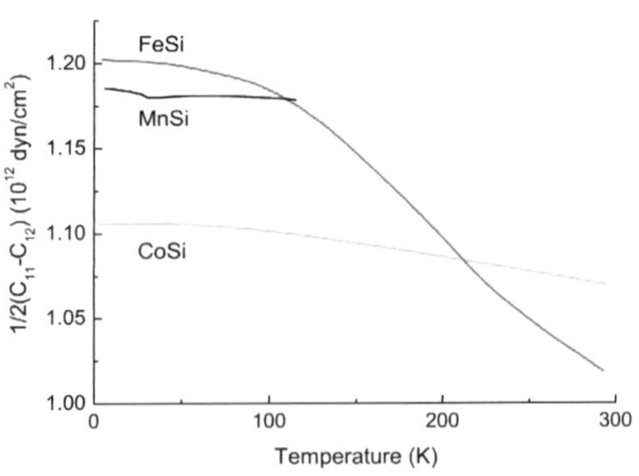

Figure 4. (Color online) Shear moduli $(c_{11} - c_{12})/2$ of MnSi, FeSi and CoSi

Returning to CoSi we are reminded that, in accordance to the Nernst theorem, temperature derivatives of physical quantities level off at low temperatures and tend to zero at $T \to 0$. As seen in Fig. 1- 3, it seems to occur at about $T/\Theta_D \approx 1/6$, where Θ_D-Debye temperature. But this is not always the case as indicated by the temperature dependence of the CoSi elastic constants featured in Fig. 5. One can see that with decreasing temperature the elastic constants of CoSi pass through maxima between 40 K and 70 K and continue to change down to 2.5 K. This may be tied to the Curie-Weiss like increase of the magnetic susceptibility of CoSi below $\sim 50K$ (see Fig. 5), associated with local moment formation.

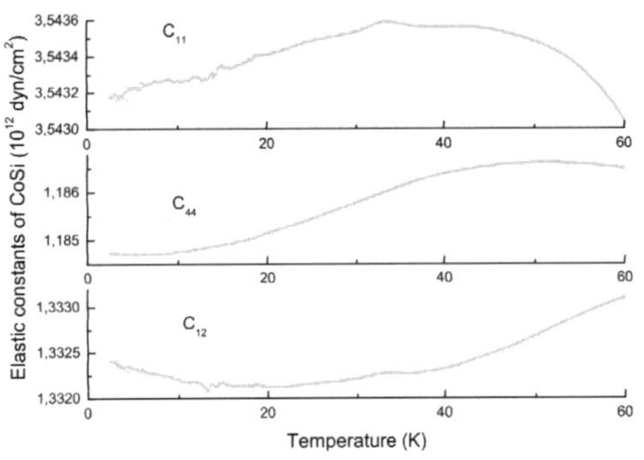

Figure 5. (Color online) Elastic constants c_{11}, c_{12} and c_{44} of CoSi (enlarged view)

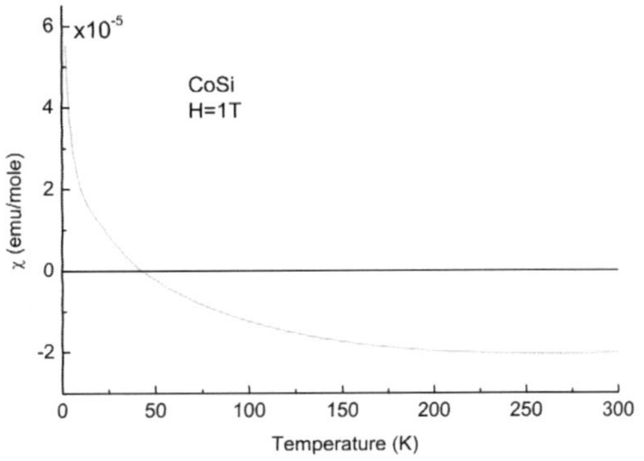

Figure 6. (Color online) Magnetic susceptibility of CoSi

The origin of these moments is probably connected with conduction electron localization at lattice defects that in turn renormalize the phonon spectra of CoSi. Clearly, this is not a complete explanation considering that FeSi, which also features a Curie-Weiss tail, does not display any irregularities in the elastic constants. Along with this unexplained behavior the description of the elastic properties of CoSi cannot be done based on a single a single energy scale like the Debye temperature. Rather an entire spectrum of characteristic energies down to

Table 1. Elastic constants of MnSi, FeSi and CoSi (a-lattice parameters, Θ_D-Debye temperature, calculated from elastic constants)

MnSi a=4.5598Å Θ_D=660 K			
C_{ij}	T=6.5	T=78	T=115
C_{11}	3.2057	3.2045	3.2047
C_{44}	1.2615	1.2582	1.2540
C_{12}	0.8523	0.8574	0.8477
FeSi a=4.483Å Θ_D=680 K			
C_{ij}	T=6.5	T=77.8	T=292.8
C_{11}	3.4626	3.4454	3.1670
C_{44}	1.3916	1.3858	1.2521
C_{12}	1.0576	1.0608	1.1298
CoSi a=4.444Å Θ_D=625 K			
C_{ij}	T=6.5	T=77.8	T=292.8
C_{11}	3.5432	3.5404	3.4529
C_{44}	1.1847	1.1857	1.1593
C_{12}	1.3323	1.3331	1.3128

~ 1 K may be required.

In conclusion, we express our deep gratitude to Dr. Joe Thompson for the technical assistance. AEP, SMS and VNK appreciate support of the Russian Foundation for Basic Research , Program of the Physics Department of RAS on Strongly Correlated Systems and Program of the Presidium of RAS on Physics of Strongly Compressed Matter. WMY and TAL wish to acknowledge research performed at Ames Laboratory. Ames Laboratory is operated for the U.S. Department of Energy by Iowa State University under Contract No. DE-AC02-07CH11358.

References

[1] Jaccarino V, Wertheim G K, Wernick J H, Walker L R, Arajs S 1967 *Phys.Rev.* **160** 476
[2] Schlesinger Z, Fisk Z, Zhang Hai-Tao, Maple M B, DiTusa J F, Aeppi G 1993 *Phys. Rev. Lett.* **71** 1748
[3] Mandrus D, Sarrao J L, Migliori A, Thompson J D, Fisk Z 1995 *Phys.Rev.* B **51** 4763
[4] Paschen S, Felder E, Chernikov M A, Degiorgi L, Schwer H, Ott H R, Young D P, Sarrao J L, Fisk Z 1997 *Phys.Rev.* B **56** 12916
[5] Fäth M, Aarts J, Menovsky A A, Nieuwenhuys G J, Mydosh J A 1998 *Phys. Rev.* B **58** 15483
[6] Amamou A, Bach P, Gautier F, Robert C 1972 *J. Phys. Chem. Solids* **33** 1697
[7] Petrova A E, Stishov S M 2009 *J.Phys.: Condens. Matter* **21** 196001
[8] Zinoveva G P, Andreeva L P, Geld P V 1974 *Phys.Stat.Sol.* (a) **23** 711
[9] Sarrao J L, Mandrus D, Migliori A, Fisk Z, Bucher E 1994 *Physica* B **199-200** 478
[10] Marklund K, Larson M, Byström S, Lindqvist T 1974 *Physica Scripta* **9** 47
[11] Stishov S M, Petrova A E, Khasanov S, Panova G Kh, Shikov A A, Lashley J C, Wu D, Lograsso T A 2008 *J. Phys.Condens. Matter* **20** 235222
[12] Arita M, Shimada K, Takeda Y, Nakatake M, Namatame H, Taniguchi M, Negishi H, Oguchi T, Saitoh T, Fujimori A, Kanomata T 2008 *Phys.Rev.* B **77** 205117
[13] Klein M, Zur D, Menzel D, Schoenes J, Doll K, Röder J, Reinert F 2008 *Phys. Rev. Lett.* **101** 046406

Journal of Physics: Conference Series **273** (2011) 012057 doi:10.1088/1742-6596/273/1/012057

A possible Fermi-surface deformation in mixed-valence systems

Takashi Sugibayashi and Hiroaki Kusunose

Department of Physics, Ehime University, Matsuyama 790-8577, Japan

E-mail: takashi@cmt.phys.sci.ehime-u.ac.jp

Abstract. The possibility of a significant Fermi-surface deformation as a result of valence crossover in mixed-valence systems is investigated theoretically. The Anderson lattice model on a square lattice with an additional local repulsion between c and f electrons is used to set up a system that shows a sudden valence crossover upon making small changes in the parameters. Analyzing the model using the dynamical mean-field theory demonstrates a significant change of the Fermi surface across the valence crossover. Thus, the structure of the irreducible susceptibility in the momentum-frequency space changes completely, which results in a significant change of effective interactions among quasiparticles. It also smears out high-energy Hubbard satellite bands in the weakly correlated regime, which provides a direct experimental test for identifying a sudden valence crossover.

1. Introduction

Recently, the mixed-valence systems such as Yb- and Sm-based compounds have attracted much attention[1, 2]. These systems with almost localized f electrons exhibit a variety of unusual phenomena such as various types of anisotropic superconductivities and the quantum criticality. The so-called heavy-fermion physics has been investigated extensively so far, and we have learned that such unusual phenomena are essentially caused by a weak hybridization between the localized and the itinerant orbitals. As the hybridization or the energy level of the localized orbital increases, the f-electron valence tends to fluctuate, enhancing a local charge fluctuation.

Recent research has indicated that a sudden change in the resistivity exponent, the specific-heat coefficient, and the T^2-resistivity coefficient occurs under external or chemical pressures in $CeCu_2(Si,Ge)_2$[3] and $CeCu_2Si_2$ compounds[4]. It has been suggested that a close relationship may exist these phenomena and the emergence of superconductivity of these compounds because of the strongly enhanced valence fluctuation[5], although there have been no direct observations till date. To describe theoretically a situation in which the f-electron valence changes significantly by slightly changing the parameters, the Anderson lattice model with an additional local repulsion U_{cf} among c- and f electrons has been used as a minimal model[6], as it allows control over the extent of the valence crossover along with f-electron energy ϵ_f. It has been experimentally found that at an ambient pressure, a heavy quasiparticle band exists with multiple Fermi surfaces, as measured by angle-resolved photoemission spectroscopy (ARPES)[7], and these measurements are reasonably consistent with the band-structure calculations[8].

When the valence crossover occurs, it is natural to expect an overall electronic structure to be reconstructed since the heavy-fermion state itself is formed on the basis of the strong electron

Published under licence by IOP Publishing Ltd

correlations, which suddenly becomes inactive across the valence crossover. A sudden decrease in the electron correlation causes a significant change in the single-particle spectrum in which the Hubbard satellite bands are smeared out in the high-energy region, and in the low-energy region a considerable changes in the quasiparticle dispersion and in the structure of the effective interaction among quasiparticles are expected. The latter phenomenon vitally affects the nature of superconductivity. Since all the experiments were performed at finite temperatures, the degree of the change in the electronic structure depends on a relationship between the renormalized Fermi energy and the temperature. Motivated by these results, we investigate the possibility of a significant Fermi-surface deformation on the sudden valence crossover at finite temperatures, and demonstrate that such a change indeed occurs.

2. Model and Method

The extended Anderson lattice model is given by

$$H = \sum_{\boldsymbol{k}\sigma} \epsilon_{\boldsymbol{k}} c^{\dagger}_{\boldsymbol{k}\sigma} c_{\boldsymbol{k}\sigma} + \epsilon_f \sum_{\boldsymbol{k}\sigma} f^{\dagger}_{\boldsymbol{k}\sigma} f_{\boldsymbol{k}\sigma} + v \sum_{\boldsymbol{k}\sigma} (c^{\dagger}_{\boldsymbol{k}\sigma} f_{\boldsymbol{k}\sigma} + H.c.) + U \sum_{\sigma} n_{if\uparrow} n_{if\downarrow} + U_{cf} \sum_{i\sigma\sigma'} n_{ic\sigma} n_{if\sigma'}, \quad (1)$$

where the last term is the additional local repulsion between c and f electrons and the other notations are standard. We adopt the nearest-neighbor dispersion on a square lattice for the c band, i.e., $\epsilon_{\boldsymbol{k}} = -2t(\cos k_x + \cos k_y)$, and we fix the parameters as $t = 0.25$, $v = 0.2$, $U = 3.0$, the total electron density $n_{\rm tot} = 1.75$, and the temperature $T = 0.01$. The values of ϵ_f and the U_{cf} are left as control parameters for the valence crossover.

Because the valence crossover or the first-order transition is essentially determined by an averaged energetics in the local site, the main role of U_{cf} can be qualitatively simulated by the mean-field decoupling, i.e., $U_{cf} \sum_{\sigma,\sigma'} n_{ci\sigma} n_{fi\sigma'} \rightarrow U_{cf} \sum_{\sigma} (\langle n_f \rangle n_{ci\sigma} + \langle n_c \rangle n_{fi\sigma})$. The effect of the U_{cf} is absorbed in the effective levels of c and f electrons and the resultant standard Anderson lattice model is solved by the dynamical mean-field theory (DMFT)[9], in which the modified iterative perturbation theory (mIPT)[10] is used as an impurity solver. We have confirmed that the finite-temperature "Luttinger sum-rule," $-(1/\pi) \sum_{\boldsymbol{k}} \int d\omega\, f(\omega) {\rm Im}\, {\rm Tr}[\hat{G}(\boldsymbol{k},\omega)(\partial\hat{\Sigma}(\boldsymbol{k},\omega)/\partial\omega)] = 0$, holds within 5% accuracy for all the parameter sets used in this study.

3. Results

To determine the parameters by which the sudden valence crossover occurs, Figure 1 shows the ϵ_f dependence of the f-electron density, n_f (the left panel) and the corresponding mass-enhancement factor, $z_f^{-1} = 1 - \partial\Sigma_f(\omega)/\partial\omega|_{\omega=0}$ (the right panel), for the fixed U_{cf}. As U_{cf} increases, the valence crossover becomes steeper, and above $U_{cf} \sim 0.7$ the crossover changes to the first-order valence transition at about $\epsilon_f = -0.35$. Following the valence crossover/transition, the corresponding z_f^{-1} decreases sharply towards the value of the weak-correlation regime (unity) from the extremely enhanced one as ϵ_f increases. These behaviors are qualitatively consistent with those found in previous works[11], although the parameters yielding the steep valence crossover themselves are different attributed to the mean-field treatment of U_{cf}. In the figure, we define the three regimes of the valence for $U_{cf} = 0.6$ as (A) the heavily renormalized regime ($\epsilon_f = -0.45$), (B) the critically upper-valence regime ($\epsilon_f = -0.35$), and (C) the critically lower-valence regime ($\epsilon_f = -0.30$). In the following section, we consider the change in the electronic structures for the three different regimes.

Figure 2 represents the single-particle spectrum defined by $A_{c,f}(\boldsymbol{k},\omega) = -1/\pi {\rm Im} G_{c,f}(\boldsymbol{k},\omega)$ at the three different regimes. In regime A, $A_f(\boldsymbol{k},0)$ and $A_c(\boldsymbol{k},0)$ have strong peaks roughly along the $(\pi,0)$-$(0,\pi)$ line, which corresponds to the "Fermi surface" at the finite temperatures accessible to the ARPES measurements. In other words, the f-electron forms a hole Fermi

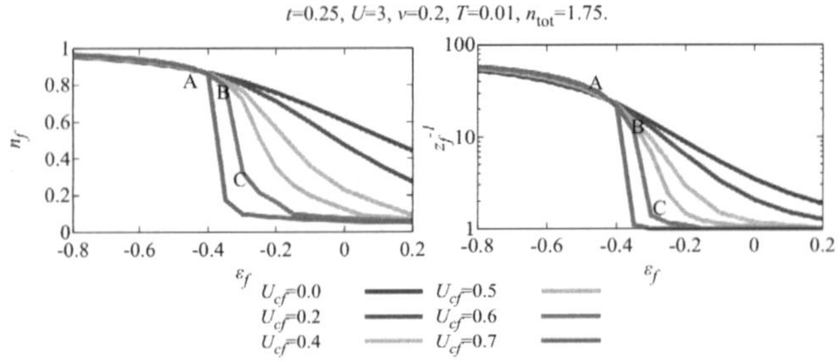

Figure 1. The ϵ_f dependences of n_f (left panel) and z_f^{-1} for several values of U_{cf}. The indices A, B and C represent three different regimes in the valence crossover.

surface around (π, π), while the c-electron forms an electron Fermi surface around $(0, 0)$. In regime B (center panel), the Fermi surface forms a slight concave bend around (π, π). The shrink of the area around (π, π) corresponds to the decrease of n_f and the increase of n_c, respectively. In regime C, the Fermi surfaces in both components change significantly as a result of the sudden valence crossover. In other words, the weight of the quasiparticle contributing to the Fermi surfaces is redistributed across the valence crossover. The Fermi surface shown in regime C is significantly different from that in regime B as a consequence of the sudden valence crossover. This significant change should be observed on both sides of the sudden valence crossover whenever the f-electron valence change considerably. Note that in regime C, the electron correlations are essentially weak, so that the Hubbard satellite peaks disappear in the high-energy single-particle spectrum as well. Using the same values of the f-electron density as in regimes B and C, the fluctuation exchange approach[12] gives a shape similar to that of the Fermi surfaces. This observation indicates that the significant change of the Fermi surface is ascribed to the sudden valence crossover.

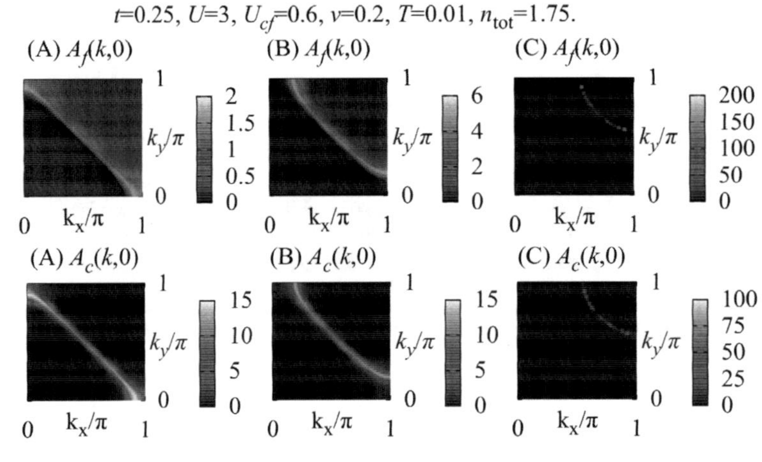

Figure 2. The single-particle excitation spectrum for c and f electrons at three different regimes in the valence crossover. Fermi surface in regime C is significantly different from that in regime B as a consequence of the sudden valence crossover.

Finally, to understand the residual interactions among quasiparticles, we show the structure

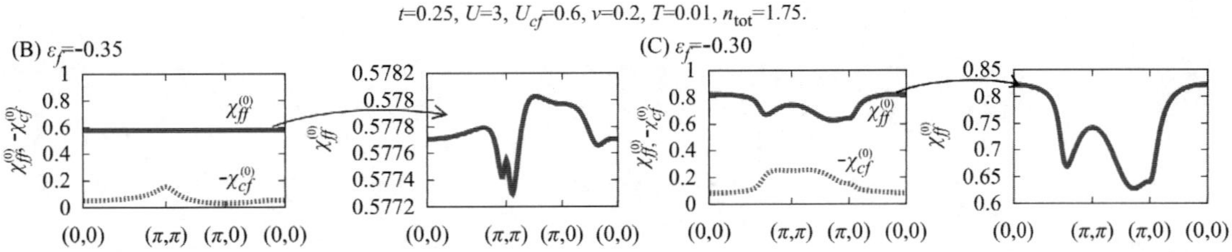

Figure 3. Structure of the irreducible susceptibilities, $\chi_{ff}^{(0)}(\boldsymbol{k},0)$ (solid line) and $\chi_{cf}^{(0)}(\boldsymbol{k},0)$ (dashed line) at regime B (left two panels) and at regime C (right two panels).

of the irreducible susceptibilities in Fig. 3. In regime B (left two panels), the very flat dispersion appears whose structure is reflected by the heavily renormalized quasiparticle bands. Across the valence crossover, the f-electron susceptibility becomes much more dispersive in regime C, reaching a maximum at $(0,0)$, which is expected from the small Fermi surface pockets around (π,π) (right panel in Fig. 2). The significant change in the structure of the irreducible susceptibility affects the residual interaction among quasiparticles, which must be taken into account to consider the stability of the superconductivity.

4. Summary

We have investigated the possibility of a significant change of the Fermi-surface topology and the corresponding structure of the effective interactions across the sudden valence crossover caused by a slight change in the parameters of the extended Anderson lattice model. Since the effective electron correlation for the renormalized quasiparticle in the heavy-fermion systems sensitively depends on the f-electron valence, the sudden change of the valence indeed results in the reformation of the quasiparticles and their $(\boldsymbol{k},\omega)$-structure in the irreducible susceptibility, yielding essentially weak-coupling behaviors.

Acknowledgments

This work was supported by a Grant-in-Aid for Scientific Research on Innovative Areas "Heavy Electrons" (No. 20102008) of The Ministry of Education, Culture, Sports, Science and Technology, Japan.

References

[1] Buschow K 1979 *Rep. Prog. Phys.* **42** 1373
[2] Batlogg B, Kaldis E, Schlegel A and Wachter P 1976 *Phys. Rev.* B **14** 5053
[3] Yuan H Q, Grosche F M, Deppe M, Geibel C, Sparn G and Steglich F 2003 *Science* **302** 2104
[4] Holmes A T, Jaccard D and Miyake K 2004 *Phys. Rev.* B **69** 024508
[5] Miyake K 2007 *J. Phys.: Condens. Mat.* **19** 125201
[6] Onishi Y and Mikaye K 2000 *Physica* B **281-282** 191; Onishi Y and Miyake Y 2000 *J. Phys. Soc. Jpn.* 043710
[7] Fujimori S, Okane T, Okamoto J, Mamiya K, Muramatsu Y, Fujimori A, Harima H, Aoki D, Ikeda S, Shishido H, Tokiwa Y, Haga Y and Ōnuki Y 2003 *Phys. Rev.* B **67** 144507; Fujimori S, Fujimori A, Shimada K, Narimura T, Kobayashi K, Namatame H, Taniguchi M, Harima H, Shishido H, Ikeda S, Aoki D, Tokiwa Y, Haga Y, and Ōnuki Y 2006 *Phys. Rev.* B **73** 224517
[8] Haga Y, Inada Y, Harima H, Oikawa K, Murakawa M, Nakawaki H, Tokiwa Y, Aoki D, Shishido H, Ikeda S, Watanabe N and Ōnuki Y 2001 *Phys. Rev.* B **63** 060503
[9] Georges A, Kotliar G, Krauth W and Rozenberg M J 1996 *Rev. Mod. Phys.* **68** 13
[10] Kajueter H and Kotliar G 1996 *Phys. Rev. Lett.* **77** 131
[11] Saiga Y, Sugibayashi T and Hirashima D S 2008 *J. Phys. Soc. Jpn.* **77** 114710
[12] Sugibayashi T, Saiga Y and Hirashima D S 2008 *J. Phys. Soc. Jpn.* **77** 024716

Journal of Physics: Conference Series **273** (2011) 012058 doi:10.1088/1742-6596/273/1/012058

Effect of pressure on transport properties of $CeIrIn_5$

Y Takaesu[1], N Aso[1,] , Y Tamaki[1], M Hedo[1], T Nakama[1], K Uchima[2], Y Ishikawa[3], K Deguchi[3], N K Sato[3]

[1]Faculty of Science, University of the Ryukyus, Nishihara, Okinawa 903-0213, Japan
[2]General Education, Okinawa Christian Junior College, Nishihara, Okinawa 903-0207, Japan
[3]Graduate School of Science, Nagoya University, Nagoya, Aichi 464-8601, Japan

E-mail: aso@phys.u-ryukyu.ac.jp

Abstract. Electrical resistivity and thermoelectric power S of a heavy-fermion superconductor $CeIrIn_5$ have been measured at temperatures from 2.0 K to 300 K under hydrostatic pressures up to 2.2 GPa. The thermoelectric power S exhibits a large positive value up to 90 V/K, which is characteristic of heavy-fermion systems. S also shows a sharp maximum in its temperature dependence and its maximum temperature $T_{S,\ \mathrm{max}}$ increases with pressure, while its maximum value is constant independent of pressure. These experimental results strongly indicate that the Kondo temperature of $CeIrIn_5$ increases by applying the pressure.

1. Introduction

The interplay between magnetism and superconductivity (SC) remains an interesting topic in condensed matter physics. It has long been believed that these two long-range ordered states would be antagonistic because of their competitive nature, i.e., the exclusion of magnetic induction by Meissner e ect and internal elds generated by magnetic orderings. According to investigations on heavy-fermion superconductors, however, antiferromagnetism (AFM) can coexist with SC, because the periodicity of AFM is usually much shorter than the SC coherence length, and hence the internal elds seen by itinerant electrons are canceled out. Actually, there is a growing list of heavy-fermion compounds the exhibiting coexistence of AFM and SC at ambient or high pressure [1, 2], however many questions remain to be resolved.

Members of the $CeTIn_5$ (T = Rh, Ir and Co) family with a tetragonal $HoCoGa_5$-type crystal structure [3], are ideal for studying the correlation between AFM and SC, because its transition temperatures (Neel temperature T_N and superconducting transition temperature T_c) can be controlled by the application of an external pressure and/or the chemical doping. At ambient pressure, $CeIrIn_5$ is a superconductor with T_c = 0.4 K and the electronic speci c heat = 700 mJ/(K^2mol) [4]. As the pressure increases, T_c passes through a maximum of 1.0 K around 2.5 GPa and then vanishes in the vicinity of 6 GPa [5]. Nuclear quadrupole resonance experiments [6] suggested that SC state has the line nodes of a superconducting energy gap, con rmed by the thermal conductivity measurements [7, 8].

Nowadays the measurements of S have been carried out extensively to study the ground state of the strongly correlated electrons systems and gave an important information concerning to the electronic density of state near the Fermi level. No measurements of S, however, have been

Published under licence by IOP Publishing Ltd

performed on a heavy-fermion superconductor $CeIrIn_5$. In this paper, we report S results of $CeIrIn_5$ for the rst time.

2. Experimental

The single crystals of $CeIrIn_5$ were grown by the In- ux method of combining stoichiometric amounts of Ce and Ir with excess In. They were cut by spark erosion to a size of about $1 \times 1 \times 3$ mm^3. The measurements of electrical resistivity ρ and thermoelectric power S under pressure were carried out simultaneously by using the standard four-probe dc method and the di erential method with a seesaw heating procedure, respectively. The thermoelectric power measurement technique of so-called "seesaw heating method" using two gradient heaters on both sides of the sample has been developed for the high sensitive and stable measurement under a strong magnetic eld [9]. For the S measurements under pressure, we employed the modi ed "seesaw heating method" with one gradient heater due to the limitation of the sample space [10]. Chromel-constantan thermocouples were used as the probes for thermoelectric power measurements because of the small temperature dependence of the pressure e ects [11]. The junctions of the thermocouples were made by spot welding, and glued directly to the sample by using Ag paste. With this experimental setup, both electrical current I and heat ow Q were applied in the same direction of a-axis of the crystals. The crystals were cooled down to 2.0 K by using a conventional liquid-^{4}He pumping cryostat. A clamp-type NiCrAl/CuBe hybrid piston cylinder pressure cell with a pressure transmitting media of Daphne oil 7373 [12] was utilized for producing hydrostatic pressures up to 2.2 GPa at low temperature.

3. Results and Discussion

Figure 1 shows the temperature dependence of the a-axis electrical resistivity ρ of $CeIrIn_5$ at the temperature range between 2.0 K and 300 K under pressure up to 1.5 GPa. The ρ of $LaIrIn_5$ at ambient pressure [13] has been subtracted as the phonon backgrounds. One can recognize that ρ exhibits the typical dense Kondo behavior: on cooling from room temperature at ambient pressure, the ρ increases and shows a maximum of the coherence peak at $T_{\rho,\ \mathrm{max}} = 42.2$ K, and then goes down into the coherent Kondo state. Inelastic neutron scattering measurements of $CeIrIn_5$ show the Kondo temperature T_{K} is 56 K [14], which is comparable to $T_{\rho,\ \mathrm{max}}$. One can also nd that the coherence peak moves to higher temperatures up to 86.4 K for 1.5 GPa with increasing P.

In the Fermi liquid metals, ρ varies as $\rho = \rho_0 + AT^n$ ($n = 2$) at low temperatures. In the case of $CeIrIn_5$ n is estimated to be 1.0 at ambient pressure, indicating the so-called non-Fermi liquid behavior. The n value, shown in the inset of Fig. 1, increases linearly with pressure, exhibiting the gradual recovery of Fermi liquid behavior. It should be noted that the overall features in the ρ results are in excellent agreement with the previous results [4, 15].

Figure 2 illustrated the temperature dependence of the a-axis thermoelectric power S of $CeIrIn_5$. Compared with a simple metal, the S value is found to be very large up to 90 μV/K, which is characteristic of heavy-fermion systems [16]. All the S data presented in Fig. 2 have positive values and then there is a pronounced maximum in each pressure data. Cox and Grewe calculated the thermoelectric power of the Anderson lattice model by using the Average T-matrix Approximation [17]. They showed not only a development of a sharp peak near Fermi level in the electronic density of state but also a maximum in the temperature dependence of the thermoelectric power at $T = T_{\mathrm{K}}/3$. We should remark that the temperature dependence of the S results is qualitatively similar to their theory.

The maximum peak at $T_{S,\ \mathrm{max}}$ moves to higher temperatures from 26 K (at 0 GPa) to 62 K (at 2.2 GPa) by applying the pressure, while its maximum value remains close to the ambient pressure value. The inset of Fig. 2 shows the thermoelectric power normalized to its maximum value, S/S_{max}, is plotted against the temperature normalized to its maximum peak

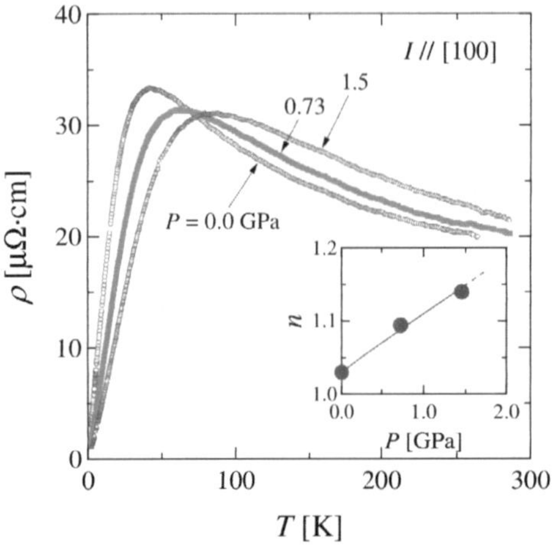

Figure 1. The temperature dependence of the a-axis electrical resistivity of $CeIrIn_5$ after subtraction of the phonon backgrounds. The inset shows the pressure dependence of n obtained from the low temperature electrical resistivity. See text for details.

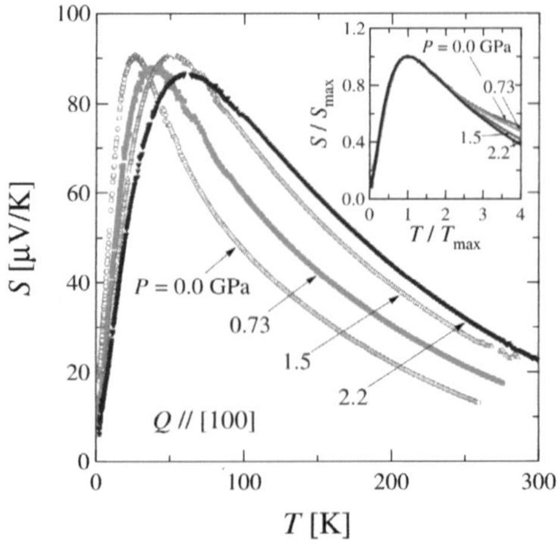

Figure 2. The temperature dependence of the a-axis thermoelectric power of $CeIrIn_5$ as a function of P. The inset shows the pressure dependence of the normalized thermoelectric power vs the normalized temperature. See text for details.

temperature, $T/T_{S,\ \mathrm{max}}$. One can clearly recognize its pressure-independent behavior below $T/T_{S,\ \mathrm{max}}$ 2.2, which strongly illustrates that the thermoelectric power at temperature near $T_{S,\ \mathrm{max}}$ is determined by the one energy parameter. Obtained values for $T_{\rho,\ \mathrm{max}}$, n, $T_{S,\ \mathrm{max}}$, and $T_{\rho,\ \mathrm{max}}/T_{S,\ \mathrm{max}}$ as a function of pressure are summarized in Table 1. In particular we stress that the ratio of $T_{\rho,\ \mathrm{max}}/T_{S,\ \mathrm{max}}$ are found to be a pressure-independent constant value around 1.6. These experimental ndings strongly support that the increase of $T_{S,\ \mathrm{max}}$ is due to that of T_{K} by applying the pressure.

In the low temperature region below $T/T_{S,\ \mathrm{max}}$ 0.2, there tends to have slight di erence in the pressure dependence of S/S_{max}, We suppose that these di erence might have the correlation with the quantum criticality of this system. To clarify this phenomena in detail, we need to perform the S measurements at lower temperature as a function of pressure.

In summary, we measure electrical resistivity ρ and thermoelectric power S of a heavy-fermion superconductor $CeIrIn_5$ at temperatures from 2.0 K to 300 K under hydrostatic pressures up to 2.2 GPa. S exhibits a large positive value up to 90 μV/K, which is characteristic of heavy-fermion systems. S also shows a sharp maximum in its temperature dependence and then its maximum temperature $T_{S,\ \mathrm{max}}$ increases with pressure, while its maximum value is constant independent of pressure. These experimental results indicate that the Kondo temperature of $CeIrIn_5$ increases by applying the pressure.

Acknowledgements
This work was supported by a Grant-in-Aid for Scienti c Research on Innovative Areas "Heavy Electrons" (No. 21102519) of The Ministry of Education, Culture, Sports, Science, and Technology, Japan.

Table 1. Values for $T_{\rho,\ \max}$, n, $T_{S,\ \max}$, and $T_{\rho,\ \max}/T_{S,\ \max}$.

P [GPa]	$T_{\rho,\ \max}$ [K]	n	$T_{S,\ \max}$ [K]	$T_{\rho,\ \max}/T_{S,\ \max}$
0.0	42.2	1.03	26.0	1.62
0.73	59.8	1.09	38.3	1.57
1.5	86.4	1.14	51.6	1.67
2.2	—	—	61.6	—

References

[1] Mathur N D, Grosche F M, Julian S R, Walker I R, Freye D M, Haselwimmer R K W, and Lonzarich G G 1998 *Nature* **394** 39
[2] Bauer E, Hilscher G, Michor H, Paul Ch, Scheidt E W, Gribanov A, Seropegin Yu, Noël H, Sigrist M, and Rogl P 2004 *Phys. Rev. Lett.* **92** 027003
[3] Moshopoulou E G, Fisk Z, Sarrao J L, and Thompson J D 2001 *J. Solid State Chem.* **158** 25
[4] Petrovic C, Movshovich R, Jaime M, Pagliuso P G, Hundley M F, Sarrao J L, Fisk Z and Thompson J D 2001 *Europhys Lett.* **53** 354
[5] Muramatsu T, Kobayashi T C, Shimizu K, Amaya K, Aoki D, Haga Y, Onuki Y 2003 *Physica C* **388-389** 539
[6] Kohori Y, Yamato Y, Iwamoto Y, and Kohara T 2000 *Eur. Phys. J. B* **18** 601
[7] Shakeripour H, Tanatar M A, Li S Y, Petrovic C, and Taillefer L 2007 *Phys. Rev. Lett.* **99** 187004
[8] Kasahara Y, Iwasawa T, Shimizu Y, Shishido H, Shibauchi T, Vekhter I, and Matsuda Y 2008 *Phys. Rev. Lett.* **100** 207003
[9] Resel R, Gratz E, Burkov A T, Nakama T, Higa M and Yagasaki K 1996 *Rev. Sci. Instrum.* **67** 1970
[10] Hedo M, Nakamura D, Takaesu Y, Fujiwara T, Uchima K, Yagasaki K and Nakama T 2010 *J. Phys.: Conf. Series* **215** 012186
[11] Choi E S, Kang H, Jo Y J and Kang W 2002 *Rev. Sci, Instrum.* **73** 2999
[12] Yokogawa K, Murata K, Yoshino H and Aoyama S 2007 *Jpn. J. Appl. Phys.* **46** 3636
[13] Shishido H, Settai R, Aoki D, Ikeda S, Nakawaki H, Nakamura N, Iizuka T, Inada Y, Sugiyama K, Takeuchi T, Kindo K, Kobayashi T C, Haga Y, Harima H, Aoki Y, Namiki T, Sato H and Onuki Y 2002 *J. Phys. Soc. Jpn.* **71** 162
[14] Christianson A D, Bauer E D, Lawrence J M, Riseborough P S, Moreno N O, Pagliuso P G, Sarrao J L, Thompson J D, Goremychkin E A, Trouw F R, Hehlen M P, and McQueeney R J 2004 *Phys. Rev. B* **70** 134505.
[15] Nakajima Y, Shishido H, Nakai H, Shibauchi T, Hedo M, Uwatoko Y, Matsumoto T, Settai R, Onuki Y, Kontani H, and Matsuda Y 2008 *Phys. Rev. B* **77** 214504
[16] Behnia K, Jaccard D and Flouquet J 2004 *J. Phys.: Condens. Matter* **16** 5187
[17] Cox D L and Grewe N 1988 *Z. Phys. B* **71** 321

Metamagnetic behavior in a heavy fermion compound $YbCo_2Zn_{20}$

Tetsuya Takeuchi[1], Masahiro Ohya[2], Shingo Yoshiuchi[2], Masaki Matsushita[2], Fuminori Honda[2], Rikio Settai[2], and Yoshichika Ōnuki[2]

[1] Low Temperature Center, Osaka University, Toyonaka, Osaka 560-0043, Japan
[2] Graduate School of Science, Osaka University, Toyonaka, Osaka 560-0043, Japan

E-mail: takeuchi@cqst.osaka-u.ac.jp

Abstract. We studied an electronic state at low temperatures by measuring the specific heat and ac-susceptibility under magnetic fields for a heavy fermion compound $YbCo_2Zn_{20}$. The magnetic specific heat in the form of $C_{\rm mag}/T$ increases steeply and exhibits an extremely large value 8 J/(K^2·mol) below 0.2 K, which corresponds to an electronic specific heat coefficient with a very low Kondo temperature of 0.2-1 K. The magnetic entropy increases rapidly at higher temperatures up to 10 K and reaches $R\ln 8$ at 50-60 K, which corresponds to a small crystalline-electric-field (CEF) splitting energy. In addition, we found a metamagnetic anomaly at $H_{\rm m} = 5.7$ kOe below the characteristic temperature $T_{\chi_{\rm max}} = 0.32$ K, where the temperature dependence of the ac-susceptibility shows a broad peak. Interestingly, the metamagnetic behavior was observed as the double peak at 4.0 and 7.5 kOe in the magnetic field dependence of the specific heat C/T at 95 mK.

1. Introduction

Most of the cerium or uranium compounds order antiferromagnetically at low temperatures on the basis of the Ruderman-Kittel-Kasuya-Yosida (RKKY) interaction. On the other hand, some cerium or uranium compounds, such as $CeCu_6$, $CeRu_2Si_2$ and UPt_3, exhibit no long-range magnetic ordering [1]. The magnetic susceptibility of these compounds increases with decreasing temperature, following the Curie-Weiss law at high temperatures, and shows a maximum at a characteristic temperature $T_{\chi_{\rm max}}$. Below $T_{\chi_{\rm max}}$, the susceptibility decreases weakly or becomes temperature-independent, and the f-electron system is changed into a new electronic state, namely the heavy fermion state. The almost localized f-electrons at high temperatures become itinerant at temperatures lower than $T_{\chi_{\rm max}}$ via the many-body Kondo effect. The crossover from "localized state" to "itinerant state" occurs at a characteristic temperature $T_{\chi_{\rm max}}$, which approximately corresponds to the Kondo temperature $T_{\rm K}$.

One of the characteristic features in the heavy fermion compounds is a metamagnetic behavior or an abrupt nonlinear increase of magnetization at the magnetic field $H_{\rm m}$ below $T_{\chi_{\rm max}}$. The metamagnetic behavior appears at $H_{\rm m} = 77$ kOe in $CeRu_2Si_2$ [2], for example. Recently, a new family of Yb-based heavy fermion compounds YbT_2Zn_{20} (T = Fe, Co, Ru, Rh, Os, and Ir) with the cubic $CeCr_2Al_{20}$-type ($Fd\bar{3}m$ space group) structure has been reported [3]. Among them, we found the metamagnetic behavior at $H_{\rm m} \simeq 100$ kOe in $YbIr_2Zn_{20}$ [4]. The metamagnetic anomaly is accompanied by a large volume contraction $\Delta V/V \simeq -3 \times 10^{-4}$ at 120 kOe [5]. The

Published under licence by IOP Publishing Ltd

metamagnetic behavior is expected to occur even in Yb-based compounds on the basis of the hole-electron analogy between $4f^{13}$-Yb^{3+} and $4f^1$-Ce^{3+} electronic configuration.

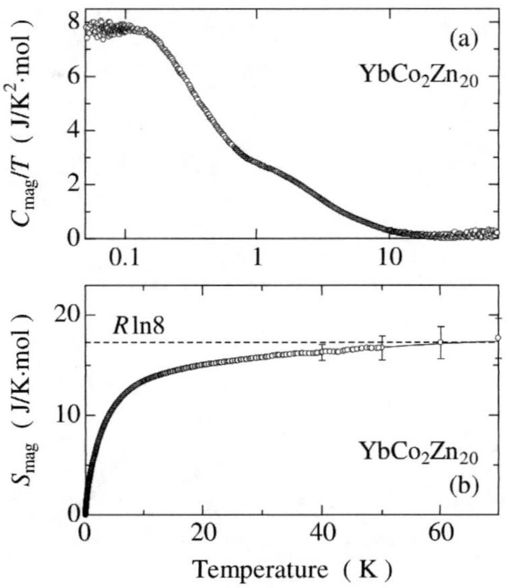

Figure 1. Temperature dependence of (a) the magnetic specific heat in the form of C_{mag}/T in a logarithmic temperature scale and (b) the magnetic entropy S_{mag} of $YbCo_2Zn_{20}$.

We continued our experimental study for a similar compound $YbCo_2Zn_{20}$. The following characteristic features were already found in the previous works [3, 6]. The Kondo temperature in $YbCo_2Zn_{20}$ is expected to be much smaller than that in $YbIr_2Zn_{20}$ with $T_{\chi_{max}} = 7.4$ K because the magnetic susceptibility in $YbCo_2Zn_{20}$ increases continuously with decreasing temperature, following the Curie-Weiss law, down to the measured lowest temperature 2 K [3]. The corresponding electronic specific heat coefficient is extremely large; $\gamma \simeq 8$ J/(K^2·mol) at low temperatures. The pressure experiment was also carried out, and the non-magnetic electronic state was found to be changed into a magnetically ordered state at pressures larger than the quantum critical pressure $P_c \simeq 1$ GPa [6]. In this work, we studied the low-temperature electronic state in single crystalline $YbCo_2Zn_{20}$ by measuring the specific heat and ac-susceptibility under magnetic fields.

2. Experimental results and discussion

Single crystals of $YbCo_2Zn_{20}$ were grown by the Zn-self flux method as in $YbIr_2Zn_{20}$ [4]. Figures 1(a) and 1(b) show temperature dependences of the magnetic specific heat in the form of C_{mag}/T in a logarithmic temperature scale and the magnetic entropy S_{mag}, respectively. Here the magnetic specific heat C_{mag} was obtained by subtracting the temperature dependence of the specific heat of $LuCo_2Zn_{20}$ from that of $YbCo_2Zn_{20}$: $C_{mag} = C(YbCo_2Zn_{20}) - C(LuCo_2Zn_{20})$. The temperature dependence of C_{mag}/T increases steeply below about 1 K and saturates below 0.2 K. An extremely large value $C_{mag}/T \simeq 8$ J/K^2·mol corresponds to the γ value, which is based on a very low Kondo temperature of 0.2-1 K. These results are consistent with the previous works [3]. S_{mag}, which was obtained by integrating C_{mag}/T over temperature, increases very rapidly at higher temperatures up to 10 K and reaches $R \ln 8$ at 50-60 K, which corresponds to the crystalline-electric-field (CEF) splitting energy of Yb^{3+} state. The present small CEF splitting energy in $YbCo_2Zn_{20}$ is compared with the rather large CEF splitting energy in $YbIr_2Zn_{20}$: 0 (doublet)-43.5 (doublet)-106 K (quartet) [5].

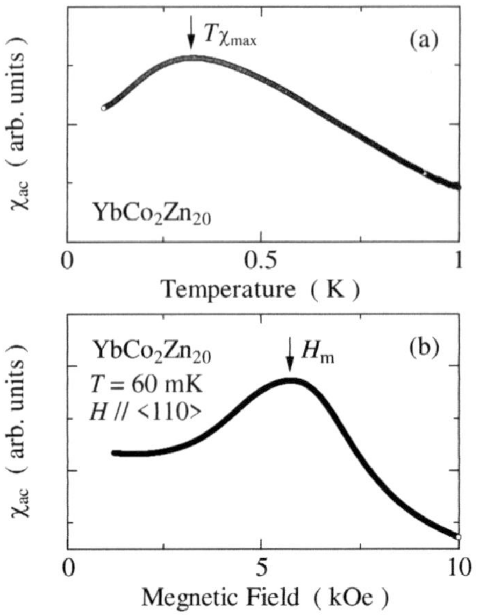

Figure 2. (a) Temperature dependence of the ac magnetic susceptibility χ_{ac} at $H = 0$ and (b) the magnetic field dependence of χ_{ac} at 60 mK for $H \parallel \langle 110 \rangle$ in $YbCo_2Zn_{20}$.

The temperature dependence of magnetic susceptibility χ follows the Curie-Weiss law down to 2 K, with the paramagnetic Curie temperature $\Theta = -4.3$ K and the effective magnetic moment $\mu_{eff} = 4.3\ \mu_B$/Yb [3]. For further decreasing temperature, it was found that $\chi(T)$ indicates a broad peak at $T_{\chi_{max}} = 0.32$ K, as shown in Fig. 2(a). From this result, the metamagnetic behavior is expected to occur below $T_{\chi_{max}}$. In fact, we found a metamagnetic behavior at $H_m = 5.7$ kOe in the magnetic field dependence of χ_{ac} at 60 mK, as shown in Fig. 2(b). The metamagnetic behavior was observed roughly below $T_{\chi_{max}} = 0.32$ K. $T_{\chi_{max}}$ approximately corresponds to the Kondo temperature and is consistent with the value of 0.2-1 K mentioned above. In the previous paper, we proposed the relation between the metamagnetic field H_m (kOe) and the characteristic temperature $T_{\chi_{max}}$ (K) in the heavy fermion compounds as H_m (kOe) $= 15T_{\chi_{max}}$ (K) [5]. The observed values of $H_m = 5.7$ kOe and $T_{\chi_{max}} = 0.32$ K almost satisfy this relation.

The metamagnetic behavior was also studied by the specific heat measurement. Figure 3 shows the magnetic field dependence of the specific heat in the form of C/T at 95 and 200 mK for $H \parallel \langle 100 \rangle$. Here the metamagnetic field H_m, which was determined by the ac-susceptibility measurements mentioned above, is indicated by a broken vertical line. At 95 mK, C/T shows distinct two peaks at about 4.0 and 7.5 kOe, as indicated by arrows, and the magnetic field, where C/T shows a minimum, corresponds to the metamagnetic field H_m. With increasing temperature, the peak at higher(lower) magnetic field shifts to higher (lower) fields, as indicated by arrows for the data at 200 mK. It is interesting that H_m is roughly located between the double peak of $C(H)/T$. The double peak in C/T was observed below $T_{\chi_{max}} = 0.32$ K. The similar feature of $C(H)/T$ was also observed in the typical heavy fermion compound $CeRu_2Si_2$ around the metamagnetic field of 80 kOe [7].

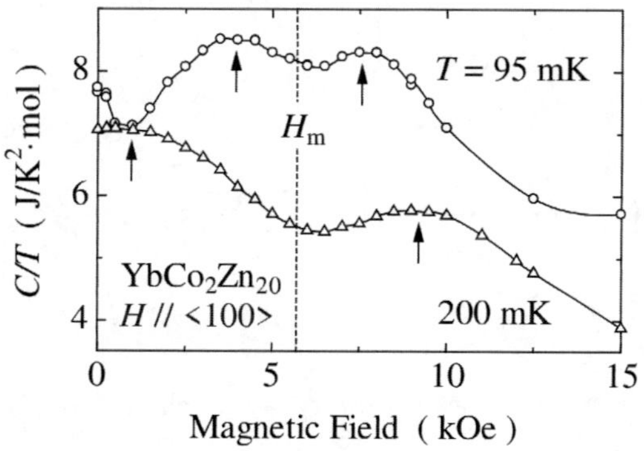

Figure 3. Magnetic field dependence of the specific heat in the form of C/T measured at 95 and 200 mK for $H \parallel \langle 100 \rangle$ in $YbCo_2Zn_{20}$.

3. Conclusion

The metamagnetic behavior was observed in the field dependence of ac-susceptibility and specific heat at $H_m = 5.7$ kOe, with the double peak in the specific heat. This is realized on the basis of a very low Kondo temperature of 0.2-1 K, where the ac-susceptibility indicates a peak at $T_{\chi_{max}} = 0.32$ K. These results indicate that the electronic state of $YbCo_2Zn_{20}$ is located in the vicinity of the quantum critical point.

Acknowledgments

This work was supported by a Grant-in-Aid for Scientific Research on Specially Promoted Research (No. 20001004), the Osaka University Global COE Program "Core Research and Engineering of Advanced Materials-Interdisciplinary Education Center for Materials Science" (No. G10) and Grant-in-Aid for Scientific Research on Innovative Areas "Heavy Electrons"(20102002, 21102511), Scientific Research (B)(21340102), Scientific Research for Young Scientist (B) (21740256) from the Ministry of Education, Culture, Sports, Science and Technology, Japan. This work (partly) was also supported by the Sasagawa Scientific Research Grant (21-231) from the Japan Science Society.

References

[1] Y. Ōnuki, R. Settai, K. Sugiyama, T. Takeuchi, T. C. Kobayashi, Y. Haga, and E. Yamamoto: J. Phys. Soc. Jpn. **73** (2004) 769.
[2] P. Haen, J. Flouquet, F. Lapierre, and P. Lejay: J. Low Temp. Phys. **67** (1987) 391.
[3] M. S. Torikachvili, S. Jia, E. D. Mun, S. T. Hannahs, R. C. Black, W. K. Neils, Dinesh Martien, S. L. Bud'ko, and P. C. Canfield: Proc. Natl. Acad. Sci. U.S.A. 104 (2007) 9960.
[4] S. Yoshiuchi, M. Toda, M. Matsushita, S. Yasui, Y. Hirose, M. Ohya, K. Katayama, F. Honda, K. Sugiyama, M. Hagiwara, K. Kindo, T. Takeuchi, E. Yamamoto, Y. Haga, R. Settai, T. Tanaka, Y. Kubo, and Y. Ōnuki: J. Phys. Soc. Jpn. **78** (2009) 123711.
[5] T. Takeuchi, S. Yasui, M. Toda, M. Matsushita, S. Yoshiuchi, M. Ohya, K. Katayama, Y. Hirose, N. Yoshitani, F. Honda, K. Sugiyama, M. Hagiwara, K. Kindo, E. Yamamoto, Y. Haga, T. Tanaka, Y. Kubo, R. Settai, and Y. Ōnuki: to be published in JPSJ.
[6] Y. Saiga, K. Matsubayashi, T. Fujiwara, M. Kosaka, S. Katano, M. Hedo, T. Matsumoto, and Y. Uwatoko: J. Phys. Soc. Jpn. **77** (2008) 053710.
[7] S. Holtmeier, P. Haen, A. Lacerda, P. Lejay, J. L. Tholence, J. Voiron, and J. Flouquet: Physica B **204** (1995) 250.

Journal of Physics: Conference Series **273** (2011) 012060 doi:10.1088/1742-6596/273/1/012060

de Haas-van Alphen effect in $SmOs_4P_{12}$

S. Tanaka[1], M. Sakoda[1], E. Matsuoka[1], H. Sugawara[1], R. Settai[2] and Y. Ōnuki[2]

[1] Department of Physics, Graduate School of Science, Kobe University, Kobe 657-8501, Japan
[2] Department of Physics, Graduate School of Science, Osaka University, Osaka 560-0043, Japan

E-mail: sugawara@crystal.kobe-u.ac.jp

Abstract. We have succeeded in growing single crystals of $SmOs_4P_{12}$ and measured the de Haas-van Alphen effect. An almost spherical Fermi surface, whose cyclotron effective masses are ranging 2.0-2.3 m_0, has been observed.

1. Introduction

The filled skutterudite compounds RT_4X_{12} (R= rare earth, T= Fe, Ru, Os, and X=pnictogen) have attracted much attention because of the wide variety of transport and magnetic properties, such as unconventional superconductivity, Kondo semiconducting behavior, metal-insulator (M-I) transition, multipole ordering, and heavy fermion (HF) behavior [1]. Among the filled skutterudites, Sm-based compounds have been most intensively investigated next to the Pr-based system, because some of them exhibit unprecedented features, such as the first Sm-based HF ferromagnetism in $SmFe_4P_{12}$ [2], the new type of multipolar ordered state in $SmRu_4P_{12}$ [3, 4], and the HF behavior robust against magnetic fields in $SmOs_4Sb_{12}$ with a ferromagnetic ground state [5]. In order to understand the mechanism of unusual properties, the information of physical properties in related compounds may be helpful.

$SmOs_4P_{12}$ was reported to be an antiferromagnetic (AFM) metal with a Néel temperature $T_N = 4.6$ K [6]. The temperature dependence of magnetic susceptibility $\chi(T)$ shows a distinct cusp at T_N, which is a typical feature of AFM transition, although the $\chi(T)$ shows upturn below ~ 3.5 K. The specific heat shows a pronounced λ-type peak at T_N that agrees with temperatures of the cusp in $\chi(T)$. The magnetic entropy estimated in zero field reaches R ln4 at T_N, which suggests that the crystalline electric field (CEF) ground state is a quartet Γ_{67} [6, 7]. Meanwhile the temperature dependence of electrical resistivity $\rho(T)$ shows Kondo like anomaly; the resistivity increases logarithmically with decreasing temperature from far above T_N, which is common feature of most Sm-based skutterudites [8].

In the previous works on $SmOs_4P_{12}$, the physical properties were investigated on the polycrystalline samples synthesized by the high-pressure technique. Recently, we have succeeded in growing single crystals of $SmOs_4P_{12}$ for the first time. In this work, we report on the measurements of the de Haas-van Alphen (dHvA) effect to deepen the understanding of electronic state.

Published under licence by IOP Publishing Ltd

2. Experiment

Single crystals of $SmOs_4P_{12}$ were grown by a tin-flux method which is basically the same as described in elsewhere. The raw materials were 3N (99.9% pure)-Sm, 4N-Os, 6N-P and 5N-Sn. The size of single crystals (~ 0.2 mm in length) are rather smaller than those of other ROs_4P_{12}. Crystal structure of the single crystal was confirmed by a back Laue X-ray diffraction method. The dHvA experiment was performed by the torque method using the micro-cantilever with the modulation field of $h \simeq 100\,\mathrm{Oe}$ (a modulation frequency of $\omega/2\pi = 11\,\mathrm{Hz}$) in a top loading dilution refrigerator with a 17 T superconducting magnet [9, 10].

3. Results and discussion

Figure 1(a) shows a typical recorder trace of the dHvA oscillations and Fig. 1(b) shows its fast Fourier transformation (FFT) spectrum at 59 mK for $H \| [001]$ in $SmOs_4P_{12}$.

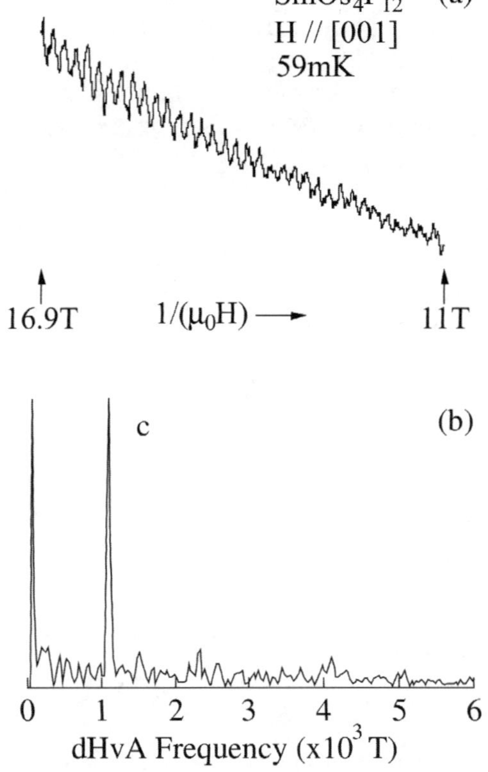

Figure 1. (a) Typical dHvA oscillations and (b) its FFT spectrum in $SmOs_4P_{12}$.

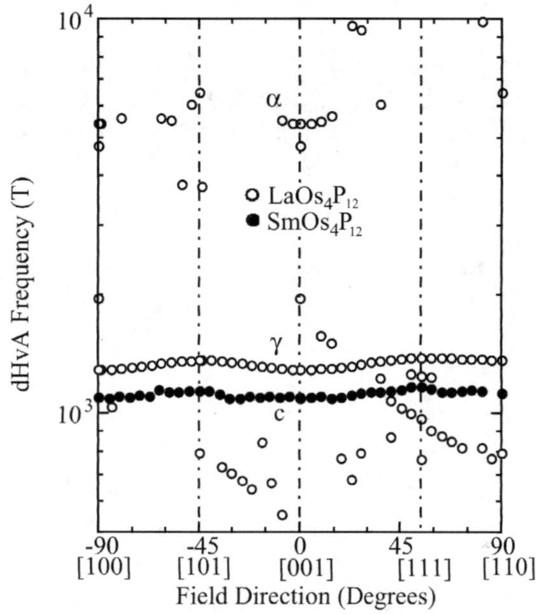

Figure 2. Comparison of the angular dependence of dHvA frequencies between $LaOs_4P_{12}$(○) and $SmOs_4P_{12}$(●). Data of $LaOs_4P_{12}$ are sited from Ref. [11].

One dHvA branch named as c with the dHvA frequency $F \sim 1.1 \times 10^3$ T was observed over the whole field angles with a slight angular dependence, indicating almost spherical Fermi surface (FS). The angular dependence is almost the same as that of γ-branch in $LaOs_4P_{12}$ except the $\sim 17\%$ smaller magnitude [11]. The γ-branch in $LaOs_4P_{12}$ is well identified as the 47th-band FS by the band structure calculation [12]. The closeness of the FS to that in $LaOs_4P_{12}$ suggests the localized nature of $4f$-electrons in $SmOs_4P_{12}$, that is consistent with the trivalent state of Sm ions suggested by the result of magnetic susceptibility [6]. However, any other branches found in $LaOs_4P_{12}$ were not observed. Such a result is the same as that in $SmFe_4P_{12}$ [13], in which

Table 1. Comparison of the dHvA frequency F and the cyclotron effective mass m_c^* between $LaOs_4P_{12}$ (γ-branch) and $SmOs_4P_{12}$ (c-branch). Data of $LaOs_4P_{12}$ are sited from Ref. [11].

	$H\|\|[101]$		$H\|\|[001]$		$H\|\|[111]$	
	$F(\times 10^3$ T)	$m_c^*(m_0)$	$F(\times 10^3$ T)	$m_c^*(m_0)$	$F(\times 10^3$ T)	$m_c^*(m_0)$
$LaOs_4P_{12}$	1.37	1.3	1.29	1.1	1.39	1.2
$SmOs_4P_{12}$	1.14	2.3	1.10	2.2	1.17	2.0

the multiply-connected 48th-band FS with heavy mass of $\sim 59m_0$ was expected. Differently from $SmFe_4P_{12}$, $SmOs_4P_{12}$ orders antiferromagnetically below 4.6 K. The $H-T$ phase diagram was determined by the specific heat measurements under the magnetic fields [6, 3], in which the AFM ordered state survives up to 8 T at 4 K. Therefore, it is inferred that the topology of FS in paramagnetic state changes in the AFM state due to the magnetic Brillouin zone. In the present dHvA experiments, we observed an evident metamagnetic anomaly at the limited filed angles ($52° \sim 61°$) around $H\|[111]$ (not shown); e.g. the metamagnetic anomaly was observed at $H_{\rm m} = 13.2$ T for field along the $H\|[111]$ at 97 mK. However, the dHvA spectrum shows no essential change below and above $H_{\rm m}$ except the enhancement of amplitude above $H_{\rm m}$, suggesting that the reconstruction of magnetic Brillouin zone is not the essential reason for undetecting 48th-band FS in $SmOs_4P_{12}$.

From the temperature dependence of the dHvA amplitude, we have determined the cyclotron effective mass m_c^* for the three principal directions as shown in Fig. 3. The m_c^* of $SmOs_4P_{12}$

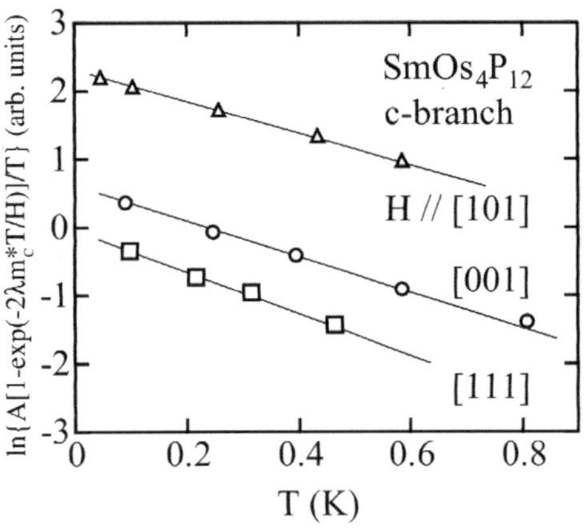

Figure 3. The semi-logarithmic plot of the dHvA amplitude A vs. temperature for c-branch in $SmOs_4P_{12}$. λ in the vertical-axis label is a constant $\lambda = 2\pi^2 ck_{\rm B}/e\hbar$.

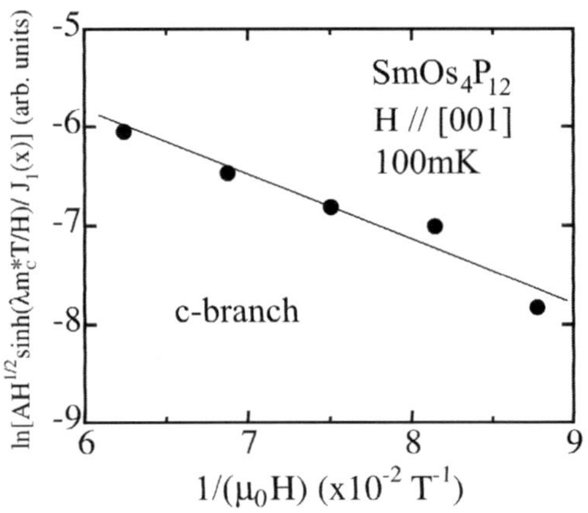

Figure 4. Dingle plot for c-branch in $SmOs_4P_{12}$. $J_1(x)$ in the vertical-axis label is the Bessel function with $x = 2\pi Fh/H^2$ due to the modulation method.

is roughly twice as large as that of $LaOs_4P_{12}$ as shown in Table 1. Such a enhancement of m_c^* is frequently observed in the magnetic materials and ascribable to the electron-magnon interaction [14]. In $LaOs_4P_{12}$ the larger FS (α-brach) with $m_c^* = 4.7m_0$, which corresponds to the open orbit of 48th-band FS, was observed around $H\|[001]$. If we simply estimate the effective mass of 48th-band FS for $H\|[001]$ in $SmOs_4P_{12}$ assuming the same mass enhancement; i.e, $(m_c^*)_{\rm Sm48th}/(m_c^*)_{\rm La48th}=(m_c^*)_{\rm Sm47th}/(m_c^*)_{\rm La47th}$, the larger mass of $m_c^* \sim 9.4m_0$ is expected in $SmOs_4P_{12}$, which may marginal value for detecting the dHvA signal in the present experimental condition.

We have also determined the Dingle temperature $T_{\mathrm{D}}(=\hbar/2\pi k_{\mathrm{B}}\tau)$, which is inversely proportional to the scattering lifetime τ, from the field dependence of dHvA amplitude as shown in Fig. 4. The Dingle temperature is T_{D} =2.0 K for c-branch from which we can estimate the mean free path $\ell = 580$Å using the relations: $S_{\mathrm{F}} = \pi k_{\mathrm{F}}^2, \hbar k_{\mathrm{F}} = m_c^* v_{\mathrm{F}}$, and $\ell = v_{\mathrm{F}}\tau$, where k_{F} is half of the caliper dimension of a circular S_{F} and v_{F} is the Fermi velocity. For the comparison, we have also estimated the Dingle temperature and mean free path in $LaOs_4P_{12}$ for the similar condition at 100 mK. The estimated values are T_{D} =0.093 K and $\ell = 27000$Å, respectvely. The mean free path in $SmOs_4P_{12}$ is roughly 1/50 of that in $LaOs_4P_{12}$, that is also the reason why the 48th-band FS could not be detected in $SmOs_4P_{12}$. Taking into account the same manner of sample preparation, such a large difference of mean free path between these compounds is not understantable only from the difference of sample quality due to the impurity and/or defect in lattice. The magnetic scattering associated with Kondo like behavior in $\rho(T)$ may be a possible origin for the smaller mean free path in $SmOs_4P_{12}$.

In summary, we have succeeded in observing the dHvA effect in $SmOs_4P_{12}$. An almost spherical FS, which is corresponding to the 47th-band hole FS, was observed. In order to understad the electric and magnetic state further, the measurements under higher magnetic fields including anisotropy are necessary.

Acknowledgments

This work was partially supported by a Grant-in-Aid for Scientific Research on Priority Areas "Skutterudite" (No. 15072206), "New Materials Science Using Regulated Nano Spaces-Strategy in Ubiquitous Elements" (No. 20045008), Innovative Areas "Heavy Electrons" (No. 20102002), and Scientifc Research (C) (No. 20540358) from the Ministry of Education, Culture, Sports, Science and Technology, Japan.

References

[1] Sato H, Sugawara H, Aoki Y and Harima H, 2009 *Handbook of Magnetic Materials* vol 18, ed K H J Buschow (Amsterdam: Elsevier) p 1.
[2] Takeda N and Ishikawa M 2003 *J. Phys. Condens. Matter* **15** L229.
[3] Matsuhira M, Hinatsu Y, Sekine C, Togashi T, Maki H, Shirotani I, Kitazawa H, Takamatsu T, Kido G 2002 *J. Phys. Soc. Jpn.* **71** (Suppl.) 237.
[4] Yoshizawa M, Nakanishi Y, Oikawa M, Sekine C, Shirotani I, Saha S R, Sugawara H, Sato H 2005 *J. Phys. Soc. Jpn.* **74** 2141.
[5] Sanada S, Aoki Y, Aoki H, Tsuchiya A, Kikuchi D, Sugawara H, Sato H 2005 *Phys. Soc. Jpn.* **74** 246.
[6] Giri R, Sekine C, Shimaya Y, Shirotani I, Matsuhira K, Doi Y, Hinatsu Y, Yokoyama M, Amitsuka H 2003 *Physica B* **329-333** 458.
[7] Matsuhira K, Doi Y, Wakeshima M, Hinatsu Y, Amitsuka H, Shimaya Y, Giri R, Sekine C, Shirotani I 2005 *J. Phys. Soc. Jpn.* **74** 1030.
[8] Sato H, Kikuchi D, Tanaka K, Aoki H, Kuwahara K, Aoki Y, Kohgi M, Sugawara H, Iwasa K 2007 *J. Magn. Magn. Mater.* **310** 188.
[9] Settai R, Kawai T, Endo T, Muranaka H, Doi Y, Ōnuki Y and Harima H 2008 *J. Phys. Soc. Jpn.* **77** (Suppl. A) 345.
[10] Kawai T, Muranaka H, Endo T, Dung N D, Doi Y, Ikeda S, Matsuda T D, Haga Y, Harima H, Settai R and Ōnuki Y 2008 *J. Phys. Soc. Jpn.* **73** 064717.
[11] Sugawara H, Iwahashi Y, Magishi K, Saito T, Koyama K, Harima H, Kikuchi D, Sato H, Endo T, Settai R, Ōnuki Y 2009 *Phys. Rev. B* **79** 035104.
[12] Harima H and Takegahara K 2008 *Physica B* **403** 906.
[13] Kikuchi D, Sugawara H, Tanaka K, Aoki H, Kobayashi M, Sanada S, Kuwahara K, Aoki Y, Shishido H, Settai R, Ōnuki Y, Harima and H, Sato H 2008 *J. Phys. Soc. Jpn.* **77** 114705.
[14] Ōnuki Y and Hasegawa A 1995 *Hand book on the Physics and Chemistry of Rare Earth,* vol 20, chap 135, ed. K A Gschneidner Jr and L Eyring (Amsterdam, Elsevier) p 1.

Journal of Physics: Conference Series **273** (2011) 012061 doi:10.1088/1742-6596/273/1/012061

Heavy Fermion Behavior in the New Antiferromagnetic Compound UIr_4Al_{15}

P H Tobash[1], B L Scott[1], V A Sidorov[1], F Ronning, K Gofryk[1], J D Thompson[1], R C Albers[1], J –X Zhu[1], M D Jones[2], E D Bauer[1]
[1]Los Alamos National Laboratory, Los Alamos, NM 87545, USA
[2]University at Buffalo, SUNY, Buffalo, NY 14260, USA

E-mail: ptobash@lanl.gov

Abstract. We report the synthesis, crystal structure, and physical properties on the new compound UIr_4Al_{15} which crystallizes with the $NdRh_4Al_{15.4}$ structure type in the tetragonal space group $P4_2/nmc$. From single crystal X-ray diffraction, the compound was found to have unit cell parameters $a = b = 9.0239(6)$ Å and $c = 15.513(2)$ Å. The compound has been synthesized from the flux growth method using an excess of Al metal. UIr_4Al_{15} undergoes antiferromagnetic order at approximately 20 K which is consistent with kink-like features in both electrical resistivity and specific heat measurements. Electrical resistivity under hydrostatic pressure shows that the antiferromagnetic ordering temperature is slightly shifted to higher temperatures which is suggestive of the very little pressure dependence and the localized nature of the U *5f* electrons. Electronic structure calculations indicated that the Fermi surface of UIr_4Al_{15} was quasi two-dimensional.

1. Introduction

Since the discovery of the first heavy fermion (HF) superconductor, $CeCu_2Si_2$ [1], the list of additional compounds showing similar properties continues to grow and is detailed very nicely in a review by Pfleiderer [2]. The search continues to be carried out to understand the origin of unconventional superconductivity in this special class of intermetallic compounds. One avenue which is proving to be significant is defining a list of necessary "ingredients" a compound should display so that superconductivity may manifest itself. Two of these qualities which seem promising are crystal systems having tetragonal symmetry and also highly layered systems. As discussed by Lonzarich and co-workers [3], tetragonal systems have a crystal symmetry which is ideal for d-wave overlap and Bauer et al. have shown the correlation of layered systems in the "115" family of compounds as a function of T_C [4]. By far, the number of tetragonal Ce-based HF superconductors dominates the known compounds so far. The isostructural URu_2Si_2 [5], which shows a superconducting transition at 1.5 K along with a hidden order transition at 17.5 K, is an example of how this crystal chemistry and physics can be extrapolated to actinide-containing compounds. This list of uranium HF superconductors remains small which means that better understanding the dual nature of the *5f* electrons in the actinides and how this unconventional superconductivity is mediated can only be carried out by discovering new U compounds. In our search for novel examples of lanthanide and actinide-based binary and ternary HF compounds, we have recently found a new tetragonal antiferromagnet, UIr_4Al_{15}. We discuss its crystal structure, electronic structure, and usual basic physical properties of magnetization, heat capacity, and electrical resistivity at both ambient pressure and applied hydrostatic pressures up to 6 GPa.

Published under licence by IOP Publishing Ltd

2. Experimental Details

The compound was prepared from its respective elements using an excess of Al. All starting elements were used as received (U: Ames, Ir: Alfa Aesar, Al: Alfa Aesar). Initially, the elements were loaded in 2 cm^3 alumina crucibles in the stoichiometric ratio 1:4:15 which were then encapsulated in evacuated quartz tubes. These vessels were then placed in a high temperature furnace that was programmed according to the profile: heated to 800 °C (rate *ca.* 100 °C/h) where it dwelled for 12 h, heated to 1050 °C (rate *ca.* 100 °C/h) where it dwelled for 12 h, cooled to 900 °C (rate *ca.* 60 °C/h) and dwelled for 1 day, and finally cooled to 700 °C (rate *ca.* 5 °C/h). At this temperature, the reactions were removed and the molten Al flux was isolated through centrifugation resulting in small irregular-shaped crystals corresponding to the title compound. The crystal had shiny facets and the majority of them measured 2-4 mm in length. In the case of the non-magnetic analogue, $ThIr_4Al_{15}$, which was prepared using an identical recipe, the crystals were typically smaller in size and usually formed aggregates or clusters of small crystallites. Single crystal X-ray and powder diffraction on the flux-grown crystals revealed that the crystal structure of the compound was tetragonal with unit cell parameters $a = b = 9.0239(6)$ Å and $c = 15.513(2)$ Å for UIr_4Al_{15}. A detailed summary of the data collection and crystallographic parameters can be found in Table 1. In some cases, small irregular-shaped crystals corresponding to Ir_2Al_9 were found in the reaction and could also be seen being attached to the crystals of UIr_4Al_{15}. The magnetization studies were done on a Quantum Design MPMS in the temperature range 350 K down to 2 K. The raw magnetization data was corrected for the holder contribution and subsequently converted to molar susceptibility. Specific heat measurements were completed on a Quantum Design PPMS with a He3 option. The electrical leads for resistivity measurements were made by spot welding 25 μm Pt wires on a polished face of the crystals. The measurements were then made using an excitation current of 300 μA. Hydrostatic pressures to 6 GPa were generated in a miniature clamped toroid-type anvil cell, equipped with an alumina-epoxy gasket. A sample was placed in a Teflon capsule 2 mm in diameter and filled with a glycerol-water mixture. Pressure was determined by a superconducting transition of Pb.

Table 1. Single crystal X-ray Diffraction data collection parameters.

Empirical formula	UIr_4Al_{15}
Formula weight	1411.53 g/mol
Collection temperature	273 K
Radiation, wavelength (λ)	0.71073
Crystal system	Tetragonal
Space group	$P4_2/nmc$
Unit cell dimensions	$a = 9.0239(6)$ Å
	$c = 15.513(2)$ Å
Unit cell volume, Z	1263.2(2) $Å^3$, 4
Density (ρ_{calc})	7.422 g cm^{-1}
Absorption coefficient (μ)	55.774 mm^{-1}
Reflections collected	11884
Unique reflections	870
GOF on F^2	1.155
Final R indices ($I>2\sigma_{(I)}$)	$R_1 = 0.0198$
	$wR_2 = 0.0413$

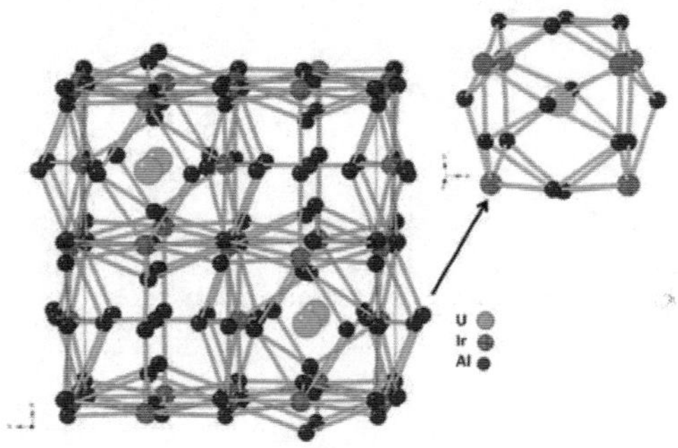

Figure 1. The crystal structure of UIr_4Al_{15} with the unit cell outlined in red. The arrow points to the isolated U atom with the coordination atoms around it.

3. Results and Discussion

The crystal structure for UIr_4Al_{15} is shown in Figure 1. From X-ray single crystal and powder diffraction analysis, it was determined that the compound crystallized with the $NdRh_4Al_{15.4}$ structure type [6] in the tetragonal space group $P4_2/nmc$. More recently, two new rare-earth quaternary phases adopting this type were reported and are described with the formula $REFe_4Al_9Si_6$ (RE = Tb, Er) [7]. Since the crystal structure is complex and has already been detailed very nicely in the aforementioned phases, only a brief description will be given here. The structure contains 10 crystallographic sites, one U, two Ir, and seven Al atoms. The structure can be envisioned as being constructed from full and empty polyhedral of U atoms located in the center of a cage made up of Ir and Al atoms. A closer view of this coordination environment is shown in Figure 1 as an isolated polyhedron. The U atom is surrounded by 20 nearest neighbours with a coordination environment similar to that found in cubic skutterudite structures.

The paramagnetic band-structure for UIr_4Al_{15} has been calculated with the augmented plane-wave Wien2k code [8] in the GGA approximation [9]. The unit cell contains four formula units and has a total of 80 atoms. The calculations were converged with on a k-point mesh of 17 X 17 X 10 using R_{mt} X K_{max} = 7 and G_{max} = 20. The muffin-tin radii were 2.5 a_0 (Bohr radius) for U, 2.46 a_0 for Ir, and 2.19 a_0 for Al. Spin-orbit coupling was included using a second-variational method [8]. The total density of states (DOS) and the Fermi surface for three representative bands that cross the Fermi surface are presented in Figure 2. The main features in the DOS are a broad Ir *d* band that is mainly between -5 eV and 0 eV (where energies are plotted relative to the Fermi energy, which is set to 0 eV), and a very narrow spin-orbit split set of U *5f* bands with bandwidths between about 0.3 to 0.5 eV and a spin-orbit splitting of approximately 0.7 eV. Lying partially

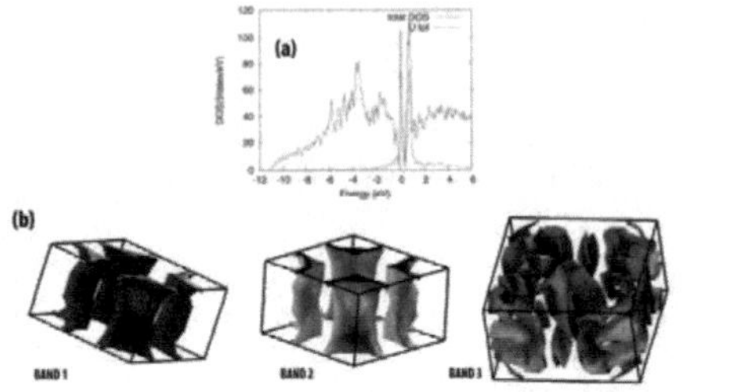

Figure 2. Spin orbit electronic structure calculation results on UIr_4Al_{15}. (a) The total density of states (DOS) plot. (b) a view of three representative Fermi surface plots for bands that cross the Fermi surface.

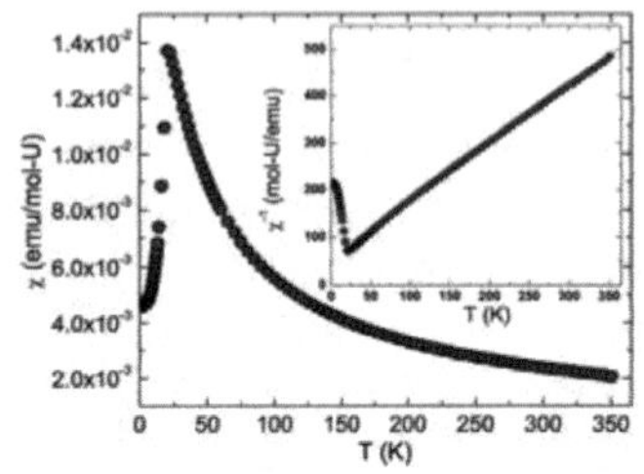

Figure 3. The magnetic susceptibility of UIr_4Al_{15} measured within the temperature range of 350 – 2 K. The top inset shows the inverse susceptibility as a function of temperature. The data was measured in a magnetic field of 0.1 T.

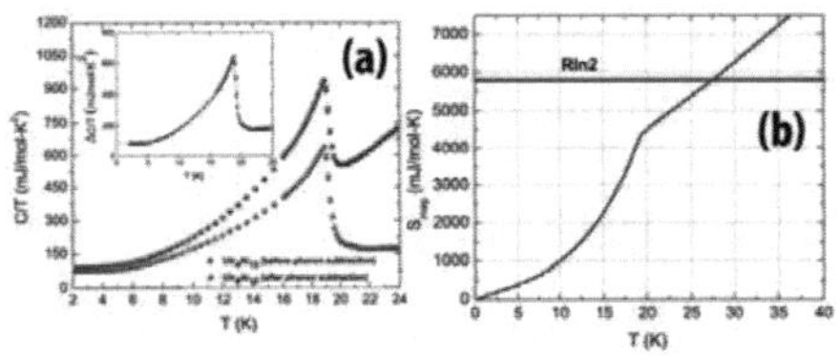

Figure 4. (a) The specific heat of UIr_4Al_{15} plotted as C/T vs T for UIr_4Al_{15}. The inset shows the subtracted data and fit to determine the gap in the magnon spectrum. (b) The magnetic entropy associated with the onset of antiferromagnetic order.

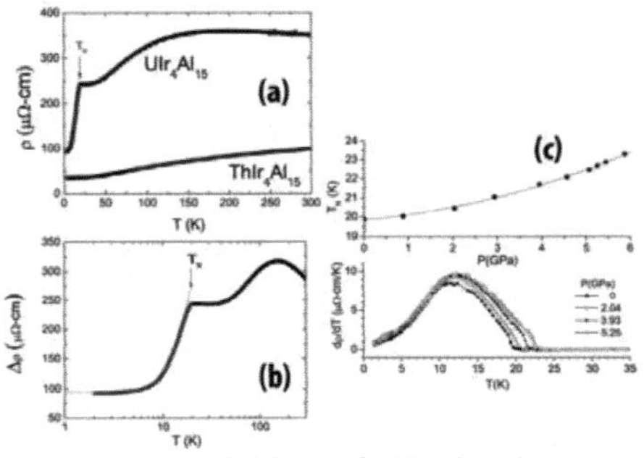

Figure 5. (a) Resistivity data for UIr_4Al_{15} and $ThIr_4Al_{15}$. (b) The resistivity data plotted on a log scale with a fit according to reference 11. (c) evolution of the antiferromagnetic ordering temperature as a function of pressure.

below and also hidden beneath the Ir-*d* and U-*f* peaks are broad Al s-p bands. The total DOS reflects the relative weighting of the atomic projected DOS by 15 Al atoms and 4 Ir atoms per every U atom (i.e., the Al and Ir contributions to the total DOS are enhanced by these atomic ratios). In the paramagnetic phase, the DOS at the Fermi energy is dominated by very narrow U-f (J=5/2) bands. The total DOS at the Fermi energy is about 115 states/(eV cell), and corresponds to a band-structure contribution to the specific heat gamma of about 68 mJ/(mol K^2). With so many atoms per unit cell and such flat U-*f* bands, the number of electronic energy bands that cross the Fermi energy is large (approximately 10 bands; although at least 4 of these bands are marginal in that the crossing occurs at the very top or very bottom of these bands, and small numerical inaccuracies could cause them to miss; in fact, increasing the mesh size from 11 X 11 X 6 to 17 X 17 X 10 increases the number of these crossing bands from 8 to 10). Some representative Fermi surface sheets are presented in Figure 2. Most of the bands have almost vertical sheets indicating a high degree of two-dimensional character.

Figure 3 displays the magnetic susceptibility $\chi(T)$ normalized per mol U atom in the formula unit. The data was measured from 350 K down to 2 K in an applied magnetic field of 0.1 T. The $\chi(T)$ data show Curie-Weiss behaviour down to *ca.* 20 K where an anomaly is observed which is of antiferromagnetic origin. A linear fit to the data, plotted as $\chi^{-1}(T)$ (shown in the inset of Figure 3), using $\chi(T) = C/(T - \theta_{CW})$, where $C = N_A^2/3k_B$ is the Curie constant and θ_{CW} is the Curie-Weiss temperature, yields an effective moment of 2.65 μB which is slightly lower for that expected for U^{3+}/U^{4+} and a negative Curie-Weiss temperature of -54 K that is consistent with an antiferromagnetic ground state.

The specific heat results are provided in Figure 4. The heat capacity was measured on flux-grown crystals of UIr_4Al_{15}. As consistent with the cusp-like feature in $\chi(T)$, C/T vs. T for UIr_4Al_{15} shows a sharp peak at *ca.* 20 K with a heat capacity reaching nearly 1 J/mol-K^2 in the ordered state as shown in Figure 4a. The nonmagnetic counterpart was also successfully synthesized and was used for suitable phonon subtraction to the specific heat. The Sommerfeld coefficient, gamma, obtained by extrapolation of the data below the ordering temperature is on the order of 86 mJ/mol-K^2, indicating moderately heavy fermion-like behaviour. The top inset in Figure 4a, shows a fit to the data in order to determine the gap ΔT in the magnon spectrum. According to the equation delta $C(T) = \gamma T + C_{mag}(T)$, where $C_{mag}(T)$ is defined by an equation taken from Perdew [9] and de Medeiros [10], the specific heat data was fit and it was determined that the value of the gap was 29 K. By taking into account the phonon contribution to the specific heat by subtraction of $ThIr_4Al_{15}$, results in the total entropy of the magnons being slightly smaller than Rln2 for UIr_4Al_{15}, suggesting a doublet ground state for U as displayed in Figure 4b.

The electrical resistivity, $\rho(T)$, measured on the same flux-grown crystals as the specific heat measurements for UIr_4Al_{15} is provided in Figure 5. The $\rho(T)$ data was typical of a metallic system and remains fairly temperature independent down to 150 K, where a broad coherence peak is observed. The kink in $\rho(T)$ is attributed to the onset of long range antiferromagnetic order. The $\rho(T)$ data were also fit using a gap equation proposed by Continentino [11] to determine the gap in the magnon spectrum. The $\rho(T)$ data were first subtracted from the electrical resistivity data for $ThIr_4Al_{15}$ and the value of the gap was determined to be 24 K, close to that determined from the specific heat data. ρ_0 was found to be 94 μΩ-cm which would result in a RRR value of nearly 3. The derivative plots in Figure 5c indicate clearly that the antiferromagnetic ordering temperature smoothly moves to higher temperature for increasing pressure. The small dependence on the applied hydrostatic pressure suggests that the U magnetic moments can be thought of as being more localized than itinerant.

Paul H. Tobash acknowledges financial support from Los Alamos National Laboratory through the Seaborg Fellowship. Work at Los Alamos was performed under the auspices of the US DOE Office of Basic Energy Sciences, Division of Materials and Engineering as was funded ni part by the Los Alamos Laboratory Directed Research and Development Program. V.A. Sidorov acknowledges support of RFBR (grant 09-02-00336).

References

[1] Steglich F, Aarts J, Bredl C D, Lieke W, Meschede D, Franz W and Schafer H 1979 *Phys. Rev. Lett.* **43** 1892
[2] Pfleiderer C 2009 *Rev. Mod. Phys.* **81** 1551
[3] Monthoux P, Pines D and Lonzarich G G 2007 *Nature* **450** 1177
[4] Bauer E D, Thompson J D, Sarrao J L, Morales L A, Wastin F, Rebizant J, Griveau J C, Javorsky P, Boulet P, Colineau E, Lander G H and Stewart G R 2004 *Phys. Rev. Lett.* **93** 147005
[5] Palstra T T M, Menovsky A A, Vandenberg J, Dirkmaat A J, Kes P H, Nieuwenhuys G J andMydosh J A 1985 *Phys. Rev. Lett.* **55** 2727
[6] Fehrmann B and Jeitschko W 2000 *J. Alloys Compd.* **298** 153
[7] Sieve B, Gray D L, Henning R, Bakas T, Schultz A J and Kanatzidis M G 2008 *Chem. Mater.* **20** 6107.
[8] Blaha P, Schwarz K, Madsen G K H, Kvasnicka D and Luitz J WIEN2k, *An Augmented Plane Wave + Local Orbitals Program for Calculating Crystal Properties*, Karlheinz Schwarz, Tech. Universität Wien, Austria, 2001, ISBN 3-9501031-1-2
[9] Perdew J P, Burke K, and Ernzerhof M, 1996 *Phys. Rev. Lett.* **77**, 3865

[10] de Medeiros S N, Continentino M A, Orlando M T D, Fontes M B, Baggio-Saitovitch E M, Rosch A and Eichler A 2000 *Physica B* **281-282** 340

[11] Continentino M A, de Medeiros S N, Orlando M T D, Fontes M B and Baggio-Saitovitch E M 2001 *Phys. Rev. B* **64** 012404

International Conference on Strongly Correlated Electron Systems (SCES 2010) IOP Publishing
Journal of Physics: Conference Series **273** (2011) 012062 doi:10.1088/1742-6596/273/1/012062

Low-temperature antiferromagnetic ordering in the heavy-fermion metal YbPd

Y. Tokiwa, S. Grünheit, H.S. Jeevan, C. Stingl, and P. Gegenwart
I. Physik. Institut, Georg-August-Universität Göttingen, D-37077 Göttingen, Germany
E-mail: ytokiwa@gwdg.de

Abstract. We investigate the low-temperature properties of YbPd by specific heat and thermal expansion measurements down to 0.1 K. This material crystallizes in a cubic CsCl-structure and four phase transitions at 125, 105, 1.9 and 0.5 K have been reported previously. The two transitions at higher temperature were suspected to be of structural origin, whereas the two low-T transitions are magnetic, confirmed by susceptibility measurements and Mössbauer spectroscopy. Our low temperature specific heat and thermal expansion data prove antiferromagnetic ordering at T_N=1.9 K and an additional first-order antiferromagnetic transition at 0.6 K. The entropy reaches approximately half of $R\ln 2$ at 5 K, confirming a doublet ground state. The enhanced value of the Sommerfeld coefficient at 0.1 K suggests a classification of YbPd as a magnetically ordered heavy-fermion metal.

The cubic system YbPd, crystallizing in the CsCl-structure, has provided puzzling experimental observations. It exhibits four transitions below room temperature at 125, 105, 1.9 and 0.5 K [1]. At the higher two transition temperatures, specific heat and thermal expansion show two symmetric peaks and two sharp spikes, respectively, implying first-order structural phase transitions, although a possible tetragonal distortion at 4.2 K is very small [2]. Furthermore, the ^{170}Yb Mössbauer measurements [2] detected no asymmetric spectrum at 4.2 K, showing Yb nuclei experiencing no electric field gradient, and thus, behaving essentially cubic-like. The most puzzling behavior is found in the phase below the transition temperature of 1.9 K. AC-susceptibility shows a kink and ^{170}Yb Mössbauer measurements confirm its magnetic origin. The Mössbauer study demonstrates that only one half of the Yb-atoms experiences the hyperfine field due to the magnetic order while the other half stays non-magnetic. We note, however, that a uniform tetragonal distortion would not produce crystallographically inequivalent Yb sites. The study concluded that the system possesses two Yb charge states with equal fraction, one magnetic, causing the magnetic transition and the other one non-magnetic Yb^{2+}. This is reminiscent of the charge order with Yb^{2+} and Yb^{3+} sites in Yb_4As_3, accompanied by a trigonal distortion along [111] [3]. Here, a trigonal distortion produces two crystallographically inequivalent Yb sites, resulting in an unequal fraction of 1:3 for Yb^{3+}:Yb^{2+}. Finally, the transition at 0.6 K in YbPd is less studied. AC-susceptibility exhibits an anomaly with a hysteresis, indicating a first order magnetic transition. In this paper, we examine the scenario of two different Yb charge states by measuring specific heat at low temperatures and obtaining its magnetic entropy. We also focus on thermal expansion to get further insight on the two phase transitions at low temperatures.

Polycrystalline samples were prepared by melting the high-purity elements with 3% excess of Yb in closed Ta-crucibles. The formation of a single phase was confirmed by X-ray

Published under licence by IOP Publishing Ltd

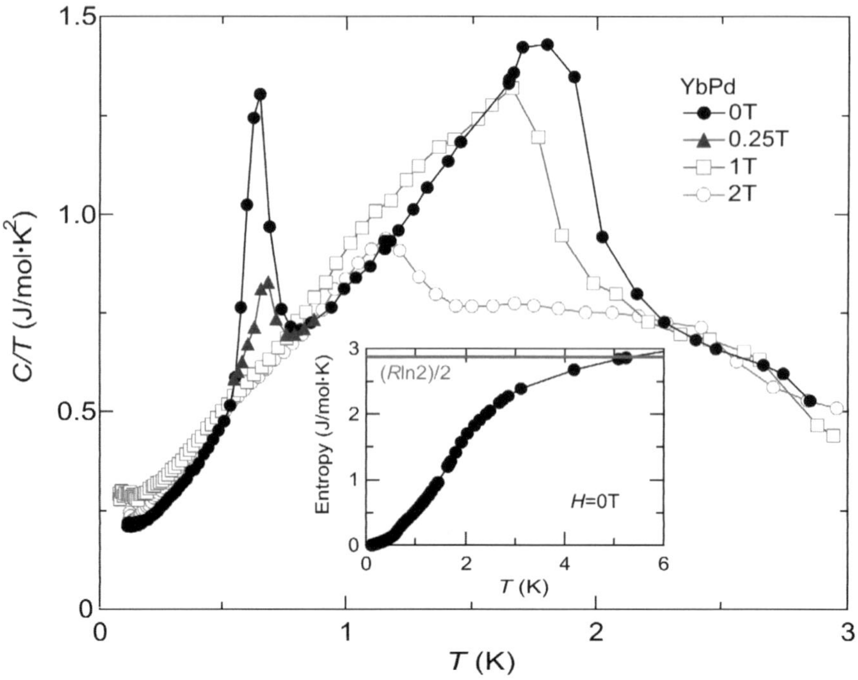

Figure 1. Specific heat divided by temperature, C/T, of YbPd as a function of temperature at different magnetic fields. Inset: magnetic entropy at zero-field.

diffraction measurements. Electrical resistivity and specific heat measurements above 4.2 K are in close agreement to Ref. [1] and reproduce the phase transitions at 130 and 110 K. The low-temperature specific heat and the thermal expansion have been determined in a dilution refrigerator by utilizing the quasiadiabatic heat pulse technique and an ultrahigh resolution capacitive dilatometer, respectively.

Figure 1 shows the specific heat of YbPd at low temperatures in different fields. The zero-field data exhibit two anomalies at 1.9 and 0.6 K, in agreement with the reported studies. The one at 1.9 K has a lambda-like shape, while the other one at 0.6 K shows a sharp and more symmetric anomaly, indicative of a first order transition. Magnetic field shifts the transition temperature of 1.9 K to lower temperatures, indicating antiferromagnetic order. The anomaly at 0.6 K is rapidly suppressed by magnetic field, however, the transition temperature stays constant. Notably, the residual C/T as $T \to 0\,K$ is enhanced, around 0.22 J/mol·K^2, which classifies YbPd as heavy fermion system. From the zero-field $C(T)/T$ data, we determined the magnetic entropy shown in the inset. Above the transition at 1.9 K it tends to saturate and reaches half of Rln2 at 5 K. This result can be understood by assuming one half of the Yb sites being in a nonmagnetic configuration. The other half contributes as heavy fermion antiferromagnet with a doublet ground state and Kondo interaction reducing the entropy below $0.5R\ln 2$ at 1.9 K.

Next, we discuss the linear thermal expansion, $\Delta L/L$, of our YbPd polycrystals. $\Delta L/L$ is negative in the entire measured temperature range up to 4 K, indicating the sample shrinking with increasing temperature. Such behavior is expected for Yb-based heavy-fermion systems and results from the negative pressure dependence of the Kondo temperature. At 1.9 K, it changes its slope, corresponding to a smeared step in thermal expansion coefficient $\alpha(T)$. $\Delta L/L$ shows a step for the transition at 0.6 K. The hysteresis between heating and cooling is due to the first

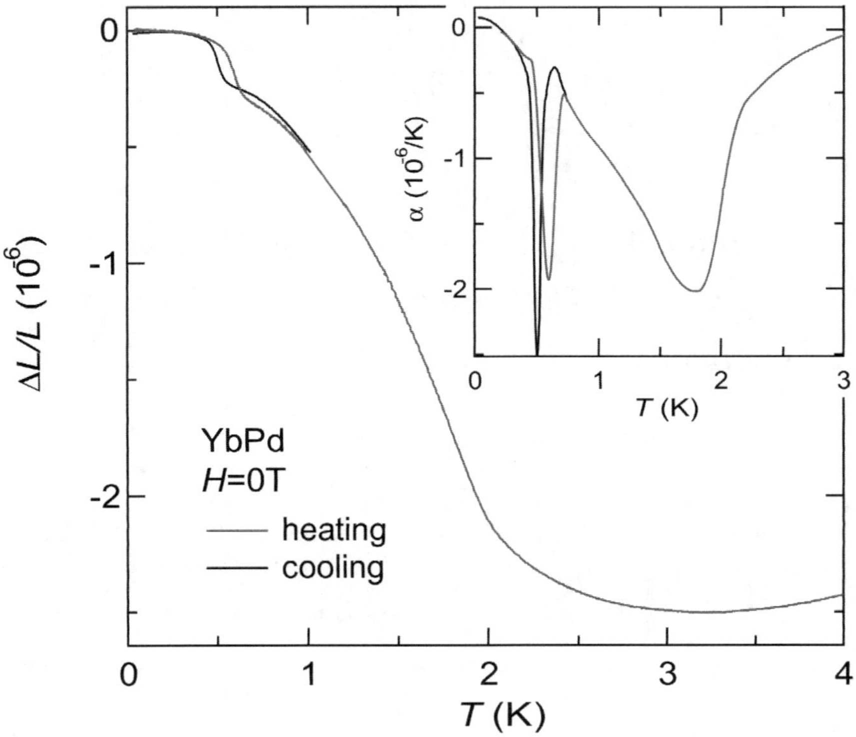

Figure 2. Thermal expansion of YbPd at zero-field. Red and black curves are for heating and cooling the sample. Inset: thermal expansion coefficient of YbPd.

order nature of the transition.

In conclusion, we have studied the low temperature phases with transitions at 1.9 and 0.6 K in YbPd by measuring specific heat and thermal expansion. The obtained magnetic entropy at 1.9 K is smaller than one half of $R\ln 2$. This confirms a doublet crystal electric field ground state, excluding the possibility of a Γ_8 quartet, proposed in Ref. [4], and also supports the interpretation of the Mössbauer spectroscopy experiments that only half of the Yb sites is in a magnetic configuration [2]. We observed a large residual Sommerfeld coefficient as $T \to 0$, indicating heavy fermion behavior. Within the picture of a 1:1 fraction of magnetic and non-magnetic Yb ions, our results suggest that the magnetic ones form a heavy fermion antiferromagnet. The magnetic transitions have also been confirmed by thermal expansion.

We thank C. Geibel for valuable conversations and acknowledge financial support by the German science foundation DFG through research unit 960 "Quantum phase transitions".

[1] Pott R, Boksch W, Leson G, Politt B, Schmidt H, Freimuth A, Keulerz K, Langen J, Neumann G, Oster F, Röhler J, Walter U, Weidner P, and Wohlleben D 1985 *Phys. Rev. Lett.* **54** 481
[2] Bonville P, Hammann J, Hodges J A, Imbert P, and Jéhanno G J 1986 *Phys. Rev. Lett.* **57** 2733
[3] Ochiai A, Suzuki T, Kasuya T 1990 *J. Phys. Soc. Jpn.* **59** 4129
[4] Walter U and Wohlleben D 1987 *Phys. Rev.* B **35** 3576

Sharp Valence Change as Origin of Drastic Change of Fermi Surface and Transport Anomalies in $CeRhIn_5$ under Pressure

Shinji Watanabe, Kazumasa Miyake
Graduate School of Engineering Science, Osaka University, Toyonaka, Osaka 560-8531, Japan

Abstract. The drastic changes of Fermi surfaces as well as transport anomalies near $P = P_c \sim 2.35$ GPa in $CeRhIn_5$ are explained theoretically from the viewpoint of sharp valence change of Ce. It is pointed out that the key mechanism is the interplay of magnetic order and Ce-valence fluctuations. It is shown that the antiferromagnetic state with "small" Fermi surfaces changes to the paramagnetic state with "large" Fermi surfaces with huge enhancement of effective mass of electrons with keeping finite c-f hybridization. This naturally explains the de Haas-van Alphen measurement and also the transport anomalies of T-linear resistivity emerging simultaneously with the residual resistivity peak at $P = P_c$ in $CeRhIn_5$.

1. Introduction

A heavy electron metal $CeRhIn_5$ has attracted much attention recently since it has been recognized to offer a new insight into the quantum critical phenomena by accumulated experimental studies. By applying pressure under the magnetic field larger than the upper critical field, the antiferromagnetic (AF) order is suppressed abruptly at $P = P_c \approx 2.35$ GPa [2]. The de Haas-van Alphen (dHvA) measurement revealed that the Fermi surfaces (FS) change from approximately the same FS as those of $LaRhIn_5$ to FS similar to those of $CeCoIn_5$ at $P = P_c$ with huge mass enhancement toward $P = P_c$. Transport anomalies such as emergence of T-linear resistivity and enhancement of residual resistivity are prominent near $P = P_c$ [3, 4, 5]. So far, it has been considered that localized-to-itinerant transition of f electrons might explain these anomalies [4]. However, this scenario encounters a serious difficulty in elucidating the experimental fact that the Sommerfeld constant at ambient pressure $P = 0$ in $CeRhIn_5$, $\gamma = 50$ mJ/molK^2, is about 10-times larger than that of $LaRhIn_5$ [2], which strongly suggests that heavy quasiparticles are formed in the AF-ordered state, showing an importance of hybridization between f and conduction electrons.

In this paper, we propose a unified explanation for resolving this outstanding puzzle in $CeRhIn_5$. First, we point out that the emergence of T-linear resistivity and residual-resistivity peak are quite similar to the observations in $CeCu_2Ge_2$ [6], $CeCu_2Si_2$ [7], and $CeCu_2(Si_xGe_{1-x})_2$ [8] at pressure $P = P_v$ located at higher pressure side than the AF-paramagnetic boundary $P = P_c$. The Cu-NQR measurement detected that the electric-field gradient starts to change at $P = 4$ GPa slightly less than $P_v \approx 5$ GPa where Ce valence changes [9]. Theoretically, it has been shown that enhanced valence fluctuations cause the T-linear resistivity [7] and residual-resistivity peak at $P = P_v$ [10]. Since in $CeRhIn_5$ such a

Published under licence by IOP Publishing Ltd

transport anomalies occur at the pressure of the AF-paramagnetic boundary, P_c seems to almost coincide with P_v, i.e., $P_c \approx P_v$. Below we demonstrate theoretically that this situation is indeed realized for realistic parameters for $CeRhIn_5$ and show that the interplay of the AF order and Ce-valence fluctuations is a key mechanism for resolving the outstanding puzzle in $CeRhIn_5$ in a unified way [11].

2. Model Calculation and Results

Let us start our discussion by introducing a minimal model which describes the essential part of the physics of $CeRhIn_5$ in the standard notation:

$$\mathcal{H} = H_{\rm c} + H_{\rm f} + H_{\rm hyb} + H_{U_{\rm fc}}, \tag{1}$$

where $H_{\rm c} = \sum_{\mathbf{k}\sigma} \varepsilon_{\mathbf{k}} c^{\dagger}_{\mathbf{k}\sigma} c_{\mathbf{k}\sigma}$ represents the conduction band, $H_{\rm f} = \varepsilon_{\rm f} \sum_{i\sigma} n^{\rm f}_{i\sigma} + U \sum_{i=1}^{N} n^{\rm f}_{i\uparrow} n^{\rm f}_{i\downarrow}$ the f level $\varepsilon_{\rm f}$ and onsite Coulomb repulsion U for f electrons, $H_{\rm hyb} = V \sum_{i\sigma} \left(f^{\dagger}_{i\sigma} c_{i\sigma} + c^{\dagger}_{i\sigma} f_{i\sigma} \right)$ the hybridization V between f and conduction electrons, and $H_{U_{\rm fc}} = U_{\rm fc} \sum_{i=1}^{N} n^{\rm f}_{i} n^{\rm c}_{i}$ the Coulomb repulsion $U_{\rm fc}$ between f and conduction electrons. The $H_{U_{\rm fc}}$ term is a key ingredient for explaining the various anomalies caused by enhanced Ce-valence fluctuations: The T-linear resistivity and residual resistivity peak have been shown theoretically on the basis of this model eq. (1) [7, 10]. To discuss $CeRhIn_5$, we analyze this model by taking into account existence of the AF order. To make a comparison with the dHvA measurement for the β_2 branch which is a two-dimensional-like cylindrical Fermi surface, we consider the AF order with the ordered vector $\mathbf{Q} = (\pi, \pi)$ on the square lattice with $\varepsilon_{\mathbf{k}} = -2t(\cos(k_x) + \cos(k_y))$ at the filling $n \equiv (\bar{n}_{\rm f} + \bar{n}_{\rm c})/2 = 0.9$ with $\bar{n}_{\rm f} \equiv \sum_{\mathbf{k}\sigma} \langle f^{\dagger}_{\mathbf{k}\sigma} f_{\mathbf{k}\sigma} \rangle / N$ and $\bar{n}_{\rm c} \equiv \sum_{\mathbf{k}\sigma} \langle c^{\dagger}_{\mathbf{k}\sigma} c_{\mathbf{k}\sigma} \rangle / N$. We take $t = 1$ as an energy unit.

To treat the effects of the AF order and Ce-valence fluctuations on equal footing, we apply the slave-boson mean-field theory [12] to eq. (1) [11]. The resultant ground-state phase diagram is shown in Fig. 1(a). The first-order valence transition (solid line with triangles) terminates at the quantum critical end point (filled circle), i.e., the QCP, and the valence crossover occurs at the dashed line with open circles. At the QCP the valence fluctuation diverges and the enhanced valence fluctuations appear even far away from the QCP, as indicated by enhanced valence susceptibility $\chi_{\rm v} \equiv -\partial \bar{n}_{\rm f} / \partial \varepsilon_{\rm f}$ in Fig. 1(b) calculated for moderate $U_{\rm fc}$'s. These are the results obtained within the paramagnetic solutions. The result when the AF order is taken into account is shown as the solid line with filled squares in Fig. 1(a). One can see that the AF-paramagnetic boundary, which is the first-order transition, almost coincides with the first-order valence-transition line and the valence-crossover line. This result is favorably compared with the observations in $CeRhIn_5$. When pressure is applied to the Ce compounds, $\varepsilon_{\rm f}$ increases. Thus, $\varepsilon_{\rm f}$ can be regarded as pressure. As shown in Fig. 1(a), the AF order is suppressed at the Ce-valence transition and crossover points, which indicates that the suppression is caused by the valence transition or enhanced valence fluctuations. This result implies that $P_c \approx P_v$ observed in $CeRhIn_5$ is reproduced by our model eq. (1). For comparison, hereafter, we show our results for $U_{\rm fc} = 0.5$, which is a moderate $U_{\rm fc}$ as expected to be a realistic parameter for $CeRhIn_5$.

Figures 2(a) and 2(b) show the contour plots of the lower hybridized band with minority spin for $\varepsilon_{\rm f} = -0.29$ and $\varepsilon_{\rm f} = -0.28$, respectively. Here, we apply the magnetic field to the system as $\mathcal{H} - h \sum_i (S^{{\rm f}z}_i + S^{{\rm c}z}_i)$ with $h = 0.005$ to simulate the field of $H = 15$ T used in the dHvA measurement [2]. We have verified that the drastic change of the Fermi surface occurs at the AF-paramagnetic-transition point $\varepsilon_{\rm f} = \varepsilon^{\rm c}_{\rm f} = -0.283$: Folded Fermi surface in Fig. 2(a) in the AF phase changes to the unfolded Fermi surface in Fig. 2(b) in the paramagnetic phase as $\varepsilon_{\rm f}$ passes across $\varepsilon^{\rm c}_{\rm f}$. We note here that the dashed line represents the Fermi surface of conduction electrons at $\bar{n}_{\rm c} = 0.8$. This is the small Fermi surface obtained by setting $V = 0$ in the model eq.(1), where f electrons with $\bar{n}_{\rm f} = 1$ are completely localized. One can see that the Fermi surface in the AF phase calculated for finite V is nearly the same as the "small" Fermi surface.

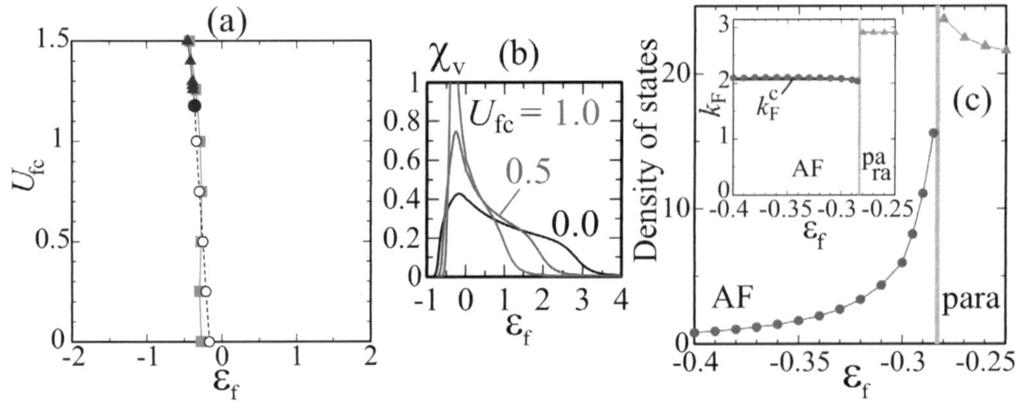

Figure 1. (color online) (a) Ground-state phase diagram in the $U_{\rm fc}$-$\varepsilon_{\rm f}$ plane for paramagnetic and AF states (see text). The first-order valence-transition line (solid line with triangles) terminates at the quantum critical end point (filled circle). Valence crossover occurs at the dashed line with open circles, at which $\chi_{\rm v}$ has a maximum, as shown in (b). The solid line with filled squares represents the boundary between the AF state and the paramagnetic state. (c) Total density of states vs $\varepsilon_{\rm f}$ for $U_{\rm fc} = 0.5$. The inset shows $k_{\rm F}$ vs $\varepsilon_{\rm f}$: In AF phase, black line represents the small Fermi surface of conduction electrons $\varepsilon_{\mathbf{k}}$ at $\bar{n}_{\rm c} = 0.8$. All results in (a)-(c) are calculated for $t = 1$, $V = 0.2$, and $U = \infty$ at $n = 0.9$.

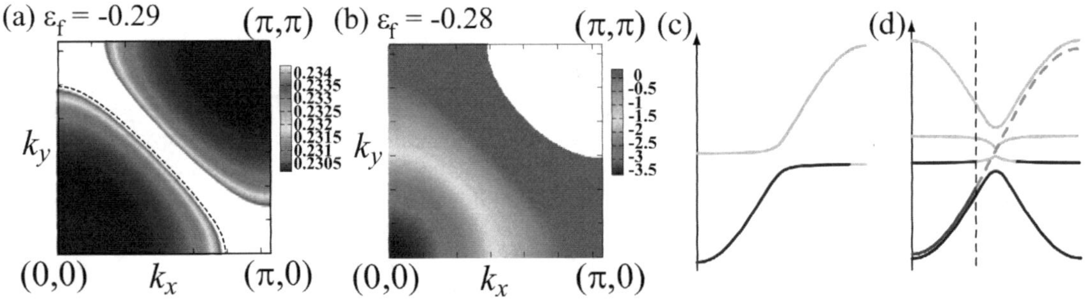

Figure 2. (color online) The contour plot of the energy band with $\downarrow$ spin located at the Fermi level μ for $t = 1$, $V = 0.2$, $U = \infty$, $U_{\rm fc} = 0.5$, and $n = 0.9$ with $h = 0.005$: (a) $\varepsilon_{\rm f} = -0.29$ and (b) $\varepsilon_{\rm f} = -0.28$. The $E_{\mathbf{k}\downarrow} > \mu$ parts are represented by white regions. In (a), the dashed line indicates the Fermi surface of the conduction band, $\varepsilon_{\mathbf{k}}$ for $\bar{n}_{\rm c} = 0.8$. Occupied (solid lines) and empty (gray lines) bands in the periodic Anderson model at $n = (\bar{n}_{\rm f} + \bar{n}_{\rm c})/2 = 0.9$ are shown in paramagnetic phase (c) and in AF-ordered phase (d). In (d), gray dashed line indicates the energy band of the conduction band, $\varepsilon_{\mathbf{k}}$ at $\bar{n}_{\rm c} = 0.8$. Solid dashed line is a guide for the eyes indicating that the Fermi surface in AF-ordered phase with finite c-f hybridization coincides with the small Fermi surface where f electrons are completely localized.

These results are naturally understood if we draw the energy bands in the paramagnetic phase and the AF phase as in Figs. 2(c) and 2(d), respectively. In Fig. 2(c), the lower hybridized band of Hamiltonian (1) is occupied as illustrated by the solid line forming the "large" Fermi surface whose volume contains f electrons, i.e., $\bar{n}_{\rm f} + \bar{n}_{\rm c}$. On the other hand, the energy band of conduction electrons is illustrated in Fig. 2(d) by the gray dashed line whose occupied part for $\bar{n}_{\rm c}$

is represented by the solid line, clearly indicating the "small" Fermi surface, in sharp contrast to Fig. 2(c). Once the AF order occurs, the lower and upper hybridized bands are folded as shown in Fig. 2(d). Since the lower-hybridized band in the magnetic Brillouin zone is fully occupied, the Fermi surface emerging at the folded lower-hybridized band shows the area occupied for the filling 0.8 among $n = (1 + 0.8)/2$, giving rise to the same Fermi surface of the conduction band at $\bar{n}_c = 0.8$. Namely, the "small" Fermi surface appears in the AF phase even though the c-f hybridization remains finite $\langle f^{\dagger}_{\mathbf{k}\sigma} c_{\mathbf{k}\sigma} \rangle \neq 0$. Thus, the mechanism of the coincidence of the Fermi surface in the AF phase and the dashed line shown in Fig. 2(a) is naturally understood.

We find that the total density of states $D(\mu) \equiv \sum_{\mathbf{k}\sigma} \delta(\mu - E_{\mathbf{k}\sigma})/(2N)$ at the Fermi level shows an enhancement toward $\varepsilon_f = \varepsilon_f^c$ as shown in Fig. 1(c). Here, μ is a chemical potential and $E_{\mathbf{k}\sigma}$ is the quasiparticle energy band with σ spin. The reason why such a mass enhancement occurs at the AF-paramagnetic boundary is explained as follows: In the AF phase, as ε_f increases, the gap between the folded lower-hybridized-band and the original lower-hybridized-band increases (see Fig. 2(d)). Then, the f-electron-dominant flat part of the folded band approaches the Fermi level, giving rise to the mass enhancement. In the paramagnetic phase, as ε_f decreases, $\bar{n}_f$ increases to reach the $\bar{n}_f = 1$ state, i.e., to approach the Kondo regime. Hence, as ε_f approaches ε_f^c, mass enhancement occurs. Our result clearly shows that the mass enhancement is caused by the band effect including the local correlations of f electrons. This explains the experimental fact that the $\sqrt{A}/m^*$ =const. scaling holds under pressure from $P = 0$ to $P = P_c$ [5], where A is the coefficient of the T^2 term in the low-temperature resistivity at $H = 15$ T and m^* is a cyclotron mass of the β_2 branch of the cylindrical Fermi surface observed for $H = 12 \sim 17$ T in $CeRhIn_5$ [2]. For $P > P_c$, the dHvA signal of the β_2 branch was not detected [2], probably because of heavy effective mass larger than $m^* \approx 90m_0$ in $CeCoIn_5$. This is also consistent with our result shown in Fig. 1(c).

The Fermi wavenumber k_F defined by the crossing point of the line connecting $\mathbf{k} = (0, 0)$ and (π, π), and the Fermi surface with minority spin at $h = 0.005$ exhibits the abrupt change at the AF-paramagnetic boundary as shown in the inset of Fig. 1(c). Thus, the drastic change of Fermi surfaces as well as the mass enhancement observed in the dHvA measurement [2] is naturally understood by our theory. We note that the density of states at $\varepsilon_f = -0.4$ is about 10 times larger than that of conduction electrons at $\bar{n}_c = 0.8$, $D_c(\mu) \equiv \sum_{\mathbf{k}} \delta(\mu - \varepsilon_{\mathbf{k}})/N = 0.092$, which is also consistent with the enhanced γ of $CeRhIn_5$ from that of $LaRhIn_5$ at $P = 0$ [2].

3. Summary

We have proposed a theory for resolving the outstanding puzzle on Fermi-surface change and transport anomalies in $CeRhIn_5$ in a unified way. The key mechanism has been shown to be the interplay of the AF order and Ce-valence fluctuations. Experimental tests to examine our proposal, in particular, direct observations of the Ce valence change under pressure and magnetic field near $P = P_c$ are greatly desired.

References

[1] Hegger H *et al.*, 2000 *Phys. Rev. Lett.* **8** 4986.
[2] Shishido H *et al.*, 2005 *J. Phys. Soc. Japan* **74** 1103.
[3] Park T *et al.*, 2006 *Nature* **440** 65.
[4] Park T *et al.*, 2008 *Nature* **456** 366.
[5] Knebel G, Aoki D, Brison J P, and Flouquet J, 2008 *J. Phys. Soc. Japan* **77** 114704.
[6] Jaccard D, Wilhelm D H, Alami-Yadri K, and Vargoz E, 1999 *Physica B* **259-261** 1.
[7] Holmes A T, Jaccard D, and Miyake K, 2004 *Phys. Rev. B* **69** 024508.
[8] Yuan H Q, Grosche F M, Deppe M, Geibel C, Sparn G, and Steglich F, 2003 *Science* **302** 2104.
[9] Fujiwara K *et al.*, 2008 *J. Phys. Soc. Japan* **77** 123711.
[10] Miyake K and Maebashi H, 2001 *J. Phys. Soc. Japan* **71** 1007.
[11] Watanabe S and Miyake K, 2010 *J. Phys. Soc. Japan* **79** 033706.
[12] Kotliar G and Ruckenstein A E, 1996 *Phys. Rev. Lett.* **57** 1362.

Theoretical reconsideration of antiferromagnetic Fermi surfaces in URu_2Su_2

Hiroshi Yamagami

Department of Physics, Faculty of Science, Kyoto Sangyo University, Kyoto 603-8555, Japan

E-mail: yamagami@cc.kyoto-su.ac.jp

Abstract. In an itinerant $5f$-band model, the antiferromagnetic (AFM) Fermi surfaces of URu_2Si_2 are reconsidered using a relativistic LAPW method within a local spin-density approximation, especially taking into account the lattice parameters dependent on pressures. The reduction of the z-coordinate of the Si sites results in the effect of flattening the Ru-Si layers of URu_2Si_2 crystal structure, thus weakening a hybridization/mixing between the U-$5f$ and Ru-$4d$ states in the band structure. Consequently the $5f$ bands around the Fermi level are more flat in the dispersion with decreasing the z-coordinate, thus producing three closed Fermi surfaces like "curing-stone", "rugby-ball " and "ball". The origins of de Haas-van Alphen branches can be qualitatively interpreted from the obtained AFM Fermi surfaces.

1. Introduction

In highly-correlated-electron system, the " hidden order " (HO) parameter of URu_2Si_2 at T_o = 17.5 K is one of reminding issues we have not yet shared a common view [1]: The bulk experiments of specific heat and Hall effect indicate that a large energy gap opens at the Fermi level (E_F), thus releasing most of the carriers below T_o [2]. Under the pressure the HO state with a tiny magnetic moment changes into a large moment antiferromagnetic (AFM) state at P_c=15 kbar [3]. However, de Haas-van Alphen (dHvA) measurements provide very limited information concerning the Fermi surface [4], and theoretically it is still hard to give a consistent understanding of the U-$5f$ states, not only in the band structure and Fermi surface, but also in the paramagnetic (PM) and ordered states [4, 5, 6].

At ambient pressure, URu_2Si_2 crystallizes in the tetragonal $ThCr_2Si_2$ crystal structure (space group $I4/mmm$) with the lattice constants a = 4.1239Å and c = 9.5817Å [7] and the z-coordinate of the Si sites z = 0.371 [8]. The ambient pressure lattice symmetry and crystal structure persist in the PM, HO and AFM phases. With increasing pressures, both the a- and c-axes reduce with the c-axis contracting slightly more than the a-axis, and additionally the z-coordinate decrease with applied pressures. At 27 kbar, the lattice parameters of the AFM phase are measured as a = 4.109Å , c = 9.515Å and z = 0.356 [9]. The reduction of z-coordinate is expected to control a c-f hybridization, thus resulting in a flattening of the $5f$ bands and a distortion of the Fermi surface volume.

Recently itinerant $5f$ bands of URu_2Si_2 around E_F was directly detected in the PM phase using soft x-ray angle-resolved photoemission spectroscopy (SXARPES) [10], and the AFM Fermi surfaces are closely related to the HO Fermi surfaces in the dHvA measurements under the pressure [11]. In an itinerant $5f$-based model, therefore, we need to reconsider the PM and

Published under licence by IOP Publishing Ltd

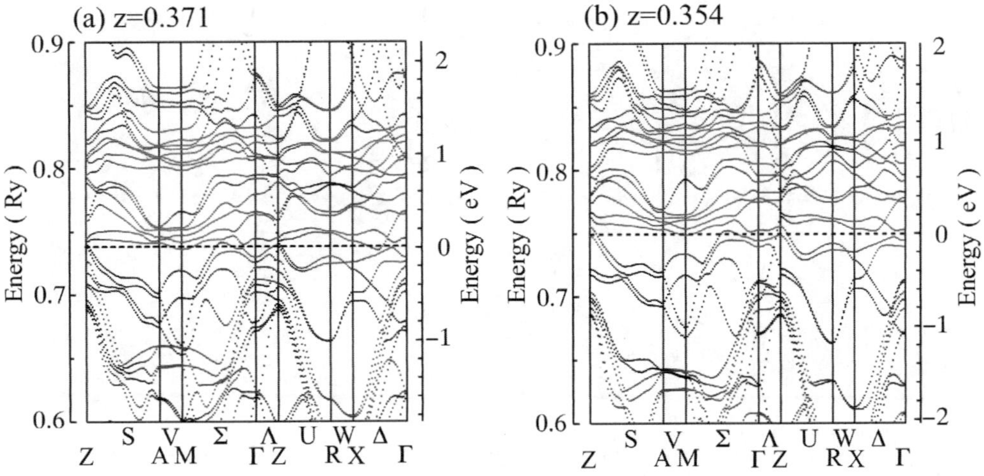

Figure 1. AFM band structures of URu_2Si_2 calculated with two z-parameters of (a) $z = 0.371$ and (b) $z = 0.354$.

AFM electronic structure theoretically. Taking into account the lattice parameters dependent on pressures [9], here, the AFM Fermi surface calculations are performed using the same relativistic LAPW (RLAPW) method [12] within a local spin-density approximation (LSDA) [13] as previously used to obtain the PM [4] and the AFM Fermi surfaces [14].

The RLAPW method is based on the spin-polarized coupled Dirac equation (SPCDE) in the LSDA. The AFM configuration is set by orienting the spin and orbital directions at each site with use of local rotation matrix. The densities and potentials are constructed in the muffin-tin approximation. In the iterative process, all-electron calculations are carried out; the $5f^3$, $6p^6$, $6d^1$ and $7s^2$ states at the U site, $4d^7$ and $5s^1$ at the Ru sites, and $3s^2$ and $3p^2$ at the Si sites are treated as valence electrons, while the other electrons are solved by SPCDE under atomic-like boundaries at each site.

2. Band structure of AFM phase dependent on z coordinate

Figure 1 shows the z-parameter dependence of the AFM band structures of URu_2Si_2 in the vicinity of E_F, which is indicated by the horizontal broken line. In Fig. 1(a) the AFM band structure calculated with $z = 0.371$ at ambient pressure, of which the z-parameter was used to create the previous AFM Fermi surfaces [14], is shown along the simple-tetragonal symmetry axes of the AFM Brillouin zone. The AFM band structure with $z = 0.354$ of the AFM phase at 27 kbar [9] is illustrated in Fig. 1(b). The U-$5f$ and the Ru-$4d$ components for each of the bands are expressed by a relative ratio from red ($5f$) to blue ($4d$) colors. The main differences of the two band structures are as follows: First, the top of the Ru-$4d$ (blue) bands of $z = 0.354$ (Fig. 1(b)) shifts lower down from E_F than that of $z = 0.371$ (Fig. 1(a)) with narrowing the bandwidth, and thus the hybridization between the U-$5f$ and the Ru-$4d$ states becomes much weaker. Secondly, the U-$5f$ (red) bands just around E_F along the Σ and Δ axis are partially more flattened with decreasing the z-parameter, and especially we can see that a gap opens along the Δ axis.

3. Fermi surfaces of AFM phase and de Haas-van Alphen frequencies

As seen in Fig.1, two bands indexed as band36 and band37 keep crossing E_F in both the band structures to make up the Fermi surfaces, where the band index is determined by numbering

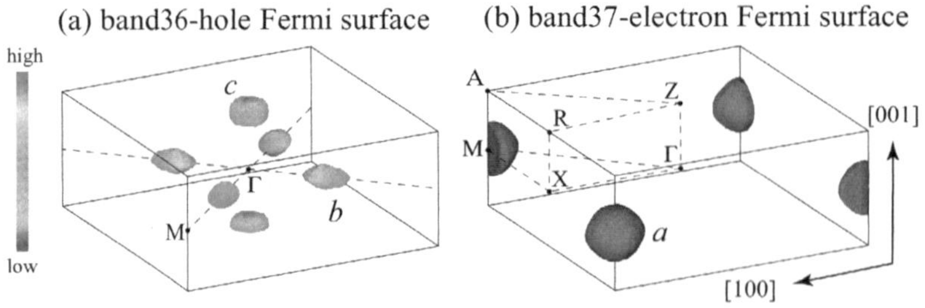

Figure 2. (a) Hole Fermi surface for band36 and (b) electron Fermi surface of band37 obtained from AFM band structure of $z = 0.354$ in URu_2Si_2.

from the bottom of the valence bands. A few small parts of the previous Fermi surfaces [14], however, vanish for the cause of changing the band dispersions around E_F. Figures 2(a) and (b) show the hole Fermi surface of the band36 and electron Fermi surface of the band37, respectively, which are obtained from the AFM band structure of $z = 0.354$ in Fig. 1(b). The hole Fermi surfaces are composed of a "curling-stone"-shaped part around the Z point and four "rugby-ball"-like parts along the Γ-M (Σ) axis. On the other hand the electron Fermi surface is formed from a big "ball" centered at the M point, of which the volume is equal to that of all the hole Fermi surfaces because of the electron number being even in URu_2Si_2. Accordingly it can be said that the AFM electronic state of URu_2Si_2 is a semi-metallic with the small electron and hole carriers. The electronic specific heat coefficient γ is estimated as 1.45 mJ/(K^2·mole) from the total density of states.

The color expression of Fig. 2 shows the intensity of $5f$ contribution on the Fermi surfaces according to a color indicator, where the highest contribution is red and the opposite is blue. As understood from the colors on the Fermi surfaces, all the hole Fermi surfaces have an anisotropic $5f$ distribution orienting higher near the Γ point, while the electron Fermi surface are constructed from almost pure $5f$ bands. Generally the intensity of the $5f$ states on the Fermi surfaces has a certain amount of relationship with a size of local band gradient around E_F, or the cyclotron effective mass (CEM) in any direction. Therefore the CEM of the curling-stone and rugby-ball Fermi surfaces behaves as a monotonically decreasing function with respect to the field angles from the [001] to the [100] directions, while the big-ball CEM is large with a small angular dependence.

In Fig. 3 we show the angular dependence of dHvA frequencies obtained from the big ball, the rugby-ball and the curling-stone Fermi surfaces, corresponding to theoretical branches a, b and c, respectively. For comparison, the experimental dHvA frequencies measured in the HO phase are depicted by open circles [4], since the dHvA frequencies of the AFM phase are similar to those of the HO phase. Although there appear quantitative differences between the theoretical and the experimental branches, it can be suggested that the α, β and γ branches are derived from the big ball, the rugby-ball and the curling-stone Fermi surfaces, respectively. Moreover it should be emphasized that the theoretical branches a, b and c remain even in changing the lattice parameters.

4. Discussion and conclusion

Under pressures the reduction of the z-parameter results in the effect of flattening the Ru-Si layers of URu_2Si_2 crystal structure [9]. In the band structure the dispersion of the $5f$ bands around E_F gets more flat, because the hybridization/mixing between the U-$5f$ and Ru-$4d$ states becomes weaker. From the AFM calculations with $z = 0.354$, it is found out that the local spin moment at the U site is $-0.39\mu_B$ and the corresponding orbital moment is $0.93\mu_B$. Therefore

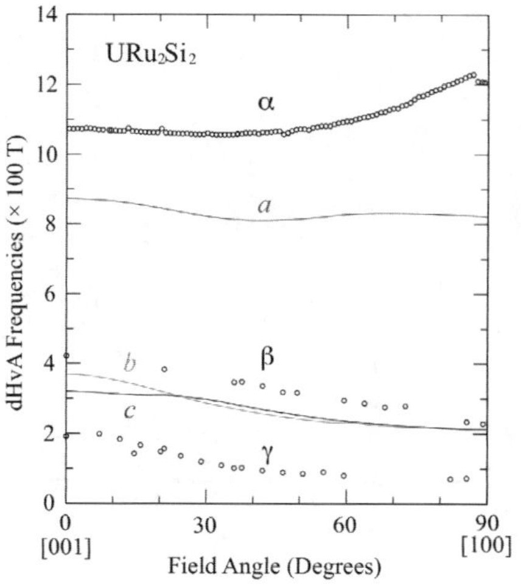

Figure 3. Angular dependence of de Haas-van Alphen frequencies from [001] to [100] directions of applied field. The theoretical branches a, b and c by colored lines originate from the big ball, the rugby-ball and the curling-stone Fermi surfaces, respectively. The observed branches α, β and γ are shown by open circles [4].

the total magnetic moment is obtained as $0.54\mu_B$ in comparable agreement with $0.4\mu_B$ of the AFM phase in neutron-scattering experiment [3]. Surprisingly, the local magnetic moment at the Ru sites is polarized with $0.36\mu_B$ in the calculation. As a next step to clarify the HO Fermi surface, we will discuss the SXARPES [10] with the PM Fermi surface calculated in the itinerant $5f$ band picture.

Acknowledgments

I would like to thank J. Flouquet and D. Aoki for providing valuable information and suggestions. This works was financially supported by a Grant-in-Aid for Scientific Research (C) (No. 22540374) and Grant-in-Aid for Scientific Research on Innovative Areas "Heavy Electrons" (No. 20102003) from the Ministry of Education, Culture, Sports, Science and Technology, Japan.

References

[1] Shah N, Chandra P, Coleman P and Mydosh J A 2000 *Phys. Rev.* B **61** 564
[2] Behnia K, Bel R, Kasahara Y, Nakajima Y, Jin H, Aubin H, Izawa K, Matsuda Y, Flouquet J, Haga Y, Ōnuki Y and Lejay P 2005 *Phys. Rev. Lett.* **94** 156405
[3] Amitsuka H, Sato M, Metoki N, Yokoyama M, Kuwahara K, Sakakibara T, Morimoto H, Kawarazaki S, Miyako Y and Mydosh J A 1999 *Phys. Rev. Lett.* **83** 5114
[4] Ohkuni H, Inada Y, Tokiwa Y, Sakurai K, Settai R, Honma T, Haga Y, Yamamoto E, Ōnuki Y, Yamagami H, Takahashi S and Yanagisawa S 1999 *Phil. Mag.* B **79** 1045
[5] Elgazzar S, Rusz J, Oppeneer P M and Mydosh J A 2009 *Nat. Mater.* **8** 337
[6] Haule K and Kotliar G 2009 *Nat. Phys.* **5** 796
[7] Palstra T T M, Menovsky A A, van den Berg J, Dirkmaat A J, Kes P H, Nieuwenhuys G J and Mydosh 1985 *Phys. Rev. Lett.* **55** 2727
[8] Cordier G, Czech E, Schäfer H, and Woll P 1985 *J. Less-Common Met.* **110** 327
[9] Jeffries J R, Butch N P, Hamlin J J, Sinogeikin S V, Evans W J and Maple M B 2010 *cond-mat.str-el* arXiv:1002.2245v1
[10] Kawasaki I, Fujimori S.-i, Takeda Y, Okane T, Yasui A, Saitoh Y, Yamagami H, Haga Y, Yamamoto E and Ōnuki Y. to be submitted to 2010 *J. Phys.: Conf. Series* (SCES2010)
[11] Flouquet J *private commun.*
[12] Yamagami H 2000 *Phys. Rev.* B **61** 2426
[13] von Barth U and Hedin L 1972 *J. Phys.* C **5** 1629
[14] Yamagami H and Hamada N 2000 *Physica* B **284-288** 14155

Journal of Physics: Conference Series **273** (2011) 012065 doi:10.1088/1742-6596/273/1/012065

Elastic Constants of Filled Skutterudite $LaRu_4As_{12}$

T Yanagisawa[1], H Saitoh[2], H Hidaka[2], H Amitsuka[2], T Cichorek[3], Z Henkie[3] and M B. Maple[4]

[1] Creative Research Institution, Hokkaido University, Sapporo 001-0021, Japan
[2] Department of Physics, Hokkaido University, Sapporo 060-0810, Japan
[3] Institute of Low Temperature and Structure Research, Polish Academy of Sciences, 50-950 Wroclaw, Poland
[4] Department of Physics and Institute for Pure and Applied Physical Sciences, University of California San Diego, La Jolla 92093, USA

E-mail: `tatsuya@cris.hokudai.ac.jp`

Abstract. Ultrasonic measurements with frequencies varying between 73 and 160 MHz were performed for the $LaRu_4As_{12}$ single crystal. This As-based filled skutterudite compound is a BCS-type superconductor with T_{SC} as high as 10.3 K. In order to examine a presence of off-center rattling motion in $LaRu_4As_{12}$, we measured elastic constants $C_{L111} = (C_{11} + 2C_{12} + 4C_{44})/3$ and $C_{T111} = (C_{11} - C_{12} + 4C_{44})/3$, which are related to longitudinal and transverse sound waves propagating along the [111] direction, respectively. No ultrasonic dispersion was found in the whole temperature range from 1.4 to 300 K.

1. Introduction

Off-center rattling motion of ion in the cage structured metallic compound attracts a lot of attention because possible interaction between local phonon of the rattler and conduction electrons may lead to several exotic phenomena like: a large magnetic-field-independent Sommerfeld coefficient of the electronic specific heat γ of $SmOs_4Sb_{12}$, and unconventional superconductivity of $PrOs_4Sb_{12}$ [2] and KOs_2O_6. [1, 2, 3] In order to understand low-energy excitation caused by the rattling effect, numerous efforts have been made including ultrasonic experiments. [5, 6, 7]. Recently, an ultrasonic dispersion (UD) and a low-temperature softening due to the off-center rattling and tunneling were found in $LaOs_4Sb_{12}$ and $LaFe_4Sb_{12}$. [4, 8] Intriguingly, the UD were not detected in $LaRu_4Sb_{12}$. [9] Arsenic based filled skutterudite compounds are much less explored, also by means of ultrasound, because significant progress in their crystal growth has been obtained very recently. $LaRu_4As_{12}$ was firstly investigated by ultrasound in the present work. This filled-skutterudite compound shows superconducting transition at $T_{SC} = 10.3$ K. [10] The specific heat data show a relatively large ratio $\Delta C/\gamma T_{SC} \sim 1.74$. [11] Additionally, these features are accompanied by the upper critical field slightly larger than 10 T, as concluded form the electrical resistivity measurements. [12]

2. Experimental Details

Single crystals of $LaRu_4As_{12}$ were grown by a molten metal flux method at high temperatures and pressures, as reported elsewhere. [13] After removing the flux by distillation, the single crystal of As-based filled skutterudite typically has a truncated-octahedron-shape with facets

Published under licence by IOP Publishing Ltd

that are perpendicular to the [111] directions. (See schematic picture in Table 1.) The orientation of crystal was checked by using x-ray Laue diffractions. We have used a single crystal with length of $L = 0.4207$ mm for the present ultrasonic measurement. Piezoelectric $LiNbO_3$ wafer of 36°Y-cut and X-cut with 100 μm thickness were used for generating and detecting longitudinal and transverse ultrasonic waves, respectively. A phase comparator using double balanced mixer can detect the relative change of sound velocity down to 1.4 K by using ^{4}He refrigerator. In the present measurements, the absolute value of the sound velocity could not be estimated because the sample thickness was not enough. Since the UD with upturn due to rattling had been observed in Γ_{23} symmetry-related elastic constants of $R\mathrm{Os_4Sb_{12}}$ (R = La, Pr, Nd) and both Γ_{23} and Γ_4 symmetry modes of $R\mathrm{Fe_4Sb_{12}}$ (R = La, Pr), the both ultrasonic modes should be analyzed spectroscopically in order to search the rattling in the filled skutterudites. In the present research, frequency dependence of the elastic constants C_{L111} and C_{T111} has been investigated on $\mathrm{LaRu_4As_{12}}$. As shown in Table 1, C_{T111} consists both Γ_{23} and Γ_4 symmetry related ultrasonic modes, while C_{L111} consists Γ_1 and Γ_4 symmetry modes in $\mathrm{T_h}$ point group. Here, $C_{\mathrm{B}} = (C_{11} + 2C_{12})/3$ is known as bulk modulus, which is associated with volume change. [14]

Table 1. Elastic Wave, Type, Elastic Constant and Symmetry [14]

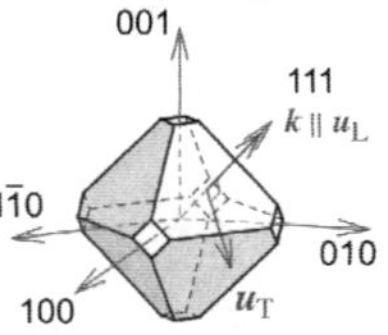

Propagation	Polarization	Type	Elastic Constant	Symmetry
$k \parallel [111]$	$u_{\mathrm{L}} \parallel [111]$	Longitudinal	$C_{\mathrm{L111}} = \frac{C_{11}+2C_{12}}{3} + \frac{4C_{44}}{3}$	$\Gamma_1 \bigoplus \Gamma_4$
	$u_{\mathrm{T}} \perp [111]$	Transverse	$C_{\mathrm{T111}} = \frac{2}{3}\frac{C_{11}-C_{12}}{2} + \frac{C_{44}}{3}$	$\Gamma_{23} \bigoplus \Gamma_4$

3. Results and Discussions

Figure 1 and figure 2 show relative change of elastic constants; the longitudinal C_{L111} and the transverse C_{T111}, of $\mathrm{LaRu_4As_{12}}$, respectively. Both elastic constants C_{L111} and C_{T111} increase monotonically and gradually level off with decreasing temperature. None of clear elastic anomaly was found in the C_{L111} and C_{T111} within the present temperature and frequency ranges of the ultrasonic measurements. In other words, we can conclude that a characteristic upturn of UD due to rattling, which had been observed in $R\mathrm{Os_4Sb_{12}}$ (R = La, Pr, Nd) at 30 − 40 K, will not be present in $\mathrm{LaRu_4As_{12}}$. One can analyze the temperature dependence of the elastic constants of La ($4f^0$)-system, without crystalline electric field effect and/or other quantum effects, by using phenomenological expressions [15];

$$C_{ij} = C_{ij}^0 - \frac{s}{e^{t/T} - 1} \quad (1)$$

where C_{ij}^0, s and t are obtained as fitting parameter for experimental results. For the present results, the fitting function well reproduces the data within the high temperature region of 100 K $< T <$ 300 K, however, in the low temperature region ($T <$ 100 K), the data shows small deviation from the fits indicated by the red and blue solid lines in Fig. 1 and 2, respectively. Following parameters are roughly obtained from the fits; (s, t) = (0.0278, 267.55 K) for C_{L111} at 160 MHz, and (0.0243, 294.78 K) for C_{T111} at 73 MHz. The parameter t is comparable to the Debye temperature of $\mathrm{LaRu_4As_{12}}$ Θ = 233 − 355 K. [7, 8] The inset of Fig. 1 and 2 show a close up of the low temperature data. Vertical dotted lines in the both figures indicate superconducting temperature T_{SC} = 10.3 K. No anomaly was observed at T_{SC} in C_{L111} and

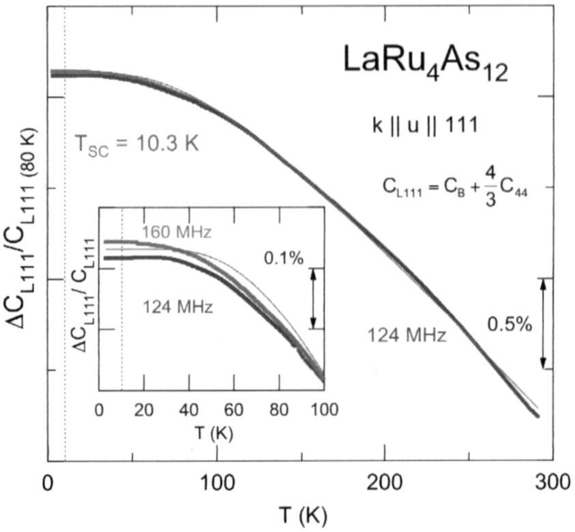

Figure 1. Relative change of elastic constant C_{L111} of $LaRu_4As_{12}$ as a function of temperature at frequencies of 124 MHz and 160 MHz. The inset displays an enlarged view below 80 K. $T_{\mathrm{SC}} = 10.4$ K is indicated by the vertical dotted lines. Solid line (red) indicates calculated elastic constant (see text).

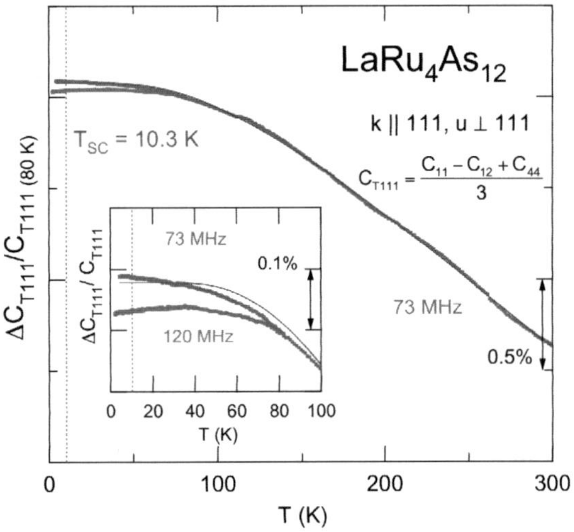

Figure 2. Relative change of elastic constant C_{T111} of $LaRu_4As_{12}$ as a function of temperature at frequencies of 73 MHz and 120 MHz. The inset displays an enlarged view below 80 K. T_{SC} is indicated by the vertical dotted lines. Solid line (blue) indicates calculated elastic constant (see text).

C_{T111}, which is similar behavior to a BCS superconductor $LaOs_4Sb_{12}$. [4] Below 80 K, small deviations between different measurement frequencies were observed with a magnitude of 10^{-5} for $\Delta C_{\mathrm{L111}}/C_{\mathrm{L111}}$. According to the measurement with more low frequency of 8 – 52 MHz (not shown), an origin of the deviation is considered as due to an extrinsic effect. For example, it would be expected that a poor directionality of ultrasonic pulse waves in the small ($L \sim 0.4207$ mm) single crystal in the present measurements even at frequencies of ~100 MHz causes indirectly reflections of the ultrasonic echoes and mixing of transverse mode to the longitudinal mode and vice versa.

On the other hand, the transverse $\Delta C_{\mathrm{T111}}/C_{\mathrm{T111}}$ also shows a deviation of 10^{-4} between 120 MHz and 73 MHz. It also seems to be due to an extrinsic origin, however, C_{T111} exhibits an opposite tendency of the frequency compared to C_{L111}, *i.e.*, C_{T111} with lower frequency of 73 MHz increases while higher frequency of 120 MHz gradually decreases at low temperatures. Since the $LaRu_4As_{12}$ does not have $4f$ electron, the tiny softening in C_{T111} could not be originated in crystalline electric field effect due to orbital degrees of freedom. At first glance, the low-temperature softening of the transverse elastic constant C_{T111} may resemble the softening due to the quantum tunneling ($\propto 1/T$), which had been observed in $LaOs_4Sb_{12}$ and $LaFe_4Sb_{12}$

at temperatures where the thermally-activated rattling is frozen out. [4, 5] However, in the $LaRu_4As_{12}$ case, the softening does not follow the $1/T$ dependence and the characteristic UD was not also found up to 300 K. In order to clarify whether the softening is intrinsic or not, further low-temperature experiments under magnetic fields and in different frequency range are in progress. Finally, we shortly summarize the results of ultrasonic studies on various La-based filled skutterudites. Ultrasonic dispersion was only found in $LaOs_4Sb_{12}$ and $LaFe_4Sb_{12}$, while the UD was not observed in $LaRu_4Sb_{12}$, $LaRu_4As_{12}$ (present work), $LaRu_4P_{12}$, and $LaFe_4P_{12}$. [16] Although, ultrasonic investigations of $LaOs_4P_{12}$, $LaOs_4As_{12}$ and $LaFe_4As_{12}$ have not been reported yet, no ultrasonic dispersion was detected in $PrOs_4As_{12}$ and $PrFe_4As_{12}$. [14, 17] Intriguingly, the only two compounds, which shows ultrasonic dispersion as well as $LaOs_4P_{12}$ (the UD has not been confirmed) exhibit negative hyperfine coupling constant in the La-NMR signal [18]. A microscopic linkage between ultrasonic and NMR measurements may be a key to understand the off-center rattling motion.

This work was supported by Grant-in-Aid for Scientific Research on Innovative Areas "Heavy Electrons" (No. 21102501) of The Ministry of Education, Culture, Sports, Science, and Technology, Japan, by U.S. Department of Energy Grant DE FG02-04ER46105 and U.S. National Science Foundation Grant NSF DMR 0802478 and by Polish Ministry of Science and Higher Education Grant NN 202 4129 33. One of the authors (T.Y.) was supported by Hokkaido Univ. Leader Development System in the Basic Interdisciplinary Research Areas.

References

[1] Sanada S, Aoki Y, Aoki H, Tsuchiya A, Kikuchi D, Sugawara H and Sato H 2005 *J. Phys. Soc. Jpn* **74** 246.
[2] Bauer E.D, Frederick N. A, Ho, P.-C, Zapf V. S and Maple M. B 2002 *Phys. Rev. B* **65** 100506.
[3] Hiroi Z, Yonezawa S, Nagao Y, and Yamaura J 2007 *Phys. Rev. B* **76** 014523.
[4] Nemoto Y, Yanagisawa T, Yasumoto Y, Kobayashi H, Yamaguchi A, Tsuduku S, Goto T, Takeda N, Ochiai A, Sugawara H, Sato H and Kitazawa H 2008 *J. Phys. Soc. Jpn.* **77** Suppl. A 153.
[5] Nemoto Y, Yamaguchi T, Horino T, Akatsu M, Yanagisawa T, Goto T, Suzuki O, Dönni A and Komatsubara T 2003 *Phys. Rev. B* **68** 184109.
[6] Goto T, Nemoto Y, Yamaguchi T, Akatsu M, Yanagisawa T, Suzuki O and Kitazawa K 2004 *Phys. Rev. B* **70** 184126.
[7] Goto T, Nemoto Y, Yamaguchi T, Akatsu M, Yanagisawa T, Suzuki, Dönni A and Komatsubara T 2004 *Phys. Rev. B* **69** 180511.
[8] Ishii I, Fujita T, Mori I, Sugawara H, Yoshizawa M, Takegahara K, and Suzuki T 2009 *J. Phys. Soc. Jpn.* **78** 084601.
[9] Kumagai T, Nakanishi Y, Sugawara H, Sato H and Yoshizawa M 2003 *Physica B* **329-333** 471.
[10] Shirotani I, Uchiumi T, Ohno K, Sekine C, Nakazawa Y, Kanoda K, Todo S, Yagi T 1997 *Phys. Rev. B* **56** 7866.
[11] Namiki T, Sekine C, Matsuhira K, Wakeshima M and Shirotani T 2008 *J. Phys. Soc. Jpn.* **77** 336.
[12] L. Bochenk at el. in preparation.
[13] Henkie Z, Maple M. B, Pietraszko A, Wawryk R, Cichorek T, Baumbach R. E, Yuhasz W. M, and Ho P.-C 2008 *J. Phys. Soc. Jpn.* **77** Suppl. A 128.
[14] Yanagisawa T, Yasumoto Y, Nemoto Y, Goto T, Yuhasz W. M, Ho P.-C, Henkie Z, Pietraszko A, and Maple M. B 2008 *J. Phys. Soc. Jpn.* **77** Suppl. A 225.
[15] Lüthi B, *Physical Acoustics in the Solid State* (Springer, Heidelberg, 2005) Vol. 148.
[16] *News Letter of Emergency of Heavy Electrons and Their Ordering* 2010 Vol. 2, No. 2 pp. 13 (Japanese), and private communications with Dr. Y. Nakanishi.
[17] Sayles T. A, Yuhasz W. M, Paglione J, Yanagisawa T, Jeffries J. R, Maple M. B, Henkie Z, Pietraszko A, Cichorek T, Wawryk R, Nemoto Y and Goto T 2008 *Phys. Rev. B* **77** 144432.
[18] Nakai Y, Ishida K, Sugawara H, Kikuchi D and Sato H 2008 *Phys. Rev. B* **77** 041101.

Journal of Physics: Conference Series **273** (2011) 012066 doi:10.1088/1742-6596/273/1/012066

A predictive standard model for heavy electron systems

Yi-feng Yang[1,2], N J Curro[3], Z Fisk[4], D Pines[3], J D Thompson[1]

[1] Los Alamos National Laboratory, Los Alamos, NM 87545, USA
[2] Institute of Physics, Chinese Academy of Sciences, Beijing 100190, China
[3] Department of Physics, University of California, Davis, CA 95616, USA
[4] Department of Physics and Astronomy, University of California, Irvine, CA 92697, USA

E-mail: yifengyyf@gmail.com

Abstract. We propose a predictive standard model for heavy electron systems based on a detailed phenomenological two-fluid description of existing experimental data. It leads to a new phase diagram that replaces the Doniach picture, describes the emergent anomalous scaling behavior of the heavy electron (Kondo) liquid measured below the lattice coherence temperature, T^*, seen by many different experimental probes, that marks the onset of collective hybridization, and enables one to obtain important information on quantum criticality and the superconducting/antiferromagnetic states at low temperatures. Because T^* is $\sim J^2\rho/2$, the nearest neighbor RKKY interaction, a knowledge of the single-ion Kondo coupling, J, to the background conduction electron density of states, ρ, makes it possible to predict Kondo liquid behavior, and to estimate its maximum superconducting transition temperature in both existing and newly discovered heavy electron families.

1. Introduction

Although we have yet to obtain a microscopic solution of the Kondo lattice problem that serves as a model for heavy electron materials, thanks to a recent phenomenological analysis [1, 2, 3, 4] of decades of experimental efforts on heavy electron behavior, we now have an experiment-based model for their behavior that explains existing experiments and has predictive power in the exploration of new families of heavy electron materials. Our model provides a unified explanation of the emergence of the heavy electron Kondo liquid that is responsible for the anomalies seen in the Knight shift, magneto-transport, and other experiments, leads to a new phase diagram that replaces the Doniach picture, and has the potential for becoming the standard model of heavy electron physics. It contains several basic aspects:

- There exists a unique temperature scale T^* for each heavy electron material; it marks the emergence of a new state of matter, a coherent heavy electron (Kondo) liquid that is produced by the collective hybridization of the local moments with the conduction electron sea, and exhibits universal behavior that scales with T^*[1].
- T^* is the nearest neighbor RKKY (Ruderman-Kittel-Kasuya-Yosida) local moment interaction; for the local moments it separates their high-T weakly interacting phase from their low-T entangled phase. It is given by $\sim 0.45J^2\rho$, where J is the single-ion Kondo coupling to the background conduction electron density of states ρ [2].
- The entangled phase below T^* is described by a two-fluid model [1, 5, 6] in which an order parameter, $f(T/T^*)$, characterizes the reduction of the overall system entropy brought about by the transfer of spectral weight from the f-electron local moments to the itinerant heavy electron Kondo liquid formed out of the conduction electron sea.

Published under licence by IOP Publishing Ltd

- The quasiparticles of the Kondo liquid exhibit coherent non-Landau Fermi liquid behavior; their spin susceptibility and specific heat vary as $\ln(T^*/T)$ down to a crossover temperature T_0.
- Their behavior below T_0 can lead to unconventional superconductivity produced by quantum critical fluctuations [4], or an antiferromagntic ordering of the local moments in which Kondo liquid relocalization occurs as a precursor [7].

The position of a number of heavy electron materials on our proposed new phase diagram is given in Fig. 1, where we see that $J\rho \sim 0.15$ roughly separates the antiferromagnets and superconductors. In striking contrast to the conventional Doniach diagram which postulates a competition between collective behavior produced by RKKY interaction and single-ion behavior characterized by the single-ion Kondo temperature, T_K, the latter quantity is found to be much smaller than T^* for almost all the materials considered there. In Fig. 2 we plot the variation of T^*, T_0, T_K, and the ordering temperatures, T_N and T_c as a function of pressure for the heavy electron material $CeRhIn_5$. There one can see how the competition between localization (antiferromagnetic order) and itinerancy (superconductivity) for Kondo liquid quasiparticles plays out as $J\rho$ increases.

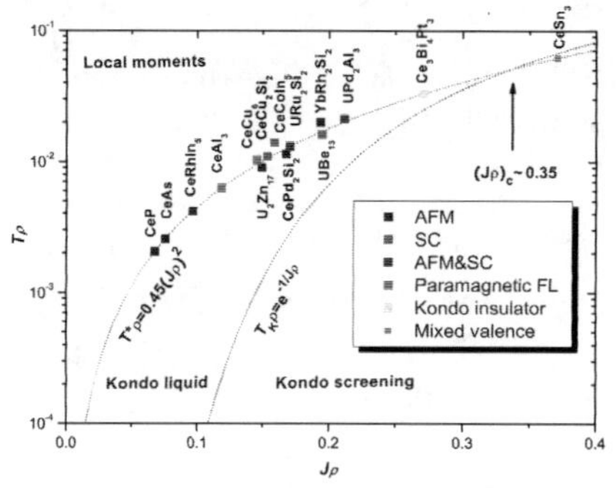

Figure 1. General phase diagram. J is the local Kondo coupling and ρ is the conduction electron density of states. T^* marks the onset of coherence, spin correlations and anomalies.

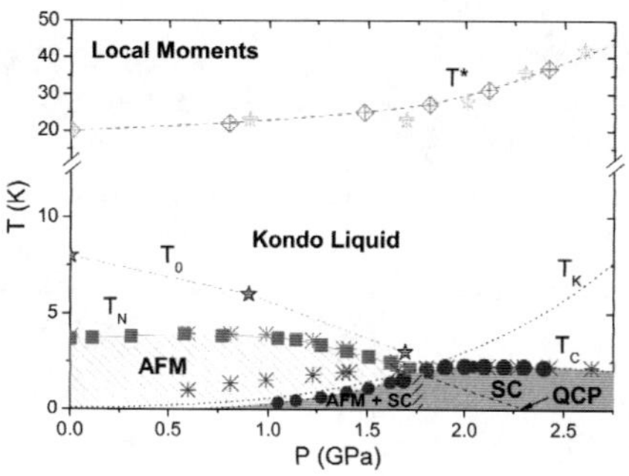

Figure 2. New phase diagram of $CeRhIn_5$ [8]. T^* is from resistivity peak and Hall anomaly. T_0 is from Hall anomaly [9]. T_K is estimated from $Ce_{1-x}La_xRhIn_5$.

2. The physical origin and determination of the fundamental temperature scale T^*

The Kondo coupling J between a localized f-moment and conduction electron spins gives rise to two competing characteristic temperature scales in a Kondo lattice:

$$k_B T_K = \rho^{-1} e^{-1/J\rho}, \quad (1)$$

$$k_B T_{RKKY} = cJ^2\rho, \quad (2)$$

where ρ, the density of states of conduction electrons, can be determined by the Sommerfeld coefficient γ of the corresponding nonmagnetic compound (e.g., $LaCoIn_5$ for $CeCoIn_5$), $\gamma = \pi^2 k_B^2 \rho/3$. After a correction to account for change of chemical pressure upon doping, the value of J for the Kondo lattice material can be determined by measurements of T_K in the diluted compound (e.g., $Ce_{1-x}La_xCoIn_5$), where it is given by $\pi^2 k_B^2/3\gamma \ln(\pi^2 k_B/3\gamma T_K)$. An analysis [2] of dozens of heavy electron materials and the pressure data of $CeRhIn_5$ showed that in all cases of interest, T_K was quite small compared to T^*, so that it was natural to inquire whether T^* might be given by the RKKY interaction, Eq. (2), with a near-universal value of c. As may be seen in Fig. 1, this is indeed the case, with a best fit analysis yielding $c \sim 0.45$, so that

$$k_B T^* = 0.45 J^2 \rho. \quad (3)$$

3. The emergent heavy electron liquid

Although T^* marks the onset of coherent collective hybridization, the process is not instantaneous; the Kondo liquid emerges over a broad temperature range from T^* to a much lower cut-off temperature, T_0, and throughout this temperature region, one therefore expects two components: the emergent Kondo non-Landau Fermi liquid composed of conduction electrons made heavy by their collective hybridization with the local moments and a local moment spin liquid formed by the substantial fraction of local moments that have not hybridized.

One is thus led to postulate a two-fluid model [5] with an order parameter $f(T/T^*)$ that denotes the fraction of the f-moments that participate in the collective hybridization; both $f(T/T^*)$ and the scaling properties of the Kondo liquid can be determined directly from experiments displaying the various heavy electron anomalies. Consider, for example, Knight shift measurements [6]. Since the f-moments are fully localized above T^*, the Knight shift of a given probe nucleus measures directly their susceptibility. Below T^*, it probes both the local moment and Kondo liquid contributions to the total magnetic susceptibility, χ. But since it couples differently to each, it will no longer be proportional to the total susceptibility, hence the appearance of the anomaly seen beginning at T^*. The two-fluid model expressions are thus [1]:

$$\chi = f(T)\chi_h + [1 - f(T)]\chi_l, \tag{4}$$

$$K = K_0 + Af(T)\chi_h + B[1 - f(T)]\chi_l, \tag{5}$$

where χ_h and χ_l are the intrinsic susceptibility of the heavy electrons and the unhybridized local moments, respectively. A and B are their hyperfine couplings. Since the Knight shift anomaly below T^* is given by, $K_a = K - K_0 - B\chi = (A - B)f(T)\chi_h$, we see that it acts as a direct probe of the emergent Kondo liquid susceptibility. A careful analysis of the Knight shift and other experiments on $CeCoIn_5$ shows that $f(T/T^*) = f_0(1 - T/T^*)^{3/2}$, while if one requires that the intrinsic entropy of the Kondo liquid approaches the free moment entropy $R\ln 2$ at T^*,

$$\chi_h = R_W \frac{R\ln 2}{2T^*}\left(1 + \ln\frac{T^*}{T}\right), \tag{6}$$

where R_W is the Wilson ratio. Knight shift measurements on other heavy electron materials and fits to the anomalous behavior seen in magneto-transport, point contact spectroscopy, and Raman spectroscopy show that these expressions are universal [1]. A comparison to the recent state-of-the-art DMFT calculations [10] shows good agreement with the Kondo liquid quasi-particle density of states determined from Eq. (6), and demonstràtes that that these contain the feedback needed to achieve collective hybridization.

4. Superconducting condensation and antiferromagnetic relocalization

The low temperature cut-off at T_0 in the measured scaling behavior of the Kondo liquid marks the onset of new physical processes that take precedence over its growth due to collective hybridization. The two-fluid model continues to apply in this region, and provides important information about both the unconventional superconducting states and the onset of antiferromagnetism observed in heavy electron materials. Knight shift measurements are especially informative, as these provide direct information on changes in the Kondo liquid as the material becomes superconducting or antiferromagnetic. An analysis of $CeCoIn_5$ confirms the universal behavior of the Kondo liquid [4]: it follows two-fluid prediction above T_0 ($\sim T_c$) and, upon becoming superconducting, follows the BCS d-wave prediction below T_c, independent of the probe nuclei. Important information on the pairing mechanism for the measured d-wave superconductivity can be obtained from the NQR spin-lattice relaxation rate T_1^{-1} by using the two-fluid formula, $T_1^{-1} = f(T)T_{1h}^{-1} + [1 - f(T)]T_{1l}^{-1}$, where the local moment T_{1l}^{-1} is given by the high temperature fit. A simple two-fluid analysis reveals quantum critical fluctuations in the Kondo liquid, and yields [4],

$$T_{1h}T \propto (T + T_{QCP}), \tag{7}$$

where T_{QCP} is the distance from the quantum critical point.

A similar two-fluid analysis applied to heavy electron antiferromagnets $CeRhIn_5$ and $CePt_2In_7$ has recently led to the discovery that at a temperature T_0 well above the Néel temperature, the Kondo liquid begins to relocalize, with a corresponding decrease in $f(T)$, confirming the important role played by the Kondo liquid in bringing about the antiferromagnetic order of both it and the residual local moments present at T_N.

5. Toward a standard predictive model of heavy electron materials

By analogy to the idea of a standard model based on a well-defined microscopic theory, we have proposed here an experiment-based standard phenomenological model for heavy electron materials that derives from the successful systematic examination of existing experiments on a broad spectrum of heavy electron materials. Its predictions for heavy electron emergence at T^* accompanied by anomalous scaling behavior can easily be checked for any newly-discovered heavy electron family, and have recently been confirmed by experiments on the 127 family.

Of particular interest are the correlations in our model between measured values of $J\rho$ and the appearance of antiferromagnetism and superconductivity at low temperatures. As we have noted, materials with $J\rho < 0.15$ all appear to be antiferromagnets, but can easily be made into superconductors by applying pressure to increase $J\rho$, as Fig. 2 demonstrates vividly for the Rh-115 family. Moreover, since the superconducting transition temperature seems to scale with T^*, a strategy for finding a higher T_c family of heavy electron materials might begin by searching for materials with large values of T^*, a strategy that proved successful in the discovery of the PuGa members of the 115 family. We note this can be achieved either by finding materials with values of $J\rho$ that significantly exceed 0.15, or, with $J\rho \sim 0.15$, finding materials that possess a comparatively low value of ρ, or equivalently a large conduction electron bandwidth, and a correspondingly high value of J. Moreover, to the extent that T_c scales with T^*, for a fixed value of $J\rho$, one would have T_c scaling with J or the conduction electron bandwidth $W \sim \rho^{-1}$.

Our model is based on phenomenology, and represents a summary of our efforts to understand what experiment is telling us about the physics of Kondo lattice systems. It is incomplete, in that it poses the questions to be answered by microscopic theory but does not answer them. The remarkably successful calculations for CeIrIn5 using DMFT represent a significant first step toward finding those answers [10]. As noted above, this is not a single site calculation; it yields a quantitative account of collective hybridization in this material by incorporating the feedback of the hybridization at one Kondo lattice site on the hybridization at its neighboring site. But what is needed are transparent microscopic calculations, beginning with perturbation theory, that point the way to the eventual renormalization group analysis needed to provide the desired microscopic theory of the collective hybridization that experiment suggests is brought about by the nearest neighbor coupling between the localized spins.

Acknowledgments

Y.Y. thanks H.-O. Lee and T. Park for experimental data of $CeRhIn_5$. Y.Y. and D.P. acknowledge ICAM support. Z.F. acknowledges support through NSF grant DMR-0801253. Work at Los Alamos was performed under the auspices of the US Department of Energy.

References

[1] Yang Y-F and Pines D 2008 *Phys. Rev. Lett.* **100** 096404
[2] Yang Y-F, Fisk Z, Lee H-O, Thompson J D and Pines D 2008 *Nature* **454** 611
[3] Yang Y-F 2009 *Phys. Rev. B* **79** 241107(R)
[4] Yang Y-F, Urbano R R, Curro N J, Pines D and Bauer E D 2009 *Phys. Rev. Lett.* **103** 197004
[5] Nakatsuji S, Pines D and Fisk Z 2004 *Phys. Rev. Lett.* **92** 016401
[6] Curro N J, Young B-L, Schmalian J and Pines D 2004 *Phys. Rev. B* **70** 235117
[7] apRoberts-Warren N, Dioguardi A P, Shockley A C, Lin C H, Crocker J, Klavins P, Pines D, Yang Y-F and Curro N J, unpublished.
[8] Park T and Thompson J D 2009 *New J. Phys.* **11** 055062
[9] Nakajima Y, Shishido H, Nakai H, Shibauchi T, Behnia K, Izawa K, Hedo M, Uwatoko Y, Matsumoto T, Settai R, Onuki Y, Kontani H and Matsuda Y 2007 *J. Phys. Soc. Jpn.* **76** 024703
[10] Shim J H, Haule K and Kotliar G 2007 *Science* **318** 1615

Journal of Physics: Conference Series **273** (2011) 012067 doi:10.1088/1742-6596/273/1/012067

Soft X-ray angle-resolved photoemission study of $YbCu_2Ge_2$

A. Yasui[1], S.-I. Fujimori[1], I. Kawasaki[1], T. Okane[1], Y. Takeda[1], Y. Saitoh[1], H. Yamagami[1,2], A. Sekiyama[3], R. Settai[4], T. D. Matsuda[5], Y. Haga[5], and Y. Ōnuki[4,5]

[1]Condensed Matter Science Division, Japan Atomic Energy Agency, SPring-8, Sayo, Hyogo 679-5148, Japan

[2]Department of Physics, Kyoto Sangyo University, Kyoto 603-8555, Japan

[3]Division of Materials Physics, Graduate School of Engineering Science, Osaka University, Toyonaka, Osaka 560-8531, Japan

[4]Department of Physics, Graduate School of Science, Osaka University, Toyonaka, Osaka 560-0043, Japan

[5]Advanced Science Research Center, Japan Atomic Energy Agency, Tokai, Ibaraki 319-1195, Japan

E-mail: a-yasui@spring8.or.jp

Abstract. Soft X-ray photoemission spectroscopies have been performed on $YbCu_2Ge_2$, which is thought to be a Yb divalent system, to investigate its bulk electronic structure including Yb $4f$ electrons. Small but finite Yb^{3+} multiplet peaks were found at 6-12 eV in the angle-integrated spectrum, suggesting that it is a valence fluctuation compound. An "anti-crossing" Yb $4f$ bands with conduction bands were observed in the angle-resolved spectrum. This is the evidence that the Yb $4f$ electrons form a band state through the hybridization with Ge $4p$ and Cu $3d$ electrons.

1. Introduction

Yb-based compounds have been attracting much attention due to their interesting physical properties [1, 2, 3]. As a first step toward a unified understanding of these properties, it is important to examine the electronic structure of the valence bands containing the Yb $4f$ electrons. An angle-resolved photoemission spectroscopy (ARPES) is a powerful experimental tool to directly observe the valence bands and the Fermi surfaces. Although many angle-integrated photoemission spectroscopies (AIPES) have been performed for various Yb compounds, there are only a few ARPES studies for $YbRh_2Si_2$ and $YbIr_2Si_2$ [4]. The previous ARPES studies have utilized the incident photon of 21.2 to 110 eV. However, those ARPES measurements are surface sensitive and obtained spectra have considerably large contributions of the surface Yb^{2+} state. On the other hands, in the bulk-sensitive soft X-ray ARPES (SX-ARPES) study, the surface Yb^{2+} contribution is suppressed and thus the spectra can be directly compared with the band dispersion and the Fermi surface predicted from band structure calculations. For example, Fermi surfaces obtained by the recent SX-ARPES for $CeRu_2Si_2$ can be explained by a band structure calculation based on a local-density-approximation (LDA) [7], while they are quite different from those obtained by the previous ARPES with the incident photon of $\sim$ 100 eV [8].

Published under licence by IOP Publishing Ltd

$YbCu_2Ge_2$ crystallizes in the body-centered tetragonal $ThCr_2Si_2$ crystal structure with one formula unit per primitive cell. The NMR study [5], and the X-ray absorption spectroscopy study [6] suggested that it is a divalent compound. Dung *et al.* carried out various transport and thermodynamic measurements and a de Haas-van Alphen (dHvA) experiment on this compound [9]. They clarified that $YbCu_2Ge_2$ shows the Pauli paramagnetic behavior with the small electronic specific coefficient γ of 9-10 mJ/(mol K^2). This small γ value indicates that $4f$ bands are located far from the Fermi energy E_F. In addition, the result of the dHvA experiment was explained by "Yb^{2+} model" which was performed by a calculation within such a situation. However, from those previous experiments, it has been not yet known the information of Yb $4f$ electrons, for example, where $4f$ bands are located and whether the $4f$ electrons hybridize with the conduction electrons (c-f hybridization) or not. It is important to clarify those kind of the information for various valent compounds to understand the formation of the heavy fermion state in the Yb system. Additionally, in the Yb divalent system, the Yb atom often seems to be Yb^{2+} ion. That means that the Yb $4f$ bands should be flat in the reciprocal space and be separated with the other conduction band in the language of the band picture. Nevertheless, such band structures can not be explained from the band calculation. In order to clarify the bulk valence-band structure, we have carried out the SX-AIPES and -ARPES measurements for $YbCu_2Ge_2$.

2. Experiment and calculation

Single crystals of $YbCu_2Ge_2$ were grown by the Sn-flux method [9]. AIPES and ARPES measurements were performed at soft X-ray beamline BL23SU in SPring-8. We have used the incident photon energy of $h\nu = 450$ eV which corresponds to the high-symmetry plane including the Γ point by assuming the inner potential value of 12 eV. The energy resolution was 90 meV for both AIPES and ARPES measurements. Clean sample surfaces were obtained by *in situ* cleaving samples along [001] surface. The sample temperature was kept at 20 K because $YbCu_2Ge_2$ does not show the phase transition over all temperature region [9] and the phonon broadening is negligible at the temperature. The valence band structure of $YbCu_2Ge_2$ was calculated by using a first-principles LDA calculation with the relativistic linearized-augmented-plane-wave (RLAPW) method [10]. This calculation is based on the Dirac one electron equation, so that all the relativistic effects including the spin-orbit (SO) interaction is directly taken into account. In this calculation, we used lattice constants and each atomic position determined by the X-ray diffraction measurement [9].

3. Results and discussion

3.1. Angle-integrated spectrum

Figure 1(a) shows the AIPES spectrum of the valence band of $YbCu_2Ge_2$. The Yb $4f$ states and the Cu $3d$ states are mainly observed because their photoionization cross sections (I) are larger than other states in the photon energy of $h\nu = 450$ eV ($I_{\rm Yb} : I_{\rm Cu} : I_{\rm other} \approx 1 : 1/6 : 1/200$). Two sharp peaks centered at 0.12 eV and 1.42 eV are the SO split bands of the bulk Yb^{2+} $4f$ states with the splitting energy of 1.3 eV. In this SX-AIPES spectrum, the surface Yb^{2+} $4f$ states which should be observed at the higher binding energy side of the bulk Yb^{2+} $4f$ peak were hardly observed. The broad peaks centered at 4 eV are originated from the Cu $3d$ state. Contrary to our expectation, we found small but clear Yb^{3+} multiplet peaks at 6-12 eV, which are well reproduced by atomic calculations [11]. Because of no phase transition over temperatures and only one Yb site in an unit cell, that indicates that $YbCu_2Ge_2$ is a valence fluctuation compound. The existence of the Yb^{3+} multiplet feature in the AIPES spectrum was not predictable from the transport and thermodynamic measurements [9], the NMR one [5], and the X-ray absorption spectroscopy [6]. In these measurements, it may be difficult to separate Yb^{2+} and Yb^{3+} components in valence fluctuation compounds. In the photoemission

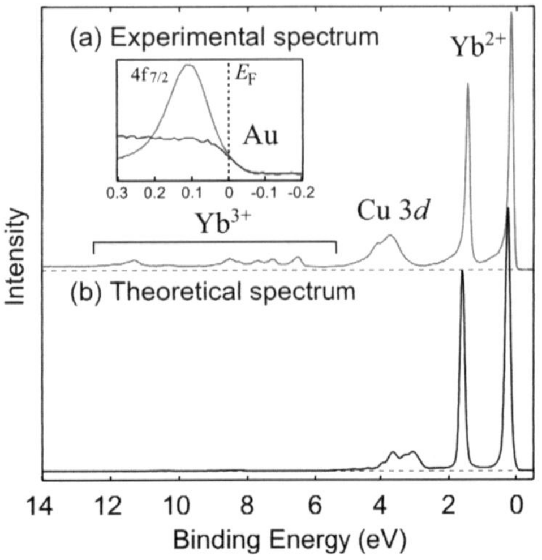

Figure 1. (a) Experimental AIPES spectrum and (b) theoretical one for $YbCu_2Ge_2$. Near-E_F region together with the spectrum of Au (blue line) is shown in the inset of (a). The theoretical spectrum is obtained by the RLAPW calculation. It should be noted that the experimental spectrum (a) is the momentum-integrated one in the $\boldsymbol{k}$ range covering larger than the first Brillouin zone (BZ) in the k_x-k_y plane including the Γ point because of using the single crystal in this experiment. On the other hand, the theoretical one (b) was obtained by the integration over the first BZ.

measurements, Yb^{2+} and Yb^{3+} states are well separated ($\sim$ 5 eV). It should be noted that the Yb^{3+} multiplets are not caused by the surface oxidation because O 1s peaks were not observed during this measurement. We exhibit the near-E_F region together with a spectrum of Au in the inset of Fig. 1(a). From the band picture, the existence of the Yb^{3+} multiplets indicates that the tail of the Yb^{2+} $4f_{7/2}$ peak should cross E_F. However the clear Fermi edge was not observed, suggesting that the Yb $4f$ states do not mainly contribute to the Fermi surfaces. That may be a reason why $YbCu_2Ge_2$ seems to be the divalent compound on the previous experimental studies.

In Fig. 1(b), we show a theoretical spectrum simulated from partial density-of-states obtained by the RLAPW calculation, taking into account the photoionization cross sections of each atomic orbital, the energy resolution, and Fermi-Dirac function. The theoretical spectrum is in a qualitative agreement with the experimental one, except for the presence of the Yb^{3+} multiplets. The peak position of the Yb^{2+} $4f$ doublet is almost the same value as the experimental spectrum, while the Cu $3d$ band lies at the lower binding energy side than the experimental one.

3.2. Angle-resolved spectrum

In order to know the c-f hybridization effects in the valence bands, we have performed the ARPES measurements. The ARPES spectra along the X$-\Sigma-$X line is shown in Figs. 2(a) and (b). Figure 2(a) depicts the spectra normalized with the integrated intensity of the energy-distribution-curve (EDC) at each k-point. In the EDC spectra, two flat peaks with high intensity are Yb^{2+} $4f$ bands and their peak tops denote by dashed line in Figs. 2(a) and (b). It should be noted that the $4f_{7/2}$-band locates a little away from E_F as shown in Fig. 1(a). The Cu $3d$ band locates at 3-5 eV with large dispersion and the flat-band centered at 6.5 eV is a part of Yb^{3+} multiplet peaks. The sinusoidal structure lying on the $4f_{5/2}$-band is conduction bands. We made a second derivative (SD) along the momentum axis to show the peak position of the conduction bands more clearly as shown in Fig. 2(b). In the SD spectra, the intensity of dispersive bands are enhanced, while that of flat bands are weakened. An anti-crossing feature of the Yb $4f_{5/2}$ bands with the conduction bands is clearly observed at the middle point of X$-\Sigma$. This is the evidence for the hybridization between the Yb $4f$ electrons and the conduction electrons in the bulk state. From the calculated result, the intersection part of the sinusoidal conduction bands with Yb $4f_{5/2}$ bands is the hybridized band mainly due to the Ge $4p$ and Cu $3d$ electrons.

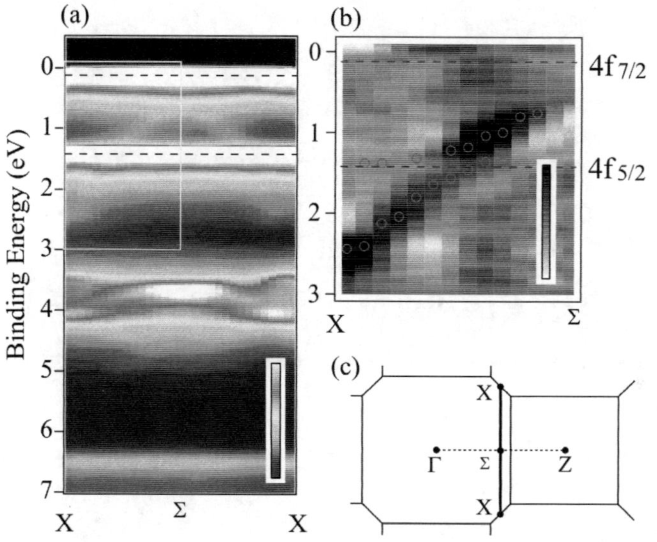

Figure 2. (a,b) ARPES spectra along the X−Σ−X line. The EDC spectra are shown in (a), while (b) shows the negative second derivatives spectra of MDC spectra in the yellow rectangular area of (a). The broken lines indicate peak positions of Yb $4f$ bands. The open circles represent peak positions of hybridized bands between Yb $4f_{5/2}$ bands and conduction bands in (b). (c) Brillouin zone and measured $\boldsymbol{k}$ range (bold line) in (a) and (b).

4. Conclusion

We have performed the soft X-ray photoemission spectroscopy to investigate the bulk band structure of $YbCu_2Ge_2$. $YbCu_2Ge_2$ has finite Yb^{3+} components, indicating that it is a valence fluctuation compound. That is not expected from the previous experimental studies. The anti-crossing feature of Yb $4f$ bands with conduction bands shows that the Yb $4f$ electrons form the band states through the hybridization with the conduction electron. The $4f$ bands should cross E_F somewhere and make Fermi surfaces because the Yb^{3+} multiplets exist in our spectra. In order to find the evidence of Yb $4f$ bands crossing E_F, we have carried out ARPES measurements on the high symmetry plane and analyzed the experimental data.

Acknowledgements

This work was performed under the Proposal No. 2009B3831 at SPring-8 BL23SU and supported by a Grant-in-Aid for Scientific Research on Innovative Areas "Heavy Electrons" (No. 20102003) from the Ministry of Education, Culture, Sports, Science and Technology, Japan.

References

[1] Stewart G R 1984 *Rev. Mod. Phys.* **56** 755
[2] Stewart G R 2001 *Rev. Mod. Phys.* **73** 797
[3] Stewart G R 2006 *Rev. Mod. Phys.* **78** 743
[4] For expample, Danzenbächer S, Kucherenko Yu, Laubschat C, Vyalikh D V, Hossain Z, Geibel C, Zhou X J, Yang W L, Mannella N, Hussain Z, Shen Z-X and Molodtsov S L 2006 *Phys. Rev. Lett.* **96** 106402
[5] Sampathkumaran E V, Gupta L C and Vijayaraghavan R 1979 *J. Phys.* C **12** 4323
[6] Padalia B D, Hatwar T K and Ghatikar M N 1983 *J. Phys.* C **16** 1537
[7] Okane T, Ohkochi T, Takeda Y, Fujimori S -I, Yasui A, Saitoh Y, Yamagami H, Fujimori A, Matsumoto Y, Sugi M, Kimura N, Komatsubara T and Aoki H 2009 *Phys. Rev. Lett.* **102** 216401
[8] Denlinger J D, Gweonb G -H, Allen J W, Olson C G, Maple M B, Sarrao J L, Armstrong P E, Fisk Z and Yamagami H 2001 *J. Electron Spectrosc. Relat. Phenom.* **117** 347
[9] Dung N D, Matsuda T D, Haga Y, Ikeda S, Yamamoto E, Ishikura T, Endo T, Tatsuoka S, Aoki Y, Sato H, Takeuchi T, Settai R, Harima H and Ōnuki Y 2009 *J. Phys. Soc. Jpn.* **78** 084711
[10] Yamagami H 1998 *J. Phys. Soc. Jpn.* **67** 3176
[11] Gerken F1983 *J. Phys.* F **13** 703

Superconducting pairing of interacting electrons: implications from the two-impurity Anderson model

Lijun Zhu and Jian-Xin Zhu

Theoretical Division and Center for Nonlinear Studies, Los Alamos National Laboratory, Los Alamos, New Mexico 87545, USA

E-mail: ljzhu@lanl.gov

Abstract.
We study the non-local superconducting pairing of two interacting Anderson impurities, which has an instability near the quantum critical point from the competition between the Kondo effect and an antiferromagnetic inter-impurity spin exchange interaction. As revealed by the dynamics over the whole energy range, the superconducting pairing fluctuations acquire considerable strength from an energy scale much higher than the characteristic spin fluctuation scale while the low energy behaviors follow those of the staggered spin susceptibility. We argue that the glue to the superconducting pairing is not the spin fluctuations, but rather the effective Coulomb interaction. On the other hand, critical spin fluctuations in the vicinity of quantum criticality are also crucial to a superconducting pairing instability, by preventing a Fermi liquid fixed point being reached to keep the superconducting pairing fluctuations finite at low energies. A superconducting order, to reduce the accumulated entropy carried by the critical degrees of freedom, may arise favorably from this instability.

1. Introduction

In conventional superconductors, the electron pairing arises from the mediation of lattice vibrations [1]. In cuprate superconductors, the pairing mechanism for superconductivity has been hotly debated as to whether the pairing occurs via the coupling between fermionic quasiparticles and bosonic modes, or whether a bosonic glue is really necessary [2]. In the former type of mechanism, the scenario of electronic coupling to spin fluctuations is the leading contender [3]. However, in high-T_c cuprates, the antiferromagnetic Néel state is terminated far before the unconventional superconductivity emerges as the hole doping varies. As such, the relation between a magnetic quantum criticality point and high-T_c superconductivity has not been developed. Furthermore, the final identification of the mechanism is also complicated by the non-negligible electron-phonon coupling as revealed in various experiments [4]. In this aspect, heavy fermion materials have been providing test grounds for novel mechanism for superconductivity [5]. In these systems, the interplay between magnetism and superconductivity has been observed unambiguously. The generic phase diagram shows that, when a tuning parameter such as doping, magnetic field, and pressure is varied, the magnetic state is either interrupted by or in close vicinity to the appearance of a superconducting state. Therefore, they are the most ideal systems for the study on unconventional superconductivity associated with strong electron correlations in general and the quantum criticality in specific.

Published under licence by IOP Publishing Ltd

To address the relation between superconductivity and magnetic quantum criticality, we consider here a two-impurity Anderson model, which is a minimal setting for antiferromagnetism and non-local superconducting pairing. It is also the quantum impurity/cluster model for the periodical Anderson lattice model or the Hubbard model within the cluster dynamical mean-field theoretical approach [6]. This model captures the essence of the competition between the local Kondo coupling between spins of the localized electrons to a conduction band, and the non-local spin exchange interaction between spins of localized electrons, e.g., Ruderman-Kittel-Kasuya-Yosida (RKKY) interaction in heavy fermion materials. Due to this competition, a quantum phase transition occurs between a Kondo resonance state and an inter-impurity spin singlet state [7, 8, 9, 10, 11, 12, 13, 14], and multiple energy scales can show up [15]. In this paper, we discuss the superconducting pairing between two Anderson orbitals and its mechanism.

2. Model

The Hamiltonian for the two-impurity Anderson model can be written as

$$H = \sum_{\mathbf{k}\sigma} \epsilon_{\mathbf{k}} c^\dagger_{\mathbf{k}\sigma} c_{\mathbf{k}\sigma} + \sum_{\mathbf{k}\sigma i} (V_{\mathbf{k}} e^{i\mathbf{k}\cdot\mathbf{r}_i} c^\dagger_{\mathbf{k}\sigma} f_{i\sigma} + \text{h.c.}) + \sum_{i\sigma} (\epsilon_f f^\dagger_{i\sigma} f_{i\sigma} + \frac{U}{2} n_{fi\sigma} n_{fi\bar{\sigma}}) + I\mathbf{S}_1 \cdot \mathbf{S}_2 \,,\quad (1)$$

which describes two interacting local orbitals $f_{i\sigma}$ $(i = 1, 2)$, with energy level ϵ_f and onsite Coulomb repulsion U, in hybridization with a non-interacting conduction electron band $c_{\mathbf{k}\sigma}$ with the strength $V_{\mathbf{k}}$ at each impurity site $\mathbf{r}_i$. In terms of the even or odd parity combinations $f_{p=(e,o),\sigma} = (f_{1\sigma} \pm f_{2\sigma})/\sqrt{2}$, the fluctuations due to conduction electrons can be represented by two separate baths with the hybridization functions $\Gamma_{e,o}(\omega) = -\frac{1}{2}\text{Im}[\sum_{\mathbf{k}} V^2_{\mathbf{k}} |e^{i\mathbf{k}\cdot\mathbf{r}_1} \pm e^{i\mathbf{k}\cdot\mathbf{r}_2}|^2/(\omega - \epsilon_{\mathbf{k}} + i0^+)]$. In a realistic system RKKY interaction is perturbatively generated. Here we consider a case $\Gamma_{e,o}(\omega) = \Gamma_0$ for $|\omega| < D$ (Γ_0 is a constant and D the half bandwidth of the conduction electrons), for which the generated RKKY interaction vanishes, and we tune the intersite spin interaction by a direct RKKY term $I\mathbf{S}_1 \cdot \mathbf{S}_2$.

We focus on a non-local superconducting pairing correlation function $\chi_{sc} = \langle\langle \hat{\Delta}^\dagger; \hat{\Delta} \rangle\rangle$ with $\hat{\Delta}^\dagger = (f^\dagger_{1\uparrow} f^\dagger_{2\downarrow} + f^\dagger_{2\uparrow} f^\dagger_{1\downarrow})/\sqrt{2}$, i.e., in a singlet configuration. To illustrate its relation to spin and charge fluctuations, we also provide results on the uniform and staggered charge susceptibilities $\chi_{c(u,a)} = \langle\langle n_1 \pm n_2; n_1 \pm n_2 \rangle\rangle/2$, the uniform and staggered spin susceptibilities $\chi_{u,a} = \langle\langle S_{1z} \pm S_{2z}; S_{1z} \pm S_{2z} \rangle\rangle/2$, the current fluctuation χ_{curr} with $\hat{\Delta}^\dagger_{curr} = i\sum_\sigma (f^\dagger_{1\sigma} f_{2\sigma} - f^\dagger_{2\sigma} f_{1\sigma})/\sqrt{2}$, as well as the single-particle properties including the spectral function $A_f = -(1/\pi)\text{Im}G_f$ and the interaction part of the self energy $-\text{Im}\Sigma_f$. These quantities are calculated from the complete-Fock-space numerical renormalization group method (CFS-NRG) [16] while a two-particle Green's function method [17] is adopted for calculating the self energy.

3. Correlation fluctuations and the associated energy scales

The results for these dynamical quantities for different values of I are shown in Fig. 1. Some of the calculation details can be found in Ref. [15]. With the tuning of an antiferromagnetic I, there is a change of the ground state from the Kondo resonance state $A_f(\omega = 0) \approx 1/(\pi\Gamma_0)$ [e.g., Fig. 1(a-b)], to the inter-impurity spin singlet state with a pseudogap $A_f(\omega) \sim \omega^2$ [e.g., Fig. 1(d)]. Though the spectral weight has a sudden change at the Fermi energy, this transition is continuous as the effective (local) Fermi liquid (FL) temperature T_F (both competing states belong to FL fixed points) vanishes uniformly at the critical point $I_c \approx 2.3T_K$ [Fig. 1(c)], where T_K is the single-impurity Kondo temperature. χ_u and χ_a have different behaviors. While $\chi''_u(\omega)$ can be characterized by one monotonically increasing energy scale T_{sf}, $\chi''_a(\omega)$ in general has a flattened peak spanning a finite energy range from T_F to T_{sf}. (T_F and T_{sf} correspond to T_L and T_H in Ref. [15]). T_{sf} therefore characterizes a spin fluctuation scale, which takes the value between T_K and I. Between T_F and T_{sf}, there are still finite spectral weights but these

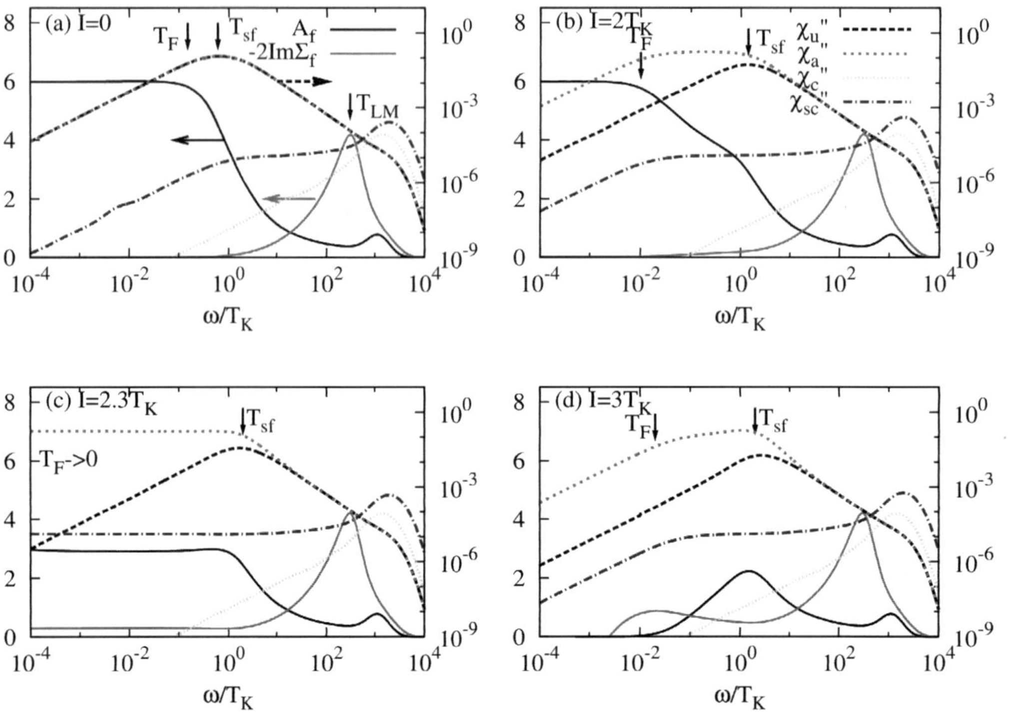

Figure 1. Dynamical properties of the two-impurity Anderson model for various values of the inter-impurity spin exchange interaction I: (a) $I = 0$, (b) $I < I_c$, (c) $I \approx I_c$ and (d) $I > I_c$. The spectral functions A_f (in unit of D), and the imaginary part of the self-energy $-\text{Im}\Sigma_f$ (in unit of $1/D$) are shown in log-linear scale (left y-axis), while the imaginary parts of the uniform spin susceptibility χ''_u, the staggered spin susceptibility χ''_a, the uniform and staggered charge susceptibilities (the same in this case) χ''_c, and the superconducting pairing fluctuation χ''_{sc} (also the half of the current fluctuation χ_{curr} due to the axial charge symmetry), all in unit of $1/T_K$, are shown in log-log scale (right y-axis). Different energy scales, such as the Fermi liquid scale T_F, the spin-fluctuation scale T_{sf}, and the local moment scale T_{LM} are labeled. Here $\Gamma_{e,o} = 0.045\pi D$ and $\epsilon_f = -U/2 = -D \approx -10^3 T_K$. In this case, A_f and $-\text{Im}\Sigma_f$ are found to be the same for different parities and spins.

quasiparticles develop non-Fermi liquid behaviors, which relate to partially screened Kondo resonances [15].

We now examine other correlation functions. For the charge fluctuations, we find that they play little role in the low energy physics. They have resonance peaks around $U/2$ and decrease as $\sim \omega$ at low energies from a scale T_{LM}. Here, the uniform and staggered charge susceptibilities are found to be the same and remain unchanged with the tuning of I. (However, in the particle-hole asymmetric case, it is found that the uniform charge susceptibility also diverges at the quantum critical point [18, 19]). For the superconducting (SC) pairing correlation function $\chi_{sc}(\omega)$, we find the low energy part of $\chi''_{sc}(\omega)$ follows the behavior of $\chi''_a(\omega)$, as $\sim \omega$ for $\omega < T_F$ and as $\sim C_{sc}$ for $\omega > T_F$, where C_{sc} is relatively a constant. At the critical point, as $T_F \to 0$, $\chi'_{sc}(0)$ becomes divergent, leading to a SC pairing instability. However, the high energy part doesn't follow the spin fluctuations: it has a resonance peak at U and the SC fluctuation softening happens from

T_{LM}, which is much larger than T_{sf}, down to T_F, or $\chi'_{sc}(0) \approx C_{sc} \ln(T_{LM}/T_F)$.

T_{LM} is the also the energy where $-\text{Im}\Sigma(\omega)$ reaches a maximal value. In the single-impurity Anderson model [20], the renormalization group (RG) flow starts from the free orbital fixed point, rather than directly from the local moment fixed point as in the Kondo model. The free orbital fixed point is unstable against any interaction U, or the hybridization with the conduction electrons Γ. U_{eff} and Γ_{eff} increase at first, approaching the local moment fixed point, which is characterized by $U_{eff} \to \infty$ and a small Kondo coupling $J_{K,eff} = \Gamma_{eff}/U_{eff}$, and then U_{eff} decreases while Γ_{eff} keeps increasing, reaching the strong coupling fixed point $J_{K,eff} \to \infty$ (due to the marginal nature of J_K, such change is slow and only becomes abrupt near T_K). T_{LM} therefore characterizes the proximity to the local moment fixed point in the RG flow.

4. On the superconducting pairing

The fact that the SC pairing fluctuations become significant before either T_K or I, and in addition their weight remains relatively unchanged at $\omega = T_{sf}$ for finite Is, implies that the SC pairing might not require the spin fluctuations as the glue. The central issue is therefore to understand the SC pairing fluctuation softening beginning at T_{LM}. In terms of the even and odd parity orbitals, the interaction part of the original Hamiltonian can be rewritten as

$$H_{f,int} = (U/2 - 3I/8)(n_{e\uparrow}n_{e\downarrow} + n_{o\uparrow}n_{o\downarrow}) + (U/2 + 3I/8)\mathbf{J}_e \cdot \mathbf{J}_o - (U/2 - I/8)\mathbf{S}_e \cdot \mathbf{S}_o, \quad (2)$$

where $\mathbf{J}$ and $\mathbf{S}$ are axial charge [7] and spin of the even and odd orbitals. For $J_p^+ = f_{p\uparrow}^\dagger f_{p\downarrow}^\dagger$, it can be easily shown that the transverse "staggered" and "uniform" axial charge susceptibilities, $\langle\langle J_e^+ \mp J_o^+; J_e^- \mp J_o^- \rangle\rangle$ correspond to the non-local singlet SC pairing correlation function and the local SC pairing correlation function $\chi_{loc}^{sc} = \langle\langle \sum_i f_{i\uparrow}^\dagger f_{i\downarrow}^\dagger; \sum_i f_{i\downarrow} f_{i\uparrow} \rangle\rangle$. Due to the axial charge rotational symmetry, they are the same as twice of the correlation functions of $J_e^z \mp J_o^z = (n_e \mp n_o)/2$, which relate to the current fluctuations and uniform charge susceptibility, respectively (explicitly verified by calculations). With an "antiferromagnetic" axial charge exchange interaction $I_{ac} = U + 3I/4$, the staggered component has negative energy while the uniform component has positive energy, i.e, the non-local SC pairing is energetically favored against the local SC pairing. Such axial charge fluctuations are also subjected to the coupling to the conduction electron bath. In Fig. 2, we show the behaviors of χ''_{sc} and $(\chi_{loc}^{sc})''$ for different values of the hybridization constant Γ_0 (U is fixed). We observe that the non-local SC pairing fluctuations are always stronger than the local ones, indicating the "antiferromagnetic" nature. Besides, the enhancement (or the softening) of χ''_{sc} is only achieved by a relative large value of Γ_0. While a detailed analysis of this behavior is desired, we can associate it with an "axial charge Kondo" effect, by making an analog between the axial charge and the spin. We expect a "Kondo"-like exchange interaction between the axial charges of the local orbitals and those of the conduction electrons in the form $J_{ac}\mathbf{J}_p \cdot \mathbf{J}_{c,p}$, and as expected, $J_{ac} \sim J_K$. Such an interaction competes with I_{ac} in the same fashion as the competition between the Kondo coupling and RKKY interaction in the spin channel. Indeed, we find the behaviors of χ''_{sc} in Fig. 2 share similarities to the spin susceptibilities when the ratio between T_K and I is varied. In the spin dynamics, the softening of χ''_a is due to the equivalent strength of local Kondo splin-flip scattering and inter-impurity spin singlet to triplet scattering. The same type of physics is expected for the axial charge fluctuations: the axial charge fluctuations with conduction electrons compete with the axial charge fluctuations between non-local ("singlet") and local ("triplet") SC pairs.

On the other hand, the low energy edge of the SC pairing fluctuation softening is T_F, rather than T_{sf} for finite Is. This indicates that the SC pairing fluctuations are only suppressed when the strong coupling fixed point, or a FL fixed point is reached. This can be understood as that the effective Kondo couplings in the spin and axial charge channels are the same. When the strong coupling fixed point is reached, $J_{K,eff} \to \infty$, the axial charge for each parity forms an "axial

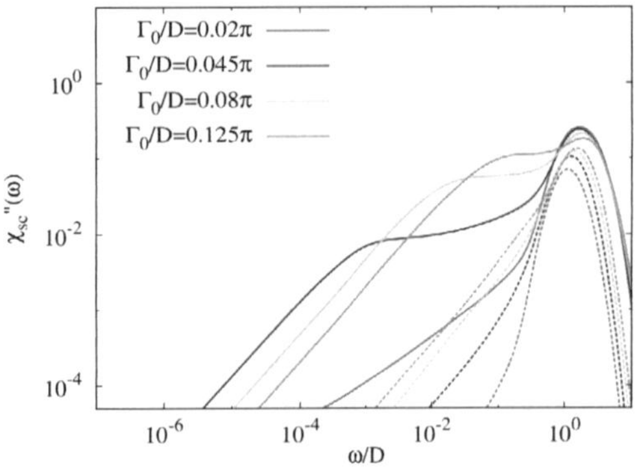

Figure 2. The non-local singlet (χ_{sc}, solid lines) and the local (χ_{loc}^{sc}, dashed lines) superconducting pairing fluctuations for different values of $\Gamma_{e,o}(\omega) = \Gamma_0$. U is fixed as $2D$, as in Fig. 1, and $I = 0$.

charge Kondo singlet" with those of the conduction electrons. The axial charge fluctuations between even and odd parities, i.e., the SC pairing fluctuations are suppressed. While the spin exchange interaction I (antiferromagnetic in our study, but we expect similar physics for the triplet pairing with ferromagnetic I) can create an "antiferromagnetic" interaction (I_{ac}) in the axial charge channel as well [Cf. Eq.(2)], but in general, $I \ll U$. This also explains that $\chi_{sc}''(\omega)$ has little change at $\omega = I$. The effect of I in this picture is rather to compete with the Kondo effect and keep $J_{K,eff}$ finite at low energies. This implies that the SC pairing can only keep its strength to low energies when the Fermi liquid fixed point is avoided, i.e., near the quantum critical point. This can also be evidenced from a low energy effective Hamiltonian in terms of Fermi liquid parameters [7], where $I_{ac,eff}$ is found to be divergent at the quantum critical point.

In the traditional theory of superconductivity, the formation of Cooper pairs requires an attractive interaction, direct or generated with certain glues. Our results rather imply that such an effective attractive interaction in the non-local singlet channel is originated from the onsite repulsive interaction. The SC pairing fluctuations at high energies may simply be two-particle excitation fluctuations. The "antiferromagnetic" nature of I_{ac}, which appears as an attractive interaction in the site basis $-(I_{ac}/2)(n_{1\uparrow}n_{2\downarrow} + n_{2\uparrow}n_{1\downarrow})$, may be due to the repulsive nature of local Coulomb interaction, i.e., the non-local two-particle excitations are favored against local pairs. The central issue revealed by this study is that such two-particle excitations can be preserved to low energies. This is achieved by an "axial charge Kondo effect" to enhance pair fluctuations with coupling to conduction electrons. Such an enhancement maintains further to the Fermi energy when the spin dynamics are critical near a quantum critical point. We notice that the spin fluctuations are the dominating fluctuations at low energies. In other words, the weights of the low energy states are dominated by the single-occupancy states (in the site basis). The SC pairing fluctuations are subjected to the weight of the double or zero occupancy configurations in the low energy states, which is $\sim (\Gamma/U)^2$. Therefore, the effective SC pairing energy scale is $I_{ac,eff} = U(\Gamma/U)^2 = \Gamma^2/U$ instead of U. The dominating excitations would still be single-particle excitations. However, the single-particle excitations at quantum criticality have non-Fermi liquid behaviors. In the two-impurity model, they are formulated as Majorana Fermions [10, 11, 13]. The origin for Majorana Fermions is also related to the finite $J_{K,eff}$, which can only lead to partial-screened Kondo resonances. They carry finite residue entropies [10, 11, 13]. A superconducting order, by forming Cooper pairs which is a lower entropy state, becomes favorable to these single-particle excitations. We notice that the entropy accumulation is also a generic thermodynamic feature of quantum criticality [21].

In summary, from the dynamical property of the superconducting pairing correlation function in the two-impurity Anderson model, we learn that the superconducting pairing doesn't necessarily need spin fluctuations, or bosonic modes as the glue. It can be originated from the effective Coulomb interaction. However, such pairing strength at low energies only remains finite when spin fluctuations are critical to avoid a Fermi liquid fixed point. This is manifested as being in the vicinity of certain quantum critical points. This provides us a new picture on the origin of unconventional superconductivity in strongly correlated electron systems. It will be interesting to check how this idea can be applied to the lattice systems.

Acknowledgments

We would like to thank Qimiao Si, Joe D. Thompson and Chandra M. Varma for helpful discussions. This work was supported by the U.S. DOE through the LDRD Program at LANL.

[1] Parks R D (ed.) *Superconductivity* (Dekker, New York, 1969)
[2] Anderson P W 2007 *Science* **316** 1705
[3] Maier T A, Macridin A, Jarrell M and Scalapino D J 2007 *Phys. Rev. B* **76** 144516
[4] Cuk T, Baumberger F, Lu D H, Ingle N, Zhou X J, Eisaki E, Kaneko N, Hussain Z, Devereaux T, Nagaosa N, Shen Z-X 2004 *Phys. Rev. Lett.* **93** 117003
[5] For a review, see, e.g., Stewart G R 1984 *Rev. Mod. Phys.* **56** 755; 2001 *Rev. Mod. Phys.* **73** 797
[6] Sun P and Kotliar G 2005 *Phys. Rev. Lett.* **95** 016402; De Leo L, Civelli M and Kotliar G 2008 *Phys. Rev. B* **77** 075107
[7] Jones B A and Varma C M 1987 *Phys. Rev. Lett.* **58** 843; Jones B A, Varma C M and Wilkins J W 1988 *Phys. Rev. Lett.* **61** 125; Jones B A and Varma C M 1989 *Phys. Rev. B* **40** 324
[8] Sakai O and Shimizu Y, 1991 *J. Phys. Soc. Jpn.* **61** 2333; 2348. They have also calculated similar correlation functions but no discussion on the superconducting pairing is given. Also they statement on the staggered charge susceptibility is not correct.
[9] Jones B A, Kotliar B G and Millis A J 1989 *Phys. Rev. B* **39** 3415
[10] Affleck I and Ludwig A W W 1992 *Phys. Rev. Lett.* **68** 1046; Affleck I, Ludwig A W W and Jones B A 1995 *Phys. Rev. B* **52** 9528
[11] Sire C, Varma C M and Krishnamurthy H R 1993 *Phys. Rev. B* **48** 13833
[12] Fye R M 1994 *Phys. Rev. Lett.* **72** 916
[13] Gan J 1995 *Phys. Rev. B* **51** 8287
[14] Silva J B, Lima W L C, Oliveira W C, Mello J L N, Oliveira L N and Wilkins J W 1996 *Phys. Rev. Lett.* **76** 275
[15] Zhu L and Zhu J-X 2010 *Preprint* arXiv:1005.5154
[16] Anders F B and Schiller A 2006 *Phys. Rev. B* **74** 245113
[17] Bulla R 1997 *J. Phys.: Condens. Matter* **9** 10463
[18] Zhu L, Varma C M and Zhu J.-X. (unpublished)
[19] Hattori K and Miyake K 2010 *J. Phys. Soc. Jpn.* **79** 073702
[20] Krishna-murthy H R, Wilkins J W, and Wilson K G 1980 *Phys. Rev. B* **21** 1003
[21] Zhu L, Garst M, Rosch A and Si Q 2003 *Phys. Rev. Lett.* **91** 066404

A Kondo cluster-glass model for spin glass Cerium alloys

F.M. Zimmer[1], S.G. Magalhaes[2], B. Coqblin[3]

[1] Departamento de Fisica, Universidade Federal de Santa Maria, Santa Maria, RS, Brazil
[2] Instituto de Fisica, Universidade Federal Fluminense, Niterói, Rio de Janeiro, Brazil
[3] L.P.S., CNRS UMR 8502, Université Paris-Sud, 91405-Orsay, France

Abstract.
There are clear indications that the presence of disorder in Ce alloys, such as Ce(Ni,Cu) or Ce(Pd,Rh), is responsible for the existence of a cluster spin glass state which changes continuously into inhomogeneous ferromagnetism at low temperatures. We present a study of the competition between magnetism and Kondo effect in a cluster-glass model composed by a random inter-cluster interaction term and an intra-cluster one, which contains an intra-site Kondo interaction J_k and an inter-site ferromagnetic one J_0. The random interaction is given by the van Hemmen type of randomness which allows to solve the problem without the use of the replica method. The inter-cluster term is solved within the cluster mean-field theory and the remaining intra-cluster interactions can be treated by exact diagonalization. Results show the behavior of the cluster glass order parameter and the Kondo correlation function for several sizes of the clusters, J_k, J_0 and values of the ferromagnetic inter-cluster average interaction I_0. Particularly, for small J_k, the magnetic solution is strongly dependent on I_0 and J_0 and a Kondo cluster-glass or a mixed phase can be obtained, while, for large J_k, the Kondo effect is still dominant, both in good agreement with experiment in Ce(Ni,Cu) or Ce(Pd,Rh) alloys.

Recent experimental results for $CePd_{1-x}Rh_x$ [1] and $CeNi_{1-x}Cu_x$ [2] alloys show a complex scenario coming from the competition between the RKKY and Kondo interaction in the presence of disorder displaying, for instance, a cluster glass state (CSG). In the particular case of $CeNi_{1-x}Cu_x$, the small doping region ($x < 0.2$) is dominated by the Kondo effect. However, for the intermediate doping regime ($0.3 < x < 0.6$), short range ferromagnetic correlations induce the formation of clusters that become frozen in a CSG order when the temperature decreases, the CSG is replaced by a ferromagnetic phase at very low temperatures.

From a theoretical point of view, several models have been analyzed in order to account the experimental results described above. The models consider the Kondo-Ising Lattice (KIL) model with an additional Ising intersite interaction between disordered localized spins [3, 4]. Despite of the relative success to reproduce the experimental phase diagram of the $CeNi_{1-x}Cu_x$ in Ref. [4], this approach has a shortcoming that the phase diagram can be dependent of the particular replica solution scheme [5]. Quite recently, a van Hemmen (vH) like interaction [6] for the Ising term has recently been studied in the KIL, which allows to treat the disorder without the use of replica method [7]. Furthermore, this model gives results in which a Kondo phase is found for large J_k (the strength of Kondo coupling). For smaller J_k values, a spin glass phase, a mixed spin glass-ferromagnetic one and finally a ferromagnetic (FE) phase are found by decreasing temperature. Theses results are in agreement with the experimental ones of the $CeNi_{1-x}Cu_x$, but nevertheless it considers canonical spins.

Published under licence by IOP Publishing Ltd

Journal of Physics: Conference Series **273** (2011) 012069 doi:10.1088/1742-6596/273/1/012069

In the present work, the competition between Kondo effect and cluster glass is studied in the following Kondo Lattice Cluster Glass (KLCG) model (see Ref. [8] and references therein) given by:

$$\hat{H} = \sum_{a}^{N_{cl}}\sum_{i,j}^{n_s} t_{ij}^{aa} d_{i\sigma a}^{\dagger} d_{j\sigma a} + J_k \sum_{a=1}^{N_{cl}}\sum_{i=1}^{n_s}\left(\hat{S}_{ia}^{+}\hat{s}_{ia}^{-} + \hat{S}_{ia}^{-}\hat{s}_{ia}^{+}\right) - J_0 \sum_{a=1}^{N_{cl}}\sum_{(i,j)}^{n_s} \hat{S}_{ia}^{z}\hat{S}_{ja}^{z} - \sum_{a<b}^{N_{cl}} J_{ab}\hat{S}_{a}^{z}\hat{S}_{b}^{z} \quad (1)$$

where N_{cl} is the number of clusters and n_s is the number of sites in each cluster. The indices (a, b) and (i, j) indicate clusters and sites inside a cluster, respectively. The three first terms represent intra-cluster interactions and the last one is an inter-cluster disordered coupling, in which J_{ab} is a random variable given as in the vH model: $J_{ab} = \frac{J}{N_{cl}}(\xi_a\eta_b + \eta_a\xi_b) + \frac{I_0}{N_{cl}}$. In this interaction, I_0 is the ferromagnetic average inter-cluster interaction while ξ_a and η_a are random variables that follow the bimodal distribution: $P(x) = \frac{1}{2}[\delta(x-1) + \delta(x+1)]$.

In Eq. (1), the hopping t_{ij}^{aa} is only inside the cluster, the second term is the Kondo interaction and the third term is a ferromagnetic intra-cluster interaction. The spin operators are defined as: $\hat{S}_{ia}^{+} = f_{ia\uparrow}^{\dagger} f_{ia\downarrow}$, $\hat{S}_{ia}^{-} = (\hat{S}_{ia}^{+})^{\dagger}$, $\hat{s}_{ia}^{-} = d_{ia\downarrow}^{\dagger} d_{ia\uparrow}$, $\hat{s}_{ia}^{+} = (\hat{s}_{ia}^{-})^{\dagger}$ and $\hat{S}_{a}^{z} = \sum_{i=1}^{n_s}\hat{S}_{ia}^{z} = \sum_{i=1}^{n_s}\sum_{\sigma=\uparrow\downarrow}\sigma f_{i\sigma a}^{\dagger} f_{i\sigma a}$, where $f^{\dagger}$ (f) is a localized fermion creation (annihilation) operator and $d^{\dagger}$ (d) is a conduction fermion creation (annihilation) operator.

To write the partition function Z of equation (1), the inter-cluster disorder can be treated in a mean field approach without the use of replica method. For that, the order parameters $q_1 = 1/N_{cl}\sum\langle\xi_a\hat{S}_a^z/n_s\rangle$, $q_2 = 1/N_{cl}\sum\langle\eta_a\hat{S}_a^z/n_s\rangle$ and $m = 1/N_{cl}\sum\langle\hat{S}_a^z/n_s\rangle$ (magnetization) are introduced in Z. Particularly, the saddle-point approximations are used to obtain q_1, q_2 and m. As a result of this calculation, one can obtain an effective one-cluster problem with the free energy per cluster given by:

$$\beta F = \frac{\beta I_0 m^2}{2} + \beta J q^2 - \frac{1}{n_s}\langle\langle \ln Tr T e^{-\beta\hat{H}_{eff}}\rangle\rangle \quad (2)$$

where $\langle\langle\cdots\rangle\rangle$ means the average over ξ and η, $\beta = 1/T$ (T is the temperature), $q = q_1 = q_2$ and:

$$\hat{H}_{eff} = \sum_{i,j,\sigma}^{n_s} t_{ij}\hat{d}_{i\sigma}^{\dagger}\hat{d}_{j\sigma} + J_k\sum_{i=1}^{n_s}\left[\hat{S}_i^{+}\hat{s}_i^{-} + \hat{S}_i^{-}\hat{s}_i^{+}\right] - J_0\sum_{i<j}^{n_s}\hat{S}_i^z\hat{S}_j^z - 2h(m,q)\sum_{j=1}^{n_s}\hat{S}_j^z \quad (3)$$

with the internal field $h(m, q) = I_0 m + Jq(\xi + \eta)$.

Numerical results are obtained by exact diagonalization of the effective Hamiltonian given by (3), where it is used a basis of states that satisfies the following criteria: each site is exactly occupied by one localized fermion (only two states per site: spin up or down), while four possible conduction fermion occupancies at each site are allowed (sites unoccupied, single occupied and double occupied). The intracluster hopping term ($t_{ij} = 5J$) and size of cluster n_s (=4 or 5) are constants. The intracluster interaction between localized fermions J_0 is ferromagnetic.

The behavior of order parameters as a function of temperature is exhibited in Fig. (1) for $J_0/J = 1$, different values of J_k and I_0. In particular, Fig. (1-a) shows results for $I_0 = 0$, where the cluster spin glass order parameter q becomes greater than zero at the freezing temperature T_f. For increasing J_k, T_f and q are decreased and the correlation $\lambda = \frac{1}{ns}\sum_i^{ns} < \hat{S}_i\hat{s}_i >$ has nonzero values ($\hat{s}_i = \sum_\sigma \sigma d_{i\sigma}^{\dagger} d_{i\sigma}$). Figure (1-b) shows results when $I_0/J = 0.1$. In this situation, the magnetization m also appears at T_c for temperatures lower than T_f ($T_c < T_f$), which characterizes the mixed phase CSG+FE with both $q > 0$ and $m > 0$. For $J_k/J = 2$ (see dashed lines of Fig. (1)), the critical temperatures decrease while λ increases, which suggests that the suppression of the magnetic order is a result of the enhancement of Kondo effect.

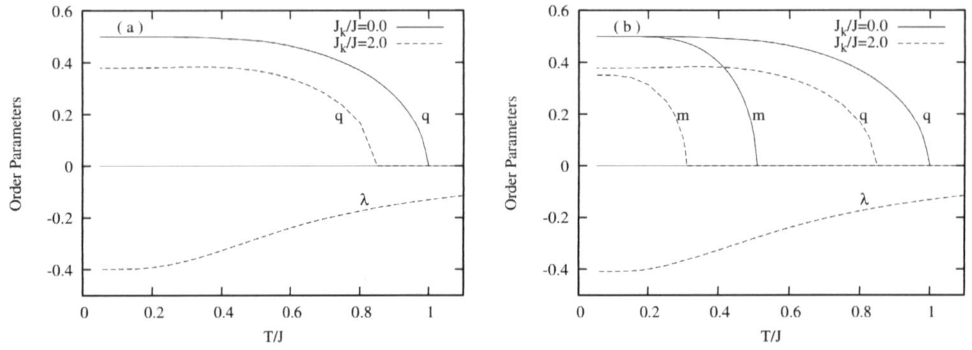

Figure 1. Order parameters versus temperature for $n_s = 4$ and two values of $J_k/J = 0.$ and 2.0. Panels (a) and (b) present results for $I_0/J = 0$ and $I_0/J = 0.1$, respectively.

The effects of J_k on the order parameters become more evident in Fig. (2), where the behaviors of q, m and λ versus J_k are exhibited for $J_0/J = 1$ and two temperatures: $T/J = 0.4$ and $T/J = 0.2$. For instance, the CSG order parameter decreases towards zero when J_k increases. On the contrary, the correlation λ enhances with J_k and it assumes higher intensities at low temperatures in the region where $q = 0$. In this figure one can also analyze the role of I_0. When $I_0 > 0$ it is possible to find the order parameter m (see Fig. (2-b)) that is also decreased by the increase of λ. Therefore, the interaction J_k is able to destroy the CSG and the mixed phase at the same time that the correlation λ is enhanced. This behavior can be associated with the Kondo interaction that could dominate at higher values of J_k.

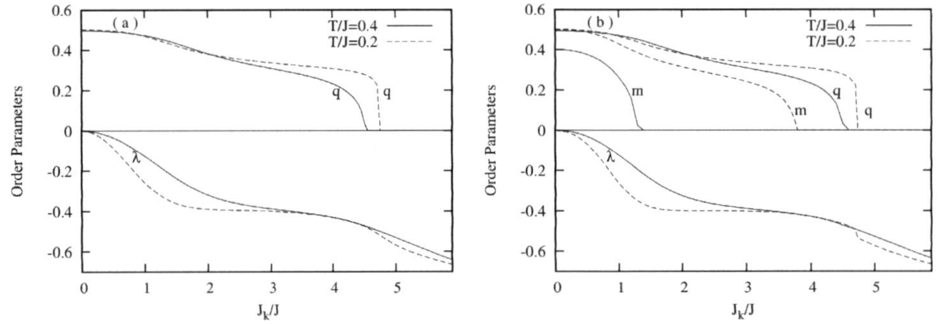

Figure 2. Order parameters versus J_k for $n_s = 4$ and two values of T/J= 0.4 and 0.2. Panels (a) and (b) present results for $I_0/J = 0$ and $I_0/J = 0.1$, respectively.

The short range intra-cluster interaction J_0 also has an important role in this problem. For example, Fig. (3) exhibits a comparison between results for order parameters with $J_0/J = 1$ (full lines) and $J_0/J = 1.5$ (dashed lines) with a constant $I_0/J = 0.1$. This shows that the increase of J_0 enhances the range of T (Fig. (3-a)) and J_k (Fig. (3-b)) where the magnetic orders occur.

Figure (4) shows results for $n_s = 5$, $J_0/J = 1$ and $I_0/J = 0.1$, where the correlation λ is also enhanced with J_k (see Fig. (4-b)). However, the mixed phase region is increased by the cluster size when the results of $n_s = 5$ are compared with those ones for $n_s = 4$. Therefore, the cluster size is relevant to the magnetic properties of the KLCG model.

In conclusion, a Kondo lattice cluster model with an inter-cluster magnetic disordered spin

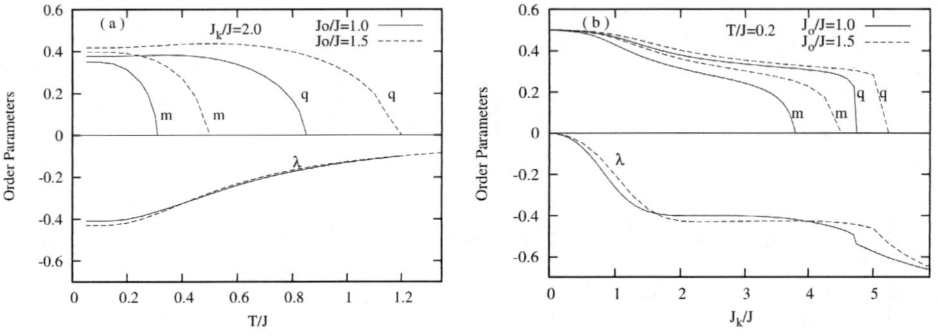

Figure 3. Panel (a) presents q, m and λ as a function of T for $J_k/J = 2$. Panel (b) shows q, m and λ versus J_k for $T/J = 0.2$. These results are for $n_s = 4$, $I_0/J = 0.1$ and two values of J_0/J= 1 and 1.5.

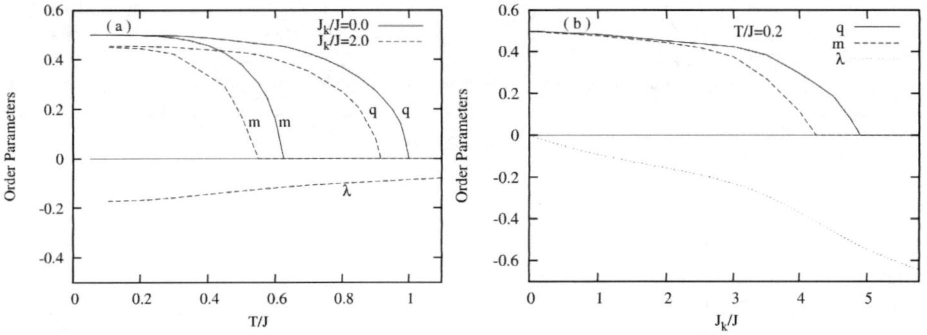

Figure 4. Panel (a) presents q, m and λ as a function of T/J for $J_k/J = 0$ and 2. Panel (b) presents q, m and λ versus J_k/J for $T/J = 0.2$. These results are for $n_s = 5$ and $I_0/J = 0.1$.

glass interaction is studied. The results show a competition between the Kondo effect and the magnetic orders, cluster spin glass (CSG) and mixed phase (CSG+FE). These magnetic orders are strongly dependent on the ferromagnetic inter-cluster interaction I_0 and the Kondo exchange interaction J_k. For example, the CSG phase can be found when J_k is small, but it can present a transition to the mixed phase at lower temperatures if I_0 assumes small values. However the Kondo effect becomes dominant for large J_k.

Acknowledgments

FMZ and SGM acknowledge the Brazilian agency CNPq for financial support.

References

[1] T. Westerkamp, M. Deppe, R. Kuchler, M. Brando, C. Geibel, P. Gegenwart, A. P. Pikul, F. Steglich 2009 *Phys. Rev. Lett.* **102** 206404.
[2] N. Marcano, J. C. Gomez Sal, J. I. Espeso, L. Fernandez Barquin, C. Paulsen 2007 *Phys. Rev. B* **76** 224419.
[3] S. G. Magalhaes, A. A. Schmidt, A. Theumann, B. Coqblin 2002 *Eur. Phys. J. B.* **30** 419..
[4] S. G. Magalhaes, F. M. Zimmer, P. R. Krebs, B. Coqblin 2006 *Phys. Rev. B* **74** 014427.
[5] S. G. Magalhaes, F. M. Zimmer, B. Coqblin 2006 *Physica B* **378-380** 131.
[6] J. L. van Hemmen 1982 *Phys. Rev. Lett.* **49** 409.
[7] S. G. Magalhaes, F. M. Zimmer, B. Coqblin 2010 *Phys. Rev. B* **81** 094424.
[8] F. M. Zimmer, S. G. Magalhaes, B. Coqblin 2009 *Physica B* **404** 2972.

Journal of Physics: Conference Series **273** (2011) 012070 doi:10.1088/1742-6596/273/1/012070

Concomitant Larkin-Ovchinikov states in polarized atomic gases

C. J. Bolech,[1,2] **L. O. Baksmaty,**[2] **H. Lu**[2] **and H. Pu**[2]

[1]Department of Physics, University of Cincinnati, Ohio 45221-0011, USA

[2]Physics & Astronomy Department, Rice University, Houston, Texas 77005, USA

Abstract. Fermionic superfluids continue to be a topic of current interest and debate in condensed matter and sub-nuclear physics. Recent experiments with laser-cooled atomic gas clouds of fermions seek to explore this physics and point out that discrepancies might arise due to different confining geometries. We reported on realistic numerical simulations with up to 100'000 atoms (a number comparable to that in the actual experiments) that shed light into what might be going on in the laboratory realizations.

1. Introduction

Recent developments in laser cooling allow experimentalists to trap clouds of atomic gases and, in tandem with evaporative cooling techniques, reach the limit of quantum degeneracy. This capability in combination with the use of Feshbach resonances [1], allows to produce systems in which the interactions and the confining potentials are controllable. Thus, new vistas open up to experimentally produce systems with interesting quantum behavior and, in particular, exotic superfluid phases. In particular, there exist strong motivations to seek for the realization of states similar to the long-sought FFLO (Ferrel-Fulde-Larkin-Ovchinnikov) states, in which fermions pair up with non-zero center-of-mass momentum [2, 3]. This unusual superconducting states are not seen in standard metals, but are thought to be realized in the context of organic conductors [4] and heavy fermions [5]. Not only that, they are also probably important for the understanding of the expected superfluid states of nuclear and quark matter [6].

Recent state-of-the-art experiments studying the effects of spin imbalance on pairing in unitary Fermi gases encountered mutual qualitative and quantitative discrepancies which we interpret to be a function of the confining geometry. To substantiate this idea we presented, and report in this proceeding, numerical results based on solutions of the Bogoliubov-de Gennes (BdG) equations for the actual experimental setups. In particular, we do not resort to the use of a local density approximation or LDA (extensively used in studying trapped gases, though its violation has been documented in some of the afore mentioned experiments in elongated 3d atomic clouds [7] due to the presence of surface/interface effects [8, 9]).

2. Modeling

We model the system of trapped fermionic atoms starting from the following one-body Hamiltonian:

$$H_\sigma = -\frac{\hbar^2}{2m}\left(\nabla_r^2 + \frac{\partial^2}{\partial z^2}\right) + V(r,z) \quad (1)$$

Published under licence by IOP Publishing Ltd

[where $V(r,z) = \frac{m}{2}(\omega_\perp^2 r^2 + \omega_z^2 z^2)$ is the trapping potential; m is the atomic mass and $\omega_{\perp,z}$ are the respective trapping frequencies] and taking into account fermion-fermion interactions using a fully spatial-dependent self-consistent paring approximation:

$$\begin{pmatrix} H_\uparrow - \mu_\uparrow & \Delta(r,z) \\ \Delta(r,z) & -H_\downarrow + \mu_\downarrow \end{pmatrix} \begin{pmatrix} u_j \\ v_j \end{pmatrix} = E_j \begin{pmatrix} u_j \\ v_j \end{pmatrix} . \tag{2}$$

One has to find all the solutions of this eigensystem and the superconducting order parameter needs to be defined self-consistently via $\Delta(r,z) \propto \sum_j u_j(r,z) v_j^*(r,z) f(E_j)$, with f the Fermi function. The self-consistency is reached via an iterative solution of the full problem.

3. Discussion and Results

We are concerned here with the fate of the superfluid state in the case of a polarized gas. Following the experiments, we fix $N = N_\uparrow + N_\downarrow$ (total particle number) and $P = \frac{N_\uparrow - N_\downarrow}{N}$ (global polarization). We take P in the 30%-40% range and try to have N as large as possible to approach the experimental number of about a quarter of a million atoms. We find that the nature of the solution evolves as the number of atoms is increased (see Fig. 1). While for a small number the solution is unique and displays a trap-induced phase separation with a superfluid core and polarized wings, as the number increases (and approaches the experimental values) the most stable solution turns out to be close to the LDA prediction with a uniform superfluid phase. New solutions are also found which are close in energy to the lowest one (within our error estimates) and that display an intermediate axial layer with an oscillating order parameter as in the Larkin-Ovchinikov ansatz solution. This solution becomes more stable and the LO region better defined as the number of particles augments. Also, as the aspect ratio of the trapping potential increases, the energy barrier between the different solutions seem to get larger. For large number of particles, which solution one converges to depends strongly on the initial ansatz, which suggests that which state is realized in the experiments would depend heavily on the initial/boundary conditions and cooling procedure.

We carried out numerical solutions of the BdG equations with up to 100'000 atoms and we obtained column-density profiles in good agreement with the experimental ones. A detailed report of the comparison between theory and experiments will be presented elsewhere [10]. Our results indicate that the confining geometry plays an important role in deciding what type of solution is stabilized. This is consistent with the scenario from one dimension, since in that limit it is known that one can stabilize a polarized liquid with paired states and many characteristics akin to an FFLO superfluid state (see Ref. [11] for an example and more references; see also Ref. [12] for preliminary experimental results).

4. Conclusions

To summarize, we have studied attractive spin-imbalanced ultracold atomic clouds of fermions in 3d elongated geometries. Our results are consistent with the possible experimental observation of polarized superfluid states. One auspicious aspect of the results worth commenting on is the radial alignment of the nodes of the FFLO phase. A fact which is promising for the prospects of experimental detection and characterization, because it implies that the Larkin-Ovchinikov state could yield a measurable signal in the density profile. It also suggests that for an array of 1d tubes, such as are being used in current experiments [12], if the inter-tube coupling is substantial then the nodes in the order parameter at each tube (and thus the peaks in the density difference between the two species) are likely to be radially aligned among tubes and yield a measurable signal 'easy' to access in the experiments.

Our approach is well suited to explore interesting crossover effects, which are challenging to capture with other theoretical methods. Moreover, our inhomogeneous numerical scheme

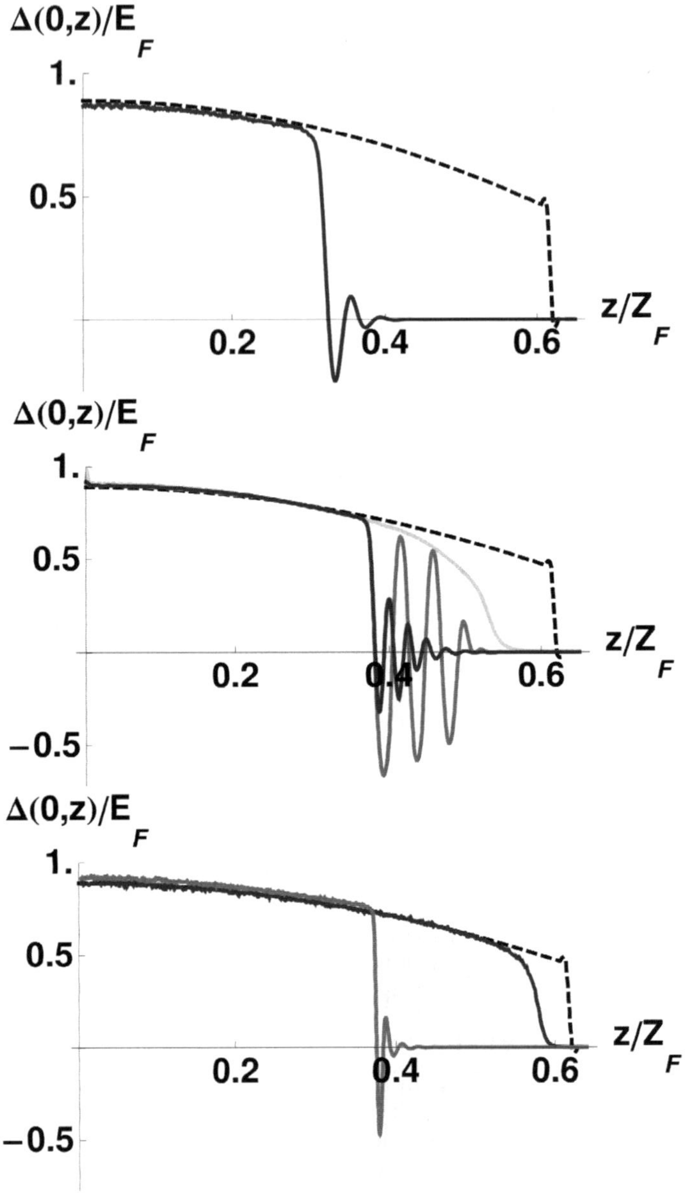

Figure 1. Axial profiles of the order parameter at the center of an elongated trap for different numbers of trapped atoms (from top to bottom: 200, 1'100 and 10'000, respectively). The dotted line represents the local-density solution and is given for comparison. The different solid lines correspond to different self-consistent solutions of the BdG equations corresponding to different free-energy local minima. The colors correlate with free energy so that the blue line indicates the most stable solution and so forth (see the discussion in the text).

could be applied to answer questions about dynamics that are currently still open. Efforts in this direction are underway. Such calculations would further address the possible experimental signatures of the inhomogeneous, partially-polarized, superfluid phases of attractive atomic Fermi gases near the unitary limit. The synergy between theory and experiments is likely to produce new insights into the physical properties of polarized superfluid matter with potentially far reaching implications.

Acknowledgments

We would like to thank the conference organizers for giving us the opportunity to present this work during the 2010 edition of the International Conference on Strongly Correlated Electron Systems (SCES2010). This work was supported in part by DARPA-ARO (grant no. W911NF-07-1-0464) and by the W. M. Keck and Welch foundations (C-1669 and C-1681). Part of this work was performed at NERSC, Navy DSRC and the ARSC.

References

[1] Feshbach H 1958 *Ann. Phys.* **5** 357
[2] Fulde P and Ferrell R A 1964 *Phys. Rev.* **135** A550
[3] Larkin A I and Ovchinnikov Y N 1964 *Zh. Eksp. Teor. Fiz.* **47** 1136 ; Sov. Phys. JETP **20**, 762 (1965)
[4] Singleton J *et al* 2000 *J. Phys.: Condens. Matter* **12** L641 ; Yonezawa S *et al* 2008 Phys. Rev. Lett. **100**, 117002. See also : Bolech C J and Giamarchi T 2004 Phys. Rev. Lett. **92** 127001 ; Bolech C J and Giamarchi T 2005 Phys. Rev. B **71** 024517
[5] Radovan H A *et al* 2003 *Nature* **425** 51 ; Miclea C F *et al* 2006 Phys. Rev. Lett. **96** 117001
[6] Sedrakian A, Clark J W and Alford M G (eds) 2006 *Pairing in Fermionic Systems: Basic Concepts and Modern Applications* (Singapore: World Scientific)
[7] Partridge G B *et al* 2006 *Science* **311** 503 ; Partridge G B *et al.* 2006 *Phys. Rev. Lett.* **97**, 190407
[8] De Silva T N and Mueller E J 2006 *Phys. Rev. Lett.* **97** 070402
[9] Imambekov A, Bolech C J, Lukin M and Demler E 2006 *Phys. Rev. A* **74** 053626 ; See also: Baksmaty L O *et al.* 2010 *arXiv:1003.4488*
[10] Baksmaty L O, Lu H, Bolech C J and Pu H 2010 *arXiv:1003.4488*
[11] Kakashvili P and Bolech C J 2009 *Phys. Rev. A* **79** 041603(R)
[12] Liao Y, Rittner A S C, Paprotta T, Li W, Partridge G B, Hulet R G, Baur S K and Mueller E J 2009 *arXiv:0912.0092*

Journal of Physics: Conference Series **273** (2011) 012071 doi:10.1088/1742-6596/273/1/012071

Effect of Phase Fluctuations on the Superconducting Properties of Strongly Disordered 3D NbN Thin Films

Madhavi Chand[1*], Mintu Mondal[1], Anand Kamlapure[1], Garima Saraswat[1], Archana Mishra[2], John Jesudasan[1], Vivas C. Bagwe[1], Sanjeev Kumar[1], Vikram Tripathi[3], Lara Benfatto[4], and Pratap Raychaudhuri[1]

[1]Department of Condensed Matter Physics and Materials Science, Tata Institute of Fundamental Research, Homi Bhabha Road, Colaba, Mumbai 400005, India
[2] IIC, Indian Institute of Technology Roorkee, Roorkee, Uttarakhand 247667, India
[3]Department of Condensed Theoretical Physics, Tata Institute of Fundamental Research, Homi Bhabha Road, Colaba, Mumbai 400005, India
[4] INFM-CNR Statistical Mechanics and Complexity Center, University of Rome "La Sapienza," P.le A. Moro 5, 00185 Rome, Italy

*Email: chand@tifr.res.in

Abstract. We present transport, Hall effect, electronic tunnelling and penetration depth studies in 3D homogeneously disordered epitaxial NbN thin films with disorder ranging from the moderately clean limit (k_Fl~10.12) to the very dirty limit (k_Fl~1.24). The superconducting transition temperature (T_c) decreases from ~17K to less than 350mK with increasing disorder. The T_c and conductivity at the lowest temperature (σ_0) both asymptotically approach zero as $k_Fl \rightarrow 1$, indicating a coincidence of the metal-insulator transition (MIT) and the superconductor-insulator transition (SIT). Close to critical disorder there is spatial inhomogeneity in the superconducting density of states (DOS) and the superconducting state is governed by quantum phase fluctuations. This results in suppression of the superfluid density (n_s) and a pseudogap state where the resistance is no longer zero but the energy gap (Δ) remains finite.

Introduction

The disorder driven superconductor-insulator and metal-insulator transitions have been studied in a number of systems recently and the physics that governs the behaviour of superconducting and electronic properties close to these transitions is a topic of fundamental interest today.

It has been observed in studies on doped materials like Nb_xSi_{1-x}[1] and Au_xSi_{1-x}[2] that superconductivity vanishes on the metallic side of the MIT. Theories suggest [3,4] that in systems with homogenous disorder, superconductivity is suppressed primarily by increased effectiveness of Coulomb repulsion due to loss of screening and is therefore likely to vanish in the metallic regime.

On the other hand, in granular systems, superconductivity is found to persist on the insulating side of the MIT. Examples of materials studied are granular Al [5] and InO [6]. However, most of the experiments have been done on 2D systems where thickness is used to control the sheet resistance. This therefore poses an open question for 3D systems regarding whether or not superconductivity would persist on the insulating side of the MIT. It has been predicted [7] for strong localization that in the absence of *el-el* repulsion, superconductivity is possible below the mobility edge as long as

Published under licence by IOP Publishing Ltd

$\rho\Delta l_{loc}^{D} \geq 1$. Here ρ is the DOS at Fermi level, Δ is the superconducting energy gap, l_{loc} is the localization length and D is the dimension. This is tantamount to saying that superconductivity will survive as long as Δ is larger than the level spacing for particles confined within l_{loc}.

Stoichiometric NbN is a conventional BCS superconductor with a relatively high T_c of 17K. However, disorder can be introduced without external doping so the properties are not dependent on composition. At the same time, the films are 3D and disorder does not depend on thickness. Disorder is characterized by the Ioffe Regel parameter k_Fl ($k_F \equiv$ Fermi wave vector and $l \equiv$ mean free path), obtained experimentally from Hall effect and resistivity measurements at 285K using free electron theory [8,9]. The existence of strong interactions has been established in a temperature dependent study of Hall effect wherein it is clear that localization theories alone or even weak interaction theories are inadequate in explaining the normal state behaviour of this system[9]. In this paper we explore the superconductor-insulator transition in the context of the metal-insulator transition and study the superconducting properties in the region of critical disorder.

Experimental Details

NbN films were grown epitaxially on (100) oriented single crystalline MgO substrates using reactive D.C. magnetron sputtering of a Nb target in an Ar/N_2 gas mixture. Films with k_Fl ranging from 1.24 to 10.12 were deposited by controlling the sputtering power and Ar:N_2 ratio.

The temperature dependence of resistivity (ρ) was measured using a standard A.C. four probe technique. To obtain k_Fl experimentally, the Hall coefficient was measured on samples patterned in a Hall bar geometry by sweeping a magnetic field from -12T to 12T and then folding over the data to cancel the resistive component. Tunneling measurements were performed on NbN/insulator (oxide)/Ag planar tunnel junctions similar to those studied in [10]. The scanning tunnelling spectroscopy (STS) is done in a home-made low temperature STM. Details of the setup can be found in Ref [11]. The penetration depth (λ) as a function of temperature was measured down to 350mK using a "two coil" mutual inductance technique operating at 60 kHz. The advantage of this technique is that it allows measurement of the absolute value of λ over the entire temperature range up to T_c without any prior assumption about the dependence of λ on T. An explanation of this technique can be found in Ref [11].

Results and Analysis

Figure 1 (a) shows the temperature dependence of conductivity for films with different k_Fl. We can see that the positive temperature coefficient of conductivity gets stronger as we increase the disorder. A detailed analysis of this temperature dependence along with the corresponding temperature dependence of Hall coefficient can be found in Ref [9].

In the case of the most disordered sample (k_Fl ~1.24), where the zero resistance regime could not be reached, we verified that the upturn in conductivity is due to superconductivity by measuring the σ-T in presence of a magnetic field which shows the superconductivity getting quenched at 5T (inset of Figure 1(b)). Since neither the sign of the temperature coefficient of resistance nor the absolute value of k_Fl can be used to exactly identify the metal insulator transition, the determining quantity used is the zero temperature conductivity. Therefore, we plot the lowest temperature conductivity (σ_0, measured at the peak value just above T_c) as a function of k_Fl (Figure 1(b)). We can see that within experimental error, as $k_Fl \rightarrow 1$, $\sigma_0 \rightarrow 0$ confirming that $k_Fl \rightarrow 1$ does indeed correspond to the metal-insulator transition.

On the same plot we look at T_c as a function of k_Fl and observe that T_c is suppressed as k_Fl decreases and tends to zero as $k_Fl \rightarrow 1$ indicating the coincidence of the metal-insulator and superconductor-insulator transitions at a single quantum critical point.

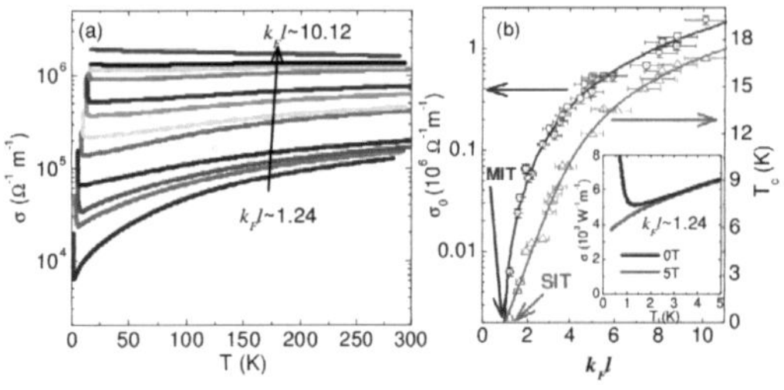

Figure 1: (a) σ *vs.* T for films of k_Fl ~ 10.12, 8.82, 8.13, 8.01, 5.5, 4.98, 3.65, 3.27, 2.21, 1.68, 1.58 and 1.24. (b) Minimum σ vs. k_Fl as well as T_c vs k_Fl. Inset of (b): zero field and in-field σ *vs.* T for the most disordered film with k_Fl~1.24.

Now, with the dual quantum critical point established, it is interesting to explore the properties of the superconducting state close to the critical region.

Figure 2(a) shows a histogram of the variation in Δ (defined as half the difference between the coherence peak positions) measured via STS at different points on a highly disordered sample (T_c~4.1K) over a 150X30nm region. The observed spatial inhomogeneity demonstrates as a spatial distribution in the values of Δ. Such a spontaneous segregation has been predicted in numerical studies [12] as well. The existence of this inhomogeneity brings into question the degree of phase stiffness in the superconducting state. In figure 2(b), the evolution of the DOS with temperature obtained by taking a spatial average of the local DOS shows that the gap feature persists even above the temperature at which resistance goes to zero. This can be understood in the following way: In most mean field theories, the electron-pairing and long range phase coherence occur at the same temperature, T_c. This means that the global phase coherence and Δ vanish at the same temperature, due mainly to the suppression of the gap with temperature. However, it has been shown [13] that for systems with low conductivity and small n_s, the phase ordering temperature is reduced significantly and becomes comparable to the pairing temperature. Therefore, the 'pseudogap' in figure 2(b) indicates that in this system, the superconducting state is possibly destroyed by phase fluctuations.

The phase fluctuations scenario is corroborated by measurement of n_s (shown in terms of $1/\lambda^2(0)$ in figure 2(c)) at 350mK for a series of samples with different disorder which is compared with the value predicted from BCS: $1/\lambda_{BCS}^2(0) = \pi\mu_0\Delta(0)\sigma_0/\hbar$ where μ_0 is the permeability of free space. We observe that for samples with $T_c \geq 6$K, the experimental and theoretical trends match closely, but for samples with $T_c < 6$K there is strong suppression in the measured n_s compared to the value predicted

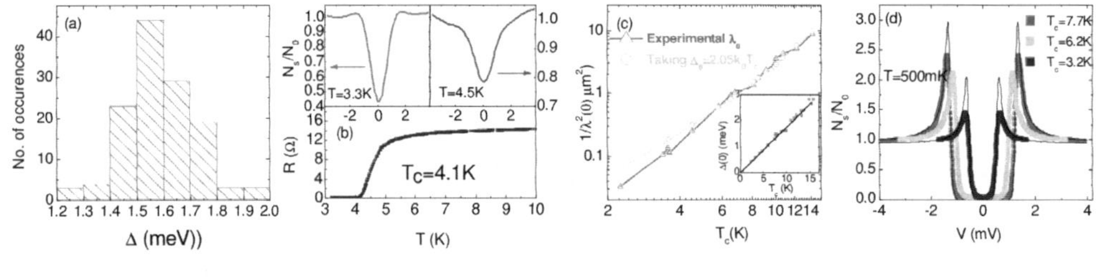

Figure 2: (a) Histogram of observed energy gap obtained from measurement of local density of states. The value is taken as the average of the positions of the coherence peaks on either side of zero bias; (b) (top)averaged conductance curves measured at 3.3K and 4.5K for a sample of T_c=4.1K and (bottom) resistance vs. temperature for the same sample ; (c) experimental and theoretical values of $1/\lambda^2(0)$ for different samples, (inset: superconducting energy gap (Δ) vs. T_c for different samples (magenta squares) along with a linear fit extrapolated to zero (black line)); (d) Conductance curves measured at 500mK for samples of different T_c. The black lines correspond to an attempted BCS fit to the density of states.

from BCS. In this calculation, $\Delta(0)$ is taken to be $2.05k_BT_c$, corresponding to a strong coupling superconductor, which is obtained from a linear fitting of Δ vs. T_c for samples with less disorder i.e. T_c>8K (inset of figure 2(c)).

In a strongly interacting system like ours, we expect that number fluctuations, arising from the number phase uncertainty relation when there is long range phase coherence, would have a large Coulomb energy cost making it energetically favourable to have large quantum phase fluctuations in the system. However, in spite of large suppression of n_s there will still be a phase coherent ground state as long as the conductivity at T_c is greater than the quantum conductivity [13] i.e. $\sigma/\sigma_Q>1$ where $\sigma_Q=(1/3)(2e)^2/ha$. Here we take the characteristic length of phase fluctuations 'a' to be ~ ξ which is the measured coherence length. We have found that this condition is true even for our most disordered samples with T_c~2K where the ratio is ~3.7.

In addition to the spontaneous inhomogeneity, pseudogap feature and reduced n_s, we observe that the coherence peaks in the superconducting DOS are suppressed. In figure 2(d) we plot conductance curves for samples of different disorder measured at ~500mK using planar tunnel junctions. We find that the coherence peaks are strongly suppressed compared to what is expected from the BCS DOS: $N(E) = E/\sqrt{E^2 - \Delta^2}$.

Conclusions

We have measured resistivity, Hall effect, electron tunnelling and penetration depth for a series of 3-dimensional homogenous NbN thin films of varying disorder. We observe that the metal-insulator and superconductor-insulator transitions coincide at a single quantum critical point. The superconducting state is governed by quantum phase fluctuations arising due to spontaneous inhomogeneity. The phase fluctuations result in a pseudogap region, suppression of the superfluid density and suppression of the coherence peaks in the DOS.

Acknowledgements

We would like to thank Mr. S.P. Pai for technical help with fabrication of the STM and the penetration depth set-up.

References

[1] Bishop B J, Spencer E G and Dynes R C 1985 *Solid State Electronics*, **28** 73
[2] Furubayashi T, Nishida N, Yamaguchi M, Morigaki K and Ishimoto H 1985 *Solid State Commun.*, **55** 513
[3] Finkelstein A M 1994 *Physica* B, **197** 636-648
[4] Anderson P W, Muttalib K A and Ramakrishnan T V 1983 *Phys. Rev.* B **28** 117
[5] Chui T, Deutscher G, Lidenfeld P and Mclean W L, *Phys. Rev.* B 1981 **23** 6172
[6] Shahar D and Ovadyahu Z 1992 *Phys. Rev.* B **46** 10917
[7] Ma M and Lee P 1985 *Phys. Rev.* B **32** 5658
[8] Chockalingam S P, Chand M, Jesudasan J, Tripathi V and Raychaudhuri P 2008 *Phys. Rev.* B **77** 214503
[9] Chand M, Mishra A, Xiong Y M, Kamlapure A, Chockalingam , Jesudasan J, Bagwe V C, Mondal M, Adams P W, Tripathi V and Raychaudhuri P *Phys. Rev.* B **80**, 134514
[10] Chockalingam S P, Chand M, Kamlapure A, Jesudasan J, Mishra A, Tripathi V and Raychaudhuri P 2009 *Phys. Rev.* B **79** 094509
[11] Kamplapure A, Mondal M, Chand M, Mishra A, Jesudasan A, Bagwe V C, Benfatto L, Tripathi V and Raychaudhuri P 2010 *Appl. Phys. Lett.* **96**, 072509
[12] Ghosal A, Randeria M and Trivedi N 2001 *Phys. Rev.* B **65** 014510
[13] Emery V J and Kivelson S A 1995 *Phys. Rev. Lett.* **74** 3253

Journal of Physics: Conference Series **273** (2011) 012072 doi:10.1088/1742-6596/273/1/012072

Current and spin correlations in Fulde-Ferrell-Larkin-Ovchinnikov state of a s-wave superconductor

P Dey[1], **S Basu**[1] **and R Kishore**[2]

[1] Department of Physics, Indian Institute of Technology Guwahati, Guwahati, Assam 781039, India

[2] Instituto Nacional de Pesquisas Espaciais, CP 515, S.J. Campos, SP, 12245-970, Brazil

E-mail: poulumi@iitg.ernet.in, saurabh@iitg.ernet.in, kishore@las.inpe.br

Abstract. In an attempt to characterize the Fulde-Ferrell-Larkin-Ovchinnikov (FFLO) phase in a two dimensional s-wave superconductor, we investigate current-current and spin-spin correlations and compare them with those in the BCS phase. The existence of the FFLO phase is demonstrated using a weakly attractive fermionic Hubbard model in presence of a harmonic trap and a population imbalance, caused by a magnetic field. The lower and upper values of the magnetic field, between which the order parameter shows spatial modulations (owing to a finite center of mass momentum pairing), mark the extent of the FFLO phase. In this regime, the current-current correlations show spatial modulation similar to the behaviour of the order parameter, while the longitudinal spin-spin correlations show modulations with a period half of that of the order parameter in the absence of trap and a bimodal distribution in the presence of the harmonic confinement, thereby underlining a dissimilarity in behaviour between the spin and charge sectors.

1. Introduction

The study of superconductivity in presence of magnetic field commenced nearly half a century ago with the works of Clogston and Chandrasekhar [1]. More abounding implications of the presence of an external magnetic field is elucidated by Fulde and Ferrel and by Larkin and Ovichinnikov [2] where a possibility of finite momentum pairing between the different participation species of electrons is explored. Backed by weak and ambiguous evidences in various condensed matter systems, such as the heavy fermion compounds, organic superconductors and very recently pnictides, the progress made in the field of trapped ultracold atomic gases has raised the possibility of the realization of the FFLO phase [3].

In this paper, we shall consider a population imbalanced atomic superfluid in presence of a harmonic trap which mimics the optical and various other trapping effects which find a mention in cold atom literature. Further we shall focus on the effect of the trap on properties of the ground state by reviewing different correlation functions that characterize the state.

Published under licence by IOP Publishing Ltd

2. Model and Formalism

The mean field decoupling of the interaction term in the attractive Hubbard model yields the effective Hamiltonian of the form,

$$\mathcal{H}_{eff} = \sum_{ij,\sigma} \mathcal{H}_{ij\sigma}(c^\dagger_{i\sigma}c_{j\sigma} + \text{H.c.}) + \sum_i \left[\Delta_i c^\dagger_{i\uparrow}c^\dagger_{i\downarrow} - \Delta^*_i c_{i\uparrow}c_{i\downarrow}\right] \tag{1}$$

here $\Delta_i = -|U|\langle c_{i\downarrow}c_{i\uparrow}\rangle$ is the gap parameter for the fermionic superfluid. $\mathcal{H}_{ij\sigma} = -t\delta_{i\pm 1j} + (V_i - \mu - U\delta n_{i\bar{\sigma}} + \sigma h)\delta_{ij}$ where $\delta n_{i\bar{\sigma}} = n_{i\bar{\sigma}} - 1/2$ with $\langle n_{i\sigma}\rangle = \langle c^\dagger_{i\sigma}c_{i\sigma}\rangle$ and $\bar{\sigma} = -\sigma$. Here V_i is assumed to be of the form $V_i = V_0(r_i - r_0)^2$ where V_0 is the strength of the trapping potential. All of U, h, μ and V_0 are expressed in units of hopping strength, t. Eq. (1) is hence diagonalized using Bogoliubov transformation which yields the gap parameter and density (and hence magnetization, $m_i = \langle n_{i\uparrow}\rangle - \langle n_{i\downarrow}\rangle$, see Eq. (3)) in terms of the eigenvectors $u_n(\mathbf{r}_i)$ and $v_n(\mathbf{r}_i)$ at a temperature T as,

$$\Delta_i = -|U|\sum_n [u_n(\mathbf{r}_i)v^*_n(\mathbf{r}_i)f(E_{n\uparrow}) - u_n(\mathbf{r}_i)v^*_n(\mathbf{r}_i)f(-E_{n\downarrow})] \tag{2}$$

$$\langle n_{i\sigma}\rangle = \sum_n \left[|u_n(\mathbf{r}_i)|^2 f(E_{n\sigma}) + |v_n(\mathbf{r}_i)|^2 f(-E_{n\bar{\sigma}})\right] \tag{3}$$

where $f(E_{n\sigma})$ is the Fermi distribution function. Δ_i and $\langle n_{i\sigma}\rangle$ are obtained self-consistently from Eq. (2) and Eq. (3) at each lattice site.

3. Results

We first present a schematic diagram of the BCS and FFLO phases, both in the absence and presence of trap in Fig. 1. The FFLO phase is marked by a lower and a upper critical magnetic field values, h_{c_1} and h_{c_2} respectively and lie intermediate to the BCS and normal phases. These boundaries are obtained from the behaviour of the gap parameter obtained solving the first of Eq. (2). Fig. 1 shows that the values for h_{c_1} and h_{c_2} are $0.35t$ and $0.5t$ in the absence of trap whereas it shifts to $h_{c_1} = 0.25t$ and $h_{c_2} = 0.4t$ when the trapping potential is switched on for $|U| = 2.5t$ and $\mu = -0.5t$.

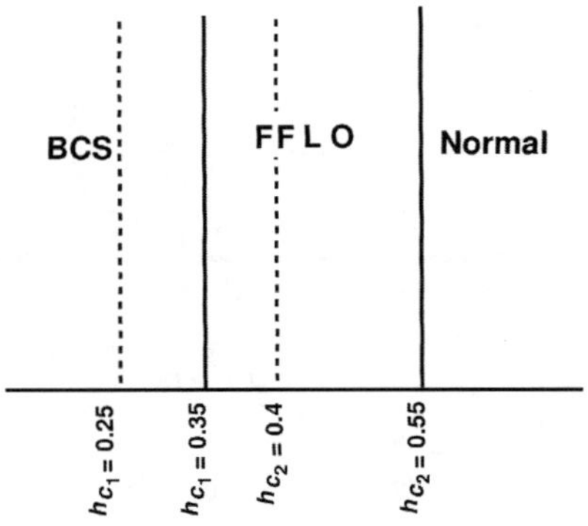

Figure 1. A schematic representation of the FFLO phase is shown as a function of the magnetic field, h both in absence and presence of the trap for $|U| = 2.5t$ and $\mu = -0.5t$ (same for all plots). The boundaries of the FFLO phase are marked by h_{c_1} and h_{c_2} in the absence (solid line) and presence (broken lines) of the trap of strength, $V_0 = 0.016t$. Same numerical value of V_0 is chosen for Figs. 2-5.

We now present our results for the gap parameter, Δ_i both with and without trap. It may be noted that Δ_i is homogeneous (BCS) for low values of magnetic fields (population imbalance being small) in the absence of trap, whereas the trapping effects lead to a maximum of Δ_i at the trap center, hence decreases and finally vanishes away from the center (see Fig. 2 (a) and (b)). Such contrasting behaviour is caused by the trapping effects of the harmonic potential that results in the accumulation of atomic densities around the center thereby leading to such an inhomogeneous distribution of Δ_i in real space. Next we compare Δ_i for the modulating (FFLO) phase. Δ_i modulates along one direction (x-axis) of the lattice without the trap while it modulates along the radial direction in the presence of trap, as shown in Figs. 2 (c) and (d). That is, far away from the trap center, the gap parameter undergoes a sign change confirming the presence of the FFLO phase.

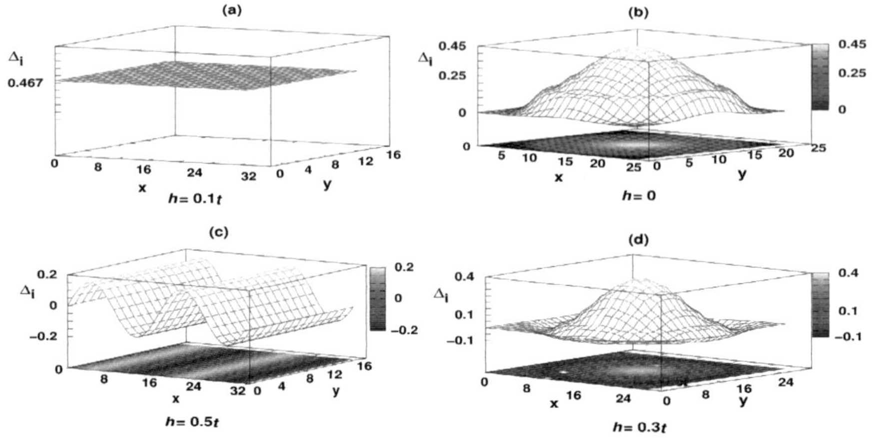

Figure 2. (Colour online) Local pairing amplitude, Δ_i in BCS ((a) and (b)) and modulating FFLO ((c) and (d)) phases are shown. Figs. (a) and (c) are for cases without trap while (b) and (d) are in presence of trap. The system size is 32×16 in the absence of trap and 24×24 when the trapping potential is present and are same for all our results.

We proceed to discuss results on the longitudinal spin-spin correlations e.g. the local magnetization, m_i ($= 2\langle S_i^z \rangle$) presented in Fig. 3. While m_i for the BCS phase, is zero understandably (being non-magnetic due to lack of unpaired particles) across the lattice (see Fig. 3(a) and (b)), it modulates with the period half of that of Δ_i in the FFLO phase when trapping effects are not included (Fig. 3(c)). The large values of magnetization occurs at nodal lines with broken pairs, whereas lattice sites with finite Δ_i correspond to weak magnetization, thereby leading to a phase difference between these two quantities. The magnetization profile in the presence of trap, exhibits a bimodal structure (as shown in Fig. 3(d)) since the magnetization is non-zero around the ring like nodal line. In other words, a small number of particles are squeezed into the inner core of the harmonic trap due to pair formation in the superfluid phase, while the unpaired (majority) carriers are pushed out to outside of the core. It may be worth mentioning that the transverse spin-spin correlations ($S_i^{\pm}$) are the all the while zero.

Finally, we analyze the behaviour of the paramagnetic current-current correlations $\Lambda_{ij}^{\alpha\beta} = \langle j_\alpha(r_i) j_\beta(r_j) \rangle$ in an environment of harmonic confinement where j_α is the component of the paramagnetic current density operator at site r_i given by $j_\alpha(r_i) = it \sum_\sigma \left[c_{i+\alpha,\sigma}^\dagger c_{i,\sigma} - c_{i,\sigma}^\dagger c_{i+\alpha,\sigma} \right]$. Here r_i and r_j denote lattice sites. Λ_{ij}^{xx} (with $\alpha = \beta = x$) rapidly decays to zero in the BCS phase, while it modulates across the lattice and remains finite in the FFLO phase, irrespective of the value of the trapping potential as shown in Fig. 4 and 5. The response of the system to an external perturbation is more in the FFLO phase than that in the BCS phase and thus in turn reiterates the presence of stronger superconducting correlations in the homogeneous BCS phase. The likely reason for lower Λ_{ij}^{xx} in the presence of trap must be because of the reduced mobility of the charge carriers due to localization effects.

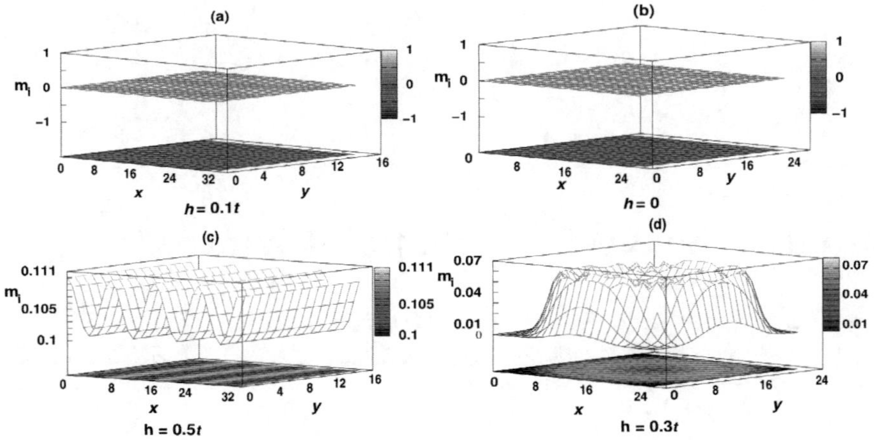

Figure 3. (Colour online) Local magnetization, m_i is shown in BCS (a) and (b) and FFLO (c) and (d) phases both in the absence and presence of the confining potential.

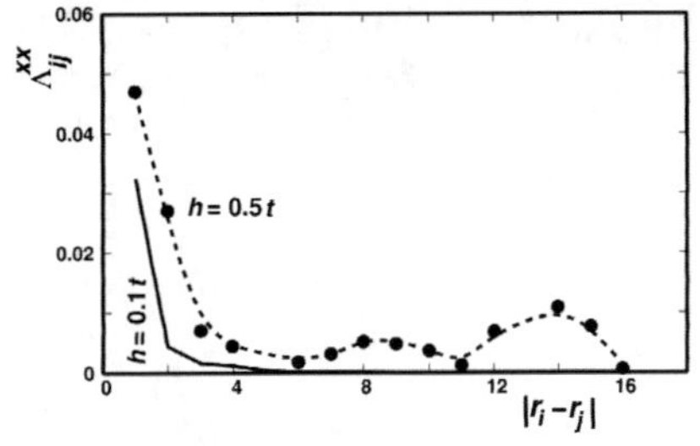

Figure 4. Λ_{ij}^{xx} with no trap.

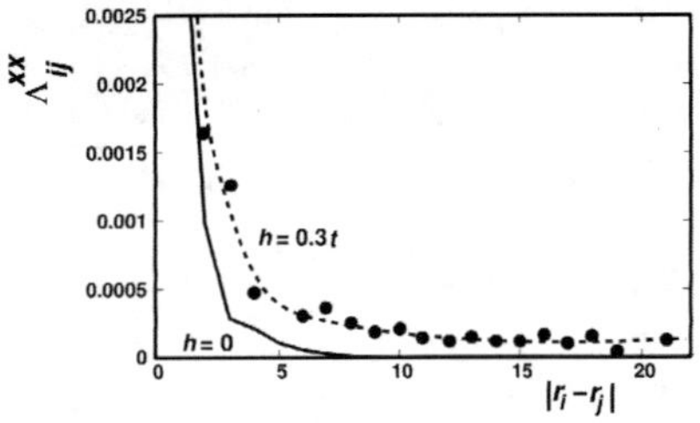

Figure 5. Λ_{ij}^{xx} with trap.

4. Conclusions

In summary, we have explored the ground state of a population imbalanced Fermi gas on a two dimensional lattice in presence of a harmonic confinement. The current-current and spin-spin correlations obtained behave differently in the BCS and FFLO phases for both with and without trap. While the current-current correlations show spatial modulation similar to the behaviour of the order parameter, the longitudinal spin-spin correlations (obtained in the form of magnetization) show modulations whose period is half of that of the order parameter. We hope that experimental realization of such correlations will be possible in near future for bosonic systems, however computing them in fermionic systems is going to be difficult.

5. Acknowledgments

We thank CSIR and DST, India under the Grants - F.No:09/731(0048)/2007-EMR-I, No.03(1097)/07/EMR-II and SR/S2/CMP-23/2009.

References

[1] Clogston A M 1962 *Phys. Rev. Lett.* **9** 266; Chandrasekhar B S 1962 *Appl. Phys. Lett.* **1** 7
[2] Fulde P and Ferrell 1964 *Phys. Rev.* **135** A550; Larkin A I and Ovichinnikov Y N 1964 *Zh. Eksp. Teor. Fiz.* **47** 1136
[3] Sheehy D E and Radzihovsky L 2006 *Phys. Rev. Lett* **96** 060401

Tunneling limit of heavy-fermion point contacts

Mikael Fogelström[1] and Matthias J. Graf[2]

[1] Department of Microtechnology and Nanoscience, Chalmers, S-412 96 Göteborg, Sweden
[2] Theoretical Division, Los Alamos National Laboratory, Los Alamos, New Mexico 87545, USA

Abstract. We present results for a multichannel tunneling model that describes point-contact spectra between a metallic tip and a superconducting heavy-fermion system. We calculate tunneling spectra both in the normal and superconducting state. In point-contact and scanning tunneling spectroscopy many heavy-fermion materials, like $CeCoIn_5$, exhibit an asymmetric differential conductance, dI/dV, combined with a strongly suppressed Andreev reflection signal in the superconducting state. For Andreev reflection to occur a junction has to be in the highly transparent limit. Here we focus on the opposite limit, namely that of low transparency leading to BCS-like dI/dV curves. We discuss the consequences of a multichannel tunneling model for $CeCoIn_5$ assuming itinerant electron bands and localized f electrons.

1. Introduction

Point contact spectroscopy (PCS) and scanning tunneling spectroscopy (STS) have been widely used to study the electronic properties of heavy-fermion superconductors (HFS). A general problem in this field has been the interpretation of tunneling data, which show asymmetric conductances in the normal state and significant deviations from the standard BTK formalism (Blonder-Tinkham-Klapwijk) [1]. In the past, the BTK formalism has been very successful in describing tunneling conductances between metal tips and conventional superconductors, while resulting in unphysical parameterizations of the HFS tunneling conductances.

Here we present results for a multichannel tunneling model between a metallic tip and a heavy-fermion superconductor. These results are discussed with respect to the anomalous properties observed in $CeCoIn_5$ [2, 3, 4, 5, 6], but are readily applied to other heavy fermions (HF) like $CeCu_2Si_2$, URu_2Si_2 and many others [7, 8, 9, 10].

2. Tunneling model

We model the HF material by two itinerant bands and additional localized surface states, which may be caused by broken f-electron bonds at the surface due to the broken translation symmetry,

$$\mathcal{H}_{HF} = \sum_{\alpha;k,\sigma} \mathcal{E}_\alpha(k) c^\dagger_{\alpha;k\sigma} c_{\alpha;k\sigma} + E_0 \sum_{i\sigma} f^\dagger_{i\sigma} f_{i\sigma}. \quad (1)$$

The heavy-fermion Hamiltonian $\mathcal{H}_{HF}$ represents two bands of itinerant conduction electrons with band index $\alpha \in \{\text{light}, \text{heavy}\}$ and localized electrons near the surface with site index i. The operators $c^\dagger_{\alpha;k\sigma}$ ($c_{\alpha;k\sigma}$) create (destroy) an itinerant electron with momentum k and spin σ in band α, while operators $f^\dagger_{i\sigma}$ ($f_{i\sigma}$) create (destroy) an f electron at site i with spin σ. $\mathcal{E}_\alpha(k)$ are the respective electronic dispersions and E_0 is the energy level of the localized f electrons.

Published under licence by IOP Publishing Ltd

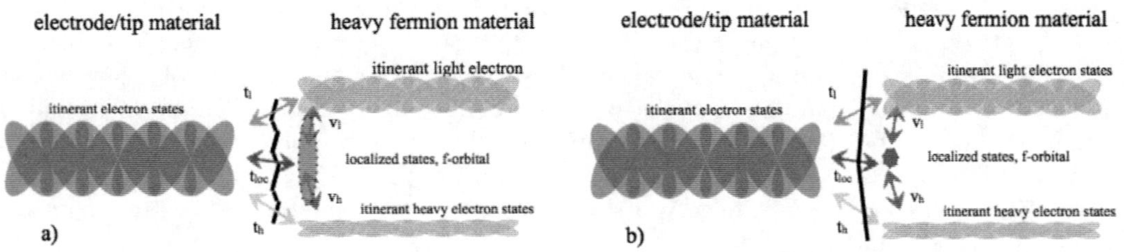

Figure 1. (Color online) A cartoon of the tunneling processes from the tip of the point contact to the heavy-fermion material (localized and itinerant electrons), which are necessary to account for the measured asymmetry in point-contact junction conductances and reduced Andreev reflection signals. In (a) the localized state appears as a broad resonance at the interface while in (b) the localized state forms a sharp surface state, which acts as a resonant tunneling center.

A simple description of a tunneling experiment is comprised of Hamiltonians for the heavy-fermion material, the counter electrode, and the transfer or tunneling processes between them: $\mathcal{H} = \mathcal{H}_{\rm HF} + \mathcal{H}_{\rm electrode} + \mathcal{H}_{\rm T}$. The counter electrode is given by normal conduction electrons

$$\mathcal{H}_{\rm electrode} = \sum_{k,\sigma} \mathcal{E}_e(k) e^{\dagger}_{k\sigma} e_{k\sigma}, \tag{2}$$

and the tunneling Hamiltonian describes all possible transfers

$$\mathcal{H}_{\rm T} = \sum_{\alpha:k,\sigma;k'\sigma'} \left[t^{\alpha}_{k,\sigma;k'\sigma'} e^{\dagger}_{k\sigma} c_{\alpha;k'\sigma'} + t^{\alpha}_{k,\sigma;k'\sigma'} c^{\dagger}_{\alpha;k\sigma} e_{k'\sigma'} \right] + \sum_{k,\sigma;\sigma'} \left[t^{loc}_{k,\sigma;\sigma'} e^{\dagger}_{k\sigma} f_{i\sigma'} + t^{loc}_{k,\sigma;\sigma'} f^{\dagger}_{i\sigma'} e_{k\sigma} \right]. \tag{3}$$

In addition to the standard overlap integrals t_α between the conduction band in the point contact and itinerant heavy-fermion bands there is a finite overlap, t_{loc}, from the point contact to the localized states in the HF. We also account for weak coupling between the localized surface electrons and itinerant electrons through scattering terms v_α (see Fig. 1). In general, to get a Fano resonance in the conductance one needs interference between different tunneling paths [11]. The resulting differential conductance calculated from this model $\mathcal{H}_{\rm T}$ will have an asymmetric Fano line shape. Figure 1 shows the processes that are active in tunneling between a metallic point contact and the HF material. Here we extend the picture of conduction through individual quantum channels to a tunneling model to account for point contacts on a HF material. When deriving the general tunneling expression, we consider strong overlap between electron states in the contact and the HF compound and thus go beyond the strict tunneling limit.

We calculate the tunneling current through a quantum channel by employing the standard non-equilibrium Green's function technique [12, 13, 14, 15]. To further simplify our calculations, we make several assumptions: (1) The itinerant microscopic Green's functions are described by quasiclassical Green's functions near the Fermi energy. (2) It is essential to keep the full energy dependence of the localized Green's function. We assume a single localized level at energy $\varepsilon = E_0$. (3) Only the heavy electrons undergo a superconducting transition at $T = T_c$, while the light electrons remain uncondensed. For details of this model and formalism see Ref. [16].

3. Tunneling conductance

For a multiband tunneling model the differential conductance of a single quantum channel was derived in Ref. [16]. It leads to a generalized Fano expression for the conductance

$$\frac{dI}{dV}(V) = \mathcal{D}\frac{e^2}{\hbar}\frac{1}{T} \int_{-\infty}^{\infty} \frac{d\varepsilon}{2\pi} \frac{|q\Gamma + \varepsilon - \tilde{E}_0|^2}{\Gamma^2 + (\varepsilon - \tilde{E}_0)^2} \cosh^{-2}\left[\frac{\varepsilon - eV}{2T}\right]. \tag{4}$$

In Eq. (4) $\mathcal{D}$ is the transparency of the junction, $\tilde{E}_0$ is the tunneling-renormalized position of the localized energy relative to the Fermi level, Γ is the half-width of the resonance, and $q = q_F + i\, q_{\mathcal{B}}$ with $q_{\mathcal{B}} = \mathcal{B}/\Gamma$. The conventional Fano parameter q_F controls the resonance shape. The additional parameter $q_{\mathcal{B}}$ is present for multiband models only, when tunneling through a resonant localized state couples differently to the HF conduction bands (see below). The term $\mathcal{B}$ adds a Lorentzian to the conventional Fano resonance.

For notational convenience, we introduce the following parameterization of the microscopic parameters $t_{\{h,l,loc\}}$ and $v_{\{h,l\}}$

$$\begin{array}{l} t_h = \tilde{t}\sin\alpha\sqrt{\eta_h}\cos\theta_t \\ t_l = \tilde{t}\sin\alpha\sqrt{\eta_l}\sin\theta_t \end{array}, \quad \begin{array}{l} v_h = \tilde{v}\sqrt{\eta_0\,\eta_h}\cos\theta_v \\ v_l = \tilde{v}\sqrt{\eta_0\,\eta_l}\sin\theta_v \end{array}, \quad t_{loc} = \tilde{t}\sqrt{\eta_0}\cos\alpha, \tag{5}$$

where $\tilde{t} = t\sqrt{\mathcal{N}_c\mathcal{N}_{HF}}$, $\tilde{v} = v\,\mathcal{N}_{HF}$ are the effective tunneling elements with the density of states at the Fermi level $\mathcal{N}_{\{HF,c\}}$ in the heavy fermion (contact) material. The factor η_0 is the fraction of localized surface states and $\eta_{\{h,l\}}$ give the relative fraction of heavy and light electrons. The angles θ_t and θ_v give the relative overlap integrals between the direct tunneling and the hybridization matrix elements. For simplicity, we assume $\theta_t = \theta_v \equiv \theta$, which results in $\mathcal{B} \equiv 0$ [16]. Finally, the angle α quantifies the relative proportion of tunneling into a localized state relative to direct tunneling. The four non-zero phenomenological model parameters introduced in Eq. (4) now depend on the microscopic parameters $(\tilde{t}, \tilde{v}, E_0, \alpha, \theta)$ and the three fractions $\eta_{\{0,h,l\}}$.

3.1. Tunneling limit in the normal state

In the tunneling limit, $\tilde{t}^2 \ll 1$, we keep the coupling term $\tilde{v}$ between localized and itinerant states at arbitrary strength and obtain

$$\mathcal{D} \approx 4\eta_i\tilde{t}^2\sin^2\alpha, \tag{6}$$

$$\tilde{E}_0 \approx E_0 - \eta_0\eta_i\tilde{v}\tilde{t}^2\sin 2\alpha, \tag{7}$$

$$\Gamma \approx \eta_0[\eta_i\tilde{v}^2 + \tilde{t}^2(\cos^2\alpha - \eta_i^2\tilde{v}^2\sin^2\alpha)], \tag{8}$$

$$q_F \approx -\frac{\cot\alpha}{\eta_i\tilde{v}}[1 - \frac{\tilde{t}^2}{n\tilde{v}^2}(\cos^2\alpha + \eta_i^2\tilde{v}^2\sin^2\alpha)]. \tag{9}$$

Here $\eta_i = \eta_h\cos^2\theta + \eta_l\sin^2\theta$ is the itinerant fraction. It quantifies the relative weight of direct tunneling into light and heavy bands. The parameter $\tilde{v}$ is the effective hybridization matrix element at the surface, which is temperature independent.

In this limit the point-contact spectra measure a dI/dV curve with a Fano line shape and allow us to directly relate the model parameters to the bulk state of the HF material. At temperatures low compared to the coherence temperature of the Kondo lattice, $T \ll T_{coh}$, the HF state can be described as a renormalized Fermi liquid [17, 18]. The density of state factors can be related to the ratio of the bare mass m with the effective mass m^* as $\eta_h \approx (1 - m/m^*)$ and $\eta_l \approx m/m^*$ [17]. The temperature dependence of m/m^* can be obtained, for example, from experiment. For qualitative purposes, we model it as $m/m^* \sim (T/T_{coh})^p$, with $p > 0$. The factor η_0 describes the fraction of localized states and its physical meaning and value are under debate, see for example the discussion by Yang [19, 20].

We can gain some physical insight by considering three special scenarios on how the measured peak width Γ of the Fano resonance should depend on temperature if it was entirely due to the effective mass. For simplicity, we keep only leading order tunneling terms and drop terms proportional to $\tilde{t}^2$:

(1) The localized states or moments at the interface are not connected to the HF physics of the bulk. In this case, the fraction η_0, giving the number of surface states, is temperature

independent. We obtain

$$\Gamma(T) \approx \eta_0 \frac{\tilde{v}^2}{2}(1+\cos 2\theta(1-2\frac{m}{m^*})), \tag{10}$$

$$q_F(T) \approx -\frac{2\cot\alpha}{\tilde{v}\left[1+\cos 2\theta(1-2\frac{m}{m^*})\right]}, \tag{11}$$

where $\Gamma(T)$ decreases with increasing temperature, while $q_F(T)$ increases, which is inconsistent with PCS measurements for $CeCoIn_5$, see Ref. [16].
(2) Localized states or moments at the interface originating from the Ce 4f states at energy E_0, *which are the same as in the bulk of* $CeCoIn_5$. In this case, it is reasonable to assume that η_0 is proportional to the number of unscreened moments, i.e., $\eta_0(T) \sim \eta_0\, m/m^*$. We obtain

$$\Gamma(T) \approx \eta_0 \frac{m}{m^*}\frac{\tilde{v}^2}{2}(1+\cos 2\theta(1-2\frac{m}{m^*})), \tag{12}$$

$$q_F(T) \approx -\frac{2\cot\alpha}{\tilde{v}\left[1+\cos 2\theta(1-2\frac{m}{m^*})\right]}, \tag{13}$$

where both $\Gamma(T)$ and $q_F(T)$ increase with T for $T \ll T_{coh}$. The increase of $\Gamma(T)$ is consistent with the PCS measurements, while that of $q_F(T)$ is not [16].
(3) The dominant temperature behavior of $\Gamma(T)$ *is caused by inelastic scattering processes like spin fluctuations.* Resistivity experiments on $CeCoIn_5$ show that $\rho(T)$ may be described by a self-consistent spin-fluctuation theory [21]. Similarly, spin-lattice relaxation rates $1/T_1 \sim T^{1.3}$ suggest the importance of spin fluctuations [22, 23, 24]. In a spin-fluctuation scenario the inelastic broadening increases as $\Gamma(T) \sim T^p$, where $p > 1$, whereas to leading order $q_F(T)$ is temperature independent. Therefore, broadening due to electrons scattering off spin fluctuations is a reasonable interpretation for the temperature dependence of $\Gamma(T)$ found in $CeCoIn_5$ [16] and more recently in URu_2Si_2 [9, 10]. We conclude that the self-consistent incorporation of inelastic scattering processes into our calculations, similar to Ref. [25], is an important next step to account for the observed line broadening.

3.2. Tunneling limit in the superconducting state

It is necessary to study tunneling in the superconducting state, where Andreev reflection processes modify the conductance, because one can extract additional information about the microscopic origin of the Fano conductance. The unknown relative ratio of tunneling into heavy or light bands can be determined as the dI/dV depends in a non-linear way on θ. At very low temperatures, $T < T_c \ll T_{coh}$, one can assume that m/m^* has saturated to its low-T limit $\sim 1-\eta_{HF}$. Thus we write for η_i defined in the previous section $\eta_i \sim \eta_{HF}\cos^2\theta + (1-\eta_{HF})\sin^2\theta$ and study the superconducting conductance as a function of θ and fixing the numerical value of η_{HF} to 0.9. For simplicity, we assume that only heavy electrons become superconducting, while light electrons remain uncondensed. In our model, we consider both an s-wave and d-wave superconducting order parameter. For the d-wave case we show results for two principal orientations. The orientation of the HF crystal lattice relative to the interface normal is either antinodal (0°) or nodal (45°) leading to strikingly different tunneling conductances important for order parameter spectroscopy of the nodes.

In Fig. 2 we plot the conductances in the superconducting state calculated in the tunneling limit by setting $\mathcal{D} = 0.005$. The Fano parameters are chosen to fit experimentally observed dI/dV curves just above $T_c = 2.3$ K in $CeCoIn_5$. Hence we set $E_0 = 2.01$ meV, $\Gamma = 13.0$ meV, and $q_F = -2.16$ in agreement with PCS measurements with a Au-tip on $CeCoIn_5$, see Fig. 3 in Ref. [16]. Since the dI/dV curve is only weakly dependent on the tunneling angle α in the superconducting state, we set in our calculations $\alpha = \pi/20$, i.e., stronger tunneling

Journal of Physics: Conference Series **273** (2011) 012073 doi:10.1088/1742-6596/273/1/012073

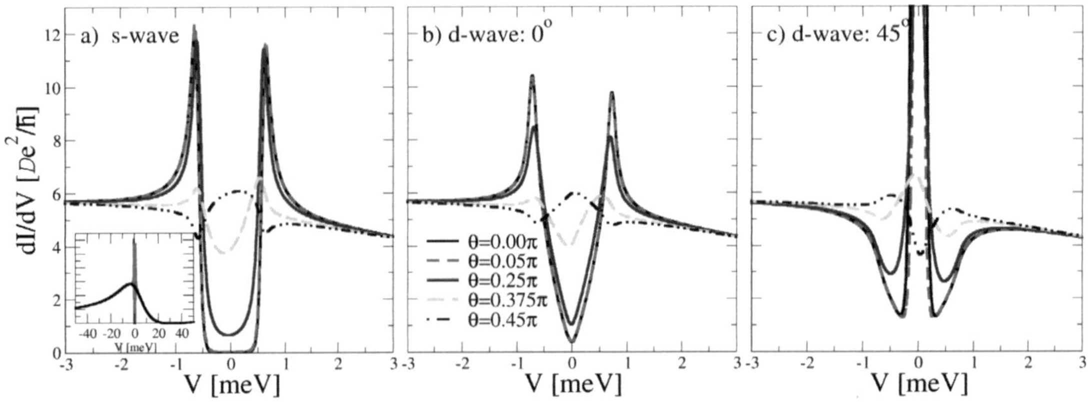

Figure 2. (Color online) Conductances for an HFS in the tunneling limit. The parameters of the curves are $\mathcal{D} = 0.005$, $E_0 = 2.01\,\text{meV}, \Gamma = 13.0\,\text{meV}, q_F = -2.16, \alpha = \pi/20$, and $T = 0.41$ K with $T_c = 2.3$ K. In panels (a)-(c) we vary the heavy-to-light tunneling parameter θ. The conductances are calculated both for an s-wave and d-wave superconducting state, setting $\Delta = 0.6\,\text{meV}$. As shown, the superconducting dI/dV is very sensitive to the co-tunneling into two bands. The inset in panel (a) shows the Fano shaped conductance in the normal state at $T > T_c$ with a superimposed conductance curve computed at $T < T_c$.

through the localized state compared to the itinerant bands. In the case of the ideal one-band superconducting tunneling model, i.e. $\theta = 0$, we recover the known dI/dV for all three cases considered. In Fig. 2(a) we see the usual BCS superconducting density of states on top of a Fano background. In the inset of panel (a) we show the voltage dependence of the dI/dV over a large voltage range. In panels (b) and (c) the case of $\theta = 0$ reproduces the V-shaped dI/dV for the 0°-junction (panel b), while for the 45°-junction we see the hallmark zero-bias peak (panel c). Note that for tunneling into the heavy band only, $\theta = 0$, the condensate suppresses the normal state dI/dV by 100% for voltages below the gap, $-\Delta < \text{eV} < \Delta$.

Incorporating the additional possibility of tunneling into a band of uncondensed light electrons has two main effects on the dI/dV characteristics. First, we see that the coherence peaks in the dI/dV curves are reduced from a 100% effect for small θ to a ~10%-effect, when $\theta > \pi/4$. Second, the dI/dV curves are qualitatively changed in shape from their ideal one-band $\theta = 0$ appearance. In panel (a) we see that the sub-gap conductance below Δ can even be increased above its normal-state value, resembling a high-transmission sub-gap conductance but with a strongly suppressed Andreev reflection signal. Next, turning to the d-wave conductances we see that in general, for a fixed set of microscopic parameters, there is a strong dependence on the junction orientation. This property may be used as a smoking gun to identify a d-wave symmetry. However, if we assume that PCS junctions are made on surfaces with different crystal orientations and that these PCS junctions have very different characteristics with different sets of microscopic parameters, then examining the curves in panels (b) and (c) may not be as distinctive as expected. In this situation one can find cases where the zero-bias peak is strongly suppressed making the distinction between principal nodal/antinodal orientations less striking. In fact, it may even be difficult to convincingly discriminate between an s-wave and d-wave superconducting state based on a few tunneling curves alone.

4. Conclusions

We presented results for a multichannel tunneling model for a superconducting heavy-fermion material. On the large voltage scale the derived conductances have the features of Fano-like dI/dV characteristics. This allows us to extract the microscopic model parameters that describe the relevant tunneling processes. Assuming a modified Fermi-liquid state at low temperatures and a tunneling contact in the tunneling limit, i.e., low transparency, we find for the normal state of $CeCoIn_5$ that the extracted Fano parameters are most likely describing localized surface states and not directly probing the formation of the heavy-fermion state and the temperature behavior of the effective mass. In the superconducting state, the calculated dI/dV curves demonstrate that additional microscopic information can be obtained about tunneling into heavy vs. light bands, which is not possible from studying the normal state alone. Finally, we investigated how the dI/dV characteristics depend on the pairing symmetry of the superconducting state. We found that in order to reliably extract information from tunneling experiments about the symmetry of the superconducting order parameter detailed modeling of the tunneling processes is required, which goes beyond a conventional BTK analysis.

Acknowledgments

We benefited from discussions with T. Löfwander, W. K. Park. L. H. Greene, G. Goll, A. V. Balatsky, Y. Dubi and P. Wölfle. M. F. was supported by the Swedish Research Council. M. J. G. was supported in parts by the U.S. DOE at Los Alamos National Laboratory under contract No. DE-AC52-06NA25396 and the Office of Science for BES.

References

[1] Blonder G E, Tinkham M and Klapwijk T M 1982 *Phys. Rev. B* **25** 4515
[2] Goll G 2005 *Adv. Sol. State Phys.* **45** 213
[3] Goll G, Brugger T, Marz M, Kontermann S, v. Löhneysen H, Sayles T, Maple M B 2006 *Physica B* **378-380** 665
[4] Park W K, Sarrao J L, Thompson J D, Greene L H 2008 *Phys. Rev. Lett.* **100** 177001
[5] Park W K, Bauer E D, Sarrao J L, Thompson J D, Greene L H 2009 *J. Phys: Conf. Ser.* **150** 052207
[6] Ernst S, Wirth S, Steglich F, Fizk Z, Sarrao J L, Thompson J D 2010 *preprint* arXiv:1002.2878
[7] Groschke R A, Gloos K, Geibel C, Ekino T, Steglich F 1996 *Czechoslovak J. Phys.* **46** Suppl. 2 797
[8] Steglich F, Ahlheim U, Rauchschwalbe U, Spille H 1987 *Physica B* **148** 6
[9] Schmidt A R, Hamidian M H, Wahl P, Meier F, Balatsky A V, Garrett J D, Williams T J, Luke G M, Davis J C 2010 *Nature (London)* **465** 570
[10] Aynajian P, da Silvia Neto E H, Parker C V, Huang Y, Pasupathy A, Mydosh J, Yazdani A 2010 *PNAS (USA)* **107** (23) 10383
[11] Fano U 1961 *Phys. Rev.* **124** 1866
[12] Schrieffer J R and Wilkins J W 1963 *Phys. Rev. Lett.* **10** 17
[13] Caroli C, Combescot R, Nozieres P, Saint-James D 1971 *J. Phys. C: Solid State Physics* **4** 916
[14] Cuevas J C, Martín-Rodero A and Yeyati A L 1996 *Phys. Rev. B* **54** 7366
[15] Cuevas J C and Fogelström M 2001 Phys. Rev. B **64** 104502
[16] Fogelström M, Park W K, Greene L H, Goll G and Graf M J 2010 *Rev. Phys. B* **82** 014527
[17] Hewson A C 1993 *The Kondo Problem to Heavy Fermions* (Cambridge Univ. Press)
[18] von Löhneysen H, Rosch A, Vojta M, Wölfle P 2007 *Rev. Mod. Phys.* **79** 1015
[19] Yang Y-F and Pines D 2008 *Phys. Rev. Lett.* **100** 096404
[20] Yang Y-F 2009 *Phys. Rev. B* **79** 241107(R)
[21] Bianchi A, Movshovich R, Vekhter I, Pagliuso P G, Sarrao J L 2003 *Phys. Rev. Lett.* **91** 257001
[22] Kawasaki Y, Kawasaki S, Yashima M, Mito T, Zheng G Q, Kitaoka Y, Shishido H, Settai R, Haga Y, Onuki Y 2003 *J. Phys. Soc. Jpn.* **72** 2308
[23] Curro N J, Sarrao J L, Thompson J D, Pagliuso P G, Kos S, Abanov A, Pines D 2003 *Phys. Rev. Lett.* **90** 227202
[24] Curro N J and Pines D 2007 *J. Phys. Chem. of Solids* **68** 2028
[25] Löfwander T and Fogelström M 2005 *Phys. Rev. Lett.* **95** 107006

Journal of Physics: Conference Series **273** (2011) 012074 doi:10.1088/1742-6596/273/1/012074

Superconductivity in $Rh_{17}S_{15}$ and $Pd_{17}Se_{15}$: A comparative study

Naren H.R., Arumugam Tamizhavel and Ramakrishan S.

Tata Institute of Fundamental Research, Mumbai, India

E-mail: nareni@tifr.res.in

Abstract. The presence of strongly correlated superconductivity in $Rh_{17}S_{15}$ has been recently established. In this work we estimate normal and superconducting parameters of a single crystal of $Rh_{17}S_{15}$ and compare them with those of a polycrystalline sample of $Pd_{17}Se_{15}$ which is reported here for the first time to be a superconductor below 2.2 K. The presence of superconductivity in two iso-structural (space group Pm3m) compounds provides an opportunity to understand the strongly correlated superconductivity in $Rh_{17}S_{15}$. We see that large unit volume of $Pd_{17}Se_{15}$ and the large separation of Pd-Pd atoms in its structure as compared those in a pure Pd metal lead to the absence of strong correlations in this compound. Finally $Pd_{17}Se_{15}$ can be characterized as an intermediate type-II superconductor as opposed to the strongly correlated superconductivity in $Rh_{17}S_{15}$.

1. Introduction

Recently there has been intense activity in the study of transition metal sulphides and selenides due to a variety of their properties such as, strongly correlated superconductivity, spin and charge density waves. In our effort to understand the physical properties of noble metal chalcogenides, we have studied cubic $Rh_{17}S_{15}$ [1, 2] which is considered as an excellent catalyst by chemists [3, 4]. $Rh_{17}S_{15}$ (known as Miassite), a mineral mainly found near the Miass river in Russia, and $Pd_{17}Se_{15}$ (known as Palladseite), a mineral mainly found in Brazil, are iso-structural cubic compounds (space group Pm3m) whose crystal structure is shown in the Fig. 1. They have a large unit cell with 64 atoms in each (two formula units) and Rh (or Pd) occur in four inequivalent crystallographic sites.

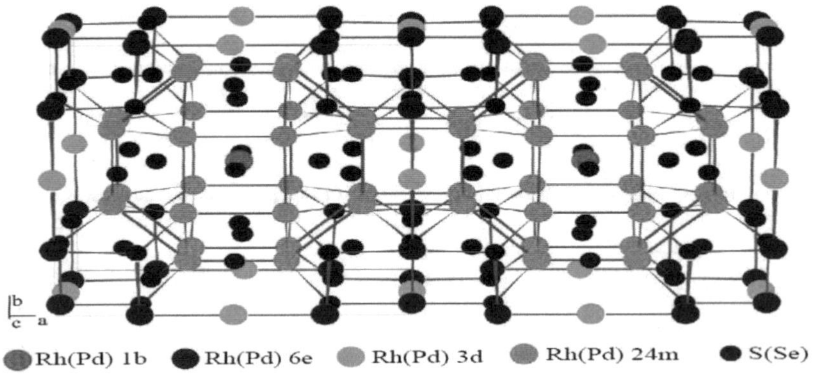

Figure 1. The crystal structure of $Rh_{17}S_{15}$ which consists of two formula units with 64 atoms in the unit cell.

Published under licence by IOP Publishing Ltd

Some of us had reported the observation of strongly correlated superconductivity (T_c = 5.4 K) in $Rh_{17}S_{15}$ [1, 2] via magnetization, resistivity, heat capacity and Hall coefficient measurements. The compound was remarkable for its unusually high upper critical field ($H_{c2}(T)$) values (around 13 T at 3 K) well above the Pauli paramagnetic limiting field (9.9 T) [5, 6]. The observation of T^2 dependence of resistivity, enhanced susceptibility, moderate Sommerfeld coefficient(γ), large value of the upper critical field and large heat capacity jump ($\Delta C/\gamma T_C$=2.5) suggest that $Rh_{17}S_{15}$ is a strongly correlated system. Or estimates of 2 for Wilson's coefficient ($\pi^2\ k_B^2\ \chi(0)/3\mu_B^2\ \gamma$) and 5 x 10^{-5} for the Kadawaki-Woods ratio (A/γ^2) further substantiate our claim that $Rh_{17}S_{15}$ is a strongly correlated system. One way to achieve a moderate density of low-energy fermionic excitations (as seen by the appreciable value of γ) is from the large density of states of the narrow 4d band of Rh at the Fermi level [5, 6, 7]. This is supported by the structure [8] where some of the Rh atoms (Rh atoms at the 3d and 6e positions) are only 0.258 nm apart (the nearest Rh-Rh distance in Rh element is 0.269 nm). In the iso-structural $Pd_{17}Se_{15}$, however, the shortest Pd-Pd distance (0.278 nm) is more than that in the elemental Pd (0.275 nm) [9]. In this work, we will establish that $Pd_{17}Se_{15}$ is a conventional superconductor below 2.2 K. The existence of $Pd_{17}Se_{15}$ and its superconductivity provide us a way to understand the strongly correlated superconductivity in $Rh_{17}S_{15}$.

2. The Samples

Both compounds were prepared by heating stoichiometric compositions of the two elements at the rate of 8 C/hr to a temperature of 1150 C in an alumina crucible. The alumina crucible was sealed in an evacuated quartz tube. The reacted mixture is annealed at 1080 C for two days and then cooled down firstly at 8 C/hr till 600 C and further at 30 C/hr. In this method we obtained a single crystal of $Rh_{17}S_{15}$ with a residual resistivity ratio (RRR) of 10 and a polycrystalline sample of $Pd_{17}Se_{15}$ with an RRR of 7. The samples were characterized using powder XRD and Laue spectrometers. The single crystal of $Rh_{17}S_{15}$ showed clear Laue spots indentifying the space group Pm3m. The lattice constants were 9.912 Å for $Rh_{17}S_{15}$ and 10.607 Å for $Pd_{17}Se_{15}$.

3. Results and Discussion

3.1. Superconducting and Normal state properties of $Rh_{17}S_{15}$

Fig. 2 shows a plot of the heat capacity(C_p) versus temperature(T) for the single crystal of $Rh_{17}S_{15}$ in the absence of applied magnetic field from 2 to 15 K.

A fit to the equation

$$C_p\ =\ \gamma\, T +\ \beta\, T^3\ +\ \delta\, T^5 \tag{1}$$

is also shown in the same figure where γ is due to the electronic contribution, β is due to the lattice contribution and δ is the contribution due to anharmonicity. The estimated values of γ, β and δ are 110 mJ/mol-K^2, 0.63 mJ/mol-K^4 and 1.5 μJ/mol-K^6 respectively. The significant value of δ suggests a complex phonon structure that could arise from the involved crystal structure of this compound. From the β value, one can estimate the Debye temperature θ_D to be 456 K. Using the value of θ_D and T_c (T_c = 5.27 K as obtained by an equal entropy estimate at the superconducting transition), we can estimate the electron-phonon coupling parameter, λ, from McMillan's formula [10] to be 0.58 which puts $Rh_{17}S_{15}$ as an intermediate electron-phonon coupling superconductor. However, the large heat capacity jump of ΔC = 2.18 J/mol K gives a $\Delta C/\gamma T_c$ of 2.5 which suggests that the observed superconductivity is in the strong coupling regime. This contradiction suggests that the system is an unconventional superconductor. We can also estimate the density of states at the Fermi level N*(0) using the relation, $N^*(0)\ =\ 0.4248\ \gamma\ states/eV - formula\ unit$ where γ is expressed in mJ/mol K^2. We get a value of 46 states/eV-formula unit for N*(0).

Fig. 3 shows the temperature dependence of the upper critical field measurement from 70 mK to 5.4 K as estimated from dc and ac magnetization, heat capacity and resistance measurements.

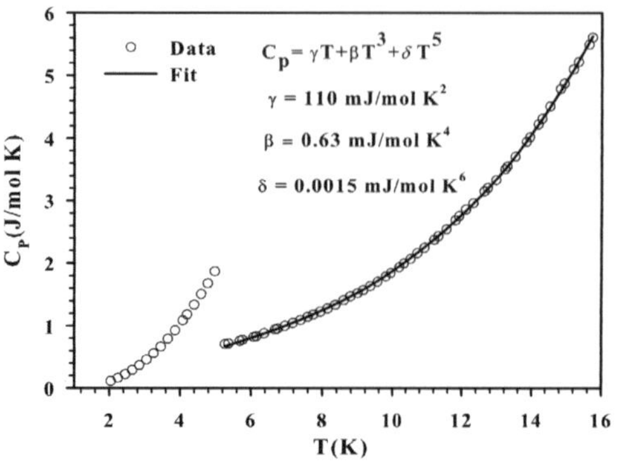

Figure 2. A plot of the heat capacity(C_p) vs temperature(T) from 2 to 15 K of a single crystal of $Rh_{17}S_{15}$. The solid line is a fit to the expression which is described in the text.

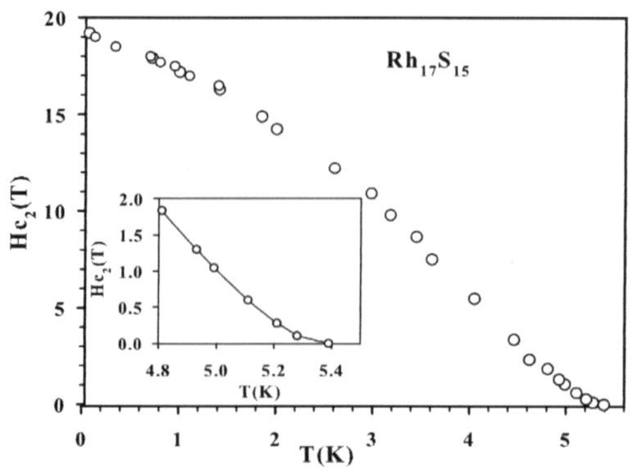

Figure 3. The temperature dependence of the upper critical field (H_{c2}) of $Rh_{17}S_{15}$ down to 70 mK. The inset shows the temperature dependence of H_{c2} near the superconducting transition.

The very low temperature data (below 2 K) have been included from a recent study [6]. From this one can estimate the value of the upper critical field at H_{c2} at 0 K to be 20 T which is in agreement with the recent single crystal study of Settai *al* [5]. The slope (dH_{c2}/dT) near the superconducting transition is estimated to be 3.6 T/K which is smaller than the value quoted by Settai *al* [5]. Using this, one can an estimate a value of 13.3 T for $H_{c2}(0)$ from the WHH theory expression[11],

$$H_{c2}(0) \;=\; 0.693\; T_c\; (dH_{c2}/dT)_{T_c} \tag{2}$$

This value (13.3 T) is much smaller than the observed value of 20 T. Moreover, the Pauli paramagnetic limiting field ($H_p(0) = 1.84 \times T_c$) is only 9.9 T which means that the system is not Pauli limited implying unconventional superconductivity in $Rh_{17}S_{15}$. We have also estimated the penetration depth λ_0 as 7000 Å from the μSR measurements performed at the PSI, Switzerland recently [12]. From the estimation of the upper critical field $H_{c2}(0)$, one can calculate the coherence length (ξ_0) which is given by, $\xi_0^2 \;=\; \phi_0/2\pi H_{c2}(0)$. This gives a value of 40 Å for ξ_0. Using λ_0 and ξ_0, this we can calculate the lower critical field H_{c1} using the relation,

$$Hc_1(0) \;=\; \frac{\phi_0}{4\pi\lambda_0^2}\, ln(\kappa_0) \tag{3}$$

where $\kappa_0 = \lambda_0/\xi_0$. The value of the lower critical field is 1.7 mT which makes $Rh_{17}S_{15}$ as an extreme(κ=175) type-II superconductor.

3.2. Normal and superconducting state properties of $Pd_{17}Se_{15}$ and its comparison with those of $Rh_{17}S_{15}$

Fig. 4 shows the temperature dependence of the resistivity (ρ(T)) of a polycrystalline sample of $Pd_{17}Se_{15}$ from 1.6 to 300 K as compared with that of the single crystal of $Rh_{17}S_{15}$. The inset (a) shows the superconducting transitions in these compounds (T_c = 2.2 K for $Pd_{17}Se_{15}$). A T^2 fit to the low temperature data ($\rho \;=\; \rho_0 \;+\; A\,T^2$) of $Pd_{17}Se_{15}$ gives a value of 43.2 $\mu\Omega$ cm for the residual resistivity and a value of 0.014 $\mu\Omega$ cm/K^2 for the coefficient A. The value of A for $Pd_{17}Se_{15}$ is comparable to that of $Rh_{17}S_{15}$ [1].

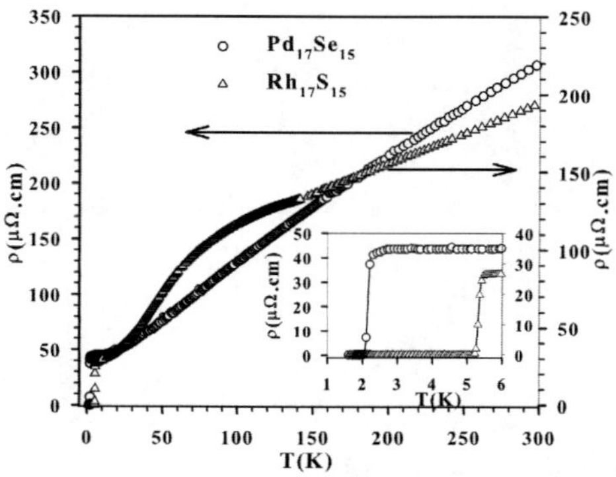

Figure 4. Comparison of the temperature dependence of the resistivity of the single crystal $Rh_{17}S_{15}$ with that of the polycrystalline $Pd_{17}Se_{15}$.

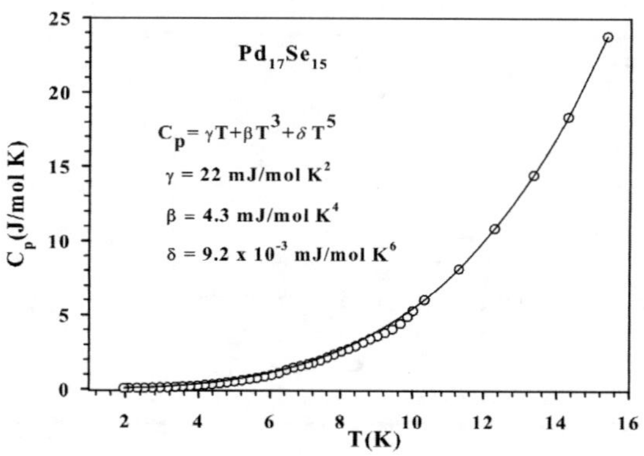

Figure 5. A plot of the heat capacity(C_p) vs temperature T from 2 to 15 K of a polycrystalline $Pd_{17}Se_{15}$. The solid line is a fit to an expression which is described in the text.

The two curves are distinctly different in the appearance of a knee-like feature at around 60 K in $Rh_{17}S_{15}$ which is completely absent in $Pd_{17}Se_{15}$. We do not, at present, understand the reason for the knee-like feature. However, we note that there is a change of sign of majority charge carriers occurs around 60 K in $Rh_{17}S_{15}$ which was seen in the Hall measurements reported earlier [1]. Though the absolute value of the resistivity of both samples are comparable at 300 K, there is a distinct possibility that the strongly correlated behavior observed in $Rh_{17}S_{15}$ could be related to this feature.

Fig. 5 shows a plot of C_p versus T for the polycrystalline $Pd_{17}Se_{15}$ in the absence of applied magnetic field from 2 to 15 K. Our range of C_p measurement, unfortunately cannot capture the transition region well. The solid line in the same figure is a fit to the equation (1) described earlier. The fit yields a γ of 22 mJ/mol-K^2, β of 4.3 mJ/mol-K^4 and δ of 9.2 μJ/mol-K^6. From the value of β we obtain a Debye temperature (θ_D) of 233.3 K which indicates that $Pd_{17}Se_{15}$ is a much softer lattice as compared with $Rh_{17}S_{15}$. However, the anharmonicity term δ is much larger for $Pd_{17}Se_{15}$ as compared to that of $Rh_{17}S_{15}$. More importantly, the enhanced density of states, $N^*(0)$, as calculated from the earlier expression is around 9.35 states/eV-formula unit is smaller than that of $Rh_{17}S_{15}$ by a factor of 5 indicating that the correlations are much weaker in $Pd_{17}Se_{15}$. From the values of γ and θ_D, we have also estimated a value of 0.54 for the electron-phonon coupling constant λ (from [10]) for $Pd_{17}Se_{15}$ which is very similar to the λ value of $Rh_{17}S_{15}$. We believe that $Pd_{17}Se_{15}$ is truly a conventional and intermediate type-II superconductor where Mcmillan's equation gives a good estimate of λ. However, $Rh_{17}S_{15}$ is a strongly correlated unconventional superconductor where one cannot use Mcmillan's equation to get the correct estimate of λ. The reduction in the density of states in $Pd_{17}Se_{15}$ as compared to that of $Rh_{17}S_{15}$ is also reflected in the temperature dependence of normal state magnetic susceptibility as shown in the Fig. 6.

Since none of Rh, Pd, S or Se carry a magnetic moment we expect to see a temperature independent susceptibility essentially sum of Pauli paramagnetism, Landau diamagnetism and core diamagnetism. However, although the values of susceptibility (χ(T)) are small, the data display a distinct temperature dependence for $Rh_{17}S_{15}$. There is a gradual increase in susceptibility as one goes to lower temperatures until the superconducting transition. The weak

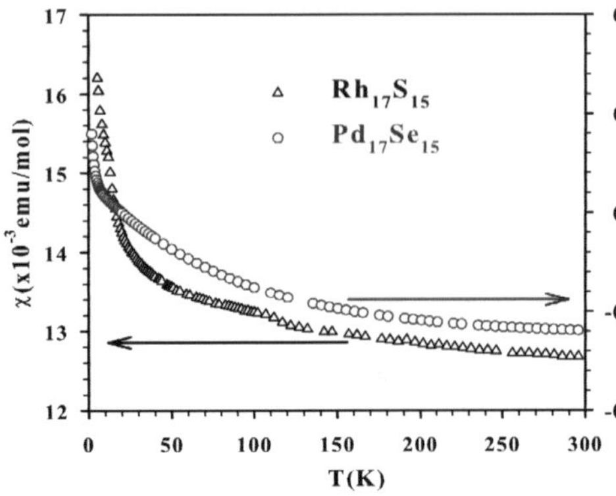

Figure 6. The temperature dependence of the susceptibility of $Pd_{17}Se_{15}$ and $Rh_{17}S_{15}$ in a field 0.1 T.

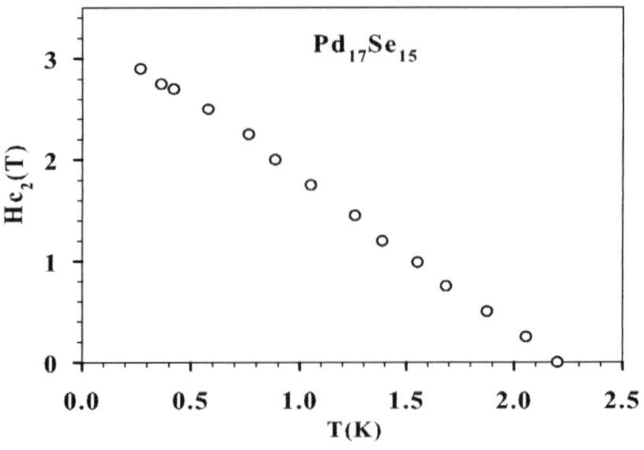

Figure 7. The temperature dependence of the upper critical field (H_{c2}) of $Pd_{17}Se_{15}$ down to 270 mK.

temperature dependence of $\chi(T)$ could arise due to a sharp density of states at the Fermi level which leads to the temperature dependence of the Pauli spin susceptibility. We believe that the mechanism is similar to the observed in V_3Si [13]. This conjecture is recently supported by the observation of temperature dependent Knight shift from the ^{103}Rh-NMR experiment [7]. In fact, the value of susceptibility per Rhodium atom in $Rh_{17}S_{15}$ is of the same order as that of susceptibility per Vanadium atom in V_3Si [13]. However the $\chi(T)$ of $Pd_{17}Se_{15}$ is diamagnetic and shows very weak temperature dependence and its $\chi(T)$ is much smaller that of $Rh_{17}S_{15}$ indicating that the density states is much smaller which is confirmed by the heat capacity data.

Finally, we show the temperature dependence of the upper critical field (H_{c2}) of $Pd_{17}Se_{15}$ from 270 mK to 2.2 K in Fig. 7. The curvature of H_{c2} near T_c is more like a conventional superconductor as opposed the one seen in $Rh_{17}S_{15}$ [1]. However, both show a tendency to saturate at lower temperatures. The saturation is better seen in $Rh_{17}S_{15}$ because we have data upto 70 mK whereas, we have presented data only upto 270 mK in $Pd_{17}Se_{15}$. The $H_{c2}(0)$ is estimated to be 3.3 T by an extrapolation which is nearly 6 times smaller than that of $Rh_{17}S_{15}$. The BCS coherence length ξ_0 is 100 Å as estimated from the $H_{c2}(0)$(3.3 T). The slope (dH_{c2}/dT) near T_c is 1.5 T/K which yields a value of 2.3 T for $H_{c2}(0)$ from the WHH theory. However, the Pauli paramagnetic limiting field ($H_p(0)$) is estimated to have a value of 4 T which suggests that that orbital critical field is Pauli limited in contrast to the situation in $Rh_{17}S_{15}$.

4. Conclusion

Table 1 contains the values of various estimated parameters for both $Rh_{17}S_{15}$ and $Pd_{17}Se_{15}$. Earlier, [1], we have conjectured that the strong correlations in $Rh_{17}S_{15}$ could be due to a high density of states of Rh-4d band at the Fermi level arising due to strong Rh-Rh interactions. On the other hand $Pd_{17}Se_{15}$ has a large unit cell volume and Pd-Pd distances are larger than that of pure Pd metal. Hence, we see the absence of strong correlations from its small values of γ and the normal state susceptibility ($\chi(T)$). More over, the superconducting state properties of $Pd_{17}Se_{15}$ can be explained by the intermediate type-II superconductivity (λ=0.54) with its upper critical field (3.3 T) Pauli limited as in conventional superconductors unlike the case of $Rh_{17}S_{15}$. Finally, our comparative study supports our conjecture that strong correlations seen

sample name	ρ_{300K} $\mu\Omega$ cm	γ in mJ/mol-K^2	θ_D K	χ_{300K} 10^{-3} emu/mol-Oe	T_c K	ξ_0 Å	λ Å	H_p (T)
$Rh_{17}S_{15}$	195	110.0	424	1.07	5.4	40.0	7500	9.9
$Pd_{17}Se_{15}$	310	22.0	234	-0.25	2.2	100.0	-	4.0

Table 1. Normal and superconducting state properties of $Pd_{17}Se_{15}$ and $Rh_{17}S_{15}$.

in $Rh_{17}S_{15}$ comes from the strong Rh-Rh interactions which lead a narrow 4-d band at the Fermi level. It will be interesting to investigate other superconducting noble metal chalcogenides for strong correlations due to narrow 4d and 5d bands at the Fermi level [3, 14].

[1] H.R. Naren *et al* 2008 *Phys. Rev. Lett.* **100** 026404.
[2] H. R. Naren *et al* 2009 *J. Phys.: Conf. Ser.* **150** 052183.
[3] Sandip Dey and Vimal K Jain 2004 *Platinum Metals Review (Published by Johnson Matthey, Hatton Garden, London)* **48** 16.
[4] Oswaldo Diguez and Nicola Marzari 2009 *Phys. Rev. B* **80** 214115.
[5] Rikio Settai, Keisuke Katayama, Hiroshi Muranaka, Tetsuya Takeuchi, Arumugam Thamizhavel, Ilya Sheikin and Yoshisicko Onuki 2010 *Physics and Chemistry of Solids* **71** 700.
[6] M. Uhlarz, O. Ignatchik, J. Wosnitza, H. R. Naren, A. Thamizhavel and S. Ramakrishnan 2010 *J. Low Temp. Phys.* **159** 176.
[7] T. Koyama, K. Kanda, K. Ueda, T. Mito, T. Kohara and H. Nakamura, 2010 *J. Phys.: Conf. Ser.* **200** 012095.
[8] S. Geller 1962 *Acta Cryst.* **15** 1198.
[9] S. Geller 1962 *Acta Cryst.* **15** 713.
[10] W.L. McMillan 1968 *Phys. Rev.* **167** 331.
[11] N.R. Werthamer, E. Helfand and P. C. Hohenberg 1966 *Phys. Rev.* **147** 295.
[12] H.R. Naren *et al to be published.*
[13] J. Labbe 1967 *Phys. Rev.* **158** 647.
[14] S. Nagata and T. Atake 1999 *Journal of Thermal Analysis and Calorimetry* **57** 807.

Journal of Physics: Conference Series **273** (2011) 012075 doi:10.1088/1742-6596/273/1/012075

Coexistence of d-wave superconductivity and antiferromagnetism induced by paramagnetic depairing

Yuhki Hatakeyama and Ryusuke Ikeda
Department of Physics, Kyoto University, Kyoto 606-8502, Japan
E-mail: ryusuke@scphys.kyoto-u.ac.jp

Abstract. It is shown theoretically that, in the superconducting state with $d_{x^2-y^2}$-pairing, a strong Pauli paramagnetic depairing (PD) induces not only the modulated Fulde-Ferrell-Larkin-Ovchinnikov (FFLO) superconducting state but also an incommensurate antiferromagnetic (AFM) (or spin-density wave) ordering with **Q**-vector nearly parallel to a gap node. In this mechanism of field-induced coexistence of the d-wave superconducting and AFM orders, a pair-density wave does not have to be assumed. It is argued that this is the common origin of *both* the coexistent FFLO and AFM phases of $CeCoIn_5$ *and* the AFM quantum critical behavior around the superconducting $H_{c2}(0)$ seen in several unconventional superconductors.

1. Introduction

At present, we have two intriguing issues on a magnetic order or fluctuation occurring close to $H_{c2}(0)$ in d-wave paired superconducting states. One is the antiferromagnetic (AFM) quantum critical behavior reflected in transport measurements around $H_{c2}(0)$ of heavy fermion superconductors $CeCoIn_5$ [1-3], pressured $CeRhIn_5$ [4], $NpPd_5Al_2$ [5], and Tl-based cuprates [6]. The other is the AFM order [7] in the high field and low temperature (HFLT) phase [8] of $CeCoIn_5$ which has been identified, based on measurements [9,10] and theoretical explanations [11,12] of elastic properties and doping effects, with a Fulde-Ferrell-Larkin-Ovchinnikov (FFLO) state [8,13]. It is notable that the materials listed above have a couple of common features, such as the d-wave pairing and a strong Pauli paramagnetic depairing (PD).

In this report, we point out that, in *nodal* d-wave superconductors, a field-induced enhancement of PD tends to induce an AFM ordering in the superconducting phase just below $H_{c2}(0)$. Conventionally, the AFM ordering is suppressed by the superconducting ordering in zero field [14], and, even in nonzero fields, the quasiparticle damping effect brought by the AFM fluctuation usually suppresses a relatively weak PD effect, suggesting a competition between the two orderings. It is found, however, that a strong PD rather favors coexistence of a d-wave superconductivity and an *incommensurate* AFM order, leading to an enhancement of AFM ordering or fluctuation just below $H_{c2}(0)$.

2. Model and Calculation

We work in a BCS-like electronic Hamiltonian [14] with the superconducting energy gap Δ and the AFM moment $\mathbf{m}$. By treating Δ at the mean field level and $m \equiv |\mathbf{m}|$ as a fluctuation,

Published under licence by IOP Publishing Ltd

respectively, the free energy describing the two possible orderings in zero field is described by

$$\begin{aligned}\mathcal{F}(\mathbf{H}=0) &= \int d^3r\, g^{-1}|\Delta(\mathbf{r})|^2 - T\ln \mathrm{Tr}_{c,c^\dagger,m}\exp[-(H_{\Delta\, m}-\mu N)/T],\\ H_{\Delta\, m}-\mu N &= \sum_{\mathbf{q}} \frac{1}{U}|\mathbf{m}(\mathbf{q})|^2 + \sum_{\mathbf{k},\alpha,\beta} \hat{c}^\dagger_{\mathbf{k},\alpha}\,\xi(\mathbf{k})\,\delta_{\alpha,\beta}\,\hat{c}_{\mathbf{k},\beta}\\ &- \sum_{\mathbf{q},\alpha,\beta}\Big(\Delta(\mathbf{q})\hat{\Psi}^\dagger(\mathbf{q}) + m_\nu(\mathbf{q})\,\hat{S}^\dagger_\nu(\mathbf{q}) + \mathrm{h.c.}\Big),\end{aligned} \tag{1}$$

where $\hat{\Psi}(\mathbf{q}) = -\mathrm{i}(\sigma_y)_{\alpha,\beta}\sum_{\mathbf{k}} w_{\mathbf{k}}\,\hat{c}_{-\mathbf{k}+\mathbf{q}/2,\alpha}\hat{c}_{\mathbf{k}+\mathbf{q}/2,\beta}/2$, $\hat{S}_\nu(\mathbf{q}) = (\sigma_\nu)_{\alpha,\beta}\sum_{\mathbf{k}}\hat{c}^\dagger_{\mathbf{k}-\mathbf{q},\alpha}\hat{c}_{\mathbf{k}+\mathbf{Q},\beta}/2$, $\hat{c}^\dagger_{\mathbf{k},\alpha}$ creates a quasiparticle with spin index α and momentum $\mathbf{k}$, σ_ν are the Pauli matrices, μ is the chemical potential, and the positive parameters g and U are the attractive and repulsive interaction strengths leading to the superconducting and AFM orderings, respectively. The gap function $w_{\mathbf{k}}$ satisfies $w_{\mathbf{k}+\mathbf{Q}} = -w_{\mathbf{k}}$ for the $d_{x^2-y^2}$-pairing state, and the dispersion $\xi(\mathbf{k})$ satisfies $\xi(\mathbf{k}) = -\xi(\mathbf{k}+\mathbf{Q}) + T_c\delta_{\mathrm{IC}}$, where $\mathbf{Q}$ is the *commensurate* AFM modulation wavevector and is $(\pi,\ \pi)$ for the $d_{x^2-y^2}$-pairing. A small deviation from the perfect nesting is measured by a small parameter δ_{IC} for a nearly-free electron model. If the tight-binding model with dispersion $\xi(\mathbf{k}) = -t_1(\cos(k_xa)+\cos(k_ya)) - t_2\cos(k_xa)\cos(k_ya) - \mu$ and the lattice constant a is directly used in examining an AFM ordering, the corresponding incommensurability is measured by the second term of the above $\xi(\mathbf{k})$. In a nonzero field ($\mathbf{H}\neq 0$) of our interest, the Zeeman term $\gamma_B H(\sigma_z)_{\alpha,\beta}$ needs to be added to $\xi(\mathbf{k})\delta_{\alpha,\beta}$. At least in the case with a continuous H_{c2}-transition like Fig.2 (a) below, the orbital depairing needs to be incorporated through the familiar quasiclassical treatment on the quasiparticle Green's function [13].

To see the position of the AFM ordering, it is convenient to examine the Gaussian AFM fluctuation term $\mathcal{F}_m$ in the free energy $\mathcal{F}$, $\mathcal{F}_m = T\sum_\Omega \ln\det[\,U^{-1}\delta_{\mathbf{q},\mathbf{q}'} - \chi_{\mathbf{q},\mathbf{q}'}(\Omega)\,]$, where

$$\chi_{\mathbf{q},\mathbf{q}'}(\Omega) = \int_0^{T^{-1}} d\tau\,\langle\, T_\tau\,\hat{S}^\dagger_\nu(\mathbf{q};\tau)\,\hat{S}_\nu(\mathbf{q}';0)\,\rangle\, e^{i\Omega\tau}, \tag{2}$$

and $\hat{S}_\nu(\mathbf{q};\tau)$ denotes $\hat{S}_\nu(\mathbf{q})$ at imaginary time τ. For the moment, we focus on the Pauli limit with no orbital depairing and with uniform Δ in which $\chi_{\mathbf{q},\mathbf{q}'}(\Omega) = [\chi^{(\mathrm{n})}(\mathbf{q},\Omega) + \chi^{(\mathrm{an})}(\mathbf{q},\Omega)]\delta_{\mathbf{q},\mathbf{q}'}$, and $\mathcal{F}_m = -T\sum_{\mathbf{q},\Omega}\ln X(\mathbf{q},\Omega)$, where $X^{-1}(\mathbf{q},\Omega) = U^{-1} - \chi^{(\mathrm{n})}(\mathbf{q},\Omega) - \chi^{(\mathrm{an})}(\mathbf{q},\Omega)$. A second order AFM ordering occurs when $X_0^{-1} = X^{-1}(0,0) = 0$. The $\mathrm{O}(|\Delta|^2)$ terms in $\chi^{(\mathrm{n})}$ and $\chi^{(\mathrm{an})}$ are expressed by Fig.1 (a) and (b), respectively. They have been studied previously [14] in $H=0$ case, where $\chi_s(\Delta) \equiv \chi^{(\mathrm{n})}(0,0) - \chi^{(\mathrm{n})}(0,0)|_{\Delta=0} + \chi^{(\mathrm{an})}(0,0)$ taking the form

$$\chi_s(\Delta) = T\int\frac{d^3p}{(2\pi)^3}\sum_{\varepsilon,\sigma}\Big[2w_{\mathbf{p}}^2\,(\mathcal{G}_{\varepsilon,\sigma}(\mathbf{p}))^2\,\Delta^*\mathcal{G}_{-\varepsilon,-\sigma}(-\mathbf{p})\,\Delta\,\mathcal{G}_{\varepsilon,\bar{\sigma}}(\mathbf{p}+\mathbf{Q})$$

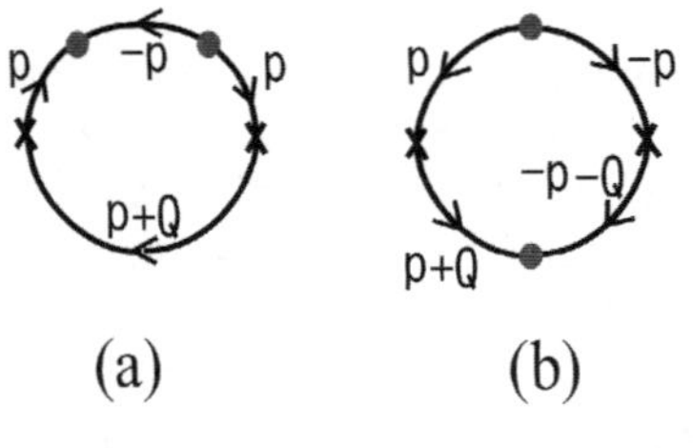

Figure 1. Diagrams describing (a) $\chi^{(n)}$ and (b) $\chi^{(an)}$ up to $\mathrm{O}(|\Delta|^2)$, where the cross denotes the particle-hole vertex on the AFM fluctuation, while the filled circle implies the particle-particle vertex on Δ or Δ^*.

$$- \quad w_{\mathbf{p}} w_{\mathbf{p}+\mathbf{Q}}\, \mathcal{G}_{\varepsilon,\sigma}(\mathbf{p})\, \Delta^* \mathcal{G}_{-\varepsilon,-\sigma}(-\mathbf{p}) \mathcal{G}_{\varepsilon,\overline{\sigma}}(\mathbf{p}+\mathbf{Q}) \Delta \mathcal{G}_{-\varepsilon,-\overline{\sigma}}(-\mathbf{p}-\mathbf{Q}) \Big] \qquad (3)$$

behaves like T^{-2} in $T \to 0$ limit and is negative so that the AFM ordering is suppressed by superconductivity [14]. In eq.(3), $\mathcal{G}_{\varepsilon,\sigma}(\mathbf{p}) = (\mathrm{i}\varepsilon - \xi_{\mathbf{p}} + \gamma_B H \sigma)^{-1}$ denotes the quasiparticle Green's function defined in the normal state. The last term implies $\chi^{(\mathrm{an})}(0,0)$.

To explain effects of strong PD, let us first explain the $\mathbf{m} \parallel \mathbf{H}$ case in which $\overline{\sigma} = \sigma$. In this case, the two terms in eq.(3) are found to take the same form as the coefficient of the $\mathrm{O}(|\Delta|^4)$ term of the superconducting Ginzburg-Landau (GL) free energy and thus, change their sign upon cooling [13]. Hence, $\chi_s(\Delta)$ becomes positive for stronger PD, leading to a lower $\mathcal{F}_m$, i.e., an *enhancement* of the AFM ordering in the superconducting phase. As well as the corresponding PD-induced sign-change of the $\mathrm{O}(|\Delta|^4)$ term which leads to the first order H_{c2}-transition [13], the PD-induced positive χ_s is also unaffected by inclusion of the orbital depairing.

In $\mathbf{m} \perp \mathbf{H}$ where $\overline{\sigma} = -\sigma$, a different type of PD-induced AFM ordering occurs in a d-wave pairing case with a gap node along $\mathbf{Q}$ where $w_{\mathbf{p}+\mathbf{Q}} = -w_{\mathbf{p}}$: In this case, the first term of eq.(3) arising from $\chi^{(\mathrm{n})}(0,0)$ remains negative as in zero field case and becomes $-N(0)|\Delta|^2/[2(\gamma_{\mathrm{B}}H)^2]$ in $T \to 0$ limit with no PD-induced sign change, where $N(0)$ is the normal density of states. Instead, the last term of eq.(3) implying $\chi^{(\mathrm{an})}(0,0)$ and thus, χ_s are divergent like $N(0)[\,|\Delta|/(\gamma_{\mathrm{B}}H)\,]^2\,|\ln[\mathrm{Max}(t, |\delta_{\mathrm{IC}}|)]|$ in $T \to 0$ limit while keeping their positive signs owing to the relation $w_{\mathbf{p}+\mathbf{Q}} \cdot w_{\mathbf{p}} < 0$, where $t = T/T_c$. This divergence is *unaffected* by including the orbital depairing. That is, in the $d_{x^2-y^2}$-wave case with $\mathbf{Q} = (\pi, \pi)$, the AFM order tends to occur upon cooling in $\mathbf{m} \perp \mathbf{H}$. In contrast, $\chi^{(\mathrm{an})}(0,0)$ is also negative in the d_{xy}-wave case with the same $\mathbf{Q}$ satisfying $w_{\mathbf{p}} w_{\mathbf{p}+\mathbf{Q}} > 0$ so that the AFM ordering is suppressed with increasing H.

3. Examples of Phase Diagrams

In this section, examples of the resulting low temperature phase diagram near $H_{c2}(0)$ will be presented. In the BCS-like model (1) and up to the $\mathrm{O}(|\Delta|^2)$ terms (see Fig.1), the H_{c2}-transition, i.e., the mean field superconducting transition in $H \neq 0$, is of second order even at lower temperatures for $\alpha \equiv \gamma_B H_{\mathrm{orb}}(0)/(2\pi T_c) \leq 0.3$ (see Fig.2(a)), where α is nothing but the Maki parameter except a difference in the numerical factor, while it becomes of first order for larger $\alpha \simeq 1.1$ [13]. It is reasonable to expect the former to correspond to the case of $\mathrm{CeRhIn_5}$ under a pressure [4]. Figure 2(a) is one of the phase diagrams in such a case, where the Neel temperature T_{N} in the normal state with perfect nesting or U was assumed to be the only parameter measuring the pressure dependence. The actual AFM transition temperature in $H > H_{c2}(0)$ for $T_{\mathrm{N}}/T_c = 0.02$ and 0.35 are zero and less than $0.35T_c$, respectively, because of the finite $\delta_{\mathrm{IC}} \simeq 0.6$ used in the calculation. Reflecting the AFM ordering enhanced by PD, the decrease of T_{N}, corresponding to an increase of pressure, results in the shrinkage of the AFM phase just below $H_{c2}(0)$, which reduces to an apparent AFM quantum critical point by a further increase of pressure.

Figure 2(b) is the corresponding result in the tight binding model in the Pauli limit with no orbital depairing (vortices). The H_{c2}-transition is of first order in the temperature range shown there. Due to the discontinuous nature of the H_{c2}-transition, an apparent AFM quantum critical point is estimated, in $h = H/H_{c2}(0) > 1$, to lie at a lower field than $H_{c2}(0)$ in spite of the PD-induced AFM ordering just below $H_{c2}(0)$. This is consistent with the observations in $\mathrm{CeCoIn_5}$ [2,7]. We note that the anomalous doping effect in $\mathrm{CeCoIn_5}$ [10] cannot be explained without a spatial modulation of $|\Delta|$ in the HFLT phase [12], implying that both the AFM and FFLO orders coexist in the HFLT phase of $\mathrm{CeCoIn_5}$. Calculation results in the case including the FFLO structure will be reported elsewhere [15].

We note that, in the present theory explaining the AFM order just below $H_{c2}(0)$ in $\mathrm{CeCoIn_5}$

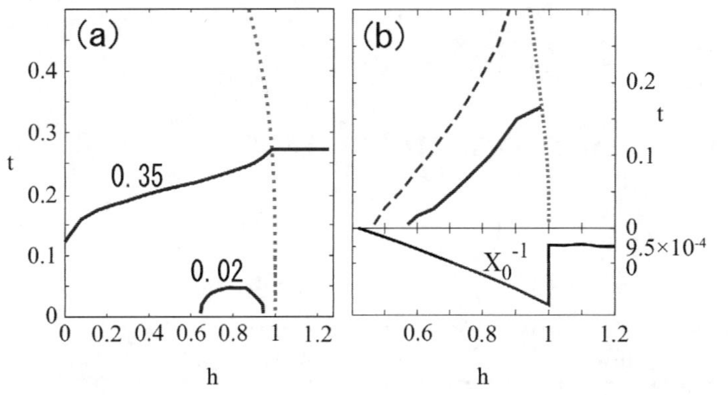

Figure 2. Typical t $(= T/T_c)$ v.s. $h = H/H_{c2}(0)$ phase diagrams (a) following from the use of $\alpha = 0.3$ leading to a second order H_{c2}-transition even in low t limit and (b) in the Pauli limit with a first order H_{c2}-transition in low t, respectively. In both figures, an AFM phase can occur below a solid curve on which $X_0^{-1} = 0$, and each nearly vertical dotted curve is the corresponding $H_{c2}(T)$-curve. Note that, in Fig.2(a), two AFM phase boundaries for $T_N/T_c = 0.35$ and 0.02 are shown in a single figure. Figure 2 (b) was obtained from the corresponding calculation in the tight binding model with the parameters $U = 33$, $t_1 = 100$, and $t_2 = 0.25$ in the unit of T_c and by taking account of the *full* Δ dependence with no limitation to the $O(|\Delta|^2)$ term. The dashed curve denotes the possible upper limit of the AFM transition temperatures at which $\chi_s = 0$. The lower panel of (b) is the h-dependence of X_0^{-1} at $t = 0.05$.

with strong PD, the assumption [16] of an additional pairing channel (pair-density wave) is unnecessary, and that both of the AFM order [7] and other observations, such as the anomalous doping effect [10], in the HFLT phase of $CeCoIn_5$ are explained consistently if the FFLO modulation in the HFLT phase is assumed.

This work was financially supported by Grant-in-Aid for Scientific Research [No. 20102008 and 21540360] from MEXT, Japan.

References

[1] Paglione J et al. 2003 *Phys. Rev. Lett.* **91** 246405.
[2] Singh S et al., 2007 *Phys. Rev. Lett.* **98** 057001.
[3] Kasahara Y et al., 2005 *Phys. Rev.* B **72** 214515(R).
[4] Park T et al., 2006 *Nature* **440** 65 and arXiv: 0910.2287.
[5] Honda F et al., 2008 *J. Phys. Soc. Jpn. Suppl.* A **77** 339.
[6] Shibauchi T et al., 2008 *Proc. Natl. Acad. Sci. USA* **105**, 7120 (arXiv:0805.2215).
[7] Kenzelmann M et al., 2008 *Science* **321** 1652.
[8] Bianchi A D et al., 2003 *Phys. Rev. Lett.* **91** 187004.
[9] Watanabe T et al., 2004 *Phys. Rev.* B **70** 020506(R).
[10] Tokiwa Y et al., *Phys. Rev. Lett.* **101** 037001 (2008).
[11] Ikeda R, 2007 *Phys. Rev.* B **76** 054517.
[12] Ikeda R, 2010 *Phys. Rev. B* **81** 060510(R).
[13] Adachi H and Ikeda R, 2003 *Phys. Rev.* B **68** 184510.
[14] Konno R and Ueda K, 1989 *Phys. Rev.* B **40** 4329 ; Kato M and Machida K, 1988 *Phys. Rev.* B **37** 1510.
[15] R. Ikeda, Y. Hatakeyama, and K. Aoyama, arXiv:1003.0309, v2 and in preparation.
[16] Aperis A et al., arXiv:0902.0553.

Journal of Physics: Conference Series **273** (2011) 012076 doi:10.1088/1742-6596/273/1/012076

Pressure effect on the superconductivity of β-pyrochlore oxides

T. Isono[1], D. Iguchi[1], Y. Machida[1], K. Izawa[1], B. Salce[2], J. Flouquet[2], H. Ogusu[3], J. Yamaura[3], and Z. Hiroi[3]

[1] Department of Physics, Tokyo Institute of Technology, Meguro, Tokyo 152-8551, Japan
[2] SPSMS, CEA-Grenoble, 38054 Grenoble Cedex 9, France
[3] Institute for Solid State Physics, University of Tokyo, Kashiwa, Chiba 277-8581, Japan

E-mail: isono.t.aa@m.titech.ac.jp

Abstract. We report the results of the specific heat measurements for β-pyrochlore oxide superconductor $RbOs_2O_6$ with superconducting transition temperature $T_c \sim 6.3$ K in magnetic field and under pressure. We find that T_c rises up to 9.3 K by applying the pressure concomitant with an increase of an initial slope of the upper critical field H_{c2} and a jump of the specific heat at T_c. From the analysis of our data, the effective mass of conduction electrons and the electron-phonon coupling appear to increase under the pressure.

1. Introduction

Conventional superconductivity in simple metals and metal alloys is well described by the BCS theory based on attractive force between electrons through the exchange of phonons. Recently, much attention has been directed toward clarifying the issue that superconducting properties in some cage compounds are not simply explained by the conventional BCS theory [1, 2].

β-pyrochlore oxide superconductors AOs_2O_6 (A = K, Rb, and Cs) are one of the cage compounds in which anomalous phonons are observed as an anharmonic oscillation of an A ion, called "rattling" motion, located inside a large atomic cage [3]. Intriguingly, the rattling motion participates the superconducting properties [2, 4]. For instance, the superconducting transition temperature T_c of AOs_2O_6 rises from 3.3 K for $CsOs_2O_6$ to 9.6 K for KOs_2O_6 with increasing the magnitude of the rattling motion and the electron-phonon coupling [5], while the density of state decreases from $CsOs_2O_6$ to KOs_2O_6 [6]. Apparently, this trend is opposite to what is expected from the weak-coupling BCS theory.

In this paper, we report the high-pressure specific heat measurements on a single-crystalline $RbOs_2O_6$ under the magnetic field in order to further explore the role of the rattling motion on the superconductivity . The pressure is one of a powerful tool for this purpose because it directly affects on the rattling motion by modifying the crystal structure. We discuss implications for the pressure variation of T_c in terms of the rattling motion.

2. Experimental

High-quality single crystals of $RbOs_2O_6$ were synthesized by a solid-state reaction at 470 – 520 °C for 24 hours, as reported previously [2]. As-grown sample was cut into a size of $\sim 0.20 \times 0.20 \times 0.05$ mm^3 for the measurements. The specific heat measurements on two different samples

Published under licence by IOP Publishing Ltd

were performed by an ac-calorimetry method, which is a good measure to evaluate a relative change in a specific heat. In ac-calorimetry, a thermal response of a sample T_{ac} with respect to an input ac-heat P_0 provides information of the specific heat C: $T_{ac} = \frac{P_0}{\kappa + i\omega C}$ [7], where ω and κ are a frequency of the input ac-heat and a thermal conductivity between the sample and bath, respectively. Moreover, we can estimate the specific heat by measuring a phase difference ϕ between the thermal response and the input ac-heat in addition to T_{ac}: $C = \frac{P_0}{\omega T_{ac}} \sin\phi$ [8]. In our measurements, the ac-heating power was produced by a laser modulated at the frequency $\omega \sim 3 - 6$ kHz by an acousto-optic elements. T_{ac} was determined by a measured voltage V_{ac} of a thermocouple Au/AuFe (0.07 %) and its thermoelectric power S: $T_{ac} = V_{ac}/S$. Pressure was generated by using a diamond anvil cell filled up by argon as a pressure medium, providing high hydrostatic pressure environment. The applied pressure was determined by a ruby fluorescence around 4.2 K. The measurements were performed in the temperature down to 2 K, under the pressure up to 9 GPa, and in magnetic fields up to 7 T.

3. Results and discussion

First, we show the temperature variation of the specific heat divided by temperature C/T for $RbOs_2O_6$ under the several pressures in Figure 1. The data are shifted upward for clarity. At ambient pressure, the specific heat exhibits a clear anomaly associated with the superconducting transition at $T_c \sim 6.3$ K. By applying the pressure, T_c monotonically increases up to ~ 9 K. In addition to the superconducting anomaly, a shoulder like anomaly appears around 7.2 K at 5.8 GPa, as denoted by the upward arrows. Interestingly, this anomaly becomes sharp with increasing the pressure, concomitant with a sudden disappearance of the superconducting anomaly at 6.0 GPa. It has been found that this second anomaly is also attributed to the superconducting transition from the electrical resistivity measurements [9]. With further increasing the pressure, an additional anomaly appears at ~ 3 K above 7.1 GPa as denoted by a dashed arrow, indicating the presence of a high-pressure phase. The nature of this high-pressure phase is unclear.

From the high-pressure specific heat measurements, we construct a thermodynamic temperature vs. pressure phase diagram of $RbOs_2O_6$ as shown in Fig. 3(a). We found the following remarkable features in the phase diagram; (1) T_c increases up to 9.3 K with increasing the pressure, while a slope of dT_c/dP levels off above 3 GPa. (2) The superconducting phase (SC1) unexpectedly vanishes and a secondary superconducting phase (SC2) appears at the critical pressure $P^* \sim 6.0$ GPa. (3) In addition to the SC2, another high-pressure phase emerges above 7.1 GPa.

In order to clarify the superconducting properties of the SC1, we performed the high-

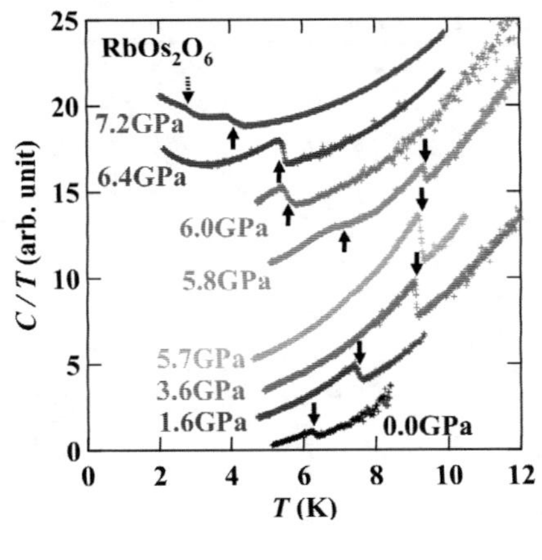

Figure 1. Temperature dependence of the specific heat divided by temperature C/T for $RbOs_2O_6$ under the several pressures. The downward and upward arrows denote the superconducting transition temperature of the two different superconducting phase, respectively. The dashed arrows represent the transition temperature of the high-pressure phase. The data are shifted upward for clarity.

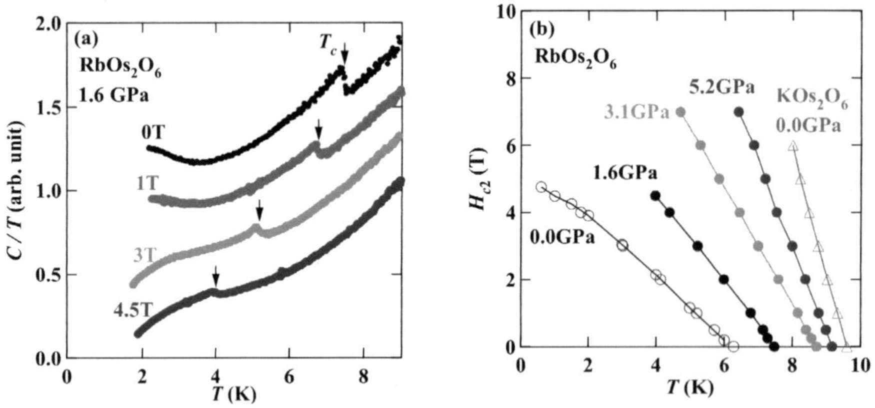

Figure 2. (a) Temperature dependence of the specific heat divided by temperature C/T for $RbOs_2O_6$ under the pressure of 1.6 GPa in the various magnetic fields. The allows indicate T_c. The data are shifted upward for clarity. (b) Upper critical field $H_{c2}(T)$ as a function of temperature at various pressures. The present results are shown by the solid circles. The open circles and triangles denote $H_{c2}(T)$ for $RbOs_2O_6$ [2] and KOs_2O_6 [10] at ambient pressure, respectively.

pressure specific heat measurements in the magnetic fields. Figure 2 (a) shows the temperature dependence of the specific heat divided by temperature C/T under the pressure of 1.6 GPa in the several magnetic fields. The data are shifted upward for clarity. As it can be seen, the superconducting anomalies become smaller and shift to lower temperatures. We plot the temperature dependence of the upper critical field $H_{c2}(T)$ under the various pressures in Fig. 2 (b). Here, T_c is determined by the entropy conserving construction. At ambient pressure, $H_{c2}(T)$ shows a linear increase at low fields, but has a concave-upward curvature at high fields [2]. It should be noted that an initial slope of $H_{c2}(T)$ at T_c ($-\mathrm{d}H_{c2}/\mathrm{d}T\,|_{T=T_c}$) monotonically increases with the pressure from 0.9 T/K at ambient pressure to 2.5(2) T/K at 5.2 GPa. Moreover, with increasing the pressure, the initial slope at 5.2 GPa is close to the one for KOs_2O_6 at ambient pressure. From the initial slope, we estimate the effective mass m^* of the conduction electrons through a relation $-\mathrm{d}H_{c2}/\mathrm{d}T\,|_{T=T_c} \propto m^{*2}$ as shown in Fig. 3(b). Interestingly, m^* monotonously increases with the pressure in the SC1 phase.

To further elucidate the superconducting properties of the SC1 and a role of the rattling motion, we examine the electron-phonon coupling from a jump of the specific heat $\Delta C/\gamma T_c$ at T_c. Although the qualitative estimate of $\Delta C/\gamma T_c$ is not straightforward by the ac-calorimetry method, we evaluate the relative change of $\Delta C/\gamma T_c$ against the pressure as follows. First, we multiply constant values to each $C(T)$ curve, so as to fall on an identical curve above T_c. Then, we obtain the relative change of the specific heat ΔC. Next, we substitute m^* to $\Delta C/\gamma T_c$, which is determined above, instead of the Sommerfeld coefficient γ because of the proportionality between m^* and γ. Figure 3 (c) displays the pressure dependence of $\Delta C/m^*T_c$ in the SC1. It is clearly seen that $\Delta C/m^*T_c$ shows a steep increase up to 3 GPa and a weak pressure dependence above 3 GPa. Notice that the pressure variation of $\Delta C/m^*T_c$ resembles that of T_c, suggesting that T_c is governed by the electron-phonon coupling under the pressure. Here, a question is raised the issue of what is a leading source providing the enhancement of the electron-phonon coupling. One possible answer is the rattling motion of the Rb ions, which might become stronger by the applied pressure. This is because one can expect the stronger electron-phonon coupling due to the enhanced rattling motion from the analogy to the variation of T_c in the series of AOs_2O_6.

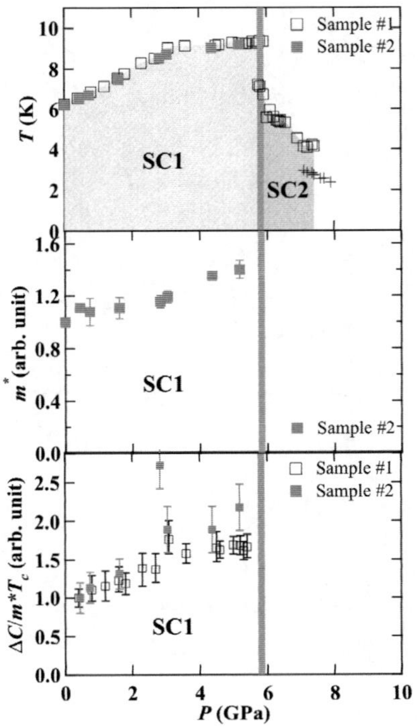

Figure 3. (a) Temperature vs. pressure phase diagram of $RbOs_2O_6$. Solid and open squares denote the superconducting transition temperature. Cross symbols show the transition temperature of the high-pressure phase. Pressure variation of (b) the normalized effective mass m^* of the conducting electrons and (c) the jump of the specific heat $\Delta C/m^*T_c$ for $RbOs_2O_6$. Both data are normalized at 0.4 GPa. The solid line is a guide to the eye.

In summary, we constructed the thermodynamic temperature vs. pressure phase diagram for $RbOs_2O_6$ from the high-pressure specific heat measurements. We found the three different phases under the pressure; the former two are attributed to the superconducting phases. In the lower pressure superconducting phase, the effective mass of the conduction electrons and the electron-phonon coupling appear to increase by applying the pressure. These findings point to an intriguing possibility of a pressure-induced enhancement of the rattling motion of the Rb ions.

Acknowledgements

We thank Marie-José Blanchard for technical assistance and K. Hattori for fruitful disscussion. This work is supported by Grant-in-Aid for Scientific Research on Innovative Areas "Heavy Electrons" (No. 20102006) from the Ministry of Education, Culture, Sports, Science and Technology (MEXT) and by Global Center of Excellence Program from MEXT, Japan through the Nanoscience and Quantum Physics Project of the Tokyo Institute of Technology.

References

[1] Bauer E D, Frederick N A, Ho P C, Zapf V S, and Maple M B 2002 *Phys. Rev.* B **65** 100506R
[2] Nagao Y, Yamaura J, Ogusu H, Okamoto Y, and Hiroi Z 2009 *J. Phys. Soc. Jpn.* **78** 064702
[3] Kuneš J, Jeong T, and Pickett W E 2004 *Phys. Rev.* B **70** 174510
[4] Hattori K, and Tsunetsugu H 2010 *Phys. Rev.* B **81** 134503
[5] Hiroi Z, Yamaura J, Yonezawa S, and Harima H 2007 *Physica* C **460-462** 20
[6] Sainz R, and Freeman A J 2005 *Phys. Rev.* B **72** 024522
[7] Sullivan P F, and Seidel G 1968 *Phys. Rev.* **173** 679
[8] Derr J, Knebel G, Lapertot G, Salce B, Méasson M-A, and Flouquet J 2006 *J. Phys.: Condens. Matt.* **18** 2089
[9] Ogusu H, Takeshita N, Yamaura J, and Hiroi Z *Phsica* C in press
[10] Hiroi Z, Yonezawa S, Nagao Y, and Yamaura J 2007 *Phys. Rev.* B **76** 014523

International Conference on Strongly Correlated Electron Systems (SCES 2010) IOP Publishing
Journal of Physics: Conference Series **273** (2011) 012077 doi:10.1088/1742-6596/273/1/012077

Low-temperature thermal conductivity of the noncentrosymmetric superconductor $LaRhSi_3$

N. Kurita[1,2], C. F. Miclea[1], C. Putzke[3], G. Seyfarth[3,4], C. Capan[4], A. Bianchi[3], Z. Fisk[4], J. D. Thompson[1], R. Movshovich[1]

[1] Los Alamos National Laboratory, Los Alamos, New Mexico 87545, USA
[2] National Institute for Materials Science, Tsukuba, Ibaraki 305-0003, Japan
[3] Department de Physique, Universite de Montreal, Montreal, Canada H3C 3J7
[4] Department of Physics and Astronomy, University of California, Irvine, California 92697, USA

E-mail: `KURITA.Nobuyuki@nims.go.jp`

Abstract. We report on low-temperature thermal conductivity of the noncentrosymmetric superconductor $LaRhSi_3$ ($T_c = 2.3\,K$), which is a paramagnetic analog of the pressure-induced superconductor $CeRhSi_3$. In the normal state (either in zero field above T_c, or in magnetic field above H_{c2}), the thermal conductivity κ is mostly due to electrons, and shows a nearly T-linear dependence. In the superconducting state, κ decreases exponentially below T_c, and crosses over to $\kappa \propto T^2$ behavior at $T \ll T_c$. The temperature dependence of thermal conductivity of $LaRhSi_3$ is similar to that of a number of conventional s-wave superconductors. The field dependence of the residual linear term κ_0/T as a function of magnetic field suggests that $LaRhSi_3$ has no nodes in the superconducting gap.

1. Introduction

The discovery of the heavy fermion superconductor $CePt_3Si$ [1] has triggered intensive research efforts aimed to understand how the lack of crystal inversion symmetry affects the superconducting (SC) properties. The lack of an inversion center, a key symmetry for Cooper pairing, is mainly reflected in the electronic properties through the antisymmetric spin-orbit coupling (ASOC). This has an important influence on the electronic band structure therefore influencing both, the normal and the superconducting state. Several unusual properties appear in noncentrosymmetric superconductors (NCS) depending in particular on the specific form of the spin-orbit coupling as well as on the pairing symmetry. A Cooper pairing model with a two-component order parameter composed of spin-singlet and spin-triplet pairing components has been suggested [2, 3] in order to account for the unusual physical properties.

Interestingly, all of the NCS superconductors discovered so far among the f-electron systems, with the exception of $CePt_3Si$ [1], exhibit superconductivity only under applied pressure, p [4–8]. Among the remarkable characteristics of these compounds, probably associated with ASOC, are the large upper critical field H_{c2} exceeding the Pauli-Clogston paramagnetic limit, H_p and its anisotropy, as reported for $CeTSi_3$ ($T = Rh, Ir$) [9, 10] and $CeTGe_3$ ($T = Co, Ir$) [8, 11]. Therefore, paramagnetic depairing is not the main pair braking mechanism in these compounds as expected for a significant triplet component of the order parameter. In the presence of magnetic f-electrons, however, the SC paring mechanism could be coupled to a magnetic instability,

Published under licence by IOP Publishing Ltd

which makes the role of crystal inversion symmetry unclear. However, noncentrosymmetric superconductivity extends also outside the heavy fermion class. Among the NCS transition-metal compounds, where electrons are not strongly correlated, Li_2Pt_3B was reported to be unconventional with line nodes and spin-triplet pairing [12, 13], whereas a number of others fall into the class of s-wave superconductors [14–16]. At present, there is no clear indication of whether the strong ASOC alone can result in unconventional superconductivity.

A NCS superconductor $LaRhSi_3$ ($T_c = 2.3\,K$) without magnetic f-electron provides another opportunity to examine the effect of the violation of the inversion symmetry on the superconducting state, and to compare its behavior to that of the p-induced f-electron superconductor $CeRhSi_3$ [5, 9]. Here, we report on thermal conductivity measurements on a single crystalline $LaRhSi_3$. Thermal conductivity is a bulk probe of low energy delocalized excitations and a good tool to probe the SC gap symmetry.

2. Experimental details

Single crystals of $LaRhSi_3$ with tetragonal $BaNiSn_3$-type structure were grown by a Sn-flux method. Bulk superconductivity with $T_c = 2.3\,K$ was confirmed by a specific heat jump and a sharp resistance drop. No signature of Sn inclusions have been revealed by resistivity investigations. Thermal conductivity was measured down to 60 mK on a $1 \times 0.1 \times 0.1\,mm^3$ parallelepiped crystal by a standard one-heater and two-thermometers technique. Heat current q was flowing in a direction parallel to [100]. Thermal links from the sample to the heater, thermometers, and the bath were provided by Pt wires spot-welded to the sample. Thermal isolation of the heater and RuO_2 thermometers from the support frame was achieved by using superconducting NbTi filaments. Electrical resistivity was measured with the electrical current $J \parallel [100]$, using the same crystal and as electrical contacts the same Pt wires used for the thermal conductivity measurement. Magnetic field $H \parallel [001]$ ensures that $H \perp q$.

3. Results and Discussion

Figure 1 shows the temperature dependence of thermal conductivity $\kappa(T)$ of $LaRhSi_3$ in zero field and in the normal state at $H = 40\,mT$ well above the upper critical field $H_{c2}(0) = 20\,mT$. The heat current and the magnetic field were applied perpendicular to each other and in the directions $q \parallel [100]$ and $H \parallel [001]$. For the non-magnetic compound $LaRhSi_3$, we expect that the measured κ can be expressed as $\kappa = \kappa_e + \kappa_{ph}$, where κ_e and κ_{ph} are contributions from the delocalized thermal excitations of electrons and phonons, respectively. In zero field above T_c, $\kappa(T)$ shows close to T-linear behavior, which extends to the lowest temperature of our investigation when the superconductivity is suppressed at $H = 40\,mT$. In the normal state, κ_e can be calculated via the Wiedemann-Franz (WF) law, which correlates thermal and charge conductivity in the normal state, as $\kappa_e = L_0 T/\rho$, with $L_0 = 2.44 \times 10^{-8}\,W\Omega/K^2$ the Lorenz number and ρ the measured resistivity. In $LaRhSi_3$, the estimated κ_e, indicated by open triangles, accounts almost completely for experimentally measured thermal conductivity in the normal state. Therefore, we can conclude that the normal state thermal conductivity κ_n is dominated by electrons ($\kappa_n \approx \kappa_e$). In the superconducting state, κ follows a BCS-like expression $\exp(-a T_c/T)$ with $a = 1.4$ (dotted curve) down to 0.7 K. Below 0.7 K ($T \ll T_c$), κ deviates from the exponential form, and exhibits phonon-like $\kappa \propto T^2$ behavior down to the lowest temperature. The overall temperature dependence of κ in $LaRhSi_3$ is qualitatively comparable to that of typical s-wave superconductors [17]. As an example, thermal conductivity of Aluminum ($T_c = 1.2\,K$, $\kappa_n \approx \kappa_e \propto T^1$, $a = 1.5$, and $\kappa \propto T^{2.4}$ at $T \ll T_c$) [18] is shown in the inset of Fig. 1.

In conventional s-wave superconductors, $\kappa(T)$, dominated by electrons in the normal state, exponentially decreases upon opening the SC gap, and becomes negligible compared to the phonon contribution κ_{ph} at $T \ll T_c$. Here, we estimate the upper limit of κ_{ph} of $LaRhSi_3$ based on the scattering off the sample boundaries. Using the phonon specific heat

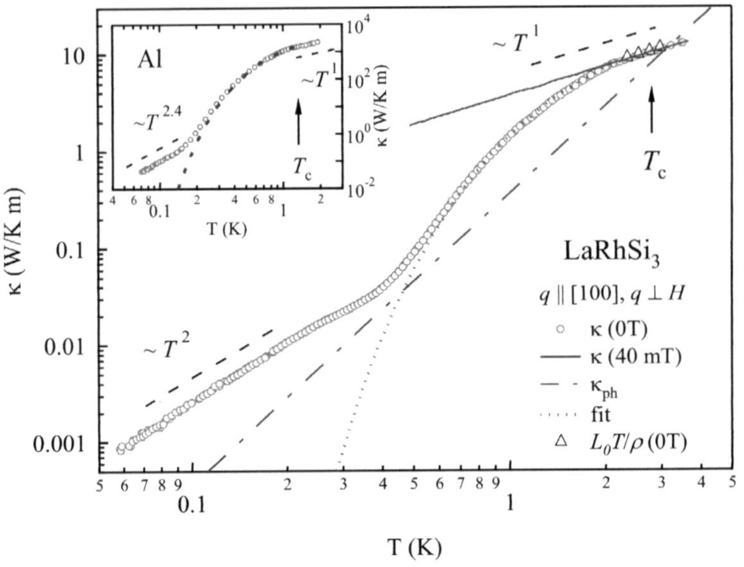

Figure 1. (Color online) Temperature dependence of the thermal conductivity $\kappa(T)$ of $LaRhSi_3$ for heat current $q \parallel [100]$ in zero field and at $40\,\mathrm{mT} \gg H_{c2}$ for $H \perp [100]$. The arrow indicates $T_c = 2.3\,\mathrm{K}$, determined from the specific heat and resistivity measurements. Open triangles represent the electronic conductivity in the normal state, obtained via the Wiedemann-Franz law from resistivity. A dotted line shows a fit to κ with $\kappa = C\exp(-a\,T_c/T)$ [17]. A dashed-dotted line corresponds to the phonon conductivity estimate for boundary scattering. Dashed lines are guides to eye, representing several $\propto T^n$ variations. The inset shows $\kappa(T)$ of Al ($T_c = 1.2\,\mathrm{K}$) [18].

coefficient $\beta = 2.68\,\mathrm{J/K^4\,m^3}$, mean phonon velocity $\langle v \rangle = 3570\,\mathrm{m/s}$ [19] and the phonon mean free path $l_{ph} = 1.13 \times 10^{-4}\,\mathrm{m}$, $\kappa_{ph} = \frac{1}{3}\beta T^3 \langle v \rangle\, l_{ph} = 0.36 \times 10^{-4} \times T^3\,\mathrm{W/K\,m}$ was obtained, and is indicated by a dashed-dotted line in Fig. 1. The estimated κ_{ph} is relatively close to the experimental κ at 0.4 K, although the experimental value ($n = 2$) is markedly different from theoretical one ($n = 3$). The reduction of the exponent value could be due to specular reflection of the phonon assuming that the faces of a crystal are smooth [20]. Another possibility is phonons scattering off either grain boundaries or electrons which are expected to give $n = 2$ [17], or combination of boundary scattering, phonon, and electron scattering. Further study should be performed to disentangle this issue.

Figure 2 displays the normalized residual linear term of thermal conductivity $(\kappa_0/T)/(\kappa_n/T)$ of $LaRhSi_3$ for $H \perp q$ as a function of the normalized field H/H_{c2}. κ_0/T was obtained by $T = 0$ extrapolation of the low-temperature κ/T data using an expression $\kappa/T = \kappa_0/T + bT^2$ at each field (not shown). κ_n/T is the normal state value estimated at $H_{c2} = 20\,\mathrm{mT}$. In s-wave superconductors, at low field above H_{c1}, quasiparticles are mostly localized around the vortex cores and cannot carry heat. When the applied field increases and the intervortex spacing is decreased, and quasiparticles begin to move between the vortices. This is the main reason why κ_0/T exhibits a sharp increase toward κ_n/T when field approaches H_{c2}, as can be seen for Nb data [21]. On the other hand, a remarkably rapid increase of κ_0/T at low field for nodal superconductor $Tl_2Ba_2CuO_{6+\delta}$ (Tl-2201) [22] comes from the excitation of nodal quasiparticles via Volovik effect [23]. A roughly field independent κ_0/T in $LaRhSi_3$ for fields below $\approx 0.65\,H_{c2}$ and a sharp rise above this field is most likely due to a weak Type II nature of superconductivity in this compound, with $H_{c1} \approx 0.65\,H_{c2}$. The overall field dependence of the residual linear term

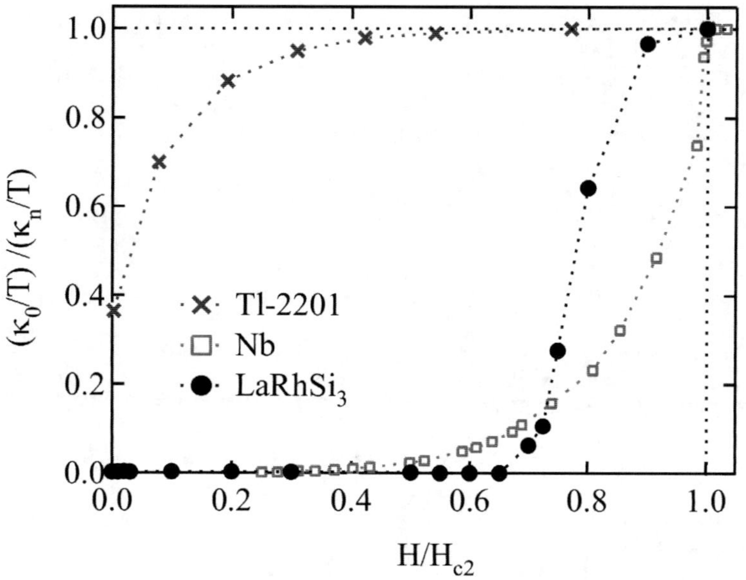

Figure 2. (Color online) Normalized residual linear term $(\kappa_0/T)/(\kappa_n/T)$ versus normalized field H/H_{c2} of $LaRhSi_3$ for $H \perp q$. κ_n/T is the normal state value extrapolated to $T \to 0$ at $H_{c2} = 20\,mT$. For comparison, data for several superconductors with different gap structures are also displayed: Nb (clean, fully gapped *s*-wave) [21], Tl-2201 (*d*-wave with line nodes) [22].

suggests that the superconducting gap in full open in $LaRh_3Si$

A similar approach of comparing the SC gap structures in isostructural compounds without and with *f*-electrons, was taken via penetration depth measurements on $LaPt_3Si$ and $CePt_3Si$ [24], respectively. The results indicated that $LaPt_3Si$ is a conventional *s*-wave superconductor, in contrast to the unconventional superconductivity with line nodes found in $CePt_3Si$ [24–26].

4. Conclusion

To conclude, we have performed a low-temperature magneto-thermal conductivity measurement of a noncentrosymmetric superconductor $LaRhSi_3$. A roughly field-independent residual thermal conductivity at $H < 0.65\,H_{c2}$ is consistent with $LaRhSi_3$ being a weakly Type II superconductor with $H_{c1} \approx 0.65\,H_{c2}$ at $T = 0$. The field evolution of the residual T-linear term suggests that no nodes are present in the superconducting gap. The temperature dependence of thermal conductivity in $LaRhSi_3$ at zero field is similar in many aspects to that of conventional *s*-wave superconductors. Further studies are required to ascertain whether superconductivity in $LaRhSi_3$ is unconventional.

References

[1] Bauer E, Hilscher G, Michor H, Paul Ch, Scheidt E W, Gribanov A, Seropegin Yu, Noël H, Sigrist M and Rogl P 2004 *Phys. Rev. Lett.* **92** 027003
[2] Frigeri P A, Agterberg D F, Koga A and Sigrist M 2004 *Phys. Rev. Lett.* **92** 097001
[3] Hayashi N, Wakabayashi K, Frigeri P A and Sigrist M 2006 *Phys. Rev. B* **73** 024504
[4] Akazawa T, Hidaka H, Fujiwara T, Kobayashi T C, Yamamoto E, Haga Y, Settai R and Ōnuki Y 2004 *J. Phys.: Condens. Matter* **16** L29.
[5] Kimura N, Ito K, Saitoh K, Umeda Y and Aoki H 2005 *Phys. Rev. Lett.* **95** 247004

[6] Sugitani I, Okuda Y, Shishido H, Yamada T, Thamizhavel A, Yamamoto E, Matsuda T D, Haga Y, Takeuchi T, Settai R and Ōnuki Y 2006 *J. Phys. Soc. Jpn.* **75** 043703
[7] Settai R, Okuda Y, Sugitani I, Ōnuki Yi, Matsuda T D, Haga Y and Harima H 2007 *Int. J. Mod. Phys. B* **21** 3238
[8] Honda F, Bonalde I, Yoshiuchi S, Hirose Y, Nakamura T, Shimizu K, Settai R and Ōnuki Y 2009 *Physica C* to be published
[9] Kimura N, Ito K, Aoki H, Uji S and Terashima T 2007 *Phys. Rev. Lett.* **98** 197001
[10] Settai R, Miyauchi Y, Takeuchi T, Levy F, Sheikin I and Ōnuki Y 2008 *J. Phys. Soc. Jpn.* **77** 073705
[11] Méasson M -A, Muranaka H, Kawai T, Ota Y, Sugiyama K, Hagiwara M, Kindo K, Takeuchi T, Shimizu K, Honda F, Settai R and Ōnuki Y 2008 *J. Phys. Soc. Jpn.* **78** 124713
[12] Yuan H Q, Agterberg D F, Hayashi N, Badica P, Vandervelde D, Togano K, Sigrist M and Salamon M B 2006 *Phys. Rev. Lett.* **97** 017006
[13] Nishiyama M, Inada Y and Zheng G -q 2007 *Phys. Rev. Lett.* **98** 047002
[14] Klimczuk T, Ronning F, Sidorov V, Cava R J and Thompson J D 2007 *Phys. Rev. Lett.* **99** 257004
[15] Harada A, Tamura N, Mukuda H, Kitaoka Y, Wakui K, Akutagawa S and Akimitsu J 2009 *J. Phys. Soc. Jpn.* **78** 025003
[16] Bauer E, Khan R T, Michor H and Royanian E 2009 *Phys. Rev. B* **80** 064504
[17] See, for example, Berman R 1976 *Thermal conduction in Solids (Oxford Univ. Press, Oxford)*
[18] Movshovich R, Hubbard M A, Salamon M B, Balatsky A V, Yoshizaki R, Sarrao J L and Jaime M 1998 *Phys. Rev. Lett.* **80** 1968.
[19] Miclea C F *et. al.* unpublished
[20] Pohl R O and Stritzker B 1982 *Phys. Rev. B* **25** 3608
[21] Lowell J and Sousa J B 1970 *J. Low. Temp. Phys.* **3** 65
[22] Proust C, Boaknin E, Hill P W, Taillefer L and Mackenzie A P 2002 *Phys. Rev. Lett.* **89** 147003
[23] Volovik G E 1995 *JETP Lett.* **58** 469
[24] Ribeiro R L, Bonalde I, Haga Y, Settai R and Ōnuki Y 2009 *J. Phys. Soc. Jpn.* **78** 115002
[25] Bonalde I, Brämer-Escamilla W and Bauer E 2005 *Phys. Rev. Lett.* **94** 207002
[26] Izawa K, Kasahara Y, Matsuda Y, Behnia K, Yasuda T, Settai R and Ōnuki Y 2005 *Phys. Rev. Lett.* **94** 197002

International Conference on Strongly Correlated Electron Systems (SCES 2010) IOP Publishing
Journal of Physics: Conference Series **273** (2011) 012078 doi:10.1088/1742-6596/273/1/012078

Platinum metal silicides and germanides: superconductivity in non-centrosymmetric intermetallics

K. Miliyanchuk[1], F. Kneidinger[2], C. Blaas-Schenner[1], D. Reith[1], R. Podloucky[1], P. Rogl[1], T. Khan[2], L. Salamakha[2], G. Hilscher[2], H. Michor[2], E. Bauer[2], A. D. Hillier[3]

[1] Institute of Physical Chemistry, University of Vienna, A-1090 Wien, Austria
[2] Institute of Solid State Physics, Vienna University of Technology, A-1040 Wien, Austria
[3] ISIS facility, STFC, Rutherford Appleton Laboratory, Chilton, Oxfordshire, OX11 0QX, U.K.

E-mail: `bauer@ifp.tuwien.ac.at`

Abstract.
The family of intermetallics crystallizing in the $BaAl_4$ type is rich in sub-groups with different ordered variants. Novel ternary compounds $EpTMX_3$ (Ep = Ba,Sr; TM = Pt,Pd; X = Si,Ge) crystallizes in the body-centred tetragonal $BaNiSn_3$ structure (space group $I4mm$) which does not possess a centre of inversion. Superconductivity in terms of a fully gapped s-wave type occurs for $SrPdGe_3$ at $T_c = 1.49$ K and $SrPtGe_3$ at $T_c = 1.0$ K.

1. Introduction

Intermetallics with the $BaAl_4$ type are rich in sub-groups with different ordered variants. Novel ternary compounds $EpTMX_3$ (Ep = Sr,Ba; TM = Pd,Pt; X = Si,Ge) crystallizing in the body-centred tetragonal $BaNiSn_3$ structure with $c/a \approx 2.2$ and exhibit superconductivity below 3 K. The $BaNiSn_3$ structure, space group $I4mm$, does not posses a centre of inversion with respect to its c-axis (see e.g., Ref. [1]). This causes an antisymmetric spin-orbit coupling (ASOC) and gives rise to a splitting of electronic bands which destroys the spin-degeneracy of the bands. The strength of this effect is proportional to Z^2 (Z is the atomic number); hence, the ASOC may be "tuned" by increasing the element masses in the respective materials. Theoretically, the splitting of bands leads to a mixture of spin-singlet and spin-triplet components in the superconducting condensate [2]. In this study, the normal and the superconducting state of $EpTMX_3$ will be characterised experimentally, and we make use of density functional theory (DFT) calculations to provide electronic density of states (DOS) at the Fermi energy (E_F) as well as the splitting of bands along high-symmetric directions around E_F. We find that the different atomic masses in the compounds modify the ASOC and the band splitting is modified as well.

2. Experimental and computational details

$BaPtSi_3$, $SrPdGe_3$, and $SrPtGe_3$ were prepared by arc-melting on a water-cooled copper hearth in Ti-gettered argon from elemental ingots with a minimum purity of 99.9 mass%. All samples

Published under licence by IOP Publishing Ltd

were sealed in quartz tubes and annealed at 900°C for 100 h before quenching in cold water. X-ray powder diffraction and other measurements were carried out with standard techniques [5].

The DFT calculations were done by using the Vienna *ab initio* simulation package (VASP) utilizing the projector augmented wave construction for the pseudopotentials (Ref. [6]). The potential for Sr comprises 10 valence states including the 4s and 4p states. The Pt potential was constructed for 10 valence states, whereas for Pd also the 4p semicore states are included resulting in 16 valence states in total. For Ge 14 valence states are taken into account incuding the 3d states. The electronic many-body interactions were described within the local density approximation. All structural parameters were optimized with an error smaller than 1.5% as compared to experiment.

3. Results and discussion

3.1. $BaPtSi_3$

Recently, $BaPtSi_3$ as an example of an non-centrosymmetric superconductor was studied. [1] A superconducting phase transition was observed at 2.25 K in combination with an upper critical field at $T = 0$ K of ≈ 0.05 T. Based on a BCS-like heat capacity for $T < T_c$ a simple s-wave superconducting state was suggested. Relativistic DFT calculations revealed only a rather small splitting of bands at E_F and Fermi surface nesting features were found for two characteristic double-sets of bands [1]. Hence, superconductivity seems to appear in an almost undisturbed BCS state. To microscopically proof this conclusion, we have carried out low temperature muon spin resonance (μSR) spectroscopy.

In this type of experiment, a field orthogonal to the muon polarisation is applied with the sample temperature above the superconducting transition temperature and then the sample is cooled. The muon response as a function of applied field and temperature provides accurate determination of the magnetic penetration depth, λ, coherence length, ξ and also the superfluid density, n_s. The results of the muon experiments are shown in Fig. 1. The inset of Fig. 1 shows the temperature dependence of the muon depolarisation rate, σ_s, and precession frequency. The precession frequency shows a characteristic drop at T_c indicating that a vortex state has been established and that we have a type II superconductor. The σ_s value shows an increase at T_c which plateaus at low temperatures. The temperature dependence of σ_s has been fitted using the N-fluid model, $\sigma_s = \sigma(0)(1-(T/T_C)^N)$, which give a reasonable description to the data. The results of the fit are $\sigma(0) = 1.10(2)\ \mu s^{-1}$, $T_c = 1.75(2)$ K and $N = 2.0(1)$. This is strongly suggestive, and confirms the heat capacity results that $BaPtSi_3$ is an s-wave BCS superconductor. The reduction in T_c is due to the applied field being close to H_{c2}. In Fig. 1, σ_s shows a decrease as the applied field is increased, because the flux lines are overlapping. The data can be well described assuming a hexagonal flux line lattice [3, 4], which is usually valid for $\kappa = \lambda/\xi > 5$,

$$\sigma_s = 4.83 \times 10^4 \left(1 - \frac{\mu_0 H}{\mu_0 H_{c2}}\right) \left(1 + 1.21 \left(1 - \frac{\mu_0 H}{\mu_0 H_{c2}}\right)^{0.5}\right)^3 \lambda^{-2}. \quad (1)$$

Here we find $\lambda = 188(10)$ nm and $\mu_0 H_{c2} = 64(3)$ mT yielding $\xi = 72(5)$ nm. This gives $\kappa = 2.6$. Although this value is less than that which is normally acceptable for this model, we can see the fit to the data is good and our results are in good agreement with those reported in Ref. [1].

3.2. $SrPdGe_3$ *and* $SrPtGe_3$

The crystal structure data were determined by X-ray Rietveld refinement and found to be isotypic to the body-centred tetragonal $BaNiSn_3$ type. The lattice parameters observed are $a = 0.44677(4)$ nm and $c = 1.02863(10)$ nm ($SrPdGe_3$) and $a = 0.44859(3)$ nm and $c = 1.01364(8)$ nm ($SrPtGe_3$).

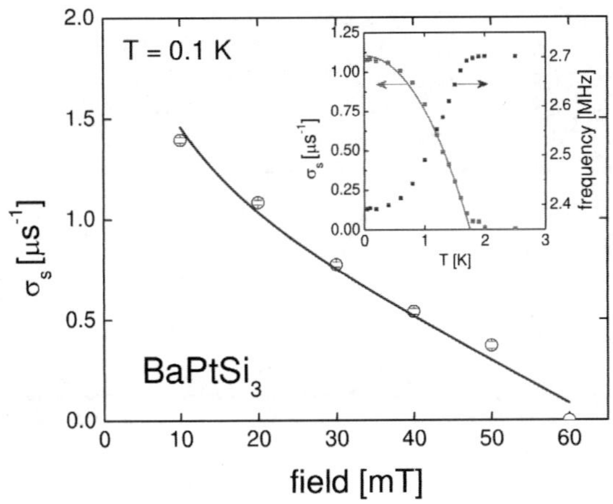

Figure 1. The field dependence of σ at 0.1 K. The insert shows the temperature dependence of σ and the precession frequency at 0.03 T. In both figures the solid line is a fit to the data as described in the text.

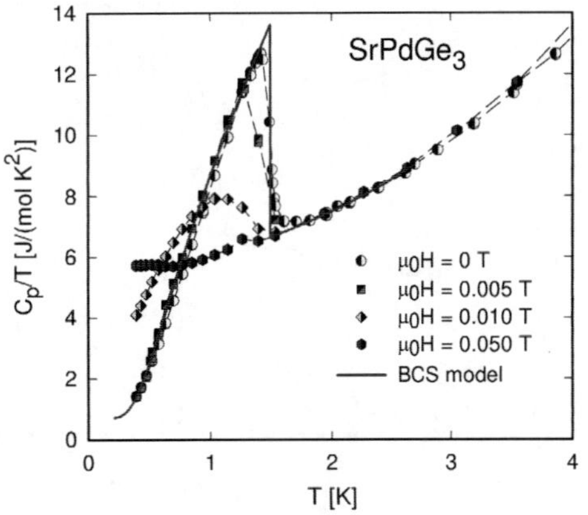

Figure 2. (a) Temperature dependent specific heat C_p of $SrPdGe_3$ for various values of applied field. The solid line represents the heat capacity of an s-wave BCS superconductor.

In order to characterise superconductivity occurring in $EpMGe_3$, we have carried out resistivity measurements as a function of temperature, field and pressure and we have studied the heat capacity at various externally applied magnetic fields. The resistivity in both cases shows metallic behaviour and a superconducting transition at 1.5 K ($SrPdGe_3$) and 1 K ($SrPtGe_3$). The application of external pressure reduces the T_c of $SrPdGe_3$ by about 0.2 K at 20 kbar. Such a behaviour is typical for the overwhelming number of BCS superconductors. External magnetic fields rapidly oppresses superconductivity, which vanishes in both samples already at fields below a tenth of a Tesla. The fact that $SrPdGe_3$ exhibits the larger transition temperature compared to the Pt based system is in line with the lighter element Pd and follows also the trend of typical BCS superconductors.

The temperature dependent heat capacity C_p of $SrPdGe_3$ and $SrPtGe_3$ is displayed in Figs. 2,3 as C_p/T vs. T. These measurements clearly reveal bulk superconductivity in both compounds, excluding impurities as the origin of the phase transitions around at 1.5 and 1.0 K, respectively. The application of a magnetic field suppresses superconductivity for fields of the order of 0.1 T, in agreement with the resistivity data. To derive information regarding the superconducting order parameter, the BCS model calculation by Mühlschlegel is used to describe $C_p(T)$ for $T < T_c$. This model assumes s-wave pairing and a fully opened gap in the electronic density of states (DOS) at the Fermi energy E_F. The BCS model does not contain any adjustable parameter, except the superconducting transition temperature T_c and the Sommerfeld coefficient γ_n, implying that the temperature dependent heat capacity is a universal function, applicable to each BCS-like superconductor. This model reveals a jump of the heat capacity at $T = T_c$ with $[\Delta C/(\gamma_n T_c)] = 1.43$. An adjustment of the transition temperature reveals $T_c = 1.49$ K for $SrPdGe_3$ and $T_c = 1.0$ K for $SrPtGe_3$. The agreement between the BCS model and the present experimental data is convincing assuming $\gamma = 5$ and 4 mJ/molK^2, for the Pd and Pt system, respectively and adding an impurity term of the order of 0.6 mJ/molK^2. The latter might be a consequence of errors in the addenda calibration of the calorimeter (PPMS, Quantum Design). The excellent agreement observed is considered as an indication of standard s-wave BCS superconductivity in both, $SrPdGe_3$ and $SrPtGe_3$.

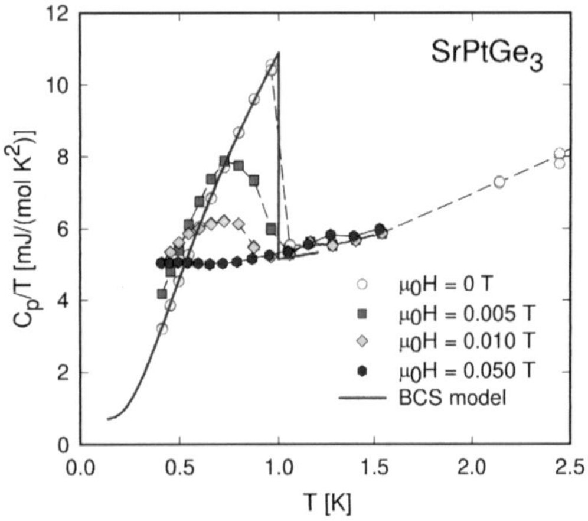

Figure 3. Temperature dependent specific heat C_p of $SrPtGe_3$ for various values of applied field. The solid line represents the heat capacity of an s-wave BCS superconductor.

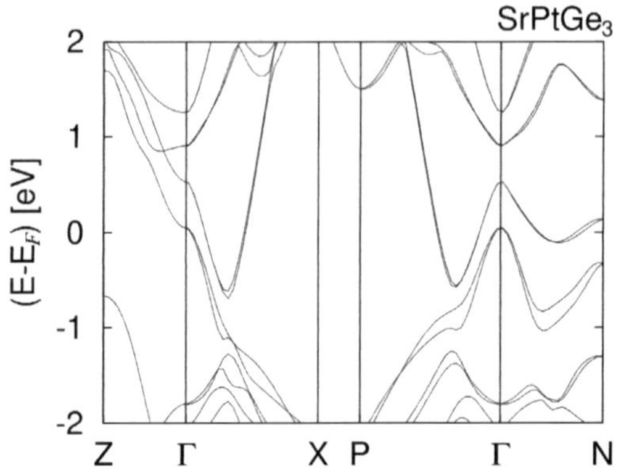

Figure 4. Electronic band structure of $SrPtGe_3$ for high symmetry directions of the body-centred tetragonal crystal structure including spin-orbit interaction.

The very low values for $\mu_0 H_{c2}(0)$ (well below the Pauli limiting field of $\mu_0 H > 1$ T) and the exponential temperature dependence of C_p make unconventional superconductivity in $SrPdGe_3$ and $SrPtGe_3$ rather unlikely. A further indication that the lack of inversion symmetry is of negligible importance for superconductivity can be gained from the DFT relativistic band structure, which includes spin-orbit coupling (SO), as shown for $SrPtGe_3$ in Fig. 4. SO causes a sizeable splitting of degenerate bands, e.g., at the Γ-point of about 0.5 eV. Furthermore, the remaining degeneracy becomes lifted by the electric field gradient owing to the absence of inversion symmetry (compare e.g., the bands along $\Gamma - N$). Both splitting effects are absent in non-SO calculation. The splitting of bands due to the non-inversion symmetry, however, remains small, in particular at E_F. This might be an indication that a mixing of spin-singlet and spin-triplet components in the superconducting condensate is weak and spin-singlet Cooper pairs dominate. The electronic band structure of $SrPdGe_3$ is rather similar, but the splitting effects are less pronounced because of the smaller mass of Pd in comparison to Pt.

In summary, we have shown that body-centred tetragonal $SrPdGe_3$ and $SrPtGe_3$ are new superconductors without inversion symmetry with transition temperatures around 1.5 and 1 K, respectively. The absence of strong correlations among electrons and the minor splitting of bands favours a fully gapped s-wave superconducting state. A microscopic evidence for a simple superconducting state was derived for $BaPtSi_3$ from μSR studies.

Work supported by the Austrian Science Foundation, project P22295. This research project has been supported by the European Commission under the 7th Framework Programme through the "Research Infrastructures" action of the "Capacities" Programme, Contract No: CP-CSA INFRA-2008-1.1.1 Number 226507-NMI3

References

[1] Bauer E *et al.*, 2009 *Phys. Rev. B* **80** 064504
[2] Gor'kov L P and Rashba E I 2001 *Phys. Rev. Lett.* **87** 037004
[3] Brandt E H 1988 *Phys. Rev. B* **37** 2349
[4] Brandt E H 2003 *Phys. Rev. B* **68** 054506
[5] Bauer E *et al.*, 2002 *Phys. Rev. B* **66** 214421
[6] Kresse G and J. Furthmüller 1996 *Phys. Rev. B* **54** 11169; Kresse G and Joubert D *Phys. Rev. B* **59** 1758

Journal of Physics: Conference Series **273** (2011) 012079 doi:10.1088/1742-6596/273/1/012079

(P, T) Phase Diagram of Clathrate Superconductor $Ba_{24}Ge_{100}$

Atsushi Miyake[1], **Tomoko Kagayama**[1], **Katsuya Shimizu**[1], **Jun Tang**[2], **Katsumi Tanigaki**[2]

[1] KYOKUGEN, Osaka University, Toyonaka, Japan,
[2] WPI/Graduate School of Science, Tohoku University, Sendai, Japan

E-mail: amiyake@cqst.osaka-u.ac.jp

Abstract. Precise electrical resistivity and ac calorimetry measurements under pressures have revealed (P, T) phase diagram of superconducting clathrate $Ba_{24}Ge_{100}$. Following conventional superconducting theory, the pressure dependence of the density of state at Fermi energy, $N(E_F)$, estimated by the electronic specific heat C/T has been discussed with the $T_c(P)$. Up to P= 2.1 GPa, where $T_c(P)$ starts to saturate, C/T increases with T_c in agreement with the theory. On further compression, C/T is strongly suppressed while T_c slightly decreases. Across $P_c \sim 2.9$ GPa, where structural deforming transitions at $T_{s1(s2)}$ are suppressed, pressure variation of C/T becomes smaller. In addition to the enhancement of the $N(E_F)$, other changes in crystal and/or electrical structure may play important roles to pressure dependence of T_c in $Ba_{24}Ge_{100}$.

Oversized-cage compounds including atoms or molecules, such as skutterudites, pyrochlores and clathates, have been attracted much attention for not only scientific interests but their thermoelectric potential. Clathrate $Ba_{24}Ge_{100}$ is a superconductor below T_c = 240 mK at ambient pressure [1]. Well above T_c, two successive structure deforming phase transitions have been observed at T_{s1} = 215 K and T_{s2} = 180 K accompanied by a pronounced thermal hysteresis [2]. The two-step increases in the electrical resistivity ρ on cooling suggests the reduction of the electronic density of state at Fermi energy, $N(E_F)$, at low temperature. It is also indicated by the soft X-ray spectroscopy [3]. With increasing pressure, T_{s1} and T_{s2} are shifted to lower temperature, but T_c increases remarkably to 3.8 K at P_c = 2.8 GPa where the structural transitions are suppressed [1]. From the $\rho(P)$ at low temperature, $dH_{c2}/dT|_{T_c}$ [4] and NMR results [5], the obtained $N(E_F)$ is strongly enhanced under pressure, and thus, the superconductivity in $Ba_{24}Ge_{100}$ are concluded as a conventional one, namely T_c varies with $N(E_F)$. Because of the few number of precise experiments around P_c, it is still not revealed the detail of the change of electronic state, e.g. how $N(E_F)$ and crystal structure varies across P_c. Here, we have studied the pressure temperature phase diagram of $Ba_{24}Ge_{100}$ via the simultaneous ρ and ac calorimetry C_{ac} measurements under pressure with special focus on the vicinity of the structural instability at P_c. The $N(E_F)$ estimated by the electronic specific heat coefficient C/T increases up to $P \sim 2.1$ GPa, where $T_c(P)$ starts to saturate. Interestingly, C/T strongly decreases on further compression, although $T_c(P)$ keeps constant or slightly decreases. These results indicate that the modification of crystal structure and/or Fermi surface, which does not strongly affect T_c, seems to occur around P_c.

The simultaneous ρ and C_{ac} have been measured as a function of temperature between 1.8 K

Published under licence by IOP Publishing Ltd

and room temperature and pressures up to 4 GPa. Two Au-AuFe (0.07%) thermocouples (TC) are welded to the both ends of sample, and two Au wires are also attached in between them for ρ measurement. For C_{ac} measurement, a modulating heat was applied by current through the sample or TC, and the oscilating temperature was measured by another TC. Pressure was applied by a diamond anvil pressure cell filled with glycerin as a pressure medium and estimated by a ruby fluorescence method at low temperature. Additional signal on C_{ac} from pressure medium shows a clear liquid to glass transition anomaly. The pressure dependence of the anomaly indicates the glycerin solidifies around P= 4 GPa at room temperature, in good agreement with the previously reported phase diagram of glycerin [6]. From these facts, at least room temperature and the pressure region discussing here, pressure is tuned and applied homogeneously.

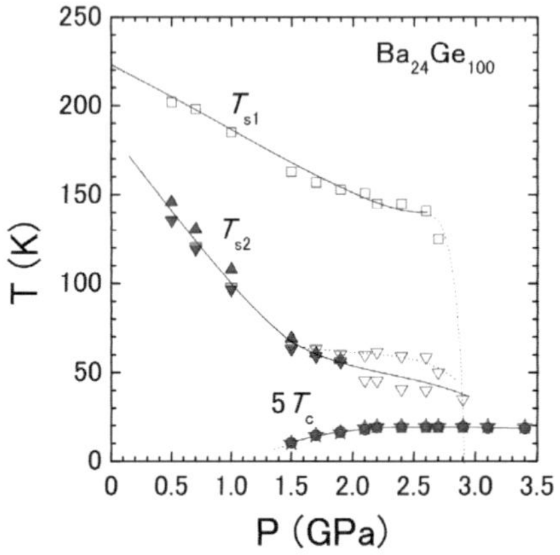

Figure 1. Pressure-temperature phase diagram of $Ba_{24}Ge_{100}$. Open and closed symbols are phase boundaries determined by ρ and C_{ac}, respectively. For T_{s2}, the downward and upward triangles are obtained on cooling and warming processes. T_{c} is multiplied by 5 for clear view. The lines are guides for eyes.

Fig. 1 shows pressure temperature phase diagram obtained by temperature dependence of ρ and C_{ac}. It is almost consistent with previous report [1]. A clear anomaly around T_{s2} with thermal hysteresis (downward and upward triangles) is observed in both curves of $\rho(T)$ and $C_{\text{ac}}(T)$, indicating a first-order phase transition. With increasing pressure, $T_{\text{s1(s2)}}$ is shifted to lower temperature, but T_{c} is enhanced drastically accompanied with the reduction of ρ at low temperature, as seen in Fig. 2(a). $\rho(T)$ around T_{s2} above 2 GPa shows two step anomalies as plotted in Fig. 1, although any C_{ac} anomalies are not detected. Because of a strong smearing of the $\rho(T)$ and C_{ac} anomalies above 2 GPa, it is not unambiguously determined P_{c}. However, our obtained $P_{\text{c}} \sim 2.9$ GPa is very close to a previously reported value of $P_{\text{c}} = 2.8$ GPa [1].

In our experimental region above T = 1.8 K, the superconducting transition was observed above P = 1.5 GPa as shown in Fig. 2. The $\rho(T)$ at low temperature has a large value and weakly depends on temperature reflecting the random scattering from disordered phase [2, 7]. A clear λ-like anomaly in specific heat at T_{c} under pressure is firstly observed, indicating that the superconductivity in $Ba_{24}Ge_{100}$ is bulk nature even in the higher temperature superconducting region. T_{c} is defined as a midpoint of the resistive drop and of the C_{ac} jump and is plotted in Fig. 1 and 3. T_{c} strongly increases with pressure up to $P \sim 2.2$ GPa and slightly varies on further compression.

As argued by the previous experiments [4, 5], we will discuss how the $N(E_{\text{F}})$ varies with T_{c}. Assuming that the pressure dependence of specific heat of pressure medium glycerin at

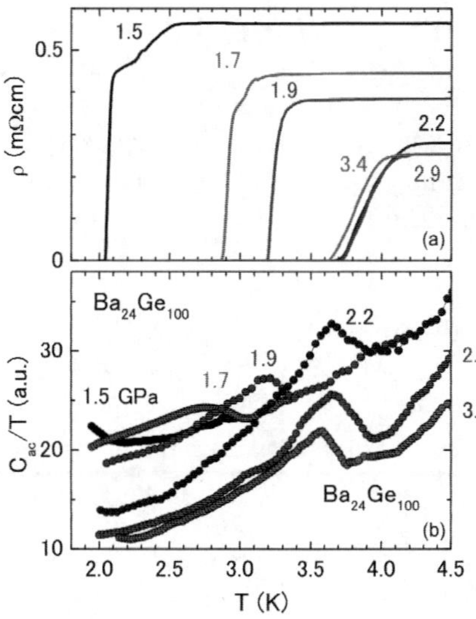

Figure 2. Temperature dependence of (a)ρ and (b)C_{ac}/T of $Ba_{24}Ge_{100}$ under several pressures.

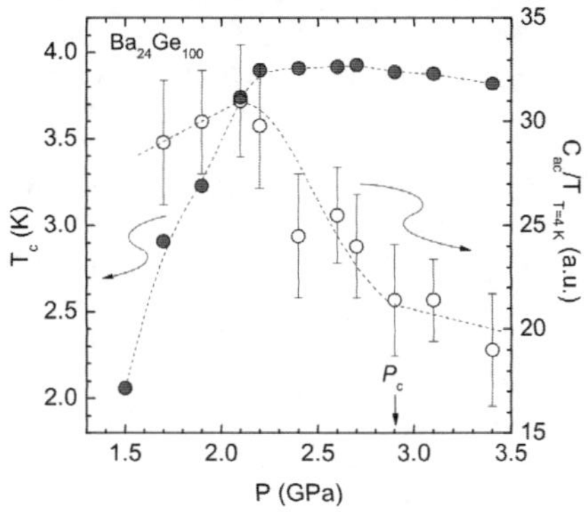

Figure 3. Pressure dependence of T_c (closed circle) and the normal state C_{ac}/T at 4 K (open circle) corresponding to the electronic specific heat coefficient.

low temperature is negligibly small, we estimated the electronic specific heat at normal state as the C_{ac}/T at 4 K. Pressure dependence of $C_{ac}/T(\propto N(E_F))$ as well as T_c are represented in Fig. 3. C_{ac}/T increases accompanied with the increase of T_c up to P ~2.1 GPa. It is in good agreement with the previous suggestions that the $N(E_F)$ is strongly enhanced with T_c [4, 5]. It is in agreement with the conventional BCS theoty that T_c varies with the $N(E_F)$ as follows, $T_c \propto \Theta_D \exp[-1/N(E_F)V]$, where Θ_D is the Debye temperature, V is the electron-phonon coupling constant. Pressure collapse of the structural transitions may reduce the random scattering leading to a drastic enhancement of T_c. At higher pressures than P = 2.2 GPa, T_c weakly depends on pressure, although the $N(E_F)$ is strongly suppressed. Above $P_c \sim$ 2.9 GPa, where the structural phase transitions at $T_{s1(s2)}$ are suppressed, the C/T seems to be weak pressure dependence. Between 2.2 GPa and P_c, the low and high pressure phase may coexist arising from the first-order transition at T_{s2}. Moreover, Raman scattering experiment revealed pressure-induced structural phase transition at room temperature and P=3.2 GPa, which seems to be due to the structural distortion combined with the enhanced guest-host interactions [8]. Up to 2.2 GPa, the pressure evolution of T_c seems strongly correlated the enhancement of $N(E_F)$. Above 2.2 GPa, a probable intermediated phase makes it difficult to simply discuss the relation between T_c and C_{ac}/T. However, it seems not rule out the conventional superconducting scenario. For first-order phase transition, the volume discontinuity across P_c is expected. It is important to take into account for the pressure variation of the volume to discuss the density of state. Thermal expansion measurements under pressure can directly determine pressure variance of the volume and is desired.

In conclusion, we have thermodynamically determined pressure temperature phase diagram of $Ba_{24}Ge_{100}$ through the precise electrical resistivity and specific heat measurements. We revealed that the $N(E_F)$ increases with T_c up to $P \sim 2.2$ GPa and decreases at higher pressure while T_c doesn't vary. In addition to the enhancement of the $N(E_F)$, other changes in crystal and/or electrical structure play important roles to the strong pressure variation of T_c.

This work was partially supported by Osaka University Global COE Program (Core Research and Engineering of Advanced Materials-Interdisciplinary Education Center for Material Science) and Scientific Research on Priority Areas (New Materials Science Using Regulated NanoSpaces-Strategy in Ubiquitous Elements) from the Ministry of Education, Culture, Sports, Science and Technology of Japan.

References

[1] Grosche F M, Yuan H Q, Carrillo-Cabrera W, Paschen S, Langhammer C, Kromer F, Sparn G, Baenitz M, Grin Y and Steglich F 2001 *Phys. Rev. Lett.* **87** 247003

[2] Carrillo-Cabrera W, Borrmann H, Paschen S, Baenitz M, Steglich F and Grin Y 2005 *Journal of Solid State Chemistry* **178** 715

[3] Rachi T, Kitajima M, Kobayashi K, Guo F, Nakano T, Ikemoto Y, Kobayashi K and Tanigaki K 2005 *The Journal of Chemical Physics* **123** 074503

[4] Yuan H Q, Grosche F M, Carrillo-Cabrera W, Pacheco V, Sparn G, Baenitz M, Schwarz U, Grin Y and Steglich F 2004 *Phys. Rev. B* **70** 174512

[5] Kanetake F, Harada A, Mukuda H, Kitaoka Y, Rachi T, Tanigaki K, Itoh K M and Haller E E 2009 *Journal of the Physical Society of Japan* **78** 104710

[6] Cook R L, H E King J, Herbst C A and Herschbach D R 1994 *The Journal of Chemical Physics* **100** 5178

[7] Zerec I, Carrillo-Cabrera W, Voevodin V, Sichelschmidt J, Steglich F, Grin Y, Yaresko A and Kimura S 2005 *Phys. Rev. B* **72** 045122

[8] Shimizu H, Fukushima T, Kume T, Sasaki S, Fukuoka H and Yamanaka S 2007 *Journal of Applied Physics* **101** 113531

Anisotropy of magnetic susceptibility of URu_2Si_2 under pressure

Gaku Motoyama, Hideki Sakai, Akira Yamaguchi,
Akihiko Sumiyama, Yasukage Oda

Graduate School of Material Science, University of Hyogo,
Kamigori-cho, Ako-gun, Hyogo 678-1297, Japan

E-mail: motoyama@sci.u-hyogo.ac.jp

Abstract. We have carried out measurements of magnetic susceptibility on a single crystal of the heavy fermion compound URu_2Si_2 under pressure up to 0.7 GPa. The external magnetic fields were applied parallel to *a*- and *c*-axes. Hidden order transition at T_0 and first-order phase transition between hidden ordered phase and large-staggered-moment antiferromagnetic phase at T_M are observable in both measurements. Temperature dependence of magnetic susceptibility of *a*- and *c*-axes exhibits a kink and a drop at T_0 and T_M, respectively. However, the magnitude of the variation on *a*-axis at T_0 and T_M were about one tenth of those on *c*-axis, respectively. We conclude that the HO phase has a similar anisotropy to that of AFM phase.

1. Introduction

URu_2Si_2 is a heavy fermion superconductor with a superconducting transition temperature $T_c \sim 1.5$ K under ambient pressure[1]. Furthermore the compound undergoes a successive phase transition(a hidden order transition) at $T_0 \sim 17.5$ K. Since the discovery of a development of the antiferromagnetic(AFM) staggered moment with increasing pressure[2], many investigations under pressure have been carried out so far. Nuclear magnetic resonance(NMR) experiments suggested that the sample consisted of the AFM and hidden order(HO) regions below T_0; the spatially inhomogeneous AFM volume fraction develops by applying pressure[3]. Thermal expansion measurements revealed an existence of the first order transition between HO and AFM phases at T_M and the pressure(P) dependence of T_M[4]. The simultaneous neutron scattering and thermal expansion measurements were carried out in order to detect the successive phase diagram through the eyes of microscopic and macroscopic[5]. Moreover, there are some studies of superconducting phase; the superconductivity exists in only HO phase but not in AFM phase[6], and two distinct superconducting gap structures with different nodal topology are suggested[7]. However, a question for the nature of hidden order parameter still remains. There is a large anisotropy for URu_2Si_2 on magnetization under ambient pressure. It would be important to reveal the anisotropy of magnetization for understanding the ordered ground states. Detailed measurements of temperature(T) dependence of magnetic susceptibility for *a*-axis($\chi_a(T)$) under pressure have never been reported, because $\chi_a(T)$ is too small. Here we report about anisotropy of magnetic susceptibility of URu_2Si_2 under pressure.

Published under licence by IOP Publishing Ltd

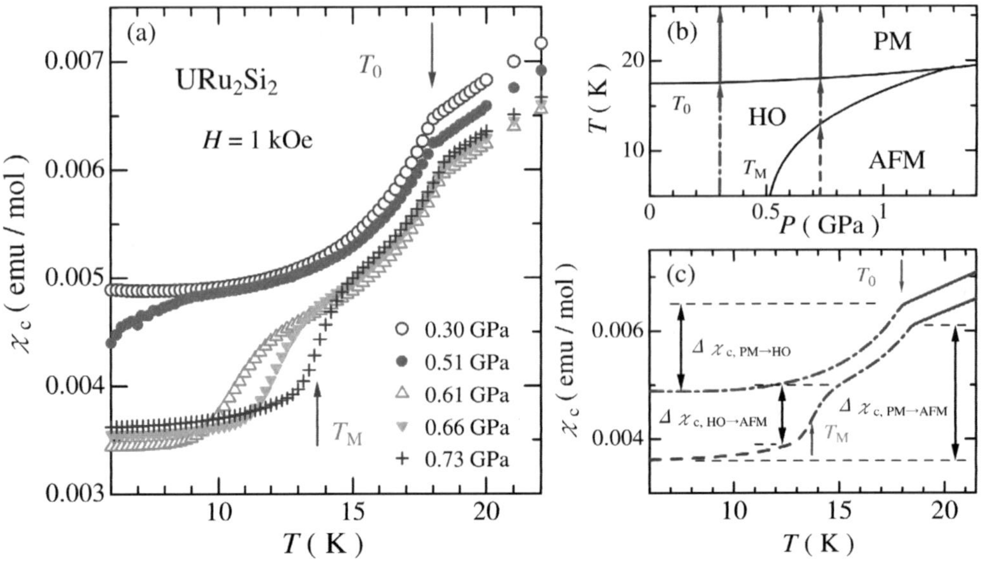

Figure 1. (a)Temperature dependences of the magnetic susceptibility applying a magnetic field of 1 kOe parallel to the c-axis at 0.30, 0.51, 0.61, 0.66, and 0.73 GPa. Down-pointing and up-pointing arrows indicate T_0 at 0.30 GPa and T_M at 0.73, respectively. (b)Schematic P-T phase diagrum of URu_2Si_2. If $\chi_c(T)$ is carried out at 0.30 GPa and 0.73 GPa, $\chi_c(T)$ at 0.30 GPa crosses T_0, and $\chi_c(T)$ at 0.73 GPa crosses T_M and T_0, respectively. (c)Selected data of P = 0.30 and 0.73 GPa is plotted to illustrate the $\Delta\chi_{c,PM\rightarrow HO}$, $\Delta\chi_{c,HO\rightarrow AFM}$, and $\Delta\chi_{c,PM\rightarrow AFM}$

2. Experiments

We first synthesized a polycrystalline material by melting a stoichiometric amount of constituent elements, each of which has the purity of 99.9 %, 99.99 %, and 99.9999 % for natural U, Ru, and Si, respectively. Then we grew a single crystal by the Czochralski pulling method from the polycrystal under high purity argon atmosphere. The obtained single crystal was cut into several small pieces. Typical size of the sample is about $\sim$ 2 mm^3. Before magnetic measurements, thermal expansion measurements were carried out to choose the sample which showed a distinct anomaly at T_M. The anomaly of T_M exhibits a sample quality because of its strong sample dependence. We measured the dc magnetic susceptibility by using commercial SQUID magnetometer(MPMS). The pressure was generated by means of a copper-beryllium clamp-type cylinder with a piston made of zirconia. Pressure transmitting medium was a Dafune7373. We determined the pressure by measuring the superconducting transition temperature of an indium.

3. Results

We performed the measurements of $\chi_a(T)$ and $\chi_c(T)$ in several magnetic fields(H) of 1, 2, 5, and 10 kOe applied parallel to the a and c-axes. There were little H dependence on the whole our measurements of $\chi_a(T)$ and $\chi_c(T)$, we present the results of H = 1 kOe in figure 1 and 2. First, we focus on the results of the $\chi_c(T)$. Our results of $\chi_c(T)$ under pressure agree to the previous work, which reported the T dependence of the T derivative of the dc magnetization, $dM/dT(T)$[6]. We observed the anomalies of phase transition at T_0 and T_M, respectively. $\chi_c(T)$ of 0.30 GPa decreases with decreasing T down to T_0, then it shows a sharp kink at T_0. After rapid decreasing below T_0, the decreasing rate becomes slower and exhibits little T dependence.

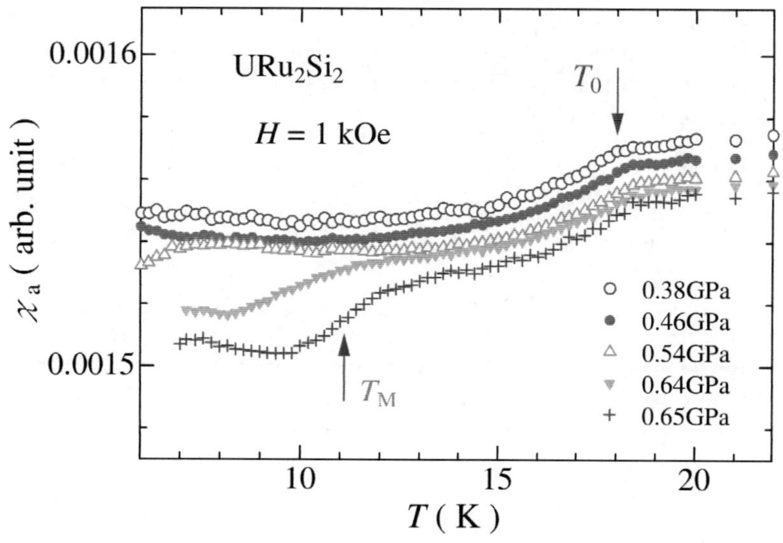

Figure 2. Temperature dependences of the magnetic susceptibility applying an magnetic field of 1 kOe parallel to the a-axis at 0.38, 0.46, 0.54, 0.64, and 0.65 GPa. Our experimental accuracy is equivalent to $\sim$ 0.0002 emu/mol because of the pressure cell. Therefore, the magnitude of the magnetic susceptibility parallel to the a-axis cannot be argued. The data of the magnetic susceptibility parallel to the a-axis are slightly shifted downward, for clarity. Down-pointing and up-pointing arrows indicate T_0 at 0.38 GPa and T_M at 0.65 GPa, respectively.

It slightly shows a paramagnetic behavior at lower temperature(not shown). When P increases to 0.51 GPa, the anomaly of T_M appears. It is a stair-like anomaly with a transition width of $\sim$ 3 K. The width becomes sharper at higher pressure as seen in the previous experiments of thermal expansion and electrical resistivity[4, 8]. More noteworthy is a magnitude of the magnetic susceptibility of the each phase. In the paramagnetic(PM) phase above T_0, χ_c(20 K) shows a trend to decrease gradually with increasing P. However, the detail could not be described, because a variation of that was comparable to our experimental accuracy. In order to compare with the decreases of the χ_c at the two phase transitions, we defined $\Delta\chi_{c,PM\to HO}$ at $P <$ 0.5 GPa and $\Delta\chi_{c,PM\to AFM}$ at $P >$ 0.5 GPa as $\chi_c(T_0)$ - $\chi_{c,min}$ at the each pressure, respectively, and $\Delta\chi_{c,HO\to AFM}$ as $\chi_c(T_{M,onset})$ - $\chi_c(T_{M,offset})$. $\Delta\chi_{c,HO\to AFM}$ shows no P dependence up to 0.73 GPa. $\Delta\chi_{c,PM\to HO}$ at $P >$ 0.5 GPa can be obtained from $\Delta\chi_{c,PM\to AFM}$ - $\Delta\chi_{c,HO\to AFM}$. If the χ_c of the AFM phase is compared to that of the HO phase, there is a significant decrease from HO to AFM. $\Delta\chi_{c,PM\to HO}$ and $\Delta\chi_{c,PM\to AFM}$ are 0.0015 and 0.0025 emu/mol, which are 25 and 40 % of the $\chi_c(T_0)$, respectively. P dependence of $\Delta\chi_{c,PM\to HO}$ and $\Delta\chi_{c,PM\to AFM}$ are small.

Next, we discuss about $\chi_a(T)$ shown in Fig. 2. First of all, there is little doubt that magnitude of $\chi_a(T)$ keeps very small with increasing P. The variations of $\chi_a(T)$ are also small. These anisotropic behaviors remain at higher pressures. According to the neutron scattering experiments, the staggered moment of ordered state orients to the c-axis[2], which is consistent with our results of large anisotropy of magnetic susceptibility in the AFM phase. On the other hand, a kink and a drop anomalies of $\chi_a(T)$ are found to be similar to those of the $\chi_c(T)$, as see in fig. 1(a). The ratio of $\Delta\chi_{a,PM\to HO}$ to $\Delta\chi_{a,PM\to AFM}$ almost agrees with that of $\Delta\chi_{c,PM\to HO}$ to $\Delta\chi_{c,PM\to AFM}$. If the measured sample has a mis-configuration of $\sim$ 5 degrees from the a-axis, these anomalies which was observed on our results of $\chi_a(T)$ below T_0 may be explained as a contribution from $\chi_c(T)$. However $\chi_a(T)$ shows no broad peak around 60 K. The broad peak is also characteristic behavior for $\chi_c(T)$. Therefore we consider that $\chi_a(T)$ exhibits quite small anomalies compared to that of $\chi_c(T)$.

4. Discussions

Our present result of the magnitude of the $\chi(T)$ indicate that the HO phase has a similar anisotropy to that of AFM phase. At least in AFM phase, the ordered local moment was observed. However, heavy fermion liquid states is achieved just above T_0; specific heat shows the linear dependence as plotted C/T vs. T^2[1, 9], nuclear spin-lattice relaxation time$(1/T_1)$ follows the Korringa law[10]. Heavy fermion liquid behavior remains even at high pressure[11], namely, it is achieved just above antiferromagnetic phase transition temperature. We need to account for the experimental results with seemingly unreasonable behaviors. Although the hidden order parameter which is exhibited below second order phase transition at T_0 is still open question, we revealed an anisotropy of magnetic susceptibility of both ordered phases with useful hints to solve the hidden order parameter.

5. Acknowledgments

We thank T. Kohara, Y. Takahashi, and Y. Hasegawa for their useful discussions. This work was partially supported by the Grant-in-Aid from the Ministry of Education, Culture, Sports, Science and Technology, Japan.

References

[1] Palstra T T M, Menovsky A A, van den Berg J, Dirkmaat A J, Kes P H, Nieuwenhuys G J, and Mydosh J A: Phys. Rev. Lett. **55** (1985) 2727.
[2] Amitsuka H, Sato M, Metoki N, Yokoyama M, Kuwahara K, Sakakibara T, Morimoto H, Kawarazaki S, Miyako Y, and Mydosh J A: Phys. Rev. Lett. **83** (1999) 5114.
[3] Matsuda K, Kohori Y, Kohara T, Kuwahara K, and Amitsuka H: Phys. Rev. Lett. **87** (2001) 087203.
[4] Motoyama G, Nishioka T, and Sato N K: Phys. Rev. Lett. **90** (2003) 166402.
[5] Villaume A, Bourdarot F, Hassinger E, Raymond S, Taufour V, Aoki D, and Flouquet J: Phys. Rev. B. **78** (2008) 012504.
[6] Uemura S, Motoyama G, Oda Y, Nishioka T, and Sato N K: J. Phys. Soc. Jpn. **74** (2005) 2667.
[7] Kasahara Y, Iwasawa T, Shishido H, Conczykowski M, Behnia K, Haga Y, Matsuda T D, Onuki Y, Shibauchi T, and Matsuda Y: Phys. Rev. Lett. **99** (2007) 116402.
[8] Motoyama G, Yokoyama N, Sumiyama A, and Oda Y: J. Phys. Soc. Jpn. **77** (2008) 123710.
[9] Maple M B, Chen J W, Dalichaouch Y, Kohara T, Rossel C, Torikachvili M S, McElfresh M W, and Thompson J D: Phys. Rev. Lett. **56** (1986) 185.
[10] Kohori Y, Matsuda K, and Kohara T: J. Phys. Soc. Jpn. **65** (1996) 1083.
[11] Matsuda K, Kohori Y, Kohara T, Amitsuka H, Kuwahara K, and Matsumoto T: J. Phys.: Condens. Matter **15** (2003) 2363.

International Conference on Strongly Correlated Electron Systems (SCES 2010) IOP Publishing
Journal of Physics: Conference Series **273** (2011) 012081 doi:10.1088/1742-6596/273/1/012081

Anomalous low-field diamagnetic response in ultraclean URu_2Si_2 superconductor

R Okazaki[1], M Shimozawa[1], H Shishido[1,2], M Konczykowski[3], Y Haga[4], T D Matsuda[4], E Yamamoto[4], Y Onuki[4,5], Y Yanase[6], T Shibauchi[1] and Y Matsuda[1]

[1]Department of Physics, Kyoto University, Kyoto 606-8502, Japan
[2]Research Center for Low Temperature and Materials Sciences, Kyoto University, Kyoto 606-8501, Japan
[3]Laboratoire des Solides Irradiés, CNRS-UMR 7642 & CEA/DSM/DRECAM, Ecole Polytechnique, 91128, Palaiseau, France
[4]Advanced Science Research Center, Japan Atomic Energy Agency, Tokai 319-1195, Japan
[5]Graduate School of Science, Osaka University, Toyonaka, Osaka 560-0043, Japan
[6]Department of Physics, Niigata University, Niigata 950-2181, Japan

E-mail: okazaki@scphys.kyoto-u.ac.jp

Abstract. The lower critical field H_{c1} in ultraclean URu_2Si_2 single crystals ($T_c = 1.4$ K) with residual resistivity ratio exceeding 700 is precisely determined by the positional dependence of the local magnetic induction measured down to 55 mK. We show that the whole $H_{c1}(T)$ for $\boldsymbol{H} \parallel a$ can be explained by the two gaps with line and point nodes. In contrast, for $\boldsymbol{H} \parallel c$ we find a distinct kink in the slope of H_{c1} at 1.2 K, which cannot be accounted for by the two gaps. This outstanding anomaly demonstrates a highly unusual superconducting state embedded in the hidden order state of URu_2Si_2 possibly related to the time reversal symmetry breaking.

1. Introduction

The heavy-fermion compound URu_2Si_2 has mystified researchers since the superconductivity occurs deep inside the mysterious 'hidden order' state (whose transition temperature is $T_h =$ 17.5 K) [1, 2]. Below T_h, most of the carriers (~90%) disappears owing to the formation of the electronic excitation gap at a large portion of the Fermi surface, resulting in a semimetallic electronic structure, and the remaining small number of carriers undergo a transition into the exotic superconducting state below $T_c = 1.4$ K [3].

Here, to shed further light on the unusual superconducting properties of URu_2Si_2, we measured low-field diamagnetic response of ultraclean single crystals by using miniature Hall sensors, from which the lower critical field H_{c1} is accurately determined down to very low temperatures [4]. We find an anomalous kink structure in the temperature dependence of H_{c1} for $\boldsymbol{H} \parallel c$, which has not been reported in the samples with low residual resistivity ratio (RRR) [5]. This anomaly suggests that the superconducting state of ultraclean URu_2Si_2 is highly unusual, and possible implications of these findings are discussed.

Published under licence by IOP Publishing Ltd

2. Experimental

Experiments have been performed on URu_2Si_2 single crystals (with typical dimensions of $2.3 \times 0.75 \times 0.15\,mm^3$) with very large RRR value exceeding 700 [6]. Reliable determination of H_{c1} is a difficult task in the presence of the flux pinning and magnetic relaxation. To avoid this difficulty, we use a method by determining H_{c1} as the field H_p at which first flux penetration occurs from the edge of the crystal [7]. For this purpose, we used a miniature Hall-sensor array tailored in a GaAs/AlGaAs heterostructure with each active area of $\sim 5 \times 5\,\mu m^2$ (the center-to-center distance of neighboring sensors is $20\,\mu m$) [8]. The crystal is directly placed on top of the array, as illustrated in the inset of Fig. 1(c).

3. Results and discussion

Figures 1(a) and (b) display the field dependence of local induction B_{edge} measured by the Hall sensor near the edge for $\boldsymbol{H} \parallel a$ and $\boldsymbol{H} \parallel c$, respectively. The flux penetration fields H_p shown by the arrows are clearly resolved by the deviation from the Meissner state ($B_{edge} = 0$). Figure 1(c) displays the positional dependence of H_p. H_p increases with increasing the distance d from the edge at low temperatures. However, H_p is position-independent close to the edge ($d \leq 40\ \mu m$) even at low temperatures. These results demonstrate that we can accurately determine the lower critical field from H_p measured by sensors near the edge.

Figures 2 depict the temperature dependence of H_{c1}^a and H_{c1}^c, where H_{c1}^a and H_{c1}^c are the lower critical field for $\boldsymbol{H} \parallel a$ and $\boldsymbol{H} \parallel c$, respectively. The absolute value of H_{c1} is obtained by taking into account the demagnetization effect. For a platelet sample, H_{c1} is given by $H_{c1} = H_p/\tanh\sqrt{0.36b/a}$, where a and b are the width and the thickness of the crystal, respectively [9]. In the present crystal, H_{c1} are evaluated as $H_{c1}^c = 3.82H_p$ and $H_{c1}^a = 1.15H_p$. As seen in Figs. 2, $H_{c1}^c(T)$ exhibits a distinct change in the slope with a kink at $T_Q \simeq 1.2\,K$ and $H_{c1}^a(T)$ exhibits a cusp behavior at $T_p \simeq 1.0$ K. At T_p the specific heat measurements have reported no anomalies in the crystals with $RRR > 100$. However, a peak structure in the specific

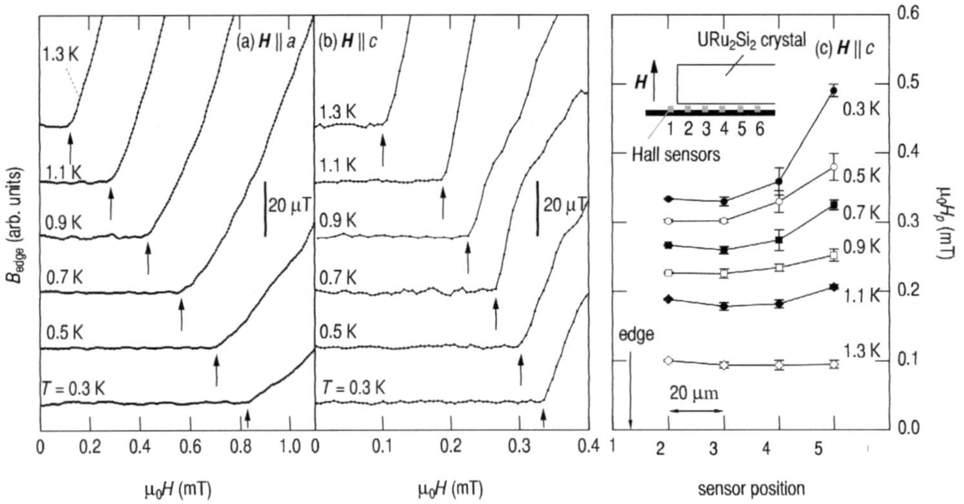

Figure 1. (a) Local magnetic induction B_{edge} measured by the sensor at the edge of the crystal as a function of H for $\boldsymbol{H} \parallel a$. The arrows indicate the flux penetration field H_p. Each data is vertically shifted for clarity. (b) The same plot for $\boldsymbol{H} \parallel c$. (c) Positional dependence of H_p. Inset: Schematic illustration of the experimental setup. Sensor 1 locates just outside of the crystal.

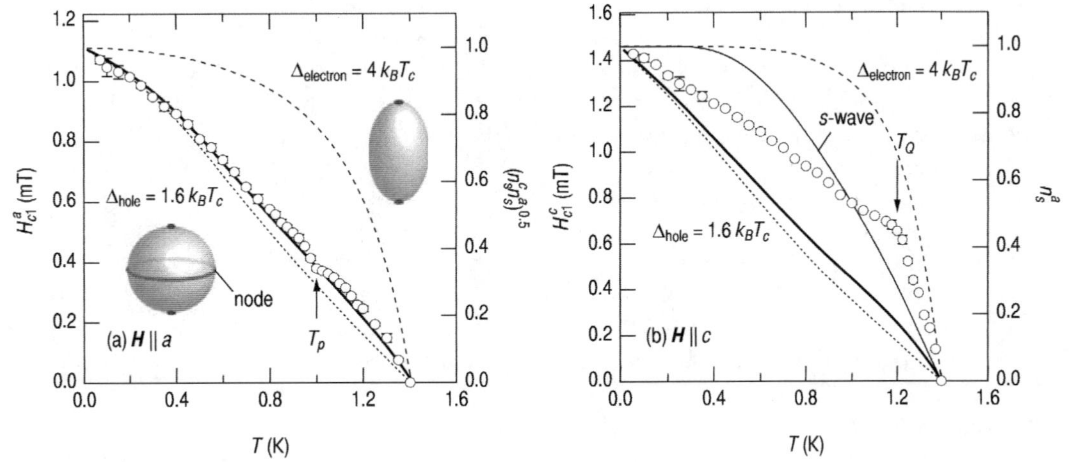

Figure 2. Temperature dependence of the lower critical fields H_{c1} for (a) $\boldsymbol{H}\parallel a$ and (b) $\boldsymbol{H}\parallel c$ (open symbols). Arrows indicate a kink anomaly at $T_Q = 1.2\,\text{K}$ and a cusp behavior at $T_p \simeq 1\,\text{K}$. Normalized superfluid densities (a) $\sqrt{n^a_s n^c_s}$ and (b) n^a_s are also plotted. The dotted and dashed lines indicate the superfluid densities for the hole band with $\Delta_h = 1.6k_BT_c$ and the electron band with $\Delta_e = 4.0k_BT_c$, respectively, which are obtained by assuming the illustrated nodal topologies suggested by the thermal conductivity measurements [3]. The solid thick and thin lines represent the results of the two-gap fitting and the calculation for s-wave gap, respectively.

heat has been reported in the vicinity of T_p for crystals with low RRR values, possibly due to the inhomogeneous distribution of T_c [6]. Therefore we cannot rule out a possibility that the anomaly at T_p in H^a_{c1} is due to a tiny portion with low-T_c phase in the crystal, though no specific heat anomaly is observed in the present crystal at T_p. As for the kink at T_Q, there is no evidence for the low-T_c phase at this temperature. So we infer that the pronounced kink in the slope of H^c_{c1} found in the ultraclean crystals is intrinsic in the superconducting state of URu_2Si_2.

We next discuss the temperature dependence of H_{c1} in terms of the two-gap model since it has been shown that below T_h the nearly perfect compensation (equal number of electrons and holes) is realized in URu_2Si_2 [3]. In the two-gap model, the in-plane and out-of-plane superfluid density normalized by their values at $T = 0\,\text{K}$, n^a_s and n^c_s, respectively, can be written as $n^i_s(T) = x^i n^i_h(T) + (1 - x^i) n^i_e(T)$ where n^i_h and n^i_e are the normalized superfluid density of hole and electron bands, respectively, and x^i is the ratio of the electron and hole mass given by $x^i = X^i_h/(X^i_h + X^i_e)$, $X^i_{h(e)} = \int dS_{h(e)} (v^i_{F,h(e)})^2/|\boldsymbol{v}_{F,h(e)}|$, where $v^i_{F,h(e)}$ is i $(= a, c)$ component of the Fermi velocity of holes (electrons) $\boldsymbol{v}_{F,h(e)}$ [10]. The hole mass determined by the dHvA measurements is nearly isotropic $m^a_h \approx m^c_h \approx 13m_0$ [11], where m^a_h (m^c_h) is the mass of the hole along the a (c) axis and m_0 is the free electron mass. The heavy electron mass is anisotropic and is estimated to be $m^a_e \approx 85m_0$ and $m^c_e \approx 305m_0$ from the large Sommerfeld coefficient in the heat capacity ($\gamma \sim 80$ mJ/K^{2}mol) [2] and the anisotropy of H_{c2} [12]. Then we obtain $x^a \approx 0.87$ and $x^c \approx 0.95$. Using these parameters we calculate the superfluid density by assuming the nodal topologies $\Delta_h(\boldsymbol{k}, T) = \Delta_h(T) \times 2\sin\theta\cos\theta$ with point and horizontal line nodes for hole bands and $\Delta_e(\boldsymbol{k}, T) = \Delta_e(T) \times \sin\theta$ with c-axis point nodes for electron bands, which have been suggested by the thermal conductivity measurements [3].

The lower critical fields $H^{a,c}_{c1}(T)$ are related to the superfluid density as $H^c_{c1} \propto n^a_s$ and $H^a_{c1} \propto \sqrt{n^a_s n^c_s}$. First we try to fit $H^a_{c1}(T)$, and the best fit is obtained by $\Delta_h(0) = 1.6k_BT_c$ and $\Delta_e(0) = 4.0k_BT_c$. As shown by the thick line in Fig. 2(a), the fitting result well reproduces the overall temperature dependence of $H_{c1}^{a}(T)$. We note that this $\Delta_h(0)$ value is close to $\Delta_h(0) \sim \hbar v_{F,h}/\pi\xi_h \sim 1.5k_BT_c$, which is obtained from $v_{F,h} \sim 2 \times 10^4$ m/s [11] and $\xi_h \sim 25$ nm. Here $\xi_h = (\Phi_0/2\pi\mu_0 H_{c2}^{h})^{0.5}$ is the coherence length of the hole band estimated from the "virtual upper critical field" of the hole band $\mu_0 H_{c2}^{h} \sim 0.5$ T [3]. In Fig. 2(b), $H_{c1}^{c}(T)$ calculated by using the same $\Delta_h(0)$ and $\Delta_e(0)$ values is plotted by the thick line. In sharp contrast to $H_{c1}^{a}(T)$, the calculation strongly deviates from the data. We tried to fit the data by assuming various $\Delta_h(0)$ and $\Delta_e(0)$ values, but could not reproduce the data, particularly the anomaly at T_Q.

At the present stage of the study, the origin of the anomaly at T_Q is not clear. A possible explanation for this may be given by a peculiar vortex dynamics associated with chiral domains due to the multi-component superconducting order parameter with broken time reversal symmetry (TRS). As suggested in Ref. 13, magnetic fields penetrate inside through the domain wall even in the Meissner state. In fact, the superconducting order parameter with broken TRS having a form $k_z(k_x \pm ik_y)$ has been proposed in URu_2Si_2 at low temperatures [3]. The order parameters in the forms k_zk_x and k_zk_y may have slightly different transition temperatures due to the hidden order, and then the time reversal symmetry will be broken below the lower temperature. This order parameter with broken TRS can produce domain walls along the c axis, which may give rise to a peculiar reduction of penetration field only for $\boldsymbol{H} \parallel c$.

4. Summary

In summary, we have determined accurately the lower critical fields of ultraclean URu_2Si_2. We find several distinct features, which has never been reported in crystals with low RRR values. Particularly, the remarkable kink structure in the temperature dependence of H_{c1} observed only for $\boldsymbol{H} \parallel c$ cannot be simply accounted for by the multiband nature. The observed unprecedented behavior demonstrates a highly unusual superconducting state of URu_2Si_2, which requires further theoretical and experimental investigations.

Acknowledgments

We thank D.F. Agterberg, E.H. Brandt, H. Ikeda, K. Machida and M. Sigrist for helpful discussions and V. Mosser for providing Hall sensors. This work was supported by Grant-in-Aid for the Global COE program "The Next Generation of Physics, Spun from Universality and Emergence", Grant-in-Aid for Scientific Research on Innovative Areas "Heavy Electrons" (No. 20102006) from MEXT of Japan, and KAKENHI from JSPS.

References

[1] Palstra T T M, Menovsky A A, van den Berg J, Dirkmaat A J, Kes P H, Nieuwenhuys G J and Mydosh J A 1985 *Phys. Rev. Lett.* **55** 2727
[2] Maple M B, Chen J W, Dalichaouch Y, Kohara T, Rossel C, Torikachvili M S, McElfresh M W and Thompson J D 1986 *Phys. Rev. Lett.* **56** 185
[3] Kasahara Y, Iwasawa T, Shishido H, Shibauchi T, Behnia K, Haga Y, Matsuda T D, Onuki Y, Sigrist M and Matsuda Y 2007 *Phys. Rev. Lett.* **99** 116402
[4] Okazaki R *et al Preprint* arXiv:0909.1075 (to be published in *J. Phys. Soc. Jpn.*)
[5] Wüchner S, Keller N, Tholence J L and Flouquet J 1993 *Solid State Commun.* **85** 355
[6] Matsuda T D, Aoki D, Ikeda S, Yamamoto E, Haga Y, Ohkuni H, Settai R and Onuki Y 2008 *J. Phys. Soc. Jpn.* **77** Suppl. A 362
[7] Okazaki R *et al* 2009 *Phys. Rev. B* **79** 064520
[8] Shibauchi T, Konczykowski M, van der Beek C J, Okazaki R, Matsuda Y, Yamaura J, Nagao Y and Hiroi Z 2007 *Phys. Rev. Lett.* **99** 257001
[9] Brandt E H 1999 *Phys. Rev. B* **60** 11939
[10] Prozorov R and Giannetta R W 2006 *Supercond. Sci. Technol.* **19** R41
[11] Ohkuni H *et al* 1999 *Philo. Mag.* B **79** 1045
[12] Okazaki R, Kasahara Y, Shishido H, Konczykowski M, Behnia K, Haga Y, Matsuda T D, Onuki Y, Shibauchi T and Matsuda Y 2008 *Phys. Rev. Lett.* **100** 037004
[13] Ichioka M, Matsunaga Y and Machida K 2005 *Phys. Rev. B* **71** 172510

Anomalous pressure effect on the superconducting transition temperatures and critical fields of $Ho_{1-x}Y_x Ni_2B_2C$

Gendo Oomi[1,2], Yukio Sakikawa[2], Mio Ohmura[2], Masashi Ohashi[3], P.C.Canfield[4] and B.K.Cho[5]
[1]Department of Education and Creation Engineering, Kurume Institute of Technology, Kamitsu-machi, Kurume, Fukuoka 830-0052, Japan
[2] Department of Physics, Kyushu University, Fukuoka 812-8581, Japan
[3] Faculty of Environmental Design, Kanazawa University, Kanazawa 920-1192, Japan
[4] Ames Labo.,Iowa State Univ., Ames, IA 50011,USA
[5] Center for Frontier Materials and Development of Materials Science and Engineering, Gwangju Institute of Science and Technology, Gwangju, South Korea

E-mail: geomi@cc.kurume-it.ac.jp

Abstract. The superconducting transition temperatures T_C and upper critical fields H_{C2} of borocarbides $Ho_{1-x}Y_x Ni_2B_2C$ have been measured under high pressure up to 3 GPa. It is found that T_C of these compounds decreases with increasing pressure but the pressure derivatives have a maximum near $x = 0.7$,where the magnetic ordering temperature T_N becomes 0. The pressure coefficients of H_{C2} show also a maximum near x=0.7. The possible origins of these anomalies are briefly discussed in connection with the instability of antiferromagnetism.

1. Introduction

The borocarbides RNi_2B_2C (R: rare earth element) have been studied extensively because the antiferromagnetism(AFM) coexists with superconductivity(SC)[1]. It is a typical example to explore a new origin of superconductivity. The structure of this material is a variant of the $ThCr_2Si_2$ type, which consists of alternating layers of RC planes and Ni_2B_2 slabs. Among these compounds, the most interesting material is R=Ho, in which the superconductivity is observed around T_C = 9K, reenters normal conducting state near 5K due to antiferromagnetic ordering around T_N = 5K and then shows a reentrant superconducting behavior below 5K. The interplay of AFM and SC is the most significant in this material because of the large value of T_N/T_C of about 0.63[2]. It has been revealed that as pressure increases, T_C decreases but T_N increases[3,4].

In the system of $Ho_{1-x}Y_x Ni_2B_2C$ the ratio is easily controlled by changing x because the sample of x = 1 is superconductor with T_C =16K and the T_N decreases with x followed by a disappearance of AFM around x = 0.6 [5]. The effect of pressure on the T_C reflects the stability of superconductivity: the value of pressure coefficient of T_C of x = 0 is extremely larger than that of x = 1[6,7], which indicates that the interplay of AFM and SC is closely related to the magnitude of pressure coefficient.

In the present work we attempted to measure the effect of pressure on the T_C and the upper critical field H_{C2} of $Ho_{1-x}Y_x Ni_2B_2C$ by measuring the electrical resistivity under high pressure and high magnetic field in order to clarify the interplay of AFM and SC in this system. Anomalous pressure effect on the T_C and H_{C2} near x=0.7 is found and the possible origins will be briefly discussed.

Published under licence by IOP Publishing Ltd

2. Experimental procedure

Single crystals were grown by a Ni_2B flux method. The details of sample preparation and characterization were reported previously [8]. High pressure was generated by using a piston-cylinder method up to 3 GPa. The pressure was kept constant throughout the experiment by controlling the load of hydraulic press. The electrical resistance was measured by using standard four-probe method and the current was in the *ab* plane. The magnetic field was generated by using a superconducting magnet and applied parallel to the *ab* plane. The details of present high pressure apparatus were published elsewhere [9].

3. Results and discussion

3.1. Effect of pressure on the T_C

Figure 1 shows the T_C as a function of Y concentration x at ambient pressure. In this figure the T_N is also shown for comparison [5]. It is found that the T_C increases almost linearly with x while T_N decreases with x and seems to disappear around $x = 0.7$. This result indicates that the SC competes with AFM, i.e., the SC is stabilized with increasing x although the AFM becomes unstable. The T_C of all samples decreases with increasing pressure but the decreasing rate depends on x: the value of $\partial T_C/\partial P$ of $x = 0$ is about -0.4K/GPa while that of x=1 is almost 0 [4,7].

Figure 2 shows the pressure dependence of T_C as a function of x. It is found that the magnitude of $-\partial T_C/\partial P$ decreases with increasing x below $x = 0.5$ but increases quickly above that followed by a peak around $x = 0.7$, i.e., $-\partial T_C/\partial P$ does not show a linear dependence against x such as T_C. As mentioned above, the value of x showing the peak agrees well with that where the AFM disappears. In other words, the peak observed in the concentration dependence of Y is closely related with the disappearance of AFM.

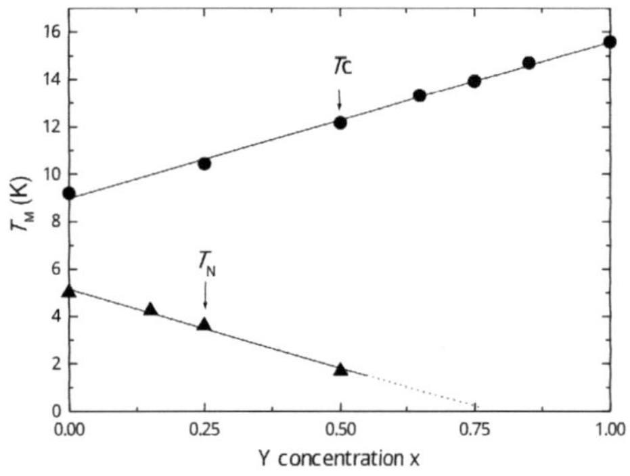

Figure 1. T_C and T_N at ambient pressure as a function of x. T_N was referred from ref.5.

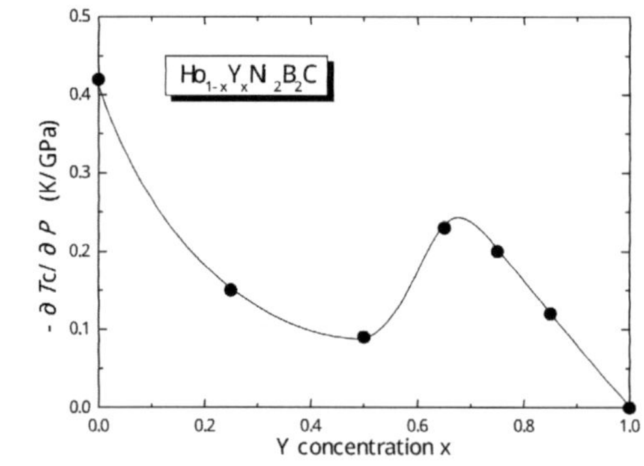

Figure 2. Pressure derivatives of T_C as a function of x.

3.2 Effect of pressure on the upper critical fields H_{C2}

Figure 3 shows the upper critical field H_{C2} at 4.2K as a function of x at ambient pressure. The H_{C2} of $x = 0$ at 4.2K is 0.2T and that of x=1 is larger than 5T, which is not shown in Fig.3. H_{C2} increases with x, i.e., the sign of $\partial H_{C2}/\partial x$ is positive. The increasing rate of H_{C2} changes between x=0.6 and 0.7: the value of $\partial H_{C2}/\partial x$ above 0.6 is about 5.5 times larger than that below 0.5. The concentration $x = 0.6$ showing the anomaly mentioned above is nearly the same as that showing the disappearance of AFM as illustrated in Fig.1.

Next we consider the pressure dependence of H_{C2}. It was reported [3,4,7] that the H_{C2} of $x = 0$ at 7K decreases with pressure having a rate, $\partial H_{C2}/\partial P$=-0.14 T/GPa, which is different from that of the re-entrant SC phase below 5K[2,6,10]. Figure 4 shows the values of $\partial H_{C2}/\partial P$ at 4.2K as a function of x

(>0.2). It is found that the value increases with increasing x and shows a peak around x = 0.7, where the AFM disappears as shown in Fig.1.

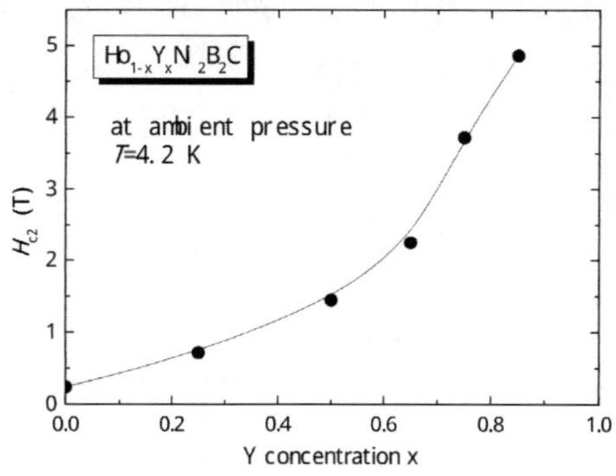

Figure 3. The upper critical field H_{C2} at 4.2K as a function of x.

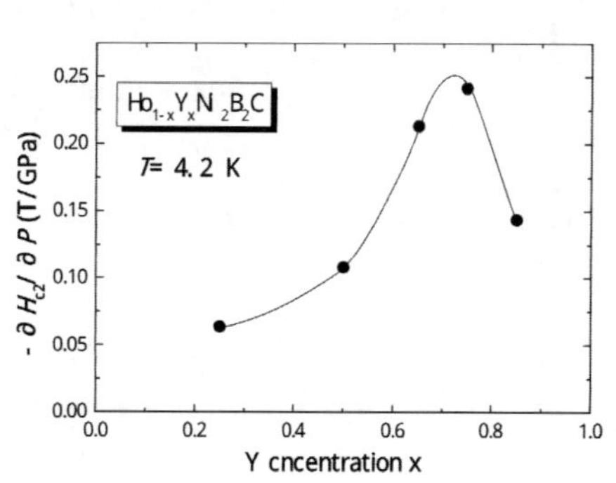

Figure 4. Pressure derivatives of H_{C2} at 4.2K.

From these results, it is suggested that the instability of AFM state is closely related with the pressure dependence of T_C and H_{C2} as was mentioned in 3.1. Near x = 0.7, a large antiferromagnetic fluctuation may be induced at low temperature. The application of pressure is expected to suppress the fluctuation effect, which gives rise to a large pressure effect on the electronic and magnetic properties in this system. This consideration explains qualitatively the anomalies near x = 0.7 observed in the present work.

Finally we comment the lattice properties of this system. It should be noted that the x dependence of the unit cell volume and the compressibility in $Ho_{1-x}Y_x Ni_2B_2C$ show an anomaly near x=0.7, which suggests that some kind of lattice instability exists near 0.7[11]. The peak in the pressure derivatives of T_C and H_{C2} may be partly due to this anomaly. We pointed out previously the close relation between superconductivity and the compressibility in high T_c superconductors and UGe_2[12,13,14]. However in order to clarify this point, more information about lattice properties of this system such as thermal expansion measurements under high pressure are highly desired.

4. Coclusions

In this work we have measured the electrical resistance of ternary borocarbides $Ho_{1-x}Y_x Ni_2B_2C$ under high pressure and magnetic fields. The main results are summarized as follows.

1) The superconducting transition temperatures of this system decrease with increasing pressure. The pressure derivatives show a peak around x = 0.7.
2) The upper critical fields H_{C2} increase with x while the increasing rate changes around x = 0.7.
3) The pressure dependence of H_{C2} shows a peak around x = 0.7 same as that of T_C.
4) These results indicate that the antiferromagnetic fluctuation plays an important role in the behaviour of the superconducting properties under high pressure.

References

[1] Canfield P C, Gammel P L, Bishop D J, 1998 Phys. Today **51** 40
[2] Oomi G, Masaoka D, Kagayama T, Kuroda N, Cho B K, Canfield P C, 2003 Physica C **388-389** 177
[3] Uwatoko Y, Oomi G, Canfield P C, Cho B K, 1996 Physica B **216**, 329
[4] Akiyama H, Kaji S, Oomi G, Cho B K, Canfield P C, 2006 J.Alloys and Compd. **408-412** 226
[5] Eversmann K, Handstein A, Fuchs G, Cao L, Müller K -H, 1996 Physica C **266**, 27
[6] Oomi G, Matsuda N, Kagayama T, Cho B K, Canfield P C, 2003 Intern.J.Mod.Phys.B **17** 3664

[7] Oomi G, Takeya H, Kadowaki K, 1998 Rev.High Pressure Sci.Technol. **7** 592
[8] Canfield P C, Cho B K, Johnston D C, Finnemore D K, Hundley M F, 1994 Physica C **230**, 397
[9] Honda F, Kaji S, Ohashi M, Oomi G, Eto T, Kagayama T, 2002 J. Phys.: Condens. Matters **14** 11501
[10] Oomi G, Kagayama T, 1998 Materia Japan **37** 397 (in Japanese)
[11] Sakigawa Y, 2006 Master Thesis, Department of Physics, Kyushu University
[12] Suenaga K, Oomi G, 1991 J. Phys. Soc. Jpn **60** 1189
[13] Kagayama T, Oomi G, 1999 J. Materials Processing Technology **85** 229
[14] Oomi G, Ohashi M, Honda F, Haga Y, Onuki Y, 2003 J. Phys.: Condens. Matters **15** S2039

Possible indicators for low dimensional superconductivity in the quasi-1D carbide Sc_3CoC_4

E–W Scheidt, C Hauf, F Reiner, G Eickerling, and W Scherer
CPM, Institut für Physik, Universität Augsburg, 86159 Augsburg, Germany
E-mail: Ernst-Wilhelm.Scheidt@physik.uni-augsburg.de

Abstract. The transition metal carbide Sc_3CoC_4 consists of a quasi-one-dimensional (1D) structure with $[CoC_4]_\infty$ polyanionic chains embedded in a scandium matrix. At ambient temperatures Sc_3CoC_4 displays metallic behavior. At lower temperatures, however, charge density wave formation has been observed around 143 K which is followed by a structural phase transition at 72 K. Below $T_c^{onset} = 4.5$ K the polycrystalline sample becomes superconductive. From $H_{c1}(0)$ and $H_{c2}(0)$ values we could estimate the London penetration depth ($\lambda_L \cong 9750$ Å) and the Ginsburg-Landau (GL) coherence length ($\xi_{GL} \cong 187$ Å). The resulting GL-parameter ($\kappa \cong 52$) classifies Sc_3CoC_4 as a type II superconductor. Here we compare the puzzling superconducting features of Sc_3CoC_4, such as the unusual temperature dependence i) of the specific heat anomaly and ii) of the upper critical field $H_{c2}(T)$ at T_c, and iii) the magnetic hysteresis curve, with various related low dimensional superconductors: e.g., the quasi-1D superconductor $(SN)_x$ or the 2D transition-metal dichalcogenides. Our results identify Sc_3CoC_4 as a new candidate for a quasi-1D superconductor.

1. Introduction

Low dimensional superconductivity is a vital and controversial research topic in solid sate physics and chemistry. However, beside the well-established examples for quasi-one-dimensional (1D) superconductors, e. g., polysulfur nitride $(SN)_x$ [1], the organic Bechgaard salts [2] or the transition-metal trichalcogenides [3], there are only a few systems known which allow us to study the nature of the phenomenon in greater detail. Recently, we observed superconductivity in Sc_3CoC_4 which might represent a new benchmark system of a quasi-1D superconductor [4]. Sc_3CoC_4 consists of quasi-one-dimensional $[CoC_4]_\infty$ polyanionic chains embedded in a scandium matrix [5]. From specific heat, magnetic susceptibility, resistivity and X-ray diffraction measurements a charge density wave around 143 K and a structural phase transition below 72 K were identified [4]. Therefore, it is very natural to ask if the superconductivity of this quasi-1D carbide is also of low dimensional character. In the absence of a single crystal, we compare in this paper i) structural features and physical properties like ii) the specific heat anomaly of the superconducting transition, iii) the magnetization behavior, and iv) the temperature dependence of the upper critical field with the established low dimensional superconductors like polysulfur nitride $(SN)_x$, the Bechgaard salts, high-T_c-cuprates, and the dichalcogenide $NbSe_2$ [7].

2. Structural details

High resolution single crystal X-ray diffraction studies at room temperature provide precise structural parameters of Sc_3CoC_4 (space group Immm) [5]: the Co atoms are coordinated by four

Published under licence by IOP Publishing Ltd

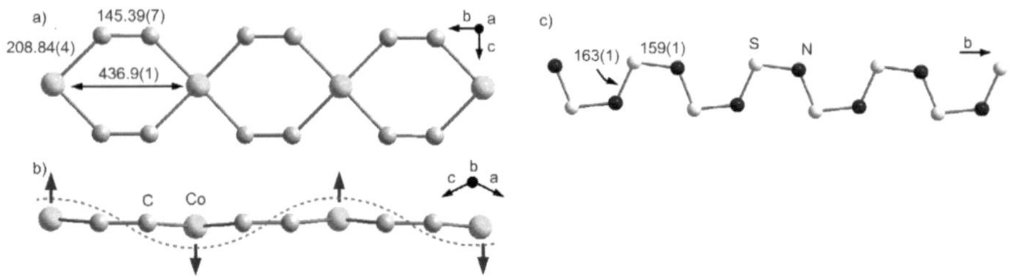

Figure 1. The $[CoC_4]_\infty$ ribbons of Sc_3CoC_4 a) in the crystallographic b,c plane and b) in the low temperature modification (side view) [4]. c) The $[SN]_\infty$ ribbons of the polysulfur nitride $(SN)_x$ showing a zigzag pattern with alternating short and long S-N distances [8]. Selected bond distances are given in pm.

C_2 pairs in an almost square planar manner, establishing one-dimensional infinite $[CoC_4]_\infty$ chains as pictured in Fig. 1a. Below the structural phase transition at 72 K Sc_3CoC_4 is characterized by an alternating displacement of the Co atoms above and below the $[CoC_4]_\infty$ ribbons (Fig. 1b, space group $C\,2/\mathrm{m}$). Such a zigzag chain deformation pattern is also observed in quasi-1D superconductors like polysulfur nitride $(SN)_x$ (Fig. 1c)[8], or the organic Bechgaard salts [9]. Therefore, the zigzag distortion of the $[CoC_4]_\infty$ ribbons might be a structural prerequisite of the low-dimensional superconductivity, even though the zigzag pattern in Sc_3CoC_4 is only weakly pronounced in comparison to $(SN)_x$. However, for the non superconducting analogues Sc_3NiC_4 and Sc_3FeC_4 no structural phase transition was observed above 2 K [4].

3. Experimental results and discussion

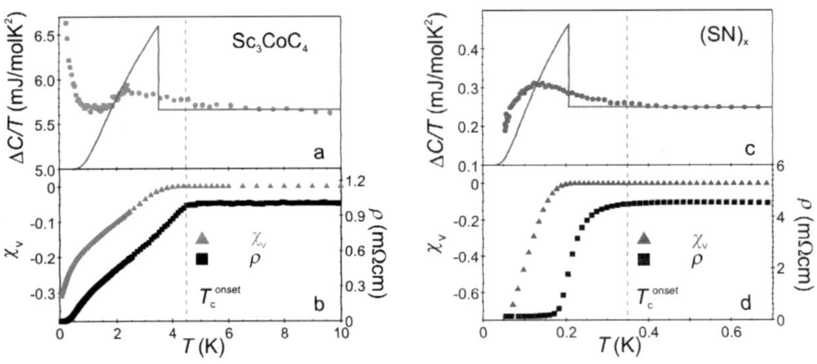

Figure 2. Temperature dependence of a) the electronic contribution of the specific heat divided by temperature, $\Delta C/T$, and b) the volume susceptibility, χ_V, at $H = 1$ Oe (left scale; triangles) and the electrical resistivity, ρ, (right scale; squares) for Sc_3CoC_4 in comparison with the temperature dependence of the same physical properties, c) $\Delta C/T$ and d) χ_V, ρ for the quasi-1D superconductor $(SN)_x$. The solid lines represent the superconducting transition as it is predicted by the BCS model [10].

The low temperature electronic specific heat of Sc_3CoC_4, shown as $\Delta C/T$ vs. T plot in Fig. 2a, reveals a broad superconducting transition below $T_c^{onset} = 4.5$ K which was determined by resistivity measurements (Fig. 2b). Additionally, this transition is indicated by the magnetic susceptibility showing a lowering of the magnetic response just below 4.4 K. These results and the fact, that the volume susceptibility at 50 mK approaches -0.3, identifies Sc_3CoC_4 as a bulk superconductor (Fig. 2b). The steep increase of $\Delta C/T$ below 1 K is mainly due to hyperfine interactions caused by the nuclear quadrupole moments of the Sc and Co atoms.

For the superconducting transition in Sc_3CoC_4 we calculated the expected ideal BCS curve of a weak coupled superconductor, taking into account the entropy balance between the normal and superconducting state (solid line in Fig. 2a) [10]. This calculation indicates, that only a small part ($\Delta\gamma_n = 0.65 \pm 0.05$ mJ/mol K^2) of the total electronic specific heat $\gamma_n = 5.7$ mJ/molK^2, which was graphically determined from a C/T vs. T^2 plot, can be involved in the development of the superconducting state. Therefore, the bulk superconductivity may be related solely to small areas of the Fermi-surface. Furthermore, the overall shape of the specific heat anomaly differs substantially from the BCS behavior, missing the characteristic pronounced jump at T_c.

The unusual broad shape of this anomaly and the fact, that only a small fraction (12 % in Sc_3CoC_4) of the band electrons contribute to the superconducting state is also observed in the well known quasi-1D superconductor polysulfur nitride $(SN)_x$ (Fig. 2c). Our specific heat measurement on a single crystal $(SN)_x$ sample indicates that 60 % of the total electronic specific heat ($\gamma_n = 0.25$ mJ/molK^2) is involved in the superconducting pairing process. The volume susceptibility at 70 mK approaches only a value of -0.66 of the full diamagnetic response (Fig. 2d). These analogies suggest that the unusual broad shape of the specific heat anomaly and the small percentage of band electrons contributing to the superconducting state may be characteristic features for the presence of quasi-one-dimensional superconductivity in Sc_3CoC_4.

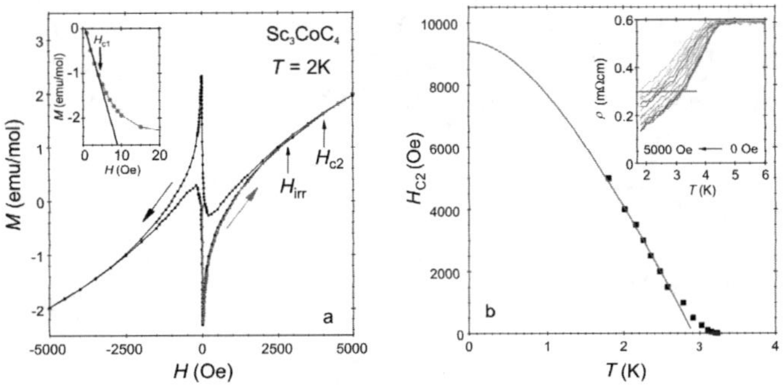

Figure 3. a) Magnetic hysteresis loop of Sc_3CoC_4 at 2 K. The underlying pronounced slope (M/H) and the deviation from the linear behavior of the magnetization is possibly due to the Pauli susceptibility of the normal conducting electrons and of a small impurity concentration of elementary Co, respectively. To determine H_{c1}, the insert displays an expanded region of the initial magnetization curve up to 20 Oe. b) Phase diagram of the upper critical fields $H_{c2}(T)$. The solid line corresponds to a WHH-fit [12]. $T_c(H_{c2})$ is obtained from the 50 % values of the normal state resistivity (solid line in the insert).

A further hint for low-dimensional superconductivity in Sc_3CoC_4 presents the magnetic hysteresis loop (Fig. 3a). This magnetization curve clearly indicates a type-II superconducting behavior. The characteristic dip near zero field crossings and the high reversibility between the irreversibility field ($H_{irr}(2K) = 2800 \pm 400$ Oe) and the upper critical field ($H_{c2}(2K) = 4000 \pm 200$) is also found in layered cuprate superconductors, but only at higher temperatures [6]. In Sc_3CoC_4 and also in the 2D highly anisotropic dichalcogenide superconductors like $NbSe_2$ [7, 11] such a reversible behavior above a threshold field occurs at very low temperatures and therefore might not only be due to granularity [6] but also due to the existence of low-dimensionality.

From $H_{c1} = 4.4 \pm 0.2$ Oe at 2 K, as determined by the deviation from the initial linearity of the magnetization curve (insert in Fig. 3a), the lower critical field $H_{c1}(0)$ is estimated to be 7 ± 0.4 Oe. $H_{c2}(0) = 94000 \pm 500$ is calculated from the WHH-theory [12], with a Maki parameter of $\alpha = 0.25$ estimated from the slope of the upper critical field at T_c and the spin-orbit parameter λ_{so} assumed to be zero. By combining the upper and lower critical field values with the Ginsburg Landau (GL) expressions for isotropic superconductors ($H_{c2} = \Phi_0/(2\pi\xi_{GL}^2)$; $H_{c1} = (\Phi_0/(4\pi\lambda_L^2))(0.08 + ln\kappa)$) we obtained the London penetration depth of $\lambda_L = 9750 \pm 300$ Å and the GL-coherence length of $\xi_{GL} = 187 \pm 5$ Å. The resulting GL parameter $\kappa = 52$ classifies Sc_3CoC_4 as a type-II superconductor.

A further interesting effect is the upturn of $H_{c2}(T)$ at temperatures close to T_c (Fig. 3b). Beside other explanations, e.g., sample imperfections, this effect is mainly attributed to a reduction of the dimensionality. This upturn is also found in the quasi-1D superconductors $(SN)_x$ [13], the Bechgaard salts [14] and as R. A. Klemm summarised: "seems to be a general feature of many, but not all highly anisotropic materials" [15].

4. Summary

Resistivity, magnetization and specific heat measurements clearly support the presence of bulk superconductivity in the quasi-one-dimensional transition metal carbide Sc_3CoC_4. Several characteristic features might identify Sc_3CoC_4 as a promising system in the rare class of quasi-one-dimensional superconductors: i) the structural zigzag chain, ii) the broad shape of the specific heat anomaly, iii) the small percentage of band electrons contributing to the superconductivity state, iv) the partially reversibility of the magnetic hysteresis loop, and v) the upturn of the $H_{c2}(T)$ at temperatures near T_c. However, future work on single crystalline samples of Sc_3CoC_4 is essential to analyze in detail the electronic anisotropy in both, the normal and the superconducting state.

5. Acknowledgments

We acknowledge valuable discussion with K. Lüders. This work was supported by the Deutsche Forschungsgemeinschaft (SPP1178). We thank J. Passmore for providing a $(SN)_x$ sample.

6. References

[1] Lou L F 1989*J. Appl. Phys.* **66**, 979
[2] Jérome D and Schulz H J 1982 *Adv. Phys.* **31** 299
[3] Srivastava S and Avasthi B N 1992 *J. Materials Science* **27** 3693
[4] Scherer W, Hauf C, Presnitz M, Scheidt E–W, Eickerling G, Eyert V, Hoffmann R–D, Rodewald U, Hammerschmidt A, Vogt C and Pöttgen R 2010 *Angew. Chem. Int. Ed.* **49** 1578
[5] Rohrmoser B, Eickerling G, Presnitz M, Scherer W, Eyert V, Hoffmann R–D, Rodewald U, Vogt C and Pöttgen R 2007 *J. Am. Chem. Soc.* **129** 9356
[6] Senoussi S, Oussena M and Hadjoudj S 1988 *J. Appl. Phys.* **63** 4176
[7] Soto F, Berger H, Cabo L, Carballeira C, Mosqueira J, Pavuna D and Vidal F 2007 *Phys. Rev.* B **75** 094509
[8] Cohen M J, Garito A F, Heeger A J, MacDiarmid A G, Mikulski C M, Saran M S and Kleppinger J 1976 *J. Am. Chem. Soc.* **98** 3844
[9] Thorup N, Rindorf G, Soling H and Bechgaard B 1981 *Acta. Cryst.* **B37** 1236
[10] Mühlschlegel B 1959 *Z. Phys.* **155** 313
[11] Sonier J E, Hundley M F, Thompson J D and Brill J W 1999 *Phys. Rev. Lett.* **82** 4914
[12] Werthamer N R, Helfand E and Hohenberg P C 1966 *Phys. Rev.* **147** 295
[13] Azevedo L J, Clark W G, Deutscher G, Greene R L, Street G B and Sutert L J 1976 *Solid State Comm.* **19** 197
[14] Lee I J, Naughton M J, Danner G M and Chaikin P M 1997 *Phys. Rev. Lett.* **78** 3555
[15] Klemm A R, 1985 *Electronic Properties of Inorganic Quasi-One-Dimensional Compounds Part I*, (Eds.: Reidel D) (Publishing Company, Dordrecht/Boston/Lancaster) 205

Magnetization in the Superconducting Mixed State of the Heavy-Fermion Compound UBe_{13}

Yusei Shimizu[1,†], Yoichi Ikeda[1,5], Takumi Wakabayashi[1], Kenichi Tenya[2], Yoshinori Haga[3], Hiroyuki Hidaka[1], Tatsuya Yanagisawa[4], and Hiroshi Amitsuka[1]

[1]Graduate School of Science, Hokkaido University, Sapporo 060-0810, Japan.
[2]Faculty of Education, Shinshu University, Nagano 385-8544, Japan.
[3]Advanced Science Research Center, Japan Atomic Energy Agency, Tokai 319-1184, Japan.
[4]Creative Research Institution, Hokkaido University, Sapporo 001-0021, Japan.
[5]Present Address: Graduate School of Natural Science and Technology, Okayama University, Okayama 700-8530, Japan.

E-mail: †`yusei@phys.sci.hokudai.ac.jp`

Abstract.
Static dc magnetization measurements in the superconducting mixed state of a single crystal UBe_{13} along the [110] axis were performed by means of a capacitance Faraday-force method down to 0.24 K. Below the upper critical eld B_{c2}, not only a peak effect but also an additional broad anomaly have been observed in magnetization curves. We report superconducting phase diagram of UBe_{13} obtained by our magnetization measurements, including these anomalies.

1. Introduction

The heavy-fermion compound UBe_{13} (cubic, $O_h^6(Fm3c)$) has attracted much attention since the discovery of non-BCS type superconductivity (SC) 27 years ago [1]. The SC transition of this compound takes place at about 0.85 K with a large specific-heat jump C, which is the same order of the electronic specific heat $C \sim 1$ J/Kmol observed just above T_c. This indicates that the heavy electrons form the Cooper pairs which condense into the SC state. In order to understand the unconventional SC of UBe_{13}, it is very important to clarify the symmetry of the Cooper pairs and investigate the SC phase diagram precisely. First, as for the SC symmetry, an early specific-heat study made by Ott *et al.* [2] has suggested a p-wave Anderson-Brinkman-Morel (ABM) state from T^3 behavior of the specific heat well below T_c, because it is similar to the behavior observed in the ABM state of ^{3}He superfluidity. NMR spin-relaxation rates [3] and ultrasonic velocities [4] have also suggested that the SC gap has a nodal structure. These results indicate that there is a line node in the SC gap, whereas the ABM state has point nodes. In addition, it remains also unclear whether the spin state of a Cooper pair is singlet or triplet. A recent $^+$-Knight shift study [5] in $B||[001]$ has suggested that the spin susceptibility decreases below T_c, indicating a spin-singlet pairing or a spin-triplet pairing with a strong spin-orbit coupling. On the other hand, a recent NMR study [6] has suggested that the Knight shift for $B||[001]$ shows no decrease below T_c, which indicates the spin-triplet pairing. In this way, no clear experimental consensus has been obtained about not only the SC gap structure but also the Cooper pairing state in UBe_{13}. Second, it is noted that the SC phase diagram of UBe_{13} is quite

Published under licence by IOP Publishing Ltd

different from that of the conventional BCS superconductors. One of the unusual properties is the upper critical field curve $B_{c2}(T)$ that exhibits a clear upturn at low temperatures [7]. Moreover, low-T specific-heat and thermal-expansion measurements by Kromer *et al.* show a new anomaly at T_L below T_c for pure UBe_{13} in zero field. In field scans of $C(B, T = \text{const.})$, this anomaly corresponds to $B^*(T)$. They have proposed that this anomaly is a precursor of another phase transition observed in the $U_{1-x}Th_xBe_{13}$ system at T_{c2} [8], because it seems that $T_L(x)$ smoothly merges into $T_{c2}(x)$ on a T-x phase diagram [9, 10].

In order to obtain further information about the unusual SC of UBe_{13}, we have performed low-temperature dc magnetization measurements in the SC mixed state of UBe_{13} using a high-quality single crystal.

2. Experimental Procedure

A single crystal of UBe_{13} was grown by an Al-flux method, and its weight was 6.6 mg. The SC transition temperature is about 0.8 K, which is defined by the peak-top temperature of the specific-heat jump for zero field. The dc magnetization measurements along [110] were performed by using a capacitive Faraday-force magnetometer in a ^{3}He-^{4}He dilution refrigerator [11] at temperatures down to 0.24 K and in fields up to 6.9 T. Magnetization processes were measured in a field gradient of 500-900 Oe/cm applied in addition to the uniform magnetic field.

3. Results and Discussion

Figure 1 shows the magnetization process of UBe_{13} observed for $B||[110]$ at 0.24 K up to 6.9 T. Below $\sim$ 2 T, a clear irreversibility attributed to a vortex flux pinning is observed. On the other hand, the irreversibility decreases with increasing field above $\sim$ 2 T. The smallness of this hysteresis in high field indicates a high-quality of the used sample, since flux-pinning centers are impurities or lattice defects in general. Here, we define B_{c2} as the field where the irreversibility vanishes completely. The linear component above B_{c2} is the normal state magnetization M_n. Just below B_{c2}, a peak effect, which is occasionally seen in type-II superconductors, is observed. We define the onset of the peak as B_{peak}. Figure 2 shows hysteresis magnetization curves M measured at various temperatures: $M=(M_{dec}-M_{inc})/2$, where M_{dec} and M_{inc} are the magnetization of decreasing- and increasing-field processes, respectively. As temperature

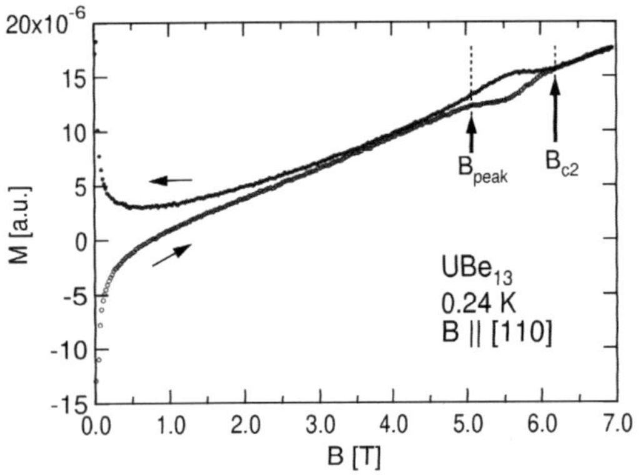

Figure 1. Magnetization curves of the single crystalline UBe_{13} at 0.24 K ($B||$ [110]). The upper critical field B_{c2} is pointed by the arrow. B_{peak} denotes the onset of the peak effect.

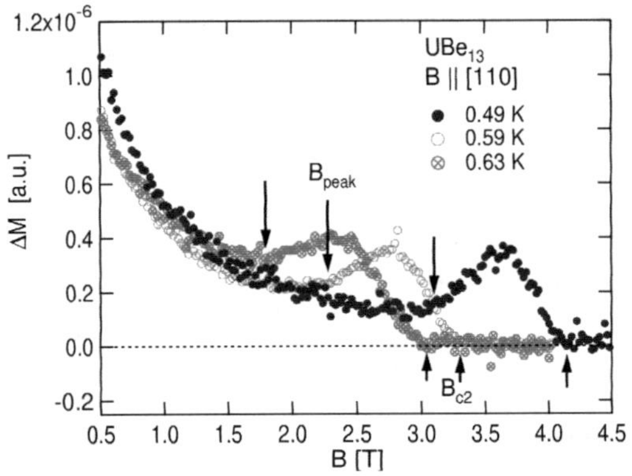

Figure 2. The hysteresis magnetization M of UBe_{13} in various temperatures ($B||[110]$). Down and up arrows indicate B_{peak} and B_{c2}, respectively.

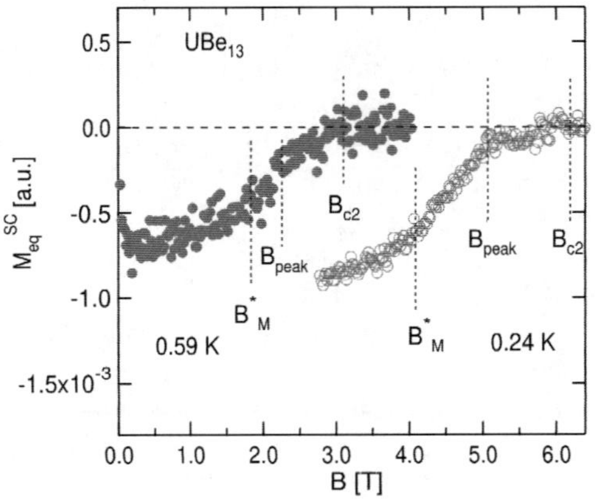

Figure 3. SC diamagnetic magnetization M_{eq}^{SC} of UBe_{13} at 0.24 K and 0.59 K ($B||[110]$). Below B_{peak}, the broad anomaly B_M^* is observed at each temperatures.

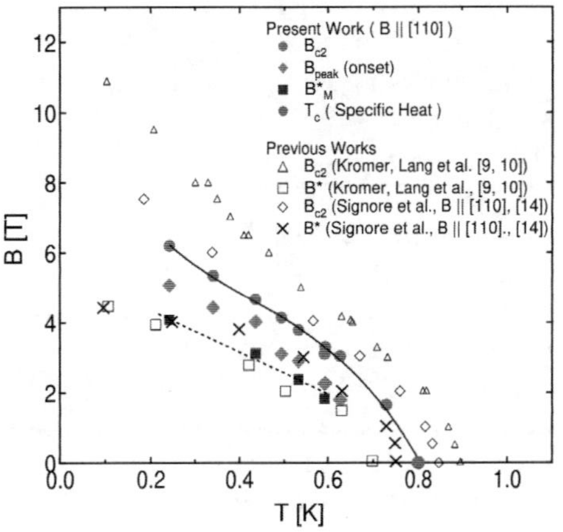

Figure 4. SC phase diagram of UBe_{13} for $B||[110]$, including our specific-heat results and the previous studies [9, 10, 14]

.

increases, the peak becomes indistinctive due to the interference with a large hysteresis in the low-field region. The peak effect has been observed also in other heavy-fermion superconductors such as UPt_3 and $CeCoIn_5$ [12, 13]. In particular, UPt_3 shows the peak effect just below B_{c2}, which is similar to the behavior of UBe_{13}.

Since the hysteresis of the magnetization is sufficiently small in the high-field region, we can approximately obtain the thermal equilibrium magnetization M_{eq} in the mixed state by averaging the increasing- and decreasing-field processes. The M_{eq} consists of the normal state component M_n and the diamagnetic magnetization component M_{eq}^{SC}: $M_{eq} = M_n + M_{eq}^{SC}$. We estimated M_n in the mixed state by extrapolating the normal state magnetization above B_{c2} to the orgin, assuming the linear region: $M_n = \chi_n B$. Figure 3 shows the M_{eq}^{SC} obtained at 0.24 K and 0.59 K. For the M_{eq}^{SC} at each temperatures, there are two bends below B_{c2} (at ~ 4 T and ~ 5 T, for 0.24 K). First, we note that M_{eq}^{SC} bends at a field clearly lower than B_{peak}. This anomaly is broad but distinctive. We define this broad anomaly as B_M^*. At nearly the same field ~ 4 T, the specific heat and thermal expansion measurements [9, 10] have suggested that there is an anomaly at B^* below B_{c2}, as mentioned in Introduction. Ac-susceptibility (ac-χ) also shows an anomaly at ~ 4 T for $T \sim 0.25$ K [14]; we call this anomaly B_{ac}^* in order to distinguish it from B^* obtained by Kromer *et al.* Second, we note another bend observed in M_{eq}^{SC} near B_{peak} (at ~ 5 T for 0.24 K). This behavior, however, does not necessarily indicate an occurrence of a phase transition: if there is a slight deference in the magnitude of the magnetization peaks between increasing- and decreasing-field processes, then the bend may appear, except for a phase transition. Therefore, this bend at B_{peak} in M_{eq}^{SC} may not be intrinsic.

Figure 4 represents the obtained SC phase diagram of UBe_{13} for $B||[110]$. In the Figure 4, the previouslly reported data, [9, 10, 14], are also plotted for comparison. It has been reported that there is no anisotropy in B^* [9]. Therefore, although magnetic field direction is not shown in refs. [9] and [10], we may compare B_M^* obtained by our measurements with the data of refs. [9] and [10]. It can be seen that the obtained $B_M^*(T)$ data are in good agreement with the $B^*(T)$ values . Next, we compare our result with the result of ref. [14]. Below ~ 0.3 K, B_{ac}^* is consistent with B_M^*. However, above ~ 0.4 K, the bahavior of B_{ac}^* disagrees with that of B_M^*. As seen in Figure 4, B_M^* gets closer to B_{peak} above ~ 0.5 K. In fact, above ~ 0.6 K, we could not distinguish $B^*_{\rm M}$ from $B_{\rm peak}$. Therefore, this disagreement may be caused by the difficulty of distinguishing $B^*_{\rm M}$ from $B_{\rm peak}$ above $\sim$ 0.5 K, and $B^*_{\rm ac}$ above $\sim$ 0.5 K may correspond to $B_{\rm peak}$. Anyway, all works have suggested that there is at least an anomaly in the SC phase diagram in the same field range $B \lesssim 4.5$ T.

On the other hand, the scale of $B_{\rm c2}$ obtained from our measurements disagree with that in [9, 10] and [14]. Here, we can guess that there is little anisotropy of $B_{\rm c2}$, because of the lack of anisotropy at least between $B||[100]$ and $B||[110]$, which has been suggested from the ac-χ mearsurements [14]. In particular, although $B^*_{\rm M}$ is almost consistent with B^* obtained by Kromer *et al.*, the scale of the $B_{\rm c2}(T)$ curve obtained for the present sample is obviously smaller than that of refs. [9] and [10]. If $B^*_{\rm M}$ and B^* are due to the same orgin, the above comparison suggests that the scale of B^* is independent of that of $B_{\rm c2}$. This may indicate that the anomaly at B^* originates from some phenomenon which is not basically associated with the SC. As for the origin of B^*, the following two cases may be considered: (1) a some magnetic correlation which coexists with the SC [9]; (2) a flux-pinning mechanism [14]. If B^* is orginated from the case (1), the scale of B^* may not necessarily depend on that of $B_{\rm c2}$. In contrast, if B^* is orginated from the case (2), tempearature dependence of B^* would follow $B_{\rm c2}(T)$ as seen in the behavior of $B_{\rm peak}(T)$. Therefore, the proposal that B^* is orginated from the magnetic correlation coexisting with the SC might be constent with our results. In order to uncover the relationship between B^* and the unusual SC of UBe_{13}, further studies such as a check of the sample-quality dependence will be needed.

4. Summary

The superconducting phase diagram of a single-crystalline UBe_{13} for $B||[110]$ has been obtained by means of the static dc magnetization measurements down to 0.24 K. In addition to the peak effect just below the upper critical field $B_{\rm c2}$, we found a broad anomaly at $B^*_{\rm M}$, which is clearly below $B_{\rm peak}$. In contrast to the disagreement of the B_{c2} curves, the observed $B^*_{\rm M}$ values are nearly comparable to the fields where specific heat and thermal expansion show an anomaly.

Acknowledgements

This work was supported by Grant-in-Aid for Scientific Research B (19340086) and S (20224015) from the Ministry of Education, Culture, Sports, Science and Technology (MEXT), Japan.

References

[1] Ott H R, Rudigier H, Fisk Z, Smith J L 1983 *Phys. Rev. Lett.* **50** 1595
[2] Ott H R, Rudigier H, Rice T M, Ueda K, Fisk Z, Smith J L 1984 *Phys. Rev. Lett.* **52** 1915
[3] MacLaughlin D E, Tien C, Clark W G, Lan M D, Fisk Z, Smith J L, Ott H R 1984 *Phys. Rev. Lett.* **53** 1833
[4] Golding B, Bishop D J, Batlogg B, Haemmerle W H, Fisk Z, Smith J L, Ott H R 1985 *Phys. Rev. Lett.* **55** 2479
[5] Sonier J E, Heffner R H, Morris G D, MacLaughlin D E, Bernal O O, Cooley J, Smith J L, Thompson J D 2003 *Physica B* **326** 414
[6] Tou H, Tsugawa N, Sera M, Haga Y, Onuki Y 2007 *J. Magn. Magn. Mat.* **310** 706
[7] Rauchschwalbe U, Ahlheim U, Steglich F, Rainer D, Franse J J M 1985 *Z. Phys. B-Cond. Matt* **60** 379
[8] Ott H R, Rudigier H, Felder E, Fisk Z, Smith J L 1986 *Phy. Rev. B.* **33** 126
[9] Kromer F, Helfrich R, Lang M, Steglich F, Langhammer, Bach A, Michels T, Kim J S, Stewart G R 1998 *Phys. Rev. Lett.* **81** 4476
[10] Lang M, Helfrich R, Kromer F, Langhammer C, Steglich F, Stewart G R, Kim J S 1999 *Physica B* **259-261** 608
[11] Sakakibara T, Mitamura H, Tayama T, Amitsuka H 1994 *Jpn. J. Appl. Phys.* **33** 5067
[12] Tenya K, Ikeda M, Tayama T, Sakakibara T, Yamamoto E, Maezawa K, Kimura N, Settai R, Ōnuki Y 1996 *Phys. Rev. Lett.* **77** 3193
[13] Tayama T, Harita A, Sakakibara T, Haga Y, Shishido H, Settai R, Ōnuki Y 2002 *Phys. Rev. B.* **65** 180504
[14] Signore P J C, Andraka B, Stewart G R, Meisel M W 1995 *Phys. Rev. B.* **52** 10315

Journal of Physics: Conference Series **273** (2011) 012085 doi:10.1088/1742-6596/273/1/012085

Neutron scattering studies of the lattice expansion in a ferromagnetic superconductor UGe_2 under pressure.

D A Sokolov[1], R Ritz[2], C Pfleiderer[2], T Keller[3,4] and A D Huxley[1]

[1] Scottish Universities Physics Alliance, School of Physics, The University of Edinburgh, Edinburgh EH9 3JZ, UK

[2] Physik-Department E21, Technische Universität München, D-85748 Garching, Germany

[3] Max-Planck-Institute fur Festköperforchung, Heisenbergstr. 1, D-70569 Stuttgart, Germany

[4] Forschungsneutronenquelle Heinz Maier-Leibnitz (FRMII), Technische Universität München, D-85748 Garching, Germany

E-mail: dsokolov@staffmail.ed.ac.uk

Abstract. We report high-resolution measurements of the lattice constants of UGe_2 under pressure probed by a novel technique which utilises Larmor precession of polarised neutrons and surpasses the resolution of conventional scattering methods by two orders of magnitude. We confirm the presence of sharp anomalies in the lattice parameters at both the Curie and crossover temperatures at ambient pressure. We find that for pressure of 9.3 kbar the anomaly at the Curie temperature shifts to lower temperature in agreement with the known phase diagram. At 9.3 kbar, the pressure corresponding to an onset of superconductivity, the lattice expansion at the ferromagnetic transition is much stronger than at ambient pressure. The results indicate a complex evolution of the electronic structure of UGe_2 with pressure and suggest that magneto-elastic coupling is strengthened at the pressures at which superconductivity appears; magneto-elastic coupling therefore may play an important role in stabilising superconductivity.

1. Introduction

UGe_2 is the first material in which bulk superconductivity co-exists rather than competes with ferromagnetism[1, 2]. The schematic pressure-temperature phase diagram of UGe_2 is shown in Fig. 1. At ambient pressure UGe_2 orders at $T_C \approx 53$ K. Within the ferromagnetic state a crossover temperature $T_X \approx 30$ K separates two ferromagnetic states, FM1 (high pressure) and FM2 (low pressure) characterised by different values of the ordered moment. At low pressures the line of T_C is of second order, whereas T_X is a crossover. At pressure higher than ≈ 11.5 kbar the T_X line is a first order transition with a critical end point for this transition line at ≈ 11.5 kbar[2]. The electronic specific heat coefficient γ increases from ≈ 50 mJ/mole·K^2 to ≈ 100 mJ/mole·K^2 when approaching the FM2 to FM1 transition at zero temperature[3] suggesting that the unconventional superconductivity in UGe_2 arises from the strong spin fluctuations associated with the critical end point of T_X.

Since the ferromagnetism in UGe_2 is suppressed with pressure we expect a strong anomaly in the thermal expansion at T_C and also near the pressure driven transition between FM1 and FM2 states. Use of a novel neutron scattering method, Larmor diffraction allows us to measure

Published under licence by IOP Publishing Ltd

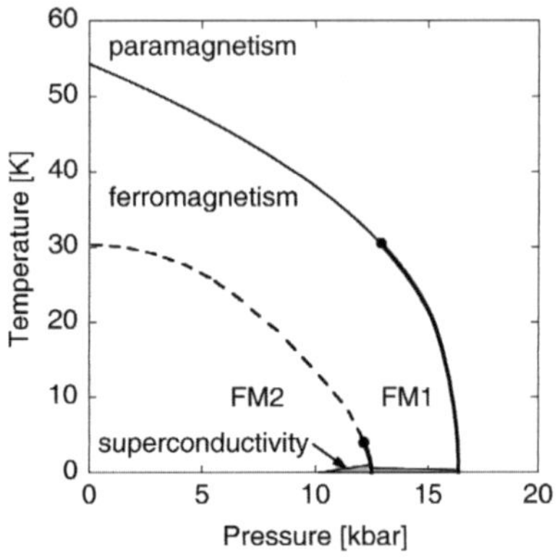

Figure 1. Schematic pressure-temperature phase diagram of UGe_2. Thin (thick) lines represent the second (first) order phase transitions. The dashed line marks the crossover regime and the dots mark the critical end points. The superconducting region is a dark area at the bottom.

the thermal expansion of two lattice constants under hydrostatic pressure conditions with the resolution of $\Delta d/d \sim 1\times10^{-6}$[4, 5]. Recently the potential of Larmor diffraction to study the thermal expansion and the distribution of the lattice constants of non-ferromagnetic systems has been demonstrated[6, 7] but our study is the first study of a ferromagnetic state.

2. Experiment

Single crystals of UGe_2 were grown using the Czochralski technique under a protective atmosphere of purified Ar to suppress the vapour pressure of Ge. Single crystals were oriented using white beam X-ray back scattering "Laue method" and sparkcut to fit into a pressure cell. The mosaic of the samples used was ≈1 degree (FWHM). Samples cut off the single crystals used in the neutron scattering measurements showed residual resistivity ratio, RRR≈70-120 and T_C=52.6 K indicating the good quality of the crystals. Pressure of up to 12 kbar was applied using Cu:Be piston-cylinder pressure cells with Fluorinert as the pressure transmitting medium. The pressure cells were the same used as used in previous studies[6, 7]. Neutron Larmor diffraction measurements were performed with the TRISP spectrometer at FRMII(Germany); details of the method are given in Refs.[6, 7]. The temperature and pressure dependence of all three lattice constants of the orthorhombic structure of UGe_2 were measured at (200), (060) and (002) positions in reciprocal space. The lattice expansion was measured on heating between 3 K and 75 K. The pressure was estimated at low temperatures from the measured temperature of the paramagnetic to ferromagnetic transition determined from the feature in the intensity of the (040) peak and known phase diagram of UGe_2[2, 8]. A demagnetisation procedure was applied to ensure equal populations of oppositely polarised ferromagnetic domains and to prevent depolarisation of the diffracted neutrons by the ordered moment of the sample.

3. Results and Discussion

At ambient pressure the lattice expansion in UGe_2 is quite anisotropic, Fig. 2. The lattice contracts on cooling as evidenced by the positive slope of $\Delta d/d$ at $T>T_C$. The temperature dependence of $\Delta d/d$ shows a sharp anomaly at T_C for *b* and *c* axes but merely a change of slope for the *a* axis, whereas the crossover regime in the ferromagnetic state at ≈30 K is marked by broad anomalies along *a* and *b* axes but not along *c* axis in a good quantitative agreement with results of strain gauge measurements[9]. The relative length along the *b* axis increases by $\approx5\times10^{-5}$ on cooling through T_X, while the change along *a* axis is of the opposite sign and of roughly the same magnitude. The linear thermal expansion coefficients α were obtained by differentiating $\Delta d/d$ with respect to T (not shown). At the ferromagnetic transition α along *a* increases by $\approx1\times10^{-6}$ K^{-1} in agreement with strain gauge and dilatometry measurements[9, 10].

The crossover region T_X manifests itself as a broad feature along *b* axis near 30 K. Observation of rather large ($\Delta d/d \approx 5\times10^{-5}$) feature at T_X indicates that the electronic structure of UGe_2 changes strongly when crossing between FM1 and FM2.

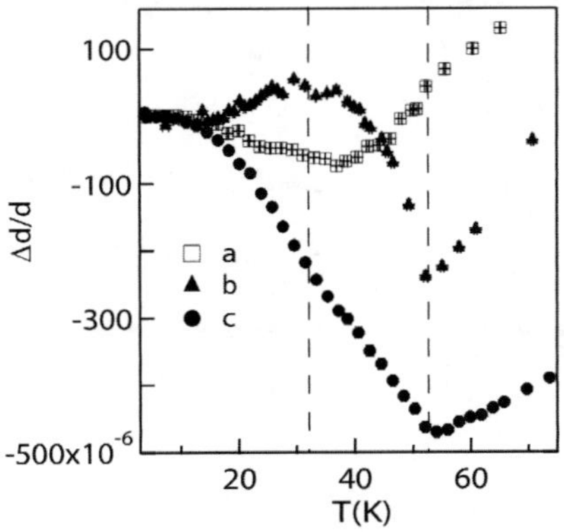

Figure 2. Temperature dependence of the linear lattice expansion along *a*, *b*, and *c* axes at ambient pressure. $\Delta d/d$ is relative to the lowest temperature measured (3 K). Error bars are the size of markers. Vertical dashed lines mark T_X and T_C.

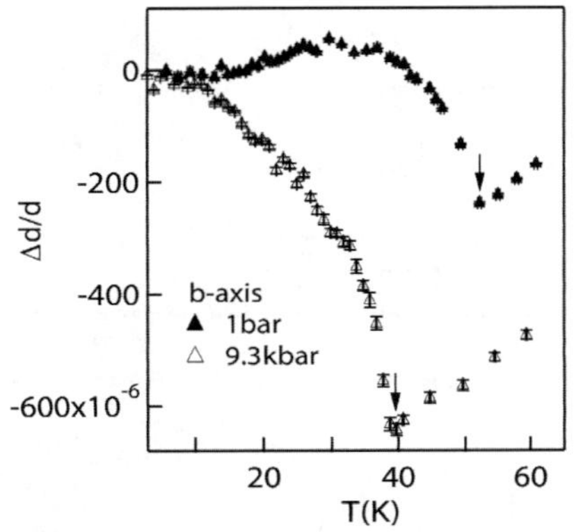

Figure 3. Temperature dependence of the linear lattice expansion along the *b* axis. $\Delta d/d$ is relative to the lowest temperature measured (3 K). Error bars are smaller than the size of the markers where not shown explicitly. Arrows mark T_C.

The results of the high pressure neutron Larmor diffraction measurements are shown in Fig. 3 for the *b* axis. As the pressure increases the anomaly at T_C moves to lower temperatures in agreement with the pressure-temperature phase diagram. Unexpectedly we find that the change in value of $\Delta d/d$ between 3 K and the transition temperature increases under pressure by a factor of $\sim$3, indicating strong magneto-elastic coupling in UGe_2 near the pressure at which superconductivity is induced. A broad maximum corresponding to T_X is no longer visible in the data at 9.3 kbar. This increase is bigger than that of γ and is consistent with an increase in the Grüneisen parameter as predicted approaching a quantum critical point[11]. Based on ambient pressure results we expected to see a stronger signature of T_X at higher pressure, however it appears that the transition at T_C dominates the lattice expansion.

4. Conclusions

We have measured the lattice expansion of UGe_2 under pressure and shown that the lattice expansion increases near the pressure at which the superconductivity is induced. Our results indicate that the magnetoelastic coupling in UGe_2 is strong and its electronic structure is quite sensitive to the variation of the lattice parameters. These are to our knowledge the first measurements that show Larmor diffraction can be applied in the ferromagnetic state.

5. Acknowledgments

We would like to thank P. Böni for pointing out that Larmor diffraction should be possible in Ising ferromagnets. Support by the team of FRMII is gratefully acknowledged. This research project has been supported by the European Commision under the 7th Framework Programme through the Research Infrastructures action of the Capacities Programme, Contract No:CP-CSA INFRA-2008-1.1.1 Number 226507-NMI3. D. A. S. and A. D. H. acknowledge support by a grant from EPSRC, UK (EP/E063985/1). R.R and C.P. acknowledge financial support in the framework of the research unit FOR 960 on Quantum Phase Transitions by the German Science Foundation DFG.

References

[1] Saxena S S, Agarwal P, Ahilan K, Grosche F M, Haselwimmer R K W, Steiner M J, Pugh E, Walker I R, Julian S R, Monthoux P, Lonzarich G G, Huxley A, Sheikin I, Braithwaite D and Flouquet J 2000 *Nature* B **406** 587
[2] Huxley A, Sheikin I, Ressouche E, Kernavanois N, Braithwaite D, Calemczuk R and Flouquet J 2001 *Phys. Rev.* B **63** 144519
[3] Tateiwa N, Kobayashi T C, Amaya K, Haga Y, Settai R and Ônuki Y 2004 *Phys. Rev.* B **69** 180513
[4] Keller T 2002 *Appl. Phys.* A **74** (suppl.) s332
[5] Rekveldt M T, Keller T and Golub R 2001 *Europhys. Lett.* A **54** 342
[6] Pfleiderer C, Böni P, Keller T, Rößler U K, Rosch A 2007 *Science* **316** 1871
[7] Niklowitz P G, Pfleiderer C, Keller T, Vojta M, Huang Y -K and Mydosh J A 2010 *Phys. Rev. Lett.* **104** 106406
[8] Pfleiderer C and Huxley A D 2002 *Phys. Rev. Lett.* **89** 147005
[9] Nishimura K, Oomi G, Yun S W and Ônuki Y 1994 *J. Alloys Comp.* **213/214** 383
[10] Hardy F, Meingast C, Taufour V, Flouquet J, v Löhneysen H, Fisher R A, Phillips N E, Huxley A and Lashley J C 2009 *Phys. Rev.* B **80** 174521
[11] Zhu L, Garst M, Rosch A and Si Q 2003 *Phys. Rev. Lett.* **91** 066404

Journal of Physics: Conference Series **273** (2011) 012086 doi:10.1088/1742-6596/273/1/012086

Magnetic-field modulation of the Josephson effect between polycrystalline $CeCu_2Si_2$ and Al

A Sumiyama, N Miyakawa, Y Ushida, G Motoyama, A Yamaguchi and Y Oda

Graduate school of Material Science, University of Hyogo, Ako-gun, Hyogo 678-1297, Japan

E-mail: sumiyama@sci.u-hyogo.ac.jp

Abstract. Josephson critical current I_c between a polycrystalline $CeCu_2Si_2$ and Al has been measured for the junctions on a $CeCu_2Si_2$ surface. A dominant maximum peak at zero magnetic field is observed in the magnetic field H dependence of I_c, suggesting that the Josephson currents flowing from variously-oriented crystallites of $CeCu_2Si_2$ are in phase at $H = 0$. If $CeCu_2Si_2$ is in the $d_{x^2-y^2}$-wave state, in which the sign of the order parameter changes according to the crystallographic orientation, an s-wave component induced on the surface is considered as a possible origin of the s-wave-like behavior of $I_c(H)$ patterns.

$CeCu_2Si_2$, which is the first heavy-fermion superconductor discovered in 1979[1], has attracted much attention in the recent years, since the superconducting transition occurs in the vicinity of a quantum critical point. As expected in superconductors close to an antiferromagnetic instability, a variety of physical properties in $CeCu_2Si_2$ suggest the d-wave superconductivity; the $d_{x^2-y^2}$ state is proposed to be the most reasonable candidate[2, 3]. Recently, another superconducting state has been found in the pressure-temperature phase diagram of $CeCu_2(Si_{1-x}Ge_x)_2$ and ascribed to a different pairing mechanism, because it occurs far away from the threshold of magnetism[4].

The Josephson effect between a conventional (s-wave) and an unconventional (non s-wave) superconductor gives direct information about the order parameter of the unconventional superconductor. For the high-T_c cuprates, the magnetic field dependence of Josephson critical current I_c, has given conclusive evidence for the $d_{x^2-y^2}$-wave superconductivity[5]. The Josephson effect of $CeCu_2Si_2$ has been investigated in point contacts between $CeCu_2Si_2$ and a conventional superconductor by Poppe *et al*[6]. They observed I_c values between 15 and 80 % of the Ambegaokar-Baratoff limit. In our previous paper we have compared the Josephson effect between S(Nb)-N(Cu)-S′($CeCu_2Si_2$) and S(Nb)-N(Cu)-S(Nb) junctions and found no significant suppression of I_c due to the unconventional superconductivity of $CeCu_2Si_2$[7]. In this paper, the magnetic field dependence of Josephson critical currents that come from multiple crystallites of polycrystalline $CeCu_2Si_2$ has been investigated and the existence of s-wave order parameter is discussed.

A polycrystalline sample of nominal composition $CeCu_{2.2}Si_2$ was prepared by argon-arc melting the corresponding amounts of the elements. The sample was annealed in Ar at 900°C for 5 days. Figure 1(a) shows the orientation imaging micrograph of the polished surface obtained by EBSD (Electron Backscatter Diffraction) patterns. The colors indicate the crystallographic orientation normal to the surface. Typical crystallites are thin long ones in a width of about 30

Published under licence by IOP Publishing Ltd

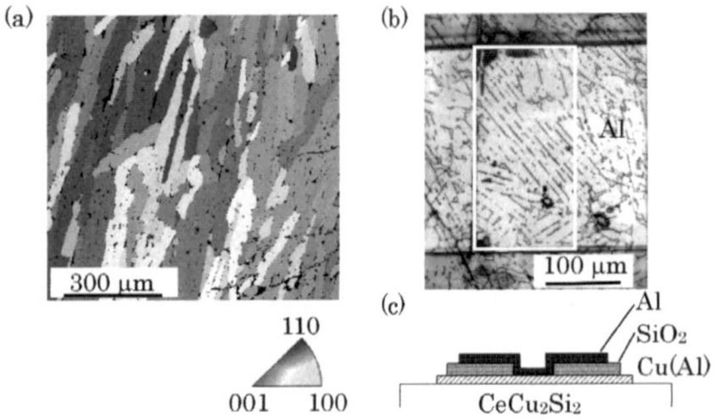

Figure 1. (a) Orientation imaging micrograph of $CeCu_2Si_2$ surface. The colors indicate the crystallographic orientation normal to the surface. (b) Optical micrograph of J-1. The rectangle indicates the junction area where the Josephson current flows in the normal direction. (c) Cross-sectional view of the Josephson junction.

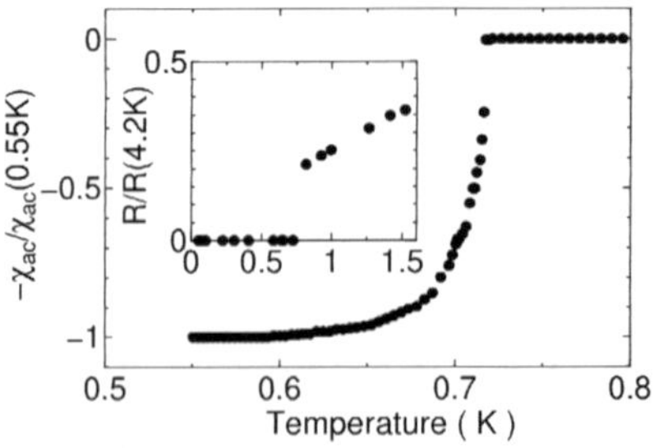

Figure 2. Temperature dependence of ac susceptibility of $CeCu_2Si_2$ showing the superconducting transition. Inset: Resistance vs. temperature.

μm; crystallites in different longitudinal directions tend to show different crystal faces.

The polycrystal was cut to slices of 1 mm thickness to use as a substrate. The surface was then polished with diamond polish down to a grain size of 1 μm, resulting in a flat mirror-like surface. In order to inspect the grain structure of the junction area, two of six junctions, denoted as J-1 and J-2, were made on the substrate which was lightly etched in dilute nitric acid to produce a microstructural contrast. The other four junctions (J-3, 4, 5, 6) were made on the mirror-like surface.

The $CeCu_2Si_2$ (S) substrate was set in a sputtering apparatus and a SNS′ junction was prepared; after the surface was sputter cleaned by Ar ion, Cu doped with 5 wt% Al (N: normal metal), SiO_2 and Al (S′: *s*-wave superconductor) were deposited successively using rf-sputtering technique, as shown in Figs. 1(b) and (c). Even after the deposition, it is still visible that the junction area contains many crystallites in different directions. The thickness of normal metal d_N was 1.6 μm with the exception of J-2 with d_N=1.2 μm. The Josephson effect was measured by a SQUID picovoltmeter. The details of the measuring technique have been described in our previous paper[8].

Figure 2 shows the superconducting transition of the $CeCu_2Si_2$ polycrystal observed by ac susceptibility and resistance measurements. As the temperature is decreased, the resistance shows a gradual decrease and then drops to zero at $T_c \sim 0.72$ K. Below T_c the diamagnetic susceptibility appears, rapidly increases and then shows a perfect diamagnetism. The high T_c value above 0.7 K ensures the good quality of the present sample.

The temperature dependence of critical current density J_c is shown for the six junctions in Fig. 3, where J_c is the critical current per unit area of the junction. The inset is the Shapiro steps appeared in the $I-V$ curve, when ac current is superposed on dc current. Since $h\nu/2e$ is 20 pV at a frequency of 10 kHz, the observed steps indicate that the Josephson current flows by the lowest-order (second-order) tunneling process.

Although $CeCu_2Si_2$ is superconducting below $T_c \sim 0.7$ K, the Josephson current becomes measurable well below T_c and increases exponentially at low temperatures, because the use of the "dirty" normal layer Cu(Al) decreases the Josephson coupling between the two superconductors. The weak coupling simplifies the analysis of the Josephson effect as follows.

International Conference on Strongly Correlated Electron Systems (SCES 2010) IOP Publishing
Journal of Physics: Conference Series **273** (2011) 012086 doi:10.1088/1742-6596/273/1/012086

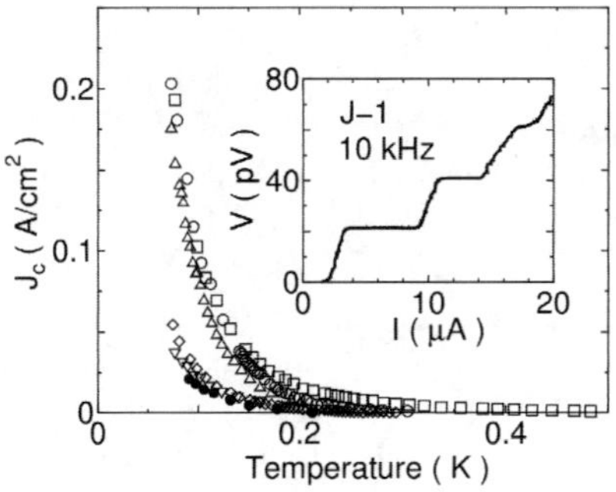

Figure 3. Temperature dependence of Josephson critical current density J_c for J-1(●), J-2(◇), J-3(□), J-4(○), J-5(△) and J-6(▽). Inset: Conventional Shapiro steps observed for J-1.

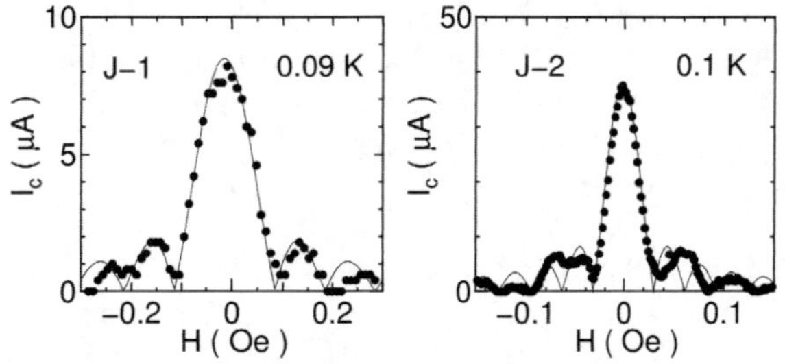

Figure 4. Magnetic field dependence of Josephson critical current I_c for J-1 and J-2. The solid line represents a fit by a Fraunhofer diffraction pattern.

If the Josephson penetration depth λ_J, which decreases with an increase in J_c, becomes smaller than the width w of the junction, the junction is self-field limited; the current distribution is not uniform and the effective current carrying width becomes of the order of λ_J. In the present junction, λ_J is about 0.5 mm even for $J_c \sim 0.06$ A/cm^2; λ_J is larger than the junction width and the current flow is uniform throughout the junction area. This ensures that the present results reflect the overall junction area.

The magnetic field dependence of I_c indicates the quality of the junctions, as shown in Fig. 4. If we refer to a coordinate system with the z axis along the current direction and apply an external magnetic field H in the y direction, the total Josephson current I through the junction is given by

$$I = \int\int dxdy J_c(x,y) \sin(\frac{2\pi d}{\Phi_0} Hx + \psi_0), \tag{1}$$

where $J_c(x,y)$ is the local Josephson critical current density, Φ_0 is the flux quantum, and ψ_0 is the phase difference between two superconductors; d is given by $d = d_N + \lambda_S + \lambda_{S'}$, where λ_S and $\lambda_{S'}$ are the penetration depth of the superconductors S and S′, respectively[9]. In case that the junction is uniform, that is, $J_c(x,y)$ is a constant, the maximum value I_c of I shows a Fraunhofer diffraction pattern;

$$I_c(H) = I_{c0} \left| \frac{\sin(\pi H/\Delta H)}{\pi H/\Delta H} \right|, \tag{2}$$

where I_{c0} and ΔH are constants and ΔH corresponds to a flux quantum threading through the junction. Even if $J_c(x,y)$ varies spatially, I shows a global maximum $I_c = \int\int dxdy J_c(x,y)$ at zero magnetic field, unless the phase difference also varies spatially.

The solid lines in the figures indicate the Fraunhofer pattern calculated using the global maximum value of I_c as I_{c0} and the H interval between the global maximum and the nearest local minimum as ΔH. The fact that J-1 shows a better fit to the pattern indicates that the Josephson current flows more uniformly as compared with J-2. The other four junctions also showed a modified Fraunhofer pattern like J-2; a global maximum of I_c near $H = 0$ was commonly observed. Similar Fraunhofer-like patterns have already been reported for point contacts on $CeCu_2Si_2$, although the size and shape of the junction area that affect $I_c(H)$ patterns were not clear[6].

If the spatial variation of the phase difference $\tilde{\psi}(x, y)$ exists ($\psi_0 \rightarrow \psi_0 + \tilde{\psi}(x, y)$), the peak at zero magnetic field is expected to become less dominant. Such a magnetic field dependence has been reported for the $d_{x^2-y^2}$ superconductors; an oscillating pattern is observed for the junctions on the (110) face of $YBa_2Cu_3O_{7-\delta}$[10] and on the (001) face of $CeCoIn_5$[11], because the Josephson current from + and - lobes of the order parameter should cancel out for [110] and [001] directions, while facets, pits and projections on the surface cause a contribution of the Josephson current reflecting a change of the relative sign of the order parameter ($\tilde{\psi}(x, y) = 0$ or π).

If $CeCu_2Si_2$ is also a $d_{x^2-y^2}$ superconductor, an oscillating $I_c(H)$ pattern should be observed, as long as the junction area contain crystallites that face both + and - lobes of the order parameter to the surface. Although the distribution of the crystallographic orientation for each junction area was not determined, the junction area was chosen so that it may consist of crystallites in various directions, as shown in Fig. 1(b). Since such a microstructure leads to the contributions of different crystallographic orientations, the Fraunhofer-like pattern in the present investigation apparently contradicts the $d_{x^2-y^2}$ superconductivity. One possible explanation is a surface-induced s-wave component[12]. Although such a component also explains the large I_c values observed for $CeCu_2Si_2$[6, 7], it will be a future issue why $CeCu_2Si_2$ behaves differently from other $d_{x^2-y^2}$ superconductors. It should be noted that unlike $CeCoIn_5$ $CeIrIn_5$ has also shown such an s-wave-like behavior[11, 13].

In conclusion, we have investigated the Josephson critical current I_c between $CeCu_2Si_2$ and a conventional superconductor by fabricating junctions on a $CeCu_2Si_2$ polycrystal. The magnetic field dependence of I_c has shown that a global maximum at zero magnetic field is dominant as compared with other peaks. This result may suggest the existence of an s-wave component on the surface of the multiple crystallites.

Acknowledgments

This work was supported by a Grant-in-Aid from the Ministry of Education, Culture, Sports, Science and Technology (MEXT), Japan.

References

[1] Steglich F, Aarts J, Bredl C D, Lieke W, Meschede D, Franz W and Schäfer H 1979 *Phys. Rev. Lett.* **43** 1892
[2] Eremin I, Zwicknagl G, Thalmeier P and Fulde P 2008 *Phys. Rev. Lett.* **101** 187001
[3] Lengyel E, Nicklas M, Jeevan H S, Sparn G, Geibel C, Steglich F, Yoshioka Y and Miyake K 2009 *Phys. Rev B* **80** 140513
[4] Yuan H Q, Grosche F M, Deppe M, Geibel C, Sparn G and Steglich F 2003 *Science* **302** 2104
[5] Wollman D A, Van Harlingen D J, Giapintzakis J and Ginsberg D M 1995 *Phys. Rev. Lett.* **74** 797
[6] Poppe U 1985 *J. Magn. Magn. Mater.* **52** 157
[7] Koyama T, Sumiyama A, Nakagawa M and Oda Y 1998 *J. Phys. Soc. Jpn.* **67** 1797
[8] Sumiyama A, Hata R, Oda Y, Kimura N, Yamamoto E, Haga Y and Ōnuki Y 2005 *Phys. Rev. B* **72** 174507
[9] Barone A and Paternò G 1982 *Physics and Applications of the Josephson Effect* (New York: John Wiley and Sons) p 71
[10] Neils W K and Van Harlingen D J 2002 *Phys. Rev. Lett.* **88** 047001
[11] Sumiyama A, Tsuji Y, Ikeda N, Oda Y, Shishido H, Settai R and Ōnuki Y *to be published in Physica C*
[12] Mößle M and Kleiner R 1999 *Phys. Rev. B* **59** 4486
[13] Sumiyama A, Tsuji Y, Oda Y, Shishido H, Settai R and Ōnuki Y 2007 *J. Mag. Mag. Mater.* **310** 599

Journal of Physics: Conference Series **273** (2011) 012087 doi:10.1088/1742-6596/273/1/012087

High-pressure electrical resistivity measurement on heavy fermion superconductor URu_2Si_2 using super clean crystal

Naoyuki Tateiwa[1], Tatsuma D. Matsuda[1], Yoshinori Haga[1], Etsuji Yamamoto[1], and Yoshichika Ōnuki[1,2]

[1]Advanced Science Research Center, Japan Atomic Energy Agency, Tokai, Ibaraki 319-1195, Japan

[2]Graduate School of Science, Osaka University, Toyonaka, Osaka 560-0043, Japan

E-mail: tateiwa.naoyuki@jaea.go.jp

Abstract. We have measured the electrical resistivity of URu_2Si_2 under high pressure using a high quality single crystal sample. The pressure phase diagram was re-investigated. With increasing pressure, the transition temperature T_0 from the paramagnetic to the "hidden order" states increases and the superconducting transition temperature T_{sc} decreases simply. The superconducting transition temperature shows the applied electrical current dependence and the width of the transition temperature $\Delta T_{sc}/T_{sc}$ strongly increases above 0.94 GPa, suggesting the filamentary superconducting state above the pressure. It is suggested that the first order phase boundary of P_x lies between 0.75 and 0.94 GPa. We analyzed the temperature dependence of the electrical resistivity ρ_a for the current along the a-axis with a generalized power law $\rho = \rho_0 + AT^n$ in the low temperature regions from T_{sc} to 2.75, 3.0, and 3.4 K. Below P_x, a power law exponent n smaller than 2 is required to represent the data and the values of n are generally about 1.5 ± 0.1. Above P_x, n increases with increasing pressure and the value is about 2.0 at 1.51 GPa. It is suggested that the derivation of the exponent n from the Fermi liquid theory is intrinsic to the hidden order state.

1. Introduction

URu_2Si_2 is a heavy-fermion superconductor with a superconducting transition temperature $T_{sc} \sim 1.5$ K at ambient pressure [1]. Furthermore, the compound undergoes a successive phase transition at $T_0 \sim 17.5$ K, and the ordered state coexists with the superconductivity. Only a tiny staggered moment of about $\sim 0.03\ \mu_B$/U was observed in the neutron diffraction experiments below T_0 and there were strong sample dependences on the value of the magnetic moment. Although intensive theoretical and experimental studies have been carried out, the nature of the ordered state has not been still explained consistently and the state is known as "hidden order" (HO) [2]. The applying pressure on the compound induces a first order phase transition from the HO to a large moment antiferromagnetic (LAFM) state at $P_x = 0.5$ GPa [2, 3, 4, 5, 6].

Recently, Zhu *et al.* reported that the resistivity along the a-axis ρ_a exhibits a non-Fermi liquid behavior ($\rho_a \propto T$) in the low temperature limit, while that along the c-axis ρ_c shows a typical Fermi liquid behavior ($\rho_c \propto T^2$) [7]. This behavior was interpreted theoretically based on a model assuming the quadrupolar ordering of the HO state [8, 9]. We succeeded in growing

Published under licence by IOP Publishing Ltd

high-quality single crystals of URu_2Si_2 and investigated the sample dependence of the electrical resistivity [10, 11]. The T-linear dependence of ρ was confirmed in a sample with a middle-class quality when ρ between $T_{sc} < T < 2.8$ K was fitted with a relation $\rho = \rho_0 + AT^n$. With increasing sample quality, the value of n increases and finally approaches to about 1.5. There have been several studies on the electrical properties of URu_2Si_2 under high pressure [2, 12, 13]. Since the electric state of this compound depends the sample quality, it is needed to investigate the electrical properties using a high quality single crystal. In this paper, we show the result of the electrical resistivity measurement on URu_2Si_2 using the high quality single crystal.

2. Experimental

We have grown the high quality single crystal of URu_2Si_2 by the Czochralski pulling method. The details of the sample preparation is given in the ref. 10. The electrical resistivity under high pressure was measured by the ac four terminal method in a ^{3}He cryostat. The pressure was generated by a hybrid CuBe/MP35 piston cylinder-type high-pressure cell. The Daphne 7474 oil was used as a pressure-transmitting medium [14, 15]. It is very difficult to estimate the residual resistivity ratio RRR ($= \rho_{RT}/\rho_{0K}$) because the residual resistivity ρ_{0K} becomes a negative value if the resistivity just above T_{sc} is simply extrapolated to 0 K. The value of RRR was estimated as about 300 using the resistivity just above T_{sc} (RRR $= \rho_{RT}/\rho_{T_{sc}}$). It seems that the residual resistivity is very small and the real RRR value exceeds 1000 [11].

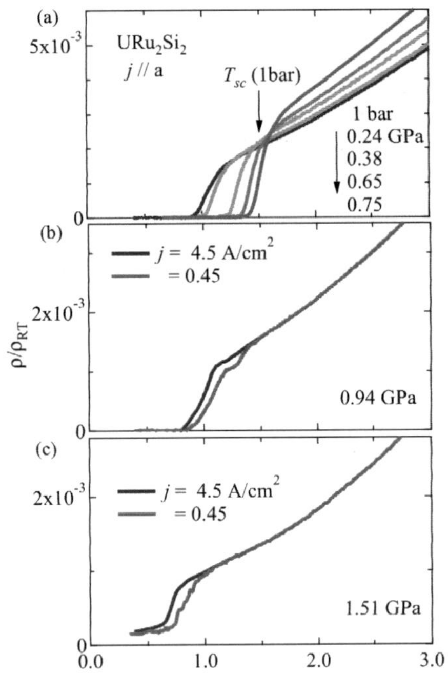

Figure 1. Temperature dependence of the electrical resistivity in URu_2Si_2 at 1 bar, 0.24, 0.38, 0.65, 0.75,0.94 and 1.51 GPa

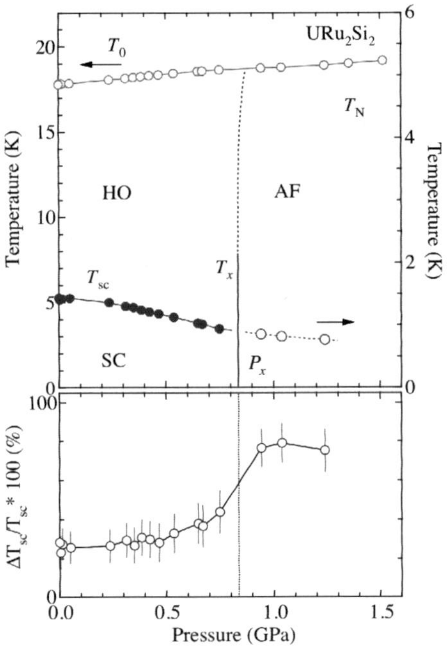

Figure 2. (upper panel)Pressure phase diagram in URu_2Si_2 determined the present resistivity measurement. (lower panel)Pressure dependence of the width on the superconducting transition temperature $\Delta T_{sc}/T_{sc}$

3. Results and discussions

Figure 1 (a) shows the temperature dependence of the electrical resistivity ρ_a for the current along the a-axis at 1 bar, 0.24, 0.38, 0.65 and 0.75 GPa. At 1 bar, a clear superconducting transition temperature was observed at $T_{sc} = 1.43$ K. The transition temperature decreases with increasing pressure. Figure 1 (b) and (c) show the temperature dependence of ρ_a at 0.94 and 1.51 GPa, respectively. Above 0.94 GPa, the superconducting transition temperature depends on the applied electrical current. The zero resistivity was not observed at 1.35 and 1.51 GPa. These results suggest that the superconducting state in the sample is not the bulk-type but filamentary above 0.91 GPa.

The upper panel of Figure 2 shows the pressure phase diagram of URu_2Si_2 determined by the present resistivity measurement. There are three phase lines in URu_2Si_2: $T_0(P)$ of the transition from the paramagnetic (PM) to the HO, $T_x(P)$ from the HO to the LAFM state and $T_N(P)$ from the PM to HO states [2, 6]. It was almost established by the previous studies with the ac magnetic susceptibility and heat capacity that the bulk-superconducting state exists only below P_x and the superconducting transition in the resistivity observed above P_x is due to a filamentary superconducting state in a sample [2, 5, 6]. In this study, the superconducting transition temperature shows the applied electrical current dependence and the width of the transition temperature $\Delta T_{sc}/T_{sc}$ increases largely above 0.94 GPa as shown in the lower panel of Figure 2. It is supposed that the value of P_x lies between 0.75 and 0.94 GPa in the present study.

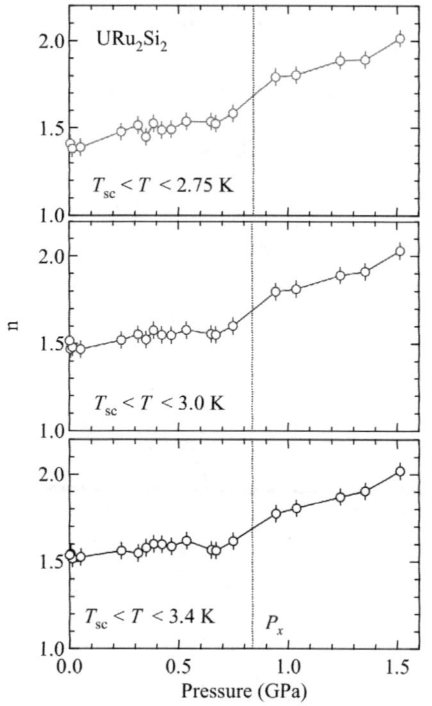

Figure 3. Pressure dependences of the resistivity exponent n obtained by fittings a generalized power low to the low temperature resistivity in the three temperature regions in URu_2Si_2

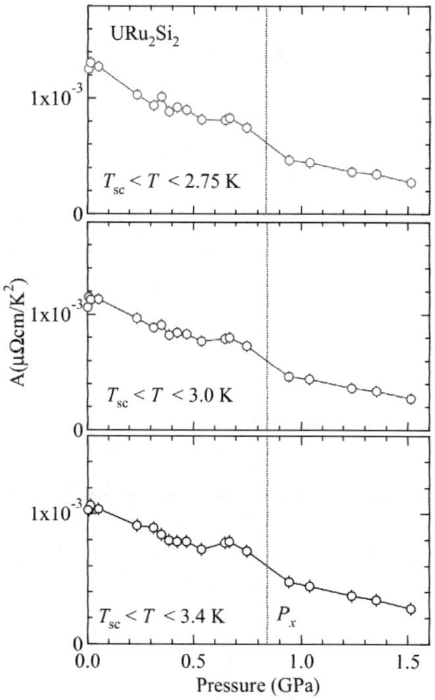

Figure 4. Pressure dependences of A obtained by fittings a generalized power low to the low temperature resistivity in the three temperature regions in URu_2Si_2

Next we briefly discuss the phase boundary of $T_x(P)$. There has been a controversy whether the line of $T_x(P)$ ends in a critical end point or intersects with the other two lines of $T_0(P)$ and $T_N(P)$ at a tricritical point of P^* [2, 4, 6]. This is an important point concerning the symmetry of the order parameter of the HO [16]. Since the present study shows only the resistivity data, we make a rough speculation assuming that the tricritical point exists. The reference 13 reported a small anomaly appeared around the phase boundary of $T_x(P)$ in the temperature derivative of the electrical resistivity $\partial\rho_a/\partial T$. In our study, there is no anomaly in the temperature dependences of $\partial\rho_a/\partial T$ from 0.94 to 1.51 GPa (data not shown). The ref. 2 pointed out that the width of the transition at T_0 in the resistivity and ac heat capacity increases above P^*. In our study, the width of the transition at T_0, estimated from T_{min}-T_0 where T_{min} is the temperature where the resistivity shows a minimum value just above T_0, increases strongly above 0.94 GPa (data not shown). One speculation is that the phase boundary line of $T_x(P)$ runs upward in a very steep gradient and P^* is located below 0.94 GPa as shown in dotted line. This should be checked by a future study with other type of measurements such as a thermal expansion or heat capacity.

We analyzed the temperature dependence of the electrical resistivity ρ_a with a generalized power law $\rho = \rho_0 + AT^n$ in the temperature regions from T_{sc} to 2.75, 3.0, and 3.4 K. Below P_x, a power law exponent n smaller than 2 is required to represent the data in the temperature ranges. The pressure dependences of n and A are shown in Figure 3 and 4, respectively. Below P_x, n shows a very weak pressure dependence and the values of n are generally about 1.5 $\pm$ 0.1. Above P_x, n increases with increasing pressure and the value of n is about 2.0 at 1.51 GPa. These results suggest that the derivation of the exponent n from the Fermi liquid theory is intrinsic to the hidden order state. The value of A simply decreases with increasing pressure. There is no anomalous behavior in the pressure dependences of A.

Finally, we discuss the exponent n of the antiferromagnetic state above P_x. As pointed out in the ref. 2, the appearance of the filament superconductivity above P_x suggests the residual HO phase, origin of the superconductivity, still exists in the antiferromagnetic phase as impurities. Generally, the behavior of the resistivity is strongly influenced by impurities. With increasing pressure above P_x, the superconducting transition temperature becomes more sensitive to the applied electrical current and the zero resistivity was not observed at 1.3 and 1.5 GPa, suggesting the decrease of the residual HO phase in the sample at the higher pressure region. If we estimate the resistivity exponent n from those at 1.35 and 1.51 GPa, the value of n in the antiferromagnetic state is estimated as close to 2.

4. Acknowledgments

This work was financially supported by a Grant-in-Aid for Scientific Research on Innovative Areas "Heavy Electrons (No. 20102002), Scientific Research S (No. 20224015), C (No. 21540373, 22234567), Specially Promoted Research (No. 20001004) and Osaka University Global COE program (G10) from the Ministry of Education, Culture, Sports, Science and Technology (MEXT) and Japan Society of the Promotion of Science (JSPS).

References

[1] Broholm C, Kjems J K, Buyers W J L, Matthews P, Palstra T T M, Mnovsky A A, and Mydosh J A, 1987 *Phys. Rev. Lett.* **58** 1467

[2] for a recent review: Hassinger E, Knebel G, Izawa K, Lejay P, Salce B, and Flouquet J, 2008 *Phys. Rev. B* **77** 115117

[3] Amitsuka H, Sato M, Metoki N, Yokoyama M, Kuwahara K, Sakakibara T, Moromoto H, Karawazaki S, Miyake Y and Mydosh J A, 1999 *Phys. Rev. Lett.* **83** 5114

[4] Motoyama G, Nishioka T, and Sato N K, 2003 *Phys. Rev. Lett.* **90** 166402

[5] Uemura S, Motoyama G, Oda, Nishioka T and Sato N K, 2005 *J. Phys. Soc. Jpn.* **75** 2667

[6] Amitsuka H, Matsuda K, Kawasaki I, Tenya K, Yokoyama M, Sekine C, Tateiwa N, Kobayashi T C, Kawarazaki S, and Yoshizawa H, 2007 *J. Magn. Magn. Mater.* **310** 214
[7] Zhu Z, Hassinger E, Xu Z, Aoki D, Flouquet J, and Behnia K, 2009 *Phys. Rev. B* **80** 172501
[8] Harima H, Miyake K and Flouquet, 2010 *J. Phys. Soc. Jpn.* **79** 033705
[9] Miyake K, and Flouquet J, 2010 *J. Phys. Soc. Jpn.* **79** 035001
[10] Matsuda T D, Aoki D, Ikeda S, Yamamoto E, Haga Y, Ohkuni H, Settai R and Ōnuki Y, 2008 *J. Phys. Soc. Jpn.* **77** 362
[11] Matsuda T D, Aoki D, Hassinger E, Knebel G, Flouquet J, in preparation
[12] Oomi G, Kagayama T, Ōnuki Y, Komatsubara T, 1994 *Physica B* **199-200** 148
[13] Motoyama G, Yokoyama N, Sumiyama A, and Oda Y, 2008 *J. Phys. Soc. Jpn.* **77** 123710
[14] Murata K, Yokogawa K, Yoshino H, Klotz S, Munsch P, Irizawa A, Nishiyama M, Iizuka K, Nanba T, Okada T, Shirage Y, and Aoyama S, 2008 *Rev. Sci. Instrum.* **79** 085101
[15] Tateiwa N, and Haga Y, 2009 *Rev. Sci. Instrum.* **80** 123901
[16] Mineev V P, and Zhitomisrsky M E, 2005 *Phys. Rev. B* **72** 0144323

Pressure studies on the superconductor Mo_3Sb_7

V. H. Tran[1], R. T. Khan[2], P. Wiśniewski[1], E. Bauer[2]

[1] Institute of Low Temperature and Structure Research, Polish Academy of Sciences, P. O. Box 1410, 50-950 Wrocław, Poland

[2] Institute of Solid State Physics, Vienna University of Technology, A-1040 Wien, Austria

E-mail: v.h.tran@int.pan.wroc.pl

Abstract. The electrical resistivity ($\rho(T)$) and magnetization ($M(T)$) of the superconductor Mo_3Sb_7 (T_c = 2.2 K) has been measured under pressures up to 22 kbar and at temperatures from 0.4 to 80 K. Hydrostatic pressure is found to show up the spin-density wave state below T_{SDW}, which is revealed by a sharp jump in $\rho(T)$ and a maximum in $M(T)$ dependencies. With increasing pressure T_{SDW} decreases while T_c increases. The pressure dependencies of T_c and T_{SDW} can be understood on the basis of competition of the SDW and the superconducting states in this compound.

1. Introduction

Mo_3Sb_7 with the lattice parameter $a = 0.9591$ nm crystallizes in the cubic Ir_3Ge_7-type structure (space group Im-3m). This compound has been reported to enter into a superconducting state below T_c = 2.2 K [1, 2]. Moreover, specific heat and muon spin rotation experiments [2, 3] indicated that the effective mass of the charge carriers is enhanced ($\sim$ 16-18 m_0), if compared to that of conventional BCS superconductors with a weak electron-phonon coupling. At present, Mo_3Sb_7 is concluded being nonmagnetic due to a spin dimerization below $T^* = 50$ K, which is reflected by anomalies in the heat capacity, magnetic susceptibility [4], muon spin rotation [5, 6, 7], and magnetic excitations in inelastic neutron scattering [5]. The crystal structure of this compound is certainly favorable for spin dimerization, because of a distinct difference between the Mo-Mo intradimer distance ($\sim$ 0.3 nm) and Mo-Mo interdimer distance ($\sim$ 0.46 nm). In general, the strength of magnetic interactions in a given compound can be modified by applying external hydrostatic pressure. Furthermore, taking into account the fact that the effective mass of superconducting carriers in Mo_3Sb_7 is sizable, approaching that of heavy-fermion superconductors, one would expect a similar response of its electronic state to applied pressure. In heavy-fermion superconductors, magnetic instabilities or magnetic fluctuations are responsible for enhancing the effective mass of the superconducting quasiparticles, and the pairing mechanism in unconventional superconductors is most probably associated with magnetism [8, 9, 10, 11]. In view of these arguments, a considerable pressure response on the magnetic state is expected in Mo_3Sb_7. Therefore, the electrical resistivity and magnetization of Mo_3Sb_7 was studied under pressure. In the following, we present pressure-dependent properties of Mo_3Sb_7. Results of our study revealed that a spin-density-wave exists in parallel with superconductivity in Mo_3Sb_7.

Published under licence by IOP Publishing Ltd

2. Experimental

Mo_3Sb_7 was prepared from Mo and Sb (purity 99.95% from Alfa Aesar) by solid-state reaction. Synthesis, purity characterization and determination of crystal parameters of the sample have been carried out in a similar technique as given in Ref. [12]. Electrical resistivity under pressure up to 22 kbar was measured in the temperature range 0.3-80 K by means of a standard four-probe technique, using a ^{3}He cryostat with a Cryogenics Ltd. 12 T magnet. Hydrostatic pressure was generated by a piston-cylinder cell using Daphne oil as the pressure-transmitting medium. The resistivity data were taken on cooling at a rate below 0.1 K/min. The magnetization was measured at a pressure of 7 kbar in several magnetic fields up to 0.07 T using a nonmagnetic pressure cell with a SQUID magnetometer (Quantum Design MPMS-5). High purity Pb was used as pressure indicator. The width of the superconducting transition of Pb did not exceed 10 mK, corresponding to an uncertainty of measured pressure $\pm$ 0.2 kbar.

3. Results

The electrical resistivity of Mo_3Sb_7 at temperatures from 3 to 80 K for various pressures up to 22 kbar is presented in Fig. 1 a. Above about 7 K, Mo_3Sb_7 behaves metallic, without substantial changes upon a change of pressure. For temperatures around T^*, a shoulder in $\rho(T)$ becomes unveiled. We have analyzed the data around T^* by taking the temperature derivative of the resistivity $d\rho(T)/dT$; results are plotted in the inset of Fig. 1 a. The maximum of $d\rho(T)/dT$, associated with the spin dimerization transition, decreases with increasing pressure. The most important finding of this work, however, is the observation of a new phase transition as evidenced by a sharp anomaly of the 4.5 kbar resistivity at $T_{SDW} = 6.6$ K, indicated by an arrow in Fig. 1 a, and it is shown in more detail in Fig. 1 b. Nevertheless, a very tiny anomaly around 6.7 K can be recognized for the data collected at 0.2 kbar (not shown here). It is reasonable to ascribe the upturn in the resistivity to the occurrence of a density wave phase, alike the spin-density wave in URu_2Si_2 [13], and $BaFe_{2-x}Co_xAs_2$ [14], or the charge-density wave (CDW) in $Lu_5Ir_4Si_{10}$ [15] and $Lu_5Rh_4Si_{10}$ [16]. We attribute the resistivity upturn in Mo_3Sb_7 to the opening of an energy gap at some portions of the Fermi surface associated with an antiferromagnetic state.

To corroborate the magnetic origin of the transition at T_{SDW} the magnetization was measured in low magnetic fields up to 0.02 T and under pressure of 6 kbar (Fig. 2 a). While no anomalies occur at ambient pressure [4], the magnetization under applied pressure and at 0.01 T exhibits a transition at 6.8 K. In a manner typical for antiferromagnets, the application of higher magnetic fields shifts the transition to lower temperatures and in a field of 0.07 T a clear maximum suggesting an antiferromagnetic-type interaction appears at 2.8 K. According to the band structure calculations [4], the Fermi surface is nested. Such a property may trigger a SDW ordering.

Fig. 2b shows low-temperature resistivity $\rho(T)$ data under pressure. The resistivity at 0.2 kbar is characterized by a superconducting transition at 2.15 K, in agreement with the previous report for ambient pressure.[12] With growing pressure, the critical temperature increases at an initial rate $dT_c/dP \sim 0.02$ K/kbar, and at maximum applied pressure of 22 kbar T_c reaches a value of 2.37 K. Simultaneously, the width of the critical transition ΔT_c decreases, from 0.2 K at 0.2 kbar to 0.1 K at 22 K, indicating that a more homogeneous superconducting state is realized under pressure. The observations of an increase of T_c and of a gradually sharpening of the phase transition with increasing pressure (see Fig. 2 b) imply that superconductivity is favored by pressure.

4. Summary

In summary, electrical resistivity and magnetization measurements were carried out under hydrostatic pressure for Mo_3Sb_7. The application of pressure increases the superconducting phase transition temperature T_c. Also, pressure reveals a transition from a metallic to a SDW phase

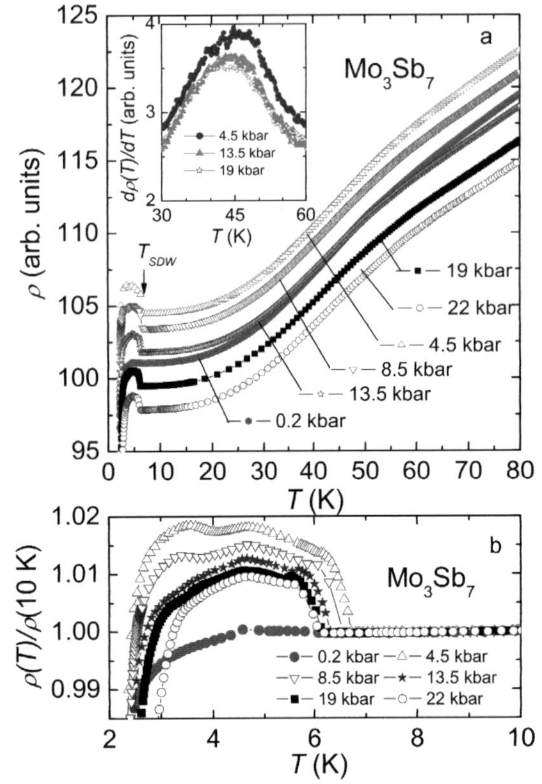

Figure 1. (color online) Electrical resistivity of Mo_3Sb_7 for different pressures. a) Electrical resistivity of Mo_3Sb_7 at temperatures up to 80 K. The inset shows temperature derivative of the resistivity vs. temperature around T^*. b) The resistivity data around T_{SDW} are normalized at $T = 10$ K.

above 4.5 kbar. From the pressure response of T_{SDW} and T_c, we suggest that superconducting Mo_3Sb_7 is an interesting example, where superconductivity competes with a SDW phase. The increasing T_c in presence of antiferromagnetic interactions implies that superconductivity in Mo_3Sb_7 is most likely mediated by magnetic fluctuations. Studies of either NMR or neutron scattering under pressure should bring new insights to the spin-density-wave phase scenario or possibly "hidden order" in this binary compound.

V.H.T. and P.W. would like to thank the Ministry of Science and Higher Education in Poland for financial support under grant N202 082/0449. R.T.K. and E.B. acknowledge support by the Austrian FWF, P18054 and COST P16.

References

[1] Z. Bukowski, D. Badurski, A. J. Stępień-Damm and R. Troć, 2002 *Solid State Commun.* **123** 283.
[2] V. H. Tran, W. Miiller and Z. Bukowski 2008 Acta Materialia, **56** 5694.
[3] V. H. Tran, A. D. Hillier, D. T. Adroja and Z. Bukowski, 2008 Phys. Rev. B **78** 172505.
[4] V. H. Tran, W. Miiller and Z. Bukowski 2008 Phys. Rev. Lett. **100** 137004.

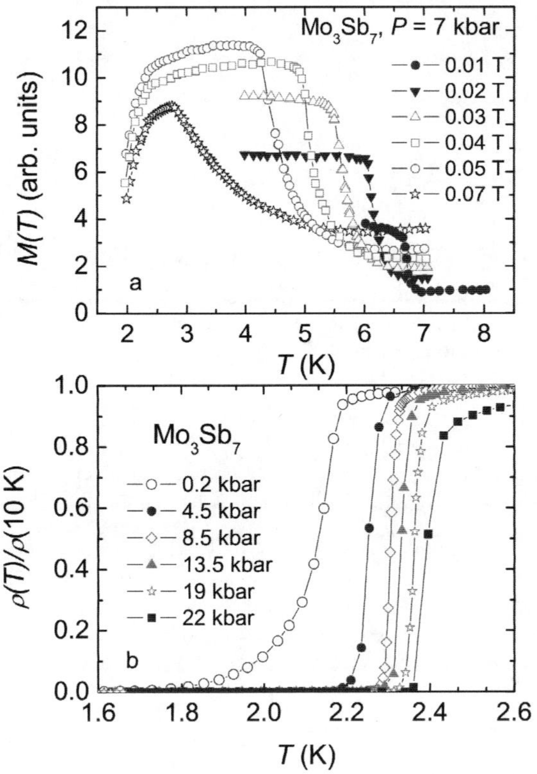

Figure 2. (color online) a) dc-magnetization of Mo_3Sb_7 at 7 kbar measured in the temperature range 2 - 8 K for several magnetic fields. b) Normalized electrical resistivity of Mo_3Sb_7 in the temperature range 1.6 - 2.6 K for different pressures.

[5] V. H. Tran, A. D. Hillier, D. T. Adroja and Z. Bukowski, submitted to J. Phys.: Condens. Matter.
[6] T. Koyama, H. Yamashita,Y. Takahashi, T. Kohara, I. Watanabe, Y. Tabata and H. Nakamura 2008 Phys. Rev. Lett. **101** 126404.
[7] Y. Tabata, T. Koyama, T. Kohara, I. Watanabe and H. Nakamura 2009 Physica B **404** 746.
[8] N. D. Mathur, F. M. Grosche, S. R. Julian, I. R. Walker, D. M. Freye, R. K. W. Haselwimmer and G. G. Lonzarich 1998 Nature **394** 39.
[9] N. K. Sato, N. Aso, K. Miyake, R. Shiina, P. Thalmeier, G. Varelogiannis, C. Geibel, F. Steglich, P. Fulde and T. Komatsubara 2001 Nature **410** 340.
[10] T. Moriya and K. Ueda, 2003 Rep. Prog. Phys. **66** 1299.
[11] P. Coleman and A. Schofield 2005 Nature **433** 226.
[12] V. H. Tran and Z. Bukowski 2008 Acta Phys. Pol. A **114** 67.
[13] T. T. M. Palstra, A. A. Menovsky and J. A. Mydosh 1986 Phys. Rev. B **33** 6527.
[14] K. Ahilan, J. Balasubramaniam, F. L. Ning, T Imai, A. S. Sefat, R. Jin, M. A. McGuire, B. C. Sales and D. Mandrus 2008 J. Phys.: Condens. Matter **20** 472201.
[15] R. N. Shelton, L. S. Hausermann-Berg, P. Klavins, H. D. Yang, M. S. Anderson and C. A. Swenson 1986 Phys. Rev. B **34** 4590.
[16] H. D. Yang, P. Klavins and R. N. Shelton 1991 Phys. Rev. B **43** 7676.

Superconductivity in $Sr(Pd_{1-x}Ni_x)_2Ge_2$[1]

C D Yang[a], H C Hsu[a], W Y Tseng[a], H C Chen[a], H C Ku[b], M N Ou[c], Y Y Chen[c], and Y Y Hsu[a2]

[a] Department of Physics, National Taiwan Normal University, Taipei 11677, Taiwan
[b] Department of Physics, National Tsing Hua University, Hsinchu 30013, Taiwan
[c] Institute of Physics, Academia Sinica, Taipei 11529, Taiwan

E-mail: hcku@phys.nthu.edu.tw

Abstract. The systematic variations of lattice structure and superconducting transition temperature in $Sr(Pd_{1-x}Ni_x)_2Ge_2$ are reported. The X-ray diffraction patterns can be labeled by tetragonal *I4/mmm* (no. 139) space group with only minor impurities, which indicates the substitution Ni into Pd sites is successful and formed pure phase. The tetragonal lattice parameter *a* decreases and *c* increases monotonically with Ni concentration x, while the unit cell volume decreases. A systematic, gradual change of superconducting transition temperature T_c increases slightly to 3.2 K for very small Ni doping of x $\sim$ 0.1 then slowly decreases to 2.1 K for x = 0.5. Further investigation for $SrNi_2Ge_2$ by low temperature resistivity indicated a $T_{c,mid}$ of 0.87 K suggesting that superconductivity exists through the whole Ni substitution levels.

1. Introduction

The appearance of superconductivity in iron-based pnictide compounds has attracted much attention, providing both a new aspect in understanding the physics of unconventional, non-BCS mechanism, superconductivity and a big new family of superconducting materials of fundamental and technological interest. Superconducting transition temperature $T_c = 26$ K was first reported in $LaO_{1-x}F_xFeAs$, later was raised to the highest T_c in these materials $\sim$ 55 K in $SmFeAsO_{1-x}F_x$ [1,2]. Oxygen-free compounds with the $ThCr_2Si_2$-type structure also exhibit superconductivity with a maximum $T_c \simeq 37$ K in $(Ba_{1-x}K_x)Fe_2As_2$ (122 system) [3]. Because of the common structural element of Fe-As tetrahedral layer, these two systems are regarded as members of the same family and share a common mechanism of superconductivity. Due to the far better possibility to grow high quality and big crystals, a lot of research efforts were focused on the 122 systems for its superconducting properties and possible mechanisms. However, the importance of the 122-type compounds is not only for its easiness of high-quality crystals, but also for its wide variety of physical properties reported in a huge range of different compounds and compositions.

More recently the $ThCr_2Si_2$-type compound $SrPd_2Ge_2$ was reported to be superconducting with $T_c = 3.04$ K [4]. However, the reported property of the closely related compound $SrNi_2Ge_2$ was not superconducting down to 2 K with neither magnetic orderings [5]. To further explore the understandings of superconductivity in the 122-type structure, it is important whether superconductivity of $SrPd_2Ge_2$ is a singular existence or belonged to a series of superconductors with tunable physical variables related to the superconductivity. In this report, systematic

[1] This work was supported by NSC of Taiwan under grand no. NSC-97-2112-M-003-001-MY3.
[2] To whom any correspondence should be addressed.

Published under licence by IOP Publishing Ltd

Ni doped compounds $Sr(Pd_{1-x}Ni_x)_2Ge_2$ were synthesized and investigated for structural and superconducting properties. The existence of superconductivity and systematic variations of T_c with Ni concentration x were observed for the entire system.

2. Experimental

The intermetallic compounds $Sr(Pd_{1-x}Ni_x)_2Ge_2$ were prepared by two-step arc melting under Ar gas atmosphere on a water-cooled copper hearth. First the start materials having higher melting points, Pd wire (99.99%), Ni foil (99.99%) and Ge grains (99.9999%) were weighed by stoichiometric ratio and arc melted several times with intermediated turning over of the melted button. After that, Sr rock (99.5% pure) was added and carefully melted together with the previous button for several times to minimize the loss of Sr. The weight loss of the resultant samples were within a few percent. Structural analysis was performed by powder X-ray diffraction with a Rigaku Rotaflex 18-kW rotating anode diffractometer with graphite monochromatized Cu-K_α radiation in 2θ range of 15°-65°. The magnetic susceptibility measurements were performed by a QUANTUM DESIGN μ-metal shielded 1-T $MPMS_2$ superconducting quantum interference device (SQUID) magnetometer at temperature range of 2-5 K. Low temperature Resistivity was done by standard four-probe technique in a ^3He refrigerator down to 0.4 K.

3. Results and discussion

The powder X-ray diffraction patterns of the as-melted samples $Sr(Pd_{1-x}Ni_x)_2Ge_2$ ($0 \leq x \leq 0.5$) were shown collectively in Figure 1 with each pattern shifted for clearance. The diffraction patterns can be well indexed by tetragonal $ThCr_2Si_2$-type structure as indicated in the figure. The minor impurities observed in some samples were marked by asterisks. The lattice constants were then obtained by least-square analysis on the indexed patterns which were in good agreement with the reported values of $SrPd_2Ge_2$ and $SrNi_2Ge_2$ [4,5]. The tetragonal a-axis contracted with the Ni doping monotonically from 0.4422 nm for x = 0 to 0.4292 nm for x = 0.5, then finally to 0.4178 nm for x = 1, while the c-axis increase from 1.0131 nm (x = 0) through 1.0179 nm (x = 0.5) to 1.0252 nm (x = 1). The obtained unit cell volume shrinks almost linearly with the Ni concentration x from 0.19799 nm^3 for x = 0 to 0.17922 nm^3 for x =1 which corresponds to V $\sim$ 10% decrease. On the contrary, the c/a values also increase nearly linearly from 2.29 (x = 0) to 2.45 (x = 1). The systematic variation of the tetragonal lattice constants a and c as well as only randomly observed minor impurities indicate that Ni can be substituted into the Pd-site of the $Sr(Pd_{1-x}Ni_x)_2Ge_2$ system without any solubility limit.

The low temperature, T $\leq$ 4 K, molar magnetic susceptibility $\chi_m(T)$ for $Sr(Pd_{1-x}Ni_x)_2Ge_2$ system with 10 G applied field B_a in zero-field-cooled (ZFC) mode was shown collectively in Figure 2. Meissner effect was observed clearly with a large diamagnetic signal indicates bulk

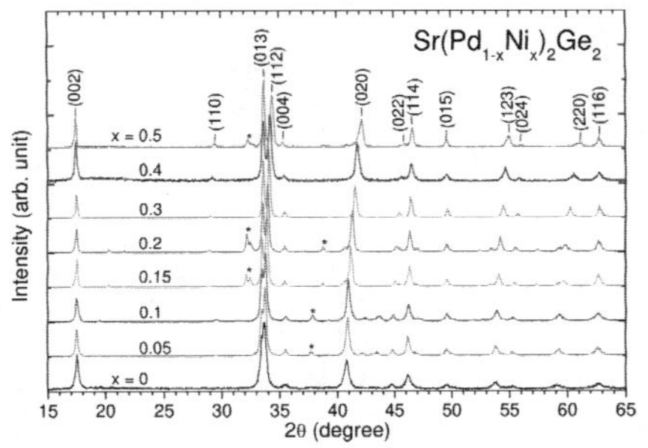

Figure 1. Collective X-ray diffraction patterns of the $Sr(Pd_{1-x}Ni_x)_2Ge_2$ system. The peaks were indexed by tetragonal $ThCr_2Si_2$-type structure and the lattice parameters were obtained from least-square fitting. The asterisks mark the peaks arising from the impurity.

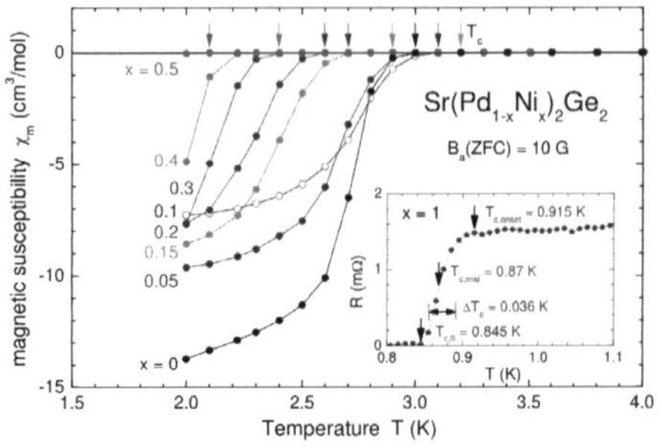

Figure 2. Low temperature molar magnetic susceptibility χ_m(T) of $Sr(Pd_{1-x}Ni_x)_2Ge_2$ at 10 G applied field. The T_c's were indicated by arrows at the onset of Meissner diamagnetic signal. Inset: The superconducting transition observed in $SrNi_2Ge_2$ by resistance.

superconductivity of all compounds observed. The superconducting transition temperature T_c of each Ni doping concentration was denoted by arrows at the onset of diamagnetic signal. A systematic change of T_c with the Ni doping was observed. In the inset, low temperature (T $\leq$ 1.1 K) resistance R(T) of the x = 1 $SrNi_2Ge_2$ compound was shown. Superconductivity was confirmed by the rapid decrease of resistance below $T_{c,onset}$ = 0.915 K which dropped to zero at $T_{c,0}$ = 0.845 K with a transition mid-point and width of $T_{c,mid}$ = 0.87 K and ΔT_c = 0.036 K. The observation of T_c varied through out the whole Ni substitution levels indicates the existence of superconductivity in all compounds of the system.

Figure 3 is the phase diagram of superconducting transition temperature as a function of Ni substitution level, T_c-x, for $Sr(Pd_{1-x}Ni_x)_2Ge_2$ system. As the Ni concentration increases, T_c is first raised slightly from 3 K for $SrPd_2Ge_2$ to 3.2 K for x = 0.1, then gradually lowered down to 2.1 K for x = 0.5, $SrPdNiGe_2$. By plotting the T_c of $SrNi_2Ge_2$ into the phase diagram, it is more clear that the superconductivity exists through out the system and follows a systematic change. Combining information from both the structural, lattice parameters changed almost linearly, and superconducting T_c has a slight maximum at x = 0.1, it is clearly indicated that Ni-doped $Sr(Pd_{1-x}Ni_x)_2Ge_2$ is a superconducting system whose T_c is believed to be tuned by the electronic structure changed with x due to structural changes and element substitution effects.

4. Conclusions

Compounds of $Sr(Pd_{1-x}Ni_x)_2Ge_2$ system were successfully synthesized in good quality. Systematic variations of tetragonal lattice parameters, a and c, as well as superconducting T_c with Ni concentration x was reported. The finite T_c was observed for the whole system which increases from 3 K (x = 0) to a slightly higher maximum 3.2 K (x = 0.1) then drops down monotonically to 0.87 K for x = 1.

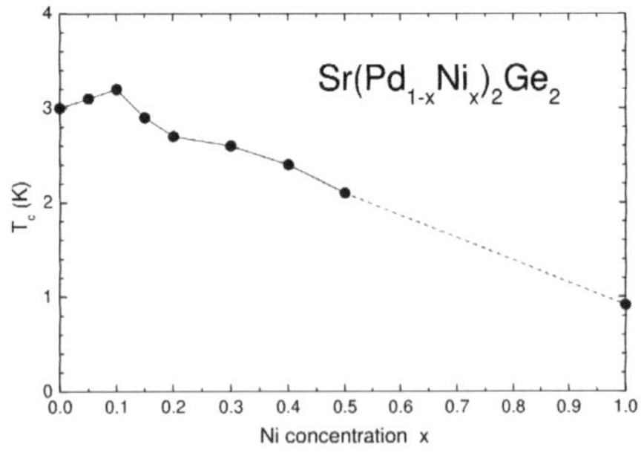

Figure 3. Systematic variation of the superconducting transition temperature T_c of the $Sr(Pd_{1-x}Ni_x)_2Ge_2$ as a function of Ni concentration x. The dashed line is draw as a guide to the eyes.

References

[1] Y Kamihara, T Watanabe, M Hirano, H Hosono, *J. Am. Chem. Soc.* 2008, **130**, 3296.
[2] Z A Ren, W Lu, J. Yang, W Yi, X L Shen, Z C Li, G C Che, X L Dong,L L Sun, F Zhou, Z X Zhao, *Chin. Phys. Lett.* 2008, **25**, 2215.
[3] Marianne Rotter, Marcus Tegel, and Dirk Johrendt *Phys. Rev. Lett.* 2008 **101** 107006.
[4] H Fujii,and A Sato *Phys. Rev.* B 2009 **79** 224522.
[5] H Fujii,and A Sato *J. Alloy. Compd.* 2009 **487** 198.

Journal of Physics: Conference Series **273** (2011) 012090 doi:10.1088/1742-6596/273/1/012090

Electric transport measurements on $La_{0.7}Ca_{0.3}MnO_3/YBa_2Cu_3O_{7-\delta}$ heterostructures

T. Hu[1], H. Xiao[1], C. Visani[2], J. Santamaria[2], and C. C. Almasan[1]

Abstract. Charge transport measurements were performed on $La_{0.7}Ca_{0.3}MnO_3/YBa_2Cu_3O_{7-\delta}$ (LCMO/YBCO) trilayer samples. They reveal that the out-of-plane stray field gives rise to flux vortices whose dissipation produces resistance peaks at the coercive fields of the top and bottom LCMO layers. A superconducting surface spin valve effect, similar to the one present in bilayer samples, was also observed in trilayers. These results were observed in the LCMO/YBCO trilayers below the superconducting transition temperature of the trilayers and are not present in the normal state, which shows the typical two fold magnetoresistance behavior.

[1]Department of Physics, Kent State University, Kent, Ohio, 44242, USA
[2]GFMC, Departamento Fisica Aplicada III, Universidad Complutense de Madrid, Madrid, 28040, Spain

1. Introduction

The superconductor/ferromagnet heterostructures have attracted a lot of research interest since (1) novel physics arises at the interface as the result of the interaction between superconductivity and magnetism and (2) they have potential application in the developement of superconductive magnetoresistance memory devices. A few examples of the novel physics present in these materials are long range proximity effect [1], domain wall superconductivity, spin imbalance effect, and induced triplet superconductivity [2]. The principle behind a giant magnetoresistance (GMR) memory device is the spin valve effect present in ferromagnet/nonmagnetic-spacer/ferromagnet (F/N/F) structures. GMR depends on the relative orientation of the magnetic moments of the top and bottom F layers [3]. A surface spin valve effect was reported in F/N structures [4] as a result of the fact that the interface magnetization differ significantly from the bulk magnetization of the F layer. It is of great interest to know if a similar spin valve effect can be observed when replacing the N layer with a superconducting S layer.

Our previous work on $La_{0.7}Ca_{0.3}MnO_3/YBa_2Cu_3O_{7-\delta}$ (LCMO/YBCO) *bilayers* has shown presence of resistance peaks [5], due to stray field, and superconducting surface spin valve effect [6], due to the fact that the magnetization of the LCMO/YBCO interface is significantly different from bulk LCMO magnetization. Here we present charge transport measurements on LCMO/YBCO *trilayer* performed in order to study their spin dependent transport properties.

1.1. Experimental Details

LCMO/YBCO trilayers were grown on (100)-oriented $SrTiO_3$ single crystals. A buffer layer of $PrBa_2Cu_3O_{7-\delta}$ (PBCO) (6 u.c.) was deposited between the substrate and the first LCMO layer in order for the two LCMO layers to be magnetically similar. (The PBCO and YBCO are very

Published under licence by IOP Publishing Ltd

similar in unit cell size). The LCMO/YBCO interfaces are sharp and perfectly coherent [1]. The details of sample preparation can be found elsewhere [7]. The thickness of the LCMO layer in these trilayer samples is 40 u.c., while the thickness of the superconducting layer is 4 u.c. (4.8 nm), 9 u.c. (10.8 nm), or 12 u.c. (14.4 nm); the superconducting transition temperature T_c (defined as 10% drop in resistance) for the three different trilayers studied is 30.0, 84.5, and 90.0 K, respectively. All samples have an area of $1\times$ 0.5 cm^2.

A four-contact method was used to measure the resistance R of the trilayers in an applied current I of 100 μA, with $I||[100]$ crystallographic direction. The angle φ dependent R was measured by rotating the magnetic field H within the ab-plane of the sample, with φ measured from the [010] direction of the LCMO layer. The contribution to $R(\varphi)$ from a small out-of-plane misalignment of H has been subtracted. Details of the misalignment produced by a one-axis rotator and its subtraction are given elsewhere [8].

2. Results and Discussions

We performed $R(H)$ measurements ($H \parallel [100]$) on trilayers with three different thicknesses of the YBCO layer, namely, $d_s^{tri} = 4$, 9, and 12 u.c. Fig. 1(a) shows the normalized resistance data $R(H)/R(1000\text{ Oe}) - 1$ vs H for increasing and decreasing H, measured at $T < T_c$ of trilayers. With increasing/decreasing H, two resistance peaks are present in $R(H)$ of the samples with $d_s^{tri} = 9$ and 12 u.c, while they become one broader peak for the $d_s^{tri} = 4$ u.c. trilayer. For example, the position of the first and second peak are $H_{p1} = 280$ Oe and $H_{p2} = 170$ Oe, respectively, for the $d_s^{tri} = 12$ u.c. trilayer.

The H_{p1} peak has the same position as in bilayers [Fig. 1 (b)]. Hence, it corresponds to the coercive field (zero magnetization and largest stray field) of the bilayer and it is a result of the dissipation of flux vortices generated by the out-of-plane stray fields present in the domain walls, as previously observed in bilayers [5]. This stray field is produced by the stresses in the twins of LCMO layer induced by the cubic-to-tetragonal transition in the $SrTiO_3$ substrate [9].

Note that, as the thickness of the YBCO layer increases, the ratio of the amplitude of the first to second peak decreases. This result is consistent with a decrease of the effect of the substrate on the top LCMO layer with increasing the thickness of the YBCO layer, such that the first peak eventually would vanish when d_s^{tri} is large enough. However, the effect of the substrate on the bottom LCMO layer is always present and should not change from sample to sample. So, we conclude that the peaks at H_{p1} and H_{p2} are a result of increased vortex dissipation due to the largest stray field present at the coercive field of the top and bottom LCMO layer, respectively.

We note that recent work on trilayers without buffer layer shows that spin dependent scattering effect (in the antiparallel orientation of the net magnetic moments of the top and bottom layer) dominates their resistance, producing much larger magnetoresistance values [10]. The buffer layer in the present samples makes the two LCMO layers magnetically similar, thus giving rise to a different physics than in trilayers without buffer layers, if a good antiferromagnetic state (antiparallel LCMO moments) is never established; i.e., the physics revealed here is vortex dissipation as a result of the out-of-plane stray field produced by the stresses in the twins in the LCMO layer induced by the cubic-to-tetragonal transition in the $SrTiO_3$ substrate.

Figure 2(a) is a plot of the normalized resistance $R^{[100]}(\varphi)/R^{[100]}_{min} - 1$ vs the angle φ, measured at 5 K $< T_c$ in 500 and 1000 Oe on a trilayer sample with $d_s^{tri} = 4$ u.c. The φ dependent MR data show maxima, corresponding to the maximum stray field for these H values (maximum number of domain walls), at 90, 180, or 270°. This is consistent with the fact that the hard axes for the LCMO/YBCO are the [010] and [100] crystallographic directions since the maximum number of domain walls takes place when the induced magnetization M is along the hard axis, while the easy axes are in the diagonal directions. Also, since this 4-fold symmetry is similar with what we have recently reported in bilayer samples [6], it shows that the magnetic coupling between top and bottom LCMO layers at these H values does not determine the overall dissipation, but

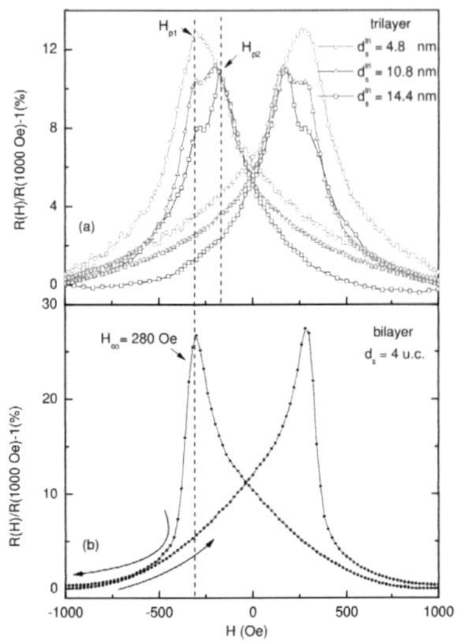

Figure 1. (a) Normalized resistance $R(H)/R(1000\ \text{Oe}) - 1$ vs. magnetic field H for trilayer samples with thickness d_s^{tri} of YBCO layer of 4, 9, and 12 unit cell measured at $T < T_c$. (b) $R(H)/R(\ \text{Oe}) - 1$ vs. H for LCMO/YBCO bilayer with thickness d_s of YBCO layer of 4 u.c. measured at a temperature T of 45 K ($< T_c$).

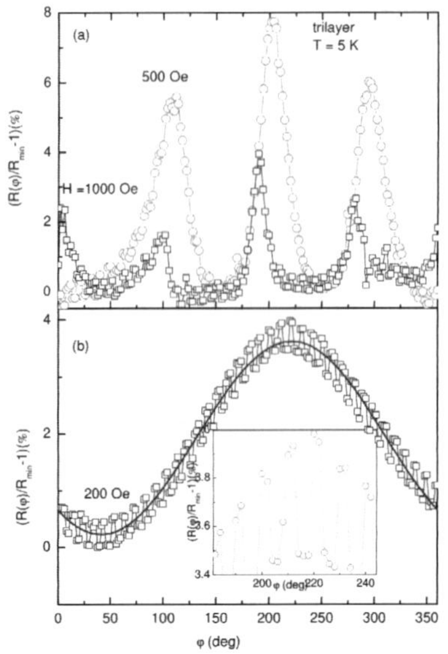

Figure 2. (a) Normalized resistance $R^{[100]}(\varphi)/R^{[100]}_{min} - 1$ vs. angle φ measured at 5 K for trilayers with $d_s^{tri} = 4$ u.c. The current is applied along the [100] crystallographic direction. (a) $H = 500, 1000$ Oe. (b) $H = 200$ Oe. Inset: the enlarged plot of the oscillation.

that these peaks in dissipation are a result of vortices produced by the out-of-plane stray field and their magnitude reflects the vortex - stripe domain structure interaction [6].

Note that the resistance peaks are not located exactly at the hard axes and also that the peaks are closer to the hard axis for the 1000 Oe data than for the 500 Oe data. We have also shown on bilayers [6] that the larger the applied magnetic field, the smaller this deviation is. The reason is the following. There are two contributions to the total free energy of the system, namely, the Zeeman energy, which tends to align M along H, and the magnetocrystal anisotropy energy which forces M to be along the easy axis. For a large magnetic field, e.g. 1000 Oe, the magnetization is close to saturation, hence the Zeeman energy dominates and the peaks take place at the hard axes. However, for a smaller field, e.g. 500 Oe, the contribution of the magnetocrystal anisotropy energy can not be neglected anymore and, therefore, M laggs behind H. This gives the small deviation of the resistance peaks from the hard axes.

Recent measurements of in-plane angular-dependent resistance of LCMO/YBCO bilayers at low H values have shown a one fold symmetry, associated with the surface spin valve effect [6]. A surface spin valve effect was initially reported within a few atomic layers at a F/N interface [4]. This phenomenon is a result of the fact that the ferromagnetic spins at the interface differ significantly from the magnetic character of the spins inside the F layer, and they can act as current or field-driven spin valves with respect to the bulk magnetization in the ferromagnetic layer. There is evidence that the magnetization of the LCMO layer at the LCMO/YBCO interface is significantly different from the bulk magnetization inside the LCMO layer [11, 12, 13]; namely, the magnetic coupling near the interface is much weaker than that of the bulk [14]. Hence the direction of the interface magnetization can be tuned by the rotation of a small magnetic field without changing the direction of the bulk magnetization in the LCMO layers. Therefore, a superconducting surface spin valve effect is produced in the LCMO/YBCO *bilayers* [6] in which the magnetoresistance depends on the relative angle between the magnetization at the LCMO/YBCO interface and the bulk magnetization of the LCMO layer. We searched for such a phenomenon in LCMO/YBCO trilayers at small applied magnetic fields.

Figure 2(b) is a plot of the normalized resistance $R^{[100]}(\varphi)/R^{[100]}_{min} - 1$ vs φ for the trilayers measured in a low H of 200 Oe using the following protocol. A magnetic field is applied along the [110] direction (easy axis) and is increased up to +2000 Oe to saturate and pin the bulk magnetization along the [110] direction. Then the magnetic field is decreased to 200 Oe and rotated within the ab-plane while measuring the resistance. The magnetoresistance curve of Fig. 2(b) is well fitted with $R^{[100]}(\varphi)/R^{[100]}_{min} - 1 = 0.035 sin^2[(\varphi + 41.4^o)/2]$ as shown by the solid curve of the figure. This one-fold symmetry is similar to what we observed in bilayer samples. This indicates that the interface spin valve effect is responsible for this behavior. It is worthwhile to comment that in trilayers with no buffer layer, where a good antiferromagnetic state can be established, a similar result is found but with much larger magnetoresistance values arising from the rotation of the magnetization of one of the layers (with a small anisotropy field), while the magnetization of the other LCMO layer (with a larger anisotropy field) is kept frozen at these small values of the rotating field [10]. Furthermore, note that in addition to the one-fold symmetry curve of Fig. 2(b), there are additional periodic oscillations [see Fig. 2(b) and its inset]. Further experiments are needed to explore the origin of these periodic oscillations.

3. Summary

In summary, charge transport measurements were performed on YBCO/LCMO trilayer heterostructures with a buffer layer between the substrate and the first LCMO layer. The vortices induced by the stray field give rise to two resistance peaks located at the coercive fields of the top and bottom LCMO layers. A superconducting spin valve behavior is observed in trilayers at low fields, which indicates that the interface effect dominate the charge dissipation.

Acknowledgments

This research was supported by the National Science Foundation under Grant No. DMR-0705959 at KSU and MCYT MAT 2008-06517 at U. Complutense de Madrid.

References

[1] Peña V, Sefrioui Z, Arias D, Leon C, Santamaria J, Varela M, Pennycook S J and Martinez J L 2004 *Phys. Rev. B* **69** 224502
[2] Hu T, Xiao H, Visani C, Sefrioui Z, Santamaria J and Almasan C C 2009 *Phys. Rev. B* **80** 060506(R)
[3] Baibich M N, Broto J M, Fert A, Dau F N V, Petroff F, Etienne P, Creuzet G, Friederich A and Chazelas J 1998 *Phys. Rev. Lett.* **61** 2472
[4] Yanson I K, Naidyuk Y G, Fisun V V, Konovalenko A, Balkashin O P, Triputen L Y and Korenivski V 2007 *Nano Lett.* **7** 927
[5] Hu T, Xiao H, Visani C, Sefrioui Z, Santamaria J and Almasan C C 2008 *Physica B* **403** 1167
[6] Hu T, Xiao H, Visani C, Sefrioui Z, Santamaria J and Almasan C C 2010 *Phys. Rev. B submitted*
[7] Peña V, Visani C, Garcia-Barriocanal J, Arias D, Sefrioui Z, Leon C and Santamaria J 2006 *Phys. Rev. B* **73** 104513
[8] Xiao H, Hu T, Almasan C C, Sayles T A and Maple M B 2008 *Phys. Rev. B* **78** 014510
[9] Vlasko-Vlasov V K, Lin Y K, Miller D J, Welp U, Crabtree G and Nikitenko V I 2000 *Phys. Rev. B* **84** 2239
[10] Visani C, Nemes N M, Rocci M, Sefrioui Z, Leon C, te Velthuis S G E, Hoffmann A, Fitzsimmons M R, Simon F, Feher T, Garcia-Hernandez M and Santamaria J 2010 *Phys. Rev. B* **81** 094512 (2010)
[11] Stahn J, Chakhalian J, Niedermayer C, Hoppler J, Gutberlet T, Voigt J, Treubel F, Habermeier H U, Cristiani G, Keimer B and Bernhard C 2005 *Phys. Rev. B* **71** 140509(R)
[12] Chakhalian J, Freeland J W, Srajer G, Strempfer J, Khaliullin G, Cezar J C, Charlton T, Dalgliesh R, Bernhard C, Cristiani G, Habermeier H U and Keimer B 2006 *Nature Phys.* **2** 244–248
[13] Santamaria J 2006 *Nature Phys.* **2** 229–230
[14] Luo W, Pennycook S J and Pantelides S T 2008 *Phys. Rev. Lett.* **101** 247204

Relationship between band populations and band structure in the three-band Hubbard model

A Avella[1,2,3], F Mancini[1,3], F P Mancini[1,3,4] and E Plekhanov[1,3]

[1] Dipartimento di Fisica "E.R. Caianiello", Università degli Studi di Salerno, 84084 Fisciano (SA), Italy
[2] CNR-SPIN, UoS di Salerno, 84084 Fisciano (SA), Italy
[3] Unità CNISM di Salerno, Università degli Studi di Salerno, 84084 Fisciano (SA), Italy
[4] Sezione INFN di Perugia, 06123 Perugia, Italy

E-mail: fpmancini@physics.unisa.it

Abstract. We study the two-dimensional three-band Hubbard model by means of a four-pole approximation within the Composite Operator Method framework. The model has been solved by considering as basic composite field a four-component spinor field, which includes the p field, the two Hubbard operators for the d field, and a composite operator describing the p field dressed by the spin excitations of the d field. This solution correctly reproduces many results of numerical simulations. In this manuscript, we investigate the relationship between p- and d- populations and the band structure of the model, by varying the on-site potential, the charge-transfer gap, the doping and the hopping integral between the p orbitals.

1. Introduction

It is believed that the two-dimensional three-band Hubbard model [1] can capture sufficient physics to describe real oxide cuprate materials. The model mimics the CuO_2 layers, with one Cu and two O atoms per unit cell, characterized by a strong intra-atomic Coulomb repulsion at Cu sites and by a very high correlation between the $3d$ electrons of copper in the atomic orbital $3d_{x^2-y^2}$ and the $2p$ electrons of oxygen in the atomic orbitals $2p_x$ and $2p_y$. It is not difficult to show that the antibonding degree of freedom of the p-electrons can be counted out through a Bogoliubov transformation [2]. As a result, if one neglects the Coulomb repulsion at O-sites and among the p- and d-electrons, the Hamiltonian of the three-band Hubbard model reads as:

$$
\begin{aligned}
H = & \sum_i [(\varepsilon_d - \mu) d^\dagger(i) d(i) + (\varepsilon_p - \mu) p^\dagger(i) p(i) + U n_\uparrow^d(i) n_\downarrow^d(i)] \\
& + 2 \sum_i \left\{ t_{pp} p^\dagger(i) p^\lambda(i) + t_{pd} [p^{\gamma\dagger}(i) d(i) + d^\dagger(i) p^\gamma(i)] \right\} .
\end{aligned}
\tag{1}
$$

U describes the Coulomb repulsion at Cu sites. The short range hopping integrals t_{pd} and t_{pp} describe the transfer between d and bonding p orbitals, and within the bonding p orbital, respectively. $p(i)$ ($p^\dagger(i)$) and $d(i)$ ($d^\dagger(i)$) are annihilation (creation) operators of p- and d-electrons in the spinor notation, satisfying canonical anticommutation relations. We use the Heisenberg picture: $i = (\mathbf{i}, t)$ where $\mathbf{i}$ stands for the Bravais vector $\mathbf{R}_i$ in the CuO_2 plane. μ is the chemical potential; ε_d and ε_p are the energy levels; $n_\sigma^d(i) = d_\sigma^\dagger(i) d_\sigma(i)$ is the

Published under licence by IOP Publishing Ltd

charge density operator of d-electrons with spin σ. For a generic operator $\Phi(i)$, we define $\Phi^{\kappa}(i) = \sum_j \kappa_{ij}\Phi(j)$, where κ_{ij} is a generic projector. The Fourier transforms of the projectors relevant to our study are: $\gamma(\mathbf{k}) = \sqrt{1-\alpha(\mathbf{k})}$, $\lambda(\mathbf{k}) = [1-2\alpha(\mathbf{k})+\beta(\mathbf{k})]/[1-\alpha(\mathbf{k})]$, where $\alpha(\mathbf{k}) = [\cos(k_x a)+\cos(k_y a)]/2$, and $\beta(\mathbf{k}) = \cos(k_x a)\cos(k_y a)$, with a the lattice constant. α and β correspond to first- and second (diagonal)-neighbor projectors, whereas γ and λ have a more complex structure in result of the bonding - anti-bonding transformation. In the framework of the composite operator method [3], the Hamiltonian (1) can been solved by considering as basic composite field a four-component spinor field: $\psi^{\dagger}(i) = \left(p^{\dagger}(i), \xi^{\dagger}(i), \eta^{\dagger}(i), p_s^{\dagger}(i)\right)$; $\xi(i) = [1-n^d(i)]d(i)$ and $\eta(i) = n^d(i)d(i)$ are the Hubbard projection operators, and $p_s(i) = \sigma_k n_k^d(i)p^{\gamma}(i) - 6c\xi(i)/(2-n_d) - 6b\eta(i)/n_d$. σ_k are the Pauli matrices, and $n_k^d(i) = d^{\dagger}(i)\sigma_k d(i)$ is the spin density operator of d-electrons. b and c are given by: $b = \langle p^{\gamma}(i)\eta^{\dagger}(i)\rangle$, $c = \langle p^{\gamma}(i)\xi^{\dagger}(i)\rangle$, where $\langle\cdots\rangle$ stands for the thermal average in the grand canonical ensemble. As it is shown in Ref. [4], this choice of the composite field $\psi(i)$ correctly reproduces many results of numerical simulations [5]. In the pole approximation [3], the basic field $\psi(i)$ satisfies the equation of motion

$$\left[i\frac{\partial}{\partial t} - \varepsilon(-i\nabla)\right]\psi(i) = 0, \tag{2}$$

where the 4×4 energy matrix $\varepsilon(\mathbf{i},\mathbf{j})$ is given by $\varepsilon(\mathbf{i},\mathbf{j}) = \sum_{\mathbf{l}} m(\mathbf{i},\mathbf{j})I^{-1}(\mathbf{l},\mathbf{j})$. The matrices $I(\mathbf{i},\mathbf{j})$ and $m(\mathbf{i},\mathbf{j})$ are defined as: $I(\mathbf{i},\mathbf{j}) = \langle\{\psi(\mathbf{i},t),\psi^{\dagger}(\mathbf{j},t)\}\rangle$ and $m(\mathbf{i},\mathbf{j}) = \langle\{i\partial\psi(\mathbf{i},t)/\partial t\},\psi^{\dagger}(\mathbf{j},t)\}\rangle$, respectively. The Fourier transforms of these matrices not only depend on the parameters model, but also on a finite set of unknown parameters, namely:

$$\begin{aligned} n_d &= \langle n^d(i)\rangle, & \chi_s^{\alpha} &= \langle n_k^d(i)n_k^{d\alpha}(i)\rangle/3, \\ a_s &= 4\langle p^{\gamma}(i)p_s^{\dagger}(i)\rangle - 6\langle n_{\uparrow}^d(i)n_{\downarrow}^d(i)\rangle, & \chi_s^{\beta} &= \langle n_k^d(i)n_k^{d\beta}(i)\rangle/3, \end{aligned} \tag{3}$$

and μ, b, c, plus two more parameters $m_{44}^{(0)}$ and $m_{44}^{(1)}$, whose lengthy expression is not reported here. By means of the equation of motion (2), the retarded Green's function (GF) $G(i,j) = \theta(t_i - t_j)\langle\{\psi(i),\psi^{\dagger}(j)\}\rangle$ and the correlation function (CF) $C(i,j) = \langle\psi(i)\psi^{\dagger}(j)\rangle$ can be easily computed:

$$\begin{aligned} G(i,j) &= \frac{ia^2}{(2\pi)^3}\sum_{n=1}^{4}\int_{\Omega_B} d^2k\, d\omega\, e^{i\mathbf{k}\cdot(\mathbf{R}_i-\mathbf{R}_j)-i\omega(t_i-t_j)}\frac{\sigma^{(n)}(\mathbf{k})}{\omega - E_n(\mathbf{k}) + i\delta}, \\ C(i,j) &= \frac{a^2}{2(2\pi)^2}\sum_{n=1}^{4}\int_{\Omega_B} d^2k e^{i\mathbf{k}\cdot(\mathbf{R}_i-\mathbf{R}_j)-iE_n(\mathbf{k})(t_i-t_j)}[1+T_n(\mathbf{k})]\sigma^{(n)}(\mathbf{k}). \end{aligned} \tag{4}$$

In the above equations, $E_n(\mathbf{k})$ are the eigenvalues of the energy matrix $\varepsilon(\mathbf{k}) = m(\mathbf{k})I^{-1}(\mathbf{k})$ and $T_n(\mathbf{k}) = \tanh[\beta E_n(\mathbf{k})/2]$. The spectral matrices $\sigma^{(n)}(\mathbf{k})$ are given by: $\sigma_{ab}^{(n)}(\mathbf{k}) = \sum_{c=1}^{4}\Omega_{an}(\mathbf{k})\Omega_{nc}^{-1}(\mathbf{k})I_{cb}(\mathbf{k})$; $\Omega(\mathbf{k})$ is the matrix whose columns are the eigenvectors of $\varepsilon(\mathbf{k})$. The parameters appearing in $I(\mathbf{k})$ and $m(\mathbf{k})$, and therefore in the GF and CF, have to be self-consistently determined. Some of the unknown parameters are expressed as CFs of operators belonging to the basis, and can be determined by means of the following self-consistent equations:

$$\begin{aligned} b &= C_{13}^{\gamma}, & n_d &= 2[1 - C_{22} - C_{33}], \\ c &= C_{12}^{\gamma}, & a_s &= 4(C_{14}^{\gamma} + 3I_{22}^{-1}c^2 + 3I_{33}^{-1}b^2) - 6(n_d/2 - C_{33}). \end{aligned} \tag{5}$$

In the above equations, we defined $C_{ab} = \langle\psi_a(i)\psi_b^{\dagger}(i)\rangle$ and $C_{ab}^{\kappa} = \langle\psi_a^{\kappa}(i)\psi_b^{\dagger}(i)\rangle$, where κ is a generic projector. Algebra constraints [3] provide us with four more equations

$$\begin{aligned} n_T &= n_d + 2(1 - C_{11}), & C_{23} &= 0, \\ C_{24} &= 3C_{12}^{\gamma} - 3cI_{22}^{-1}C_{22}, & C_{34} &= -3bI_{33}^{-1}C_{33}, \end{aligned} \tag{6}$$

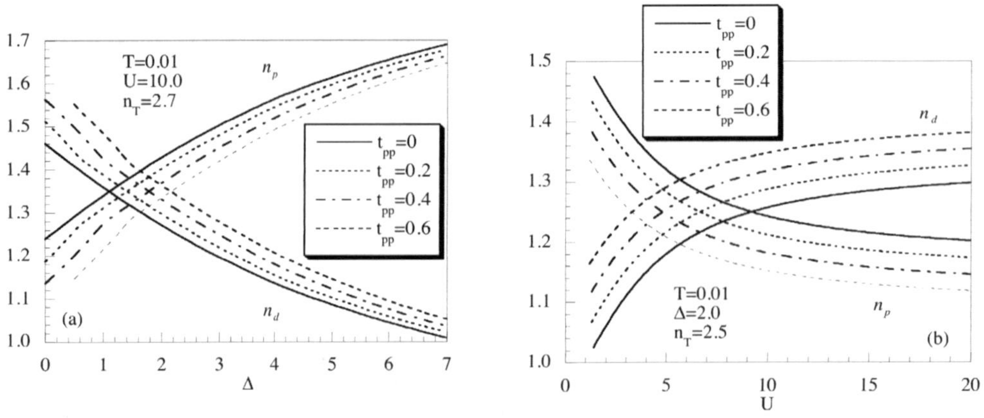

Figure 1. (a) The number of electrons per site n_d and n_p as a function of the charge-transfer gap Δ, for $U = 10$, $n_T = 2.7$, $T = 0.01$ and several values of t_{pp}. (b) The number of electrons per site n_d and n_p as a function of U for $n_T = 2.5$, $\Delta = 2$, $T = 0.01$ and several values of t_{pp}.

allowing us to determine the parameters μ, χ_s^α, $m_{44}^{(0)}$ and $m_{44}^{(1)}$. Here, $n_T = \langle d^\dagger(i)d(i)\rangle + \langle p^\dagger(i)p(i)\rangle = n_d + n_p$ is the number of electrons per site in the CuO_2 layer. By using a decoupling procedure, the last parameter χ_s^β can be determined by means of the CFs as

$$\chi_s^\beta = -2(C_{22}^\beta + 2C_{23}^\beta + C_{33}^\beta)^2. \quad (7)$$

The set of coupled equations (5)-(7) allows us to self-consistently compute the unknown parameters, and, consequently, the GFs and CFs: the physical properties of the system can now be investigated.

2. Results and discussion

In this short manuscript, we focus our attention on the dependence of the relative populations of p- and d- electrons on varying all model parameters (U, $\Delta = U - (\varepsilon_p - \varepsilon_d)$, and t_{pp}) and the total number of electrons per site n_T. In what follows, we will take t_{pd} as energy unit. As it is shown in Ref. [6], the band structure is constituted by four bands: the Hubbard subbands, upper and lower, the one stemming from the bare bonding p level (hereafter the p-dominant band), and the Zhang-Rice singlet band (ZRSB) [7]. In the explored range of parameters, we find that the upper Hubbard subband is located above the ZRSB, which is centered on the chemical potential, whereas the p-dominant and the lower Hubbard subband are below the ZRSB, in this order. In Fig. 1a, we report n_p and n_d as functions of the charge-transfer gap Δ for $U = 10$, $n_T = 2.7$, $T = 0.01$ and $t_{pp} = 0, 0.2, 0.4, 0.6$. On increasing Δ, at a fixed value of the filling between 2 and 3, the upper Hubbard subband (mainly d in content) moves upwards and the d-electron fraction in the p-dominant band and ZRSB diminishes steadily down to zero. n_d goes to 1 at large, but smaller than U, values of Δ since only the lower Hubbard subband (almost exclusively d in content) remains filled with d-electrons.

In Fig. 1b, we report the p- (n_p) and d- (n_d) electron filling as functions of the on-site Coulomb repulsion between d-electrons U for $\Delta = 2$, $n_T = 2.5$, $T = 0.01$ and $t_{pp} = 0, 0.2, 0.4, 0.6$. Very counter-intuitively, the increase of U at fixed Δ leads to an increase of the d fraction in the ZRSB. This quite anomalous behavior can be understood considering that diminishing the ratio between Δ and U by increasing U has the same net effect as to invert the process described in the previous case where the ratio was actually increased by increasing Δ.

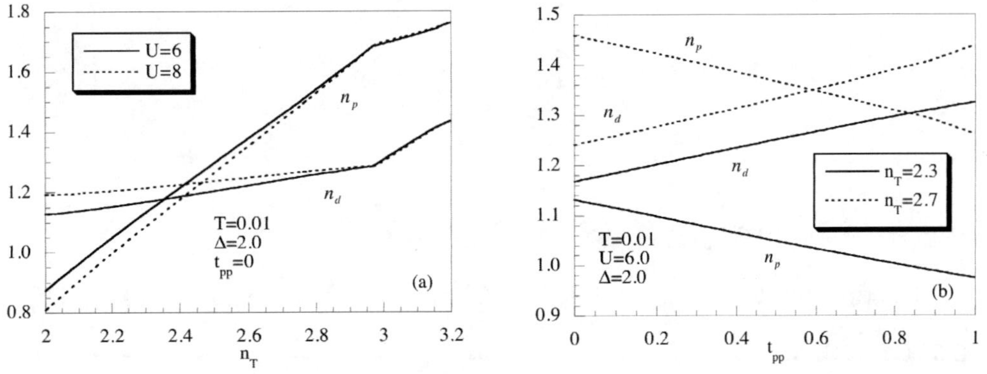

Figure 2. (a) The number of electrons per site n_d and n_p as a function of the total number of electrons, for $t_{pp} = 0$, $\Delta = 2$, $T = 0.01$ and two different values of U. (b) The number of electrons per site n_d and n_p as a function of the hopping integral t_{pp} for $U = 6$, $\Delta = 2$, $T = 0.01$ and two different values of the total number of electrons.

In Fig. 2a, we report n_p and n_d as functions of the total filling n_T for $\Delta = 2$, $T = 0.01$, $t_{pp} = 0$, and $U = 6, 8$. For these ratios of Δ over U and range of the filling, the d-electrons populate only the lower Hubbard subband and partially the ZRSB. Upon increasing the total filling, only the p fraction of the ZRSB steadily increases, while the d one stays practically constant or just increases very slowly. Only when the total filling overcomes $n_T = 3$, the d- electrons start effectively populating the upper Hubbard subband with a significant speed (dn_d/dn_T) comparable to the speed at which the p- electrons fill the ZRSB. In Fig. 2b, we report n_p and n_d as functions of the transfer integral t_{pp} for $U = 6$, $\Delta = 2$, $T = 0.01$, and $n_T = 2.3, 2.7$. On increasing t_{pp} from zero, both the p-dominant band and the ZRSB start to warp along the anti-diagonal $((\pi, 0) \to (0, \pi))$ [6] with a significant increase in the kinetic energy for the sole p component of the band (the band develops crossings along the anti-diagonal). Accordingly, at fixed total filling n_T, the p fraction decreases in favor of the d fraction which takes over as main component of the ZRSB. This behavior is strongly dependent on the ratio Δ/U.

3. Conclusions

In summary, by means of a four-pole approximation, within the Composite Operator Method, we have studied the 2D three-band Hubbard model and, in particular, how the relative populations of p- and d- electrons change on varying all model parameters (U, Δ and t_{pp}) and total filling n_T. The interplay between on-site d-electron repulsion U and the charge-transfer gap Δ can lead to very anomalous behaviors, often very counter-intuitive, that can be understood on the basis of the relative populations of the single-particle bands of the system.

References

[1] Emery V and Reiter G Phys. Rev. B **38** (1988) 4547.
[2] Matsumoto H and Tachiki M Suppl. Prog. Theor. Phys. **101**, (1990) 353.
[3] Mancini F and Avella A Adv. Phys. **53**, (2004) 537.
[4] Fiorentino V, Mancini F, Žaşinas E and Barabanov A F 2001 Phys. Rev. B **64** 214515.
[5] Dopf G, Muramatsu A and Hanke W 1990 Phys. Rev. B **41** 9264; Scalettar R T, Scalapino D J, Sugar R L and White S R 1991 Phys. Rev. B **44** 770; Dopf G *et al.* 1992 Phys. Rev. Lett. **68** 2082.
[6] Avella A, Mancini F, Mancini F P and Plekhanov E 2010 preprint unisa/cond-mat/001052010.
[7] Zhang F C and Rice T M 1988 Phys. Rev. B **37** 3759.

Journal of Physics: Conference Series **273** (2011) 012092 doi:10.1088/1742-6596/273/1/012092

Electronic Structure of $PrFeAsO_{1-\delta}$: An Investigation Using X-ray Absorption and Emission Spectroscopy

B. FREELON,[1] YUHAO LIU,[1] Y.S. LIU,[2] C. R. ROTUNDU,[3] S. D. WILSON,[3] J.H. GUO,[4] JENGLUNG CHEN,[2,4] W. YANG,[4] C. L. CHANG,[2] P.A. GLANS,[4] P. M. SHIRAGE,[5] A. IYO,[5] and R. J. BIRGENEAU [1,3]

[1] Department of Physics, University of California, Berkeley, CA 94720
[2] Department of Physics, Tamkang University, Tamsui, Taiwan 250
[3] Materials Science Division, Lawrence Berkeley National Laboratory, Berkeley, CA 94720
[4] Advanced Light Source, Lawrence Berkeley National Laboratory, Berkeley, CA 94720
[5] National Institute of Advanced Industrial Science and Technology, Tsukuba, Ibaraki 305-8568, Japan

E-mail: BKFreelon@lbl.gov

Abstract. Soft X-ray absorption and emission spectroscopy was used to determine the electronic density of states of oxygen vacated PrFeAsO. The measured oxygen and iron spectra are in good agreement with mean field theory and local density approximations. Iron-arsenic hybridized spectral features reveal that Fe-As bonding is involved in the process of electron addition near the Fermi-level. X-ray emission and absorption spectra shows an increase in the Fe density of states at the Fermi-level as the doping decreases. This doping dependent electronic behavior indicates the possibility of a magnetic instability in the undoped compound.

1. Introduction

The recently discovered Fe-based rare-earth (RE)-pnictide high-temperature superconductors have attracted a great deal of attention.[1] These compounds form several classes including materials in which RE-substitution (RE = La, Ce, Pr, Nd, Sm, Gd) in $RE\text{FeAsO}_{1-x}\text{F}_x$ or $RE\text{FeAsO}_{1-\delta}$ has yielded superconducting transition temperatures (T_c) as high as 55 K in the case of RE = Sm.[2] Early reports suggested that the parent phase might be either a non-magnetic metal or an antiferromagnetic (AF) semi-metal[3] while experiments confirm the latter. It now accepted that the FeAs materials are not in a Mott insulating phase[4] but there is inconclusiveness concerning the degree of electronic correlations.[5] The basic magnetic nature of the parent phase is also debated, although neutron scattering shows convincingly an AF phase below temperatures typically of the order of 140 K. Therefore, it is important to study, experimentally, the electronic behavior in comparison with the cuprates and to reconcile the electronic properties with magnetic information.

The superconducting state is reached in REFeAsO materials by electron doping or hole doping the FeAs planes. Electron-doped FeAs superconductors have been fabricated in at least three ways. All of the doping steps *e.g.* fluorine-substitution for oxygen, oxygen deficiency, and

Published under licence by IOP Publishing Ltd

the substitution of Th^{4+} for RE^{3+} occur in the charge reservoir *RE*-O layer by substitution and induce electron carriers. The effect of electron-doping on the electronic structure has been studied theoretically and experimentally but, so far, the conclusions are indefinite. Local density approximation (LDA), dynamical mean field theory (DMFT), and spin-dependent LDA (LDSA) calculations have given differing results on the question of whether or not the FeAs family of superconductors are strongly correlated. Haule *et al.*,[6] using DMFT, concluded that the materials are strongly correlated. In contrast, Kurmaev *et al.*,[7] and Yang *et al.*,[8] based on Fe $3d$ electronic structure measurements by x-ray absorption and emission spectroscopy, suggested the absence of strong electronic correlations. Angle-resolved[9] and integrated[10] photoemission spectroscopy results from the first reported iron-based superconductor, LaFeOP, as well as $LaFeAsO_{1-x}F_x$, support proposed itinerant magnetism ground state models of these materials. This claim is at variance with the strongly correlated electron picture that emphasizes an antiferromagnetic (AF) ground state Mott physics interpretation.

We report measurements of the electronic structure in the set of materials $PrFeAsO_{1-\delta}$ with $\delta = 0.15$, 0.30 and 0.35. Soft x-ray absorption and emission spectroscopy (XAS and XES) are used to extract element specific density of states (DOS) spectra that can be directly compared to theoretical calculations of the electronic structure of the PrFeAsO materials. XES of the Fermi-level crossing, Fe $3d$ states show a single, narrow dominant peak having an As hybridization shoulder in agreement with other experimental[7] and theoretical[11] results. The Fe conduction band onset shifts to higher energy as a function of electron doping (oxygen vacancy). From Fe XES data, we provide an estimate of the Fe $3d$ bandwidth W. An increase in the Fe electronic density of states at the Fermi-level is demonstrated for less doped samples. This electronic trend indicates an approach toward a magnetic instability in the parent phase of $PrFeAsO_{1-\delta}$.

2. Experimental Procedure

$PrFeAsO_{1-\delta}$ polycrystalline samples were synthesized at high pressure and high temperature using a cubic-anvil high-pressure apparatus.[12-14] The nominal oxygen deficiencies at the start of the synthesis were $\delta = 0.15$, 0.30, and 0.35 in $PrFeAsO_{1-\delta}$. Neutron scattering analysis[15] of $NdFeAsO_{1-\delta}$ indicated that the oxygen deficiencies were smaller than those of the starting materials. It is likely that a similar condition exists for $PrFeAsO_{1-\delta}$. This was likely due to the oxidation during the synthesis process of those materials and $PrFeAsO_{1-\delta}$. *RE*FeAsO are quaternary equiatomic compounds that have a rather simple structure of alternating FeAs and *RE*-O layers with eight atoms in a tetragonal unit cell of space group $P4/nmm$ at 300 K.

Table 1. The lattice parameters (T_c) of the $PrFeAsO_{1-\delta}$ samples.

δ	$T_c(K)$	$a(\AA)$	$c(\AA)$
0.15	3	3.9866	8.6046
0.30	43	3.9770	8.5869
0.35	48	3.9627	8.5721

Table 1 shows that the a-parameter systematically decreases upon increasing the nominal δ, which means that the oxygen deficiencies of the samples systematically increased though, as stated above, there is some ambiguity in the absolute value of the oxygen deficiency parameter δ. The T_c of the $PrFeAsO_{1-\delta}$ samples were measured, using a SQUID magnetometer, under a magnetic field of 5 Oe after zero-field cooling from sufficiently far above T_c.

X-ray absorption and emission spectroscopy was performed at the Advanced Light Source. XES intensity is proportional to the occupied partial density of states (DOS) of the valence band while XAS intensity is proportional to the unoccupied partial DOS associated with the conduction band. O $1s$ XAS and O $K\alpha$-edge XES were used to investigate the electronic structure of the oxygen-associated bands. O $1s$ XAS entails the $1s \rightarrow 2p$ transition and O $K\alpha$ XES records the intensity resulting from the O $2p \rightarrow 1s$ transition. Transition metal (TM) edge x-ray spectroscopy is highly sensitive to the chemical environment of the TM atoms; therefore, we employed Fe $2p$ XAS and Fe $L_{2,3}$-edge XES. In an Fe $L_{2,3}$-edge XAS, Fe $2p$ electrons are excited into empty Fe $3d$ states. Fe $L_{2,3}$ XES profiles were obtained by collecting x-ray emission resulting from the Fe $3d \rightarrow 2p$ transition. All samples were filed immediately before being placed in the ultra-high vacuum synchrotron end-station. All measurements were performed at room-temperature. The beamlines 7.0.1 and 8.0.1 undulators and monochromators produced intense x-ray beams with spot-sizes on the order of 100 microns and energy resolutions of 0.2 eV(0.2 eV) and 0.5 eV(0.6 eV) for oxygen(iron) x-ray absorption and emission, respectively. X-ray absorption spectra were measured in the total fluorescence yield (TFY) mode with a resolving power of $E/\Delta E = 5000$. All spectra were measured with the incident radiation impinging at 20° with respect to the sample surface. The x-ray emission spectrometer, mounted in the synchrotron orbit plane, used a grazing-incidence, two-dimensional multi-channel plate X-ray emission detector and a resolution of 0.35 eV.

3. Results and Discussion

Figure 1 shows the total fluorescence yield (TFY) mode of O $1s$ XAS from $PrFeAsO_{1-\delta}$ (δ = 0.15, 0.30, 0.35). The O XAS profiles, normalized by setting the maximum peak of each spectrum equal to unity, are plotted in a collapsed fashion showing the main absorption lines **A**, **B**, and **C**. The O $1s$ XAS data are consistent with numerous LDA band structure calculations.[16] A weak feature at 1 eV, together with more prominent O spectral band features at about 3 eV and 6 eV above E_F. The pre-peak labeled **A** is primarily due to the hybridization of Fe $3d$, arsenic and oxygen states according to LDA[6,11,17-19] calculations. The main absorption edge **B** is due to the hybridization of Pr $5d$ states and oxygen. A band, **C**, of features at 537 eV forms due to strong hybridization bonds of oxygen with Pr states. Oxygen LDA band structure calculations show the presence of O $2p$ DOS in the range 1 to 2 eV above the Fermi-level where the bonding of Fe $3d$ [*c.f.* Fig. 2(a) of Ref. 17] and As $4p$ is the strongest. LDA predicts the spectral weight of the oxygen $2p$ DOS to be significantly less than that of the Fe or As DOS at low lying energies. The presence of very weak oxygen states near the Fermi-level is not surprising given that the O ions are separated from the FeAs layer. The small O DOS that exists at low energies may be explained in the following way: O $2p$ orbitals can hybridize with the Pr through Pr $5d$ states. Pr $5d$ states, in turn, hybridize with the empty As $5s$ and relatively long-range As $4p$ orbitals. Fe $3d$ electrons form covalent sp^3 bonds with As $4p$ states in the superconducting layer. This complicated hybridization scheme, based on long-range oxygen intermixing, is consistent with band theory. In Ref [17] Sawatzky *et al.* stressed that the As $4p$-Fe interaction is weaker than the planar hybridization found in the cuprates. Reported band structure calculations indicate that the O $2p$ hybridization in *RE*-FeAsO materials is even weaker than the Fe-As bonding, but nevertheless, it is clearly present in our experimental and theoretical results just above and below E_F. The long-range intermixing of the oxygen may play a role in the FeAs layer doping. While a doping mechanism based on weak oxygen intermixing might be plausible, one clearly must also include the electrostatic influence of the doped *RE* layers on the metallic FeAs layers.

The oxygen $1s$ absorption spectra of $PrFeAsO_{1-\delta}$ show a strong similarity to that of $RE\text{FeAsO}_{1-x}F_x$.[20] The overall O $1s$ XAS spectra do not shift significantly with doping δ. We do not observe an onset shift that might reflect a change in the chemical potential; however, Kroll *et al.*[20] reported a trend in F-doped LaFeAsO such that onset of the O $1s$ XAS pre-peak undergoes small shifts toward higher energies with increased electron doping. The O 1*s* XAS onset shift was interpreted as being due to the change in O Madelung potential and the change in the chemical potential due to electron doping. Higher energy absorption features **B** and **C** do not change energy as doping varies (inset Fig. 1.). This agrees with results from $LaFeAsO_{1-x}F_x$ where the higher energy peaks **B** and **C** were stationary with doping and the pre-peak **A** was observed to change position only by 100 meV as x was varied from $x = 0.0$ to 0.1.[20]

In Fig. 2 the XA and XE spectra are plotted to show the occupied and unoccupied oxygen partial density of states (PDOS). The O $K\alpha$ XES contains four major peaks **I-IV**. **I** and **II** are due primarily to Pr, Fe and O hybridization. These features match those in the oxygen spectral weight (DOS) suggested by LSDA+U for PrFeAsO.[18] A dual peak structure occurs in the energy range 4.0 to 5.0 eV below E_F. Several FeAs LDA studies predict this structure, we attribute these peaks to be due do to an oxygen band splitting induced by the Pr crystal field.[21] **III** and **IV** are thought to originate from the oxygen hybridization with Fe-d and As-4p states.[6] LDA calculations by Sawatzky *et al.* [17] support this spectral assignment [*c.f.* Fig. 2(a) of Ref. 17] by showing As DOS as the strongest spectral contribution located $\sim$ 2 - 3 eV below E_F. Arsenic 4p orbitals, possessing a longer range than O 2p orbitals, interact primarily with Fe states; therefore, As 4p orbitals, while bonding to Fe, simultaneously interact with the *RE*-layer, and thereby, may participate in the transfer of charge carriers to the superconducting layer by two observations i) LDA results show that the As DOS dominates the oxygen spectral weight near the Fermi-level and ii) DMFT, LDA and LSDA+U calculations suggest that the As contributions are distributed over a broad energy range. **III** and **IV** both appear to be due to Fe, As, and oxygen hybridization, but in this energy region, according to theoretical calculations, As dominates the near-E_F O spectral weight. This fact may pose challenges in extracting doping dependent trends from O valence band spectra.

The Fe $L_{2,3}$ x-ray absorption spectrum for $PrFeAsO_{1-\delta}$ ($\delta = 0.15$) is shown in Fig. 3(a). The XA process yields a direct measurement of the unoccupied (hole) states of the Fe 3d band. X-rays tuned to the $L_{2,3}$-edge excited Fe 2p electrons ($2p^6 3d^6$) into empty Fe 3d states ($2p^5 3d^7$). Two major Fe $L_{2,3}$ spin-orbit split absorption features result at 707.6 eV (Fe $2p_{3/2} \rightarrow$Fe 3d) and 720 eV (Fe $2p_{1/2} \rightarrow$Fe 3d). The Fe XA spectral profile, Fig. 3(a), of $PrFeAsO_{0.85}$ closely matches that of pure Fe-metal.[8] The shoulder appearing at the high energy tail of the Fe L_3 peak is thought to be due to covalent sp^3 bonds between Fe 3d and As 4p states. [22] Fe XAS is a very sensitive tool for determining the chemical environment, electronic structure and valence of Fe ions in compounds.[23,24] The single-peak structure of the Fe L_3 line in addition to the S. O. splitting of 12.4 eV indicates that the Fe ion of $PrFeAsO_{1-\delta}$ is in a 2+ valence (high spin) state. [8,25] Fig. 3(b) displays non-resonant Fe L-edge XE spectra obtained with 740 eV incident photons. The L-edge XE profile of transition metals is composed of two major peaks due to the $3d \rightarrow 2p_{3/2}$ and $3d \rightarrow 2p_{1/2}$ de-excitation transitions. These XES peaks occur at 705 eV (L_3) and $\sim$717 eV (L_2) and are separated in energy by the S.O. splitting of the Fe $2p_{j=1/2,3/2}$ states.

The XA spectrum in Fig. 3(a) was reduced from raw data, in Fig. 4(a), taken from the samples $PrFeAsO_{1-\delta}$ ($\delta = 0.15$, 0.30 and 0.35). Each raw $PrFeAsO_{1-\delta}$ spectrum consists of the Fe L_3 mainline and an extraneous shoulder on the high-energy tail. The additional spectral feature originates from the crystal field splitting of a small, unreacted portion of one of the starting materials[14] Fe_2O_3. Of the two major Fe_2O_3 XAS peaks **A*** and **B***, only **B*** appears prominently in the $PrFeAsO_{1-\delta}$ data. The relative intensity of **A*** and **B*** is well-known for Fe_2O_3 but it is not manifested in the L_3 band of the raw $PrFeAsO_{1-\delta}$ spectra indicating that Fe_2O_3 has a non-dominant concentration.

In order to further explore this point, the experimental spectra of Fig. 4(a) were fitted to determine the Fe_2O_3 spectral contribution to the $PrFeAsO_{1-\delta}$ XA profile. A superposition of Gaussian functions, 1 - 4, was used in a non-linear least squares fit (gray line in Figure 4(b)) to estimate the contribution of four major characteristics (see vertical hash marks) of the raw $PrFeAsO_{1-\delta}$ spectra. A fit of pure Fe_2O_3 determined the relative magnitudes and widths of the spectral components of the unreacted material. The Fe_2O_3 peaks (see Figure 4(c)) were scaled to the extraneous component, 2, of the raw $PrFeAsO_{1-\delta}$ Fe XAS data. This step ensured that the appropriate magnitude of the Fe_2O_3 spectral contribution was attributed to the raw XAS $PrFeAsO_{1-\delta}$ profiles. The outlined fitting program assumed[26] the profile of raw $PrFeAsO_{1-\delta}$ to be a linear superposition of Fe_2O_3 and the actual $PrFeAsO_{1-\delta}$ spectral intensity. Accordingly, Fe_2O_3 was isolated from the raw data to obtain the resulting $PrFeAsO_{0.70}$ spectral line shape in Fig. 4(d). Similar spectra obtained for $PrFeAsO_{1-\delta}$ ($\delta = 0.15$ and 0.35) were found to be in excellent agreement with XAS reported[7,8,20,22,27] for other iron-arsenide materials. For consistency, this analysis approach was applied to the oxygen XA spectra. We investigated the presence of Fe_2O_3 in the raw $PrFeAsO_{1-\delta}$ oxygen XAS data. Fe_2O_3 has two crystal-field split pre-peaks (at 530 eV) below the O 1s edge.[28,29] An Fe_2O_3 signature is absent from the O 1s XAS pre-peak region of the raw $PrFeAsO_{1-\delta}$ spectra (Figure 1). The close similarity of the raw $PrFeAsO_{1-\delta}$ O 1s pre-peaks in all of the samples rules out the presence of random oxygenation and provides direct evidence that any extraneous oxidation due to Fe_2O_3 is negligibly small[30] in the oxygen spectra. The intensity minimum[29] in the O 1s XA profile of Fe_2O_3 at 532 eV results in an insignificant contribution to the XAS of $PrFeAsO_{1-\delta}$ at this energy. Thus, the small amount of unreacted Fe_2O_3 has a negligible effect on the oxygen XA spectra.

$PrFeAsO_{1-\delta}$ possesses an XE edge intensity ratio $I(L_2)/I(L_3)$ (*c.f.* Ref. [31]) value that is identical to that of other FeAs materials such as $CaFe_2As_2$[27] and LiFeAs.[27] Kurmaev *et al.*,[27] in the case of other iron arsenides, noted the proximity of their $I(L_2)/I(L_3)$ values to that of Fe-metal.

Figure 5 shows a non-resonant VB Fe L_3 XE spectral profile that possesses a single dominant peak and a low-energy shoulder in agreement with other studies.[7] The main peak is due primarily to Fe 3d bands[7,32-33] and we interpret the shoulder near 704 eV to originate from the hybridization of Fe 3d and As 4p states.[6] This assignment is supported by LDSA+U calculations which show that the spectral weight $\sim$ 3.0 eV below E_F derives from both Fe and arsenic states.[16] The Fe XES can be used to ascertain information about the bandwidth of the Fe 3d states. A measurement of the XES full-width at half maximum (FWHM) yields a width of 3.1 eV; however, this value is core-hole lifetime and multiplet broadened and cannot be taken as a direct measurement of the Fe 3d bandwidth. The width 2 eV of the dominant portion of the Fe L_3 XE peak can serve as an upper limit of the Fe 3d bandwidth W. In other words, the Fe 3d width should be no greater than 2 eV. A range of reported values for the Coulomb parameter U, from 0.8 eV[34] for a basis consisting of only Fe 3d orbitals to 2 eV[17-18,35] for a As p - Fe 3d mixing model,[35-36] is consistent with our O and iron spectra. Recently, Wang *et al.*[8] argued that $U \sim 2$ eV based on theoretical calculations that matched x-ray spectroscopy in the limiting condition. These values imply that the magnitude of U is similar to or less than that of the estimated Fe 3d bandwidth. Importantly, the relative magnitude of U and W can implicate the extent of electron correlations in $PrFeAsO_{1-\delta}$. In the present case, $U/W \sim 1$ suggests that $PrFeAsO_{1-\delta}$ system is in an intermediate state between the metallic and insulating phases. [6,17]

In Fig. 6(a) we present a magnified view of the low-energy tail of the Fe L_3 XAS peaks of $PrFeAsO_{1-\delta}$ ($\delta = 0.15$, 0.30 and 0.35). The peak onset (see the arrow), defined as 10 percent of the absorption maximum, shifts as a function of doping. The onset's upward shift of $\sim$ 900 meV to higher energies is due to the addition of electrons at Fe metallic sites. Equivalently, the onset shifts represent a reduction in the XAS intensity (hole DOS) near the Fermi-level. Figure 6(b) shows normalized XA spectra and Fe L-edge XES. All the doped $PrFeAsO_{1-\delta}$ samples are arranged in composite intensity profiles. If $PrFeAsO_{1-\delta}$ is taken to be metallic, the Fermi-level E_F can be approximated as the intersection of the Fe XE and XA spectral profiles. Fe 2p XA and Fe L-edge XE spectral intersection points are reduced relative to the valence and conduction band maxima. In addition, these points shift by ~ 200 meV (Fig. 6(b))on account of the doping δ. A closer examination of this shift is afforded by the left inset of Fig. 6(b) where the arrows indicate the positions of the points of intersection of the XA and XE spectra. The Fe L_3 XAS main-line shifts by $\sim$0.15 eV upward with doping from $\delta = 0.15$ to 0.35; therefore, the Fe conduction band (CB) maximum and low-energy electronic structure *i.e.*, the onset, undergoes an observable change as a result of doping. Evolution of the Fe L_3 peak onset as δ goes from 0.15 to 0.35 indicates the addition of doped electrons to the low-energy Fe states. The removal of low energy (near-E_F) unoccupied spectral weight is equivalent to electron addition per Fe atom.

The changes in the Fe DOS near the Fermi-level are accompanied by a systematic increase in the hole DOS at the Fe-As hybridization peak (shoulder). Spectral changes in the high energy shoulder of the Fe L_3-edge XA peak suggest that, concomitant with the addition of electrons to states near E_F, doping induces the removal of electrons from covalent bonding states associated with Fe and As atoms. This suggests the DOS at the Fermi-level $D(E_F)$, (right inset of Fig. 6(b)) increases as the doping decreases, giving a signal of a possible approach toward a parent phase antiferromagnetic instability.[37-38] The trend also suggests that the parent compound is in the itinerant magnetism regime. The mechanism of the magnetic instability in FeAs parent materials is not yet known. While the spin-density wave (SDW) in LaFeAsO was argued to originate from the Fermi-surface nesting[39] of electron and hole pockets, an alternative proposal argued that the super exchange interaction of Fe ions mediated through the off-plane As atoms was important.[32,40-41] The doping behavior, presented here, of the Fe L_3 XAS shoulder is consistent with the notion that the Fe-As bonding might play an important role in the occurrence of the magnetic instability of the parent phase. Therefore, XES offers a method to access the link between the electronic structure at the Fermi-level and the magnetic behavior in FeAs superconductors.

4. Conclusion

Oxygen and iron soft X-ray emission and absorption data has been used to probe the structure of the low-lying electronic states of $PrFeAsO_{1-\delta}$ as a function of doping. The results indicate that the oxygen density of states of $PrFeAsO_{1-\delta}$ is well hybridized and is consistent with a band structure models for this material. The O $2p$ states are shown to be involved in long-range mixing, into the FeAs hybridized states, which might facilitate electron transfer between the layers. This long-range, non-planar mechanism is consistent with band structural calculations as well as the report of three-dimensional superconductivity and the magnetism in some FeAs materials. The oxygen XAS and XES are well-described by LDA + small-U, LSDA and, DMFT calculations; however, only LDA + small-U matches the observed low energy O DOS just above and below E_F. The $PrFeAsO_{1-\delta}$ occupied and unoccupied O bands show behavior that is similar to that observed in the fluorine-doped material REFeAs$O_{1-x}F_x$. However, the energy shifts in the onset of the Fe unoccupied band, as given by the Fe XAS, is consistent with electron doping. The Fermi-level density of states $D(E_F)$ increases as the parent phase is approached. This behavior reveals the tendency toward itinerant magnetism with decreasing doping. We believe that the experimental spectra given here are robust and should guide future theoretical models and understanding.

Acknowledgments

This work was supported by the U.S. Department of Energy, under Contract No. DE-AC02-05CH11231 and Office of Basic Energy Sciences U.S. DOE under Contract No. DE-AC03-76SF008. P.M.S. is grateful to the Japan Society for the Promotion of Science for the JSPS postdoctoral fellowship.

Journal of Physics: Conference Series **273** (2011) 012092 doi:10.1088/1742-6596/273/1/012092

References

[1] Y. Kamihara, T.Watanabe, M. Hirano, and H. Hosono: *J. Am. Chem. Soc.* **130** (2008) 3296.

[2] Ren Zhi-An, Lu Wei, Yang Jie, Yi Wei, Shen Xiao-Li, Zheng-Cai, Che Guang-Can, Dong Xiao-Li, Sun Li-Ling, Zhou Fang and Zhao Zhong-Xian: Chinese Physics Letters 25 (2008) 2215; Zhi-An Ren, Guang-Can Che, Xiao-Li Dong, Jie Yang, Wei Lu, Wei Yi, Xiao-Li Shen, Zheng-Cai Li, Li-Ling Sun, Fang Zhou and Zhong-Xian Zhao: *Europhys. Lett.* **83** (2008) 17002.

[3] F. Ma and Z.-Y. Lu: *Phys. Rev. B* **78** (2008) 033111.

[4] Gianluca Giovannetti, Sanjeev Kumar and Jeroen van den Brink: *Physica B* **403** (2008) 3653.

[5] I. I. Mazin, . I. Mazin, M. D. Johannes, L. Boeri, K. Koepernik, and D. J. Singh: *Phys. Rev. B* **78** (2008) 0851104.

[6] Haule, J. H. Shim, and G. Kotliar: *Phys. Rev. Lett.* **100** (2008) 226402.

[7] E. Z. Kurmaev, J. A. McLeod, A. Buling, N. A. Skorikov, A. Moewes, M. Neumann, M. A. Korotin, Yu. A. Izyumov, N. Ni, and P. C. Canfield: *Phys. Rev. B* **80** (2009) 054508.

[8] W. L. Yang, A. P. Sorini, C-C. Chen, B. Moritz, W.-S. Lee, F. Vernay, P. Olalde-Velasco, J. D. Denlinger, B. Delley, J.-H. Chu, J. G. Analytis, I. R. Fisher, J. Yang, W. Lu, Z. X. Zhao, J. van den Brink, Z. Hussain, Z.-X. Shen, and T. P. Devereaux: *Phys. Rev. B* **80** (2009) 014508.

[9] D. H. Lu, M. Yi, S.-K. Mo, A. S. Erickson, J. Analytis, J.-H. Chu, D. J. Singh, Z. Hussain, T. H. Geballe, I. R. Fisher and Z.-X. Shen: *Nature* **455** (2008) 81.

[10] D.R. Garcia, C. Jozwiak, C.G. Hwang,A. Fedorov, S.M. Hanrahan,S.D. Wilson, C.R. Rotundu, B.K. Freelon, R.J. Birgeneau, E. Bourret-Courchesne and A. Lanzara: *Phys. Rev. B* **78** (2008) 245119.

[11] C. Cao, P. J. Hirschfeld, and H.-P. Cheng: *Phys. Rev. B* **77** (2008) 220506.

[12] K. Miyazawa, K. Kihou, P. M. Shirage, C. H. Lee, H. Kito, H. Eisaki, and A. Iyo, J.: *Phys. Soc. Jpn.* **78** (2008) in press.

[13] K. Hashimoto, T. Shibauchi, T. Kato, K. Ikada, R. Okazaki, H. Shishido, M. Ishikado, H. Kito, A. Iyo, H. Eisaki, S. Shamoto, Y. Matsuda: *Phys. Rev. Lett.* **102** (2009) 017002.

[14] P. M. Shirage, K. Miyazawa, M. Ishikado, K. Kihou, C. H. Lee, N. Takeshita, H. Matsuhata, R. Kumai, Y. Tomioka, T. Ito, H. Kito, H. Eisaki, S. Shamoto and A. Iyo: *Physica C* **469** (2009) 355.

[15] C.H. Lee, A. Iyo, H. Eisaki, H. Kito, M.T. Fernandez-Diaz, T. Ito, K. Kihou, H. Matsuhata, M. Braden, and K. Yamada: *J. Phys. Soc. Jpn.* **77** (2008) 083704.

[16] D. J. Singh and M.-H. Du: *Phys. Rev. Lett.* **100** (2008) 237003.

[17] G. A. Sawatzky, I. S. Elfimov, J. van den Brink, and J. Zaanen: *Europhys. Lett.* **86** (2009) 17006.

[18] I. A. Nekrasov, Z. V. Pchelkina, and M. V. Sadovskii: *JETP Lett.* **87** (2008) 620.

[19] F. Ma and Z.-Y. Lu: *Phys. Rev. B* **78** (2008) 033111.

[20] T. Kroll, S. Bonhommeau, T. Kachel, H.A. Duerr, J. Werner, G. Behr, A.Koitzsch, R. Huebel, S. Leger, R. Schoenfelder, A. Ariffin, R. Manzke, F.M.F. de Groot, J. Fink, H. Eschrig, B. Buechner, M. Knupfer: *Phys. Rev. B* **78** (2008) 220502.

[21] D. J. Singh: *private communication* (2009).

[22] F. Bondino, E. Magnano, M. Malvestuto, F. Parmigiani, M. A. McGuire, A. S. Sefat, B. C. Sales, R. Jin, D. Mandrus, E.W. Plummer, D. J. Singh, and N. Mannella: *Phys. Rev. Lett* **101** (2008) 267001.

[23] R. D. Leapman and L. A. Grunes: *Phys. Rev. Lett.* **45** (1980) 397.

[24] B. T. Thole and G. van der Laan: *Phys. Rev. B 38* (1988) 3158.

[25] Frank M. F. de Groot, Pieter Glatzel, Uwe Bergmann, Peter A. van Aken, Raul A. Barrea, Stephan Klemme, Michael Hvecker, Axel Knop-Gericke, Willem M. Heijboer, and Bert M. Weckhuysen: *J. Phys. Chem. B* **109** (2005) 20751.

[26] D. Nolle, E. Goering, T. Tietze, G. Schtz, A. Figuerola and L. Manna: *New. J. Phys.* **11** (2009) 033034.

[27] E. Z. Kurmaev, J. A. McLeod, N. A. Skorikov, L. D. Finkelstein, A. Moewes, Yu A Izyumov and S. Clarke: *J. Phys. Condens. Matter* **21** (2009) 345701.

[28] F. M. F. de Groot, M. Grioni, J. C. Fuggle, J. Ghijsen, G. A. Sawatzky and H. Petersen: *Phys. Rev. B* 40 (1989) 5715.

[29] J. A. Liero, G. Drager and W. Czolbe: *X-ray Spectroscopy* **22** (1993) 272.

[30] A recent report by Nie et. al., *arXiv:0912.4539v1*, suggests that oxidation of Fe-based samples or starting materials might contribute to variations in the intensity of the O 1s XAS pre-peak if the exposure to O_2 is sufficient. The lack of dual iron-oxide pre-peaks and the absence of change in the pre-peak intensity of the PrFeAsO samples supports our claim that the unreacted Fe_2O_3 has a negligible effect on the oxygen XA spectra.

[31] B. Freelon, Y.S. Liu, Y.H. Liu, C. R. Rotundu, S. D. Wilosn, J. Guo, J.L. Chen, W. Yang, C. Chang, Per. Anders. Glans, P. Shirage, A. Iyo, and Robert. J. Birgeneau: *J. Phys. Jap. Soc.* **In press** (2010)

[32] K. Schwarz, A. Neckel and J. Nordgren: *J. Phys. F: Metal Phys.* **9** (1979) 2509.

[33] G. van der Laan and I. W. Kirkman: *J. Phys. Condens. Matter* **4** (1992) 4189.

[34] V. I. Anisimov, Dm. M. Korotin, A. V. Kozhevnikov, J. Kunes, A. O. Shorikov, S. L. Skornyakov and S. V. Streltsov: *J. Phys. Condens. Matter* **21** (2009) 075602.
[35] Jiansheng Wu, Philip Phillips, and H. Castro Neto: *Phys. Rev. Lett.* **101** (2008) 126401.
[36] V. Cvetkovic and Z. Tesanovic: *Europhys. Lett.* **85** (2009) 37002.
[37] R. Skomski: *Simple Models of Magnetism* (Oxford University Press, Oxford, 2008) pg.121.
[38] I. I. Mazin, D. J. Singh, M. D. Johannes and M.H. Du: *Phys. Rev. Lett.* **101** (2008) 057003.
[39] J. Dong, H. J. Zhang, G. Xu, Z. Li, G. Li, W. Z. Hu, D. Wu, G. F. Chen, X. Dai, J. L. Luo, Z. Fang and N. L. Wang: *Europhys. Lett.* **83** (2008) 27006.
[40] T. Yildrim: *Phys. Rev. Lett.* **101** (2008) 057010.
[41] Q. Si and E. Abrahams: *Phys. Rev. Lett.* **101** (2008) 076401.

Figure Captions Freelon *et. al.*,

Fig. 1. (a) Oxygen 1s x-ray absorption spectra of $PrFeAsO_{1-\delta}$ ($\delta = 0.15$, 0.30 and 0.35 taken in total fluorescence yield (TFY) mode. Feature **A** is the oxygen pre-peak due to O-Fe 3d and As 4p hybridization; **B** is an O-Pr 5d hybridization peak. Features **A**, **B** and **C** (see inset) do not shift with respect to oxygen vacancy concentration.

Fig. 2. Normalized O 1s XA and O $K\alpha$ XE spectra of $PrFeAsO_{1-\delta}$ ($\delta = 0.15$, 0.30 and 0.35). The XES(XAS) reflect the occupied(unoccupied) oxygen partial density of states. The major spectral features **I-IV**, **A**, **B** and **C** are discussed in the text.

Fig. 3. Fe L-edge X-ray absorption and emission spectra of $PrFeAsO_{0.85}$ $T_c = 3$ K. (a) the total electron yield (TEY) mode Fe 2p XA intensity plotted as a function of photon energy. (b)The non-resonant Fe L-edge XE intensity profile collected at an incident energy of 740 eV.

Fig. 4. The Fe TFY L_3 XAS band of (a) raw $PrFeAsO_{1-\delta}$ ($\delta = 0.15$, 0.30 and 0.35). Both raw $PrFeAsO_{0.70}$ (b) and Fe_2O_3 (c) Fe L_3 XA spectra are shown with fitted curves (gray lines) based on Gaussian functions (see text). (d) $PrFeAsO_{0.70}$ Fe L_3 XA spectrum.

Fig. 5. A Comparison of Fe 2p XAS and Fe $L_{2,3}$ XES data of FeO, $PrFeAsO_{0.85}$ and Fe-metal. The ratio $I(L_2)/I(L_3)$ of the integrated XES main peak intensities of samples FeO, Fe-metal and $PrFeAsO_{0.85}$.

Fig. 6. Enhanced views of the normalized Fe 2p XA and Fe $L_{2,3}$ XE spectra of $PrFeAsO_{1-\delta}$ ($\delta = 0.15$, 0.30 and 0.35). (a) The onset of the leading edge of the Fe L_3 XA edge undergoes a shift as a function of oxygen vacancy doping. (b) The XE and XA spectra are juxtaposed in order to depict the Fe band occupied and unoccupied electronic bandstructure versus doping. XE spectra, collected at an incident photon energy of 740 eV, do not change with doping. Fe 2p XA spectra show that the conduction band maximum does not shift with electron doping. A shift in the chemical potential is revealed (inset) as a change in the normalized intensity of the XE and XA spectral intersection.

Figure 1, Freelon *et al.*

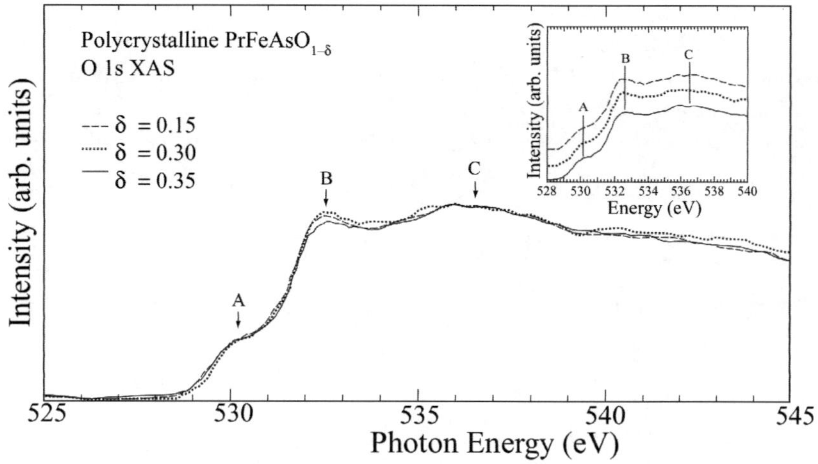

Figure 1. (a) Oxygen 1*s* x-ray absorption spectra of $PrFeAsO_{1-\delta}$ (δ = 0.15, 0.30 and 0.35 taken in total fluorescence yield (TFY) mode. Feature **A** is the oxygen pre-peak due to O-Fe 3*d* and As 4*p* hybridization; **B** is an O-Pr 5*d* hybridization peak. Features **A**, **B** and **C** (see inset) do not shift with respect to oxygen vacancy concentration.

Figure 2, Freelon *et al.*

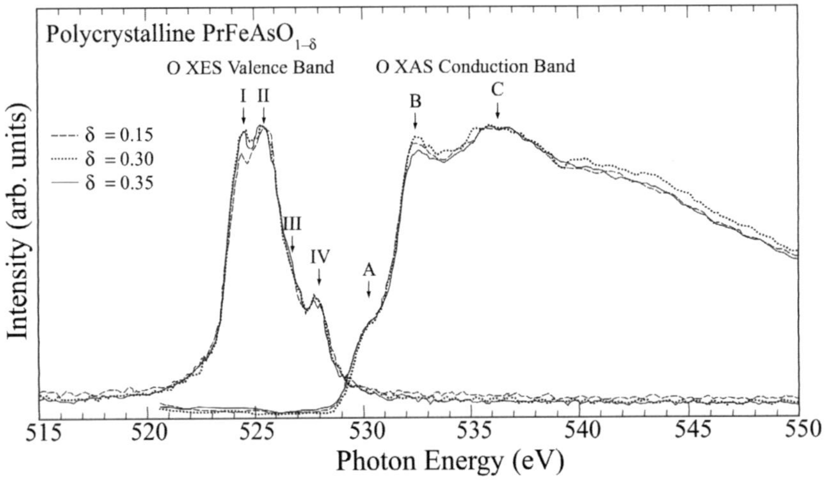

Figure 2. Normalized O 1s XA and O $K\alpha$ XE spectra of $PrFeAsO_{1-\delta}$ ($\delta = 0.15$, 0.30 and 0.35). The XES(XAS) reflect the occupied(unoccupied) oxygen partial density of states. The major spectral features **I-IV**, **A**, **B** and **C** are discussed in the text.

Figure 3, Freelon *et al.*

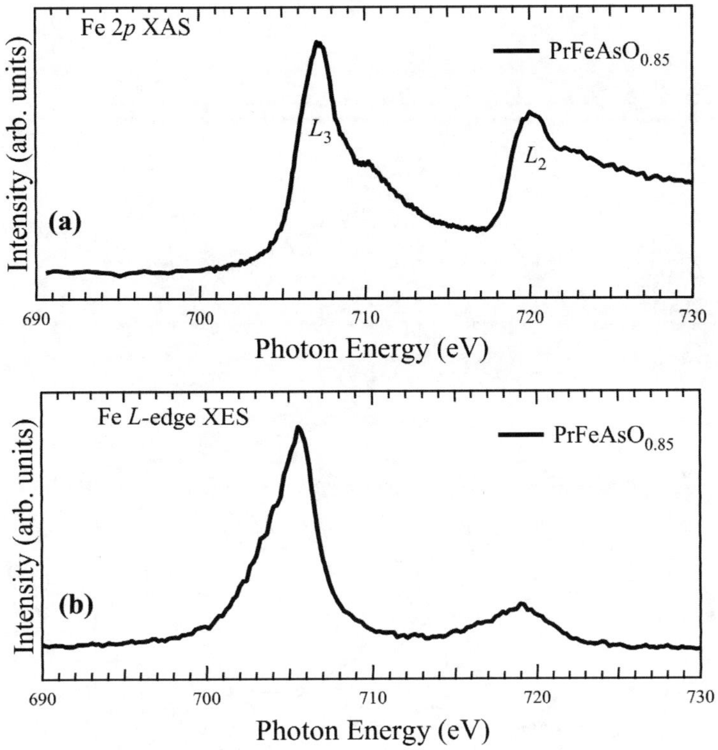

Figure 3. Fe *L*-edge X-ray absorption and emission spectra of $PrFeAsO_{0.85}$ $T_c = 3$ K. (a) the total electron yield (TEY) mode Fe $2p$ XA intensity plotted as a function of photon energy. (b)The non-resonant Fe *L*-edge XE intensity profile collected at an incident energy of 740 eV.

Figure 4, Freelon, *et. al.*

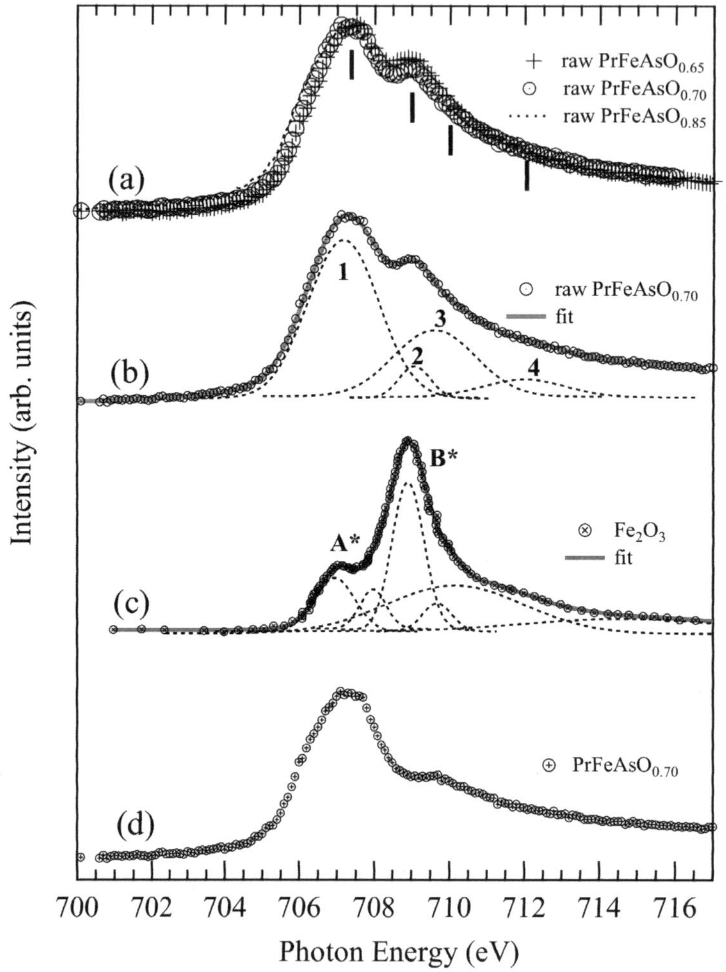

Figure 4. The Fe TFY L_3 XAS band of (a) raw $PrFeAsO_{1-\delta}$ ($\delta = 0.15$, 0.30 and 0.35). Both raw $PrFeAsO_{0.70}$ (b) and Fe_2O_3 (c) Fe L_3 XA spectra are shown with fitted curves (gray lines) based on Gaussian functions (see text). (d) $PrFeAsO_{0.70}$ Fe L_3 XA spectrum.

Figure 5, Freelon ***et al.***

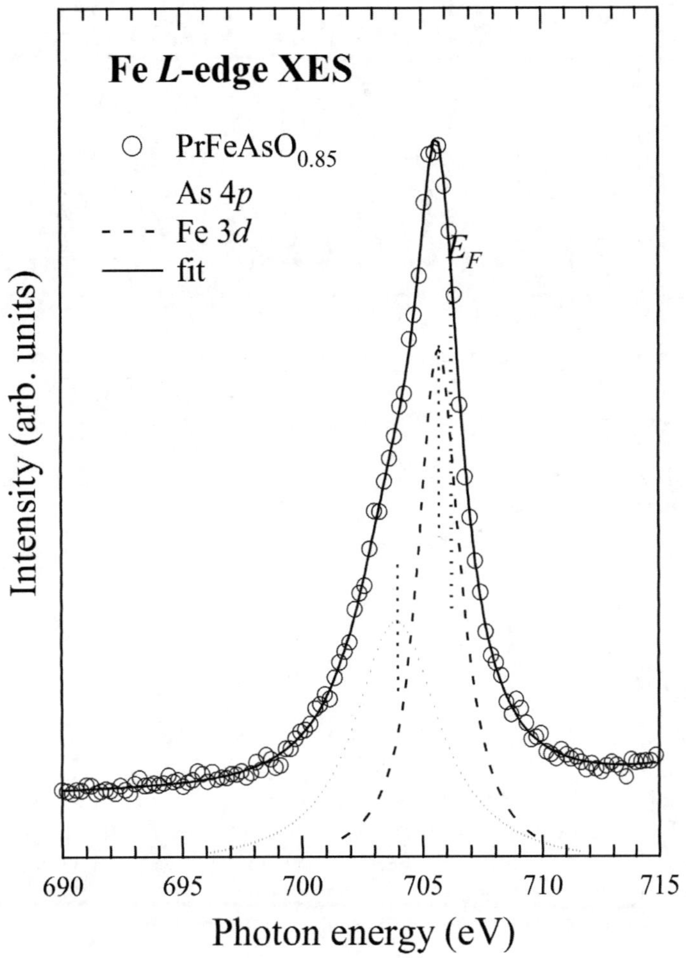

Figure 5. A Comparison of Fe $2p$ XAS and Fe $L_{2,3}$ XES data of FeO, $PrFeAsO_{0.85}$ and Fe-metal. The ratio $I(L_2)/I(L_3)$ of the integrated XES main peak intensities of samples FeO, Fe-metal and $PrFeAsO_{0.85}$.

Figure 6, Freelon *et al.*

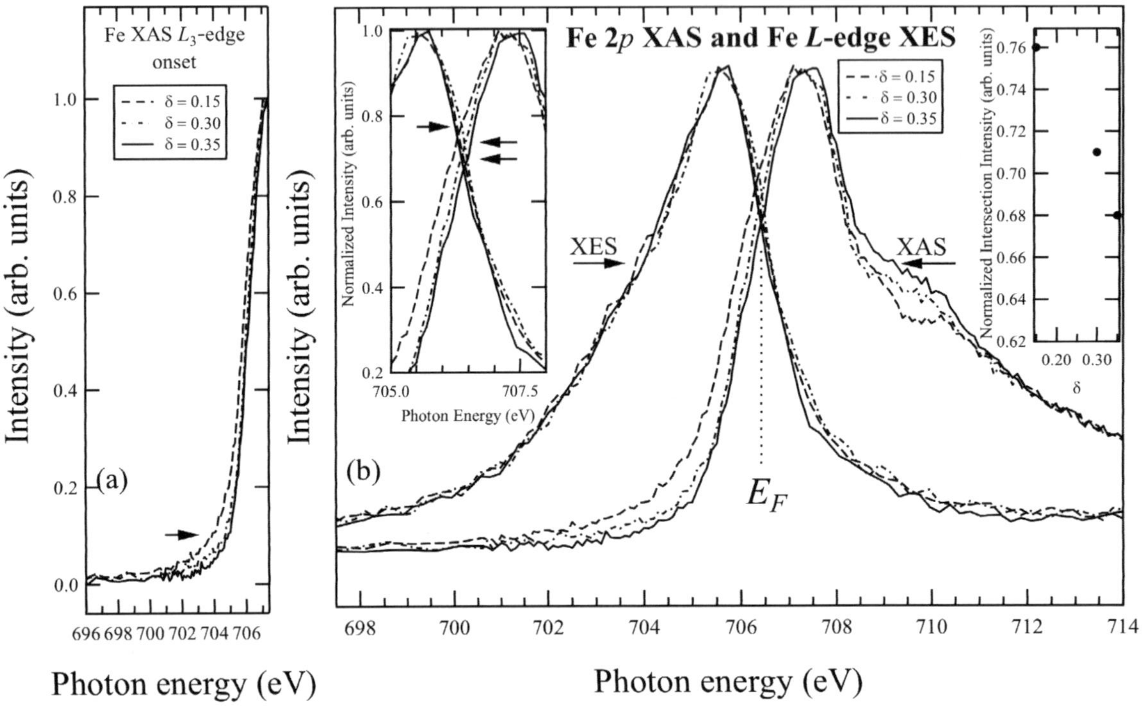

Figure 6. Enhanced views of the normalized Fe $2p$ XA and Fe $L_{2,3}$ XE spectra of $PrFeAsO_{1-\delta}$ (δ = 0.15, 0.30 and 0.35). (a) The onset of the leading edge of the Fe L_3 XA edge undergoes a shift as a function of oxygen vacancy doping. (b) The XE and XA spectra are juxtaposed in order to depict the Fe band occupied and unoccupied electronic bandstructure versus doping. XE spectra, collected at an incident photon energy of 740 eV, do not change with doping. Fe $2p$ XA spectra show that the conduction band maximum does not shift with electron doping. A shift in the chemical potential is revealed (inset) as a change in the normalized intensity of the XE and XA spectral intersection.

Journal of Physics: Conference Series **273** (2011) 012093 doi:10.1088/1742-6596/273/1/012093

La doping effects in the coupling between localized and itinerant electronic states in $EuFe_2As_2$ probed by Eu^{2+} ESR.

FA Garcia[1], EM Bittar [1], C Adriano [1], TM Garitezi [1],C Rettori[1], PG Pagliuso[1]

[1]Instituto de Física "Gleb Wataghin", UNICAMP, Campinas-SP, 13083-970, Brasil.

E-mail: `fgarcia@ifi.unicamp.br`

Abstract. In this work we present an Electron Spin Resonance (ESR) study on La-doped $EuFe_2As_2$ aiming to further understand the coupling of localized spin states and itinerant electrons in these systems. The temperature dependence of the Eu^{2+} ESR linewidth, g-value and intensity was investigated for both monocrystalline and powdered samples. Our results confirm the suggestion of previous ESR experiments on the pure compound [1], that the SDW transition dramatically changes the coupling between localized and itinerant states. In addition, we found a peculiar dome like shape in the Eu^{2+} ESR linewidth temperature dependence for temperatures between T_{SDW} and T_N.

1. INTRODUCTION

The discovery of superconductivity in Fe-based pnictides [2] gave rise to a series of experimental and theoretical work on this field. As previously speculated in the case of cuprates, magnetic fluctuations appears to be of great importance on the onset and stabilization of the superconducting state. It makes the study of related magnetic non-superconducting compounds an important tool to understand the role and interplay of these fluctuations with superconductivity.

Among these compounds, $EuFe_2As_2$ calls attention of the community since it can be tuned to a superconducting state upon doping on any of the Eu, Fe and As sites [3, 4, 5] and also by applying pressure, which leads to the appearance of a re-entrant superconducting state [6]. Its electronic and magnetic properties are the subject of several recent studies [7, 8].

Electron Spin Resonance (ESR) is a very sensitive microscopic probe to local spin fluctuations and magnetic interactions [9]. Moreover, it can also extract, in some cases, information about the superconducting state of unconventional superconductors as it has been shown recently for Fe-based pnictides superconductors [3, 10].

This work discuss the La doping effects on the coupling between the localized Eu^{2+} and itinerant electronic states as probed by ESR. In general, our results are in qualitative agreement with a previous work done on the pure compound [1]. Some peculiarities, however, exist and are reported in this paper, being the most interesting a conspicuous dome-like shape in the ESR linewidth in the temperature interval between T_N and T_{SDW}.

Published under licence by IOP Publishing Ltd

Journal of Physics: Conference Series **273** (2011) 012093 doi:10.1088/1742-6596/273/1/012093

2. METHOD AND RESULTS

Single crystals of $Eu_{0.6}La_{0.4}Fe_2As_2$ were grown in a molten Sn flux according to the method described in Ref. [11]. The crystal structure (space group I4/mmm) and phase purity were checked by X-ray powder diffraction. The La content is the stoichiometric nominal value used in the synthesis. The ESR experiments were carried on both powdered and single crystal samples. The powdered sample was obtained from crushed single crystals with high grain size homogeneity and for the single crystal experiment a platelet ($\approx 0.5mm \times 0.2mm \times 0.01mm$) shaped sample was chosen.

The ESR spectra were taken in a X (9.48 GHz)-band Bruker spectrometer using appropriate resonators coupled to a T-controller of helium gas flux system for $4.2 \leq T \leq 300$ K. For the single crystals, the ESR lines usually showed a Dysonian lineshape. This asymmetric lineshape reflects the skin depth effects due to the penetration of the microwave on a metallic surfface, in a regime that the skin depth is smaller than the sample size. For the powdered sample, a Lorentzian lineshape was observed. At low temperatures, close to the reported T_N for the parent compound, substantial lineshape deformation was observed.

Figures 1a and 1b show some representative spectra for the powdered and single crystal samples, respectively. All the experimental data were fitted to a Dysonian lineshape to extract the ESR linewidth (ΔH), resonance field (converted to g-value), intensity (double integration of the resonance spectrum) and the absorption/dispersion relation (D/A , that express the lineshape asymmetry, being 0 for a Lorentzian lineshape.) In figure 1a we show a comparison between the fitting and experimental results and a very good agreement is achieved for all T-interval. Figure 1b shows that above the magnetic transition ($T_N \approx 20$K) no significant anisotropy was found in the spectra.

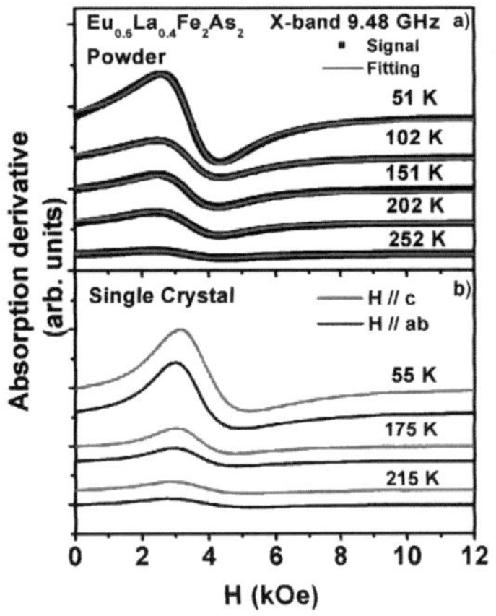

Figure 1. X-Band Eu^{2+} ESR spectra for a a) powdered sample, with a comparison between the fitting and experimental results and b) single crystal sample, in some selected temperature region, comparing spectra taken for different orientations of the applied magnetic field.

Figure 2 presents the ESR linewidth as a function of temperature. At high temperatures ($T > 200$K), the linewidth increases linearly with temperature in a Korringa-like relaxation with $b = 3(1)$ Oe/K (powdered sample), a value significantly lower than the one previous reported of ≈ 8 Oe/K [1]. Below this temperature, this relaxation ceases and the linewidth start to increase until reaching a maximum at $T \approx 100$K . Then it decreases again down to $T \approx 25$K, where magnetic fluctuations become important resulting in strong anisotropy (single crystal) and broadening (powder) of the resonance lines. The inset in this figure, shows that, within experimental error, the g-value is isotropic and temperature independent for $T > 25$K. The g-value of 1.96(9) is consistent with a Eu^{2+} g-value and this same value was reported earlier

[1]. The described dome-like shape in the linewidth is a robust result that was not reported in this previous work.

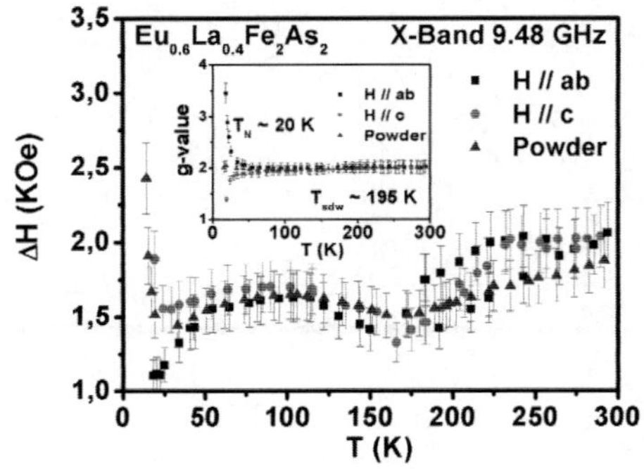

Figure 2. X-Band ESR linewidth behavior in the temperature interval $4.2 \le T \le 300$K for all experiments. In the inset, the calculated g-values are presented as a function of temperature.

Figure 3 presents the normalized ESR intensity for temperatures well above ($T > 50$K) the magnetic transition at $T \approx 25$K. The experimental results follow closely a Curie law confirming that our resonance arises from a localized electronic state. The absorption/dispersion relation is plotted in the inset. Away from the magnetic transition, this quantity is closely constant, with a variation smaller than 10%. Both results are in contrast to those previously reported for the pure compound [1].

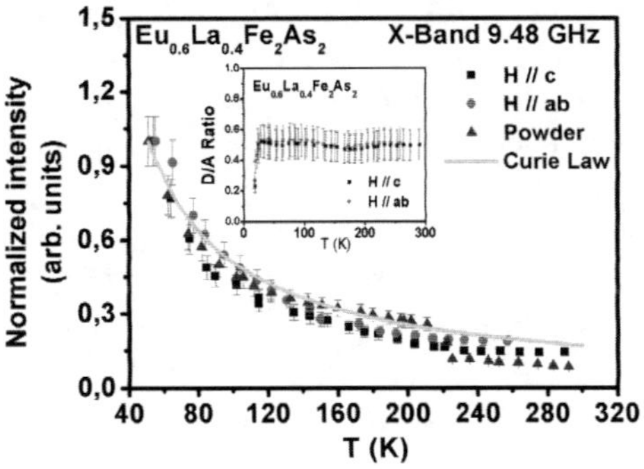

Figure 3. X-Band ESR normalized intensity as a function of temperature. The intensities are normalized for $T > 50$K since due to the proximity to the magnetic transition at $T \approx 25$K, and also low temperature spin-spin interaction, the ESR intensity may deviate from the expected Curie behavior. The inset shows that the absorption/dispersion ratio is nearly constant in a broad T-interval.

3. DISCUSSION

The key aspects of the magnetic ordering phenomena in $EuFe_2As_2$ and similar compounds are reflected in changes of the ESR parameters. In particular, appearance of SDW ordering is reported to be related to changes in the linewidth ($EuFe_2As_2$ [1], $EuFe_{2-x}Co_xAs_2$ [3]) g-value ($EuFe_{2-x}Co_xAs_2$ [3]) and intensity ($EuFe_{2-x}Co_xAs_2$ [3]) behavior.

In the linewidth, the reported changes are the appearance of anisotropic linewidth and modification in the relaxation regime, where the linewidth is closely constant below T_{SDW}. It is usually thought that the SDW transition leads to nesting of the Fermi surface that greatly reduces the exchange interaction of the conduction electrons with the localized Eu^{2+} states. This exchange interaction dominates the spin dynamics above T_{SDW} leading to the observed Korringa relaxation. All these experiments were done on concentrated Eu^{2+} systems.

Qualitatively, it is expected that the La-doping dilutes the Eu^{2+}ions reducing the localized spin-spin interaction. At the same time, it may also affect the local density of itinerant electronic states at a given Eu^{2+} site and/or change the localized-itinerant spin J_{fs} exchange coupling. Localized spin-spin interaction leads to deformation and broadening of the resonance line. Inspection of figures 1(a)-(b) and the inset of figure 3 indicates that, for all temperature interval (above $\approx$ 25K), the broadening effects of the Eu^{2+} spin-spin interaction remains strong but no significant lineshape deformation appears. This is in contrast to the previous results where lineshape deformation was observed.

On figure 2, the reduction of the Korringa relaxation at high temperatures may be attributed to one, or both, of the cited putative *La*-doping effects. This is so because its value is direct connected to the values of the density of states and J_{fs} ($b = \frac{\pi k_B}{g\mu_B} J_{fs}^2 \eta_F^2$). The regime between $25 \lesssim T \lesssim 195$K, however, is unique and different from the previous report.

An anisotropy was found in the pure compound and was attributed to CF effects reflecting the low symmetry of the system. These effects were absent at high temperatures due to interaction with the conduction electrons, which was turned off at T_{SDW}. Figure 2 shows that, within experimental error, the linewidth and g-values are the same for the field applied perpendicular and parallel to the plane of the sample. This leads us to believe that the anisotropy observed for the pure compound is not due to CF effects but to localized Eu^{2+} spin-spin interaction, which is weakly suppressed in our case due to the La-doping. This point is based upon the fact that the CF effect is a single ion effect which would remain in the Eu^{2+} ion in the La-doped sample.

Another interesting distinct behavior of our doped system, is that in $EuFe_2As_2$ [1] (and also $EuFe_{2-x}Co_xAs_2$ [3]) the linewidth is nearly constant in this T-interval. In our case, a conspicuous dome like shape appears in this region. The strange behavior of ESR linewidth in the temperature region below T_{SDW} is usually ascribed to intricate Fermi surface nesting characteristics, and may resemble the coherence peak usually found in NMR data near a gap opening transition. A systematic study as a function of La-doping may elucidate this issue.

For $EuFe_2As_2$, a correction of the intensity values due to change of the lineshape (expressed by change in the A/D ration), was needed to adjust the intensity to a Curie law. Figure 3 shows that this is not necessary in our case. Our intensity naturally follows a Curie law above the low-T magnetic transition. We claim that it is further evidence that localized Eu^{2+} spin-spin interaction is affecting the lineshape analysis in the results for $EuFe_2As_2$.

In conclusion, ESR proved to be an important tool to further investigate magnetic and electronic ordering properties of Fe-based pnictide compounds. We expect that a systematic ESR study with different La-doping, and possible other dopings at both the Fe and As sites, may reveal interesting new physics of these materials.

ACKNOWLEDGMENTS

This work was supported by FAPESP and CNPq, Brazil.

4. REFERENCES

[1] E. Dengler, J. Deseinhofer, H.-A. Krug von Nidda, Seunghyun Khin, J. S. Kim, Kee Hoon Kim, F. Casper, C. Felser, and A. Loidl, 2010, *Phys Rev B* **81**, 024406.
[2] Kamihara Y, Watanabe T, Hirano M , Hosono H, 2008 *J.Ame.ChemSoc.* **130**, 11.
[3] H. S. Jeevan, *et al*, 2008, *Phys Rev B* **78**, 092406.
[4] Z. Ren, *et al,* 2009, *Phys Rev Letters* **102**, 137002.
[5] J. J. Ying, *et al,* 2010, *Phys Rev B* **81**, 052503.
[6] C. F. Miclea, *et al*, 2009, *Phys Rev B* **79**, 212509.
[7] Y. Xiao, *et al*, 2009, *Phys Rev B* **80**, 174424.
[8] B. Zhou, *et al*, 2010, *arXiv* : 1001.4537*v*1.
[9] A. Abragam and B. Bleaney, *EPR of Transition Ions*, (Clarendon Press, Oxford, 1970).
[10] Pascher N, *et al*, 2010, *arXiv* : 1001.1302*v*1.
[11] Z. Fisk and J. P. Remeika, *Handbook on the Physics and Chemistry of Rare Earths* (Elsevier, North-Holland, Amsterdam, 1989), Vol. 12, p. 53.

International Conference on Strongly Correlated Electron Systems (SCES 2010) IOP Publishing
Journal of Physics: Conference Series **273** (2011) 012094 doi:10.1088/1742-6596/273/1/012094

Effect of annealing on the specific heat of optimally doped $Ba(Fe_{0.92}Co_{0.08})_2As_2$

K. Gofryk[1], A. S. Sefat[2], M. A. McGuire[2], B. C. Sales[2], D. Mandrus[2], T. Imai[3], J. D. Thompson[1], E. D. Bauer[1], F. Ronning[1]

[1]Condensed Matter and Thermal Physics, Los Alamos National Laboratory, Los Alamos, New Mexico 87545, USA

[2]Materials Science and Technology Division, Oak Ridge National Laboratory, Oak Ridge, Tennessee 37831, USA

[3]Department of Physics and Astronomy, McMaster University, Hamilton, Ontario L8S4M1, Canada

E-mail: gofryk@lanl.gov

Abstract. We report the temperature dependence of the low-temperature specific heat down to 400 mK of the electron-doped $Ba(Fe_{0.92}Co_{0.08})_2As_2$ superconductors. We have measured two samples extracted from the same batch: first sample has been measured just after preparation with no additional heat treatment. The sample shows T_c=20 K, residual specific heat γ_0=3.6 mJ/mol K^2 and a Schottky-like contribution at low temperatures. A second sample has been annealed at 800 ^oC for two weeks and shows T_c=25 K and γ_0=1.4 mJ/mol K^2. By subtracting the lattice specific heat, from pure $BaFe_2As_2$, the temperature dependence of the electronic specific heat has been obtained and studied. For both samples the temperature dependence of $C_{el}(T)$ clearly indicate the presence of low-energy excitations in the system. Their specific heat data cannot be described by single clean *s*- or *d*-wave models and the data requires an anisotropic gap scenario which may or may not have nodes.

1. Introduction

The discovery of superconductivity in FeAs-based materials[1, 2] has re-opened new interest in superconductivity research. Unfortunately, despite a large theoretical[3] and experimental[4] effort the nature of the superconductivity in these materials including the pairing mechanism and the symmetry of the order parameter remains unclear.

One way to address the nature of the superconductivity is to study the low-temperature heat capacity. The measurements of the temperature dependence of the specific heat and its magnetic filed response in superconducting state give an important information about the symmetry of the order parameter[5, 6, 7, 8, 9, 10]. For optimally Co doped material[11] such a analysis suggest the presence of an anisotropic gap structure which may or may not have nodes (see Ref.[5, 6, 7, 8]). Moreover, the specific heat displays significant residual specific heat possibly coming from the non-superconducting fraction that is present in the system[6]. Thus, the heat capacity at very low temperatures is very sensitive on the quality of the crystals studied and it is very often dominated by a Schottky-like contribution[12]. This could make the the gap structure analysis quite complex.

Published under licence by IOP Publishing Ltd

In this paper, we present results of our specific heat studies of optimally doped $Ba(Fe_{0.92}Co_{0.08})_2As_2$ samples. We have measured two samples extracted from the same batch: sample one has been measured just after growing and without any additional heat treatment. The sample shows T_c=20 K, residual specific heat γ_0=3.6 mJ/mol K^2 and a Schottky-like contribution at low temperatures. Sample two has been annealed at 800 ^oC for two weeks with T_c=25 K and γ_0=1.4 mJ/mol K^2. For this material no Schottky term is observed (see Fig.1 inset). Then using similar approach as previously reported in Ref.[5, 6] we extract and study the full electronic T-dependence for these two samples.

2. Experimental details

Single crystals were grown out of FeAs flux with a typical size of about 2×1.5×0.2 mm^3[11]. They crystalize in well-formed plates with the [001] direction perpendicular to the plane of the crystals. The doping level was determined by microprobe analysis. After synthesis a second piece of the sample was annealed for two weeks at 800 ^oC while the first one has not been heat treated. The heat capacity was measured down to 400 mK using a thermal relaxation method implemented in a Quantum Design PPMS-9 device.

3. Results and discussion

Fig.1 shows the temperature dependence of the specific heat of the annealed and non-annealed $Ba(Fe_{0.92}Co_{0.08})_2As_2$ samples. As may be seen the annealing process has increased T_c quite remarkably from 20 K for the non-annealed sample up to 25 K for the annealed one. It is worth noting the transition temperature of about 25 K is one of the highest reported within the Co doped Ba122 family. Above T_c the $C_p(T)$ curves indicate very similar temperature dependence of the phonon part in these two materials as expected. As may be seen from the inset of Fig.1 the annealing strongly suppress the residual specific heat γ_0 from 3.6 to 1.4 mJ/mol K^2 respectively for the non-annealed and annealed sample. Moreover, no Schottky-like anomaly is observed for

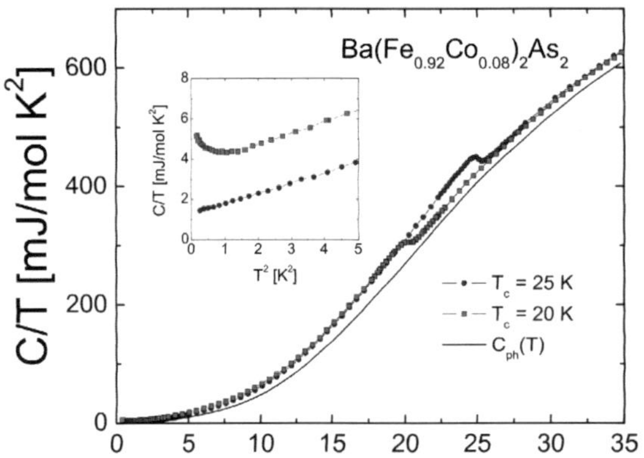

Figure 1. (Color online) The temperature dependence of the heat capacity of $Ba(Fe_{0.92}Co_{0.08})_2As_2$ samples. Circles: annealed sample with T_c = 25 K. Squares: non-annealed sample with T_c = 20 K. The solid line describes the normal state specific heat (see text). Inset: low-temperature specific heat of these two materials.

the annealed sample. The considerably increase of T_c, decrease of γ_0 and lack of Schottky-type contribution at low temperature indicate significant improvement of the quality of the single crystals after the heat treatment.

In order to obtain the electronic part of the specific heat, we have used the same approach used previously for the optimal and doped $Ba(Fe_{1-x}Co_x)_2As_2$ samples (see Ref.[5, 6]). We have assumed that the phonon contribution of the specific heat is independent of doping and we use the lattice specific heat obtained from the parent compound. We separate the lattice contribution to the specific heat of the pure Ba122 compound as $C_{ph} = C^{BaFe_2As_2}$ - $\gamma_{el}^{BaFe_2As_2}T$ where $\gamma_{el}^{BaFe_2As_2}$ is the T→0 intercept of C/T of $BaFe_2As_2$ as shown by a solid line in Fig.1. So, the electronic specific heat is simply determined by $C_{el}(T)^{Ba(Fe_{1-x}Co_x)_2As_2} = C_p(T)^{Ba(Fe_{1-x}Co_x)_2As_2} - C(T)_{ph}$. The obtained normal state $\gamma(T)$ from above T_c provide perfect entropy balance for the electronic specific heat in the superconducting state below T_c for both, non-annealed[5] and annealed samples.

Fig.2 shows the electronic part of the specific heat of $Ba(Fe_{1-x}Co_x)_2As_2$ samples. For the non-annealed sample a small Schottky contribution below 1 K has also been subtracted. For both samples pronounced peaks are observed at the transition temperature. For the annealed sample the specific heat exhibits a jump $\Delta C/T_c = 33.6$ mJ/mol K^2 at T_c. Taking $\gamma_n = 22$ mJ/mol K^2 the ratio $\Delta C/T_c\gamma = 1.53$ is slightly larger then the weak-coupling BCS value 1.43. It is also larger then $\Delta C/T_c\gamma = 1.34$ obtained for the non-annealed sample[5].

Below T_c the C_{el}/T data in Fig.2 were analyzed by the BCS α-model:

$$C_{BCS} = t\frac{d}{dt}\int_0^\infty dy\left(-\frac{6\gamma\Delta_0}{k_B\pi}\right)[flnf + (1-f)\,ln\,(1-f)] \tag{1}$$

where $t = \frac{T}{T_c}$, f is the Fermi function $f = \frac{1}{e^{\frac{E}{k_BT}}+1}$, $E = \sqrt{\epsilon^2+\Delta^2}$ and $y = \frac{\epsilon}{\Delta}$ (see Ref.[13]) plus residual normal state component.

As has been shown before a single s-wave gap scenario cannot describe the specific heat data

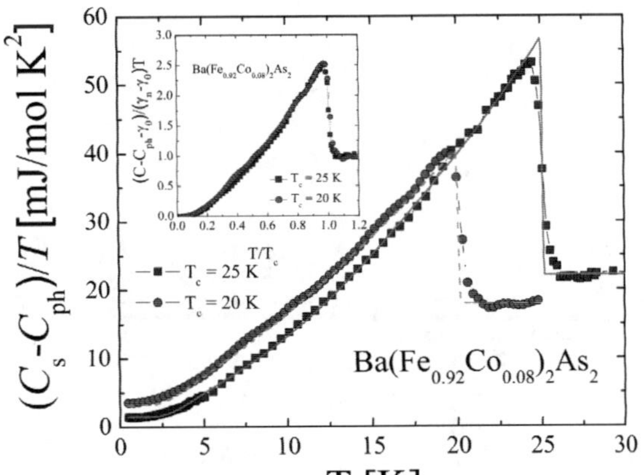

Figure 2. (Color online) The temperature dependence of the non-lattice part of the specific heat of $Ba(Fe_{0.92}Co_{0.08})_2As_2$ samples. Solid and dashed lines are fits of the BCS α-model (see text). Inset: normalized temperature dependence of the superconducting part of the specific heat.

in Co-doped Ba122 material[5, 7, 6, 8]. It clearly indicates the presence of an anisotropic gap structure in this material. Moreover, the specific heat data of optimally doped Ba122 cannot be accounted by using a clean d-wave gap aproach either[5, 7, 8]. In order to sufficiently describe the specific heat data of annealed $Ba(Fe_{0.92}Co_{0.08})_2As_2$ two separate s-wave gaps (solid line) have been used as shown in Fig.2. This parametrization is not unique but provides a reasonable estimate of the dominant energy scales in the gap structure. It gives parameters: $\Delta_1 = 1.83$ meV, $\Delta_2 = 4.76$ meV and $A = 0.72$. The values of the gap are larger then the ones observed in non-annealed materials[5, 7] $\Delta_1 = 1.65$ meV, $\Delta_2 = 3.75$ meV and $A = 0.62$ (see dashed line in Fig.2). It is consistent with the lower transition temperature observed in the non-annealed materials. Inset of Fig.2 presents the electronic part of the specific heat of the superconducting portion of the $Ba(Fe_{0.92}Co_{0.08})_2As_2$ samples. It has been obtained by subtracting the residual linear term γ_0, together with a small Schottky contribution below 1 K for the non-annealed sample, and normalized by (γ_n-γ_0). Interestingly, these two curves match, indicating no significant change of the gap structure in these two samples.

4. Summary and conclusions

In summary, using the low-temperature specific heat we explore details of the superconducting state of $Ba(Fe_{0.92}Co_{0.08})_2As_2$ with $T_c = 25$ and 20 K. By subtracting the phonon specific heat the temperature dependence of the electronic specific heat has been studied. The temperature variation of C_{el} as well as its field dependence (not shown) cannot be described by a single isotropic s-wave gap, indicating the presence of an anisotropic gap structure in the system. The temperature variation of the electronic specific heat below T_c may be well described by the presence of two superconducting gaps, pointing to a complex gap structure in the system. It has been shown recently that for optimally Co-doped $BaFe_2As_2$ samples a minimum of two superconducting gaps are necessary to describe the temperature dependence of the electronic specific heat[5, 7, 6, 8]. Moreover, for these two optimally doped samples with different quality the details of the superconducting state do not change significantly. It is in agreement with the lack of doping dependence in $Ba(Fe_{1-x}Co_x)_2As_2$[6], indicating that the gap structure does not change significantly as a function of doping[6]. This does not address the possibility that accidental nodes may or may not to be present and could be sample, doping, and/or material dependent.

Work at Los Alamos National Laboratory was performed under the auspices of the U.S. Department of Energy, Office of Science and supported in part by the Los Alamos LDRD program. Research at Oak Ridge National Laboratory is sponsored by the Division of Material Sciences and Engineering Office of Basic Energy Sciences.

References

[1] Y. Kamihara, T. Watanabe, M. Hirano, and H. Hosono, J. Am. Chem. Soc. **130**, 3296 (2008).
[2] X. H. Chen, T. Wu, G. Wu, R. H. Liu, H. Chen, and D. F. Fang, Nature **453**, 761 (2008).
[3] I. I Mazin and J. Schmalian, Physica C **469**, 614 (2009).
[4] K. Ishida, Y. Nakai, and H. Hosono, J. Phys. Soc. Jpn. **78**, 062001 (2009).
[5] K. Gofryk, A. S. Sefat, E. D. Bauer, M. A. McGuire, B. C. Sales, D. Mandrus, J. D. Thompson, and F. Ronning, New J. Phys. **12**, 023006 (2010).
[6] K. Gofryk, A. S. Sefat, M. A. McGuire, B. C. Sales, D. Mandrus, J. D. Thompson, E. D. Bauer, and F. Ronning, Phys. Rev. B **81**, 184518 (2010).
[7] F. Hardy, T. Wolf, R. A. Fisher, R. Eder, P. Schweiss, P. Adelmann, H. v. Löehneysen, C. Meingast, Phys. Rev. B **81**, 060501 (2010).
[8] B. Zeng, Ga. Mu, B. Shen, P. Cheng, H. Luo, H. Yang, L. Shan, C. Ren, Hai-Hu Wen, arXiv:1006.2785v1
[9] J. S. Kim, P. J. Hirschfeld, G. R. Stewart, S. Kasahara, T. Shibauchi, T. Terashima, and Y. Matsuda, Phys. Rev. B **81**, 214507 (2010)
[10] P. Popovich, A. V. Boris, O. V. Dolgov, A. A. Golubov, D. L. Sun, C. T. Lin, R. K. Kremer, B. Keimer, arXiv:1001.1074v2

[11] A. S. Sefat, R. Jin, M. A. McGuire, B. C. Sales, D. J. Singh, and D. Mandrus, Phys. Rev. Lett. **101**, 117004 (2008).
[12] J. S. Kim, E. G. Kim, and G. R. Stewart, J. Phys.: Condens. Matter **21**, 252201 (2009).
[13] M. Tinkham, Introduction to Superconductivity (McGraw-Hill, New York, 1975).

New intermetallic MIrP (M=Ti, Zr, Nb, Mo) and MgRuP compounds related with MoM'P (M'=Ni and Ru) superconductor

Hijiri Kito, Akira Iyo and Toshimi Wada

Superconducting Materials Group, Nanoelectronics Research Institute (NeRI) , National Institute of Advanced Industrial Science and Technology (AIST), Tsukuba Central 2, 1-1-1, Umezono, Tsukuba, Ibaraki, Japan, 305-8568

E-mail: h.kito@aist.go.jp

Abstract. Using a cubic-anvil high-pressure apparatus, ternary iridium phosphides MIrP (M=Ti, Zr, Nb, Mo) and MgRuP have been prepared by reaction of stoichiometric amounts of each metal and phosphide powders at around 2 Gpa and above 1523 K for the first time. The structure of these compounds prepared at high-pressure has been characterized by X-ray powder diffraction. Diffraction lines of these compounds are assigned by the index of the Co_2Si-type structure. The electrical resistivity and the d.c magnetic susceptibility of MIrP (M=Ti, Zr, Nb, Mo) have measured at low temperatures. Unfortunately, no superconducting transition for MIrP (M=Ti, Zr, Nb, Mo) and MgRuP are observed down to 2 K.

1. Introduction

Many ternary systems MRuP, where M represent 3*d*, 4*d* and 5*d* transition metals, have been studied crystallographically [1-5]. These compounds primarily exhibit hexagonal Fe_2P-type or orthorhombic Co_2Si (ε-TiNiSi)-type structures. For the hexagonal Fe_2P-type structure is built from neary rhombohedral $M_IRu_{II}P$ subcells stacked [5] to form columns containing M_I-Ru_{II} zig-zag chains. M_I and M_{II} represent the crystallographically in equivalent sites for the metal atoms coordinated by a tetrahedron and tetragonal pyramid of P anions, respectively, with M_I being typically the smaller and more electronegative of metal atoms [6]. Especially, the ZrRuP and HfRuP systems show superconductivity and superconducting transition temperature (T_c) of ternary phosphides with the hexagonal Fe_2P-type structure is considerably high: T_c is about 13K for ZrRuP and HfRuP [1, 8]. Several phosphide superconductors with T_c above 15K have also been found; MoNiP with the Fe_2P-type structure [9] and MoRuP with the Co_2Si-type structure have T_c of about 15.5K [9, 10] and other interesting superconductors of phosphides have also been found; T_c of $LaRu_4P_{12}$ with skutterudite ($CoAs_3$-type) structure is about 7K [11, 12].

Here, if Ru atom in MoRuP can be substituted by Ir atom or Mo atom in MoRuP can be substituted by Mg atoms, superconductivity also may be expected. We have searched in ternary phosphide; M–Ir–P system (M=Ti, Zr, Nb and Mo) and MgRuP have prepared at high temperature and high pressures for the first time. For Mo–Ir–P system, we had previously reported [13]. In this paper, the physical properties of MIrP (M=Ti, Zr, Nb and Mo) and MgRuP compounds are reported.

Published under licence by IOP Publishing Ltd

2. Experiment and results

Polycrystalline sample of the ternary metal phosphides MIrP (M=Ti, Zr, Nb and Mo) and MgRuP were prepared by high-pressure synthesis method using a cubic-anvil high-pressure apparatus (RIKEN CAP-07). The starting materials were mixed stoichiometric amounts of each powder metal and P for MIrP and MgRuP and ground by an agate mortar in a glove box filled with dry nitrogen gas. The starting material is put into a crucible made of BN. The crucible with a graphite heater is inserted into the pyrophyllite cube. The samples were synthesized by heating the mixtures in BN crucibles under a pressure of about 2 GPa at 1523-1873 K for 1.0 h.

These obtained compounds were characterized for the phase purity and lattice parameters by powder X-ray diffraction using CuKα radiation (Rigaku RINT 1100, 40 kV, 40 mA). Figure 1 shows the X-ray powder diffraction pattern for MIrP (M=Ti, Zr, Nb and Mo). The diffraction lines in MIrP (M=Ti, Zr, Nb and Mo) are assigned by the index of Co_2Si-type structure. The X-ray profiles of MIrP are similar to those of orthorhombic MoRuP [10]. These new compounds with the Co_2Si type structure have a/c and $(a+c)/b$ ratios very similar to those of other ternary transition metal phosphides with that structure; so all belong to the Co_2Si branch of the anti-$PbCl_2$ type [14]. Figure 2 shows the variation of the lattice parameters as a function of metallic radii for the six coordination. For larger size of the metallic radii M, the lattice parameter a and c expand and lattice parameter b shrink, respectively. Figure 3 shows the X-ray powder diffraction pattern for MgRuP. The diffraction lines in MgRuP are also assigned by the index of Co_2Si-type structure. The lattice constants of MgRuP prepared at high-pressure are also given as follows; a=6.444Å, b=3.835Å and c=6.970Å.

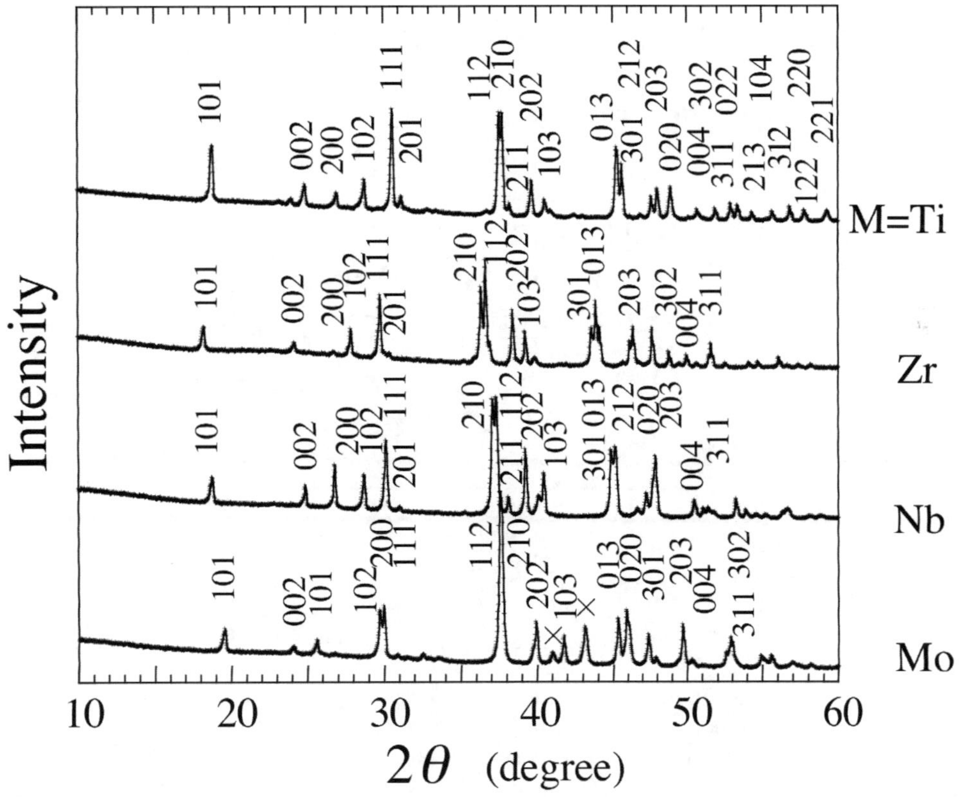

Figure 1. X-ray powder diffraction of MIrP (M=Ti, Zr, Nb and Mo).

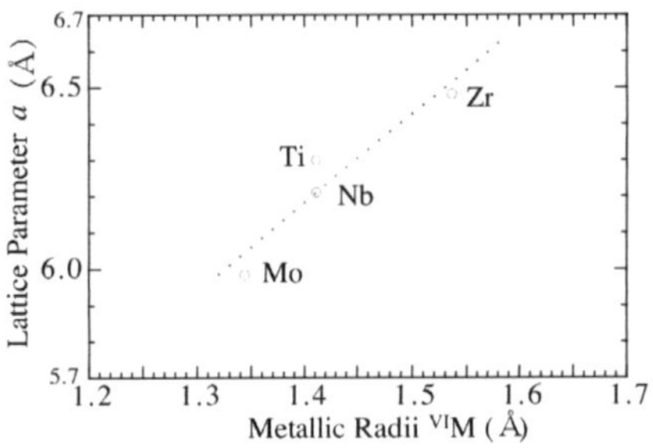

Figure 2(a). Lattice parameter *a* as a function of metallic radii M.

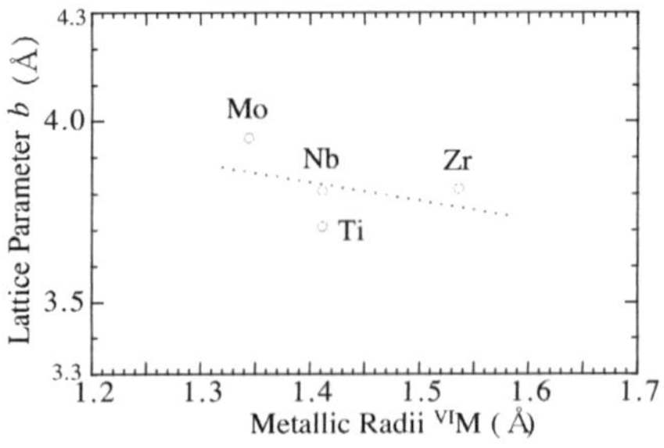

Figure 2(b). Lattice parameter *b* as a function of metallic radii M.

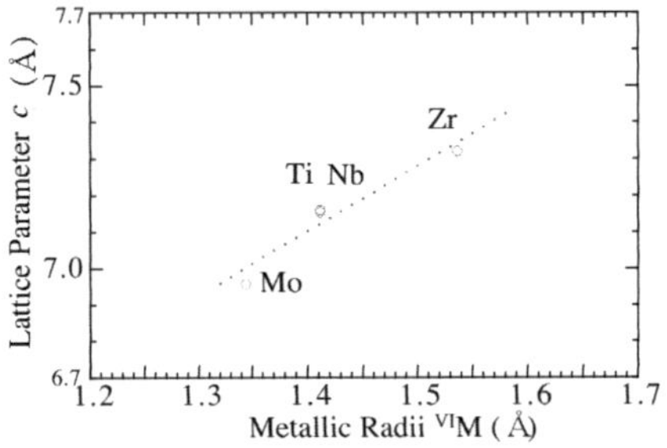

Figure 2(c). Lattice parameter *c* as a function of metallic radii M.

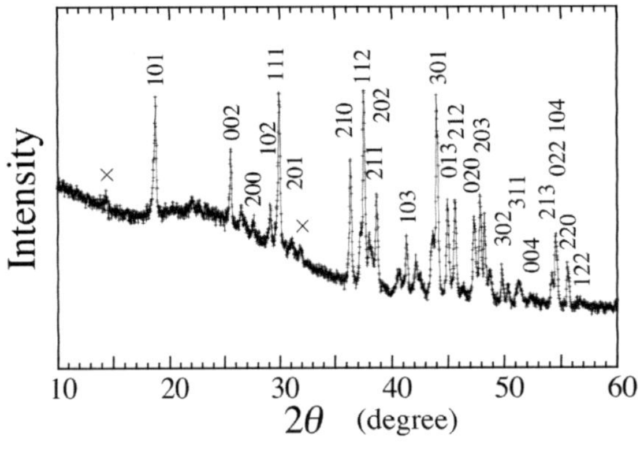

Figure 3. X-ray powder diffraction of MgRuP.

Electrical resistivity measurements were carried out using a conventional DC four-probe method. Gold lead was attached to polycrystalline samples with silver filled by epoxy and measurements were performed at low temperatures. The resistivity for MIrP (M=Ti, Zr, Nb and Mo) decreases with decreasing temperature. MIrP (M=Ti, Zr, Nb and Mo) behaves as a metal. At low temperature, large residual resistivity was observed for MIrP (M= Nb). This suggests that the purity of the specimen is still low. The superconductivity is not observed down to 5 K.

The DC magnetic susceptibility at H=20Oe of the polycrystalline MIrP (M=Ti, Zr, Nb and Mo) and MgRuP samples were measured in the range of 2–300K with a Quantum Design MPMS SQUID magnetometer. No superconducting transition is observed down to 2 K. The Susceptibility of the MIrP (M=Ti, Zr, Nb and Mo) is essentially temperature independent in the region of 2-300 K. These show Pauli paramagnetism and are consistent with metallic behavior, which are expected from the temperature dependence of the resistivity.

These MIrP (M=Ti, Zr, Nb and Mo) and MgRuP may be possible to show the superconductivity under high pressure or at very low temperatures.

This work was supported by a Grant-in-Aid for Specially Promoted Research (21540375) from the Ministry of Education, Culture, Sports, Science and Technology (MEXT), Japan.

References

[1] H. Barz, H.C. Ku, G.P. Meisner, Z. Fisk and B.T. Mattias 1980 *Proc. Natl. Acad. Sci. USA* **77** 3132.

[2] R. Müller, R.N. Shelton, J.W. Richardson, JR. and R.A. Jacobson 1983 *J. Less-common Metals* **92** 177.

[3] G.P. Meisner and H.C. Ku 1983 *Apll. Phys.* A **31** 201.

[4] G.P. Meisner and H.C. Ku, in G.K. Shenoy, B.D. Dunlap, F.Y. Fradin (eds.) 1981, *Ternary Superconductors*, North-Holland, New York p.255.

[5] R. Fruchart, A. Roger and J.P. Sénateur 1969 *J. Appl. Phys.* **40** 1250.

[6] J. Roy-Montreuil, B. Deyris, A. Michel, R. Fruchart, J.P. Sénateur and D. Boursier 1979 *Ann. Chim. Fr.* **4** 411.

[7] G.R. Stewart, G.P. Meisner and H.C. Ku, in W. Buckel, W.Weber (eds.) 1982, *Superconductivity in d- and f-band metals*, Kernforschungszentrum, Karlsruhe p.331.

[8] I. Shirotani, N. Ichihashi, K. Nozawa, M. Kinoshita, T. Yagi, K. Suzuki and T. Enoki 1983 *Jpn. J. Appl. Phys. Suppl.* **32(3)** 695.

[9] I. Shirotani, M. Takaya, I. Kaneko, C. Sekine and T. Yagi 2000 *Solid State Commun* **116** 683.

[10] W. Wong-Ng, W.Y. Ching, Y.-N. Xu, J.A. Kaduk and I. Shirotani, L. Swartzendruber 2003 *Phys. Rev.* B **67** 144523

[11] G.P. Meisner 1981 *Physica* B **108** 763.

[12] I. Shirotani, T. Adachi, K. Tachi, S. Todo, K. Nozawa, T. Yagi and M. Kinoshita 1996 *J.Phys.Chem. Solids* **57** 211.

[13] H. Kitô, T. Wada and H. Abe 2006 *Physica* B **378-380** 1146.

[14] W. Jeitschko 1968 *Acta Cryst.* B **24** 930.

Journal of Physics: Conference Series **273** (2011) 012096 doi:10.1088/1742-6596/273/1/012096

Pressure and K doping induced superconductivity in $BaFe_2As_2$

T Yamazaki[1], N Takeshita[2,3], R Kobayashi[1], T Saito[1], Y Yamada[1], K Kondo[1], M Hirano[1], H Fukazawa[2,3], Y Kohori[2,3], K Kihou[2,3], C H Lee[2,3], H Kito[2,3], A Iyo[2,3], and H Eisaki[2,3]

[1]Department of Physics, Graduate School of Science, Chiba University, Chiba 263-8522, Japan
[2]National Institute of Advanced Industrial Science and Technology, Tsukuba 305-8562, Japan
[3]JST, TRIP, Chiyoda-ku, Tokyo 102-0075, Japan

E-mail: `kohori@physics.s.chiba-u.ac.jp`

Abstract. We have measured electrical resistivity under pressure P and ^{75}As NMR on single crystal and poly crystalline $BaFe_2As_2$ and $(Ba_{1-x}K_x)Fe_2As_2$ to elucidate mechanism of antiferromagnetic/superconducting phase transition in $BaFe_2As_2$. This system is an antiferromagnet, AF, with a transition temperature of 141 K, which shows superconductivity by applying P or by a carrier doping of K for Ba. From high P resistivity measurements, we understood that the transition was highly sensitive to the P homogeneity. Small uniaxial stress strongly suppresses AF order and stabilizes superconductivity at much lower pressure. ^{75}As NMR study in $(Ba_{1-x}K_x)Fe_2As_2$ shows existence of large AF spin fluctuations in tetragonal phase having a superconducting ground state. The AF spin fluctuations decrease gradually with increasing K concentration.

1. Introduction

The discovery of superconductivity in F-doped LaFeAsO at 26 K by Kamihara et al. heralded a new era of superconductivity based on Iron [1]. The common feature of the new superconductors is existence of Iron pnictide layers, which is analogous to CuO_2 planes of the cuprate superconductors. After the extensive investigations, the high temperature superconductivity with the transition temperature T_c of 56 K was observed in rare earth iron oxypnictides (1-1-1-1 compound)[2,3,4]. Members of superconducting iron family were observed one after another in the oxygen free compounds. Among them, $AEFe_2As_2$ (AE =Ba, Ca and Sr) (1-2-2 compound) occupies a singular position since the high quality large single crystals are grown. The 1-2-2 compound has a first order antiferromagnetic AF transition on cooling at moderately high temperature, which always accompanied phase transition from tetragonal to orthorhombic structure. In the case of $BaFe_2As_2$, crystal structure/magnetic phase transition occurs at 141 K [5]. The carrier doping of K for Ba, or Co for Fe strongly suppresses the transition in $BaFe_2As_2$. The highest T_c of 38 K is obtained in K doped $BaFe_2As_2$ in a paramagnetic phase located adjacent to AF state [6]. Here, we report ^{75}As NMR measurement of hole-doped $Ba_{1-x}K_xFe_2As_2$ having a maximum T_c ($x = 0.4$) and end member ($x = 1$) with T_c of 3.5 K. The spin-lattice relaxation rate $1/T_1$ was measured both in normal and superconducting SC state. Another method to suppress AF order in $BaFe_2As_2$ is applying pressure P. Many confusing results were reported under pressure using high quality single crystals. In order to solve this

Published under licence by IOP Publishing Ltd

confusion, electrical resistivity was measured under pressure in different hydrostaticity with single crystals of same batch.

2. Experimental

Single crystals of $BaFe_2As_2$ were grown by the FeAs self-flux method, and polycrystalline $Ba_{1-x}K_xFe_2As_2$ samples were synthesized by a high T and high-pressure method. X-ray diffraction analysis revealed that all the samples have a single phase. The ^{75}As NMR experiments were carried out using phase-coherent pulsed NMR spectrometer and a superconducting magnet with a maximum magnetic field of 8 T. ^{75}As NMR was performed with a field oriented powder whose crystal c- axis aligned perpendicular to external field. For KFe_2As_2 ($x = 1$), ^{75}As NQR measurements were performed around 12.4 MHz down to 70 mK using a ^{3}He-^{4}He dilution refrigerator. $1/T_1$ was obtained from a recovery of nuclear magnetization after a saturation pulse. Electrical resistivity was measured in $BaFe_2As_2$ under high pressure using a cubic anvil apparatus and modified Bridgman anvil cell, which provided very hydrostatic pressure and quasi-hydrostatic pressure, respectively.

3. Results and Discussion

T dependence of $1/T_1$ obtained in parent $BaFe_2As_2$ ($x = 0$), $Ba_{0.6}K0_{0.4}Fe_2As_2$ ($x = 0.4$) and KFe_2As_2 ($x = 1$) were shown in figure 1. The value of $1/T_1T$ for $x = 0.4$ is nearly the same as that for $x = 0$ in the normal state. The $1/T_1T$ varies proportionally to $1/T$ above 40 K for $x = 0.4$, which indicates existence of large two-dimensional AF spin fluctuations [7]. The behavior of $1/T_1T$ observed in hole doped system is different from that of the electron doped system like F- doped LaFeAsO or $Ba(Fe_{1-x}Co_x)_2As_2$ in which spin fluctuations are strongly suppressed with increasing electron doping [8.9]. Doping dependence of spin fluctuations is well explained by FLEX approximation [10].

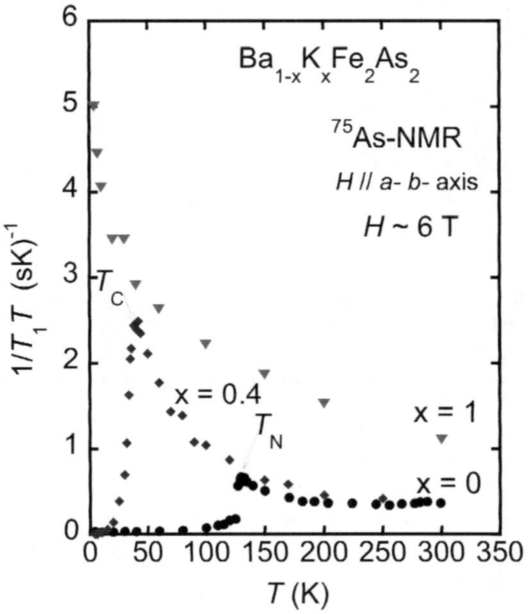

Figure 1. T dependence of $1/T_1T$ for x =0, 0.4 and 1.

The maximum T_c of the hole doped system appears close to quantum critical point. With further increasing carrier by K doping, the value of $1/T_1T$ becomes large, which reflects enhanced density of states at Fermi level. But, T dependence of $1/T_1T$ becomes weak. The AF spin

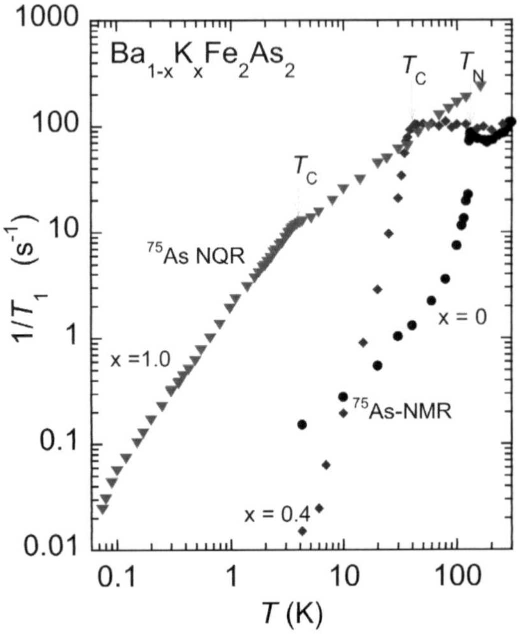

Figure 2. T dependence of $1/T_1$ is plotted in a Log-Log scale to enlarge low T region.

fluctuations decrease gradually with increasing K concentration due to the change of band structure with hole doping. The T dependence of $1/T_1$ is plotted in figure 2 against T in a Log - Log scale to enlarge low temperature superconducting region. For $x = 0.4$, $1/T_1$ decreases rapidly on cooling without a coherence peak just below T_c, and varies close to T^3 at lower temperatures. Similar behavior in the SC state is reported in many NMR studies of iron pnictides. Since T dependence of other physical properties like field penetration depth showed existence of a fully-gapped multiple gap [11], we could not conclude line node superconductivity from T^3 behavior of $1/T_1$. Indeed, our $1/T_1$ could be interpreted with a multiple gap structure with fully-gapped $s\pm$ model. In order to explain by simple s_{++} model, a intense damping of quasiparticles,[12,13] which is strong enough to suppress the coherent peak below T_c, is required. At present, $s\pm$ model seems most probable to account for data [14,15]. The T dependence of $1/T_1$ for $x = 1$ is much weaker than that for $x = 0.4$ in the SC state. The simple calculation has shown existence of small gap, about 1/10 of larger gap, successfully explains the T dependence. $1/T_1$ and specific heat data are consistently explained by a multiple gap with nodal-line gap structure [16]. Further study is required to conclude the gap structure of KFe_2As_2.

Electrical resistivity was measured in $BaFe_2As_2$ obtained from same batch under high pressure using both a cubic anvil apparatus and modified Bridgman anvil cell. The measurement using a cubic anvil apparatus provides highly hydrostatic pressure. In this condition, crystal structure and AF transition decreases robustly and no superconductivity was observed up to 10 GPa. The measurement using modified Bridgman anvil cell, which contains liquid Fluorinerts pressure transmitting medium, provides quasi-hydrostatic pressure since a solidification of Fluorinert above 1.2 GPa yields small pressure inhomogeneity. Figure 3 shows $\rho(T)$ obtained in this condition. The lower T anomaly was observed at 30 K above 2.5 GPa. The rapid drop of resistivity shows appearance of superconductivity. Non-zero resistivity below the transition indicates that the induced superconductivity is filamentary. This anomaly around 30 K was observed by many groups. The superconductivity in $BaFe_2As_2$ is very sensitive to a small uniaxial stress. We expect that the crystal structure transition from tetragonal to orthorhombic is sensitive to uniaxial stress and superconductivity appears only in tetragonal phase.

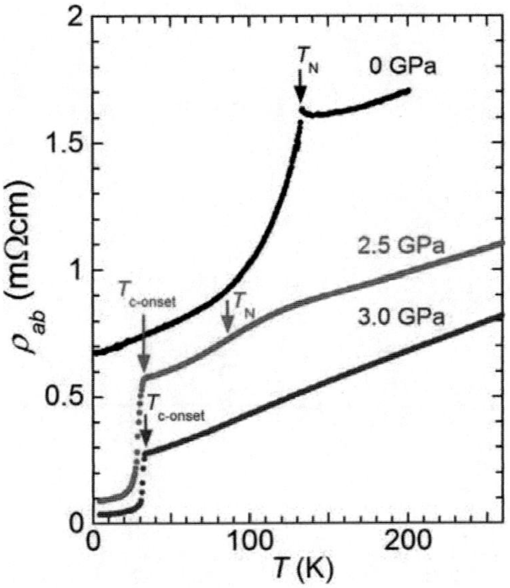

Figure 3. Pressure depedence of $\rho(T)$. Filamentary superconductivity appears above 2.5 GPa.

4. Conclusion

$BaFe_2As_2$ shows superconductivity by applying pressure or by a carrier doping of K for Ba. From high pressure resistivity measurements, we understood that the transition was highly sensitive to the pressure homogeneity. Small uniaxial stress strongly suppresses AF order and stabilizes superconductivity at much lower pressure. ^{75}As NMR study in $(Ba_{1-x}K_x)Fe_2As_2$ shows existence of large AF spin fluctuations in tetragonal and paramagnetic phase having a superconducting ground state. The AF spin fluctuations decrease gradually with increasing K concentration due to the change of band structure.

Acknowledgments

This work is supported by a Grant-in-Aid for Scientific Research from MEXT and JSPS, Global COE program of Chiba University.

[1] Kamihara Y, Watanabe T, Hirano M, Hosono H 2008 *J. Am. Chem. Soc.* **130** 3296
[2] Kito H, Eisaki H, Iyo Y J. Phys. Soc Jpn 2008 **77** 063707
[3] Ren Z - A, Yang J, Lu Yi W, Shen X L, Li Z C, Che G C, Dong X L, Sun L L, Zhao Z -X 2008 *Europhs. Lett.* **82** 57002
[4] Wang C, Li L, Chi S, Zhu Z, Ren Z, Li Y, Wang Y, Lin X, Luo Y, Shuai S, Xu X, Guanghan Cao G, Zhu'an Xu Z 2008 *Europhs. Lett.* **83** 67006
[5] Rotter M, Tegel M, Johrendt D 2008 *Phys. Rev. Lett.* **101** 107006
[6] Rotter M, Tegel M, Johrendt D 2008 *Phys. Rev.* B **78** 020503
[7] Moriya T 1991 *J. Mag. Mag. Matter.* **100** 261
[8] Nakai Y, Kitagawa S, Ishida K, Kamihara Y, Hirano M 2009 *Phys. Rev.* B **79** 212506
[9] Ning F L, Ahilan K, Imai T, Sefat A S, McGuire M A, Sales B C, Mandrus D, Cheng P, Shen B, Wen H -H 2010 *Phys. Rev. Lett.* **104** 037001
[10] Ikeda H, Arita R, Kuneš J *arXive*:1002.4471
[11] Shibauchi T, Hashimoto K, Okazaki R, and Matsuda Y 2009 *Physica* C **469** 590
[12] Ohsugi S, Kitaoka Y, Kyogaku M, Ishida K, Asayama K, Ohtani T 1992 *J. Phys. Soc. Jpn.* **61** 3054
[13] Allen P B, Rainer D 1991 *Nature* **349** 396
[14] Mazin I I, Singh D J, Johannes M D, and Du M H 2008 *Phys. Rev. Lett.* **101** 057003
[15] Kuroki K, Onari S, Arita R, Usui H, Tanaka Y, Kontani H, Aoki H, 2008 *Phys. Rev. Lett.* **101** 087004
[16] Fukazawa H, Yamazaki T, Kondo K, Kohori Y, Takeshita N, Shirage P M, Kihou K, Miyazawa K, Kito H, Eisaki H, Iyo A 2009 *J. Phys. Soc Jpn* **78** 033704

Fully gapped superconductivity in Ni-pnictide superconductors $BaNi_2As_2$ and $SrNi_2P_2$

N Kurita[1,2], F Ronning[1], C F Miclea[1], Y Tokiwa[1,3], E D Bauer[1], A Subedi[4,5], D J Singh[4], H Sakai[1,6], J D Thompson[1] and R Movshovich[1]

[1] Los Alamos National Laboratory, Los Alamos, New Mexico 87545, USA
[2] National Institute for Materials Science, Tsukuba, Ibaraki 305-0003, Japan
[3] I. Physikalisches Institut, Georg-August-Universität Göttingen, D-37077 Göttingen, Germany
[4] Materials Science and Technology Division, Oak Ridge National Laboratory, Oak Ridge, Tennessee 37831-6114, USA
[5] Department of Physics and Astronomy, University of Tennessee, Knoxville, Tennessee 37996-1200, USA
[6] Advanced Science Research Center, Japan Atomic Energy Agency, Ibaraki 319-1195, Japan

E-mail: KURITA.Nobuyuki@nims.go.jp

Abstract. We have performed low-temperature specific heat C and thermal conductivity κ measurements on the Ni-pnictide superconductors $BaNi_2As_2$ ($T_c = 0.7$ K) and $SrNi_2P_2$ ($T_c = 1.4$ K). The temperature dependences $C(T)$ and $\kappa(T)$ of the two compounds are similar to the results of a number of s-wave superconductors. Furthermore, the concave field responses of the residual κ for $BaNi_2As_2$ rules out the presence of nodes on the Fermi surfaces. We postulate that fully gapped superconductivity could be universal for Ni-pnictide superconductors. Specific heat data on $Ba_{0.6}La_{0.4}Ni_2As_2$ shows a mild suppression of T_c and H_{c2} relative to $BaNi_2As_2$.

1. Introduction

Since the discovery of superconductivity in LaFeAs(O,F) [1], a number of studies have been performed for understanding the origin of high-T_c in the Fe-pnictide superconductors (SCs). The structure of superconducting gap symmetry, closely related with the paring mechanism, is one of the most important issues to resolve. However, it remains still controversial whether superconductivity in Fe-pnictides is conventional, or unconventional with or without node(s) [2]. On the other hand, Ni-counterparts with the same crystal structure also superconduct [3]. Until now, it was reported that some Ni-pnictides had several properties in common with Fe-pnictides, for example, in structural transitions from tetragonal to lower symmetry [4, 5]. In contrast, there are crucial differences including the magnitude of T_c which does not exceed 5 K in any Ni-pnictide, the absence of magnetism, and a more three dimensional structure of the Fermi surface in Ni-pnictides [3, 5]. Therefore, it can be expected that identifying the SC gap symmetry of Ni-pnictides would yield a key information for that of Fe-pnictides.

Here, we present the results of low-temperature thermal transport measurements, as established powerful tools for probing low energy excited quasiparticles, on $BaNi_2As_2$

Published under licence by IOP Publishing Ltd

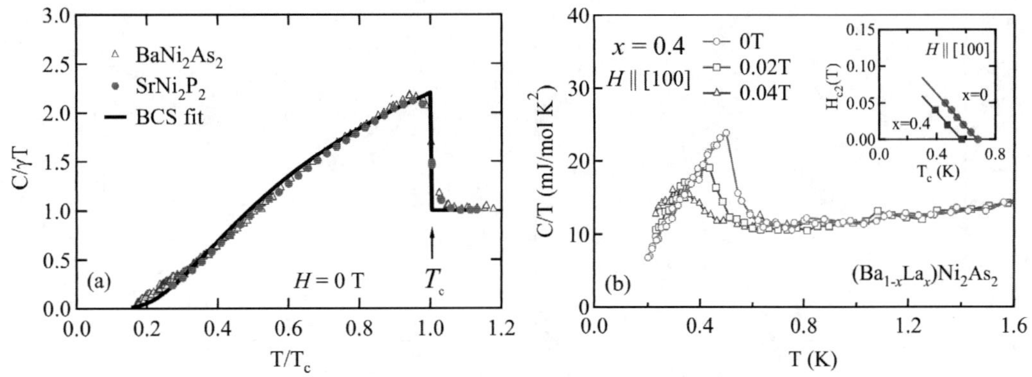

Figure 1. (Color online) (a) Scaled C/T vs T of $BaNi_2As_2$ and $SrNi_2P_2$ in zero field. Solid curve represents a BCS fit with the parameters found in Ref. [6]. (b) C/T vs T for $Ba_{1-x}La_xNi_2As_2$ ($x = 0.4$) in several fields for $H \parallel ab$. The inset shows H_{c2} vs T_c in $x = 0$ [6] and 0.4 for $H \parallel ab$. The solid lines represent a least-squares fit to the data.

($T_c = 0.7$ K) and $SrNi_2P_2$ ($T_c = 1.4$ K) which exhibit structural transitions at 130 K [4] and at 325 K [5], respectively.

2. Experimental details

Single crystals of $Ba_{1-x}La_xNi_2As_2$ ($x = 0$ and 0.4 nominal composition) and $SrNi_2P_2$ were grown in Pb-flux [4] and Sn-flux [5], respectively. Thermal conductivity was measured in a dilution refrigerator on plate-like crystals by a standard one-heater and two-thermometers technique with a heat current $q \parallel [100]$. Electrical resistivity was measured for electrical current $J \parallel [100]$, using the same crystal with the same electrical contacts as for the thermal conductivity measurements. Specific heat was measured in a dilution refrigerator by a standard heat-pulse method for $(Ba,La)Ni_2As_2$ and in a PPMS (Quantum Design) by a relaxation method for $SrNi_2P_2$. Magnetic field was applied for $H \parallel [100]$. In the thermal conductivity study, we measured two samples from different growths denoted as #1 and #2.

3. Results and Discussion

Figure 1(a) shows the scaled C/T as a function of T/T_c of $BaNi_2As_2$ and $SrNi_2P_2$ in zero field. A nuclear quadrupole contribution arising from As was already subtracted in $BaNi_2As_2$. A theoretical curve based on weak coupling BCS [6] can be fit to nearly identical data sets for the two compounds. This C/T behavior is in strong contrast to the results of nodal SCs such as Sr_2RuO_4 [7], which has a smaller jump at T_c as well as much more low energy excitations visible at low temperatures. Figure 1(b) shows C/T vs T of La-doped $Ba_{1-x}La_xNi_2As_2$ ($x = 0.4$) in several fields for $H \parallel ab$. The inset displays the upper critical field H_{c2} vs T_c for $x = 0$ [6] and $x = 0.4$, determined by the midpoint of the jump in C/T. The initial slopes -0.23 T/K for $x = 0$ and -0.22 T/K for $x = 0.4$ give $H_{c2}(0) = 0.11$ T and 0.088 T, respectively, using $H_{c2}(0) = -0.7\,T_c dH_{c2}/dT_c$ [8]. The reduction of T_c and $H_{c2}(0)$ by doping in $BaNi_2As_2$ is opposite to the enhancement of the two values in LaNiAsO by Sr- or F-doping [9, 10]. It should be noted that the similar enhacement of T_c and $H_{c2}(0)$ with doping was widely confirmed in Fe-pnictides [2].

Figure 2 shows the temperature dependence of thermal conductivity $\kappa(T)$ of (a) $BaNi_2As_2$ (#1 [6] and #2) and (b) $SrNi_2P_2$. Heat current q and magnetic field H were applied in the directions $H \parallel q \parallel [100]$. For $BaNi_2As_2$(#1), κ in zero field above T_c exhibits approximately

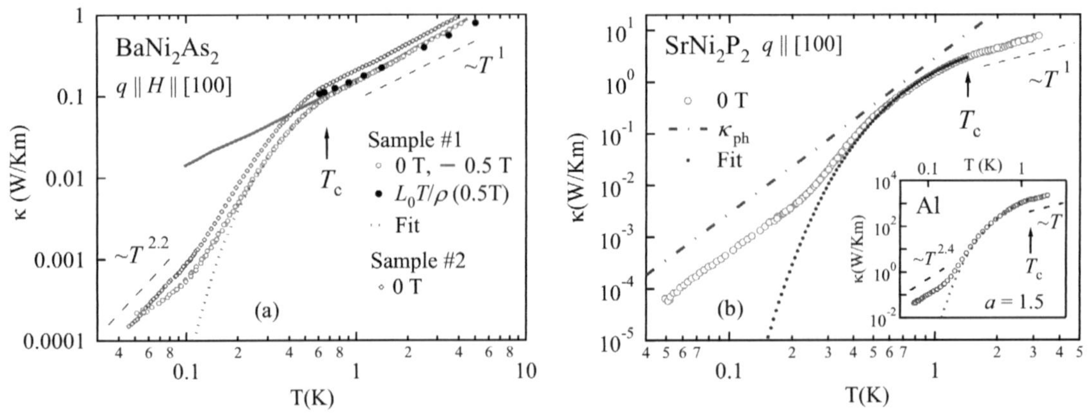

Figure 2. (Color online) κ vs T in (a) $BaNi_2As_2$(#1 [6], #2) and (b) $SrNi_2P_2$ for heat current $q \parallel [100]$. Dotted lines show a fit to $\kappa(T)$ based on BCS theory defined as $\kappa = C\exp(-aT_c/T)$. Solid symbols in (a) represent the electronic thermal conductivity $\kappa_e = L_0T/\rho$ with $L_0 = 2.45 \times 10^{-8}\,\mathrm{W\Omega/K^2}$ for 0.5 T ($\gg H_{c2}$), derived from resistivity data using the Wiedemann-Franz law. The inset of (b) exhibits κ versus T of Al with $T_c = 1.2$ K [14]. A dashed-dotted line ($\propto T^3$) is the upper limit for the phonon conductivity based on scattering from the crystal surfaces.

T-linear behavior and continues to lower temperature in the normal state at 0.5 T ($\gg H_{c2}$). Using the Wiedemann-Franz law, we estimate the electronic conductivity κ_e in the normal state through $\kappa_e = L_0T/\rho$ with $L_0 = 2.44 \times 10^{-8}\,\mathrm{W\Omega/K^2}$ and the measured resistivity (ρ) data at 0.5 T as indicated by the solid symbols in Fig. 2(a). The close agreement of the estimated κ_e with the experimental κ indicates that the normal state thermal conductivity κ_n can be attributed mostly to electrons rather than phonons. In the superconducting state just below T_c, κ exhibits a BCS-like expression $\exp(-aT_c/T)$ as indicated by the dotted curves for $BaNi_2As_2$ and $SrNi_2P_2$. It is worth noting that $a = 1.3$ ($BaNi_2As_2$) and 1.5 ($SrNi_2P_2$) are comparable to the values 1.3-1.5 obtained in conventional s-wave SCs Al, Sn and Zn [11]. At $T \ll T_c$, κ does not decrease as rapidly as expected from exponential behavior. Similar deviations from an exponential variation were found in a number of s-wave superconductors usually with a form of $\kappa \propto T^\alpha$ and are commonly attributed with phonon contributions. In fact, $\kappa \propto T^\alpha$ can be clearly seen in $BaNi_2As_2$ (#2) and $SrNi_2P_2$ with $\alpha = 2.2$ and 3, respectively although an anomalous behavior was found in $BaNi_2As_2$ (#1). In particular, $\alpha = 3$ in $SrNi_2P_2$ is equal to the value expected in a theoretical phonon conductivity $\kappa_{ph} \propto T^3$, as indicated by a dashed-dotted line [12] in Fig. 2(b), based on a crystal-boundary scattering model [11]. However, there is no consistent explanation why the exponent α ranges between 2 and 3 in many SCs. Note that $\kappa_{ph} \propto T^{2.2}$ was reported in $BaFe_2As_2$ which is a parent compound of Fe-pnictide SCs [13]. The $\kappa(T)$ behaviors in $BaNi_2As_2$ and $SrNi_2P_2$ discussed above are well consistent with the results of a number of s-wave superconductors including Al with $T_c = 1.2$ K [14] as can be seen in the inset of Fig. 2(b)

Figure 3 (a) shows low-temperature κ/T vs T^2 of $BaNi_2As_2$ in several fields $H \parallel q \parallel [100]$, not presented in Ref [6]. The rise in κ/T can only be attributed to κ_e, as κ_{ph} may only go down in magnetic field due to additional scattering from vortices in the mixed state. The straight lines are fits to $\kappa/T = \kappa_0/T + bT^2$, where κ_0/T is the residual term extrapolated to $T = 0$ K at each field. Figure 3(b) displays the normalized κ_0/T of $BaNi_2As_2$ for #1 ($H \perp q$ [6] and $H \parallel q$) and #2 ($H \perp q$), with respect to H/H_{c2}. In s-wave SCs, quasiparticles localized around the vortex

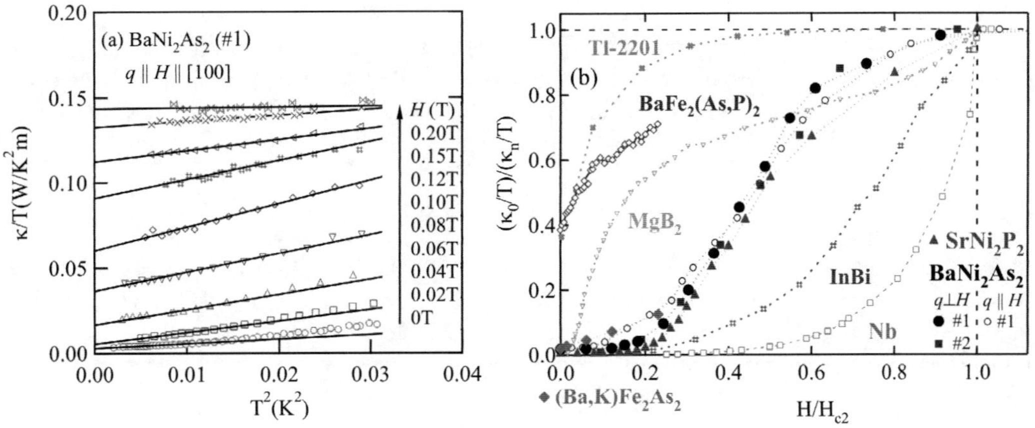

Figure 3. (Color online) (a) low-temperature κ/T vs T^2 of $BaNi_2As_2$ in several fields $H \parallel q \parallel [100]$. (b) Scaled residual linear term $(\kappa_0/T)/(\kappa_n/T)$ vs H/H_{c2} of $BaNi_2As_2$ for #1 ($H \perp q$ [6] and $H \parallel q$) and #2 ($H \parallel q$), and $SrNi_2P_2$ ($H \perp q$). Data for Nb (clean, fully gapped s-wave) [15], InBi (dirty, fully gapped s-wave) [18], MgB_2 (multi-band gap) [19], and Tl-2201 (d-wave with line nodes) [16] as well as Fe-pnictide SCs (Ba,K)Fe_2As_2 [20] and $BaFe_2(As,P)_2$ [21] are shown for comparison.

cores at low field ($> H_{c1}$) cannot carry heat until the intervortex distance gets so narrowed with field that quasiparticles can conduct between the vortices. This is the main reason why κ_0/T exhibits an exponential-like increase toward H_{c2} as seen for Nb data [15]. In contrast, for nodal SCs such as $Tl_2Ba_2CuO_{6+\delta}$ (Tl-2201) [16], κ_0/T exhibits a rapid enhancement in the low field region because of the excitation of nodal quasiparticles by the Volovik effect [17]. Therefore, the concave field dependences of κ_0/T at low field in $BaNi_2As_2$ (#1 and #2) and $SrNi_2P_2$ are clear evidence for fully gapped superconductivity.

The rapid increase of $\kappa(H)$ for $BaNi_2As_2$ and $SrNi_2P_2$ more closely resembles the dirty s-wave SCs such as InBi [18], than the clean case (i.e., Nb [15]), which is consistent with the specific heat analysis in $BaNi_2As_2$ [6]. The shoulder-like anomaly close to H_{c2} may reflect a spread in H_{c2} as reflected by the width of the heat capacity anomaly in field close to H_{c2} [6]. On the other hand, the shoulder might originate from multiband superconductivity although the anomaly is much less dramatic than in MgB_2 [19]. Further studies should be performed on cleaner crystals to resolve this issue.

As to Fe-pnictide SCs, there is no consistent conclusion for the superconducting pairing symmetry as can be inferred from fully gapped (Ba,K)Fe_2As_2 [20] and nodal $BaFe_2(As,P)_2$ [21], both of which have 30 K-class high-T_c. It is suggested that the strongly k-dependent gap function in Fe-pnictides could give rise to nodes on the Fermi surface [22–24]. First principles calculations show that the Fermi surface of the Ni-pnictide compounds is much more complicated than that of the Fe-pnictide compounds because of the two additional valence electrons in Ni relative to Fe [25, 26]. Accordingly, it is difficult to find any nodal plane which does not intersect the Fermi surface in the Ni-pnictide compounds. This leads to the conclusion that the superconducting pairing symmetry of $BaNi_2As_2$ and $SrNi_2P_2$ is most likely conventional s-wave rather than sign-reversing $\pm s$-wave as suggested in Fe-pnictide SCs [27, 28].

4. Conclusion

To conclude, we have investigated the structure of the superconducting gap symmetry of $BaNi_2As_2$ and $SrNi_2P_2$ through low-temperature thermal transport measurements. The concave field variations of κ_0/T in the low field region have revealed that $BaNi_2As_2$ and $SrNi_2P_2$ are fully gapped SCs. This conclusion is also supported by the low-temperature dependences of κ and C.

Acknowledgment

We would like to thank I. Vekhter, M. Graf and S.-H. Baek for useful discussions. Work at Los Alamos National Laboratory was performed under the auspices of the US Department of Energy. Work at Oak Ridge was supported by the DOE, Division of Materials Sciences and Engineering.

References

[1] Kamihara Y, T. Watanabe, M. Hirano, and H. Hosono 2008 *J. Am. Chem. Soc.* **130** 3296
[2] K. Ishida, Y. Nakai, H. Hosono, 2009 *J. Phys. Soc. Jpn.* **78** 062001 and references therein.
[3] Ronning F, Bauer E D, Park T, Kurita N, Klimczuk T, Movshovich R, Sefat A S, Mandrus D and Thompson J D 2009 *Physica C* **469** 396 and references therein.
[4] Ronning F, Kurita N, Bauer E D , Scott B L, Park T, Klimczuk T, Movshovich R and Thompson J D 2008 *J. Phys.: Condens. Matter* **20** 342203
[5] Ronning F, Bauer E D, Park T, Baek S -H, Sakai H, and Thompson J D 2009 *Phys. Rev. B* **79** 134507
[6] Kurita N, Ronning F, Tokiwa Y, Bauer E D, Subedi A, Singh D J, Thompson J D and Movshovich R 2009 *Phys. Rev. Lett.* **102** 147004
[7] Suzuki M, Tanatar M A, Kikugawa N, Mao Z Q, Maeno Y and Ishiguro T 2002 *Phys. Rev. Lett.* **88** 227004
[8] Werthamer N R, Helfand E and Hohenberg P C 1966 *Phys. Rev.* **147** 295
[9] Fang L, Yang H, Cheng P, Zhu X, Mu G and Wen H -H 2008 *Phys. Rev. B* **78** 104528
[10] Li Z, Chen G F, Dong J, Li G, Hu W Z, Wu D, Su S K, Zheng P, Xiang T, Wang N L and Luo J L 2008 *Phys. Rev. B* **78** 060504(R)
[11] Berman R 1976 *Thermal conduction in Solids* (Oxford Univ. Press, Oxford) and references therein
[12] Kurita N, Ronning F, Miclea C F, Bauer E D, Thompson J D, Gofryk K, and Movshovich R 2010 in preparation
[13] Kurita N, Ronning F, Miclea C F, Bauer, E D, Thompson J D, Sefat A S, McGuire M A, Sales B C, Mandrus D and Movshovich R 2009 *Phys. Rev. B* **79** 214439
[14] Movshovich R, Hubbard M A, Salamon M B, Balatsky A V, Yoshizaki R, Sarrao J L and Jaime M 1998 *Phys. Rev. Lett.* **80** 1968.
[15] Lowell J and Sousa J B 1970 *J. Low. Temp. Phys.* **3** 65
[16] Proust C, Boaknin E, Hill P W, Taillefer L and Mackenzie A P 2002 *Phys. Rev. Lett.* **89** 147003
[17] Volovik G E 1995 *JETP Lett.* **58** 469
[18] Willis J and Ginsberg D 1976 *Phys. Rev. B* **14** 1916
[19] Sologubenko A V, Jun J, Kazakov S M, Karpinski J and Ott H R 2002 *Phys. Rev. B* **66** 014504
[20] Luo X G, Tanatar M A, Reid J Ph, Shakeripour H, Doiron-Leyraud N, Ni N, Bud'ko S L, Canfield P C, Luo H, Wang Z, Wen H H, Prozorov R and Taillefer L 2009 *Phys. Rev. B* **80** 140503
[21] Hashimoto K, Yamashita M, Kasahara S, Senshu Y, Nakata N, Tonegawa S, Ikada K, Serafin A, Carrington A, Terashima T, Ikeda H, Shibauchi T and Matsuda Y 2010 *Phys. Rev. B* **81** 220501
[22] Machida Y, Tomokuni K, Isono T, Izawa K, Nakajima Y and Tamegai T 2009 *J. Phys. Soc. Jpn.* **78** 073705
[23] Dong J K, Zhou S Y, Guan T Y, Zhang H, Dai Y F, Qiu X , Wang X F, He Y, Chen X H and Li S Y 2010 *Phys. Rev. Lett.* **104** 087005
[24] Reid J Ph, Tanatar M A, Luo X G, Shakeripour H, Doiron-Leyraud N, Ni N, Bud'ko S L, Canfield P C, Prozorov R and Taillefer L 2010 *cond-mat arXiv:*1004.3804
[25] Subedi A and Singh D J 2008 *Phys. Rev. B* **78** 132511
[26] Singh D J 2008 *Phys. Rev. B* **78** 094511
[27] Mazin I I, Singh D J, Johannes M D and Du M H 2008 *Phys. Rev. Lett.* **101** 057003
[28] Kuroki K, Usui H, Onari S, Arita R and Aoki H 2009 *Phys. Rev. B* **79** 224511

International Conference on Strongly Correlated Electron Systems (SCES 2010) IOP Publishing
Journal of Physics: Conference Series **273** (2011) 012098 doi:10.1088/1742-6596/273/1/012098

High-Pressure Electrical Resistivity Measurements of $EuFe_2As_2$ Single Crystals

N Kurita[1,2], M Kimata[1,2], K Kodama[1,3], A Harada[1], M Tomita[1], H S Suzuki[1], T Matsumoto[1], K Murata[4], S Uji[1,2,3] and T Terashima[1,2]

[1]National Institute for Materials Science, Tsukuba, Ibaraki 305-0003, Japan
[2]JST, Transformative Research-Project on Iron Pnictides (TRIP), Chiyoda, Tokyo 102-0075, Japan
[3]Graduate School of Pure and Applied Sciences, University of Tsukuba, Ibaraki 305-0003, Japan
[4]Division of Molecular Materials Science, Graduate School of Science, Osaka City University, Osaka 558-8585, Japan

E-mail: KURITA.Nobuyuki@nims.go.jp

Abstract.
High-pressure electrical resistivity measurements up to 3.0 GPa have been performed on $EuFe_2As_2$ single crystals with residual resistivity ratios $RRR = 7$ and 15. At ambient pressure, a magnetic / structural transition related to FeAs-layers is observed at $T_0 = 190$ K and 194 K for samples with $RRR = 7$ and 15, respectively. Application of hydrostatic pressure suppresses T_0, and then induces similar superconducting behavior in the samples with different RRR values. However, the critical pressure 2.7 GPa, where $T_0 \rightarrow 0$, for the samples with $RRR = 15$ is slightly but distinctly larger than 2.5 GPa for the samples with $RRR = 7$.

1. Introduction

Since the discovery of superconductivity in LaFeAs(O,F) with $T_c = 26$ K [1], a family of Fe-pnictide superconductors has attracted much attention. In particular, $A\mathrm{Fe_2As_2}$ (A = Ca, Sr, Ba, Eu, etc.) with a tetragonal $ThCr_2Si_2$-type structure has been intensively studied because of the availability of stoichiometric single crystals with high quality. It turned out that, in Fe-pnictide compounds, the superconducting (SC) ground state could appear in accordance with the suppression of a magnetic/structural transition by doping [2]. In the phase diagrams, it is argued that the superconductivity could coexist and/or compete with the antiferromagnetism [3, 4]. However, a random potential introduced by doping could smear the intrinsic SC properties. For understanding the origin of the high-T_c superconductivity with T_c up to 55 K [5], it is of considerable importance to probe the systematic change of ground states using high-quality single crystals. An alternative way to tune the ground state is to apply hydrostatic pressure (P). For instance, recent high-P ac-susceptibility and resistivity measurements have revealed that $A\mathrm{Fe_2As_2}$ (A = Sr, Eu) exhibits P-induced bulk superconductivity by suppressing the magnetic/structural transition [6–10]. Meanwhile, superconductivity under hydrostatic P is absent in $CaFe_2As_2$ [11–14], and remains a controversial issue in $BaFe_2As_2$ [8, 15–17]

Among the $A\mathrm{Fe_2As_2}$ series, $EuFe_2As_2$ is quite unique because the localized Eu^{2+} moments order antiferromagnetically at $T_N \sim 20$ K, in addition to the magnetic/structural transition related to FeAs-layers at $T_0 \sim 190$ K [18–21]. Interestingly, the magnetic order of Eu^{2+} moments

Published under licence by IOP Publishing Ltd

can be detected even in the SC state induced by doping or application of pressure, which could be a main reason for the novel reentrant-SC-like behavior [9, 10, 22–25].

Here, we report the results of high-P electrical resistivity measurements in $EuFe_2As_2$ using newly grown single crystals with a residual resistivity ratio (RRR) as high as 15. At ambient P, the magnetic/structural transition occurs at a higher temperature of $T_0 = 194\,K$, compared with 190 K for single crystals with $RRR = 7$. Consequently, it is found that the higher quality single crystal requires higher-P to suppress T_0, and to induce the SC ground state in $EuFe_2As_2$.

2. Experimental Details

Single crystals of $EuFe_2As_2$ were grown by Bridgman method from a stoichiometric mixture of the constituent elements. In this study, we examined several crystals from two different batches with residual resistivity ratios $RRR = 7$ and 15, where RRR is defined as $_{300K}/\ _{4K}$. Single crystals measured in Ref [10] were taken from a batch with $RRR = 7$. High-pressure resistivity measurements of samples with $RRR = 7$ and 15 have been performed simultaneously up to 3.0 GPa using a hybrid-type piston cylinder pressure device [26]. The resistivity was measured by the four-probe method with an ac current $I = 0.3$ mA in the ab-plane. To generate hydrostatic pressure, Daphne 7474 (Idemitsu Kosan) oil, which remains in a liquid state up to 3.7 GPa at room temperature [27], was used as a pressure-transmitting medium. Samples were cooled down in Oxford ^{4}He system, slowly with an average rate of 0.5 K/min. Applied pressure was estimated at 4.2 K from the resistance change of a calibrated Manganin wire [28].

3. Results and Discussions

Figure 1 shows the temperature (T) dependence of electrical resistivity scaled at 300 K ($/\ _{300K}$) in $EuFe_2As_2$ single crystals with $RRR = 7$ and 15, where RRR is determined as $_{300K}/\ _{4K}$. The measurement was performed in zero field at ambient pressure outside a pressure device with current direction $I \parallel ab$. To our knowledge, $RRR = 15$ is the largest value in $EuFe_2As_2$ single crystals [9, 21, 29]. Overall T-variations of the resistivity in the samples with $RRR = 7$ and 15

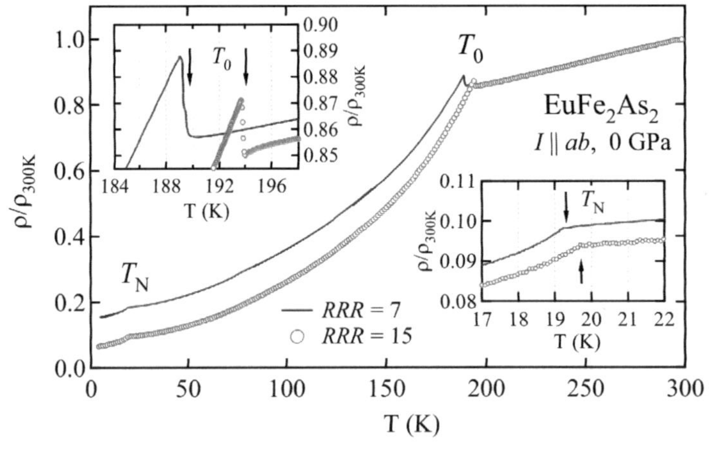

Figure 1. (Color online) The scaled electrical resistivity $/\ _{300K}$ versus temperature in $EuFe_2As_2$ single crystals with $RRR = 7$ and 15. The measurement was carried out in zero-field at ambient pressure with the current direction $I \parallel ab$. Upper left and lower right insets represent the expanded views around $T = T_0$ and T_N, respectively. The data for the sample with $RRR = 7$ in the lower right inset is arbitrarily shifted in vertical direction for clarity.

are qualitatively similar to each other, and are consistent with previous results [9, 20, 21, 29]. It is worthwhile to mention that, as shown in the upper left inset, a magnetic/structural transition temperature $T_0 = 194$ K for the sample with $RRR = 15$ is higher than $T_0 = 190$ K for the sample with $RRR = 7$. This would be the reason why samples with $RRR = 15$ needs higher pressure (P) to suppress T_0, as will be discussed below. The Néel temperature T_N of the localized Eu^{2+} moments for the sample with $RRR = 15$ is slightly higher than the value for the sample with $RRR = 7$, as can be seen in the lower right inset.

Next, we turn to the pressure effect on the electrical resistivity for the samples with $RRR = 7$ (Fig. 2) and 15 (Fig. 3), which are simultaneously measured in the same pressure device. With increasing P, the resistivity peak related to the magnetic/structural transition is suppressed to a lower temperature in both samples as shown in Figs. 2(a) and 3(a). For the sample with $RRR = 7$, a reminiscence of the peak is clearly recognized at 2.38 GPa around 100 K, and faintly visible at 2.46 GPa around 70 K as shown in Fig. 2(a). At 2.55 GPa, there is no detectable anomaly, which implies that the critical pressure P_c, where $T_0 \to 0$, may be about 2.5 GPa. For the sample with $RRR = 15$, P_c would be ~ 2.7 GPa since the resistivity hump is slightly recognized at 2.69 GPa, but undetectable at 2.77 GPa as shown in Fig. 3(a). It is of interest that the resistivity follows nearly T-linear behavior above T_c at 2.55 and 2.77 GPa ($P \sim P_c$) for samples with $RRR = 7$ and 15, respectively, as guided by a dashed line. A similar T-variation of resistivity was also reported in several optimally-doped Fe-pnictide superconductors [4, 30–32]. For the sample with $RRR = 7$, a resistivity upturn and a small maximum, as indicated by an arrow in Figs. 2(b), in the broad SC transition below 31 K are observed at $P = 2.38$ GPa ($< P_c$). It suggests that the superconductivity is suppressed by the magnetic order of the Eu^{2+} moments; consequently, reentrant-SC-like behavior appears. A similar behavior is also slightly seen for the sample with $RRR = 15$ (Figs. 3(b)), but more smeared out. At $P > P_c$, resistivity exhibits sharp SC transitions to zero-resistivity with $T_c \sim 30$ K for both samples. With increasing P, the SC

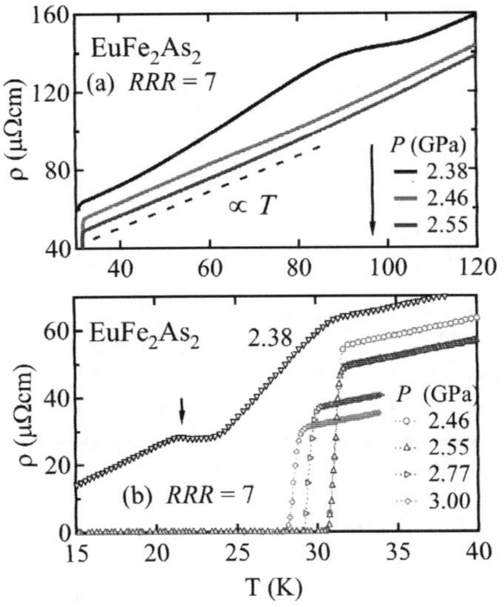

Figure 2. vs T of a $EuFe_2As_2$ single crystal ($RRR = 7$) up to 3.0 GPa in the temperature ranges (a) 30 – 120 K and (b) 15 – 40 K. An arrow in (b) indicates an anomaly attributed to T_N

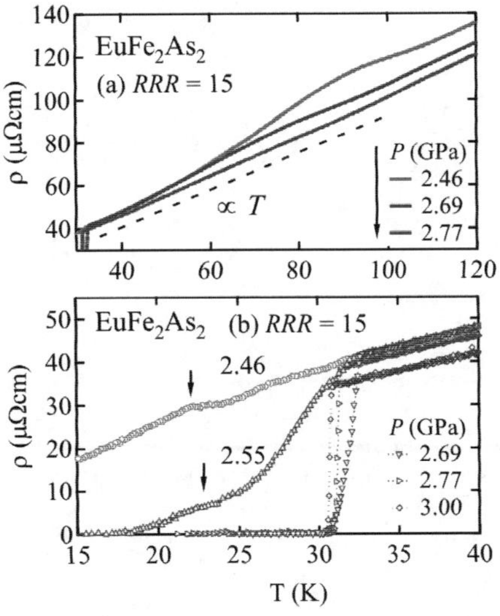

Figure 3. vs T of a $EuFe_2As_2$ single crystal ($RRR = 15$) up to 3.0 GPa in the temperature ranges (a) 30 – 120 K and (b) 15 – 40 K. Arrows in (b) indicate anomalies attributed to T_N.

transitions persist up to 3.00 GPa although the T_c continuously decreases. Thus, the P-variation of the resistive behavior between the samples with different quality is qualitatively similar to each other, and is consistent with the previous result [10]. However, $P_c \sim 2.7$ GPa for the sample with $RRR = 15$ is slightly but distinctly larger than ~ 2.5 GPa for the sample with $RRR = 7$, which may be as a consequence of the larger value of T_0 for the higher-quality sample at ambient-P. We have repeated similar high-P resistivity measurements using several single crystals, and confirmed that the observed difference in the magnitude of T_0 and P_c between the samples with $RRR = 7$ and 15 is beyond the error of the pressure estimation (± 2-3 10^{-2} GPa) [28]. Another meaningful issue, which probably relates to the sample quality, is the width of a SC transition ΔT_c. The minimum values of ΔT_c are 1 K and 0.8 K for samples with $RRR = 7$ and 15, respectively. These facts suggest that the higher-quality single crystals have larger values of T_0 and P_c as well as a sharper SC transition in $EuFe_2As_2$.

Until now, there has been no report concerning the quantum oscillation in $EuFe_2As_2$, despite the importance for understanding the Fermi surface topology and mass renormalization. In fact, we have already tried de Haas-van Alphen (dHvA) measurements of $EuFe_2As_2$ using the samples with $RRR = 7$ at 0.6 K with fields up to 35 T, but could not detect any dHvA oscillation. Given that quantum oscillations were successfully detected in $SrFe_2As_2$ ($RRR \sim 8$) [33] and $BaFe_2As_2$($RRR = 10$) [34], it is worthwhile to perform the dHvA measurement of $EuFe_2As_2$ using the newly grown single crystals with $RRR = 15$.

4. Conclusions

We have performed high-pressure electrical resistivity measurements up to 3.0 GPa in $EuFe_2As_2$ single crystals with $RRR = 7$ and 15. At ambient pressure, a magnetic/structural transition occurred at $T_0 = 190$ K and 194 K for the samples with $RRR = 7$ and 15, respectively. Although P-induced superconductivity was confirmed in the samples with different RRR values, the critical pressure $P_c \sim 2.7$ GPa for the samples with $RRR = 15$ was slightly but distinctly larger than ~ 2.5 GPa for the samples with $RRR = 7$.

References

[1] Kamihara Y, Watanabe T, Hirano M and Hosono H 2008 *J. Am. Chem. Soc.* **130** 3296
[2] Ishida K, Nakai Y and Hosono H 2009 *J. Phys. Soc. Jpn.* **78** 062001 and references therein.
[3] Chen H, Ren Y, Qiu Y, Bao W, Liu R H, Wu G, Wu T, Xie Y L, Wang X F, Huang Q and Chen X H 2009 *Europhys. Lett.* **85** 17006
[4] Wang X F, Wu T, Wu G, Liu R H, Chen H, Xie Y L and Chen X H 2009 *New J. Phys.* **11** 045003
[5] Ren Z A, Lu W, Yang J, Yi W, Shen X L, Li Z C, Che G C, Dong X L, Sun L L, Zhou F and Zhou Z X 2008 *Chin. Phys. Lett.* **25** 2215
[6] Alireza P L, Ko Y T C, Gillett J, Petrone C M, Cole J M, Lonzarich G G and Sebastian S E 2009 *J. Phys.: Condens. Matter* **21** 012208
[7] Kotegawa H, Sugawara H and Tou H 2009 *J. Phys. Soc. Jpn.* **78** 013709
[8] Matsubayashi K, Katayama N, Ohgushi K, Yamada A, Munakata K, Matsumoto T and Uwatoko Y 2009 *J. Phys. Soc. Jpn.* **78** 073706
[9] Miclea C F, Nicklas M, Jeevan H S, Kasinathan D, Hossain Z, Rosner H, Gegenwart P, Geibel C and Steglich F 2009 *Phys. Rev. B* **79** 212509
[10] Terashima T, Kimata M, Satsukawa H, Harada A, Hazama K, Uji S, Suzuki H S, Matsumoto T and Murata K 2009 *J. Phys. Soc. Jpn.* **78** 083701
[11] Torikachvili M S, Bud'ko S L, Ni N and Canfield P C 2008 *Phys. Rev. Lett.* **101** 057006.
[12] Park T, Park E, Lee H, Klimczuk T, Bauer E D, Ronning F and Thompson J D 2009 *J. Phys.: Condens. Matter* **20** 322204
[13] Lee H, Park E, Park T, Sidorov V A, Ronning F, Bauer E D and Thompson J D 2009 *Phys. Rev. B* **80** 024519
[14] Yu W, Aczel A A, Williams T J, Bud'ko S L, Ni N, Canfield P C and Luke G M 2009 *Phys. Rev. B* **79** 020511
[15] Ishikawa F, Eguchi N, Kodama M, Fujimaki K, Einaga M, Ohmura A, Nakayama A, Mitsuda A and Yamada Y 2009 *Phys. Rev. B* **79** 172506

[16] Colombier E, Bud'ko S L, Ni N and Canfield P C 2009 *Phys. Rev. B* **80** 224518
[17] Yamazaki T, Takeshita N, Kobayashi R, Fukazawa H, Kohori Y, Kihou K, Lee C.-H., Kito H, Iyo A and Eisaki H 2010 *Phys. Rev. B* **81** 224511
[18] Raffius H, Mörsen E, Mosel B D, Müller Warmuth W, Jeitschko W, Terbüchte L and Vomhof T 1993 *J. Phys. Chem. Solids* **54** 135
[19] Tegel M, Rotter M, Weiss V, Schappacher F M, Pottgen R and Johrendt D 2008 *J. Phys.: Condens. Matter* **20** 452201
[20] Ren Z, Zhu Z W, Jiang S, Xu X F, Tao Q, Wang C, Feng C M, Cao G H and Xu Z A 2008 *Phys. Rev. B* **78** 052501
[21] Jeevan H S, Hossain Z, Kasinathan D, Rosner H, Geibel C and Gegenwart P 2008 *Phys. Rev. B* **78** 052502
[22] Ren Z, Lin X, Tao Q, Jiang S, Zhu Z, Wang C, Cao G and Xu Z 2008 *Phys. Rev. B* **78** 092406.
[23] Jiang S, Xing H, Xuan G, Ren Z, Wang C, Xu Z and Cao G 2009 *Phys. Rev. B* **80** 184514.
[24] Ren Z, Tao Q, Jiang S, Feng C, Wang C, Dai J, Cao G and Xu Z 2009 *Phys. Rev. Lett.* **102** 137002.
[25] Zheng Q J, He Y, Wu T, Wu G, Chen H, Ying J J, Liu R H, Wang X F, Xie Y L, Yan Y J, Li Q J and Chen X H 2009 arXiv:0907.5547v1
[26] Uwatoko Y, Hedo M, Kurita N, Koeda M, Abliz M and Matsumoto T 2003 *Physica C* **329-333** 1658
[27] Murata K, Yokogawa K, Yoshino H, Klotz S, Munsch P, Irizawa A, Nishiyama M, Iizuka K, Nanba T, Okada T, Shiraga Y and Aoyama S 2008 *Rev. Sci. Instrum.* **79** 085101
[28] Terashima T, Tomita M Kimata M, Satsukawa H, Harada A, Hazama K, Uji S, Suzuki H S, Matsumoto T and Murata K 2009 *J. Phys. Soc. Jpn.* **78** 118001
[29] Mitsuda A, Matoba T, Wada H, Ishikawa F and Yamada Y 2010 *J. Phys. Soc. Jpn.* **79** 073704
[30] Liu R H, Wu G, Wu T, Fang D F, Chen H, Li S Y, Liu K, Xie Y L, Wang X F, Yang R L, Ding L, He C, Feng D L and Chen X H 2008 *Phys. Rev. Lett.* **101** 087001
[31] Gooch M, Bing L, Lorenz B, Guloy A M and Chu C W 2009 *Phys. Rev. B* **79** 104504
[32] Jiang S, Xing H, Xuan G, Wang C, Ren Z, Feng C, Dai J, Xu Z and Cao G 2009 *J. Phys.: Condens. Matter* **21** 382203
[33] Sebastian S E, Gillett J, Harrison N, Lau P H C, Singh D J, Mielke C H and Lonzarich G G 2008 *J. Phys.: Condens. Matter* **20** 422203
[34] Analytis J G, McDonald R D, Chu J -H, Riggs S C, Bangura A F, Kucharczyk C, Johannes M and Fisher I R 2009 *Phys. Rev. B* **80** 064507

Journal of Physics: Conference Series **273** (2011) 012099 doi:10.1088/1742-6596/273/1/012099

Coherence effect in multiband superconductors: A suggestion of an experiment to determine the correct symmetry of the order parameter

Keisuke Masuda and Susumu Kurihara

Department of Physics, Waseda University, Okubo, Shinjuku, Tokyo 169-8555, Japan

E-mail: masuda@kh.phys.waseda.ac.jp

Abstract. An experimental method to determine the pairing symmetry in iron pnictides is theoretically proposed. We focus on two kinds of s-wave states, s_{+-}- and s_{++}-wave states, which are possible candidates for the pairing states of iron pnictides. For each state, we calculate both the density and spin response functions by using the two-band BCS model within the one-loop approximation. As a consequence, a clear difference is seen between the s_{+-}- and s_{++}-wave states in the temperature dependence of the response functions at nesting vector **Q**, which connects the hole and electron Fermi surfaces. We mention that this difference originates from the coherence effect in multiband superconductors. We suggest that the pairing symmetry could be identified by observing the temperature dependence of both the density and spin structure factors at low energies and at the nesting vector **Q** in neutron scattering measurements. Furthermore, we discuss the influence of impurities on the coherence effect in multiband superconductors.

For unconventional superconductors, determining the pairing symmetry is one of the main subjects of both the experimental and theoretical researches because the pairing symmetry is closely related to the mechanism of superconductivity. Although two years have passed since the discovery of iron pnictides [1], the pairing symmetry in this material has not been determined completely. In many previous studies, it has been shown that a fully gapped s-wave state with sign reversal (s_{+-}-wave state) [see Fig. 1(a)] is the most promising candidate for the pairing symmetry in iron pnictides [2]. However, in recent research, Onari and Kontani pointed out that a fully gapped s-wave state without sign reversal (s_{++}-wave state) [see Fig. 1(b)] is more likely than the s_{+-}-wave state [3]. In experiments such as ARPES [4] and penetration depth

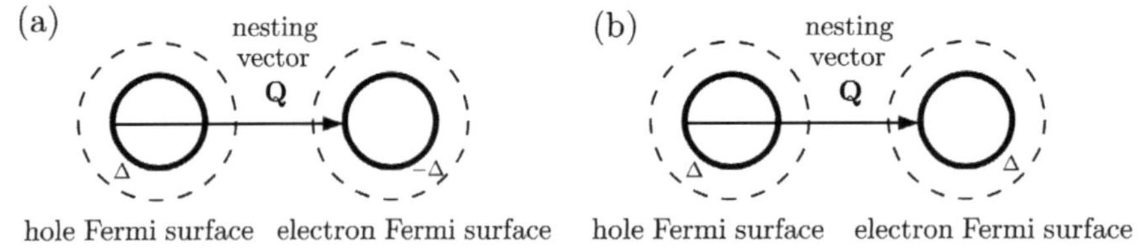

Figure 1. Schematic representation of the s_{+-}- (a) and s_{++}- (b) wave state. The thick solid lines represent the Fermi surfaces, and the arrow indicates the nesting vector **Q**.

Published under licence by IOP Publishing Ltd

measurements [5], multiband and fully gapped s-wave natures have been confirmed. However, the information about the relative phase between the gap on the hole Fermi surface and that on the electron Fermi surface has not been fully obtained [6]. Therefore, an experiment to capture this relative phase is much needed. In this work, we theoretically suggest an experimental method to determine the pairing state of iron pnictieds. Our idea is to use the temeprature dependence of both the density and spin responses which depends on the relative phase between the gaps. First, we introduce the following two-band BCS Hamiltonian $\mathcal{H}$, which we use in our calculation:

$$\mathcal{H} = \mathcal{H}_0 + \mathcal{H}', \tag{1}$$

$$\mathcal{H}_0 = \sum_{\mathbf{k},\sigma,\nu=\pm} \xi_\nu(\mathbf{k})\gamma^\dagger_{\nu\sigma}(\mathbf{k})\gamma_{\nu\sigma}(\mathbf{k}), \tag{2}$$

$$\mathcal{H}' = -\sum_{\mathbf{k},\nu=\pm}\left[\Delta_\nu\gamma^\dagger_{\nu\uparrow}(\mathbf{k})\gamma^\dagger_{\nu\downarrow}(-\mathbf{k}) + \Delta_\nu\gamma_{\nu\downarrow}(-\mathbf{k})\gamma_{\nu\uparrow}(\mathbf{k})\right], \tag{3}$$

where $\gamma_{\nu\sigma}(\mathbf{k})$ is the annihilation operator with spin σ and wave vector $\mathbf{k}$ in the band ν, and $\xi_\nu(\mathbf{k})$ is the band energy obtained from the two-orbital tight binding model proposed by Raghu *et al.* [7]. Throughout this work the parameters of Ref. 7 are used: $t_1 = -1.0, t_2 = 1.3, t_3 = t_4 = -0.85$, and $\mu = 1.45$ (all in units of $|t_1|$). Δ_ν is the superconducting gap of the band ν, and we use the interpolation formula $\Delta_\nu \simeq \Delta^{(0)}_\nu \tanh\left(1.74\sqrt{\frac{T_c}{T}-1}\right)$ in numerical calculations. In our two-band BCS model, the condition $\Delta_+ = -\Delta_-$ ($\Delta_+ = \Delta_-$) corresponds to the s_{+-}-wave (s_{++}-wave) case.

Next, we calculate the response functions for our model. According to the Kubo formula, the density response function $D(\mathbf{q}, i\omega_m)$ and the spin response function $\chi^{+-}(\mathbf{q}, i\omega_m)$ are given by

$$D(\mathbf{q}, i\omega_m) = -\sum_{rt}\int_0^\beta d\tau e^{i\omega_m\tau}\langle T_\tau\rho_r(\mathbf{q},\tau)\rho_t(-\mathbf{q},0)\rangle, \tag{4}$$

$$\chi^{+-}(\mathbf{q}, i\omega_m) = -\sum_{rt}\int_0^\beta d\tau e^{i\omega_m\tau}\langle T_\tau S^+_r(\mathbf{q},\tau)S^-_t(-\mathbf{q},0)\rangle. \tag{5}$$

Here, $\rho_r(\mathbf{q},\tau) = \sum_{\mathbf{k}\sigma} d^\dagger_{r\sigma}(\mathbf{k}+\mathbf{q},\tau)d_{r\sigma}(\mathbf{k},\tau)$, and $S^+_r(\mathbf{q},\tau) = (S^-_r(-\mathbf{q},\tau))^\dagger = \sum_{\mathbf{k}} d^\dagger_{r\uparrow}(\mathbf{k}+\mathbf{q},\tau)d_{r\downarrow}(\mathbf{k},\tau)$, where $d_{r\sigma}(\mathbf{k})$ annihilates an electron with spin σ and wave vector $\mathbf{k}$ in the orbital r. Note that the orbital operator $d_{r\sigma}(\mathbf{k})$ is related to the band operator $\gamma_{\nu\sigma}(\mathbf{k})$ by the unitary transformation $d_{r\sigma}(\mathbf{k}) = \sum_{\nu=\pm} a^r_\nu(\mathbf{k})\gamma_{\nu\sigma}(\mathbf{k})$, where $a^r_\nu(\mathbf{k})$ is the matrix element between the orbital r and the band ν. The response functions (4) and (5) are transformed into the equivalent functions in the band representation and are calculated within the one-loop approximation. Then, the final expressions are obtained by substituting the normal and anomalous Green's functions derived from the above Hamiltonian $\mathcal{H}$ into the response functions. The details of the calculations are presented in Ref. 8. In the following we show the numerical results of our calculations. In Fig. 2(a), we show the temperature dependence of the density and spin response functions at $\omega \ll |\Delta^{(0)}_\nu| + |\Delta^{(0)}_{\nu'}|$ and at the nesting vector $\mathbf{Q} = (\pi, 0)$ for the s_{+-}-wave state. The density response shows a peak just below T_c. In contrast, the spin response decreases monotonically as temperature deacreses. This opposite behavior of the density and spin responses is the dominant feature of the "coherence effect" [9] in superconductors and can be understood as follows. The response function mainly consists of two parts: one is the density of states (DOS) and the other is the coherence factor. As is well known, in the superconducting state, the DOS has a peak at the gap energy [see the case of $I = 0$ in Fig. 3(a)]. For

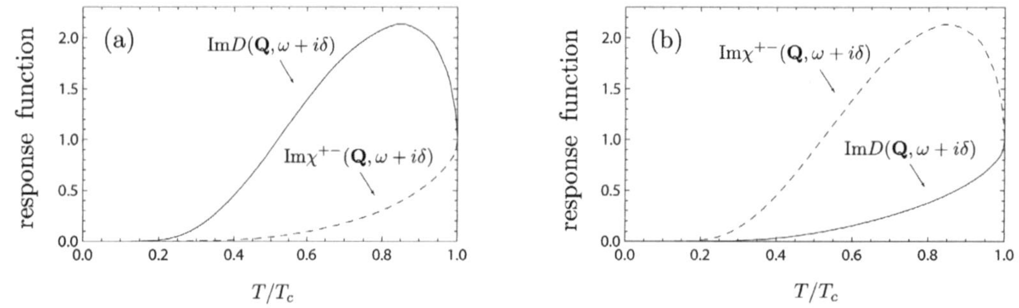

Figure 2. Temperature dependence of the density and spin response functions at low energies ($\omega \ll |\Delta_+^{(0)}| + |\Delta_-^{(0)}|$) for s_{+-}- (a) and s_{++}- (b) wave states normalized by its value at $T = T_c$ with $|\Delta_+^{(0)}| = |\Delta_-^{(0)}| = 0.05$, $T_c = 0.03$, $\omega = 0.001$, $\mathbf{Q} = (\pi, 0)$, and $\delta = 0.0001$. The solid (dashed) line represents the imaginary part of the density (spin) response function.

$\omega \ll |\Delta_\nu^{(0)}| + |\Delta_{\nu'}^{(0)}|$ the dominant coherence factor is given in the following form [10]:

$$\frac{1}{2}\left\{\left(1+\frac{\xi_{\nu'}(\mathbf{k})}{E_{\nu'}(\mathbf{k})}\right)\left(1+\frac{\xi_\nu(\mathbf{k}+\mathbf{q})}{E_\nu(\mathbf{k}+\mathbf{q})}\right)\mp\frac{\Delta_{\nu'}\Delta_\nu}{E_{\nu'}(\mathbf{k})E_\nu(\mathbf{k}+\mathbf{q})}\right\}, \tag{6}$$

where $E_\nu(\mathbf{k}) = \sqrt{\xi_\nu^2(\mathbf{k}) + \Delta_\nu^2}$ and the upper (lower) sign corresponds to the case of the density (spin) response. In the s_{+-}-wave state, the product of the gaps $\Delta_{\nu'}\Delta_\nu$ becomes negative at $\mathbf{q} = \mathbf{Q}$. Therefore, the minus (plus) sign in front of $\Delta_{\nu'}\Delta_\nu/E_{\nu'}(\mathbf{k})E_\nu(\mathbf{k}+\mathbf{q})$ in (6) is effectively reversed, and the coherence factor in the density (spin) response simply behaves as one (zero) at the Fermi level. As a result, in the density (spin) response, the coherence factor keeps (cancels) the peak in the DOS and the peak in the response appears (disappears). In Fig. 2(b), we show the temperature dependence of the density and spin response functions for the s_{++}-wave state. In this state, the behavior of both the responses is just the opposite to the case of s_{+-}-wave. By noting that $\Delta_{\nu'}\Delta_\nu > 0$ in the s_{++}-wave state, this behavior of both the responses can be understood in the same way as the s_{+-}-wave case.

In most iron pnictides, superconductivity is induced by chemical doping. From this perspective, studying the influence of impurities on the coherence effect is of importance. We analyze the effect of orbital-diagonal local impurities [3] on the DOS of our two-band BCS model. In Fig. 3(a) [Fig. 3(b)], we show the DOS in the s_{+-}-wave (s_{++}-wave) state. In the s_{+-}-wave state, the peak in the DOS is eliminated by realistic impurity potentials. However, in the s_{++}-wave state, the peak in the DOS is hardly changed by impurities. If impurities eliminate the peak in the DOS, both the density and spin response functions decrease monotonically as temperature decreases. Hence, it is found that in the dirty s_{+-}-wave state both the response functions have no peak just below T_c and show monotonic decrease as temperature decreases.

By using the results obtained above, let us discuss the experiment determines the pairing state in iron pnictides. For the determination of the pairing state, measuring the temperature dependence of both the density and spin responses (density and spin structure factors in actual experiments) at $\omega \ll |\Delta_\nu^{(0)}| + |\Delta_{\nu'}^{(0)}|$ and at the nesting vector $\mathbf{Q}$ in neutron scattering measurements is of necessity. According to our calculations, the following three possibilities can be considered: (i) The density response shows a peak just below T_c, and the spin response decreases monotonically as temperature decreases. (ii) The density response shows a monotonic decrease as temperature decreases and the spin response exhibits a peak just below T_c. (iii)

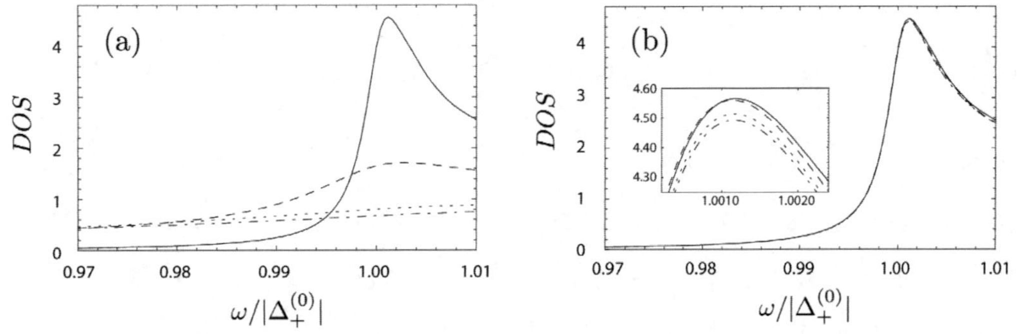

Figure 3. The DOS in the s_{+-}- (a) and s_{++}- (b) wave states at $T = 0$ for $n_{imp} = 0.01$, $|\Delta_+^{(0)}| = |\Delta_-^{(0)}| = 0.05$, $T_c = 0.03$, and $\delta = 0.0001$. $I = 0$ (solid line), $I = 0.5$ (dashed line), $I = 4.5$ (dotted line), and $I = \infty$ (dotted-dashed line). 1024×1024 **k**-meshes are used in the numerical calculations.

Both the responses show monotonic decrease with decreasing temperature. In the first case, we can conclude that the clean s_{+-}-wave state is realized in this material. In the second case, it is concluded that the s_{++}-wave state with or without impurities is realized. In the last case, we can conclude that the dirty s_{+-}-wave state is realized.

In summary, a proposal for an experiment to determine the pairing state in iron pnictides has been made. We have analyzed both the density and spin responses at the nesting vector **Q** for the s_{+-}- and s_{++}-wave states, and have found a clear difference between these two states. An explanation of this differnece has been given by using the coherence effect in superconductors. In addition, we have considered the influence of impurities on such a coherence effect. Our suggestion is that the pairing state in iron pnictides is identified by measuring the temperature dependence of both the density and spin responses at low energies ($\omega \ll |\Delta_\nu^{(0)}| + |\Delta_{\nu'}^{(0)}|$) and at the nesting vector **Q** in neutron scattering measurements.

References

[1] Kamihara Y, Watanabe T, Masahiro H and Hosono H 2008 *J. Am. Chem. Soc.* **130** 3296
[2] Mazin I I, Singh D J, Johannes M D and Du M H 2008 *Phys. Rev. Lett.* **101** 057003; Kuroki K, Onari S, Arita R, Usui H, Tanaka Y, Kontani H and Aoki H 2008 *Phys. Rev. Lett.* **101** 087004
[3] Onari S and Kontani H 2009 *Phys. Rev. Lett.* **103** 177001
[4] Ding H *et al* 2008 *Europhys. Lett.* **83** 47001
[5] Hashimoto K *et al* 2009 *Phys. Rev. Lett.* **102** 017002
[6] Very recently in scanning tunneling microscopy experiments a result supporting the s_{+-}-wave state was reported; Hanaguri T, Niitaka S, Kuroki K and Takagi H 2010 *Science* **328** 474
[7] Raghu S, Qi X L, Liu C X, Scalapino D J and Zhang S C 2008 *Phys. Rev.* B **77** 220503(R)
[8] Masuda K and Kurihara S 2009 *Preprint* arXiv:0911.0528 (to appear in J. Phys. Soc. Jpn.)
[9] Cooper L N 1992 in *Nobel Lectures in Physics 1971-1980* (Singapore: World Scientific)
[10] Note that in this work we focus on the region $\omega \ll |\Delta_\nu^{(0)}| + |\Delta_{\nu'}^{(0)}|$ and $0 \le T \le T_c$, not the region $\omega \approx |\Delta_\nu^{(0)}| + |\Delta_{\nu'}^{(0)}|$ and $T \ll T_c$. For $\omega \approx |\Delta_\nu^{(0)}| + |\Delta_{\nu'}^{(0)}|$ and $T \ll T_c$, the dominant coherence factor is different from (6) (for details, see Ref.8). The resonance peak in inelastic neutron scattering spectrum [11] is discussed in this region [12].
[11] Christianson A D *et al* 2008 *Nature* (London) **456** 930
[12] Maier T A and Scalapino D J 2008 *Phys. Rev.* B **78** 020514; Korshunov M M and Eremin I 2008 *Phys. Rev.* B **78** 140509(R); Onari S, Kontani H and Sato M 2010 *Phys. Rev.* B **81** 060504(R)

Competition between Fe-based superconductivity and antiferromagnetism of Eu^{2+} in $Eu_{1-x}Ca_xFe_2As_2$

A. Mitsuda[1], S. Seike[1], T. Matoba[1], H. Wada[1], F. Ishikawa[2] and Y. Yamada[3]

[1] Department of Physics, Kyushu University, 6-10-1 Hakozaki, Higashi-ku, Fukuoka 812-8581, Japan
[2] Graduate School of Science and Technology, Niigata University, Niigata 950-2181, Japan
[3] Department of Physics, Niigata University, Niigata 950-2181, Japan

E-mail: 3da@phys.kyushu-u.ac.jp

Abstract. Single crystals of $Eu_{1-x}Ca_xFe_2As_2$ (x = 0.25, 0.50 and 0.75) are prepared. With increasing x, the Néel temperature of Eu^{2+} decreases and disappears at $x = 0.75$, whereas the spin density wave (SDW) transition temperature, T_{SDW}, is almost unchanged. For $x = 0.50$, the pressure-induced superconductivity emerges below 20 K and the superconductivity is slightly suppressed by the antiferromagnetic order (AFM) of Eu^{2+}. Compared with $EuFe_2As_2$, for $x = 0.50$, the superconductivity seems to overcome the AFM, which is ascribed to the weaker AFM (or lower T_N) of Eu^{2+}.

1. Introduction

Since Kamihara et al. discovered the new high-T_c superconductor, $LaFeAsO_{1-x}F_x$ ($0.05 \leq x \leq 0.12$) [1], a large number of researches on Fe-based superconductors have been carried out across the world. $EuFe_2As_2$, which crystallizes in the tetragonal $ThCr_2Si_2$ type structure and has FeAs layers similar to those in LaFeAsO, exhibits an antiferromagnetic order (AFM) of Eu^2 at $T_N = 19$ K and a Fe-based spin density wave (SDW) transition at 191 K [2]. Substitution of K^+ for Eu^{2+}, which corresponds to hole doping, suppresses the SDW transition and induces the Fe-based superconductivity of $T_c = 32$ K [3]. The application of external pressure [4, 5], substitution of Co for Fe (electron doping) [6], and isovalent substitution of P for As (chemical pressure) [7] also induce the Fe-based superconductivity almost simultaneously with the suppression of the SDW. Similar behavior is also observed in AFe_2As_2 (A=Ba, Sr, Ca). However, $EuFe_2As_2$ exhibits the AFM of Eu^{2+} with a large magnetic moment of $7\mu_B$, which is destructive to the superconductivity. Actually, the electrical resistivity under pressure of 2 GPa $\leq P \leq$ 2.5 GPa presents a sharp drop, which is reminiscent of the superconductivity, but does not reaches zero at the lowest temperature, which suggests the complete formation of the superconducting state is interrupted by the ordering of Eu^{2+} [4, 5]. He et al. observed a bulk superconductivity with zero resistivity in magnetically weakened $Eu_{1-x}Sr_xFe_{2-y}Co_yAs_2$ [8]. Unfortunately, since $SrFe_{2-y}Co_yAs_2$ is also a superconductor of $T_c \sim 20$ K, it is difficult to judge whether the observed superconductivity with zero resistivity results from the weakening of the magnetism of Eu^{2+}. We have observed a pressure-induced bulk superconductivity with zero resistivity in $Eu_{0.5}Ca_{0.5}Fe_2As_2$, where magnetism of Eu^{2+} is weakened [9]. Though $CaFe_2As_2$ also exhibits a

Published under licence by IOP Publishing Ltd

superconductivity under pressures, it is distinguishable due to substantially lower T_c value than that of $EuFe_2As_2$. In the present paper, we manifest more detailed properties under ambient pressure of $Eu_{1-x}Ca_xFe_2As_2$ and competition between the Fe-based superconductivity and the AFM of Eu^{2+} for $x = 0.50$.

2. Experimental

The single crystals of $Eu_{1-x}Ca_xFe_2As_2$ ($x = 0.25, 0.50, 0.75$) were grown by a tin flux method. A mixture of constituent elements in a ratio of Eu : Ca : Fe : As : Sn = $1-x : x : 2 : 2 : 48$, which was put into an alumina crucible, was heated as described in Ref. [9]. We obtained many pieces of plate-like single crystals with typical dimensions of $\sim 3 \times 3 \times 0.1$ mm^3. The powder X-ray diffraction (XRD) at room temperature confirms that the samples are a single phase with $ThCr_2Si_2$ type. Figure 1 shows the lattice constants, a and c, determined from the XRD pattern, as a function of Ca concentration x. Both a and c decrease linearly with increasing x, which indicates the samples obeys Vegard's law. The energy dispersive X-ray (EDX) spectroscopy

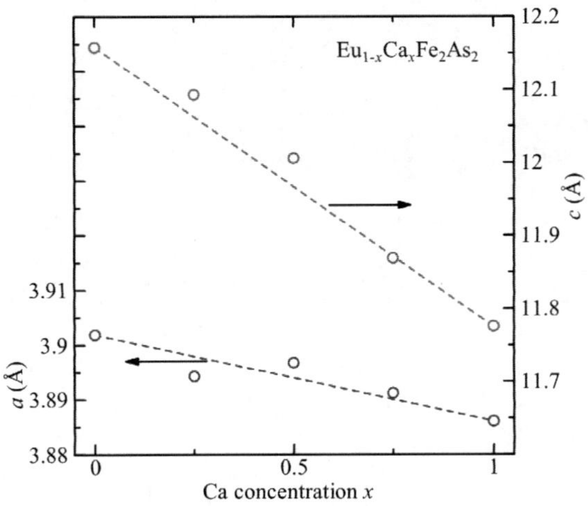

Figure 1. The lattice constants, a and c, as a function of Ca concentration x in $Eu_{1-x}Ca_xFe_2As_2$. Both a and c nearly obey Vegard's law.

demonstrates that the composition of the samples is almost Eu : Ca : Fe : As = $1-x : x : 2 : 2$. Magnetization measurement was performed by a SQUID magnetometer (MPMS; Quantum Design) in the temperature range between 2 and 300 K. Only for $x = 0.50$, the magnetization was measured under pressure of 1.4 GPa. The pressure was generated by using a piston-cylinder-type pressure cell made of CuBe alloy. A few pieces of the randomly directed samples and a tin manometer were placed into a Teflon cell filled with a pressure-transmitting medium consisting of a mixture of two types of Flourinert in a ratio of FC70 : FC77 = 1 : 1.

3. Results and Discussions

Figure 2 shows temperature dependence of the magnetic susceptibility, χ, measured in a field of 50 Oe under ambient pressure. There is no evidence of the superconductivity under ambient pressure in $Eu_{1-x}Ca_xFe_2As_2$. The clear anomaly, which is associated with the AFM of Eu^{2+}, is observed for $x = 0.25$ and 0.50. On the other hand, for $x = 0.75$, the susceptibility continues to increase down to 2 K with decreasing temperature, and there is no anomaly in the χ–T curve. The Néel temperature,T_N, decreases monotonously with increasing Ca concentration x. In our previous report, the T_N value of $x = 0.50$ in a field of 0.5 T was estimated to be ~ 4 K, which is substantially lower than the present result ($T_N \sim 9$ K). Since it is reported in $EuFe_2As_2$ that the Néel temperature shifts lower with increasing magnetic field up to 1 T [10], similar phenomenon would occur also in $Eu_{1-x}Ca_xFe_2As_2$.

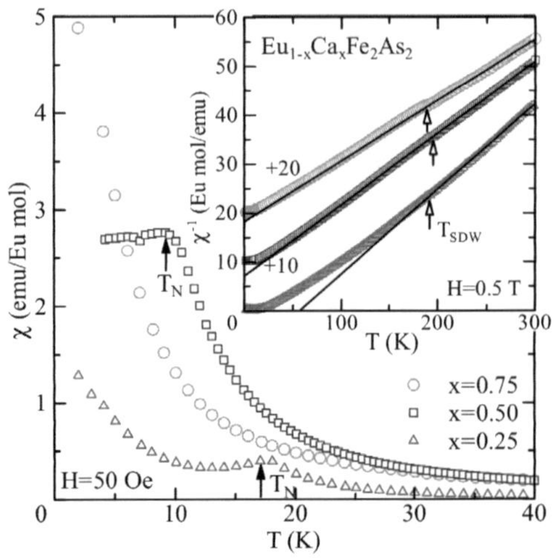

Figure 2. The temperature dependence of magnetic susceptibility measured in a field of 50 Oe under ambient pressure of $Eu_{1-x}Ca_xFe_2As_2$ with $x = 0.25$, 0.50, and 0.75. The clear anomalies corresponding to the Néel temperature (solid arrows) are observed for $x = 0.75$ and 0.50. The inset shows the temperature dependence of the inverse susceptibility measured in a field of 0.5 T. The shallow kinks corresponding to the SDW transition (open arrows) are observed. The solid lines indicate the Curie-Weiss law above $T_{\rm SDW}$.

The inset of Fig. 2 depicts temperature dependence of the inverse susceptibility, χ^{-1}, measured in a field of 0.5 T. We found a kink in the χ^{-1}–T curves of all studied samples, which corresponds to the SDW transition. In contrast to $T_{\rm N}$, $T_{\rm SDW}$, which is close to 190 K, is almost independent of x. This result coincides with no evidence of the superconductivity. Above $T_{\rm SDW}$, the χ^{-1} exhibits T-dependence, which indicates the Curie-Weiss behavior. The effective moments $\mu_{\rm eff}$ are estimated to be $6.78\mu_{\rm B}$, $7.40\mu_{\rm B}$ and $8.03\mu_{\rm B}$ for $x = 0.25$, 0.50 and 0.75, respectively. The $\mu_{\rm eff}$ values of $x = 0.50$ and 0.75 are roughly close to the theoretical value of Eu^{2+}, $7.94\mu_{\rm B}$. But that of $x = 0.25$ is a little smaller. The Weiss temperature $\Theta_{\rm p}$ is positive, which suggests a ferromagnetic exchange interaction between the Eu^{2+} moments though the AFM of Eu^{2+} is confirmed by resonant X-ray [11] and neutron diffraction [12]. This may be connected with the ferromagnetic behavior that appears in a magnetic field of 1 T [10].

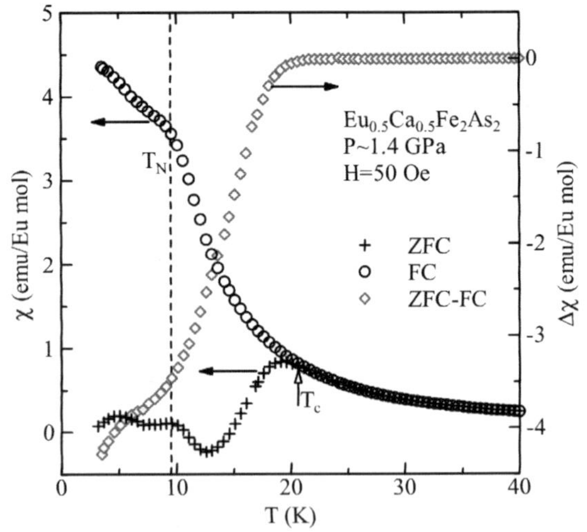

Figure 3. Temperature dependence of the magnetic susceptibility of $x = 0.50$ under pressure of 1.4 GPa measured in a field of 50 Oe after zero field cooling (ZFC) (cross) and field cooling (FC) (open circle). The open diamond data (ZFC−FC) are obtained by subtracting the FC data from the ZFC data. The ZFC−FC data exhibits a clear diamagnetic signal below $T_{\rm c} =$ 21 K and slight collapse of the superconductivity at round $T_{\rm N}$.

Figure 3 demonstrates temperature dependence of the magnetic susceptibility of $Eu_{0.5}Ca_{0.5}Fe_2As_2$ under pressure of 1.4 GPa measured in a field of 50 Oe after zero field cooling (ZFC) and field cooling (FC). Contrary to the results under ambient pressure, the ZFC data decreases significantly below 20 K with decreasing temperature. Consequently, there is a large difference in χ between ZFC and FC below 20 K. This result indicates the pressure-induced bulk superconductivity, which is in agreement with our previous results [9]. In addition, an anomaly associated with the AFM of Eu^{2+} is also observed at $T_N = 9$ K in both ZFC and FC data, which could not be found in the electrical resistivity measurement [9]. To extract the superconducting properties, the FC data is subtracted from the ZFC data, as plotted in Fig. 3. A clear diamagnetic signal is seen in the $\Delta\chi$-T curve and the diamagnetic behavior is slightly suppressed at around T_N, which indicates the superconductivity competes with the AFM. Compared with $EuFe_2As_2$, however, the superconductivity seems to overcome the AFM for $x = 0.50$, which would result from weakening the AFM of Eu^{2+}.

4. Conclusion

We could grow single crystals of $Eu_{1-x}Ca_xFe_2As_2$ with $x = 0.25$, 0.50 and 0.75, and have presented the magnetic susceptibility under ambient and high pressures. Under ambient pressure, with increasing x, T_N decreases and disappears at $x = 0.75$, whereas T_{SDW} is almost unchanged. For $x = 0.50$, under pressure of 1.4 GPa, the pressure-induced superconductivity emerges below 20 K and the superconductivity is slightly suppressed by the AFM of Eu^{2+}. Compared with $EuFe_2As_2$, the superconductivity seems to overcome the AFM for $x = 0.50$, which is ascribed to weakening AFM.

Acknowledgments

The authors thank Dr. M. Watanabe at the Center of Advanced Instrumental Analysis, Kyushu University for checking our samples by SEM-EDX.

References

[1] Kamihara Y, Watanabe T, Hirano M and Hosono H 2008 *J. Am. Chem. Soc.* **130** 3296
[2] Ren Z, Zhu Z, Jiang S, Xu X, Tao Q, Wang C, Feng C, Cao G and Xu Z 2008 *Phys. Rev.* B **78** 052501
[3] Jeevan H S, Hossain Z, Kasinathan D, Rosner H, Geibel C and Gegenwart P 2008 *Phys. Rev.* B **78** 092406
[4] Miclea C F, Nicklas M, Jeevan H S, Kasinathan D, Hossain Z, Rosner H, Gegenwart P, Geibel C and Steglich F 2009 *Phys. Rev.* B **79** 212509
[5] Terashima T, Kimata M, Satsukawa H, Harada A, Hazama K, Uji S, Suzuki H S, Matsumoto T and Murata K 2009 *J. Phys. Soc. Jpn.* **78** 083701
[6] Jiang S, Xing H, Xuan G, Ren Z, Wang C, Xu Z and Cao G 2009 *Phys. Rev.* B **80** 184514
[7] Ren Z, Tao Q, Jiang S, Feng C, Wang C, Dai J, Cao G and Xu Z 2009 *Phys. Rev. Lett.* **102** 137002
[8] He Y, Wu T, Wu G, Zheng Q J, Liu Y Z, Chen H, Ying J J, Liu R H, Wang X F, Xie Y L, Yan Y J, Dong J K, Li S Y and Chen X H 2010 *J. Phys. : Condens. Matter* **22** 235701
[9] Mitsuda A, Matoba T, Ishikawa F, Yamada Y and Wada H 2010 *J. Phys. Soc. Jpn.* to be published.
[10] Wu D, Barišić N, Drichko N, Kaiser S, Faridian A, Dressel M, Jiang S, Ren Z, Li L J, Cao G H, Xu Z A, Jeevan H S and Gegenwart P 2009 *Phys. Rev.* B **79** 2009
[11] Herrero-Martín J, Scagnoli V, Mazzoli C, Su Y, Mittal R, Xiao Y, Brueckel T, Kumar N, Dhar S K, Thamizhavel A and Paolasini L 2009 *Phys. Rev.* B **80** 134411
[12] Xiao Y, Su Y, Meven M, Mittal R, Kumar C M N, Chatterji T, Price S, Persson J, Kumar N, Dhar S K, Thamizhavel A and Brueckel T 2009 *Phys. Rev.* B **80** 174424

Competition of local-moment ferromagnetism and superconductivity in Co-substituted $EuFe_2As_2$

M Nicklas, M Kumar[1], E Lengyel, W Schnelle and A Leithe-Jasper
Max Planck Institute for Chemical Physics of Solids, Nöthnitzer Str. 40, 01187 Dresden, Germany
E-mail: nicklas@cpfs.mpg.de (Michael Nicklas)

Abstract. In contrast to $SrFe_2As_2$, where only the iron possesses a magnetic moment, in $EuFe_2As_2$ an additional large, local magnetic moment is carried by Eu^{2+}. Like $SrFe_2As_2$, $EuFe_2As_2$ exhibits a spin-density wave transition at high temperatures, but in addition, the magnetic moments of the Eu^{2+} order at around 20 K. The interplay of pressure-induced superconductivity and the Eu^{2+} order leads to a behavior which is reminiscent of re-entrant superconductivity as it was observed, for example, in the ternary Chevrel phases or in the rare-earth nickel borocarbides. Here, we study the delicate interplay of the ordering of the Eu^{2+} moments and superconductivity in $EuFe_{1.9}Co_{0.1}As_2$, where application of external pressure makes it possible to sensitively tune the ratio of the magnetic (T_C) and the superconducting ($T_{c,onset}$) critical temperatures. We find that superconductivity disappears once $T_C > T_{c,onset}$.

1. Introduction

The discovery of high-temperature superconductivity in the iron-based superconductors has stimulated an enormous interest in the study of this new class of materials. One peculiar finding is the interplay of the local 4f moments of the Eu^{2+} ions and superconductivity in $EuFe_2As_2$ under pressure, which is reminiscent of re-entrant superconductivity [1, 2]. Like the (A=Ca, Sr, Ba) members of the AFe_2As_2 family, $EuFe_2As_2$ exhibits a spin-density wave (SDW) transition around $T_0 = 190$ K related to the Fe_2As_2 layers, but in addition the magnetic moments of the localized Eu^{2+} order at $T_N = 19$ K [3, 4]. $EuFe_2As_2$ has a similar crystallographic and electronic structure compared to that of $SrFe_2As_2$ [3]. Therefore, $SrFe_2As_2$ can be considered a non-f homolog of $EuFe_2As_2$. In $SrFe_{2-x}Co_xAs_2$ Co-substitution in the Fe_2As_2 layer stabilizes a superconducting (SC) phase ($0.2 \lesssim x \lesssim 0.4$) leading to the expectation that Co-substitution in $EuFe_2As_2$ does the same, which in the meantime has been confirmed experimentally [6, 7].

In this paper we present a pressure study on $EuFe_{1.9}Co_{0.1}As_2$ by means of electrical-resistivity (ρ) and ac-susceptibility (χ_{ac}) measurements on polycrystalline samples. In $EuFe_{1.9}Co_{0.1}As_2$ in our accessible pressure range we can sensitively tune the magnetic ordering temperature T_C from $T_C < T_{c,onset}$ to $T_C > T_{c,onset}$ allowing us to study the peculiar interplay of 4f magnetism and superconductivity.

2. Experimental Details

The polycrystalline samples of $EuFe_{2-x}Co_xAs_2$ were synthesized by sintering stoichiometric amounts of the precursors EuAs, Fe_2As and Co_2As. The use of precursors minimizes the

[1] Present address: Institute for Solid State Research, IFW Dresden, D-01171 Dresden, Germany

Published under licence by IOP Publishing Ltd

elemental impurities of As and Fe. X-ray diffraction measurements confirmed the $BaAl_4$ type structure (space group I4/mmm) for all the samples. Electrical resistance and ac susceptibility were measured using an LR700 resistance/mutual inductance bridge (Linear Research). A miniature compensated coil system placed inside the pressure cell was utilized for the χ_{ac} experiments. Temperatures down to 1.8 K were reached using a physical property measurement system (PPMS, Quantum Design). Pressures up to 3 GPa have been achieved in a double-layer piston-cylinder type pressure cell with silicone fluid as pressure transmitting medium. The SC transition of Pb, which served as a pressure gauge, remained sharp at all pressures indicating good hydrostatic conditions.

3. Experimental Results

Figure 1 shows the resistance normalized to the value at room temperature, $R/R_{300K}(T)$, for $EuFe_{2-x}Co_xAs_2$ ($x = 0$, 0.1, 0.2, and 0.3). In $EuFe_2As_2$ $R/R_{300K}(T)$ exhibits two distinct features at $T_0 = 195$ K and $T_N = 12.5$ K corresponding to the SDW instability and the ordering of the localized Eu^{2+} moments, respectively, in good agreement with literature [3, 4]. Upon Co substitution T_0 decreases rapidly to 103 K for $x = 0.1$. At $x = 0.2$ no signature of the SDW transition is visible in the $\rho(T)$ data anymore. $T_{N,C}$ exhibits a weak concentration dependence. In the concentration range between $x \approx 0.1 \div 0.2$ the type of magnetic order possibly changes from antiferromagnetic (AFM) to ferromagnetic (FM). In $EuFe_{1.8}Co_{0.2}As_2$ a sharp drop of the resistance below ≈ 10 K indicates the onset of superconductivity. This observation is consistent with studies on single crystals [6]. However, there a higher T_c and reentrant-like SC behavior was reported. For $x \leq 0.1$ and $x \geq 0.3$ no signature of superconductivity is visible in our data evidencing a more narrow SC region than in the non-f homolog system $SrFe_{2-x}Co_xAs_2$ [5].

In the following, we will focus on the results of our pressure study on $EuFe_{1.9}Co_{0.1}As_2$. $\rho(T)$ for different pressures is shown in Fig. 2. Upon applying pressure the anomaly indicating the SDW transition at T_0 broadens and shifts rapidly to lower temperatures. At 1.61 GPa no feature related to T_0 can be resolved in $\rho(T)$ anymore suggesting that the SDW order is already suppressed at this pressure. At low temperatures, displayed in Fig. 3a, a kink in $\rho(T)$ at 14.8 K and 16.9 K indicates the ordering of the Eu^{2+} moments at 0 and 0.43 GPa, respectively [9]. At 1.01 GPa a first signature of a SC transition becomes evident: $\rho(T)$ drops sharply below $T_{c,onset} = 24.2$ K followed by a small maximum indicating the magnetic ordering of the Eu^{2+} and a further decrease in $\rho(T)$ on lowering the temperature. This behavior is similar to that

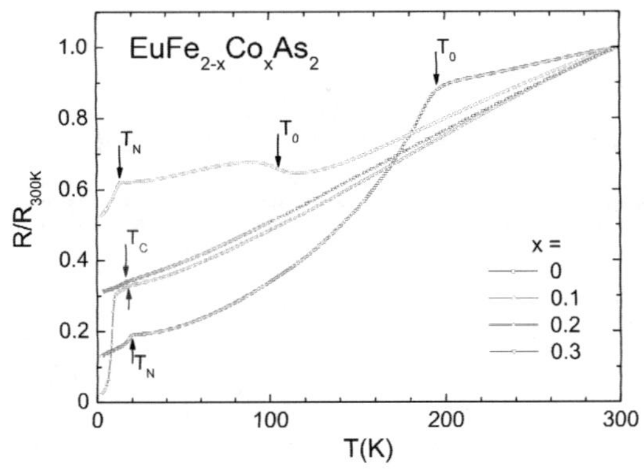

Figure 1. Electrical resistivity normalized to the value at room temperature (R/R_{300K}) for $EuFe_{2-x}Co_xAs_2$.

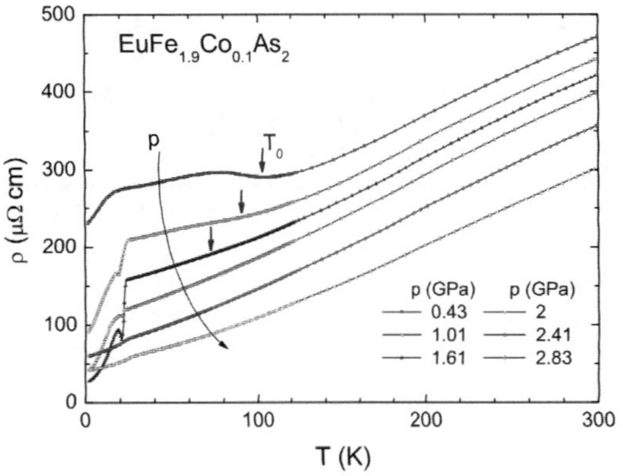

Figure 2. $\rho(T)$ of $EuFe_{1.9}Co_{0.1}As_2$ at different external pressures.

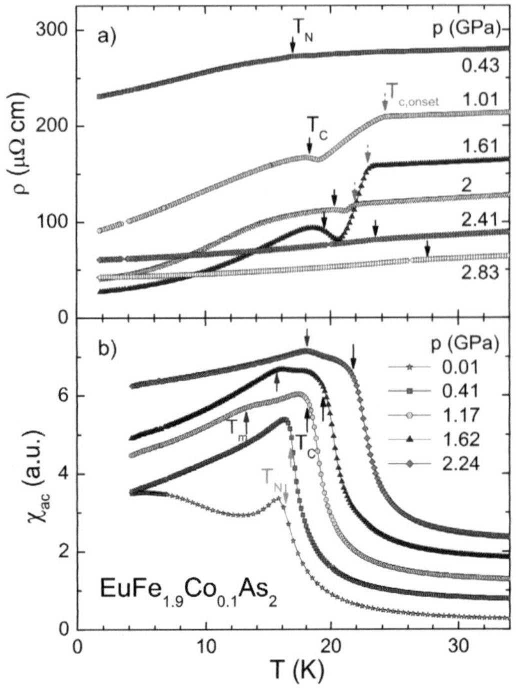

Figure 3. (a) Low-temperature resistivity and (b) ac susceptibility of $EuFe_{1.9}Co_{0.1}As_2$ at different external pressures. For clarity the $\chi_{ac}(T)$ data at different pressures have been shifted with respect to each other.

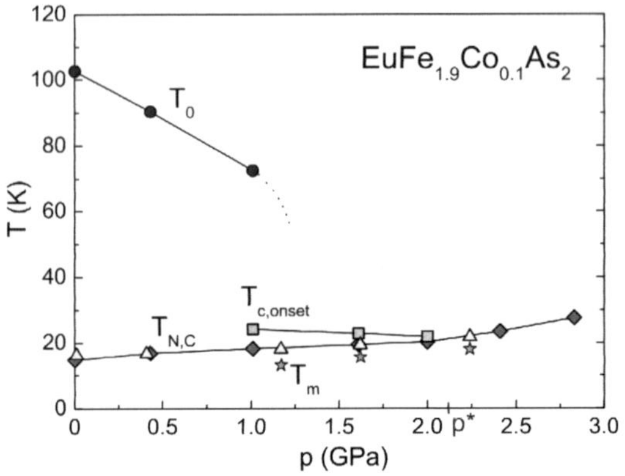

Figure 4. Temperature-pressure phase diagram of $EuFe_{1.9}Co_{0.1}As_2$.

observed previously, but at higher pressures in $EuFe_2As_2$ [1]. The drop in $\rho(T)$ as well as the following maximum are much more pronounced at $p = 1.61$ GPa, but are hardly visible anymore at $p = 2$ GPa. At even higher pressures no indication of superconductivity is present in the resistivity data anymore.

The results of the $\chi_{ac}(T)$ experiments are presented in Fig. 3b. The $\chi_{ac}(T)$ curve at the lowest pressure ($p = 0.01$ GPa) and the curves at higher pressures ($p \geq 1.01$ GPa) are qualitatively different. While the shape of $\chi_{ac}(T)$ at $p = 0.01$ GPa is reminiscent of an AFM phase transition, the shape at $p \geq 1.01$ GPa is typical for a FM transition. A change of the magnetic groundstate from AFM to FM has been found in $EuFe_2As_2$ on P-doping on the As-site too [10]. P doping corresponds to the application of positive chemical pressure, therefore, our finding is not surprising and in agreement with the doping studies. On increasing pressure the anomaly in $\chi_{ac}(T)$ at $T_{N,C}$ shifts to higher temperatures [9]. At pressures $p \geq 1.01$ GPa a second feature appears in $\chi_{ac}(T)$ below T_C at T_m [9]. We attribute this kink to a change in the magnetic structure from FM to AFM. No feature in $\rho(T)$ is visible at T_m. A detailed study of the magnetic properties will be presented elsewhere. It is important to note that we do not observe a clear signature of superconductivity in our χ_{ac} results.

In our study we observe a less robust superconductivity than reported in literature [6, 7]. This might be related to different sample preparation procedures or to internal strain in the samples. An important influence of internal/external residual strain on the observed physical properties appears also in our experiments. After releasing the pressure, we still find reentrant SC behavior in $\rho(T)$ similar to the data at 1.61 GPa (not shown).

The $T-p$ phase diagram in Fig. 4 summarizes our results. While $T_c(p)$ decreases only slightly with increasing pressure $T_{N,C}(p)$ increases strongly by a factor of nearly 2 from 14.8 K at ambient pressure to 27.6 K at 2.83 GPa, the highest pressure of our experiment. We find a good agreement between $T_{N,C}(p)$ obtained from the $\rho(T)$ and $\chi_{ac}(T)$ experiments. The strong increase of $T_{N,C}(p)$ with pressure has not been reported in $EuFe_2As_2$ [1]. There, the magnetic ordering temperature is nearly independent of pressure. The different pressure dependencies of $T_{c,onset}(p)$ and $T_{N,C}(p)$ in $EuFe_{1.9}Co_{0.1}As_2$ lead to a crossing of both temperature lines $[T_{c,onset}(p) = T_C(p)]$ in the $T-p$ phase diagram at a pressure p^* slightly higher than 2 GPa. Above p^* once $T_C > T_c$ no indication of any SC transition is visible in the resistivity data. Already at 2 GPa where $T_c = 21.9$ K and $T_C = 20.3$ K get rather close. The feature associated to the SC transition is considerably reduced compared to the previous pressure.

4. Discussion and Summary

In summary our results show that in the substitution series $EuFe_{2-x}Co_xAs_2$ the SDW transition temperature T_0 is suppressed rapidly upon increasing Co concentration, while the ordering temperature of the Eu^{2+} moments is almost constant. For $x = 0.2$ we observe the onset of superconductivity in $\rho(T)$ at ≈ 10 K. For $x < 0.2$ and $x > 0.2$ no indication of superconductivity is present. The strong suppression of T_0 as function of the Co concentration and the appearance of superconductivity in the $T-x$ phase diagram is consistent with the expectation from the comparison with the homolog non-f Co-substitution series $SrFe_{2-x}Co_xAs_2$ [5]. However, there superconductivity is stable in a broader concentration range. The presence of only a narrow SC regime in the Eu system might be related to the presence of the local moment Eu^{2+} magnetism.

Pressure rapidly suppresses the SDW instability at T_0 in $EuFe_{1.9}Co_{0.1}As_2$. In the same pressure region where the signature of T_0 is lost in $\rho(T)$ the onset of superconductivity is observed. At low temperatures our results suggest a strong detrimental effect of the Eu^{2+} magnetism on the superconductivity. In an intermediate pressure range (1.01 GPa $\leq T \leq$ 2 GPa) the system is on the verge to a SC state, indicated by the drop in the resistivity below $T_{c,onset}$. However, the formation of the SC state is inhibited by the formation of long range magnetic order of the Eu^{2+} moments. A similar result has been reported previously in undoped $EuFe_2As_2$ [1]. The important difference in the present study on $EuFe_{1.9}Co_{0.1}As_2$ is that upon increasing pressure $T_{N,C}(p)$ increases while $T_{c,onset}(p)$ decreases and, therefore, $T_C(p)$ becomes greater than $T_{c,onset}(p)$ above a certain pressure p^*. Once $T_C(p) > T_{c,onset}(p)$ no signature of superconductivity is found anymore. We take this as an evidence for the strong detrimental effect of the Eu^{2+} magnetism on the the formation of the SC state. As our ac-susceptibility results indicate Eu^{2+} orders ferromagnetically under pressure. Thus, we speculate that the internal magnetic fields due to the ferromagnetically ordered Eu^{2+} moments suppress the superconductivity in the iron-arsenide layers.

Acknowledgements

We gratefully acknowledge the Deutsche Forschungsgemeinschaft (DFG) for financial support through SPP 1458.

References

[1] Miclea C F *et al.* 2009 *Phys. Rev. B* **79** 212509.
[2] Terashima T *et al.* 2009 *J. Phys. Soc. Jpn.* **78** 083701.
[3] Jeevan H S, Hossain Z, Kasinathan D, Rosner H, Geibel C and Gegenwart P 2008 *Phys. Rev. B* **78** 052502.
[4] Ren Z, Zhu Z, Jiang S, Xu X., Tao Q., Wang C, Feng C, Cao G and Xu Z 2008 *Phys. Rev. B* **78** 052501.
[5] Leithe-Jasper A, Schnelle W, Geibel C and Rosner H 2008 *Phys. Rev. Lett.* **101** 207004.
[6] Jiang S, Xing H, Xuan G, Ren Z, Wang C, Xu Z and Cao G 2009 *Phys. Rev. B* **80** 184514.
[7] Zheng Q J *et al.* 2009 *arXiv*:0907.5547 (to be published).
[8] Kumar M *et al.* 2008 *Phys. Rev. B* **78** 184516.
[9] $T_{N,C}$ and $T_{c,onset}$ were defined from $\rho(T)$ by taking the position of the corresponding minimum in $d^2\rho(T)/dT^2$. $T_{N,C}$ and T_m from $\chi_{ac}(T)$ were determined accordingly.
[10] Ren Z, Tao Q, Jiang S, Feng C, Wang C, Dai J, Cao G and Xu Z 2009 *Phys. Rev. Lett.* **102** 137002.

Relation between superconductivity and tetragonal phase stabilized by uniaxial pressure in $CaFe_2As_2$

K. Prokeš [1], A. Kreyssig [2,3], B. Ouladdiaf [4], D. K. Pratt [2,3], N. Ni [2,3], S. L. Bud'ko [2,3], P. C. Canfield [2,3], R. J. McQueeney [2,3], A. I. Goldman [2,3] and D. N. Argyriou [1]

[1] Helmholtz-Zentrum Berlin, M-I1, Hahn-meitner-Platz 1, 141 09 Berlin, Germany
[2] Ames Laboratory, U.S. DOE, Iowa State University, Ames, Iowa 50011, USA
[3] Dep. of Physics and Astronomy, Iowa State University, Ames, Iowa 50011, USA
[4] Institut Laue-Langevin, 38042 Grenoble Cedex, France

prokes@helmholtz-berlin.de

Abstract. We report on neutron diffraction experiments on high quality $CaFe_2As_2$ single crystals performed under uniaxial pressure and combined with in-situ electrical resistance measurements. We conclude that the superconductivity is associated with the high-temperature tetragonal phase that is stabilized down to the lowest temperatures by a small uniaxial pressure of at least 0.075 GPa applied along the c-axis.

1. Introduction

Despite a century of intensive efforts in the field of superconductivity (SC), this phenomenon remains one of the most active research areas of the condensed matter physics. Especially large families of superconducting compounds allow us to draw more general conclusions regarding the relation between the crystal structure (and underlined magnetic properties) on one side and superconductivity on the other. Much that we understand about unconventional high temperature superconductivity comes from our knowledge of the cuprate family [1]. The parent phase of those materials is an antiferromagnetic (AF) Mott insulator and when doped chemically with holes, ordered magnetism is suppressed and SC emerges. In the newly discovered *R*FeAsO (*R* = rare earth) [2] and $AEFe_2As_2$ (*AE* = Ca, Sr, Ba) [3] families, the situation is quite similar. SC emerges when the non-SC parent compounds, are chemically doped much in the same fashion as in the cuprates. These compounds undergo a structural and *AF* transitions that are, at least in some instances, intimately linked [4-5]. Both, structural transition and magnetism is suppressed with doping and eventually the SC appears at low temperatures. It seems that both, AF and SC coexist, in some cases. Surprisingly, unlikely the cuprates, the application of pressure on these parent compounds leads to SC as high as ~50 K [6-7]. There have been several studies to unravel the question which phase in $AEFe_2As_2$ compounds when subjected to pressure hosts SC. At ambient pressure, $CaFe_2As_2$ undergoes a first order transition from a high temperature tetragonal (*T*) phase ($ThCr_2Si_2$ structure) to a structure with orthorhombic (*O*) symmetry at T_{TO} = 172 K connected with an *AF* transition [5]. Upon the application of modest pressures, using liquid media clamping cells, the structural and *AF* transition is rapidly suppressed and *SC* was observed for $P > 0.23$ GPa and $T_{SC} \sim 12$ K [7]. *SC* has also been observed in electrical resistance measurements of samples under uniaxial pressure [8]. In that material initial bulk property

Published under licence by IOP Publishing Ltd

measurements under less than ideal hydrostatic conditions showed superconductivity above 0.1 GPa, while measurements close to ideal hydrostatic pressure showed no evidence of *SC*. The later result is consistent with a previously known non-magnetic phase *T* phase with a significantly reduced unit cell volume, the so-called collapsed tetragonal (*cT*) phase. In this contribution we report on our recent neutron diffraction experiments on high quality $CaFe_2As_2$ single crystals performed under uniaxial pressure combined with in-situ electrical resistance measurements.

2. Experimental

Several high quality single crystals of $CaFe_2As_2$ with dimensions of 2-3mm x 3-4mm x 0.2mm (mass between 8 and 12 mg) were grown out of a Sn flux. The crystals were carefully polished to produce parallel surfaces perpendicular to the *c*-axis and checked by neutron Laue technique that confirmed the good quality of samples. Crystals were clamped between two ZrO_2 cylindrical pistons that comprise a small uniaxial pressure cell. A force up to 1 kN was applied on the sample and the corresponding pressure calculated from the calibrated displacement of the clamping screws and the measured sample cross section. Several different single crystals at several pressures between ambient pressure and 0.3 GPa were studied.

The neutron diffraction experiments were performed at two different diffractometers: on the E4 double-axis diffractometer at the Helmholtz-Zentrum Berlin using a neutron wavelength of λ=2.45 Å using a standard cryostat and on the D10 diffractometer at the Institute Laue-Langevin (ILL) with a wavelength of 2.36 Å using a four-circle closed cycle refrigerator. In both cases, two-dimensional area detectors that provide a diffraction image over a range of scattering angles (2θ) as the sample is rocked over a specified angular range (ω) were used. Pyrolytic graphite filters were

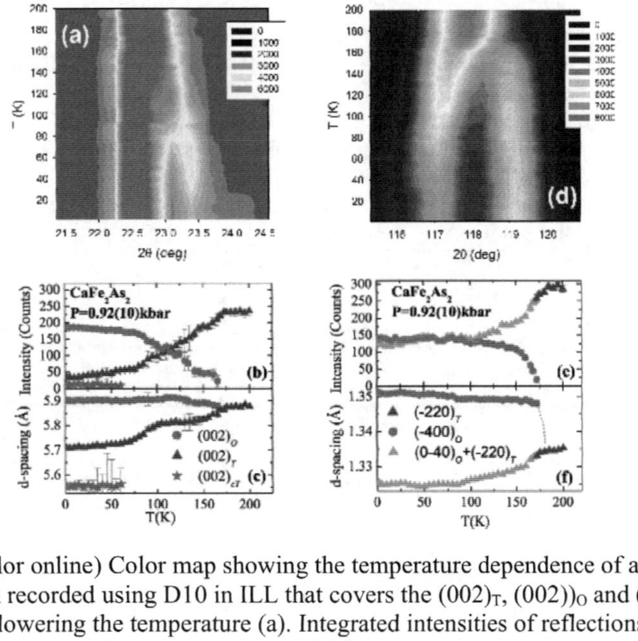

Figure 1. (Color online) Color map showing the temperature dependence of a portion of the diffraction pattern recorded using D10 in ILL that covers the $(002)_T$, $(002))_O$ and $(002)_{cT}$ reflections of $CaFe_2As_2$ with lowering the temperature (a). Integrated intensities of reflections shown in (a) as a function of the temperature (b) and d-spacings (c). Color map showing the temperature dependence of the diffraction range covering the $(-220)_T$ reflection that splits at lower temperatures to $(-400)_O$ and $(0-40)_O$ (with remnants of the $(-2\ 20)_T$ reflections is shown in panel (d). Fitted integrated intensities and d-spacings obtained from data shown in (d) are shown in (e) and (f), respectively.

employed in both sets of measurements to reduce the higher harmonic contents to less that 10^{-4} of the primary beam. During the E4 measurements we have measured in-situ electrical resistance using a standard 2-point or 4-point ac methods.

3. Results

As an example, in Fig. 1a we show the temperature dependence of the neutron diffracted intensity recorded in the vicinity of the (002) Bragg reflections at the D10 instrument with decreasing temperature. The nominal pressure applied in this case on a 5.3 mg $CaFe_2As_2$ single crystal was 0.092 GPa. It is seen that the pressure stabilizes the *T* phase to temperatures far below $T_{TO} \sim 170$ K and induces the *cT* phase below $T_{TcT} \sim 70$ K. The corresponding fitted integrated intensities and calculated d-spacings are shown in Figs. 1b and 1c, respectively. The $(002)_O$ reflection grows much faster between T_{TO} and T_{TcT} than below T_{TcT}. Color map showing the temperature dependence of the diffracted signal recorded under the same conditions around the $(-220)_T$ Bragg reflection is shown in Fig. 1d. The *T-O* transition is clearly visible, although the $(-220)_T$ reflection does not split below T_{TO} into $(-400)_O$ and $(0\text{-}40)_O$ completely. Part of the sample volume retains the *T* symmetry. This is reflected in a different width of the two split reflections and different temperature dependence of the reflection positions. Resulting fitted integrated intensities and d-spacings are shown in Figs. 1e and 1f, respectivelly. The influence of the *T* and *O* phase mixing in the low-temperature limit eventually reduces and both split reflections tend to have similar intensities and also the width. Due to strong anisotropic absorption of the pressure cell it is impossible to determine all the crystallographic details of the stabilized *T'* phase. However, smooth temperature dependence of the lattice parameters and Bragg reflection intensities give a good reason to conclude that this stabilized *T'* phase is simply the very same T phase that is stable at high temperatures in samples without influence of pressure. Although we have performed several types of reciprocal scans, no signal that could be ascribed to magnetic origin could be found. We conclude therefore that the T' phase does not order magnetically.

Considering also other pressure runs at different crystals we can conclude that at pressures below a starting pressure of 0.06 GPa, the *cT* phase is absent at all temperatures and the *T* phase is observed only over a narrow range of temperatures below T_{TO}. Upon increasing the pressure, we observe that the relative fractions of the stabilized tetragonal (*T*) and *cT* phases increase at the expense of the *O* phase. A finite fraction of the *T* phase extends down to the lowest temperatures measured for uniaxial pressures greater than 0.075 GPa. At even higher applied pressure, the *cT* phase appears at progressively higher temperatures and its weight fraction increases together with that of the

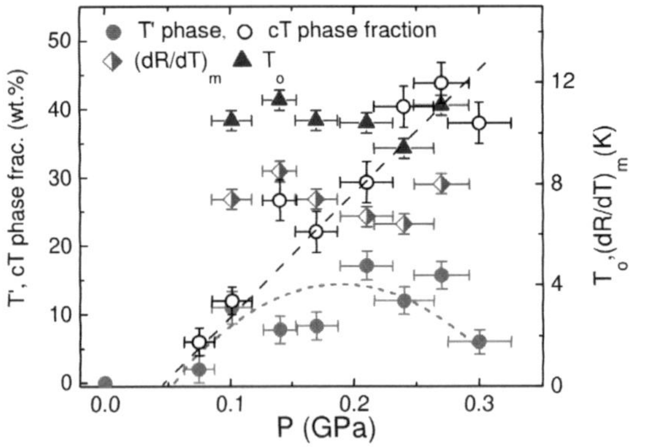

Figure 2. Triangles and half-filled diamonds show the pressure dependence of the onset of the SC transition and the temperature with the maximal dR/dT value of $CaFe_2As_2$. Filled and open circles show the weight fraction of the pressure stabilized phase *T'* and the collapsed tetragonal phase cT as a function of pressure, respectively. Dotted lines are guides to the eye.

T' phase at the expense of the *O* phase. However, at pressures higher than ~ 0.2 GPa the weight fraction of the *T'* phase does not grow anymore although the fraction of the *cT* phase continues to increase on the expense of the *O* phase. This trend is documented in Fig. 2, in which we show by full full (open) points the pressure dependence of the weight fraction of the stabilized *T'* (collapsed tetragonal cT) phase deduced from neutron diffraction data recorded at 2 K. Unfortunately, we have not been able to determine the upper limit of the T phase existence due to limitations of our pressure cell. In the same picture we show the pressure dependence of the onset of the *SC* transition T_0 deduced from the in-situ measured electrical resistance and of the temperature with the maximal temperature derivative of the electrical resistance $(dR/dT)_m$. Clearly, the *T* phase appears to be stable in a part of the sample in the very same pressure range at which also the *SC* exists. The latter data are in good agreement with electrical resistivity curves obtained under uniaxial pressure by Torikachvili et al. [8].

4. Conclusions

We find that in a narrow range of uniaxial pressure applied along the c axis of $CaFe_2As_2$ the high-temperature tetragonal phase is stabilized down to the lowest temperatures. This tetragonal phase differs from the so-called collapsed tetragonal phase that is not found in hydrostatic measurements and appears also to be present in our measurements. Mapping out its stability in temperature and pressure and using in-situ resistivity measurements we argue that this non-magnetic, pressure-stabilized tetragonal phase is the likely host of superconductivity in $CaFe_2As_2$ under pressure.

Acknowledgments

This work was supported in part by the DFG project SPP 1458. We also gratefully acknowledge the HZB and ILL for the allocated beamtime and their support during experiments. The work at the Ames Laboratory was supported by the U.S. DOE under Contract No. DEAC02- 07CH11358.

References

[1] E. W. Carlson, V. J. Emery, S. A. Kivelson, and D. Orgad, in "The Physics of Conventional and Unconventional Superconductors, Vol. II " edited by K.H. Bennemann and J.B. Ketterson, Springer-Verlag (2004).
[2] H. Yanagi, T. Kamiya and H. Hosono, J. Am. Chem. Soc. Y. Kamihara, H. Hiramatsu, M. Hirano, R. Kawamura, **128**, 10012 (2006).
[3] M. Rotter, M. Tegel and D. Johrendt, Phys. Rev. Lett.**101**, 107006 (2008).
[4] Y. Ishida, F. Nakai and H. Hosono, J. Phys. Soc. Jpn. **78**, 062001 (2009)
P. C. Canfield, S. L. Bud'ko, N. Ni, A. Kreyssig, A. I. Goldman, R. J. McQueeney, M. S. Torikachvili, D. N. Argyriou, G. Luke and W. Yu, Physica C **469**, 404 (2009).
C. Krellner, N. Caroca-Canales, A. Jesche, H. Rosner, A. Ormeci and C. Geibel, Phys. Rev. B **78**, 100504(R) (2008).
[5] A. I. Goldman, D. N. Argyriou, B. Ouladdiaf, T. Chatterji, A. Kreyssig, S. Nandi, N. Ni, S. L. Bud'ko, P. C. Canfield and R. J. McQueeney, Phys. Rev. B **78**, 100506(R) (2008).
[6] H. Takahashi, K. Igawa, K. Arii, Y. Kamihara, M. Hirano and H. Hosono, Nature **453**, 376 (2008).
[7] M. S. Torikachvili, S. L. Bud'ko, N. Ni and P. C. Canfield, Phys. Rev. Lett. **101**, 057006 (2008).
[8] M. S. Torikachvili, S. L. Bud'ko, N. Ni, P. C. Canfield and S. T. Hannahs, Phys. Rev. B **80**, 014521 (2009).
[9] W. Yu, A. A. Aczel, T. J. Williams, S. L. Bud'ko, N. Ni, P. C. Canfield and G. M. Luke, Phys. Rev. B **79**, 020511(R) (2009).
[10] K. Prokeš, A. Kreyssig, B. Ouladdiaf, D. K. Pratt, N. Ni, S. L. Bud'ko, P. C. Canfield, R. J. McQueeney, D. N. Argyriou, and A. I. Goldman, Phys. Rev. B **81**, 180506 (2010)

Journal of Physics: Conference Series **273** (2011) 012103 doi:10.1088/1742-6596/273/1/012103

Heat capacity of $Ba_{1-x}K_xFe_2As_2$, x=0 and 0.41

C. R. Rotundu[1], B. Freelon[2], S. D. Wilson[3], G. Pinuellas[2], A. Kim[2], E. Bourret-Courchesne[4], N. E. Phillips[1,5], and R. J. Birgeneau[1,2,6]

[1] Materials Sciences Division, Lawrence Berkeley National Laboratory, Berkeley, CA 94720, USA

[2] Department of Physics, University of California, Berkeley, CA 94720, USA

[3] Department of Physics, Boston College, Chestnut Hill, MA 02467, USA

[4] Life Sciences Division, Lawrence Berkeley National Laboratory, Berkeley, CA 94720, USA

[5] Department of Chemistry, University of California, Berkeley, CA 94720, USA

[6] Department of Materials Science and Engineering, University of California, Berkeley, CA 94720, USA

E-mail: CRRotundu@lbl.gov (Costel R. Rotundu)

Abstract. Heat-capacity measurements on exceptionally high-quality samples of $Ba_{1-x}K_xFe_2As_2$ give the values of a number of parameters relevant to the electronic structure, the structural/magnetic transition in the undoped parent compound, and the superconducting transition in the near optimally doped, x = 0.41 sample. In $BaFe_2As_2$ the changes in the lattice and magnetic structure appear as a single, sharp first-order transition at 140 K; the Sommerfeld coefficient is $\gamma = 5.1$ mJ K^{-2} mol^{-1}. For the superconducting, x = 0.41, sample, the residual density of electron states in the superconducting states is essentially zero. The specific-heat anomaly at T_c suggests extreme strong coupling, with $\Delta(0)/k_BT_c \approx 2.6$, and a normal-state Sommerfeld coefficient $\gamma_n \approx 45$ mJ K^{-2} mol^{-1}.

1. Introduction

The structural/magnetic transition in the 122 series of Fe pnictide superconductos, e.g. $BaFe_2As_2$, takes various forms in different samples and different measurements, and has been described as both first and second order. Superconductivity is produced by doping with electrons, e.g., by substitution of Co or Ni on the Fe sites, or holes, e.g., by substitution of K or Na on the Ba sites. The superconducting samples have also shown a great variety of behavior, leading to different conclusions about the symmetry of the order parameter. Complications in the analysis of the low-temperature data associated with the presence of paramagnetic impurities are one obvious factor in these discrepancies, and the presence of a residual density of electron states, "non superconducting material" is another. Since the hole-doped materials are more difficult to prepare, most measurements have been made on the electron-doped materials. There is good reason, however, to think that there may be significant differences between the two and for that reason measurements on the hole-doped materials are of interest. Here we report annealing studies on $BaFe_2As_2$ that are relevant to the order of the structural/magnetic transition, and measurements on a sample of $Ba_{0.59}K_{0.41}Fe_2As_2$ that has an unusually low concentration of paramagnetic centers and essentially zero residual density of states in the superconducting state.

Published under licence by IOP Publishing Ltd

2. Experimental Details

Measurements on the parent sample were made on a 10.3 mg piece from a larger crystal grown by a modified self-flux method, on which we reported neutron results earlier [1]. All the measurements were made on the same sample after successive annealing periods of time. Potassium doped samples were prepared by standard self-flux (FeAs) method [2]. The stoichiometry and the potassium doping was checked by both inductively coupled plasma (ICP) and electron microprobe analysis WDS (wavelength-dispersive X-ray spectroscopy). Measurements on both samples were made in Quantum Design apparatus, heat-capacity, resistivity and AC-magnetization measurements in a PPMS, and DC-magnetization in a MPMS.

3. Results and Discussion

3.1. $BaFe_2As_2$: the annealing study

The heat capacity of the as-grown $BaFe_2As_2$ sample is shown in Fig. 1. The structural/magnetic transition occurs at 135.4 K, with an entropy of 0.84 J K^{-1} mol^{-1}. Annealing the sample for 30 days at 700°C sharpened the transition and shifted it to 140.2 K (as shown in the inset) with essentially no change in the entropy For both the as-grown and annealed samples the transition was slightly hysteretic (as shown for the as-grown sample in Fig. 2) and unchanged in a magnetic field of μ_0H = 14 T (as shown for the annealed sample in Fig. 2).

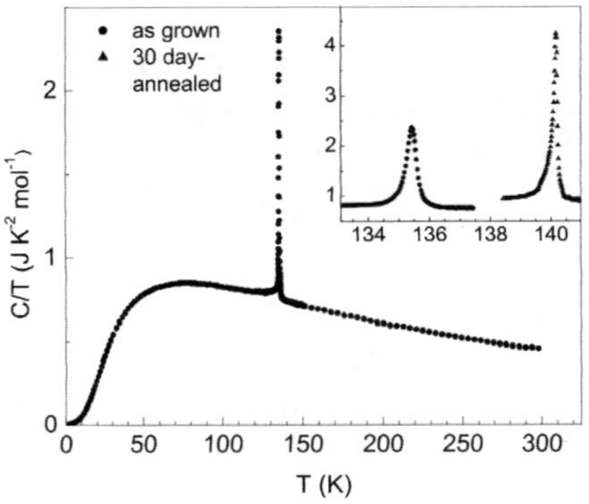

Figure 1. The specific heat of the as-grown sample of $BaFe_2As_2$. The anomaly at the structural/magnetic transition is compared with that of the annealed sample in the inset.

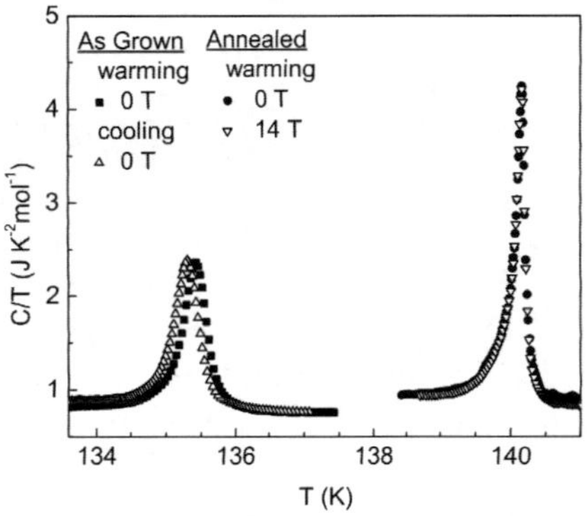

Figure 2. Examples of the effects of magnetic field and hysteresis on the specific heat at the structural/magnetic transition in $BaFe_2As_2$.

The heat capacity of the as-grown sample below 10 K and in fields to μ_0H = 14 T is shown in Fig. 3. To the accuracy of the measurements the heat capacity is independent of field, which shows that the concentration of paramagnetic impurities is well below 10^{-3} mol mol^{-1}. The data in all fields were fit simultaneously with $C = \gamma T + B_3T^3 + B_5T^5 + B_7T^7$, to obtain γ = 5.11 mJ K^{-2} mol^{-1}, B_3 = 0.412 mJ K^{-4} mol^{-1}, B_5 = 8.27 x 10^{-5} mJ K^{-6} mol^{-1} and B_7 = 6.22 x 10^{-8} mJ K^{-2} mol^{-1}. The experimental data are compared with the resulting fit in Fig. 4, where the data in fields other than zero are shifted in C/T by increments of 5 mJ K^{-2} mol^{-1} for clarity, but in each case the common fitting expression is shown.

3.2. The hole-doped $Ba_{0.59}K_{0.41}Fe_2As_2$

The zero-field heat capacity of the $Ba_{0.59}K_{0.41}Fe_2As_2$ sample is shown in Fig. 4. One notable feature is the absence of structure in the heat capacity near 71 K. FeAs, which is a common

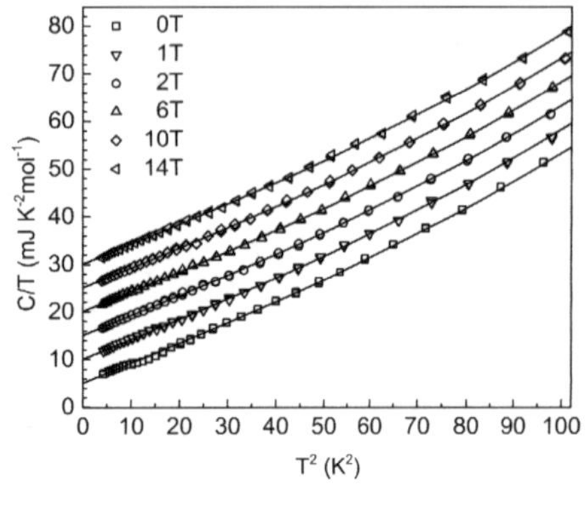

Figure 3. The low-temperature specific heat of $BaFe_2As_2$. The data in magnetic fields are displaced by progressive increments of 5 mJ K^{-2} mol^{-1} for clarity. For each field the curve represents a simultaneous fit to all the data.

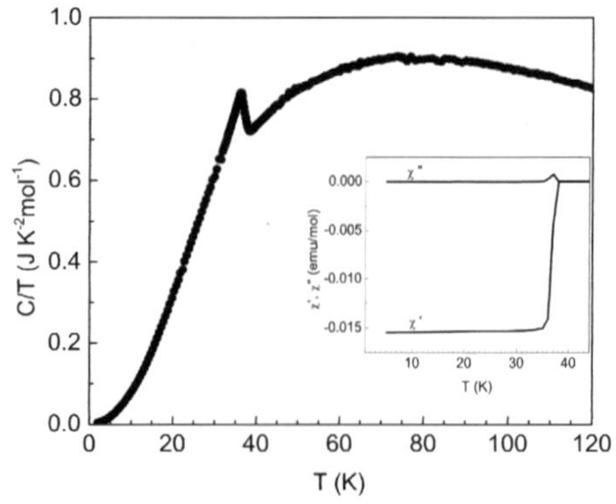

Figure 4. The specific heat of $Ba_{0.59}K_{0.41}Fe_2As_2$. The feature in the susceptibility at the transition to the superconducting state is shown in the inset.

impurity in superconducting samples made by doping $BaFe_2As_2$, has a sharp peak in the heat capacity at that temperature, and, apparently, a significant Sommerfeld coefficient [3]. A sample of $Ba_{0.55}K_{0.45}Fe_2As_2$ that showed the 71-K FeAs feature in the heat capacity had a substantially larger residual density of states in the superconducting state and a smaller heat capacity anomaly at the superconducting transition than reported here for the $Ba_{0.59}K_{0.41}Fe_2As_2$ sample. It seems probable that some of the discrepancies between the results reported here and other measurements are also associated with the presence of FeAs in the other samples. The feature near 37 K is the transition to the superconducting state. The corresponding feature in the susceptibility is shown in the inset. The heat-capacity anomalies at T_c for fields to $\mu_0H = 14$ T

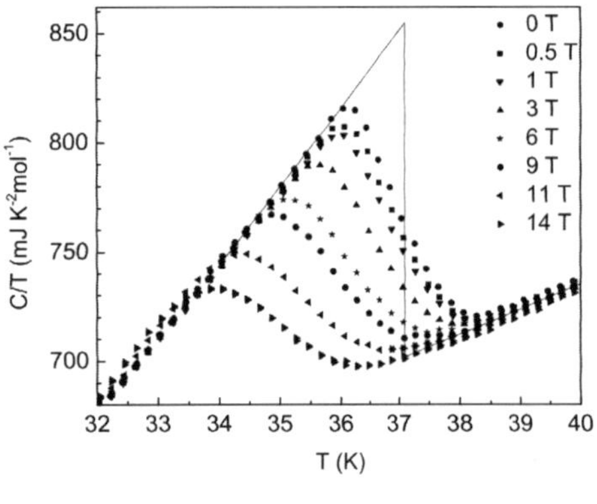

Figure 5. The specific heat of $Ba_{0.59}K_{0.41}Fe_2As_2$ in the vicinity of the transition to the superconducting state.

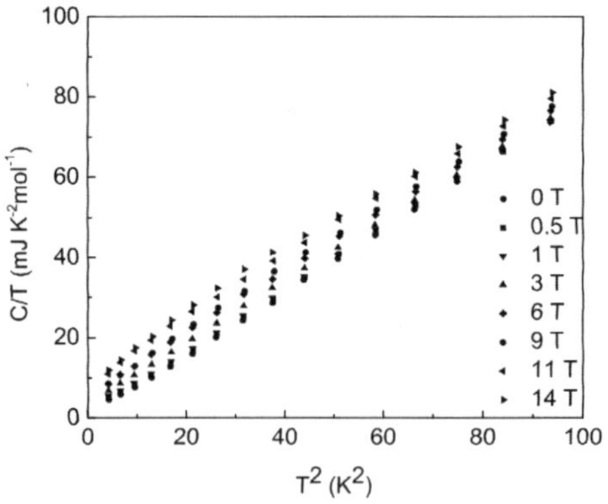

Figure 6. Specific heat of $Ba_{0.59}K_{0.41}Fe_2As_2$ at low temperatures.

are shown in Fig. 5. The straight-line, entropy-conserving construction on the zero-field data gives $T_c = 37.2$ K and $\Delta C(T_c)/T_c = 153$ mJ K^{-2} mol^{-1}. The shape of the zero-field anomaly suggests strong coupling. The assumption of a single gap in the superconducting state, and comparison with the α model [4] gives estimates of the normal-state Sommerfeld coefficient, γ_n, and the coupling strength, as measured by the parameter $\alpha = \Delta(0)/k_B T_c$: (The α model gives the ratios of the discontinuities in C and dC/dT to γ_n.) The straight-line construction in the figure gives $\gamma_n = 45$ mJ K^{-2} mol^{-1}, and $\alpha = 2.61$, substantially larger than the BCS weak-coupling value, 1.764.

The low-temperature data are shown in Fig. 6 as C/T vs T^2. The small upturns in the 0-, 0.5-, and 1-T data at the lowest temperatures, and the negative curvature in the 9-, 11-, and 14-T data indicate the presence of paramagnetic impurities. The 0-, 0.5-, and 1-T data below 6 K were fit separately with the high-temperature tail of a two-level Schottky anomaly and terms for the vortex and lattice contributions, $\gamma_v(H)T$ and B_3T^3, respectively. For all three fields the fits gave 2 x 10^{-3} mol mol^{-1} for the concentration of paramagnetic centers and $B_3 = 0.79$ mJ K^{-4} mol^{-1}; the value of $\gamma_v(H)$ was significant in magnitude and positive only for $\mu_0 H = 1$ T, $\gamma_v(1) = 0.23$ mJ K^{-2} mol^{-1}. Thus, both the concentration of paramagnetic centers and the residual density of states are remarkably low.

Acknowledgements
This work was supported by the Director, Office of Science, Office of Basic Energy Sciences, U.S. Department of Energy, under Contract No. DE-AC02-05CH11231 and Office of Basic Energy Sciences US DOE DE-AC03- 76SF008.

References

[1] Wilson S D, Yamani Z, Rotundu C R, Freelon B, Bourret-Courchesne E, and Birgeneau R J 2009 *Phys. Rev.* B **79** 184519.
[2] Luo H, Wang Z, Yang H, Cheng P, Zhu X, and Wen H-H 2008, *Supercond. Sci. Technol.* **21** 125014.
[3] Gonzalez-Alvarez D, Grønvold F, Falk B, Westrum E F Jr., Blachnik R, and Kudermann G 1989, *J. Chem. Thermodynamics* **21** 363.
[4] Padamsee H, Neighbor J E, and Shiffman C A 1973 *J. Low Temp. Phys.* **12** 387.

Uniform chemical pressure effect in solid solutions $Ba_{1-x}Sr_xFe_2As_2$ and $Sr_{1-x}Ca_xFe_2As_2$

S. R. Saha, K. Kirshenbaum, N. P. Butch, and J. Paglione
Center for Nanophysics and Advanced Materials, Department of Physics, University of Maryland, College Park, MD 20742, USA.

P. Y. Zavalij
Department of Chemistry and Biochemistry, University of Maryland, College Park, MD 20742
E-mail: `paglione@umd.edu`

Abstract.
The effect of alkaline earth substitution on structural parameters was studied in high-quality single crystals of $Ba_{1-x}Sr_xFe_2As_2$ and $Sr_{1-x}Ca_xFe_2As_2$ grown by the self-flux method. The results of single-crystal and powder x-ray diffraction measurements suggest a continuous monotonic decrease of both a- and c-axis lattice parameters, the c/a tetragonal ratio, and the unit cell volume with decreasing alkaline earth atomic radius as expected by Vegard's law. As a result, the system experiences a continuously increasing chemical pressure effect in traversing the phase diagram from $x = 0$ in $Ba_{1-x}Sr_xFe_2As_2$ to $x = 1$ in $Sr_{1-x}Ca_xFe_2As_2$.

The recent discovery of high-temperature superconductivity in iron-based compounds has attracted much interest. The parent phases of these compounds generally show antiferromagnetic order that onsets between 130 K and 200 K, with superconductivity emerging when the antiferromagnetic order of the parent compounds is suppressed [1, 2, 3, 4, 5]. This proximity of magnetic and superconducting order parameters is widely thought to be a key argument for an unconventional pairing mechanism, likely mediated by spin fluctuations [6, 7] similar to the cuprates [8, 9]. But in strong contrast to the copper oxides, superconductivity in iron arsenides can be induced without changing the carrier concentration, either by applying external pressure [10, 11] or by isovalent chemical substitution. The highest T_c achieved so far in these materials is ~ 55 K in $SmO_{1-x}F_xFeAs$ [5] and (Sr,Ca)FeAsF [12, 13]. Oxygen-free FeAs-based compounds with the $ThCr_2Si_2$-type (122) structure also exhibit superconductivity induced by chemical substitution of alkali or transition metal ions [3, 14, 15, 16], the application of large pressures [11, 17, 18, 19], or lattice strain [20], with transition temperatures as high as ~ 37 K.

For the 122 phase, superconductivity has been induced by substituting Fe with not only $3d$-transition metals such as Co and Ni, but also some of the $4d$- and $5d$-transition metals. Recently, Ru, Ir, and Pt substitution for Fe were also shown to induce superconductivity in $SrFe_2As_2$ and $BaFe_2As_2$ [21, 22, 23]. Superconductivity with $T_c \sim 31$ K has also been shown to occur by isovalent substitution of P for As [24]. This gives the opportunity to tune magnetic character without nominally changing charge carrier concentrations, for instance making the interpretation of transport coefficients much simpler than in the case of charge doping.

Published under licence by IOP Publishing Ltd

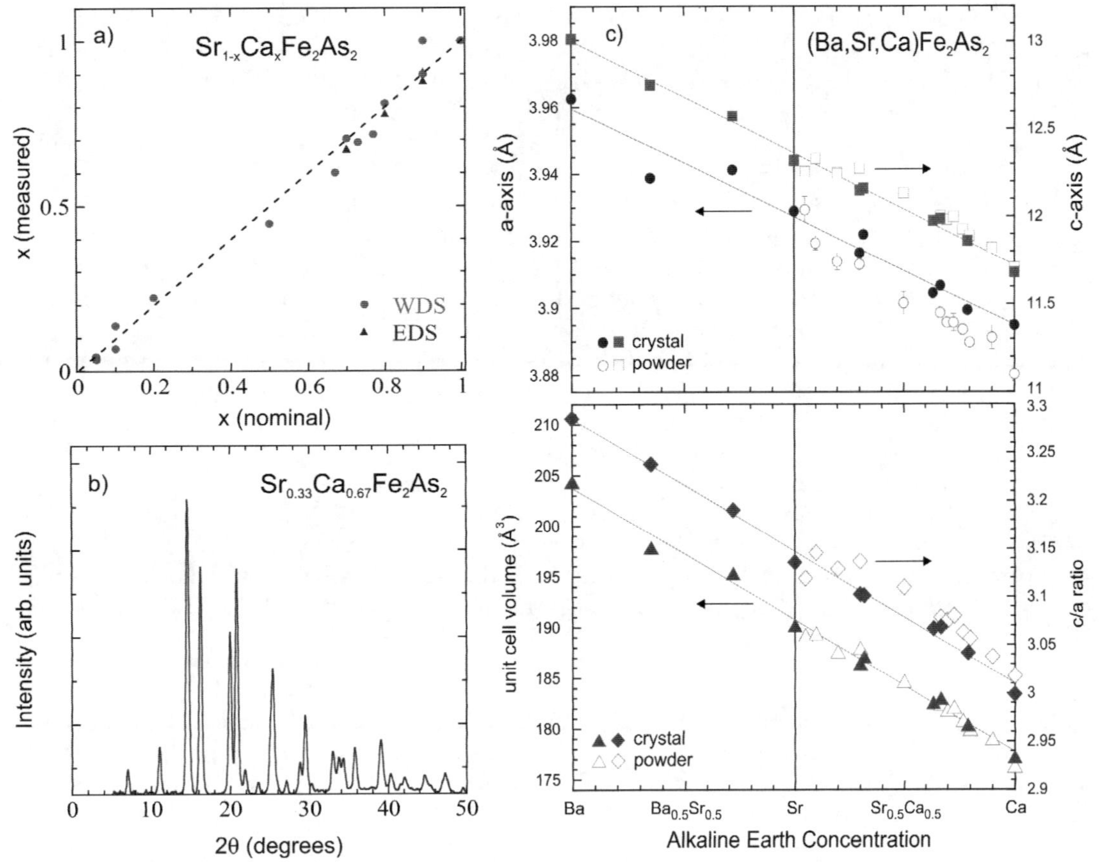

Figure 1. (a) Measured Ca concentration of $Sr_{1-x}Ca_xFe_2As_2$ single-crystal samples as a function of nominal concentration x, as determined by wavelength dispersive x-ray spectroscopy (data points represent an average value of 8 scanned points for each concentration). Some of the specimens are also confirmed by energy dispersive X-ray spectroscopy (EDS). The dotted line is a guide to eye which traces x(measured)=x(nominal). (b) Typical x-ray powder diffraction pattern, shown for sample $Sr_{0.33}Ca_{0.67}Fe_2As_2$, obtained by using Mo-$K_\alpha$ radiation. The main peaks can be indexed with a tetragonal structure and there are no impurity phases detected within experimental accuracy. (c) Upper panel: Variation of the a- and c-axis lattice constants as a function of alkaline earth substitution in the series $Ba_{1-x}Sr_xFe_2As_2$ (left half) and $Sr_{1-x}Ca_xFe_2As_2$ (right half) as determined from single crystal x-ray diffraction measurements at 250 K of single-crystal samples. Corresponding c/a ratio and the unit cell volume are plotted in the lower panel. In both panels, solid symbols indicate data acquired using single-crystal specimens and open symbols represent data determined by powder x-ray diffraction.

In order to investigate the possibility of applying uniform chemical pressure in a continuous manner, we have synthesized the series of solid solutions $Ba_{1-x}Sr_xFe_2As_2$ and $Sr_{1-x}Ca_xFe_2As_2$ by substituting isovalent alkaline earth atoms, and investigated the evolution of the crystal structure by high-resolution powder and single-crystal x-ray diffraction. Here we present our preliminary results that suggest the unit cell of the Ba-Sr-Ca substitution series experiences a monotonic uniform chemical pressure as a function of alkaline earth atomic radius.

Single-crystal samples of $Ba_{1-x}Sr_xFe_2As_2$ and $Sr_{1-x}Ca_xFe_2As_2$ were grown using the FeAs self-flux method [20]. Fe was first separately pre-reacted with As via solid-state reaction of Fe (99.999%) powder with As (99.99%) powders in a quartz tube of partial atmospheric pressure of Ar. The precursor materials were mixed with elemental Sr (99.95%) with either Ba (99.95%) or Ca (99.95%) in the ratio 4:1$-x$: x, placed in an alumina crucible and sealed in a quartz tube under partial Ar pressure. The mixture was heated to 1150°C, slow-cooled to a lower temperature and then quenched to room temperature. Typical dimensions of as-grown single crystal specimen are ~100 μm thickness and up to 5 mm width. Chemical analysis was performed using both energy- and wavelength-dispersive x-ray spectroscopy (WDS).

Both EDS and WDS analysis of all $Ba_{1-x}Sr_xFe_2As_2$ and $Sr_{1-x}Ca_xFe_2As_2$ samples showed the proper 1:2:2 stoichiometry in all specimens reported herein, with no indication of impurity phases. Figure 1(a) compares the nominal alkaline earth concentration x in $Sr_{1-x}Ca_xFe_2As_2$ crystals with that measured by WDS and EDS analysis, using an average value determined from 8 different spots on each specimen. As shown by the dotted line guide, the actual concentrations found by WDS are equal to the nominal values of x to within experimental error, indicating homogeneous substitution in this series of solid solutions.

Diffraction patterns were obtained by both powder and single-crystal x-ray diffraction and Rietfeld refinement (SHELXS-97) to $I4/mmm$ structure. Powder x-ray diffraction was performed at 250 K using a Smart Apex2 diffractometer with Mo-K_α radiation and a graphite monochromator. Figure 1(b) shows a typical x-ray diffraction pattern obtained from a single-crystal sample of $Sr_{0.3}Ca_{0.7}Fe_2As_2$. All of the main peaks can be indexed to the $ThCr_2Si_2$ structure, with no impurity phases detected.

Table 1. Crystallographic data for $SrFe_2As_2$ and $Sr_{0.3}Ca_{0.7}Fe_2As_2$ determined by single-crystal x-ray diffraction at 250 K. The tetragonal structure was solved and refined using the SHELXS-97 software, yielding lattice constants with residual factor R= 1.36% and 1.95% for $SrFe_2As_2$ and $Sr_{0.33}Ca_{0.67}Fe_2As_2$, respectively.

	$SrFe_2As_2$	$Sr_{0.33}Ca_{0.67}Fe_2As_2$
Temperature	250 K	250 K
Space group	I4/mmm	I4/mmm
a(Å)=b(Å)	3.9289(3)	3.9066(8)
c(Å)	12.3172(12)	11.988(5)
V(Å^3)	190.17(4)	182.95(9)
Z	2	2
Density(g/cm^3)	6.098	6.045
Atomic parameters:		
Sr/Ca	2a(0,0,0)	2a(0,0,0)
Fe	4d(1/2,0,1/4)	4d(1/2,0,1/4)
As	4e(0,0,z)	4e(0,0,z)
	z=0.36035(5)	z=0.36423(7)
Atomic displacement parameters U_{eq} (Å^2):		
Sr1/Ca1	0.0108(2)	0.0116(5)
Fe1	0.0096(2)	0.0125(3)
As1	0.00964(17)	0.0119(2)
Bond lengths and angles:		
Sr/Ca-As (Å)	3.2677(4) × 8	3.2062(7) × 8
Fe-As (Å)	2.3890(4) × 4	2.3855(7) × 4
Fe-Fe (Å)	2.7782(2) × 4	2.7624(6) × 4
As-Fe-As(deg)	110.63(3) × 2	109.94(4) × 4
	108.896(14) × 4	109.24(2) × 4
Fe-As-Fe(deg)	71.105(14) × 4	70.76(2) × 4

Table 1 shows the crystallographic parameters determined by single-crystal x-ray-diffraction at 250 K in $Sr_{0.3}Ca_{0.7}Fe_2As_2$. A Bruker Smart Apex2 diffractometer with Mo-K_α radiation, a graphite monochromator with monocarp collimator, and a CCD area detector were used for this experiment. The structure was refined with SHELXL-97 software using 1033 measured reflections of which 115 were unique and 108 observed. The final residuals were R_1= 1.36% and 1.96% for the observed data and wR_2= 3.31% and 4.52% for all data for $SrFe_2As_2$ and $Sr_{0.3}Ca_{0.7}Fe_2As_2$ respectively. Sr and Ca atoms were found to reside in the same site with a refined Ca:Sr = 0.33(1):0.67(1), giving the exact formula $Sr_{0.33}Ca_{0.67}Fe_2As_2$ from x-ray analysis.

Figure 1(c) presents the variation of the a- and c-axis lattice constants (upper panel), the unit cell volume and the tetragonal c/a ratio (lower panel) with Ba-Sr and Sr-Ca concentrations determined from refinements of the single crystal x-ray diffraction data for $Ba_{1-x}Sr_xFe_2As_2$ and $Sr_{1-x}Ca_xFe_2As_2$ crystals taken at 250 K. Within experimental accuracy, the a- and c-axis lattice constants, the c/a ratio, and the unit cell volume all show a monotonic linear decrease with alkaline earth substitution in the continuous series from $BaFe_2As_2$ to $SrFe_2As_2$ to $CaFe_2As_2$. This fact indicates that the whole (Ba,Sr,Ca)Fe_2As_2 series progression experiences a uniform chemical pressure effect due to the reduction of the cation size that follows Vegard's law, as expected for the decreasing ionic radii of Ba, Sr and Ca, respectively.

The lattice parameters of $Ba_{1-x}Sr_xFe_2As_2$ obtained in our experiments are consistent with the data reported in a recent study [25, 26], which found a systematic increase of T_0 with increasing Sr content and no superconductivity. On the other hand, substitution of arsenic for the smaller phosphorus atoms, also instituting a chemical pressure effect, induces superconductivity in $BaFe_2As_{2-x}P_x$ [25]. Thus, a pressure-volume effect is clearly an oversimplified explanation for superconductivity in $BaFe_2As_{2-x}P_x$. In the future, it will be interesting to investigate the evolution of superconductivity combining both the chemical pressure effect of alkaline earth substitution studied here and another tuning parameter that induces superconductivity in order to investigate the role of lattice density in these phenomena.

References

[1] Kamihara Y, Watanabe T, Hirano M., and Hosono H., J. Am. Chem. Soc. 130, 3296 (2008).
[2] Cruz de la C *et al.*, 2008 *Nature* **453**, 899.
[3] Rotter M, Tegel M, and Johrendt D, 2008 *Phys. Rev. Lett.* **101** 107006.
[4] Saha S R, Butch N P, Kirshenbaum K, and Paglione Johnpierre, 2009 *Phys. Rev. B* **79** 224519.
[5] Ren Z -A *et al.*, 2008 *Chin. Phys. Lett.* **25** 2215.
[6] Mazin I I, Schmalian J, 2009 *Physica C* **469** 614.
[7] Christianson A D *et al.*, 2008 *Nature* **456**, 930.
[8] Shirane G *et al.*, 1987 *Phys. Rev. Lett.* **59** 1613.
[9] Keimer B *et al.*, 1992 *Phys. Rev. B* **46**, 14034.
[10] Okada H *et al.*, 2008 *J. Phys. Soc. Jpn.* **77** 113712.
[11] Alireza P L *et al.*, 2009 *J. Phys.: Condens. Matt.* **21**, 012208.
[12] Zhu X *et al.*, 2009 *Europhys. Lett.* **85** 17011.
[13] Cheng P *et al.*, 2009 *Euro Phys. Lett.* **85** 67003.
[14] Sasmal K *et al.*, 2008 *Phys. Rev. Lett.* **101** 107007.
[15] Sefat A S, Jin R, McGuire M A, Sales B C, Singh D J, and Mandrus D, 2008 *Phys. Rev. Lett.* **101** 117004.
[16] Leithe-Jasper A, Schnelle W, Geibel C, and Rosner H, 2008 *Phys. Rev. Lett.* **101** 207004.
[17] Torikachvili M S, Bud'ko S L, Ni N, and Canfield P C, 2008 *Phys. Rev. Lett.* **101** 057006.
[18] Park T *et al.*, 2008 *J. Phys.: Condens. Matt.* **20** 322204.
[19] Kumar M *et al.*, 2008 *Phys. Rev. B* **78** 184516.
[20] Saha S R, Butch N P, Kirshenbaum K, and Paglione J., 2009 *Phys. Rev. Lett.*, **103** 037005.
[21] Schnelle W *et al.*, 2009 *Phys. Rev. B* **79** 214516.
[22] Han F *et al.*, 2009 *Phys. Rev. B.* **80** 024506.
[23] Saha S R *et al.*, 2010 *J. Phys. Condens. Matt.*, **22** 072204.
[24] Kasahara S *et al.*, 2010 *Phys. Rev. B* **81** 184519.
[25] Wang Z, Yang H, Ma C, Tian H, Shi H, Lu J, Zeng L, Li J, 2009 *J. Phys. Condens. Matt.*, **21** 495701.
[26] Rotter M, Hieke C, Johrendt D, 2010 *Phys. Rev. B* **82** 014513.

Sample dependence of superconductivity for V_3Si under high pressure

Shigeki Tanaka, Atsushi Miyake, Tomoko Kagayama, Katsuya Shimizu

KYOKUGEN, Osaka University, Osaka, Japan

E-mail: tanaka@djebel.mp.es.osaka-u.ac.jp

Abstract. V_3Si is one of the superconducting A15 compounds and has relatively high T_c = 17 K at ambient pressure. The superconductivity of V_3Si and the pressure effect of that were strongly dependent on the sample. The pressure dependence of T_c was considered to be in uenced by the composition and the crystal disorder.

1. Introduction

In many superconductors, a structural phase transition seems to play the important roles in their appearance of superconductivity, but it is still unclear. The difficulty may be arisen from the difference between their transition temperature, for example, the superconducting transition occurs at much lower temperature than the structural phase transition temperature [1].

V_3Si crystallizes in the A15 crystal structure with a cubic unit cell. The Si atoms form a bcc lattice and each cube face is bisected by orthogonal V chains. The importance of these chains is often emphasized to understand qualitatively the generally high critical temperatures of A15 compounds. The superconductivity of V_3Si, which is one of the A15 compounds had been already observed at relatively high T_c =17 K [2]. That superconductivity also appears near the lattice instability, martensitic transition at 21 K [3,4]. V_3Si is expected to be the good sample to study the relationship between a superconductivity and a structural instability. However, there are two types of crystals in V_3Si, transforming (T) and nontransforming (NT) ones and the difference of these two crystals is not yet understood. In previous works, the pressure dependence of T_c and T_M for both singlecrystalline samples were investigated up to 3 GPa by Chu *et al* [3,4]. These studies reported that the pressure dependence of T_c for T and NT samples was different above 2.5 GPa. T_c of T sample monotonically increased and of NT sample suddenly decreased above 2.5 GPa. We investigated the pressure dependence of T_c for NT polycrystalline sample and the pressure dependence of T_c was different from the reports of Chu *et al* [5].

In this work, we focus on the sample dependence of the pressure effect of the superconductivity for V_3Si.

2. Experimental details

Two $V_{1-x}Si_x$ samples used in this work were prepared by arc melting the commercial V_3Si polycrystalline sample (purity: 99.9%). The compositions of them were controlled by changing the arc melting time. The melting point of V_3Si is over 1900 °C and the vapor pressure of silicon around this temperature is 10 times higher than that of vanadium. The longer the arc melting time is extended, the more silicon evaporates. One sample grown by melting for about 20 minutes was the same with the sample used in our previous work could be Si-poor sample [5]. The other sample was made by melting for several minutes. The x-ray fluorescence analysis and the powder x-ray diffraction analysis were done for both samples to investigate the composition and the crystal structure. T_c of them at ambient pressure were determined by the electrical resistivity measurement. The specific heat was measured for the sample arc melted for 3 minutes by means of the ac-calorimetric method. Two Au/AuFe (0.07%) thermocouples were used as the heater and the thermometer, respectively. Pressure was applied using a diamond anvil cell (DAC) with glycerol as a pressure transmitting medium under pressures up to 2.5 GPa.

3. Experimental results

From the results of the x-ray fluorescence analysis, the compositions of two samples were $V_{1-x}Si_x$ at $x = 0.15$ (Sample 1) and 0.25 (Sample 2). The atomic Si content x for Sample 1 was the average of the total composition, V_3Si and the parasitic component which was revealed by the x-ray diffraction analysis (mentioned below). According to V-Si binary phase diagram, V_3Si can exist in the composition range of $0.22 < x < 0.25$ [6]. Sample 1 was considered to be the Si-poor crystal within the composition range.

The x-ray diffraction (XRD) patterns showed that Sample 1 was the mixture of V_3Si and the other structural parasitic component (Fig 1). Any stoichiometric compounds in V-Si binary phase diagram doesn't reproduce the experimental diffraction pattern. The XRD pattern of

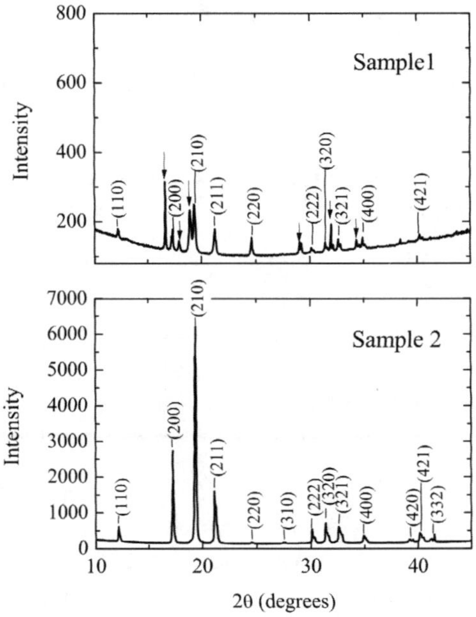

Figure 1. X-ray diffraction pattern of V_3Si; Sample 1 (Si-poor), Sample 2 (stoichiometric).

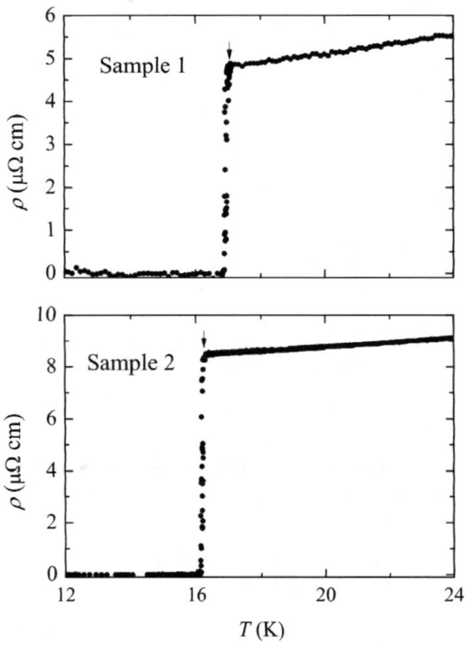

Figure 2. Temperature dependence of the electrical resistivity of V_3Si; Sample 1 (Si-poor), Sample 2 (stoichiometric).

Sample 2 shows that Sample 2 doesn't contain any impurities. The lattice parameters of V_3Si were 4.738 Å and 4.741 Å for Sample 1, 2, respectively. The results were consistent with the previous report that the lattice parameter of $V_{1-x}Si_x$ was independent of the composition around $x = 0.25$ [7].

Figure 2 shows the temperature dependence of the electrical resistivity for two samples. The residual resistance ratio were RRR $= \rho$ (300 K) / ρ (18 K) $= 20, 10$ for Sample 1, 2, respectively. It is possible that RRR of Sample 1 is determined mainly by the parastic components. It is unclear what caused the relatively low T_c of Sample 2. The relationship between T_c and RRR has not been understood. And T_c is independent of the composition near stoichiometry in the previous report [7]. It is considered that there are some other important factors that determine T_c of V_3Si. Table 1 is the overview of Sample 1, 2.

Table 1. Atomic Si content, lattice parameter (a), electrical resistivity (ρ) at room temperature, and T_c of Sample 1, 2.

	atomic Si content	a (Å)	ρ ($\mu\Omega$·cm)	T_c (K)
Sample 1	15%	4.738	98	17.1
Sample 2	25%	4.741	89	16.3

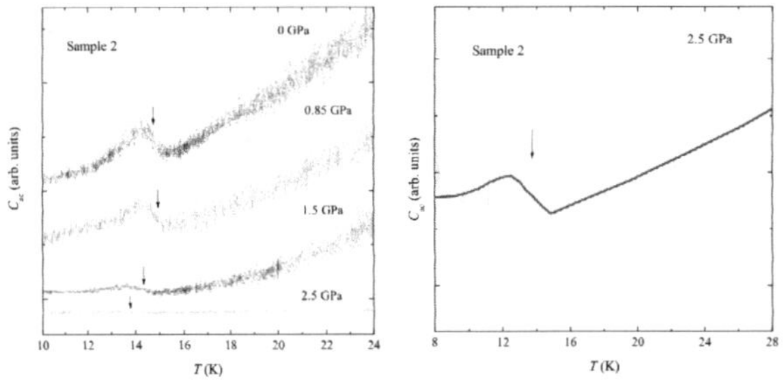

Figure 3. Temperature dependence of the specific heat measured by the ac-calorimetry of V_3Si (stoichiometric Sample 2) under pressures. The arrows show T_c.

Figure 3 shows the temperature dependence of the specific heat for Sample 2 at several pressures. T_c was determined from the half height of the jumps. The jumps of the specific heat becomes smaller with applying pressure. T_c was lower than that determined by the electrical resistivity measurement at ambient pressure and decreased with applying pressure up to 2.5 GPa. The discrepancy of T_c is not caused by the sample heating in the ac-calorimetric measurement. We confirmed that the sample temperature changed little by the application of heat at a certain temperature around T_c. The relatively low RRR and the broad superconducting transitions are considered to be influenced by the sample inhomogeneity. And it is possible that the low T_c and the discrepancy of T_c for Sample 2 are also caused by the same reason and the quality of the sample is rather poor. Figure 4 shows the pressure dependence of T_c for V_3Si NT crystals. The pressure dependence of T_c are positive and negative for Sample 1, 2 respectively, and different from the reports of Chu *et al* [4]. T_c of Sample 1 determined by the electrical resistivity measurement (not shown here) are consistent with the results of the ac-suscepetibility measurement for Sample 1 [5]. The difference of the pressure dependence of T_c is expected to be influenced by each sample's composition. But we couldn't find the synthesis process and how Chu's NT sample was characterized. To reveal the relationship between the

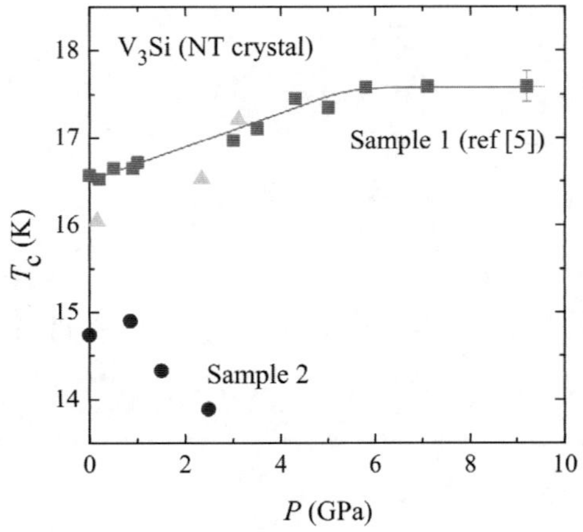

Figure 4. Pressure dependence of T_c for V_3Si NT samples. The square, triangle and circle symbols indicate T_c determined by the ac-susceptibility [5], the electrical resistivity and the ac-calorimetric measurements, respectively.

composition and the pressure effect is important to study the superconsuctivity of V_3Si and the other A15 compounds.

There are still unsolved problems of superconductivity under high pressure in A15 compounds including V_3Si. The sample growth of V_3Si single crystal and further experiments (e.g. thermal expansion measurement) for the sample under high pressure are planned to investigate the mechanism of the pressure dependence of T_c for V_3Si.

4. References

[1] J. Q. Yan, A. Kreyssig, S. Nandi, N. Ni, S. L. Bud ' ko, A. Kracher, R. J. McQueeney, R. W. McCallum, T. A. Lograsso, A. I. Goldman, and P. C. Can eld 2008 *Phys. Rev. B.* **78** 024516.
[2] Hardy. G and Hulm. J 1953 *Phys. Rev.* **89** 884.
[3] C. W. Chu and L. R. Testardi 1974 *Phys. Rev. Lett.* **32** 766.
[4] C. W. Chu and V. Diatschenko 1978 *Phys. Rev. Lett.* **41** 572.
[5] S. Tanaka, A. Miyake, B. Salce, D. Braithwaite, T. Kagayama, K. Shimizu 2010 *J. Phys. Conf. Ser.* **200** 012202-012204.
[6] J. F. Smith 1985 *Bull. Alloy Phase Diagrams.* **6** 3.
[7] A. Junod, J.-L. Staudenmann, J. Muller, and P. Spitzli 1971 *Journal of Low Temperature Physics.* **5** 25.

Journal of Physics: Conference Series **273** (2011) 012106 doi:10.1088/1742-6596/273/1/012106

Theory of the time-resolved spectral function of high-temperature superconductors

Jianmin Tao and Jian-Xin Zhu

Theoretical Division & CNLS, Los Alamos National Laboratory, Los Alamos, New Mexico 87545, USA

Abstract. We have recently developed a three-temperature model to simulate the time evolution of the electron and phonon temperatures in high-temperature superconductors. This is the first model that is valid in the superconducting state. Based on this model, we calculated the time-resolved spectral function via the double-time Green's functions. Our calculations show that the theory not only reveals the familiar dip-hump structure in the normal state, but also a new signature of the electron-phonon coupling in the superconducting state. We further show that this signature can only be observed when the phonons are pumped directly. In this proceedeing, we provide detailed derivation of our theory.

1. Introduction

The physics underlying pairing in high-temperature superconconductors (HTSC) remains a topic of strong interest. The small isotope effect of optimally doped samples and the very nature of the d-wave pairing suggested the importance of a spin-based pairing interaction. Indeed the dip-hump structure of the density of states observed in electron tunneling spectroscopy [1] could be explained with the electronic coupling to spin fluctuations [2]. Due to the improvement of sample quality and momentum resolution, similar structures have been observed in angle-resolved photoemission spectroscopy (ARPES) [3]. These features have been equally well explained [4] with the electron-phonon (el-ph) coupling. This suggests that some phonon modes that are strongly coupled to electrons may be of relevance. To better understand the physics behind the superconductivity, a good knoweledge of the el-ph coupling is essential.

However, the usual ARPES only provides the information of the el-ph coupling in thermally populated excitation. Accessing the thermally inaccessible states provides the ability to uncover novel signatures of the el-ph coupling, which may not be captured by the standard ARPES. For this reason, time-resolved (TR) ARPES has been developed to offer the capability to capture the single-particle (frequency domain) and collective (time domain) information, thus making it possible to directly probe the link between the collective modes and single-particle states. In this setting, either electrons or phonons (i.e., lattice vibrational modes) can be selectively excited with an ultrafast laser pulse. This new technique has been applied to the studies of transient electronic structure in Mott insulators [5, 6] and HTSC [7] with optical pump. The TR-ARPES with the capability of directly pumping vibrational or phonon mode has been developed to study the el-phonon coupling in manganites [8], though not yet in the cuprates.

To study the dynamics of the el-ph coupling in HTSC, We have recently proposed [9] a theory of the time-resolved spectral function, which is valid in the superconducting state where

Published under licence by IOP Publishing Ltd

the quasiparticle excitation is well defined [2]. The theory consists of two parts: (i) a three-temperature model is proposed to simulate the time dependence of the electron and phonon temperatures, and (ii) time-resolved spectral function up to second order in the coupling constant is calculated with the double-time Green's function approach. To illustrate the theory, we applied it to the optimally doped cuprates to study the time evolution of the el-ph interaction between electrons and phonons of the out-of-plane and out-of-phase buckling mode [10]. Numerical calculations show that, when phonons are pumped directly, a new signature of the el-ph coupling can be observed in the superconducting state and then immediately disappears when the system evolves into the normal state with the increase of the electronic temperature. This enhancement of the el-ph coupling in the superconducting state suggests that the role of the el-ph coupling cannot be ignored when we explore the physics behind high-temperature superconductivity. Here we will give a detailed dicussion on the derivation of this theory.

2. Three-temperature model

Our three-temperature model [9] is based on the two assumptions: (i) The system can be divided into three subsystems, namely, electrons, hot phonons, and cold lattice, and (ii) the energy transfer within each subsystem is much faster than that between different subsystems. The first assumption requires three temperatures to characterize the system, while the second guarantees that a local equilibrium within each subsystem can be reached quickly after pumping. According to the energy conservation law, the rate equations for the electron, hot phonon, cold lattice temperatures [7] can be written as

$$C_e\frac{\partial T_e}{\partial \tau} = \frac{\partial E_e}{\partial \tau} + P_e, \tag{1}$$

$$C_{ph}\frac{\partial T_{ph}}{\partial \tau} = -\frac{\partial E_e}{\partial \tau} + P_{ph} - \frac{C_{ph}(T_{ph}-T_l)}{\tau_\beta}, \tag{2}$$

$$C_l\frac{\partial T_l}{\partial \tau} = \frac{C_{ph}(T_{ph}-T_l)}{\tau_\beta}, \tag{3}$$

where P_e is the power for pumping electrons, P_{ph} is the power for pumping hot phonons, and τ_β is the time of the anharmonic decay of hot phonons. The specific heat of electrons can be calculated from the Boltzmann entropy $S_e = -2k_B\sum_{\mathbf{k}}\{[1-f(E_{\mathbf{k}})]\ln[1-f(E_{\mathbf{k}})]+f(E_{\mathbf{k}})\ln f(E_{\mathbf{k}})\}$ by $C_e = T_e\partial S_e/\partial T_e$, while the specific heat of hot phonons for *one*-vibrational mode can be calculated directly from the internal energy of phonons [which is given by the third term of Eq. (6)] by $C_{ph} = \hbar\Omega\partial N(\Omega)/\partial T_{ph}|_{\Omega=\Omega_0}$. In these expressions, $E_{\mathbf{k}} = \sqrt{\xi_{\mathbf{k}}^2+\Delta_{\mathbf{k}}^2}$ is the superconducting-state quasiparticle energy, $f(E_{\mathbf{k}}) = 1/(e^{\beta_e E_{\mathbf{k}}}+1)$ is the Fermi distribution function of electrons, and $N(\Omega_0) = 1/(e^{\beta_{ph}\Omega_0}-1)$ is the Bose-Einstein distribution function of phonons, with $\beta_e = 1/k_BT_e$ and $\beta_{ph} = 1/k_BT_{ph}$ being the temperature parameters of electrons and phonons. Simple algebra leads to the results

$$C_e = \beta_e k_B\sum_{\mathbf{k}}\left[-\frac{\partial f(E_{\mathbf{k}})}{\partial E_{\mathbf{k}}}\right]\left(2E_{\mathbf{k}}^2+\beta_e\Delta_k\frac{\partial\Delta_k}{\partial\beta_e}\right), \tag{4}$$

$$C_{ph} = \frac{k_B}{4}(\hbar\Omega_0\beta_{ph})^2\left[\coth^2\left(\frac{\hbar\Omega_0\beta_{ph}}{2}\right)-1\right]. \tag{5}$$

Now we calculate the energy exchange rate $\partial E_e/\partial\tau$ due to el-ph scattering. For this purpose, we consider a two-dimensional superconductor exposed to a time-dependent laser field. The model Hamiltonian for a vibrational mode ν can be written as

$$\begin{aligned} H &= \sum_{\mathbf{k}\sigma} \xi_{\mathbf{k}} c^\dagger_{\mathbf{k}\sigma} c_{\mathbf{k}\sigma} + \sum_{\mathbf{k}} (\Delta_{\mathbf{k}} c^\dagger_{\mathbf{k}\uparrow} c^\dagger_{-\mathbf{k}\downarrow} + \text{h.c.}) + \sum_{\mathbf{q}} \hbar\Omega_{\nu\mathbf{q}} \left(b^\dagger_{\nu\mathbf{q}} b_{\nu\mathbf{q}} + \frac{1}{2} \right) \\ &+ \frac{1}{\sqrt{N_L}} \sum_{\mathbf{kq}\sigma} g_\nu(\mathbf{k},\mathbf{q}) c^\dagger_{\mathbf{k}+\mathbf{q},\sigma} c_{\mathbf{k}\sigma} A_{\nu\mathbf{q}} + H_{\text{field}}(\tau), \end{aligned} \tag{6}$$

where $c^\dagger_{\mathbf{k}\sigma}$ ($b^\dagger_{\nu\mathbf{q}}$) and $c_{\mathbf{k}\sigma}$ ($b_{\nu\mathbf{q}}$) are the creation and annihilation operators for an electron with momentum $\mathbf{k}$ and spin σ (phonon with momentum $\mathbf{q}$ and vibrational mode ν), $A_{\nu\mathbf{q}} = b^\dagger_{\nu,-\mathbf{q}} + b_{\nu\mathbf{q}}$, $\xi_{\mathbf{k}}$ is the normal-state energy dispersion [11], μ the chemical potential, $\Delta_{\mathbf{k}}$ the gap function [12], and g_ν the coupling matrix. To simplify the calculation, we perform the Bogoliubov-de Gennes transformation [13] $c_{\mathbf{k}\uparrow} = u_{\mathbf{k}}\alpha_{\mathbf{k}} - v_{\mathbf{k}}\beta^\dagger_{\mathbf{k}}$ and $c_{-\mathbf{k}\downarrow} = u_{\mathbf{k}}\beta_{\mathbf{k}} + v_{\mathbf{k}}\alpha^\dagger_{\mathbf{k}}$, where $u^2_{\mathbf{k}} = (1+\xi_{\mathbf{k}}/E_{\mathbf{k}})/2$ and $v^2_{\mathbf{k}} = (1-\xi_{\mathbf{k}}/E_{\mathbf{k}})/2$, and obtain the diagonalized expression for the electronic quasiparticle energy, $E_e = \sum_{\mathbf{k}} E_{\mathbf{k}}(\langle\alpha^\dagger_{\mathbf{k}}\alpha_{\mathbf{k}}\rangle - \langle\beta_{\mathbf{k}}\beta^\dagger_{\mathbf{k}}\rangle)$. The time evolution of $\langle\alpha^\dagger_{\mathbf{k}}\alpha_{\mathbf{k}}\rangle$ and $\langle\beta_{\mathbf{k}}\beta^\dagger_{\mathbf{k}}\rangle$ can be calculated using the equations-of-motion approach. The results, correct to second order in g_ν, are given by

$$\frac{\partial\langle\alpha^\dagger_{\mathbf{k}}\alpha_{\mathbf{k}}\rangle}{\partial\tau} = \frac{2\pi}{N_L}\sum_{\mathbf{q}} g^2_\nu (u_{\mathbf{k}}u_{\mathbf{k}-\mathbf{q}} - v_{\mathbf{k}}v_{\mathbf{k}-\mathbf{q}})^2 (\delta_2 e^{\beta_e\Omega_0} - \delta_1)\left[e^{(\beta_{ph}-\beta_e)\Omega_0} - 1\right](1-f_{\mathbf{k}}) f_{\mathbf{k}-\mathbf{q}} N_{\Omega_0} \tag{7}$$

and $\partial\langle\beta_{\mathbf{k}}\beta^\dagger_{\mathbf{k}}\rangle/\partial\tau = -\partial\langle\alpha^\dagger_{\mathbf{k}}\alpha_{\mathbf{k}}\rangle/\partial\tau$, where $f_{\mathbf{k}} = f(E_{\mathbf{k}})$, $N_{\Omega_0} = N(\Omega_0)$, $\delta_1 = \delta(E_{\mathbf{k}-\mathbf{q}} - E_{\mathbf{k}} + \Omega_0)$, and $\delta_2 = \delta(E_{\mathbf{k}-\mathbf{q}} - E_{\mathbf{k}} - \Omega_0)$. Here we have specifically set $\Omega_\nu = \Omega_0$. Differentiation of both sides of the expression for E_e with respect to τ leads to

$$\frac{\partial E_e}{\partial\tau} = \frac{4\pi}{N_L}\sum_{\mathbf{kq}} g^2_\nu (u_{\mathbf{k}}u_{\mathbf{k}-\mathbf{q}} - v_{\mathbf{k}}v_{\mathbf{k}-\mathbf{q}})^2 \delta(E_{\mathbf{k}-\mathbf{q}} - E_{\mathbf{k}} - \Omega_0)\Omega_0\left[e^{(\beta_{ph}-\beta_e)\Omega_0} - 1\right] f_{\mathbf{k}}(1-f_{\mathbf{k}-\mathbf{q}}) N_{\Omega_0} \tag{8}$$

Equations (1)-(5) and (8) constitute our three-temperature model. The original version of this model was phenomenologically proposed [7] as an extension of the two-temperature model [14] for the normal state. Compared to the original version, the present model has the additional ingredients: (i) it incorporates the detailed band structure, (ii) it is valid for the superconducting state and will essentially reduce to the original version [7] in the normal state, (iii) it includes the anistropic effect on the el-ph coupling, and (iv) it offers the capability of selectively pumping electron and phonon degrees of freedom.

Finally we point out that in the calculation of the equations of motion for $\langle\alpha^\dagger_{\mathbf{k}}\alpha_{\mathbf{k}}\rangle$ and $\langle\beta_{\mathbf{k}}\beta^\dagger_{\mathbf{k}}\rangle$, two cross terms $\langle\alpha^\dagger_{\mathbf{k}}\beta^\dagger_{\mathbf{k}}\rangle$ and $\langle\beta_{\mathbf{k}}\alpha_{\mathbf{k}}\rangle$, which are coupled to the former two direct terms, have been neglected. These two terms describe the formation and breaking of the Cooper pairs arising from the time-dependent perturbation. At the initial state, they vanish. This negligence should not change the main physics discussed here, because they do not contribute to the energy directly.

3. Time-resolved spectral function

The time-resolved spectral function is defined as $A(\mathbf{k},\omega) \equiv -\frac{2}{\pi}\,\text{Im}\mathcal{G}_{11}(\mathbf{k},\omega)$. Here $\mathcal{G}_{11}$ is the one-one component of the retarded Green's function $\hat{\mathcal{G}}(\mathbf{k},\omega)$, which is related to the self-energy by $\hat{\mathcal{G}}^{-1}(\mathbf{k},\omega) = \hat{\mathcal{G}}^{-1}_0(\mathbf{k},\omega) - \hat{\Sigma}(\mathbf{k},\omega)$, with $\hat{\mathcal{G}}^{-1}_0(\mathbf{k},\omega) = \omega\hat{\sigma}_0 - \Delta_{\mathbf{k}}\hat{\sigma}_1 - \xi_{\mathbf{k}}\hat{\sigma}_3$. $\hat{\sigma}_{0,1,2,3}$ are the unit and Pauli matrices. As is known [4], in the equilibrium state with $T_e = T_{ph}$, the self-energy can be evaluated more conveniently within the imaginary-time Green's function approach, in which a key step is to convert the Bose-Einstein distribution to the Fermi distribution, $n_B(i\omega_n \pm E_{\mathbf{k}-\mathbf{q}}) = -n_F(\pm E_{\mathbf{k}-\mathbf{q}})$ (with $\omega_n = (2n+1)\pi T$, $T = T_e = T_{ph}$). For the current situation, the temperatures of electrons and hot phonons are no longer tied to each other and the above conversions are not valid any more. To avoid this restriction, here we apply the double-time Green's function approach to calculate $\hat{\Sigma}(\mathbf{k},\omega)$. The result is

$$\begin{aligned}\hat{\Sigma}(\mathbf{k},\omega) &= \frac{1}{N_L}\sum_{\mathbf{q}} |g_\nu(\mathbf{k},-\mathbf{q})|^2\{[(\omega-\Omega_0)\Phi_1 - (\omega+\Omega_0)\Phi_2 + \Omega_0(\Phi_3+\Phi_4)]\hat{\sigma}_0 \\ &+ [2E_{\mathbf{k}-\mathbf{q}}(\Phi_1-\Phi_3) + 2\Omega_0(\Phi_3-\Phi_4)]\hat{\sigma}_1 + [\xi_{\mathbf{k}}(\Phi_1-\Phi_2) + (\Omega_0\xi_{\mathbf{k}}/E_{\mathbf{k}})(\Phi_3-\Phi_4)]\hat{\sigma}_3\} \quad (9)\end{aligned}$$

where $\Phi_1 = N(\Omega_0)/[(\omega-\Omega_0+E_{\mathbf{k}-\mathbf{q}})(\omega-\Omega_0-E_{\mathbf{k}-\mathbf{q}})]$, $\Phi_2 = N(-\Omega_0)/[(\omega+\Omega_0+E_{\mathbf{k}-\mathbf{q}})(\omega+\Omega_0-E_{\mathbf{k}-\mathbf{q}})]$, $\Phi_3 = f(-E_{\mathbf{k}-\mathbf{q}})/[(\omega+\Omega_0-E_{\mathbf{k}-\mathbf{q}})(\omega-\Omega_0-E_{\mathbf{k}-\mathbf{q}})]$, and $\Phi_4 = f(E_{\mathbf{k}-\mathbf{q}})/[(\omega+\Omega_0+E_{\mathbf{k}-\mathbf{q}})(\omega-\Omega_0+E_{\mathbf{k}-\mathbf{q}})]$. In our calculation, a bare $\hat{\mathcal{G}}_0$ is used. Since the coupling constant for the cuprates is small, self-consistency effect on the self-energy is small and is thus neglected in this work. Notice that the dynamics due to the perturbative Hamiltomian (6) enters the spectral function through the three-temperature model. From Eq. (9), we see that the dip-hum structure of the density of states, $\rho(\omega) = \frac{1}{N_L}\sum_{\mathbf{k}} A(\mathbf{k},\omega)$, arise from the pole structure [15] of the self-energy $\hat{\Sigma}(\mathbf{k},\omega)$. Because of the inverse relationship between the Green's function and the self-energy, the larger the self-energy is, the smaller the Green's function and so is the spectral function, and vice visa.

In conclusion, we have explained in detail a theory of the time-resolved spectral function we recently proposed to study the time-evolution of the signature of the el-ph coupling in HTSC materials. The theory is valid in the superconducting state. Our study shows that in the superconducting state, the el-ph coupling signature gets enhanced, suggesting the relevance of the el-ph coupling to the superconductivity. In the viewpoint that the electronic spin fluctuations arise from the strong correlation within the electronic DoF itself, and if one can assume that the spin fluctuations will ride on the electrons and thus its effective temperature will be tied to the electronic temperature, direct pumping of hot phonons will provide a unique way to differentiating the bosonic modes being of electronic or phononic origin, to which electronic quasiparticles are strongly coupled.

`Acknowledgments` This work was supported by the National Nuclear Security Administration of the U.S. DOE at LANL under Contract No. DE-AC52-06NA25396, the U.S. DOE Office of Science, and the LDRD Program at LANL.

[1] Huang Q, Zasadzinski J F, Gray K E, Liu J Z, and H. Claus H 1989 *Phys. Rev.* B **40**, 9366
[2] Eschrig M and Norman M R 2000 *Phys. Rev. Lett.* **85**, 3261
[3] Damascelli A, Hussain Z, and Shen Z X 2003 *Rev. Mod. Phys.* **75**, 473
[4] Devereaux T P, Cuk T, Shen Z X, and Nagaosa N 2004 *Phys. Rev. Lett.* **93**, 117004
[5] Perfetti L, Loukakos P A, Lisowski M, Bovensiepen U, Berger H, Biermann S, Cornaglia P S, Georges A, and Wolf M 2006 *Phys. Rev. Lett.* **97**, 067402
[6] Schmitt F, Kirchmann P S, Bovensiepen U, Moore R G, Rettig L, Krenz M, Chu J H, Ru N, Perfetti L, Lu D H, Wolf M, Fisher I R, Shen Z X 2008 *Science* **321**, 1649
[7] Perfetti L, Loukakos P A, Lisowski M, Bovensiepen U, Eisaki H, and Wolf M 2007 *Phys. Rev. Lett.* **99**, 197001
[8] Rini M, Tobey R, Dean N, Itatani J, Tomioka Y, Tokura Y, Schoenlein R W, and Cavalleri A 2007 *Nature* **449**, 72
[9] Tao J and Zhu J X, 2010 *Phys. Rev.* B (to appear).
[10] Zhu J X, Balatsky A V, Devereaux T P, Si Q, Lee J, McElroy K, and Davis J C 2006 *Phys. Rev.* B **73**, 014511
[11] Norman M R, Randeria M, Ding H, and Campuzano J C 1995 *Phys. Rev.* B **52**, 615.
[12] Gross F, Chandrasekhar B S, Einzel D, Andres K, Hirschfeld P J, Ott H R, Beuers J, Fisk Z, and Smith J L 1986 *Z. Phys.* B **64**, 175
[13] de Gennesi P G 1966 *Superconductivity of Metals and Alloys* (Benjamin, New York).
[14] P. B. Allen P B 1987 *Phys. Rev. Lett.* **59**, 1460
[15] Lee J, Fujita K, McElroy K, Slezak J A, Wang M, Aiura Y, Bando H, Ishikado M, Masui T, Zhu J X, Balatsky A V, Eisaki H, Uchida S, and Davis J C 2006 *Nature* **442**, 546

Competing orders in underdoped $(Ba_{1-x}K_x)Fe_2As_2$

R. R. Urbano[1], E. L. Green[1], W. G. Moulton[1], A. P. Reyes[1], P. L. Kuhns[1], E. M. Bittar[2], C. Adriano[2], P. G. Pagliuso[2]

[1] National High Magnetic Field Laboratory, FSU, Tallahassee, Florida 32306-4005, USA.
[2] Instituto de Física "Gleb Wataghin", Unicamp, 13083-970, Campinas-São Paulo, Brazil.

E-mail: urbano@magnet.fsu.edu

Abstract. We report ^{75}As Nuclear Magnetic Resonance (NMR) measurements in the high-T_c superconductor $(Ba_{1-x}K_x)Fe_2As_2$ in the underdoped regime. A structural transition at $T_S \simeq$ 110 K is followed by an antiferromagnetic (AFM) order at $T_N \simeq 102$ K for our $x = 0.16$ single crystal [1]. Superconductivity (SC) also appears at $T_c \simeq 20$ K. We find that the ordered Fe moment $\langle S \rangle$ is reduced upon hole-doping. Both spectrum analysis and relaxation measurements indicate that pinned vortices are present below T_c and SC is coexisting with AFM fluctuations.

The newly discovered Fe-based superconductors have stimulated a great interest due to their high-T_c and rich phase diagrams [2, 3], with similarities to the high-T_c cuprates. In the $\mathbf{A}(Fe_{1-x}T_x)_2As_2$ family ($\mathbf{A}$ = Ca, Sr or Ba and T = transition metal) the simultaneous structural/magnetic phase transition that occurs at high temperature in the undoped material splits and is suppressed by chemical doping. Also, superconductivity commonly emerges sometimes coexisting with the AFM state [4, 5, 6]. Recently, similar features have been reported for the hole-doped $(Ba_{1-x}K_x)Fe_2As_2$ compound [1]. An important and controversial issue is the competition and/or coexistence of AFM and superconductivity in the underdoped regime. Despite the enormous theoretical and experimental effort [7] the nature of the SC in these materials including the pairing mechanism and the symmetry of the order parameter remain unknown. Local probe studies have found evidence for phase separation into AFM domains and SC [8]. Moreover, the experimental results reported so far are often contradictory, reflecting large sample and doping dependences [7].

In this work we carried out ^{75}As NMR measurements to further explore the effect of hole- (*out of plane*) doping in the electronic and magnetic properties of $BaFe_2As_2$. Particular attention is devoted to the interplay between AFM fluctuations and SC. We found a reduced ordered Fe moment $\langle S \rangle$ in the underdoped regime, consistent with the suppression of the AFM order with the appearance of SC. Based on the relaxation experiments we conclude that these two broken symmetries are coexisting microscopically and must be related.

Single crystals of $(Ba_{1-x}K_x)Fe_2As_2$ were grown out of Sn-flux. Detailed bulk characterization is described elsewhere [1]. A high quality single crystal was mounted on a NMR probe equipped with a goniometer allowing a fine alignment of the crystallographic axes with H_{ext}. The field-swept ^{75}As NMR spectra ($I = 3/2$; $\gamma/2\pi = 7.2919$ MHz/T) were obtained by summing the Fourier transform of the spin-echo signal stepwise.

Fig. 1(a) and 1(b) present the field-swept ^{75}As-NMR spectra at constant frequency $\nu =$ 62.238 MHz with $H \perp c$ and $H \parallel c$, respectively. The spectra at 120 K represent the normal paramagnetic (PM) state while the broad spectra at 4.2 K belong to the coexisting AFM+SC

Published under licence by IOP Publishing Ltd

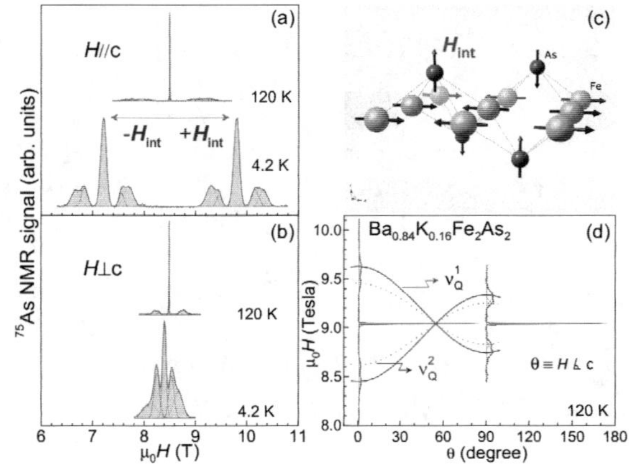

Figure 1. ^{75}As NMR spectra of $Ba_{0.84}K_{0.16}Fe_2As_2$ at 120 K and 4.2 K for (a) $H\|c$ and (b) $H\perp c$. The spectra present three main peaks arising from $m = 3/2$, 1/2 and -1/2 transitions ($I = 3/2$ nucleus). Below T_N, $H_{\rm int}\|c$ split the NMR spectra for $H\|c$ and shift the unsplit spectra for $H\perp c$. Solid lines are Gaussian fits of the spectra; (c) Schematic representation of the $H_{\rm int}\|c$ at the As sites and the ordered Fe moments in the ab-plane; (d) Fits of the angular variation of the satellite lines with Eq. 1 as described in the text.

states. In the PM state, a sharp central transition ($+1/2 \leftrightarrow -1/2$) is observed along with two sets of broader satellite lines ($\pm 3/2 \leftrightarrow \pm 1/2$) split by the quadrupolar interaction of the As nuclei with the local electric field gradient (EFG). In Fig. 1(d) we show the angular variation of the ^{75}As spectrum with $H \simeq 9$ T ($\nu = 66.225$ MHz) at 120 K. The solid (dotted) lines represent the NMR frequencies ν given by Eq. 1

$$\nu_{m\leftrightarrow m-1} = \mu_0\ ^{75}\gamma(1-K^i)H_{\rm ext} + \frac{\nu_{zz}}{2}(m-1/2)(3\ cos^2\theta - 1 + \eta\ sin^2\theta\ cos2\phi) + \quad (1)$$
$$+\ 2^{nd} \text{ order correction}$$

where the first term is the effective field $H_{\rm eff}$ and K^i is the knight shift along the i-axis. The second term in Eq. 1 is the 1st order quadrupolar shift for the three allowed transitions ($\Delta m = \pm 1$). The asymmetry parameter of the EFG is defined as $\eta = (\nu_{xx} - \nu_{yy})/\nu_{zz}$. In the PM state, $\nu_{zz} = \nu_Q$ and $\eta = 0$ due to the high symmetry of the tetragonal phase ($\nu_{xx} = \nu_{yy} = \nu_{zz}/2$). From the fits in Fig. 1(d) we obtained two distinct quadrupole frequencies $\nu_Q^1 \simeq 4.30(1)$ MHz, $\nu_Q^1 \simeq 3.06(1)$ MHz. This means that alloying Ba and K not only induces a substantial distribution of the EFG resulting on the broadening of the satellite lines but also creates two distinct sets of quadrupole splitting. The latter is proportional to the EFG at the As, V_{zz}, through the relation: $\nu_Q = eQV_{zz}/2\sqrt{1+\eta^2/3}$, where Q is the nuclear quadrupole moment of ^{75}As. V_{zz} arises from the hybridization between the As-4p and the Fe-3d orbitals (V_{zz}^{intra}) with an added contribution from any non-cubic distribution of surrounding ions (V_{zz}^{inter}). These ν_Q's are larger than $\nu_Q = 2.21$ MHz observed in the undoped compound [9] and confirm previous observations that ν_Q increases with doping [10, 11]. It also suggests that this sample has no chemically (macroscopically) segregated inhomogeneities [8]. The twinned structural domains are only observed in the orthorhombic ordered phase [9] and is another source for the line broadening observed in the AFM ordered phase.

Below $T_N \simeq 102$ K, static internal hyperfine fields $H_{\rm int}$ develop along the c-axis at the As sites generated by the commensurate AF ordered Fe moments (see Fig. 1(c)). These $H_{\rm int}$ split the ^{75}As-NMR spectra doubling the number of resonance lines when $H\|c$ since the resonance field is given by $H_{\rm eff} = H_0 \pm H_{\rm int}$ (Fig. 1(a)). Alternatively, because of $H_{\rm int}\|c$, when $H\perp c$ the resonance field is given by the magnitude of the vector sum of the mutually orthogonal external and internal fields, $H_{\rm eff} = \sqrt{H^2 + H_{\rm int}^2}$, causing the shift of the unsplit ^{75}As-NMR spectra towards lower fields as displayed in Fig. 1(b) [1].

The T dependence of $H_{\rm int}$ determined with $H_{\rm ext}^{\perp c} \simeq 9$ T is shown in Fig. 2(a). The solid (red) line is the best fit to $M(T) = M_0[1 - (T/T_N)^\alpha]$ with $M_0 = 1.052(7)$, $T_N = 102$ K and

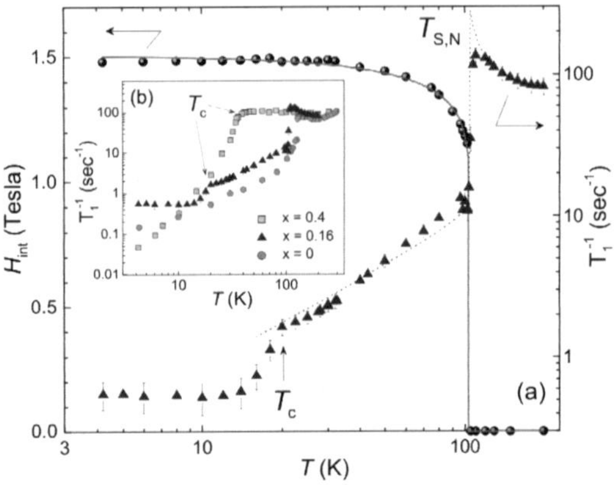

Figure 2. (a) T dependence of $H_{\rm int}$ (sphere, left axis) and T_1^{-1} (triangle, right axis) with $H_{\rm ext}^{\perp c} \simeq 9$ T. The dotted (solid) lines are fits to the magnetic order parameter $M(T)$ (SCR theory) as described in the text. (b) T_1^{-1} of $(Ba_{1-x}K_x)Fe_2As_2$ for $x = 0$, 0.4 (from Ref. [12]) and the sample studied here ($x = 0.16$). The $1/T_1$ obtained for our sample seems to be a mixture of $1/T_1$ of both the pure magnetic $BaFe_2As_2$ and the pure superconducting $(Ba_{0.6}K_{0.4})Fe_2As_2$.

$\alpha = 0.08(1)$. The transition at T_N is almost 1st-order-like with the curvature of $M(T)$ much steeper than obtained with $\alpha = 0.5$, the value expected for a 3-D mean field. The reduced value of the critical exponent α is related to the bi-dimensionality of the magnetic fluctuations [13]. From the direct measurement of the magnetic order parameter $H_{\rm int}$ ($M(T)$) along with the hyperfine coupling constant $A_{hf} \simeq 3.93$ T/μ_B obtained from the Clogston-Jaccarino plot (not shown) we were able to estimate the ordered Fe moment $\langle S\rangle \simeq 0.37\mu_B/Fe$ for our sample, since $\langle S\rangle \equiv H_{\rm int}/A_{hf}$. The smaller $\langle S\rangle$ compared to that observed for $BaFe_2As_2$ ($\langle S\rangle \simeq 0.87\mu_B/Fe$) is consistent with the suppression of T_N and the emergence of SC upon increasing K-doping [1].

In Fig. 2, we also present the T dependence of $1/T_1$ measured at the central transition. The values of the spin-lattice relaxation time T_1 were obtained after saturating the nuclear magnetization $M(t)$ and fitting the recovery with the appropriate solutions to the Master equation. $1/T_1$ exhibits a divergence at $T_N \simeq 102$ K. Its T dependence can be roughly fitted to the self-consistent renormalization (SCR) theory for weak itinerant AFM [13, 14] given by Eq. 2 with the fitting parameters a = -0.112(6) (s K^{-1}), b = 5.12(4)(s^{-1} K$^{-1/2}$) and c = 0.090(1) (s K)$^{-1}$. However, the fast suppression of $1/T_1$ just below T_N can not be well reproduced by Eq. 2 which implies that the nearby structural (tetragonal-orthorhombic) transition may be affecting the spin-fluctuations around T_N.

$$1/T_1 = \begin{cases} aT + bT/\sqrt{T - T_N} & , T > T_N \\ cT/M(T) & , T < T_N \end{cases} \qquad (2)$$

We carried out the $1/T_1$ measurements using the standard saturation recovery technique but we noted that the signal can be fully inverted after an inversion pulse in the PM phase. However, the full inversion of the line is no longer observed for $T \leq T_N$ when the ^{74}As spectrum becomes broader due to the distribution of static hyperfine fields in the AFM order. A distribution of $1/T_1$ is also observed below $T_c \simeq 20$ K. This is not a surprising result once the vortices induced by $H_{\rm ext}$ are expected to cause such behavior.

Further insight into the role of the vortices below T_c is obtained from $1/T_2$ vs. T shown in Fig. 3(a). The spin-spin relaxation time T_2 were obtained by measuring the decay of the echo intensity as a function of $2\tau_e$ using the typical Hahn-echo sequence: $\pi/2$ - τ_e - π - τ_e - *echo signal*. A single component relaxation function such as Gaussian was not sufficient to fit the relaxation data properly. This is not an issue for high T but, below $T_c \simeq 20$ K, the decays are clearly exponential and consist of both components at intermediate T, as inferred from Fig. 3(c). In order to separate out these components, we fit the data to the function $M(t) = M(0)[\delta e^{-t/T_{2e}} + (1-\delta)e^{-(t/T_{2g})^2}]$ where T_{2g} and T_{2e} are the Gaussian and Lorentzian relaxation times, respectively. We find that T_{2e} has an onset coincident with T_c and associate

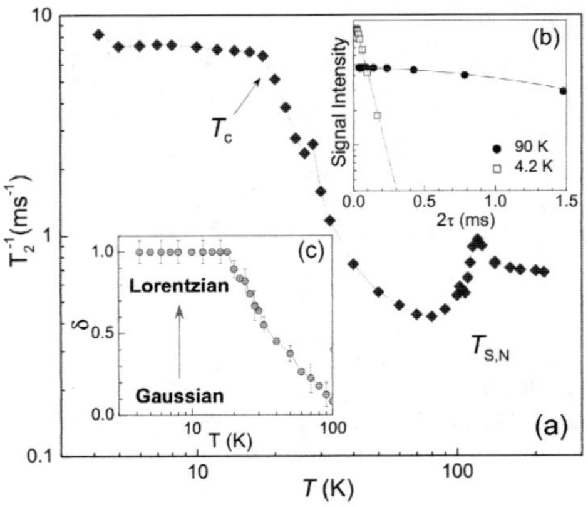

Figure 3. (a) Spin-spin relaxation rate T_2^{-1} at $\simeq$ 9 T derived from fitting the Hahn-echo intensity to the equation $M(t) = M(0)[\delta e^{-t/T_{2e}} + (1-\delta)e^{-(t/T_{2g})^2}]$ where T_{2g} and T_{2e} are the Gaussian and Lorentzian (exponential) relaxation times, respectively; (b) Spin-echo intensity vs. time at 90 K and 4.2 K and respective fits; (c) Mixing parameter δ below 100 K indicating that the vortex dynamics dominate the spin-spin relaxation process with a Lorentzian spectral density below T_c.

the emergence of this new relaxation mechanism with the presence of pinned vortices whose dynamics produce a z-component field fluctuations with a Lorentzian spectral density [15]. The increase of δ, and thus the Lorentzian component, is consistent with the drop in $1/T_1$ at T_c. This observation independently confirms the microscopic coexistence of frozen vortices and AFM order in the SC state contradicting recent reports [8].

In summary, we have used ^{75}As NMR experiments to investigate the interplay between AFM and SC in the underdoped $(Ba_{1-x}K_x)Fe_2As_2$. We found a reduced ordered Fe moment $\langle S\rangle \simeq 0.37\mu_B/Fe$ for our sample confirming that AFM order is suppressed upon doping, probably to give way to SC in these materials. This means that these two broken symmetries must be intimately related. We also conclude that the AFM order uniformly coexists with SC in underdoped compositions, at least on a lattice parameter length scale, given that both $1/T_1$ and $1/T_2$ show the anomalies associated with the AFM order (T_N) and SC (T_c) probed locally by the same ^{75}As nuclei.

Acknowledgments

Work at NHMFL was performed under the auspices of the NSF through the Cooperative Agreement No. DMR-0654118 and the State of Florida. Work at Unicamp was supported by Fapesp, CNPq and Capes.

References

[1] Urbano R R, Green E L, Moulton W G *et al.* 2010 arXiv:1005.3718v1 [cond-mat.supr-con]
[2] Kamihara Y, Watanabe T, Hirano M and Hosono H 2008 *J. Am. Chem. Soc.* **130** 3296-97
[3] Rotter M, Tegel M and Johrendt D 2008 *Phys. Rev. Lett.* **101** 107006
[4] Ni N, Tillman M E, Yan J -Q *et al.* 2008 *Phys. Rev.* B **78** 214515
[5] Pratt D K, Tian W and Kreyssig A *et al.* 2009 *Phys. Rev. Lett.* **103** 087001
[6] Lester C, Chu J -H, Analytis J G *et al.* 2009 *Phys. Rev.* B **79** 144523
[7] Johnston D C 2010 arXiv:1005.4392v1 [cond-mat.supr-con]
[8] Park J T, Inosov D S, Niedermayer Ch *et al.* 2009 *Phys. Rev. Lett.* **102** 117006
[9] Kitagawa K, Katayama N, Ohgushi K *et al.* 2008 *J. Phys. Soc. Jpn* **77** 114709
[10] Julien M H, Mayaffre H, Horvatic M *et al.* 2009 *Europhys. Lett.* **87** 37001
[11] Mukuda H, Terasaki N, Yashima M *et al.* 2009 *Physica* C **469** 559
[12] Fukazawa H, Yamazaki T, Kondo K *et al.* 2009 *J. Phys. Soc. Jpn.* **78** 033704
[13] Nakai Y, Ishida K, Kamihara Y, Hirano M and Hosono H, 2008 *J. Phys. Soc. Jpn.* **78** 033704
[14] Moriya T 1985 *Spin Fluctuations in Itinerant Electron Magnetism* (Berlin, Springer-Verlag)
[15] Bachman H N, Reyes A P, Mitrovic V F *et al.* 1998 *Phys. Rev. Lett.* **80** 1726

Journal of Physics: Conference Series **273** (2011) 012108 doi:10.1088/1742-6596/273/1/012108

Magnetic relaxation studies on $FeTe_{0.60}Se_{0.40}$ superconductor

C S Yadav and P L Paulose
Dept. of Condensed Matter Physics and Material Science
Tata institute of Fundamental Research Mumbai-400005 (India)

E mail: csyadav@tifr.res.in

Abstract: Among the Iron chalcogenide superconductors, $FeTe_{0.60}Se_{0.40}$ shows highest transition temperature of 14.5K. The Magnetic relaxation studies have been performed on the high quality $FeTe_{0.60}Se_{0.40}$ single crystal. The time dependence of the magnetization (M) exhibits thermally activated flux motion and the magnetization relaxation rate 'S' increases with the temperature. The activation energy U(j) has been analyzed using the Maley's method and correlation between the scaled activation energy U and magnetization M suggests the applicability of the collective pinning model and crossover from the single vortex regime to the collective bundle pinning regime in this compound.

1. Introduction

Ever since the advent of superconductivity in the Fe based compound in 2008, there have been frenetic research in the field and the transition temperature (T_c) reached to 56K for iron pnictide material. The FeAs layers are considered to be main structure for the superconductivity in $LaO_{1-x}F_xFeAs$ (1111), $Ba_{1-x}K_xFe_2As_2$ (122), and LiFeAs (111) compounds [1-3]. The discovery of superconductivity in the simpler but structurally similar Iron chalcogenides systems (FeSe, $FeTe_{1-x}Se_x$, known as '11'system) gave an impetus to the activities for understanding the mechanism of the superconductivity in these pnictide superconductors [4].

The tetragonal compound FeSe shows superconductivity at 8K, which on doping with Te increases to 15K for $FeTe_{0.5}Se_{0.5}$. Also with the application of 7GPa pressure, the superconductivity rises to 37K for FeSe [5]. The connection between T_c and the pressure is related to the enhancement of spin fluctuations and modulation of electronic properties [6] due to evolution of the inter-layer Se-Fe-Se separations in the absence of pressure [5, 6]. The absence of a coherence peak in NMR measurements on polycrystalline FeSe compound suggests the unconventional nature of superconductivity [7]. Muon spin rotation study of the penetration depth in $FeSe_x$ was consistent with either anisotropic s-wave or a two gap s-wave pairing [8].

In these iron pnictide and chalcogenide superconductors, the conducting FeAs/FeSe layers are responsible for the superconductivity, which looks very similar to the case of the HTSC cuprate materials. For cuprates superconductors, due to the high anisotropy, short coherence length, and high operation temperature, the vortex motion and fluctuations are quite strong, leading to the small pinning energy and the single vortex to vortex bundles pinning collectively by many small pinning centers. Magnetic relaxation is a valuable tool for determining vortex related parameters of a superconductor. The relaxation rate depends on the pinning parameters as well as the structure of the Abrikosov vortices and vortex lattice [9]. In the present study we have performed magnetic relaxation measurements to bring out the dynamics of vortex in the $FeTe_{0.60}Se_{0.40}$ compound.

Published under licence by IOP Publishing Ltd

2. Experimental:

The single crystals of the $FeTe_{0.60}Se_{0.40}$ were prepared by the chemical reaction of the high purity elements (Fe powder of 99.999% purity Te powder of 99.99% purity and Se powder of 99.98% purity), in stoichiometric proportion, inside the evacuated quartz tube. The thoroughly grinded mixture was slowly heated to 950^0C at the rate 50^0C/hrs and kept for 12 hours before cooling down to 400^0C at the rate of 6^0C/hrs, and then furnace cooled to the room temperature. The crystals were found to grow along the ab-basal plane. The detailed sample preparation method and the quality of the crystal are reported elsewhere [10]. Magnetization measurements have been performed on the high quality single crystal of $FeTe_{0.60}Se_{0.40}$.

The magnetization relaxation studies were performed using a SQUID (Superconducting Quantum Interference Device) magnetometer. The approximate dimension of the sample were $4 \times 2\times 1$ mm^3, with the long axis parallel to the magnetic field (H//ab). We used a scan length of 4 cm, and for each scan 32 data points were taken. This relatively short scan length minimizes the magnetic field variation during the sample movement. For the relaxation measurement, applied magnetic field was increased to 1T and then reduced to zero. The decay rate of remnant magnetization M_{rem} was then recorded.

3. Results and Discussion:

The vortices in type II superconductors, moves over the characteristic pinning energy U(j,T,B) upon thermal activation with an average velocity given by $v = v_0\, exp(-U(j,T,B)/k_BT)$, where '$v$' is the attempt velocity of vortex movement, and j is the critical current density. When a magnetic field is applied, i.e. when there gradient in the magnetic flux density in the superconductor, the Lorentz force reduces the pinning barrier and magnetic flux start penetrating in the material, and magnetization begins to decay from the critical state where $J = J_c$ and U(j) =0 [9]. Using the Anderson Kim model, the time dependence of the magnetic moment is given by $M(t,T) = M_0(T)\left[1 - \frac{kT}{U}\ln\left(\frac{t}{t_0}\right)\right]$, where $M_0(T)$ is the magnetization at t=0, and t_0 is the hopping time (10^{-12} s < t_0 <10^{-6} s), Anderson Kim model assumes a linear dependence of U on J [9,11].

The time decay of the magnetization (-M) measured at different temperatures, in 1T magnetic field along the ab-plane is plotted in log-log scale in Figure 1. The logarithmic decay of the non-equilibrium magnetization (-M) is as expected from the thermally activated motion of the flux creep. The temperature dependence of the normalized relaxation rate S = |d ln M/d ln t| obtained from this data, is shown in the figure 2. The relaxation rate increases linearly with the increase in temperature but shows a tendency of slower relaxation at higher temperatures. The extrapolation of the relaxation rate S gives a finite value of S at T=0K which may be due to either quantum tunneling of vortices or the multi gap nature of the compound like for the MgB_2 [12]. We have calculated the characteristic pinning energy using the expression$T/(S(T,B)) = U_c(T,B)/k_B + \mu(T,B)CT$, where C~ln($2v_0$B/l(dH/dt)) is a weakly temperature dependent parameter [13]. By extrapolating the curve T/S vs. T down to zero temperature, we got the value of the activation energy $U_c(0)/k_B$ = 9K at 1Tesla field. This value is small compared to that for the $SmFeAsO_{0.9}F_{0.1}$ (40K), $PrFeAsO_{0.9}F_{0.1}$ (75K), $Ba(Fe_{0.93}Co_{0.07})_2As_2$ single crystal (75K), YBCO film (100-400K), and MgB_2 (3000K) [13-16]. The parameter C can be determined from the curve '–d ln M/dln t' vs S/T and we find C = 0.306. The slope of the T/S vs T gives the value of μC = 0.201 at low temperatures in 1T field. This gives the μ = 0.67. This value is much less than the Anderson-Kim's criterion of μ = 1 for the thermally activated flux creep motion.

Another important feature that characterizes the vortex dynamics is the dependence of activation energy U on the critical current J and hence on magnetization 'M'. Models based on the vortex glass and

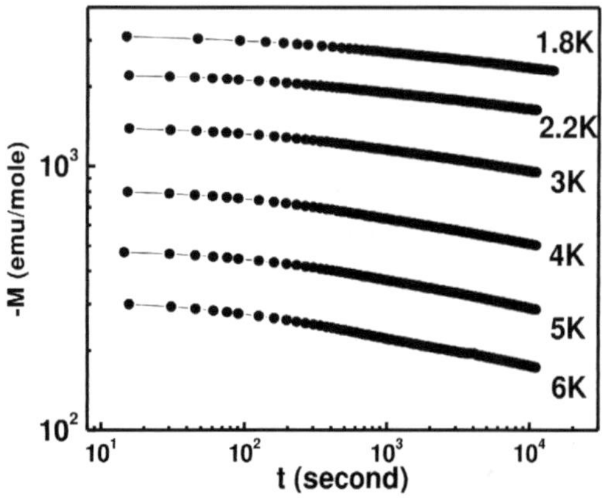

Figure 1. The log-log plot of magnetization '–M (emu/mole)' versus time 't (second)' at various temperature measured in the H=1T along the ab-plane of the crystal).

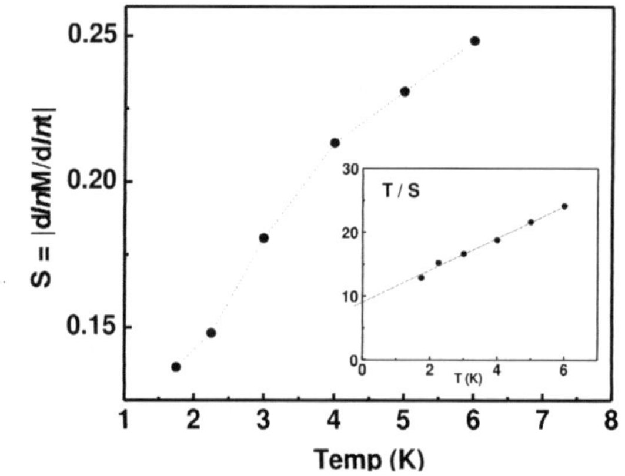

Figure 2. The temperature dependence of the normalized relaxation rate S (= |*d ln M/d ln t* |) measured at 1T field (H//ab). The inset shows the T/S vs T curve, intercept of which gives the value of $U_c(0) = 9K$.

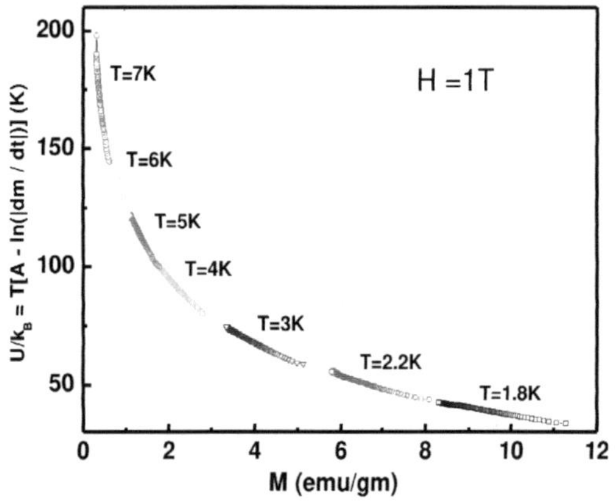

Figure 3. Magnetization dependence of the scaled flux creep activation energy U(K) (= *T[A–|d M / d ln t)]*), calculated using Maley's techniques with parameter A=14.

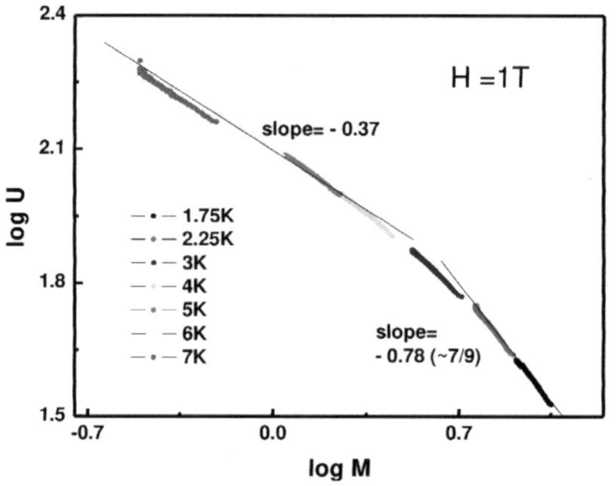

Figure 4. The plot of activation energy versus M on a double logarithmic scale. The solid lines are drawn to show the slopes of the curve.

collective pinning-collective flux creep theories have derived an inverse power law expression for U(J) that can be generally written as $U(J) = U(0)\ (J_c/J)^{\mu}$, where U_c is the characteristic pinning energy and the exponent μ determine the nature of the pinning barrier. In order to meet the boundary condition $U(j)=0$ at $J=J_c$, this expression can be modified as $U(J) = U_c\ [\ (J_c/J)^{\mu} - 1]$. The exponent μ = -1 in the above equation describes the Anderson Kim barrier linear in J/J_c, whereas the positive value of μ leads to a non linear U-J dependence and describe the collective creep barrier. The collective pinning-collective flux creep model considers the current dependence of the flux bundle size, and the different regimes of the bundle size lead to different power law behavior [9,11,17]

We used Maley's method to analyze the activation energy *U(j)*. This model proposes that the magnetic relaxation data at different temperatures can be scaled by the general relation $U/k_B = -T\,[\,ln\,(dM/dt) - A\,]$, where 'A' is the constant that depends on the average hopping velocity only [18]. The figure 3 shows the magnetization dependence of the of the flux creep activation energy, where all the curves at different temperatures seems to fall on a single universal curve for A = 14. However the double log-log plot of U(j) versus M shows two different slops, leading to the two different power law regime depending on the value of j (figure 4). The slopes of the low temperature and high temperature region are found to be 0.78 (μ=7/9) and 0.37 (μ~1/3) respectively. The low temperature value of the slope 0.78 (μ = 7/9), match well with the μ=7/9, which corresponds to the flux creep caused by the large vortex bundles. In the collective flux creep model for the three dimensional superconductors, where the size of bundle is more than the penetration depth of the system, the activation energy has, $U(j) \sim j^{-7/9}$ dependence, where *J* is the critical current density [14,17]. It is noted here that the $FeTe_{0.60}Se_{0.40}$ crystal shows the fish tail effect in the magnetic isotherm curve [10]. This fishtail behavior can be understood as due to cross over in the vortex dynamics with increasing magnetic field, when the system undergo from single vortex regime to the collective bundle pinning regime or the occurrence of the first order phase transition from an ordered elastically pinned low vortex phase to a high field disordered phase.

4. Conclusion:

We have measured the magnetic relaxation on the $FeTe_{0.60}Se_{0.40}$ crystal. The logarithmic decay of the non-equilibrium magnetization is as expected from the thermally activated motion of the flux creep. The activation energy is found to follow $U(j) \sim j^{-7/9}$ dependence on the activation energy suggesting the size of the magnetic flux bundle to be greater than the penetration depth. This behavior along with the fish tail effect observed in the magnetic isotherms shows the crossover form the single vortex regime to the collective bundle pinning regime.

References:

[1] Kamihara Y, Hiramtsu H, Hirano M, Kawamura R, Yanagi H, Kamiya T, and Hosono H (2006) *J.Am.Chem.Soc* **128** 10012; (2008) J.Am.Chem.Soc **130** 3296
[2] Wang C, Li L, Chi S, Ahu Z, Ren Z, Li Y, Wang Y, Lin X, Luo Y, Jiang S, Xu X, Cao G, and Lu Z (2008) *Euro. Phys. Letts.* **83** 67006
[3] Rotter M, Tegel M, Johrendt D, (2008) *Phys Rev. Lett.* **101** 026403
[4] Hsu F C, Luo J Y, Yeh K W, Chen T K, Huang T W, Wu P M, Lee Y C, Huang Y L , Chu Y Y, Yan D C, and Wu M K; (2008) *Proc. Natl. Acad. Sci. USA*; **105** 14262
[5] Margadonna S, Takabayashi Y, Ohishi Y, Mizuguchi Y, Takano Y, Kagavama Y, Nakagawa T, Takata M, and Prassides K, (2009) *Phys Rev B* **80**, 064506.
[6] Masaki S, Kotegawa H, Hara Y, You H, Murato K, Mizuguchi Y, and Takano Y, (2009) *J Phys Soc Jpn* **78**, 063704
[7] Kotegawa H, Masaki S, Awai S, Tou H, Mizuguchi Y, and Takano Y, (2008) *J Phys Soc Jpn* **77**, 113703
[8] Khasanov R, Conder R, Pmjakushina E, Amato E, Baines C, Bukowsaki Z, Karpinski J, Katrych S, Klauss H H, Luetkens H, Shegelaya S, and Zhigaldo N D, (2008) *Phys Rev B*, **78**, 220510(R)
[9] Yeshurun Y, Malozemoff A P, and Shaulov A, (1996) *Rev. Mod. Phys.* **68**, 911
[10] Yadav C S and Paulose P L (2009) *New Journal of Physics*, **11**, 103046
[11] Anderson P W, and Kim Y B, (1964) *Rev Mod Phys* **36**, 39
[12] Wen H H, Li S L, Zhao Z W, Jin H, Ni Y M, Kang Y N, Kim H J, Choi E M, and Lee S I, (2001) *Phys Rev B* **64**, 134505.
[13] Huan Yang, Cong Ren, Lei Shan, and Hai-Hu Wen, (2008) *Phys. Rev. B* **78**, 092504
[14] Bhoi D, Mandal P, Chaudhari P, Dash S, and Banerjee A, arXiv:1002.0208v1
[15] Prozorov R, Ni N, Tanatar M A, Kogan V G, Gordon R T, Martin C, Blomberg E C, Prommapan P, Yan J Q, Budko S L, and Canifield P C, (2008) *Phys. Rev. B* **78**, 224506
[16] Wen H H, Li S L, Zhao Z W, Jin Z, Ni Y M, Ren Z A, Che G C, and Zhao Z X, (2001) *Physica C* 363, 170
[17] Fei'gelman M V, Geshkenbein B V, Larkin A I, and Vinokur V M, (1989) *Phys Rev Lett* **63**, 2303
[18] Maley M P, Willis J O, Lessure H, and McHenry M E, (1990) *Phys Rev Lett* 42, 2639

Anisotropic structural and magnetic properties of the field-aligned superconducting system $SmFeAsO_{1-x}F_x$ ($x = 0$, 0.1, 0.2, 0.25 and 0.3)

Y B You[1], T K Hsiao[1], B C Chang[1], M F Tai[1], Y Y Hsu[2], H C Ku[1], Z Wei[3], K Q Ruan[3] and X G Li[3]

[1] Department of Physics, National Tsing Hua University, Hsinchu 30013, Taiwan
[2] Department of Physics, National Taiwan Normal University, Taipei 10677, Taiwan
[3] Department of Physics, University of Science and Technology of China, Hefei 230026, China

E-mail: `ypyou@phys.nthu.edu.tw`

Abstract. Anisotropic structural and magnetic properties of the field-aligned superconducting system $SmFeAsO_{1-x}F_x$ ($x = 0$, 0.1, 0.2, 0.25 and 0.3) are reported. Due to the Fe spin-orbital related anisotropic exchange coupling, all the tetragonal microcrystalline powders in epoxy were aligned at room temperature using the field-rotation method where the tetragonal *ab*-plane is parallel to the magnetic alignment field B_a of 0.9 T and the *c*-axis parallels to the rotating axis. Anisotropic magnetic properties are studied through low temperature magnetic measurements along the *c*-axis and paralleled to the *ab*-plane of aligned samples in both zero-field-cooled (ZFC) and field-cooled (FC) modes. The under-doped compound ($x = 0.1$) is not superconducting with an antiferromagnetic Néel temperature $T_N \sim 40$ K, while the two optimum-doped compounds ($x = 0.2$ and 0.25) show high superconducting transition temperatures T_c of 49K and 50K, respectively. The variation of anisotropic structural and magnetic properties for this system are discussed and compared with the previously reported 52 K anisotropic superconductor $Sm_{0.95}La_{0.05}FeAsO_{0.85}F_{0.15}$.

1. Introduction

High-T_c superconducting iron-based RFeAs$O_{1-x}F_x$ (rare earth R = La, Ce, Pr, Nd, or Sm) system with transition temperature up to 55 K were reported since 2008[1][3][6]. The ZrCuAsSi-type tetragonal structure (space group $P4/nmm$) is a layered structure where the metallic FeAs layer is separated by the insulating $RO_{1-x}F_x$ layer. The parent compound LaFeAsO is a normal semi-metal which shows a spin density wave (SDW) type antiferromagnetic order below 150 K due to Fermi surface nesting with a tetragonal to orthorhombic structural transition[2]. Electron doping to the FeAs layer through F^--substitution in the O^{2-} site or O^{2-}-deficiency suppresses both the magnetic order and the structural distortion in favor of superconductivity[1][3]. On the other hand, hole doping through Sr^{2+}-substitution in the R^{3+} site gives similar effect[4]. Therefore, similar to the high-T_c cuprate systems, the superconductivity in these iron-based compounds occurs in close proximity to a long range antiferromagnetic ground state.

Since the FeAs layer is the superconducting layer of the RFeAs$O_{1-x}F_x$ system, studies on the anisotropic magnetic properties and their relation to the structural transition are crucial to understanding this new iron-based system. However, high-quality single crystal is difficult

Published under licence by IOP Publishing Ltd

to acquire, a much simpler way using the field-rotation alignment method presents a practical mean to study the anisotropic magnetic and structural transition properties[5].

In this report, $SmFeAsO_{1-x}F_x$ (x = 0, 0.1, 0.2, 0.25, 0.3) powders are field-aligned in epoxy at room temperature and magnetic susceptibility were measured along the *ab*-plane and *c*-axis in a 10-G external magnetic field. Lattice parameters are determined from x-ray diffraction data and systematic variations on lattice aspect ratio c/a and cell volume $V = a^2c$ are presented.

2. Experimental

The polycrystalline $SmLaFeAsO_{1-x}F_x$ powder samples were prepared by the conventional solid state reaction method. First the SmAs powder was prepared by reacting Sm and As powders at 650°C for about 5 hours in an evacuated and sealed quartz tube, and then fine powders of SmAs, FeAs, Fe, LaF_3, Fe_2O_3 were mixed together according to the stoichiometric ratio of $SmLaFeAsO_{1-x}F_x$, grounded thoroughly, and then pressed into pellets. The pellets were further wrapped in Ta foil to be sealed in an evacuated quartz tube, and annealed at 1,160°C for about 50 hours.

The $SmLaFeAsO_{1-x}F_x$ were further grounded into powders with average grain sizes ranged between 1 to 10 μm, and were mixed with epoxy in a quartz tube ($\phi = 8$ mm) with a powder-to-epoxy ratio of 1 to 5. The quartz tube filled with powder-in-epoxy was immediately placed in a 0.9-T magnetic field at room temperature and rotated at a speed of 10 rpm for 4 hours (time for epoxy curing) with the rotation axis perpendicular to the 0.9-T applied magnetic field such that the tetragonal *ab*-plane is parallel to the applied magnetic field and *c*-axis is parallel to the rotation axis[5]. The cured cylindrical compounds were sliced cross-wise into a round button to prepare a surface for x-ray diffraction study on the *ab*-plane, and then sliced perpendicularly to the button surface to prepare another surface for x-ray study along *c*-axis.

Magnetic susceptibility data were collected with a Quantum Design 1-T μ-metal shielded $MPMS_2$ or a 7-T MPMS superconducting quantum interference device (SQUID) magnetometer from 2 K to 300 K.

3. Results and Discussions

The diffraction patterns of both random and aligned powders were compared to confirm proper orientations for further magnetic measurements. Enhanced $(hk0)$ diffraction lines $(h + k = 2n)$ were expected and observed for surfaces parallelled to the *ab*-plane, and enhanced $(00l)$ diffraction lines, for surface parallelled to the *c*-axis. The lack of the forbidden line (210) is consistent with the space group $P4/nmm$. The relative intensity of the (102) line which may be due to intrinsic weak magnetic anisotropy, low aligned field (0.9 T), imperfect powder preparation or alignment procedure gives estimated degrees of alignment of 70-80% for $x = 0.25$ and 0.3 samples and of 80-90% for $x = 0.1$ and 0.2 samples.

For field-aligned compounds with F content $x = 0.2$, 0.25, and 0.3, molar magnetic susceptibility χ_{mol} of aligned powder in an applied magnetic field $B_a = 10$ G are measured and both field-cooled (denoted by filled markers) and zero-field-cooled (denoted by opened markers) data are shown collectively in Fig. 1. χ_c (denoted by round markers) is the molar magnetic susceptibility measured along *c*-axis and, χ_{ab} (denoted by square markers), along the tetragonal *ab*-plane. Anisotropic relation $\chi_c > \chi_{ab}$ are obvious for all field-aligned compounds below superconducting transition temperature T_c. Previously reported $Sm_{0.95}La_{0.05}FeAsO_{0.85}F_{0.15}$ (T_c = 52 K)[5] is also shown as a reference to $x = 0.15$. The field-aligned $SmLaFeAsO_{0.9}F_{0.1}$ compound is not superconducting with an antiferromagnetic Néel temperature $T_N \sim 40$ K.

Superconducting transition temperature T_c were determined from the magnetic measurement data since conducting regions of aligned-powder-in-epoxy specimens are dispersed and disconnected. For near-optimum-doped $x = 0.2$ compound, both FC and ZFC data show a

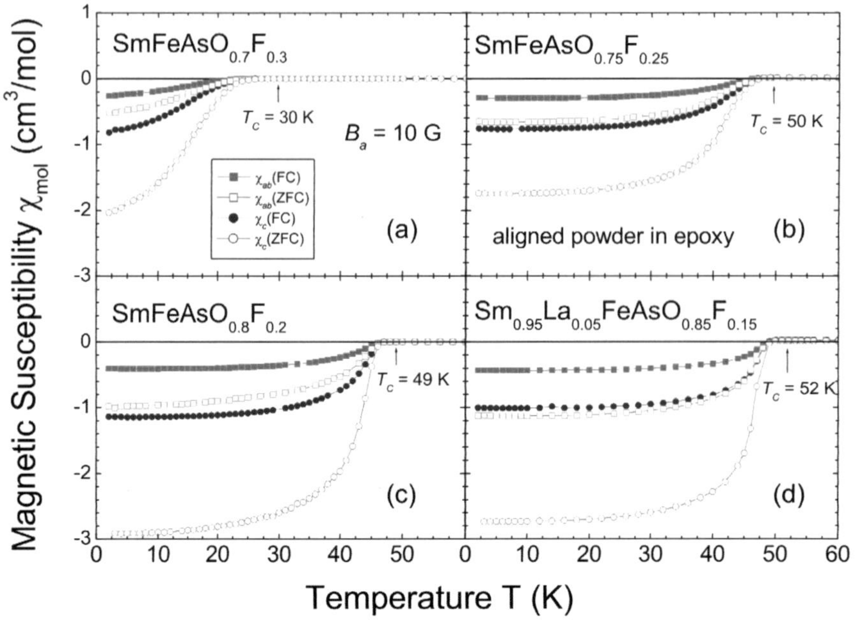

Figure 1. Anisotropic molar magnetic susceptibility χ_{mol} in an applied magnetic field $B_a = 10$ G of $SmFeAsO_{1-x}F_x$ ($x = 0.3$ (a), 0.25 (b), 0.2 (c)) and $Sm_{0.95}La_{0.05}FeAsO_{0.85}F_{0.15}$ (d) aligned powders. χ_c (circle markers) is the molar magnetic susceptibility measured along c axis and, χ_{ab} (square markers), along ab plane. Both field-cooled (FC, filled markers) and zero-field-cooled (ZFC, opened markers) data are shown.

sharp transition at $T_c = 49$ K. For optimum-doped $x = 0.25$ compound, a similar transition is at $T_c = 50$ K. For over-doped $x = 0.3$ specimen, a transition at $T_c = 30$ K was observed.

Large ZFC intragrain Meissner shielding signals were observed with $\chi_c \sim -3$ cm^3/mol and $\chi_{ab} \sim -1$ cm^3/mol for F content $x = 0.2$ and $\chi_c \sim -2$ cm^3/mol and $\chi_{ab} \sim -0.5$ cm^3/mol for F content $x = 0.25$ and 0.3 at $T < \frac{2}{3}T_c$.

Further analysis of anisotropic diamagnetic parameter $\gamma = \chi_c/\chi_{ab}$ of the field-aligned $SmFeAsO_{1-x}F_x$ ($x = 0.2$, 0.25, 0.3) system in an applied magnetic field B_a reveals that the anisotropic diamagnetic parameter γ are between 2.4 to 4 (Fig. 2). At $T \sim T_c$, the FC flux-trapped signals of similar γ magnitude are observed. Previously reported $Sm_{0.95}La_{0.05}FeAsO_{0.85}F_{0.15}$ is shown as a reference to $x = 0.15$ which has a ZFC anisotropic diamagnetic parameter γ of 2.4 and FC flux-trapped signal anisotropic ratio γ of 2.4.

Superconducting transition temperatures T_c are observed to increase with decreasing ZFC diamagnetic parameter γ. Starting from $T_c = 30$ K for $\gamma = 4$ and $x = 0.3$, T_c increases to 49 K for $\gamma = 3$ and $x = 0.2$, a small increase of T_c to 50 K for $\gamma = 2.7$ and $x = 0.25$, and finally, $T_c =$ 52 K for $\gamma = 2.4$ and a reference to $x = 0.15$ of previously reported $Sm_{0.95}La_{0.05}FeAsO_{0.85}F_{0.15}$.

Figure 3 plots lattice aspect ratios c/a and the cell volumes against the F content of $SmFeAsO_{1-x}F_x$ ($x = 0$, 0.1, 0.2, 0.25, and 0.3). The cell volume $V = a^2c$ decreases linearly while the lattice aspect ratio c/a increases with F content. Since the under-doped $x = 0.1$ compound is antiferromagnetic while the previously reported $Sm_{0.95}La_{0.05}FeAsO_{0.85}F_{0.15}$ (as a reference to $x = 0.15$ compound) is a 52 K superconductor, we suggested an antiferromagnetic and superconducting phase boundary at $x = 0.12$.

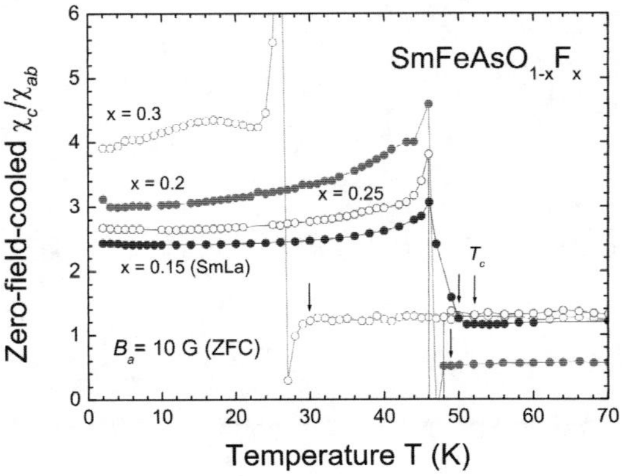

Figure 2. ZFC molar magnetic susceptibility ratio χ_c/χ_{ab} in an applied magnetic field B_a = 10 G of $SmFeAsO_{1-x}F_x$ (x = 0.2, 0.25, 0.3) aligned powders. Previously reported $Sm_{0.95}La_{0.05}FeAsO_{0.85}F_{0.15}$ is shown as a reference to $x = 0.15$. At $T < \frac{2}{3}T_c$, χ_c/χ_{ab} are between 2 to 4 among this system. At $T \sim T_c$, dynamic flux penetrations are obvious.

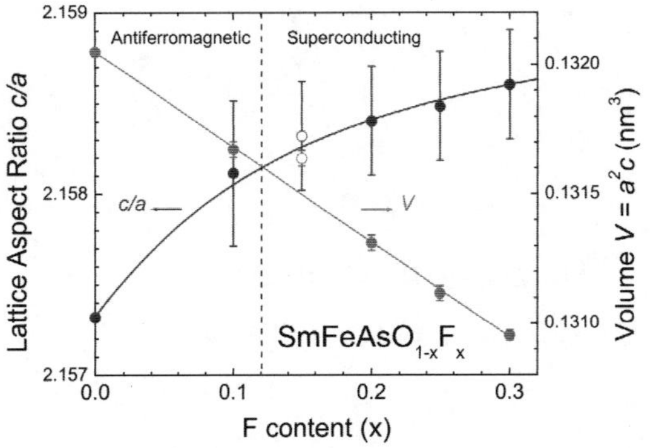

Figure 3. Lattice aspect ratios c/a increase with respect to the F content of $SmFeAsO_{1-x}F_x$ (x = 0, 0.1, 0.2, 0.25, and 0.3). The cell volume $V = a^2c$ is decreased linearly with respect to F content. Previously reported $Sm_{0.95}La_{0.05}FeAsO_{0.85}F_{0.15}$ is shown as a reference to x = 0.15 (open circle markers). We suggested an antiferromagnetic and superconducting phase boundary at x = 0.12 (dashed line).

4. Conclusion

Low temperature magnetic measurements along the c-axis and ab-plane shows anisotropic relation $\chi_c > \chi_{ab}$ for both FC and ZFC modes were observed for all field-aligned specimens below superconducting transition temperature T_c. The under-doped compound (x = 0.1) is not superconducting with an antiferromagnetic Néel temperature $T_N \sim$ 40 K, while the two optimum-doped compounds (x = 0.2 and 0.25) show high superconducting transition temperatures T_c of 49K and 50K, respectively. The over-doped compound (x = 0.3) reveals a low superconducting temperature T_c of 30 K.

The isotropic diamagnetic parameter γ are constant and between 2.4 to 4 among this system at $T < \frac{2}{3}T_c$. In this system, superconducting transition temperatures T_c were observed to raise with decreasing γ values with the smallest value $\gamma = 2.4$ of the previously reported 52 K superconductor $Sm_{0.95}La_{0.05}FeAsO_{0.85}F_{0.15}$.

Acknowledgments

Funding for this work provided by NSC95-2112-M-007-006-MY3, 97-2112-M-003-001-MY3, and 98-2112-M-007-013-MY3.

References

[1] Kamihara Y, Watanabe T, Hirano M and Hasono H 2008 *J. Am. Chem. Soc.* **130** 3296
[2] de la Cruz C, Huang Q, Lynn J W, Li J Y, Ratcliff W II, Zarestky J L, Mook H A, Chen G F, Luo J L, Wang N L and Dai P C 2008 *Nature* **453** 899
[3] Ren Z A, Che G C, Dong X L, Yang J, Lu W, Yi W, Shen X L, Li Z C, Sun L L, Zhou F and Zhao Z X 2008 *EPL* **83** 17002
[4] Wen H H, Mu G, Fang L, Yang H and Zhu X 2008 *EPL* **82** 17009
[5] Chang B C, Hsu C H, Hsu Y Y, Wei Z, Ruan K G, Li X G and Ku H C 2008 *EPL* **84** 67014
[6] Chen X H, Wu T, Wu G, Liu R H, Chen H and Fang D F 2008 *Nature* **453** 761

Journal of Physics: Conference Series **273** (2011) 012110 doi:10.1088/1742-6596/273/1/012110

The magnetoresistance and Hall effect in CeFeAsO: a high magnetic field study

H Q Yuan[1,*], L Jiao[1], J Singleton[2], F F Balakirev[2], G F Chen[3], J L Luo[3], N L Wang[3]

[1] Department of Physics, Zhejiang University, Hangzhou, Zhejiang 310027, China
[2] NHMFL, Los Alamos National Laboratory, MS E536, Los Alamos, NM 87545, USA
[3] Beijing National Laboratory for Condensed Matter Physics, Institute of Physics, Chinese Academy of Science, Beijing 10080, China

[*] E-mail: hqyuan@zju.edu.cn

Abstract. The longitudinal electrical resistance and the transverse Hall resistance of CeFeAsO are simultaneously measured up to a magnetic field of 45T using the facilities of pulsed magnetic field at Los Alamos. Distinct behaviour is observed in both the magnetoresistance $R_{xx}(\mu_0H)$ and the Hall resistance $R_{xy}(\mu_0H)$ while crossing the structural phase transition at $T_s \approx 150K$. At temperatures above T_s, little magnetoresistance is observed and the Hall resistivity follows linear field dependence. Upon cooling down the system below T_s, large magnetoresistance develops and the Hall resistivity deviates from the linear field dependence. Furthermore, we found that the transition at T_s is extremely robust against the external magnetic field. We argue that the magnetic state in CeFeAsO is unlikely a conventional type of spin-density-wave (SDW).

1. Introduction

Following the discovery of superconductivity in $La(O_{1-x}F_x)FeAs$ [1], variant families of the iron-based superconductors have been successively found in the last two years. Resembling the high T_c cuprates, all these iron pnictides crystallize in a layered crystal structure. Superconductivity occurs upon suppressing the magnetic transition via either doping [2] or applying pressure [3], which provided another important example for studying the interplay of superconductivity and magnetism. On the other hand, the iron pnictides also exhibit many distinct features from the high T_c cuprates. For example, superconductivity in iron pnictides is nearly isotropic even though they are layered crystals [4,5], and their parent compounds are usually bad metal instead of a Mott insulator [6]. A comparison of these two types of superconductors might provide important insights on the puzzles of high T_c superconductivity.

In order to study the interplay of superconductivity, magnetism and structural distortion and, therefore, the mechanism of superconductivity in iron pnictides, it is crucial to elucidate the nature of the magnetic state/structural phase transition in the parent compounds. Variant approaches, either based on Fermi surface nesting [7,8] or started from the proximity to a Mott-insulator [9], have been proposed to describe the magnetic states, but no consensus has been reached. A fundamental problem concerns on the itinerancy of the 3d-electrons, which degree can be measured against the robustness of the magnetic transition to external magnetic field. Measurements in a pulsed magnetic field are,

Published under licence by IOP Publishing Ltd

therefore, highly desired. In iron pnictides, intensive efforts have been devoted to the superconducting properties using pulsed magnetic field [4, 5], but little attention has been drawn into the magnetic state.

In this paper, we report the measurements of the electrical resistance $R_{xx}(\mu_0 H)$ and the Hall resistivity $R_{xy}(\mu_0 H)$ of CeFeAsO up to 45T. We found that both $R_{xx}(\mu_0 H)$ and $R_{xy}(\mu_0 H)$ demonstrate distinct behaviour in the magnetic and paramagnetic state, respectively. The observations of extreme robustness of the magnetic transition/structural transition to the applied magnetic field might indicate that the magnetic state is unlikely a conventional type of spin-density-wave state.

2. Experimental methods

Polycrystalline samples of CeFeAsO are synthesized by solid state reaction method using CeAs, CeO_2, Fe_2As as starting materials [10]. The raw materials were thoroughly mixed and pressed into pellets, which were then wrapped with Ta foil and sealed in an evacuate quartz tube. The materials were then annealed at 1150 °C for 50 h. The resulting samples were characterized by powder x-ray diffraction (XRD) at room temperature, which identifies a single phase with a tetragonal ZrCuSiAs-type structure (space group *P4/nmm*).

The longitudinal resistivity and the transverse Hall resistivity were simultaneously measured with a typical 5-probe method using the facilities of pulsed magnetic fields in the National High Magnetic Field Laboratory at Los Alamos. In order to eliminate the effects of contact asymmetries, forward- and reverse-field shots were made at the same temperatures for Hall resistance measurements. The data traces were recorded on a digitizer using a custom designed high-resolution, low-noise synchronous lock-in technique. The temperature dependence of the resistivity at zero magnetic field was measured with a Lakeshore resistance bridge. To minimize the self-heating effect in a pulsed magnetic field, samples with typical size of about 2mm×0.5mm×0.02mm were cut off from the large batch for resistivity measurement.

3. Results and discussion

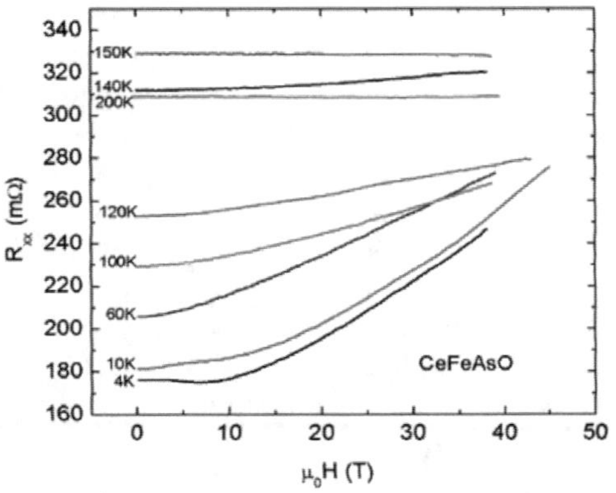

Figure 1: The field dependence of the electrical resistance $R_{xx}(\mu_0 H)$ for CeFeAsO at variant temperatures of T=4K,10K,60K,100K,120K,140K,150K, and 200K, respectively.

Figure 1 shows the electrical resistance $R_{xx}(\mu_0 H)$ of CeFeAsO at variant temperatures. It is noted that CeFeAsO undergoes a structural phase transition around $T_s \approx 150K$, followed by an antiferromagnetic transition around $T_N \approx 140K$ as derived from the measurements of neutron scattering [11]. The difference between T_s and T_N might become smaller with improving sample quality [12]. From figure 1, one can see that the magnetoresistance is negligible at temperatures above the structural phase transition (T>150K), but becomes significant once cooling down below T_s. At low temperature (e.g., T=4K), a large magnetoresistance is observed.

THE GRAPHIC ON THIS PAGE WAS NOT AVAILABLE ON THE E-MEDIA VERSION OF THIS PAPER.

Figure 2. Temperature dependence of the electrical resistance $R_{xx}(T)$ at variant magnetic fields for CeFeAsO. The solid line is the electrical resistivity measured at zero field.

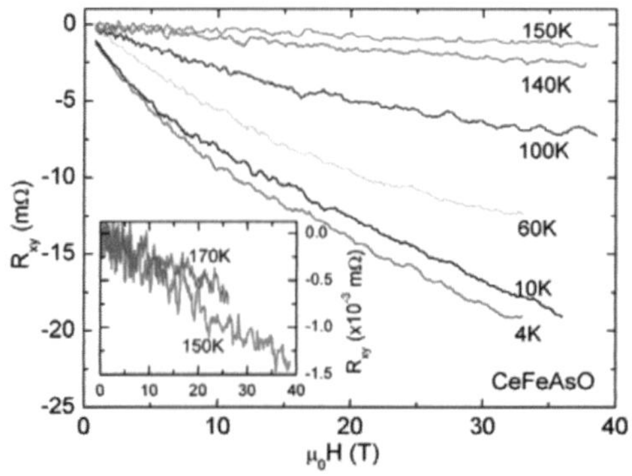

Figure 3. The Hall resistance $R_{xy}(\mu_0H)$ of CeFeAsO at variant temperatures. The inset shows $R_{xy}(\mu_0H)$ at T=150K and 170K, respectively.

To show the evolution of the magnetic/structural transitions with magnetic field, the electrical resistance is replotted as a function of temperature in figure 2. Note that the thick line is the resistive curve measured with the Lakeshore resistance bridge while cooling down the sample at zero field, and the symbols represent the data measured at variant temperatures as shown in figure 1. Obviously, the data collected at variant shots (empty squares) are consistent with those measured at zero field. From figure 2, one can see that the electrical resistance, R_{xx}, is largely enhanced at low temperatures upon applying a magnetic field, but the magnetoresistance seems to always vanish around 150K at which a phase transition from tetragonal to orthorhombic structure has been identified [11,12]. Such behaviour is similar to what we observed in other 1111- and 111-series of iron pnictides [13]. The disappearance of magnetoresistance at the structural phase transition, which still requires more detailed measurements, would suggest that spin fluctuations or short range magnetic order might already develop at the structural phase transition, prior to the magnetic transition. This indicates the importance of magnetic interactions on the subsequent occurrence of the structural and magnetic phase transitions in CeFeAsO. Remarkably, the structural phase transition is independent of the magnetic field up to 40T. Since the structural transition is likely driven by magnetic interaction as argued above, the extreme robustness of the structural transition to magnetic field demonstrates that the magnetic state in CeFeAsO cannot be a conventional type of spin-density-wave state. Otherwise, the magnetic/structural transitions would be suppressed by applying a sufficiently large magnetic field. A magnetic field induced shoulder-like transition is observed with further lowering temperature which nature remains unclear.

Figure 3 shows the field dependence of the Hall resistance at variant temperatures. It is found that the behaviour of the Hall resistance $R_{xy}(\mu_0H)$ also undergoes a dramatic change at the structural phase transition (T_s=150K). At temperatures above T_s, the Hall resistance follows linear field dependence, but deviates from such linear behaviour as the system is cooled below T_s. Band structure calculations show two types of charge carriers (holes and electrons) in iron pnictides [14]. According to the two-band model [15], the linear field dependence of $R_{xy}(\mu_0H)$ might hint a good balance of electrons and holes above T_s. The deviation of the Hall resistivity from the linear field dependence below T_c indicates that the charge carriers or Fermi surface is significantly modified in the magnetic state. It is noted that the initial slope of the Hall resistance undergoes a sharper decrease while crossing the structural phase transition. One possibility for such a decrease might be attributed to the loss of electron carriers at T_s.

The above experimental findings, including the magnetoresistance and the Hall resistance, strongly indicate that the electronic properties of CeFeAsO undergo substantial changes at the structural/magnetic transitions, which is likely driven by magnetic interactions. To elucidate the nature of the magnetic transition and its interplay with the structural transition, it is desired to measure these physical properties in the interesting temperature range ($T_N<T<T_S$) in more detail and to extend the measurements to higher magnetic fields.

4. Conclusion

To summarize, we have found that both the magnetoresistance and the Hall resistance of CeFeAsO show clear changes at the structural phase transition. Independent of the applied magnetic field, the magnetoresistance seems to always vanish at the structural transition, indicating a close relation between the structural distortion and the magnetic ordering, which might originate from the magnetic interactions. The steeper slopes of the Hall resistivity $R_{xy}(\mu_0 H)$ and its deviation from the linear field dependence at temperatures below T_s indicate that the magnetic/structural transitions might lead to a change of the electronic structure.

Acknowledgements
We acknowledge very helpful discussions with J. P. Hu, T. Xiao, Z. Y. Weng and M. B. Salamon. This work was supported by NSFC (No. 10874146 and No.10934005), the National Basic Research Program of China (No. 2009CB929104), the PCSIRT of the Ministry of Education of China, Zhejiang Provincial Natural Science Foundation of China and the Fundamental Research Funds for the Central Universities. Work at NHMFL-LANL is performed under the auspices of the National Science Foundation, Department of Energy and State of Florida.

References:
[1] Kamihara Y, Watanabe T, Hirano M and Hosono H 2008 *J Am Chem Soc.* **130** 3296.
[2] Ishida K, Nakai Y and Hosono H 2009 *J Phys Soc Jpn.* **78** 62001.
[3] Park T, Park E, Lee H, Klimczuk T, Bauer E D, Ronning F and Thompson J D 2008 *Journal of Physics: Condensed Matter.* **20** 322204.
[4] Yuan H Q, Singleton J, Balakirev F F, Baily S A, Chen G F, Luo J L and Wang N L 2009 *Nature.* **457** 565.
[5] Fang M H, Yang J H, Balakirev F F, Kohama Y, Singleton J, Qian B, Mao Z Q, Wang H D and Yuan H Q 2010 *Phys. Rev. B.* **81** 20509.
[6] Sadovskii M V 2008 *Physics-Uspekhi.* **51** 1201.
[7] Dong J, Zhang H J, Xu G, Li Z, Li G, Hu W Z, Wu D, Chen G F, Dai X, Luo J L, Fang Z and Wang N L 2008 *EPL.* **83** 27006.
[8] Mazin I I, Singh D J, Johannes M D and Du M H 2008 *Phys. Rev. Lett.* **101** 57003.
[9] Si Q and Abrahams E 2008 *Phys. Rev. Lett.* **101** 76401.
[10] Chen G F, Li Z, Wu D, Li G, Hu W Z, Dong J, Zheng P, Luo J L and Wang N L 2008 *Phys Rev Lett.* **100** 247002.
[11] Zhao J, Huang Q, de la Cruz C, Li S L, Lynn J W, Chen Y, Green M A, Chen G F, Li G, Li Z, Luo J L, Wang N L and Dai P C 2008 *Nature Materials.* **7** 953.
[12] Jesche A, Krellner C, de Souza M, Lang M and Geibel C 2010 *Phys. Rev. B.* **81** 134525.
[13] H. Q. Yuan, unpublished results.
[14] Qazilbash M M, Hamlin J J, Baumbach R E, Zhang L, Singh D J, Maple M B and Basov D N 2009 *Nature Physics.* **5** 647.
[15] Ashcroft N W and Mermin N D 1976 *Solid State Physics,* (Philadelphia: Saunders College) .

Journal of Physics: Conference Series **273** (2011) 012111 doi:10.1088/1742-6596/273/1/012111

Magnetoelastic quantum oscillations in GdSb to 55 T

R Daou[1,2], A Haase[1,2], M Doerr[3], M Rotter[4], F Weickert[1,2], M Nicklas[1] and F Steglich[1]

[1] Max Planck Institute for Chemical Physics of Solids, D-01187 Dresden, Germany.
[2] Hochfeld-Magnetlabor Dresden, FZ Dresden-Rossendorf e.V., D-01314 Dresden, Germany.
[3] Technische Universität Dresden, Institut für Festkörperphysik, D-01062 Dresden, Germany.
[4] University of Oxford, Dept. of Physics, Clarendon Lab., Parks Road, OX1 3PU Oxford, UK.

E-mail: Ramzy.Daou@cpfs.mpg.de

Abstract. The rare-earth monopnictide materials include many examples of strongly correlated electron materials with low carrier density and magnetic ground states. The semimetal GdSb orders antiferromagnetically below 23.4 K, and at low temperature a fully polarised ferromagnetic state is reached for applied magnetic fields greater than the saturation field, H_s, of 33.5 T. Magnetostriction measurements in pulsed magnetic fields up to 55 T show that the magnetoelastic coupling to the Gd^{3+} moment system changes sign on crossing H_s. We also observe large-amplitude quantum oscillations of the sample length.

1. Introduction

The Gd monopnictides GdX (X = N, P, As, Sb, Bi) have attracted particular attention [1, 2] because the ground state and fundamental magnetic interactions can be studied easily. All of them adopt the f.c.c. NaCl structure. Since Gd^{3+} with $S = 7/2$ has no orbital angular momentum, they are pure spin systems and therefore no single-ion anisotropy is expected. Most of the magnetic properties can be explained by a Heisenberg molecular field model. The two-ion coupling leads to antiferromagnetic ordering with a Néel temperature, T_N, of 23.4 K for GdSb [2]. The standard model of rare-earth magnetism predicts that the magnetic field required to fully align the Gd^{3+} moments, $H_s = \frac{3k_B(T_N-\theta_W)}{g_J\mu_B(J+1)}$ where k_B is the Boltzmann constant, $\theta_W = -31.3$ K [3] is the Weiss temperature, μ_B is the Bohr magneton and $g_J = 2$ is the g-factor. From these values we predict $H_s = 27$ T, which is close to the observed value of 33.5 T.

The progression of magnetic phases is first from a simple antiferromagnetic (AFM) ground state into a spin-flop state at low applied field (~ 0.5 T). The canting angle between spins is then smoothly decreased as the field is increased until a critical value, H_c, is reached. At higher fields the response is paramagnetic (PM). Thus at intermediate temperatures $T < T_N$, $H_c < H_s$.

The electrical transport properties of the Gd-monopnictides are also of interest as they are examples of clean, nearly compensated low carrier-concentration magnetic materials. While the rest of the series are antiferromagnetic semimetals, GdN is found to be a half-metallic, ferromagnetic semiconductor [4] with potential applications in spintronics. Bandstructure calculations [5, 6] find semimetallic ground states and can reproduce the increasing trend of H_s with anion mass by incorporating small nearest- and next-nearest-neighbour interactions.

Because of the high saturation field, GdSb has not been well investigated experimentally. Previous measurements of magnetization and magnetostriction [3, 7] in GdSb did not determine

Published under licence by IOP Publishing Ltd

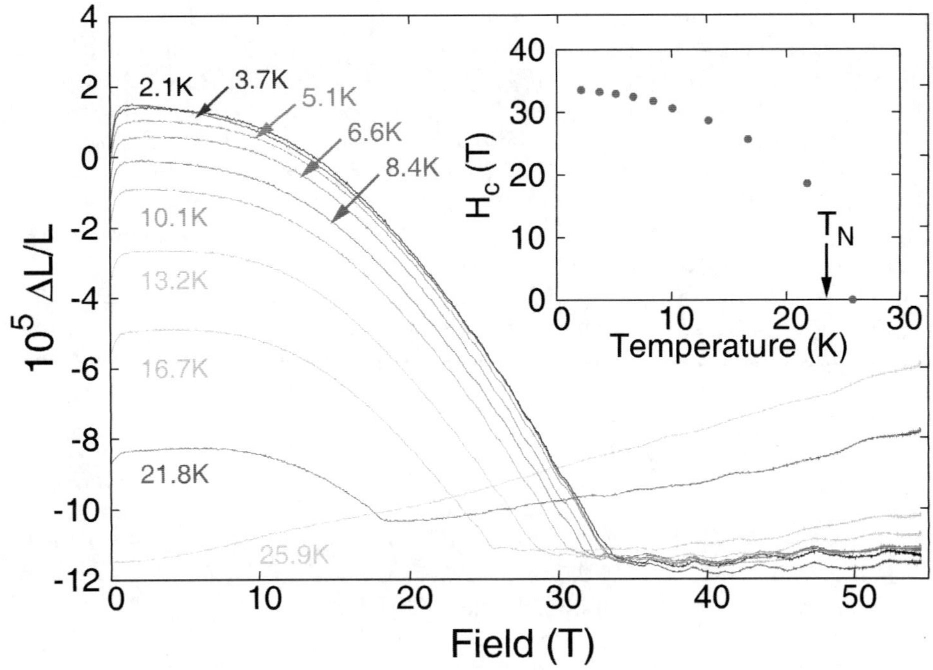

Figure 1. The linear magnetostriction of GdSb at various temperatures, relative to the value at 2.1 K and 0 T. The curves are offset by the relative thermal expansion of the sample. The data shown was taken during the upsweep of the pulsed magnetic field; the downsweeps match well. *Inset:* The critical field H_c as a function of temperature.

the magnetic phase diagram or examine Fermi surface properties in the high field state. Here we investigate the magnetoelastic behaviour in GdSb in both the antiferromagnetic and polarised paramagnetic states by using pulsed magnetic fields up to 55 T. We extract the magnetic phase diagram and succeed in observing quantum oscillations in the magnetostriction for the first time in pulsed magnetic fields.

2. Methods

The single crystal of GdSb used in this study was prepared at AMES laboratory, Iowa (ED088/3). The crystal dimensions along principal axes were $1.4 \times 1.6 \times 3.0\,\mathrm{mm}^3$ and its mass was 23 mg.

Measurements of the longitudinal magnetostriction were performed using a fibre optic strain gauge glued to the sample with cyanoacrylate epoxy. Details of the method can be found in Ref. [8]. This technique is suitable for the challenging environment of pulsed magnetic fields, and it offers a resolution for the relative length change of the sample, $\Delta L/L$, of about 10^{-7} ($\Delta L \sim 3\,\text{Å}$ for the current sample). While superconducting magnets allow better measurements, this is the highest resolution so far achieved in pulsed fields. The capacitor-driven pulsed magnet was energised to a maximum field of 55 T in a rise time of 33 ms. All of the data shown is taken from the upsweeps, but the data from the downsweeps match very well.

3. Results

In Figure 1 we show the linear magnetostriction of GdSb at several temperatures between 2.1 K and 26 K. Both magnetic field and the measured elongation were aligned along a principal axis of the crystal. Below 1 T the spin flop transition can be seen as a rapid increase in sample length.

In the spin-flop phase the sample length decreases smoothly as the moments rotate to align with the field. The critical field H_c is sharply defined by a kink in the data at high fields for $T < T_N$, and the inset to Figure 1 shows the magnetic phase diagram described by these points. In the paramagnetic phase above T_N, the sample length increases monotonically with field. At high fields and low temperatures, quantum oscillations in the sample length can be observed.

The relative thermal expansion is negative in the antiferromagnetic and spin-flop phases, but positive in the polarised paramagnetic phase. This can be explained by a change in sign of the strain dependence of the exchange interaction between antiferro- and ferromagnetic configurations, as expected, for example, for the RKKY mechanism.

In Figure 2 we have subtracted the background magnetostriction to focus on the quantum oscillations in the strain using fourth order polynomial fits for the low- and high-field data that are constrained to meet at H_c. Below H_c, only one quantum oscillation frequency can be resolved. However, above H_c there is a clear change in the periodicity of the oscillations that cannot be well fitted by a single frequency.

Bandstructure calculations predict the presence of several small ellipsoidal Fermi surface sheets, with electron pockets at the X-point and hole pockets at the Γ-point[5, 6]. The exact number and size of these pockets is somewhat sensitive to the position of the Fermi level. A calculation in the ferromagnetic state [9] shows that strong spin-splitting is expected in all sheets. The expected doubling of the unit cell size to reflect the periodicity of magnetic order in the AFM state also has consequences for the electronic structure [10]. The pockets at the Γ-point are so small, however, that they are not expected to be changed by the unfolding of the bands.

The temperature dependence of the amplitude of the observed frequencies is reasonably consistent with the Lifshitz-Kosevich formula [11], resulting in effective masses in the range 0.3–0.6 m_e, similar to those reported in Ref. [2]. We also estimate the Dingle temperature for the F_0 frequency to be (17 ± 3) K from the field dependence of the 2.1 K data for $H < H_c$. The Dingle temperatures for F_{1-4} are harder to extract as there are only a few periods.

There are two possible explanations for the change in the frequency spectrum across H_c. In the first scenario, the onset of AFM order causes a reconstruction of the Fermi surface, as has been observed in NdB_6 [10]. In the second, the Fermi surface remains unchanged through the AFM-PM transition but the frequency spectrum becomes more complex as a result of Zeeman splitting, back-projection effects (see following paragraph) and Dingle damping, which strongly damps higher frequencies at low fields.

Since the magnetisation increases linearly up to H_c [3], we would expect only smooth changes in the quantum oscillation spectrum as a function of magnetic field when the Fermi surface is spin-split. However, a smooth change in the Fermi surface can appear as an abrupt change in the quantum oscillation frequency due to the back projection effect [12], an effective Doppler shift of the observed frequency, F_{obs}, with respect to the true frequency, F_{true}, such that $F_{obs} = F_{true}(B) - B\frac{dF_{true}(B)}{dB}$. Hence, if F_{true} changes only linearly with B up to H_c, no change would be seen in F_{obs} until H_c is crossed.

4. Conclusion

We have measured the magnetostriction in GdSb up to 55 T with a new high resolution technique. The improved resolution allows us to resolve quantum oscillations in the magnetostriction both above and below the critical field. We observe a change in the quantum oscillation spectrum coincident with the switching of the sense of the magnetoelastic coupling to the local moment system. This may either be a consequence of magnetic reconstruction of the Fermi surface, or of Zeeman splitting largely hidden by the subtleties of quantum oscillation measurements. It is difficult, given the signal-to-noise ratio and small number of oscillations involved, to differentiate between these two scenarios; higher resolution quantum oscillation measurements are required.

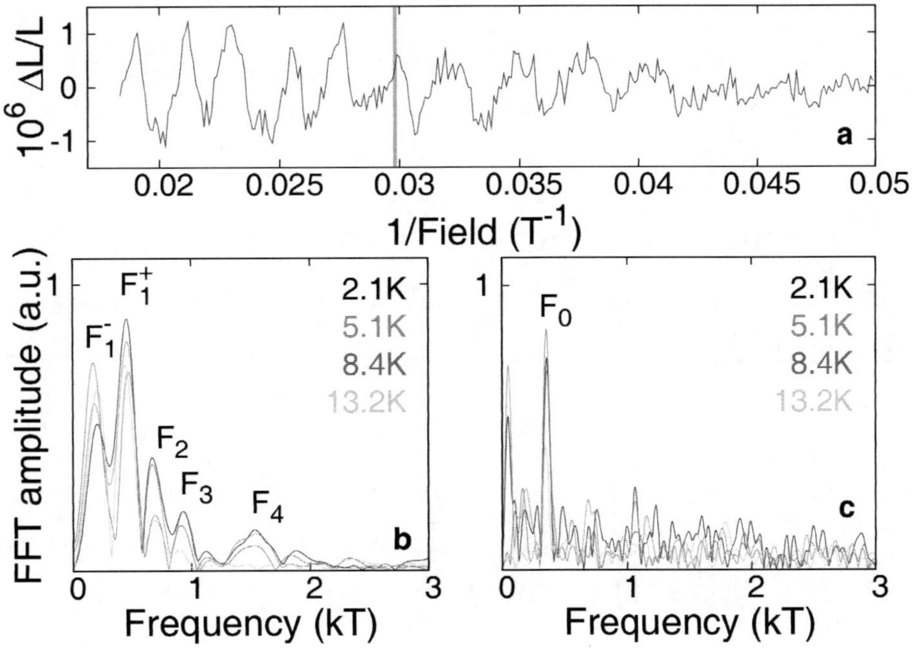

Figure 2. *a)* The oscillatory magnetostriction of GdSb at 2.1 K as a function of inverse field. The saturation field H_s is indicated by the vertical grey line. *b)* Fourier transform of the data for the fixed field range 34–54 T for several temperatures for several temperatures using a Hann window to eliminate sidebands. Several frequencies F_{1-4} are present. F_1^- is likely to be an artifact as its amplitude is highly dependent on the order of polynomial background subtraction that is used. *c)* Fourier transforms for the field range between 15 T and $H_c(T)$. Only one frequency is clearly present, $F_0 = 360 \pm 20$ T. The change in frequency spectrum on crossing H_c is discussed in the text.

Acknowledgments
We thank Andreas Kreyssig for providing us with the GdSb single crystal. This study was partly supported by EuroMagNetII under EU contract 228043 and by the MPG Research Initiative "Materials Science and Condensed Matter Research at the Hochfeld-Magnetlabor Dresden."

References

[1] Sugawara K, Huang C Y, Huber D L 1977 *J. Low Temp. Phys.* **26** 525.
[2] Yoshida M, Koyama K, Shirakawa M, Ochiai A, Watanabe K, Motokawa M 2003 *J. Phys. Soc. Japan* **72** 1763.
[3] Li D X, Haga Y, Shida H, Suzuki T, Kwon Y S, Kida G 1997 *J. Phys.: Cond.* **9**, 10777.
[4] Duan C-G, Sabiryanov R F, Liu J, Mei W N, Dowben P A, Hardy J R 2005 *Phys. Rev. Lett.* **94** 237201.
[5] Hasegawa A, Yanase A 1977 *J. Phys. Soc. Jpn* **42** 492.
[6] Larson P, Lambrecht W R L 2006 *Phys. Rev. B* **74** 085108.
[7] Doerr M *et al* 2006 *J. Phys.: Conf. Series* **51** 561.
[8] Daou R, Weickert F, Nicklas M, Steglich F, Haase A, Doerr M 2010 *Rev. Sci. Instrum.* **81** 033909.
[9] Duan C G et al 2007 *J. Phys.: Condens. Matter* **19** 315220.
[10] R G Goodrich, N Harrison, Z Fisk 2006 *Phys. Rev. Lett.* **97** 146404.
[11] Chandrasekhar B S and Fawcett E 1971 *Adv. Phys.* **20** 775.
[12] J M van Ruitenbeek *et al* 1982 *J. Phys. F: Met. Phys.* **12** 2919

Journal of Physics: Conference Series **273** (2011) 012112 doi:10.1088/1742-6596/273/1/012112

Electrical transport properties of ternary phosphides RRu_2P_2 (R = La, Ce, Pr and Eu) with $ThCr_2Si_2$ type crystal structure

T Fujiwara[1], K Kanto[1], K Matsubayashi[2], Y Uwatoko[2] and T Shigeoka[1]

[1] Graduate School of Science and Engineering, Yamaguchi University, Yamaguchi 753-8512, Japan.

[2] Institute for Solid State Physics, University of Tokyo, Kashiwa, Chiba 277-8581, Japan.

E-mail: tetsuyaf@yamaguchi-u.ac.jp

Abstract. It was found the growth method of single crystals of ternary lanthanide transition metal phosphide RRu_2P_2 (R = La, Ce, Pr and Eu) with enough large volume to measure the macroscopic physical properties, such as magnetization, electrical resistivity and specific heat. As a result, we succeeded in the measurements of electrical resistivity with a sufficiently good accuracy and reveal an experimental evidence of the superconducting transition in $LaRu_2P_2$ for the first time. On the basis of the results of electrical resistivity measurements we discuss that how P-P bonding state does affect on the electrical transport properties of RRu_2P_2 compounds.

1. Introduction

Hundreds of ternary intermetallic compounds RT_2X_2 (R = Rare earth metals, T = Transition metals, X = Si, Ge, P, As etc) with $ThCr_2Si_2$-type tetragonal structure have been synthesized since the middle of 1960s. Among them, silicides and germanides, RT_2Si_2 and RT_2Ge_2, have been studied most intensively and it is well known that they show various physical properties such as magnetism due to localized spins or/and itinerant electrons, multiple magnetic phase transition, valence fluctuation, valence transition, heavy fermion and superconductivity [1, 2, 3]. On the other hand, little is known about macroscopic physical properties, such as magnetization, electrical resistivity and specific heat, for the lanthanide-transition metal-phosphides RT_2P_2, except some compounds [4, 5]. Physical properties of these compounds have been mainly studied from only the points of view of structural properties by powder X-ray diffraction experiments or microscopic ones by either Mössbauer spectroscopies or neutron diffraction studies so far [6, 7, 8]. According to powder X-ray diffraction studies, RT_2P_2 compounds are classified broadly two isostructural phases characterized by two extremely different P-P interatomic distances along the [001] direction, d_{P-P}, within the RT_2P_2 series: In ones there is a large P-P distance ($d_{P-P} \geq 2.8$ Å) with a non-bonding P-P state; in others it is much shorter ($d_{P-P} \sim 2.3$ Å) forming a P-P single bond state. Band-structure calculations on some RT_2P_2 compounds indicate that there exist significant difference in the electronic structures of the compounds between these P-P bonding states [9]. Therefore, it can be sufficiently expected that the isostructural phase transition bring about a novel and interesting physical phenomena

Published under licence by IOP Publishing Ltd

accompanied with a drastic change of the physical properties. However, as mentioned above, systematic investigations of the physical properties in RT_2P_2 compounds have not been carried out yet because of a difficulty of producing single phase bulk sample. Recently, we have succeeded in growth of single crystals of some RT_2P_2 compounds. In this paper, we will reveal the electrical transport properties of RRu_2P_2 (R = La, Ce, Pr and Eu).

2. Experimental procedure

Single crystals of RT_2P_2 (R = La, Pr, Ce and Eu) were grown out a Sn flux. In the case of La, Ce and Pr compounds, RRu_x binary alloys with Ru contents x of 0.54 ~ 0.70 were prepared by arc-melting under Ar atmosphere as precursors. Then, appropriate amount of RRu_x, P and Sn were put inside a Ta crucible and those were sealed into an evacuated quartz ampoule. In the case of Eu compound, appropriate amount of Eu, Ru, P and Sn were mixed in a quartz crucible sealed into the evacuated quartz ampoule. Subsequently, the ampoules were heated up to 1100 deg C and kept for 5 hours, then, those were gradually cooled down to 700 deg C with cooling rate of 2 deg C per hour. Finally, molten flux was separated by centrifugation. Typical volume of single crystals of Eu and other compounds obtained in this work are $2 \times 2 \times 0.5$ mm^3 and $0.7 \times 0.7 \times 0.1$ mm^3, respectively. This is enough large to measure the electrical resistivity. In the latter case, the largest crystals sized up to $\sim 1 \times \sim 1 \times 0.2$ mm^3. The chemical composition of obtained crystals was verified to be R : Ru : P = 1 : 2 : 2 by EPMA using energy dispersive spectrometer. Electrical resistivity measurements were performed usign a standard a.c. four-probe method with excitation currents along the [100] direction of tetragonal structure.

3. Results and discussion

Figure 1 and 2 show the temeprature dependences of the electrical resistivities along the [100] direction in $CeRu_2P_2$ and $PrRu_2P_2$, respectively. $CeRu_2P_2$ does not exhibit any anomaly due to magnetic ordering down to 0.36 K. A broad maximum appears around $T \sim 180$ K in the $\rho - T$ curve, indicating that Ce ions are valence fluctuated. On the other hand, we can see an anomaly at 25 K in the $\rho - T$ curve of $PrRu_2P_2$, which correspond to the magnetic ordering temperature. Recent studies of RRu_2P_2 compounds by Mizusaki *et al* [5], $PrRu_2P_2$ shows an antiferromagnetism below the Néel temperature, $T_N = 25$ K. The behavior of the single crystal $PrRu_2P_2$ obtained in this work is in good agreement with that observed from the magnetic susceptibility measurements by them [5]. However, a residual resistance ratio estimated in this work is much larger than the value from their work. It seems that $PrRu_2P_2$ sigle crystal obtained in this work is of high quality.

Figure 3 exhibits the resistivity parallel to the [100] direction as a function of temperature in $EuRu_2P_2$. An anomaly is clearly recognized at 29 K. According to Jeitschko *et al*'s studies [5], $EuRu_2P_2$ has relatively larger lattice spacing than that expected from the lanthanide contraction. This suggests that Eu ions are in the divalent state, Eu^{2+}. In consequence, the anomaly at 29 K must be attributed to the magnetic ordering of magnetic moments in Eu^{2+} ions. The magnetization measurements and Mössbauer spectroscopy or X-ray absorption spectroscopy measurements of this compound are required to clarify the origin of this anomaly.

Figure 4 gives the resistivity parallel to the [100] direction in $LaRu_2P_2$ compound as a function of temperature and the inset shows that at temperature between 2 and 6 K. As clearly seen from the figure, a superconducting transition occurs at low temperature. Although Jeitschko *et al* [7] describes that the $LaRu_2P_2$ shows the superconducting transition at $T_c = 4.1$ K, no evidence has been released so far. Accordingly, we can notice that this result is the first published experimental evidence. Superconducting critical temperatures, T_c, were determined to be 4.4 K by estimeting the onset temperature of the transition. The properties of the superconductivity of $LaRu_2P_2$ were investigated by using few single crystals. The experimental results of AC

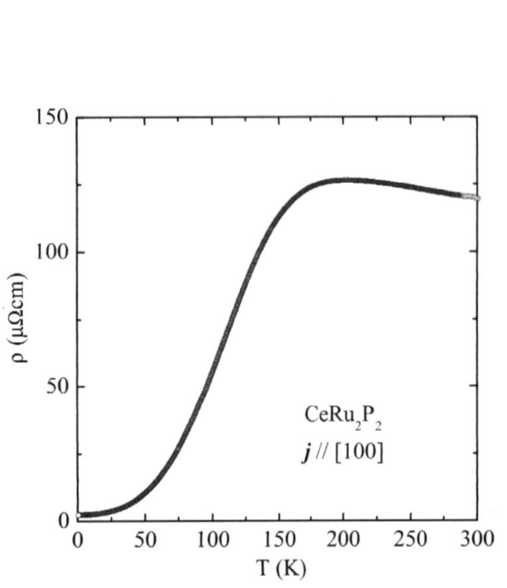

Figure 1. Temperature dependence of electrical resistivity along the [100] direction in $CeRu_2P_2$.

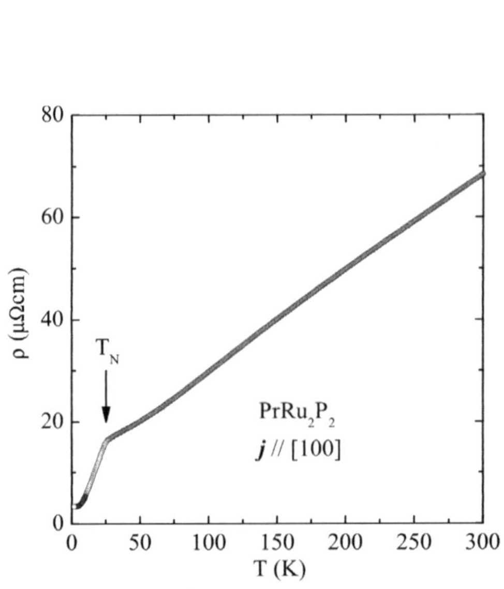

Figure 2. Temperature dependence of electrical resistivity along the [100] direction in $PrRu_2P_2$.

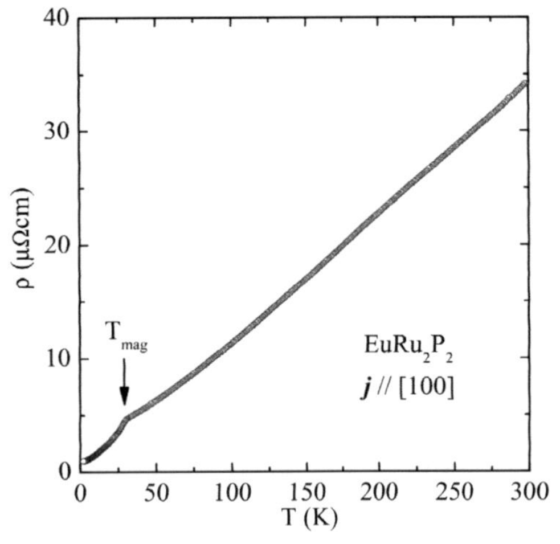

Figure 3. Temperature dependence of electrical resistivity along the [100] direction of $EuRu_2P_2$.

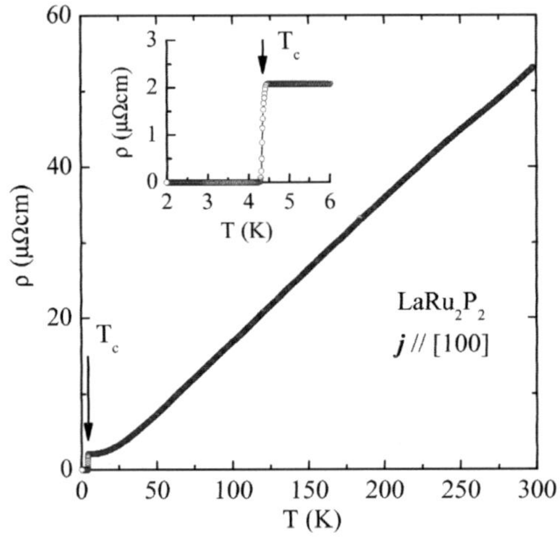

Figure 4. Temperature dependence of electrical resistivity along the [100] direction in $LaRu_2P_2$.

susceptibilities, DC magnetization and specific heat measurements will be published elsewhere in very near future.

The overall temperature dependences of the resistivities show metallic except that observed above $\sim$ 200 K in $CeRu_2P_2$. We notice that the resistivities of RRu_2P_2 compounds are getting smaller and smaller in the order of Ce, Pr, La and Eu compounds. On the contrary, the P-P interatomic distances along [001] direction d_{P-P} of RRu_2P_2 compounds are getting larger and larger in the same order [7]. In addition, d_{P-P} of $EuRu_2P_2$ is above 3 Å, so that the P-P single bond is not formed in this compounds [9]. Hence, it would be appear that the P-P single bond is not formed in $LaRu_2P_2$ as well, because the lattice spacing of the $LaRu_2P_2$ is as sufficiently large as that of Eu compound [7]. Compared to in the P-P non-bonding state, in the P-P single bond state, P $2p$-electrons are additionally doped to the system. The increase of the number of doped electrons may bring about the significant reduction of the electrical conductivity of the ternary lanthanide transition metal phosphide RT_2P_2. In the $CeRu_2P_2$ compound, due to that the valence of Ce ions are fluctuating, the electrons is supplied to the system from the Ce ions more additionally than in the Pr compound. As a consequence, the electrical resistivity in the Ce compound might be enhanced remarkably. Indeed, this hypothesis is not inconsistent with that both $LaRu_2P_2$ and $EuRu_2P_2$ are good conductor comparison with the Ce and Pr compounds. Moreover, this also yields an understanding for the fact that the resistivity of the $EuRu_2P_2$ further reduces compared to that of La compound. In short, the divalent Eu ions tend to supply less electrons to the system than R^{3+} ions. Therefore, we can conclude that both the P-P bonding state and the valence state of the R ions affect on the transport properties as well as the magnetic characteristics of ternary lanthanide transition metal phosphides RT_2P_2 [11].

References

[1] Szytula A and Leciejewicz J 1989 *Handbook on the Physics and Chemistry of Rare Earths* vol 12, ed K A Gschneidner Jr and L Erwin (Amsterdam: Elsevier) p 133
[2] Steglich F, Aarts J, Bredl C D, Lieke W, Meschede D, Franz W and Schäfer 1979 *Phys. Rev. Lett.* **43** 1892
[3] Sampathkumaran E V, Gupta L C, Vijayaraghavan R, Gopalakrishnan K V, Pillay R G and Devare H G 1981 *J. Phys.* C **14** L237
[4] Reehuis R and Jeitschko W 1990 *J. Phys. Chem. Solids* **51** 961
[5] Mizusaki S, Adachi Y, Ohnishi T, Taniguchi T, Natata Y, Ozawa T C, Noko Y and Samata H 2009 *J. Alloys. Comp* **468** 28
[6] Marchand R and Jeitschko W 1978 *J. Solid. State. Chem* **24** 351
[7] Jeitschko W, Glaum R and Boonk L 1987 *J. Solid. State. Chem* **69** 93
[8] Mösen E, Mosel B D, Müller-Warmuth W, Reehuis M and Jeithschko W 1988 *J. Phys. Chem. Solids* **49** 785
[9] Hffmann R and Zheng C 1985 *J. Phys. Chem* **89** 4175
[10] Huhnt C, Schlabitz W, Wurth A, Mewis A and Reehuis M 1998 *Physica* B **252** 44
[11] Chefki M, Abd-Elmeguid M M, Micklitz H, Huhnt C and Schlabitz W 1998 *Phys. Rev. Lett* **80** 802

International Conference on Strongly Correlated Electron Systems (SCES 2010) IOP Publishing
Journal of Physics: Conference Series **273** (2011) 012113 doi:10.1088/1742-6596/273/1/012113

SC-to-AFM transition in $CeCo(In_{1-x}Cd_x)_5$: de Haas-van Alphen Measurements

R. G. Goodrich[1], C. Capan[2], A. D. Bianchi[2], Z. Fisk[2], J. F. DiTusa[3] I. Vekhter[3], D. P. Young[3], L. Balicas[4], Y-J Yo[4], T. Murphy[4] J. Y. Cho[5], and J. Y. Chan[5]

[1.] Department of Physics, George Washington University, Washington, DC 20052

[2.] Department of Physics and Astronomy, University of California Irvine, Irvine, CA 95616

[3.] Department of Physics and Astronomy, Louisiana State University, Baton Rouge, LA 70803

[4.] National High Magnetic Field Laboratory, Florida State University, Tallahassee, FL 32310

[5.] Department of Chemistry, Louisiana State University, Baton Rouge, LA 70803

E-mail: phgood@gwu.edu

Abstract. The results of de-Haas-van-Alphen (dHvA) measurements on Cd doped $CeCoIn_5$ in the high magnetic field paramagnetic phase are reported. The replacement of trivalent In by divalent Cd is known to induce an antiferromagnetic order coexisting with superconductivity in this heavy fermion superconductor. We find a small but systematic change in the dHvA frequencies with Cd doping, reflecting the chemical potential shift due to the removal of conduction electrons. The frequencies and effective masses are close to those found in the undoped compound. We observe no abrupt change in the electron Fermi surface (FS) volume in the high field paramagnetic phase for $x > x_c$ corresponding to the onset of antiferromagnetic ordering at zero magnetic field in $CeCo(In_{1-x}Cd_x)_5$. Our results show that no significant change of the Fermi surface volume occurs to cause the AFM onset inside the SC state in $CeCoIn_5$ with Cd doping. Only slight changes causing the Band 15 electron FS to be more cylindrical are observed making nesting more likely.

1. Introduction

$CeCoIn_5$ is a heavy fermion superconductor in which Cooper pairs are formed out of a non-Fermi liquid metallic state. The ability to grow sizable, high quality, single crystals enables detailed investigations of the effect of chemical doping in this and other 115 compounds. While Sn substitution for In has been found to suppress T_c without revealing any incipient magnetism[1], Cd replacing In induces an antiferromagnetic ground state in $CeCoIn_5$[2]. Because Sn and Cd are next to In in the periodic table, their substitutions for In result in electron and hole doping, respectively. The effect of Cd doping is quite unusual in that it takes a very small amount of Cd in $CeCoIn_5$ to induce a change in the ground state from paramagnetic to antiferromagnetic (AFM) that can be reversed with the application of pressure[2]. How Cd induces long range AFM order with a large ordered magnetic moment[3] ($0.7\mu B$ per Ce) in $CeCoIn_5$ remains an open question. One possible mechanism is the formation of antiferromagnetic droplets at the Cd sites, as was inferred from NMR measurements[3]. Long range AFM order occurs once the density of such droplets reaches a percolation threshold for nearest neighbor• interactions. However, the density of Cd necessary to induce ordering is well below the percolation threshold. Thus the ordering at such a small Cd concentration requires interactions having a longer interaction range. Since the ordered moments are likely local moments on the Ce sites,

Published under licence by IOP Publishing Ltd

one way to account for the reversibility of Cd doping with pressure is for the change in carrier density and disorder caused by Cd substitution to affect the valence of nearby Ce atoms. Application of pressure to metals with localized f-electrons (corresponding to Ce^{3+}) tends to increase their hybridization with the conduction band and promote a mixed valence regime. Another possibility is that the AFM state is due to a Fermi Surface (FS) instability, which is the well known explanation in the case of elemental Cr. Recent neutron scattering results on Cd doped $CeCoIn_5$[4] have demonstrated that the AFM ordering has a wavevector, Q, of (1/2,1/2,1/2) suggesting that if the AFM is nesting-driven, the nesting wavevector is commensurate with the lattice. This is a plausible, but unusual, situation that occurs, for example, in Mn doped Cr[5]. One possibility in the case of Cd doping is that in the doping region intermediate between local and itinerant magnetism the magnetic ordering is driven by the unscreened component of the spin. The improved near-nesting with Cd doping could lower the energy necessary to open the gap for itinerant electrons at the magnetic Brillouin Zone boundary and this would enabling the long-range order to appear. One of the most surprising features of this ordered state is that the ordering wavevector[6] coincides with the wavevector at which an inelastic neutron scattering resonance[7] is observed in the superconducting state of pure $CeCoIn_5$. The origin of this resonance has been attributed to AFM magnons[8]. A similar AFM state can be induced with Rh substitution for Co in $CeCoIn_5$ for Rh concentrations greater than ≅ 25%. Here, the AFM wavevector is identical to that found for Cd doped $CeCoIn_5$, which coexists with superconductivity for Rh concentrations of less than ≅ 60%. For larger Rh concentrations the AFM wavevector becomes (1/2,1/2,~ 0.3) and superconductivity is suppressed. The change in wavevector and the loss of superconductivity at Rh concentrations above 60% suggest that the AFM states in Rh (for $x_{Rh} < 0.6$) and Cd doped $CeCoIn_5$ have a different mechanism from the incommensurate antiferromagnetic order found in $CeRhIn_5$. In order to understand more fully the superconducting state in $CeCoIn_5$ and the relevance of the coexisting AFM, the character of the AFM state will require further investigations. In this paper, we report one such investigation by specifically exploring the evolution of the FS in the high magnetic field paramagnetic phase of $CeCoIn_5$ as a function of Cd doping via de Haas van Alphen (dHvA) measurements.

II. Experimental Details

The single crystals of $CeCo(In_{1-x}Cd_x)_5$ used in these measurements were grown from high purity starting materials, as described elsewhere[9]. The values of x we give are the nominal flux concentrations and it has been found that the actual Cd concentration is consistently 10% of the nominal flux concentration across the entire range of flux compositions[9]. The lattice parameters were determined by using both powder and single crystal X-ray diffraction and the lattice constants were found to change by < 0.1 % across the doping range indicating that FS volume changes due to changes in the BZ dimensions are insignificant. Extensive measurements previously have shown that the Cd atoms are distributed uniformly throughout the samples at each concentration[10].

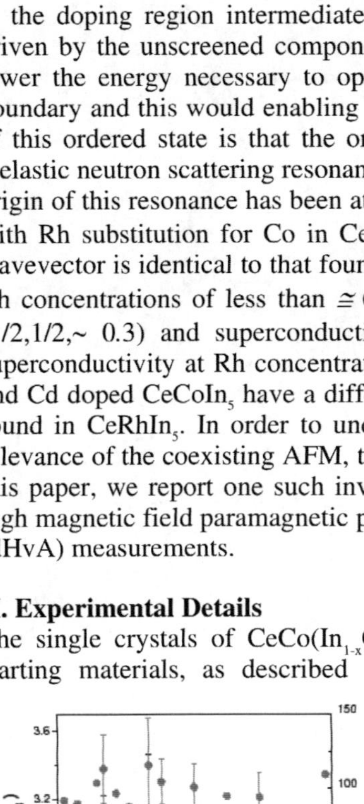

Fig. 1 Curie-Weiss moment μ_{eff} (•) and Curie-Weiss Temperature (°) vs. nominal Cd concentration

The evolution of the FS of the same crystals used for the single crystal X-ray diffraction measurements was investigated via the dHvA oscillations measured using a torque magnetometer[11]. Single crystals were mounted on a Cu-Be cantilever in a dilution refrigerator equipped with a sample rotator. The dHvA measurements were made at the National High Magnetic Field Laboratory in magnetic fields of up to 45 T, and for temperatures down to 0.05 K. The susceptibility of single crystals of $CeCo(In_{1-x}Cd_x)_5$ was measured at H = 0.1 T applied perpendicular to [001] axis for temperatures ranging from 1.8 to 400 K using a Superconducting Quantum Interference Device (SQUID) magnetometer. For the dHvA data analysis it is assumed that H • B and no demagnetizing factor corrections were made, since we estimate from the susceptibility measurements the magnetization of $CeCo(In_{1-x}Cd_x)_5$ to be ~0.4% of the applied field. The in-plane susceptibility (measured down to 1.8 K) is extrapolated with a power law fit to $\chi_\perp = 0.0143$ emu/mol to T = 50 mK in 5% Cd, which corresponds to $\chi_{\|}\psi = 0.0286$ emu/mol for H || [001] (with a magnetic anisotropy of 2) and to a volume magnetization of 4 M = 0.0126 T at an applied field of 35 T.

III. Experimental Results

Figure 2 shows the magnetization as a function of applied field for the 5% and 7.5% Cd samples. In the 7.5% sample both the superconducting transition H_{c2} and the AFM to paramagnetic transition, H_M, can be seen, while in the 5% sample only H_{c2} is observed. All of the dHvA signals reported here are from the paramagnetic phase at fields greater than either H_{C2} or H_M. In Fig. 3 the dHvA signal in the 5% sample at base temperature is shown. Fourier transforms (FTs) of data such as this were performed to obtain the dHvA frequencies (proportional to 1/H) in all of the samples. The angular dependence of the frequencies is shown in Figs. 4 and 5 for the 2.5% and 7% Cd samples. The three frequencies, F3, F4, and F5 arise from the Band 15 undulating cylindrical shaped electron sheets located at the corners of the BZ, see Fig. 7.

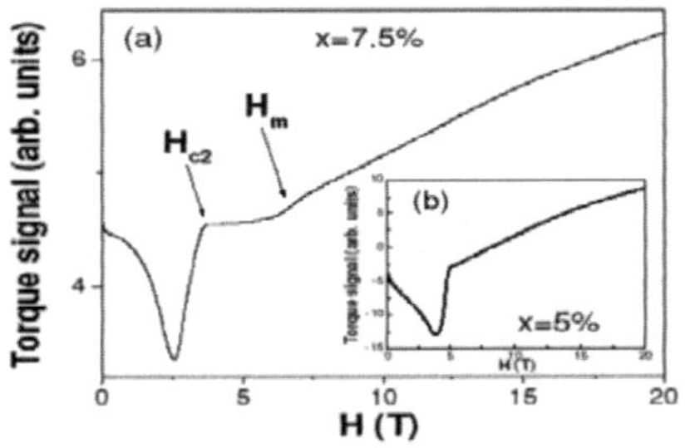

Fig. 2 Torque signal,

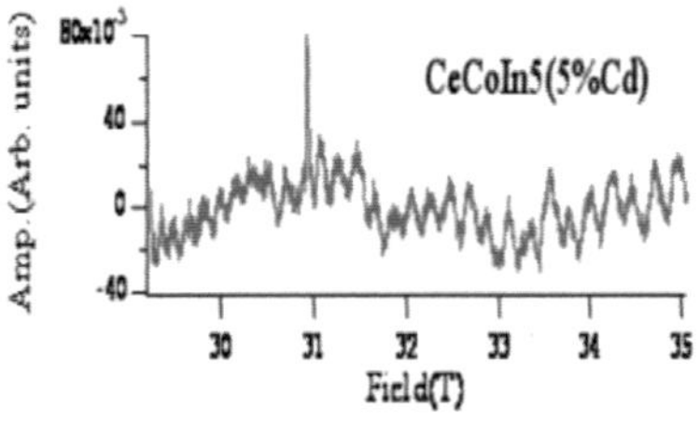

Fig. 3 Example of raw data

As can be seen from the data from the two Cd concentrations, the Band 15 surface becomes more cylindrical as the amount of Cd is increased causing the F3, F4, and F5 frequencies to merge into one frequency. The frequencies for H along [001] are given in the Table below.

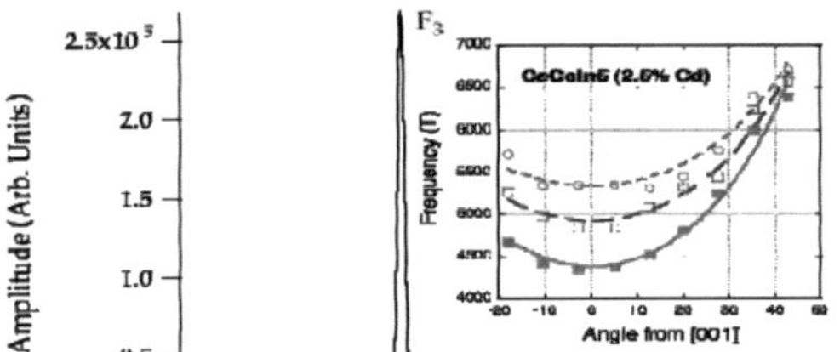

Fig. 4 FFT and angular dependence of the F_3, F_4 and F_5 frequencies in the 2.5% Cd sample.

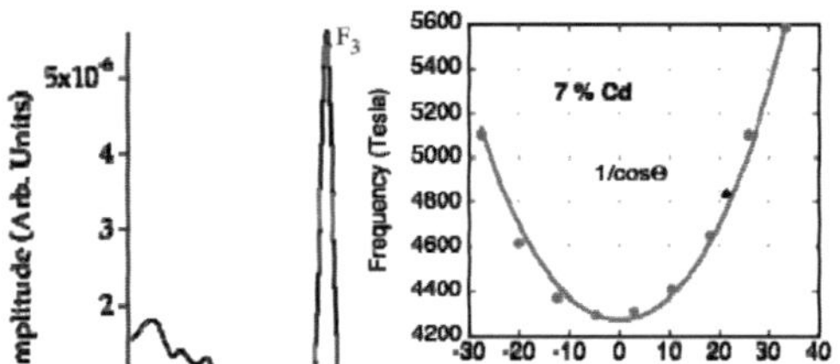

Fig. 5 FFT and angular dependence of the F_3, Frequency in the 7% sample.

III. Conclusions

We find a small but systematic change in the dHvA frequencies with Cd doping in $CeCo(In_{1-x}Cd_x)_5$ reflecting the chemical potential shift due to the removal of conduction electrons. The frequencies and effective masses of the Band 15 electron portions of the FS are close to those found in the undoped compound (see Fig.7)[12]. We note that in the doped compounds we only observe signals from the Band 15 electron sheet. We observe no abrupt change in the Band 15 electron Fermi surface volume in the high field paramagnetic state for $x > x_c$ corresponding to the onset of antiferromagnetic ordering that occurs at H = 0 in $CeCo(In_{1-x}Cd_x)_5$. Our results show that no significant change of the Fermi surface volume in the paramagnetic state occurs to cause the AFM onset inside the SC state in $CeCoIn_5$ with Cd doping. Only slight changes causing the Band 15 electron FS to be more cylindrical (note that a 1/cosθ angular dependence indicates a cylindrical surface) are observed making nesting more likely. In another paper [13] we show that the changes that occur with Cd doping in $LaCoIn_5$ and $CeCoIn_5$ compounds suggests that Cd does not induce any valence fluctuation on neighboring Ce atoms, as originally assumed in order to explain the reversible tuning. The high field Fermi surface in antiferromagnetic $CeCo(In_{1-x}Cd_x)_5$ is large, i. e. of a size similar to the FS of $CeCoIn_5$, in contrast to the small, f electron localized, FS found in $CeRhIn_5$, suggesting a separate mechanism for the antiferromagnetism. It follows that the commensurate AFM order in $CeCo(In_{1-x}Cd_x)_5$ could be due to an itinerant, SDW-type, mechanism which relies on FS nesting. This would make the current situation similar to the example of SDW ordering that is commensurate with the underlying lattice in the elemental antiferromagnet Cr which, although incommensurate when pure, evolves to a commensurate state with small amounts of Mn doping[5].

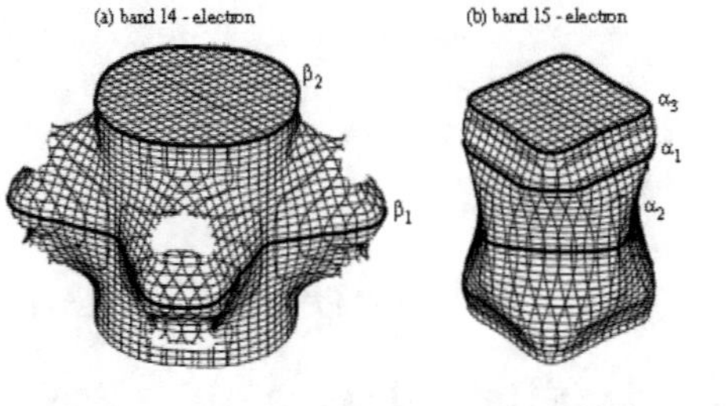

Fig. 6 Electron Fermi surfaces in $CeCoIn_5$. The α_1, α_2, and α_3 orbits are our F3, F4, and F5 frequencies.

Table of dHvA frequencies and m*

	D. Hall et. al. PRB, **64,** 212508 (2001)		Cd 2.5 %		Cd 4.8 %		Cd 7.0 %	
	F(T)	m*	*F*(T)	m*	*F*(T)	m*	*F*(T)	m*
F_1	267		215	13.8±3.0	350			
F_2	411	8.7±2.2						
F_3	4566	11.2±0.2	4390	8.5±0.8	4347	12.6±1.2	4278	9.4±0.7
F_4	5161	20.3±0.7	4928	14.9±1.4				
F_5	5401		5353	16.7±2.0				
F_6	7535		8651	15.2±--				

References

[1] Bauer E D, Capan C, Ronning F, et. al., (2005) *Phys. Rev. Lett.* **94** 047001.
[2] Pham L D, Park T, Maquilon S, Thompson J D and Fisk Z, *Phys. Rev. Lett.* (2006) **97** 056404.
[3] Urbano R R,. Young B-L, Curro N J, et. al., (2007) *Phys. Rev. Lett.* **99**, 146402.
[4] Nicklas M, Stockert O, Park T, et. al., (2007) *Phys. Rev. B* **76**, 052401.
[5] Fawcett E E, Alberts H L, Galkin V Y, et. al., (1994), *Rev. Mod. Phys.* **66**, 25.
[6] Nicklas M, Stockert O, Park T, et. al., *Phys. Rev. B* **76**, 052401 (2007).
[7] Singh S, Capan C, Nicklas M, et. al., (2007) *Phys. Rev. Lett.* **98** 057001.
[8] Chubukov A V and Gor'kov L P, (2008) *Phys. Rev. Lett.* **101**, 147004.
[9] Petrovic C, Pagliuso P G, Hundley M F, et. al., (2001) *J. Phys. : Condens. Matter* **13**, L337.
[10] Kenzelmann M, Strassle T, Niedermayer C, et al., (2008) *Science* **321** 1652, and Ref. 2.
[11] Goodrich R G, Harrison N, Vuillemin J J, et. al., (1998) *Phys. Rev. B* **58** 14896.
[12] Settai R, Shishido H, Ikeda S, Murakawa Y, et. al., (2001) *J. Phys.: Condens. Matter* **13**, L627.
[13] Capan C, et. al., arXive.1003.272v1 (2010).

Journal of Physics: Conference Series **273** (2011) 012114 doi:10.1088/1742-6596/273/1/012114

Magnetic and dielectric properties of multiferroic $Tb_{0.5}Eu_{0.5}MnO_3$[1]

H C Hsu[a], C D Yang[a], W Y Tseng[a], H C Ku[b] and Y Y Hsu[a2]

[a] Department of Physics, National Taiwan Normal University, Taipei 11677, Taiwan
[b] Department of Physics, National Tsing Hua University, Hsinchu 30013, Taiwan

E-mail: yyhsu@phy.ntnu.edu.tw

Abstract. The structural, magnetic and dielectric properties of multiferroic compound $Tb_{0.5}Eu_{0.5}MnO_3$ are reported near the boundary of the multiferroic phase of the Eu-doped $Tb_{1-x}Eu_xMnO_3$ system. The Mn^{3+} moments order in an incommensurate antiferromagnetic manner below $T_N = 50$ K, followed by a spin canting-like response around 25 K. Below T_N, extra hysteresis loops were observed for applied fields B_a around 1-2 T, with decreasing magnetic moment for increasing magnetic fields. The dielectric constant ε' showed a peak at 28 K, which was attributed to the ferroelectric transition in analogy with the parent compound $TbMnO_3$. Structural analysis indicates the Mn-O2-Mn bond angle within the ab-plane is 145.9° suggesting that there exists a critical angle in the perovskite manganate multiferroic systems.

1. Introduction

Since the discovery of multiferroic behavior in small radius rare-earth $RMnO_3$ compounds, the origin and detailed behaviors of these systems have attracted much interest [1]. Although the ferroelectric transition temperature T_C in these compounds are not high enough for application purposes, the strong electromagnetic coupling and the novel magnetic interaction origin of the ferroelectricity make a new era for research in strongly correlated systems[2]. In addition to the investigations of pure rare earth $RMnO_3$ systems, the substituted compounds $Tb_{1-x}Gd_xMnO_3$ show a systematic variation from para-electric canted-antiferromagnetism (PE-CAF) to coexistence of ferroelectricity and antiferromagnetism, know as multiferroic behavior, as the system goes from the Gd to the Tb rich side and the ab-plane Mn-O2-Mn angle decreases [3]. The ferroelectric transition starts to be observed in this Tb-Gd system close to the Gd end at a Mn-O2-Mn angle ~146°, or an averaged rare earth ionic radius $\langle r_R \rangle \sim r_{Gd}$. However, the complicated spontaneous polarization re-orientation from a- to c-axis with increasing Tb concentration suggests that the detailed mechanism could be much more complicated [4]. On the other hand, similar phase diagrams were reported for $Eu_{1-x}Y_xMnO_3$ with multiferroic behavior starting at a similar $\langle r_R \rangle$ value but with very different polarization re-orientation behavior. Moreover, the phase diagrams reported by different groups only agree qualitatively with each other [5,6], which reflects the complicated nature of the interplay between magnetic moments and electric polarizations.

[1] This work was supported by NSC of Taiwan under grand no. NSC-97-2112-M-003-001-MY3.
[2] To whom any correspondence should be addressed.

Published under licence by IOP Publishing Ltd

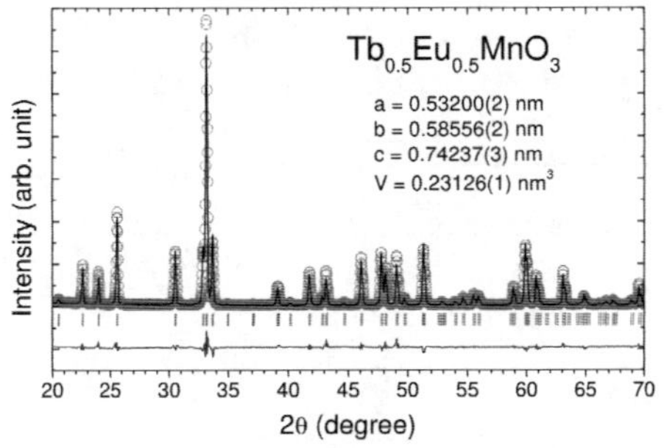

Figure 1. The x-ray diffraction data (open circles), Rietveld refinement and the deviation from data (black and blue lines respectively) of $Tb_{0.5}Eu_{0.5}MnO_3$. The vertical bars represent allowed reflections of the *Pbnm* structure.

Since the phase diagrams of Eu-Y and Tb-Gd systems only qualitatively agree with each other, new similar systems are always welcomed for further investigations and comparisons. To study the multiferroic mechanisms especially near the boundary of its appearance, the $Tb_{1-x}Eu_xMnO_3$ system was proposed and studied for its wider averaged rare earth radius $\langle r_{Tb-Eu} \rangle$, which can cover the averaged $\langle r_{Tb-Gd} \rangle$ value of the $Tb_{1-x}Gd_xMnO_3$ system, and avoids the possible drawbacks raised by too small Y^{3+} ionic size. Here, we report detailed investigations of lattice structure, magnetic susceptibility and dielectric measurements for $Tb_{0.5}Eu_{0.5}MnO_3$, which is near the multiferroic boundary of the Eu-doped system $Tb_{1-x}Eu_xMnO_3$.

2. Experimental

Polycrystalline $Tb_{0.5}Eu_{0.5}MnO_3$ samples were prepared by the solid state reaction method [7]. High-purity oxide ingredient powders were mixed in stoichiometric ratios then calcined at 900°C in air for 30 hours with several intermediate regrindings for homogeneity. After adding 3% wt of polyvinyl alcohol (PVA) to ensure high packing density, the pressed pellets were sintered at 1380°C in flowing Ar atmosphere for 24 hours followed by slow cooling to room temperature. The structure of the sintered compounds were analyzed by Rietveld refinement with a FullPROF program on powder X-ray diffraction patterns collected by a Rigaku Rotaflex 18-kW rotating anode diffractometer with graphite monochromatized Cu-K_α radiation in 2θ range of 20°-70° [8]. The magnetic measurements were carried out with a Quantum Design 7-T MPMS superconducting quantum interference device (SQUID) magnetometer from 5-300 K. The dielectric permittivity was measured in a parallel-plate capacitor geometry by using an impedance analyzer (Hewlett-Packard, 4194A) with a closed-cycle cryocooler from 13-300 K.

3. Results and discussions

The structural characterization of the sintered $Tb_{0.5}Eu_{0.5}MnO_3$ sample by powder x-ray diffraction is shown in Figure 1, which can be well indexed by an orthorhombic structure with space group *Pbnm* (No. 62) indicating high purity of the sample. The obtained lattice constants a = 0.53200(2) nm, b = 0.58556(2) nm, c = 0.74237(3) nm and unit cell volume V = 0.23126(1) nm^3 is in good agreement with the expected values extrapolated from the reported stoichiometric $TbMnO_3$ and $EuMnO_3$ compounds indicating high sample quality with no, or at least low, vacancies due to off-stoichiometry of oxygen content [7,9].

The refined atomic positions (atom sites of *Pbnm* is shown in the parenthesis) of orthorhombic $Tb_{0.5}Eu_{0.5}MnO_3$ are listed in Table 1. The deduced MnO_6 octahedra in the perovskite structure elongates along and rotates 13.6° about the c-axis then tilts 17.5° away from c-axis resulting a buckled Mn-O network in the ab-plane with an Mn-O2-Mn angle of 145.9°. The large deviation of the Mn-O2-Mn angle from ideal 180° thus frustrates the magnetic exchange interactions between the Mn moments, which order incommensurately below the Néel temperature T_N.

Table 1. Refined atomic positions (atom sites denoted in the parenthesis) of orthorhombic $Tb_{0.5}Eu_{0.5}MnO_3$ with space group *Pbnm* and lattice constants a = 0.53200(2) nm, b = 0.58556(2) nm, c = 0.74237(3) nm.

Atom	x	y	z	B_{iso}	occupation
Tb/Eu (4c)	-0.0176(2)	0.0795(2)	1/4	0.9(2)	0.5/0.5
Mn (4b)	1/2	0	0	0.5(3)	1
O1 (4c)	0.102(1)	0.462(1)	1/4	0.8(9)	1
O2 (8d)	0.715(1)	0.334(1)	0.0497(8)	0.3(7)	1
R_p = 9.18%, R_{wp} = 10.7%, R_{exp} = 2.96%, R_{Bragg} = 5.23%					

Figure 2a shows the zero-field-cooled (ZFC) and field-cooled (FC) molar magnetic susceptibility χ_m(T) and $1/\chi_m$ for $Tb_{0.5}Eu_{0.5}MnO_3$. At elevated temperatures, the susceptibility follows a Curie-Weiss (CW) law as clearly revealed by the linear $1/\chi_m$ with an effective paramagnetic moment $\mu_{eff} = 8.70\ \mu_B$, which is attributed mainly to paramagnetic Tb^{3+} and Mn^{3+} moments with minor contributions from Eu^{3+} low-lying multiplets excitations at these temperatures. Since the sample orders antiferromagnetically with an incommensurate magnetic moment arrangement below T_N, there is no clear peak or inflection point of the χ_m(T) curve at T_N. Since χ_m(T > 60 K) follows a Curie-Weiss law, the deviation point was used to determine the Néel temperature as in the literature [10]. The obtained T_N = 50 K is similar to the reported values for $EuMnO_3$ and $TbMnO_3$ and similar to doped systems (e.g., $Eu_{1-x}Y_xMnO_3$ and $Tb_{1-x}Gd_xMnO_3$) [4,5]. As the temperature is lowered further, ZFC and FC curves of χ_m(T) deviate at an irreversibility point T_{irr} = 38 K followed by canted-antiferromagnetic-like (CAF-like) behavior below T_{CAF} = 25 K, which is the peak for ZFC and the inflection point for FC curves. No trace of Tb^{3+} ordering was observed in the temperature range investigated.

The magnetic hysteresis loops $M(B_a)$ of $Tb_{0.5}Eu_{0.5}MnO_3$ in the temperature range 5-30 K are shown in Figure 2b. At T = 5 K, a saturation behavior of $M(B_a)$ is observed with a saturation moment ~1.9 μ_B per formula. This is far smaller than the expected saturation

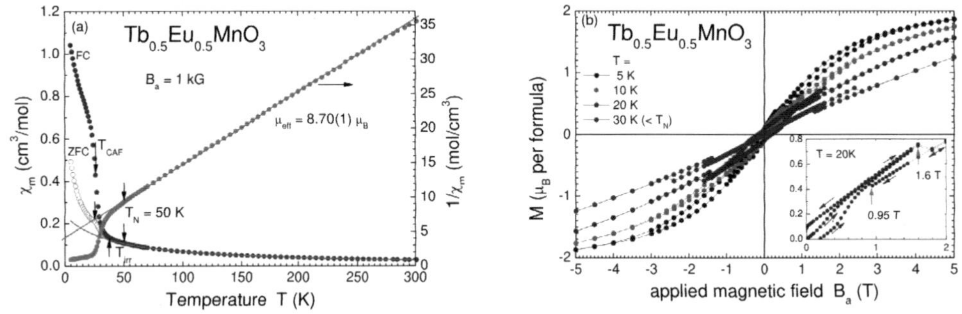

Figure 2. (a) Temperature dependence of the molar magnetic susceptibility χ_m(T) and its inverse, $1/\chi_m$, for $Tb_{0.5}Eu_{0.5}MnO_3$ with an applied field B_a = 1 kG. (b) The magnetic hysteresis $M(B_a)$ for temperature between 5 to 30 K. The inset is a closer look at the extra loop at T = 20 K. Thin arrows label the loop direction, where the start and end of the loop are indicated by vertical arrows.

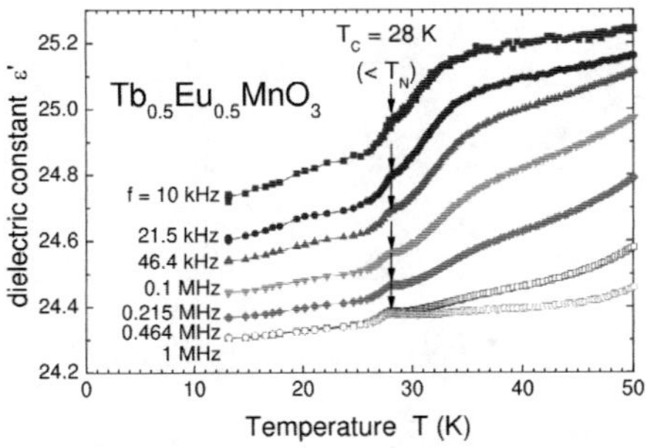

Figure 3. Temperature-dependence of the real part of the dielectric constant ε'(T) of $Tb_{0.5}Eu_{0.5}MnO_3$ at frequencies from 10 kHz to 1 MHz. The dielectric anomaly near T = 28 K is marked and attributed to the ferroelectric transition temperature T_C.

moment calculated from the ionic contributions, which suggests further transitions induced at higher applied field. In addition, meta-magnetic loops were observed at 1-2 T. The inset of Figure 2b is a closer look at the loop at T = 20 K. As B_a increases, unlike general meta-magnetic behavior, the magnetic moment M(B_a) experienced an abrupt decrease of ~0.1 μ_B around B_a = 1.6 T and an abrupt increase of ~0.06 μ_B with descending B_a at 0.95 T. Since the change of the magnetic moment tends to increase the magnetic potential energy -M·B_a, which is in contrast to general moment changes, it is believed to be accompanied by other transitions that lower the total energy even further. These extra loops were not reported previously probably due to the small change in moment and typical large applied field increment in this field regime. Since it is important to clarify the effect of inhomogeneous field of the SQUID magnetometer on these loops further investigations by other means are planed and in progress.

Figure 3 shows the low-temperature dependence of the real part dielectric constant ε'(T) of $Tb_{0.5}Eu_{0.5}MnO_3$ at frequencies of 10 kHz to 1 MHz. The observed ε'(T) is slightly frequency dependent, which is attributed to the capacitance contributed from the grain boundaries of the polycrystalline sample. A small peak in ε' can be observed at 28 K. Since the peak position is frequency-independent and can be systematically traced with different Eu doping to $TbMnO_3$, we attributed this to the ferroelectric (FE) transition temperature T_C [11].

Comparing the anomaly peak strength with $TbMnO_3$ and the similar doping systems, $Tb_{1-x}Gd_xMnO_3$ and $Eu_{1-x}Y_xMnO_3$, we conclude that $Tb_{0.5}Eu_{0.5}MnO_3$ is close to the multiferroic boundary. The observation of CAF-like behavior below $T_{CAF} < T_C$ and no second ε' peak suggest possible coexistence of the FE and CAF phases. On the other hand, the Mn-O2-Mn angle is very close to the reported value in Tb-Gd system, indicating that there is a critical angle for the appearance of multiferroic behavior in the perovskite manganate compounds.

References

[1] T Kimura, T Goto, H Shintanl, K Ishizaka, T Arima and Y Tokura 2003 *Nature* **426** 55.
[2] S W Cheong and M Mostovoy 2007 *Nature Mater.* **6** 13.
[3] T Goto, T Kimura, G Lawes, A P Ramirez and Y Tokura 2004 *Phys. Rev. Lett.* **92** 257201.
[4] T Goto, Y Yamasaki, H Watanabe, T Kimura and Y Tokura 2005 *Phys. Rev.* B **72** 220403.
[5] J Hemberger, F Schrettle, A Pimenov, P Lunkenheimer, V Yu Ivanov, A A Mukhin, A M Balbashov and A Loidl 2007 *Phys. Rev.* B **75** 035118.
[6] Y Yamasaki, S Miyasaka, T Goto, H Sagayama, T Arima and Y Tokura 2007 *Phys. Rev.* B **76** 184418.
[7] B Dabrowski, S Kolesnik, A Baszczuk, O Chmaissem, T Maxwell and J Mais 2005 *J. Solid State Chem.* **178** 629.
[8] J Rodrífuez-Carvajal 1993 *Physica* B **192** 55.
[9] J A Alonso, M J Martnez-Lope, M T Casais and M T Fernndez-Daz 2000 *Inorg. Chem.* **39** 23.
[10] J Blasco, C Ritter, J García, J M de Teresa, J Pérez-Cacho and M R Ibarra 2000 *Phys. Rev.* B **62** 5609-5618.
[11] H C Hsu, Y Y Hsu et al. in preparation.

Journal of Physics: Conference Series **273** (2011) 012115 doi:10.1088/1742-6596/273/1/012115

New intermetallic MIrP (M=Ti, Zr, Nb, Mo) and MgRuP compounds related with MoM'P (M'=Ni and Ru) superconductor

Hijiri Kito, Akira Iyo and Toshimi Wada

Superconducting Materials Group, Nanoelectronics Research Institute (NeRI) , National Institute of Advanced Industrial Science and Technology (AIST), Tsukuba Central 2, 1-1-1, Umezono, Tsukuba, Ibaraki, Japan, 305-8568

E-mail: h.kito@aist.go.jp

Abstract. Using a cubic-anvil high-pressure apparatus, ternary iridium phosphides MIrP (M=Ti, Zr, Nb, Mo) and MgRuP have been prepared by reaction of stoichiometric amounts of each metal and phosphide powders at around 2 Gpa and above 1523 K for the first time. The structure of these compounds prepared at high-pressure has been characterized by X-ray powder diffraction. Diffraction lines of these compounds are assigned by the index of the Co_2Si-type structure. The electrical resistivity and the d.c magnetic susceptibility of MIrP (M=Ti, Zr, Nb, Mo) have measured at low temperatures. Unfortunately, no superconducting transition for MIrP (M=Ti, Zr, Nb, Mo) and MgRuP are observed down to 2 K.

1. Introduction

Many ternary systems MRuP, where M represent 3*d*, 4*d* and 5*d* transition metals, have been studied crystallographically [1-5]. These compounds primarily exhibit hexagonal Fe_2P-type or orthorhombic Co_2Si (ε-TiNiSi)-type structures. For the hexagonal Fe_2P-type structure is built from neary rhombohedral $M_IRu_{II}P$ subcells stacked [5] to form columns containing M_I-Ru_{II} zig-zag chains. M_I and M_{II} represent the crystallographically in equivalent sites for the metal atoms coordinated by a tetrahedron and tetragonal pyramid of P anions, respectively, with M_I being typically the smaller and more electronegative of metal atoms [6]. Especially, the ZrRuP and HfRuP systems show superconductivity and superconducting transition temperature (T_c) of ternary phosphides with the hexagonal Fe_2P-type structure is considerably high: T_c is about 13K for ZrRuP and HfRuP [1, 8]. Several phosphide superconductors with T_c above 15K have also been found; MoNiP with the Fe_2P-type structure [9] and MoRuP with the Co_2Si-type structure have T_c of about 15.5K [9, 10] and other interesting superconductors of phosphides have also been found; T_c of $LaRu_4P_{12}$ with skutterudite ($CoAs_3$-type) structure is about 7K [11, 12].

Here, if Ru atom in MoRuP can be substituted by Ir atom or Mo atom in MoRuP can be substituted by Mg atoms, superconductivity also may be expected. We have searched in ternary phosphide; M–Ir–P system (M=Ti, Zr, Nb and Mo) and MgRuP have prepared at high temperature and high pressures for the first time. For Mo–Ir–P system, we had previously reported [13]. In this paper, the physical properties of MIrP (M=Ti, Zr, Nb and Mo) and MgRuP compounds are reported.

Published under licence by IOP Publishing Ltd

2. Experiment and results

Polycrystalline sample of the ternary metal phosphides MIrP (M=Ti, Zr, Nb and Mo) and MgRuP were prepared by high-pressure synthesis method using a cubic-anvil high-pressure apparatus (RIKEN CAP-07). The starting materials were mixed stoichiometric amounts of each powder metal and P for MIrP and MgRuP and ground by an agate mortar in a glove box filled with dry nitrogen gas. The starting material is put into a crucible made of BN. The crucible with a graphite heater is inserted into the pyrophyllite cube. The samples were synthesized by heating the mixtures in BN crucibles under a pressure of about 2 GPa at 1523-1873 K for 1.0 h.

These obtained compounds were characterized for the phase purity and lattice parameters by powder X-ray diffraction using CuKα radiation (Rigaku RINT 1100, 40 kV, 40 mA). Figure 1 shows the X-ray powder diffraction pattern for MIrP (M=Ti, Zr, Nb and Mo). The diffraction lines in MIrP (M=Ti, Zr, Nb and Mo) are assigned by the index of Co_2Si-type structure. The X-ray profiles of MIrP are similar to those of orthorhombic MoRuP [10]. These new compounds with the Co_2Si type structure have a/c and $(a+c)/b$ ratios very similar to those of other ternary transition metal phosphides with that structure; so all belong to the Co_2Si branch of the anti-$PbCl_2$ type [14]. Figure 2 shows the variation of the lattice parameters as a function of metallic radii for the six coordination. For larger size of the metallic radii M, the lattice parameter a and c expand and lattice parameter b shrink, respectively. Figure 3 shows the X-ray powder diffraction pattern for MgRuP. The diffraction lines in MgRuP are also assigned by the index of Co_2Si-type structure. The lattice constants of MgRuP prepared at high-pressure are also given as follows; a=6.444Å, b=3.835Å and c=6.970Å.

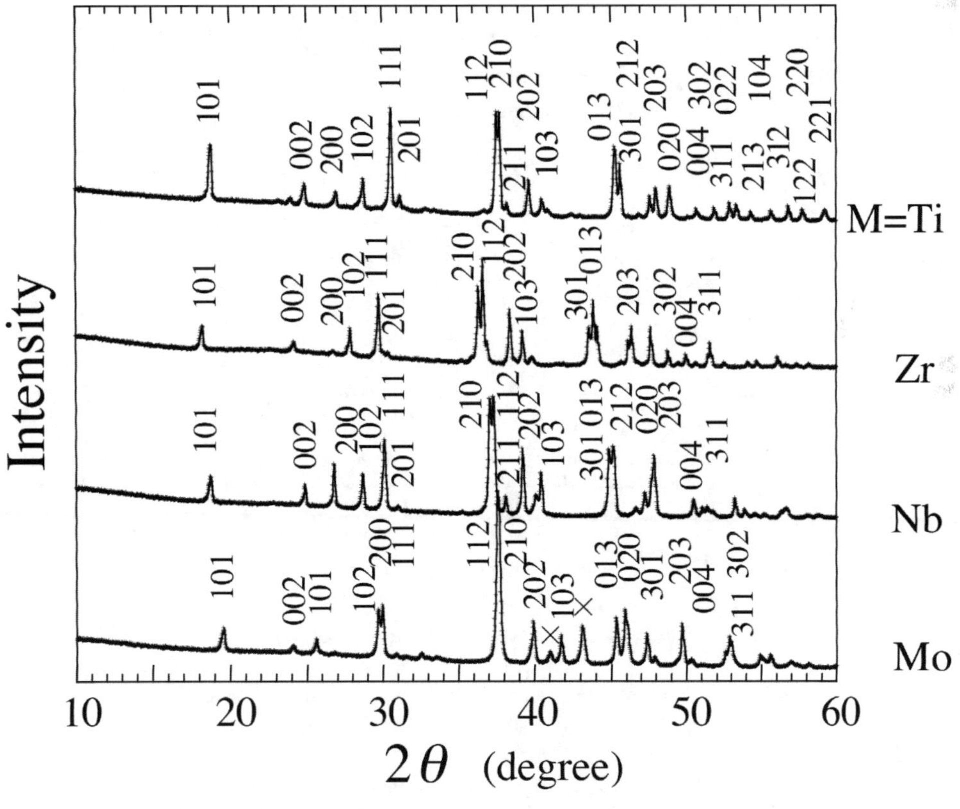

Figure 1. X-ray powder diffraction of MIrP (M=Ti, Zr, Nb and Mo).

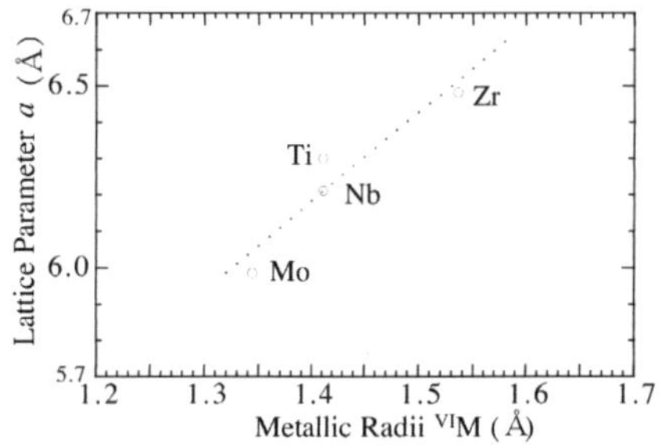

Figure 2(a). Lattice parameter *a* as a function of metallic radii M.

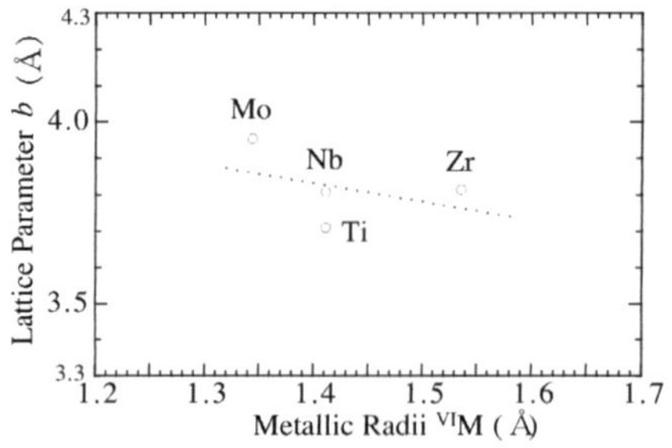

Figure 2(b). Lattice parameter *b* as a function of metallic radii M.

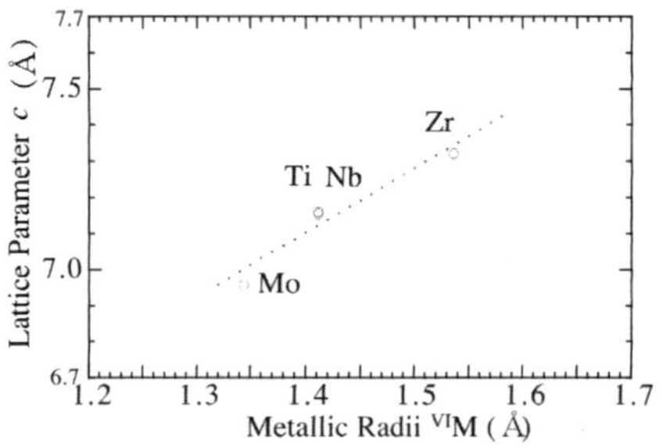

Figure 2(c). Lattice parameter *c* as a function of metallic radii M.

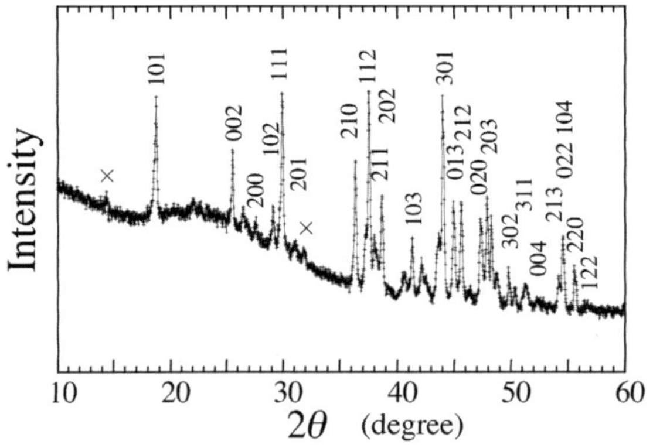

Figure 3. X-ray powder diffraction of MgRuP.

Electrical resistivity measurements were carried out using a conventional DC four-probe method. Gold lead was attached to polycrystalline samples with silver filled by epoxy and measurements were performed at low temperatures. The resistivity for MIrP (M=Ti, Zr, Nb and Mo) decreases with decreasing temperature. MIrP (M=Ti, Zr, Nb and Mo) behaves as a metal. At low temperature, large residual resistivity was observed for MIrP (M= Nb). This suggests that the purity of the specimen is still low. The superconductivity is not observed down to 5 K.

The DC magnetic susceptibility at H=20Oe of the polycrystalline MIrP (M=Ti, Zr, Nb and Mo) and MgRuP samples were measured in the range of 2–300K with a Quantum Design MPMS SQUID magnetometer. No superconducting transition is observed down to 2 K. The Susceptibility of the MIrP (M=Ti, Zr, Nb and Mo) is essentially temperature independent in the region of 2-300 K. These show Pauli paramagnetism and are consistent with metallic behavior, which are expected from the temperature dependence of the resistivity.

These MIrP (M=Ti, Zr, Nb and Mo) and MgRuP may be possible to show the superconductivity under high pressure or at very low temperatures.

This work was supported by a Grant-in-Aid for Specially Promoted Research (21540375) from the Ministry of Education, Culture, Sports, Science and Technology (MEXT), Japan.

References

[1] H. Barz, H.C. Ku, G.P. Meisner, Z. Fisk and B.T. Mattias 1980 *Proc. Natl. Acad. Sci. USA* **77** 3132.

[2] R. Müller, R.N. Shelton, J.W. Richardson, JR. and R.A. Jacobson 1983 *J. Less-common Metals* **92** 177.

[3] G.P. Meisner and H.C. Ku 1983 *Apll. Phys.* A **31** 201.

[4] G.P. Meisner and H.C. Ku, in G.K. Shenoy, B.D. Dunlap, F.Y. Fradin (eds.) 1981, *Ternary Superconductors*, North-Holland, New York p.255.

[5] R. Fruchart, A. Roger and J.P. Sénateur 1969 *J. Appl. Phys.* **40** 1250.

[6] J. Roy-Montreuil, B. Deyris, A. Michel, R. Fruchart, J.P. Sénateur and D. Boursier 1979 *Ann. Chim. Fr.* **4** 411.

[7] G.R. Stewart, G.P. Meisner and H.C. Ku, in W. Buckel, W.Weber (eds.) 1982, *Superconductivity in d- and f-band metals*, Kernforschungszentrum, Karlsruhe p.331.

[8] I. Shirotani, N. Ichihashi, K. Nozawa, M. Kinoshita, T. Yagi, K. Suzuki and T. Enoki 1983 *Jpn. J. Appl. Phys. Suppl.* **32(3)** 695.

[9] I. Shirotani, M. Takaya, I. Kaneko, C. Sekine and T. Yagi 2000 *Solid State Commun* **116** 683.

[10] W. Wong-Ng, W.Y. Ching, Y.-N. Xu, J.A. Kaduk and I. Shirotani, L. Swartzendruber 2003 *Phys. Rev.* B **67** 144523

[11] G.P. Meisner 1981 *Physica* B **108** 763.

[12] I. Shirotani, T. Adachi, K. Tachi, S. Todo, K. Nozawa, T. Yagi and M. Kinoshita 1996 *J.Phys.Chem. Solids* **57** 211.

[13] H. Kitô, T. Wada and H. Abe 2006 *Physica* B **378-380** 1146.

[14] W. Jeitschko 1968 *Acta Cryst.* B **24** 930.

Journal of Physics: Conference Series **273** (2011) 012116 doi:10.1088/1742-6596/273/1/012116

Superfluid gap formation in a fermionic optical lattice with spin imbalanced populations

Akihisa Koga[1], **Philipp Werner**[2]

[1]Department of Physics, Tokyo Institute of Technology, Tokyo 152-8551, Japan
[2]Theoretische Physik, ETH Zurich, 8093 Zürich, Switzerland

E-mail: koga@phys.titech.ac.jp

Abstract. We investigate the attractive Hubbard model in infinite spatial dimensions at quarter filling. By combining dynamical mean-field theory with continuous-time quantum Monte Carlo simulations in the Nambu formalism, we directly deal with the superfluid phase in the population imbalanced system. We discuss the low energy properties in the polarized superfluid state and the pseudogap behavior in the vicinity of the critical temperature.

1. Introduction

Recently, ultracold fermions have attracted much interest. One of the most interesting topics in the field is the superfluid state [1], where Cooper pairs composed of distinct components condense at low temperatures. Owing to the high controllability of the interaction strength and the particle number, interesting phenomena have been observed such as the BCS-BEC crossover [2, 3, 4] and the superfluid state in imbalanced systems [5, 6]. These observations stimulate further experimental and theoretical investigations on fermionic systems.

For optical lattice systems, which are formed by loading ultracold fermions into a periodic potential, various types of superfluid states have been proposed [7, 8, 9]. One of the possible candidates is the polarized superfluid state, which is naively expected to be realized in higher dimensional systems with imbalanced populations [10]. In our previous paper [11], we have investigated the attractive Hubbard model in infinite spatial dimensions to clarify how the polarized superfluid state is realized at low temperatures. However, dynamical properties were not yet studied in detail in the intermediate correlation and temperature region. In particular, the question whether or not a pseudogap structure appears in the density of states above the critical temperature, has not been answered. To clarify this point, we study the formation of the superfluid gap in the system with and without spin-imbalanced populations.

The paper is organized as follows. In §2, we introduce the model Hamiltonian and briefly summarize the DMFT framework. In §3, we study the attractive Hubbard model at quarter filling. By combining dynamical mean-field theory (DMFT) [12, 13, 14, 15] with the continuous time quantum Monte Carlo (CTQMC) method [16], we discuss how the polarized superfluid state is realized and the superfluid gap appears in the density of states at low temperatures. A brief summary is given in §4.

Published under licence by IOP Publishing Ltd

2. Model and Method

We consider a correlated fermion system with attractive interactions, which may be described by the Hubbard Hamiltonian,

$$\hat{\mathcal{H}} = \sum_{(i,j),\sigma} \left[-t - (\mu + h\sigma)\,\delta_{ij}\right] c_{i\sigma}^{\dagger} c_{j\sigma} - U \sum_{i} n_{i\uparrow} n_{i\downarrow}, \tag{1}$$

where $c_{i\sigma}$ ($c_{i\sigma}^{\dagger}$) is an annihilation (creation) operator of a fermion on the ith site with spin σ, and $n_{i\sigma} = c_{i\sigma}^{\dagger} c_{i\sigma}$. U is the onsite attractive interaction, t is the transfer integral between sites, μ is the chemical potential, and h is the magnetic field, which controls the spin imbalanced populations in the system.

To study the attractive Hubbard model in infinite spatial dimensions [10, 11, 17, 18, 19, 20], we make use of DMFT [12, 13, 14, 15]. In DMFT, the original lattice model is mapped to an effective impurity model, which accurately takes into account local particle correlations. The lattice Green's function is obtained via a self-consistency condition imposed on the impurity problem. This treatment is formally exact in infinite spatial dimensions, and DMFT has successfully been applied to strongly correlated fermion systems. When the superfluid state is directly treated in the framework of DMFT, the Green's function should be described by a 2×2 matrix. The self-consistency condition for the model [21] is given as,

$$\hat{G}_0^{-1}(i\omega_n) = (i\omega_n + h)\,\hat{\sigma}_0 + \mu\hat{\sigma}_z - \left(\frac{D}{4}\right)^2 \hat{\sigma}_z \hat{G}(i\omega_n)\hat{\sigma}_z, \tag{2}$$

where $\hat{G}_0(i\omega_n)[\hat{G}(i\omega_n)]$ is the non-interacting (interacting) Green's function for the impurity model, $\omega_n[= (2n+1)\pi T]$ is the Matsubara frequency, T is the temperature, $\hat{\sigma}_0$ is the identity matrix, and $\hat{\sigma}_z$ is the z-component of the Pauli matrix. Here, we have used the semi-circular density of states, $\rho_0(x) = 2/\pi D\sqrt{1-(x/D)^2}$, where D is the half bandwidth. In the following, we set D as a unit of the energy.

When low energy properties in strongly correlated systems are discussed in the framework of DMFT, an impurity solver is necessary to obtain the Green's function for the effective impurity model. Here we use the recently developed CTQMC technique, which has successfully been applied to the Hubbard model [11, 22, 23], the periodic Anderson model [24], the Kondo lattice model [25] and the Holstein-Hubbard model [26]. In this paper, by using a CTQMC method in the continuous-time auxiliary field formulation [27] and extended to the Nambu formalism as an impurity solver, we discuss how the polarized superfluid state is realized in the system. The details of the method have been explained in Ref. [11].

3. Results

We study the superfluid state in the Hubbard model at quarter filling to discuss how the gap structure appears in the density of states. A quarter filled system is chosen to study the attractive Hubbard model at a generic filling. The pair potential $\Delta(= \langle c_{i\uparrow} c_{i\downarrow}\rangle)$ and the density of states $\rho_\sigma(\omega)$ in the case $U = 2$ are shown in Fig. 1. When $h = 0.0$, as temperature is decreased, the pair potential is induced at $T_c \sim 0.095$, as shown in Fig. 1 (a). This implies that a phase transition occurs from the normal metallic state to the superfluid state. A superfluid gap characteristic of this state clearly appears in the density of states at low temperatures, as shown in Fig. 1 (b). When the magnetic field is switched on, pairing correlations are suppressed and the normal metallic state becomes stable. In fact, the superfluid order parameter is decreased and the critical temperature is shifted to $T_c \sim 0.077$ when $h = 0.3$, as shown in Fig. 1 (a). When a large magnetic field is applied, the polarized superfluid state is no longer stable even at very low temperatures. This instability has been discussed in Ref. [11]. The density of states for

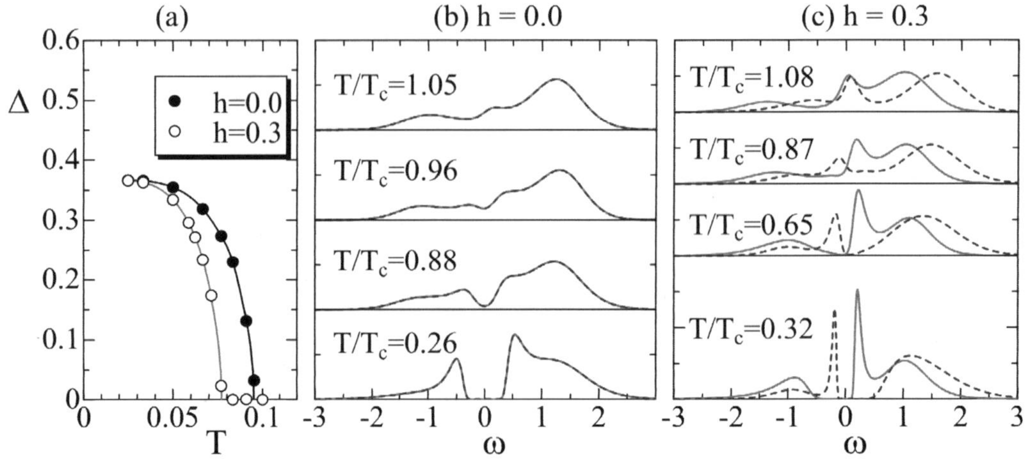

Figure 1. (a) Pair potential as a function of the temperature when $h = 0.0$ and $h = 0.3$. Solid and dashed lines represent the spectral functions for fermions with up (down) spin when $h = 0.0$ (b) and $h = 0.3$ (c).

each spin in the $h = 0.3$ case is shown in Fig. 1 (c). It is found that the density of states around the Fermi level decreases, with decrease of temperature from T_c, similar to the case with $h = 0.0$. These results suggest that the superfluid gap appears only at $T < T_c$. Therefore, we conclude that no pseudogap behavior is found at $T > T_c$, at least, in the quarter filled Hubbard model with the interation $U = 2$. However, there still exists an open question whether or not a pseudogap behavior appears in other parameter regions such as the weak or strong coupling region and other fillings. Further systematic calculations are necessary to clarify this point, and will be discussed elsewhere.

4. Summary

We have investigated the attractive Hubbard model in infinite spatial dimensions, by combining dynamical mean-field theory with continuous-time quantum Monte Carlo simulations based on the Nambu formalism. We have calculated the superfluid order parameter and the density of states to discuss how the polarized superfluid state is realized at low temperatures. It was found that no pseudogap behavior appears above the critical temperature.

Acknowledgment

Parts of the computations were done on TSUBAME Grid Cluster at the Global Scientific Information and Computing Center of the Tokyo Institute of Technology. This work was partly supported by the Grant-in-Aid for Scientific Research 20740194 (A.K.) and the Global COE Program "Nanoscience and Quantum Physics" from the Ministry of Education, Culture, Sports, Science and Technology (MEXT) of Japan. PW acknowledges support from SNF Grant PP002-118866.

References

[1] Regal C A, Greiner M, and Jin D S 2004 *Phys. Rev. Lett.* **92** 040403
[2] Jochim S, Bartenstein M, Altmeyer A, Hendl G, Riedl S, Chin C, Denschlag J H, and Grimm R 2003 *Science* **302** 2101

[3] Zwierlein M W, Stan C A, Schunck C H, Raupach S M F, Gupta S, Hadzibabic Z, and Ketterle W 2003 *Phys. Rev. Lett.* **91** 250401
[4] Bourdel T, Khaykovich L, Cubizolles J, Zhang J, Chevy F, Teichmann M, Tarruell L, Kokkelmans S J J M F, and Salomon C 2004 *Phys. Rev. Lett.* **93** 050401
[5] Zwierlein M W, Schirotzek A, Shunck C H, and Ketterle W 2006 *Science* **311** 492
[6] Partridge G B, Li W, Kamar R I, Liao Y, and Hulet R G 2006 *Science* **311** 503
[7] Pour F K, Rigol M, Wessel S, and Muramatsu A, 2007 *Phys. Rev. B* **75** 161104; Xianlong G, Rizzi M, Polini M, Fazio R, Tosi M P, Campo Jr. V L, and Capelle K 2007 *Phys. Rev. Lett.* **98** 030404; Tezuka M and Ueda M 2008 *Phys. Rev. Lett.* **100** 110403; Machida M, Yamada S, Okumura M, Ohashi Y, and Matsumoto H 2008 *Phys. Rev. A* **77** 053614
[8] Koponen T K, Paananen T, Martikainen J P, and Törmä P 2007 *Phys. Rev. Lett.* **99** 120403; Chen Y, Wang Z D, Zhang F C, and Ting C S 2009 *Phys. Rev. B* **79** 054512; Tamaki H, Miyake K, and Ohashi Y 2009 *J. Phys. Soc. Jpn.* **78** 073001
[9] Koga A, Higashiyama T, Inaba K, Suga S, and Kawakami N 2008 *J. Phys. Soc. Jpn.* **77** 073602; 2009 *Phys. Rev. A* **79** 013607; Fujihara Y, Koga A, and Kawakami N, 2009 *Phys. Rev. A* **79** 013610
[10] Dao T L, Ferrero M, Georges A, Capone M, and Parcollet O 2008 *Phys. Rev. Lett.* **101** 236405
[11] Koga A and Werner P 2010 *J. Phys. Soc. Jpn.* **79** 064401
[12] Metzner W and Vollhardt D 1989 *Phys. Rev. Lett.* **62** 324
[13] Müller-Hartmann E 1989 *Z. Phys. B* **74** 507
[14] Georges A, Kotliar G, Krauth W and Rozenberg M J 1996 *Rev. Mod. Phys.* **68** 13
[15] Pruschke T, Jarrell M, and Freericks J K 1995 *Adv. Phys.* **42** 187
[16] Rubtsov A N, Savkin V V and Lichtenstein A I 2005 *Phys. Rev. B* **72** 035122
[17] Keller M, Metzner W, and Schollwöck U 2001 *Phys. Rev. Lett.* **86** 4612
[18] Garg A, Krishnamurthy H R, and Randeria M 2005 *Phys. Rev. B* **72** 024517
[19] Toschi A, Capone M, and Castellani C 2005 *Phys. Rev. B* **72** 235118
[20] Bauer J, Hewson A C, and Dupuis N 2009 *Phys. Rev. B* **79** 214518; Bauer J and Hewson A C 2009 *Europhys. Lett.* **85** 27001
[21] Georges A, Kotliar G, and Krauth W 1993 *Z. Phys. B* **92** 313
[22] Werner P, Comanac A, de'Medici L, Troyer M, and Millis A J 2006 *Phys. Rev. Lett.* **97** 076405; Werner P and Millis A J 2007 *Phys. Rev. B* **75** 085108
[23] Werner P and Millis A J 2006 *Phys. Rev. B* **74** 155107; 2007 *Phys. Rev. Lett.* **99** 126405; Werner P, Gull E, and Millis A J 2009 *Phys. Rev. B* **79** 115119
[24] Luitz D J and Assaad F F 2010 *Phys. Rev. B* **81** 024509
[25] Otsuki J, Kusunose H, Werner P, and Kuramoto Y 2007 *J. Phys. Soc. Jpn.* **76** 114707; Otsuki J, Kusunose H, and Kuramoto Y 2009 *Phys. Rev. Lett.* **102** 017202
[26] Assaad F F and Lang T C 2007 *Phys. Rev. B* **76** 035116; Werner P and Millis A J 2007 *Phys. Rev. Lett.* **99** 146404
[27] Gull E, Werner P, Parcollet O and Troyer M 2008 *Europhys. Lett.* **82** 57003

Journal of Physics: Conference Series **273** (2011) 012117 doi:10.1088/1742-6596/273/1/012117

Comparison study of magnetic ordering for Fe-free and Fe-doped $LiMn_2O_4$ spinel oxide

Yang Li [1,2], Boyu Ma [2], Aihua Wang [3], Ning Chen [4], Lihua Liu [1], Yang Liu [4], Weipeng Wang [2], Xiaoxiang Li [2], Guohui Cao [2], Xingqiao Ma [2], Jun Lu [5]

1) Department of Engineering Science and Materials, University of Puerto Rico at Mayaguez, PR 00681, USA, 2) Department of physics, University of Science and Technology Beijing, 100083, China, 3) Department of Physics, Capital Normal University, Beijing 100037, China, 4) School of Material Science and Engineering, University of Science and Technology Beijing, 100083, China, 5) National High Magnetic Field Lab, Florida State University, Tallahassee, FL 32310, USA

Corresponding author: YL Email: yang.li@upr.edu

Abstract. The structural and physical properties of $LiMn_{2-x}Fe_xO_4$ (x = 0 to 0.5) were investigated. Contrasting $LiMn_2O_4$ with $LiMn_{1.5}Fe_{0.5}O_4$, we observed a remarkable difference in the magnetic frustration. The $LiMn_2O_4$ sample has a magnetic ordering transition at 61 K while Fe-doped $LiMn_{1.5}Fe_{0.5}O_4$ has an antiferromagnetic ordering transition with a Neel temperature at T_N = 34 K. Our result showed a larger value of the frustration index ($f=|\ _w|/T_C$) for pure $LiMn_2O_4$ than for $LiMn_{1.5}Fe_{0.5}O_4$. Fe-doping results in a decrease in the degree of frustration. Fe dopants occupying positions of Mn ion can break the original moment equilibrium so as to suppress the magnetic frustration in $LiMn_2O_4$.

1. Introduction

$LiMn_2O_4$ presents a Verwey-like phase transition at T_V ~ 285 K [1,2], which has attracted great interest. Above T_V the manganese formally exists in the Mn^{4+} ($t_{2g}^3e_g^0$) and Jahn-Teller active Mn^{3+} ($t_{2g}^3e_g^1$) configurations. The distribution of cations can be represented by $(Li^+)[Mn^{3+}Mn^{4+}]O_4^{2-}$. The Li-O coordination is a tetrahedral site (the A-site) and the Mn-O coordination is an octahedral site (the B-site). The significant physical characteristics are the mixed oxidation states of Mn^{3+} and Mn^{4+}.

Resistivity and differential scanning calorimetry (DSC) measurements indicated that the Verwey-like transition is first order, which leads to a hysteresis of ~15 K upon cooling and then heating to room temperature [2]. Electrical conduction above the transition temperature can be described by a model of electron hopping with an activation energy of E_h = 0.16 eV [3], where a dynamic Jahn-Teller distortion in the environment around Mn^{3+} is sufficient to impede the motion of the e_g electrons. The presence of a Verwey transition suggests that the e_g electrons become partially localized with distinguishable Mn^{3+} and Mn^{4+} cations and environments, consistent with the observed distortion away from cubic symmetry [4]. The magnetic susceptibility of $LiMn_2O_4$ shows a field-cooled and zero-field-cooled irreversibility at 65 K. The neutron diffraction results indicate that the short-range order develops above 100 K. Upon cooling to 60 K a long-range magnetic order occurs [4]. The extended x-ray-absorption fine-structure spectrum also agree with the neutron-scattering results [5].

Published under licence by IOP Publishing Ltd

In spinel $LiMn_2O_4$, Mn atoms have a special magnetic sublattice in which there is a network of apex sharing Mn-Mn tetrahedra [4,6]. Four moments at the apexes of a Mn-Mn tetrahedron cannot align to be simultaneously antiparallel to each other. Because the connectivity of sharing only one vertex between tetrahedra is insufficient to define a unique spin structure, various magnetic properties relative to magnetic moment frustration are expected to occur in this compound. The magnetic properties extensively depend on the precise balance among various exchange pathways in the system. In order to investigate further the mechanism of magnetic ordering, we performed a comparison study of magnetic ordering for Fe-free and Fe-doped $LiMn_2O_4$ spinel oxides. We observed that Fe dopants substituting for Mn ions heavily suppress the magnetic frustration in $LiMn_2O_4$.

2. Experiment

A series of samples of $LiMn_{2-x}Fe_xO_4$ (x=0, 0.1, 0.2, 0.3, 0.4 and 0.5) were prepared by the solid state reaction. Appropriate amount of Li_2CO_3, MnO_2 and Fe_2O_3 were mixed and sintered at 750 °C for 24 h. The precursors of Li-Mn-Fe-O were ground and mixed fully, then pressed into a pellet, and annealed at 860 °C for 100 hours. The pellets were annealed under oxygen flowing atmosphere. Samples were analyzed by X-ray diffraction (XRD) and refinement using the general structure analysis software (GSAS) package. The energy dispersive spectroscopy (EDS) compositional analysis was carried out using an SEM-based electron microprobe system. Magnetic measurements were performed on a SQUID. The resistivity and specific heat were measured in a Physical Property Measurement System.

3. Results and discussions

X-ray diffraction structural analysis for $LiMn_{2-x}Fe_xO_4$ (x=0 to 0.5) showed characteristic reflections of the spinel structure. All samples exhibited single-phase X-ray patterns. The refined lattice parameters of $LiMn_{2-x}Fe_xO_4$ (x=0 – 0.5) exhibit an increasing trend with x due to the larger atomic size of Fe than that of Mn, which suggests that the Fe atoms enter $LiMn_2O_4$ lattice. In addition, as a result of the refinement, we find that Fe preferentially occupies the Mn sites. SEM observations showed that samples are uniform, and big grains showed spinel morphigraphics. The EDS compositional analyses were performed on particles selected by the electron diffraction. The quantification results are consistent with the sample compositions of $LiMn_{2-x}Fe_xO_4$ (x= 0 - 0.5).

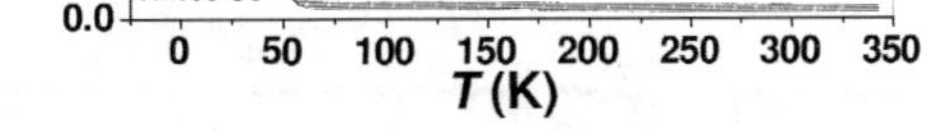

Figure 1. $LiMn_2O_4$ *dc* susceptibility. Inset: Hysteresis loop of $LiMn_2O_4$ at T = 2 K. Inset left: Magnetization vs. field. Inset right: coercivity H_C= 0.55 T.

The *dc* susceptibility of $LiMn_2O_4$ was measured in a 100 Oe magnetic field. The susceptibility data can be fitted by the Curie-Weiss law $_m= [p^2/3k_B(T - {}_w)+ {}_d]$ for T >230K to yield the effective moment p_{eff} =4.16±0.08 μ_B per Mn, which shows Mn ions being in the high-spin configuration. The Weiss temperature $_w$ = - 276±10 K, and $_d$ = - 0.0004(8) emu/mole, which denotes a temperature-independent magnetic background including diamagnetism. Due to our different sample preparation conditions from literatures, this effective moment p_{eff} =4.16 μ_B is smaller than the reported effective moment 4.4 μ_B [4], which corresponds to an average Mn moment based on the free-ion Mn^{4+} moment (3.87 μ_B) and the free Mn^{3+} moment (4.90 μ_B) in the sample with Mn^{3+}/Mn^{4+} =1:1 [4]. An abrupt jump of the susceptibility was not observed around 285 K corresponding to Verwey transition temperature. The electrical resistivity measurement (not shown here) also shows no irreversibility around this temperature range. We might attribute the disappearance of Verway-like transition to the proportionate reducing of Jahn-Teller-active Mn^{3+} in our sample. A rough estimation for the average Mn valence is +3.75 based on the effective moment (p_{eff} =4.16 μ_B), i.e., the ratio of Mn^{3+}/Mn^{4+} ion is 1:3. This departure from the typical ratio 1:1 of $LiMn_2O_4$ would cause excess oxygen in the lattice. In addition, the extra oxygen also may result from partial substitution of Mn for Li lattice sites so as to form site inversion at the *A*-sites. The rich oxygen environment during sample preparation is propitious to excess oxygen entering the lattice structure. We suggest that the mismatching of Mn^{3+}/Mn^{4+} results in suppression of Verwey transition. However, more direct experimental proof is still needed becauce XRD and EDS are not sensitive to the change of oxygen content below 5%. Moreover, the negative value of Weiss temperature $_w$= –275 K suggests antiferromagnetic coupling among Mn moments. Upon cooling, a magnetic ordering transition starts at the irrevesibility temperature T_{irr}=61 K. A difference between the susceptibility of the susceptibility in ZFC (zero-field-cooled) and FC (field-cooled) procedures occurs, which corresponds to a transition from a short-range order to a long-range magnetic order [4]. The T_{irr} is smaller by ~ 4 K than that of the $LiMn_2O_4$ sample with Mn^{3+}/Mn^{4+} = 1:1 [4]. We reconcile the discrepancy in the irreversibility T_{irr} by a higher oxidation of Mn in our samples. It is clear that the Mn^{3+} cations play an important role in this magnetic transition.

The magnetic transition temperature T_C (=46 K) was determined from a large maximum in the slope of d $_m$ $/dT$ at the critical point. We obtained a large value f = 6 of the frustration index ($f=|\theta_w|/T_C$), i.e. the large negative value of the Weiss temperature $|\theta_w|$ with respect to a lower magnetic ordering temperature T_C. The depression of the ordering temperature (T_C) demonstrates that the $LiMn_2O_4$ system is frustrated. In Fig. 2, the cusp of C_p/T indicated by an arrow coincides with the irreversibility temperature T_{irr} in the $_m$ –T curve in Fig. 1. The measurement of specific heat (C_p) for this sample also verifies a phase transition. An approximate straight-line C_p/T = + T^2 fit was obtained at low temperature (T< 10 K), with electron and phonon contributions = 14.7 mJ/K^2 and = 1.2 mJ/K^4, respectively.

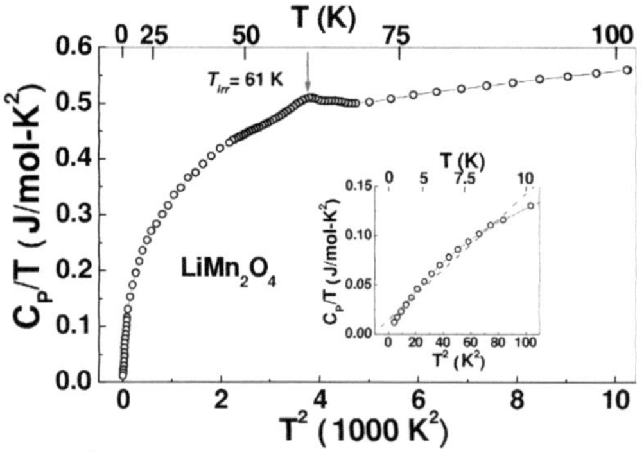

Figure 2. C_p/T vs. T^2 curve of $LiMn_2O_4$. Inset: C/T with the fitting curve $\gamma+\beta T^2$.

In the inset of Fig. 1, the *M-H* loop shows ferromagnetic hysteresis behavior at 2 K, in which the coercivity is 5.5 kOe. However, full saturation is not reached even at 7 T, where the magnetization corresponds to 0.17 μ_B/Mn at 2 K. This value is considerably lower than expected, given the value p_{eff}= 4.16 μ_B/Mn obtained from the Curie-Weiss fit. This suggests a noncollinear spin configuration. Considering the ferromagnetic hysteresis in the ZFC/FC curves but along with a very small saturated moment, which indicates some degree of canted antiferromagnetism, so in this magnetic system actually there is a ferrimagnetism between the Mn^{3+} and Mn^{4+} ions. In view of the spinel structure of $LiMn_2O_4$, where the coupling occurs the moments of Mn-Mn atoms with tetrahedral coordination, the distribution of magnetic moments of Mn atom shows a magnetic frustration behavior. In this Mn moment system, there is no simple preferential moment orientation. The noncollinear spins result in a very small net magnetization. In spite of the high magnetic field, the moment arrangement cannot be aligned, which could explain the lack of saturation for $LiMn_2O_4$ at 2 K and 7 Tesla.

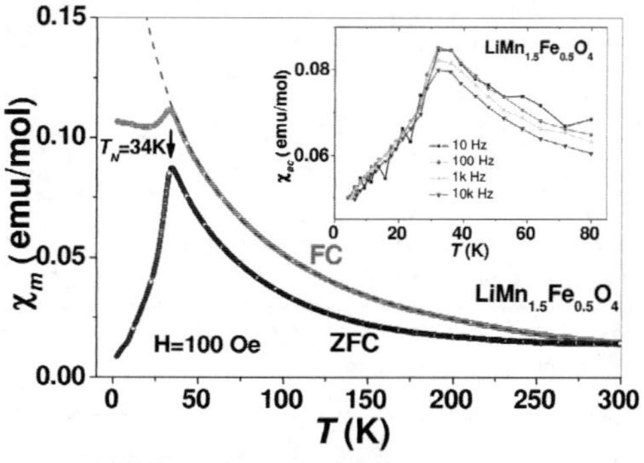

Figure 3. *dc* susceptibility of $LiMn_{1.5}Fe_{0.5}O_4$. Inset: *ac* susceptipilities of $LiMn_{1.5}Fe_{0.5}O_4$ with different frequencies.

The *dc*-susceptibility of $LiMn_{1.5}Fe_{0.5}O_4$ are shown in Fig. 3. There is no Verwey transition in the Fe-doped sample, which is consistent with resistivity measurements. A fit to $_m$= [$p^2/3k_B(T$ - $_w$)+ $_d$] above 50 K yeilds p =5.71 μ_B per 3*d* atom ($Fe_{0.25}Mn_{0.75}$), a Weiss temperature of $_w$ = - 31 K, and $_d$ = - 0.0035 emu/mole. The value of the p_{eff} is consistent with Fe^{3+} ion being in the high-spin configuration. Based on the fact of the free-ion moments of Mn^{3+}, Mn^{4+} and Fe^{3+} are 4.90, 3.87 and 5.92 μ_B, respectively, the value of p_{eff} =5.71 μ_B is larger than the average value of free-ion moments. Thus, Fe doping gives rise to an enhancement of the magnetic correlations among 3*d* ions. The negative value of the Weiss temperature $_w$ = - 31 K indicates antiferromagnetic correlation among moments. As shown in Fig. 3, a cusp in the susceptibility manifests the characteristic of antiferromagnetic ordering with T_N=34 K for $LiMn_{1.5}Fe_{0.5}O_4$. In addition, the frequency-dependent *ac* susceptibility show that no shift occurs in the peak for the frequencies between 10 Hz and 10 kHz, which excludes the occurrence of a spin-glass-like transition. For $LiMn_{1.5}Fe_{0.5}O_4$, the frustration index f=| $_w$|/T_N ~1 which is much smaller than that of the Fe-free sample (f = 6). A smaller frustration index implies that Fe-doping results in a decrease on the degree of frustration. Fe dopants give rise to an antiferromagnetic ordering in the frustrated $LiMn_2O_4$. The experimental results show that the magnetic properties are very sensitive to Fe-doping because the ground state degeneracy can be removed by introducing magnetic Fe sites. In the Fe-doped sample, due to the difference between Fe and Mn ion moments, the Fe ions occupying positions of Mn ion can break the original moment equilibrium so as to suppress the magnetic frustration in $LiMn_2O_4$.

This work was supported by National Science Foundation under Grant No. DMR-0821284, National Natural Science Foundation of Beijing (Grant No. 1072007), NASA PRSGC IDEAS-ER (Grant Nos. NNX10AM80H and NNX07AO30A-EPSCoR).

References

[1] Yamada A, Miura K, Hinokuma K, and Tanaka M, 1995 *J. Electrochem. Soc.* 142, 2149.
[2] Yamada A and Tanaka M, 1995 *Mater. Res. Bull.* 30, 715.
[3] Shimakawa Y, Numata T, Tabuchi J J, 1997 *J. Solid State Chem.*131, 138.
[4] Wills A S, Raju N P, and Greedan J E, 1999 Chem. Mater. 11, 1510.
[5] Paolone A, Castellano C, et al., 2003 Phys. Rev. B68, 014108.
[6] Greedan J E, Wiebe C R, Wills A S, Stewart J R, 2002 *Phys. Rev.* B65 184424.

Journal of Physics: Conference Series **273** (2011) 012118 doi:10.1088/1742-6596/273/1/012118

Magnetic order in the filled skutterudites $R\text{Pt}_4\text{Ge}_{12}$ (R = Nd, Eu)

M Nicklas, R Gumeniuk, W Schnelle, H Rosner, A Leithe-Jasper, F Steglich and Yu Grin

Max Planck Institute for Chemical Physics of Solids, Nöthnitzerstr. 40, 01187 Dresden, Germany

E-mail: `nicklas@cpfs.mpg.de (Michael Nicklas)`

Abstract. Rare-earth metal filled skutterudites $R\text{Pt}_4\text{Ge}_{12}$ with R=La-Nd, and Eu exhibit a variety of different ground states, e.g., conventional and unconventional superconductivity in $\text{LaPt}_4\text{Ge}_{12}$ and $\text{PrPt}_4\text{Ge}_{12}$, respectively, and intermediate valence behavior in $\text{CePt}_4\text{Ge}_{12}$. In this work we investigate the magnetic state of $\text{NdPt}_4\text{Ge}_{12}$ and $\text{EuPt}_4\text{Ge}_{12}$ by specific heat, dc-susceptibility and magnetization. $\text{NdPt}_4\text{Ge}_{12}$ shows two magnetic phase transitions at $T_{N1} = 0.67$ K and $T_{N2} = 0.58$ K, while $\text{EuPt}_4\text{Ge}_{12}$ displays a complex magnetic phase diagram below the magnetic ordering temperature of 1.78 K. The specific heat indicates that in $\text{NdPt}_4\text{Ge}_{12}$ the crystalline electric field (CEF) ground state of the Nd^{3+} is a quartet and that, as expected, in $\text{EuPt}_4\text{Ge}_{12}$ the Eu^{2+} state is fully degenerate.

1. Introduction

Filled skutterudites have become a topic of considerable interest with respect to basic and applied solid state sciences. This is particularly due to a variety of physical properties which can be intimately related to the underlying structural chemistry [1, 2, 3, 4, 5]. Their stoichiometry can be rationalized with the chemical formula MT_4X_{12}, with M being an electropositive element like alkali, alkaline-earth, early rare-earth, actinide, or thallium metal, T standing for a transition metal of the iron- or cobalt-group, and X representing a pnictogen element as they are phosphorus, arsenic, or antimony. They all crystallize with the cubic $\text{LaFe}_4\text{P}_{12}$ [6] structure where cation M stabilizes the T_4X_{12} host structure. Recently it has been shown by us and others [7, 8] that the transition metal is not restricted to the iron or cobalt group, but can also be the noble metal platinum, which together with germanium acts as the framework forming elements stabilized by the alkaline-earth metals Sr and Ba. Moreover, we have discovered a whole new family of rare-earth metal (R) based filled skutterudites $R\text{Pt}_4\text{Ge}_{12}$ with R = La-Nd, and Eu [8]. In turn, this series of compounds has been extended to actinide based $A\text{Pt}_4\text{Ge}_{12}$ (A= Th, U) phases [9, 10]. $M\text{Pt}_4\text{Ge}_{12}$ compounds (M = Sr, Ba, La, Pr) are superconductors with T_c up to 8.3 K [8]. $\text{PrPt}_4\text{Ge}_{12}$ is an unconventional superconductor with nodes in the superconducting energy gap [11]. Experimental and theoretical analysis of the electronic structure and chemical bonding revealed deep-lying Pt $5d$ states which only partially form covalent bands with the Ge $4p$ electrons. Consequently, the states at the Fermi level, which are relevant for superconductivity of these compounds, can be assigned to originate mainly from Ge $4p$ electrons [12]. Here, we report on the low-temperature magnetic properties of $R\text{Pt}_4\text{Ge}_{12}$ (R = Nd, Eu) studied by specific heat, dc-susceptibility, and magnetization experiments.

Published under licence by IOP Publishing Ltd

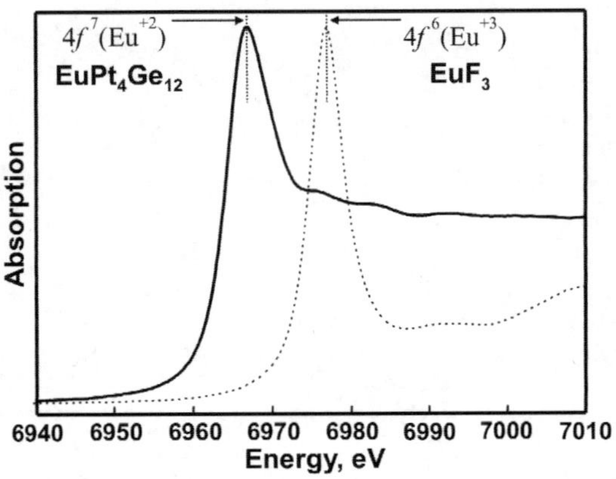

Figure 1. XAS of $EuPt_4Ge_{12}$ and EuF_3 at the Eu $L_{\rm III}$ edge. The XAS were recorded at room temperature at the EXAFS beamline A1 of HASYLAB at DESY using the four-crystal mode of the Si(111) monochromator and determined in transmission mode using powdered samples (~10 mg) which were diluted with B_4C and embedded in paraffin wax.

2. Experimental Details

Samples were synthesized as described elsewhere by arc-melting from the elements followed by annealing procedures [8]. We carried out low temperature ($0.35\text{K} \le T \le 10\text{K}$) specific-heat ($C$) experiments using the ^{3}He option of a PPMS (Quantum Design). The magnetization (M) in the temperature range $0.48\text{K} \le T \le 1.9\text{K}$ was measured using a SQUID magnetometer (MPMS, Quantum Design) equipped with a ^{3}He option (iQuantum). X-ray absorption spectroscopy at the Eu $L_{\rm III}$ edge (6977 eV) was used to determine valence of this rare-earth metal in the structures of new filled skutterudite.

3. Experimental Results

3.1. $EuPt_4Ge_{12}$

The X-ray absorption spectrum (XAS) for $EuPt_4Ge_{12}$ shown in Fig. 1 indicates that Eu is in the magnetic $4f^7$ (Eu^{2+}) configuration. No admixture of the $4f^6$ state is visible in the data. In Eu^{2+} only the spin component contributes to the total angular momentum of the ground multiplet $^8S_{7/2}$ and, therefore, distinct crystalline electric field effects are expected to be absent.

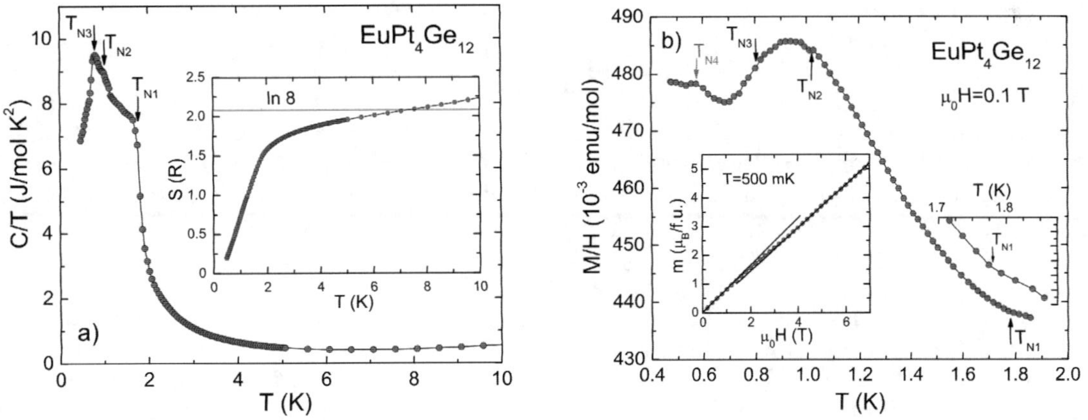

Figure 2. Specific heat as $C(T)/T$ (a) and dc-susceptibility $M(T)/H$ (b) of $EuPt_4Ge_{12}$. The arrows in (a) indicate the phase transitions. The inset of (a) shows the entropy $S(T)$. The black arrows in (b) correspond to the transition temperatures from specific heat, while the red arrow indicates an additional anomaly in $M(T)/H$. The inset of (b) presents $m(H)$ at $T \approx 500$ mK.

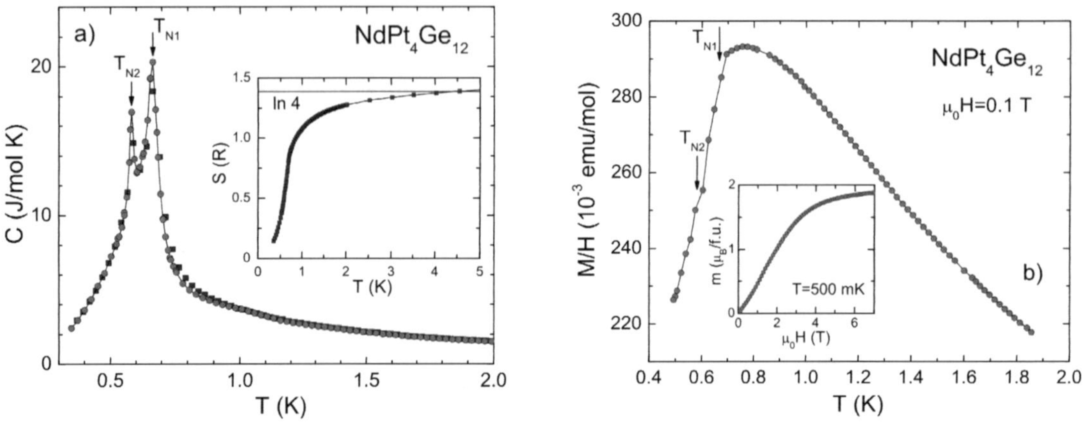

Figure 3. Specific heat as $C(T)/T$ (a) and dc-susceptibility $M(T)/H$ (b) of $NdPt_4Ge_{12}$. For $C(T)/T$ the results from two different measurements are presented. The inset of (a) shows the entropy $S(T)$ determined form one of the $C(T)$ experiments. The inset of (b) displays $m(H)$ at $T \approx 500$ mK. The arrows in both panels correspond to the transition temperatures determined from specific heat.

Accordingly, a fully degenerated $J = 7/2$ ground state multiplet is anticipated. The specific heat, $C(T)/T$, of $EuPt_4Ge_{12}$ is displayed in Fig. 2a. An analysis of $C(T)/T$ finds -as expected- a magnetic entropy of $R\ln 8$ (inset of Fig. 2a) at about 7 K.

$C(T)/T$ exhibits several anomalies at low temperatures. In contrast to previous works, reporting only one antiferromagnetic transition around $T_N \approx 1.7$, we can clearly identify at least three different anomalies, the first at $T_{N1} = 1.78$ K, at a slightly higher temperature than previously reported [8, 14], and at $T_{N2} = 1.03$ K and $T_{N3} = 0.8$ K. To further elucidate the nature of the phase transitions we conducted dc-susceptibility ($M(T)/T$) measurements. The general shape of $M(T)/H$ is dominated by a strong increase upon decreasing temperature and a pronounced maximum just below 1 K. Below the maximum $M(T)/H$ decreases again and saturates toward lower temperatures. This overall shape is typical for antiferromagnetic ordering with a Néel temperature slightly below temperature where the maximum in $M(T)/H$ is observed. However, $C(T)$ exhibits a strong anomaly at a nearly twice as large temperature $T_{N1} = 1.78$ K. Surprisingly, the transition at T_{N1} indicated by an arrow in Fig. 2b, is hardly visible in $M(T)/H$. Only in a magnification (right inset of Fig. 2b) a kink is visible at T_{N1} hinting at a magnetic character of the transition. This transition has been also previously identified in resistivity data, while no indications for additional phase transitions at lower temperatures have been reported [14]. At T_{N2} a small but clearly visible feature is evident in both, $C(T)/T$ and $M(T)/H$, while the most pronounced feature in $M(T)/H$ can be associated with T_{N3}. At T_{N3} $C(T)/T$ shows a well developed peak. In $M(T)/H$ one additional phase transition can be recognized at T_{N4}. On a closer look a change in slope is visible in $C(T)/T$ at this temperature. The magnetization, $M(H)$, at $T \approx 0.5$ K is proportional to the magnetic field in the whole measurement range up to 7 T. We do not observe any tendency toward saturation since only ≈ 5 μ_B are attained at our highest field. Only at about 2.5 T a tiny step hints at a subtle reorientation of the magnetic moments. Our results indicate complex magnetic ordering phenomena in $EuPt_4Ge_{12}$ which ask for further investigations.

3.2. $NdPt_4Ge_{12}$

The specific heat, $C(T)/T$, of $NdPt_4Ge_{12}$ displayed in Fig. 3a shows two distinct peaks at $T_{N1} = 0.67$ K and $T_{N2} = 0.58$ K indicating two magnetic phase transitions. A previous resistivity study reported a rapid decrease of $\rho(T)$ below ~0.7 K [13], which is in good agreement with T_{N1}. However, like in the case of $EuPt_4Ge_{12}$, no hint at the second phase transition has been observed in $\rho(T)$ [13]. The dc-susceptibility, $M(T)/T$, shows the typical temperature dependence of an antiferromagnetic material (see Fig. 3): with decreasing temperature $M(T)/T$ first continuously increases, exhibiting a maximum and then sharply drops on further decreasing temperature. As usually observed in an antiferromagnet the transition temperature (T_{N1}) coincides with the onset temperature of the drop in the susceptibility. At the second phase transition at T_{N2} only a small kink, possibly indicating a reorientation of the magnetic moments, is evident in $M(T)/T$. The magnetization at $T \approx 0.5$ K deviates from a linear behavior above 4 T, but no saturation is reached up to 7 T. At 7 T $m = 1.88\ \mu_B$/f.u..

An analysis of the magnetic entropy points at a quartet ground state in the CEF level scheme. Already at ~4 K the complete entropy corresponding to a quartet ground state, $R\ln 4$, is recovered, as displayed in the inset of Fig. 3a. In a CEF analysis of susceptibility and magnetization data of $NdPt_4Ge_{12}$ indeed a $\Gamma_{67}^{(2)}$ quartet ground state has been favored against a Γ_5 doublet [13]. In addition to our entropy analysis also the observed magnetic moment at 7 T and 0.5 K, $m = 1.88\ \mu_B$/f.u., is significantly larger compared to the expected saturation moment in case of a Γ_5 doublet of 1.33 μ_B/f.u.. Therefore, our data strongly supports the proposed $\Gamma_{67}^{(2)}$ quartet ground state [13].

4. Summary

In summary, $EuPt_4Ge_{12}$ and $NdPt_4Ge_{12}$ order magnetically below 1.78 K and 0.67 K, respectively. In $EuPt_4Ge_{12}$, Eu is in a stable magnetic Eu^{2+} configuration. The analysis of the specific heat at low temperatures yields an entropy of $R\ln 8$ around 6 K consistent with a fully degenerate $J = 7/2$ multiplet. We could identify four anomalies in specific heat and dc-susceptibility evidencing a complex magnetic phase-diagram. At the lowest obtained temperature ($T \approx 0.5$ K) we do not observe any saturation of the magnetic moment in fields up to 7 T. $NdPt_4Ge_{12}$ shows one additional magnetic transition inside the magnetically ordered state which is possibly related to a re-arrangement of the magnetic moments. The analysis of the magnetic entropy points to a quartet CEF ground state. To clarify the nature of the magnetic order in $EuPt_4Ge_{12}$ and $NdPt_4Ge_{12}$ further studies on single crystals would be desirable.

References

[1] Sales B C, Mandrus D and Williams R K (1996) *Science* **272** 1325.
[2] Uher C (2001) *Semicond, Semimetals* **69** 139.
[3] Chen G, Dresselhaus M S, Dresselhaus G, Fleurial J-P and Caillat T (2003) *Int. Materials Rev.* **48** 45.
[4] Uher C in *Chemistry, Physics, and Materials Science of Thermoelectric Materials: beyond Bismuth Telluride*, eds. Kanatzidis M G, Hogan T P and Mahanti S D, Spinger 2008.
[5] Sales B C 2007 *Int. J. Appl. Ceram. Technol* **4** 291.
[6] Jeitschko W and Braun D (1977) *Acta Cryst B* **33** 3401.
[7] Bauer E *et al.* (2007) *Phys. Rev. Lett.* **99** 217001.
[8] Gumeniuk R, Schnelle W, Rosner H, Nicklas M, Leithe-Jasper A and Grin Yu (2008) *Phys. Rev. Lett.* **100** 017002.
[9] Kaczorowski D and Tran V H (2008) *Phys. Rev. B* **77** 180504R.
[10] Bauer E *et al.* (2008) *Phys. Rev. B* **78** 064516.
[11] Maisuradze A *et al.* (2009) *Phys. Rev. Lett.* **103** 147002.
[12] Rosner H *et al.* (2009) *Phys. Rev. B* **80** 075113.
[13] Toda M *et al.* (2008) *J. Phys. Soc. Jpn.* **77** 124702.
[14] Grytsiv A *et al.* (2008) *J. Phys. Soc. Jpn. Suppl. A* **77** 121.

Journal of Physics: Conference Series **273** (2011) 012119 doi:10.1088/1742-6596/273/1/012119

Sample quality and magnetic properties of $Ca_{1-x}Na_xV_2O_4$

Hiroya Sakurai[1], **Hiroyuki Takeya**[2], **and Kazuo Hirata**[2]

[1] National Institute for Materials Science, 1-1 Namiki, Tsukuba 305-0044, Japan
[2] National Institute for Materials Science, 1-2-1 Sengen, Tsukuba 305-0047, Japan

E-mail: sakurai.hiroya@nims.go.jp

Abstract. $Ca_{1-x}Na_xV_2O_4$ shows a variety of physical properties depending on temperature, magnetic field and the composition. We made the $x = 1/6$, $1/3$ and $1/2$ samples under a high pressure of 6 GPa and grew single crystal of CaV_2O_4 at ambient pressure using floating-zone technique. The samples of the solid-solution system were found to be of high quality. On the other hand, antiferromagnetic transition temperature, T_N, of the single crystal was $T_N = 66$ K, which is significantly lower than that of the raw materials ($T_N = 78$ K) as reported previously. This is probably because Ca content of the single crystal is slightly smaller than that of the powder. Electron probe microanalysis indicated that the true composition of the single crystal is $Ca_{0.99}V_2O_4$.

1. Introduction

AV_2O_4 (A = Ca or Na) crystallizes in calcium ferrite type structure, which is composed of the double chains of VO_6 octahedra, as shown in Fig. 1. Because low-dimensional electron correlations and geometrical frustration are expected in the compounds, much effort has recently been made to clarify their physical properties. The double chains in CaV_2O_4, which is electrically insulating, can be identified as Haldane chains at a high temperature and as spin ladders at a low temperature, and it finally settles in antiferromagnetic (AF) ordered state below $T_N = 51 - 78$ K [1, 2, 3, 4]. In the magnetic behavior, the orbital degree of freedom plays an important role [2, 3, 4]. On the other hand, NaV_2O_4 shows AF ordering at $T_N = 140$ K maintaining metallic conductivity [5]. Photoemission spectroscopy measurement revealed the suppression of the spectral density of state and the power law behavior, as expected for the Tomonaga-Luttinger liquid, indicating one-dimensional nature of the compound [6]. Due to neutron diffraction measurements, incommensurate spin-density-wave state with magnetic moment perpendicular to the chains is thought to be the most probable magnetic ground state [7]. Interestingly, the AF phase can be divided into, at least, three subphases of AF1, AF2, and AF3, depending on the temperature and the magnetic field [8]. AF1 phase covers the small area of high temperature and low field, AF2 phase is stabilized under a high field in a wide temperature range below T_N, and AF3 phase, including the ground state, is seen at a low-temperature and low-field region. Magnetization, electrical resistivity and specific heat measurements indicated that, when Na ions are substituted with Ca ions, the T_N value decreases and AF1 phase becomes dominant as seen in Fig. 2. Below a certain Na content, the compound exhibits metal-insulator transition at the temperature where the magnetic susceptibility shows a peak. As the Na content

Published under licence by IOP Publishing Ltd

further decreases, the metal-insulator transition becomes obscure and finally disappears [8]. This complex phase diagram has been confirmed by muon spin rotation/relaxation (μSR) [9, 10], ^{51}V and ^{23}Na nuclear magnetic resonance (NMR) [11, 12], and optical conductivity measurements [13]. Pressure effect was also studied [14]; T_N and T_{n1}, the temperature of the phase boundary between AF1 and AF2, decrease with increasing pressure under the magnetic field of 1 T at the suppression rates of 5.2 K/GPa and 4.2 K/GPa, respectively, which is very consistent with μSR results at 1.7 GPa [9, 10]. However, it is strange that AF3 phase becomes stable at 1.7 GPa at 3 K for $Ca_{1/6}Na_{5/6}V_2O_4$, in which at ambient pressure only AF1 phase was observed down to 1.8 K [9, 10], although for NaV_2O_4 T_{n2}, the temperature of the phase boundary between AF2 and AF3, decreases more rapidly with increasing pressure above 0.4 GPa under 1 T than T_N or T_{n1}.

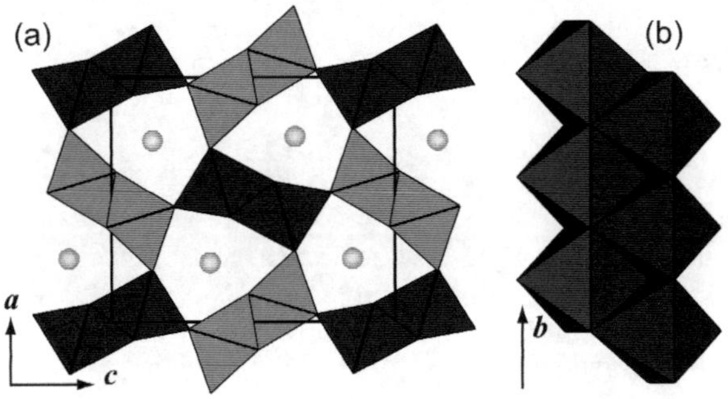

Figure 1. Structure of AV_2O_4 (A = Ca or Na) (a) and the double chain (b) The octahedra and the circle represent VO_6 octahedra and A, respectively.

2. Experimental

Powder CaV_2O_4 was obtained by reduction reaction of CaV_2O_6, which was in advance made from the stoichiometric mixture of $CaCO_3$ and V_2O_5 at 700°C in air. The reaction was performed in flowing H_2 gas at 1000°C for 6 hours. Single crystal of CaV_2O_4 was grown from the powder by conventional floating-zone method in flowing 5% H_2/Ar gas. White powder appeared very thinly on the lining of the quart tube of the floating-zone furnace. The crystal was annealed in flowing H_2 gas at 1000°C for 2 hours and then cooled slowly in the furnace.

Powder samples of $Ca_{1-x}Na_xV_2O_4$ (x = 1/6, 1/3, and 1/2) were synthesized from the stoichiometric mixtures of CaO, V_2O_3 and $Na_4V_2O_7$. CaO was prepared by decomposition of $CaCO_3$ at 1050°C in flowing Ar gas. V_2O_3 was made from V_2O_5 by being heated in flowing H_2 gas at 600°C for 4 hours and then at 800°C for 2 hours. $Na_4V_2O_7$ was synthesized from the stoichiometric mixture of Na_2CO_3 and V_2O_5 in air at 550°C . Because CaO and $Na_4V_2O_7$ are reactive with water and/or carbon dioxide in air, they were put in a dry box filled with Ar immediately after they were synthesized. Weighing, mixing and sealing in Au capsule were performed in the dry box. The samples were pressed in a belt-type press under 6 GPa, and heated at 1300°C for 1 hour. The pressure was released after the temperature was quenched to room temperature.

X-ray diffraction (XRD) measurements were performed on a commercial diffractmeter, X'pert, Panalatica, using Cu $K\alpha$ radiation. Ca contents in the CaV_2O_4 samples were determined by electron probe microanalysis (EPMA). Four pieces of single crystals taken from different parts of the crystal rod, as well as sintered pellet of the powder sample, were used for the analysis. Ca content, y in $Ca_yV_2O_4$, was determined as the mean of the values obtained from more than five spots on each sample, to be y = 1.00 for the powder sample and y = 0.99 for each single crystal. Magnetic data were corrected under 1 T using a commercial magnetmeter, MPMS-XL, Quantum Design.

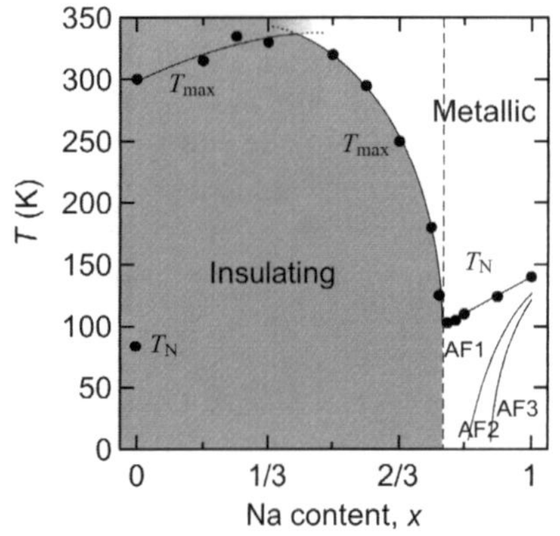

Figure 2. Phase diagram of $Ca_{1-x}Na_xV_2O_4$ at magnetic field of 1 T.

3. Results and Discussion

XRD patterns of the samples are shown in Fig. 3. In the pattern of the $x = 1/6$, $1/3$, or $1/2$ sample, a peak from impurity is detectable at $2\theta = 48.3^\circ$. No other impurity peak was distinguishable from the peaks coming form $Ca_{1-x}Na_xV_2O_4$. The impurity is likely $CaVO_3$ because Varga et al. observed it in their samples [15]. We did not find V_2O_3 or NaV_6O_{11}, which they found in their samples. It should be emphasized that our previous experiments [7, 8, 9, 10, 11, 12, 13] were performed using as high quality samples as these.

For $x = 1/6$ and $1/3$ samples, the width of some peaks, for example the peak at $2\theta \sim 36^\circ$, is a little larger than that for $x = 0$ or $1/2$, which is likely due to the phase separation observed by μSR and NMR [9, 10, 11]. Since the phase separation seems imperfect as suggested by the gradual change in the normalized weak transverse-field asymmetry [9, 10], it is possibly suppressed by more rapid thermal quenching and/or by quenching from higher temperature.

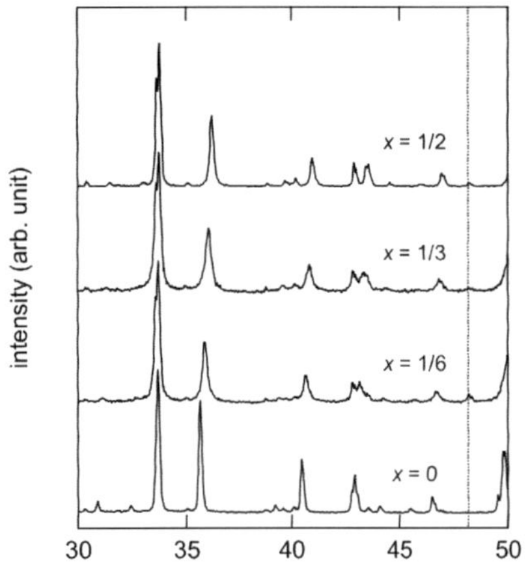

Figure 3. XRD patterns of $Ca_{1-x}Na_xV_2O_4$ samples. The dotted line shows the position of the second largest peak of $CaVO_3$, whose intensity is half of that of the main peak.

Magnetic susceptibility of CaV_2O_4 was shown in Fig. 4. The T_N values were estimated from the peaks of the differentiated curves to be $T_N = 78$ K for the powder sample and 66 K for the single crystal. The difference in the T_N value is most likely due to the difference in Ca content. Probably, the compound can accommodate such a amount of Ca defects ($\sim 1\%$) at around the melting point although it cannot below about 1200°C . This idea can explain why T_N

of the as-grown crystal is the lowest and why V_2O_3 inclusion appeared in the crystals annealed at 1200°C in flowing 5% H_2/He gas [2]. The partial lack of Ca ions increases the V valence as in the case of the substitution of Ca with Na ions to lower T_N, and annealing enhances T_N because the Ca defects gather to form CaV_2O_4 with full Ca ions. V_2O_3 would be made during the annealing process because VO_2 is reduced by hydrogen gas.

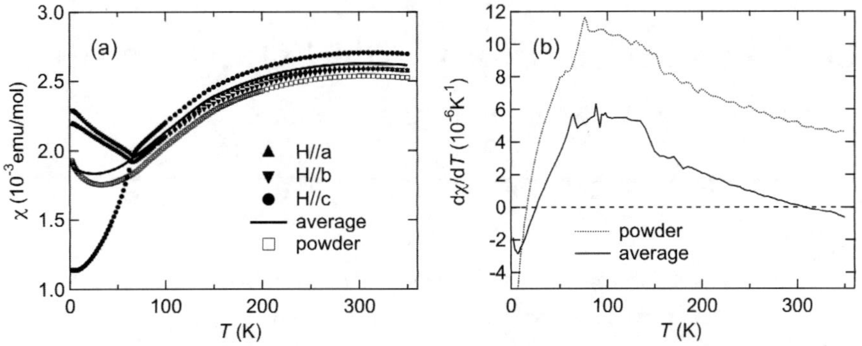

Figure 4. Temperature dependences of magnetic susceptibility of CaV_2O_4 (a) and their differentiated curves (b). The average curve was calculated as the mean of the three kinds of susceptibility.

Acknowledgments

Special thanks to T. Taniguchi for his experimental support for the high-pressure synthesis and to K. Kosuda for EPMA measurements. This study is supported partially by Grant-in-Aid for Scientific Research (A) from JSPS (22246083), and by FIRST, JSPS.

References

[1] Zong X, Suh B J, Nizai A, Yan J Q, Schlagel D L, Lograsso T A, and Johnston D C 2008 *Phys. Rev B* **77** 014412
[2] Niazi A, Bud'ko S L, Schlagel D L, Yan J Q, Lograsso T A, Kreyssig A, Das S, Nandi S, Goldman A I, Honecker A, McCallum R W, Reehuis M, Pieper O, Lake B and Johnston D C 2009 *Phys. Rev. B* **79** 104432
[3] Pieper O, Lake B, Daoud-Aladine A, Reehuis M, Prokeš K, Klemke B, Kiefer K, Yan J Q, Niazi A, Johnston D C and Honecker A 2009 *Phys. Rev. B* **79** 180409
[4] Chern G-W and Perkins N 2009 *Phys. Rev. B* **80** 220405
[5] Yamaura K, Arai M, Sato A, Karki A B, Young D P, Movshovich R, Okamoto S, Mandrus D and Takayama-Muromachi E 2007 *Phys. Rev. Lett.* **99** 196601
[6] Qian T, Nakayama K, Sun Y, Arakane T, Sato T, Takahashi T, Yamaura K and Takayama-Muromachi E 2009 *J. Phys. Soc. Jpn.* **78** 024709
[7] Nozaki H, Sugiyama J, Månsson M, Harada M, Pomjakushin V, Sikolenko V, Cervellino A, Roessli B and Sakurai H 2010 *Phys. Rev. B* **81** 100410
[8] Sakurai H 2008 *Phys. Rev. B* **78** 094410
[9] Sugiyama J, Ikedo Y, Goko T, Ansaldo E J, Brewer J H, Russo P L, Chow K H and Sakurai H 2008 *Phys. Rev. B* **78** 224406
[10] Sugiyama J, Ikedo Y, Russo P L, Goko T, Ansaldo E J, Brewer J H, Chow K H and Sakurai H 2009 *Physica B* **404** 789
[11] Takeda H, Itoh M and Sakurai H 2010 *J. Phys.: Conf. Ser.* **200** 012200
[12] Takeda H, Itoh M and Sakurai H 2010 *65th Annual Meeting of the Physical Society of Japan*
[13] Irizawa A, Nanba T and Sakurai H 2008 *2008 Annual Meeting of the Physical Society of Japan*
[14] Tseng K F, Ho C J, Chou C C, Huang C L, Sakurai H and Yang H D 2010 *J. Phys.: Conf. Ser.* **200** 012211
[15] Varga T, Mitchell J F, Yamaura K, Mandrus D G and Wang J 2009 *Solid State Sci.* **11** 694

International Conference on Strongly Correlated Electron Systems (SCES 2010) IOP Publishing
Journal of Physics: Conference Series **273** (2011) 012120 doi:10.1088/1742-6596/273/1/012120

Magnetic properties of new lled skutterudite compounds $GdFe_4As_{12}$ and $TbFe_4As_{12}$

C. Sekine [1], K. Ito [1], K. Matsui [1] and T. Yagi [2]

[1] Graduate School of Engineering, Muroran Institute of Technology, 27-1 Mizumoto-cho, Muroran, Hokkaido 050-8585, Japan
[2] Institute for Solid State Physics, University of Tokyo, Kashiwa, Chiba 277-8581, Japan

E-mail: `sekine@mmm.muroran-it.ac.jp`

Abstract. New As-based lled skutterudite compounds $GdFe_4As_{12}$ and $TbFe_4As_{12}$ have been synthesized under high pressure. The samples have a cubic structure with lattice parameters a=8.3024 Å, and 8.2961 Å, respectively. Both compounds show metallic behavior and a pronounced kink at low temperature in the temperature dependence of resistivity. Magnetic measurements revel that $GdFe_4As_{12}$ and $TbFe_4As_{12}$ exhibit a ferromagnetic-like phase transition at 56K and 38K, respectively. The unexpectedly high transition temperatures could be related to the interaction of Gd (or Tb) 4f moments with Fe 3d moments.

1. Introduction

The filled skutterudite compounds, particularly P- and Sb-based compounds, have attracted much attention of their wide variety of strongly correlated electron behaviors, such as unconventional superconductivity[1], non-Fermi liquid behavior[2, 3], anomalous metal–insulator transition[4], and multipole ordering[5]. As-based filled skutterudite compounds also should be important for systematically investigating skutterudite systems. In fact, several compounds exhibit exotic properties, including superconductivity in $Ln\mathrm{Ru}_4\mathrm{As}_{12}$ (Ln=La, Pr)[6], canted ferromagnetic or ferrimagnetic phase transition in $EuFe_4As_{12}$[7]. However, only preliminary studies have been conducted on the properties of As-based compounds because the compounds are quite difficult to prepare. High-pressure synthesis using pressure above 4GPa is a powerful technique for preparing As-based skutterudite compounds. We have suceeded in synthesizing samples of new As-based filled skutterudite compounds $GdFe_4As_{12}$ and $TbFe_4As_{12}$ using the high-pressure synthesis technique. In this paper, we report the magnetic properties of $GdFe_4As_{12}$ and $TbFe_4As_{12}$ for the first time.

2. Experimental

Polycrystalline $GdFe_4As_{12}$ and $TbFe_4As_{12}$ were prepared at high temperatures and high pressures using a wedge-type cubic anvil high-pressure apparatus. The sample cell assembly for the preparation of the As-based filled skutterudite compounds is similar to that used for the high-pressure synthesis of P-based filled skutterudite compounds[8]. The compounds were prepared by reacting stoichiometric amounts of 3N (99.9% pure)-Gd, Tb chips, 4N-Fe, and 6N-As powders at 4 GPa. The reaction temperatures were between 840 and 850°C. The prepared samples were characterized by powder x-ray diffraction using $\mathrm{CoK}_{\alpha 1}$ radiation and silicon as

Published under licence by IOP Publishing Ltd

Journal of Physics: Conference Series **273** (2011) 012120 doi:10.1088/1742-6596/273/1/012120

a standard. Resistivity was measured by a standard dc four-probe method. Magnetization and dc magnetic susceptibility were measured with a Quantum Design MPMS superconducting quantum interference device magnetometer.

3. Results and discussion

Figure 1 shows an x-ray diffraction pattern of the new filled skutterudite $GdFe_4As_{12}$ prepared at high pressures. Most of the observed diffraction lines were indexable using the skutterudite structure though small amount of impurity phases ($FeAs_2$, metallic As) were detected. The lattice constant determined by a least-squares fit to the data was 8.3024Å. The x-ray diffraction pattern of the new compound $TbFe_4As_{12}$ also indicates that the main phase consists of the filled skutterudite with less than 5 wt% impurity phases of $FeAs_2$ or metallic As. The lattice constant of $TbFe_4As_{12}$ was determined in the same manner and found to be 8.2961Å.

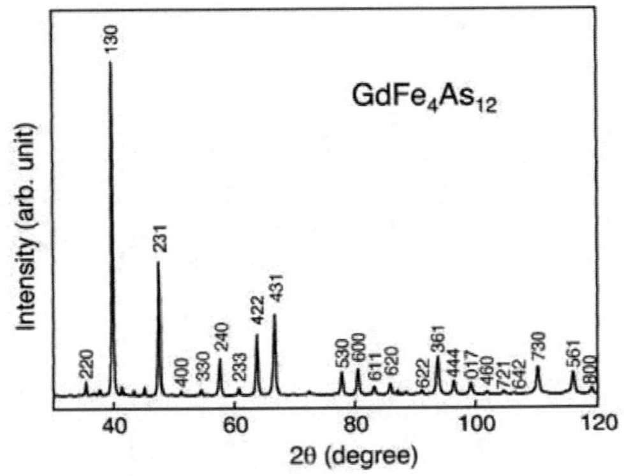

Figure 1. X-ray diffraction pattern of the new filled skutterudite compound $GdFe_4As_{12}$.

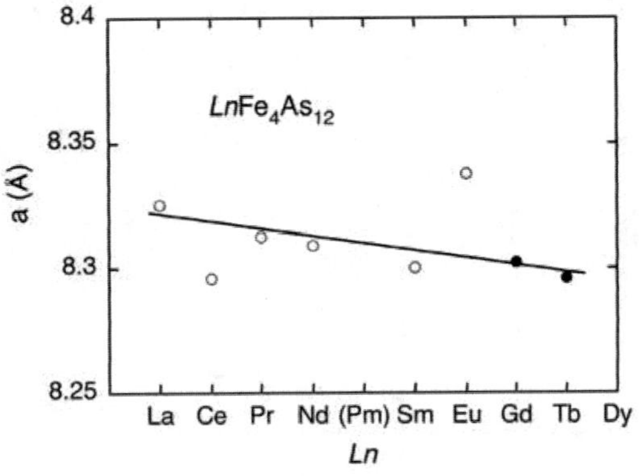

Figure 2. Lattice constants of filled skutterudite arsenides $LnFe_4As_{12}$ (Ln=La, Ce, Pr, Nd, Sm, Eu, Gd, and Tb.)

Figure 2 plots the lattice constants of the new filled skutterudite compounds $GdFe_4As_{12}$ and $TbFe_4As_{12}$ along with the values reported earlier for the filled skutterudite arsenides $LnFe_4As_{12}$[7]. Starting with La, there is a typical reduction in lattice constant due to the contraction of the trivalent ionic radii of the lanthanides with increasing atomic number. This indicates that Gd and Tb ions are in a trivalent state. The small lattice constants of Ce compounds are due to strong c-f hybridization. The large lattice constants of the Eu compounds suggest that Eu ions are in a divalent or mixed-valence state. The same tendency of the lattice constant was observed in P-based filled skutterudite compounds [9].

Figure 3 shows the temperature dependence of the magnetic susceptibility χ and the inverse magnetic susceptibility χ^{-1} at 1T for $GdFe_4As_{12}$. The inset of Fig. 3 shows the temperature dependence of χ below 100K at 0.01T. A large increase in $\chi(T)$ below 56K indicates a ferromagetic-like transition. Figure 4 shows the temperature dependence of χ and χ^{-1} at 1T for $TbFe_4As_{12}$. The inset of Fig. 4 shows the low temperature region of χ at 0.05T. $TbFe_4As_{12}$ also exhibits a large increase in $\chi(T)$ below 38K indicating ferromagnetism. $\chi(T)$ of $GdFe_4As_{12}$ and $TbFe_4As_{12}$ follows the Curie–Weiss law above 150K. Least-squares fits yield effective moments of 8.09 μ_B/Gd and 9.76 μ_B/Tb for $GdFe_4As_{12}$ and $TbFe_4As_{12}$, respectively. These effective moments are close to the expected values for trivalent ions of μ_{eff}= 7.94 μ_B/Gd and 9.72μ_B/Tb.

Magnetization measurements at 2 K on $GdFe_4As_{12}$ and $TbFe_4As_{12}$ reveal a feature due to spontaneous magnetization (Fig. 5). The magnetization increases with increasing fields, reaching 6μ_B/Gd and 4.6μ_B/Tb at 5 T. The saturation magnetization should reach 7 μ_B/Gd

Journal of Physics: Conference Series **273** (2011) 012120 doi:10.1088/1742-6596/273/1/012120

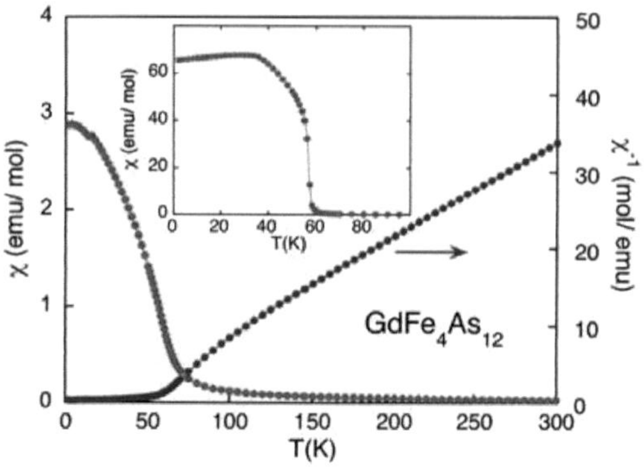

Figure 3. Magnetic susceptibility χ and inverse magnetic susceptibility χ^{-1} vs temperature at 1T for $GdFe_4As_{12}$. The inset shows the low-temperature region of χ at 0.01T.

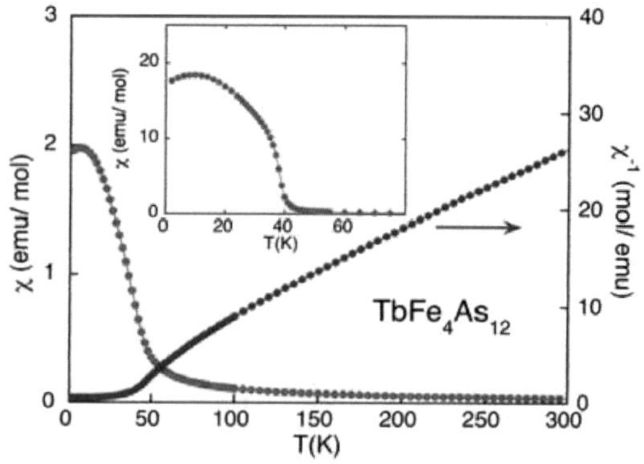

Figure 4. Magnetic susceptibility χ and inverse magnetic susceptibility χ^{-1} vs temperature at 1T for $TbFe_4As_{12}$. The inset shows the low-temperature region of χ at 0.05T.

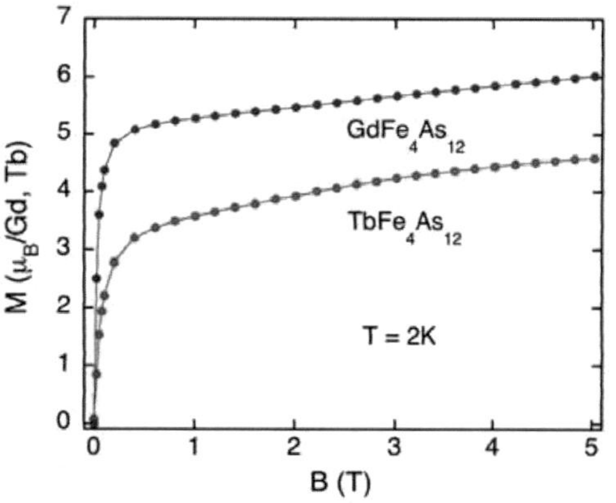

Figure 5. Magnetization for $GdFe_4As_{12}$ and $TbFe_4As_{12}$ at 2 K.

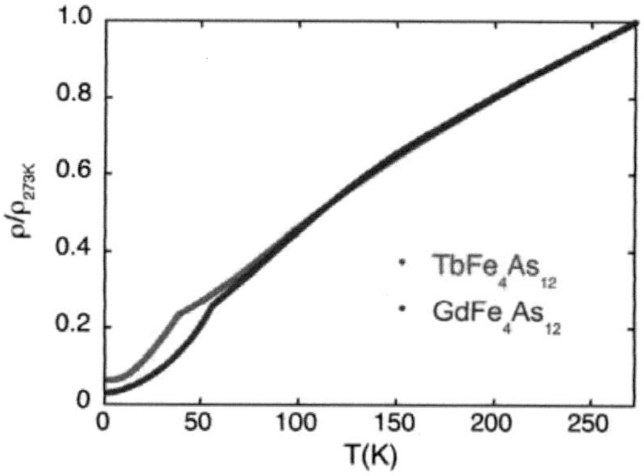

Figure 6. Temperature dependence of electrical resistivity ρ/ρ_{273K} for $GdFe_4As_{12}$ and $TbFe_4As_{12}$ normalized at 273 K.

and $9\mu_B$/Tb. The significantly small experimental values of the magnetization and relatively high transition temperatures suggest that these compounds have ferrimagnetic or canted ferromagnetic ground state. The nonlinear, s-like shape of χ^{-1} also supports this scenario. $EuFe_4Sb_{12}$ with high transition temperature T_C of 85K has been found to be ferrimagnetic in which Fe and Eu moments are ordered with antiferromagnetic coupling[12].

Figure 6 shows the temperature dependence of the electrical resistivity $\rho(T)/\rho_{273K}$ for $GdFe_4As_{12}$ and $TbFe_4As_{12}$ normalized at 273K. The overall behaviors of these compounds indicate metallic states. The $\rho(T)$'s of these compounds show a pronounced kink around T_C=56 and 38K, corresponding to the ferromagnetic-like transitions.

4. Summary

We have studied the physical properties of new filled skutterudite compounds $GdFe_4As_{12}$ and $TbFe_4As_{12}$. Magnetic measurements reveal that $GdFe_4As_{12}$ and $TbFe_4As_{12}$ exhibit a ferromagnetic-like transition at 56K and 38K. The magnetic properties of $GdFe_4As_{12}$ and

Table 1. Transition temperatures to the ordered state T_{TR} (T_C: ferromagnetic order, T_N: antiferromagnetic order), effective magnetic moments μ_{eff}, and Weiss temperatures θ_p of LnT_4X_{12} (Ln=Gd, Tb, T= Fe, Ru, and Os; X=P, and As).

Compound	T_{TR}(K)	$\mu_{eff}(\mu_B)$	θ_p(K)	Ref. No.
$GdFe_4P_{12}$	T_C: 23	7.90	17.5	9
$GdRu_4P_{12}$	T_N: 22	8.04	23	10
$GdOs_4P_{12}$	T_C: 5	8.54	2.9	8
$GdFe_4As_{12}$	T_C: 56	8.09	23	This study
$TbFe_4P_{12}$	T_C: 10	9.48	10	11
$TbRu_4P_{12}$	T_N: 20	9.76	8	10
$TbOs_4P_{12}$	No	10.86	3	9
$TbFe_4As_{12}$	T_C: 38	9.76	16	This study

$TbFe_4As_{12}$ are summarized in Table 1 along with those of filled skutterudite phosphides with Gd and Tb. Fe and Os compounds show a ferromagnetic transition except for $TbOs_4P_{12}$, which remains paramagnetic down to 2K. On the other hand, Ru compounds show an antiferromagnetic transition. The positive Weiss temperatures for these compounds suggests that basically there are ferromagnetic correlations in this system. In the absence of other effects, e.g., Fermi surface instability or multipole interaction, these compounds are expected to show ferromagnetic ordering. The unexpectedly high transition temperatures of $GdFe_4As_{12}$ and $TbFe_4As_{12}$ could be due to the interaction between Gd (or Tb) 4f moments and Fe 3d moments. To elucidate this point, element-specific magnetic measurements are needed.

Acknowledgments

This work was supported by a Grant-in-Aid for Scientific Research on Innovative Areas "Heavy Electrons" (No. 20102004) of The Ministry of Education, Culture, Sports, Science, and Technology, Japan and a Grant-in Aid for Scientific Research (C) (No. 20540339) of the Japan Society for the Promotion of Science.

References

[1] Aoki Y, Tayama T, Sakakibara T, Kuwahara K, Iwasa K, Kohgi M, Higemoto W, MacLaughlin D E, Sugawara H and Sato H 2007 *J. Phys. Soc. Jpn.* **76** 051006
[2] Takeda N and Ishikawa M 2000 *J. Phys. Soc. Jpn.* **69** 868
[3] Yamamoto A, Wada S, Shirotani I and Sekine C 2006 *J. Phys. Soc. Jpn.* **75** 063703
[4] Sekine C, Uchiumi T, Shirotani I and Yagi T 1997 *Phys. Rev. Lett.* **79** 3218
[5] Yoshizawa M, Nakanishi Y, Oikawa M, Sekine C, Shirotani I, Saha S R, Sugawara H and Sato H 2005 *J. Phys. Soc. Jpn.* **74** 2141
[6] Shirotani I, Uchiumi T, Ohno K, Sekine C, Nakazawa Y, Kanoda K, Todo S and Yagi 1997 *Phys. Rev.* B **56** 7866
[7] Sekine C, Akahira K, Ito K and Yagi T 2009 *J. Phys. Soc. Jpn.* **78** 093707
[8] Kihou K, Shirotani I, Shimaya Y, Sekine C and Yagi T 2004 *Mater. Res. Bull.* **39** 317
[9] Sekine C, Ando H, Sugiuchi Y, Shirotani I, Matsuhira K and Wakeshima M 2008 *J. Phys. Soc. Jpn.* **77** Suppl. A 135
[10] Sekine C, Uchiumi T, Shirotani I, Matsuhira K, Sakakibara T, Goto T and Yagi T 2000 *Phys. Rev.* B **62** 11581
[11] Shirotani I, Takeda K, Sekine C, Hayashi J, Nakata R, Kihou K, Ohishi Y and Yagi T 2006 *Z. Naturforsch.* **61b** 1471
[12] Krishnamurthy V V, Lang J C, Haskel D, Keavney D J, Srajer G, Robertson J L, Sales B C, Mandrus D G, Singh D J and Bilc D I 2007 *Phys. Rev. Lett.* **98** 126403

Ref.: Ms. No. SCES127

Magnetic properties of new filled skutterudite compounds $GdFe_4As_{12}$ and $TbFe_4As_{12}$

Journal of Physics: Conference Series

Thank you very much for positive evaluation of our study.

>P.1, Line 26, Abstract: Authors should specify that the metallic behavior and the kink is observed
>in the temperature dependence of resistivity

1) We have added the phrase "in the temperature dependence of resistivity" to the third line sentence in Abstract.

> P3: Fig. 3 and Fig. 4 Captions: Indicate the magnetic field for the main figures.

2) We have added the phrase "at 1T" in the figure captions.

> Please indicate in text whether the samples are single- or poly- crystalline.

3) We have added the word "Polycrystalline" to the first sentence in the chapter of "Experimental".

>The nonlinear, s-like shape of the inverse susceptibility also provides support for ferrimagnetism or
>canted ferromagnetism. Adding a note to this effect near line 46 (page 3) would bolster your claim
>concerning the nature of the ordered state.

4) We have added the sentence "The nonlinear, s-like shape of χ-1 also supports this scenario." near line 46 (page 3).

With the best wish

The authors

Novel magnetic behaviour of $GdPd_2Si_2$ single crystal

T. Shigeoka[1], T. Fujiwara[1], K. Kanto[1], Y. H. Zhang[1], Y. Uwatoko[2]

[1]Graduate School of Science and Engineering, Yamaguchi University, Yamaguchi 753-8512, Japan.
[2]Institute for Solid State Physics, University of Tokyo, Kashiwa-shi Chiba 277-8581, Japan.

shigeoka@yamaguchi-u.ac.jp

Abstract. Magnetic and specific heat measurements have been performed on a $GdPd_2Si_2$ single crystal compound having the tetragonal $ThCr_2Si_2$-type crystal structure. Specific heart shows two λ-type anomalies at T_N=16.7 K, T_1=12.3 K and a small hump with hysteresis at T_2=6.8 K; there are three phases below T_N. The magnetic entropy of Rln6 is released below T_N. The temperature dependence of magnetic susceptibility shows a very peculiar behaviour. With decreasing temperature, the magnetic susceptibility along the [001] direction increases through a small kink at T_N down to T_1 where a clear cusp is seen, decreases rapidly below T_N, then shows an upturn and a clear kink at T_2 followed by a gradual increase. This behaviour examines at various magnetic field along [001] and [100]. The B_{001}-T and B_{100}-T magnetic phase diagrams have been constructed. The field-induced phase appears near by the 100] magnetic field, B_{100}.

1. Introduction

The ternary compounds having the tetragonal $ThCr_2Si_2$-type crystal structure have been extensively studied for last three decades because of a large variety of physical properties [1]. A large number of these compounds show an antiferromagnetic order and are Ising-like systems with the easy magnetization axis of the [001] direction (c-axis). Much less study has been made on compounds with the easy magnetization direction in the basal plane. The behavior of magnetic moments having the basal plane anisotropy is interesting. Magnetic study has been planning on the compounds series RPd_2Si_2 (R=rare earth) which has an anisotropy of easy direction in the basal plane [1, 2]. The compound $GdPd_2Si_2$ without crystal field effects is taken as the first target. The compound shows antiferromagnetic order below 13 K or 16.5 K and a metamagnetic behavior [3, 4]. The report suggests Gd magnetic moments lie in the basal plane. The detail magnetic behavior has been, however, unknown yet. Magnetic behavior on a $GdPd_2Si_2$ single crystal has been studied by magnetic and specific measurements.

2. Experimental

The single crystal has been grown by the tetra-arc Czochralski method. The single phase nature has been confirmed by X-ray powder diffraction. Good quality of the single crystal has been confirmed and crystallographic orientations were determined by back Laue method. The crystal was fixed on a plastic plate so that the desired direction is perpendicular to the plate within experimental angle

Published under licence by IOP Publishing Ltd

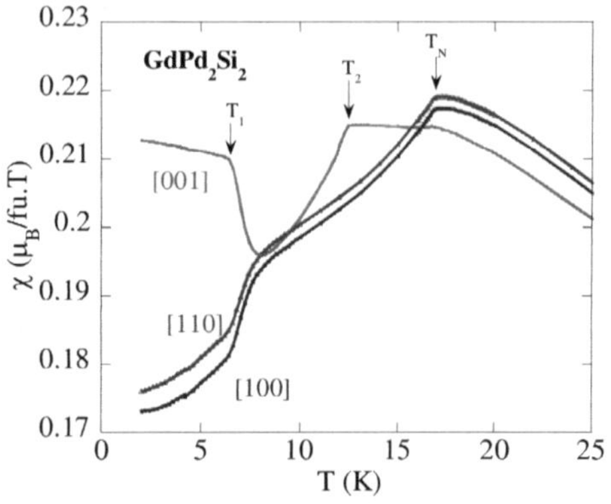

Figure 1 Temperature dependence of magnetic susceptibilities along the main symmetry axes on the $GdPd_2Si_2$ single crystal

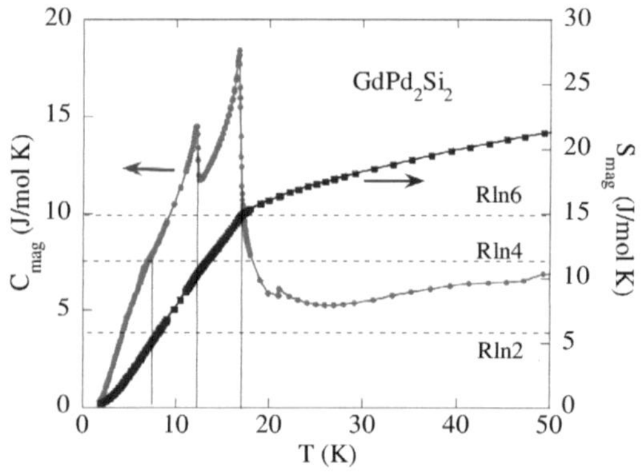

Figure 2 Thermal variation of magnetic specific heat and entropy

accuracy of one degree. Then it is used for magnetic measurements which have been carried out by use of a PPMS (Quantum Design). The specific heat has been performed by a relaxation method also using the PPMS.

3. Results and discussion

Figure 1 shows temperature dependences of magnetic susceptibilities along the symmetry axes in a tetragonal cell on the $GdPd_2Si_2$ single crystal. The dependencies are very peculiar. With decreasing temperature, the [001] susceptibility increases down to T_2=12.6 K via a small kink at T_N=16.7 K, then decreases steeply, goes through a deep valley, increases rapidly and shows a clear kink at T_1=6.8 K followed by almost constant value. There are three anomalies at T_N, T_2 and T_1, indicating phase transitions. On the other hand, the [100] and [110] susceptibilities in the basal plane are almost identical. They increase up to the maximum value at T_N and decrease with a rapid change around T_1. No clear anomaly is seen around T_2. These low temperature behaviours indicate that the magnetic moments are in the basal plane in the ground state and their direction may change for intermediate temperature. This is very strange. The direction of a Gd moment, magnetic anisotropy Gd compounds, is insensitive for the crystal field effects. The dipole-dipole interaction plays a dominant role to determine the moment direction. In this crystal structure, the maximum electric field gradient is the [001] direction [2]. This leads generally the magnetic moments direction of [001]. Other interactions so that lead frustration of magnetic moments, should be considered.

The magnetic susceptibilities above 50 K follow the Curie-Weiss law. The curves become linear and parallel with each other for high temperature. The effective magnetic moment μ_{eff} is 8.1 μ_B which is slightly larger than one of Gd^{3+} free ion. The paramagnetic Curie temperature Θ_p is -45.3 K, indicating a dominant interaction is antiferromagnetic.

The thermal variations of magnetic specific heat C_{mag} and entropy S_{mag} of $GdPd_2Si_2$ are shown in Fig.2. The magnetic contribution of specific heat is deduced from subtracting the specific heat of nonmagnetic compound YPd_2Si_2 from total specific heat of $GdPd_2Si_2$. There are three anomalies: two λ-type ones at T_N and T_2, and a small hump with a hysteresis loop at T_1. The existence of three phases below T_N is confirmed. The phase transitions are the second-order ones at T_N and T_2, and the first-order one at T_1. The magnetic entropy is obtained from numerical integration of C_{mag}/T vs. T. It is evidenced from the figure the entropy of Rln6, Rln4 and Rln2 is released below T_1, T_2 and T_N, respectively. This indicates that the ground state of is a pseudo-sextet at least.

The temperature dependences of the [001] magnetization/magnetic field: $M_{[001]}/B$, at various applied fields are shown in Fig. 3. The transition temperature of T_N is almost constant for this applied field

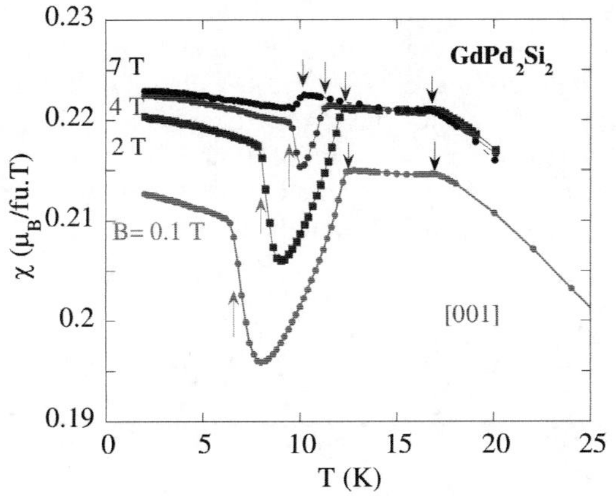

Figure 3 Thermal variation of *M/B* along the [001] direction under various magnetic fields.

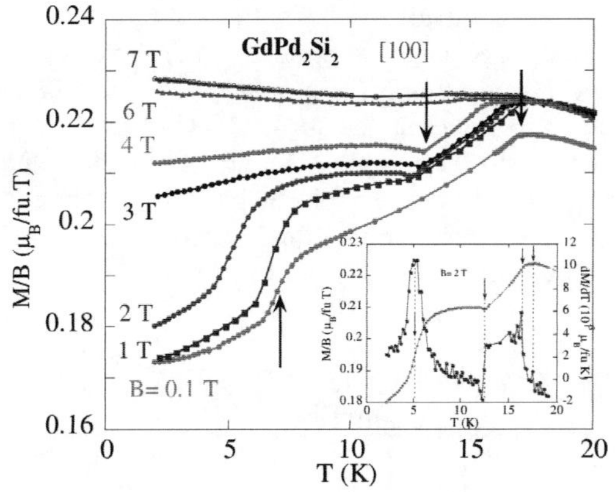

Figure 4 Thermal variation of *M/B* along the [100] direction under various magnetic fields. The inset shows thermal variation of *M/B* under 2T and d*M*/d*T*.

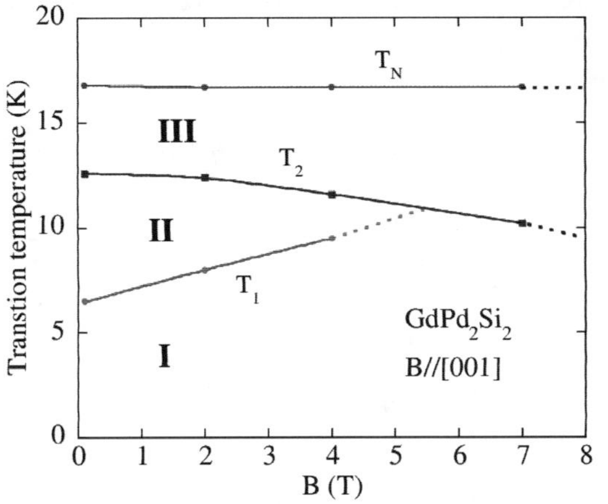

Figure 5 B_{001}-T magnetic phase diagram.

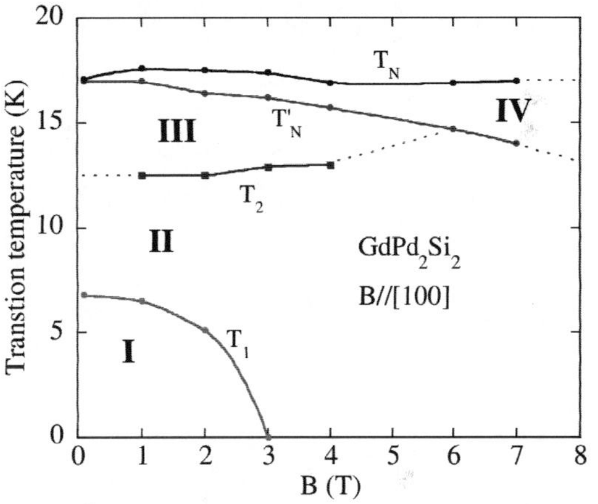

Figure 6 B_{100}-T magnetic phase diagram.

range. The transition temperature T_2 decreases with increasing magnetic field while T_1 increases and disappears around 5 T. Three ordered phases exist, which are referred to as the phase I, II and III for $T<T_1$, $T_1<T<T_2$ and $T_1<T<T_N$, respectively. On the other hand, the variation of the transition temperatures on the [100] applied magnetic field is different from one of [001], which is evidenced from Fig.4. The field-induced transition, the phase IV, appears below and near T_N. We call it T_N' here after. The transition points T_1 is determined from a peak temperature in d*M*/d*T* vs. *T* curve shown in the inset of Fig. 5. The T_2 and T_N' are from a rapid change and T_N is corresponding to the temperature where d*M*/d*T* is across zero at high temperature. The T_N increases slightly in low field, which is a strange behaviour, and becomes almost constant for higher field. This behaviour is not explained by a usual antiferromagnetic interaction only, suggesting an existence of other interactions. The field-induced transition temperature decreases with increasing field. The temperature of T_2 increases gradually and disappears around 6 T. The T_1 decreases and disappears around 3 T. From these results, the $B_{[001]}$-T and $B_{[001]}$-T magnetic phase diagrams have been constructed and are shown in Fig.5 and Fig.6, respectively.

The comparison of both diagrams is interesting. The magnetic behaviour is anisotropic nevertheless a Gd compound. There are three magnetic phases the phase, I, II and III for $T<T_1$, $T_1<T<T_2$ and $T_1<T<T_N$, in non magnetic field. A field-induced phase, the phase IV, appears only by magnetic field applied to the [100] direction. The [100] field makes the phase II be stable and the phase I and III instable. In contrast, the phase I and III become stable by the [001] magnetic field and the phase II becomes instable. This strange behaviour suggests a new mechanism to determine magnetic moment direction. In order to explain this, magnetic structures should be determined.

References

[1] Gignoux D, Schmitt D,1997, *Handbook of Magnetic. Materials.* Vol.**10,** ed. Buschow K H J (Amsterdam:Elsevier) 239

[2] Czjzek C, Oestreich V, Schmitt H, Latek K, Tomala K, 1989 *J .Magn. Magn. Mater* **79** 42

[3] Yakinttos J.K, Gamari-Seale H, 1982 *Z. Phys.* B **48** 251

[4] Duijn H.G.M, 2000, *The Institutional Repositonaryof the University of Amsterdam 109.*

International Conference on Strongly Correlated Electron Systems (SCES 2010) IOP Publishing
Journal of Physics: Conference Series **273** (2011) 012122 doi:10.1088/1742-6596/273/1/012122

Single crystal growth and physical properties of ternary uranium compounds UM_2Al_{10} (M=Fe, Ru and Os)

Takashi Sugai[1, 2], Yoshinori Haga[1, 2], Tatsuma D Matsuda[2], Etsuji Yamamoto[2], Naoyuki Tateiwa[2], Fuminori Honda[3], Rikio Settai[3] and Yoshichika Ōnuki[1, 3]

[1]Advanced Science Research Center, Japan Atomic Energy Agency, Tokai, Ibaraki 319-1195, Japan
[2]Graduate school of Science, Tohoku University, Sendai, Miyagi 980-8578, Japan
[3]Graduate school of Science, Osaka University, Toyonaka, Osaka 560-0043, Japan

E-mail: sugai.takashi@jaea.go.jp

Abstract. We investigated the ternary uranium compounds UM_2Al_{10} (M = Fe, Ru and Os) with the orthorhombic $YbFe_2Al_{10}$-type crystal structure. URu_2Al_{10} and UOs_2Al_{10} are new compounds grown for the first time. These compounds show no ordering despite the Curie - Weiss behavior at high temperatures. The temperature dependence of the electrical resistivity, the specific heat and the magnetic susceptibility indicate that UM_2Al_{10} (M = Fe, Ru and Os) has non-magnetic ground states.

1. Introduction

Uranium based intermetallic compounds show a variety of physical properties including the heavy fermion state and various magnetic/multipolar orderings where the $5f$ electrons play principal roles. Because the $5f$ electrons have a large spatial extent compared to $4f$ electron, the hybridization with the ligand sites is significant, leading to those different behaviors.

UFe_2Al_{10} with the orthorhombic $YbFe_2Al_{10}$-type crystal structure is one of the rare examples of uranium compounds which have a paramagnetic ground state. It exhibits Curie - Weiss behavior in the temperature dependence of the magnetic susceptibility $\chi(T)$ at high temperatures above 150 K[1, 2]. However, no magnetic ordering is observed down to low temperatures. On the other hand, CeM_2Al_{10} (M=Fe, Ru and Os) with the same structure are reported to show valence instability or a possible magnetic phase transition, indicating that the hybridization between f-states and conduction bands is significant in this crystal structure[3, 4].

In this paper we report the physical properties of UFe_2Al_{10} and related compounds on the single crystalline samples.

2. Sample preparation and characterization

Single crystals of UM_2Al_{10} (M : Fe, Ru and Os) were grown by the Al self-flux method. The starting materials were U (99.98 %), Fe (99.9975 %), Ru (99.99 %), Os (99.99 %) and Al (99.999 %). Samples were prepared in the typical atomic ratio U:M:Al = 1:2:40. The elements were

Published under licence by IOP Publishing Ltd

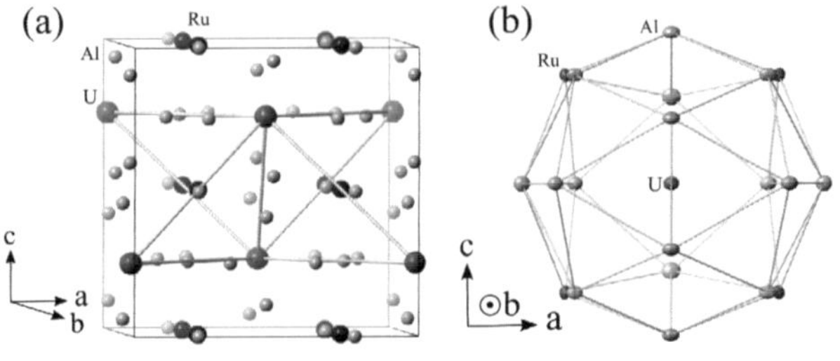

Figure 1. (a)The crystal structure for URu_2Al_{10}. (b)Local atomic arrangements around the uranium site.

placed in an alumina crucible. The crucible was sealed in an evacuated quartz tube. The tube was heated up to 1050 °C and was kept for a few days, followed by a slow cooling down to 700 °C. The excess Al flux was removed by the centrifuge. The chemical composition and homogeneity of the samples were checked by the electron probe micro analyses. No impurity phase was observed within accuracy of measurements. The crystal structure was determined by the single crystal X-ray diffraction. The all Bragg reflections were successfully indexed for the orthorhombic $YbFe_2Al_{10}$-type crystal structure. In addition to the reported UFe_2Al_{10}, we succeeded in synthesizing URu_2Al_{10} and UOs_2Al_{10}.

UM_2Al_{10} compounds have the orthorhombic $YbFe_2Al_{10}$-type (Space Group: $Cmcm$, No.63) crystal structure(Figure 1(a)). The lattice parameters of UM_2Al_{10} are listed in Table 1. The lattice parameters for a and c-axis are similar, while that for b-axis is slightly longer. As shown in Figure 1(b), uranium atom occupies the $4c$ Wycoff position(0 y 1/4) and is surrounded by Al and the transition metal. The local arrangement around the uranium site has a quasi four-fold symmetry. The large interatomic distance between the uranium atoms $\sim$ 5.2 Å implies that the direct hybridization between uranium sites is not significant. The crystallographic parameters determined in the present study are in accordance with the previous report[1].

Table 1. Lattice parameters of UM_2Al_{10} (M:Fe, Ru and Os) determined in the present study.

	a (Å)	b (Å)	c (Å)	
UFe_2Al_{10}	8.9195(6)	10.1998(8)	9.0135(3)	
UFe_2Al_{10}	8.9146(3)	10.1986(3)	9.0114(3)	ref. [1]
URu_2Al_{10}	9.0350(6)	10.2486(8)	9.1265(5)	
UOs_2Al_{10}	9.0559(6)	10.2626(7)	9.1377(8)	

3. Magnetic susceptibility

Figure 2(a) and 2(b) show the temperature dependence of magnetic susceptibility measured on a piece of single crystal for UFe_2Al_{10} and URu_2Al_{10}, respectively. The Magnetic field was applied to [001], [100] and [010] directions. For $H \parallel [001]$ and [100], a distinct Curie -Weiss behavior was observed for both compounds. Among the two directions, [001] shows larger susceptibility. On the other hand, magnetic susceptibility for [010] is small and almost independent of temperature. Such magnetic anisotropy is consistent with the quasi four-fold symmetry with the principal axis along the [010] as mentioned above. From the Curie - Weiss law above 150 K, the effective magnetic moments are 3.14 μ_B/U for $H \parallel [001]$ and 3.13 μ_B/U for $H \parallel [100]$ for UFe_2Al_{10} and 3.07 μ_B/U for $H \parallel [001]$ and 3.50 μ_B/U for $H \parallel [100]$ for URu_2Al_{10}. These values are close to those of free uranium ion. As temperature decreases, the susceptibility of UFe_2Al_{10} becomes smaller than that of Curie - Weiss law. With further decreasing temperature, the susceptibility

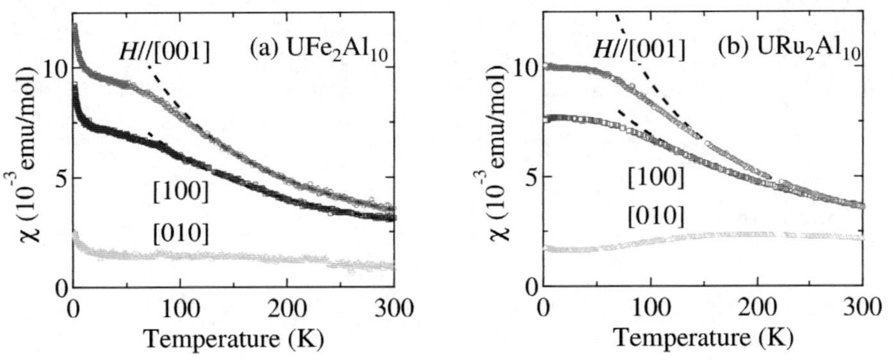

Figure 2. (a)The temperature dependence of electrical resistivity for UFe_2Al_{10}. (b)The temperature dependence of electrical resistivity for UFe_2Al_{10}. The broken lines are the result of Curie - Weiss fitting.

increases again below 20 K, consistent with the previous report. The origin of the steep increase of the magnetic susceptibility below 20 K is not clear at present. However the isotropic behavior contrasts with the anisotropic bulk susceptibility. Furthermore such low-temperature increase is not observed for URu_2Al_{10} which shows lower residual resistivity than that for UFe_2Al_{10}. It is therefore supposed that the present behavior should be attributed to magnetic impurities. For both compounds, there is no signature corresponding to magnetic ordering down to 2 K. It was also confirmed by the specific heat measurements down to 0.4 K (not shown). The deviation from the Curie -Weiss law is reminiscent of the crystalline field effects or the heavy fermion formation with the chacteristic energy scale of $\sim$ 100 K

Our preliminary measurements on a tiny crystal of UOs_2Al_{10} also show the similar susceptibility behavior. The similar paramagnetic effective moments on these compounds suggest that the magnetic properties are governed by $5f$ electrons of uranium and the contribution from d-electrons is negligible. Absence of the Curie-Weiss behavior in the isostructural $LaFe_2Al_{10}$ also supports this idea[3].

4. Electrical resistivity

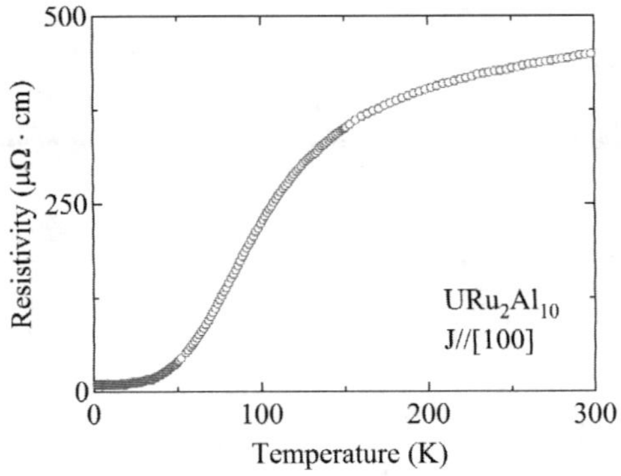

Figure 3. The temperature dependence of electrical resistivity for URu_2Al_{10}.

Figure 3 shows the temperature dependence of the electrical resistivity for URu_2Al_{10} measured with the dc four-probe method. The current was applied along [100]. As temperature decreases from 300 K to 150 K, the electrical resistivity gradually decreases. Then it rapidly decreases below 150 K and approaches to a constant value. The residual resistivity ratio ρ_{RT}/ρ_0 is 46, indicating a high quality sample. Below 6 K, the resistivity behavior is characterized by the usual AT^2 term and the exponential term with the energy gap of about 100 K. The small A coefficient 0.0013 $\mu\Omega$·cm/K^2 is indicative of a weak electron correlation, which is consistent with the small Sommerfeld coefficient $\gamma \sim 22$ mJ/K^2·mol.

5. de Haas-van Alphen effect

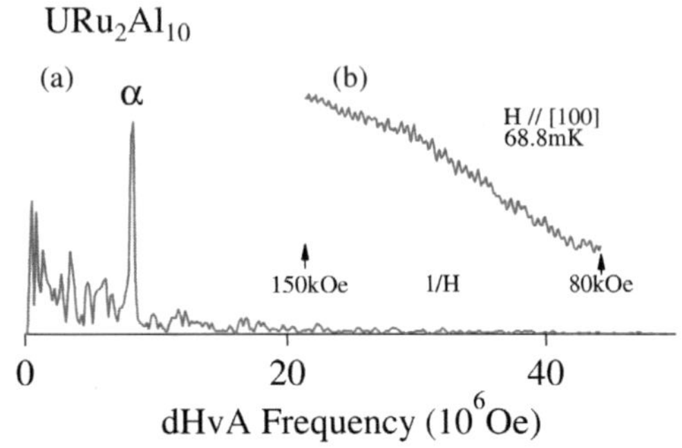

Figure 4. The dHvA oscillation (a) and corresponding FFT spectrum (b) for URu_2Al_{10}.

To clarify the contribution of 5f electron to the Fermi surface, we measured de Haas-van Alphen(dHvA) effect. We succeeded in observing dHvA oscillation for URu_2Al_{10}. Figure 4 shows the typical dHvA oscillation and corresponding FFT spectrum. One branch(α) was observed. The α branch is almost independent of magnetic field angle, corresponding to a spherical Fermi surface. From the cyclotron effective mass m_c^* of α branch 2.3 m_0, the specific heat coefficient is estimated as 2.4 mJ/K^2·mol which is one order of magnitude smaller than value estimated from specific heat. This discrepancy indicates the presence of other Fermi surfaces compared to α branch.

6. Summary

We succeeded in growing single crystals of new compounds UM_2Al_{10}(M=Fe, Ru and Os). The saturating of magnetic susceptibility, the small γ - value, the small cyclotron effective mass and the exponential behavior of resistivity might be accounted for by a crystalline field effect with the characteristic energy scale of $\sim$ 100 K.

References

[1] Noël H, Gonçalves AP and Waerenborgh JC 2004 *Intermetallics* **12** 189-194.
[2] Niemann N and Jeitschko W 1995 *Z Kristallogr* **210** 338-341.
[3] Muro Y, Motoya K, Saiga Y and Takabatake T 2009 *J. Phys. Soc. Jpn.* **78** 083707.
[4] Nishioka T, Kawamura Y, Takesaka T, Kobayashi R, Kato H, Matsumura M, Kodama K, Matsubayashi K and Uwatoko Y 2009 *J. Phys. Soc. Jpn.* **78** 123705.

Magnetic properties of geometrically frustrated polymorphic crystals of $Cu_{4-x}Mg_x(OH)_6Cl_2$

Shaoyan Chu

Center for Materials Science and Engineering, Massachusetts Institute of Technology
77 Massachusetts Avenue, Cambridge, MA 02139, USA

E-mail: sc79@mit.edu

Abstract. Geometric frustration can suppress magnetic long-range order (LRO) in either ferromagnetic (FM) [1-3] or antiferromagnetic (AFM) systems with Kagomé structure. Crystals of $Cu_{4-x}Mg_x(OH)_6Cl_2$, magnesium-containing paratacamite and herbertsmithite ($x < 1$ and $x \sim 1$, $R\bar{3}m$) and haydeeite ($x \sim 1$, $P\bar{3}m1$), have been grown by the hydrothermal technique and characterized by single crystal X-ray diffraction. The 2D S =1/2 polymorphs can be classified into the two types of magnets based on the susceptibility at high temperature. Temperature and dc field-dependence of magnetization of the magnesium-containing phase are quite similar to those of isostructural single crystals of $Cu_{4-x}Zn_x(OH)_6Cl_2$ ($R\bar{3}m$), zinc-containing paratacamite (or herbertsmithite). However, for haydeeite, though a ferromagnetic long range ordering at 20.6 K can be predicted by the Curie-Weiss law, the magnetic phase transition is absent until 4.3 K, indicating a high degree of frustration ($|\theta|/T_c \sim 5$). Its ac frequency and dc field-dependence of spin freezing temperature are analogous to those of spin ice and/or spin glass.

Strong geometric frustration usually arises in systems where the spins occupy the sites of a Kagomé structure. An AFM nearest-neighbor (NN) exchange in these systems may lead to a spin liquid state or valance-bond-solid state. However a FM exchange with strong single-ion anisotropy (Ising model) is assumed to attribute to the spin ice ground state in pyrochlore oxides of $M_2Ti_2O_7$ (M = Dy or Ho) where the Kagomé plane is perpendicular to the <111> axes [1-3]. It is commonly believed that the most promising candidate for exhibiting spin liquid state is the 2D spin-1/2 Kagomé lattice because quantum fluctuations are particularly strong for low spin systems; in contrast, significant super-exchange and dipolar interaction of spins with large moments play an important role in spin ice magnets rather than quantum fluctuations. Studies on isostructural systems and polymorphs are essential for verifying this belief due to their comparable magnetic properties. The report here involves novel polymorphs of $Cu_{4-x}Mg_x (OH)_6Cl_2$. Though arrangements of Cu^{2+}cations in the 2D Kagomé nets of the polymorphs are quite similar, their magnetic properties are strikingly different.

Crystals of $Cu_{4-x}Mg_x(OH)_6Cl_2$ were grown under hydrothermal conditions [4]. As an analogue of the Zn-containing atacamite family, the Mg-containing family has four members. Clinoatacamite $Cu_2(OH)_3Cl$ is the x = 0 end compound of the family [5]. This parent phase with $P2_1/n$ structure has a deformed Kagomé lattice of Cu^{2+} (S = ½). Two ideal x = 1 end compounds exist as polymorphs. One is haydeeite [6], an isostructural mineral of kapellasite ($P\bar{3}m1$). The crystal structure of haydeeite is based on brucite-like sheets parallel to (0001) built from edge-sharing angle-distorted $Mg(OH)_6$ and Jahn-Teller distorted CuO_4Cl_2 octahedra. The sheets containing perfect Kagomé lattices of Cu^{2+} are stacked directly on each other (*...AAA...* stacking). Unit-cell parameters of a haydeeite of $Cu_{2.85}Mg_{1.15}(OH)_6Cl_2$ at 100 K are a = b = 6.2605(7) Å and c = 5.7218(7) Å. Another ideal x =1 end member is an isostructural compound of herbertsmithite ($R\bar{3}m$). An ideal compound of $Cu_3Mg (OH)_6Cl_2$ with $R\bar{3}m$ structure can be named Mg-containing herbertsmithite (called Mg-herbertsmithite for short) and has not been reported in mineralogy. Its higher symmetry structure is caused by the substitution Mg for all of Cu^{2+} in the interlayer of the parent phase, clinoatacamite. Unit-cell parameters of Mg-herbertsmithite, $Cu_{3.04}Mg_{0.96}(OH)_6Cl_2$, at T = 100 K are a = b = 6.8337(10) Å and c = 13.965(2) Å. If the cations in the interlayer are partially replaced by Mg^{2+}, Mg-paratacamite appears as the fourth

Published under licence by IOP Publishing Ltd

phase, an analogue of Zn-paratacamite (R$\bar{3}$m or R$\bar{3}$) [5]. This phase is a closely related to Mg-herbertsmithite (x = 1) in structure and does not show significant changes in Kagomé lattice parameters when the concentration of Mg varies [6]. It is known that, with increasing number of Mg occupying the interlayer sites, LRO temperature of Mg-paratacamites is decreased. For this reason, one of the samples used here for magnetic measurements is a collection of single crystals of Mg-paratacamite (x = 0.31) with ~1 mm diameter-size, This sample has a LRO transition at 4.7 K which is comparable with that of well-known antiferromagnets of clinoatacamite and Zn-paratacamite. Another sample is a collection of 0.1 mm diameter-sized crystals of Haydeeite, $Cu_{2.85}Mg_{1.15}(OH)_6Cl_2$. Idealized structure of the partial unit-cell of the R$\bar{3}$m phase and P$\bar{3}$m1 phase are shown in the inset of Figure 1.

A SQUID magnetometer (MPMS2, Quantum Design Inc) was used to measure the dc magnetization, M, as well as the real part, M'_{ac}, and imaginary part, M''_{ac}, of the ac magnetization, M_{ac}, with a drive amplitude of ~0.3 mT. All of the samples were randomly oriented with the applied field.

The main panel of Figure 1 shows the temperature-dependence of inverse dc susceptibility $1/\chi$, where χ = M/H, of clinoatacamite, Zn-paratacamite (x = 0.34), Mg-paratacamite (x = 0.31) and haydeeite (x = 1.15) at an applied field of H = 20 mT. Absence of any temperature hysteresis above 20 K indicates a paramagnetic (PM) character for all of the samples at high temperatures. The Curie-Weiss law, $\chi = C/(T - \theta)$ is a suitable model for the data within the temperature range of 200 K - 340 K. Fitting a straight line to each set of data yields the Curie constant, C = 0.483 and 0.441 (cm^3·K/mol.Cu) for the Mg-paratacamite and haydeeite. Typical g-factors of 2.27 and 2.17 for Cu^{2+} (S = ½) in the two polymorphs can be determined respectively. The Curie-Weiss temperature, θ, of the Mg-paratacamite (x = 0.31) is -187 K, indicating an AFM character. This value is nearly equal to that of Zn-paratacamite (x = 0.34), θ = -220 K. However, in contrast with these paratacamites, haydeeite would be predicted to have a FM ground state below θ = 20.6 K. Absence of LRO phase and spin freezing until a critical temperature T_c ~ 4.3 K indicates a high degree of frustration ($|\theta|/T_c$ ~ 5) for this system. Surprisingly, theoretical calculation [7] and phenomenological analysis [8] cannot deduce a FM NN interaction because the O-H distance and the Cu-O-Cu angle between two neighbor plaquettes reported here are 0.827 Å and 104.92 °. Therefore, FM character of haydeeite may be attributed to the dipole-dipole interaction and next-nearest-neighbor (NNN) interaction and/or interaction along the diagonal of the Kagomé lattice. This situation is probably the same as that in the spin ice materials of pyrochlore oxides implying haydeeite is an excellent candidate for demonstrating a quantum 2D spin ice [9], as long as single-ion anisotropy of Cu^{2+} (S = ½) is involved in the frustration.

To reveal a difference in the magnetization process between Mg-paratacamite and haydeeite, one

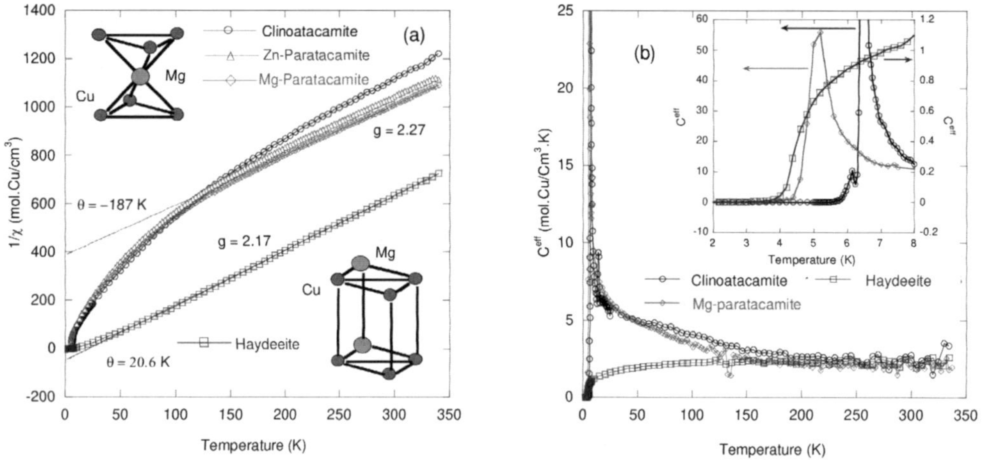

Figure 1. (Color online) (a) Temperature dependences of $1/\chi$ for clinoatacamite, Zn-paratacamite (x = 0.34), Mg-paratacamite (x = 0.31) and Haydeeite (x = 1.15). Insert: structures of partial unit-cell of R$\bar{3}$m phase (top) and P$\bar{3}$m1 phase (low). (b) Main panel: temperature dependence of $C^{eff} = d(1/\chi)/dT$. Insert: same data at low temperature

can define an effective Curie constant $C^{eff} = d(1/\chi)/dT$. This parameter is independent of the demagnetization factor, D, because of an intrinsic susceptibility $\chi_{int} = \chi/(1-4\pi D\chi)$. Temperature dependences of C^{eff} for clinoatacamite, Mg-paratacamites, and haydeeite are presented in Figure 1-(b). A separation between the C^{eff}-T curves starts at ~ 200 K and increases at lower temperatures. The C^{eff}-T curves below 8 K presented in inset of Figure 1-(b) show different profiles: multi-peaked for clinoatacamite, single-peaked for Mg-paratacamite, and monotonically decreasing for haydeeite. These differences become trivial at T < 4 K. However, as shown in Figure 2-(a), the M-T curve of haydeeite is flat below a well-defined freezing temperature, T_f = 4.3 K, implying that spins freeze out. It is extremely different from that of clinoatacamite and Mg-paratacamite where a typical PM-FM phase transition occurs.

ZFC (zero field cooling) and FC (field cooling) measurements are used to detect the irreversibility of the magnetization process. As shown in Figure 2-(a), upon cooling, the FC curves of Mg-paratacamite and clinoatacamite increase monotonically, but the ZFC curves have a maximum at temperatures below their individual Curie points. For haydeeite, the ZFC magnetization is only slightly smaller than that of FC at same temperatures. Such a tiny difference in magnetization is further detected by minor M-H loops at an isothermal condition and lower field of up to 10 mT. Two loops shown in inset of Figure 2-(b) were performed after a zero field cooling. As an evidence of blocking phenomenon, magnetization (relative change of magnetization, ΔM/M) of haydeeite at a higher temperature, 3 K, has a larger (smaller) value as compared with that at a lower temperature, 1.8 K.

Hysteresis loops using Mg-paratacamite and haydeeite were measured at T = 2 K and a field of up to 5 T. As shown in the main panel of Figure 2-(b), an average magnetic moment for each Cu^{2+} in the $R\bar{3}m$ phase at 5 T is only 0.16 μ_B. This value and the slope of M vs. H at high field are comparable with that of Zn-paratacamite (x = 0.34) in the same conditions. The weak magnetism is caused by a canting spin configuration in these systems. In contrast, haydeeite is mostly saturated at a field of ~ 1 T and has a saturation moment of ~1.0 μ_B/Cu^{2+}. It is worth noting that at high field, the crystals might re-orient their magnetic easy axes along the applied field direction during the saturation process.

Nevertheless here is no lattice of corner-sharing tetrahedron in the structure and no significant random disorder of Cu^{2+} in the single crystal, haydeeite display a time-dependent magnetization reminiscent of what is observed in spin ice materials and/or spin glasses. Main panels of Figure 3-(a) and (b) show temperature dependences of ac magnetization of Mg-paratacamite and haydeeite at zero field. A sharp decline of M'_{ac} (or M''_{ac}) for Mg-paratacamite at LRO state indicates that the spins' dynamic response is slowed such that it cannot respond to the time-varying drive field, which implies

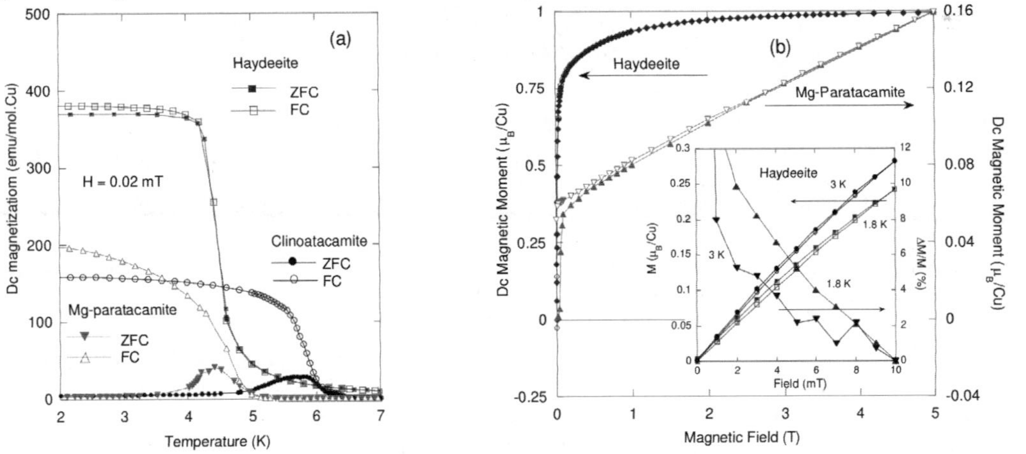

Figure 2. (Color online) (a) Temperature-dependence of dc magnetization for clinoatacamite Mg-paratacamite and haydeeite. (b) Hysteresis loops of Mg-paratacamite (♦ and ◊) and haydeeite (▲ and Δ) at T = 2 K. Insert: relative changes of magnetizations (▲ and ▼) drawn from hysteresis loops for haydeeite at temperatures of 3 K (○ and ●) and 1.8 K (□ and ■). The fields swept up to 10 mT.

that upon cooling the spins gradually become firmly locked. For haydeeite, the M'_{ac}-T curve shows a slight decline at temperatures below T_f, while the M''_{ac}-T curve has a multistep structure. Beside a steep increase of M''_{ac} at T_f, a broad peak at a higher temperature can be detected on the M''_{ac}-T curve. The frequency-dependence of the peak temperature shown in the insert of Figure 3-(b) is quite similar to that of spin glass and/or spin ice systems. When a dc magnetic field was applied, the broad peak was suppressed. However, another maximum of M'_{ac} (and M''_{ac}) near the T_f appeared. As shown in the insert of Figure 3-(a), the temperature at which the maximum appears increases with increasing field. The field-dependence of the peak temperature is in striking contrast with the AT line in spin glasses, where a magnetic field suppresses its freezing temperature [10]. It is not clear now whether the nature of the enhanced T_f here is same as that in the 3D spin ice system [4]. One of explanations for the observation is that the application of a filed enhances a short range ordering (which decreases the spin degree of freedom), consequentially suppressing the ac magnetization for all temperatures below T_f through spin locking. This assumption is consistent with the observation of the magnetization process shown in Figure 2-(b).

This work was supported primarily by the MRSEC Program of the NSF under award number DMR – 0819762 and DOE under Grant No. DE-FG02-04ER46134. The author acknowledges helpful discussions with Patrick Lee, Daniel G. Nocera, Young S. Lee and Peter Müller.

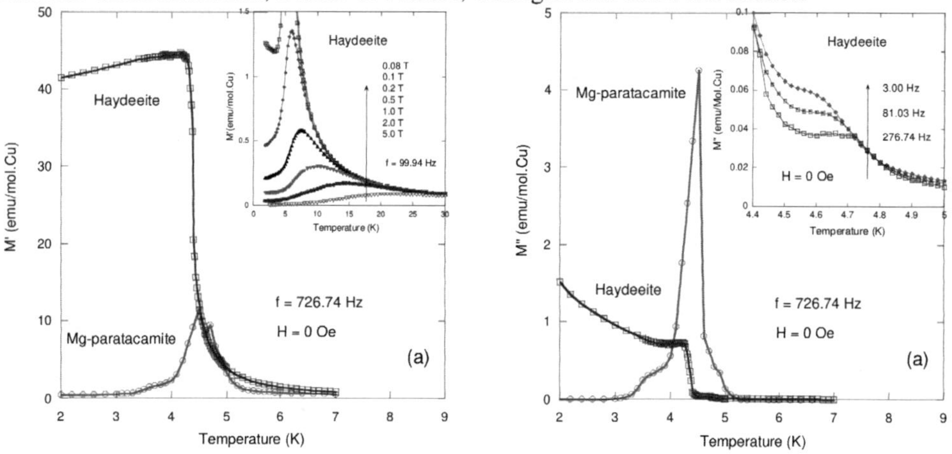

Figure 3. (Color online) Main panels of (a) and (b) show M'_{ac}-T and M''_{ac}-T curves for Mg-paratacamite and haydeeite at a zero field (~ 0.1 mT). Insert of (a): M'_{ac}-T curves at varied dc fields. Inset of (b): frequency-dependence of M''_{ac} at a zero field.

References

[1] Harris M J, Bramwell S T, McMorrow D F, Zeiske T and Godfrey K W 1997 *Phys. Rev. Lett.* **79** 2554
[2] Matsuhira K, Hinatsu Y, Tenya K and Sakakibara T 2000 *J. Phys. Condens. Matter* **12** L649
[3] Snyder J, Slusky J S, Cava R J and Schiffer P 2001 *Nature* **413**, 48
[4] Chu S-Y, et.al. 2010 *J. Am. Chem. Soc.* **132** 5570
[5] Krause W, Bernhardt H J, Braithwaite R S W, Kolitsch U and Pritchard R 2006 *Mineral. Mag.* **70** 329
[6] Malcherek T and Schlüer J 2007 *Acta Cryst.* B **63**, 157
[7] Janson O, Richter J and Rosner H 2008 *Phys. Rev. Lett.* **101** 106403
[8] Crawford V H, Richardson H W, Wasson J R, Hodgson D J and Hatfield W E 1976 *Inorg. Chem.* **15** 2107
[9] Wills A S, Ballou R and Lacroix C 2002 *Phys. Rev.* B **66** 144407
[10] de Almeida J R L and Thouless D J 1978 *J. Phys. A: Math. Gen.* **11** 983

Long-time variation of magnetic structure in rare-earth intermetallic compounds

Kiyoichiro Motoya[1], Taketo Moyoshi[1] and Toru Shigeoka[2]

[1]Department of Physics, Faculty of Science and Technology, Tokyo University of Science, Noda 278-8510, Japan

[2]Graduate School of Science and Engineering, Yamaguchi University, Yamaguchi 753-8512, Japan

E-mail: motoya@ph.noda.tus.ac.jp

Abstract. Long-time variations of magnetic structure have been studied by means of time-resolved neutron scattering measurements in two kinds of non-diluted uniform magnets, $PrCo_2Si_2$ and $TbNi_2Si_2$. These materials show commensurate (C) to incommensurate (IC) magnetic transitions below Neel temperature. When a sample is cooled to the low temperature C-phase, magnetic Bragg scattering signals from the C and the IC phases coexist and then their intensity vary with time. These characteristics are similar to those of the previously studied material $CeIr_3Si_2$. However, the temperature ranges in which clear time variations exist are much narrower in the present materials.

1. Introduction

Recently, long-time variation of magnetic structure has been observed in a non-diluted uniform magnet $CeIr_3Si_2$ [1]. This material shows successive magnetic transitions in zero magnetic field and multi-step metamagnetic transitions at relatively low magnetic field. These observations suggest that the frustrating magnetic interactions cause the long-time variation of magnetic structure in a material without random magnetic interactions. We have searched for other materials which show similar long-time variation of magnetic structure. We have found two intermetallic compounds $PrCo_2Si_2$ and $TbNi_2Si_2$.

Magnetic properties of intermetallic compounds REM_2X_2, where RE, M and X stand for a rare earth metal, a transition metal and silicon or germanium, respectively, have been extensively studied since 1980's [2]. These materials crystallize in the body centered tetragonal structure (space group $I4/mmm$), with RE, M and X atoms occupying the $2a$, $4d$ and $4e$ positions, respectively. $PrCo_2Si_2$, $TbNi_2Si_2$ and some of other REM_2X_2 compounds show similar characteristics found in $CeIr_3Si_2$, namely the successive magnetic transitions and multi-step metamagnetic transitions at relatively low magnetic field.

Neutron diffraction study of $PrCo_2Si_2$ using a single crystal sample clarified the magnetic structure [3]. Three different magnetic phases were found below T_N=30 K. For all these phases, the magnetic moments of Pr atoms ($\sim$3.2 $_{\rm B}$ at 4.2K) on the same (001) plane align ferromagnetically parallel or antiparallel to the c-axis. Therefore, the magnetic propagation vector is given as $\mathbf{k} = (0, 0, k)[2\pi/c]$. For $T < T_1$=9 K, the sequence of ferromagnetic planes is +-+- which corresponds to the propagation vector of k=1 (C-phase). It transforms to the incommensurate structure with k=0.926 at T_1 (IC_1-phase). In this phase, the presence of the

Published under licence by IOP Publishing Ltd

third harmonics of Bragg peaks indicates a square-wave structure. Above T_2=17 K, another incommensurate structure with k=0.777 develops and persists up to T_N (IC_2-phase). High-field magnetization measurements along the c-axis at 4.2 K showed four steps of increase at fields 1.2, 3.8, 6.7 and 12.2 T.

Neutron diffraction studies of $TbNi_2Si_2$ were made by several groups [2, 4, 5, 6]. It was shown that a simple antiferromagnetic structure with a +-+- sequence of ferromagnetic (110) planes appears below T_t=7.9 K (C-phase). Above T_t, an incommensurate structure with $\mathbf{k} = (1/2 + \quad, 1/2 - \quad, 0)[2\pi/a]$, where $\quad = 0.074$ appears and persists up to T_N=14.6 K (IC-phase). The magnetization value measured at T=1.3 K with the field parallel to the c-axis jumps at five critical fields of 1.5, 2.2, 2.4, 4.4 and 4.6 T. Above 4.6 T it reaches the saturation value of 8.8 $_B$/f.u., which is in good agreement with the results of the neutron experiments.

Based on these previous works, we made macroscopic and neutron scattering experiments to detect time variations of magnetic structure after rapid change of temperature across the magnetic transition temperatures. In this report, we present the results of time-resolved neutron scattering experiments of these two materials.

2. Experimental

Single crystal samples of $PrCo_2Si_2$ and $TbNi_2Si_2$ were cut from the ingots grown by the Czochralski pulling method and used in the previous works [3, 6]. Magnetization measurements detected no appreciable aging effect in these samples. The mass of each sample is 0.21 g and 0.34 g, respectively. Neutron scattering experiments of $PrCo_2Si_2$ were conducted using the 4G triple-axis spectrometer installed at the JRR-3M reactor of JAEA-Tokai. Measurements were made by a double-axis mode using pyrolytic graphite crystals for a monochromator and a filter. The wave-length of the incident neutron and the horizontal beam collimation are 2.35 Å and $40' - 40' - 40'$, respectively. The sample was mounted with the [100] direction vertical and the magnetic Bragg peaks in the $a^* - c^*$ plane were measured.

Neutron scattering experiments of $TbNi_2Si_2$ were made using the T1-1 triple-axis spectrometer installed at the thermal guide of the JRR-3M reactor. Measurements were made using pyrolytic graphite crystals for a monochromator and a filter. The wave-length of the incident neutron and the horizontal beam collimation are 2.46 Å and guide - $40' - 40'$, respectively. The sample was mounted with the [001] direction vertical and the magnetic Bragg peaks in the $a^* - b^*$ plane were measured. A closed-cycle refrigerator was used for each experiment.

3. Results and discussion

Figure 1 shows the temperature variations of the amplitude of three kinds of magnetic Bragg diffractions. The measurements were made with heating and cooling processes. In each process, scans across the peaks were started after the sample temperature reached the target temperature same as usual measurements. The time required for the measurement at one temperature was typically 20 min. The result is basically same as the previous one [3], which showed the presence of three magnetic phases. In the present measurements, clear thermal hysteresis has been observed at the boundary between the C and IC_1 phases.

Figure 2 shows the time evolutions of neutron scattering patterns from (a) the C-phase and (b) the IC_1-phase of $PrCo_2Si_2$ at T=10 K after cooled from 11 K. The amplitude of the C-phase signal increases with time whereas the amplitude of the IC_1-phase signal decreases. Neither the center position nor the width of each signal changes with time.

Time variation measurements of these signals were made at various temperatures. Figure 3 shows the results at representative temperatures. Clear time variations of magnetic scattering intensity exist in the narrow temperature range around T_1. The time variations of the signal amplitude can not be expressed by a simple exponential function. We tentatively analyzed the

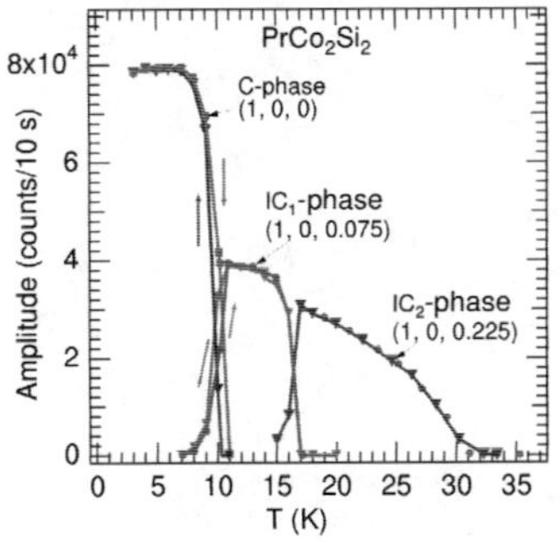

Figure 1. Temperature variations of the amplitude of three kinds of magnetic Bragg diffractions of $PrCo_2Si_2$.

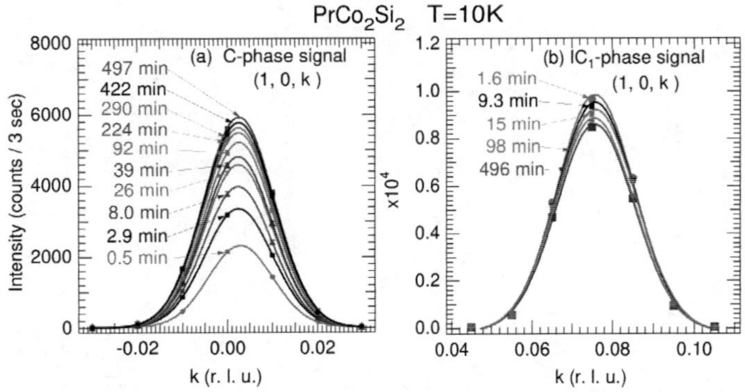

Figure 2. Time evolutions of neutron scattering patterns from (a) the C-phase and (b) the IC_1-phase of $PrCo_2Si_2$ measured at 10 K after cooled from 11 K. Curves represent the results of the least squares fittings described in the text.

observed time variation as a sum of two exponential functions having different characteristic times. The curves in Fig. 3 show the results of the fittings.

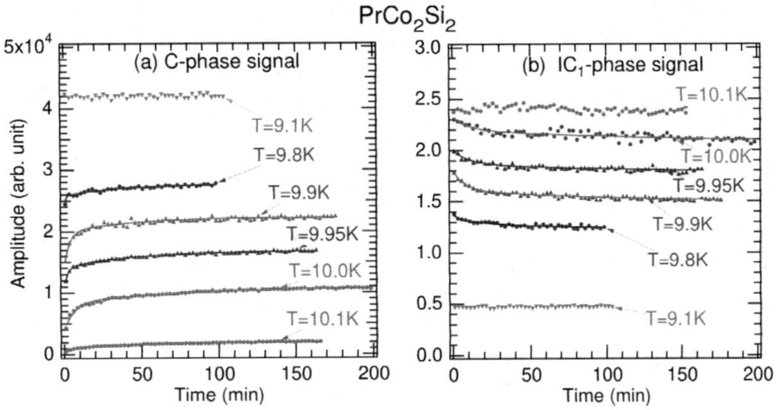

Figure 3. Time variations of the amplitude of neutron scattering signals from (a) the C-phase and (b) the IC_1-phase of $PrCo_2Si_2$ at various temperatures. Curves represent the results of the fittings described in the text.

Figure 4 shows the temperature variations of the amplitude of two kinds of magnetic Bragg diffraction signals from $TbNi_2Si_2$. New features of the temperature variation behavior have been observed in addition to the previous result [6]. Firstly, large thermal hysteresis has been observed in wide temperature region. Secondly, the amplitude of the C-phase signal at low temperature varies with the cooling speed. Based on these observations, we made time variation measurements of neutron diffraction signals.

Figure 5 shows the time variations of the amplitudes of (a) the C-phase and (b) the IC-phase signals at representative temperatures. The observed time variation behavior is basically same as the case of $PrCo_2Si_2$.

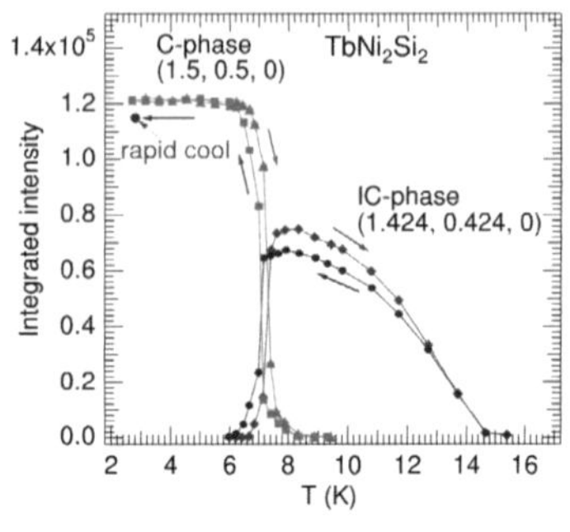

Figure 4. Temperature variations of the amplitude of two kinds of magnetic Bragg diffractions of $TbNi_2Si_2$.

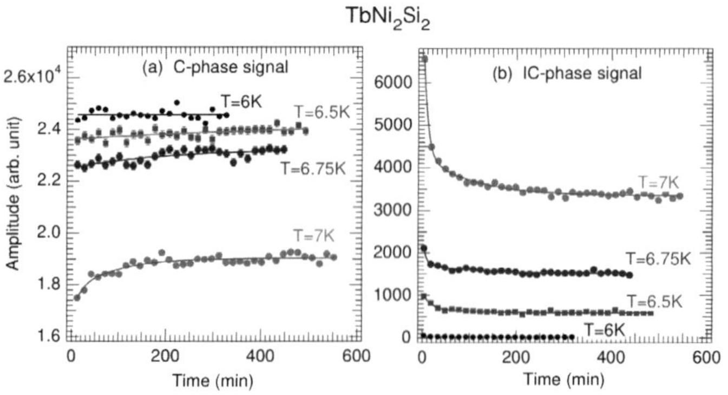

Figure 5. Time variations of the amplitude of neutron scattering signals from (a) the C-phase and (b) the IC-phase of $TbNi_2Si_2$ at various temperatures.

We have shown that magnetic Bragg peaks corresponding to different types of magnetic structures coexist near the transition temperatures and the amplitudes of these peaks vary with time after the change of sample temperature in both materials. We note that only the amplitude of Bragg peaks vary with time. Neither the position nor the line width showed appreciable time variation. These results indicate that we have observed the change of the volume fractions of two different magnetic phases same as the case of previously studied material $CeIr_3Si_2$ [1]. It should be noted that in the present materials time variations have been observed only in narrow temperature range around the commensurate to incommensurate transition temperatures.

Acknowledgments

We thank the members of the Neutron Scattering Laboratory, Institute for Solid State Physics, the University of Tokyo for supporting our experiments. This work is partly supported by a Grant-in-Aid for Scientific Research from the Ministry of Education, Culture, Sports, Science, and Technology.

References

[1] Motoya K, Muro Y and Takabatake T 2010 *J. Phys.:Conf. Series* **200** 032048
[2] Barandiaran J M, Gignoux D, Schmitt D, Gomez Sal J C, and Rodriguez Fernandez J 1987 *J. Mag. Mag. Mater.* **69** 61
[3] Shigeoka T, Iwata N, Fujii H, Okamoto T and Hashimoto Y 1987 *J. Mag. Mag. Mater.* **70** 239
[4] Nguyen V N, Tcheou F, Rossat-Mignod J and Ballestracci R 1983 *Solid State Commun.* **45** 209
[5] Blanco J A, Gignoux D, Schmitt D and Vettier C 1991*J. Mag. Mag. Mater.* **97** 4
[6] Shigeoka T, Fujii H, Nishi M, Uwatoko Y, Takabatake T, Oguro I, Motoya K, Iwata N and Ito Y 1992 *J. Phys. Soc. Jpn.* **61** 4559

International Conference on Strongly Correlated Electron Systems (SCES 2010) IOP Publishing
Journal of Physics: Conference Series **273** (2011) 012125 doi:10.1088/1742-6596/273/1/012125

Long-time variation of magnetic structure in a geometrically frustrated magnet $Ca_3Co_2O_6$

Taketo Moyoshi, Rui Takahashi and Kiyoichiro Motoya

Department of Physics, Tokyo University of Science, 2641 Yamazaki, Noda 278-8510, Japan

E-mail: moyoshi@ph.noda.tus.ac.jp

Abstract. Magnetization and neutron scattering measurements have been made on a single crystal of geometrically frustrated compound $Ca_3Co_2O_6$ which shows successive magnetic transitions at $T_{c1}\sim25$ K and $T_{c2}\sim13$ K and multi-step metamagnetic transitions in low magnetic field. In this compound without random magnetic interactions, a long-time variation of magnetic structure has been detected. The magnetic Bragg peaks corresponding to an incommensurate antiferromagnetic structure observed at (1 0 $\pm\delta$) reciprocal lattice positions after cooled below T_{c2} move toward (1 0 0) with time. The characteristic time for these variations follows the Arrhenius law with an activation energy E_a/k_B=57 K.

1. Introduction

In spin-glass systems, long-time variations of magnetic properties have been regarded as a characteristic behavior due to the multi valley structure of the free energy which comes from random magnetic interactions. Therefore, in a system without randomness, we have not expected to observe long-time variation of magnetic properties. Recently, long-time variations of magnetic structures have been detected in $CeIr_3Si_2$ which has no randomness or imperfections [1]. This compound shows multi-step metamagnetic transitions in low magnetic field, which indicates the presence of frustrating magnetic interactions. We examined several compounds which show multi-step metamagnetic transitions in low magnetic field. We have observed a long-time variation of magnetic structure in a frustrated magnet $Ca_3Co_2O_6$.

$Ca_3Co_2O_6$ has a hexagonal structure composed of Co_2O_6 chains running along the c-axis and these chains form a triangular lattice in the ab-plane [2]. The chains are made of alternating CoO_6 octahedra (CoI sites) and CoO_6 trigonal prisms (CoII sites). The first report of the magnetic property of this material was presented by Kageyama *et al.* [3]. Magnetization measurements of polycrystalline sample revealed the existence of two magnetic transitions at $T_{c1}\sim25$ K and $T_{c2}\sim13$ K and multistep metamagnetic transitions in low magnetic field. They also measured magnetization of an oriented polycrystalline sample along the chain direction. These measurements showed the easy-axis anisotropic behavior along the c-axis [4]. These properties were confirmed by the measurements using a single crystal sample by Maignan *et al.* [5]. Hardy *et al.* reinvestigated magnetization and determined the saturation magnetization value as 4.8 μ_B per formula unit [6]. Aasland *et al.* made a neutron scattering experiment using powder sample and determined that the magnetic moments align ferromagnetically along the c-axis with the antiferromagnetic coupling within the ab-plane [7]. Based on the resonant X-ray and neutron

Published under licence by IOP Publishing Ltd

scattering studies, Agrestini *et al.* presented a model of incommensurate magnetic structure [8, 9]. In this model, the amplitude of magnetic moments varies sinusoidally along the c-axis.

These previous works prompted us to search $Ca_3Co_2O_6$ for a long-time variation of magnetic structure. We made time variation measurements of magnetization and neutron diffraction patterns on a single crystal sample. We have detected the long time variation of magnetic structure. We report the time variation behavior of magnetic structure and present a new magnetic structure model in the low temperature phase.

2. Experimental

A single crystal of $Ca_3Co_2O_6$ was prepared by the flux method. The size of the single crystal is ~1.5 mm × ~1.5 mm × ~10 mm with the longest direction parallel to the c-axis. The lattice parameters at room temperature are a=9.079 Å and c=10.381 Å (hexagonal unit cell) and the mosaic spread is ~0.3° (full width at half maximum, FWHM).

Time variations of magnetization (M) value were measured using a superconducting quantum interference device (SQUID) magnetometer (Quantum Design model MPMS). A magnetic field (H) was applied parallel to the c-axis. Measurements were made with zero field cooled (ZFC) and field cooled (FC) processes. For each measurement, the sample was cooled from T=30 K ($>T_{c1}$) to the target temperature. A typical time needed to reach the target temperature was ~7 min. Then magnetization was repeatedly measured as a function of time (t).

Neutron scattering experiments were carried out using the triple-axis spectrometers, 4G and 5G, installed at the JRR-3M of JAEA in Tokai. The crystal was set with the [100] and [001] axes in the scattering plane. The temperature was controlled using a closed cycle refrigerator. Time variations of scattering patterns were measured by repeating the scans along the (1, 0, l) direction after rapidly cooled from T=30 K. The time required to cool the sample was ~5 min. For the analysis of a magnetic structure, intensities of several magnetic reflections were measured by 2θ-θ scans and l-scans through the magnetic Bragg peaks at 10.5 K. To avoid the influence of the time variation of magnetic structure, measurements were started 5 hours after the sample was cooled to 10.5 K.

3. Results and Discussion

Temperature dependence of magnetization showed two magnetic transitions at T_{c1}=25 K and T_{c2}=13 K. Magnetization curve at T=1.9 K showed multi-step metamagnetic transitions at H_{c1}=0.1 T, H_{c2}=1.2 T, H_{c3}=2.4 T, H_{c4}=3.6 T and H_{c5}=6.0 T. These values are consistent with the previous reports [4-6]. Saturation magnetization was ~5 μ_B per formula unit which is close to the value reported by Hardy *et al.* [7].

We show time (t) dependence of magnetization measured with ZFC process at H=0.005 T in Fig. 1. A remarkable increase of magnetization with time was observed for 7 K$<T<$10.5 K. However, magnetization at 16 K is independent of t.

Figure 2 shows the neutron scattering profiles measured along the (1, 0, l) reciprocal lattice line measured at representative elapsed times after cooled to T=8 K. Magnetic scattering peaks were observed at incommensurate positions (1, 0, $\pm\delta$). The peak positions move toward δ=0 with time. Observed profiles were fitted by the sum of two Gaussian functions which has peaks at (1, 0, δ) and (1, 0, $-\delta$). The curves in Fig. 2 show the results of the fittings. The time dependence of δ at representative temperatures are shown in Fig. 3. The values at 8 K and 11.5 K significantly decrease with t, while the time variations of δ at 15 K and 20 K are much smaller than those at $T<T_{c2}$. These results are consistent with the time dependences of magnetization. These δ-t curves can be represented by the function $\delta=a+b\exp(-t/t^*)$, where t^* is the characteristic time. The value of δ at $t=\infty$ takes a nearly constant value ~0.0045 (r.l.u.) below T_{c2} and increases with increasing temperature above T_{c2}. Figure 4 shows the Arrhenius plot of t^* ($\log(1/t^*)$ vs.

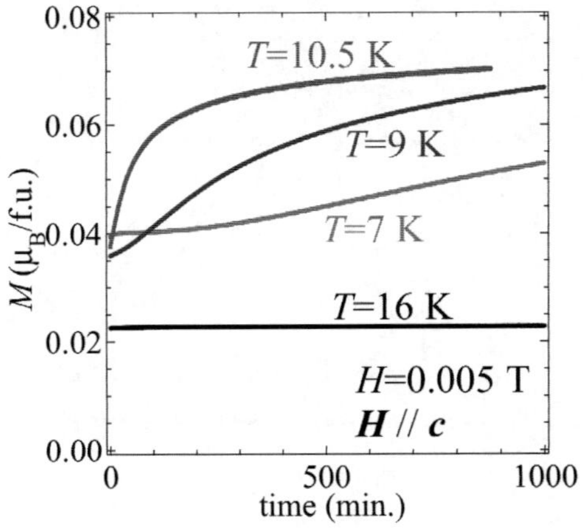

Figure 1. Time dependence of magnetization measured at representative temperatures.

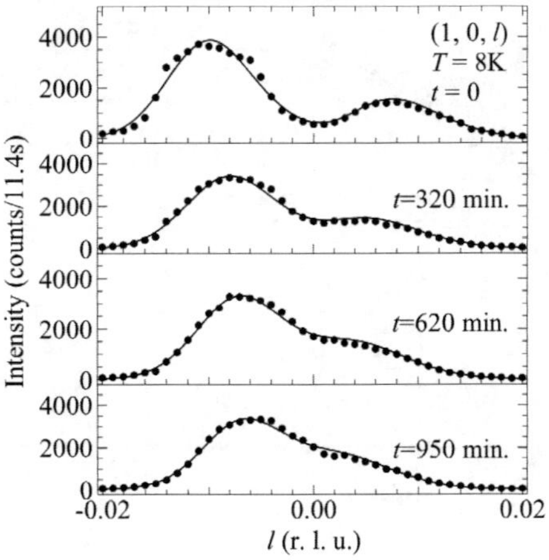

Figure 2. Neutron scattering profiles along (1, 0, l) at T=8 K at representative times. Solid lines show the fittings described in the text.

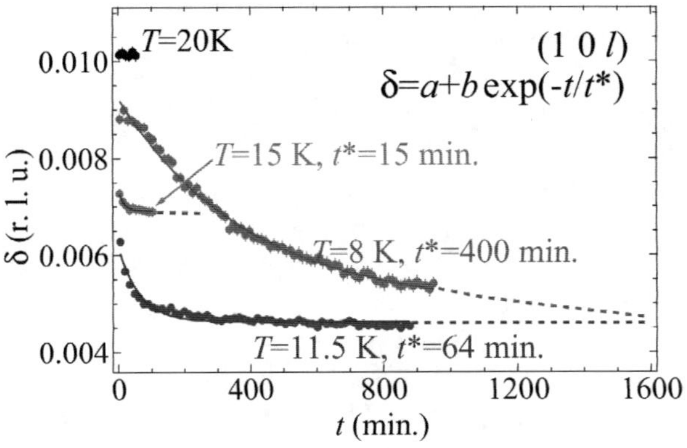

Figure 3. Time variations of magnetic peak positions δ. Curves show the fittings described in the text. t^* means the characteristic time of variation.

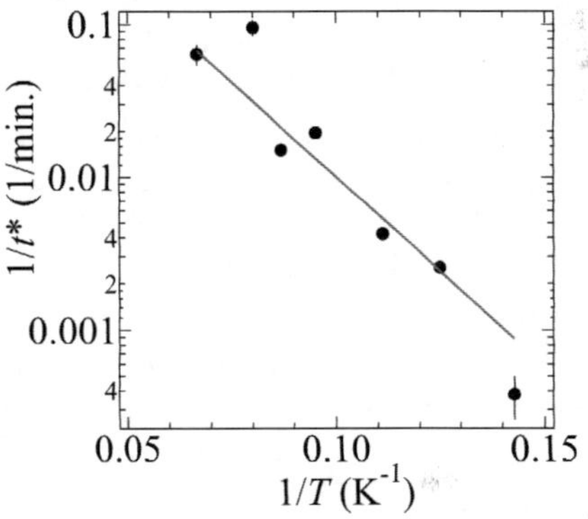

Figure 4. Arrhenius plot of $1/t^*$.

$1/T$ plot). Temperature variation of t^* follows the Arrhenius law as shown by the straight line and the activation energy E_a has been determined as E_a/k_B=57 K.

Now we move to the magnetic structure. We have shown that for $T<T_{c2}$, δ converges on nearly constant value at $t=\infty$. Therefore, we believe that the magnetic structure determined based on the measurements following the procedure described in §2 represents the stable (*i.e.* $t=\infty$) structure at any temperatures below T_{c2}.

The results are summarized as follows.

1) The (0 0 l) magnetic reflections were not observed, which means all the magnetic moments align along the c-axis.
2) All magnetic reflections were observed at the incommensurate positions (h 0 $l\pm\delta$), where $l=2n$ and $\delta\sim0.0045$ (r. l. u.).

We propose a magnetic structure model consistent with the present results shown in Fig.

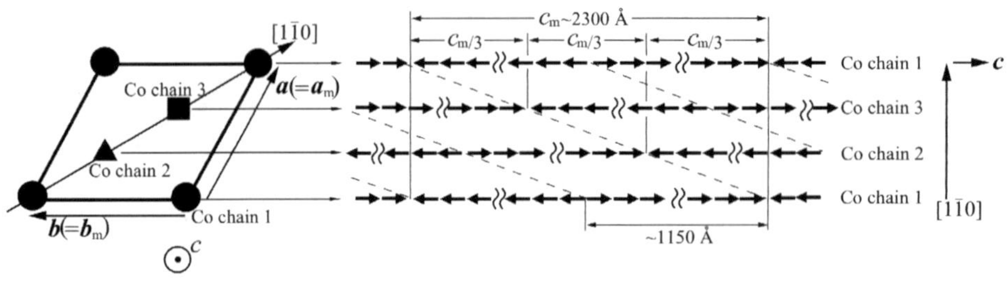

Figure 5. Magnetic structure model of $Ca_3Co_2O_6$.

5. The left-side and the right-side parts show the projections along the c-axis and the b-axis, respectively. In this figure, a_m, b_m and c_m are the lattice parameters of the magnetic unit cell. The magnetic moments of CoII sites form a series of $/+++\cdots+++/---\cdots---/+++\cdots+++/$ arrangement along the c-axis as shown by the right-side part. The magnetic period along the c-axis is given by $c_m=c/\delta=10.4$ Å$/0.0045$ (r.l.u.)$\sim$2300 Å and the length of ferromagnetically coupled regions is $c_m/2$. The boundary of the + and − moments regions shift along the c-axis regularly to make three kinds of chains (Co chain 1, 2, 3) as we proceed along the [1$\bar{1}$0] direction. The shift of the boundary is shown by the dashed lines. The calculation based on this model with the magnetic moments value of 5.3 μ_B for CoII ions (0 μ_B for CoI) gives a reasonable agreement with the observed intensities. The calculation based on the sinusoidally modulated model [9] combined with the present measurements requires the maximum magnetic moment value of 6.5 μ_B for CoII ions, which is much larger for the Co^{3+} ions. Details of the magnetic structure analysis will be published in a separate paper [10].

Based on this magnetic structure model, the observed shift of the magnetic Bragg peak positions corresponds to the time variation of the length of the ferromagnetically coupled regions ($c_m/2$) in a Co chain. Therefore, the time variation of the magnetic structure is caused by the motion of the domain boundary (the boundary between + and − moments regions). In $CeIr_3Si_2$ the time variation of magnetic structure is caused by the change of the volume fractions of the commensurate and the incommensurate phases, where the period of the incommensurate phase does not change with time. In the present material, the time variation is caused by the continuous change of the magnetic period. In this sense, we have found a new type of time variation of magnetic structure in $Ca_3Co_2O_6$.

We thank the members of the Neutron Scattering Laboratory, Institute for Solid State Physics, the University of Tokyo for supporting our experiments. This work is partly supported by a Grant-in-Aid for Scientific Research from the Ministry of Education, Culture, Sports, Science, and Technology.

References

[1] Motoya K, Muro Y and Takabatake T 2010 *J. Phys. Conf. Series* **200** 032048
[2] Fjellvåg H, Gulbrandsen E, Aasland S, Olsen A and Hauback B C 1996 *J. Solid State Chem.* **124** 190
[3] Kageyama H, Yoshimura K, Kosuge K, Mitamura H and Goto T 1997 *J. Phys. Soc. Jpn.* **66** 1607
[4] Kageyama H, Yoshimura K, Kosuge K, Azuma M, Takano M, Mitamura H and Goto T 1997 *J. Phys. Soc. Jpn.* **66** 3996
[5] Maignan A, Michel C, Masset A C, Martin C and Raveau B 2000 *Eur. Phys. J.* B **15** 657
[6] Hardy V, Lees M R, Petrenko O A, Paul D McK, Flahaut D, Hebert S and Maignan A 2004 *Phys Rev.* B **70** 064424
[7] Aasland S, Fjellvåg H and Hauback B 1997 *Solid State Commun.* **101** 187
[8] Agrestini S, Mazzoli C, Bombardi A and Lees M R 2008 *Phys Rev.* B **77** 140403
[9] Agrestini S, Chapon L C, Daund-Aladine A, Schefer J, Gukasov A, Mazzoli C, Lees M R and Petrenko O A 2008 *Phys. Rev. Leters* **101** 097207
[10] Moyoshi T and Motoya K *J. Phys. Soc. Jpn.* to be submitted

Journal of Physics: Conference Series **273** (2011) 012126 doi:10.1088/1742-6596/273/1/012126

Comparison of a Shastry-Sutherland lattice stability in Ce_2Pd_2Sn as a function of field and doping

Julian G. Sereni

Low Temperature Div., CAB-CNEA and CONICET, 8400 S.C. de Bariloche, Argentina

E-mail: jsereni@cab.cnea.gov.ar

Abstract. A comparative study of the stability of a Shastry-Sutherland lattice (SSL) in Ce_2Pd_2Sn under different physical conditions is presented. Applied magnetic field suppresses the SSL of stoichiometric Ce_2Pd_2Sn in a magnetic critical point at $T_{cr}(B) = 4.2$ K and $B_{cr} = 0.13$ T, whereas 25% of Ni does it at $T_{cr}(Ni) = 3.4$ K. Electronic concentration variation, driven by the increase of Pd (holes) concentration in $Ce_2Pd_{2+y}In_{1-y}$, decreases the magnetic transition down to $T_M = 2.8$ K in the limit of the alloy solubility, i.e. $y = 0.4$. The existence of a $M(B)$ plateau in SSL predicted by theory and the crossing of those isotherms, previously observed in the model compound $SrCu_2(BO_3)_2$ are analyzed.

1. Introduction

The Shastry-Sutherland lattice (SSL) can be described as a 2D square lattice built up from a network of dimers disposed orthogonally to each other [1]. Those dimers form from an antiferromagnetic AF coupling J between magnetic nearest-neighbors nn atoms, which interact among them through a J' coupling. Such a configuration is topologically equivalent to the 2D square lattice Heisenberg model [1] which has an exact dimer ground state. The model compound for this type of magnetic lattice is $SrCu_2(BO_3)_2$ [2], though it was recently found to form also in some intermetallic compounds with the Mo_2FeB_2-type of structure [3].

In this contribution we present a comparative study of the SSL stability in Ce_2T_2X intermetallic compounds, where T = transition metal and X = p-metal. For that purpose we have applied different control parameters, such as: i) magnetic field, ii) T = transition-metal substitution to induce a structural pressure and iii) the variation of electronic concentration (i.e. the chemical potential).

2. Experimental and results

The studied samples are polycrystalline, prepared by arc melting the appropriate amount of components. The respective annealing procedures and X-ray characterization were described elsewhere, Ref.[4] for Ce_2Pd_2Sn and [5] for $Ce_2Pd_{2+y}In_{1-y}$ family. Specific heat was measured using the heat pulse technique in a semi-adiabatic He-3 calorimeter in the range between 0.5 and 20K, at zero and applied magnetic field up to 4T. DC-magnetization measurements were carried out using a standard SQUID magnetometer operating between 2 and 300K, and as a function of field up to 5T.

Published under licence by IOP Publishing Ltd

3. Magnetic field effect in Ce_2Pd_2Sn

The effect of magnetic field was investigated in stoichiometric and alloyed samples. However, for simplicity we only analyze here the results on stoichiometric Ce_2Pd_2Sn which is the best example showing the main characteristics of SSL under magnetic field. In this compound, the formation of a SSL in Ce_2Pd_2Sn was observed within a limited range of temperature, between an (second order) AF transition at $T_M = 4.9$ K and a (first order) ferromagnetic FM one at $T_C = 2.1$ K [6], having as the upper limit a correlated paramagnetic phase and a FM one at low temperature. In Ce_2Pd_2Sn this exotic phase builds up from FM-dimers formed by Ce nn atoms, which are coupled by J. An AF J' coupling between those dimers drives the formation of the SSL, realized as a quasi 2D square lattice of effective spin $S_{eff} = 1$ below $T_M = 4.9$ K. Since the SSL only holds within a short range of temperature it is strongly affected by external magnetic field, including its suppression at a magnetic critical point: $B_{cr} = 0.13$ T and $T_{cr} = 4.1$ K [7].

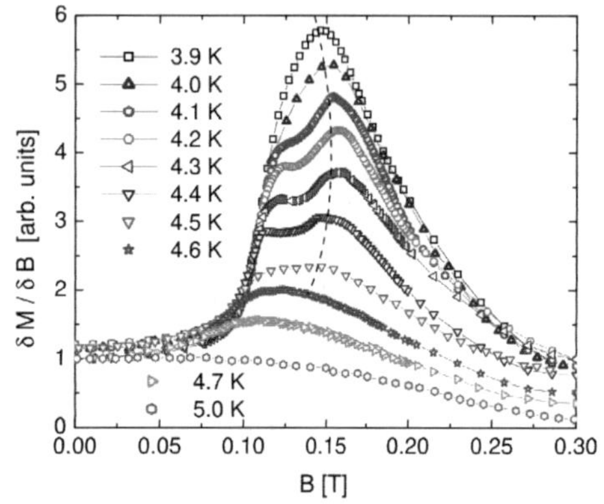

Figure 1. Detail of the $M(B)$ derivative of Ce_2Pd_2Sn within the $3.9 \leq T \leq 5$ K range showing the satellite maximum (dashed curve) around the critical field.

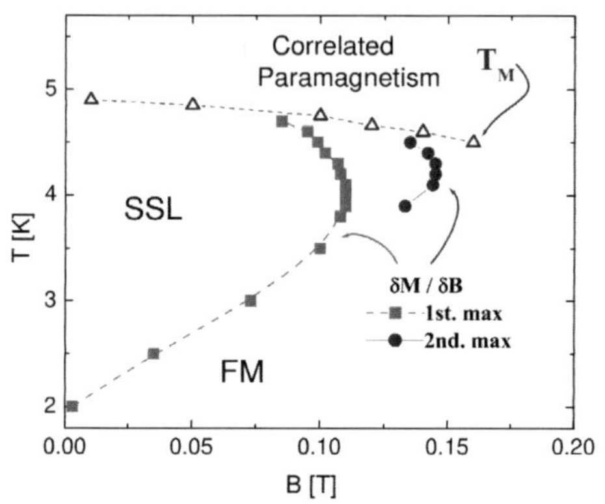

Figure 2. Phase diagram of Ce_2Pd_2Sn. The phase boundaries are extracted from the cusp of $M(T)$ at $T_M(B)$ $\triangle$, the maximum of $\partial M \partial B \mid_T$ ■ and its satellite ● .

Isothermal $M(B)$ measurements show some peculiar features characteristic of the SSL, which were previously observed in the model compound $SrCu_2(BO_3)_2$ [2, 8], such as quantized magnetization plateaux at fractional values of the saturation moment M_{sat} (c.f. 1/4 and 1/8 M_{sat}) and the crossing of magnetization isotherms as a function of B at certain temperatures. One of those plateaux is revealed incipiently in Ce_2Pd_2Sn as a satellite maximum in the $\partial M/\partial B \mid_T$ derivative, see Fig. 1. In the magnetic phase diagram presented in Fig. 2, the low temperature phase boundary (related to T_C) was extracted from the maximum of $\partial M/\partial B(B) \mid_T$ for $2 \leq T \leq 5$ K (only shown for $T > 3.9$ K in Fig. 2). The upper phase boundary (related to $T_M(B)$) was extracted from the cusp of $M(T)$ [7]. Around the critical region the satellite maximum of $\partial M/\partial B \mid_T$ indicates the field at which an incipient $M(B)$ plateau occurs, with a value $\approx 1/4 M_{sat}$. Such a plateau is related with the crossing of the $M(B)$ isotherms [7] since, within the $4 \leq T \leq 4.6$ K range, one has $\partial M/\partial T \mid_B = 0$. From Maxwell relations this implies that $\partial S/\partial B \mid_T = 0$ and, consequently, that the entropy S is locked under field variation in an isothermal process. Therefore, a plateau is expected in those $M(B)$ isotherms. Since the existence of 'crossing points' in thermodynamical parameters carries important implications in the related properties, they were identified as 'isosbestic points' in the literature [9].

4. Structural pressure in Ni doped $Ce_2(Pd_{1-x}Ni_x)_2Sn$

The structural pressure was induced by doping the Pd lattice with smaller isoelectronic Ni atoms. In this case a weakening of the Ce magnetic moment is expected due to the Kondo screening of the $4f$ Ce states. In $Ce_2(Pd_{1-x}Ni_x)_2Sn$ that effect is observed as a reduction of the upper ordering temperature T_M as a function of Ni concentration $x(Ni)$, see Fig. 3. On the contrary, the lower transition T_C to the FM phase increases with $x(Ni)$ as a consequence of the reduction of the temperature range of the SSL stability. This behavior can be explained by the weakening of the J and J' couplings due to the increasing Kondo screening. Those couplings may be eventually overcome by the inter-Ce planes coupling in the 'c' crystalline direction J_C, which for $x = 0.25$ practically inhibits the formation of the SSL phase.

The arising Kondo screening is confirmed by i) the increasing γ coefficient (extracted from a $C_m/T(T \to 0)$ extrapolation), which increases proportionally to $x(Ni)$ and ii) the weakening of the saturation magnetic moment at high field (not shown). In Fig. 4 we present Ni dependent phase diagram which extrapolates to a critical concentration $x_{cr}(Ni) \approx 0.27$.

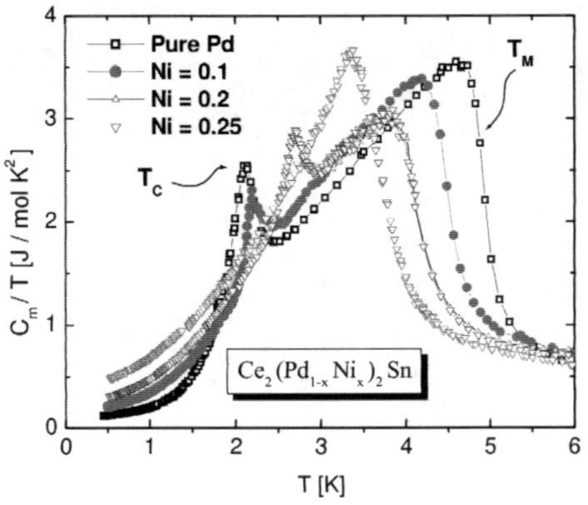

Figure 3. Thermal dependence of specific heat showing upper (T_M) and lower (T_C) transitions. The magnetic contribution C_m was evaluated after La_2Pd_2Sn phonon subtraction.

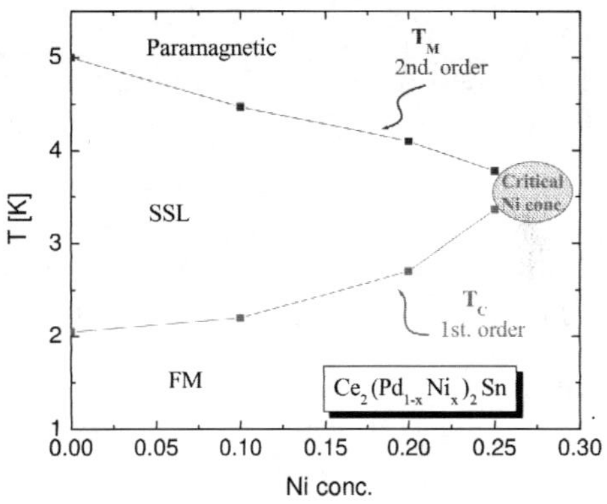

Figure 4. Ni doping dependencies of T_M and T_C, which extrapolate to a critical point at $T_{cr}(Ni) \approx 3.5\,K$ for $x_{cr}(Ni) \approx 0.27$, where the SSL is suppressed.

5. Electron concentration effect in $Ce_2Pd_{2+y}In_{1-y}$

The change of electronic concentration and consequently the chemical potential, is realized in the isotypic compound $Ce_2Pd_{2+y}In_{1-y}$. In that compound one takes profit of its extended range of solubility, between $0.1 \leq y \leq 0.4$ [5], where the actual Ce atomic concentration is 1.95 instead of 2. The electronic variation is induced by enriching the Pd (holes) concentration and by the replacement of $Sn[5s^2 5p^2]$ by $In[5s^2 5p^1]$.

In Fig. 5 we present the magnetic contribution to the specific heat which shows that the transition T_M decreases upon increasing $y(Pd)$. In these system no T_C transition is observed but a kink (at $T = T^*$) in the $y(Pd) = 2.20$ and 0.25 alloys, which transforms into a shoulder for higher Pd concentrations. Coincidentally, only the former alloys show the SSL signs, i.e. the crossing of their $M(B)$ isotherms and a satellite maximum in $\partial S/\partial B \mid_T = 0$ like in Ce_2Pd_2 Sn. Those properties vanish for $y(Pd) = 2.35$ and 0.24 concentrations indicating that the SSL is strongly affected by the Pd/In modification which also increases atomic disorder.

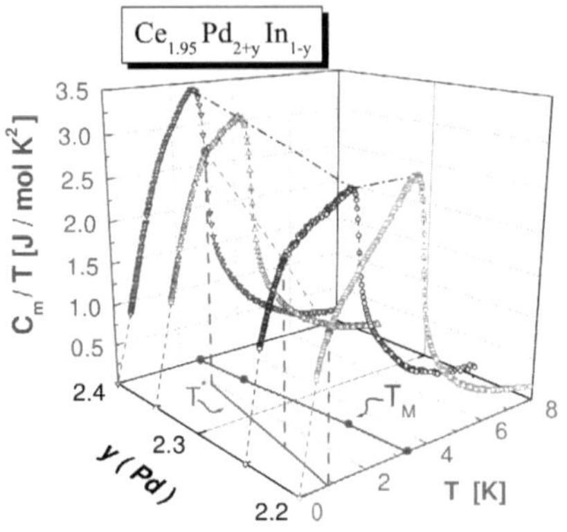

Figure 5. Magnetic specific heat dependence on temperature and $y(Pd)$ concentration. Basal plane: projection of the $y(Pd)$ dependence of T_M and T^*, see the text.

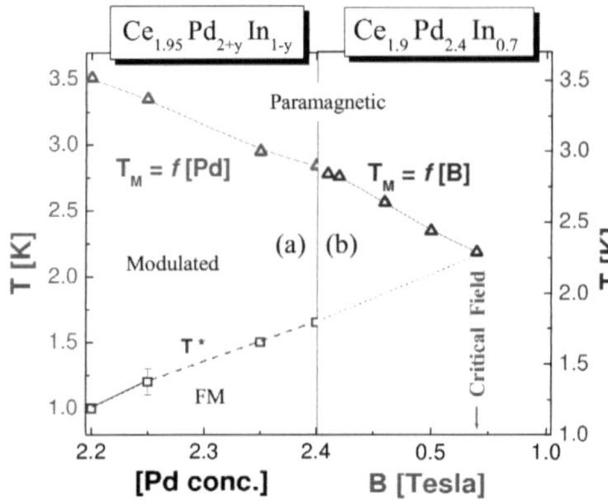

Figure 6. (a) Electronic, $y(Pd)$ holes, dependence of the T_M phase boundary and T^*. (b) Field dependence for the richest $y(Pd) = 0.24$ alloy. Dot line: extrapolation of T^* to the critical field.

Fig. 6 shows the magnetic phase diagram including both T_M and $T^*(Pd)$ dependencies. In $Ce_{1.9}Pd_{2.4}In_{0.7}$, T_M was traced down to lower temperature by applying magnetic field up to a critical value $B_{cr}(y = 0.4) = 0.7\,T$ where $T_{cr}(y = 0.4) = 2.3\,K$. Beyond that point, a field induced FM polarization takes over. Both, $y(Pd)$ concentration and field dependencies of T_M are compared in Fig. 6 by properly scaling both parameters, c.f. Fig. 6a and b respectively.

6. Conclusions

We have compared the stability of the SSL in the Ce_2T_2X family of compounds (T= Pd, Ni and X= Sn, In) driving different control parameters, c.f. magnetic field, structural pressure and electron concentration. We find that the SSL is suppressed in stoichiometric Ce_2Pd_2Sn by a moderate field ($B_{cr} = 0.13\,T$), by arising the Kondo screening of Ce moments with $x_{cr}(Ni) \approx 0.27$ doping, and by increasing electron (holes) concentration up to $y(Pd) = 2.25$ in $Ce_{1.95}Pd_{2+y}In_{1-y}$. Comparatively, a stronger magnetic field is required to induce FM polarization in indeede compounds than in the stannides, indicating a different relation between the on-plane J and J' coupling parameters strength respect to the inter-plane J_C one.

Acknowledgments
The author thanks G. Schmerber and M. Giovannini for providing the samples, M.G.- Berisso for collaborating with the measurements and H. von Löhneysen for motivating discussions.

References

[1] S. Miyahara and K. Ueda 1999 *Phys. Rev. Lett.* **82** 3701.
[2] H. Kageyama et al. 1999 *Phys. Rev. Lett.* **82** 3168.
[3] M. N. Peron et al. 1993, *J. Alloys & Comps.* **201** 203.
[4] A. Braghta et al. 2008 *J. Magn. Magn. Mater.* **320** 1141.
[5] M. Giovannini et al. 2000 *Phys. Rev. B* **61** 4044.
[6] J. G. Sereni et al. 2009 *Phys. Rev. B* **80** 024428.
[7] J. G. Sereni et al. 2010 *Phys. Rev. B* **81** 184429.
[8] G. A. Jorge et al., 2005 *Phys. Rev. B* **71** 092403.
[9] D. Wollhardt 1997, *Phys. Rev. Lett.* **78** 1307.

Journal of Physics: Conference Series **273** (2011) 012127 doi:10.1088/1742-6596/273/1/012127

Component-separated magnetic transition in $HoRh_2Si_2$ single crystal

T. Shigeoka[1], T. Fujiwara[1], K. Munakata[2], K. Matsubayashi[2], Y. Uwatoko[2]

[1]Graduate School of Science and Engineering, Yamaguchi University, Yamaguchi 753-8512, Japan.
[2]Institute for Solid State Physics, University of Tokyo, Kashiwa-shi Chiba 277-8581, Japan.

shigeoka@yamaguchi-u.ac.jp

Abstract. Magnetic susceptibility and specific heat measurements have been carried out on a $HoRh_2Si_2$ single crystal compound. Specific heat shows two λ-like anomalies at T_{N1}=29.1 K and T_{N2}=12.0 K, and an additional sharp anomaly indicating the first order transition at *T*t=27.3 K. The temperature dependence of susceptibility along the [001] direction shows a clear cusp at T_{N1} but a tiny kink at T_{N2} and *T*t. In contrast, those in the *ab*-plane show a clear cusp at T_{N2}. $HoRh_2Si_2$ shows a successive component-separated magnetic transition coming from a new type frustration not geometrical one.

1. Introduction

A frustration, which often comes from a geometrical ion arrangement, leads various unusual physical properties. "Successive component-separated magnetic transition", which is unique successive transition characterized by independent ordering of the *c*- and *ab*-component of magnetic moments at different temperatures, is one of the interesting characters. The compounds $CsNiCl_3$ and DyB_4 are well known as compounds which show this type of transition [1, 2]. In $CsNiCl_3$, it is suggested that the transition is due to competing exchange interaction and weak Ising anisotropy in the antiferromagnetic triangular lattice [3]. In a Shastry-Sutherland lattice DyB_4, geometrical quadrupolar frustration in addition to geometrical spin frustration play an important role [2]. In most cases, geometrical frustration is important. Recently, N. Sanada et al. have reported that the tetragonal compound TbCoGa5 shows successive component-separated magnetic transitions [4]. The magnetic, specific heart and elastic constant behavior indicate the presence of a strong frustration in the interactions in $TbCoGa_5$. The geometrical frustration is not valid for $TbCoGa_5$ because of the simple tetragonal arrangement of Tb atoms in this compound. The new class of frustration effects where quadrupolar degrees of freedom play an important role has been proposed. They suggest that this type transitions appear commonly in the rare earth compounds with a degeneracy of quadrupolar degrees of freedom below the highest magnetic transition temperatures. Further study on compounds which show a similar behavior to one of $TbCoGa_5$ should be required.

The ternary compound $HoRh_2Si_2$ having the tetragonal $ThCr_2Si_2$-type structure shows an antiferromagnetic ordering below T_N=27 K and an additional transition at T_N'=11 K [5, 6]. In the temperature dependence of magnetic susceptibility, two clear peaks have been observed. The neutron diffraction study have reported that the antiferromagnetic structure is a simple one; propagation vector

Published under licence by IOP Publishing Ltd

k=(0,0,1) with magnetic moments tilted from the c-axis by 28° at low temperatures and have given no information on the transition at T_N'. Takano et al have proposed that the transition at T_N' is corresponding to the temperature where magnetic moments become parallel to the c-axis with increasing temperature [7]. Although the details in the transition at T_N' have been unknown yet, multiple interactions should play a role to decline the magnetic moments. So we consider the occurrence of frustration is possible. Then $HoRh_2Si_2$ single crystals have been grown and magnetic study has been performed on the single crystal compounds.

2. Experimental

The single crystal $HoRh_2Si_2$ has been grown by the tetra-arc Czochralski method. The single phase nature has been confirmed by X-ray powder diffraction. Good quality of the single crystal has been confirmed and crystallographic orientations were determined by back Laue method. The crystal was fixed on a plastic plate so that the desired direction is perpendicular to the plate within experimental angle accuracy of one degree. Then it is used for magnetic measurements which have been carried out by use of a PPMS and SQUID magnetometer, MPMS, (Quantum Design). The specific heat has been performed by a relaxation method also using the PPMS.

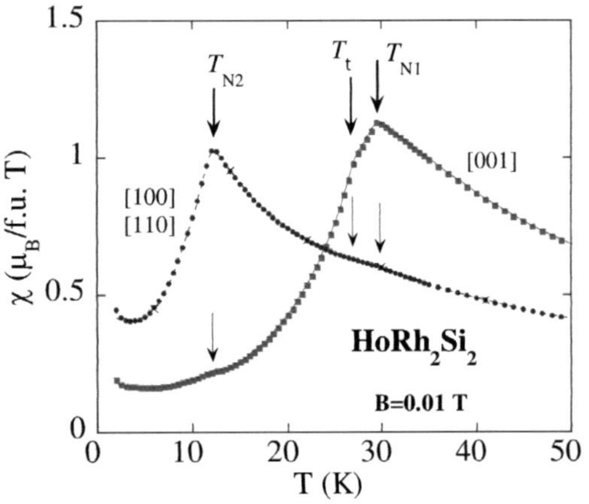

Figure 1 Magnetic susceptibility along the main symmetry axes of a tetragonal cell on the $HoRh_2Si_2$ single crystal.

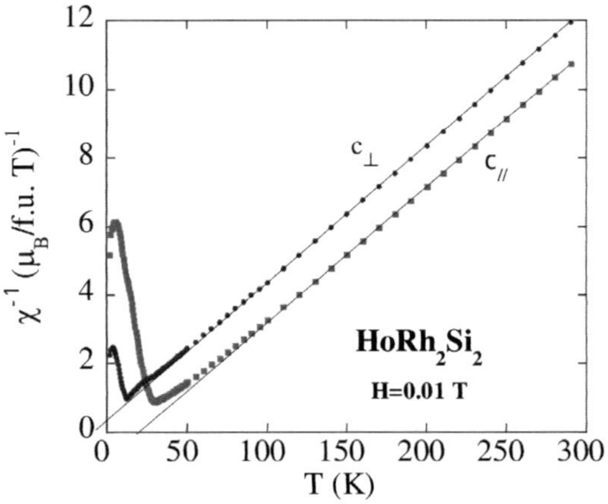

Figure 2 Inverse magnetic susceptibility along the c-axis and the directions in the basal plane

3. Results and discussion

The temperature dependence of magnetic susceptibility along the main symmetry axes for a tetragonal cell is shown in Fig.1. A magnetic anisotropy between the c-axis and directions in the basal plane is evidenced from the figure while no anisotropy is in the basal plane; the susceptibility along the [100] direction is identical to the one along [110]. The [001] magnetic susceptibility $\chi_{//}$ shows a clear maximum at T_{N1}=29.5 K and a small hump at T_{N2}=12.4 K. A small kink at T_t=27.2 K is also seen. In contrast, susceptibilities $\chi_\perp$for directions in the basal plane, $B\perp$[001], show a small hump at T_{N1} and a clear maximum at T_{N2}. Any anomaly is hardly seen around T_t. These results suggest that *c*- and *ab*-components of magnetic moments of the Ho^{3+} ions order independently at T_{N1} and T_{N2}, respectively. This type of transition is called "successive component-separated magnetic transitions". Similar behaviours have been reported in $TbCoGa_5$ [4] and DyB_4 [2]. The existence of three phases below T_{N1} is evidenced from the result. We call hereafter these the phase I (T_{N1}>T> T_t), the phase II (T_t>T> T_{N2}), the phase III (T_{N2}>T).

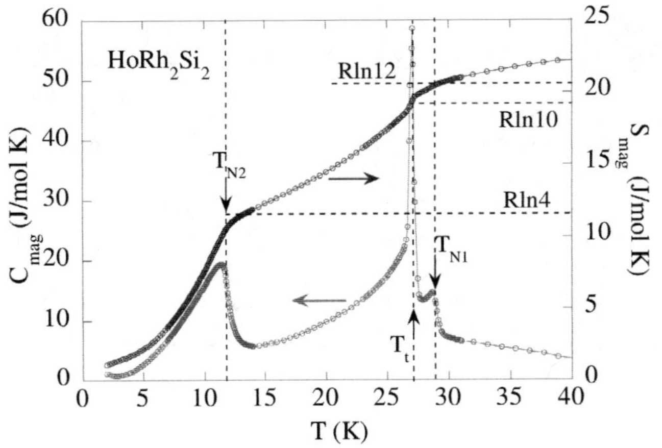

Figure 3 Temperature dependence of magnetic specific heat C_{mag} and magnetic entropy S_{mag} on the $HoRh_2Si_2$ single crystal.

The susceptibilities along the all directions obey the Curie-Weiss law above 100 K as shown in Fig. 2. The χ^{-1}-T curves become linear and are almost parallel with each other at high temperatures. The effective magnetic moment estimated is 10.60 μ_B which is in good agreement with the Ho^{3+} ion magnetic moment (10.6 μ_B). This shows that 4*f* electrons in $HoRh_2Si_2$ are well localized and Rh ion is nonmagnetic. The paramagnetic Curie temperatures along the c-axis and directions in the basal plane are $\Theta_{//}$=19.3 K and $\Theta_{\perp}$=-9.3 K, indicating that dominant interactions along the c-axis and in the basal plane are ferromagnetic and antiferromagnetic, respectively.

The thermal variations of magnetic specific heat C_{mag} and entropy S_{mag} on the $HoRh_2Si_2$ single crystal are shown in Fig.3. The magnetic contribution of specific heat is deduced from subtracting the specific heat of a phonon contribution from total specific heat of $HoRh_2Si_2$. The phonon contribution is calculated from the Debye function with the Debye temperature Θ_D=313 K for $HoRh_2Si_2$ which is estimated from Θ_D =330 K for $LaRh_2Si_2$ [6]. The specific heat shows two λ-type anomalies at T_{N1} and T_{N2}, and a sharp peak at T_t. This dependency is almost agreement with one reported by Sekizawa et al. [6], but they did not recognize the transition at T_{N1}. They determined the Neel temperature by the peak temperature of T_t. The existence of three phases below T_{N1} is confirmed. The each phase transitions at T_{N1} and T_{N2} is of the second-order one, and the first-order one at T_t. The magnetic entropy is obtained from numerical integration of C_{mag}/T vs. T. It is evidenced from the figure the entropy reaches Rln4 at T_{N2}. This indicates that the ground state is a pseudo-quartet at least. The entropy of Rln12 (=20.6 J $mol^{-1}K^{-1}$) which is released below T_{N1} is smaller than Rln(2*J*+1)=23.5 J $mol^{-1}K^{-1}$ for *J*=8 multiplet of Ho^{3+} ion. The overall splitting of the crystalline electric field levels is slightly larger than T_{N1}.

The neutron diffraction study on the single crystal and powdered sample has been performed. The antiferromagnetic structure of the phase III at low temperatures below T_{N2} is a simple one; the propagation vector k=(0,0,1) with magnetic moments tilting from the c-axis by about 30° at 4 K. The tilting angle becomes smaller with increasing temperature and almost zero at T_{N2}. For the phase II, ab-component of magnetic moments disappears and diffuse scatterings, indicating paramagnetic ab-component, appear. The phase I has an incommensurate structure. The detailed analysis of neutron result is now in progress which will be published soon elsewhere.

These results of $HoRh_2Si_2$ are quite similar to those of $TbCoGa_5$. Ho atoms in this compound are of the simple tetragonal arrangement. Then no geometrical frustration is considered. So we believe $HoRh_2Si_2$ is a candidate for a new type frustration. Further study such as elastic constant, magnetic and neutron measurements on this single crystal is now in progress.

References

[1] Achiwa A 1969 *J. Phys. Soc. Jpn.* **27** 561.
[2] Watanuki R, Sato G, Suzuki K, Ishihara M, Yanagisawa T, Nemoto Y, Goto T 2005 *J. Phys. Soc. Jpn.* **74** 2169.
[3] Kadowaki H, Ubukoshi K, Hirakawa K 1986 *J. Phys. Soc. Jpn.* **56** 751.
[4] Sanada N, Watanuki R, Suzuki K, Akatsu M, Sakakibara T 2009 *J. Phys. Soc. Jpn.* **78** 073709.
[5] Felner I, Nowik I 1984 *J. Phys. Chem. Solids* **45** 419.
[6] Sekizawa K, Takano Y, Takigami H, Takahashi Y 1987 *J. Less-Common Metals* **127** 99.
[7] Takano Y, Ohhata K, Sekizawa K 1987 *J. Magn. Magn. Mater.* **66** 187.

Journal of Physics: Conference Series **273** (2011) 012128 doi:10.1088/1742-6596/273/1/012128

Anisotropic hyperfine coupling in vanadium oxides

Y. Shimizu[1,2], H. Takeda[2], M. Tanaka[2], M. Itoh[2], S. Niitaka[3], H. Takagi[3], M. Isobe[4], T. Yamauchi[4], J. Yamaura[4], Y. Ueda[4]

[1] Institute for Andvance Research, Nagoya University, Furo-cho, Chikusa-ku, Nagoya 464-8601, Japan

[2] Department of Physics, Nagoya University, Furo-cho, Chikusa-ku, Nagoya 464-8602, Japan

[3] RIKEN, Wako, Saitama 351-0198, Japan

[4] ISSP, University of Tokyo, Kashiwa, Chiba 277-8581, Japan

E-mail: yasuhiro@iar.nagoya-u.ac.jp

Abstract. We have investigated orbital occupancies in metallic states of the vanadium oxides by ^{51}V NMR measurements on the single crystals of V_2O_3, $K_2V_8O_{16}$, and LiV_2O_4, having ground states of an antiferromagnetic Mott insulator, a spin singlet insulator, and a heavy-mass Fermi liquid, respectively. The Knight shift K vs the magnetic susceptibility plots for the principal axes give the axially anisotropic and isotropic parts of the hyperfine coupling constant, $A_{\rm ax}$ and $A_{\rm iso}$. In V_2O_3, $A_{\rm ax}$ diminishes owing to the equivalent contribution of three-hold t_{2g} orbitals to the spin susceptibility. The orbital mixing is also significant in $K_2V_8O_{16}$ having a small $A_{\rm ax}/A_{\rm iso} \sim 0.1$, which is expected to have $0.5 - 0.6$ for $3d_{xy}^1$. In LiV_2O_4, $A_{\rm ax}/A_{\rm iso}$ has a large negative value, as expected in the a_{1g} orbital. The result suggests the presence of the localized spin hidden in the itinerancy, giving the high-entropy state of LiV_2O_4 at low temperatures.

1. Introduction

The attractive quantum effects to release geometrical frustration are forming a spin-singlet cluster [1] and a Fermi liquid [2]. In transition metal oxides, orbital degrees of freedom play a significant role in stabilizing these states. The hollandite $A_2M_8O_{16}$ with a zigzag chain [3, 4] and the spinel AM_2O_4 with a pyrochlore lattice [2, 5] actually exhibit various ground states depending on the orbital occupancy.

Nuclear magnetic resonance (NMR) can probe the orbital state via the dipole hyperfine interaction with electron spins [6]. The method has been applied to Mott insulators with the orbital order [7, 8] and the rare-earth heavy fermion system with octupole order [9]. In this paper, we investigate the orbital state from the ^{51}V Knight shift tensor K in the frustrated vanadium oxides, $K_2V_8O_{16}$ and LiV_2O_4, in comparison with the conventional material V_2O_3 [10, 11]. We also discuss the relation to anomalous nature in the metallic state.

2. Experimental

^{51}V NMR measurements were performed on a single crystal of $K_2V_8O_{16}$, LiV_2O_4 and V_2O_3. The sample preparation methods were described elsewhere [2, 3, 10]. The frequency-sweep ^{51}V NMR spectra were obtained by the Fourier transformation of spin echo signals in a constant magnetic field H_0. The ^{51}V Knight shift K was defined as a relative shift from a ^{51}V resonance frequency of an aqueous solution of $NaVO_3$.

Published under licence by IOP Publishing Ltd

3. Results and discussion

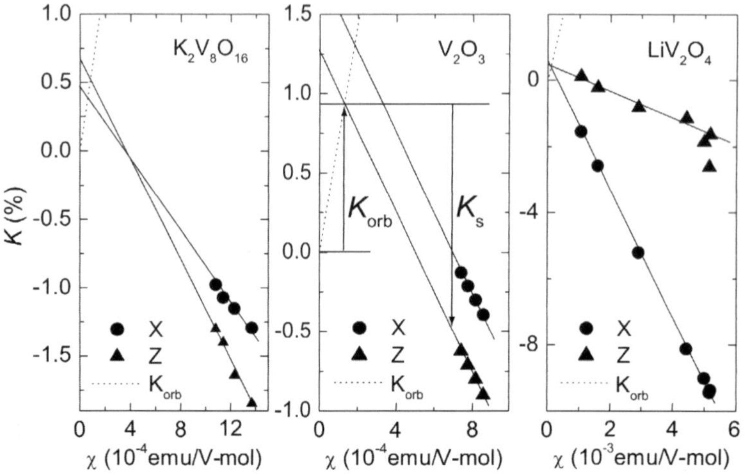

Figure 1. ^{51}V Knight shift K vs magnetic susceptibility χ plots for $K_2V_8O_{16}$, V_2O_3, and LiV_2O_4.

The hyperfine coupling tensors related to spin susceptibilities, A_s, of $K_2V_8O_{16}$, LiV_2O_4 and V_2O_3 are obtained from the linearity of the K versus magnetic susceptibility χ plots with temperature T as an implicit parameter, as shown in Fig. 1. Here the T-independent orbital component of K, $K_{\rm orb}$, was evaluated by using the orbital hyperfine coupling constant $A_{\rm orb} = 2\mu_{\rm B}\langle r^{-3}\rangle$, as shown in the dotted line in Fig. 1, where $\mu_{\rm B}$ is the Bohr magneton and $\langle r^{-3}\rangle$ is the average value of r^{-3} for the $3d$ orbital. For the t_{2g} system with the doublet and singlet levels, A_s can be decomposed into an isotropic part $A^{\rm s}_{\rm iso} = (A^s_X + A^s_Y + A^s_Z)/3$ and an axially symmetric part $A^{\rm s}_{\rm ax} = (2A^s_Z - A^s_X - A^s_Y)/6$ [12];

$$A_{\rm s} = A^{\rm s}_{\rm iso} + A^{\rm s}_{\rm ax}(3\cos^2\theta - 1), \tag{1}$$

where θ is the angle between H_0 and the Z axis. Table 1 shows $A^{\rm s}_{\rm iso}$ and $A^{\rm s}_{\rm ax}$ for various materials. $A^{\rm s}_{\rm iso}$ mainly arising from the core polarization is expressed as $-\kappa\mu_{\rm B}\langle r^{-3}\rangle$ by using a dimensionless parameter κ, while $A^{\rm s}_{\rm ax}$ reflects the anisotropy of the dipole hyperfine coupling tensor $\mathsf{A}^{\rm s}_{\rm dip}$ due to the anisotropic orbital. In a d_{xy} case, the maximum and minimum components of $\mathsf{A}^{\rm s}_{\rm dip}$ are respectively calculated as [6],

$$A_\perp = \frac{2}{7}\mu_{\rm B}\langle r^{-3}\rangle, A_{||} = -\frac{4}{7}\mu_{\rm B}\langle r^{-3}\rangle, \tag{2}$$

leading to $A^{\rm s}_{\rm ax} = -\frac{2}{7}\mu_{\rm B}\langle r^{-3}\rangle$. The unknown $\langle r^{-3}\rangle$ is canceled out by taking a ratio $A^{\rm s}_{\rm ax}/A^s_{\rm iso} = \frac{2}{7\kappa}$ ~ 0.57 for $\kappa = 0.5$. In $K_2V_8O_{16}$, the t_{2g} triplet is split into the higher d_{xy} singlet and the lower d_{yz}-d_{zx} doublet. The positive $A^{\rm s}_{\rm ax}$ shows that $\mathsf{A}^{\rm s}_{\rm dip}$ has a negative Z component, which indicates that the d_{xy} contribution to $A^{\rm s}_{\rm ax}$ is larger than the d_{yz}-d_{zx} one. The $A^{\rm s}_{\rm ax}$ value is, however, much smaller than those of NaV_2O_5 and CaV_2O_5 with a d_{xy} orbital [13, 14]. It suggests that 1.25 electrons per one V atom are nearly equally distributed in d_{xy} and d_{yz}-d_{zx} in $K_2V_8O_{16}$ due to the strong Hund's coupling. This is consistent with the orbital occupancies calculated

Table 1. The isotropic and axial parts of the hyperfine coupling constants, A^s_{iso} and A^s_{ax}, the ratio A^s_{iso}/A^s_{ax}, the number of $3d$ electrons, the ground states, and the orbital state for the various V oxides.

Material	A^s_{iso}	A^s_{ax}	A^s_{ax}/A^s_{iso}	$3d^n$	ground state	orbital
β-$Na_{0.33}V_2O_5$	-	-	0.50	0.17	AF	d_{xy}
CaV_2O_5	−117	−60	0.51	1	singlet	d_{xy}
$K_2V_8O_{16}$	−79	−9	0.11	1.25	singlet	$d_{xy} > d_{yz,zx}$
LiV_2O_4	80	−32	−0.4	1.5	metal	$a_{1g} > e'_g$
V_2O_3	−130	−1.7	0.01	2	AF	$a_{1g} \sim e'_g$
$Lu_2V_2O_7$	−97	58	−0.6	1	ferro	a_{1g}

by using a Wien 2k GGA+U package [15]. A slight predominance of d_{xy} is attributed to the quasi-one-dimensional band with a smaller bandwidth than those of $d_{yz,zx}$.

In the presence of the trigonal distortion, the t_{2g} triplet is lifted into the a_{1g} singlet and the e'_g doublet. The electron distribution on a_{1g} gives the minimum and maximum components of A^s_{dip} as,

$$A_\perp = -\frac{2}{7}\mu_B \langle r^{-3}\rangle, A_\parallel = \frac{4}{7}\mu_B \langle r^{-3}\rangle. \tag{3}$$

The sign of the ratio $A^s_{ax}/A^s_{iso} = -\frac{2}{7\kappa}$ changes compared to that of d_{xy}. As shown in Table 1, V_2O_3 has the vanishing A^s_{ax}, indicating that t_{2g} levels are equivalently occupied by two electrons. Namely, the ratio of the orbital occupancy is $a_{1g} : e'_g = 1 : 2$, in good agreement with the result of the polarized x-ray absorption measurement [16]. In contrast, we observed the highly anisotropic A_s in LiV_2O_4 with $A^s_{ax}/A^s_{iso} \sim 0.3$-$0.5$, which tends to decrease at low temperatures. It means that the a_{1g} orbital is predominant in the spin susceptibility, similarly to the ferromagnetic Mott insulator $Lu_2V_2O_7$ having the a_{1g} orbital order [8].

Based on the obtained orbital state, we discuss the origin of the anomalous χ of the metallic state in these vanadium oxides. The three vanadium oxides, $K_2V_8O_{16}$, V_2O_3, and LiV_2O_4, do not have the Pauli paramagnetic χ but the Curie-Weiss χ in spite of the metallic conductivity [2, 3, 11], with the Weiss temperature Θ of −290 K [4], −600 K [11] and −37 K [2], respectively. In a localized picture, Θ is a measure of the energy scale of the antiferromagnetic exchange interaction, where the short range correlation starts to develop. In an itinerant picture, an increase in the spin susceptibility can be interpreted as the enhancement of the density of states or the presence of the ferromagnetic correlation, which may be unlikely because the numbers of effective Bohr magneton obtained from the Curie constant are close to the full moment values. Furthermore, the resistivity exceeds the Mott-Ioffe-Regel limit at high temperatures in these materials [2, 3, 10], and the optical conductivity does not follow the Drude model in LiV_2O_4 [17]. The localized picture is also unlikely because a development of antiferromagnetic correlation has not been observed in the nuclear spin-lattice relaxation rate [2, 4] and the inelastic neutron scattering [18]. Thus the metallic states in these materials are not understood within the conventional Heisenberg or Fermi liquid picture. Compared with $|\Theta|$ and $|A^s_{ax}/A^s_{iso}|$, the materials with the larger $|\Theta|$ tend to have smaller $|A^s_{ax}/A^s_{iso}|$. It suggests the good correlation between the orbital mixing ratio and the electron hoping probability between the V sites. In this context, the electrons are nearly localized in LiV_2O_4, located close to the metal-insulator transition despite the metallic conductivity down low temperatures. The stabilization of the a_{1g} orbital in LiV_2O_4, which favors the ferromagnetic interaction as observed in $Lu_2V_2O_7$

[8], is ascribed to a reduction of the magnetic frustration. Despite the development of the antiferromagnetic correlation and the Fermi liquid coherence at low temperatures [2, 18], the magnetic order is absent. It suggests that a frustrated spin liquid state in the localized a_{1g} orbitals gives the remaining large entropy down to low temperatures.

4. Conclusion

The anisotropy of the Knight shift tensor was presented to elucidate the orbital state in the metallic frustrated systems $K_2V_8O_{16}$ and LiV_2O_4, having the spin singlet and Fermi liquid ground states, respectively. The negative and positive values of the axially anisotropic part of the hyperfine coupling interaction indicate the predominance of d_{xy} and a_{1g} orbitals in the spin susceptibility of $K_2V_8O_{16}$ and LiV_2O_4, respectively. This is in contrast to the vanishing $A^{\mathrm{s}}_{\mathrm{ax}}$ value in V_2O_3. Thus NMR was shown to be a useful probe to elucidate the orbital state of the metallic transition metal oxides.

Acknowledgments

We also thank S. Inoue for technical assistance. This work was financially supported by a Grant-in-Aid for Scientific Research on Priority Areas "Novel State of Matter Induced by Frustration" (No. 22014006) and Special Coordination Funds for Promoting Science and Technology from the MEXT, and also by Grants-in-Aid for Scientific Research (Grants No. 18104008, 19340097, and 19740224) from the JSPS.

References

[1] Schmidt M, Ratcliff II W, Radaelli P G, Refson K, Harrison N M and Cheong S W, 2004 *Phys. Rev. Lett.* **92** 056402; P. G. Radaelli, Y. Horibe, M. J. Gutmann, H. Ishibashi, C. H. Chen, R. M. Ibberson, Y. Koyama, Y-S. Hor, V. Kiryukhin and S-W. Cheong, 2001 *Nature* **416** 155
[2] Kondo S 1997 *Phys. Rev. Lett.* **78** 3729; Urano C, Nohara M, Kondo S, Sakai F, Takagi H, Shiraki T and Okubo T 2000 *Phys. Rev. Lett.* **85** 1052
[3] Isobe M, Koichi S, Kouno N, Yamaura J, Yamauchi T, Ueda H, Gotou H, Yagi T and Ueda Y 2006 *J. Phys. Soc. Jpn.* **75** 073801
[4] Shimizu Y, Okai K, Itoh M, Isobe M, Yamauchi T, Yamaura J and Ueda Y unpublished (submitted to *Phys. Rev. B*).
[5] Shimizu Y, Tanaka M, Itoh M and Katsufuji T 2008 *Phys. Rev. B* **78** 144423
[6] Abragam A and Bleaney B 1970 *Electron Paramagnetic Resonance of Transition Ions* Oxford University Press, London
[7] Khoi L D and Veillet P 1975 *Phys. Rev. B* **11** 4128
[8] Kiyama T, Shiraoka T, Itoh M, Kano L, Ichikawa H and Akimitsu J 2006 *Phys. Rev. B* **73** 184422
[9] Sakai M, Shiina R, Shiba H and Thalmeier P, 1999 *J. Phys. Soc. Jpn.* **68** 1364
[10] McWhan D B, Menth A, Remeika J P, Brinkman W F and Rice T M 1973 *Phys. Rev. B* **7** 1920
[11] Takigawa M, Ahrens E T and Ueda Y 1996 *Phys. Rev. Lett.* **76** 283
[12] C. P. Slichter, 1990 *Principles of the Magnetic Resonance* 3rd. ed., Springer-Verlag, Berlin
[13] Ohama T, Yasuoka H, Isobe M and Ueda Y 1997 *J. Phys. Soc. Jpn.* **66** 3008
[14] Ohama T, Isobe M and Ueda Y 2000 *J. Phys. Soc. Jpn.* **69** 1574
[15] Sakamaki M, Horiuchi S, Konishi T and Ohta Y arXiv:0811.4338v1
[16] Park J-H, Tjeng L H, Tanaka A, Allen J W, Chen C T, Metcalf P, Honig J M, de Groot F M F and Sawatzky G A 2000 *Phys. Rev. B* **61** 11506
[17] Jönsson PE, Takenaka K, Niitaka S, Sasagawa T, Sugai S and Takagi H 2007 *Phys. Rev. Lett.* **99** 167402
[18] Lee A H, Qiu Y, Broholm C, Ueda Y and Ruch J J 2001 *Phys. Rev. Lett* **86** 5554

Journal of Physics: Conference Series **273** (2011) 012129 doi:10.1088/1742-6596/273/1/012129

Lattice dynamics in the itinerant helical magnet MnSi

P.A. Alekseev[1], E.S. Clementyev[1,2] R. Heid[3], A.S. Ivanov[4], D. Lamago[3,5], J.-M.Mignot[5], A.E. Petrova[6] and S.M.Stishov[6]

[1]ISSSPh, Russian Research Centre "Kurchatov Institute", 123182 Moscow, Russia
[2]LNS, Institute for Nuclear Research of RAS, Troitsk 142190, Moscow Region, Russia
[3]Karlsruhe Institute of Technology (KIT), Institut für Festkörperphysik, P.O.B. 3640, D-76021 Karlsruhe, Germany
[4]Institut Laue Langevin, 156X, 38042 Grenoble cedex 9, France
[5]Laboratoire Léon Brillouin, CEA-Saclay, F-91191 Gif sur Yvette Cedex, France
[6]Institute for High Pressure Physics of RAS, Troitsk 142190, Moscow Region, Russia

E-mail: `sergei@hppi.troitsk.ru`

Abstract. The phonon dispersion relations in MnSi were measured using inelastic neutron scattering. At the same time, calculations of the phonon spectra of MnSi were carried out in the framework of density functional theory in the local density and the generalized gradient approximations. The calculations match most of the phonon modes in the frequency range up to 40 meV. The phonon density of states obtained on the basis of the phonon dispersion was used to estimate the lattice contribution to the specific heat of MnSi.

MnSi is an itinerant magnet with a helical spin order below $T_c \sim 28.8K$. Its transport, magnetic and thermodynamic properties reveal giant anomalies at temperatures centered at $\sim 30K$ that is slightly above the magnetic phase transition. The magnetic phase transition at $28.8K$ is signified by sharp peaks in heat capacity, thermal expansion coefficient and temperature derivative of resistivity in a temperature interval less than $0.1K$. Elastic constants, controlling propagation of longitudinal waves display significant softening at the temperature interval, corresponding the mentioned anomalies, and discontinuities at $\sim 28.8K$ [1, 2, 3, 4, 5, 6, 7] . These observations, indicating a strong electron-phonon coupling in MnSi, call for a search of specific features in the lattice dynamics of MnSi.

In this paper, we report first measurements of the phonon dispersion in MnSi, performed by making use of the inelastic neutron scattering (INS) technique. The experimental and theoretical data allowed us to calculate the phonon density of state (PhDOS) and subsequently estimate a lattice contribution to the specific heat in MnSi.

Measurements were performed on two samples: both were nearly perfect single crystals grown by the Bridgman method and having a very small mosaic spread. One had a volume of $2cm^3$ and the other of $7.5cm^3$. Further details of crystal characterization are given in Refs. [5, 6]. MnSi has the cubic B20-type crystal structure, space group $P2_13$ with four manganese and four silicon atoms in the unit cell that gives rise the 24 phonon modes making a lattice dynamics study highly challenging.

The INS measurements were carried out on the triple-axis spectrometer IN8 at the Institute Laue Langevin (ILL), Grenoble and on the 1T1 triple axis spectrometer at the Laboratoire Léon

Published under licence by IOP Publishing Ltd

Brillouin, Saclay. The single crystals were aligned in the [110]/[001] scattering plane in order to measure phonons with the reduced wavevector q along the main crystallographic directions [001], [110] and [111]. The measurements were done at $T = 15K$, $30K$ and $150K$. The samples were placed in a conventional closed cycle refrigerator. Most of the INS data were collected around the zone centers (1,1,4), (2,2,2) and (1,1,1) using constant-Q scans in the transverse, longitudinal or mixed polarizations at the relatively low and medium energy transfer up to a maximum value of $\sim 40meV$.

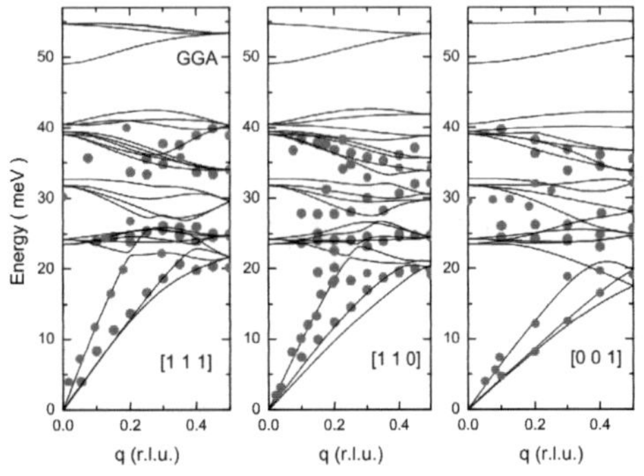

Figure 1. (Color online) Measured (circles) and calculated (solid lines) phonon frequencies in MnSi at $T = 15K$: The lines correspond to the GGA calculations at the experimental value of the lattice spacing with optimized atomic positions.

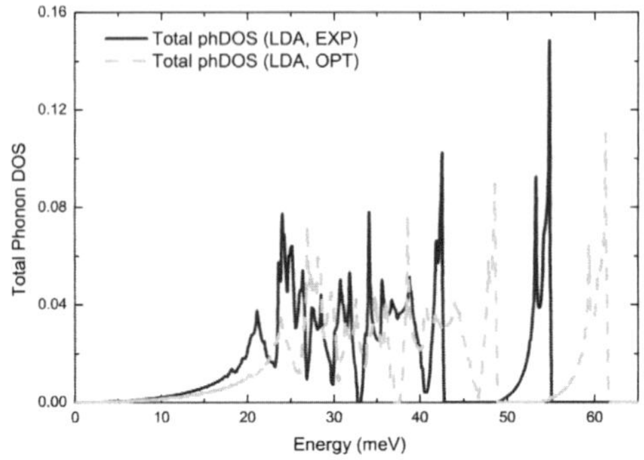

Figure 2. (Color online) The calculated total phonon density of state (PhDOS) in MnSi based on the LDA calculations using the experimental EXP(solid line) and the optimized OPT(dashed line) lattice parameters.

The phonon dispersion curves along the main crystallographic direction evaluated from these data are shown in Fig. 1.

The phonon dispersion measured at $15K$, $30K$ and $150K$ does not show any significant changes within the accuracy of the INS data. The only noticeable influence of temperature on the INS spectra is caused by the conventional Bose temperature factor for the phonon intensity. The slight changes of the phonon linewidth observed on heating lie well within experimental error. Note that no anomalous phonon mode has been observed.

The density-functional calculations were carried out within the framework of the local density approximation (LDA) and the generalized gradient approximation (GGA) using a mixed-basis pseudopotential method [8, 9, 10]. For Mn the 3s and 3p semicore states were treated as valence states. Plane waves up to a kinetic energy of 22 Ry were augmented by local functions of s, p, and d type at the Mn sites.

For both LDA and GGA, we have first performed a full optimization of the structure by minimizing the total energy with respect to the lattice parameter and the internal parameters characterizing the positions of the Mn and Si atoms in the unit cell. While the optimized internal parameters agree well with experimental data, we found for both LDA and GGA an overbinding of about 3% and 1.5%, respectively, of the lattice constant. The resulting phonon spectra turned out to be too hard when compared with the measured ones. Because for the analysis of the specific heat data we require a sufficiently accurate phonon model, we also performed additional calculations using the experimental room-temperature lattice constant of $a = 4.5598\mathring{A}$, but still optimizing the internal parameters. Additional spin-polarized calculations demonstrate that the phonon spectra are little influenced by a spin order.

All model calculations show qualitatively similar dispersion curves, which consist of weakly dispersive optic bands grouped into three sets of bands in the middle part of the spectrum, and three high-frequency bands containing mainly Si vibrations. Quantitatively, however, the spectra based on experimental (LDAexp and GGAexp) and optimized (LDAopt and GGAopt) lattice constants were different from each other regarding both acoustic and optic phonon dispersion curves. This is exemplified in Fig. 2 where the phonon density of states (PhDOS) for the two LDA models is compared. Clearly, using the optimized lattice constant results in significantly higher frequencies for practically all optic modes.

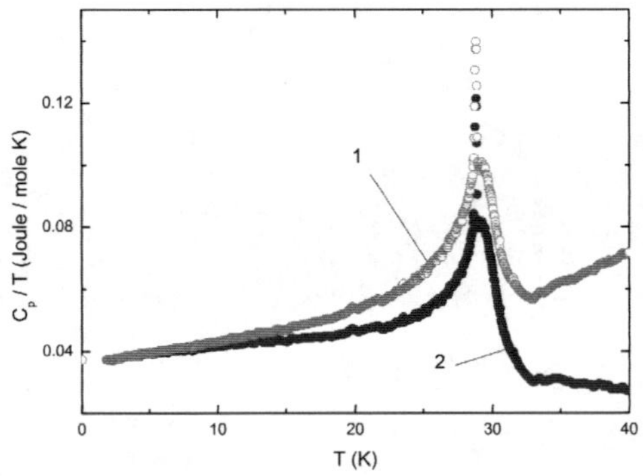

Figure 3. (Color online) Temperature dependence of the electronic and magnetic contribution to the heat capacity of MnSi devided by temperature. 1 - experimental values of Cp/T [5, 6]; 2 - electronic and magnetic contributions obtained from 1 by subtraction of the lattice part.

The best match with our neutron experiment was obtained for the models based on the experimental lattice constant (see Fig. 1). Both LDA and GGA calculations lead to very similar results. The main difference is a slightly stiffer spectrum for the GGA model, which ranges up to $55meV$ while for LDA the high-frequency cut-off is at $52.5meV$. However, the measured low energy phonon data do not allow discriminating between the two models.

Another way to judge the accuracy of the models is a comparison of the experimentally determined and calculated sound velocities. The sound velocities of MnSi were measured in the course of the study of elastic properties of MnSi with an accuracy better than 1% [7]. As is seen in Table 1, the velocities of the longitudinal waves are too high for both LDA and GGA calculations based on the fully optimized structures.

The calculated phonon density of states in MnSi allows to estimate the lattice contribution to the total heat capacity of MnSi. The GGA-exp model has been selected for the estimation of the phonon contribution to the heat capacity of MnSi material base on results of present neutron experiments. Note that the elastic continuous media approximation works purely above $\sim 10K$ in case of MnSi and can not be used to calculate the lattice heat capacity of MnSi in the required temperature range.

The curve 1 in Fig. 3 displays the experimental heat capacity of MnSi divided by temperature Cp/T . The corresponding peak is related to the magnetic phase transition in MnSi (see details in Ref. [5, 6]). After subtraction of the phonon contribution from the curve 1 one gets the electronic and magnetic contributions (curve 2).

The both curves cross the ordinate axis at the same value of $Cp/T = \gamma = 0.0365 joule/moleK^2$, which implies that the subtraction does not influence the linear term in the heat capacity [6]. On the other hand this procedure unveils a nontrivial feature in behavior of Cp/T. As is seen in Fig. 3 magnitudes of Cp/T for $T > T_c$ drop below of the corresponding

Table 1. Comparison of the sound velocity in MnSi obtained from the DFT calculations with results from ultra sound experiments at $6.5K$ [7].

Propagation and polarization	LDAopt	LDAexp	GGAopt	GGAexp	Experiment
Longitudinal waves ($\times 10^5$ cm/s)					
$V_{[100]} = [C_{11}/\rho]^{1/2}$	8.33	6.98	7.01	7.17	7.40
$V_{[110]} = [1/2(C_{11}+C_{12}+2C_{44})/\rho]^{1/2}$	8.78	7.45	8.12	7.33	7.49
$V_{[111]} = [1/3(C_{11}+2C_{12}+4C_{44})/\rho]^{1/2}]$	8.92	7.59	8.25	7.80	7.52
Transverse waves ($\times 10^5$ cm/s)					
$V_{[100]} = [C_{44}/\rho]^{1/2}]$	5.32	4.72	4.93	4.71	4.64
$V_{[110][00\bar{1}]} = [C_{44}/\rho]^{1/2}]$	5.22	4.65	4.87	3.89	4.61
$V_{[110][1\bar{1}0]} = [1/2(C_{11}-C_{12})/\rho]^{1/2}]$	4.52	3.91	4.15	4.66	4.46
$V_{[111]} = [1/3(C_{11}-C_{12}+C_{44})/\rho]^{1/2}]$	4.77	4.20	4.39	4.15	4.53

value at zero temperature. This could be treated as a consequence of evolution of the electron effective mass in MnSi with temperature. However, published experimental data show that the temperature dependence of resistivity of MnSi can be described by a single expression $\rho = \rho_0 + aT^{2+\delta}$ in the temperature range $0 - 28K$ [12,13]. That certainly indicates a lack of noticeable change of the electron effective mass in the corresponding interval of temperatures. However the same resistivity measurements show a quite different situation at temperatures above T_c, where the resistivity tends to saturate and most likely pointing to the suppression of spin fluctuations. So perhaps our observation support earlier claims of existence of a reduction of the heat capacity in weak ferromagnets as a result of interaction of the spin fluctuation modes [14, 15, 16, 17, 18].

The phonon dispersions have been measured in MnSi by means of inelastic neutron scattering experiments. The ab initio model calculations of the lattice dynamics of MnSi are in good agreement with the experimental data. The phonon density of states of MnSi was obtained on the basis of calculations and experiments that allowed an estimation of the phonon contribution to the specific heat of MnSi. So far no detectable temperature effects in the phonon spectra of MnSi at temperatures above and below the phase transition temperature $T_c = 29K$ were found within the experimental error limits.

We thank N.N. Tidden for valuable contribution at the initial stage of this study. The hospitality of the Institute Laue Langevin (ILL) in Grenoble is highly appreciated.

References

[1] Fawcett E, Maita J P , Wernick J H 1970 *Intern. J. Magnetism* **1** 29
[2] Kusaka S, Yamamoto K, Komatsubara T, Ishikawa Y 1976 *Solid State Communications* **20** 925
[3] Matsunaga M, Ishikawa Y, Nakajima T 1982 *J. Phys. Soc. Jpn.* **51** 1153
[4] Lamago D, Georgii R, Pfleiderer C, Boni P 2006 *Physica* B **385-386** 385
[5] Stishov S M, Petrova A E, Khasanov S, Panova G Kh, Shikov A A, Lashley J C, Wu D, Lograsso T A 2007 *Phys.Rev.* B *76* 052405
[6] Stishov S M, Petrova A E, Khasanov S, Panova G Kh, Shikov A A, Lashley J C, Wu D, Lograsso T A 2008 *J. Phys.Condens. Matter* **20** 235222
[7] Petrova A E, Stishov S M 2009 *J.Phys.:Condens. Matter* **21** 196001
[8] Meyer B, Elsässer C, Lecherman M, Fahnle F *FORTRAN90, a program for mixed-basis pseudopotential calculations for crystals, MPI* (für Metallforschung, Stuttgart)
[9] Heid R, Bohnen K-P 1999 *Phys. Rev.* B **60** R3709
[10] Vanderbilt D. 1985 *Phys. Rev.* B **32** 8412
[11] Ishikawa Y, Shirane G, Tarvin J A, Kohgi M 1977 *Phys. Rev.* B **16** 4956
[12] Thessieu C, Flouquet J, Lapertot G, Stepanov A N, Jaccard D 1995 *Solid State Commn.* **95** 707
[13] Petrova A E, Bauer E D, Krasnorussky V, Stishov S M 2006 *Phys. Rev.* B **74** 092401

[14] Moriya T 1985 *Spin fluctuations in itinerant electron magnetism* (Springer-Verlag)
[15] Murata K K, Doniach S 1972 *Phys.Rev.Lett.* **29** 285
[16] Callaway J 1972 *Phys. Rev.* **5** 106
[17] Makoshi K, Moriya T 1975 *J. Phys. Soc. Jpn.* **38** 10
[18] Mohn P, Hilscher G 1989 *Phys.Rev.* B **40** 9126

Journal of Physics: Conference Series **273** (2011) 012130 doi:10.1088/1742-6596/273/1/012130

Pressure effect on transport and magnetic properties of $Nd_{1-x}Tb_xCo_2$

K Uchima[1], Y Takaesu[1,2], S Yonamine[2], M Takeda[2], M Hedo[2], T Nakama[2], K Yagasaki[2], Y Uwatoko[3] and A T Burkov[4]

[1]General Education, Okinawa Christian Junior College, Nishihara, Okinawa, 903-0207, Japan
[2]Faculty of Science, University of the Ryukyus, Nishihara, Okinawa, 903-0213, Japan
[3]Institute for Solid State Physics, University of Tokyo, Kashiwa, Chiba 277-8581, Japan
[4]A. F. Ioffe Physico-Technical Institute, Russian Academy of Sciences, Sankt-Petersburg, 194021, Russia

E-mail: uchima@ocjc.ac.jp

Abstract. The electrical resistivity and magnetization of Laves phase $Nd_{1-x}Tb_xCo_2$ compounds have been investigated at temperatures from 2 K to 300 K in magnetic fields up to 10 T under pressures up to 8 GPa. The magnetic transition temperature T_C increases linearly with increasing Tb concentration x, which indicates that the 4f–3d exchange interaction energy increases linearly with increasing x. The spontaneous magnetization M_S at small x decreases linearly, vanishes around $x \approx 0.35$ and then increases linearly with increasing x. This behavior suggests that the Tb magnetic moment is antiparallel to that of Nd, while Co moment is almost independent of x.

1. Introduction

RCo_2 compounds with cubic Laves phase structure (R stays for rare earth elements) are known for magnetic instability of Co 3d-electron system [1, 2, 3]. When R is nonmagnetic (R=Y, Lu), the compounds are exchange-enhanced paramagnets exhibiting metamagnetic transition in external magnetic field of 70 T for YCo_2 [4] and about 77 T for $LuCo_2$ [5]. In the case of magnetic RCo_2 compounds, long-range magnetic order of the itinerant 3d-electron subsystem is induced by exchange field of the ordered localized magnetic moments of the rare earths. Due to antiferromagnetic spin–spin coupling between 3d and 4f moments [6], the compounds with light R elements (LR) are ferromagnetic, whereas those with heavy R (HR) are ferrimagnetic. Accordingly, $NdCo_2$ is a ferromagnet with Curie temperature of $T_C \approx 98$ K, while $TbCo_2$ is a ferrimagnet with $T_C \approx 230$ K. It has been found that electronic transport in RCo_2 compounds is strongly dependent on the state of Co 3d band and on 3d-4f interaction [3]. Particularly interesting and fruitful results have been found in quasibinary alloys of these compounds (R - R$'$)Co_2, where one can control the averaged magnitude of 4f magnetic moment and the corresponding exchange 4f-3d molecular field. The alloys of non-magnetic YCo_2 with magnetic RCo_2 have been extensively investigated, while a very few studies of the alloys with mixing of LR and HR have been performed [7, 8, 9]. It was shown in our previous paper [10] that in zero field magnetic transition temperature T_C of $Nd_{1-x}Tb_xCo_2$ increases linearly with increasing x. The magnetic transition temperature T'_C observed in external magnetic field of 10 T is higher than T_C for $x < 0.35$ and lower for $x \geq 0.35$. This result gives further support to the 3-sublattice

Published under licence by IOP Publishing Ltd

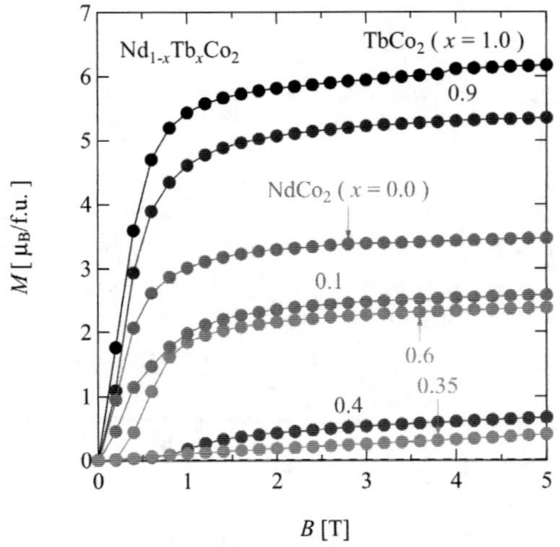

Figure 1. Field dependence of magnetization M of $Nd_{1-x}Tb_xCo_2$ at 2 K.

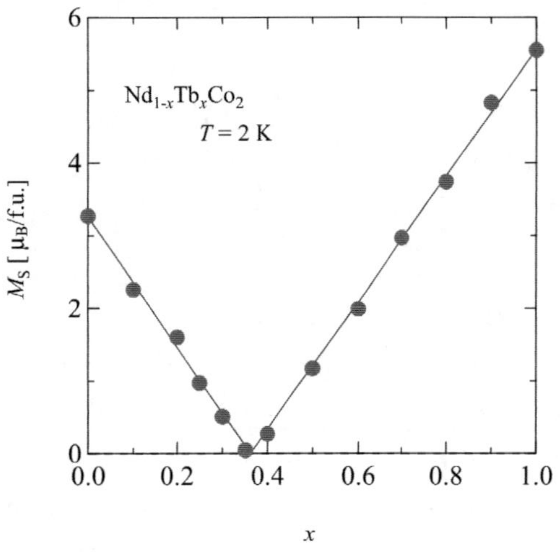

Figure 2. Concentration x dependence of M_s.

model with antiparallel alignment of Nd and Tb magnetic moments, since within this model for $x < 0.35$ the internal molecular field, acting on Co sublattice, increases in external magnetic field, while the molecular field decreases in external field for $x \geq 0.35$. In order to investigate the magnetization and the pressure effect on transport properties of $Nd_{1-x}Tb_xCo_2$, we have measured magnetization and electrical resistivity at temperatures from 2 K to 300 K under pressures up to 8 GPa and in magnetic field up to 10 T.

2. Experimental procedures

The polycrystalline samples of $Nd_{1-x}Tb_xCo_2$ compounds were prepared by melting the pure constituents(3N) in an arc furnace under a protective Ar atmosphere and were subsequently annealed in vacuum at 900 °C for one week. A ratio of R:Co = 1:1.93 for cast was chosen to avoid formation of magnetic RCo_3 phase. The cubic Laves phase structure was confirmed by X-ray diffraction measurement. The field dependence of the magnetization at 2 K was measured by a superconducting quantum interference device magnetometer. The electrical resistivity measurements were carried out by means of a standard four-probe dc method from 2 K to 300 K. A clamp-type piston cylinder pressure cell and cubic anvil cell were utilized for the measurements of ρ under pressures up to 3 GPa and up to 8 GPa, respectively.

3. Results and Discussion

The magnetization curves of $Nd_{1-x}Tb_xCo_2$ compounds at 2 K are shown in Fig. 1. The magnetization increases ferromagnetically with increasing external magnetic field, except for the compounds of $x \approx 0.35$. The spontaneous magnetization M_s at 2 K, obtained by extrapolating the higher field magnetization curve to 0 T, decreases linearly with increasing x and vanishes at $x_c \approx 0.35$. For $x > x_c$, M_s increases linearly with increasing x as shown in Fig. 2. The variation of M_s with alloy composition can be described by simple expression: $M_s = |(1-x)M_{Nd} - xM_{Tb} + 2M_{Co}|$. This behavior implies the antiparallel arrangement of Tb and Nd magnetic moments and independence of Co moment of x. The magnetic transition temperature T_C, determined from experimental temperature dependences of ρ, increases linearly with increasing x and can be expressed as: $T_C = (1-x)T_C^{NdCo_2} + xT_C^{TbCo_2}$, where $T_C^{NdCo_2}$ and

$T_C^{TbCo_2}$ are the magnetic transition temperatures of $NdCo_2$ and $TbCo_2$, respectively [10]. Within s-d model T_C of $Nd_{1-x}Tb_xCo_2$ is given by

$$T_C = \frac{N_A \mu_B^2}{3k_B}(I_{RR} + I_{Rd}^2 \chi_d)\{(1-x)\xi_{NdCo_2} + x\xi_{TbCo_2}\}, \quad (1)$$

where N_A is the Avogadro constant, I_{RR} and I_{Rd} are the respective exchange interaction coefficients between the rare earths and between the rare earth and cobalt, and χ_d is the magnetic susceptibility of Co-3d band [11]. ξ_{NdCo_2} and ξ_{TbCo_2} are de Gennes factor of $NdCo_2$ and $TbCo_2$, respectively. According to ref. [11] exchange coefficients I_{RR}, I_{Rd} and χ_d are the same in the RCo_2 series. This formula implies that T_C of $Nd_{1-x}Tb_xCo_2$ is determined by the mean magnitude of the effective spin of R atom.

Figure 3 shows the temperature dependence of ρ under pressures up to 8 GPa for the compound $Nd_{0.7}Tb_{0.3}Co_2$ in which the rare earth sublattice magnetization is almost zero. Resistivity increases with increasing temperature and shows a knee around T_C. The application of pressure leads to decrease of T_C and of the resistivity at the high temperature region, while at low temperatures the resistivity increases under pressure. The pressure variation of T_C is attributed to the pressure response of the magnetic susceptibility of Co-3d band χ_d as

$$\frac{\mathrm{d}\ln T_C}{\mathrm{d}P} = \frac{I_{Rd}^2 \chi_d}{I_{RR} + I_{Rd}^2 \chi_d} \frac{\mathrm{d}\ln\chi_d}{\mathrm{d}P}. \quad (2)$$

From the results shown in Fig. 3, we obtained the pressure derivative of T_C for $Nd_{0.7}Tb_{0.3}Co_2$, $\mathrm{d}\ln T_C/\mathrm{d}P = -0.09$, which is almost same as that for $NdCo_2$ (-0.11) and for $TbCo_2$ (-0.12) [12]. This result indicates that the parameters of Co-3d band are independent of composition of the $Nd_{1-x}Tb_xCo_2$ system.

Assuming validity of Matthiessen's rule the total electrical resistivity $\rho(T)$ for the magnetic RCo_2 compounds can be written as:

$$\rho(T) = \rho_0 + \rho_{ph}(T) + \rho_m(T) + \rho_{sf}(T).$$

Here, ρ_0, ρ_{ph}, ρ_m, and ρ_{sf} denote the contributions due to impurities, phonon, 4f magnetic moments and Co-3d spin fluctuation scatterings, respectively. Within the scope of Mott's s-d scattering model these all contributions are proportional to the 3d state density at Fermi level, so that $\rho \sim N_{\varepsilon_F}$ [13]. On the other hand, $\chi_d \sim \mu_B^2 N_{\varepsilon_F}/(1 - IN_{\varepsilon_F})$. At high temperatures, in paramagnetic state, the main effect of pressure on resistivity comes from decrease of N_{ε_F} due to increasing overlapping of 3d orbitals. Therefore, the high temperature resistivity decreases under pressure. Moreover we expect that there a linear relation between the pressure variations of T_C and ρ at high temperatures: $\mathrm{d}\ln T_C/\mathrm{d}P \sim \mathrm{d}\ln\rho/\mathrm{d}P$. Figure 4 depicts the plot of $-\Delta T_C/T_C$ vs. $-\Delta\rho(290)/\rho(290)$. As one can see, the plot of $-\Delta T_C/T_C$ vs. $-\Delta\rho(290)/\rho(290)$ shows a good linear relation, which indicates that the pressure effect on T_C and ρ at high temperatures is associated with the change of N_{ε_F}.

At low temperatures, the 3d electron system of the alloys is in the partially ordered magnetic state due to competition of ferro- and antiferromagnetic exchange with Nd and Tb ions. In this case the pressure changes the degree of static magnetic disorder and leads to the resistivity increase. The details of the resistivity behavior in (R - R$'$)Co_2 alloys with partially ordered 3d electron system are described elsewhere [14].

In summary, the magnetization and electrical resistivity of the $Nd_{1-x}Tb_xCo_2$ system were measured at temperatures from 2 K to 300 K. The spontaneous magnetization M_s at 2 K decreases linearly with increasing x, vanishing at $x_c \approx 0.35$. For $x > x_c$, M_s increases linearly with increasing x. The magnetic transition temperature T_C increases with increasing x. The

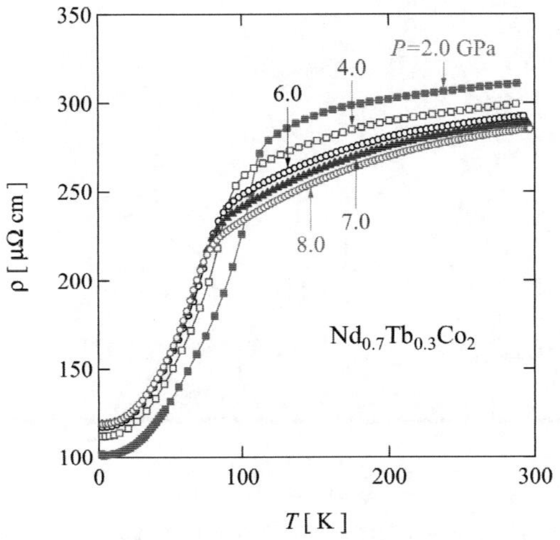

Figure 3. Temperature dependence of electrical resistivity ρ up to 8 GPa for $Nd_{0.7}Tb_{0.3}Co_2$.

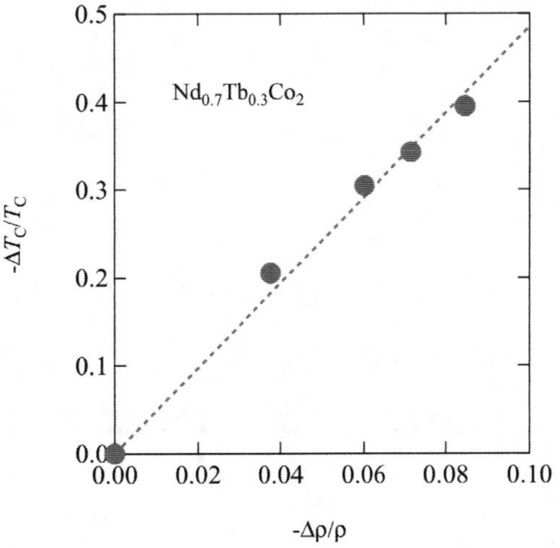

Figure 4. $-\Delta T_C/T_C$ vs. $-\Delta\rho(290)/\rho(290)$ for $Nd_{0.7}Tb_{0.3}Co_2$. The dotted line is a guide for the eye.

linear variations of M_s and T_C with x suggest the antiparallel alignment between Nd and Tb magnetic moments and almost constant value of Co moment. The decrease of T_C and of the high temperature resistivity with pressure is related to decrease of N_{ε_F} due to increasing overlapping of 3d orbitals. The pressure dependence of ρ at low temperatures is related to change of static magnetic disorder in partially ordered 3d electron system.

Acknowledgments

Part of this work was supported by a Grant-in-Aid for Scientific Research (No. 21540340) from Ministry of Education, Culture, Sports, Science and Technology of Japan, and by Russian Foundation for Basic Research under the Grant 09-02-01285a.

References

[1] Lemaire R 1966 *Cobalt* **32** 132; 1966 **33** 201
[2] Levitin R Z and Markosyan A S 1988 *Sov. Phys. - Usp.* **31** 730
[3] Gratz E, Resel R, Burkov A T, Bauer E, Markosyan A S and Galatanu A 1995 *J. Phys.: Condens. Matter* **7**6687
[4] Goto T, Fukamichi K, Sakakibara T and Komatsu H 1989 *Solid State Commun.* **72** 945
[5] Goto T, Sakakibara T, Murata K, Komatsu K and Fukamichi K 1990 *J. Magn. Magn. Mater.* **90/91** 700
[6] Cyrot M, Lavagna M 1979 *J. Appl. Phys.* **50(3)** 2333
[7] Duc N H, Brommer P E and Franse J J M 1993 *Physica* B **191** 239
[8] Ouyang Z W, Rao G H, Yang H F, Liu W F, Liu G Y and Liang J K 2002 *Appl. Phys. Lett.* **81** 97
[9] Ouyang Z W, Rao G H, Yang H F, Liu W F, Liu G Y, Feng X M and Liang J K 2003 *J. Phys.: Condens. Matter* **15** 5599
[10] Uchima K, Yonamine S, Kinjyo A, Takaesu Y, Hedo H, Nakama T, Yagasaki K and Burkov A T *J. Phys.: Conf. Ser.*,**200** (2010) 032077
[11] Bloch D and Lemaire R *Phys. Rev. B* **2** (1970) 2648-2650
[12] Hauser R, Bauer E and Gratz E *Phys. Rev. B* **57** (1998) 2904-2914
[13] Burkov A T, Gratz E, Bauer E, and Resel R *J. Alloys Compd.* **198** (1993) 117-126
[14] Burkov A T, Zyuzin A Yu, Nakama T, Yagasaki K *Phys. Rev. B* **69** (2004) 144409-1-6.

Journal of Physics: Conference Series **273** (2011) 012131 doi:10.1088/1742-6596/273/1/012131

A numerical study for two-dimensional spin 1/2 antiferromagnets: a generalization of Entanglement Perturbation Theory to two-dimensional systems

L. Wang[1,2], T. Shirakawa[1,2], H. Watanabe[1,2], and S. Yunoki[1,2]

[1] Computational Condensed Matter Physics Laboratory, RIKEN ASI, Wako, Saitama 351-0198, Japan

[2] CREST, Japan Science and Technology Agency (JST), Saitama 332-0012, Japan

E-mail: lihua.wang@riken.jp

Abstract. Two-dimensional spin 1/2 antiferromagnetic Heisenberg models are numerically studied using Entanglement Perturbation Theory, where the ground state wave function is described by a product of local matrices defined at every site, and each matrix is optimized variationally to minimize the energy. We first apply this method for the spin 1/2 antiferromagnetic Heisenberg model on the square lattices to reproduce correctly the known ground state energy and the spin structure factors. Then, we study the spin 1/2 antiferromagnetic Heisenberg model on the triangular lattice with spatially anisotropic nearest neighbor couplings, J and J', to establish the ground state phase diagram as a function of J'/J.

1. Introduction

Recently, much attention has been attracted to geometrically frustrated quantum spin systems where exotic ground states as well as nontrivial fractional excitations might appear [1]. Among them, one of the simplest systems is the spin 1/2 antiferromagnetic Heisenberg model (AFMHM) on the triangular lattice with spatially anisotropic nearest neighbor couplings J and J' (Fig. 1):

$$H = J\sum_{\langle \mathbf{i},\mathbf{j}\rangle} \vec{S}_{\mathbf{i}} \cdot \vec{S}_{\mathbf{j}} + J'\sum_{\langle \mathbf{i},\mathbf{j}\rangle'} \vec{S}_{\mathbf{i}} \cdot \vec{S}_{\mathbf{j}}. \tag{1}$$

Here $\vec{S}_{\mathbf{i}}$ is the spin operator of $|\vec{S}_{\mathbf{i}}| = 1/2$ at site $\mathbf{i}$, and $\langle \mathbf{i},\mathbf{j}\rangle$ ($\langle \mathbf{i},\mathbf{j}\rangle'$) runs over a pair of nearest neighbor spins at sites $\mathbf{i}$ and $\mathbf{j}$ along the chain (between the chains) as indicated in Fig. 1. Although this model is seemingly simple, even the ground state has not been understood except for the special coupling cases of $J'/J = 0$ (one-dimensional limit), $J'/J \approx 1$ (isotropic triangular limit), and $J'/J \gg 1$ (isotropic square limit) [2, 3, 4]. Furthermore, this model may describe the low-energy physics of various quasi-2D triangular materials such as Cs_2CuCl_4 [5], κ-(BEDT-TTF)$_2$Cu$_2$(CN)$_3$ [6], and EtMe$_3$Sb[Pd(dmit)$_2$]$_2$ [7] for which experiments have indicate a magnetically disordered Mott insulator, i.e., a possible spin liquid state, at low temperatures. The purpose of the present study is to establish the ground state phase diagram of Eq. (1) as a function of J'/J and to understand the nature of the low-energy excitations.

Published under licence by IOP Publishing Ltd

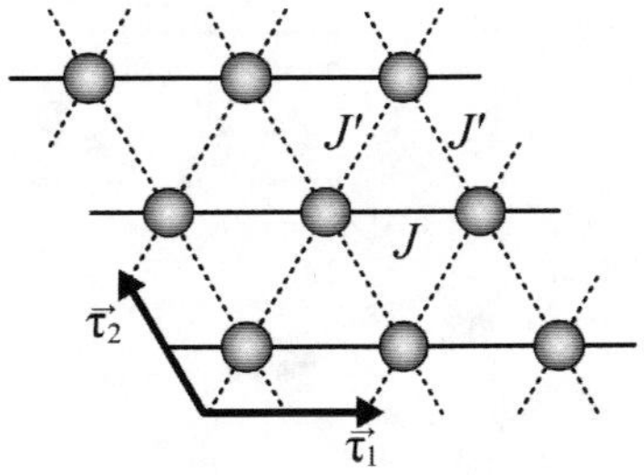

Figure 1. Spin 1/2 AFMHM on the triangle lattice with the anisotropic nearest neighbor coupling J and J' ($J'/J \geq 0$) indicated in the figure. The primitive translation vectors are denoted by $\vec{\tau}_1 = (1, 0)$ and $\vec{\tau}_2 = (-\frac{1}{2}, \frac{\sqrt{3}}{2})$. The periodic boundary conditions are assumed.

2. Numerical Method: Entanglement Perturbation Theory

To study the ground state of Eq. (1), we use Entanglement Perturbation Theory (EPT) which is originally developed in one-dimensional systems [8], and recently has been applied for the 2D Hubbard model [9]. Here, we shall extend this method to 2D spin systems.

EPT expresses the wave function $|\psi\rangle$ by a product of local square matrices as in matrix product state [10],

$$|\psi\rangle = \sum_{\sigma_1}\sum_{\sigma_2}\cdots\sum_{\sigma_N} \mathrm{Tr}\left[\xi_1^{\sigma_1}\cdot\xi_2^{\sigma_2}\cdot\cdots\cdot\xi_N^{\sigma_N}\right]|\sigma_1\rangle\otimes|\sigma_2\rangle\otimes\cdots\otimes|\sigma_N\rangle, \tag{2}$$

where $\xi_i^{\sigma_i}$ is the local wave function matrix at site i, and N is the number of sites. $\sigma_i(=\uparrow,\downarrow)$ refers to local states. Note that the trace of the matrix product for a given $\{\sigma_i\}$ is the amplitude of the corresponding Fock state in the wave function $|\psi\rangle$. The dimension p of the square matrices $\xi_i^{\sigma_i}$ controls the accuracy of the wave function $|\psi\rangle$ when $|\psi\rangle$ is used as a variational state for H, and in principle $|\psi\rangle$ with optimized $\xi_i^{\sigma_i}$ becomes exact with $p \to \infty$. The matrices $\xi_i^{\sigma_i}$ are optimized to minimize the variational energy:

$$\frac{\partial}{\partial\eta_i}\frac{\langle\psi(\eta_1,\eta_2,\cdots,\eta_N)\mid H\mid\psi(\eta_1,\eta_2,\cdots,\eta_N)\rangle}{\langle\psi(\eta_1,\eta_2,\cdots,\eta_N)\mid\psi(\eta_1,\eta_2,\cdots,\eta_N)\rangle} = 0. \tag{3}$$

Here η_i is a vector representation of $\xi_i^{\sigma_i}$. We then arrive at a generalized eigenvalue equation

$$X_i(\eta_1,\cdots,\eta_{i-1},\eta_{i+1},\cdots,\eta_N)\,\eta_i = \mu Y_i(\eta_1,\cdots,\eta_{i-1},\eta_{i+1},\cdots,\eta_N)\,\eta_i \quad (i = 1, 2\cdots, N), \tag{4}$$

where X_i and Y_i are appropriate matrices depending implicitly on $\{\eta_i\}$, and μ is the eigenvalue. The eigenfunction $\{\eta_i\}$ with the smallest μ describes the optimized ground state wave function $|\psi\rangle$. The details of how to extend EPT to 2D spin systems will be discussed elsewhere [11].

3. Results

First, we study spin 1/2 AFMHM on the square lattice with $N = 4 \times 4$ to check the validity of the method in 2D. As explained above, EPT is the variational method for the ground state $|\psi\rangle$, and the larger the dimension p of the matrices is, the more accurate $|\psi\rangle$ becomes. This is clearly seen in Fig. 2 where the energy expectation value is plotted as a function of p. The converged ground state energy per site is $-0.70175J$ with $p = 31$, which is very close to the numerically exact value $-0.70178J$ obtained by Exact Diagonalization (ED).

Next, we calculate the spin correlation functions $C(\vec{R}) = \langle\vec{S}_{\vec{R}+\vec{r}}\cdot\vec{S}_{\vec{r}}\rangle$, and the results are compared with those obtained by ED in Table.1. The excellent agreement is observed. We have

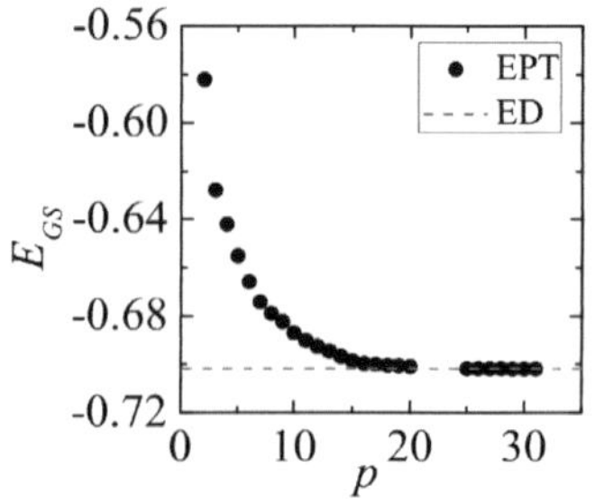

Figure 2. The expectation value of energy for spin 1/2 AFMHM on the isotropic square lattice with $N = 4\times4$ as a function of the dimension p of matrices $\xi_i^{\sigma_i}$. For comparison, the numerically exact value obtained by ED is also plotted by dashed line.

studied $C(\vec{R})$ using different lattice sizes with N up to $N = 10 \times 10$, and the results will be reported elsewhere [11]

Table 1. Spin correlation functions $C(\vec{R})$ for spin 1/2 AFMHM on the isotropic square lattice with $N = 4 \times 4$. EPT results are compared with those obtained by ED. Here $\vec{R} = i_x\hat{\mathbf{x}} + i_y\hat{\mathbf{y}}$.

(i_x,i_y)	EPT	ED
(0,1)	−0.35096	−0.35089
(1,0)	−0.35084	−0.35089
(2,0)	0.21373	0.21377
(0,2)	0.21381	0.21377
(1,2)	−0.20224	−0.20216
(2,1)	−0.20216	−0.20216
(2,2)	0.17966	0.17963

Encouraged by these results, we shall now study spin 1/2 AFMHM on the triangular lattice. First, we calculate the ground state energy per site E_{GS} for the isotropic limit $J = J'$ with $N = 6 \times 6$, and find that $p = 44$ ($E_{GS} = -0.5450J$) is enough to achieve the accuracy of about 2.7%, for which the numerically exact energy is $-0.5603J$ [2]. Next, we calculate the spin structure factor $S(\mathbf{q}) = \sum_{\mathbf{R}} \mathrm{e}^{-i\mathbf{q}\cdot\mathbf{R}} C(\mathbf{R})/N$ for $N = 6 \times 6$ and several values of J'/J. These results are shown in Fig.3. Fig.3 (b) corresponds to the isotropic limit. It is clearly seen in Fig.3 (b) that peaks appear at $\mathbf{q}^* = (4\pi/3, 0)$ and the symmetrically equivalent $\mathbf{q}$'s, indicating the classical Néel ordered spin structure. The calculated value of $S(\mathbf{q}^*)$ is 0.0946 which is about 4.4% larger than the numerically exact value (0.0904) [2]. With increasing J'/J, we find that the spin structure factor changes abruptly at around $J'/J = 1.18$ above which $S(\mathbf{q})$ has peaks at $(0, \mp 2\pi/\sqrt{3})$. These wave numbers correspond to the classical Néel ordered spin structure on the square lattice, namely, the one expected in $J'/J \to \infty$. Instead, with decreasing J'/J, we find that $S(\mathbf{q})$ shows several incommensurate peaks with a one-dimensional character for J' smaller than about 0.8, as shown in Fig.3 (a). Since the system size studied in this paper is limited, we can not exclude, at this moment, a possibility that there exists another ordered or disordered phase [3, 4].

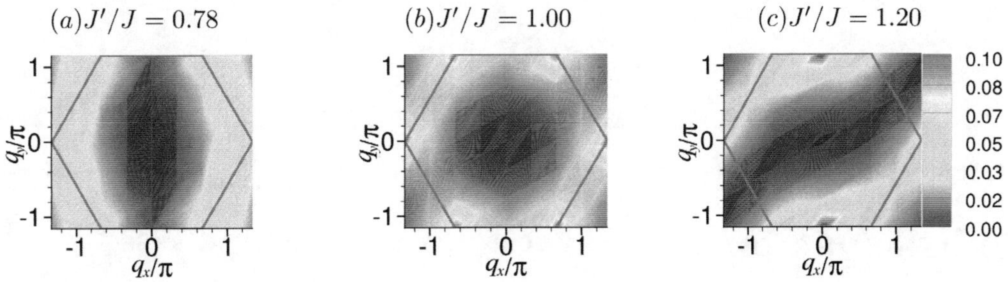

Figure 3. (color online) Contour plots of spin structure factors $S(\mathbf{q})$ for spin 1/2 AFMHM on the triangular lattice with $N = 6 \times 6$. The values of J'/J are indicated in the figures. The zone boundary of the first Brillouin zone is denoted by solid lines. $p = 29$ (a), 44 (b), and 25 (c) are used.

4. Conclusions

We have extended EPT to 2D spin systems to study the ground state properties of spin 1/2 AFMHM on the anisotropic triangular lattice. We have shown that EPT successfully reproduces the known results for spin 1/2 AFMHM on the square lattice. We then applied the method to the spatially anisotropic triangular lattice. We found that, based on our simulations for $N = 6 \times 6$, the classical Néel order with the nearest neighbor spin angle of 120° is stable for a wide range of parameter ($0.8 \lesssim J'/J \lesssim 1.2$), and that for $J'/J \gtrsim 1.2$ the spin structure becomes collinearly Néel ordered, i.e., the one expected for $J'/J \to \infty$. Although the evolution of the spin structure factors as a function of J'/J is clear, more systematic studies are necessary to determine the nature of various phases and the critical values of J'/J for the phase boundary. A study in this direction is now in progress [11].

5. Acknowledgments

The computation was performed using RIKEN Integrated Cluster of Clusters (RICC).

References

[1] Diep H T *Frustrated Spin Systems* (World Scientific, New Jersey, 2004)
[2] Bernu B, Lecheminant P, Lhuillier C, and Pierre L 1994 *Rhys. Rev.* B **50** 10048
[3] Capriotti L, Trumper A E, and Sorella S 1999 *Rhys. Rev. Lett.* **82** 3899; Yunoki S and Sorella S 2004 *Phys. Rev. Lett.* **92** 157003; Yunoki S and Sorella S 2006 *Phys. Rev.* B **74** 014408; Heidarian D, Sorella S, and Becca F 2009 *Phys. Rev.* B **80** 012404
[4] Weng M Q, Sheng D N, Weng Z Y, and Bursill R J 2006 *Rhys. Rev.* B **74** 012407
[5] Coldea R, Tennant D A, Tsvelik A M, and Tylczynski Z 2001 *Phys. Rev. Lett.* **86** 1335
[6] Shimizu Y, Miyagawa K, Kanoda K, Maesatol M, and Saito G 2003 *Phys. Rev. Lett.* **91** 107001
[7] Yamashita M, Nakata N, Senshu Y, Nagata M, Yamamoto H M, Kato R, Shibauchi T, and Matsuda Y 2010 *Science* **328** 1246
[8] Chung S G 2007 *Phys. Lett.* A **361** 396
[9] Chung S G and Ueda K 2008 *Phys. Lett.* A **372** 4845
[10] Rommer S and Ostlund S 1997 *Phys. Rev.* B **55** 2164
[11] Wang L, *et al.* 2010 to be published

International Conference on Strongly Correlated Electron Systems (SCES 2010) IOP Publishing
Journal of Physics: Conference Series **273** (2011) 012132 doi:10.1088/1742-6596/273/1/012132

Multiferroic behavior in organo-metallics

V. S. Zapf[1], F. Wolff-Fabris[1,2], M. Kenzelmann[3], F. Nasreen[4], F. Balakirev[1], Y. Chen[5] and A. Paduan-Filho[6]

[1] National High Magnetic Field Laboratory (NHMFL), Los Alamos, NM
[2] Now at Dresden Hochfeld Labor, Dresden, Germany
[3] Laboratory for Developments and Methods, Paul Scherrer Institute, Switzerland
[4] New Mexico State University, Las Cruces, NM
[5] NIST Center for Neutron Research, Gaithersburg, MD
[6] Instituto de Fisica, Universidade de Sao Paulo, Brazil

E-mail: `vzapf@lanl.gov`

Abstract. The study of multiferroic (MF) and magnetoelectric (ME) materials seeks to understand and harness the coupling between electric and magnetic polarizations. To date, research into ME and MF bulk materials has largely focused on inorganic oxides and only a relatively small number of materials have been found with significant ME couplings. We explore organo-metallics where transition metal ions are incorporated into a flexible and designable organic lattice. We present recent results on ME and MF behavior in single crystals of the organo-metallic compounds $NiCl_2$-$4SC(NH_2)_2$ and $CuCl_2$-$2(CH_3)_2SO$.

1. Introduction

Magnetoelectrics (ME) are materials in which a magnetic field H can modify the electric polarization P and conversely, an electric field E can modify the magnetization M. ME multiferroics are a subset of ME in which both M and P involve long-range order and the resulting diverging susceptibilities at the transitions lead to potentially stronger magnetoelectric coupling [1–4]. Research in this field is motivated by the promise of devices with new functionality as well as improving the speed, energy-efficiency and size of existing circuits. Most ME and multiferroics are transition-metal oxides or fluorides where the magnetoelectric interactions are mediated via superexchange through the oxide and fluoride anions. Magnetoelectric interactions can also be expected in other materials such as organo-metallic solids and very recently a few examples have been reported [5–7], thereby opening a new field of materials to exploration of these effects. Organic materials tend to possess soft lattice structures that can be easily modified by magnetic forces, leading to large magnetic field-induced changes in P. In addition, the flexibility for design of organic molecules and availability of electrically polar building blocks could allow us to construct a variety of new magnetoelectrics by design. Ferroelectricity is known to occur in a number of organic molecules [8], and was most recently discovered above room temperature in croconic acid [9]. In this paper we present two examples of magnetic field-induced electric polarizations in organo-metallic materials. In the first, $NiCl_2$-$4SC(NH_2)_2$ (DTN), thiourea ($SC(NH_2)_2$) is electrically polar and an H-induced P is likely due to magnetic forces distorting the lattice and thus the angles of these polar molecules. The second example, $CuCl_2 \cdot 2[(CH_3)_2SO$ (CDC), is a multiferroic material in which a non-collinear magnetic order occurs in magnetic fields that breaks spatial inversion symmetry (SIS),

Published under licence by IOP Publishing Ltd

thereby creating the necessary symmetry conditions for a net P to form that is closely coupled to the magnetic order and the magnetic field.

2. $NiCl_2$-$4SC(NH_2)_2$ (DTN)

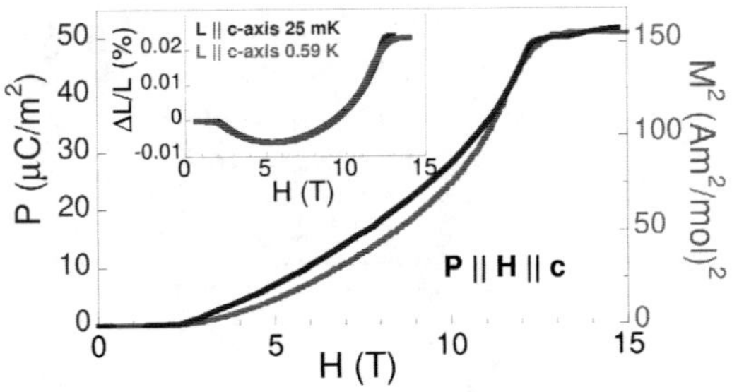

Figure 1. DTN: Comparison of the electric polarization P [7] and the square of the magnetization M^2 vs magnetic field H [10] at 0.55 K. Inset: Percentage length change ΔL of the sample along the c-axis as a function of $H||c$ relative to its zero field value [11].

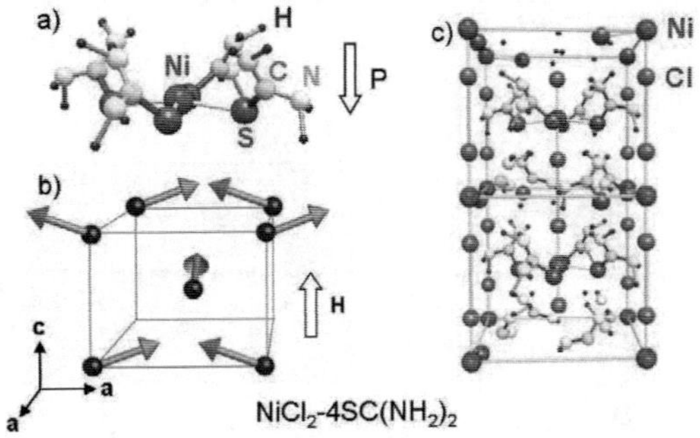

Figure 2. Crystal structure of DTN showing a) the thiourea coordination around the Ni ions and the likely direction of P, b) the Ni spins for $2 \leq H \leq 12$ T, and c) two unit cells of the crystal structure.

DTN is a magnetic field-induced antiferromagnet in which superexchange coupling between Ni spins creates long-range order in a dome-shaped region of $T-H$ space between $2 \leq H||c \leq 12$ T with a maximum T_N of 1.2 K at about 5 T [12]. Uniaxial anisotropy along the c-axis creates a zero-field spin gap of ~ 10 K between the Ni $S_z = 0$ and $S_z = 1$ states, which is closed with applied H. At 2 T, the Ni spins form an XY antiferromagnetic (AFM) structure with the spins lying perpendicular to c, and gradually cant with increasing field until they saturate at 12 T. As shown in Fig. 1, P along the c-axis is also induced above 2 T and roughly follows the square of the magnetization M^2 [7], increasing until 12 T and then saturating. The ME effect is bilinear, such that P always points along the same direction regardless of the sign of H. Empirically, we find that M^2 most closely tracks P both in its H-dependence and the fact that it is also bilinear in H. We suspect that the origin of P is the highly polar sulfur-carbon double bond in the thiourea molecules, which all tilt in the same direction along the c-axis (see the crystal structure in Fig. 2). Since the crystal is relatively soft, the magnetic forces between Ni spins can distort the lattice and we would expect the angles of the thiourea molecules to change as well. The relative change in length $\Delta L/L$ of the c and a-axis as a function of H is shown in the inset to Fig. 1. It was shown [11] that when the Ni spins are anti-aligned along the c-axis (the direction of strongest AFM coupling) the AFM bonds are satisfied and the Ni spins can reduce their energy by attracting each other, increasing the AFM coupling. This occurs between 2 and 5.5 T. However, as the Ni spins cant along the c-axis, they become increasingly aligned and thus start to repel each other, leading to a change in sign of $\Delta L/L||c$ at 5.5 T. The angles of the thiourea molecules, whose axes are tilted between c and a may not be a strict function of the c-axis magnetostriction effect. No hysteresis is observed in $P(E)$ or $P(H)$, nor do we see signatures of the 2 and 12 T transitions in the dielectric constants. Thus we conclude that only M and not P in DTN result from long-range order, thus DTN is a magneto-electric, not a multiferroic.

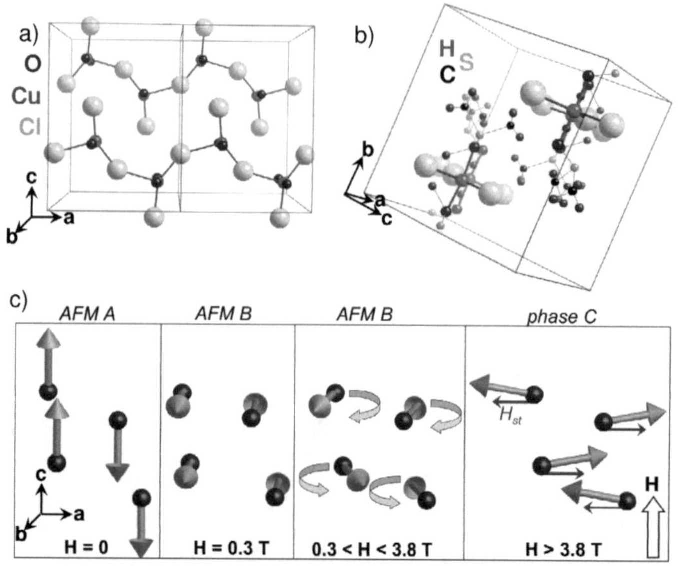

Figure 3. Crystal structure of CDC showing zig-zag Cu-Cl chains. b) Structure showing the organic molecules that mediate AFM exchange along the b-axis. c) Effect of applied H on the Cu $S = 1/2$ spins (shown as arrows) for one unit cell. The effective field-induced staggered magnetic fields H_{st} from spin-orbit couplings are indicated in the final panel [6].

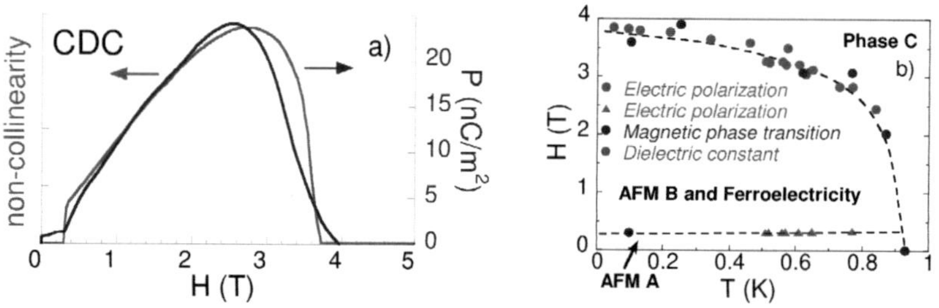

Figure 4. a) $P(H)$ of CDC compared to the non-collinearity of the spins (see ref. [6]). b) H-T phase diagram showing regions of collinear and non-collinear antiferromagnetism (AFM A and B), ferroelectricity, and the spin-orbit-induced phase (phase C) that occurs at high fields for H along [001].

3. $CuCl_2 \cdot 2[(CH_3)_2SO]$ (CDC)

CDC is one of the first examples of an organo-metallic in which multiferroic behavior is clearly established [6]. In this compound, the Cu spins form a non-collinear magnetic structure in applied H that breaks SIS and allows a net P to occur. The crystal structure and orientation of the Cu $S = 1/2$ spins are shown in Fig. 3. The details of the spin arrangements, AFM couplings, and spin-orbit interactions were unraveled via inelastic neutron diffraction and specific heat measurements [13, 14]. The non-collinearity of the Cu spins results from a competition between an AFM superexchange interaction that favors one type of AFM order (labeled AFM A in Fig. 3c), and local staggered effective magnetic fields from spin-orbit couplings that favor a different AFM spin arrangement (phase C in Fig. 3c). Since the spin-orbit couplings are H-dependent, the relative strength of these two competing interactions can be tuned with applied magnetic field. The result is that after a spin-flop transition occurs at 0.3 T, the spins rotate from the b to the a-axis and become non-collinear in the process for fields between 0.3 and 3.8 T. In this magnetic field range, the SIS-breaking magnetic structure leads us to expect a net P and indeed this can be observed in Fig. 4a. Here $P(H)$ is plotted together with a measure of the non-collinearity of the spins (actually the product of the two order parameters resulting from AFM superexchange and spin-orbit couplings) [6]. The phase diagram in Fig. 4b demonstrates that the observed region of non-collinear spins and net P exactly overlap. As described in ref. [6], this material is multiferroic since the P undergoes an actual phase transition, which can be observed asa peak in the dielectric constant at 3.8 T, and P is hysteretic and can be switched by a sequence of applied H pulses.

In summary, we show magnetoelectric and/or multiferroic behavior in two organo-metallic materials. In DTN, the polar nature of thiourea molecules is combined with Ni magnetism to create a net H-induced P of 50 $\mu C/m^2$. Thiourea itself is a potent ferroelectric with a P of 3000 $\mu C/m^2$ below 170 K. In a recent review of organic ferroelectrics [8], thiourea was identified as a promising candidates for ferroelectricity and thus for the construction of magnetoelectric materials in organic materials. In pure thiourea P is thought to originate from the slight misalignment of nearly oppositely aligned thiourea pairs. This work on DTN demonstrates that magnetism can indeed be incorporated into an organo-metallic lattice with thiourea, creating a net magnetoelectric effect.

In CDC, simultaneous phase transitions in the magnetization and electric polarization occur as a function of magnetic field, allowing a SIS-breaking structure to occur between 0.3 and 3.8 T. The H-induced suppression of both the magnetic and electric orders at 3.8 T results in a novel magnetoelectric quantum critical point. CDC is an example of using a complex non-collinear spin structure that breaks SIS to modify the lattice symmetry and allow a net P to occur. In this type of multiferroic, the non-collinear structure often result from competing or frustrating magnetic interactions (see e.g. [15] and references therein).

4. Acknowledgments

Work at the NHMFL was supported by the NSF (Grant No. DMR901624), the State of Florida, and the U.S. DOE. Work at JHU was supported by the NSF through Grant No. DMR-0306940. A.P.F. acknowledges support from CNPq (Conselho Nacional de Desenvolvimento Cientfico e Tecnolgico, Brazil).

5. References

[1] Fiebig M 2005 *J. Phys. D* **38** R123
[2] Khomskii D I 2006 *J. Magn. Magn. Mater* **306** 1
[3] Eerenstein W, Mathur N D and Scott J F 2006 *Nature* **442** 759
[4] Hill N 2000 *J. Phys. Chem. B* **104** 6694
[5] Kagawa F, Horiuchi S, Tokunaga M, Fujioka J and Tokura Y 2010 *Nature Physics* **6** 169
[6] Zapf V S, Kenzelmann M, Wolff-Fabris F, Balakirev F and Chen Y 2010 *Phys. Rev. B* **82** 060402R
[7] Zapf V S, Nasreen F, Wolff-Fabris F and Paudan-Filho A, in progress.
[8] Horiuchi S and Tokura Y 2008 *Nature Materials* **7** 357
[9] Horiuchi S et. al 2010 *Nature* **463**
[10] Paduan-Filho A, Gratens X and N F Oliveiras Jr 2004 *Phys. Rev. B* **69** 020405R
[11] Zapf V S et. al 2008 *Phys. Rev. B* **77** 020404R
[12] Zapf V S et. al 2006 *Phys. Rev. Lett.* **96** 077204
[13] Kenzelmann M, Chen Y, Broholm C, Reich D H and Qiu Y 2004 *Phys. Rev. Lett.* **93** 017204
[14] Chen Y, Stone M B, Kenzelmann M, Batista C D, Reich D H and Broholm C 2007 *Phys. Rev. B* **75** 214409
[15] Kimura T 2007 *Annu. Rev. Mater. Res.* **37** 387

Low-Temperature Elastic Properties of Non-Kramers Doublet Compound $PrMg_3$

K Araki , K Mitsumoto[1], M Akatsu[1], Y Nemoto[1], T Goto[1], H S Suzuki[2], H Tanida[3], S Takagi[4], S Yasin[5], S Zherlitsyn[5] and J Wosnitza[5]

[1] Graduate School of Science and Technology, Niigata University, Niigata 950-2181, Japan
[2] National Institute for Materials Science, 1-2-1 Sengen, Tsukuba, Ibaraki 305-0047, Japan
[3] Graduate School of Advanced Sciences of Matter, Hiroshima University, 1-3-1 Kagamiyama, Higashi-hiroshima, Hiroshima 739-8530, Japan
[4] Physics Department, Graduate School of Science, Tohoku University, Aramaki Aoba, Aoba-ku, Sendai, Miyagi 980-8578, Japan
[5] Hochfeld-Magnetlabor Dresden (HLD), Forschungszentrum Dresden-Rossendorf, 01314 Dresden, Germany

E-mail: `araki@phys.sc.niigata-u.ac.jp`

Abstract.
We have investigated low-temperature elastic properties of $PrMg_3$ with a non-Kramers Γ_3 doublet ground state by using ultrasonic measurements under high eld up to 18 T down to 20 mK. A softening of the elastic constant $(C_{11} \;\; C_{12})/2$ has been observed below 8 K due to the non-magnetic Γ_3 doublet possessing electric quadrupole O_u and O_v with Γ_3 symmetry and magnetic octupole T_{xyz} with Γ_2. This result appreciably is deviated from a theoretical tting of the quadrupole susceptibility based on the crystal-electric- eld levels below 800 mK. Furthermore, the $(C_{11} \;\; C_{12})/2$ shows strange magnetic dependence up to 18 T below 1.7 K, which are inconsistent with the theoretical results.

1. Introduction

In rare-earth compounds, crystal-electric-field (CEF) states of $4f$ electrons affects various physical properties of the system. For the f^2 configuration of Pr^{3+}, a $^3H_4$ ($J = 4$) ground state multiplet splits into a $_1$ singlet, a $_3$ doublet, $_4$ and $_5$ triplets due to the cubic CEF. In particular, a non-Kramers $_3$ doublet ground state system in Pr-based compounds is of great interest for studying strong correlated effects of the multipole because the $_3$ doublet has electric quadrupoles O_u $(=(2{J_z}^2\text{-}{J_x}^2\text{-}{J_y}^2)/\sqrt{3})$ and O_v $(={J_x}^2\text{-}{J_y}^2)$ with $_3$ representation, a magnetic octupole T_{xyz} $(=\sqrt{15}\overline{J_xJ_yJ_z}/6)$ with $_2$ and no magnetic dipoles J_x, J_y, J_z with $_4$. Here, the bar means the sum of cyclic permutations on x, y, and z. The rare-earth compounds with orbital degenerate ground state generally favor to undergo multipole ordering at low temperatures. For example, PrPtBi and $PrPb_3$ with $_3$ ground state have been known to exhibit a $_3$-type quadrupole ordering at low temperatures [1, 2, 3]. On the other hand, $PrAg_2In$ and $PrMg_3$ reveal no sign of long-range ordering at low temperatures, while these compounds show a pronounced peak with a gigantic specific heat coefficient $C/T = 6.5$ J/(mol K^2) for $PrAg_2In$ and $C/T = 2.8$ J/(mol K^2) for $PrMg_3$ at low temperatures [4, 5]. Tanida *et*

Published under licence by IOP Publishing Ltd

al. suggests that the orbital degenerate Γ_3 ground states with the multipole degrees of freedom are quenched by the hybridization with the conduction electrons. However, the mechanism that suppresses the quadrupole ordering has not yet been established.

The intermetallic compound $PrMg_3$ is crystallized in a cubic Fe_3Al-type crystal structure of space group Fm$\bar{3}$m. Previous neutron studies reported the CEF level scheme of Γ_3(0 K)-Γ_4(56 K)-Γ_1(135 K)-Γ_5(183 K) [6, 7]. The magnetic susceptibility shows unusual temperature dependence $-\log T$ ($1 \leq T \leq 10$ K) or $-\sqrt{T}$ ($0.3 \leq T \leq 2.5$ K), which indicates the hybridization effect between $4f^2$- and conduction-electrons [8]. The ultrasonic measurement is a powerful tool to investigate the Γ_3 ground state system, because the quadrupole O_v with Γ_3 representation couples to the Γ_3-type elastic strain ε_{Γ_3} corresponding to transverse $(C_{11}-C_{12})/2$ mode. We showed pre-measurements on $PrMg_3$ and found the importance of the hexadecapole-strain interaction [9]. In this paper, we report on the ultrasonic measurements under the magnetic fields up to 18 T and analysed by using the quadrupole susceptibility in order to elucidate low-temperature properties of $PrMg_3$.

2. Experiment

A single crystal of $PrMg_3$ was prepared by a Bridgman method with Mo crucible sealed in a high vacuum. We used the sample with a rectangle shape of the size of $3.4 \times 2.5 \times 2.1$ mm^3. The sound velocity has been measured with an ultrasonic apparatus based on a phase comparison method as described in Ref. [10]. The piezoelectric plates of $LiNbO_3$ with fundamental frequencies 16 $\sim$ 30 MHz were used for the ultrasonic transducers. The absolute velocity was obtained by measuring the delay time for sequence of the ultrasonic echoes. The elastic constants $C = \rho v^2$ were calculated from the sound velocity v and the mass density ρ = 3.485 g/cm^3 of $PrMg_3$ with a lattice constant a = 7.415 Å. The low-temperature measurements

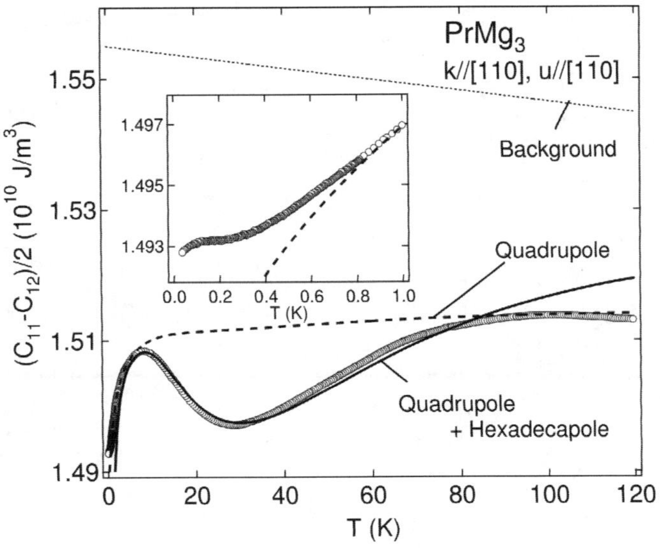

Figure 1. Elastic constant $(C_{11}-C_{12})/2$ as a function of temperature in $PrMg_3$. The dashed line is a fit with the quadrupole susceptibility with $\lambda = 0$ in the text. The solid and dotted lines show the multipole susceptibility for quadrupole and hexadecapole with $\lambda = 0.00126$, the background of $(C^0_{11}-C^0_{12})/2$. The inset shows an enlarged view below 1 K.

below 1.1 K down to 14 mK were performed using a ^{3}He-^{4}He dilution refrigerator at HLD in Dresden, Germany. The measurements above 300 mK were carried out by a ^{3}He cryostat. These systems were equipped with 20 T superconducting magnet and 12 T one, respectively.

3. Experimental results and discussions

Figure 1 shows the temperature dependence of elastic constant $(C_{11}-C_{12})/2$ in $PrMg_3$, which was measured by a transverse ultrasonic wave propagating along the $\boldsymbol{k} \parallel [110]$ direction with polarization of $\boldsymbol{u} \parallel [1\bar{1}0]$ with frequency of 218 MHz. The $(C_{11}-C_{12})/2$ shows a softening below 100 K and turns to increase below 30 K with decreasing temperature. The softening in $(C_{11}-C_{12})/2$ appears below 8 K and the transverse C_{44} shows no softening at low temperatures, not shown here. These results supports that $PrMg_3$ has the Γ_3 doublet ground state with the Γ_3-type quadrupoles O_u and O_v.

The solid and dashed lines in Fig. 1 indicate the calculated results for multipole and quadrupole susceptibility, respectively. The multipole-strain interaction can be written as $H_{ms} = \sum_{\Gamma\gamma} g_{\Gamma\gamma}(O_{\Gamma\gamma}+\lambda H_{\Gamma\gamma})\varepsilon_{\Gamma\gamma}$, where $O_{\Gamma\gamma}$ is quadrupole operator, $H_{\Gamma\gamma}$ is hexadecapole operator, $g_{\Gamma\gamma}$ is a coupling constant of multipole-strain interaction and λ is a contribution parameter for hexadecapole-strain interaction. Therefore, the temperature dependence of the elastic constants are described by the formula $C_\Gamma = C_\Gamma^0 - Ng_\Gamma^2\chi_{\Gamma\gamma}(T)/[1 - g'_\Gamma\chi_{\Gamma\gamma}(T)]$, where $\chi_{\Gamma\gamma}$ is the multipole susceptibility, g'_Γ is a coupling constant of intersite multipole interactions, C_Γ^0 is the background and N is number of Pr ions in unit volume. The dashed line based on the quadrupole susceptibility with $\lambda = 0$ well describes the softening below 8 K. The characteristic minimum around 30 K, however, is not reproduced. So we performed the analysis including contribution of the hexadecapole with $\lambda \neq 0$, where the intersite multipole interactions have

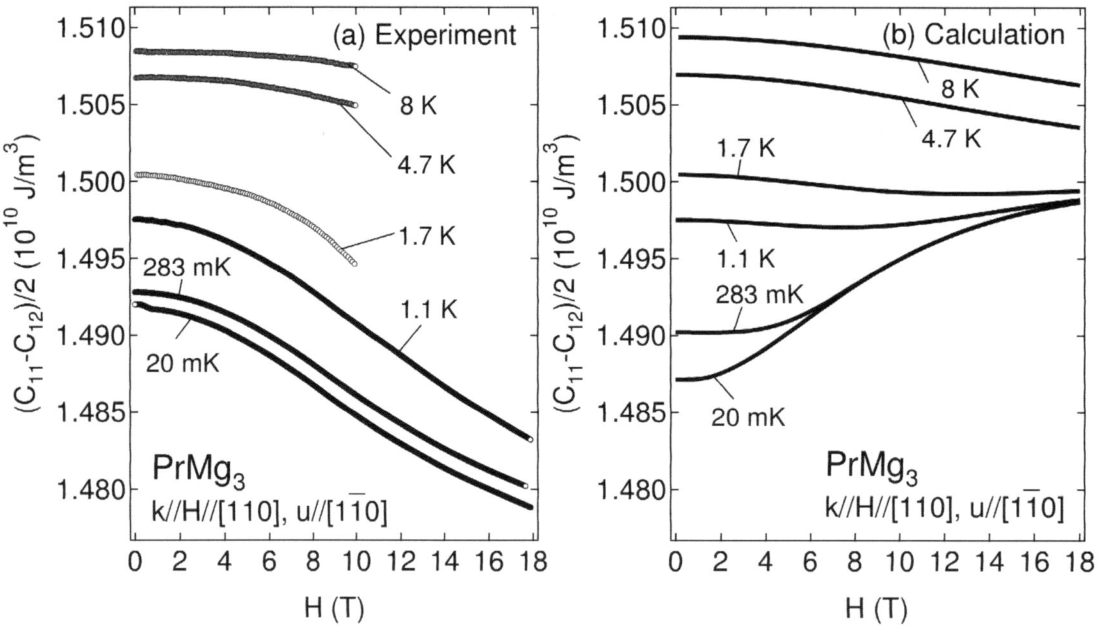

Figure 2. Magnetic field dependence of $(C_{11}-C_{12})/2$ under the field up to 18 T along the [110] direction at fixed various temperatures. (a) Experimental results. (b) Calculated results by using the quadrupole susceptibility based on the CEF levels.

little contribution. The characteristic minimum around 30 K is well described by the multipole susceptibility for quadrupole and hexadecapole. We obtained $|g_{\Gamma_3}| = 14.6$ K, $g'_{\Gamma_3} = -93$ mK for the quadrupole susceptibility and $|g_{\Gamma_3}| = 45.2$ K, $\lambda = 0.0126$ for the multipole susceptibility consisting of hexadecapoles as well as quadrupoles. As shown in the inset of Fig. 1, the $(C_{11}-C_{12})/2$ shows the softening with decreasing temperature and a shoulder anomaly around 200 mK. The results are deviated from the dashed line based on the quadrupole susceptibility below 800 mK. No evidence of phase transition was observed in the present experiment down to 20 mK. These results reproduce our previous report in Ref. [9].

The magnetic dependence of $(C_{11}-C_{12})/2$ under field along the [110] direction below 8 K in Fig. 2(a) shows a decrease and no phase transition with increasing magnetic fields up to 18 T. Figure 2(b) shows the calculated results of the quadrupole susceptibility with the $|g_{\Gamma_3}| = 45.2$ K and $g'_{\Gamma_3} = -93$ mK. The calculations of Fig. 2(b) are inconsistent with the experimental results below 1.7 K of (a).

In conclusion, the characteristic minimum around 30 K in the $(C_{11}-C_{12})/2$ was well described by the multipole susceptibility where the quadrupole and the hexadecapole are included together. At the low temperatures, the $(C_{11}-C_{12})/2$ shows the Curie-type softening due to the $_3$ ground state with the $_3$-type quadrupoles O_u and O_v and the shoulder anomaly around 200 mK. The results are deviated from the calculated result based on the quadrupole susceptibility. The magnetic dependence of $(C_{11}-C_{12})/2$ shows strange behavior which are inconsistent with the theoretical results. To elucidate the strange behavior of the $(C_{11}-C_{12})/2$ at low temperatures under the magnetic filed, ultrasonic measurements in magnetic field along the [100] and [111] direction are required.

Acknowledgments
This works was supported by a Grant-in-Aid for Specially Promoted Research (No. 18002008) "Strongly correlated quantum phases associated with charge fluctuations" of the Ministry of Education, Culture, Sports, Science and Technology, Japan. One of the authors (K. A.) appreciates the financial support of Research Fellows of the Japan Society for the Promotion of Science for Young Scientists.

References

[1] Suzuki H, Kasaya M, Miyake T, Nemoto Y and Goto T 1997 *J.Phys.Soc.Jpn.* **66** 2566-2568
[2] Bucher E, Andres K, Gossard A C, Maita J P 1972 *J.LowTemp.* **2** 322
[3] Morin P, Schmitt D, du Tremolet de Lacheisserie E 1982 *J.Magn.Magn.Mater.* **30** 257
[4] Yatskar A, Beyermann W P, Movshovich R and Can eld P C 1996 *Phys.Rev.Lett.* **77** 3637
[5] Tanida H, Suzuki H S, Takagi S, Onodera H and Tanigaki K 2006 *J.Phys.Soc.Jpn.* **75** 073705
[6] Galera R M, Murani A P and Pierre J 1981 *J.Magn.Magn.Mater.* **23** 317
[7] Andreeff A, Goremychkin E A, Griessmann H, Lippold B, Matz W, Chistyakov O and Savitskii E M 1980 *Phys.Stat.sol* **98** 283
[8] Morie T, Sakakibara T, Suzuki H S, Tanida H and Takahi S 2009 *J.Phys.Soc.Jpn.* **78** 033705
[9] Araki K, Yoshikawa T, Akatsu M Ishii I, Nemoto Y, Goto T Suzuki H S, Tanida H, Takagi S and Luthi B 2009 *PhysicaB* **404** 3187-3190
[10] Luthi B 2005 *PhysicalAcousticsintheSolidState*, Springer

Journal of Physics: Conference Series **273** (2011) 012134 doi:10.1088/1742-6596/273/1/012134

A simple model for the magnetoelectric interaction in multiferroics

Cesar J. Calderon Filho and Gaston E. Barberis

Instituto de Física "Gleb Wataghin", UNICAMP,
13083-970, Campinas, São Paulo, Brazil

E-mail: barberis@ifi.unicamp.br

Abstract. The (anti)ferromagnetic and ferroelectric transitions in some multiferroic compounds seem to be strongly correlated. Even for systems that do not show spontaneous ferroelectricity such as the $LiMPO_4$ (M = Mn, Fe, Co, Ni) compounds, the coupling between magnetic and electric degrees of freedom is evident experimentally. Here, we present a simple numerical calculation to simulate this coupling that leads to the two transitions. We assume a magnetic sublattice consisting of classical magnetic moments coupled to a separated non-magnetic sublattice consisting of classical electric dipoles. The coupling between them is realized through a phenomenological spin-lattice Hamiltonian, and the solution is obtained using the Monte Carlo technique. In the simplest version, the magnetic system is 2D Ising (anti)ferromagnetic lattice, with nearest neighbors interactions only, and the electric moments are permanent moments, coupled electrically. Within this approximation, the second order magnetic transition induces ferroelectricity in the electric dipoles. We show that these calculations can be extended to other magnetic systems, (x-y model and 3D Heisenberg) and to systems where the electric moments are created by strains, generated via spin-lattice coupling, so the model can be applied to model realistic systems such as the olivines mentioned above.

1. Introduction

Recently, there has been a revived interest in the research of multiferroics due to several new discoveries, and the possibility to use them technologically [1]. The electric and magnetic transitions are not necessarily correlated, but when it occurs - the so called *magnetoelectric effect* appears - the materials suggest possible use as devices. The LiMPO4 (M = Mn, Fe, Co, Ni) compounds generated great interest in the last times, for they are used as electrodes in batteries, but also because they present very rich physics [1-2]. The whole family is multiferroic (that is, present simultaneously magnetic and electric order), with transitions related one to the other, and their transitions show very special effects. The Ni compound undergoes a first order phase transition from collinear antiferromagnetic ground state to long-range incommensurate structure at T_N =20.8 K and at T_{IC} =21.7 K the incommensurate structure transforms to short range [2]. Inelastic and elastic neutron studies reveal anomalous spin wave in this incommensurate phase, and iron substitution (about 20%) fine tunes the phase transition and the spin-wave dispersion [2]. Many other interesting physical properties of these compounds are in the recent literature. However, just few papers are dedicated to the theoretical interpretation of the magnetoelectric effect [4].

Published under licence by IOP Publishing Ltd

We present here a simple and understandable model for the magnetoelectricity in these and similar compounds. The electromagnetic Hamiltonian is solved for very simple cases, and the solutions present similitude with the reported experiments. The model is extended to more complicated systems, and partially solved in those cases too. These results and comparisons will be published elsewhere.

2. The model

We assume that the magnetoelectric system is a set of magnetic dipoles, coupled via the exchange interaction, in a lattice with a distribution of electric charges, susceptible to change when the magnetic dipoles change their orientations. The change in orientation of the magnetic dipoles modify their environment, via spin-orbit interaction, creating local strains, and creating or orienting a set of electric dipoles in the lattice. We assume that our crystal suffers the strain in such a way that electric dipoles are oriented to a particular direction when the magnetic dipoles relax.

The model Hamiltonian used in this work follows:

$$H = H_M + H_E + H_{ME} \tag{1}$$

where H_M is the magnetic energy, H_E the electric energy and H_{ME} the magnetoelectric coupling.

2.1. *Approximations*

The first approach for a solution of eq. (1) is obtained replacing the first term in the sum of the right side by a square sublattice of Ising magnetic moments, and the electric moments in the second and third terms by random oriented classical electrical dipoles, located in a separated square sublattice. The Ising spins are coupled to their nearest neighbors only, and with periodic boundary conditions. The interaction Hamiltonian allows only nearest neighbors magnetoelectric interaction. Symmetry requires that the magnetic point group of the magnetic moment is one of the 58 Shubnikov groups that allow magnetoelectricity [1]. This forces the magnetic moments to have one electric dipole only as a nearest neighbor.

The numerical solution of the problem was done using the importance sample Monte Carlo method, looking for the minimum in energy for our system.

Thus,

$$H = -J \sum_{<i,j>} \sigma_i \sigma_j - h \sum \sigma_i - \beta \sum_{<i,j>} P_{ix} P_{jx} + \alpha \sum_i P_{ix} \tag{2}$$

is the approximated Hamiltonian, where J is the exchange coupling of the Ising magnetic spins σ, and P the electric dipoles. The symbol $< i,j >$ indicates sums over the nearest neighbors only, in both the electric and magnetic cases. The first and second term constitutes the magnetic energy, where we included the possibility of an applied or external magnetic field h. The third term represents the electric energy, proportional to the orientation of neighbor electric momenta. The interaction term, which represents a spin - lattice Hamiltonian, has been extremely simplified. As the interaction energy would depend of the value of the created dipole (or strain), it is proportional to the projection of the electric dipole in the selected direction; the value of the coupling, contained in the α parameter, which also contains the value of the magnetic moment. This very simple expression is convenient to allow the values of the coupling to be positive or negative, as the strain can add or take energy of the lattice.

2.2. Ferromagnetic case.

Our first calculation was performed in a 100×100 2D Ising ferromagnetic spins coupled to a separated 100×100 electric dipoles. The electric dipoles were oriented at random, as was the magnetic lattice, to begin with infinite temperature. The temperature was then fixed to a value, and a Monte Carlo program, where the transitions are allowed following the Metropolis technique, is iterated the time necessary to obtain thermal equilibrium of the system. Then, the converged results are used as initial condition for the following temperature. The calculation was performed reducing the temperature in each step.

The complete calculation was performed after a study of convergence of our model. As first step, we did not considered the electric and interaction energies when we looked for the minimum. This means that the model is just a 2D Ising system, moving electric dipoles together, and as expected, the magnetization follows the Ising model as was done before. The exact calculation published by Onsager allows a very good comparison, and the electric dipoles also feel a transition at the same temperature.

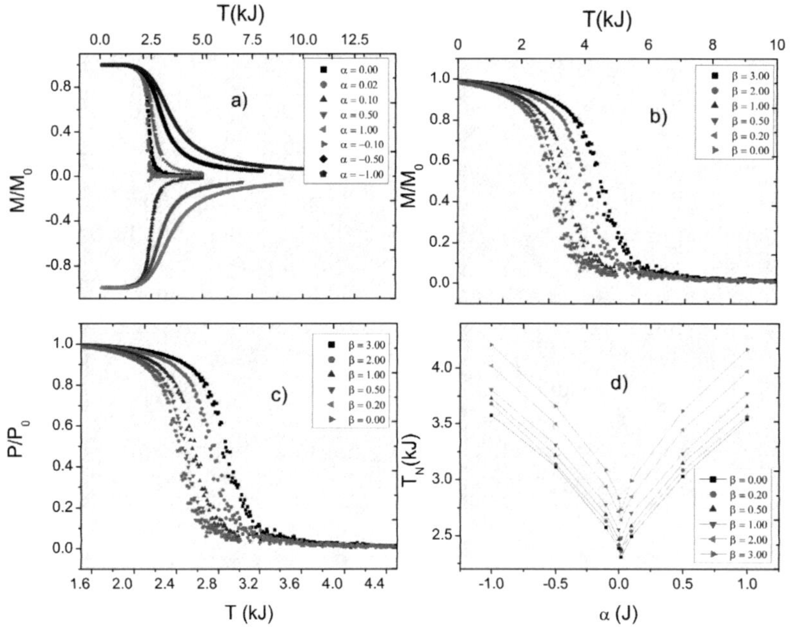

Figure 1. (Color online) The results for the ferromagnetic case. a) Magnetization as a function of T, for different vales of α; b) The same for different values of β; c) The electric polarization as function of the temperature, for different values of β; d) The complete dependence of T_c as a function on the parameters of the model.

The following step was to study the convergence when the parameters α and β in the model are different from zero. It is well known that the Ising model converges slowly near by the transition temperature, due to fluctuations, and the equal value for the energy when the system is oriented in any of the both possible directions. This is easily solved with the addition of the small external field h; however, we observed that for particular values of α, the convergency is the slowest. This can be explained by the fact that α appears in the Hamiltonian as an extra external field, in some manner. The second and last terms in the Hamiltonian can be written as

$$\alpha \sum_i P_{ix} - h \sum \sigma_i = -\sum (h - \alpha |P_{ix}|) \sigma_i \tag{3}$$

which shows that the value of $\alpha|P_{ix}|$ appears as an extra field. The mean value of $|P_{ix}|$ annulate the external field when $\alpha = 0.02$, and the convergency is the slowest for this value of α.

To make the results clear, we made calculations as function of the temperature for different values of α, when β is zero - meaning that the magnetoelectric interaction is bigger than the electric one. After that, we calculated the minimum as function of β, when α is zero. The complete calculation, with both parameters different from zero gives a clear vision of the total behavior of our model.

Fig. 1 shows the results for the ferromagnetic case. The induced electric polarization appears at the same temperature as the magnetic transition, but shows important fluctuations at temperatures higher than T_c as compared with those of the magnetic lattice.

2.3. Antiferromagnetic case

The antiferromagnetic case was treated similarly. It requires a negative value of J, but several changes in the other terms of the Hamiltonian are necessary. The magnetization of both sublattices will be coupled to the electric lattice opposed, in order to obtain the required ferroelectricity. Our results in this case are very similar to those above, and we lack here of space to show them completely. These results will be publish elsewhere, together with a more sophisticated model for the spin-lattice coupling.

3. Conclusions and perspectives.

The model presented above contains essentially the complete elements necessary to solve the magnetoelectric problem. Depending of the details in the spin-lattice (or magnetoelectric) Hamiltonian, the simulations will show different aspects of the physics involved. We are developing the approximation that corresponds with the creation of the electric moments, and the change of their values as functions of the local magnetization and strain. We believe that that approximation will simulate properly the case of the $LiMPO_4$ compounds.

3.1. Acknowledgments

We are acknowledge to Fapesp, São Paulo, Brazil, and Faepex, UNICAMP, for their financial help.

4. References

[1] Fiebig M, 2005 *J.Phys. D: Appl. Phys.* **38** R123
[2] Rivera J-P, 1996 *Ferroelectrics* **161** 147
[3] Vaknin D,Zaretsky J L , Rivera J-P and Schmid H, 2004 *Phys Rev. Lett.* **92** 207201
[4] Kosterlitz J M, Nelson D R, and Fisher M E, 1976 *Phys Rev.* B **13** 412

Journal of Physics: Conference Series **273** (2011) 012135 doi:10.1088/1742-6596/273/1/012135

Nonlinear conductivity of electronic origin in magnetic oxides: uncommon cases at E < 1000 V/cm

B.Fisher, J. Genossar, L. Patlagan, I. Feldman and G. M. Reisner

Dept. of Physics, Technion, Haifa 32000, Israel
phr06bf@physics.technion.ac.il

Abstract. We report on electronic transport measurements on polycrystalline, non-metallic samples of doped-$LaMnO_3$ in the ohmic and non-ohmic regimes. The non-ohmic regime has been investigated using d.c. as well as pulsed measurements. Large nonlinear conductivity of electronic origin, was found with pulsed currents in (nominal) $La_{0.8}Li_{0.2}MnO_3$, $LaMnO_{3+\delta}$ and in $La_{0.8}Ca_{0.2}Mn_{0.95}O_3$ at low temperatures under electric fields E< 500 V/cm. Their thermopowers exhibit maxima associated with the onset of (partial) magnetic order. For all samples, including some with perfectly linear pulsed I-V characteristic, the d.c. characteristics show, under similar fields, large nonlinearity and hysteresis, typical of self-heating in samples with activated conductivity.

1. Introduction

Nonlinear conductivity and associated phenomena (hysteresis, noise, switching, breakdown, etc) found in magnetic oxides, in particular in charge ordered (CO) manganites, have attracted considerable attention for more than a decade. Although the thermal theory of non-ohmic conductivity and associated phenomena are known for about a century, many of the recently published I-V characteristics (I-V-C) were obtained using d.c., a.c. or long-duration current pulses (of hundreds of msecs or more). Unfortunately, such measurements do not discriminate between nonlinearity of electronic origin and that due to self-heating. We measured the I-V-C of several CO oxides, using both d.c. and short-duration single current pulses[1,2] and found that in all our samples, for electric fields up to ~ 1000 V/cm, the strongly nonlinear d.c. I-V-C mask a perfectly linear or only slightly nonlinear conductivity as obtained with short current pulses. However, using the same technique we did find significant nonlinear conductivity of electronic origin (NCEO) in some non-metallic ferromagnetic oxides, most recently in $LaMnO_{3+\delta}$ ($LMO_{+\delta}$)[3]. By significant we mean that for E≤ 500 V/cm the excess current above ohmic ($\Delta J=J-J_o$) is of the order of, or larger than, the ohmic current (J_o). Even in such cases self heating takes over the NCEO at unpredictable voltages, when measured using d.c.

The excess oxygen in $LMO_{+\delta}$ appears as pairs of La-Mn vacancies[4] that contribute the finite Mn^{4+}/Mn^{3+} ratio (= 2δ/1-2δ) responsible for ferromagnetism. The disorder on the Mn lattice containing Mn vacancies causes non-metallicity. We suggestested[3] that the significant NCEO obtained in this system is due to field enhanced hopping along the Mn chains. In order to learn more about the field and temperature dependence of $\Delta J/J_o$ in insulating FM manganites and their relation to the low-field transport we extended this work to additional polycrystalline doped $LaMnO_3$ 1. Li-doped on the La site (nominal), 2. Li-doped on the Mn site (nominal) and 3. self-doped or Ca-doped with vacancies on the Mn-site. Substitution of trivalent ions (La and Mn) by monovalent ions (Li) or by vacancies is more efficient than by the common divalent ions for increasing the Mn^{4+}/Mn^{3+} ratio.The largest NCEO was obtained in $La_{0.8}Ca_{0.2}Mn_{0.95}O_3$(LCMO) and in $LMO_{+\delta}$; their relatively low resistances allowed us to extend the measurements below liquid nitrogen. It will be shown that the temperature dependence of the thermopower is correlated with the magnitude of the NCEO.

2. Sample preparation and experimental methods

The samples were prepared either by the standard solid state synthesis (SSSS)[5] or by wet chemistry[6] both followed by sintering. We initially followed the preparation protocols from the literature but

Published under licence by IOP Publishing Ltd

digressed from these protocols, increasing the number, duration and temperature of heat treatments whenever the resulting samples were not of single phase or disintegrated. The XRD patterns obtained using a Siemens diffractometer showed virtually no traces of impurity phases in the final compounds. The unit cell symmetry and lattice parameters were determined by least-square fitting in the range of $2\theta=10^{\circ} - 140^{\circ}$, collected by step scan of 0.03°, 15 secs. step. The (four-probe) d.c. and pulsed I-V-C were measured as described in Ref. 1. and their thermopower (relative to Cu) was measured as in Ref. 2. The preparation method of the investigated samples and their structural properties are summarized in columns 2- 4 of Table I and the SEM morphology of six samples is shown in Fig. 1.

Table I. Preparation and structural properties of the investigated samples and the most significant low-field and high- field electronic properties

Nominal composition	Preparation	Crystal system	Volume/ f.u. ($\mathring{A}^3$)	$T(S_{max})$ (K)	E_o(80 K) V/cm
$La_{0.8}Li_{0.2}MnO_3$ LLMO(a)	SSSS 1100° C	Orthorhombic (O)	58.87	-	> 1000
$La_{0.8}Li_{0.2}MnO_3$ LLMO(b)	SSSS 1450° C + 600 C in O_2	O close to R	59.63	185	140
$La_{0.8}Li_{0.2}MnO_3$ LLMO(c)	Wet chemistry + 1200° C	Hexagonal (H)	58.77	180	205
$LaMn_{0.9}Li_{0.1}O_3$ LMLO(a)	SSSS 1300° C	Rhombohedral (R)	58.60	-	-
$LaMn_{0.9}Li_{0.1}O_3$ LMLO(b)	Wet chemistry +1200°C	R	58.60	165	> 1000
$La_{0.9}Ca_{0.1}Mn_{0.9}Li_{0.1}O_3$ LCMLO	Wet chemistry +1200°C	Monoclinic	58.87	161	560
$LaMn_{0.95}O_3$ LMO(a)	SSSS 1450° C + 600° C in O_2	R	58.62	-	-
$LaMnO_{3+\delta}$ ($\delta\approx0.15$)[3] $LMO_{+\delta}$	SSSS 1450° C + 600° C in O_2	R	58.65	191	85
$La_{0.8}Ca_{0.2}Mn_{0.95}O_3$ LCMO	SSSS 1450° C + 600° C in O_2	O	58.95	186	105

Fig. 1. SEM images of LLMO(a), LLMO(b) and LLMO(c) (upper row) and of LCMLO, $LMO_{+\delta}$, LCMO, (second row). Note the large grains and the melted boundaries of the samples sintered at high temperatures in contrast with those sintered at low temperatures (see Table I).

3. Results and discussion

The resistivity (ρ) and the thermopower (S) as function of inverse temperature and the pulsed (symbols) and d.c. J-E characteristics of the samples at various temperatures are shown in Figs. 2 -4.

Fig. 2. (a) Semilog plot of ρ versus 1/T for $La_{0.8}Li_{0.2}MnO_3$ (a,b,c). (see Table I). (b) S versus 1/T for the three types of samples. (c) – (e) Log-log plots of J(E) for the three types of samples at various temperatures. Symbols represent pulsed measurements (NCEO) and solid lines- d.c. measurements (self-heating). Significant NCEO is observed at low T for LLMO(b) and (c). The insets in (c) and (d) represent J(E) (pulsed and d.c.) on linear scales.

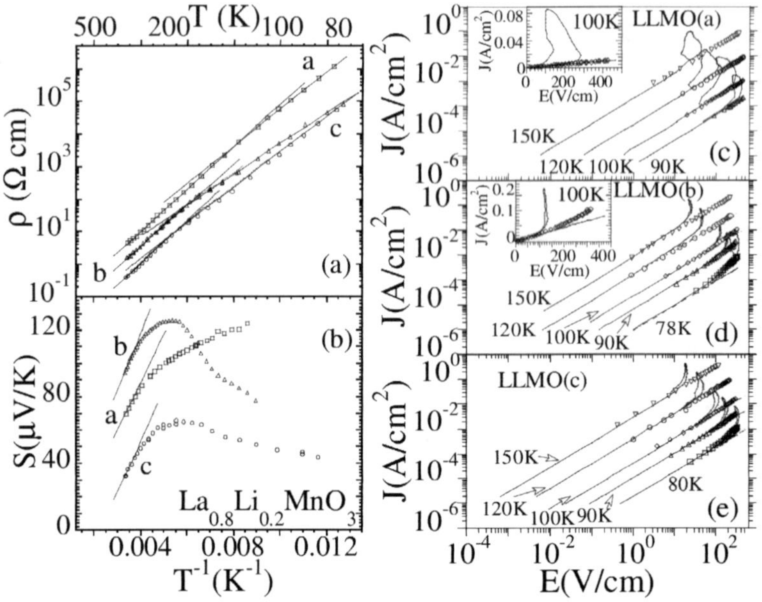

Fig. 4. (a) Semilog plot of ρ versus 1/T for $LaMn_{0.95}O_3$, $LaMnO_{3+\delta}$ and $La_{0.8}Ca_{0.2}Mn_{0.95}O_3$. (see Table I). (b) S versus 1/T for the three types of samples. (c) Semilog plot of ρ versus T. (d) – (f) Log-log plots of J(E) for the three types of samples at various temperatures. Symbols represent pulsed measurements (NCEO) and solid lines- d.c. measurements (self-heating). Large NCEO is observed at low T for $LMO_{+\delta}$ and LCMO.

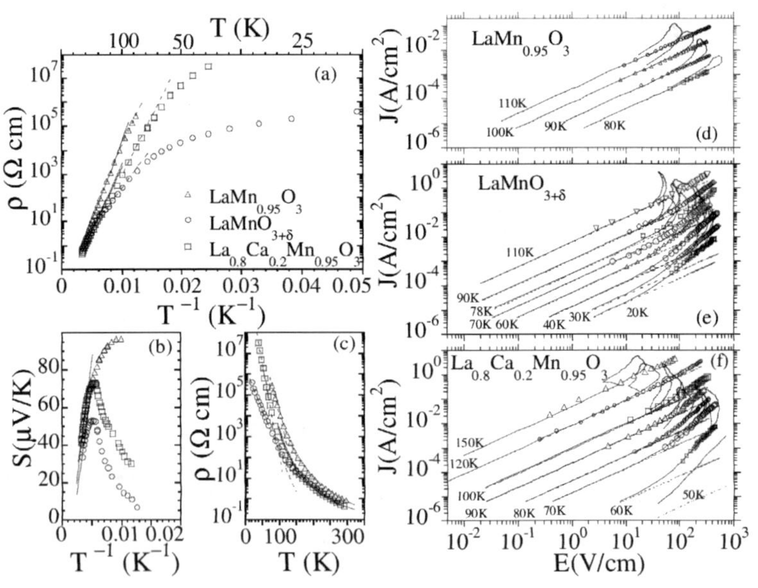

Fig. 3 containing the corresponding data for the three types of samples doped with Li on the Mn sites (see Table I) is added as a supplementary Figure. See also the Supplementary Table IS.

For all samples investigated in this work the resistivity is activated at all temperatures. All ln (ρ) vs T^{-1} plots bend from a higher to a lower activation energy at specific temperatures T_ρ, associated with the onset of ferromagnetic order.[3] Between RT and T_ρ the resistivity activation energy $\Delta_\rho(T>T_\rho)$

is lowest for LCMLO (110 meV) and highest for LMLO(a) and LCMLO (137 meV). Below T_ρ the activation energy is constant as in Fig. 2 (a) or decreases with decreasing temperature as in Figs. 3 (a) and 4 (a). Close to RT, the activation energies of the thermopower (Δ_s) are much lower (range between ~22 meV for $LMO_{+\delta}$ and ~35 meV for LLMO(b)). $\Delta_\rho >> \Delta_s$ is expected for polaron hopping. For all samples Δ_s decreases with decreasing temperature and for six samples S(T) passes through a maximum ($T(S_{max})$) at temperatures close to T_ρ. The values of $T(S_{max})$ are shown in row 5 of Table I.

The pulse J-E characteristics for LLMO (b) and (c), (Fig. 2) for $LMO_{+\delta}$ and for LCMO (Fig. 4) exhibit strong nonlinearity at low temperatures, visible even on the log-log scale. Note that although the morphology of LLMO (b) and LLMO (c) are very different they exhibit comparable NCEO. The weak nonlinearity of J(E) for LLMO(a) (Fig. 2), LMLO(b) and LCMLO (Fig. 3) is barely visible on this scale. The nonlinearity decreases with increasing temperatures and is not detectible above ~ 150 K. For LMLO(a) (Fig. 3) and LMO(a) (Fig. 4) the pulsed J(E) plots are perfectly linear at all temperatures of the measurements. For all samples, with and without NCEO, the d.c. characteristics are nonlinear and exhibit hysteresis, the typical behavior for self-heating.

Analysis of the J(E) plots shows that the dependence of the excess current density ΔJ on the electric field E can be expressed as: $\Delta J/J_o=(E/E_o)^\alpha$ where E_o and α are fitting parameters. For $LMO_{+\delta}$ and LCMO for which J-E measurements could be extended below liquid nitrogen temperature $\alpha>1$ at low temperatures (see Fig. 5(a)), where $\ln \rho/\rho_o \propto -T$ (see Fig. 3(c)). Fig. 5(b) shows the semilog plot of $E_o(T)$. $E_o(80\ K)$ is shown in the last column of Table I.

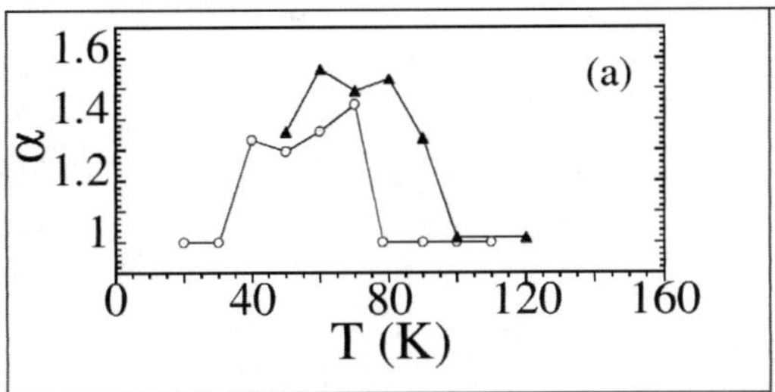

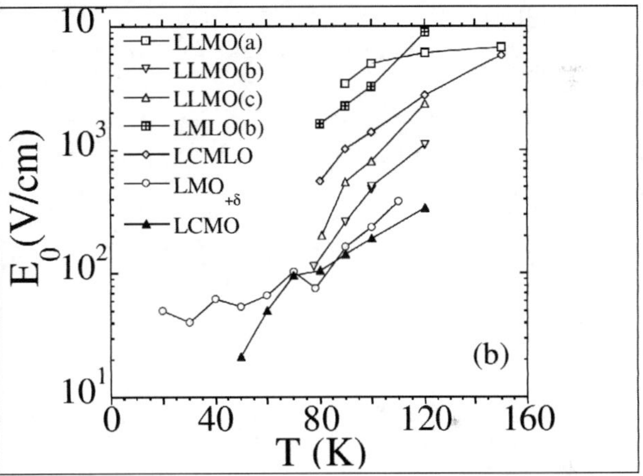

Fig. 5.Temperature dependence of fitting parameters in the relation $\Delta J/J_o=(E/E_o)^\alpha$ (a) α versus T for $LMO_{+\delta}$ and LCMO. In all other cases $\alpha=1$. (b) semilog plot of E_o versus T for all samples exhibiting NCEO.

The most significant observation of this work is revealed by the correlation between the last two columns in Table I: It is seen that measurable NCEO at fields below 500 V/cm appears in samples where S has a maximum. Moreover, the higher $T(S_{max})$ the larger the NCEO (the lower E_o).

References

[1] Fisher B, Genossar J, Chashka K B, Patlagan L and Reisner G M 2006 Appl. Phys. Lett. **88** 152103 and references therein.

[2] Fisher B, Genossar J, Patlagan L and. Reisner G M 2010 J. Magn. Mag. Matter **322**, 1239

[3] Fisher B, Genossar J, Patlagan L and. Reisner G M, 2009 Appl. Phys. Lett. **95**, 132501 and references therein.

[4] Töpfer J.and Goodenough J B, 1997 J. Solid State Chemistry **130**, 117

[5] Wang X L, Kennedy S J, Gehringer P, Lang W,. Liu H K and Dou S X, 1998 J. Appl. Phys. **83**, 7177 .

[6] Kutty T R N and Philip J, 2000 J. Phys.: Condens. Matter **12**, 7747 .

Supplementary data

Table 1S: Additional structural and electronic transport properties of the investigated samples

Nominal composition	Cell constants (Å)	T_ρ(K)	$\Delta_\rho(T\geq T_\rho)$ meV	$\Delta_\rho(T\leq T_\rho)$ meV	$T(S_{max})$ (K)	Δ_s(RT) meV
$La_{0.8}Li_{0.2}MnO_3$ LLMO(a)	5.487(2) 5.521(2) 7.774(2)	121	129	113	-	28
$La_{0.8}Li_{0.2}MnO_3$ LLMO(b)	5.520(1) 5.525(1) 7.822(1)	165	125	94	185	35
$La_{0.8}Li_{0.2}MnO_3$ LLMO(c)	5.526(1) 13.334(3)	186	130	108	180	30
$LaMn_{0.9}Li_{0.1}O_3$ LMLO(a)	5.524(1) 13.305(2)	149	137	119	-	27
$LaMn_{0.9}Li_{0.1}O_3$ LMLO(b)	5.524(1) 13.306(2)	~130	127	$12\ T^{1/2}$	165	20
$La_{0.9}Ca_{0.1}Mn_{0.9}Li_{0.1}O_3$ LCMLO♥	5.489(1) 5.497(1) 7.804(2) β=90.43°	~220	137	$11\ T^{1/2}$	161	24
$LaMn_{0.95}O_3$ LMO(a)	5.524(1) 13.310(2)	~125	136	$11\ T^{1/2}$	-	26
$LaMnO_{3+\delta}$ ($\delta\approx0.15$) $LMO_{+\delta}$	5.521(1) 13.329(2)	179	112	66♣♣	191	22
$La_{0.8}Ca_{0.2}Mn_{0.95}O_3$ LCMO	5.513(1) 5.500(1) 7.777(1)	~240	110	97♣♣	186	24

♥ The XRD spectrum of this compound contained minute unidentified peaks.
♣♣ Below 80 K for $LMO_{+\delta}$ and 90 K for LCMO $\ln(\rho/\rho_o)$.

Remarks to Table IS

- Doping on the Mn site is difficult and the blue section of Table IS represents the few successful cases, i.e. single phase and non-brittle.
- T_ρ is the temperature of the bend in the plot of ln (ρ) vs. 1/T.
- Above T_ρ, the activation energy $\Delta_\rho(T> T_\rho)$ is constant for all samples
- $\Delta_\rho(T<T_\rho)<\Delta_\rho(T>T_\rho)$ for all samples and for some of them is temperature dependent (see column 5 in the Table above).
- $T(S_{max})$ which appears in Table I was included here to show that in most cases it is close to T_ρ.
- $\Delta_S = \frac{dS}{d(T^{-1})}$ decreases with increasing temperature (see Figs. 2 (b), 3(b) and 4 (b)). Column 6 of this Table shows its value at room temperature (RT).

Fig. 3. (a) Semilog plot of ρ versus 1/T for LLMO (a,b) and LCMLO (see Table I). The inset shows the semilog plot of ρ versus $T^{-1/2}$.(b) S versus 1/T for the three types of samples.
(c) – (e) Log-log plots of J(E) for the three types of samples at various temperatures. Symbols represent pulsed measurements (NCEO) and solid lines- d.c. measurements (self-heating). Very weak NCEO is observed for LMLO(b) and LCMLO.

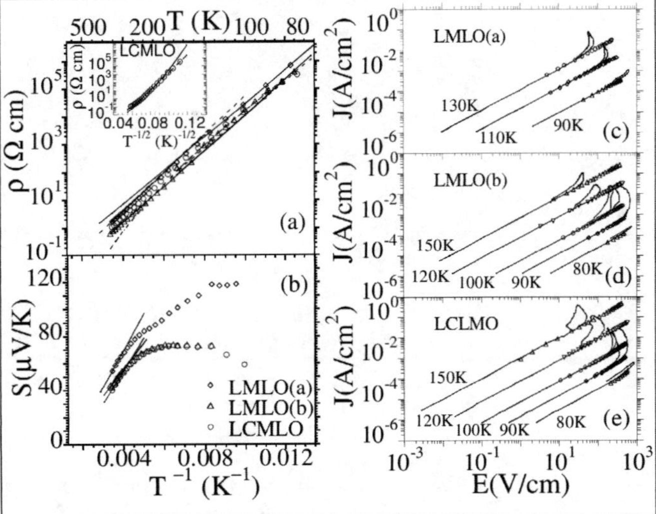

Elastic modulus of cage compound $PrRu_2Zn_{20}$

I Ishii[1], Y Suetomi[1], T K Fujita[1], T Onimaru[1], K T Matsumoto[1], Y F Inoue[1], T Takabatake[1,2] and T Suzuki[1,2,3]

[1] Department of Quantum Matter, ADSM, Hiroshima University, Higashi-Hiroshima 739-8530, Japan
[2] Institute for Advanced Materials Research, Hiroshima University, Higashi-Hiroshima 739-8530, Japan
[3] Cryogenics and Instrumental Analysis Division, N-BARD, Hiroshima University, Higashi-Hiroshima 739-8526, Japan

E-mail: ish@hiroshima-u.ac.jp

Abstract. Elastic properties of the cage compound $PrRu_2Zn_{20}$ have been investigated by means of ultrasonic measurement. The transverse modulus $(C_{11} - C_{12})/2$ shows elastic softening below 250 K and exhibits a discontinuous change in the slope at 150 K. An abrupt elastic hardening is observed below $T_s = 138$ K which is a non-magnetic phase transition temperature. The phase transition at T_s is accompanied by hysteresis indicating that the transition is of first-order. The softening above 150 K is well reproduced by Curie-Weiss-like formula for the co-operative Jahn-Teller effect with a negative quadrupole-quadrupole coupling constant. The phase transition at T_s is not caused by $4f$-electrons of Pr ions. There is a possibility that the elastic hardening below 3 K originates from the rattling motion.

1. Introduction

Cage compounds have attracted much attention in recent years because these compounds show various physical properties affected by the ionic degrees of freedom of the rattling motion, which is a large amplitude atomic motion. In ultrasonic measurements, cage compounds, such as the filled skutterudite ROs_4Sb_{12} [1, 2] and RFe_4Sb_{12} [3, 4], show the ultrasonic frequency dependences of elastic modulus and ultrasonic attenuation, which is so-called ultrasonic dispersion. Furthermore, elastic softening due to the rattling motion is observed at very low temperatures in these compounds [1, 2, 3, 4].

The cage compound RT_2Zn_{20} (R: rare earth, T = Ru, Ir, etc.) has the cubic $CeCr_2Al_{20}$-type structure (space group $Fd\bar{3}m$) with the lattice parameter of $a = 14.336$ Å at room temperature [5]. R atoms are accommodated at a single $8a$ site in the Frank-Kasper cages which consist of 16 Zn atoms. Onimaru *et al.* reported that $PrRu_2Zn_{20}$ shows a first-order phase transition at $T_s = 138$ K accompanied by structural modulation by their specific heat, electrical resistivity and magnetization experiments [6]. Temperature T dependence of the magnetic susceptibility shows no anomaly at T_s suggesting a non-magnetic phase transition. To investigate the origin of phase transition and the rattling motion in the cage compound $PrRu_2Zn_{20}$, we have measured T dependence of elastic modulus on a single crystalline sample.

Published under licence by IOP Publishing Ltd

2. Experimental

Single crystal of $PrRu_2Zn_{20}$ was grown by the Zn self-flux method. Details of the sample preparation were reported in ref. 7. The sample was shaped in a rectangular parallelepiped of $2.470 \times 2.182 \times 1.865$ mm^3. The elastic modulus $(C_{11} - C_{12})/2$, which is the transverse mode with the sound propagation direction k // [110] and the displacement direction u // $[1\bar{1}0]$, was measured as a function of T from 0.4 to 310 K using the phase comparison-type pulse echo method [8]. Detailed experimental technique was described in ref. 9. The elastic modulus C was calculated using $C = \rho v^2$ with a room-temperature mass density $\rho = 7.443$ g/cm^3, where v is the sound velocity in a sample. The frequency range was from 18 to 27MHz.

3. Results and discussion

Figure 1 shows T dependence of the transverse elastic modulus $(C_{11} - C_{12})/2$ in $PrRu_2Zn_{20}$. The modulus $(C_{11} - C_{12})/2$ increases monotonically with decreasing T above 250 K. It turns into decrease gradually below 250 K and exhibits a discontinuous change in the slope at 150 K. With further decreasing T, an abrupt elastic hardening is observed below T_s. The phase transition at T_s is accompanied by hysteresis as shown in the upper inset in Fig. 1. It suggests the first-order phase transition, being consistent with the results reported by Onimaru *et al.*. Huge ultrasonic attenuation is observed at around T_s. At low temperatures, it should be noted that $(C_{11} - C_{12})/2$ shows elastic hardening below 3 K and leveling off below 1 K as shown in the lower inset in Fig. 1. The origin of hardening below 3 K will be discussed in detail later. No elastic softening at very low temperatures is observed down to 0.4 K.

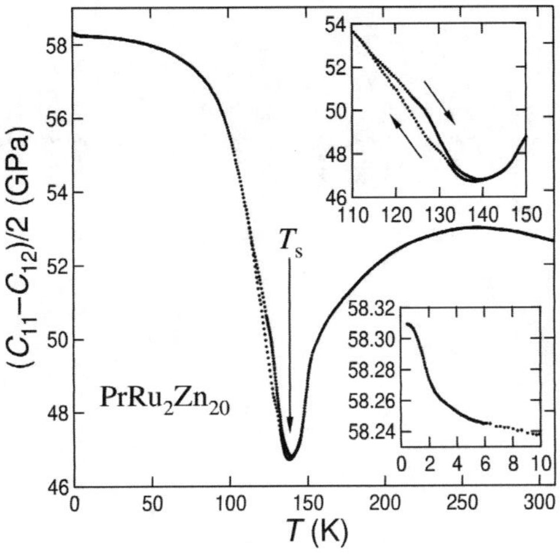

Figure 1. T dependence of transverse elastic modulus $(C_{11} - C_{12})/2$. The upper inset represents the same data in an expanded scale between 110 and 150 K. The lower inset also shows the same data between 0.4 and 10 K.

We discuss the origin of phase transition at T_s in $PrRu_2Zn_{20}$. The modulus $(C_{11} - C_{12})/2$ shows abrupt elastic hardening below T_s. On the other hand, a similar phase transition is reported at 150 K in $LaRu_2Zn_{20}$ [6]. In our preliminary ultrasonic measurements, elastic modulus in $LaRu_2Zn_{20}$ also shows the hardening below T_s without softening. These suggest that the phase transition at T_s in $PrRu_2Zn_{20}$ is not caused by $4f$-electrons of Pr ions. Onimaru *et al.* reported that the phase transition at T_s is accompanied by structural modulation. A possible candidate of the origin of phase transition is the band Jahn-Teller effect. To clarify the origin of the phase transition in $PrRu_2Zn_{20}$ and $LaRu_2Zn_{20}$, a band calculation to examine the existence of degenerated band structure at the Fermi energy is necessary.

As for the softening below 250 K, it would be caused by a strain-quadrupole and quadrupole-quadrupole interaction. We carried out the theoretical fitting based on the Curie-Weiss-like

Table 1. Fitting parameters of $(C_{11} - C_{12})/2$.

Θ (K)	$T_c - \Theta$ (K)	a (GPa)	b ($\times 10^{-2}$ GPa/K)
−1.5	41.2	74.6	−4.5

formula for the co-operative Jahn-Teller effect [10]:

$$C(T) = C_0 \frac{T - T_c}{T - \Theta} \ , \tag{1}$$

where Θ is proportional to the quadrupole-quadrupole coupling constant, $T_c - \Theta$ is a measure of Jahn-Teller energy including the strain-quadrupole coupling constant and C_0 is a back ground stiffness assumed as linear T dependence:

$$C_0 = a + bT \ . \tag{2}$$

As shown in Fig. 2, $(C_{11} - C_{12})/2$ is well reproduced above 150 K with fitting parameters listed in Table 1. The negative value of Θ suggests that there are the antiferro-quadrupolar interaction

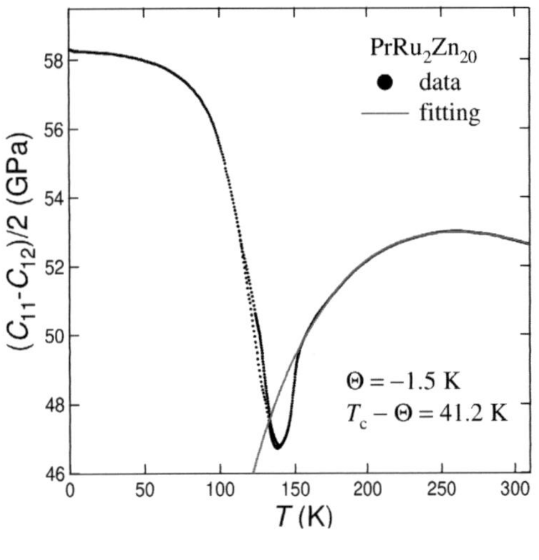

Figure 2. T dependence of $(C_{11} - C_{12})/2$. Solid curve is the result calculated with the Curie-Weiss-like formula for the co-operative Jahn-Teller effect.

among quadrupole moments at $8a$ sites. Since $(C_{11} - C_{12})/2$ shows a discontinuous change in the slope at 150 K, the theoretical curve deviates from the data. In $PrRu_2Zn_{20}$, however, no anomaly at 150 K is reported in other measurements [6]. It would be necessary to carry out further detailed measurements of physical properties, such as specific heat, because there is a possibility to be a phase transition at 150 K.

Hereafter, we discuss the elastic behaviors at low temperatures. The modulus $(C_{11} - C_{12})/2$ shows elastic hardening below 3 K and leveling off below 1 K as shown in the lower inset in Fig. 1. No anomaly has been reported around these temperatures in other measurements [6]. In cage compounds, the filled skutterudite compounds $R\mathrm{Os_4Sb_{12}}$ [1, 2] and $R\mathrm{Fe_4Sb_{12}}$ [3, 4] show elastic hardening at low temperatures, and the temperature where the hardening occurs depends on ultrasonic frequency. This ultrasonic frequency dependence of elastic modulus in cage compounds is due to the rattling motion. The modulus $(C_{11} - C_{12})/2$ in $PrRu_2Zn_{20}$ shows the characteristic hardening below 3 K. There is a possibility that it originates from the rattling motion of Pr ions. To examine whether the hardening below 3 K is ultrasonic dispersion, ultrasonic attenuation and elastic modulus measurements using the overtone of the fundamental resonance frequencies of transducers are requisite.

4. Conclusion

We have performed ultrasonic experiments on the cage compound $PrRu_2Zn_{20}$. The elastic softening is observed below 250 K in $(C_{11}-C_{12})/2$, and it exhibits a discontinuous change in the slope at 150 K. The modulus $(C_{11}-C_{12})/2$ turns into increase suddenly below T_s accompanied by hysteresis, suggesting the first-order phase transition. The softening below 250 K is caused by the strain-quadrupole and quadrupole-quadrupole interaction. The theoretical curve using the Curie-Weiss-like formula for the co-operative Jahn-Teller effect deviates from the data below 150 K, suggesting a possibility to be a phase transition at 150 K. The elastic hardening below 3 K could be ultrasonic dispersion due to the rattling motion.

5. Acknowledgments

This work was supported by a Grant-in-Aid (No. 20340093) and a Grant-in-Aid for Scientific Research on Innovative Areas "Heavy Electrons" (No. 20102005) from the Ministry of Education, Culture, Sports, Science and Technology of Japan.

References

[1] Goto T, Nemoto Y, Sakai K, Yamaguchi T, Akatsu M, Yanagisawa T, Hazama H, Onuki K, Sugawara H and Sato H 2004 *Phys. Rev. B* **69** 180511(R)

[2] Nemoto Y, Yanagisawa T, Yasumoto Y, Kobayashi H, Yamaguchi A, Tsuduku S, Goto T, Takeda N, Ochiai A, Sugawara H, Sato H and Kitazawa H 2008 *J. Phys. Soc. Jpn.* **77** Suppl. A, 153

[3] Ishii I, Fujita T, Mori I, Sugawara H, Yoshizawa M, Takegahara K and Suzuki T 2009 *J. Phys. Soc. Jpn.* **78** 084601

[4] Ishii I, Fujita T, Mori I, Sugawara H, Yoshizawa M and Suzuki T 2008 *J. Phys. Soc. Jpn.* **77** Suppl. A, 303

[5] Nasch T, Jeitschko W and Rodewald U C 1997 *Z. Naturforschung* **52B** 1023

[6] Onimaru T, Matsumoto K T, Inoue Y F, Umeo K, Saiga Y, Matsushita Y, Tamura R, Nishimoto K, Ishii I, Suzuki T and Takabatake T 2010 *J. Phys. Soc. Jpn.* **79** 033704

[7] Saiga Y, Matsubayashi K, Fujiwara T, Kosaka M, Katano S, Hedo M, Matsumoto T and Uwatoko Y 2008 *J. Phys. Soc. Jpn.* **77** 053710

[8] Moran T J and Lüthi B 1969 *Phys. Rev.* **187** 710

[9] Suzuki T 1988 Ph.D. thesis Tohoku University

[10] Lüthi B 2005 *Physical Acoustics in the Solid State* (Verlag Berlin Heidelberg: Springer) p 121

Journal of Physics: Conference Series **273** (2011) 012137 doi:10.1088/1742-6596/273/1/012137

Spontaneous strain in ferroquadrupolar phase of $TmAu_2$

M Kosaka[1], Y Shida[1], A Imanari[1], S. Katano[1], K. Ohoyama[2] and H. Onodera[3]

[1] Graduate School of Science and Engineering, Saitama University, Saitama 338-8570, Japan
[2] Institute for Materials Research, Tohoku University, Sendai 980-8577, Japan
[3] Department of Physics, Tohoku University, Sendai 980-8578, Japan

E-mail: kosaka@phy.saitama-u.ac.jp

Abstract. We present the results of a study of the spontaneous strain accompanied by a ferroquadrupolar ordering in $TmAu_2$. The strain has been investigated by means of thermal expansion, magnetostriction, and neutron diffraction measurements on a single crystalline $TmAu_2$. In the ferroquadrupolar ordered phase, the strain shows a strong enhancement by applying magnetic field. This implies that structural orthorhombic domains caused by the ferroquadrupolar ordering are transformed into a single-domain structure by a magnetic field.

1. Introduction

It is widely recognized that quadrupolar interactions play an important role in the f-electron system. When the system has a degeneracy in the crystalline electric field (CEF) ground state associated with the quadrupole moment, and the quadrupolar interaction is stronger than magnetic interactions, a quadrupolar ordering can occur in a paramagnetic phase. The type of the quadrupolar ordering is divided into two categories of a ferroquadrupolar and an antiferroquadrupolar ordering, according to the sign of its quadrupolar interaction. In the case of the ferroquadrupolar ordering, a structural phase transition occurs simultaneously, which causes a spontaneous strain in ferroquadrupolar ordered phase.

The binary thulium compound $TmAu_2$ crystallizes in the tetragonal $MoSi_2$-type structure, which shows the ferroquadrupolar and an antiferromagnetic ordering at $T_Q = 7.0$ K and $T_N = 3.2$ K, respectively [1]. At T_Q, the structural phase transition from a tetragonal to an orthorhombic phase occurs. The strain caused by the ferroquadrupolar ordering is an orthorhombic γ-symmetry-lowering mode, and then the ordering parameter has been determined to be $\langle O_2^2 \rangle$.

In this paper, we present the results of thermal expansion, magnetostriction, and neutron diffraction measurements using a single crystalline $TmAu_2$.

2. Experimental

A single crystal was prepared using the tetra-arc Czochralski method. The detail of the sample preparation was reported in our previous work [1]. Used as sample is $TmAu_2$ in the form of rectangular parallelepiped with the size of $3.20 \times 3.25 \times 3.65$ mm^3. Thermal expansion measurements were performed using the sensitive three-terminal capacitance method. The capacitive dilatometer for these measurements has been constructed with reference to the design

Published under licence by IOP Publishing Ltd

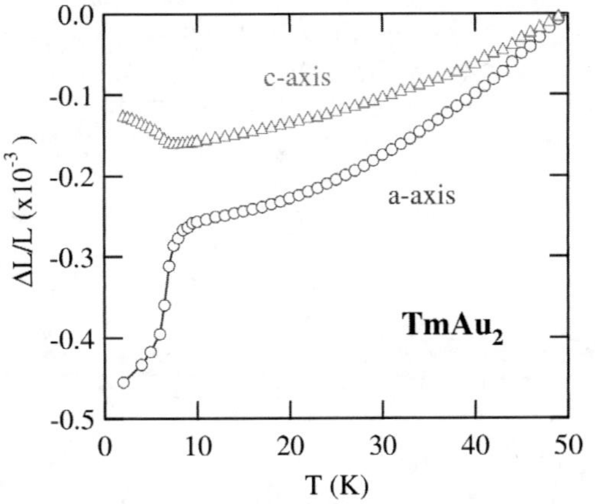

Figure 1. Temperature dependence of the thermal expansion of $TmAu_2$ along the a and c axes under zero magnetic field.

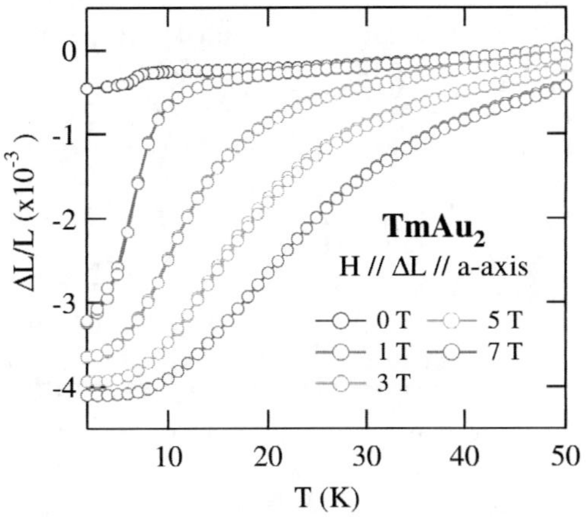

Figure 2. Temperature dependence of the thermal expansion of $TmAu_2$ along the a axis at various magnetic fields.

of the apparatus proposed by Schmiedeshoff *et al.* [2]. We have installed the dilatometer in a commercial Quantum Design (QD) physical property measurement system (PPMS). The neutron diffraction experiments were carried out using the IMR single-crystal diffractometer installed at the thermal neutron guide of the JRR-3M reactor at the Japan Atomic Energy Agency (JAEA). The diffraction patterns were recorded using neutrons of wave length 1.527 Å.

3. Results and Discussion

All thermal expansion data in this paper are normalized at 50 K under zero magnetic field. Thermal expansion measurements were performed in the process of decreasing and increasing temperature as well as magnetic field. A distinguishable hysteresis has not been observed in our measurements. Figure 1 shows the temperature dependence of the thermal expansion $\Delta L/L$ along the a and c axes at zero field. The both thermal expansion curves decrease monotonically with decreasing temperature, and show anomalies around $T_Q = 7.0$ K. The $\Delta L/L$ along the c axis turns to increasing below T_Q, whereas the $\Delta L/L$ along the a axis displays a large drop. It is well known that the crystal structure of $TmAu_2$ transforms from the paramagnetic tetragonal phase to the ferroquadrupolar ordered orthorhombic phase. The observed large change in $\Delta L/L$ along the a is considered to be in qualitative agreement with this fact. The value of the change in thermal expansion along the a axis around T_Q is estimated to be about 0.1×10^{-3}.

Figure 2 shows the temperature dependence of the $\Delta L/L$ along the a axis under several magnetic fields. Notice that, the anomaly corresponding to the ferroquadrupolar ordering develops extremely when a 1 T magnetic field is applied, the change in $\Delta L/L$ is estimated to be 2.65×10^{-3}. This behavior is considered to be associated with a formation of the orthorhombic domain structure. In this case, many orthorhombic domains are expected to occur at the structural phase transition. We previously reported that the orthorhombic distortion $2(a-b)/(a+b)$ increased gradually with lowering temperature, reaching 2.7×10^{-3} at 4.2 K [3]. This value was estimated from the temperature dependence of the lattice parameter a, which was determined by neutron powder diffraction. The magnitude of the change in $\Delta L/L$ at 1 T is almost the same as this distortion ratio. Neutron powder diffraction measurements, which are microscopic experiments, are not affected by the existence of structural domains. Therefore, the spontaneous strain at zero magnetic field must be averaged over these domains. These domains are considered to be modified into a single-domain via the coupling between the quadrupole moment and the magnetic moment induced by a magnetic field. In addition, the anomaly demonstrates a tendency to shift toward higher temperatures and to broaden with increasing magnetic field. This behavior is similar to that of the temperature dependence of the specific heat. By applying magnetic fields, the broadening of the specific heat anomaly was observed around T_Q and the beginning of the ferroquadrupolar ordering appeared to move toward higher temperatures [1].

Shown in Fig. 3 is the isothermal magnetostriction along the a and c axes at several temperatures. The initial slope of the magnetostriction along the a axis indicated with circles is drastically altered between 10 K and 5 K. On the other hand, the data along the c axis are almost constant even at 2 K. These results also indicate that the realignment of the orthorhombic multi-domain structure by applying magnetic fields.

In ordered to characterize the multi-domain structure in the ferroquadrupolar ordered phase, we have measured the splits of nuclear reflections by neutron diffraction on a single crystalline $TmAu_2$. Figure 4 and 5 show the contour map of (200) and (110) nuclear reflections at 5.2 K on the reciprocal $a^* - a^*$ plane, respectively. The indexes of h and k are determined based on these nuclear reflections in the paramagnetic phase. Below T_Q, the (200) reflection splits into four peaks, while the (110) reflection is divided into three peaks. To distribute internal stress caused by the orthorhombic distortion, the high temperature tetragonal phase appear to be converted into an assemblage of orthorhombic domains with domain walls oriented in the (110) or $(1\bar{1}0)$ planes. This plays a role in lowering the elastic energy at the structural phase transition. Switching the a and the b axes in every small domain is advantageous for relieving stress. Such situations are schematically illustrated in Fig. 6, in which the broken lines indicate the tetragonal lattice before the structural phase transition occurs. The inset shows the real lattice on the $a - a$ plane, the solid lines express the switching behavior of the orthorhombic axes with the boundary of the (110) plane. Assuming the situation described above, there should be four kinds of orthorhombic domains. The crystallographic axes of domain individuals are slightly

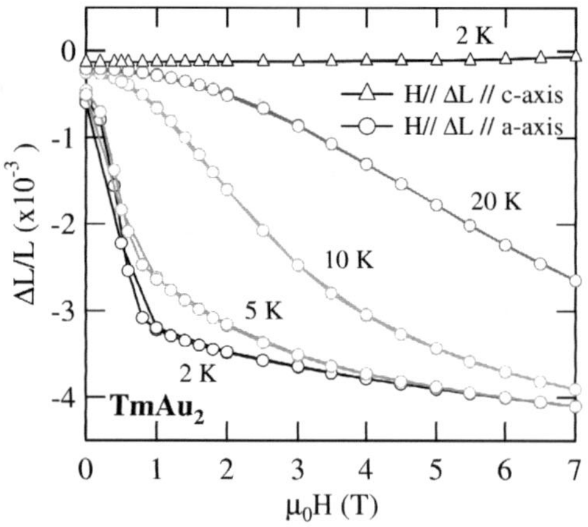

Figure 3. Magnetostriction of $TmAu_2$ along the a and c axes at various temperatures in magnetic fields applied along the same direction of $\Delta L/L$ measurements.

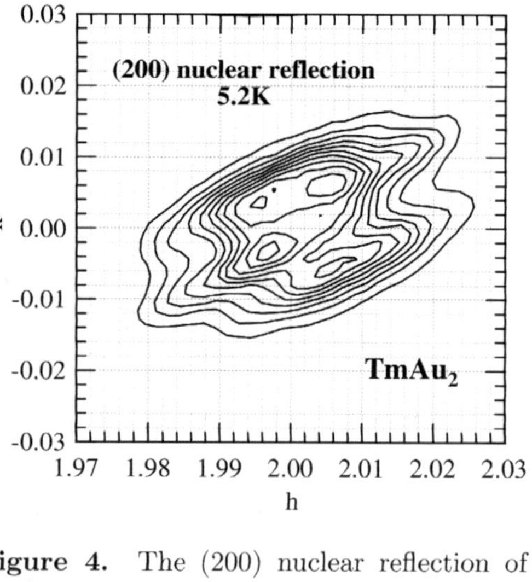

Figure 4. The (200) nuclear reflection of $TmAu_2$ at 5.2 K.

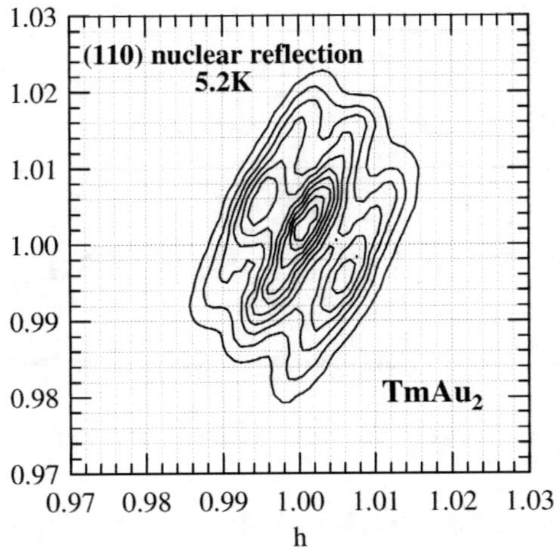

Figure 5. The (110) nuclear reflection of $TmAu_2$ at 5.2 K.

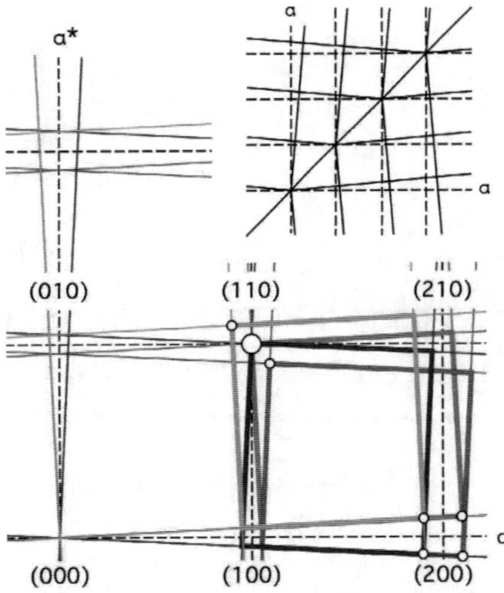

Figure 6. Schematic geometry of four orthorhombic domains in the ferroquadrupolar ordered phase on the reciprocal $a^* - a^*$ plane.

inclined from 90° each other due to the small difference between the length of a and b axes. The reciprocal lattice in Fig. 6 represents a schematic geometry of four individual orthorhombic domains. An instructive example was observed in high-T_C cuprate reported by Sueno *et al.* [4] and Saito *et al.* [5]. The origin of the similar nuclear reflection splittings was discussed in Ref. 4. They found that the twinning, which means the orthorhombic domain, occurs at a structural transition from a tetragonal phase in cooling process.

4. Conclusion

We have found that orthorhombic domains can be controlled by magnetic fields via the coupling between the quadrupole moment and the magnetic moment induced by a magnetic field. This technique makes it possible to compare the magnitude of the change in lattice constant and thermal expansion accompanied by the ferroquadrupolar ordering. The single-crystal neutron diffraction measurement reveals the existence of the multi-domain structure and its strain.

Acknowledgments

This work was partially supported by a Grant-in-Aid for Research (No. 20540344) from the Ministry of Education, Culture, Sports, Science and Technology, Japan.

References

[1] Kosaka M, Onodera H, Ohoyama K, Ohashi M, Yamaguchi Y, Nakamura S, Goto T, Kobayashi H and Ikeda S 1998 *Phys. Rev.* B **58** 6339
[2] Schmiedeshoff G M, Lounsbury A W, Luna D J, Tracy S J, Schramm A J, Tozer S W, Correa V F, Hannahs S T, Murphy T P, Palm E C, Lacerda A H, Bud'ko S L, Canfield P C, Smith J L, Lashley J C and Cooley J C 2006 *Rev. Sci. Instrum.* **77** 123907.
[3] Kamiyama T, Oikawa K, Izumi F, Kosaka M, Onodera H, Yamaguchi Y, Kasaya M and Kojima K 1997 *Physica* B **241** 376
[4] Sueno S, Nakai I, Okamura F P and Ono A 1987 *Jpn. J. Appl. Phys.* **26** L842
[5] Saito K, Nissen H U, Beeli C, Wolf T and Schauer W 1998 *Phys. Status Solidi* A **166** 861

First-order-like antiferromagnetic transition in rare-earth palladium bronze $SmPd_3S_4$

E Matsuoka[1], M Watahiki[2], M Sakoda[1], H Sugawara[1], T Sakurai[3], H Ohta[4] and H Onodera[2]

[1]Department of Physics, Graduate School of Science, Kobe University, Kobe, 657-8501, Japan

[2]Department of Physics, Graduate School of Science, Tohoku University, Sendai, 980-8578, Japan

[3]Center for Supports to Research and Education Activities, Kobe University, Kobe, 657-8501, Japan

[4]Molecular Photoscience Research Center, Kobe University, Kobe, 657-8501, Japan

E-mail: matsuoka@crystal.kobe-u.ac.jp

Abstract. Specific-heat and magnetic susceptibility measurements were performed on the single-crystalline and polycrystalline samples of $SmPd_3S_4$. The magnetic susceptibility does not exhibit a cusp at the Neél temperature T_N, which is commonly observed in antiferromagnets, but a first-order like discontinuous decrease below T_N. One of the possible origins of a first-order like transition is a simultaneous occurrence of a magnetic-dipolar and a higher multipolar orders.

1. Introduction

Rare-earth (R) palladium bronzes RPd_3S_4 crystallize into a cubic $NaPt_3O_4$-type crystal structure [1]. The R atoms form a body-centered cube and are subjected to the crystalline electric field (CEF) represented by the cubic point group T_h [1-3]. Systematic studies along the R series using polycrystalline samples have revealed that the CEF ground states of R^{3+} in RPd_3S_4 possess a multipolar degree of freedom except for R = Tm [2-4]. To clarify the effects of multipolar interaction in the RPd_3S_4 system more systematically, we have synthesized single-crystalline samples of R = Ce [5, 6], Pr [7-9], Ho [10] and Er [10] and reported various multipole orders in these compounds. In this study, the magnetic properties of single-crystalline $SmPd_3S_4$ were examined by the measurements of specific heat $C(T)$ and magnetic susceptibility M/B. The study using polycrystalline samples has reported that $SmPd_3S_4$ shows AFM ordering with the Neél temperature of 2.5 K and a nonmagnetic phase transition at 3.0 K [2].

2. Experimental method

Published under licence by IOP Publishing Ltd

Polycrystalline and single-crystalline samples of $SmPd_3S_4$ were synthesized by similar method as that reported for $PrPd_3S_4$ [7]. Single-crystalline samples of $SmPd_3S_4$ were cubic or plate-like with typical sizes and masses of about 1 mm^3 and 1 mg, respectively. The specific heat was measured by a relaxation method (Oxford Instruments, $MagLab^{HC}$) between 0.5 and 50 K in magnetic fields up to 9 T. The magnetization and magnetic susceptibility was measured using a superconducting quantum interference device (SQUID) magnetometer (Quantum Design, MPMS) between 1.8 and 300 K in magnetic fields of up to 5 T. The inset of figure 1 shows the magnetic-field dependence of the magnetization *M*(*B*) of $SmPd_3S_4$ measured at 1.8 K. For the measurement of *M*(*B*), the polycrystalline sample of 112.9 mg was used. The *M*(*B*) value of $SmPd_3S_4$ is very small: *M*(*B*) is 0.45 emu/g (0.048 $_B$/Sm) even at 5 T. This fact means that the measurement of magnetic susceptibility of single-crystalline samples is difficult since the magnetization value of 1 mg of $SmPd_3S_4$ is comparable to that of sample holder even at low-temperature and high magnetic-field condition. Therefore, the polycrystalline sample of 112.9 mg was used also for the measurement of magnetic susceptibility *M*/*B*.

3. Results and Discussion

Figure 2(a) shows the temperature dependences of the specific heat *C*(*T*) for the single-crystalline sample measured in the magnetic fields of 0 T and 9 T applied along the three principal axes. At T_N = 2.46 K, *C*(*T*) in $B = 0$ exhibits a sharp peak, that corresponds to the AFM transition reported in the previous study using polycrystalline samples [2]. Although the previous study reported the existence of a nonmagnetic phase transition at 3.0 K, the present *C*(*T*) shows no anomaly at 3.0 K. The magnetic entropy $S_{mag}(T)$ shown in figure 2(b) was calculated by integrating C_{mag}/T, where C_{mag} is the magnetic specific heat deduced by subtracting the specific heat of $LaPd_3S_4$ as a nonmagnetic contribution from that of $SmPd_3S_4$. The $S_{mag}(T)$ value exceeds *R*ln2 above 2.4 K and reaches 64% of *R*ln4 at T_N, implying that Sm ions in $SmPd_3S_4$ are trivalent and the CEF ground state of the *J*-multiplet with $J = 5/2$ is the Γ_8 quartet as reported previously [2]. It is noteworthy that the values of *M*/*B* for polycrystalline sample above 50 K shown in figure 1 are very close to those of the cubic compound SmBiPt whose AFM transition temperature is 2.2 K [11]. It was reported that the valence state of Sm ions in SmBiPt is determined to be trivalent by considering the result of quantitative analysis of magnetic susceptibility based on a second-perturbation theory [11]. The similarity between the *M*/*B* values of $SmPd_3S_4$ and those of SmBiPt above 50 K supports the trivalent state of Sm ions in $SmPd_3S_4$.

The magnetic-field dependence of T_N and that of the peak intensity of *C*(*T*) are considerably small in contrast to the remarkable field dependences of those observed in R = Ce [6], Pr [7, 8], and Dy [12]. T_N decreases slightly with increasing magnetic field, but the decreasing width is less than 0.15 K even at 9 T: T_N at 9 T are 2.37, 2.33, and 2.35 K along the [100], [110], and [111] directions, respectively. Small field dependence of T_N for $SmPd_3S_4$ is due to the small value of magnetization. For example, the magnetization value of $SmPd_3S_4$ at 1.8 K measured in the applied magnetic field of 5 T is about 1/20 of that of $CePd_3S_4$.

Figure 3 shows the temperature dependence of the magnetic susceptibility *M*/*B* measured at 1 T and the derivative of *M*/*B*. In figure 3, *C*(*T*) in the magnetic field of 1 T applied along the [100] direction is also depicted as dotted line for comparison. *M*/*B* does not exhibit a cusp at T_N, which is commonly observed in antiferromagnets, but a first-order like discontinuous decrease around T_N. In fact, T_N at 1 T that is defined by the peak temperature of *C*(*T*) (2.46 K) does not agree with the peak temperature of *M*/*B* (2.6 K) but with the peak temperature of the derivative of *M*/*B* as shown in figure 3. One of the possible origins for the appearance of the first-order-like AFM transition in $SmPd_3S_4$ may be the presence of a quadrupolar interaction as observed for TbP [13]: a quadrupolar interaction plays a dominant role for the appearance of the first-order-like AFM transition in TbP [13]. Another possible origin may be the simultaneous occurrence of a magnetic-dipolar and a higher multipolar orders as observed in $CePd_3S_4$ [6] and the tetragonal TbB_2C_2 [14]. TbB_2C_2 shows a first-order like AFM transition at the Neél temperature of 21.7 K [13]. In the case of TbB_2C_2, an AFM and octupolar orders occur simultaneously at 21.7 K and these orders can be separated by the partial substitution of Tb by

Gd [13]. If the simultaneous ordering of two kinds of order parameter occurs also in $SmPd_3S_4$, we can expect that two ordering states are separated by the partial substitution of Sm by Y or Gd as for the case of TbB_2C_2. To examine the possibility of a simultaneous transition, we are planning to synthesize substituted compounds $Sm_{1-x}Y_xPd_3S_4$ and $Sm_{1-x}Gd_xPd_3S_4$.

This work was supported by a Grant-in-Aid for Scientific Research on Innovative Areas "Heavy Electrons"(No. 21102503), Young Scientists (B) (No. 21740238) from the Ministry of Education, Culture, Sports, Science, and Technology of Japan, and Creative Scientific Research (No. 19GS1209) from Japan Society for the Promotion of Science.

References

[1] Keszler D A and Ibers J A 1983 *Inorg. Chem.* **22** 3366
[2] Abe K, Kitagawa J, Takeda N and Ishikawa M 1999 *Phys. Rev. Lett.* **83** 5366
[3] Wakeshima M and Hinatsu Y 1999 *J. Solid State Chem.* **146** 226
[4] Kitagawa J, Matsuoka E, Takeda N and Ishikawa M 1999 *Phys. Rev. B* **60** R15028.
[5] Matsuoka E, Usui D and Onodera H 2008 *Physica B* **403** 1603
[6] Matsuoka E, Usui D, Sasaki Y, Nakao H, Shida H, Ohoyama K and Onodera H 2008 *J. Phys. Soc. Jpn.* **77** 114706
[7] Matsuoka E, Usui D, Tanida H, Nakamura S, Nojima T and Onodera H 2007 *J. Phys. Soc. Jpn.* **76** 073707.
[8] Usui D, Matsuoka E, and Onodera H 2008 *Physica B* **403** 1605
[9] Matsuoka E, Usui D, Sasaki Y, Watahiki M, Iwasa K, Shida H, Ohoyama K and Onodera H 2010 *J. Phys. Soc. Jpn.* **79** 064708.
[10] Matsuoka E, Usui D, Nakamura S, Nojima T, Takagi S and Onodera H 2009 *Physica B* **404** 3297
[11] Kim M S, Sasakawa T, Suemitsu T, Bando Y, Umeo K, Iga F, Mitamura H, Goto T and Takabatake T 2001 *J. Phys. Soc. Jpn.* **70** 3650
[12] Matsuoka E, Kitagawa J, Ohoyama K, Yoshizawa H and Ishikawa M 2001 *J. Phys.: Condens. Matter* **13** 11009
[13] Morin P and Schmitt D, *Ferromagnetic Materials*, edited by K. H. J. Buschow and E. P. Wohlfarth (North-Holland, Amsterdam, 1990), Vol. 5, p. 47.
[14] Matsuoka E, Sasaki Y, Usui D, Nakagawa M and Onodera H 2006 *J. Phys. Soc. Jpn.* **75** 123707

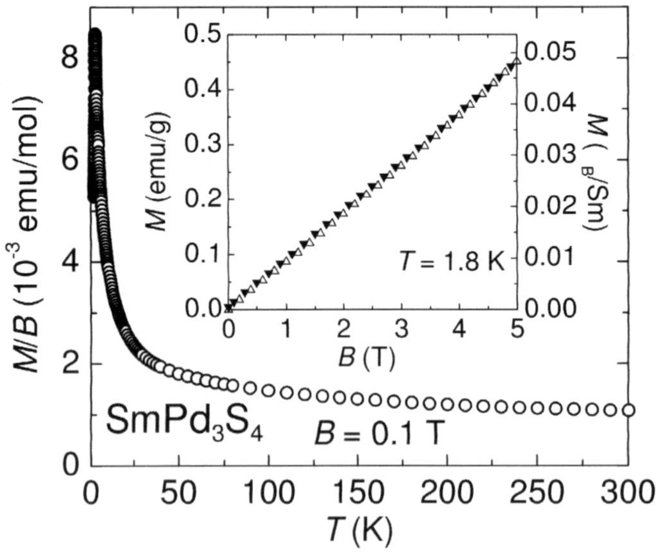

Figure 1. Temperature dependence of the magnetic susceptibility *M*/*B* of $SmPd_3S_4$. The inset shows the magnetic-field dependence of the magnetization *M*(*B*). The open and solid triangles in the inset represent the values measured with increasing and decreasing fields, respectively.

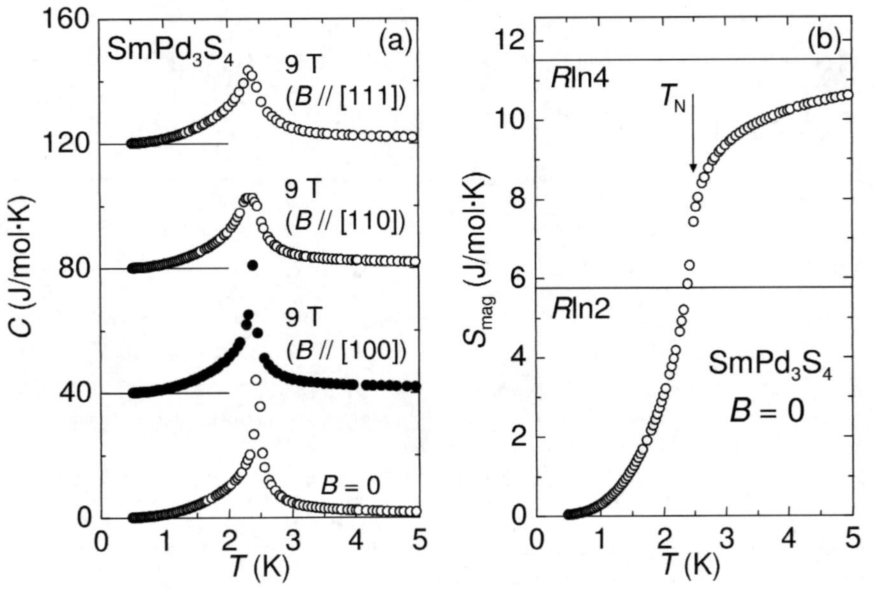

Figure 2. (a) Temperature dependence of the specific heat *C*(*T*) of $SmPd_3S_4$. (b) Temperature dependence of the magnetic entropy $S_{mag}(T)$ of $SmPd_3S_4$.

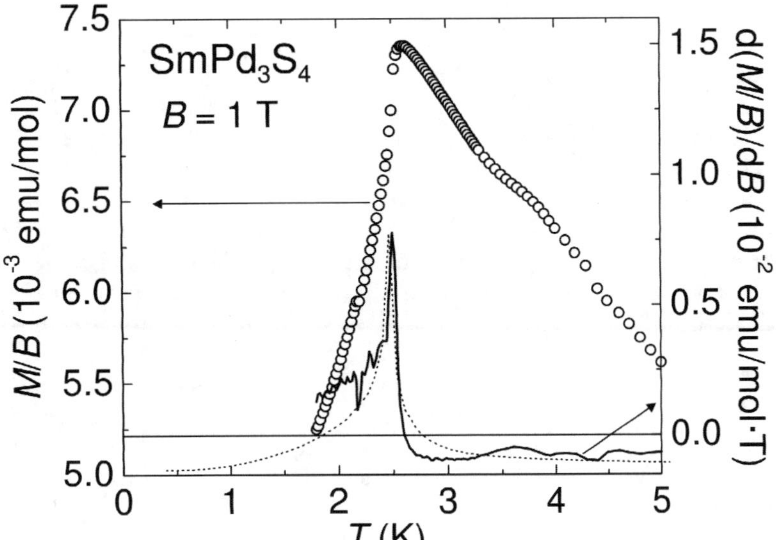

Figure 3. Temperature dependence of the magnetic susceptibility *M*/*B* and the derivative of *M*/*B* of $SmPd_3S_4$. The unit of vertical scale of the specific heat at 1 T denoted by the dotted line is arbitrary one.

NMR studies of metal-insulator transition in the spinel-type $Cu(Ir_{1-x}V_x)_2S_4$

H Niki[1], H Okuda[1], Y Okada[1], K Higa[1], M Oshiro[1], N Fukuyoshi[1], R Mahoe[1], M Yogi[1], T Nakama[1], K Yagasaki[1], S Ebisu[2] and S Nagata[2]

[1] Department of Physics, Faculty of Science, University of the Ryukyus, Nishihara, Okinawa 903-0213, Japan

[2] Muroran Institute of Materials Research, Muroran, Hokkaido 980-8577, Japan

E-mail: niki@sci.u-ryukyu.ac.jp

Abstract. In order to investigate the physical properties of a metal-insulator transition (MIT) on Ir rich side in $Cu(Ir_{1\;x}V_x)_2S_4$ from a microscopic point of view, ^{63}Cu NMR measurements for $Cu(Ir_{1\;x}V_x)_2S_4$ (x = 0.00, 0.03 and 0.05) have been carried out from 4.2 to 300 K. The temperature dependences of the line width, Knight shift and spin-lattice relaxation time (T_1) have been measured for these samples. A sudden decreases of the $1/T_1T$ and the negative contribution to the Knight shift from the core polarization of Cu conduction electrons for x = 0.00 and 0.03 samples show clearly the existence of the MIT at T_{MI} = 226 and 175 K, respectively, being attributed to the opening of a band gap below T_{MI}. However, the temperature dependences of the $1/T_1T$ and the Knight shift for x = 0.05 sample change continuously as the band gap is remarkably suppressed below T_{MI}. The density of states at Fermi level in the insulating phase is reduced to about 20 % of that in the metallic phase.

The spinel $CuIr_2S_4$ exhibits a temperature-induced metal-insulator transition (MIT) around T_{MI} = 226 K with a structural transformation from cubic to triclinic [1-3]. Simultaneously, at this MIT, a charge-ordering by Ir^{3+} and Ir^{4+} and a bond dimerization with spin-singlet state by Ir^{4+} take place [4-6]. On the other hand, CuV_2S_4 exhibits two-step anomalies at about 92 and 56 K, reflecting the CDW state [7]. The MIT and the CDW states are incompatible with each other. Therefore, a systematic study was made to investigate the concentration dependences of structural, electrical and magnetic properties for spinel-type $Cu(Ir_{1-x}V_x)_2S_4$ [8]. According to these results [5], a low substitution (x 0.05) of V for Ir leads to disappearance of the step-like sharp MIT, and similarly, a low substitution (x 0.95) of Ir for V destroys the CDW transition. It was found that both transitions are very sensitive for the change of x in the vicinity of both x = 0.00 and 1.00. In order to investigate the physical properties of the MIT on Ir rich side in $Cu(Ir_{1-x}V_x)_2S_4$ from a microscopic point of view, ^{63}Cu NMR measurements for $Cu(Ir_{1-x}V_x)_2S_4$ (x = 0.00, 0.03 and 0.05) have been carried out. The temperature dependences of the line width, Knight shift and spin-lattice relaxation time (T_1) have been measured for these samples.

The polycrystalline specimens were prepared by a solid state reaction method [8]. Mixtures of high-purity fine powders of Cu, Ir, V and S with normal stoichiometry were calcined in sealed quartz tubes at 850 °C for 10 days. Subsequently, the specimens were reground and pressed to form the pellets, and then were sintered at 850 °C for 2 days. The powdered specimens were prepared for NMR measurements. No secondary phase in the specimens was detected in the powder x-ray diffraction profile. A conventional pulse NMR spectrometer was utilized for

Published under licence by IOP Publishing Ltd

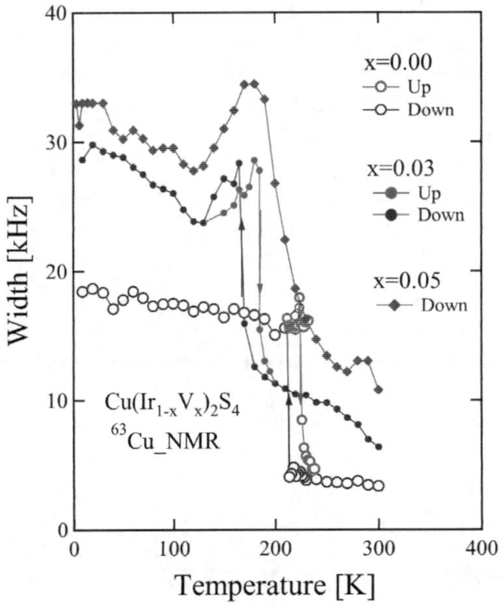

Figure 1. Temperature dependence of the line width on ^{63}Cu NMR in $Cu(Ir_{1-x}V_x)_2S_4$ ($x = 0.00$, 0.03 and 0.05).

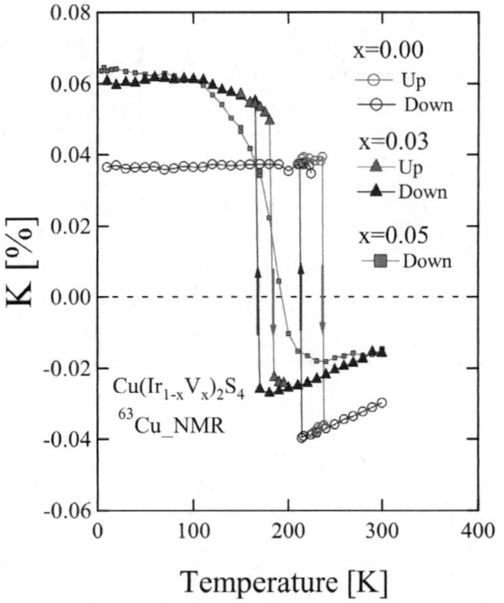

Figure 2. Temperature dependence of the Knight shift (K) on ^{63}Cu NMR in $Cu(Ir_{1-x}V_x)_2S_4$ ($x = 0.00$, 0.03 and 0.05).

the measurements of ^{63}Cu NMR. A magnetic field was applied at about 6.9 T (about 78 MHz) by a superconducting magnet with the magnetic-field homogeneity of 10^{-5}. ^{63}Cu-NMR shift was obtained with gyro-magnetic ratio of 11.285 MHz/T. The line width and peak shift of the spectrum were derived from Fourier transform of echo envelope. Spin-lattice relaxation time T_1 was measured by means of the saturation recovery methods.

The ^{63}Cu NMR in $Cu(Ir_{1-x}V_x)_2S_4$ ($x = 0.00$, 0.03 and 0.05) has measured in the temperature range between 4.2 and 300 K. A Lorentzian line has been fitted for each spectrum in the measuring temperature region. Below T_{MI}, the line width of the spectrum is broadened due to the NQR interaction and anisotropy of Knight shift caused by the structural change from cubic to triclinic symmetry. However, since both the NQR interaction and anisotropy of the Knight shift are weak, the spectrum can be analyzed as a Lorentzian line. The Cu NMR spectrum is narrower than that obtained by Sasaki et al [9] and our spectrum shape is different from that of them, which will be discussed elsewhere.

Figure 1 represents the temperature dependence of the line widths (half width of half maximum) on ^{63}Cu NMR between 4.2 and 300 K. For $x = 0.00$ sample, the line width is 3.4 kHz at 300 K and almost constant down to T_{MI}, while it shows an abrupt increase at T_{MI} (226 K) with a temperature-hysteresis. The line width is 15.1 kHz at 200 K below T_{MI} and then slightly increases with decreasing temperature. For $x = 0.03$ sample, the line width is 6.4 kHz at 300 K and then increases about twice down to T_{MI}. It abruptly increases at T_{MI} (175 K) with a temperature-hysteresis. The width becomes 23.8 kHz at 130 K below T_{MI}, and then monotonously increases with decreasing temperature. For $x = 0.05$ sample, the line width is 10.8 kHz at 300 K, increases steeper with decreasing temperature, reaches the maximum (34.5 KHz) at 180 K without the temperature hysteresis, then goes through the minimum (27.8 KHz) at 120K, and monotonously increases with decreasing temperature.

An increase of the width with increasing x, as shown in Fig. 1, is attributed to inhomogeneous broadening due to the distribution of the local field on each nuclei caused by the substitution of V for Ir. For all samples, the line widths in the metallic phase are narrow since there is no influence of the anisotropic Knight shift and nuclear quadrupole interactions because of the cubic crystal structure. To the contrary, in the insulating phase they are wider than those in the metallic phase due to the influence of the anisotropic Knight shift and nuclear quadrupole interactions owning to the triclinic crystal structure. In the $x = 0.00$ sample, NMR spectra of the metallic and insulating phases coexist during MIT, while for the $x = 0.03$ sample, the coexistence of the spectra of the metallic and insulating phases cannot be confirmed because of the wider line widths than those of the $x = 0.00$ sample.

Figure 2 shows the temperature dependence of the Knight shift (K) for ^{63}Cu NMR in $Cu(Ir_{1-x}V_x)_2S_4$ between 4.2 and 300 K. For all samples, the Knight shift in the metallic phase above T_{MI} indicates the negative contribution to the Knight shift from the core polarization of d conduction electrons. The Knight shift for $x = 0.00$ and 0.03 samples represents a sudden change to a positive side by about 0.08 % at T_{MI} with a temperature-hysteresis. For $x =$ 0.05 sample, however, it continuously changes to the positive side by about 0.08 % without a temperature-hysteresis from 200 to 100 K. These results show that the negative contribution of the Knight shift from the core polarization of the d electrons disappears due to the opening of a band gap below T_{MI}. In the insulating phase, the Knight shift for all samples is positive shift, which is mainly attributed to the orbital mechanism. The absence of the apparent temperature dependence of the Knight shift below T_{MI} indicates that the electronic state is basically nonmagnetic in the insulating phase. This means that both Cu and Ir atoms are in nonmagnetic state in the insulating phase.

Spin-lattice relaxation time T_1 has been measured from 7 to 300 K. The nuclear longitudinal magnetization recovery $f(t) = 1 - (M(t)/M_0)$ cannot be expressed by a single exponential type. The recovery $f(t)$ is well explained by $f(t) = a\exp(-t/T_{1S}) + (1 - a)\exp(-t/T_{1L})$ consisting of two components of short T_{1S} and long T_{1L}. The value of a for $x = 0.00$ sample is about 0.2 in the metallic phase and about 0.6 in the insulating phase. For both $x = 0.03$ and 0.05 samples, those of a are about 0.85 in the metallic phase and about 0.65 in the insulating phase. No secondary phase in the specimens was detected in the powder X-ray diffraction profile. Therefore, the time dependence of the recovery in the cubic metallic phase above T_{MI} should be expressed by the single exponential type. And also, even in the triclinic insulating phase below T_{MI}, the single exponential type for the recovery is expected, since the resonance line is considered to be almost saturated by the rf-pulse due to the weak NQR interaction and small anisotropy of the Knight shift as mentioned above. However, in both phases, the experimental results do not show the exponential recovery, on the contrary, the recovery can be rather represented by the two exponential type because of the distribution of T_1. The reason why T_1 is distributed in these materials is not clear.

The temperature dependence of $1/T_1T$ from 7 to 300 K is shown in Fig. 3. In the metallic phase, the $1/T_1T$ for all samples obeys the Korringa relation, which indicates $T_1T =$ constant, as expected in a paramagnetic metallic state. The values of $1/T_1T$ indicate the magnitudes of the density of states at Fermi surface. For $x = 0.00$ sample, a sudden decrease more than one order of magnitude of the $1/T_1T$ occurs at T_{MI} and the $1/T_1T$ obeys a power law of T^2 below T_{MI} , suggesting that the energy gap is quite anisotropic. On the other hand, in the case of $x =$ 0.03 sample, an enhancement of $1/T_1T$ occurs around T_{MI} and $1/T_1T$ is proportional to T below T_{MI}, which means that the band gap is suppressed compared with that for $x = 0.00$ sample. For $x = 0.05$ sample, an enhancement of $1/T_1T$ also occurs around T_{MI} and obeys Korringa relation below T_{MI}, indicating that the band gap is remarkably suppressed below T_{MI}. The density of states at Fermi level in the insulating phase is reduced to about 20 % of that in the metallic phase.

According to our results of ^{63}Cu NMR in $Cu(Ir_{1-x}V_x)_2S_4$ ($x = 0.00$, 0.03 and 0.05), the

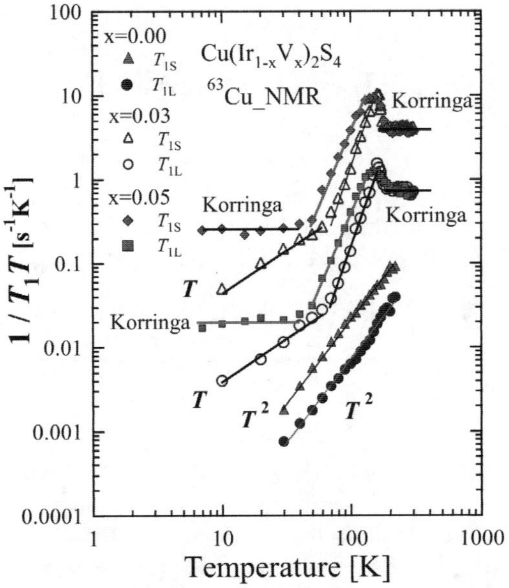

Figure 3. Temperature dependence of $1/T_1T$ on ^{63}Cu NMR in $Cu(Ir_{1-x}V_x)_2S_4$ (x = 0.00, 0.03 and 0.05).

density of states at Fermi level in the metallic phase are almost the same in all samples. The anisotropic energy gap is formed in the insulating phase for $x = 0.00$ sample. However, it is rapidly suppressed with increasing V concentration x, and the density of states at Fermi level for $x = 0.05$ sample exists in the insulating phase. Its amount is about 20 % of that in the metallic phase. The ^{63}Cu NMR results also show that the valence states of Cu site for all samples are nonmagnetic in the insulating phase. X-ray photoemission measurement [10] and calculation of the band structure [11] show that Cu atoms in $CuIr_2S_4$ are in the monovalent state of Cu^+ ion. The effective valence of Ir is suggested to be 3.5^+ in the Pauli paramagnetic state above T_{MI}. On the other hand, high resolution X-ray measurement by Radaelli *et al.* [4] concluded that $CuIr_2S_4$ undergoes the simultaneous charge-ordering and spin-dimerization transition at T_{MI}, which means that $CuIr_2S_4$ is in nonmagnetic state. This physical situation is in agreement with our NMR results.

References

[1] Nagata S, Hagino T, Seki Y and Bitoh T 1994 *Phyica* B **194-196** 1077.
[2] Furubayashi T, Msatsumoto T, Hagino T and Nagata S 1994 *J. Phys. Soc. Jpn.* **63** 3333.
[3] Kumagai K, Tuji S, Hagino T and Nagata S 1995 *Spectroscopy of Mott Insulators and Correlated Metals* ed Fujimori A and Tokura Y (Berlin: Springer) p 255.
[4] Radaelli P G, Horibe Y, Gutmaann M J, Ishibashi H, Chen C H, Ibberson R M, Koyama Y, Hor Y S, Kiryukhin V and Cheong S W 2002 *Nature* **416** 155.
[5] Kumagai K, Sasaki M, Kakuyanagi K and Nagata S 2004 *Hyperfine Interactions* **159** 181.
[6] Yagasaki K, Nakama T, Hedo M, Uwatoko Y, Shimoji Y, Notsu S, Uchima K, Msatsumoto N, Nagata S, Okada H, Fujii H, Yoshida H, Kimura H M, Yamaguchi Y and Burkov A T 2006 *J. Phys. Soc. Jpn.* **75** 074706.
[7] Fleming R M, DiSalvo F J, Cava R J and Waszczak 1981 *Phys. Rev.* **24** 2850.
[8] Kawashima Y, Horibe N, Awaka J, Yamamoto H, Ebisu S and Nagata S 2007 *Physica* B **387** 208.
[9] Sasaki M, Kumagai K, Kakuyanagi K, Nagata S, 2004 *Physica* C **408-410** 822.
[10] Matuno J, Mizokawa T, Fujimori A, Zatsepin D A, Galakhov V R, Kurmaev E Z, Kato Y and Nagata A, 1997 *Phys. Rev.* B **55** R15979.
[11] Oda T, Shirai M, Suzuki N and Motizuki K, 1995 *Condens. Matter* **7** 4443.

Journal of Physics: Conference Series **273** (2011) 012140 doi:10.1088/1742-6596/273/1/012140

Signature of phase coexistence in electron doped manganite

R. Bindu, Ganesh Adhikary and Kalobaran Maiti

Department of Condensed Matter Physics and Materials Science, Tata Institute of Fundamental Research, Homi Bhabha Road, Colaba, Mumbai - 400 005, INDIA.

E-mail: kbmaiti@tifr.res.in

Abstract.
We investigate the temperature evolution of the electronic structure of $La_{0.2}Sr_{0.8}MnO_3$ using Mn K-edge extended x-ray absorption fine structure (EXAFS) and state-of-the-art high resolution photoemission spectroscopy. This compound undergoes a transition from cubic paramagnetic metal to tetragonal C-type antiferromagnetic insulator around 265 K (T_N). EXAFS results exhibit a direct link between the temperature induced change in resistivity and the structural parameters of the MnO_6 octahedra. The high resolution photoemission spectra reveal a gradual evolution of the spectral function exhibiting a pseudo gap at 300 K to a hard gap around 205 K. The appearance of hard gap much below the transition temperature suggests coexistence of two phases in the vicinity of the phase transition temperature, which might be important in the understanding of phase separation scenario widely discussed in this class of materials.

The physics of manganites [1] are mainly governed by the interplay between the charge, spin, orbital and lattice degrees of freedom. Such intricate coupling leads to interesting properties like colossal magnetoresistance (CMR), multiferrocitiy, electronic phase separation, charge, spin and orbital ordering, etc [2]. Nowadays, the phenomenon of electronic phase separation is gaining great attention in the scientific community as it is believed to be a possible origin of CMR effect. $La_{1-x}Sr_xMnO_3$ is one such compound which shows this effect but for x<0.5 compositions. As this property arises due to coexistence of metallic/ferromagnetic and insulating/paramagnetic phases, it is equally important to understand the behavior of the insulating phase which is dominant for x> 0.5 region of the phase diagram. So the best compound which serves this purpose is $La_{0.2}Sr_{0.8}MnO_3$ as it is cubic at room temperature. In addition to this, it lies far away from the phase boundary region thereby getting rid of the complications arising due to the coexistence of different e_g electron densities. This compound undergoes various phase transitions from paramagnetic to C-type antiferromagnetic, metal to insulator, cubic to tetragonal at $\sim$ 265 K [3, 4, 5]. Another important aspect of this large band width system is the strong role played by electron phonon coupling on the transport properties. To understand the above behaviors across the region of phase transition both at the electronic and local structural level, we have carried

Published under licence by IOP Publishing Ltd

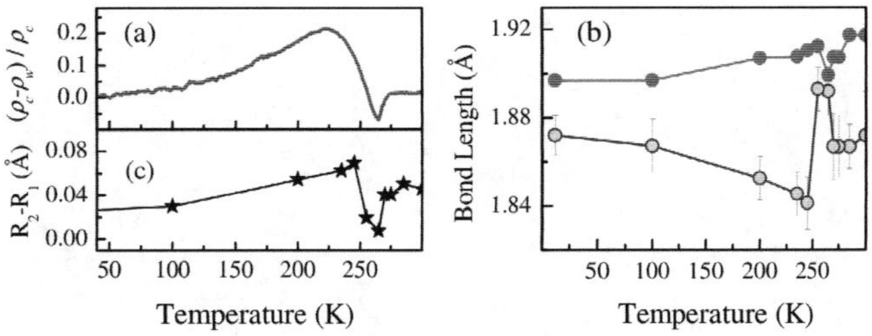

Figure 1. (a) The normalized resistivity ($(\rho_c - \rho_w)/\rho_c$) is shown as a function of temperature. (b) Temperature evolution of short (R_1)and long (R_2) Mn-O bond lengths. (c) Difference between R_1 and R_2 with temperature.

out temperature dependent ultraviolet, x-ray photoemission and Mn K-edge x-ray absorption spectroscopic experiments. Our results reveal signature of co-existence of two phases in a wide temperature range and portrays the importance of fluctuations in the Mn 3*d* orbital occupancies to understand such phase coexistence.

The sample was prepared by standard solid state route and characterized using x-ray diffraction (XRD), scanning electron microscopic, resistivity, specific heat and dc magnetization measurements [3, 4, 6]. The details of the temperature dependent Mn K-edge EXAFS measurements and photoemission spectroscopic techniques are given elsewhere [3, 4]. The band structure calculations were carried out for $La_{0.25}Sr_{0.75}MnO_3$ in the C-type antiferromagnetic phase by using a linearized muffin-tin orbital method within the atomic sphere approximation (LMTART 6.61).

The resistivity measurements exhibit hysteresis during the warming (ρ_w) and cooling (ρ_c) cycles [3]. This can be clearly seen in the normalized resistivity, $(\rho_c - \rho_w)/\rho_c$ shown in Fig (1a) that reveals a dip at ~ 265 K and a peak at ~ 225 K, which signals interesting behaviors occurring around these temperatures. Such hysteresis is indicative of first order phase transition. Since the transport property is mainly governed by the hopping of Mn 3*d* electrons among various MnO_6 octahedra, the local structure associated to MnO_6 octahedron plays an important role in deriving the electronic properties in this system.

We employed EXAFS to probe the structural change of MnO_6 octahedron, which is the ideal tool due to its element specificity. The bond lengths with respect to the Mn ions were obtained by fitting the EXAFS spectra using UWXAS 3.0 software [3]. In Fig (1b), we show the evolution of the Mn-O bondlengths as a function of temperature. We find that all the Mn-O bondlengths in the MnO_6 octahedra are not identical. The best fit corresponds to two short (R_1) and four long (R_2) Mn-O bonds [7]. R_1 and R_2 exhibit an unusual temperature dependence across the phase transition temperature. On cooling the sample from 300 K, the distortion (difference between R_1 and R_2) gradually reduces. At about 265 K, all the Mn-O bond lengths become identical. Further lowering of temperature leads the distortion to become prominent again. The difference between R_1 and R_2 is the largest at around 245 K and reduces further at lower temperatures. Interestingly, the temperature dependence of (R_1-R_2) follows the trend of the normalized

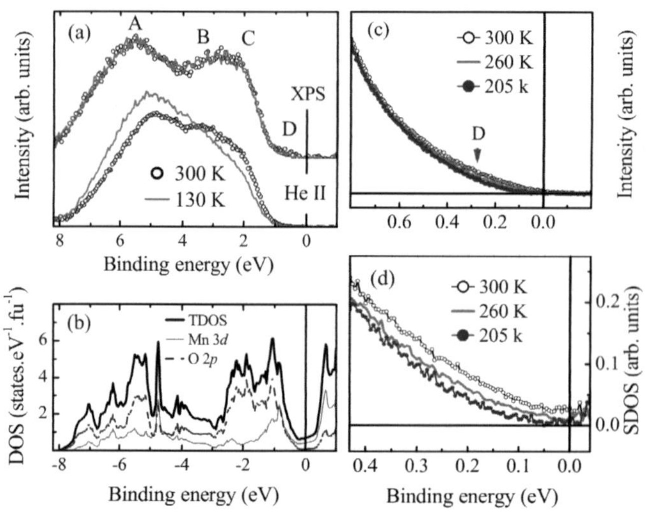

Figure 2. Temperature dependent valence band spectra of $La_{0.2}Sr_{0.8}MnO_3$ collected using (a) AlKα and He_{II} sources (b) Density of states (DOS) obtained using band structure calculations (c) High resolution He_I spectra at different temperatures. (d) Temperature variation of the Spectral density of states (SDOS).

resistivity data shown in Fig. 1(c). This suggests a direct connection between the local structural parameters and the transport properties.

In order to investigate the microscopic origin of this phenomena, we show the valence band spectra collected at different temperatures and photon energies in Fig. 2. All the spectra exhibit four distinct features marked by A, B, C and D as shown in Fig. 2(a). The relative photo-ionization cross section of Mn $3d$ states compared to O $2p$ states is dominant in Al Kα, which becomes much smaller in He II energies [8]. This helps to identify the character of these features that can be confirmed further by the results from band structure calculations [9]. The calculated density of states (DOS) are shown in Fig 2(b). From the comparison of the spectra at two different energies and the calculated results, it is evident that the peaks A and C correspond to the bonding and anti-bonding states of Mn $3d$ and O $2p$ orbitals having t_{2g} symmetry. The feature B can be attributed to the non-bonding O $2p$ states. The peak D corresponds to the e_g anti-bonding states and mainly contributes to the states at and very close to the Fermi level (ϵ_F).

The valence band spectra collected using x-ray source does not show markable temperature dependence except for the peak D. On the other hand, all the features in the He II spectra undergoes significant changes. This suggests that O $2p$ states are very sensitive to temperature and t_{2g} bands are not influenced significantly. The increase in intensity of the features at higher binding energy at lower temperatures suggests that the valence electrons are getting more and more local character as the compound enters into the C-type antiferromagnetic phase. This is also apparent by the decrease in intensity of the feature D with temperature as shown in Fig. 2(a) and 2(c) at Al $K\alpha$, He II and He I energies.

In Fig (2d), we show the temperature evolution of the spectral DOS (SDOS) at ϵ_F obtained by dividing the experimental spectra by the Fermi-Dirac distribution function. Here, all the spectra are normalized by the intensity at 1 eV binding energy. At room temperature, the DOS exhibit a pseudogap at ϵ_F. The intensity at ϵ_F gradually reduces with the decrease in temperature. No hard gap was observed at the phase transition temperature of 265 K although the compound becomes insulating. A hard gap is observed at much lower temperature ($\sim$ 205K). Such gradual evolution of the gap with temperature suggests coexistence of low temperature phase above T_N as found in other systems [10] and the high temperature phase below T_N; phase coexistence in a wide temperature range [2]. We have verified that this trend of spectral evolution is independent of the background function and normalization procedures. These results evidently indicate a relation to the observation of hysteresis in the transport measurements much below T_N.

In summary, we have studied the temperature dependence of the electronic properties and its relation to the local structural parameters of $La_{0.2}Sr_{0.8}MnO_3$ using Mn K-edge EXAFS and high resolution photoemission spectroscopy. The MnO_6 octahedra undergo interesting temperature evolution across the first order phase transition that signals the importance of fluctuations in the Mn $3d$ orbital occupancy across T_N. The high resolution photoemission spectra reveal a pseudogap in the metallic phase, which gradually evolves into a hard gap at about 205 K, much lower than the metal to insulator transition. This suggests signature of phase separation occurring in a wide temperature range in this system.

References

[1] Ramakrishnan T V, 2008 *Current Science* **95**, 1284
[2] Dagotto E, 2005 New J. Phys. **7** 67
[3] Bindu R *et al.*, 2008 *Appl. Phys. Lett.* **92**, 121906
[4] Bindu R *et al.*, 2010 *New. J. Phys.* **12**, 033026
[5] Hemberger J *et al.* 2002 *Phys. Rev. B* **66**, 094410; Chmaissem O *et al.*, 2003 *Phys. Rev. B* **67**, 094431.
[6] Bindu R, 2004 *Eur. Phys. J. B* **37**, 321.
[7] Bindu R, Pandey S K, Kumar A, Khalid S and Pimpale A V, 2005 J. Phys.: Condens. Matter **17** 6393
[8] Yeh J J and Lindau I 1985 *At. Data Nucl. Data Tables* **32**, 1
[9] Maiti K, 2006 *Phys. Rev. B* **73**, 235110; Maiti K, 2006 *Phys. Rev. B* **73**, 115119; Maiti K, 2008 *Phys. Rev. B* **77**, 212407
[10] Maiti K, Singh R S, Medicherla V R R, Rayaprol S and Sampathkumaran E V, 2005 *Phys. Rev. Lett.* **95**, 016404; Bindu R, Maiti K, Khalid S and Sampathkumaran E V, 2009 *Phys. Rev. B* **79** 094103

Theory of Unconventional Charge Ordering and Electrical Transport in Skutterudite $PrRu_4P_{12}$

Ryousuke Shiina

Department of Materials Science and Technology, Niigata University, Niigata 950-2181, JAPAN

E-mail: `shiina@eng.niigata-u.ac.jp`

Abstract. The unconventional properties of charge ordering associated with the metal-insulator transition in skutterudite $PrRu_4P_{12}$ are studied theoretically. It is demonstrated by a simpli ed model Hamiltonian that the electrical transport in the charge ordered state is governed by a large f-orbital uctuation. In particular, the mechanism of an anomalous temperature dependence of the Hall conductivity is discussed in detail. The nature of residual magnetic interaction giving magnetoresistance at low temperature is also discussed.

1. Introduction

The rare-earth filled skutterudite compounds with the general formula RT_4X_{12} exhibit a remarkable variety of phase transitions [1]. Among them, charge ordering associated with the metal-insulator transition in $PrRu_4P_{12}$ has been attracting great interest due to its unconventional properties [2]. Recent neutron scattering experiments for $PrRu_4P_{12}$ by Iwasa et al. have revealed that a strong and characteristic temperature dependence of the f-electron crystal field (CF) levels is realized in the charge order phase [3]. In the normal phase above the transition temperature $T_c = 63$K, all Pr ions are equivalent and singlet $_1$ becomes the ground state in the CF level scheme. With lowering temperature below T_c, the excitation energy to a triplet $_4$ is reduced in one of the sublattices, and level crossing with the singlet takes place at 40K. As a result, two distinct CF ground states, singlet and triplet, are realized in different sublattices at zero temperature. Apparently, this fact and other subsequent experimental results indicate importance of an interaction between conduction electrons and localized f electrons in the charge order formation. In this report, we review recent theoretical progress on the stability mechanism and electrical transport in the charge order phase of $PrRu_4P_{12}$ [4, 5, 6]. Then, we also discuss the origin of anomalous magnetic properties at low temperature.

2. Model Hamiltonian

It has been pointed out that the conduction electron state of $PrRu_4P_{12}$ in the normal phase is rather simple since it is composed of a single band with almost half-filling [7]. The so-called a_u band is well approximated by the tight-binding model for the *bcc* lattice, whose half-filling condition leads to the particle-hole symmetry and a nested Fermi surface.

Considering these facts together with the f-electron states, we introduce a simplified

Published under licence by IOP Publishing Ltd

Hamiltonian:

$$H = \sum_{k} \ {}_{k} c_{k}^{\dagger} \ c_{k} \ - J \sum_{i} n_{i} \ \ {}_{i} + H_{f}. \tag{1}$$

In the first term, $c_{k}^{\dagger}$ (c_{k}) represents conduction electron creation (annihilation) operators with momentum k and spin . We denote the band dispersion ${}_{k}$, assuming a complete nesting property ${}_{k+Q} = - {}_{k}$ for arbitrary k. In this study, we assume a parabolic density of states for the conduction band due to analytical simplicity, ${}_{0}(\ ) = \frac{2}{\pi D^2}\sqrt{D^2 - \ ^2}$, where we take $D = 1$ to be the unit of all energies. The second term in eq. (1) represents charge interaction between conduction and f electrons; $n_i = c_i^{\dagger} \ c_i$ is the number of conduction electrons at i-th lattice site. The occupation of f states is described as $= 1$ when one of the triplet states are occupied and $= 0$ when the singlet state is occupied. Note that this variable is nothing but a hexadecapole moment with the ${}_{1}$ representation in the singlet-triplet subspace [8, 9]. In addition, we take into account the f-electron part in H as

$$H_f = \ {}_{0} \sum_{i} \ {}_{i} + \sum_{ij} K_{ij} \ {}_{i} \ {}_{j}. \tag{2}$$

In the first term, ${}_{0}(> 0)$ is the bare CF potential leading to a singlet ground state in the normal phase. The second term represents an inter-site interaction of f electrons.

Throughout this paper, we shall fix the interaction parameters as $J = 0.2$, $K_{ab} = 0$ and $K \quad K_{aa} = K_{bb} = 0.035$. Hereafter, we discuss the dependences of the transport properties on the CF splitting, for which an effective value with fixed J and K is used instead of the bare value ${}_{0}$, where $= \ {}_{0} - J - 3/4K$. This is nothing but the splitting in the high-temperature limit where $= 3/4$.

3. Formulation of CPA

We consider a charge ordered state in the model Hamiltonian. Let us define order parameters in a and b sublattices, as ${}_{a(b)} = \langle \ {}_{i} \rangle$ and $n_{a(b)} = \sum \langle n_i \ \rangle$ for $i \in a(b)$, where $\langle x \rangle$ indicates the thermal average of an operator x. At zero temperature, one has saturated occupations ${}_{a} = 1$ and ${}_{b} = 0$, which result in modulating charge density as $n_a = 1 + \ q$ and $n_b = 1 - \ q$ with $q > 0$ in the case of $J > 0$. It is not difficult to determine these order parameters by the Hartree-Fock mean-field (MF) treatment.

Here, we shall study the transport properties determined by the thermal fluctuation of the order parameters. For this purpose, we employ the coherent potential approximation (CPA) which is equivalent to the dynamical mean field theory (DMFT) for given ${}_{a}$ and ${}_{b}$ [10]. In the framework of the CPA, we introduce two distinct local self-energies at a and b sites, ${}_{a}(z)$ and ${}_{b}(z)$, which contain the effect of thermal fluctuation of f orbitals as well as the static MF contribution. Then, the Green's functions of conduction electrons in the charge ordered state are derived as,

$$G_k(z) = \langle c_k \ | c_k^{\dagger} \ \rangle_z = \left(z - \ {}_{k} + \ \ - \ {}_{f}(z) - \frac{{}_{s}(z)^2}{z + \ {}_{k} + \ \ - \ {}_{f}(z)} \right)^{-1} \tag{3}$$

$$\bar{G}_k(z) = \langle c_k \ | c_{k+Q}^{\dagger} \ \rangle_z = (z + \ {}_{k} + \ \ - \ {}_{f}(z))^{-1} \ {}_{s}(z) G_k(z). \tag{4}$$

Here, ${}_{f}(z)$ and ${}_{s}(z)$ stand for the uniform and staggered self-energies ${}_{f(s)}(z) = (\ {}_{a}(z) \pm \ {}_{b}(z))/2$. Since the self-energies should be reduced to the MF values at zero temperature, a gapped quasi-particle spectrum is derived from these Green's functions.

In the spirit of the CPA, the self-energies should be determined so as to eliminate the average scattering due to local potentials. Such a condition for a site results in

$${}_{a}(z) = -J \ {}_{a} - \ {}_{a}(z) F_a(z)(J + \ {}_{a}(z)), \tag{5}$$

where $F_a(z) = 1/N \sum_k (G_k(z) + \bar{G}_k(z))$ with N being the number of lattice sites, and the corresponding equation for b site is given by a replacement $a \leftrightarrow b$. Thus, one can set up the equations for $_a(z)$ and $_b(z)$, which are coupled with each other through the expressions $F_{a(b)}(z)$. The CPA equations can be solved by a numerical iteration for input parameters, J, $_a$, and $_b$. Although the full DMFT framework requires a self-consistent treatment of $_a$ and $_b$, we here use the MF results for such f-electron parts, to study a qualitative feature of the fluctuation correction to the conduction-electron Green's function.

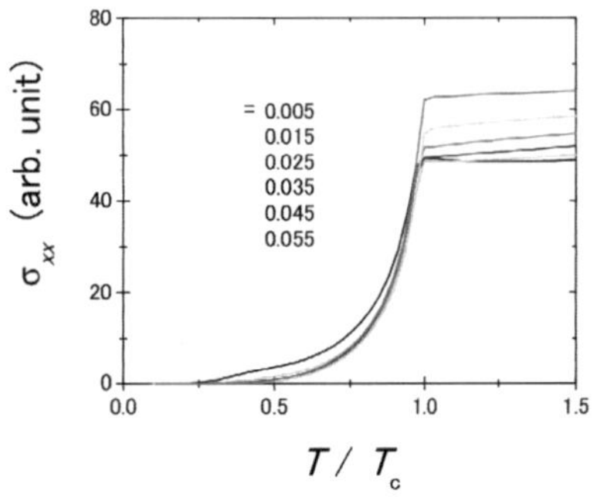

Figure 1. Temperature dependence of diagonal conductivity for various CF splittings $= 0.005 \sim 0.055$ from top to bottom in the high-temperature region.

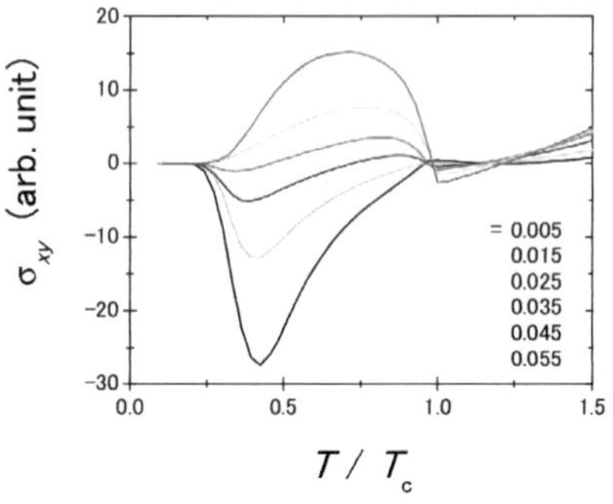

Figure 2. Temperature dependence of Hall conductivity for various CF splittings $= 0.005 \sim 0.055$ from top to bottom in the low-temperature region.

4. Transport Properties

Here, we shall study the temperature dependence of diagonal and off-diagonal conductivities in the charge order phase. These conductivities are derived by the linear response theory, in which the current vertex corrections are neglected in the framework of the DMFT [5].

The diagonal conductivity calculated by the CPA are plotted in Fig. 1 as a function of temperature for various values of . It is apparent that the charge ordering transition at T_c gives rise to a distinct gap formation showing $_{xx} \to 0$ as $T \to 0$. One can find a characteristic structure at low temperatures for larger , which should be a consequence of the densities of states renormalized due to strong f-orbital fluctuation.

Next, we study the Hall conductivity $_{xy}$, which can be induced by a simultaneous application of electric and magnetic fields along x and z axes, respectively. The numerical results derived from the CPA are presented as a function of temperature in Fig. 2. One can find that the conductivity is rather small although finite nearly above the transition temperature. It is because the particle-hole symmetry is approximately preserved around the Fermi level in the normal phase due to weak fluctuation.

It might be expected that the conductivity is suppressed for $T < T_c$ due to the gap formation, as in $_{xx}$. Contrary to this expectation, one can find that the magnitude is strongly enhanced in the intermediate-temperature region in the phase and becomes vanishing at zero temperature. It is also apparent that the sign changes as varies. These features are reasonably understood based upon the properties of the quasi-particle states. According to the CPA analysis on the density of states, the gap-edge states suffer most severely from f-orbital fluctuation, and consequently a distinct particle-hole symmetry breaking appears in the low-energy excitations.

A careful analysis has also shown that changing gives rise to changing the nature of fluctuation, resulting in switching the role of particles and holes. Thus, the unexpected enhancement and the sign change of $_{xy}$ in the charge order phase are naturally interpreted as a consequence of the fluctuation-induced particle-hole symmetry breaking.

It has been shown experimentally for $PrRu_4P_{12}$ that the Hall coefficient $_{xy}/(H\ ^{2}_{xx})$ is extremely small for $T > T_c$ while it is suddenly enhanced for $T < T_c$ [11]. In addition, it exhibits a distinct sign change in an intermediate temperature. These features of the observation are quite consistent with the present theoretical result.

5. Magnetic Properties at Low temperature

In the charge order phase, there is no indication of magnetic phase transition down to 50 mK, in spite of remaining triple degeneracy in one of the sublattices. Quite recently, we have studied this issue by analyzing the realistic hybridization process between conduction band and f orbital in both the jj-coupling and the LS-coupling schemes for f-electron states [6]. Then, we have found that a characteristic orbital dependence of the hybridization leads to a distinct difference of the interaction strengths. It is shown that the effective magnetic interaction derived from the hybridization is an order of magnitude smaller than the charge interaction. This is certainly consistent with the simplified model (1) and (2) ignoring any magnetic interaction.

To be precise, however, there is an implication of an effect of the small magnetic interaction in the transport phenomena of $PrRu_4P_{12}$. For example, the resistivity at low temperature is strongly suppressed by applying a weak magnetic field, depending on the field direction [11]. These behaviors can be explained by an idea that the field restores a disordered triplet state to the ferromagnetic metallic state. A theoretical study of such a large magnetoresistance based on this idea is now being under progress.

6. Summary

In this paper, the nature of an unconventional charge order in skutterudite $PrRu_4P_{12}$ was studied theoretically. We intoduced a simplified model Hamiltonian with relevant f-electron crystal field levels and calculated the solutions of the coherent potential approximation. It was shown that a large f-orbital fluctuation due to singlet-triplet level crossing is very important in understanding the mechanism of transport properties. A possible origin of magnetic properties at low temperature was breifly discussed.

7. Acknowledgments

The author acknowledges H. Shiba for discussions. This work was partially supported by the Grant-in-Aid for Scientific Research (No. 21540368) from JSPS and (No. 21102520) from MEXT.

References

[1] See for example, *Proceedings of Skutterudite 2007*: J. Phys. Soc. Jpn. **77** Supplement.
[2] Sekine C, *et al.* 1997 Phys. Rev. Lett. **79** 3218.
[3] Iwasa K, *et al.* 2005 Phys. Rev. B **79** 024414.
[4] Shiina R 2008 J. Phys. Soc. Jpn. **77** 083705
[5] Shiina R 2009 J. Phys. Soc. Jpn. **78** 104722.
[6] Shiina R and Shiba H 2010 J. Phys. Soc. Jpn. **79** 044704.
[7] Harima H and Takegahara K 2003 J. Phys. Condens. Matter **15** S2081.
[8] Shiina R 2004 J. Phys. Soc. Jpn. **73** 2257.
[9] Takimoto T 2006 J. Phys. Soc. Jpn. **75** 034714.
[10] Freericks J K and Zlatic V 2003 Rev. Mod. Phys. **75** 1333.
[11] Saha S R, *et al.* 2009 Rhys. Rev. B **80** 014433.

Journal of Physics: Conference Series **273** (2011) 012142 doi:10.1088/1742-6596/273/1/012142

Magnetic frustration effect in the multi-band vanadate NaV_2O_4

Hikaru Takeda[1], Masayuki Itoh[1] and Hiroya Sakurai[2]

[1]Department of Physics, Graduate School of Science, Nagoya University, Furo-cho, Chikusa-ku, Nagoya 464-8602, Japan

[2]National Institute for Materials Science, Namiki 1-1, Tsukuba, Ibaraki 305-0044, Japan

E-mail: takeda.hikaru@d.mbox.nagoya-u.ac.jp

Abstract. We have performed ^{23}Na and ^{51}V NMR measurements to study the magnetic frustration effect on the local magnetic properties of NaV_2O_4 with the double chain structure. In the paramagnetic state, we find the presence of the ferromagnetic interaction, which competes with the antiferromagnetic interaction, coming from the multi-band effect. The spin structure in the magnetically ordered state, an incommensurate helical structure which appears due to the competition between the magnetic interactions, is proposed. Thus the magnetic frustration closely related to the multi-band effect is concluded to play an essential role for the magnetic properties of NaV_2O_4.

1. Introduction

Mixed-valence vanadium oxides with geometrically frustrated lattice show fascinating physical phenomena such as heavy fermion behavior in LiV_2O_4 with the spinel structure [1]. Recently NaV_2O_4 with the mixed valence of V^{3+}:V^{4+}=1:1 has attracted attention as a candidate of the itinerant frustrated systems [2]. This vanadate crystallizes into the $CaFe_2O_4$ type structure which has a V_8O_{16} framework composed of double chains of edge sharing V(1)O_6 or V(2)O_6 octahedra. Na atoms are located in a tunnel of the framework. Then the double chain structure is expected to have a low dimensional electronic structure with the geometrical frustration.

NaV_2O_4 exhibits metallic conductivity down to 40 mK, and a transition from the paramagnetic to antiferromagnetic (AFM) states occurs at T_N=140 K [2]. Above T_N, the presence of a ferromagnetic (FM) correlation was reported from the magnetic susceptibility measurement in spite of the AFM transition. However, the origin of the FM interaction has not been clarified. Below T_N, the AFM3 phase appears at low magnetic fields in the temperature T versus magnetic field H phase diagram which includes the AFM1 and AFM2 phases [3]. The spin structures in the AF phases remain controversial in spite of several experimental studies [2, 4, 5].

In this study, we have made ^{23}Na and ^{51}V NMR measurements to investigate the magnetic frustration effect on the local magnetic properties of NaV_2O_4. We confirm the presence of the FM correlation in the paramagnetic phase and discuss the origin of the FM interaction on the basis of the multi-band effect. An incommensurate helical spin structure, which comes from the competition between the FM and AFM interactions, in the AFM3 phase is proposed.

Published under licence by IOP Publishing Ltd

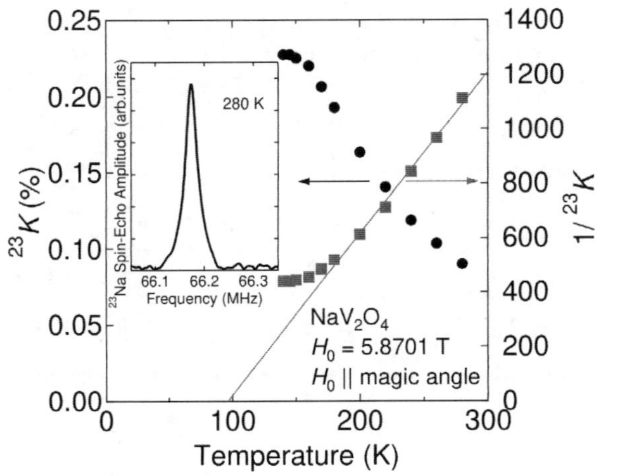

Figure 1. (Color online) Temperature dependence of the ^{23}Na Knight shift with H_0 (=5.8701T) applied along the magic angle. The solid line represents the fitted result of the data to the Curie Weiss law $1/^{23}K=(T\text{-}\Theta)/C$ with the fitting parameters C and Θ. The inset shows the ^{23}Na NMR spectrum at 280 K.

2. Experiments

Single crystals of NaV_2O_4 were prepared by the hydrothermal synthesis method in a high-pressure furnace [3]. A crystal of $0.5\times1.0\times1.5$ mm^3 was used for present NMR measurements. ^{23}Na and ^{51}V Fourier-transformed (FT) NMR spectra were measured in a magnetic field of H_0=3.0 T and 5.8701 T by using a pulsed spectrometer. The ^{23}Na Knight shift was determined as $K=(\nu_{\rm res}\text{-}\nu_0)/\nu_0$ where $\nu_{\rm res}$ and ν_0 (=66.114 MHz) are ^{23}Na resonance frequencies in the NaV_2O_4 sample and the aqueous $NaVO_3$ solution, respectively. Frequency-swept ^{51}V NMR spectra at 4.2 K were taken point by point in zero external field.

3. Results and discussion

Figure 1 shows the T dependence of the ^{23}Na Knight shift ^{23}K determined from the peak position in the ^{23}Na FT NMR spectra with H_0 (=5.8701 T) applied along the direction of the magic angle in the ab plane where the electric quadrupole splitting vanishes as shown in the inset of Fig. 1. The T dependence of ^{23}K obeys above 180 K the Curie-Weiss law $^{23}K=C/(T\text{-}\Theta)$ with C= 0.17 K and the Weiss temperature Θ=96 K. It should be noted that the positive value of Θ indicates the presence of the FM correlation. The similar behavior was also observed in the T dependence of the ^{51}V Knight shift. The origin of the FM correlation is closely related to the multi-orbital state at the Fermi level in NaV_2O_4. The mirror symmetry at the V sites leads to the splitting of the t_{2g} orbital into the doublet, d_{yz} and d_{zx}, and the singlet d_{xy}. The doublet orbital spreads to the rung direction in the double chain, whereas the singlet spreads to the leg direction. Then $3d$ electrons which occupy the doublet are considered to generate the FM interaction due to the Hund coupling, which is the origin of the FM correlation observed in the paramagnetic state, whereas $3d$ electrons on the d_{xy} orbital lead to the AFM interaction. Thus the competition between the FM and AFM interactions is expected in the double chain.

In a magnetically ordered state, spontaneous magnetic moments generally generate an internal field $\mathbf{H_n}$ at a nuclear site. Then the NMR frequency $\nu_{\rm res}$ is written as

$$\nu_{\rm res} = 2\pi\gamma_n \mid \mathbf{H_0} + \mathbf{H_n} \mid, \qquad (1)$$

where γ_n is the nuclear gyromagnetic ratio. The inset in Fig. 2 (a) shows ^{23}Na NMR spectrum with H_0 (=5.8701 T) || b at several temperatures. One central and two satellite lines split by the electric quadrupole interaction are observed above $T_{\rm N}$, whereas the characteristic spectra with six peaks appear below $T_{\rm N}$. The spectrum below $T_{\rm N}$ is composed of three broad spectra due to one central and two satellite transitions. Each broad spectrum has two peaks denoted by the arrows in Fig. 2 (a) and clearly shows a broad distribution of H_n at the Na site. The difference between the NMR frequencies denoted by the red arrows in the spectrum of the central transition $\Delta\nu_{res}$ shows the T dependence, as presented in Fig. 2 (a), which obeys the T dependence of the ordered magnetic moment.

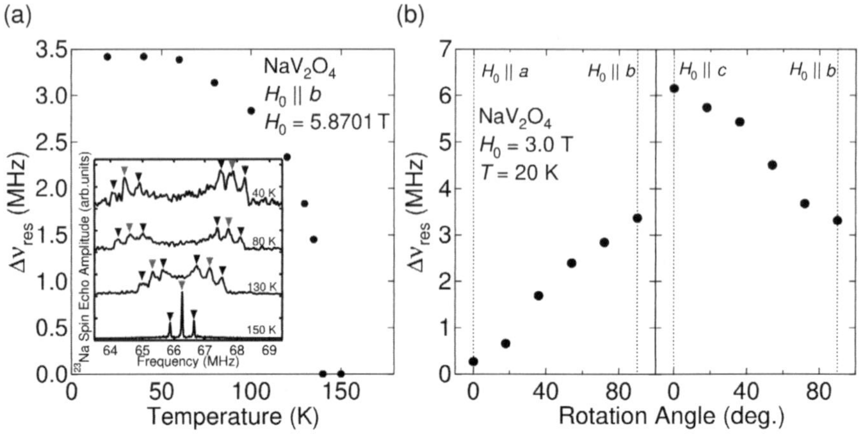

Figure 2. (Color online) (a) Temperature dependence of the difference between resonance frequencies of the peak positions denoted by the red arrows in the ^{23}Na central spectra, $\Delta\nu_{res}$, with H_0 (=5.8701 T) applied along the b axis in NaV_2O_4. The inset shows the temperature dependence of the ^{23}Na NMR spectrum with $H_0 \| b$. (b) The angular dependence of $\Delta\nu_{res}$ with H_0 (=3.0 T) rotated in the ab and bc planes at 20 K in NaV_2O_4.

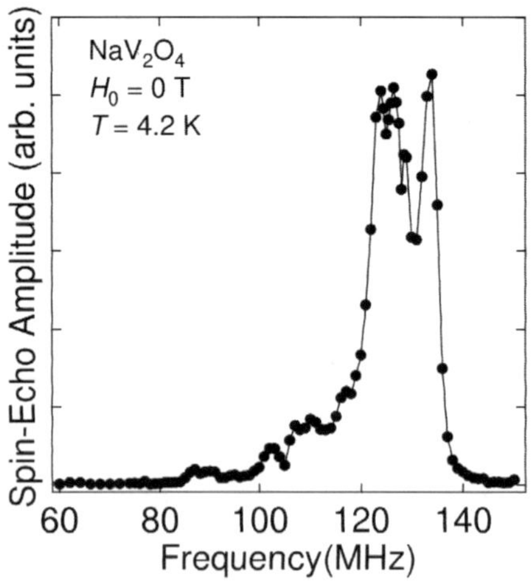

Figure 3. ^{51}V NMR spectrum under zero external field at 4.2K in NaV_2O_4.

At H_0=3.0 T and T=20 K in the AFM3 phase of NaV_2O_4, the ^{23}Na NMR spectra similar to the inset of Fig. 2 (a) were observed with H_0 applied along any direction. Figure 2 (b) shows the angular dependence of $\Delta\nu_{res}$ with H_0 rotated in the ab and bc planes. This angular dependence is governed by the spin structure via the classical dipole interaction from the magnetic moments at the V sites. The fact that $\Delta\nu_{res}$ for $H_0 \perp a$ is larger than that for $H_0 \| a$ indicates that the direction of the magnetic moments may be parallel to the bc plane, if an inter-chain ferromagnetic spin configuration is assumed. The broad spectra observed in the present NMR experiments clearly reject the collinear AF spin structures proposed in the magnetic susceptibility [2] and μSR measurements [4]. Thus we discuss in the following two models which can reproduce the broad ^{23}Na NMR spectra and the angular dependence of $\Delta\nu_{res}$. One is an incommensurate helical structure and the other is an incommensurate spin density wave (SDW) structure proposed by the neutron scattering measurement [5]. However, the SDW model cannot explain the ^{51}V NMR spectrum under zero external field at 4.2 K as shown in Fig. 3, because a spectrum spread down to zero frequency expected for the SDW model is inconsistent with the spectrum observed in the frequency range 85-140 MHz. This ^{51}V spectrum indicates an incommensurate spin structure in the AFM3 phase and supports the helical model. Thus we propose the incommensurate helical spin structure which is reasonably expected to appear by the competition between the FM and AFM interactions in the double chain.

4. Conclusion

We have performed ^{23}Na and ^{51}V NMR measurements to investigate the frustration effect on the local magnetic properties of NaV_2O_4. We confirmed that the ferromagnetic correlation exists in the paramagnetic state from the temperature dependence of the ^{23}Na Knight shift. We also discussed the origin of the ferromagnetic interaction on the basis of the $3d$ orbital state. In the magnetically ordered sate, we proposed the incommensurate helical spin structure which appears due to the competition between the ferromagnetic and antiferromagnetic interactions in the double chain.

Acknowledgement

This study was supported by the JSPS Institutional Program for Young Researcher Overseas Visits and by the Grant-in-Aid for Scientific Research on Priority Areas" Novel State of Matter Induced by Frustration" (No. 22014006) from the Ministry of Education, Culture, Sports, Science and Technology, Japan.

References

[1] Kondo S, Johnston D-C, Swenson C-A, Borsa F, Mahajan A-V, Miller L-L, Gu T, Goldman A-I, Maple M-B, Gajewski D-A, Freeman E-J, Dilley N-R, Dickey R-P, Merrin J, Kojima K, Luke G-M, Uemura Y-J, Chmaissem O and Jorgensen J-D 1997 *Phys. Rev. Lett.* **78** 3729.

[2] Yamaura K, Arai M, Sato A, Karki A-B, Young D-P, Movshovich R, Okamoto S, Mandrus D and Takayama-Muromachi E 2007 *Phys. Rev. Lett.* **99** 196601.

[3] Sakurai H 2008 *Phys. Rev.* B **78** 094410.

[4] Sugiyama J, Ikedo Y, Goko T, Ansaldo E-J, Brewer J-H, Russo P-L, Chow K-H and Sakurai H 2008 *Phys. Rev.* B **78** 224406.

[5] Nozaki H, Sugiyama J, Masson M, Harada M, Pomjakushin V, Sikolenko V, Cervellino A, Roessli B and Sakurai H 2010 *Phys. Rev.* B **81** 100410.

International Conference on Strongly Correlated Electron Systems (SCES 2010) IOP Publishing
Journal of Physics: Conference Series **273** (2011) 012143 doi:10.1088/1742-6596/273/1/012143

Variational Monte Carlo study of two-dimensional strong spin-orbit coupling system: Novel Mott insulating state in Ir oxide

Hiroshi Watanabe[1,2], **Tomonori Shirakawa**[1,2], **and Seiji Yunoki**[1,2]

[1]Computational Condensed Matter Physics Laboratory, RIKEN ASI, Saitama 351-0198, Japan
[2]CREST, Japan Science and Technology Agency (JST), Saitama 332-0012, Japan

E-mail: `h-watanabe@riken.jp`

Abstract. Motivated by the novel $5d$ Mott insulator Sr_2IrO_4, we have studied the three-orbital Hubbard model with a spin-orbit coupling (SOC) term in a two-dimensional square lattice. We have found the transition from a paramagnetic metal to an antiferromagnetic (AF) insulator at $U = U_{\rm MI}$. A large SOC, which is characteristic of $5d$ electron systems, greatly reduces $U_{\rm MI}$ and even small U can lead the system into the Mott insulator. Moreover, the Hund's coupling induces the anisotropy and stabilizes the in-plane AF order. These mechanisms are considered to be valid for Sr_2IrO_4.

1. Introduction

Recently, the layered $5d$ transition metal oxide Sr_2IrO_4 has attracted much attention as a candidate of a novel Mott insulator. In Sr_2IrO_4, t_{2g} and e_g orbitals are well separated by the large crystalline electrostatic field and the lower t_{2g} orbital is filled with five electrons, $(t_{2g})^5$. In spite of the large band width and small on-site Coulomb interaction U, Sr_2IrO_4 is an insulator with a canted antiferromagnetic (AF) order. It is proposed that the strong spin-orbit coupling (SOC), which is characteristic of $5d$ electron systems, is responsible for the insulating mechanism and even small U can lead the system into a Mott insulator [1]. The proposed scenario is as follows. The SOC splits the t_{2g} orbital into a $J_{\rm e} = 1/2$ state (twofold degenerate) and a $J_{\rm e} = 3/2$ state (fourfold degenerate). $J_{\rm e}$ denotes the "effective" total angular momentum that is represented by $-\boldsymbol{L} + \boldsymbol{S}$ [1]. When the SOC is large enough, the lower $J_{\rm e} = 3/2$ state is fully filled and the upper $J_{\rm e} = 1/2$ state becomes an active half-filled energy band. The band width of this half-filled band is much smaller than the original one without the SOC and even small U can lead the system into a Mott insulator. The picture of "$J_{\rm e} = 1/2$ Mott insulator" is supported by several experiments such as resonant x-ray scattering [2], angle resolved photoemission spectroscopy [1], and optical conductivity [1]. This novel Mott insulator has a quite different character with usual $3d$ Mott insulator where the effect of the SOC is only perturbative. However, the detailed theoretical study is still lacking except for first-principle calculations [1, 3]. With these facts in mind, we study the possibility of novel Mott insulating state and magnetic order induced by the strong SOC. The ground state properties of three-orbital Hubbard model with a SOC term are investigated with variational Monte Carlo method. We show that the large SOC works

International Conference on Strongly Correlated Electron Systems (SCES 2010) IOP Publishing
Journal of Physics: Conference Series **273** (2011) 012143 doi:10.1088/1742-6596/273/1/012143

cooperatively with U and leads the system into a novel Mott insulating state with an in-plane AF order.

2. Model and method

We study the following three-orbital Hubbard model in a two-dimensional square lattice,

$$\begin{aligned} H = & \sum_{\boldsymbol{k}\alpha} \varepsilon_\alpha(\boldsymbol{k}) c^\dagger_{\boldsymbol{k}\alpha \ } c_{\boldsymbol{k}\alpha \ } + 2\lambda \sum_i \boldsymbol{L}_i \cdot \boldsymbol{S}_i \\ & + U \sum_{i,\alpha} n_{i\alpha\uparrow} n_{i\alpha\downarrow} + \sum_{i,\alpha< \ , \ } U' n_{i\alpha \ } n_{i \ -} + (U' - J) n_{i\alpha \ } n_{i \ } \\ & + J \sum_{i,\alpha< \ } (c^\dagger_{i\alpha\uparrow} c^\dagger_{i \ \downarrow} c_{i\alpha\downarrow} c_{i \ \uparrow} + c^\dagger_{i\alpha\uparrow} c^\dagger_{i\alpha\downarrow} c_{i \ \downarrow} c_{i \ \uparrow} + \text{H.c.}). \end{aligned} \tag{1}$$

The indices α and   represent the three t_{2g}-orbitals: yz (1), zx (2), and xy (3). The first term is the band energy and the second term is the SOC. They can be written in the matrix form and shown later. The third term is the intraorbital Coulomb interaction and the fourth term is the interorbital Coulomb interactions including z-component of Hund's coupling. The last term is the spin-flip and pair-hopping interactions.

The band energy and SOC are written in the matrix form,

$$H_0 = \sum_{\boldsymbol{k}} \left(c^\dagger_{\boldsymbol{k}1 \ }, c^\dagger_{\boldsymbol{k}2 \ }, c^\dagger_{\boldsymbol{k}3-} \right) \begin{pmatrix} \varepsilon_1(\boldsymbol{k}) & \mathrm{i} \ \lambda & - \ \lambda \\ -\mathrm{i} \ \lambda & \varepsilon_2(\boldsymbol{k}) & \mathrm{i}\lambda \\ - \ \lambda & -\mathrm{i}\lambda & \varepsilon_3(\boldsymbol{k}) \end{pmatrix} \begin{pmatrix} c_{\boldsymbol{k}1 \ } \\ c_{\boldsymbol{k}2 \ } \\ c_{\boldsymbol{k}3-} \end{pmatrix}. \tag{2}$$

We can see that the SOC, λ, mixes the up- and down-spin states and the quasiparticles with pseudospins are newly formed. The total angular momentum $\boldsymbol{J}$ becomes a good quantum number and the orbital and spin angular momentum are no longer conserved quantities.

First, we construct the non-interacting tight-binding energy band to reproduce the first-principle band calculation. In Sr_2IrO_4, there is a large hybridization between xy and $x^2 - y^2$ orbitals originated from the rotation of IrO_6 octahedra and the band structure becomes quite different from the one without rotation. Figure 1 shows the energy bands with and without the rotation of octahedra and SOC. To take account of the effect of rotation, we introduce next-nearest and third-nearest hopping integrals in xy orbital. The chemical potential $\ _3$ is also introduced to take account of the tetragonal splitting. Then the energy dispersion of xy orbital is $\varepsilon_3(\boldsymbol{k}) = 2t_1(\cos k_x + \cos k_y) + 4t_2 \cos k_x \cos k_y + 2t_3(\cos 2k_x + \cos 2k_y) + \ _3$. yz and zx orbitals have almost one-dimensional character, $\varepsilon_1(\boldsymbol{k}) = 2t_5 \cos k_x + 2t_4 \cos k_y$ and $\varepsilon_2(\boldsymbol{k}) = 2t_4 \cos k_x + 2t_5 \cos k_y$ ($t_4 \ \ t_5$). These tight-binding parameters are chosen as $(t_1, t_2, t_3, t_4, t_5, \ _3, 2\lambda)$=(1.0, 0.5, 0.25, 1.028, 0.167, -0.1, 1.028) and they well reproduce the first-principle band dispersion and Fermi surface [1]. In the following, we set $t_1 = t$ as an energy unit.

Next, we consider the effect of Coulomb interaction. The following trial wave function is used in this study,

$$|\Psi\rangle = P_{\mathrm{J_c}} P_{\mathrm{G}} | \ \rangle. \tag{3}$$

$| \ \rangle$ is the one-body part and obtained by diagonalizing H_0 with renormalized tight-binding parameters, $H_0(\tilde{t}_i, \tilde{\ }_3, \tilde{\lambda}_{\alpha \ })$. Note that by the effect of Coulomb interaction and tetragonal splitting, the "effective" SOC has orbital dependence: $\lambda \rightarrow \tilde{\lambda}_{\alpha \ }$. To consider the magnetically ordered state, the AF order parameter is added to H_0. Here, we study the out-of-plane AF order (Fig. 2(a)) and in-plane AF order (Fig. 2(b)). Note that the arrows in Fig. 2 represent the pseudospins, not the original spins. Since the rotational symmetry in pseudospin

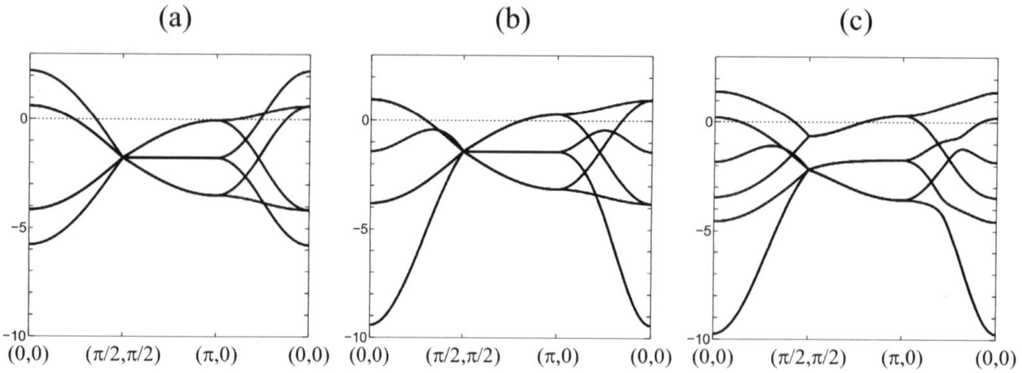

Figure 1. Non-interacting tight-binding energy band (a) without rotation of octahedra and SOC, (b) with rotation and without SOC, and (c) with rotation and SOC.

space is broken by Hund's coupling, these two states should have different energies. Then the matrix becomes 12 12 for the AF state (not shown here because of the limiting space). $P_{\mathrm{G}} = \prod_{i,\gamma} [1 - (1 - g_\gamma) |\gamma\rangle \langle\gamma|_i]$ is a Gutzwiller factor extended for the three-orbital system [4]. i is a site index and γ runs over possible spin configurations at each site. In the three-orbital system, the total number of the spin configuration at each site is $4^3 = 64$. We control the weight of each spin configuration by varying g_γ from 0 to 1. It is a natural extension of Gutzwiller factor in the single-band model. The set of $\{g_\gamma\}$ is a variational parameter and optimized so as to give the energy minimum. In this study, we classify the possible 64 patterns into 12 groups by Coulomb interaction and therefore the number of variational parameter is 12. $P_{\mathrm{J_c}} = \exp\left[-\sum_{i\neq j} v_{ij} n_i n_j\right]$ is a charge Jastrow factor that controls the long-range charge correlation. The long-range charge correlation is known to be important for describing Mott transition [5, 6]. In this study, we assume that v_{ij} depends only on the distance between charges, $v_{ij} = v(|\boldsymbol{r}_i - \boldsymbol{r}_j|)$. Since i and j run over all lattice sites, the number of variational parameter increases with the system size. For 10 10 square lattice, we consider up to 19th-neighbor correlation and therefore the number of variational parameter is 19.

The ground state energy and other physical quantities are calculated with the variational Monte Carlo method. The variational parameters mentioned above are simultaneously optimized by using the stochastic reconfiguration method [7]. This method makes it possible to optimize many parameters efficiently and stably. Once the trial wave function is determined, we can easily calculate physical quantities.

3. Numerical Results

Figure 2(c) shows the ground state phase diagram of the three-orbital Hubbard model in a two-dimensional 10 10 square lattice. The transition from the paramagnetic metal to the AF insulator occurs at $U = U_{\mathrm{MI}} \sim 3.8t$. This value is much smaller than the one in a weak SOC case. When the AF order parameter is introduced, the degeneracy along ($\pi/2$,$\pi/2$)-(π,0) is lifted and the highest band is split away by the AF gap. As shown in Figs. 1(b) and 1(c), the SOC lifts up the two branches out of six energy bands and it makes it easier to fully open the AF gap. Therefore, the larger the SOC, the easier the system becomes an AF insulator. Indeed, our calculation shows that the U_{MI} monotonically decreases with increasing the SOC, $U_{\mathrm{MI}}/t = 8.8, 5.5, 3.8, 3.1$ for $2\lambda/t = 0.2, 0.6, 1.028, 1.4$, respectively. Moreover, as U increases, the effective SOC also increases and they work cooperatively for insulating state. We expect

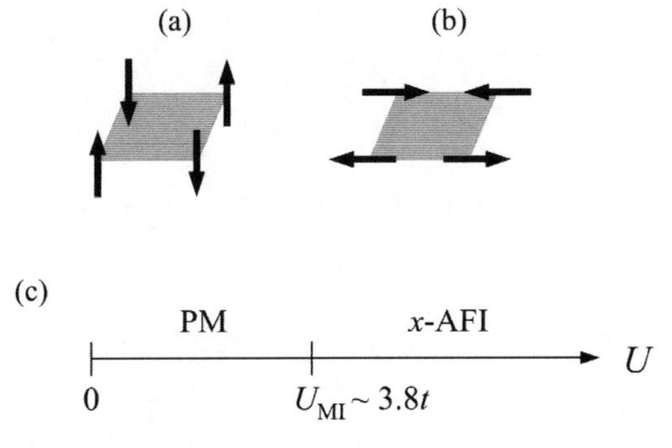

Figure 2. Schematic picture of (a) out-of-plane AF order and (b) in-plane AF order. (c) Ground state phase diagram of the three-orbital Hubbard model in a two-dimensional 10 10 square lattice with $U' = 0.7U, J = 0.15U$, and $2\lambda/t = 1.028$.

that this insulating mechanism is valid for Sr_2IrO_4. Although the Coulomb interaction U is much smaller than the band width and naively the metallic state is expected, Sr_2IrO_4 is an insulator with a canted AF order. This fact can be explained if the SOC in Sr_2IrO_4, $\lambda_{\rm Ir}$, is large enough to reduce $U_{\rm MI}$ smaller than $U_{\rm Ir}$. Indeed, $\lambda_{\rm Ir}$ is estimated as 0.4-0.5eV, much larger than $3d$ and $4d$ electron systems, and our calculation indicates $U_{\rm MI}$ ~1eV. We expect that $U_{\rm Ir} > U_{\rm MI}$ is satisfied and the "spin-orbit induced" Mott insulator is realized in Sr_2IrO_4.

In the insulating region, the in-plane AF order (x-AFI, Fig. 2(b)) is stabilized. The out-of-plane AF order (z-AFI, Fig. 2(a)) has always higher energy than x-AFI and does not appear in the phase diagram. The origin of this energy difference is a Hund's coupling. If there is no Hund's coupling, the pseudospin SU(2) symmetry is preserved and z-AFI and x-AFI are energetically degenerated. However, the Hund's coupling introduces the anisotropy and the in-plane magnetism (x-AFI) is more favored than the out-of-plane magnetism (z-AFI). This result is consistent with the study of effective strong SOC spin model by Jackeli and Khaliullin [8]. The magnetic x-ray diffraction experiment [2] also supports the in-plane magnetism in Sr_2IrO_4. To explain the canted AF order observed in the experiment, more detailed band structure and the introduction of y component of the magnetic moment is necessary. It is an interesting future problem.

4. Summary
We have studied the three-orbital Hubbard model with a SOC term in a two-dimensional square lattice. We have shown that the spin-orbit induced AF insulator is realized for $U > U_{\rm MI}$. With a large SOC, $U_{\rm MI}$ is greatly reduced and even small U can lead the system into the Mott insulator. The Hund's coupling induces the anisotropy and stabilizes the in-plane AF order.

Acknowledgments
The authors thank Y. Yanase for useful discussions. A part of this work is supported by CREST (JST). The computation in this work has been done using the RIKEN Cluster of Clusters (RICC) facility.

References
[1] Kim B J *et al* 2008 *Phys. Rev. Lett.* **101** 076402
[2] Kim B J, Ohsumi H, Komesu T, Sakai S, Morita T, Takagi H and Arima T 2009 *Science* **323** 1329
[3] Jin H, Jeong H, Ozaki T and Yu J 2009 *Phys. Rev.* B **80** 075112
[4] Bunemann J and Weber W 1998 *Phys. Rev.* B **57** 6896
[5] Capello M, Becca F, Fabrizio M, Sorella S and Tosatti E 2005 *Phys. Rev. Lett.* **94** 026406
[6] Tahara D and Imada M 2008 *J. Phys. Soc. Jpn.* **77** 093703
[7] Sorella S 2001 *Phys. Rev.* B **64** 024512
[8] Jackeli G and Khaliullin G 2009 *Phys. Rev. Lett.* **102** 017205

Pump-probe reflectivity study of ultrafast dynamics of strongly correlated 5f electrons in UO_2

Yong Q. An[1], Antoinette J. Taylor[1], Tomasz Durakiewicz[2], and George Rodriguez[1]

1. Center for Integrated Nanotechnologies, Materials Physics and Applications Division, Los Alamos National Laboratory, Los Alamos, NM 87545
2. Condensed Matter and Magnet Science, Materials Physics and Applications Division, Los Alamos National Laboratory, Los Alamos, NM 87545

E-mail: yqan@lanl.gov

Abstract. 5f electrons in the Mott insulator UO_2 produce intriguing electronic states and dynamics, such as strong correlation and f-f excitations. We have performed femtosecond pump-probe reflectivity measurements on a single crystal UO_2 at temperatures of 5-300 K to study the ultrafast dynamics of photoexcited 5f electrons. Laser pulses at 400 nm pump 5f electrons across the Mott gap, while those at 800 nm probe the pump-induced change of reflectivity. We find temperature-dependent excitation and relaxation processes and long-lived acoustic phonons, and extract picosecond risetimes and microsecond relaxation times at low temperatures. The observed slow relaxation is ascribed to the decay of Hubbard excitons formed by U^{3+}-U^{5+} pairs.

1. Introduction

The simple cubic uranium dioxide (UO_2) is a model system for the understanding of electronic transport properties, thermodynamics, and magnetism of 5f materials, and such understanding is also fundamental for strongly correlated electron physics of Mott insulators in general. For decades UO_2 has been extensively studied both experimentally [1] and computationally [2, 3], mainly due to its important role as nuclear fuel. In the earlier review of its optical properties and electronic structures by J. Schoenes [1], UO_2 was regarded as a magnetic semiconductor with a bandgap about 2 eV, without recognition of the Mott-insulating nature of its ground state. Spectroscopic techniques and theoretical modeling have made progress in accessing the specific atomic arrangement and electronic structure [3-5], suggesting UO_2 as a Mott insulator. It has an antiferromagnetic (AFM) ground state and a magnetic transition at the Neel temperature, T_N=30.8 K. Photoexcitation in a Mott insulator may involve carrier hopping between localized states, while that in a band insulator locally creates itinerant carriers.

Because of the strong electron correlation characterized by large on-site Coulomb repulsion U (~3 eV) [6], a Mott gap is formed between an occupied lower Hubbard band (LHB) and an unoccupied upper Hubbard band (UHB). In the LHB of the ground state, two electrons are localized on each U site to form a $U^{4+}(5f^{\,2})$ ion, and thus low-energy excitations across the gap are of 5f-5f character. Analogous to conductivity-inducing thermal excitation [6], photoexcitation of UO_2 causes inter-site hopping of 5f electrons, leading to carrier redistribution in the LHB and UHB, formulated as

$$2U^{4+}(5f^2) \Rightarrow U^{3+}(5f^3) + U^{5+}(5f^1). \quad (1)$$

Published under licence by IOP Publishing Ltd

A charge neutral photoexcitation produces a pair of U^{3+} and U^{5+}, which, depending on the relative strength between the on-site and nearest-site repulsion, can either form a bound state, such as a Hubbard exciton [7, 8], or remain decoupled. If a lattice deformation is incurred around $U^{3+}$and $U^{5+}$by these excitonic states, bound polaron pairs or isolated polarons may form in large or small sizes [6, 9].

In this paper, we present a pump-probe optical reflectivity study of ultrafast dynamics of photoexcited 5f electrons across the Mott gap. Such an investigation has not appeared before to our knowledge, while previous optical studies have been inadequate especially at low temperatures to explore its applications in electronics and photonics. We address ultrafast dynamics including the excitation/relaxation channels and their characteristic timescales, the binding status of U^{3+} and U^{5+}, the role of phonons and temperature, and the effect of photoexcitation on electron correlation. We find a long-lived (~1 ns) acoustic phonon in UO_2 following photoexcitation, indicating strong electron-phonon coupling. We observe a slow relaxation time of ~2 μs for photoexcited 5f electrons in the AFM region, and attribute it to the decay process of the Hubbard excitons formed by nearest-neighbor U^{3+}-U^{5+} pairs in a background of U^{4+} sites.

2. Experiment

Photoinduced reflectivity measurements were performed using a femtosecond regenerative amplifier producing 6 μJ, 80 fs pulses at 800 nm, at a repetition rate of 250 kHz. The pump pulse at 400 nm (3.1 eV) in our experiments was produced by frequency-doubling of the laser for optical excitation well above the Mott gap. The pump induced change of reflectivity, ΔR, was monitored by the probe pulse at 800 nm as a function of the delay time between pump and probe. The pump fluence ranged from 0.5-70 μJ/cm^2, while the typical probe fluence was ~5 times weaker. The fractional changes of reflectivity, $\Delta R/R$, of the probe light reflected from the sample surface were measured by a lock-in amplifier fed by the detected signal from a photodiode. The sample was a 1-mm-thick, (100) surface-oriented single crystal of UO_2 and it was mounted in a cryostat for measurements at liquid He temperatures.

3. Results and Discussion

The left panel of Fig. 1 shows the measured reflectivity transients, $\Delta R/R$, of UO_2 at several different temperatures (T) as a function of the time delay, t, between the pump and probe pulses. Each transient exhibits an initial exponential rise with a time constant, τ_{rise}, followed by a damped oscillation riding on a slow relaxation process. Each transient appears as a negative differential reflectivity, corresponding to photoinduced absorption following excitation.

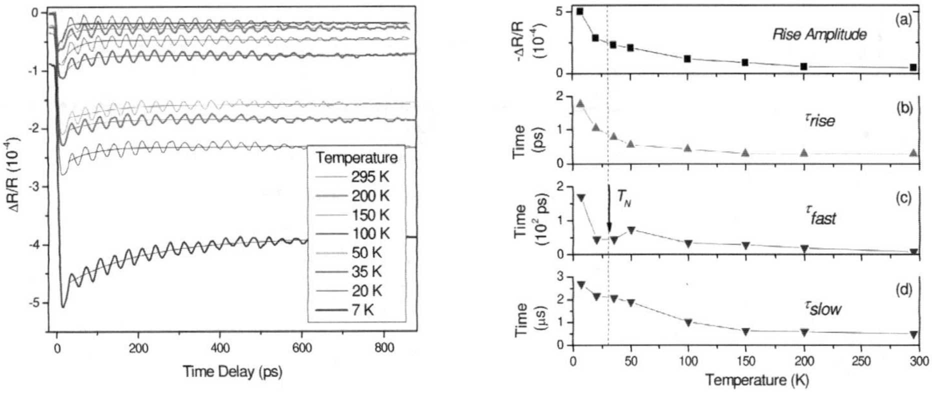

Fig. 1. Left panel: Measured reflectivity transients of UO_2 at different temperatures for a fixed pump fluence 64 μJ/cm^2. Smooth lines are exponential fits for the fast relaxation time, τ_{fast}. Right panel:

Journal of Physics: Conference Series **273** (2011) 012144 doi:10.1088/1742-6596/273/1/012144

Extracted temperature dependence of the rise amplitude, A_0 (a), risetime, τ_{rise} (b), fast relaxation time, τ_{fast} (c), and slow relaxation time, τ_{slow} (d). The vertical dashed line indicates T_N=30.8 K.

The relaxation can be represented as the sum of two decaying exponentials with respective different relaxation time constants, τ_{fast}, a fast relaxation time, and τ_{slow}, a slow relaxation time and a damped oscillation term as in Eq. (2)

$$\Delta R(t)/R = A_1 + A_2 e^{-t/\tau_{fast}} + A_3 e^{-t/\tau_{slow}} + A_4 e^{-t/\tau_{osc}} \cos(4\pi nvt/\lambda - \delta). \tag{2}$$

Here A_n are the component amplitudes, τ_{osc} the damping constant of oscillation, n the real refractive index, v the sound velocity, λ the probe wavelength, and δ the phase [10]. The right panel of Fig. 1 shows the extracted T-dependence of the rise amplitude, A_0, and time constants, τ_{rise}, τ_{fast}, and τ_{slow}, all increasing upon cooling, except for a slight drop in the vicinity of T_N for τ_{fast}. Remarkably τ_{slow} is very long, about 2 μs at low temperatures, reaching the limitation of measurement with the pulses separated by 4 μs. By assuming an exponential decay, we calculate τ_{slow} from the offsets of $\Delta R/R$ at zero time delay. A long-lived oscillation of $\Delta R/R$ is observed and its frequency of 29.4 GHz is matched to $2nv/\lambda$ [11], consistent with an acoustic phonon package (strain pulse) propagating in the sample [10].

The initial quick rise of $\Delta R/R$ originates from the combination of the change of the LHB occupation and the lattice deformation caused by the excitonic states, U^{3+} and U^{5+}. The probe pulse at 800 nm accesses the optical transition between the LHB and midgap states momentarily generated by the deformation. The ps rise process includes three steps: electron thermalization, carrier relaxation via optical phonon emission, and the decay of optical to acoustic phonons. If we assume that the pump induced density of midgap states is proportional to the electronic temperature T_e, while the initial density is proportional to T (this is based on our measured result that the reflectivity decreases linearly with T), the rise amplitude as a function of T, $A_0(T)$, of the Mott insulator follows the formalism based on the two-temperature model of a simple metal, written as [12]

$$A_0(T) \propto \sqrt{T^2 + \alpha F} - T. \tag{3}$$

Here α is a material constant and F is the pump fluence.

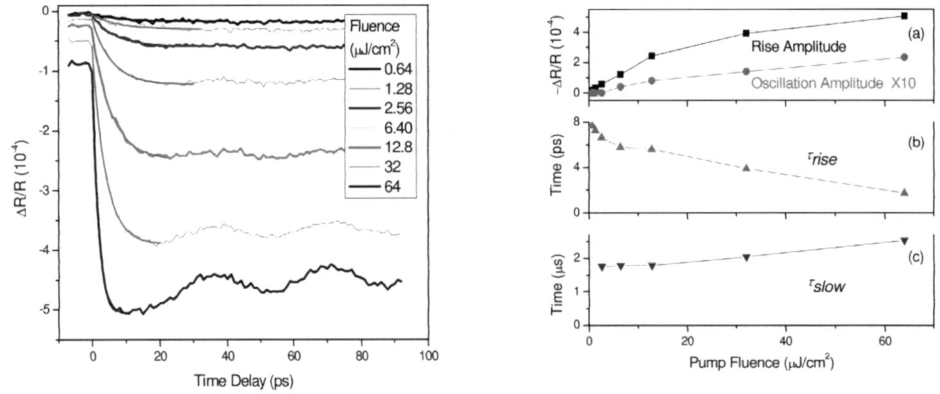

Fig. 2. Left panel: Measured reflectivity transients of UO_2 at a fixed temperature T=7 K for different pump fluences. Smooth lines are exponential fits for the risetimes. Right panel: Extracted pump fluence dependence of the rise and oscillation amplitudes, A_0 and A_4, respectively (a), risetime, τ_{rise} (b), and slow relaxation time, τ_{slow} (c).

From the T-dependence of the characteristic rise and relaxation times, we can derive ultrafast electron and lattice dynamics and perceive electron correlation effects. The $\tau_{rise}(T)$ can be fitted to the $1/\sqrt{T}$ trend, which describes a hop duration of a polaron [6] and can also be due to longer electron-lattice thermalization times at low temperatures. The $\tau_{fast}(T)$ corresponds to the reordering of electron spins after the pump disturbance. Its time scale of 100 ps at $T<T_N$ agrees with earlier identifying of magnetic ordering, though in a different oxide [13]. The observation that $\tau_{fast}(T)$ shows a drop at $\sim T_N$ and almost disappears at $T> T_N$ supports such magnetic identification. The observed acoustic phonon means lattice distortion present, which likely forms polarons. However, $\tau_{slow}(T)$ does not follow the exponential function of *1/T*, a characteristic trend for polaron hopping [6], so we exclude the relaxation channel of polaron decay. Instead, we ascribe the slow relaxation time of ~2 μs at $T<T_N$ to the decay of Hubbard excitons formed by nearest-neighbor U^{3+}-U^{5+} pairs. The U^{3+}-U^{5+} pairs form a bound state at $T<T_N$ because of magnetic ordering correlation but quickly recombine at high temperatures because of thermal and spin fluctuations.

The excitation intensity dependence of $\Delta R/R$ at low temperatures helps to reveal the relaxation mechanism, as thermal excitation is negligible when the material is in its ground state. The left panel of Fig. 2 shows the measured reflectivity transients of UO_2 at a fixed temperature T=7 K for various pump fluences, F=0.64-64 μJ/cm^2, and the extracted amplitudes and time constants are shown in the right panel of Fig. 2. Both $A_0(T)$ in Fig. 1 and $A_0(F)$ in Fig. 2 can be well fitted to Eq. (3), suggesting that our assumption of the density of midgap states is reasonable. The increasing trend of $\tau_{rise}(F)$ with decreasing F is consistent with a longer time needed for the formation of lattice deformation at low temperatures. The slight variation of $\tau_{slow}(F)$ with F excludes heating effects. Persistence of oscillations at low pump fluences indicates strong phonon emission due to inter-site photoexcitation.

4. Summary

We have studied the ultrafast dynamics of photoexcited 5f electron in the Mott insulator UO_2 using femtosecond pump-probe techniques. An unusually long relaxation time of ~2 μs in the AFM ground state UO_2 has been found and ascribed to the decay of Hubbard excitons of photoinduced U^{3+}-U^{5+} pairs. Current work includes detecting photoluminescence from UO_2 to confirm the presence of Hubbard excitons and to identify whether the decay is radiative or non-radiative. This work on strongly correlated 5f electrons will project enabling electronic applications for U compounds. In addition, such long-lived excitons in UO_2 may make the material a good candidate for the observation of a Bose condensation of excitons, as proposed in Cu_2O [7].

Acknowledgements

This work was supported by Laboratory Directed Research and Development at Los Alamos National Laboratory. Y.Q. An gratefully acknowledges helpful discussions with J. Demsar and J.-X. Zhu.

References

[1] J. Schoenes, Phys. Rep. **63**, 301 (1980).
[2] K. N. Kudin, G. E. Scuseria, and R. L. Martin, Phys. Rev. Lett. **89**, 266402 (2002).
[3] L. E. Roy, T. Durakiewicz, R. L. Martin, J. E. Peralta, G. E. Scuseria, C. G. Olson, J. J. Joyce, and E. Guziewicz, J. Comput. Chem. **29**, 2288 (2008).
[4] F. Jollet, T. Petit, S. Gota, N. Thromat, M. GautierSoyer, and A. Pasturel, J. Phys.-Condens. Mat. **9**, 9393 (1997).
[5] P. Roussel, P. Morrall, and S. J. Tull, J. Nucl. Mater. **385**, 53 (2009).
[6] J. M. Casado, J. H. Harding, and G. J. Hyland, J. Phys.-Condens. Mat. **6**, 4685 (1994).
[7] A. Mysyrowicz, D. Hulin, and A. Antonetti, Phys. Rev. Lett. **43**, 1123 (1979).
[8] S. Doniach, B. J. Roulet, and M. E. Fisher, Phys. Rev. Lett. **27**, 262 (1971).
[9] H. B. Schuttler, and T. Holstein, Phys. Rev. Lett. **51**, 2337 (1983).
[10] C. Thomsen, H. T. Grahn, H. J. Maris, and J. Tauc, Phys. Rev. B **34**, 4129 (1986).
[11] J. P. Panakkal, and J. K. Ghosh, J. Mater. Sci. Lett. **3**, 835 (1984).
[12] J. Demsar, J. L. Sarrao, and A. J. Taylor, J. Phys.-Condens. Mat. **18**, R281 (2006).
[13] H. C. Shih, T. H. Lin, C. W. Luo, J. Y. Lin, T. M. Uen, J. Y. Juang, K. H. Wu, J. M. Lee, J. M. Chen, and T. Kobayashi, Phys. Rev. B **80**, 5 (2009).

Dielectric Breakdown in the 1-D Hubbard Model

Shunsuke Kirino and Kazuo Ueda
Institute for Solid State Physics, University of Tokyo
E-mail: kirino@issp.u-tokyo.ac.jp

Abstract. Nonequilibrium transport in the one dimensional Hubbard model at half-filling is studied by the time-dependent density matrix renormalization group method. It is clearly demonstrated that dielectric breakdown of the Mott insulating phase to a nonequilibrium steady state occurs when an external voltage is larger than the charge gap. From the numerically obtained current-voltage characteristics we find that the current is suppressed by the correlation e ect when compared with that of a band insulator and that the current is scaled only by the charge gap for small voltages compared to the band width.

1. Introduction

The Hubbard model has been extensively studied by variety of methods as one of the most simpli ed model that captures important features of interacting electrons. Especially in one dimension various analytical and numerical approaches have been developed[1]. When the band is half- lled and the Coulomb energy is su ciently strong, a nite energy Δ_c is required to make a charge excitation and this is a manifestation of the Mott insulating ground state. The insulating state can be broken by applying a bias voltage larger than Δ_c[2]. This process, dielectric breakdown of a Mott insulator, may be called as *a nonequilibrium metal-insulator transition.*

From a theoretical point of view, in order to formulate a proper theory of the dielectric breakdown of the Mott insulator it is required to treat electron correlation rigorously in a nonequilibrium situation beyond the linear response theory. This di culty have prevented us to make signi cant progress in theoretical understanding of the phenomenon.

Recently the density matrix renormalization group (DMRG) method[3] was extended to time-dependent problems[4]. This new algorithm (TdDMRG) has been successfully applied to nonequilibrium problems in one spatial dimension with strong correlation, such as single quantum dot system under nite bias voltages[5] and the interacting resonant level model[6].

Oka and Aoki studied the breakdown of the Mott insulating phase of the 1-D Hubbard model caused by an external electric eld[7] using the TdDMRG method. They showed that the phenomenological Landau-Zener formula for the transition probability is consistent with the U dependence of the threshold electric eld if one replaces the band gap in the formula by the many-body charge gap Δ_c. However they only discussed the threshold eld of the breakdown and it is necessary to investigate current-voltage (I-V) characteristics beyond the breakdown to elucidate nature of nonequilibrium steady states of the 1-D Hubbard model.

In this paper the 1-D Hubbard model with a nite bias voltage is studied by the TdDMRG method. We determine the I-V characteristics for the repulsive 1-D Hubbard model at half- lling, and show that nonzero steady current appears when the bias voltage exceeds Δ_c. We

Published under licence by IOP Publishing Ltd

nd that by increasing the voltage beyond Δ_c, the steady current is scaled only by Δ_c if the voltage is small compared to the band width and the Coulomb energy U.

2. Model and Numerical Method

We consider the 1-D Hubbard model at half-filling and at $T = 0$ with an applied DC voltage. In order to study nonequilibrium steady states numerically we first calculate the ground state of the system without the voltage term by the usual DMRG method. Then the time evolution of the ground state wave function, which is governed by the Hamiltonian with the voltage term, is simulated by the TdDMRG algorithm. The desired nonequilibrium steady state is expected to appear after some transient behaviors. The model Hamiltonian to be considered is therefore written as follows:

$$H(\tau) = H_L + H_R - t' \sum_{\sigma} (c^{\dagger}_{l\sigma} c_{r\sigma} + h.c.) + \frac{eV}{2} \theta(\tau) (N_L - N_R), \tag{1}$$

$$H_{\nu} = -t \sum_{i,i+1 \in \nu, \sigma} (c^{\dagger}_{i\sigma} c_{i+1\sigma} + h.c.) + U \sum_{i \in \nu} c^{\dagger}_{i\uparrow} c^{\dagger}_{i\downarrow} c_{i\downarrow} c_{i\uparrow} \qquad (\nu = L, R), \tag{2}$$

where $L(R)$ represents the left (right) half of the system, $N_{\nu} \equiv \sum_{i \in \nu, \sigma} c^{\dagger}_{i\sigma} c_{i\sigma}$, τ is the time variable, t the hopping amplitude, U the Coulomb energy, V the bias voltage. $c_{i\sigma}$ annihilates an electron with spin σ at ith site. $l(r)$ is the index of rightmost (leftmost) site in $L(R)$ and t' is the hopping between the lth and rth sites. Here we consider only $t' = t$ case for simplicity. The time dependence of the voltage term is set as the smoothed step function $\theta(\tau) \equiv (1 + \exp[(\tau_0 - \tau)/\tau_1])^{-1}$, in order to mimic adiabatic switching-on. We fix $\tau_0 = 4h/t$ and $\tau_1 = h/t$ throughout this paper. The nonequilibrium steady states realized using the above model Eq.(1) are equivalent to the ones realized in the case of switching-on of the hopping t' instead of eV.

Results shown in this paper are obtained by the TdDMRG calculations for $L = 120$ system keeping $m = 1200$ states and using the 2nd order Suzuki-Trotter decomposition with time step $\Delta\tau = 0.05h/t$. Examples of the truncation error and the Trotter error are shown in Fig.1 for $U/t = 3$ and $eV/t = 1$ and the errors are sufficiently small.

3. Results of Current as a Function of Time

Expectation values of the current operator after the switching-on of the bias voltage

$$J(\tau) = e\langle\psi(\tau)|\dot{N}_R|\psi(\tau)\rangle \tag{3}$$

is shown in Fig.1. In each case the current shows a steady-like behavior (in some cases with an oscillation) after relaxation. After a certain time the steady-like behavior collapses. We refer to the state in the time interval where the current stays at a constant value as a quasi-steady state.

The setup of the system inevitably involves reflection of the current at the edge[5, 8, 9]. The reflected excitations eventually arrive at the center of the chain and disturb the steady current. This effect forces one to take sufficiently long system size to simulate completely relaxed steady states before the disturbances. As a by-product one can estimate velocity of the excitations carrying the transient current from the the time the quasi-steady behavior ends. For $U = 0$ excitations of the system are the particle-hole excitations in the $-2t\cos k$ band, and the reflected waves moving with the Fermi velocity, which is 2 in the units of Fig.1, appears at the time $L/2$. The velocity slightly increases from 2 by increasing U, reflecting the fact that the velocity of the charge excitations of the ground state is an increasing function of U[1].

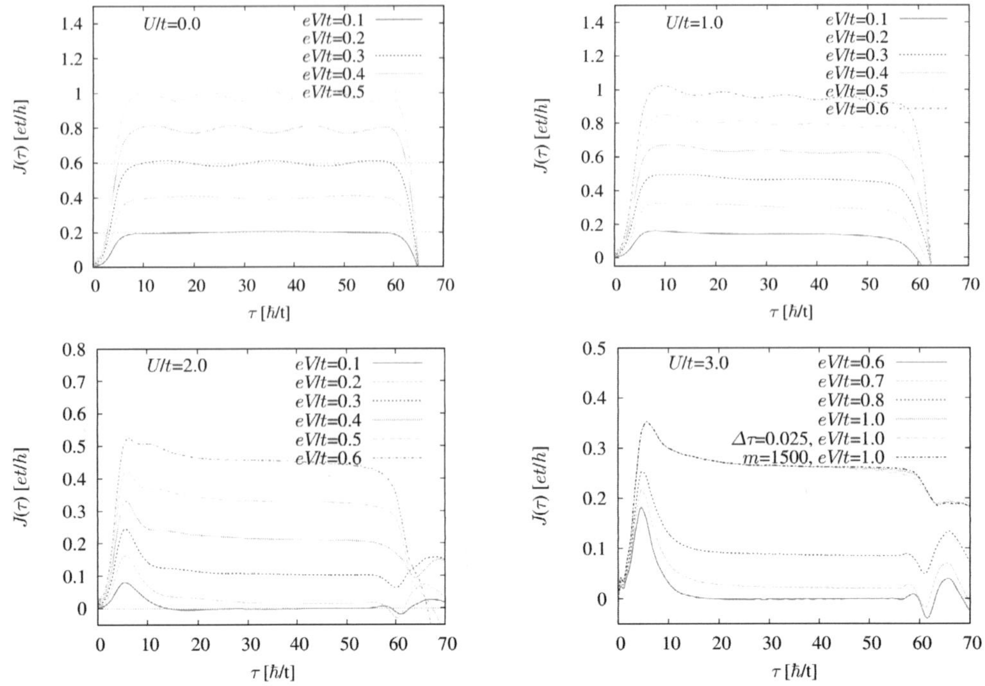

Figure 1. Expectation values of the current operator Eq.(3) at each time step of the TdDMRG calculation. Horizontal lines in the gure for $U = 0$ represent the exact values of the steady currents obtained from Keldysh formalism.

The oscillation during the quasi-steady interval which is observed when $U/t = 0$ and 1 is a nite size e ect [5, 8]. As can be seen in Fig.1 the amplitude of the oscillation is strongly suppressed by the Coulomb repulsion. This is because the repulsive interaction suppresses coherent transport in the chain. The e ects of interaction on the oscillatory behaviors provide an intriguing subject to be studied in future.

4. Results of Current-voltage Characteristics

The steady currents are calculated by averaging $J(\tau)$ inside an interval $[30, 50]$ and shown in Fig.2. For comparison we also plot exact nonlinear currents in a (noninteracting) band insulator with the same magnitude of the charge gap. The Hamiltonian for the band insulator we used is obtained by replacing the interaction term in Eq.(2) by an alternating potential $(\Delta_b/2)\sum_{j\sigma}(\ 1)^j n_{j\sigma}$. The dispersion relation of the model has a band gap Δ_b. From Fig.2 it is obvious that the current for $eV < \Delta$ is zero while for $eV > \Delta$ the current has a nite value, where Δ is Δ_c for the Mott insulator and Δ_b for the band insulator. One can see that the dielectric breakdown is successfully reproduced in our simulation. Comparison of the currents for both models shows that the correlation e ect suppresses the current not only in the Mott insulating phase but also beyond the breakdown.

In Fig.2 we also show the I-V characteristics scaled by the gap Δ. For each model data points of $J(V)$ for di erent values of Δ sit in a single curve when the voltage is small compared with the band width $D \simeq 4t$. When eV becomes comparable to D the energy dependence of the density of states of the band is not negligible, and therefore $J(V)$ deviates from the scaling curve. Likewise when eV is comparable to U, excitations which are not in uenced by the correlation e ect can

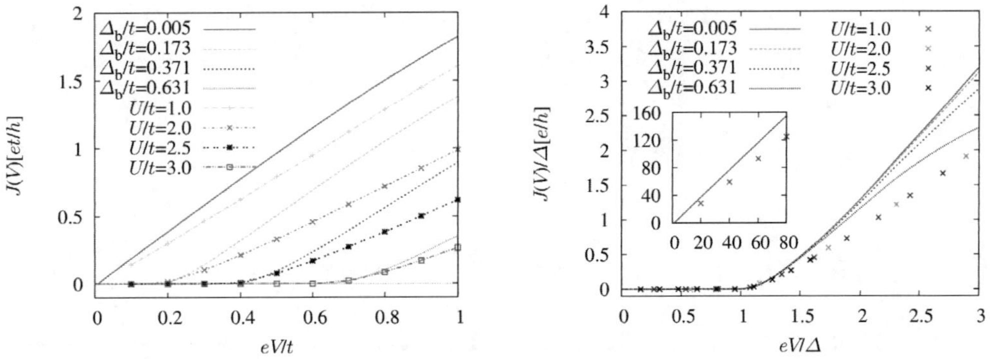

Figure 2. (left) Current-voltage characteristics of the Mott insulator and the band insulator. The values of Δ_c obtained from the exact expression[1] are $\Delta_c(U=1)=0.005$, $\Delta_c(U=2)=0.173$, $\Delta_c(U=2.5)=0.371$ and $\Delta_c(U=3)=0.631$ in units of t. The band gap Δ_b are chosen to be the same as Δ_c. (right) Current-voltage characteristics rescaled by the gap Δ_c or Δ_b.

contribute to the current. Therefore the scaling behavior appears for $eV \ll \min(D, U)$.

5. Conclusions

We have studied nonequilibrium transport in the 1-D Hubbard model at half- lling with a nite bias voltage using the TdDMRG method. We have successfully reproduced the behavior of the dielectric breakdown in the I-V characteristics and found that the correlation e ect suppresses the current compared to the noninteracting band insulator. Moreover we found that the current in the region $eV \ll \min(D, U)$ shows a universal behavior. We believe that these newly discovered nonequilibrium properties stimulate future studies of the dielectric breakdown. At the same time these reliable results demonstrate that various nonequilibrium problems in strongly correlated 1-D systems can be accurately investigated by the TdDMRG method.

Acknowledgments

The authors would like to express their gratitude to T. Fujii, M. Oshikawa, M. Sigrist, H. Tsunetsugu and N. Kawakami for helpful discussions. S. K. is supported by the Japan Society for the Promotion of Science. This work was supported by JSPS Grant-in-Aid for JSPS Fellows 21 6752, by Grant-in-Aid on Innovative Areas "Heavy Electrons" (No. 20102008) and also by Scienti c Research (C) (No. 20540347).

References

[1] see for example, F. H. L. Essler et. al., *The One-Dimensional Hubbard Model* (Cambridge, 2005).
[2] A. Asamitsu et. al., Nature (London) **388**, 50 (1997); Y. Taguchi, T. Matsumoto and Y. Tokura, Phys. Rev. B **62** 7015 (2000).
[3] S. R. White and R. M. Noack, Phys. Rev. Lett. **68** 3487 (1992).
[4] A. J. Daley et. al., J. Stat. Mech.: Theor. Exp. P04005 (2004); S. R. White and A. E. Feiguin, Phys. Rev. Lett. **93** 076401 (2004).
[5] S. Kirino et. al., J. Phys. Soc. Jpn. **77** 084704 (2008).
[6] E. Boulat, H. Saleur and P. Schmitteckert, Phys. Rev. Lett. **101** 140601 (2008).
[7] T. Oka and H. Aoki, Phys. Rev. Lett. **95** 137601 (2005); T. Oka and H. Aoki, *Lecture Notes in Physics* **762** (Springer, 2009).
[8] G. Schneider and P. Schmitteckert, cond-mat 0601389 (2006).
[9] K. A. Al-Hassanieh, A. E. Feiguin, J. A. Riera, C. A. Busser and E. Dagotto, Phys. Rev. B **73** 195304 (2006).

International Conference on Strongly Correlated Electron Systems (SCES 2010) IOP Publishing
Journal of Physics: Conference Series **273** (2011) 012146 doi:10.1088/1742-6596/273/1/012146

Quantum phases of Bose-Fermi mixtures in optical lattices

K. Noda[a], R. Peters[a], N. Kawakami[a], and Th. Pruschke[b]

[a]Department of Physics, Kyoto University, Kyoto 606-8502, Japan
[b]Institut für Theoretische Physik, Universität Göttingen, Göttingen D-37077, Germany

E-mail: noda@scphys.kyoto-u.ac.jp

Abstract. We consider a mixture of interacting bosons and fermions in optical lattices described by the Bose-Fermi Hubbard Hamiltonian. To treat bosonic degrees of freedom, we use a generalized dynamical mean field theory (GDMFT). By combining the GDMFT with the numerical renormalization group method, we revisit the zero-temperature phase diagram with particular emphasis on many-body effects in a supersolid state and discuss the origin of an anomalous peak structure emerging in the density of states for fermions.

Recently, systems of correlated atoms in optical lattices have attracted much interest[1]. In these systems, one can tune the interaction strength by the magnetic field dependence of Feshbach resonances, and thereby simulate typical models used to study quantum many-body physics. Furthermore, the discovery of Feshbach resonances in a mixture of fermions and bosons, ^{6}Li - ^{23}Na [2], ^{40}K - ^{87}Rb [3, 4, 5], and ^{6}Li - ^{87}Rb [6], creates a new route to realizing a physical system without any analog in the conventional condensed matter field. Motivated by this, we consider a mixture of interacting bosons and fermions in optical lattices described by the Bose-Fermi Hubbard Hamiltonian. Preceding studies of this system have revealed the possibility of various fascinating states, such as a supersolid state [7, 8, 9, 10, 11], where a superfluid state coexists with a density wave state, or s-wave and d-wave pairing superfluid states [12, 13, 14, 15, 16], etc. A particularly interesting point raised for the supersolid state is that an unusual peak structure emerges in the density of states (DOS) for fermions near the Fermi surface [10]. Here, we revisit this problem, and point out that the peak structure is related to a single-fermion excitation that is strongly renormalized by low-energy bosonic excitations, the origin of which is quite similar to that for heavy fermions in the Fermi liquid state.

We consider a mixture of spin-polarized fermions and bosons in an optical lattice system, which may be described by the following Bose-Fermi Hubbard Hamiltonian,

$$H = \sum_{\langle i,j \rangle}(t^b - \mu^b\delta_{ij})b_i^\dagger b_j + \sum_i \frac{U_b}{2}n_i^b(n_i^b - 1) + \sum_{\langle i,j \rangle}(t^f - \mu^f\delta_{ij})c_i^\dagger c_j + \sum_i U_{bf}n_i^b n_i^f, \tag{1}$$

where b_i (c_i) annihilates a boson (fermion) at ith site, $n_i^b = b_i^\dagger b_i$ ($n_i^f = c_i^\dagger c_i$) represents the number operator for bosons (fermions). $t^b(t^f)$ is the hopping integral between the nearest-neighbor sites, $\mu^b(\mu^f)$ the chemical potential for bosons (fermions), and U_b (U_{fb}) the on-site boson-boson (fermion-boson) interaction, respectively. $\langle i,j \rangle$ denotes summation over nearest-neighbor sites.

Published under licence by IOP Publishing Ltd

To treat bosonic degrees of freedom, we use a generalized dynamical mean field theory (GDMFT) introduced by several groups [9, 10, 17], where fermions are treated by dynamical mean field theory [18, 19] while the condensate fraction of bosons is treated by conventional static mean field theory. We perform the calculation using a semi-elliptic DOS $\rho(\omega) = 2/\pi\sqrt{1-(\omega/D)^2}$ with a half bandwidth D. Here, the half bandwidth D is given by $D = 2\sqrt{z}t^f$. In the following, we take D as the energy unit.

In order to discuss the supersolid phase with long range orders, we introduce a generalized single impurity Anderson model with a two-sublattice structure. The corresponding Hamiltonian is

$$\begin{aligned} H_{\mathrm{GSIAM}} &= \sum_{\alpha=\pm1} [-zt_b(\varphi_{\overline{\alpha}} b_\alpha^\dagger + \varphi_{\overline{\alpha}}^* b_\alpha) + \frac{U_b}{2} n_\alpha^b (n_\alpha^b - 1) - \mu_\alpha n_\alpha^b \\ &\quad -\mu_f n_\alpha^f + \sum_k \{\epsilon_k a_{k\alpha}^\dagger a_{k\alpha} + V_{k\alpha}(c_\alpha^\dagger a_{k\alpha} + \mathrm{h.c.})\} + U_{bf} n_\alpha^b n_\alpha^f], \end{aligned} \tag{2}$$

where $\alpha = \pm1$ represents the sublattice index, z the coordination number, $\varphi_\alpha = \langle b_\alpha \rangle$ the superfluid order parameter, and $V_{k\alpha}$ the hybridization. Following the study in Ref. [10], we use Wilson's numerical renormalization group (NRG) [20, 21] to solve the effective impurity model, which provides reliable results in the low energy regime. This method has already been extended to include bosonic degrees of freedom [9], which enables us to calculate the superfluid order parameter $\varphi = \langle b \rangle$ and the self energy $\Sigma(\omega)$ for fermions. Note that in the framework of our two-sublattice GDMFT, we can only treat a commensurate density wave order, so that the possibility of long-period density wave order is not addressed in this paper.

We present the results obtained under the condition $zt_b = 0.05$ and $U_b = 1.0$ with fixed filling $n^f = 1/2$ for fermions and $n^b = 5/2$, where we define $n^b = \sum_{\alpha=\pm1}\langle n_\alpha^b \rangle/2$, etc. This choice of parameter turns out to be complementary to the study in Ref. [10] ($n^f = 1/2$ and $n^b = 3/2$). By systematically calculating the DOS for fermions, $\rho(\omega)$, for several values of the fermion-boson interaction U_{fb}, we wish to elucidate the many-body nature emerging in the supersolid state. When $U_{fb} \neq 0$, there is always a finite amplitude of density-wave order parameter $\Delta N^{f(b)} (= 1/2|n_\alpha^{f(b)} - n_{\overline{\alpha}}^{f(b)}|) \neq 0$ both for fermions and bosons, whose density wave oscillations are out of phase, implying that the supersolid state can be realized when $\varphi \neq 0$. Actually, as U_{fb} is increased, bosons can become itinerant and the supersolid state is further stabilized [10]. We here focus on the supersolid region where $1/2 < \Delta N^b < 3/2$ corresponding to $1.7 \simeq U_{fb} \simeq 2.4$. In this region, the system has gapless (compressible) boson density fluctuations, which trigger the emergence of the superfluidity for bosons. This simultaneously gives rise to an unusual peak near the Fermi level in the DOS for fermions, as shown in Fig. 1. Titvinidze *et al.* [10] pointed out that the peak formation in the DOS in the supersolid phase may be related to the instability to breaking the alternating boson-fermion density wave order. We here argue that this peak

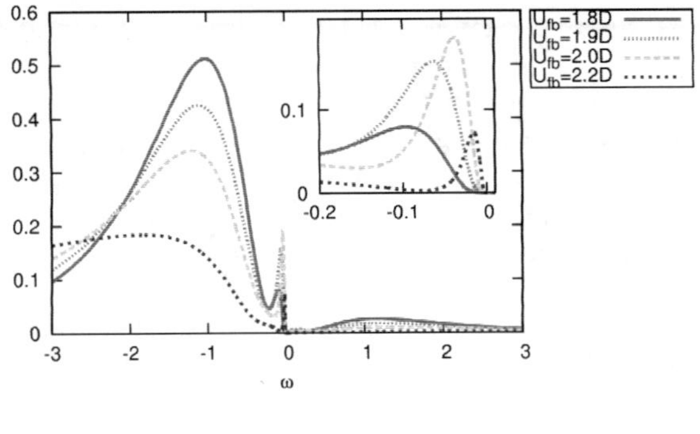

Figure 1. (Color online) Density of states in a supersolid state for $U_{fb}/D = 1.8, 1.9, 2.0, 2.2$. The inset shows the region near the Fermi surface.

Journal of Physics: Conference Series **273** (2011) 012146 doi:10.1088/1742-6596/273/1/012146

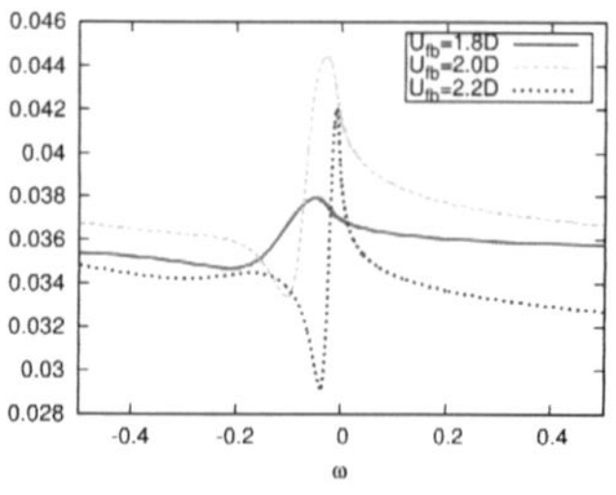

Figure 2. (Color online) Real part of the self-energy for $U_{fb}/D = 1.8, 2.0, 2.2$.

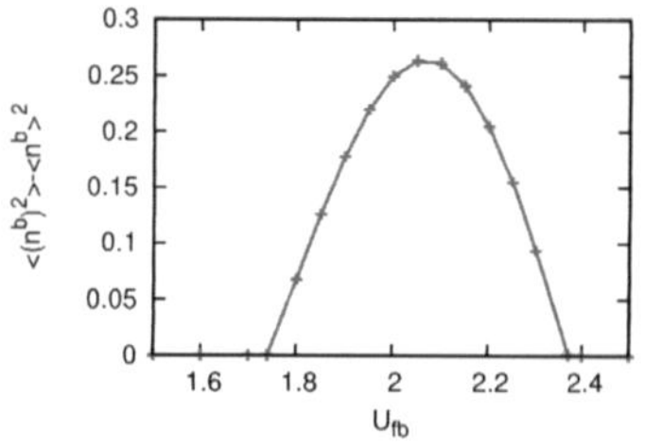

Figure 3. (Color online) Density fluctuations $(\langle (n^b)^2\rangle - \langle n^b\rangle^2)$ for bosons as a function of U_{fb}.

is caused by many-body effects characteristic of a mixture of fermions and bosons, the origin of which is analogous to the heavy-fermion formation in the Fermi liquid state. As shown in Fig. 1, a fermionic density wave order caused by the coupling to the bosonic degrees of freedom makes the DOS asymmetric with respect to $\omega = 0$ in spite of the half-filling condition. We briefly comment on the behaviour patterns of physical quantities in this region. ΔN^f (ΔN^b) monotonically increases from 0.44 to 0.49 (1.5 to 2.5). The superfluid order parameter shows a similar behaviour to that of the boson density fluctuations, as shown in Fig. 3 (explained below).

Note that there are two characteristic structures in the DOS for fermions in Fig. 1 [10], a hump in the high-energy region and a sharp peak in the low-energy region. In the insulating regime ($\Delta N^b = 1/2$ or $3/2$), there is only the high energy hump originating from a mean-field type effect due to U_{fb}. On the other hand, in the supersolid regime, there is an additional peak at low energies which may originate from gapless density fluctuations of bosons through the fermion-boson coupling term. This implies that the peak in the fermionic DOS is induced by the renormalization effect due to gapless bosonic degrees of freedom via a fermion-boson coupling. Such a renormalization effect should be dynamically generated, so that it should be due to the low-energy sector of boson excitations. In order to confirm this, we systematically investigate how the shape of the peak changes according to the gapless bosonic excitations (see Fig. 1). We also show the real part of the self-energy for fermions in Fig. 2 and the density fluctuations of bosons in Fig. 3. As the interaction U_{fb} increases from 1.7, the latter quantity increases, takes maximum values, and then decreases to zero when the system approaches another insulating state around $U_{fb} \simeq 2.4$. Accordingly, the self-energy in Fig. 2 develops a sharp structure around $\omega = 0$, giving rise to the peak in the fermionic DOS in Fig. 1. Note that the width of the peak becomes very sharp, while its weight once increases and then decreases, as seen in Fig. 1. This can be explained as follows. Around $U_{fb} \sim 1.7$, the number of effective gapless bosons is small, so that the effective interaction between bosons is not so strong, providing only the weak renormalization effect to an excited fermion. On the other hand, when $U_{fb} \simeq 2.4$, low-energy bosons are highly correlated since the system is close to the insulating phase. These results in a strong renormalization effect for fermionic excitations, giving rise to the very sharp peak structure. We have also checked that essentially the same tendency appears in the parameters regime studied in Ref. [10].

Recall that a quite similar behaviour appears ubiquitously in correlated electron systems in condensed matter physics; when few holes are doped into a Mott insulating phase of the Hubbard model (or t-J model) at half filling, the DOS acquires a sharp peak structure, which is caused by the renormalization effect due to mobile correlated electrons (or holes). In the present Fermi-Bose mixture system, the fermion sector is always insulating, so that the renormalization effect on one-particle excitations is unusual in the sense of ordinary electron systems. We note, however, that there is another massless boson degree of freedom in our case, which may be a source of the renormalization effect. In this sense, the peak formation in the DOS for fermions in the supersolid states may be regarded as a unique many-body effect inherent in mixed Fermi-Bose systems. In contrast to ordinary heavy fermion systems, the fermion sector in our model is in an insulating state, so that the DOS should vanish inside the charge excitation gap, resulting in the decrease in the DOS near the gap edge. This is why the anomalous peak, which is caused by massless bosonic excitations, is formed a little bit below the lower edge of the charge gap.

In summary, we have reexamined the ground state properties of a mixed Bose-Fermi system with particular emphasis on many-body effects in the supersolid phase. By examining the anomalous peak structure in the DOS for fermions, we have elucidated that it indeed comes from many body effects; while the high-energy hump structure in the DOS is dominated by the mean-field type effect of the fermion-boson interaction, the low-energy peak structure is induced by the density fluctuations of bosons through the fermion-boson interaction term. The formation of the low-energy peak is quite similar to that for heavy fermions in the Fermi liquid state. In the present case, the low-energy bosonic excitations provide a source of the strong renormalization.

Acknowledgements

This work is supported by the Grant-in-Aid for Scientific Research [Grant nos. 21540359, 20029013, 20102008] and the Global COE Program "The Next Generation of Physics, Spun from Universality and Emergence" from MEXT of Japan. N.K. is supported by the Japan Society for the Promotion of Science (JSPS) through its FIRST Program. R.P. acknowledges support from the Alexander von Humboldt Foundation and JSPS under Program "FY2010 JSPS Postdoctoral Fellowship for Foreign Researchers". T.P. and K.N. acknowledge support by the German Science Foundation (DFG) through the SFB 602. Part of the computations were done at the Supercomputer Center at the Institute for Solid State Physics, University of Tokyo and Yukawa Institute Computer Facility.

References

[1] Bloch I, Dalibard J and Zwerger W 2008 *Rev. Mod. Phys.* **80** 885
[2] Stan C A, Zwierlein M W, Schunck C H, Raupach S M F and Ketterle W 2004 *Phys. Rev. Lett.* **93** 143001
[3] Inouye S, Goldwin J, Olsen M L, Ticknor C, Bohn J L and Jin D S 2004 *Phys. Rev. Lett.* **93** 183201
[4] Ospelkaus S, Ospelkaus C, Humbert L, Sengstock K and Bongs K 2006 *Phys. Rev. Lett.* **97** 120403
[5] Zaccanti M, D'Errico C, Ferlaino F, Roati G, Inguscio M and Modugno G 2006 *Phys. Rev. A* **74** 041605
[6] Deh B, Marzok C, Zimmermann C and Courteille P W 2008 *Phys. Rev. A* **77** 010701
[7] Bühler H P and Blatter G 2003 *Phys. Rev. Lett.* **91** 130404
[8] Hebert F, Batrouni G G, Roy X and Rousseau V G 2008 *Phys. Rev. B* **78** 184505
[9] Titvinidze I, Snoek M and Hofstetter W 2008 *Phys. Rev. Lett.* **100** 100401
[10] Titvinidze I, Snoek M and Hofstetter W 2009 *Phys. Rev. B* **79** 144506
[11] Orth P P, Bergman D L and Hur K L 2009 *Phys. Rev. A* **80** 023624
[12] Illuminati F and Albus A 2004 *Phys. Rev. Lett.* **93** 090406
[13] Wang D W, Lukin M D and Demler E 2005 *Phys. Rev. A* **72** 051604
[14] Mathey L, Tsai S W and Neto A H C 2006 *Phys. Rev. Lett.* **97** 030601
[15] Klironomos F D and Tsai S W 2007 *Phys. Rev. Lett.* **99** 100401
[16] Zujev A, Baldwin A, Scalettar R T, Rousseau V G, Denteneer P J H and Rigol M 2008 *Phys. Rev. A* **78** 033619
[17] Byczuk K and Vollhardt D 2009 *Ann. Phys.* **18** 622
[18] Metzner W and Vollhardt D 1989 *Phys. Rev. Lett.* **62** 324
[19] Georges A, Kotliar G, Krauth W and Rozenberg M J 1996 *Rev. Mod. Phys.* **68** 13
[20] Wilson K G 1975 *Rev. Mod. Phys.* **47** 773
[21] Bulla R, Costi T A and Pruschke T 2008 *Rev. Mod. Phys.* **80** 395

Correlation-induced band suppression in the two-orbital Hubbard model

E Plekhanov[1,2], A Avella[1,2,3], F Mancini[1,2] and F P Mancini[1,3,4]

[1] Dipartimento di Fisica "E.R. Caianiello", Università degli Studi di Salerno, 84084 Fisciano (SA), Italy
[2] Unità CNISM di Salerno, Università degli Studi di Salerno, 84084 Fisciano (SA), Italy
[3] CNR-SPIN, UoS di Salerno, 84084 Fisciano (SA), Italy
[4] Sezione INFN di Perugia, 06123 Perugia, Italy

E-mail: plekhanoff@physics.unisa.it

Abstract.
The orbital degrees of freedom are of vital importance in explanation of various phenomena. Among them is the orbital-selective Mott transition (OSMT), which is thought to occur in several materials as $Ca_{2-x}Sr_xRuO_4$ and $La_{n+1}Ni_nO_{3n+1}$. OSMT is usually studied in the infinite-dimension limit, and for the time being, it is not clear if it could survive in one-dimensional (1D) case. There exist two scenarios for the OSMT: upon increasing the interaction in a two-band system i) one of the bands becomes insulating, while the other remains metallic and ii) one of the bands becomes empty, while the other may eventually undergo a single-band Mott insulator transition. In this work, we present a preliminary study of the two-orbital Hubbard model by means of Density Matrix Renormalization Group in 1D at quarter-filling, where the second scenario seems to be realized. In particular, we study the orbital densities, double occupancies and form-factors also in the case of finite inter-orbital inter-site hopping.

1. Introduction

The crucial role of the orbital degrees of freedom in explaining several intriguing phenomena such as metal-insulator transition and magnetism in doped phthalocyanines as well as the superconductivity in heavy-fermion compounds has been recently recognized. On the theoretical side, the study of multi-orbital systems is extremely complicated and has been mainly accomplished in the infinite-dimension limit by means of the Dynamical Mean-Field Theory, although, in these systems, the spatial correlations are essential.

Another fascinating phenomenon inherent to the multi-orbital systems is the orbital selective Mott transition (OSMT). In such a transition, by tuning the strength of the Coulomb repulsion, it is possible to open a gap in some band(s), while leaving the other(s) gapless. It is thought that OSMT occurs in materials as $Ca_{2-x}Sr_xRuO_4$ [1, 2] and $La_{n+1}Ni_nO_{3n+1}$ [3, 4, 5].

The simplest multi-orbital model is the two-orbital Hubbard model, whose most general Hamiltonian reads as follows:

Published under licence by IOP Publishing Ltd

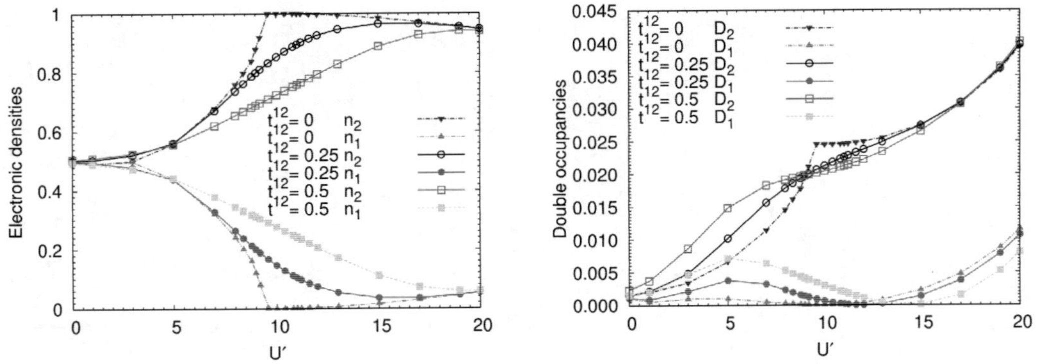

Figure 1. Left panel: U'-dependence of the average occupation number per orbital per site n_1 and n_2 at different values of inter-orbital hopping t^{12}. $U = 10$, $t^{11}/t^{22} = 0.5$, DMRG 50 sites. Right panel: The same for the average double occupancy D_1 and D_2.

$$
\begin{aligned}
H &= -\sum_{\langle ij\rangle\alpha\beta\sigma} t^{\alpha\beta}\left(c^{\dagger}_{i\alpha\sigma}c_{j\beta\sigma} + h.c.\right) + U\sum_{i\alpha} n_{i\alpha\uparrow}n_{i\alpha\downarrow} \qquad (1)\\
&+ \left(U' - \frac{J}{2}\right)\sum_i \left(n_{i1\uparrow}+n_{i1\downarrow}\right)\left(n_{i2\uparrow}+n_{i2\downarrow}\right) - 2J\sum_i\sum_{k=1}^{3} S^k_{i1}S^k_{i2} + J\sum_i\left[p_{i1}p^{\dagger}_{i2} + p_{i2}p^{\dagger}_{i1}\right].
\end{aligned}
$$

Here $t^{\alpha\beta}$ parametrizes intra- (t^{11} and t^{22}) and inter- orbital (t^{12}) hoppings between nearest-neighbor (NN) sites, U and U' are the on-site inter- and intra-orbital Coulomb repulsions, respectively, while J is the direct exchange interaction. Rotational symmetry imposes that $U = U' + 2J$. Finally, $c^{\dagger}_{i\alpha\sigma}$ creates an electron with spin σ in the α-orbital of site i, $S^k_{i\alpha} \equiv \frac{1}{2}\sum \sigma^k_{ab}c^{\dagger}_{i\alpha a}c_{i\alpha b}$ is the electron spin operator in the α-orbital of site i, being σ^k_{ab} the Pauli matrices, and the pair operator $p_{i\alpha}$ and the electron density operator $n_{i\alpha\sigma}$ are defined as follows: $n_{i\alpha\sigma} = c^{\dagger}_{i\alpha\sigma}c_{i\alpha\sigma}$, $p_{i\alpha} = c_{i\alpha\uparrow}c_{i\alpha\downarrow}$.

In the present manuscript, we use finite-size Density Matrix Renormalization Group (DMRG) method [6] to investigate the zero-temperature properties of the Hamiltonian (2). We consider chains of 50 sites with open boundary conditions. In the DMRG decimation procedure, we retain up to $M = 300$ lowest eigenstates of the density matrix, which amounts to have a truncation error on the sum of the density matrix eigenstates not larger that 10^{-5}.

2. Results

It is well known that the (OSMT) occurs at commensurate fillings ($n = 1$, $n = 2$ electrons per site). In particular, in one-dimensional (1D) case, by means of the product-wave-function renormalization-group technique, the existence of two Mott states in the Hamiltonian (2) at quarter- and half- filling for $t^{12} = 0$, $U = 10$ and $t^{11}/t^{22} = 0.5$ has been shown in Ref. [7]. It was found therein that the quarter-filling case corresponds to a complete narrow orbital suppression (NOS), while in the half-filling case there appears a true OSMT state. In the present manuscript we concentrate on the case of a quarter-filled system ($n = 1$) and explore the evolution of NOS upon introducing a finite inter-orbital hopping t^{12}. We have also studied the $t^{12} = 0$ case, already analyzed in Ref. [7], in order to compare our finite-size algorithm with their bulk one, as well as to have our own $t^{12} = 0$ reference line. In Fig. 1, we present our results for average electron

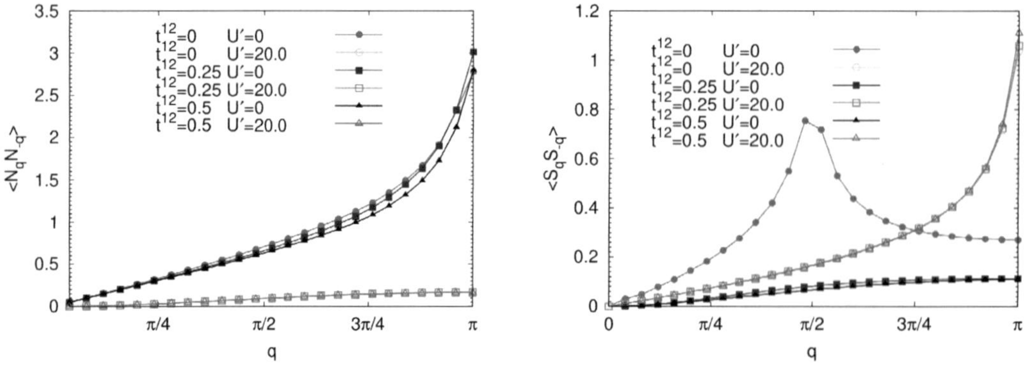

Figure 2. Fourier transforms of the density (Left panel) and spin (Right panel) static form-factors at initial ($U' = 0$) and final ($U' = 20$) points from Fig. 1 and different values of inter-orbital hopping t^{12}: $t^{12} = 0,\ 0.25,\ 0.5$. $U = 10$, $t^{11}/t^{22} = 0.5$, DMRG 50 sites.

densities and double occupancies per orbital as functions of U' at $U = 10$ and $J = (U - U')/2$. For $t^{12} = 0$, NOS occurs approximately at $U'_c = 9.4$ in qualitative agreement with Ref. [7], although in the region of $U' > U'_c$ ($J < 0$) there is a slight re-population of the otherwise empty narrow orbital. Such a re-population might be explained as the effect of the negative J Hund's exchange: the system gains more energy by populating both orbitals rather than emptying one of them. We have checked, by increasing M (not shown), that this re-population is not a DMRG artefact, and the small discrepancy between our results and those of Ref. [7] remains to be investigated. Our analysis confirm the conclusion of Ref. [7] that the quarter-filled transition is due to NOS. NOS influences inevitably the behavior of the average double occupancy per site D_i ($i = 1, 2$), as shown in the right panel of Fig.1. As U' grows for $U' < U'_c$, D_1 decreases monotonically due to the depletion of the narrow orbital. At the same time, D_2 grows due to the opposite reason and reaches the value $D_2 \approx 0.02455$ at U'_c which is very close to the Bethe-ansatz value ($D_{BA} = 0.025$ at half-filling and $U = 10$, Ref.[8]).

In order to have a deeper insight into the nature of the initial and final states involved in this transition, we have measured the static density and spin form-factors. As shown in Fig.2, in the initial state ($U' = 0$), there is a strong density correlation peaked at the wave-vector π, meaning that the density has staggered correlations. These staggered correlations are gradually suppressed at $U' > U'_c$. In such a strong-coupling regime, the wide band becomes uniformly occupied leading to an almost featureless density form-factor. The opposite picture is observed in the spin channel. At $U' = 0$ and $t^{12} = 0$, the spin form-factor exhibits a peak at $\pi/2$, implying the enhanced staggered correlations with the period of four lattice spacings, while at $t^{12} \neq 0$ the spin correlations are heavily suppressed. In the strong-coupling regime at $U' > U'_c$, only the wide orbital is active and the system resembles very much a single-band Hubbard model at half-filling. In such a case, large on-site repulsion induces strong antiferromagnetic NN correlations, as can be seen from Fig.2.

Upon switching on a finite inter-orbital hopping t^{12}, the NOS becomes incomplete, as shown in the left panel of Fig. 1. This can be understood by noting that the electrons in the narrow orbital can now reach the NN site also by making the virtual hopping through the wide orbital according to the following path: $(i, 1) \to (i + 1, 2) \to (i, 2) \to (i + 1, 1)$, where (i, α) stands for α-orbital on site i. Such an extra possibility effectively widens the narrow orbital. To better understand this widening, we have investigated the NOS when $t^{12} = 0$ and at different values of t^{11}, as shown in the left panel of Fig. 3. We have observed the full NOS for $t^{11} \lesssim 0.6$. This full

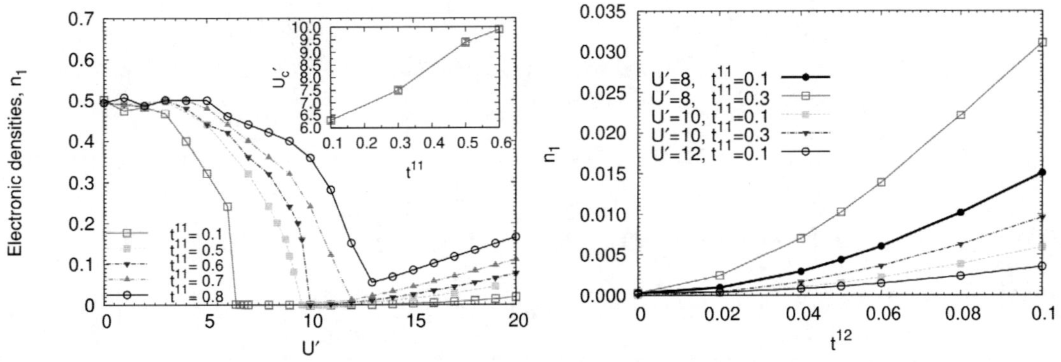

Figure 3. Left panel: U'-dependence of the narrow band occupation at $t^{12} = 0$ and different values of t^{11}. In the inset, U'_c is shown as function of t^{11}. Right panel: The narrow band occupation as function of t^{12} as the latter goes to zero at several values of U' and t^{11}.

NOS occurs at some $U'_c(t^{11})$ (shown in the inset) above which n_1 starts growing nearly linear. This situation changes drastically when $t^{12} \neq 0$. The NOS in this case is never complete even for small t^{12}, as shown in Fig. 1. In the right panel of Fig. 3 we present a study of the narrow band occupation as $t^{12} \to 0$ for different values of U' and t^{11}. We have chosen these values in such a way that there is a complete NOS at $t^{12} = 0$. It can be seen from Fig. 3 that as soon as t^{12} is finite, n_1 is never suppressed completely.

Summarizing, by using finite-system DMRG, we have studied the influence of the inter-orbital hopping t^{12} on NOS in quarter-filled two-orbital Hubbard model (2). We have found that t^{12} prevents the narrow orbital from complete suppression, by providing an additional contribution to the intra-orbital hopping. We have found that at $t^{12} = 0$, a complete NOS occurs at $U'_c(t^{11})$ for $t^{11} \lesssim 0.6$ while for greater values of t^{11} NOS is incomplete for all values of $U' \in [0, 20]$. On the contrary, our data suggest that a finite t^{12} prevents complete NOS even in case of very narrow band ($t^{11} = 0.1$). In the regime with $U' < U'_c$ we find a strong enhancement of staggered charge correlations, while in the case when $U' > U'_c$ the staggered spin correlations are favoured.

3. Acknowledgements
The numerical calculations reported in the present article were done on the CASPUR cluster (project No. 208/10).

References
[1] S. Nakatsuji and Y. Maeno, Phys. Rev. Lett. **84** (2000) 2666.
[2] S. Nakatsuji and Y. Maeno, Phys. Rev. Lett. **90** (2003) 137202.
[3] K. Sreedhar, M. McElfresh, D. Perry, D. Kim, P. Metcalf and J. M. Honig, J. Solid State Chem. **110** (1994) 208.
[4] Z. Zhang and M. Greenblatt, J. Solid State Chem. **117** (1995) 236.
[5] Y. Kobayashi, S. Taniguchi, M. Kasai, M. Sato, T. Nishioka and M. Kontani, J. Phys. Soc. Jpn. **65** (1996) 3978.
[6] S. R. White, Phys. Rev. Lett. **69** (1992) 2863.
[7] S. Miyashita, Y. Yamashita, K. Yonemitsu, A. Koga and N. Kawakami, J. Phys.: Conf. Ser. **150** (2009) 042128.
[8] F. H. L. Essler, H. Frahm, F. Göhmann, A. Klümper, and V. E. Korepin, The One-Dimensional Hubbard Model (Cambridge University Press, Cambridge, U.K., 2005).

Variational cluster approximation study of Mott transition with strong spin-orbit coupling

Tomonori Shirakawa[1,2], Hiroshi Watanabe[1,2], and Seiji Yunoki[1,2]

[1]Computational Condensed Matter Physics Laboratory, RIKEN ASI, Saitama 351-0198, Japan
[2]CREST, Japan Science and Technology Agency (JST), Saitama 332-0012, Japan

E-mail: t-shirakawa@riken.jp

Abstract. Motivated by recent experiments on Sr_2IrO_4, the ground state magnetic and electronic structures are studied theoretically for a two-dimensional three-band Hubbard model with strong spin-orbit coupling. To treat spin-orbit coupling, local Coulomb interactions, and band structure effects on the same footing, the variational cluster approximation based on the self-energy functional theory is employed. It is found that for a relatively large coupling region, the ground state is an anisotropic antiferromagnetic Mott insulator of an effective local angular momentum $J_{\rm eff} = 1/2$ with xy plane as an easy plane. This anisotropy is caused by the strong spin-orbit coupling along with the inter-orbital Hund's coupling. The momentum resolved one-particle excitations are also studied for the Mott insulating phase. It is found that the low-energy one-particle excitations consist mostly of the $J_{\rm eff} = 1/2$ state, a direct evidence of a novel $J_{\rm eff} = 1/2$ Mott insulator.

1. Introduction

Recent experiments on Sr_2IrO_4 [1] have revealed a novel Kramers doublet $J_{\rm eff} = 1/2$ Mott insulator induced by strong spin-orbit coupling (SOC) and local Coulomb interactions among Ir $5d$ electrons. Sr_2IrO_4 is a layered perovskite structure of K_2NiF_4 type [2], and each Ir ion is surrounded by six oxygens forming an octahedral cage. Because of the large crystalline electrostatic field, Ir^{4+} is the low-spin state with $(t_{2g})^5$ electron configuration. Sr_2IrO_4 is thus very similar to a $4d$ counterpart of Sr_2RhO_4 [3]. Because the local Coulomb repulsion (the band width) becomes smaller (larger) with increasing the period in the same transition metal group, it is expected that Sr_2IrO_4 should be more metallic-like than Sr_2RhO_4. However, the experimental observations have shown the opposite: while Sr_2RhO_4 is a metal [4], Sr_2IrO_4 is an insulator [1]. This seemingly counterintuitive behavior was understood [1] by noticing that the larger SOC (λ) of about 0.4 eV in Sr_2IrO_4 further splits the sixfold degenerate t_{2g} states (including spins) into twofold degenerate $J_{\rm eff} = 1/2$ states and fourfold degenerate $J_{\rm eff} = 3/2$ states. Four of the five electrons in Ir^{4+} ion fully occupy the $J_{\rm eff} = 3/2$ states, and the remaining electron partially occupy one of the $J_{\rm eff} = 1/2$ states. This half-filled $J_{\rm eff} = 1/2$ states form a band with smaller band width than the original t_{2g} band without SOC, and thus it becomes a Mott insulator once even smaller on-site Coulomb interactions than that for $4d$ electrons are considered. This picture is consistent with several experimental observations including angle resolved photoemission spectroscopy [1], optical conductivity [1], and resonant x-ray scattering [5]. However, theoretically, only the first-principles band structure calculations

Published under licence by IOP Publishing Ltd

have been reported [1, 6], and a systematic study based on many-body correlated-electron theory is highly desired.

The purpose of this paper is to study theoretically effects of the strong SOC as well as the local Coulomb interactions, including the Hund's coupling, on the ground state phase diagram and the low-lying one-particle excitations for Sr_2IrO_4. Here the variational cluster approximation based on the self-energy functional theory is used for a multi-orbital Hubbard model to treat the local static and dynamical fluctuations exactly. We found that the strong SOC along with the Hund's coupling induces an anisotropic antiferromagnetic Mott insulator of $J_{\rm eff} = 1/2$ with xy plane as an easy plane. We also found that the low-energy one-particle excitations can be described mostly by quasi-particle excitations of $J_{\rm eff} = 1/2$ characteristics.

2. Model and Method

As mentioned above, the local electronic configuration of Ir^{4+} ion in Sr_2IrO_4 is the low-spin $(t_{2g})^5$ because of the large crystal field effect, which has also been estimated much larger than the SOC and the local Coulomb interactions [6]. Therefore, we use the following effective Hubbard model for Sr_2IrO_4, consisting of three orbitals (d_{xy}, d_{yz}, and d_{zx}), defined on the square lattice:

$$H = H_{\mathbf{t}} + H_{\mathbf{U}}, \quad (1)$$

$$H_{\mathbf{t}} = \sum_{i,\alpha} E_\alpha n_{i,\alpha,\sigma} + \sum_{\mathbf{k},\alpha,\sigma} \varepsilon_\alpha(\mathbf{k}) c^\dagger_{\mathbf{k}\alpha\sigma} c_{\mathbf{k}\alpha\sigma} + 2\lambda \sum_i \mathbf{L}_i \cdot \mathbf{S}_i, \quad (2)$$

$$H_{\mathbf{U}} = U \sum_{i,\alpha} n_{i,\alpha,\uparrow} n_{i,\alpha,\downarrow} + \frac{1}{2}(U' - J) \sum_{i,\sigma} \sum_{\alpha\neq\beta} n_{i,\alpha,\sigma} n_{i,\beta,\sigma} + \frac{U'}{2} \sum_{i,\sigma} \sum_{\alpha\neq\beta} n_{i,\alpha,\sigma} n_{i,\beta,\bar{\sigma}}$$
$$-J \sum_i \sum_{\alpha\neq\beta} c^\dagger_{i,\alpha,\uparrow} c_{i,\alpha,\downarrow} c^\dagger_{i,\beta,\downarrow} c_{i,\beta,\uparrow} + J' \sum_i \sum_{\alpha\neq\beta} c^\dagger_{i,\alpha,\uparrow} c^\dagger_{i,\alpha,\downarrow} c_{i,\beta,\downarrow} c_{i,\beta,\uparrow}. \quad (3)$$

Here, $c^\dagger_{i,\alpha,\sigma}$ represents the creation operator of an electron with spin $\sigma = (\uparrow, \downarrow)$ and orbital $\alpha = (xy, yz, zx)$ on site i, and $n_{i,\alpha,\sigma} = c^\dagger_{i,\alpha,\sigma} c_{i,\alpha,\sigma}$. E_α is on-site potential for orbital α, and we set $E_{yz} = E_{zx} \neq E_{xy}$ reflecting the tetragonal symmetry of the system. The energy dispersions of xy, yz, and zx orbital are $\varepsilon_{xy}(\mathbf{k}) = -2t_1(\cos k_x + \cos k_y) - 2t_2 \cos k_x \cos k_y - 2t_3(\cos 2k_x + \cos 2k_y)$, $\varepsilon_{yz}(\mathbf{k}) = -2t_4 \cos k_y - 2t_5 \cos k_x$, and $\varepsilon_{zx}(\mathbf{k}) = -2t_4 \cos k_x - 2t_5 \cos k_y$ with $(t_2, t_3, t_4, t_5) = (0.5t_1, 0.25t_1, 1.028t_1, 0.167t_1)$ where $t_1 = 0.36$ eV, respectively. Restricting the Hilbert space to the t_{2g} orbitals, the SOC term is written as

$$2\mathbf{L}_i \cdot \mathbf{S}_i = \sum_\sigma \left(c^\dagger_{i,xy,\bar{\sigma}}, c^\dagger_{i,yz,\sigma}, c^\dagger_{i,zx,\sigma} \right) \begin{pmatrix} 0 & -s & -i \\ -s & 0 & is \\ i & -is & 0 \end{pmatrix} \begin{pmatrix} c_{i,xy,\bar{\sigma}} \\ c_{i,yz,\sigma} \\ c_{i,zx,\sigma} \end{pmatrix}, \quad (4)$$

where $s = +1$ (-1) for $\sigma = \uparrow$ $(\downarrow)$. For the local Coulomb interactions, we introduce the intra-orbital (U) and the inter-orbital (U') Coulomb interactions, the Hund's coupling J, and the pair-hopping J', with $U = U' + 2J$ and $J = J'$ [7]. The electron density is set to be 5 per site.

We adopt the variational cluster approximation (VCA) method [8] based on the self-energy functional theory (SFT) [9]. This method takes into account precisely effects of short-range static and dynamical correlations, and thus superior to simple mean field approximations. The method has been applied successfully to a wide range of strongly correlated electronic systems to study, e.g., competition of various phases in high-T_c cuprates [10, 11] and in layered organic conductors [12]. The SFT introduces a reference Hamiltonian H' with the same two-body interactions as the original Hamiltonian H but with a different one-body part ($\mathbf{t}'$), and H' may be solved numerically on a finite cluster. An approximate grand potential $\Omega\left[\mathbf{\Sigma}(\mathbf{t}')\right]$ of the original system H is given in a functional form by $\Omega\left[\mathbf{\Sigma}(\mathbf{t}')\right] = \Omega' - \mathrm{Tr}\ln\left(-\mathbf{G}_0^{-1} + \mathbf{\Sigma}(\mathbf{t}')\right) + \mathrm{Tr}\ln\left(-\mathbf{G}'^{-1}\right)$, where Ω', $\mathbf{\Sigma}(\mathbf{t}')$, and $\mathbf{G}'$ are the exact ground potential, the exact self-energy, and the exact Green function of the reference system H', respectively. $\mathbf{G}_0$ is the non-interacting Green function of the original system H. The condition $\partial\Omega\left[\mathbf{\Sigma}(\mathbf{t}')\right]/\partial\mathbf{t}'|_{\mathbf{t}'=\mathbf{t}'^*} = 0$ determines an appropriate reference system which describes the original system approximately.

In investigating broken symmetry phases, VCA treats the short-range correlations exactly within a cluster, without introducing any factorization of the interaction terms, yet the obtained Green function is still defined on the infinite lattice. The only approximation comes from the limited forms of the self-energy functional, e.g., the size of clusters and the number of variational parameters, on which the variational principle is applied. To discuss the symmetry-breaking long-range ordered states in VCA, we introduce suitably chosen fictitious Weiss fields in the set of variational parameters $\mathbf{t}'$ [8, 14, 15]. The Weiss field term associated with the magnetic order considered in this study is expressed as H'_{h_α} $(\alpha = x, z)$ where $H'_{h_z} = h'_z \sum_{i,\sigma} e^{i\mathbf{q}\cdot\mathbf{R}_i} s\gamma^\dagger_{i,\sigma}\gamma_{i,\sigma}$ and $H'_{h_x} = h'_x \sum_{i,\sigma} e^{i\mathbf{q}\cdot\mathbf{R}_i}\gamma^\dagger_{i,\sigma}\gamma_{i,\bar{\sigma}}$. Here $\gamma_{i\sigma} = (c_{i,xy,\bar{\sigma}} - sc_{i,yz,\sigma} - isc_{i,zx,\sigma})/\sqrt{3}$ is the annihilation operator of the $J_{\rm eff} = 1/2$ state at site i, $\mathbf{q} = (\pi,\pi)$ [antiferromagnetic (AF) order], and $\mathbf{R}_i$ represents the position of site i. The Weiss fields H'_{h_z} and H'_{h_x} correspond to the out-of-plane and in-plane magnetic ordering, respectively. The reference system used here consists of 2 sites and the variational parameters $\mathbf{t}'$ are the Weiss field h'_z (h'_x) and the chemical potential μ'.

3. Numerical Results

The results for $\Omega\left[\mathbf{\Sigma}(h'_\alpha)\right] - \Omega\left[\mathbf{\Sigma}(0)\right]$ with $U/t_1 = 4$ and several values of J/U are shown in Fig.1. It is found that the minimum of the grand potential appears at a finite value of h'_α, indicating that the ground state is AF Mott insulating. It is also interesting to notice that the grand potential for the in-plane AF ordering (h'_x) is smaller than that for the out-of-plane AF ordering (h'_z), and the energy difference becomes zero with $J/U \to 0$. This strongly suggests that the Hund's coupling is the origin of the magnetic anisotropy. Our finding is consistent with the recent theoretical study [13] and also in qualitative agreement with experiments [2].

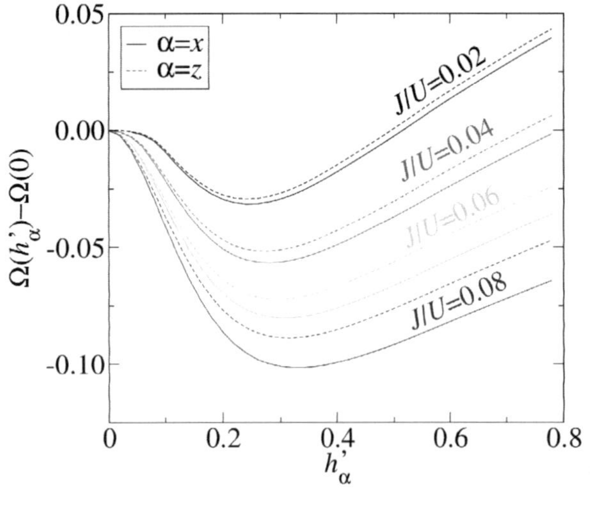

Figure 1. (Color online) The grand potential $\Omega\left[\mathbf{\Sigma}(h'_\alpha)\right] - \Omega\left[\mathbf{\Sigma}(0)\right]$ in the Mott insulating phase for several values of J/U with $U/t_1 = 4$, $\lambda/t_1 = 0.514$, and $(E_{xy} - E_{yz})/t_1 = -1.0$.

We then calculate the one-particle excitation spectra [14], and the results are shown in Fig.2. It is clearly seen that the unoccupied states are well separated from the occupied states, forming a well-defined upper Hubbard band. Projecting the spectral weight onto the $J_{\rm eff} = 1/2$ state ($\gamma_{i\sigma}$ and $\gamma^\dagger_{i\sigma}$), we find that most of the low-energy one-particle excitations, especially the upper Hubbard band, is described by quasi-particles with $J_{\rm eff} = 1/2$ characteristics, except for electron removal excitation spectra at $\mathbf{k}$'s around $(0, 0)$. These results support the picture of a novel $J_{\rm eff} = 1/2$ Mott insulator mentioned in Introduction.

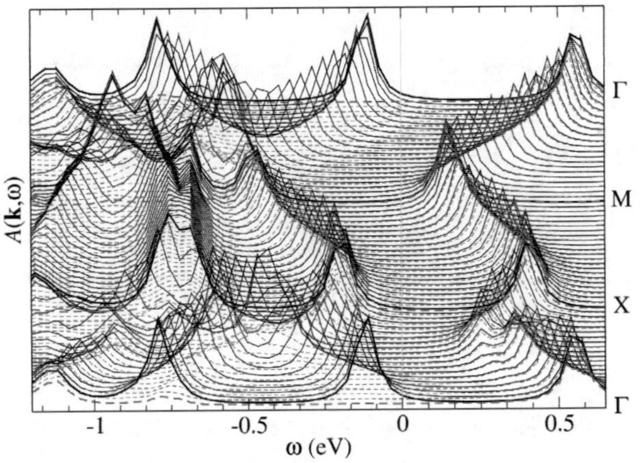

Figure 2. (Color online) Momentum resolved one-particle excitation spectra $A(\mathbf{k}.\omega)$ for $U/t_1 = 3.5$, $J/U = 0.1$, $\lambda/t_1 = 0.514$, and $(E_{xy} - E_{yz})/t_1 = -1.0$ with $t_1 = 0.36$ (eV). For comparison, projected one-particle excitation spectra onto $J_{\rm eff} = 1/2$ states are also shown in dashed lines. Here Γ, X, and M correspond to $\mathbf{q} = (0, 0)$, (π, π), and $(\pi, 0)$, respectively, in antiferromagnetic Brillouin zone. A vertical line at $\omega = 0$ corresponds to Fermi energy.

4. Summary

We applied the VCA to study the ground state magnetic and electronic structures of the three-band Hubbard model with the strong SOC for Sr_2IrO_4. We found that the Hund's coupling and the strong SOC induce an anisotropic AF Mott insulator with xy plane as an easy plane. We also found that the low-energy one-particle excitations is described mostly by quasi-particles with $J_{\rm eff} = 1/2$ characteristics.

Acknowledgments

A part of this work is supported by CREST (JST). The computation was performed using RIKEN Integrated Cluster of Clusters (RICC).

References

[1] Kim B J, Hosub J, Moon S J, Kim J-Y, Park B-G, Leem C S, Yu J, Noh T W, Kim C, Oh S-J, Park J-H, Durairaj V, Gao G and Rotenberg E 2008 *Phys. Rev. Lett.* **101** 076402
[2] Crawford M K, Subramanian M A, Harlow, Fernandez-Baca J A, Wang Z R and Johnston D C 1993 *Phys. Rev.* B **49** 9198
[3] Vogt T and Buttrey D J 1996 *J. Solid State Chem.* **123** 186
[4] Kim B J, Yu J, Koh H, Nagai I, Ikeda S I, Oh S-J and Kim C 2006 *Phys. Rev. Lett.* **97** 106401
[5] Kim B J, Ohsumi H, Komesu T, Sakai S, Morita T, Takagi H and Arima T 2009 *Science* **323** 1329
[6] Jin H, Jeong H, Ozaki T and Yu J 2009 *Phys. Rev.* B **80** 075112
[7] Kanamori J 1963 *Prog. Theor. Phys.* **30** 275
[8] Potthoff M, Aichhorn M and Dahnken C 2003 *Phys. Rev. Lett.* **91** 206402
[9] Potthoff M 2003 *Eur. Phys. J.* B **32** 429
[10] Sènèchal D, Lavertu P-L, Marois M-A and Tremblay A-M S 2005 *Phys. Rev. Lett.* **94** 156404
[11] Arrigoni E, Aichhorn M, Daghofer M and Hanke W 2009 *New. J. Phys.* **11** 055066
[12] Sahebsara P. and Sènèchal D 2006 *Phys. Rev. Lett.* **97** 257004
[13] Jackeli G and Khaliullin G 2009 *Phys. Rev. Lett.* **102** 017205
[14] Dahnken C, Aichhorn M, Hanke W, Arrigoni E and Potthoff M 2004 *Phys. Rev.* B **70** 245110
[15] Sénéchal D, Lavertu P W, Marois M-A and Tremblay A-M S 2005 *Phys. Rev. Lett.* **94** 156404

Pair-distribution function analysis of the structural valence transition in $Cp_2^*Yb(4,4'-Me_2\text{-bipy})$

C H Booth[1], E D Bauer[2], E S Bozin[3,4], S J L Billinge[3,4] and M D Walter[1]

[1]Chemical Sciences Division, Lawrence Berkeley National Laboratory, Berkeley, California 94720, USA

[2]Materials Physics and Applications Division, Los Alamos National Laboratory, Los Alamos, New Mexico 87545, USA

[3]Department of Applied Physics and Applied Mathmatics, Columbia University, New York, New York 10027, USA

[4]Condensed Matter Physics and Materials Science Department, Brookhaven National Laboratory, Upton, NY 11973, USA

E-mail: `chbooth@lbl.gov`

Abstract. The $Cp_2^*Yb(L)$ class of compounds, where Cp*=pentamethylcyclopentadienyl = C_5Me_5 and L is either a 1,4-diazabutadiene or bipy = 2,2′-bipyridine related ligand, have provided excellent analogies to the Kondo state on the nanoscale. $Cp_2^*Yb(4,4'-Me_2\text{-bipy})$ furthers this analogy by demonstrating a valence transition as the sample is cooled below 200 K. Here, pair-distribution function (PDF) analysis of x-ray powder diffraction data demonstrate that the $Cp_2^*Yb(4,4'-Me_2\text{-bipy})$ molecule is virtually unchanged through the valence transition. However, the molecule's stacking arrangement is altered through the valence transition.

In 1989, Neumann and Fulde [1] conjectured that a molecular analogue to the Kondo effect occurs in certain Ce- and Yb-based organometallic molecules. The theory of such effects (see Ref. [2] and references therein) indicates the traditional role of many-body interactions in Kondo systems is reduced to, essentially, including specific configuration interactions (CIs) to the single-electron theory using, for instance, the Complete Active Space Self-Consistent Field (CASSCF) approach. According to such methods, intermediate valence (IV) occurs in the ground state when open and closed shell singlet configurations mix, thereby lowering the combined open-shell singlet state below the triplet state [2]. For example, CASSCF calculations in $Cp_2^*Yb(\text{bipy})$ indicate that the open-shell $f_{\uparrow}^{13}\pi_{\downarrow}^{*1}$ configuration interacts with, primarily, the closed-shell $f_{\uparrow,\downarrow}^{14}\pi^{*0}$ configuration to form an open-shell singlet below the triplet configuration [2], where Cp* = pentamethylcyclopentadienyl = C_5Me_5, Me = methyl = CH_3, and bipy = 2,2′-bipyridine = $(C_5H_4N)_2$. This result is in contrast to lower-order calculations from Density Functional Theory (DFT) that indicate a triplet $f_{\uparrow}^{13}\pi_{\uparrow}^{*1}$ in the ground state in $Cp_2^*Yb(\text{bipy})$ [2]. The IV ground state in the CASSCF calculation has an f-hole occupancy of about $n_f = 0.8$, consistent with Yb L_{III}-edge x-ray absorption near-edge structure (XANES) experiments [2, 3]. In analogy to the Kondo effect, the singlet interactions are made possible both by symmetry and the more extended nature of the f-orbital in Yb and the π^* orbital in the aromatic bipy ligand, which plays the role of a quantum-confined metallic band in an intermetallic nanoparticle [4, 5]. The Kondo analogy has also worked very well for guiding many other aspects of the research

Published under licence by IOP Publishing Ltd

[2, 6]. These studies therefore have broad implications for fields such as solid-state physics, materials science, and organometallic chemistry, and may have future implications for nanoscale devices, which have spurred recent interest in understanding the Kondo effect on the nanoscale.

The Me-substituted Cp^*_2Yb(bipy) complexes demonstrate one of the more dramatic differences between bulk and nanoscale Kondo behavior in that the first excited state need not be a triplet. However, the $Cp^*_2Yb(4,4'\text{-}Me_2\text{-bipy})$ molecule (Fig. 1) also undergoes a first-order +10% valence transition when cooling below 200 K [3, 7]. Here, we address the low temperature structure of the molecule using the pair-distribution function (PDF) technique to determine whether the molecule is isomorphic through the transition, in analogy to the isomorphic valence transition in $YbInCu_4$ [8] and the α-γ transition in elemental cerium [9, 10].

Synthesis details for the $Cp^*_2Yb(4,4'\text{-}Me_2\text{-bipy})$ complex are reported in Ref. [7]. In preparation for the x-ray scattering experiments, the air-sensitive single crystals were powdered in an inert atmosphere glove box. The finely powdered sample was packed into a 5 mm diameter hole in a 1 mm thick aluminum plate, and then sealed between aluminized mylar windows. The sample was then shipped to the Advanced Photon Source (APS) at Argonne National Laboratory in a nitrogen-filled jar. The sample was removed from the jar and then placed into a displex just prior to measurement at the 6ID-D beamline of MuCAT. Measurements were made at many temperatures between 50 K and 300 K. X-ray powder-diffraction data were collected using the rapid acquisition pair distribution function method [11]. The experiments were conducted using synchrotron x-rays with a wavelength of 0.12634 Å using a circular image plate camera (Mar345) 345 mm in diameter. The camera was mounted orthogonally to the beam path with a sample-to-detector distance of 240.464 mm. Other details of data collection and analysis are as described previously [12]. The corrected total scattering structure function, $S(Q)$, was obtained using standard corrections [13] with the program PDFGETX2 [14]. Finally, the PDF was obtained by Fourier transformation of $S(Q)$ according to $G(r) = \frac{2}{\pi}\int_0^{Q\max} Q[S(Q)-1]\sin(Qr)dQ$, where Q is the magnitude of the scattering vector. A $Q_{\max} = 15.0$ Å^{-1} was used. Fig. 2 shows $G(r)$ for several selected temperatures. Fits are performed using the PDFGUI analysis package [15].

From room temperature to just above the transition, the changes in the PDF data from temperature to temperature are small, consistent with thermally-induced variations. At the transition, a clearly resolved change occurs, as illustrated in Fig. 2, where in the 10 K above the transition the data overlap well (Fig. 2a), yet in the next 10 K (Fig. 2b), the changes in the local structure are easily discernible, especially starting at about 9 Å. However, below 9 Å, the local structure is nearly unaffected. Note that the intermolecular Yb-Yb distance is about 9 Å in these crystals, while the longest intramolecular distances are about 11 Å and the shortest intermolecular distances ar about 4 Å. We can therefore conclude simply from these data and with no fitting that the main structural change occurs in the molecular stacking.

Two questions remain unanswered: (1) can we resolve an actual change in the molecule's morphology, and (2) what is the change in the packing arrangement? To answer these questions,

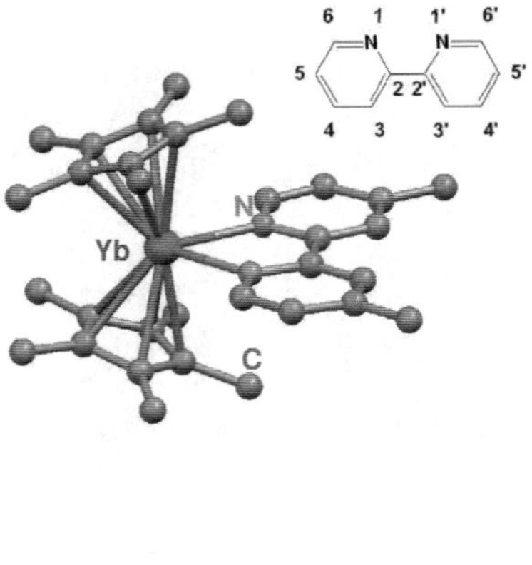

Figure 1. The $Cp^*_2Yb(4,4'\text{-}Me_2\text{-bipy})$ molecule and the position numbering scheme for bipyridine. The $Cp^*_2Yb(4,4'\text{-}Me_2\text{-bipy})$ molecule crystallizes into the $P2_1/c$ space group above 200 K.

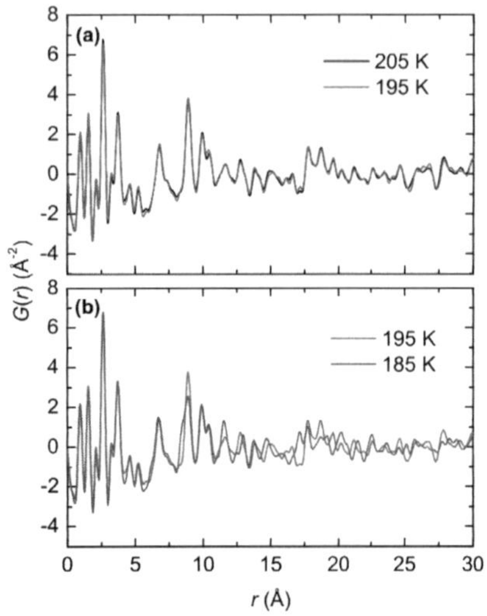

Figure 2. The PDF at two temperatures 10 K apart (a) above and (b) through the transition.

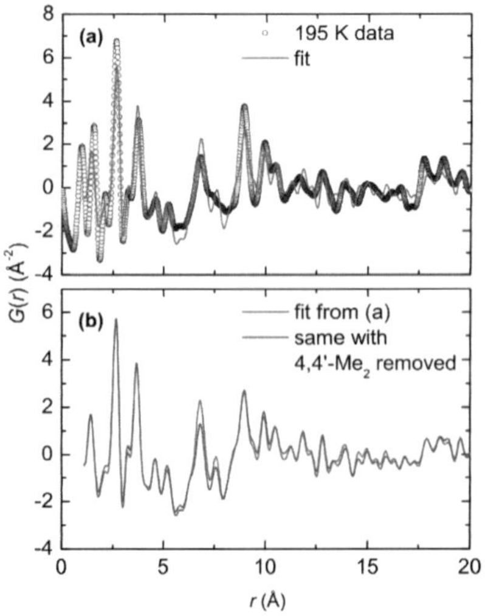

Figure 3. (a) Data and fit to the single crystal structure. (b) Fit from (a) with and without the 4,4′-Me_2 removed.

a detailed fit should be performed. Unfortunately, attempts to fit structural models to the low temperature data, even including a wide variety of constraints, did not produce satisfactory results. It is possible to fit the high-temperature-phase data to the nominal crystal structure [7] by only allowing a single mean-squared displacement parameter u^2 for each atomic species, the three lattice parameters, a scale factor, and a damping factor to vary. Such a fit is shown in Fig. 3a. Clearly, the high-temperature-phase data represent the structure from the much more accurate single-crystal diffraction results well.

Fortunately, we can demonstrate the sensitivity of the data (if not the fit) to possible changes in the molecule's morphology through the transition using these fit results. A relatively small change in the morphology, at least as far as the x-ray scattering is concerned, would be to eliminate a C atom from the molecule. In Fig. 3b, the effect of removing the Me carbons at the 4,4′ sites on the bipy is illustrated by taking the fit result shown in Fig. 3a and simply deleting these Me groups. The $G(r)$ is nearly unaffected, except at the Yb-Me distance of about 7 Å. This example, of course, illustrates the change from two carbon atoms; in any case, there is no change observed on this scale below 9 Å in the data spanning the phase transition (Fig. 2b). Consequently, we conclude that the molecule's morphology is unaffected by the valence transition. Changes in the bond lengths due to the change in valence can be observed in these data, but are more accurately presented in fits to extended x-ray absorption fine-structure (EXAFS) data described elsewhere [3].

Although the molecule has the same morphology in the high and low temperature phases, it is possible, if not likely, that the change in packing produces a different crystal symmetry than $P2_1/c$. A likely alternative is a $Pbca$ structure, such as occurs for Cp_2^*Yb(bipy), Cp_2^*Yb(5,5′-Me_2-bipy), and others [3]. As mentioned above, fits using the $Pbca$ structure are inconclusive. Attempts to fit these data using Rietveld refinement also failed to produce a unique answer, since the data can be fit using either space group if enough parameters are allowed to vary.

The molecule itself is therefore isomorphic through the valence transition, but based on the changes in packing, the most likely conclusion regarding the crystal structure is that it is not isomorphic through the transition. It remains unclear whether the valence transition is driven by the Kondo effect or not. As the low-temperature structure may well be in the *Pbca* space group, the change in structure could be driven by steric effects. The observed valence transition would then be an indication that intermolecular, Van der Waals interactions are important in these molecules, although we note that CASSCF calculations on isolated molecules obtain an n_f that is generally well within 10% of the measured values. The changing intermolecular interaction could, in any case, have an effect on the overall π^* charge density, and so the change in valence could be a manifestation of a charge-density induced change to the f-orbital coupling. This scenario would be very similar to one suggested for $YbInCu_4$, except that in that case the change in density is thought to be due to a sharp band edge near the Fermi level [16].

Another possibility is that intermolecular effects are insignificant, and the valence transition is somehow analogous to that which occurs in the α-γ transition in cerium [9, 10]. Of course, even here, there remains a controversy regarding the basic physics underlying the volume collapes in elemental cerium. The two general possibilities at present are the so-called Kondo Volume Collapse model (KVM) [17], and the Mott transition model [18]. With regard to the role of the Kondo effect, the KVM achieves a volume change by exploiting nonlinearities in the pressure/volume relationship in the Kondo impurity model [17]. Although the possibility of a Mott transition exists in cerium, it is unlikely that there is some local equivalent possiblity in $Cp^*_2Yb(4,4'\text{-}Me_2\text{-bipy})$ given the observed IV in both the high temperature and the low temperature phases. Therefore, CI-induced IV behavior appears to play an important role in generating the valence transition in $Cp^*_2Yb(4,4'\text{-}Me_2\text{-bipy})$, either through structural-induced changes in charge density or through a KVM-like mechanism.

Acknowledgments

Work at Lawrence Berkeley National Laboratory was supported by the Director, Office of Science (OS), Office of Basic Energy Sciences (OBES), of the U.S. Department of Energy (DOE) under Contract No. DE-AC02-05CH11231. Work in the Billinge group was supported by the DOE under contract No. DE-AC02-98CH10886. The APS facility at Argonne is supported by DOE Contract DE-AC02-06CH11357. Work at Los Alamos was supported by the DOE.

References

[1] Neumann C S and Fulde P 1989 *Z. Phys. B* **74** 277
[2] Booth C H, Walter M D, Kazhdan D, Hu Y J, Lukens W W, Bauer E D, Maron L, Eisenstein O and Andersen R A 2009 *J. Am. Chem. Soc.* **131** 6480
[3] Booth C H *et al.* Intermediate-valence tautomerism in decamethylytterbocene complexes of methyl-substituted bipyridines (*Preprint* `http://lise.lbl.gov/chbooth/papers/Me_bipy.pdf`)
[4] Thimm W B, Kroha J and von Delft J 1999 *Phys. Rev. Lett.* **82** 2143
[5] Schlottmann P 2002 *Phys. Rev. B* **65** 024420
[6] Booth C H, Walter M D, Daniel M, Lukens W W and Andersen R A 2005 *Phys. Rev. Lett.* **95** 267202
[7] Walter M D, Berg D J and Andersen R A 2006 *Organometallics* **25** 3228
[8] Felner I and Nowik I 1986 *Phys. Rev. B* **33** 617
[9] Lawrence J M, Riseborough R S and Parks R D 1981 *Rep. Prog. Phys.* **44** 1
[10] McMahan A K, Huscroft C, Scalettar R T and Pollock E L 1998 *J. Comput.-Aid. Mater. Des.* **5** 131
[11] Chupas P J, Qiu X, Hanson J C, Lee P L, Grey C P and Billinge S J L 2003 *J. Appl. Crystallogr.* **36** 1342
[12] Lin H, Božin E S, Billinge S J L, Androulakis J, Malliakas C D, Lin C H and Kanatzidis M G 2009 *Phys. Rev. B* **80** 045204
[13] Egami T and S J L Billinge 2003 *Underneath the Bragg peaks: Structural analysis of complex materials* (Amsterdam: Pergamon)
[14] Qiu X, Thompson J W and Billinge S J L 2004 *J. Appl. Crystallogr.* **37** 678
[15] Farrow C L, Juhás P, Liu J W, Bryndin D, Božin E S, Bloch J, Proffen T and Billinge S J L 2007 *J. Phys.: Condens. Matter* **19** 335219
[16] Figueroa E, Lawrence J M, Sarrao J L, Fisk Z, Hundley M F and Thompson J D 1998 *Solid State Commun.* **106** 347
[17] Allen J W and Martin R M 1982 *Phys. Rev. Lett.* **49** 1106
[18] Johansson B 1974 *Phil. Mag.* **30** 469

CT-QMC and Maximum Entropy Approach to a Scattering-States Formulation of Strongly Correlated Steady-State Transport

Andreas Dirks[1], Philipp Werner[2], Mark Jarrell[3], Thomas Pruschke[1]

[1] Department of Physics, University of Göttingen, D-37077 Göttingen, Germany
[2] Institut für theoretische Physik, ETH Zurich, CH-8093 Zürich, Switzerland
[3] Department of Physics and Astronomy, Louisiana State University, Baton Rouge, LA 70803, USA

E-mail: dirks@theorie.physik.uni-goettingen.de

Abstract. In a recent work [17] Han and Heary introduced a formalism approaching steady-state quantum transport through mesoscopic structures which maps the non-equilibrium problem onto a family of auxiliary equilibrium quantum impurity systems by introducing imaginary voltages. We apply continuous-time quantum Monte-Carlo solvers to obtain precise and unbiased imaginary-time data for these auxiliary models. Physical observables are obtained by an analytical continuation in both Matsubara frequency and complexified voltage using a Maximum Entropy Method (MEM). The MEM is introduced by means of a kernel function compatible with the analytical structure of the theory. While it remains a yet challenging task to obtain reliable spectral functions, this unbiased procedure seems to indicate that the formalism yields results which are compatible with those of other methods.

1. Introduction

Calculating steady-state transport properties of open quantum systems is a challenging and unsolved problem. It has been approached by perturbative methods [1–3], the time-dependent density matrix renormalization group (t-DMRG) [4, 5], real-time quantum Monte Carlo (RT-QMC) [6–9], numerical renormalization group (NRG) [10], and the functional renormalization group (fRG) [11–16]. However, none of the methods developed so far is able to give a complete and reliable description of the physical properties in all parameter regimes. More importantly, the most interesting regime, where all relevant energy scales – voltage, temperature, magnetic field, etc. – are of the same order as the low-energy scale of the Anderson Impurity Model, is usually not accessible. Therefore, the development of new or improved simulation approaches is a worthwhile and important task.

Recently, a new and rather unconventional approach to calculate the steady-state transport through interacting quantum dots or similar structures was proposed by Han and Heary [17]. Their formalism, which is based on Hershfield's density operator [18], maps the non-equilibrium steady-state of the interacting model onto an infinite set of auxiliary equilibrium systems, each characterized by some complex voltage. The complexification of the voltage bias, however, introduces a formidable new problem in the form of an analytical continuation with respect to

Published under licence by IOP Publishing Ltd

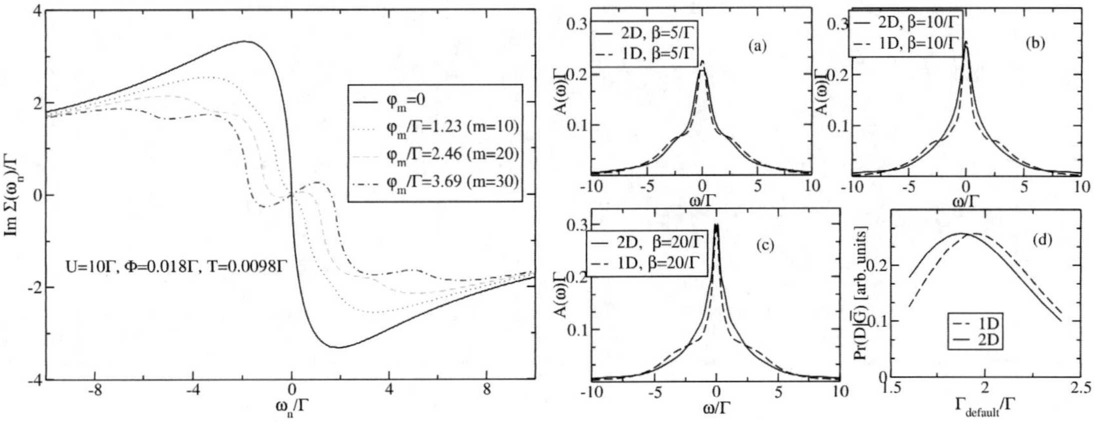

Figure 1. Monte-Carlo results (left panel) and analytically continued results (right panel). The right panel contains equilibrium results (2D) as compared to the conventional Maximum Entropy procedure for the Wick rotation (1D) for $U = 5\Gamma$, $V_G = 0$ and different temperatures.

the voltage on top of the already challenging analytic continuation from Matsubara frequencies to real frequencies.

In a recent preprint [19] we applied Continuous-Time Quantum Monte Carlo [20–22] in order to obtain high-quality data and proposed a scheme for analytical continuation using a Maximum Entropy Method [23, 24]. For the latter purpose, a linear integral equation was derived. We review the results and introduce a possible generalization of the approach.

2. Physical Model and Imaginary Voltages

We consider the single-impurity Anderson model attached to two Fermi liquid leads,

$$H = \sum_{\alpha k\sigma} \varepsilon_{\alpha k\sigma} c^\dagger_{\alpha k\sigma} c_{\alpha k\sigma} + \sum_\sigma V_G d^\dagger_\sigma d_\sigma + \sum_{\alpha k\sigma} \left(V_{\alpha k\sigma} c^\dagger_{\alpha k\sigma} d_\sigma + V^*_{\alpha k\sigma} d^\dagger_\sigma c_{\alpha k\sigma} \right) + U \left(n_\uparrow - \frac{1}{2} \right) \left(n_\downarrow - \frac{1}{2} \right),$$

where V_G is the gate voltage, U is the Coulomb repulsion, $V_{\alpha k\sigma}$ the hybridization between quantum dot and lead and $\varepsilon_{\alpha k\sigma}$ are the dispersion relations of the left ($\alpha = -1$) and right ($\alpha = +1$) leads. The two leads have temperature β and different chemical potentials μ_α, with the bias voltage $\mu_L - \mu_R =: \Phi$.

In the formalism by Han and Heary, the resulting non-equilibrium steady-state problem is mapped to the equilibrium quantum impurity problems given by the effective action for the dot electrons,

$$S_{\text{eff}} = \sum_\sigma \iint_0^\beta \mathrm{d}\tau\, \mathrm{d}\tau'\, d^\dagger_\sigma(\tau') G^{-1}_{0\sigma}(\tau', \tau) d_\sigma(\tau) + U \int_0^\beta \mathrm{d}\tau \left(d^\dagger_\downarrow(\tau) d_\downarrow(\tau) - \frac{1}{2} \right) \left(d^\dagger_\uparrow(\tau) d_\uparrow(\tau) - \frac{1}{2} \right), \tag{1}$$

where the bare Green's function $G_0(\mathrm{i}\varphi_m, \mathrm{i}\omega_n) = \sum_{\alpha=\pm 1} \frac{\Gamma_\alpha/\Gamma}{\mathrm{i}\omega_n - V_G - \frac{\alpha}{2}(\mathrm{i}\varphi_m - \Phi) + \mathrm{i}\Gamma \mathrm{sgn}(\omega_n - \alpha\varphi_m/2)}$ depends on the imaginary voltages $\varphi_m = \frac{4\pi m}{\beta}$, $m \in \mathbb{Z}$ which parametrize the set of equilibrium systems.

The action is solved using continuous-time QMC solvers [20–22]. Typical Monte-Carlo results from the interaction expansion solver are shown in the left panel of Fig. 1. High quality data (error bars of the order of the linewidth) are obtained for the non-trivial regime

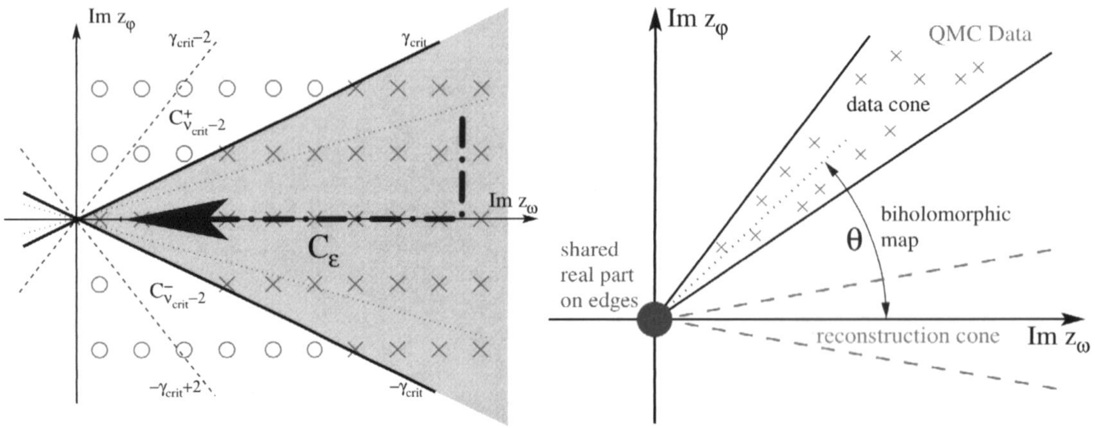

Figure 2. Analytic Structure. Green crosses (circles) denote QMC data which are (are not) taken into account.

$\Phi \approx V_G \approx \beta^{-1} \approx T_K$, where T_K is the equilibrium Kondo temperature. Analytical continuation of the data at $\Phi = 0$ yields a good agreement with results obtained from equilibrium QMC data for conventional Matsubara theory and the conventional Wick rotation inferred by a Maximum entropy method for the integral representation $G(\mathrm{i}\omega_n) = \int \mathrm{d}\varepsilon \frac{A(\varepsilon)}{\mathrm{i}\omega_n - \varepsilon}$. This is shown in the second panel of Fig. 1.

3. Analytic Continuation Procedure

The analytic continuation procedure we used [19] can be summarized as follows. The complexified Green's function $G(\mathrm{i}\varphi_m \to z_\varphi, \mathrm{i}\omega_n \to z_\omega)$ has branch cuts for certain ratios of the imaginary parts of the complex variables. Those branch cuts are denoted by lines in the left panel of Fig. 2 and occur for ratios $\mathrm{Im}\, z_\omega = \frac{\gamma}{2}\mathrm{Im}\, z_\varphi$, γ odd. The desired limiting procedure for the analytic continuation is denoted by the dash-dotted arrow. Hence, the data points within the cone C_ε are considered, where the branch cut γ_{crit} is the last relevant branch cut. This is because the series of relevant branch cuts is expected to terminate at a finite $|\gamma_{\mathrm{crit}}| < \infty$ at finite temperature and interaction. [17] Using analyticity with respect to two complex variables within the domain $T^{C_\varepsilon} = \mathbb{R}^2 + \mathrm{i}C_\varepsilon$ the following integral representation with respect to the edge of the domain can be derived [19]:

$$\mathrm{Im}\, G(\underline{z})|_{T^{C_\varepsilon}} = -\pi \int_{\mathbb{R}^2} \mathrm{d}^2 x\, \mathcal{P}_\varepsilon(\underline{z} - \underline{x}) \tilde{A}(\underline{x}), \tag{2}$$

with

$$\mathcal{P}_\varepsilon(\underline{z}) = \frac{2}{\pi^2 \varepsilon} \prod_{\mu=\pm 1} \frac{y_2 - \mu y_1/\varepsilon}{(x_2 - \mu x_1/\varepsilon)^2 + (y_2 - \mu y_1/\varepsilon)^2}. \tag{3}$$

The function $\tilde{A}(\underline{x})$ is inferred via the Maximum Entropy Method, and the dot electron spectral function is then given by $A(\omega) = \tilde{A}(\Phi, \omega)$. Data in the right panel of figure 1 were calculated with this procedure.

4. Extension of the Procedure

Experience, however, shows that the exclusion of QMC data in the other cones is suboptimal from a stochastic point of view. In the right panel of figure 2 an arbitrary cone is considered, for which the Green's function is holomorphic. We find that a crucial property of the two-variable Green's function, namely the shared real part at the point were the branch cuts cross, suffices to connect the analytic properties of all cones. This property may easily be verified for the bare Green's function. For the interacting case a rigorous proof is still missing. It may, however be motivated by an argument based on hermiticity.

We use the generalization of conformal transformations to several complex variables, called biholomorphic transformations, to perform the generalization. Namely, the formal rotation of a cone to a cone ("reconstruction cone"), for which the Kramers-Kronig relation with respect to one complex variable may be applied, is such a transformation. We define the Hilbert transformation $(\mathcal{H}f)(x,y) = \frac{1}{\pi}\mathcal{P}\int_{-\infty}^{\infty} \frac{f(x,\tilde{y})}{y-\tilde{y}}\,\mathrm{d}\tilde{y}$. and the rotation operator $\mathcal{R}_\vartheta$ which maps a function located on a cone to the cone rotated by angle ϑ using a coordinate transformation. A straightforward calculation yields data not contained in the C_ε-cone may be related to the spectral function in a linear fashion in the following way:

$$\operatorname{Im} G|_{TC_\varepsilon^\vartheta} = \pi \cdot \mathcal{P}_{\varepsilon,\vartheta}\mathcal{R}_\vartheta\mathcal{H}\mathcal{R}_\vartheta^{-1}\mathcal{H} \cdot \tilde{A} =: \mathcal{K}_{\varepsilon,\vartheta}\tilde{A}. \tag{4}$$

Here,

$$\mathcal{P}_{\varepsilon,\vartheta}((\mathrm{i}\varphi_m, \mathrm{i}\omega_n)^T - \underline{x}) = \mathcal{P}_\varepsilon(R_\vartheta^{-1}[(\mathrm{i}\varphi_m, \mathrm{i}\omega_n)^T - \underline{x}]), \tag{5}$$

with the rotation matrix $R_\vartheta^{-1} = \begin{pmatrix} \cos\vartheta & -\sin\vartheta \\ \sin\vartheta & \cos\vartheta \end{pmatrix}$. Current efforts focus on implementing this relation into the Maximum Entropy procedure in order to improve the quality of the inferred solutions.

5. Acknowledgments

We acknowledge useful conversations with Jong Han, Kurt Schönhammer, Sebastian Fuchs, Emanuel Gull, Hans Kroha, and Hartmut Monien. We acknowledge travel support by the DAAD through the PPP program and computer support by the GWDG and HLRN.

References

[1] S. Hershfield, J. H. Davies, and J. W. Wilkins, Phys. Rev. Lett. **67**, 3720 (1991).
[2] S. Hershfield, J. H. Davies, and J. W. Wilkins, Phys. Rev. B **46**, 7046 (1991).
[3] T. Fujii and K. Ueda, Phys. Rev. B **68**, 155310 (2003).
[4] F. Heidrich-Meisner, A. E. Feiguin, E. Dagotto, Phys. Rev. B **79**, 235336 (2009)
[5] P. Schmitteckert, Phys. Rev. B **70**, 121302 (2004)
[6] L. Mühlbacher and E. Rabani, Phys. Rev. Lett. **100**, 176403 (2008).
[7] S. Weiss, J. Eckel, M. Thorwart, and R. Egger, Phys. Rev. B **77**, 195316 (2008).
[8] P. Werner, T. Oka, and A. J. Millis, Phys. Rev. B **79**, 035320 (2009).
[9] M. Schiro and M. Fabrizio, Phys. Rev. B **79**, 153302 (2009).
[10] F. B. Anders, Phys. Rev. Lett. **101**, 066804 (2008).
[11] A. Rosch, J. Paaske, J. Kroha, and P. Wölfle, J. Phys. Soc. Jpn. **74**, 118 (2005).
[12] S. G. Jakobs, V. Meden, H. Schoeller, Phys. Rev. Lett. **99**, 150603 (2007).
[13] R. Gezzi, Th. Pruschke, and V. Meden, Phys. Rev. B **75**, 045324 (2007).
[14] Th. Pruschke, R. Gezzi, A. Dirks, NATO Science Series B: Electron Transport in Nanosystems, 249 (2009).
[15] H. Schoeller, and F. Reininghaus, Phys. Rev. B **80**, 045117 (2009).
[16] Th. Pruschke, A. Dirks, R. Gezzi, Physica B **404**, 3141 (2009).
[17] J. E. Han and R. J. Heary, Phys. Rev. Lett. **99**, 236808 (2007).
[18] S. Hershfield, Phys. Rev. Lett. **70**, 2134 (1993).
[19] A. Dirks, Ph. Werner, M. Jarrell, Th. Pruschke, arXiv:1002.4081v1
[20] A. N. Rubtsov, V. V. Savkin and A. I. Lichtenstein, Phys. Rev. B **72**, 035122 (2005).
[21] P. Werner, A. Comanac, L. de' Medici, M. Troyer and A. J. Millis, Phys. Rev. Lett. **97**, 076405 (2006).
[22] E. Gull, P. Werner, O. Parcollet and M. Troyer, Europhys. Lett. **82** 57003 (2008).
[23] M. Jarrell, J. E. Gubernatis, Physics Reports **269**, 133 (1996).
[24] M. Jarrell, A. Macridin, K. Mikelsons, and D.G.S.P. Doluweera, in *Lectures on the Physics of Strongly Correlated Systems XII*, AIP Conference Proc. **1014**, A. Avella and F. Mancini (Eds.), 34 (2008)

International Conference on Strongly Correlated Electron Systems (SCES 2010) IOP Publishing
Journal of Physics: Conference Series **273** (2011) 012151 doi:10.1088/1742-6596/273/1/012151

Optical conductivity and superconductivity in $LaSb_2$

J F DiTusa[1], V Guritanu[2], S Guo[1], D P Young[1], P W Adams[1], R G Goodrich[1], J Y Chan[3] and D van der Marel[2]

[1] Department of Physics and Astronomy, Louisiana State University, Baton Rouge, LA, 70803, USA

[2] Département de Physique de la Matière Condensée, Université Genève, CH-1211 Genève 4, Switzerland

[3] Department of Chemistry, Louisiana State University, Baton Rouge, LA 70803, USA

E-mail: `ditusa@phys.lsu.edu`

Abstract. We have measured the resistivity, optical conductivity, and magnetic susceptibility of $LaSb_2$ to search for clues as to the cause of the extraordinarily large linear magnetoresistance and to explore the properties of the superconducting state. We find no evidence in the optical conductivity for the formation of a charge density wave state above 20 K despite the highly layered crystal structure. In addition, only small changes to the optical reflectivity with magnetic field are observed indicating that the MR is due to scattering rate, not charge density, variations with field. Although a superconducting ground state was previously reported below a critical temperature of 0.4 K, we observe, at ambient pressure, a fragile superconducting transition with an onset at 2.5 K. In crystalline samples, we find a high degree of variability with a minority of samples displaying a full Meissner fraction below 0.2 K and fluctuations apparent up to 2.5 K. The application of pressure stabilizes the superconducting transition and reduces the anisotropy of the superconducting phase.

1. Introduction

The observation of a large linear magnetoresistance, MR, in the diantimonide, $R\mathrm{Sb}_2$ (R =La-Nd, Sm), family of compounds has stimulated recent interest because of the difficulty in modeling such behavior. These materials are layered compounds that crystallize in the orthorhombic $SmSb_2$ crystal structure demonstrated in Fig. 1[1]. This structure is characterized by La/Sb layers that alternate with flat, rectangular, sheets of Sb atoms stacked along the crystallographic c-axis. All of the $R\mathrm{Sb}_2$ compounds are metallic with $LaSb_2$ reported to be superconducting, SC, with a critical temperature, T_c, of 0.4 K[2] that is not yet fully characterized. For these reasons we investigated $LaSb_2$ to explore the mechanism for the large MR and its SC properties.

The MR of $LaSb_2$ reproduced in Fig. 2 is linear in field above 3 T[3], is highly anisotropic, and does not saturate in fields of up to 45 T. These features led to the suggestion that $LaSb_2$ is useful as a magnetoresistive sensor for high field applications[4]. Several mechanisms have been proposed over the past 35 years to explain anomalously linear MRs which were observed in disordered semiconductors[5], two-dimensional (2D) heterostructures[6], and elemental metals[7, 8]. The purported causes include the nucleation of charge density wave (CDW) distortions with MR due to quantum fluctuations about the CDW[9] ground state and/or a magnetic breakdown of the CDW gap[10], high field quantization effects[11], or singular scattering mechanisms[12]. While $LaSb_2$ does not appear to be in the extreme quantum limit at

Published under licence by IOP Publishing Ltd

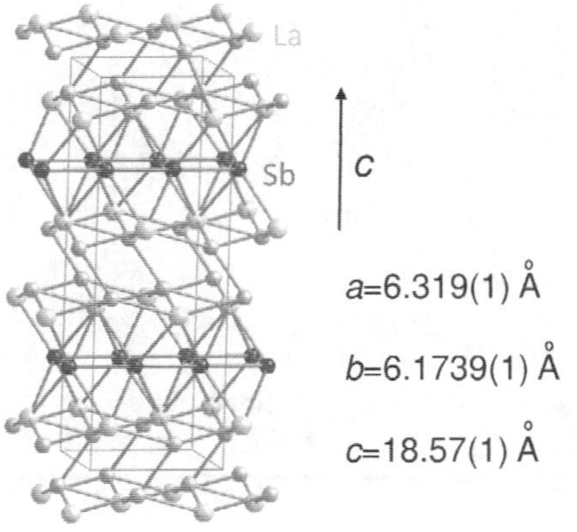

Figure 1. Lattice structure of $LaSb_2$. The layered crystal structure highlighted by flat sheets of Sb (blue) separated by the La (orange) and Sb interlayers. The unit cell is indicated by the box.

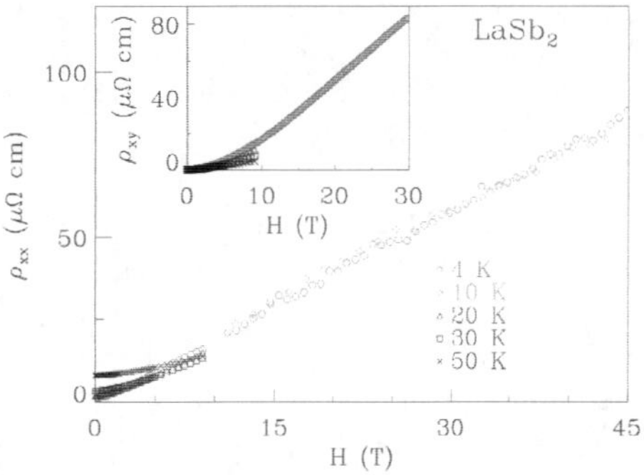

Figure 2. Magnetoresistance and Hall effect. Resistivity, ρ_{xx}, vs transverse magnetic field, H, for single crystal $LaSb_2$ at temperatures identified in the figure. Inset: Hall resistivity, ρ_{xy} vs H for the same temperatures (symbols same as in main frame).

the fields of Fig. 2, and there is no indication that conditions are favorable for strongly singular scattering, it remains possible that there is a CDW ordering. This possibility is supported by the fact that $NbSe_2$ and $TaSe_2$, two prototypical CDW systems[13], have similar crystal structures and linear MRs[14, 15]. However, measurements of the resistivity, ρ, showed no transitions or unusual temperature, T, dependencies that one usually associates with a CDW transition[3] and photoemission[16] and neutron diffraction measurements[17] did not reveal any phase transitions.

Here, we present optical ellipsometry, reflectivity, ρ, and magnetic susceptibility, χ, measurements of single crystalline $LaSb_2$. In agreement with the previous measurements outlined above, our optical measurements reveal no features indicative of CDW formation. We find an extraordinarily wide transition to an anisotropic SC phase that sharpens with pressure and which suggests that SC phase fluctuations determine T_c.

2. Experimental details

Single crystals of $LaSb_2$ were grown from high purity La and Sb by metallic flux methods that had typical dimensions 0.5 by 0.5 by 0.02 cm. The crystal structure displayed in Fig. 1 was determined from single crystal x-ray diffraction. Optical experiments were performed using spectroscopic ellipsometry at 0.75 - 3.7 eV in combination with reflectivity measurements from 7 meV to 0.85 meV. Because there is no significant difference between the 300 and 20 K reflectivity spectra in the mid-infrared energy range, the ellipsometry data were obtained at room temperature. All measurements were performed on freshly polished crystals introduced into a vacuum cryostat for the T-dependent measurements and kept in a flow of dry nitrogen for room T experiments in order to avoid the contamination of the surface. To obtain the absolute value of the reflectivity a reference gold layer was evaporated *in situ* on the sample surface. From the frequency, F, dependent reflectivity and ellipsometry data we derived the real part of the conductivity using a Kramers-Kronig consistent variational fitting procedure[18]. Reflectivity measurements were carried out in Hs of up to 7 T at a few selected Ts. ac susceptibility measurements were performed in a commercial SQUID magnetometer down to 1.75 K and in a dilution refrigerator ac χ probe for $T \geq 50$ mK. Hall effect and ρ measurements were performed on single crystals with electrical contacts made using silver epoxy employing lock-in techniques at 17 or 27 Hz at Ts between 0.05 and 300 K. Some measurements reported here were made at the National High Magnetic Field Laboratory in continuous fields of up to 45 T (35 T) for ρ (Hall effect) measurements. The crystals studied here had residual ρ ratios of between 70 and 90 for Ts between 300 and 4 K.

3. Experimental Results

The transverse MR, $\rho_{xx}(H)$, and Hall resistivity, ρ_{xy}, measured with currents parallel to the ab planes presented in Fig. 2 are similar to previous measurements[3, 4] displaying a large, highly linear MR and a large Hall constant. With the Hall carrier density, n, estimated to be 2×10^{20} cm^{-3} and using the effective mass determined from previous de Haas-van Alphen experiments[19] of $m^* = 0.2$ times the bare electron mass, we estimate the mean free path, ℓ, to be 3.5 μm for carriers moving along the ab planes. The resistivity along the c-axis, ρ_c is up to 200 times larger at 4 K as can be seen in Fig. 3.

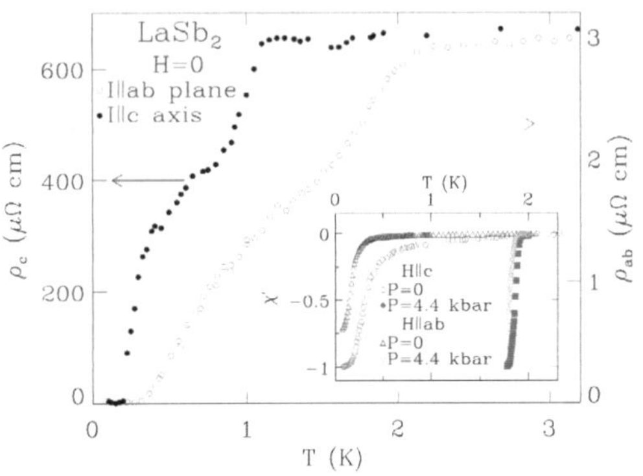

Figure 3. Resistivity and Magnetic susceptibility. The resistivity for currents parallel to the c-axis, ρ_c and the ab planes, ρ_{ab} vs temperature, T, across the superconducting transition. Inset: The real part of the ac magnetic susceptibility, χ', vs T at zero and 4.4 kbar applied hydrostatic pressure for two ac excitation magnetic field orientations.

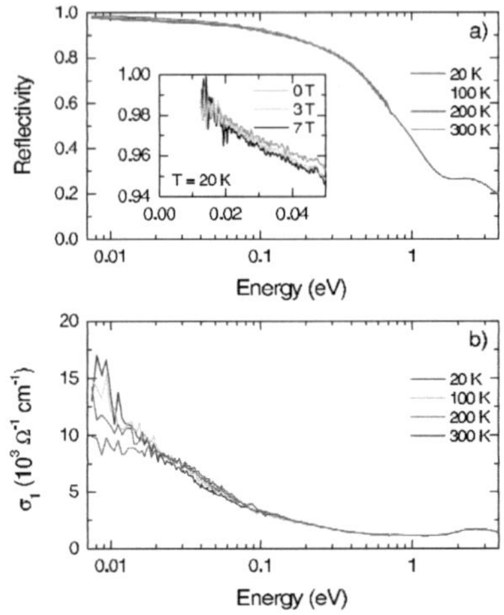

Figure 4. Optical response of LaSb$_2$. The reflectivity (a) and optical conductivity, σ_1, (b) at temperatures identified in the figure. Inset: Reflectivity at 20 K in magnetic fields of 0, 3 and 7 T.

The optical reflectivity spectra of LaSb$_2$ at several Ts is presented in Fig. 4. All spectra show a typical metallic behavior with a high value of the reflectivity at low F and a well defined plasma edge at 1.7 eV. The conductivity, σ_1, spectra obtained from the reflectivity and ellipsometry measurements are displayed in Fig. 4b. These data show a small T dependence and no interband transition appears to be present at low energy. At higher energy a peak centered at about 2.5 eV is attributed to an interband transition. At low F, σ_1 is featureless and displays no features which can be associated with the opening of a CDW gap. The normalized reflectivity versus photon energy for three applied magnetic fields at T = 20 K is shown in the inset of Fig. 4a. We see no systematic trends in these data outside of the measurement error as H is increased. The data agree with simple estimates of the field induced changes to the reflectivity based upon the MR of Fig. 2 and the Hagen-Rubens formula. Our data rule out any significant change to the electronic structure of LaSb$_2$ with magnetic field at energies above 10 meV.

The wide transition into the SC phase as probed by ρ and ac χ measurements is displayed in Fig. 3. Both of these quantities show a very wide, anisotropic, phase transition where ρ

decreases and the real part of the ac magnetic susceptibility, χ', inset, displays diamagnetism below 2.5 K. Here χ' is shown for a crystal displaying a complete Meissner fraction, $\chi' = -1$, for ac excitation fields, H_{ac}, along the c-axis for $T < 0.25$ K. The application of moderate pressure, represented in the inset to Fig. 3 by data at 4.4 kbar, significantly sharpens the transition so that by 1.7 K $\chi' = -1$ for $H_{ac} \parallel c$.

4. Discussion and Conclusions

Our measurements of σ_1 of $LaSb_2$ effectively rule out the formation of a CDW phase above 20 K showing no unusual changes from a typically metallic form. In addition, we see no significant changes in the reflectivity with magnetic field. Thus, our data suggest that the MR is a scattering induced effect which is not related to the formation of a CDW induced energy gap or variations of the electronic structure with H. In addition, our measurements reveal an unusual, anisotropic, SC transition that is very broad in T that is incomplete down to our lowest Ts in the majority of crystals. The application of P sharpens the transition dramatically, inducing a full SC transition. Similar SC phase transitions have been previously observed in layered materials such as 2H-TaS_2 where the width of the transition was associated with the competition of a CDW phase[20]. In $LaSb_2$ we have found no indication of a CDW, or other energy gap inducing, phases that compete with superconductivity either in σ_1 above 20 K nor in $\rho(T)$ at any T. This suggests a different cause for the wide SC transition we observe. Because the transition sharpens with pressure, and because of the long ℓ for charge carriers moving parallel to the ab planes, we do not believe impurity or crystal defects to be important. Instead we suggest that the SC phase in $LaSb_2$ at $P = 0$ may be phase fluctuation limited as in the underdoped high T_c SC cuprates.

4.1. Acknowledgments

JFD, DPY, and JYC acknowledge support from the NSF through DMR 084376, 0449022, and 0756281. PWA acknowledges support from the U.S. DOE through DE-FG02-07ER46420. Work at the NHMFL was performed under the auspices of the NSF and the state of Florida.

References

[1] Wang R and Steinfink H 1967 *Inorg. Chem.* **6** 1685
[2] Hulliger F and Ott H R 1977 *J. Less-common met.* **55** 103
[3] Bud'ko S L, Canfield P C, Mielke C H and Lacerda A H 1998 *Phys. Rev. B* **57** 13624
[4] Young D P, Goodrich R G, DiTusa J F, Guo S, Adams P W, Chan J Y and Hall D 2003 *Appl. Phys. Lett.* **82** 3713
[5] Husmann A, Betts J B, Boebinger G S, Migliori A, Rosenbaum T F and Saboungi L 2002 *Nature* **417** 421
[6] Stormer H L, Baldwin K W, Pfeiffer L N and West K W (1992) *Sol. St. Commun.* **84** 95
[7] Falikov L M and Smith H 1972 *Phys. Rev. Lett.* **29** 124
[8] Kapitza P L 1928 *Proc. R. Soc. London, Ser. A* **119** 358
[9] Castro Neto A H 2001 *Phys. Rev. Lett.* **86** 4382
[10] Wilson J A, DiSalvo F J and Mahajan S 1975 *Adv. Phys.* **24** 117
[11] Abrikosov A A 1999 *Phys. Rev. B* **60** 4231
[12] Young R A 1968 *Phys. Rev.* **175** 813
[13] Moncton D E, Axe J D and DiSalvo F J 1975 *Phys. Rev. Lett.* **34** 734
[14] Morris R C, Coleman R V and Bhandari R 1972 *Phys. Rev. B* **5** 895
[15] Naito M, Tanaka S and Miura N 1981 *Physics in High Magnetic Field* edited by Chikazumi S and Miura M (Springer NY) 320.
[16] Acatrinei A I, Browne D, Losovyi Y, Young D P, Moldovan M, Chan J Y, Sprunger P T and Kurtz R L 2003 *J. Phys. Cond. Mat.* **15** L511
[17] Kurtz R L Private communication.
[18] Kuzmenko A B 2–5 *Rev. Sci. Instrum.* **76** 083108
[19] Goodrich R G, Browne D A, Kurtz R L, Young D P, DiTusa J F, Adams P W and Hall D (2004) *Phys. Rev. B* **69** 125114
[20] Nagata S, Aochi T, Abe T, Ebisu S, Hagino T, Seki Y and Tsutsumi K 1992 *J. Phys. Chem. Solids* **53** 1259

Effect of elemental substitutions on ruthenates with pyrochlore-type structure

M Kato, S Harada, M Fukatsu and K Hirota

Department of Molecular Chemistry and Biochemsitry, Faculty of Science and Engineering, Doshisha University, Kyotanabe 610-0321, Japan

E-mail: makato@mail.doshisha.ac.jp

Abstract. Synthesis and physical properties of $Pb_{2-x}Ln_xRu_2O_{7-\delta}$ (Ln = Y, Eu, Lu) phase are reported. The pyrochlore ruthenates $Pb_{2-x}Ln_xRu_2O_{7-\delta}$ (Ln = Y, Eu, Lu) have been successfully synthesized by the solid-state reaction. The solid-solution compounds of $Pb_{2-x}Lu_xRu_2O_{7-\delta}$ undergo a crossover from metallic (x = 0) to insulating (x = 2) with increasing x. Owing to a localized magnetic moment on Ru site, these compounds with $x > 0.2$ also indicate the spin-glass (SG) like behaviors at low temperature ($T < T_{SG}$). From electric resistivity measurements for the Y-substituted system, the sample of x = 0.0 shows metallic behaviros and that of x = 1.0 shows insulating properties. In $Pb_{2-x}Eu_xRu_2O_{7-\delta}$ system, the solid-solution compounds show from metallic (x = 0) behaviors to insulating (x = 2) behaviors with increasing x. In a similar manner as $Pb_{2-x}Lu_xRu_2O_{7-\delta}$ system, it was found that T_{SG} decreases monotonously with decreasing x and goes to zero at x = 0.6. As a result of electric resistivity measurements, the samples with $0.4 \leq x \leq 0.8$ show the metal-insulator (MI) transition at 10-30 K where the slope of temperature dependence of electrical resistivity ( ) changes. In the metallic region, values of obey T^2-dependencies which indicate the strongly correlated electron behaviors in this system.

1. Introduction

Pyrochlore oxides constitute a large family of transition metal oxides with a chemical formula of $A_2B_2O_7$ [1–3], where A is a larger cation and B is a smaller transition metal. Both A and B ions individually form a three dimensional sub-lattices consisting of corner-shared tetrahedra. When the exchange interaction between the nearest-neighbor atoms is antiferromagnetic, the magnetic moments on the atoms in such sub-lattice are expected to be geometrically frustrated. As a result, the system shows a freezing into a spin glass state which has been found in several pyrochlores or a disordered ground state instead of a magnetically ordered state. Electric and magnetic properties of pyrochlores with B sites of $4d$ elements are especially attractive, since the $4d$ electrons generally indicate the variety of localized and itinerant behaviors [4–8] . For example, the bismuth ruthenate $Bi_2Ru_2O_7$ is metallic and Pauli paramagnetic with a nearly temperature-independent resistivity, while the rare earth ruthenates $Ln_2Ru_2O_7$ (Ln represents lanthanoids from Pr to Lu) are all semiconductors with localized magnetic moment on the ruthenium atoms. In order to clarify the transport behaviors of pyrochlore ruthenates near the boundary of the metal-insulator crossover, detailed investigations of the physical properties have been performed for the Ln-substituted system $Bi_{2\ x}Ln_xRu_2O_7$ [9, 10]. Although $Pb_2Ru_2O_{6.5}$ is known to show a Pauli paramagnetism with a metallic behavior similar to $Bi_2Ru_2O_7$, there

Published under licence by IOP Publishing Ltd

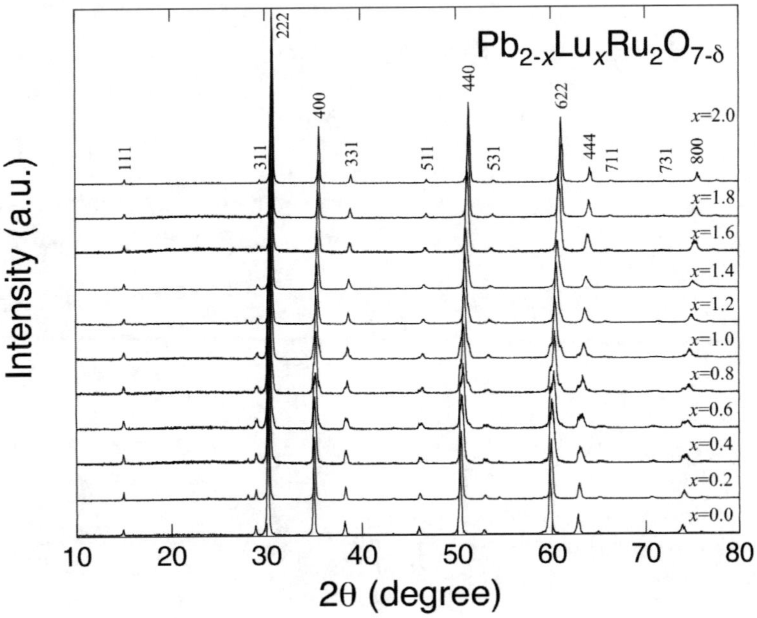

Figure 1. XRD patterns of $Pb_{2\;x}Lu_xRu_2O_7$ compounds.

have been a few reports on Ln-substituted system $Pb_{2\;x}Ln_xRu_2O_7$ because of the di culty in synthesizing a single phase [11]. In this paper, the synthesis and physical properties of $Pb_{2\;x}Ln_xRu_2O_7$ (Ln = Y, Eu, Lu) phase are reported [12].

2. Experimental

The lead-lanthanoid pyrochlores $Pb_{2\;x}Ln_xRu_2O_7$ have been synthesized by the solid-state reaction method. The raw materials of PbO, Ln_2O_3 and Ru with 99.99% purity were mixed thoroughly, pelletized and heated in air at 850 ° C for 24 hours. Then the sample was ground, pelletized and heated in air at 850 — 1200 C for 48 hours. Several samples with Pb rich composition were sealed in Pt tubes to prevent the sublimation at high temperatures. Obtained samples were characterized by X-ray di raction measurements (Rigaku, RINT-2200). Magnetic susceptibilities were measured by SQUID magnetometer (Quantum Design, MPMS-5T). Electric resistivity measurements were performed by a conventional four-terminal method.

3. Results and Discussion

Figure 1 shows XRD patterns of $Pb_{2\;x}Lu_xRu_2O_7$ samples. All the sample is successfully indexed by the pyrochlore structure. As shown in Fig. 2, lattice parameters were found to decrease linearly with increasing Lu content x. This indicates that the solid solution compounds of $Pb_{2\;x}Lu_xRu_2O_7$ were obtained as single phases. Eu and Y substituted compounds were also synthesized as single phases in the range of $0 \leq x \leq 2$.

Figure 3 shows temperature dependences of magnetic susceptibility ( ) for $Pb_{2\;x}Lu_xRu_2O_7$ compounds. $Pb_2Ru_2O_{6.5}$ compound (x = 0) shows temperature-independent Pauli paramagnetism as previously reported [13, 14]. The other end material, $Lu_2Ru_2O_7$ indicates spin-glass like behavior at $T = T_{SG}$ of about 80 K where values of for zero- eld cooling (ZFC) and eld cooling (ZFC) diverge [15]. Values of for the samples of $x > 1.0$ are well analyzed by

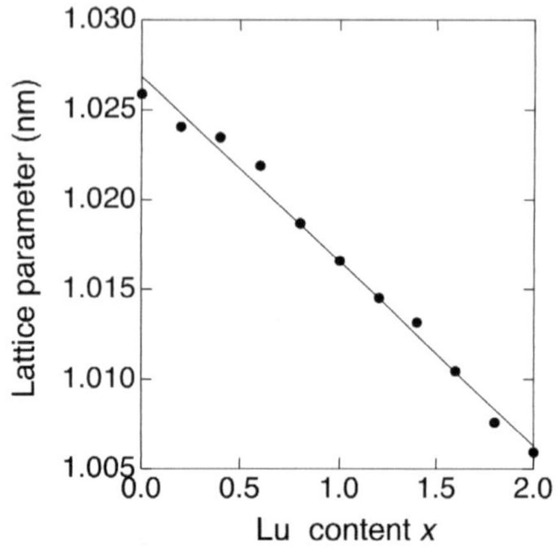

Figure 2. Lu composition dependence of lattice parameters in $Pb_{2\ x}Lu_xRu_2O_7$ compounds.

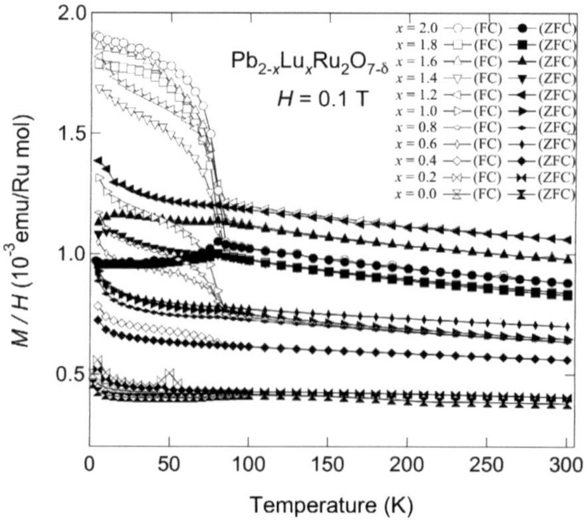

Figure 3. Temperature dependeces of magnetic susceptibility for $Pb_{2\ x}Lu_xRu_2O_7$ compounds. Open and solid marks corresponds to FC and ZFC measuraments.

Curie-Weiss law with negative values of Weiss temperatures which indicates a localized-moment character with antiferromagnetic interactions. The spin freezing temperature, T_{SG}, slightly decreases with Lu content x and disappears for $x = 0.2$. In the case of Ln = Y and Eu, similar behaviors were observed [12]. In the case of Pb-Eu system, another anomalies in $\ $-$T$ plot were observed at lower temperatures of about 25 K for samples with $0.2 < x < 1.8$. These anomalies were attributed to the magnetic order of rare-earth atoms in other compounds [13], although the ground state of Eu^{3+} is non-magnetic. These anomalies may be caused by magnetic Eu^{2+} ions due to the oxygen de ciency and/or an impurity phase.

In Fig. 4, we show temperature dependences of electric resistivity (ρ) of $Pb_{2\ x}Y_xRu_2O_7$ system. It was found that the conductivity of the system changes from metallic to semiconducting with increasing x. From a preliminary analysis, it was found that Ru-O-Ru bond angles of about 130 ° increases with decreasing x. These structural distortions can be caused by the $6s^2$ lone-pair electrons on Pb sites, leading to the band mixture with Ru $4d$ band at the Fermi surface via the framework of oxygen atoms.

Fermi liquid behaviors such as T^2-dependence of electric resistivity and linear heat-capacity coe cient ( ) are reported for a single crystal of $Pb_2Ru_2O_{6.5}$ [14]. It was found that the ratio of T^2 coe cient, A, to ^{2} of $Pb_2Ru_2O_{6.5}$ corresponds to the universal Kadowaki-Woods relation as shown in other strongly correlated transition metal oxides [16]. From this viewpoint, it is of interest to study $A/\ ^2$ of the Ln substituted system which will be reported elsewhere.

4. Summary

The lead-lanthanoid pyrochlores $Pb_{2\ x}Ln_xRu_2O_7$ (Ln = Y, Eu, Lu) have been successfully synthesized by the solid-state reaction. Obtained samples were characterized by X-ray di raction, magnetic susceptibility, and electrical resistivity measurements. Magnetic susceptibility measurements indicated that the spin freezing temperature, T_{SG}, decreases with Ln content x and goes to zero at $x = 0.2$ for Lu and Y. From electric resistivity measurements, a crossover from metallic to semiconducting behaviors was observed in the $Pb_{2\ x}Y_xRu_2O_7$

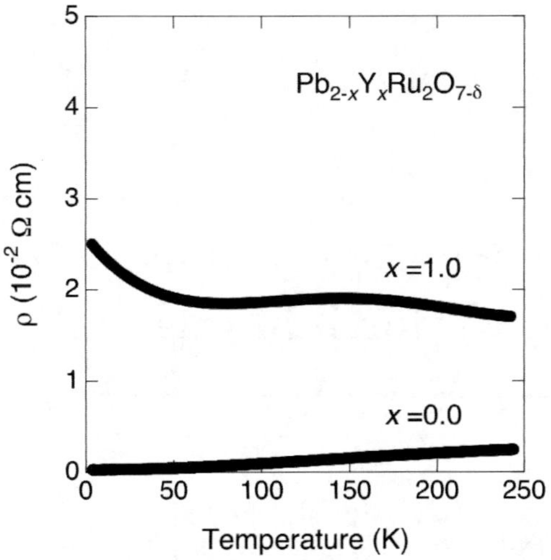

Figure 4. Temperature dependences of electric resistivity for $Pb_{2\ x}Lu_xRu_2O_7$ compounds.

system. These boundaries between metallic and insulating phases mostly correspond to values of x with $T_{SG} = 0$. Further microscopic experiments such as NMR or neutron di raction measurements should be necessary to clarify the e ect of the geometrical frustration near $x = 2.0$ on the crossover to itinerant behaviors around $x = 1.0$.

Acknowledgments
This work was partially supported by a Grant-in-Aid Scienti c Research of Japan Society for Promotion of Science (22550129), and by "Advanced Study for Integrated Particle Science and Technology," Strategic Development of Research Infrastructure for Private Universities, #S0901039, the Ministry of Education, Culture, Sports, Science and Technology (MEXT), Japan.

References

[1] Subramanian M A, Aravamudan G and Subba Rao G V 1983 *Prog. Solid State Chem.* **15**, 55
[2] Bouchard R J and Gillson J L 1971 *Mater. Res. Bull.* **6** 669
[3] Ramirez A P, Hayashi A, Cava R J, Siddharthan R and Shastry B S 1999 *Nature* **399** 333
[4] Harris M J, Bramwell S T, McMorrow D F, Zeiske T and Godfrey K W 1997 *Phys. Rev. Lett.* **79** 2554
[5] Bramwell S T and Harris M J 1998 *J. Phys.: Condens. Matter* **10** L215
[6] Hanawa M, Muraoka Y, Tayama T, Sakakibara T, Yamaura J and Hiroi Z 2001 *Phys. Rev. Lett.* **87** 187001
[7] Sakai H, Yoshimura K, Ohno H, Kato H, Kambe S, Walstedt R E, Matsuda T D, Haga Y and Onuki 2001 J. Phys.: Condens. Matter **13** L785
[8] Shimakawa Y and Kubo Y 1999 *Phys. Rev. B.* **59** 1249 (1999)
[9] Yoshii S, Murata K and Sato M 2000 *J. Phys. Soc. Jpn.* **69** 17
[10] Takeda T, Nagata M., Kobayashi H, Kanno R, Kawamoto Y, Takano M, Kamiyama T, Izumi F and Sleight A W 1998 *J. Solid State Chem.* **140** 182
[11] Yoshii S, Murata K and Sato M 2001 *J. Phys. Chem Solids.* **62** 129
[12] Kato M, Tani Y, Imamura T, Hirota K and Yoshimura K 2008 *Physica* B **403** 1315
[13] Taira N, Wakeshima M and Hinatsu Y 1999 *J. Phys: Condens. Matter.* **11** 6983
[14] Tachibana M, Shimoyama T, Harada A, Taniyama T, Itho M, Kawaji H and Atake T 2006 *Phys. Rev.* B **73** 193107
[15] Taira N, Wakeshima M and Hinatsu Y 1999 *J. Solid State Chem.* **144** 216
[16] Kadowaki K and Woods S B 1986 *Solid State Commun.* **58** 507

An Orthofermion Approach for Thermodynamic Properties of Infinite U Hubbard Model in one and two Dimensions

A. K. Mishra[a,*], **R. Kishore**[b]

[a]Institute of Mathematical Sciences, CIT Campus, Chennai 600113, India
[b]Insitituto Nacional de Pesquidas Espaciais - INPE, CP 515, S. J. Campos, SP 12245-970, Brazila

E-mail: [a]`mishra@imsc.res.in`; [b]`kishore@las.inpe.br`

Abstract. Hubbard model plays a critical role in the study of strongly correlated electron systems. The only available exact solutions of 1-D Hubbard Hamiltonian, obtained through nested Bethe ansatz, show a spin charge decoupling for infinite U when there is no possibility of double occupancy of an electron orbital state. It had been shown that these features can be directly obtained by employing orthofermion algebra. In the present communication, the thermodynamic properties, namely, specific heat and entropy are provided for a 1-D chain and 2-D square lattice using orthofermion approach. It is shown that in 1-D, our results are identical to the known exact results. We have also compared the thermodynamic properties of free fermions and orthofermions in both 1-D and 2-D. Our thermodynamic results, based on orthofermion approach, can be used to ascertain the accuracy of the approximate methods employed in the context of infinite U Hubbard model.

1. Introduction

The Hubbard model provides the simplest Hamiltonian describing a correlated electron system [1]. The U infinity limit of the Hubbard Hamiltonian has come to occupy a central role in the context of strongly correlated electron systems. It has been extensively employed in the context of high temperature superconductivity as well as for understanding the magnetic properties of correlated systems, as described by $t-J$ Hamiltonian. An important consequence of infinite intrasite electron repulsion U is that each electron orbital state can be utmost occupied by a single electron, irrespective of its spin direction.

Obtaining solutions of Hubbard Hamiltonian remains a challenging task. Till date, only exact solution of the Hubbard Hamiltonian is available for a 1-D system through the application of nested Bethe ansatz (NBA) approach [2]. Even in one dimension, although we have a fairly complete description of the ground state properties by the exact NBA solutions, there exist very few exact results at finite temperatures. Exact analytical expressions for the thermodynamic properties for $U = \infty$ have been derived by many authors in one dimension, but their methods cannot be extended to higher dimensions[3, 4, 5, 6]. However, it has been shown that orthofermion approach not only reproduces the known exact 1-D results for infinite U, but can also be used to obtain 2-D results [7].

Published under licence by IOP Publishing Ltd

In the present communication, we obtain the thermodynamic properties of noninteracting orthofermions in one and two dimensions, and compare the results with non interacting fermions having identical particle density.

2. Orthofermion algebra and nested Bethe ansatz solutions

Nested Bethe ansatz solutions exhibit nontrivial consequences in the U infinity limit. As highlighted by Fulde, a remarkable property of the U infinity NBA wavefunctions is that their spatial part correspond to that of spin-less fermions for any distribution of particles or holes on the chain [8, 9, 10]. This general feature of the NBA state vectors automatically implements the 'no double-occupancy' constraint. Unlike in the Gutzwiller projection operator approach (GPO), no additional projection operators are required. It is pertinent to point out that in the presence of holes, the NBA wave function is a superposition of spin-charge factorized wavefunctions. It may be noted that in contrast to orthofermion approach, GPO fails to reproduce the the exact energy spectrum in the U infinity limit [11]

The orthofermion algebra automatically satisfies the requirement of no double occupancy and spin charge decoupling as exhibited by the NBA solutions. It also extends the exact 1-D solutions of infinite U Hubbard model to the higher dimensions in a very simple and natural way.

3. Thermodynamic Properties

The Hamiltonian for a system of noninteracting orthofermions, in wave vector space, is given as

$$H = \sum_{\mathbf{k}\sigma} (\epsilon_k - \mu) a^\dagger_{\mathbf{k}\sigma} a_{\mathbf{k}\sigma} \tag{1}$$

where orthofermion creation and annihilation operators satisfy the algebra

$$a^\dagger_{\mathbf{k}\alpha} a^\dagger_{\mathbf{k}'\beta} + a^\dagger_{\mathbf{k}'\alpha} a^\dagger_{\mathbf{k}\beta} = \delta_{\mathbf{k}\mathbf{k}'}\delta_{\alpha\beta}; \quad a_{\mathbf{k}\alpha} a^\dagger_{\mathbf{k}'\beta} = \delta_{\alpha\beta}(\delta_{\mathbf{k}\mathbf{k}'} - \sum_{\gamma} a^\dagger_{\mathbf{k}'\alpha} a_{\mathbf{k}\beta}) \tag{2}$$

The various thermodynamic quantities can be obtained using the the orthofermion partition function

$$Z_{of} = \prod_{\mathbf{k}} (1 + e^{-\beta(\epsilon_{\mathbf{k}} - \bar{\mu})}) \tag{3}$$

where

$$\bar{\mu} = \mu + \frac{1}{\beta} \ln(2) \tag{4}$$

Since the thermal average of any operator O is given by $< O >= Tr(Oe^{-\beta H})/Z$, the distribution function $< n_{\mathbf{k}} >= \sum_\sigma a^\dagger_{\mathbf{k}\sigma} a_{\mathbf{k}\sigma}$ is given as

$$< n_{\mathbf{k}} >= -\frac{1}{\beta}\frac{\partial \ln Z}{\partial \epsilon_{\mathbf{k}}} = \frac{1}{e^{\beta(t_{\mathbf{k}} - \bar{\mu})} + 1}. \tag{5}$$

The above distribution function has been obtained earlier by Kwak and Beni [5] for the infinite U Hubbard model in absence of magnetic field.

Using the partition function (3), we get entropy S and the specific heat C_v as

$$S = -k \sum_{\mathbf{k}} [< n_{\mathbf{k}} > \ln < n_{\mathbf{k}} > + (1- < n_{\mathbf{k}} >) \ln(1- < n_{\mathbf{k}} >)] + Nk \;\; ln(2) \tag{6}$$

$$C_v = k\beta^2 [\sum_{\mathbf{k}} \epsilon^2_{\mathbf{k}} < n_{\mathbf{k}} > (1- < n_{\mathbf{k}} >) - \frac{(\sum_{\mathbf{k}} t_{\mathbf{k}} < n_{\mathbf{k}} > (1- < n_{\mathbf{k}} >))^2}{\sum_{\mathbf{k}} < n_{\mathbf{k}} > (1- < n_{\mathbf{k}} >}] \tag{7}$$

For a system of free fermions, Hamiltonian has the same form as that of orthofermions, given by eq (1), with orthofermion creation and annihilation operators being replaced by fermion creation and annihilation operator. The thermodynamic expressions (3) to (7) are of the same form except we have to replace $\bar{\mu}$ by μ and $\mathbf{k}$ by $\mathbf{k}\sigma$, and omit the second term $Nk\ ln(2)$ in the expression(6) of the entropy. In case of orthofermions this term is free spin entropy because of spin charge separation.

Both for fermions and orthofermions, the energy $\epsilon_{\mathbf{k}}$ equals to $2\Delta cos(k)$ for a 1-D system and $2\Delta(cos(k_x) + cos(k_y))$ for a two dimensional square lattice. Δ is the hopping term. Using these relations we provide the thermodynamic quantities in the next section.

4. Results and Discussions

The results corresponding to the dependence of entropy and specific heat per particle on the temperature for the average particle densities $n = 0.2$, 0.5 and 0.8, both for orthofermions and fermions, are presented in this section. It may be noted that for a 1-D system, the thermodynamic quantities obtained by orthofermion approach exactly coincide with the results of Klein, obtained through a projected Hubbard Hamiltonian [12]. One draw back of Klein approach is that the creation and annihilation operators introduced therein violate particle and spin number conservation. No such limitation exists for orthofermion creation and annihilation operators. For a better comparison of the results, plots for fermions and orthofermions appear in the same figure.

The results for entropy vrs temperature in 1-D system are plotted in FIG. 1. In contrast to entropy of fermions, which tends to zero as temperature nears zero, the entropy of orthofermions remains finite at zero temperature. For orthofermions statistics, entropy has an additional contribution of $nk\ ln(2)$, because of free spins. Therefore at low temperature, entropy of fermions are lower than orthofermions entropy, though for larger temperature this trend gets reversed.

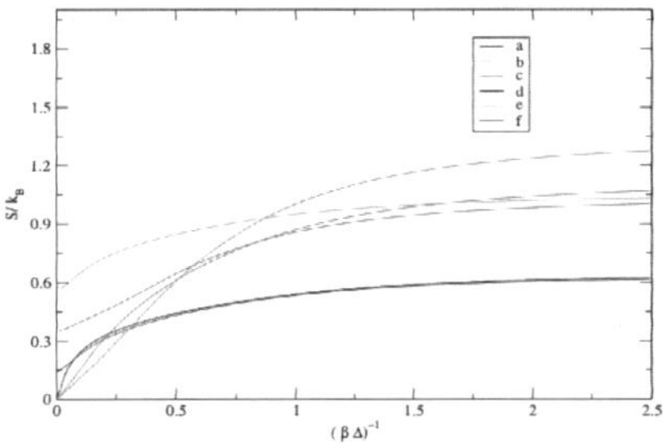

FIG. 1 Entropy of orthofermions: n = 0.2 (a), 0.5 (b), 0.8 (c); for fermions: n = 0.2 (d), 0.5 (e), 0.8 (f) in 1-D.

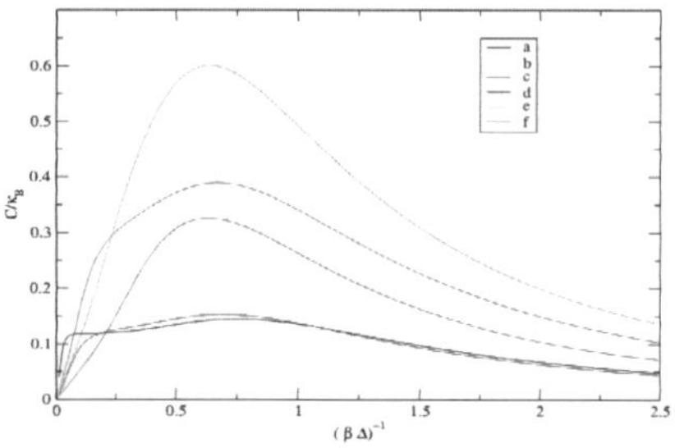

FIG. 2 Specific heat of orthofermions: n = 0.2 (a), 0.5 (b), 0.8 (c); for fermions: n = 0.2 (d), 0.5 (e), 0.8 (f) in 1-D.

Variation of specific heat with respect to temperature for different particle densities is provided in FIG. 2. The specific heat tends to zero as $T \to 0$ for fermions and orthofermions. The orthofermion specific heat exhibits a particle-hole symmetry and specific heat for a given n and $1-n$ are identical. On the other hand fermion specific heat keeps on increasing for moderate and large temperatures with increasing particle density, though at low temperature, this trend again gets reversed. Consequently, the specific heat tends to zero more steeply with decreasing temperature as the fermion density diminishes. Other important outcome of the calculations is that for not very low temperatures, specific heat for fermions remains larger than orthofermion specific heat for identical particle densities.

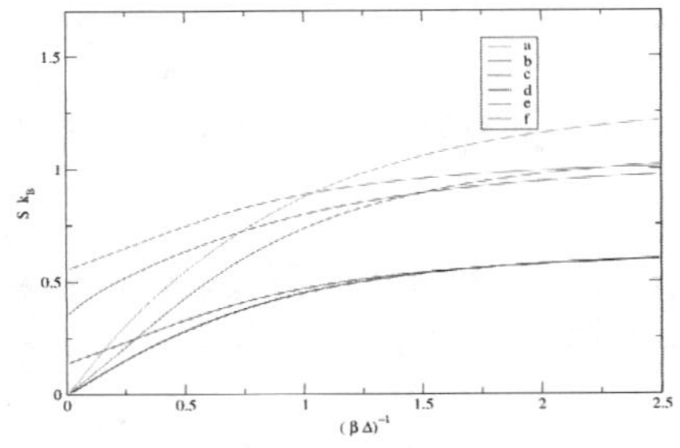

FIG. 3 Entropy of orthofermions: n = 0.2 (a), 0.5 (b), 0.8 (c); for fermions: n = 0.2 (d), 0.5 (e), 0.8 (f) in a 2-D square lattice.

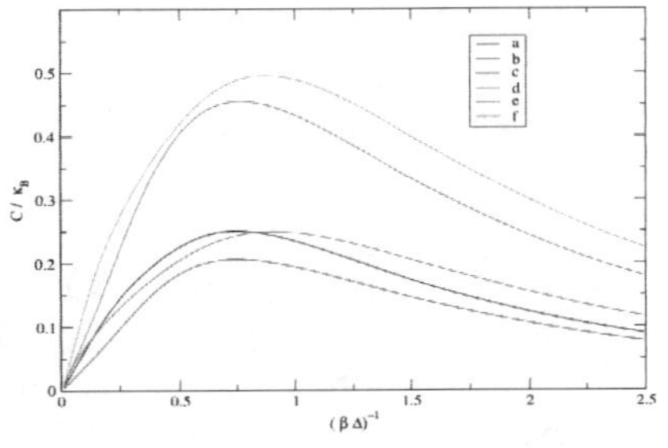

FIG. 4 Specific heat of orthofermions: n = 0.2 (a), 0.5 (b), 0.8 (c); for fermions: n = 0.2 (d), 0.5 (e), 0.8 (f) in a 2-D square lattice.

The trends noted in the thermodynamic quantities vrs temperature plots for 1-D lattice remains valid for fermions and orthofermions on a 2-D square lattice. The difference between the entropy of both kind of particles in 1-D and 2-D square lattice are not appreciable (FIG. 3). However, the variation in the specific heat at low enough temperature is now smoother (FIG. 4). The specific heat for fermions and orthofermions do not show a crossover behaviour with varying particle density in the low temperature regime for a 2-D square lattice. Also the particle-hole symmetry for orthofermions specific heat for the particle densities considered here remains valid.

5. Summary and Conclusions

Thermodynamic properties of free orthofermions on a 1-D lattice have been determined. The 1-D results coincide with the earlier known exact results in the U infinity limit of Hubbard Hamiltonian. As the earlier methods could not be extended to higher dimensions, orthofermion approach has been used to study the consequences of U infinity limit in 2-D, particularly, for a square lattice. The orthofermion results obtained here have been compared with the thermodynamic properties of noninteracting fermions in 1-D and 2-D. Besides, the similarity and differences between the the thermodynamic properties of orthofermions in 1 and 2-Dimensions have been highlighted.

References

[1] Hubbard J 1963 *Proc. Roy. Soc.*, A **276** 238
[2] Lieb E H and Wu F Y 1968*Phys. Rev.* B **20** 1445; Ogata M and Shiba H 1990 *Phys. Rev.* B **41** 2326
[3] Klein D J 1973 *Phys. Rev.* B **8** 3452
[4] Sokoloff J B 1970 *Phys. Rev.* B **2** 779
[5] Kwak J F and Beni G 1976*Phys. Rev.* B **13** 652
[6] Stafford C A 1993 *Phys. Rev.* B **44** 8430
[7] Kishore R and Mishra A K 2004, *Physica* C **408** 277; 2006 *Physica* B **378** 325
[8] P. Fulde, Electron Correlations in Molecules and Solids, Solid State Sciences, vol. 100, Springer, Berlin, 1995.
[9] M. Ogata, H. Shiba, Phys. Rev. B 41 (1990) 2326.
[10] Mishra A K 2001 *Phys. Rev.* B **63** 13405
[11] Mishra A. K. and Kishore R. 2008 *Phys. Letters* A **372** 6019
[12] Douglas J. Klein 1973 *Phys. Rev.* B **8** 3452

RF skin-depth measurement of UIrGe in high magnetic fields

Farzana Nasreen[1], Karunakar Kothapalli [1], Moaz Mohammad Altarawneh[2], Heinz Nakotte[1], Neil Harrison[2], Ekkehard Brück[3]

[1] Department of Physics, New Mexico State University, Las Cruces, NM 88003, USA
[2] National High Magnetic Field Laboratory, Pulse Field Facilty, Los Alamos National Laboratory, Los Alamos, NM 87545, USA
[3]Faculty of Applied Sciences, Delft University of Technology, 2628 CJ Delft, The Netherlands

E-mail: fnasr2@nmsu.edu

Abstract. UIrGe crystallizes in the orthorhombic TiNiSi structure and it orders antiferromagnetically at low temperatures. Previous magnetoresistance and magnetization studies had revealed multiple metamagnetic transitions between ~ 12 T and ~ 19T. Our present studies show that RF skin depth measurement offers an alternative magnetotransport probe. A proximity detector oscillator (PDO) is used to perform the contactless RF skin-depth measurements of UIrGe in pulsed magnetic fields up to 47 T in temperature range 0.57–12 K. The frequency and amplitude shifts reflect the changes in both the real and imaginary components of the conductivity. Our results confirm that the measured frequency shifts can be related to the magnetoresistance effects.

1. Introduction

Correlated *5f* electron compounds, such as uranium intermetallics, may exhibit a wide variety of exotic phenomena due to the competing localized and itinerant features of the *5f* electrons. The delicate balance between itinerancy and localization can be altered by doping, pressure and magnetic field. The magnetism in uranium intermetallics is believed to be determined by two delocalization mechanisms, one is the direct overlap of the *5f* wavefunctions, and the other is the hybridization of *5f* states with the *p* and *d* states of the ligands. Isostructural UTX (U: Uranium, T: transition element, X: *p*-electron element) compounds have been studied for quite some time in order to disentangle the effects of *5f-5f* hybridization and *5f*–ligand hybridization onto the formation of a magnetic ground state. UTX compounds crystallizing in the orthorhombic TiNiSi structure have attracted particular attention since they show a wide variety of moment configurations, whereas hexagonal ZrNiAl-type UTX compounds exhibit strong uniaxial c-axis magnetism [1].

UIrGe is one of the compounds that crystallize in the orthorhombic TiNiSi-type structure. Bulk magnetic studies revealed that the *a*-axis is the hard-magnetization direction or hard axis [2]. Temperature studies of bulk properties on powders indicated a single antiferromagnetic (AF) transition around 16.4 K, whereas in-depth single-crystal studies shows the AF order below 14.1 K and a second transition at ~ 15. 8 K [2]. Single-crystal neutron diffraction studies resulted in a non-collinear, antiferromagnetic magnetic structure at low temperatures [3, 4]. The 5f magnetic moment in UIrGe (~ $0.36\mu_B$ at 1.8 K) was found to be strongly reduced compared to the U^{3+} free-ion moment, more so than that of most other TiNiSi-type UTX compounds.

Published under licence by IOP Publishing Ltd

Measurements of field-induced transitions in small-moment systems using magnetization or magnetoresistance probes can be a challenge. Recently, we have shown that one can utilize RF skin-depth measurements for the determination of transition fields for several UTX compounds [5]. Here, we report on RF skin-depth measurements of UIrGe in high magnetic fields.

2. Experimental

A single crystal of UIrGe was grown by the modified Czochralski method using stoichiometric amounts of the constituents with a purity of at least 99.95%. Previous electrical and magnetic transport measurements [6] utilized a small bar-shaped sample of 1×2×1 mm^3 that had been cut from a larger single-crystalline ingot. The longest dimension is parallel to the *b*-axis. We used the same sample in order to perform the present radio-frequency (RF) measurements at high magnetic fields. The sample was placed in the sensor inductor coil in sample probe such that the *b*-axis is parallel to the coil axis, and then the probe was placed in a ^{3}He cryostat. The whole assembly was placed at the centre of 50-T mid-pulse magnet at the Pulse Field Facility, NHMFL, Los Alamos National Laboratory. The axis of the sensor coil and the magnetic field were aligned parallel, and thus the field was parallel to the *b*-axis of UIrGe. A proximity detector oscillator (PDO) circuit is placed at the top of the probe (almost a meter away from the centre of the magnet) and it is kept at room temperature. The main part of the PDO circuit is a commercially available integrated circuit (TDA0161), which is widely used in metal detectors. This circuit allows compensating for losses in the effective LC circuit using a principle that is very similar to a tunnel diode oscillator (TDO). The modified circuit oscillates at the resonance frequency given by

$$f = \frac{1}{2\pi\sqrt{LC}} \tag{1}$$

where *L* is the effective inductance and *C* is the capacitance of the circuit. A careful design of the PDO circuit leads to a stable resonance frequency with a phase noise of less than 50 Hz over 2 hours. The PDO used in this study is optimized to produce a resonance frequency at 10 MHz, but the stable resonance frequencies up to 28 MHz are possible [7]. A more detailed description of RF skin-depth measurements in pulse field using this PDO circuit and a comparison with RF techniques using TDOs is given in reference 7.

As the applied magnetic field or the temperature are changed for a magnetic sample, its magnetic permeability modifies the effective inductance of the whole circuit and that causes a shift in the resonance frequency, which is given by the following relation

$$\frac{\Delta f}{f_0} = -\frac{\Delta L}{2L} \tag{2}$$

The frequency shift can be directly related to the skin-depth and dynamic susceptibility of a material [8, 9]. The Drude model predicts that the skin-depth is related to complex conductivity [10], and thus a shift in resonant frequency is a measurement of magneto-transport. Here, we performed RF skin-depth studies on single-crystalline UIrGe at a frequency of 25 MHz, in temperature range from 0.57 K to 12 K and in pulsed magnetic fields from 0–47 T applied along the *b*-axis.

3. Results and discussion

Figure 1 shows the variation of frequency shift in UIrGe as a function of magnetic field applied along the *b*-axis at various fixed temperatures. The frequency strongly decreases as the field is increased. At 0.57 K, we find that the frequency at 47 T is less than 5 % of the frequency at 0 T. More interestingly, at the lowest temperature (0.57 K), we can clearly distinguish two stronger metamagnetic transitions at ~ 14.5 T and ~ 18.9 T and a third weaker transition at ~19.4 T. One can extract the temperature dependence of those metamagnetic transitions by comparison with the experimental results at elevated temperatures. All three metamagnetic transitions exhibit some small hysteresis, which persists at higher temperatures. For clarity, except at 0.57 K and 12 K variation of frequency with only up sweeps of field are shown in figure 1. The features in •*f* coincide reasonably well with similar anomalies in magnetoresistance data on the same sample [6]. However, previous *b*–axis magnetoresistance studies were limited to fields up to ~18 T, and the here observed transitions around 19 T were out to the field range.

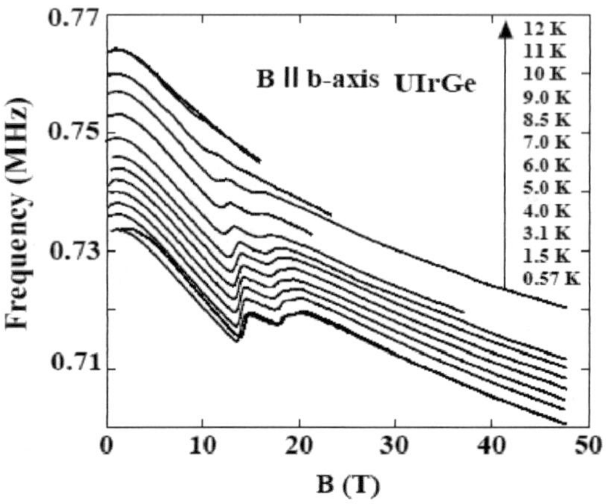

Figure 1. The field dependence of the frequency shift (•f) in the proximity detector oscillator (PDO) circuit coupled to UIrGe with applied field along the *b*-axis at different temperatures.

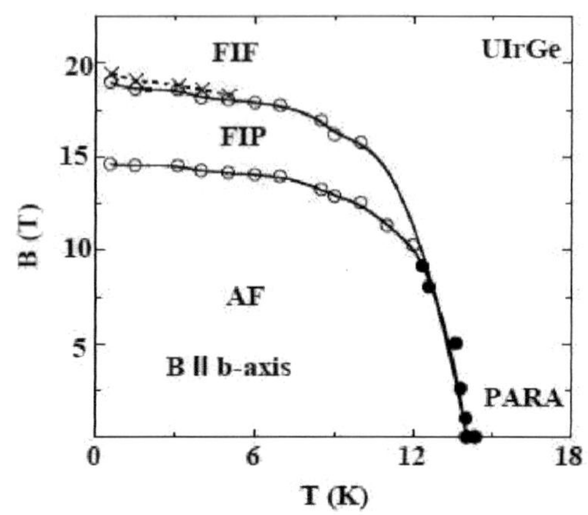

Figure 2. B-T phase diagram of UIrGe. The open circles are data points determined from RF skin-depth measurement. The solid circles are from resistivity and heat capacity [6]. The lines are guide to the eye. AF is for anti-ferromagnetic phase, FIP is for field induced phase, FIF is for field induced ferromagnetic phase and PARA is for paramagnetic phase.

We used the values determined from field-up sweeps in order to construct the phase diagram shown in figure 2. The B-T phase diagram obtained from our skin-depth studies is in reasonable agreement with the data proposed from previous bulk studies [6]. The first two metamagnetic transitions are in reasonable agreement with B-T phase diagram extracted from resistivity, magnetization and specific-heat measurements [6, 11]. However, the third weaker transition at 19.4 T at 570 mK had not been seen previously. That transition becomes weaker as the temperature is increased and starts fading at 6 K. It is possible that this third metamagnetic transition is not intrinsic, but it is due to a small misaligned UIrGe crystallite. This third weak transition contributes a very small phase as shown in figure 2. The metamagnetic transitions shift towards lower field with increasing temperature, which is consistent with an antiferromagnetic ground state. Results of Prokes *et. al* group shows that the *a*-axis component survives for field applied along the *b*-axis [4] whereas it disappears for field applied along *c*-axis [12]. In our measurements upto 47 T field along *b*-axis, there is no sign of separate transition due to the alignment of this *a*-axis component of moment in UIrGe something that had been seen in RF skin-depth measurement in UNiGe [13].

4. Conclusion

We measured the RF skin-depth for UIrGe in pulsed fields applied along the *b*-axis using PDO technique. The features in frequency with the variation of magnetic field and temperature are in good agreement with magnetoresistance data. It seems reasonable to assume that the conductivity dominates the changes in the complex-conductivity measurements since similar changes were seen in magnetoresistance data. Our B-T phase diagram is consistent with previously published magnetisation, transport and heat-capacity data.

Acknowledgements

This work was supported by a National Science Foundation Grant No. DMR 0804032. Experimental work at the National High Magnetic Field Laboratory, Pulse Field Facility at Los Alamos National Laboratory was supported by National Science Foundation, the US department of energy and the State of Florida.

References

[1] Sechovsky V and Havela L 1998 *Handbook of Magnetic Materials* **Vol. 11**, ed K.H.J. Buschow (Amsterdam: North- Holand) chapter 1
[2] Prokeš K, Tahara T, Fujita T, Goshima H, Takabatake T, Mihalik M, Menovsky A A, Fakuda S and Sakurai J 1999 *Phys. Rev. B* **60** 9532
[3] Troc R and Tran V 1988 *J. Magn. Magn. Mat.* **73** 389-397.
[4] Prokeš K, Nakotte H, Sechovsky V, Mihalik M and Andreev A V 2004 *Physica B* **350** e199-e202
[5] Alsmadi A M et. al 2009 *J. Magn. Magn. Mat.* **321** 3712–3718
[6] Chang S, Jung M H, Nakotte H, Brück E, Klaasse J C P, Mihalik M, Lacerda A H, Prokeš K, Torikachvilli M S and Schultz A J 2001 *J. Appl. Phys.* **89 No. 11** 7186-7188
[7] Altarawneh M M, Mielke C H and Brooks J S 2009 *Rev. Sci. Instrum.* **80** 066104
[8] Ohmichi E, Komatsu E and Osada T 2004 *Rev. Sci. Instrum.* **75 iss. 6** 2094-2097
[9] Vannette M D, Sefat A, Jia S, Law S A, Lapertot G, Bud'ko S L, Canfield P C, Schmalian J and Prozorov R 2008 *J. Magn. Magn. Mat.* **320** 354–363
[10] Aschroft N W and Mermin N. D 1976 *Solid State Physics* (Fort Worth: Saunders College Publishing)
[11] Chang S, Nakotte H, Alsmadi A M, Lacerda A H, Jung M H, Mihalik M, Prokes K, Klaasse, J P, Brück E and De Boer, F R 2002 *International Journal of Modern Physics B* **16 Nos. 20, 21 & 22** 3041-3044
[12] Prokeš K, Sechovsky V, De Boer, F R and Andreev A V 2008 *J. Phys.: Condens. Matter* **20** 104221
[13] Chang S, Mielke C H, Bennett M, Brück E and Nakotte H 2002 *Physica B* **312–313** 877–878.

Journal of Physics: Conference Series **273** (2011) 012155 doi:10.1088/1742-6596/273/1/012155

Thermoelectric effects in correlated quantum dots and molecules

V. Schöps[1,2], V. Zlatić[1,3] and T. A. Costi[1]

[1] Institute of Solid State Research, Research Centre Jülich, 52425 Jülich, Germany
[2] Department of Physics, University of Konstanz, 78457 Konstanz, Germany
[3] Institute of Physics, 10001 Zagreb, Croatia

E-mail: `t.costi@fz-juelich.de`

Abstract. We investigate thermoelectric properties of correlated quantum dots and molecules, described by a single level Anderson model coupled to conduction electron leads, by using Wilson's numerical renormalization group method. In the Kondo regime, the thermopower, $S(T)$, exhibits two sign changes, at temperatures $T = T_1$ and $T = T_2 > T_1$. We find that T_2 is of order the level width Γ and $T_1 > T_p \approx T_K$, where T_p is the position of the Kondo induced peak in the thermopower and T_K is the Kondo scale. No sign change is found outside the Kondo regime, or, for weak correlations, making a sign change in $S(T)$ a particularly sensitive signature of strong correlations and Kondo physics. For molecules, we investigate the effect of screening by conduction electrons on the thermoelectric transport. We find that a large screening interaction enhances the figure of merit in the Kondo and mixed valence regimes.

1. Introduction

Materials with high thermoelectric efficiency are currently under intense theoretical and experimental investigation, largely due to the prospect of applications, e.g., for the conversion of waste heat into electricity, or for refrigeration and on-chip cooling in microelectronics [1, 2]. Apart from applications, thermoelectric materials also serve as an interesting testing ground for new theoretical approaches to thermoelectric transport in solids[3, 4, 5]. Coulomb correlations and the Kondo effect give rise to large thermoelectric coefficients in strongly correlated bulk materials [6, 7, 8, 9]. They are also important in nanoscale semiconducting devices and in molecular transistors, posing interesting theoretical and experimental challenges for the understanding of electrical and thermal transport in such systems[10, 11, 12, 13, 14, 15, 16].

In this paper we address the thermoelectric properties of two related systems: a nanoscale size quantum dot exhibiting the Kondo effect, which we describe in terms of a single level Anderson impurity model with two conduction electron leads at fixed chemical potentials, and, a molecular transistor described by an extension of the above model to include a local screening of the Coulomb interaction on the molecule by the electrons in the leads. The quantum dots or molecules that we consider can be tuned from the Kondo to the mixed valence and empty orbital regimes by a gate voltage [17, 18]. Very recent experiments on nanoscale quantum dots [10] are beginning to probe the effect of Kondo correlations on the thermopower.

We use Wilson's numerical renormalization group (NRG) method [19, 20] to calculate the linear conductance, thermopower and thermal conductance as a function of temperature and gate voltage. The technique gives reliable results for transport properties in all regimes of

Published under licence by IOP Publishing Ltd

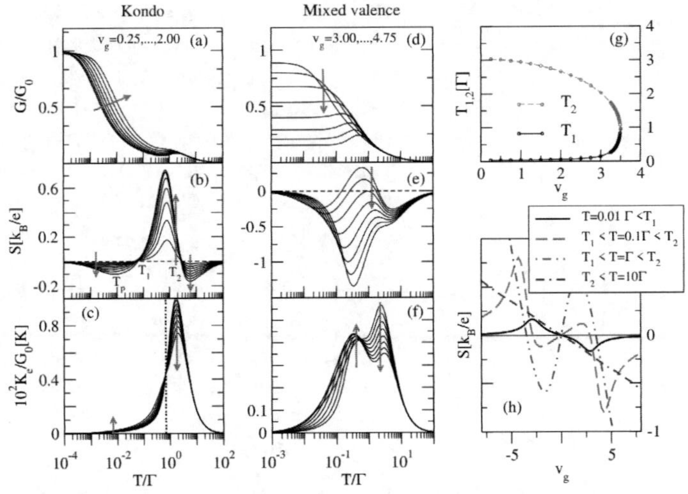

Figure 1. Conductance, G, thermopower, S, and, thermal conductance, K_e, versus T/Γ, for $U = 4\Gamma = 0.08$ and gate voltages, $v_g = (\varepsilon_d + U/2)/\Gamma$, in the Kondo (a-c) and mixed valence (d-f) regimes ($G_0 = 2e^2/h$). The v_g are indicated in top panels and increase by 0.25. Arrows show trends with increasing v_g. T_p: position of low temperature peak in S. Temperatures $T_{1,2}$ at which $S(T)$ changes sign are shown in (g) versus v_g. (h): gate voltage dependence of S at representative temperatures T relative to $T_1(0)$ and $T_2(0)$.

interest[4]. The recently introduced full density matrix (FDM) approach [21](see also [22]) is used. This allows calculations of dynamical properties at all excitation energies ω relative to the temperature T, thereby simplifying the calculation of transport properties which require knowledge of excitations, ω, above and below the temperature.

2. Model and transport properties

We describe both the nanoscale quantum dot and the molecular transistor by the single level Anderson impurity model with two conduction electron leads

$$H = \sum_{\alpha k\sigma} \epsilon_{\alpha k\sigma} c^{\dagger}_{\alpha k\sigma} c_{\alpha k\sigma} + \sum_{\sigma} \varepsilon_d d^{\dagger}_{\sigma} d_{\sigma} + U n_{d\uparrow} n_{d\downarrow} + H_{\text{screening}} + \sum_{\alpha k\sigma} t_{\alpha}(c^{\dagger}_{\alpha k\sigma} d_{\sigma} + h.c.). \quad (1)$$

Here, $\epsilon_{\alpha k\sigma}$ is the kinetic energy of electrons with wavenumber k and spin σ in lead $\alpha = (L, R)$, ε_d is the local level energy, U is the local Coulomb repulsion and t_α is the tunnel matrix element of the local level to conduction electron states in lead $\alpha = (L, R)$. $H_{\text{screening}} = U_{dc}(n_d - 1)(n_0 - 1)$, with n_d the local level occupancy and n_0 the local occupancy of the lead electrons, represents the screening interaction for the case of the molecule. We assume symmetric couplings of the dot to both leads and use the full width at half maximum, $\Gamma = 0.02$, as the energy unit throughout.

Thermoelectric transport is calculated for a situation in which a small external bias voltage, $\delta V = V_L - V_R$, and a small temperature gradient δT is applied between left and right leads following [11, 12]. To linear order, the following expressions for the electrical conductance, $G(T)$, the thermal conductance, $K_e(T)$, and the thermoelectric power, $S(T)$, are obtained

$$G(T) = e^2 I_0(T) \quad (2)$$

$$S(T) = -\frac{1}{|e|T}\frac{I_1(T)}{I_0(T)} \quad (3)$$

$$K_e(T) = \frac{1}{T}\left[I_2(T) - \frac{I_1^2(T)}{I_0(T)}\right] \quad (4)$$

where $I_n, n = 0, 1, 2$ are the transport integrals

$$I_n(T) = \frac{2}{h}\int d\omega\, \omega^n \mathcal{T}(\omega)(-\frac{\partial f}{\partial \omega}). \quad (5)$$

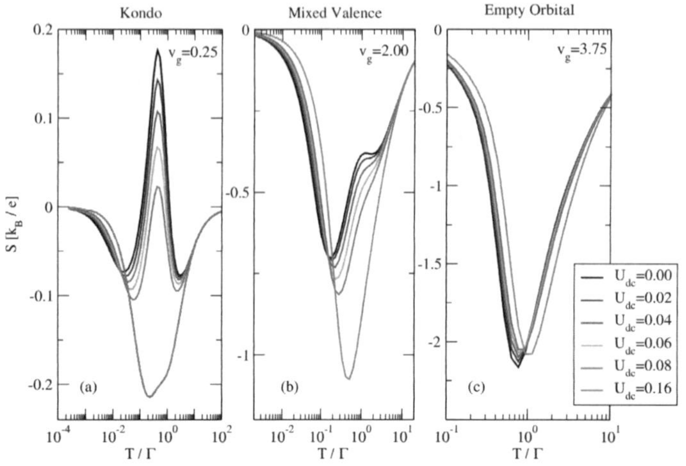

Figure 2. Temperature dependence of the thermopower, $S(T)$, for $U = 4\Gamma = 0.08$ in the Kondo, (a), mixed valence, (b), and empty orbital, (c), regimes for increasing values of the screening interaction U_{dc}. The case of large screening $U_{dc} = 0.16 > U$ is qualitatively similar to a large local electron-phonon coupling and results in destruction of Kondo correlations at finite gate voltages. Screening has a negligible effect in the empty orbital regime, but affects significantly $S(T)$ in the other regimes.

Here, e denotes the magnitude of the electronic charge and h denotes Planck's constant. The transmission function $\mathcal{T}(\omega) = \pi\Gamma A(\omega)$, where $A(\omega)$ is the d-level spectral density.

3. Results for quantum dots: $U_{dc} = 0$

The temperature dependence of the electrical conductance, $G(T)$, thermopower, $S(T)$, and electronic part of the thermal conductance, $K_{\mathrm{e}}(T)$, are shown in Fig. 1a-c in the Kondo regime and in Fig. 1d-f in the mixed valence regime [23]. The thermopower exhibits two sign changes in the Kondo regime at $T = T_1(\mathrm{v}_g)$ and $T = T_2(\mathrm{v}_g)$ and no sign changes on entering the mixed valence regime (see Fig. 1e). The gate voltage dependence of S in Fig. 1h shows three characteristic types of behaviour depending on the T relative to $T_1(0)$ as explained in [23].

4. Results for molecules: $U_{dc} \geq 0$

For molecules in metallic break junctions, we expect additional effects to be important for thermoelectric properties, e.g. local phonon modes and screening by the conduction electrons in the leads. Here, we investigate the effect of the latter, and note that the effect of a strong local electron-phonon coupling can be approximately simulated by choosing a large screening interaction $U_{dc} > U$.

For moderate screening interactions $U_{dc} < U$, we recover the expected downward renormalization of the local Coulomb repulsion $U \to U^* < U$ and an upward excitonic renormalization of the effective hybridization with increasing U_{dc} leading to a weakening of Kondo correlations. Correspondingly, transport properties are significantly affected in the Kondo and mixed valence regimes, as shown in Fig. 2 for the thermopower and in Fig. 3 for the figure of merit. For large screening interactions $U_{dc} > U$ we find a destruction of Kondo correlations for gate voltages $\mathrm{v}_g > 0$ away from particle-hole symmetry with an enhancement of the figure of merit for "Kondo" and "mixed valence" regimes (referred to the $U_{dc} = 0$ case).

5. Conclusions

The thermoelectric power of, (i), quantum dots, and, (ii), molecules in break junctions, has been investigated using a single-level Anderson model. We showed that the thermopower exhibits sign changes as a function of temperature and gate voltage. These can be used as sensitive probes of Kondo correlations. Recent measurements on Kondo correlated quantum dots [10] are in qualitative agreement with our model calculations [23]. The effect of lead electron screening on

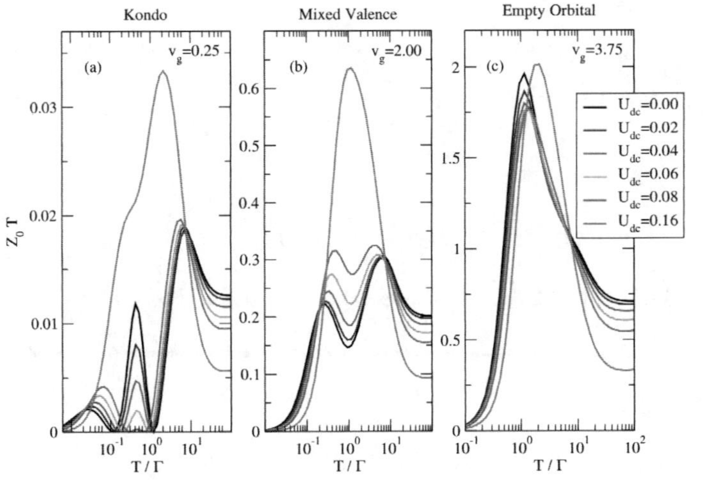

Figure 3. Temperature dependence of the "figure of merit" $Z_0T = GS^2T/K_e$ for $U = 4\Gamma = 0.08$ and increasing values of the screening interaction U_{dc} in different regimes (by reference to the $U_{dc} = 0$ case). The true figure of merit $ZT = GS^2T/(K_e + K_{ph})$ includes a material specific phonon contribution, K_{ph}, which we have neglected, but needs to be included whenever phonons dominate thermal transport. Provided all scales are less than the Debye temperature θ_D, we expect that $ZT \approx Z_0T$.

molecular transport is shown to be significant in the Kondo and mixed valence regimes, where we find that an enhanced figure of merit can result. In future, it would be interesting to include a local electron-phonon coupling of the molecular state in order to investigate the effect of this interaction on molecular transport in detail[24].

Acknowledgments

Support from the Forschungszentrum Jülich (V.Z.), the SFB 767 (V.S.) and supercomputing support from the John von Neumann Institute for Computing (Jülich) is acknowledged.

References

[1] Mahan G D 1998 *Solid State Phys.* **51** 81
[2] Kanatzidis M G 2010 *Chem. Mater.* **22** 648
[3] Zlatić V, Costi T A, Hewson A C and Coles B R 1994 *Phys. Rev.* B **48** 16152
[4] Costi T A, Hewson A C and Zlatić V 1994 *J. Phys.: Condens. Matter* **6** 2519
[5] Zlatić V and Monnier R 2005 *Phys. Rev.* B 71 165109
[6] Arita R , Kuroki R K, Held K, Lukoyanov A V, Skornyakov S and Anisimov V I 2008 *Phys. Rev.* B **78** 115121
[7] Lackner R, Bauer E and Rogl P 2006 *Physica* B **378-390** 835
[8] Bentien A, Johnsen S, Madsen G K H, Iversen B B and Steglich F 2007 *Europhys. Lett.* **80** 17008
[9] Sales B C, Mandrus D and Williams, R K 1996 *Science* **272** 1325
[10] Scheibner R, Buhmann H, Reuter D, Kiselev M N and Molenkamp L W 2005 *Phys. Rev. Lett.* **95** 176602
[11] Kim T-S and Hershfield S 2002 *Phys. Rev. Lett.* **88** 136601
[12] Dong B and Lei X L 2002 *J. Phys.: Condens. Matter* **14** 11747
[13] Roch N, Florens S, Costi T A, Wernsdorfer W and Balestro F 2009 *Phys. Rev. Lett.* **103** 197202
[14] Parks J J, Champagne A R, Costi T A, Shum W W, Pasupathy A N, Neuscamman E, Flores-Torres S, Cornaglia P S, Aligia A A, Balseiro C A, Chan G K-L, Abruña H D and Ralph D C 2010 *Science* **328** 1370
[15] Koch J, von Oppen F, Oreg Y and Sela E 2004 *Phys. Rev.* B **70** 195107
[16] Reddy P, Jang S-Y, Segalman R A and Majumdar A 2007 *Science* **315** 1568
[17] Goldhaber-Gordon D, Göres J, Kastner M A, Shtrikman H, Mahalu D and Meirav U 1998 *Phys. Rev. Lett.* **81** 5225
[18] Cronenwett S M, Osterkamp T H and Kouwenhoven L P 1998 *Science* **281** 540
[19] Wilson K G 1975 *Rev. Mod. Phys.* **47** 773
[20] Bulla R, Costi T A and Pruschke, T 2008 *Rev. Mod. Phys.* **80** 395
[21] Weichselbaum A and von Delft J 2007 *Phys. Rev. Lett.* **99** 076402
[22] Peters R, Pruschke T and Anders F B 2006 *Phys. Rev.* B **74** 245114
[23] Costi T A and Zlatić V 2010 *Phys. Rev.* B **81** 235127
[24] Leijnse M, Wegewijs M R and Flensberg K 2010 *Phys. Rev.* B **82** 045412; Entin-Wohlman O, Imry Y and Aharony A 2010 *Phys. Rev.* B **82** 115314

Journal of Physics: Conference Series **273** (2011) 012156 doi:10.1088/1742-6596/273/1/012156

Insights into phase-slip events via dynamics of current-induced transitions from a superconductive to a resistive state

Nayana Shah

Department of Physics, University of Cincinnati, Cincinnati, OH 45221-0011

E-mail: nayana.shah@uc.edu

Abstract. Phase-slip fluctuations play a central role in a variety of systems and are responsible for dissipation via a loss of coherence. By virtue of these fluctuations, quasi-one-dimensional superconductors acquire a non-zero resistance unlike the bulk superconductors in which the resistance drops to zero upon entering the superconducting phase. The present generation of experiments on superconducting nanowires are capable of characterizing non only the thermally activated phase slips in detail, but also the quantum phase slip processes as well as the crossover from thermal to quantal. The proceeding will report on the recent progress made towards achieving this goal by studying the dynamics of current-induced transitions from superconductive to resistive state and analyzing the switching statistics.

Phase slip processes–in which the relative phase between two regions of space (or space-time) changes or slips by a multiple of 2π–play an important role in a variety of theoretical and experimental settings ranging from cosmic strings in the universe to ultra-cold atomic systems in optical traps. In the Bose-Hubbard model realized using trapped atoms, the damping rate of the atomic motion was experimentally found to be induced by phase-slip processes [1]. In analogy to the phase slips in Josephson junctions or weak links in superconductors, phase slips processes are found to give rise to dissipation events in the flow of superfluid ^{4}He and ^{3}He through orifices [2, 3, 4]. To give another example, phase slips have also been studied in the context of charge and spin density wave systems [5, 6]. This proceeding aims at gaining insights into phase-slip processes by focusing on their behavior in quasi-one-dimensional superconductors [7, 8, 9, 10].

In the quasi-one-dimensional limit in which the transverse dimension is smaller than the superconducting coherence length, the superconducting order parameter has a constant amplitude and a phase that winds along the length to form a helix. In a phase-slip process [11], the amplitude of the complex-valued order parameter field fluctuates to zero (typically over a length-scale of the order of coherence length) somewhere along the wire, allowing the phase to slip by 2π. In doing so, the system goes from one metastable minimum to another by crossing a potential barrier either by thermal activation or quantum tunneling [12, 13]. Experimental observation of quantum phase slips and the nature of thermal to quantum crossover is one of the central questions that plays an essential role in determining the physics, not only in the case of quasi-one-dimensional superconductors but also for other phase-slip settings e.g. stated in the previous paragraph.

Published under licence by IOP Publishing Ltd

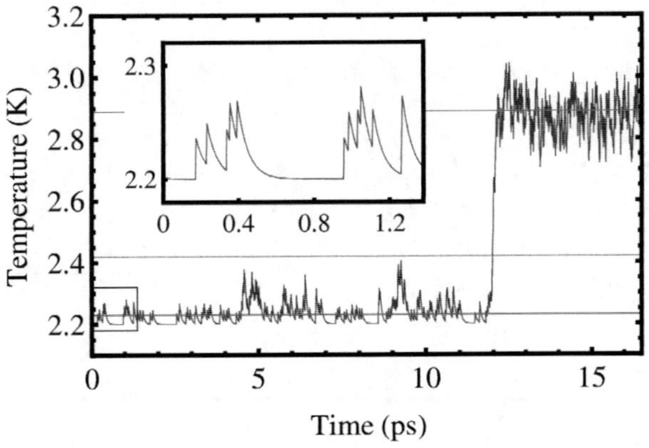

Figure 1. Trace of a typical trajectory of the temperature of the central segment (see text) of the nanowire illustrating the switching caused by thermal runaway. Overlaid on the trace are red lines showing the temperatures $T_{\rm sc}$ (lowest), $T_{\rm sp}$ (middle), and $T_{\rm rs}$ (highest) corresponding to the superconducting minimum, the saddle-point, and the resistive minimum of an effective potential in the Langevin equation. The inset shows a blow-up of the boxed region of the main figure.

The phenomenological Ginzburg-Landau (GL) formalism that describes the superconducting phase in terms of a complex order parameter field provides a natural and direct language to describe phase-slip fluctuations. This framework was used to calculate the rate of thermally activated phase slips more than four decades ago [14] and soon afterwards the time-dependent GL formalism was used to calculate the attempt frequency [15], together constituting the LAMH theory of thermally activated phase slips. The regime of validity of this theory is expected to be in the vicinity of the transition temperature, and indeed in this regime it is successful in characterizing the broad resistive transition. Some of the more recent works have used a variety of theoretical approaches to further characterize phase-slip fluctuations and their implications. For instance, the rate of thermally activated phase slips was reanalyzed using an imaginary time effective action approach [16] and microscopically calculated in the clean limit using Eilenberger equations [17]. The quantum phase slip rate was microscopically calculated [18] and the interaction between quantum phase slips was analyzed [19, 20]. A memory formalism approach was used to study the contribution of quantum phase slips to the resistivity of dirty and inhomogeneous wire [21].

The present generation of experiments on superconducting nanowires are indeed capable of accessing the regime in which quantum fluctuations play an essential role [9, 10]. Since is is difficult to unambiguously establish the existence of quantum phase slips within the linear resistance measurements due to the noise floor, recent experiments have successfully evolved in better measuring and characterizing the entire I-V characteristics for ultra-narrow wires at exceedingly lower temperatures [13, 22, 23, 24, 25, 26]. One of the very recent experiments [26] has in fact been able to measure the distribution of switching currents–the current values at which the nanowire switches from a superonductive (i.e. low resistance state) to higher resistance state–as a function of temperature. Note that it was established that the nanowires in this experiment were in an overdamped regime and did not support phase-slip centers [27]. Here I will report on this experiment and its theoretical underpinning [28, 29].

The experimental set-up consists of a superconducting nanowire fabricated using molecular templating (by using a carbon nanotube or a DNA molecule as a template) to furnish a

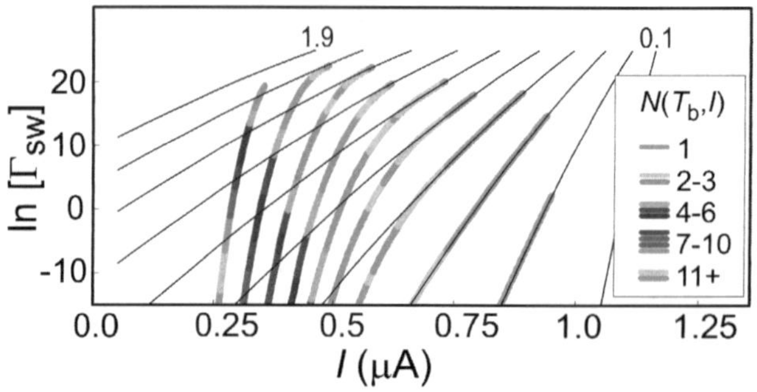

Figure 2. Logarithm of the mean switching rate, $\ln \Gamma_{\rm sw}$, as a function of current for bath temperatures $T_b = 0.1, 0.3, 0.5 ..., 1.9\,$K, obtained by assuming the rate Γ of phase slips to be given by $\Gamma_{\rm LAMH}$ i.e. the LAMH formula for the rate of thermally activated phase slips. For reference, the thinner black lines indicate the rate $\Gamma_{\rm LAMH}$.

free-standing superconducting nanowire connected to thermal baths at its two ends. If the Joule heating caused by resistive phase slip fluctuations is not overcome sufficiently rapidly by conductive cooling, it effectively reduces the depairing current, ultimately to below the applied current, thus causing switching to the highly resistive state. Although the discrete phase-slip events take place at random moments of time and are centered at random spatial locations along the wire, one can argue that it is reasonable to assume that all the phase slips and hence the Joule heating takes place within a central segment of length to which a uniform temperature T is assigned and the heat is conducted away through the end segments (within which the heat capacity is ignored).

Within this simplified model, the dynamics of the nanowire is given by a stochastic (Langevin) ordinary differential equation for the time-evolution of the temperature T of the central segment. A typical trajectory of the temperature of the central segment is shown in Figure 1. To gain intuition and illustrates the bistability, one can consider an effective potential description of the Langevin equation by replacing the sum over discrete phase slips by a phase slip rate $\Gamma(T, I)$ (the problem is then similar to that of the motion of an over-damped "particle" of position $T(t)$ at time t, moving in the fictitious potential). The switching then involves going from the low temperature 'superconducting' minimum to the hight temperature 'resistive' minimum of the potential.

As illustrated by Figure 1, the switching dynamics is dictated by the competition between discrete heating (since each phase slip heats the wire by a "quantum" of energy $hI/2e$, for a given value of the applied current I) and continuous cooling. Thus the mechanism of switching can be physically understood in terms of a thermal runaway i.e. heating by rare sequences of closely-spaced phase slips. The number of phase-slips in such a phase-slip sequence can be easily estimated by evaluating the number $N(T_{\rm b}, I)$ of phase-slips needed to cause switching in the absence of cooling.

Within this theoretical framework [28, 29] one can successfully understand the seemingly counter-intuitive broadening of the switching current distribution on lowering the bath temperature $T_{\rm b}$ observed in the experiment [26]. Alternatively one can fit the mean switching rates $\Gamma_{\rm sw}$ shown in Figure 2, to those extracted from the experiments. As one can see from the figure, $\Gamma_{\rm sw}$ deviates drastically from the rate of phase slips when the thermal runaway sequence involves multiple phase slips. The two overlap only when a single phase slip is sufficient to cause the switching. Detailed fits of the experimental data amply demonstrate that as low temperatures, it is essential to phenomenologically incorporate quantum phase slips into the phase slip rate [26, 29]. Moreover, one is able to conclude that the experiment is in fact probing individual quantum phase slips in the low temperature regime.

In conclusion, the recent study of switching dynamics is able to provide insights into the quantum phase slip fluctuations and the crossover from thermal to quantal fluctuations. The understanding of low temperature phase slip behavior is essential to establish whether superconductivity survives upon approaching zero temperature. The question of what is the state of the nanowire at zero temperature is a fundamental one and correspondingly the experimental and theoretical study of possible superconductor-metal or superconductor-insulator quantum phase transitions is an important topic of current interest [30, 31, 23, 32, 33, 34, 35, 36, 37].

Acknowledgments

I would like to gratefully acknowledge my theory collaborators, D. Pekker and P. M. Goldbart and experimental collaborators M. Sahu and A. Bezryadin.

References

[1] McKay D, White M, Pasienski M and DeMarco B 2008 *Nature* **453** 76–79
[2] Avenel O and Varoquaux E 1985 *Phys. Rev. Lett.* **55** 2704–2707
[3] Avenel O and Varoquaux E 1988 *Phys. Rev. Lett.* **60** 416–419
[4] Sato Y, Hoskinson E and Packard R E 2006 *Phys. Rev. B* **74** 144502
[5] Maki K 1995 *Physics Letters A* **202** 313 – 316 ISSN 0375-9601
[6] Glatz A and Nattermann T 2002 *Phys. Rev. Lett.* **88** 256401
[7] Skocpol W J and Tinkham M 1975 *Reports on Progress in Physics* **38** 1049–1097
[8] Tinkham M 1996 *Introduction to Superconductivity* (McGraw-Hill, New York)
[9] Arutyunov K, Golubev D and Zaikin A 2008 *Physics Reports* **464** 1 – 70 ISSN 0370-1573
[10] Bezryadin A 2008 *Journal of Physics: Condensed Matter* **20** 043202
[11] Little W A 1967 *Phys. Rev.* **156** 396–403
[12] Grabert H and Weiss U 1984 *Phys. Rev. Lett.* **53** 1787–1790
[13] Giordano N 1990 *Phys. Rev. B* **41** 6350–6365
[14] Langer J S and Ambegaokar V 1967 *Phys. Rev.* **164** 498–510
[15] McCumber D E and Halperin B I 1970 *Phys. Rev. B* **1** 1054–1070
[16] Golubev D S and Zaikin A D 2008 *Phys. Rev. B* **78** 144502
[17] Zharov A, Lopatin A, Koshelev A E and Vinokur V M 2007 *Phys. Rev. Lett.* **98** 197005
[18] Golubev D S and Zaikin A D 2001 *Phys. Rev. B* **64** 014504
[19] Zaikin A D, Golubev D S, van Otterlo A and Zimányi G T 1997 *Phys. Rev. Lett.* **78** 1552–1555
[20] Khlebnikov S 2008 *Phys. Rev. B* **78** 014512
[21] Pai G V, Shimshoni E and Andrei N 2008 *Phys. Rev. B* **77** 104528
[22] Tinkham M, Free J U, Lau C N and Markovic N 2003 *Phys. Rev. B* **68** 134515
[23] Rogachev A, Bollinger A T and Bezryadin A 2005 *Phy. Rev. Lett.* **94** 017004
[24] Altomare F, Chang A M, Melloch M R, Hong Y and Tu C W 2006 *Phys. Rev. Lett.* **97** 017001
[25] Zgirski M, Riikonen K P and Arutyunov K Y 2008 *Journal of Physics: Conference Series* **97** 012113
[26] Sahu M, Bae M H, Rogachev A, Pekker D, Wei T C, Shah N, Goldbart P M and Bezryadin A 2009 *Nature Physics* **5** 503–508
[27] Vodolazov D Y, Peeters F M, Piraux L, Matefi-Tempfli S and Michotte S 2003 *Physical Review Letters* **91** 157001 (pages 4)
[28] Shah N, Pekker D and Goldbart P M 2008 *Phys. Rev. Lett.* **101** 207001
[29] Pekker D, Shah N, Sahu M, Bezryadin A and Goldbart P M 2009 *Phys. Rev. B* **80** 214525
[30] Bezryadin A, Lau C N and Tinkham M 2000 *Nature* **404** 971
[31] Lopatin A V, Shah N and Vinokur V M 2005 *Phys. Rev. Lett.* **94** 037003
[32] Shah N and Lopatin A 2007 *Phys. Rev. B* **76** 094511
[33] Refael G, Demler E, Oreg Y and Fisher D S 2007 *Physical Review B (Condensed Matter and Materials Physics)* **75** 014522 (pages 37)
[34] Del Maestro A, Rosenow B, Shah N and Sachdev S 2008 *Phys. Rev. B* **77** 180501
[35] Del Maestro A, Rosenow B, Müller M and Sachdev S 2008 *Phys. Rev. Lett.* **101** 035701
[36] Bollinger A T, Dinsmore R C, Rogachev A and Bezryadin A 2008 *Phys. Rev. Lett.* **101** 227003
[37] Refael G, Demler E and Oreg Y 2009 *Phys. Rev. B* **79** 094524

Journal of Physics: Conference Series **273** (2011) 012157 doi:10.1088/1742-6596/273/1/012157

Concluding Remarks: Experiment from a materials perspective

Z. Fisk
Department of Physics and Astronomy, University of California, Irvine, CA 92697, USA

Abstract. The author provides some remarks regarding the current status of experiments in strongly correlated electron systems. By construction, they are biased by the author's perspectives at the time of writing.

Characterization techniques have progressed enormously since the early studies of strongly correlated electron materials, along with ever increasing sample quality and the present dominance of single crystal work in our field. Scanning tunneling microscopy that is just beginning to be turned to the study of cleaved surfaces of heavy Fermion materials and should be able to provide detailed information at the atomic level that we have never had before. Correlated electronic structure is expected to be particularly labile in the vicinity of a surface and interpreting exactly how STM measurements reflect the bulk physics will be interesting to sort out. Using the STM as an experimental knob presents interesting possibilities as well, such as locally inducing superconductivity. Ever increasing sensitivity in photo-emission experiments now approaches the energy scales of direct electronic interest. Further, the new combination of various techniques, for example neutron scattering plus x-ray absorption as presented by Severing, can further address the low temperature electronic structure in remarkable detail at the local level.

Much of the physics of interest in heavy Fermion materials occurs near a magnetic quantum critical point. Connected with this is the observation that certain physics, such as the nematic phase suggested in $Sr_3Ru_2O_7$, depends sensitively on the perfection of the material. This points to the need to characterize materials thoroughly via x-ray and related techniques. Modern light sources enable x-ray experiments never before possible: modern x-ray technique can provide structural detail that can deepen significantly our understanding of the physics and chemistry of strongly correlated electron materials.

Superconductivity has been in a certain sense the black hole organizing the galaxy of correlated electron materials. The idea that the magnetic fluctuation spectrum in dense Kondo liquids can support both antiferromagnetism and superconductivity appears consistent with the apparent general occurrence of heavy Fermion superconductivity near a magnetic quantum critical point. It is not so clear in this regard that actinide materials map exactly onto the Ce materials or why Yb heavy electron materials so far have only shown one instance of superconductivity. Why $PuCoGa_5$, which would seem to be naturally classed with the other 115 superconductors, has such a highT_c remains a mystery. A related question is whether the 2.3K maximum T_c observed so far for Ce heavy Fermion

Published under licence by IOP Publishing Ltd

superconductors might be some kind of natural limit for Ce heavy Fermions and whether some kind of local atomic physics plays significantly into correlated electron superconductivity.

The large amount of Fermi surface data that is now available for highly correlated electron materials provides important input for assessing how well the increasingly sophisticated electronic structure calculations are doing in coming to grips with electronic correlations. But what deeper lessons can we take from our knowledge of the Fermi surface? We can understand magnetic instabilities in terms of Fermi surface features but the same cannot be said for superconductivity. For example, superconductivity occurs in $CeRhIn_5$ at pressures below where its Fermi surface changes into one closely resembling that of $CeCoIn_5$. Coupling this with the observation that highly correlated electron superconductivity appears to favor certain crystal structures suggests that superconductivity might really be a chemistry problem.

The discovery of new materials continues to surprise. Superconductivity coming to terms with magnetism has been and is still a driving force in the field, and this acquires new forms. Heavy Fermion superconductivity has many aspects that bear directly on the physics of high T_c. The many faces of frustration are seen throughout correlated electron physics. Quantum criticality itself presents a type of frustration. Behind all new materials work is the motivation to understand structure/property relationships. What kind of progress does one make here? In detail, not much. In the more general sweep of 4f to 5f to transition metal compounds we might hope some patterns to emerge, but this progress is slow. The cleanest case of correspondence across this progression is with the Kondo insulator materials and perhaps this route might be generalized. These materials are a version of valence, or electron precise, compounds.

What else is interesting about intermetallics which appear to obey valence rules? We are usually dealing with semimetals, and it is interesting to note that one extreme of the cuprate and the iron pnictide materials are of this sort. At this extreme, for example with La_2CuO_4, one has a layered structure with significant charge separation between layers: the bonding differs between inter- and intra-layer. The layers appear as entities in themselves, stabilized by their 3D arrangement. Such a picture gains legitimacy from the computations described by Grin. A lot of interesting things happen when conduction electrons are present, either from band overlap or deliberate doping. In a loose way from this vantage point, we might ask what is it about URu_2Si_2 that makes it act so strangely? Perhaps we can regards RuSi as $(RuSi)^{-2}$, so that $U^{+4}(RuSi)^{-2}{}_2$ is actually a low carrier density system with some similarity to the 122 pnictides. The more general point is that there is use to not only exploring what is happening, but also why it happens in one instance and not in another.

Acknowledgments
Support from NSF-DMR-0801253 and Los Alamos National Laboratory is gratefully acknowledged.

AUTHOR INDEX

AUTHOR INDEX

AUTHOR INDEX

AUTHOR INDEX

AUTHOR INDEX

AUTHOR INDEX